INSTRUCTOR'S SOLUTIONS MANUAL

JUDITH A. PENNA

Indiana University Purdue University Indianapolis

to accompany

ALGEBRA AND TRIGONOMETRY

SECOND EDITION

AND

PRECALCULUS

SECOND EDITION

Judith A. Beecher

Indiana University Purdue University Indianapolis

Judith A. Penna

Indiana University Purdue University Indianapolis

Marvin L. Bittinger

Indiana University Purdue University Indianapolis

Boston San Francisco New York
London Toronto Sydney Tokyo Singapore Madrid
Mexico City Munich Paris Cape Town Hong Kong Montreal

Reproduced by Pearson Addison-Wesley from electronic files supplied by the author.

Publishing as Pearson Addison-Wesley, 75 Arlington Street, Boston, MA 02116

ISBN 0-321-23701-3

3 4 5 6 OPM 07 06 05

Contents

Chapter R

Basic Concepts of Algebra

Exercise Set R.1

1. Whole numbers: $\sqrt[3]{8}$, 0, 9, $\sqrt{25}$ $\quad(\sqrt[3]{8}=2,\ \sqrt{25}=5)$

2. Integers: -12, $\sqrt[3]{8}$, 0, 9, $\sqrt{25}$ $\quad(\sqrt[3]{8}=2,\ \sqrt{25}=5)$

3. Irrational numbers: $\sqrt{7}$, 5.242242224..., $-\sqrt{14}$, $\sqrt[5]{5}$, $\sqrt[3]{4}$
(Although there is a pattern in 5.242242224..., there is no repeating block of digits.)

4. Natural numbers: $\sqrt[3]{8}$, 9, $\sqrt{25}$

5. Rational numbers: -12, $5.\overline{3}$, $-\frac{7}{3}$, $\sqrt[3]{8}$, 0, -1.96, 9, $4\frac{2}{3}$, $\sqrt{25}$, $\frac{5}{7}$

6. Real numbers: All of them

7. Rational numbers but not integers: $5.\overline{3}$, $-\frac{7}{3}$, -1.96, $4\frac{2}{3}$, $\frac{5}{7}$

8. Integers but not whole numbers: -12

9. Integers but not rational numbers: -12, 0

10. Real numbers but not integers: $\sqrt{7}$, $5.\overline{3}$, $-\frac{7}{3}$, 5.242242224..., $-\sqrt{14}$, $\sqrt[5]{5}$, -1.96, $4\frac{2}{3}$, $\sqrt[3]{4}$, $\frac{5}{7}$

11. This is a closed interval, so we use brackets. Interval notation is $[-3,3]$.

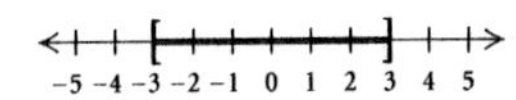

12. $(-4,4)$

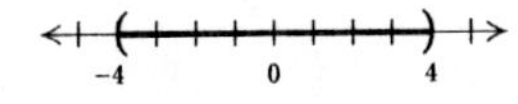

13. This is a half-open interval. We use a bracket on the left and a parenthesis on the right. Interval notation is $[-4,-1)$.

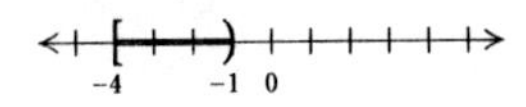

14. $(1,6]$

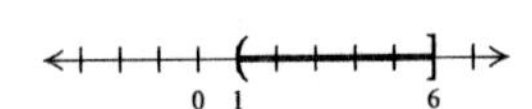

15. This interval is of unlimited extent in the negative direction, and the endpoint -2 is included. Interval notation is $(-\infty,-2]$.

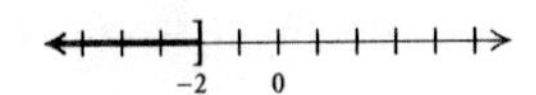

16. $(-5,\infty)$

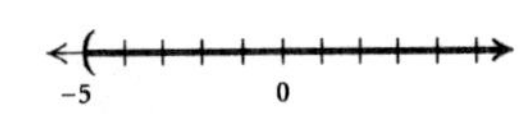

17. This interval is of unlimited extent in the positive direction, and the endpoint 3.8 is not included. Interval notation is $(3.8,\infty)$.

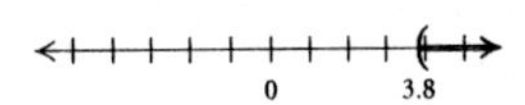

18. $[\sqrt{3},\infty)$

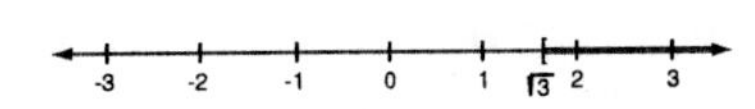

19. $\{x|7<x\}$, or $\{x|x>7\}$.
This interval is of unlimited extent in the positive direction and the endpoint 7 is not included. Interval notation is $(7,\infty)$.

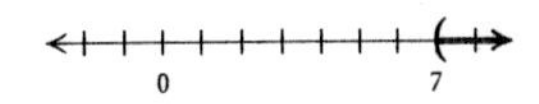

20. $(-\infty,-3)$

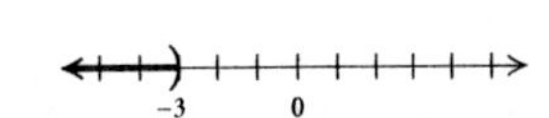

21. The endpoints 0 and 5 are not included in the interval, so we use parentheses. Interval notation is $(0,5)$.

22. $[-1,2]$

23. The endpoint -9 is included in the interval, so we use a bracket before the -9. The endpoint -4 is not included, so we use a parenthesis after the -4. Interval notation is $[-9,-4)$.

24. $(-9,-5]$

25. Both endpoints are included in the interval, so we use brackets. Interval notation is $[x,x+h]$.

26. $(x,x+h]$

27. The endpoint p is not included in the interval, so we use a parenthesis before the p. The interval is of unlimited extent in the positive direction, so we use the infinity symbol ∞. Interval notation is (p,∞).

28. $(-\infty,q]$

29. Since 6 is an element of the set of natural numbers, the statement is true.

30. True

31. Since 3.2 is not an element of the set of integers, the statement is false.

32. True

33. Since $-\frac{11}{5}$ is an element of the set of rational numbers, the statement is true.

34. False

35. Since $\sqrt{11}$ is an element of the set of real numbers, the statement is false.

36. False

37. Since 24 is an element of the set of whole numbers, the statement is false.

38. True

39. Since 1.089 is not an element of the set of irrational numbers, the statement is true.

40. True

41. Since every whole number is an integer, the statement is true.

42. False

43. Since every rational number is a real number, the statement is true.

44. True

45. Since there are real numbers that are not integers, the statement is false.

46. False

47. The sentence $6x = x6$ illustrates the commutative property of multiplication.

48. Associative property of addition

49. The sentence $-3 \cdot 1 = -3$ illustrates the multiplicative identity property.

50. Commutative property of addition

51. The sentence $5(ab) = (5a)b$ illustrates the associative property of multiplication.

52. Distributive property

53. The sentence $2(a+b) = (a+b)2$ illustrates the commutative property of multiplication.

54. Additive inverse property

55. The sentence $-6(m+n) = -6(n+m)$ illustrates the commutative property of addition.

56. Additive identity property

57. The sentence $8 \cdot \frac{1}{8} = 1$ illustrates the multiplicative inverse property.

58. Distributive property

59. The distance of -7.1 from 0 is 7.1, so $|-7.1| = 7.1$.

60. 86.2

61. The distance of 347 from 0 is 347, so $|347| = 347$.

62. 54

63. The distance of $-\sqrt{97}$ from 0 is $\sqrt{97}$, so $|-\sqrt{97}| = \sqrt{97}$.

64. $\frac{12}{19}$

65. The distance of 0 from 0 is 0, so $|0| = 0$.

66. 15

67. The distance of $\frac{5}{4}$ from 0 is $\frac{5}{4}$, so $\left|\frac{5}{4}\right| = \frac{5}{4}$.

68. $\sqrt{3}$

69. $|-5-6| = |-11| = 11$, or
$|6-(-5)| = |6+5| = |11| = 11$

70. $|0-(-2.5)| = |2.5| = 2.5$, or
$|-2.5-0| = |-2.5| = 2.5$

71. $|-2-(-8)| = |-2+8| = |6| = 6$, or
$|-8-(-2)| = |-8+2| = |-6| = 6$

72. $\left|\frac{15}{8} - \frac{23}{12}\right| = \left|\frac{45}{24} - \frac{46}{24}\right| = \left|-\frac{1}{24}\right| = \frac{1}{24}$, or
$\left|\frac{23}{12} - \frac{15}{8}\right| = \left|\frac{46}{24} - \frac{45}{24}\right| = \left|\frac{1}{24}\right| = \frac{1}{24}$

73. $|12.1-6.7| = |5.4| = 5.4$, or
$|6.7-12.1| = |-5.4| = 5.4$

74. $|-3-(-14)| = |-3+14| = |11| = 11$, or
$|-14-(-3)| = |-14+3| = |-11| = 11$

75. $\left|-\frac{3}{4} - \frac{15}{8}\right| = \left|-\frac{6}{8} - \frac{15}{8}\right| = \left|-\frac{21}{8}\right| = \frac{21}{8}$, or
$\left|\frac{15}{8} - \left(-\frac{3}{4}\right)\right| = \left|\frac{15}{8} + \frac{3}{4}\right| = \left|\frac{15}{8} + \frac{6}{8}\right| = \left|\frac{21}{8}\right| = \frac{21}{8}$

76. $|-3.4-10.2| = |-13.6| = 13.6$, or
$|10.2-(-3.4)| = |10.2+3.4| = |13.6| = 13.6$

77. $|-7-0| = |-7| = 7$, or
$|0-(-7)| = |0+7| = |7| = 7$

78. $|3-19| = |-16| = 16$, or
$|19-3| = |16| = 16$

79. Provide an example. For instance, $16 \div (8 \div 2) = 16 \div 4 = 4$, but $(16 \div 8) \div 2 = 2 \div 2 = 1$.

80. $\sqrt{a}$ is a rational number when a is the square of a rational number. That is, $\sqrt{a}$ is a rational number if there is a rational number c such that $a = c^2$.

81. Answers may vary. One such number is 0.124124412444....

82. Answers may vary. Since $-\sqrt{2.01} \approx -1.418$ and $-\sqrt{2} \approx -1.414$, one such number is -1.415.

83. Answers may vary. Since $-\frac{1}{101} = 0.\overline{0099}$ and $-\frac{1}{100} = -0.01$, one such number is -0.00999.

84. Answers may vary. One such number is $\sqrt{5.995}$.

85. Since $1^2 + 3^2 = 10$, the hypotenuse of a right triangle with legs of lengths 1 unit and 3 units has a length of $\sqrt{10}$ units.

$$c^2 = 1^2 + 3^2$$
$$c^2 = 10$$
$$c = \sqrt{10}$$

Exercise Set R.2

1. $18^0 = 1$ (For any nonzero real number, $a^0 = 1$.)

2. $\left(-\frac{4}{3}\right)^0 = 1$

3. $x^9 \cdot x^0 = x^{9+0} = x^9$

4. $a^0 \cdot a^4 = a^{0+4} = a^4$

5. $5^8 \cdot 5^{-6} = 5^{8+(-6)} = 5^2$, or 25

6. $6^2 \cdot 6^{-7} = 6^{2+(-7)} = 6^{-5}$, or $\frac{1}{6^5}$

7. $m^{-5} \cdot m^5 = m^{-5+5} = m^0 = 1$

8. $n^9 \cdot n^{-9} = n^{9+(-9)} = n^0 = 1$

9. $y^3 \cdot y^{-7} = y^{3+(-7)} = y^{-4}$, or $\frac{1}{y^4}$

10. $b^{-4} \cdot b^{12} = b^{-4+12} = b^8$

11. $7^3 \cdot 7^{-5} \cdot 7 = 7^{3+(-5)+1} = 7^{-1}$, or $\frac{1}{7}$

12. $3^6 \cdot 3^{-5} \cdot 3^4 = 3^{6+(-5)+4} = 3^5$

13. $2x^3 \cdot 3x^2 = 2 \cdot 3 \cdot x^{3+2} = 6x^5$

14. $3y^4 \cdot 4y^3 = 3 \cdot 4 \cdot y^{4+3} = 12y^7$

15. $(-3a^{-5})(5a^{-7}) = -3 \cdot 5 \cdot a^{-5+(-7)} = -15a^{-12}$, or $-\frac{15}{a^{12}}$

16. $(-6b^{-4})(2b^{-7}) = -6 \cdot 2 \cdot b^{-4+(-7)} = -12b^{-11}$, or $-\frac{12}{b^{11}}$

17. $(5a^2b)(3a^{-3}b^4) = 5 \cdot 3 \cdot a^{2+(-3)} \cdot b^{1+4} = 15a^{-1}b^5$, or $\frac{15b^5}{a}$

18. $(4xy^2)(3x^{-4}y^5) = 4 \cdot 3 \cdot x^{1+(-4)} \cdot y^{2+5} = 12x^{-3}y^7$, or $\frac{12y^7}{x^3}$

19. $(6x^{-3}y^5)(-7x^2y^{-9}) = 6(-7)x^{-3+2}y^{5+(-9)} = -42x^{-1}y^{-4}$, or $-\frac{42}{xy^4}$

20. $(8ab^7)(-7a^{-5}b^2) = 8(-7)a^{1+(-5)}b^{7+2} = -56a^{-4}b^9$, or $-\frac{56b^9}{a^4}$

21. $(2x)^3(3x)^2 = 2^3x^3 \cdot 3^2x^2 = 8 \cdot 9 \cdot x^{3+2} = 72x^5$

22. $(4y)^2(3y)^3 = 16y^2 \cdot 27y^3 = 432y^5$

23. $(-2n)^3(5n)^2 = (-2)^3n^3 \cdot 5^2n^2 = -8 \cdot 25 \cdot n^{3+2} = -200n^5$

24. $(2x)^5(3x)^2 = 2^5x^5 \cdot 3^2x^2 = 32 \cdot 9 \cdot x^{5+2} = 288x^7$

25. $\frac{b^{40}}{b^{37}} = b^{40-37} = b^3$

26. $\frac{a^{39}}{a^{32}} = a^{39-32} = a^7$

27. $\frac{x^{-5}}{x^{16}} = x^{-5-16} = x^{-21}$, or $\frac{1}{x^{21}}$

28. $\frac{y^{-24}}{y^{-21}} = y^{-24-(-21)} = y^{-24+21} = y^{-3}$, or $\frac{1}{y^3}$

29. $\frac{x^2y^{-2}}{x^{-1}y} = x^{2-(-1)}y^{-2-1} = x^3y^{-3}$, or $\frac{x^3}{y^3}$

30. $\frac{x^3y^{-3}}{x^{-1}y^2} = x^{3-(-1)}y^{-3-2} = x^4y^{-5}$, or $\frac{x^4}{y^5}$

31. $\frac{32x^{-4}y^3}{4x^{-5}y^8} = \frac{32}{4}x^{-4-(-5)}y^{3-8} = 8xy^{-5}$, or $\frac{8x}{y^5}$

32. $\frac{20a^5b^{-2}}{5a^7b^{-3}} = \frac{20}{5}a^{5-7}b^{-2-(-3)} = 4a^{-2}b$, or $\frac{4b}{a^2}$

33. $(2ab^2)^3 = 2^3a^3(b^2)^3 = 2^3a^3b^{2\cdot 3} = 8a^3b^6$

34. $(4xy^3)^2 = 4^2x^2(y^3)^2 = 16x^2y^6$

35. $(-2x^3)^5 = (-2)^5(x^3)^5 = (-2)^5x^{3\cdot 5} = -32x^{15}$

36. $(-3x^2)^4 = (-3)^4(x^2)^4 = 81x^8$

37. $(-5c^{-1}d^{-2})^{-2} = (-5)^{-2}c^{-1(-2)}d^{-2(-2)} = \frac{c^2d^4}{(-5)^2} = \frac{c^2d^4}{25}$

38. $(-4x^{-5}z^{-2})^{-3} = (-4)^{-3}(x^{-5})^{-3}(z^{-2})^{-3} = \frac{x^{15}z^6}{(-4)^3} = \frac{x^{15}z^6}{-64}$

39. $(3m^4)^3(2m^{-5})^4 = 3^3m^{12} \cdot 2^4m^{-20} =$

$27 \cdot 16m^{12+(-20)} = 432m^{-8}$, or $\dfrac{432}{m^8}$

40. $(4n^{-1})^2(2n^3)^3 = 4^2n^{-2} \cdot 2^3n^9 = 16 \cdot 8 \cdot n^{-2+9} =$

$128n^7$

41. $\left(\dfrac{2x^{-3}y^7}{z^{-1}}\right)^3 = \dfrac{(2x^{-3}y^7)^3}{(z^{-1})^3} = \dfrac{2^3x^{-9}y^{21}}{z^{-3}} =$

$\dfrac{8x^{-9}y^{21}}{z^{-3}}$, or $\dfrac{8y^{21}z^3}{x^9}$

42. $\left(\dfrac{3x^5y^{-8}}{z^{-2}}\right)^4 = \dfrac{81x^{20}y^{-32}}{z^{-8}}$, or $\dfrac{81x^{20}z^8}{y^{32}}$

43. $\left(\dfrac{24a^{10}b^{-8}c^7}{12a^6b^{-3}c^5}\right)^{-5} = (2a^4b^{-5}c^2)^{-5} = 2^{-5}a^{-20}b^{25}c^{-10}$,

or $\dfrac{b^{25}}{32a^{20}c^{10}}$

44. $\left(\dfrac{125p^{12}q^{-14}r^{22}}{25p^8q^6r^{-15}}\right)^{-4} = (5p^4q^{-20}r^{37})^{-4} =$

$5^{-4}p^{-16}q^{80}r^{-148}$, or $\dfrac{q^{80}}{625p^{16}r^{148}}$

45. Convert 405,000 to scientific notation.

We want the decimal point to be positioned between the 4 and the first 0, so we move it 5 places to the left. Since 405,000 is greater than 10, the exponent must be positive.

$$405,000 = 4.05 \times 10^5$$

46. Position the decimal point 6 places to the left, between the 1 and the 6. Since 1,670,000 is greater than 10, the exponent must be positive.

$$1,670,000 = 1.67 \times 10^6$$

47. Convert 0.00000039 to scientific notation.

We want the decimal point to be positioned between the 3 and the 9, so we move it 7 places to the right. Since 0.00000039 is a number between 0 and 1, the exponent must be negative.

$$0.00000039 = 3.9 \times 10^{-7}$$

48. Position the decimal point 4 places to the right, between the 9 and the 2. Since 0.00092 is a number between 0 and 1, the exponent must be negative.

$$0.00092 = 9.2 \times 10^{-4}$$

49. Convert 234,600,000,000 to scientific notation. We want the decimal point to be positioned between the 2 and the 3, so we move it 11 places to the left. Since 234,600,000,000 is greater than 10, the exponent must be positive.

$$234,600,000,000 = 2.346 \times 10^{11}$$

50. Position the decimal point 9 places to the left, between the 8 and the 9. Since 8,904,000,000 is greater than 10, the exponent must be positive.

$$8,904,000,000 = 8.904 \times 10^9$$

51. Convert 0.00104 to scientific notation. We want the decimal point to be positioned between the 1 and the last 0, so we move it 3 places to the right. Since 0.00104 is a number between 0 and 1, the exponent must be negative.

$$0.00104 = 1.04 \times 10^{-3}$$

52. Position the decimal point 9 places to the right, between the 5 and the 1. Since 0.00000000514 is a number between 0 and 1, the exponent must be negative.

$$0.00000000514 = 5.14 \times 10^{-9}$$

53. Convert 0.000016 to scientific notation.

We want the decimal point to be positioned between the 1 and the 6, so we move it 5 places to the right. Since 0.000016 is a number between 0 and 1, the exponent must be negative.

$$0.000016 = 1.6 \times 10^{-5}$$

54. Position the decimal point 12 places to the left, between the ones. Since 1,137,000,000,000 is greater than 10, the exponent must be positive.

$$1,137,000,000,000 = 1.137 \times 10^{12}$$

55. Convert 8.3×10^{-5} to decimal notation.

The exponent is negative, so the number is between 0 and 1. We move the decimal point 5 places to the left.

$$8.3 \times 10^{-5} = 0.000083$$

56. The exponent is positive, so the number is greater than 10. We move the decimal point 6 places to the right.

$$4.1 \times 10^6 = 4,100,000$$

57. Convert 2.07×10^7 to decimal notation.

The exponent is positive, so the number is greater than 10. We move the decimal point 7 places to the right.

$$2.07 \times 10^7 = 20,700,000$$

58. The exponent is negative, so the number is between 0 and 1. We move the decimal point 6 places to the left.

$$3.15 \times 10^{-6} = 0.00000315$$

59. Convert 3.496×10^{10} to decimal notation.

The exponent is positive, so the number is greater than 10. We move the decimal point 10 places to the right.

$$3.496 \times 10^{10} = 34,960,000,000$$

60. The exponent is positive, so the number is greater than 10. We move the decimal point 11 places to the right.

$$8.409 \times 10^{11} = 840,900,000,000$$

61. Convert 5.41×10^{-8} to decimal notation.

The exponent is negative, so the number is between 0 and 1. We move the decimal point 8 places to the left.

$$5.41 \times 10^{-8} = 0.0000000541$$

62. The exponent is negative, so the number is between 0 and 1. We move the decimal point 10 places to the left.

$$6.27 \times 10^{-10} = 0.000000000627$$

63. Convert 2.319×10^8 to decimal notation.

The exponent is positive, so the number is greater than 10. We move the decimal point 8 places to the right.

$$2.319 \times 10^8 = 231{,}900{,}000$$

64. The exponent is negative, so the number is between 0 and 1. We move the decimal point 24 places to the left.

1.67×10^{-24} g $= 0.00000000000000000000000167$ g

65.
$$\begin{aligned}&(3.1 \times 10^5)(4.5 \times 10^{-3})\\ &= (3.1 \times 4.5) \times (10^5 \times 10^{-3})\\ &= 13.95 \times 10^2 \quad \text{This is not scientific notation.}\\ &= (1.395 \times 10) \times 10^2\\ &= 1.395 \times 10^3 \quad \text{Writing scientific notation}\end{aligned}$$

66.
$$\begin{aligned}(9.1 \times 10^{-17})(8.2 \times 10^3) &= 74.62 \times 10^{-14}\\ &= (7.462 \times 10) \times 10^{-14}\\ &= 7.462 \times 10^{-13}\end{aligned}$$

67.
$$\begin{aligned}&(2.6 \times 10^{-18})(8.5 \times 10^7)\\ &= (2.6 \times 8.5) \times (10^{-18} \times 10^7)\\ &= 22.1 \times 10^{-11} \quad \text{This is not scientific notation.}\\ &= (2.21 \times 10) \times 10^{-11}\\ &= 2.21 \times 10^{-10}\end{aligned}$$

68.
$$\begin{aligned}(6.4 \times 10^{12})(3.7 \times 10^{-5}) &= 23.68 \times 10^7\\ &= (2.368 \times 10) \times 10^7\\ &= 2.368 \times 10^8\end{aligned}$$

69.
$$\begin{aligned}\frac{6.4 \times 10^{-7}}{8.0 \times 10^6} &= \frac{6.4}{8.0} \times \frac{10^{-7}}{10^6}\\ &= 0.8 \times 10^{-13} \quad \text{This is not scientific notation.}\\ &= (8 \times 10^{-1}) \times 10^{-13}\\ &= 8 \times 10^{-14} \quad \text{Writing scientific notation}\end{aligned}$$

70.
$$\begin{aligned}\frac{1.1 \times 10^{-40}}{2.0 \times 10^{-71}} &= 0.55 \times 10^{31}\\ &= (5.5 \times 10^{-1}) \times 10^{31}\\ &= 5.5 \times 10^{30}\end{aligned}$$

71.
$$\begin{aligned}&\frac{1.8 \times 10^{-3}}{7.2 \times 10^{-9}}\\ &= \frac{1.8}{7.2} \times \frac{10^{-3}}{10^{-9}}\\ &= 0.25 \times 10^6 \quad \text{This is not scientific notation.}\\ &= (2.5 \times 10^{-1}) \times 10^6\\ &= 2.5 \times 10^5\end{aligned}$$

72.
$$\begin{aligned}\frac{1.3 \times 10^4}{5.2 \times 10^{10}} &= 0.25 \times 10^{-6}\\ &= (2.5 \times 10^{-1}) \times 10^{-6}\\ &= 2.5 \times 10^{-7}\end{aligned}$$

73. The average cost per mile is the total cost divided by the number of miles.

$$\begin{aligned}&\frac{\$210 \times 10^6}{17.6}\\ &= \frac{\$210 \times 10^6}{1.76 \times 10}\\ &\approx \$119 \times 10^5\\ &\approx (\$1.19 \times 10^2) \times 10^5\\ &\approx \$1.19 \times 10^7\end{aligned}$$

The average cost per mile was about $\$1.19 \times 10^7$.

74.
$$\begin{aligned}\frac{412}{9{,}600{,}000} &= \frac{4.12 \times 10^2}{9.6 \times 10^6}\\ &\approx 0.43 \times 10^{-4}\\ &\approx (4.3 \times 10^{-1}) \times 10^{-4}\\ &\approx 4.3 \times 10^{-5} \text{ square miles}\end{aligned}$$

75. First find the number of seconds in 1 hour:

$$1 \text{ hour} = 1\,\cancel{\text{hr}} \times \frac{60\,\cancel{\text{min}}}{1\,\cancel{\text{hr}}} \times \frac{60 \text{ sec}}{1\,\cancel{\text{min}}} = 3600 \text{ sec}$$

The number of disintegrations produced in 1 hour is the number of disintegrations per second times the number of seconds in 1 hour.

$$\begin{aligned}&37 \text{ billion} \times 3600\\ &= 37{,}000{,}000{,}000 \times 3600\\ &= 3.7 \times 10^{10} \times 3.6 \times 10^3 \quad \text{Writing scientific notation}\\ &= (3.7 \times 3.6) \times (10^{10} \times 10^3)\\ &= 13.32 \times 10^{13} \quad \text{Multiplying}\\ &= (1.332 \times 10) \times 10^{13}\\ &= 1.332 \times 10^{14}\end{aligned}$$

One gram of radium produces 1.332×10^{14} disintegrations in 1 hour.

76.
$$\begin{aligned}&2\pi \times 93{,}000{,}000\\ &= 2\pi \times 9.3 \times 10^7\\ &\approx 58.43362336\\ &\approx (5.843362336 \times 10) \times 10^7\\ &= 5.843362336 \times 10^8 \text{ mi}\end{aligned}$$

77.
$$\begin{aligned}&3 \cdot 2 - 4 \cdot 2^2 + 6(3-1)\\ &= 3 \cdot 2 - 4 \cdot 2^2 + 6 \cdot 2 \quad \text{Working inside parentheses}\\ &= 3 \cdot 2 - 4 \cdot 4 + 6 \cdot 2 \quad \text{Evaluating } 2^2\\ &= 6 - 16 + 12 \quad \text{Multiplying}\\ &= -10 + 12 \quad \text{Adding in order}\\ &= 2 \quad \text{from left to right}\end{aligned}$$

78.
$$\begin{aligned} & 3[(2+4\cdot 2^2)-6(3-1)] \\ &= 3[(2+4\cdot 4)-6\cdot 2] \\ &= 3[(2+16)-6\cdot 2] \\ &= 3[18-6\cdot 2] \\ &= 3[18-12] \\ &= 3[6] \\ &= 18 \end{aligned}$$

79.
$$\begin{aligned} & 16\div 4\cdot 4\div 2\cdot 256 \\ &= 4\cdot 4\div 2\cdot 256 \quad \text{Multiplying and dividing in order from left to right} \\ &= 16\div 2\cdot 256 \\ &= 8\cdot 256 \\ &= 2048 \end{aligned}$$

80.
$$\begin{aligned} & 2^6\cdot 2^{-3}\div 2^{10}\div 2^{-8} \\ &= 2^3\div 2^{10}\div 2^{-8} \\ &= 2^{-7}\div 2^{-8} \\ &= 2 \end{aligned}$$

81.
$$\begin{aligned} & \frac{4(8-6)^2-4\cdot 3+2\cdot 8}{3^1+19^0} \\ &= \frac{4\cdot 2^2-4\cdot 3+2\cdot 8}{3+1} \quad \text{Calculating in the numerator and in the denominator} \\ &= \frac{4\cdot 4-4\cdot 3+2\cdot 8}{4} \\ &= \frac{16-12+16}{4} \\ &= \frac{4+16}{4} \\ &= \frac{20}{4} \\ &= 5 \end{aligned}$$

82.
$$\begin{aligned} & \frac{[4(8-6)^2+4](3-2\cdot 8)}{2^2(2^3+5)} \\ &= \frac{[4\cdot 2^2+4](3-16)}{2^2(8+5)} \\ &= \frac{[4\cdot 4+4](-13)}{4\cdot 13} \\ &= \frac{[16+4](-13)}{4\cdot 13} \\ &= \frac{20(-13)}{52} \\ &= \frac{-260}{52} \\ &= -5 \end{aligned}$$

83. Since interest is compounded semiannually, $n = 2$. Substitute \$2125 for P, 6.2% or 0.062 for i, 2 for n, and 5 for t in the compound interest formula.

$$\begin{aligned} A &= P\left(1+\frac{i}{n}\right)^{nt} \\ &= \$2125\left(1+\frac{0.062}{2}\right)^{2\cdot 5} \quad \text{Substituting} \\ &= \$2125(1+0.031)^{2\cdot 5} \quad \text{Dividing} \\ &= \$2125(1.031)^{2\cdot 5} \quad \text{Adding} \\ &= \$2125(1.031)^{10} \quad \text{Multiplying 2 and 5} \\ &\approx \$2125(1.357021264) \quad \text{Evaluating the exponential expression} \\ &\approx \$2883.670185 \quad \text{Multiplying} \\ &\approx \$2883.67 \quad \text{Rounding to the nearest cent} \end{aligned}$$

84. $A = \$9550\left(1+\dfrac{0.054}{2}\right)^{2\cdot 7} \approx \$13,867.23$

85. Since interest is compounded quarterly, $n = 4$. Substitute \$6700 for P, 4.5% or 0.045 for i, 4 for n, and 6 for t in the compound interest formula.

$$\begin{aligned} A &= P\left(1+\frac{i}{n}\right)^{nt} \\ &= \$6700\left(1+\frac{0.045}{4}\right)^{4\cdot 6} \quad \text{Substituting} \\ &= \$6700(1+0.01125)^{4\cdot 6} \quad \text{Dividing} \\ &= \$6700(1.01125)^{4\cdot 6} \quad \text{Adding} \\ &= \$6700(1.01125)^{24} \quad \text{Multiplying 4 and 6} \\ &\approx \$6700(1.307991226) \quad \text{Evaluating the exponential expression} \\ &\approx \$8763.541217 \quad \text{Multiplying} \\ &\approx \$8763.54 \quad \text{Rounding to the nearest cent} \end{aligned}$$

86. $A = \$4875\left(1+\dfrac{0.058}{4}\right)^{4\cdot 9} \approx \$8185.56$

87. Yes; find the results with parentheses and without them.

$4\cdot 25\div(10-5) = 4\cdot 25\div 5 = 100\div 5 = 20$,

but $4\cdot 25\div 10-5 = 100\div 10-5 = 10-5 = 5$.

88. No; x^{-2}, or $\dfrac{1}{x^2}$ is positive for all $x < 0$ and x^{-1}, or $\dfrac{1}{x}$ is negative for all $x < 0$. Partial confirmation can be obtained by inspecting the graphs of $y_1 = x^{-2}$ and $y_2 = x^{-1}$ for $x < 0$.

89. Substitute \$250 for P, 0.05 for r and 27 for t and perform the resulting computation.

$$\begin{aligned} S &= P\left[\frac{\left(1+\frac{r}{12}\right)^{12\cdot t}-1}{\frac{r}{12}}\right] \\ &= \$250\left[\frac{\left(1+\frac{0.05}{12}\right)^{12\cdot 27}-1}{\frac{0.05}{12}}\right] \\ &\approx \$170,797.30 \end{aligned}$$

90. $$t = 65 - 25 = 40$$
$$S = \$100\left[\frac{\left(1+\frac{0.04}{12}\right)^{12\cdot 40} - 1}{\frac{0.04}{12}}\right] \approx \$118,196.13$$

91. Substitute \$120,000 for S, 0.06 for r, and 18 for t and solve for P.

$$S = P\left[\frac{\left(1+\frac{r}{12}\right)^{12\cdot t} - 1}{\frac{r}{12}}\right]$$
$$\$120,000 = P\left[\frac{\left(1+\frac{0.06}{12}\right)^{12\cdot 18} - 1}{\frac{0.06}{12}}\right]$$
$$\$120,000 = P\left[\frac{(1.005)^{216} - 1}{0.05}\right]$$
$$\$120,000 \approx P(387.3532)$$
$$\$309.79 \approx P$$

92. $$t = 70 - 30 = 40$$
$$\$200,000 = P\left[\frac{\left(1+\frac{0.045}{12}\right)^{12\cdot 40} - 1}{\frac{0.045}{12}}\right]$$
$$P \approx \$149.13$$

93. $(x^t \cdot x^{3t})^2 = (x^{4t})^2 = x^{4t\cdot 2} = x^{8t}$

94. $(x^y \cdot x^{-y})^3 = (x^0)^3 = 1^3 = 1$

95. $(t^{a+x} \cdot t^{x-a})^4 = (t^{2x})^4 = t^{2x\cdot 4} = t^{8x}$

96. $$(m^{x-b} \cdot n^{x+b})^x (m^b n^{-b})^x$$
$$= (m^{x^2-bx} n^{x^2+bx})(m^{bx} n^{-bx})$$
$$= m^{x^2} n^{x^2}$$

97. $$\left[\frac{(3x^a y^b)^3}{(-3x^a y^b)^2}\right]^2 = \left[\frac{27x^{3a}y^{3b}}{9x^{2a}y^{2b}}\right]^2$$
$$= \left[3x^a y^b\right]^2$$
$$= 9x^{2a}y^{2b}$$

98. $$\left[\left(\frac{x^r}{y^t}\right)^2 \left(\frac{x^{2r}}{y^{4t}}\right)^{-2}\right]^{-3}$$
$$= \left[\left(\frac{x^{2r}}{y^{2t}}\right)\left(\frac{x^{-4r}}{y^{-8t}}\right)\right]^{-3}$$
$$= \left(\frac{x^{-2r}}{y^{-6t}}\right)^{-3}$$
$$= \frac{x^{6r}}{y^{18t}}, \text{ or } x^{6r}y^{-18t}$$

Exercise Set R.3

1. $-5y^4 + 3y^3 + 7y^2 - y - 4 =$
$-5y^4 + 3y^3 + 7y^2 + (-y) + (-4)$

Terms: $-5y^4$, $3y^3$, $7y^2$, $-y$, -4

The degree of the term of highest degree, $-5y^4$, is 4. Thus, the degree of the polynomial is 4.

2. $2m^3 - m^2 - 4m + 11 = 2m^3 + (-m^2) + (-4m) + 11$

Terms: $2m^3$, $-m^2$, $-4m$, 11

The degree of the term of highest degree, $2m^3$, is 3. Thus, the degree of the polynomial is 3.

3. $3a^4b - 7a^3b^3 + 5ab - 2 = 3a^4b + (-7a^3b^3) + 5ab + (-2)$

Terms: $3a^4b$, $-7a^3b^3$, $5ab$, -2

The degrees of the terms are 5, 6, 2, and, 0, respectively, so the degree of the polynomial is 6.

4. $6p^3q^2 - p^2q^4 - 3pq^2 + 5 = 6p^3q^2 + (-p^2q^4) + (-3pq^2) + 5$

Terms: $6p^3q^2$, $-p^2q^4$, $-3pq^2$, 5

The degrees of the terms are 5, 6, 3, and 0, respectively, so the degree of the polynomial is 6.

5. $$(5x^2y - 2xy^2 + 3xy - 5) + (-2x^2y - 3xy^2 + 4xy + 7)$$
$$= (5-2)x^2y + (-2-3)xy^2 + (3+4)xy + (-5+7)$$
$$= 3x^2y - 5xy^2 + 7xy + 2$$

6. $2x^2y - 7xy^2 + 8xy + 5$

7. $$(2x + 3y + z - 7) + (4x - 2y - z + 8) + (-3x + y - 2z - 4)$$
$$= (2+4-3)x + (3-2+1)y + (1-1-2)z + (-7+8-4)$$
$$= 3x + 2y - 2z - 3$$

8. $7x^2 + 12xy - 2x - y - 9$

9. $$(3x^2 - 2x - x^3 + 2) - (5x^2 - 8x - x^3 + 4)$$
$$= (3x^2 - 2x - x^3 + 2) + (-5x^2 + 8x + x^3 - 4)$$
$$= (3-5)x^2 + (-2+8)x + (-1+1)x^3 + (2-4)$$
$$= -2x^2 + 6x - 2$$

10. $-4x^2 + 8xy - 5y^2 + 3$

11. $$(x^4 - 3x^2 + 4x) - (3x^3 + x^2 - 5x + 3)$$
$$= (x^4 - 3x^2 + 4x) + (-3x^3 - x^2 + 5x - 3)$$
$$= x^4 - 3x^3 + (-3-1)x^2 + (4+5)x - 3$$
$$= x^4 - 3x^3 - 4x^2 + 9x - 3$$

12. $2x^4 - 5x^3 - 5x^2 + 10x - 5$

13. $(a-b)(2a^3-ab+3b^2)$
$= (a-b)(2a^3)+(a-b)(-ab)+(a-b)(3b^2)$ Using the distributive property
$= 2a^4-2a^3b-a^2b+ab^2+3ab^2-3b^3$ Using the distributive property three more times
$= 2a^4-2a^3b-a^2b+4ab^2-3b^3$ Collecting like terms

14. $(n+1)(n^2-6n-4)$
$= (n+1)(n^2)+(n+1)(-6n)+(n+1)(-4)$
$= n^3+n^2-6n^2-6n-4n-4$
$= n^3-5n^2-10n-4$

15. $(x+5)(x-3)$
$= x^2-3x+5x-15$ Using FOIL
$= x^2+2x-15$ Collecting like terms

16. $(y-4)(y+1) = y^2+y-4y-4 = y^2-3y-4$

17. $(x+6)(x+4)$
$= x^2+4x+6x+24$ Using FOIL
$= x^2+10x+24$ Collecting like terms

18. $(n-5)(n-8) = n^2-8n-5n+40 = n^2-13n+40$

19. $(2a+3)(a+5)$
$= 2a^2+10a+3a+15$ Using FOIL
$= 2a^2+13a+15$ Collecting like terms

20. $(3b+1)(b-2) = 3b^2-6b+b-2 = 3b^2-5b-2$

21. $(2x+3y)(2x+y)$
$= 4x^2+2xy+6xy+3y^2$ Using FOIL
$= 4x^2+8xy+3y^2$

22. $(2a-3b)(2a-b) = 4a^2-2ab-6ab+3b^2 = 4a^2-8ab+3b^2$

23. $(y+5)^2$
$= y^2+2\cdot y\cdot 5+5^2$ $[(A+B)^2 = A^2+2AB+B^2]$
$= y^2+10y+25$

24. $(y+7)^2 = y^2+2\cdot y\cdot 7+7^2 = y^2+14y+49$

25. $(x-4)^2$
$= x^2-2\cdot x\cdot 4+4^2$ $[(A-B)^2 = A^2-2AB+B^2]$
$= x^2-8x+16$

26. $(a-6)^2 = a^2-2\cdot a\cdot 6+6^2 = a^2-12a+36$

27. $(5x-3)^2$
$= (5x)^2-2\cdot 5x\cdot 3+3^2$ $[(A-B)^2 = A^2-2AB+B^2]$
$= 25x^2-30x+9$

28. $(3x-2)^2 = (3x)^2-2\cdot 3x\cdot 2+2^2 = 9x^2-12x+4$

29. $(2x+3y)^2$
$= (2x)^2+2(2x)(3y)+(3y)^2$ $[(A+B)^2 = A^2+2AB+B^2]$
$= 4x^2+12xy+9y^2$

30. $(5x+2y)^2 = (5x)^2+2\cdot 5x\cdot 2y+(2y)^2 = 25x^2+20xy+4y^2$

31. $(2x^2-3y)^2$
$= (2x^2)^2-2(2x^2)(3y)+(3y)^2$ $[(A-B)^2 = A^2-2AB+B^2]$
$= 4x^4-12x^2y+9y^2$

32. $(4x^2-5y)^2 = (4x^2)^2-2\cdot 4x^2\cdot 5y+(5y)^2 = 16x^4-40x^2y+25y^2$

33. $(a+3)(a-3)$
$= a^2-3^2$ $[(A+B)(A-B) = A^2-B^2]$
$= a^2-9$

34. $(b+4)(b-4) = b^2-4^2 = b^2-16$

35. $(2x-5)(2x+5)$
$= (2x)^2-5^2$ $[(A+B)(A-B) = A^2-B^2]$
$= 4x^2-25$

36. $(4y-1)(4y+1) = (4y)^2-1^2 = 16y^2-1$

37. $(3x-2y)(3x+2y)$
$= (3x)^2-(2y)^2$ $[(A-B)(A+B) = A^2-B^2]$
$= 9x^2-4y^2$

38. $(3x+5y)(3x-5y) = (3x)^2-(5y)^2 = 9x^2-25y^2$

39. $(2x+3y+4)(2x+3y-4)$
$= [(2x+3y)+4][(2x+3y)-4]$
$= (2x+3y)^2-4^2$
$= 4x^2+12xy+9y^2-16$

40. $(5x+2y+3)(5x+2y-3) = (5x+2y)^2-3^2 = 25x^2+20xy+4y^2-9$

41. $(x+1)(x-1)(x^2+1)$
$= (x^2-1)(x^2+1)$
$= x^4-1$

42. $(y-2)(y+2)(y^2+4)$
$= (y^2-4)(y^2+4)$
$= y^4-16$

43. No; if the leading coefficients of the polynomials are additive inverses, the degree of the sum is less than n. For example, the sum of the second degree polynomials x^2+x-1 and $-x^2+4$ is $x+3$, a first degree polynomial.

44. Algebraically: Choose specific values for A and B, $A \neq 0$, $B \neq 0$, and evaluate $(A+B)^2$ and A^2+B^2. For example, $(2+3)^2 = 5^2 = 25$, but $2^2+3^2 = 4+9 = 13$.

Geometrically: Show that the area of a square with side $A+B$ is not equal to A^2+B^2. See the figure below.

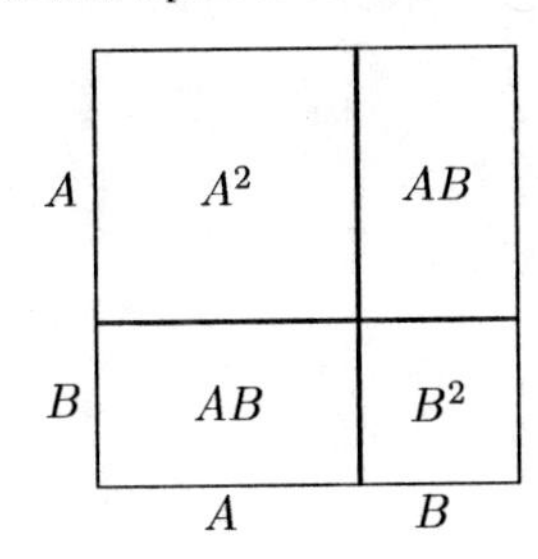

45. $(a^n+b^n)(a^n-b^n) = (a^n)^2-(b^n)^2$
$= a^{2n}-b^{2n}$

46. $(t^a+4)(t^a-7) = (t^a)^2-7t^a+4t^a-28 =$
$t^{2a}-3t^a-28$

47. $(a^n+b^n)^2 = (a^n)^2+2\cdot a^n\cdot b^n+(b^n)^2$
$= a^{2n}+2a^nb^n+b^{2n}$

48. $(x^{3m}-t^{5n})^2 = (x^{3m})^2-2\cdot x^{3m}\cdot t^{5n}+(t^{5n})^2 =$
$x^{6m}-2x^{3m}t^{5n}+t^{10n}$

49.
$$\begin{aligned}
&(x-1)(x^2+x+1)(x^3+1)\\
&=[(x-1)x^2+(x-1)x+(x-1)\cdot 1](x^3+1)\\
&=(x^3-x^2+x^2-x+x-1)(x^3+1)\\
&=(x^3-1)(x^3+1)\\
&=(x^3)^2-1^2\\
&=x^6-1
\end{aligned}$$

50.
$$\begin{aligned}
&[(2x-1)^2-1]^2\\
&=[4x^2-4x+1-1]^2\\
&=[4x^2-4x]^2\\
&=(4x^2)^2-2(4x^2)(4x)+(4x)^2\\
&=16x^4-32x^3+16x^2
\end{aligned}$$

51.
$$\begin{aligned}
&(x^{a-b})^{a+b}\\
&=x^{(a-b)(a+b)}\\
&=x^{a^2-b^2}
\end{aligned}$$

52.
$$\begin{aligned}
&(t^{m+n})^{m+n}\cdot(t^{m-n})^{m-n}\\
&=t^{m^2+2mn+n^2}\cdot t^{m^2-2mn+n^2}\\
&=t^{2m^2+2n^2}
\end{aligned}$$

53.
$$\begin{aligned}
&(a+b+c)^2\\
&=(a+b+c)(a+b+c)\\
&=(a+b+c)(a)+(a+b+c)(b)+(a+b+c)(c)\\
&=a^2+ab+ac+ab+b^2+bc+ac+bc+c^2\\
&=a^2+b^2+c^2+2ab+2ac+2bc
\end{aligned}$$

Exercise Set R.4

1. $2x-10 = 2\cdot x-2\cdot 5 = 2(x-5)$

2. $7y+42 = 7\cdot y+7\cdot 6 = 7(y+6)$

3. $3x^4-9x^2 = 3x^2\cdot x^2-3x^2\cdot 3 = 3x^2(x^2-3)$

4. $20y^2-5y^5 = 5y^2\cdot 4-5y^2\cdot y^3 = 5y^2(4-y^3)$

5. $4a^2-12a+16 = 4\cdot a^2-4\cdot 3a+4\cdot 4 = 4(a^2-3a+4)$

6. $6n^2+24n-18 = 6\cdot n^2+6\cdot 4n-6\cdot 3 = 6(n^2+4n-3)$

7. $a(b-2)+c(b-2) = (b-2)(a+c)$

8. $a(x^2-3)-2(x^2-3) = (x^2-3)(a-2)$

9.
$$\begin{aligned}
&x^3+3x^2+6x+18\\
&=x^2(x+3)+6(x+3)\\
&=(x+3)(x^2+6)
\end{aligned}$$

10.
$$\begin{aligned}
&3x^3-x^2+18x-6\\
&=x^2(3x-1)+6(3x-1)\\
&=(3x-1)(x^2+6)
\end{aligned}$$

11.
$$\begin{aligned}
&y^3-y^2+3y-3\\
&=y^2(y-1)+3(y-1)\\
&=(y-1)(y^2+3)
\end{aligned}$$

12.
$$\begin{aligned}
&y^3-y^2+2y-2\\
&=y^2(y-1)+2(y-1)\\
&=(y-1)(y^2+2)
\end{aligned}$$

13.
$$\begin{aligned}
&24x^3-36x^2+72x-108\\
&=12(2x^3-3x^2+6x-9)\\
&=12[x^2(2x-3)+3(2x-3)]\\
&=12(2x-3)(x^2+3)
\end{aligned}$$

14.
$$\begin{aligned}
&5a^3-10a^2+25a-50\\
&=5(a^3-2a^2+5a-10)\\
&=5[a^2(a-2)+5(a-2)]\\
&=5(a-2)(a^2+5)
\end{aligned}$$

15.
$$\begin{aligned}
&a^3-3a^2-2a+6\\
&=a^2(a-3)-2(a-3)\\
&=(a-3)(a^2-2)
\end{aligned}$$

16.
$$\begin{aligned}
&t^3+6t^2-2t-12\\
&=t^2(t+6)-2(t+6)\\
&=(t+6)(t^2-2)
\end{aligned}$$

17. $x^3 - x^2 - 5x + 5$
$= x^2(x-1) - 5(x-1)$
$= (x-1)(x^2-5)$

18. $x^3 - x^2 - 6x + 6$
$= x^2(x-1) - 6(x-1)$
$= (x-1)(x^2-6)$

19. $p^2 + 6p + 8$

We look for two numbers with a product of 8 and a sum of 6. By trial, we determine that they are 2 and 4.

$$p^2 + 6p + 8 = (p+2)(p+4)$$

20. Note that $(-5)(-2) = 10$ and $-5 + (-2) = -7$. Then

$$w^2 - 7w + 10 = (w-5)(w-2).$$

21. $x^2 + 8x + 12$

We look for two numbers with a product of 12 and a sum of 8. By trial, we determine that they are 2 and 6.

$$x^2 + 8x + 12 = (x+2)(x+6)$$

22. Note that $1 \cdot 5 = 5$ and $1 + 5 = 6$. Then

$$x^2 + 6x + 5 = (x+1)(x+5).$$

23. $t^2 + 8t + 15$

We look for two numbers with a product of 15 and a sum of 8. By trial, we determine that they are 3 and 5.

$$t^2 + 8t + 15 = (t+3)(t+5)$$

24. Note that $3 \cdot 9 = 27$ and $3 + 9 = 12$. Then

$$y^2 + 12y + 27 = (y+3)(y+9).$$

25. $x^2 - 6xy - 27y^2$

We look for two numbers with a product of -27 and a sum of -6. By trial, we determine that they are 3 and -9.

$$x^2 - 6xy - 27y^2 = (x+3y)(x-9y)$$

26. Note that $3(-5) = -15$ and $3 + (-5) = -2$. Then

$$t^2 - 2t - 15 = (t+3)(t-5).$$

27. $2n^2 - 20n - 48 = 2(n^2 - 10n - 24)$

Now factor $n^2 - 10n - 24$. We look for two numbers with a product of -24 and a sum of -10. By trial, we determine that they are 2 and -12. Then $n^2 - 10n - 24 = (n+2)(n-12)$. We must include the common factor, 2, to have a factorization of the original trinomial.

$$2n^2 - 20n - 48 = 2(n+2)(n-12)$$

28. $2a^2 - 2ab - 24b^2 = 2(a^2 - ab - 12b^2)$

Note that $-4 \cdot 3 = -12$ and $-4 + 3 = -1$. Then

$$2a^2 - 2ab - 24b^2 = 2(a-4b)(a+3b).$$

29. $y^4 - 4y^2 - 21 = (y^2)^2 - 4y^2 - 21$

We look for two numbers with a product of -21 and a sum of -4. By trial, we determine that they are 3 and -7.

$$y^4 - 4y^2 - 21 = (y^2+3)(y^2-7)$$

30. Note that $9(-10) = -90$ and $9 + (-10) = -1$. Then

$$m^4 - m^2 - 90 = (m^2+9)(m^2-10).$$

31. $2n^2 + 9n - 56$

We use the FOIL method.

1. There is no common factor other than 1 or -1.
2. The factorization must be of the form $(2n+\quad)(n+\quad)$.
3. Factor the constant term, -56. The possibilities are $-1 \cdot 56$, $1(-56)$, $-2 \cdot 28$, $2(-28)$, $-4 \cdot 16$, $4(-16)$, $-7 \cdot 8$, and $7(-8)$. The factors can be written in the opposite order as well: $56(-1)$, $-56 \cdot 1$, $28(-2)$, $-28 \cdot 2$, $16(-4)$, $-16 \cdot 4$, $8(-7)$, and $-8 \cdot 7$.
4. Find a pair of factors for which the sum of the outside and the inside products is the middle term, $9n$. By trial, we determine that the factorization is $(2n-7)(n+8)$.

32. $3y^2 + 7y - 20 = (3y-5)(y+4)$

33. $12x^2 + 11x + 2$

We use the grouping method.

1. There is no common factor other than 1 or -1.
2. Multiply the leading coefficient and the constant: $12 \cdot 2 = 24$.
3. Try to factor 24 so that the sum of the factors is the coefficient of the middle term, 11. The factors we want are 3 and 8.
4. Split the middle term using the numbers found in step (3):
$$11x = 3x + 8x$$
5. Factor by grouping.
$$\begin{aligned} 12x^2 + 11x + 2 &= 12x^2 + 3x + 8x + 2 \\ &= 3x(4x+1) + 2(4x+1) \\ &= (4x+1)(3x+2) \end{aligned}$$

34. $6x^2 - 7x - 20 = (3x+4)(2x-5)$

35. $4x^2 + 15x + 9$

We use the FOIL method.

1. There is no common factor other than 1 or -1.
2. The factorization must be of the form $(4x+\quad)(x+\quad)$ or $(2x+\quad)(2x+\quad)$.
3. Factor the constant term, 9. The possibilities are $1 \cdot 9$, $-1(-9)$, $3 \cdot 3$, and $-3(-3)$. The first two pairs of factors can be written in the opposite order as well: $9 \cdot 1$, $-9(-1)$.
4. Find a pair of factors for which the sum of the outside and the inside products is the middle term, $15x$. By trial, we determine that the factorization is $(4x+3)(x+3)$.

36. $2y^2 + 7y + 6 = (2y+3)(y+2)$

37. $2y^2 + y - 6$

We use the grouping method.

1. There is no common factor other than 1 or -1.
2. Multiply the leading coefficient and the constant: $2(-6) = -12$.
3. Try to factor -12 so that the sum of the factors is the coefficient of the middle term, 1. The factors we want are 4 and -3.
4. Split the middle term using the numbers found in step (3):

$$y = 4y - 3y$$

5. Factor by grouping.

$$\begin{aligned} 2y^2 + y - 6 &= 2y^2 + 4y - 3y - 6 \\ &= 2y(y+2) - 3(y+2) \\ &= (y+2)(2y-3) \end{aligned}$$

38. $20p^2 - 23p + 6 = (4p-3)(5p-2)$

39. $6a^2 - 29ab + 28b^2$

We use the FOIL method.

1. There is no common factor other than 1 or -1.
2. The factorization must be of the form $(6x+ \quad)(x+ \quad)$ or $(3x+ \quad)(2x+ \quad)$.
3. Factor the coefficient of the last term, 28. The possibilities are $1 \cdot 28$, $-1(-28)$, $2 \cdot 14$, $-2(-14)$, $4 \cdot 7$, and $-4(-7)$. The factors can be written in the opposite order as well: $28 \cdot 1$, $-28(-1)$, $14 \cdot 2$, $-14(-2)$, $7 \cdot 4$, and $-7(-4)$.
4. Find a pair of factors for which the sum of the outside and the inside products is the middle term, -29. Observe that the second term of each binomial factor will contain a factor of b. By trial, we determine that the factorization is $(3a-4b)(2a-7b)$.

40. $10m^2 + 7mn - 12n^2 = (5m-4n)(2m+3n)$

41. $12a^2 - 4a - 16$

We will use the grouping method.

1. Factor out the common factor, 4.

$$12a^2 - 4a - 16 = 4(3a^2 - a - 4)$$

2. Now consider $3a^2 - a - 4$. Multiply the leading coefficient and the constant: $3(-4) = -12$.
3. Try to factor -12 so that the sum of the factors is the coefficient of the middle term, -1. The factors we want are -4 and 3.
4. Split the middle term using the numbers found in step (3):

$$-a = -4a + 3a$$

5. Factor by grouping.

$$\begin{aligned} 3a^2 - a - 4 &= 3a^2 - 4a + 3a - 4 \\ &= a(3a-4) + (3a-4) \\ &= (3a-4)(a+1) \end{aligned}$$

We must include the common factor to get a factorization of the original trinomial.

$$12a^2 - 4a - 16 = 4(3a-4)(a+1)$$

42. $12a^2 - 14a - 20 = 2(6a^2 - 7a - 10) = 2(6a+5)(a-2)$

43. $$\begin{aligned} m^2 - 4 &= m^2 - 2^2 \\ &= (m+2)(m-2) \end{aligned}$$

44. $z^2 - 81 = (z+9)(z-9)$

45. $$\begin{aligned} 9x^2 - 25 &= (3x)^2 - 5^2 \\ &= (3x+5)(3x-5) \end{aligned}$$

46. $16x^2 - 9 = (4x-3)(4x+3)$

47. $6x^2 - 6y^2 = 6(x^2 - y^2) = 6(x+y)(x-y)$

48. $8a^2 - 8b^2 = 8(a^2 - b^2) = 8(a+b)(a-b)$

49. $$\begin{aligned} 4xy^4 - 4xz^2 &= 4x(y^4 - z^2) \\ &= 4x[(y^2)^2 - z^2] \\ &= 4x(y^2+z)(y^2-z) \end{aligned}$$

50. $5x^2y - 5yz^4 = 5y(x^2 - z^4) = 5y(x+z^2)(x-z^2)$

51. $$\begin{aligned} 7pq^4 - 7py^4 &= 7p(q^4 - y^4) \\ &= 7p[(q^2)^2 - (y^2)^2] \\ &= 7p(q^2+y^2)(q^2-y^2) \\ &= 7p(q^2+y^2)(q+y)(q-y) \end{aligned}$$

52. $$\begin{aligned} 25ab^4 - 25az^4 &= 25a(b^4 - z^4) \\ &= 25a(b^2+z^2)(b^2-z^2) \\ &= 25a(b^2+z^2)(b+z)(b-z) \end{aligned}$$

53. $$\begin{aligned} y^2 - 6y + 9 &= y^2 - 2 \cdot y \cdot 3 + 3^2 \\ &= (y-3)^2 \end{aligned}$$

54. $x^2 + 8x + 16 = (x+4)^2$

55. $$\begin{aligned} 4z^2 + 12z + 9 &= (2z)^2 + 2 \cdot 2z \cdot 3 + 3^2 \\ &= (2z+3)^2 \end{aligned}$$

56. $9z^2 - 12z + 4 = (3z-2)^2$

57. $$\begin{aligned} 1 - 8x + 16x^2 &= 1^2 - 2 \cdot 1 \cdot 4x + (4x)^2 \\ &= (1-4x)^2 \end{aligned}$$

58. $1 + 10x + 25x^2 = (1+5x)^2$

59. $$\begin{aligned} &a^3 + 24a^2 + 144a \\ &= a(a^2 + 24a + 144) \\ &= a(a^2 + 2 \cdot a \cdot 12 + 12^2) \\ &= a(a+12)^2 \end{aligned}$$

60. $y^3 - 18y^2 + 81y = y(y^2 - 18y + 81) = y(y-9)^2$

61. $$\begin{aligned} &4p^2 - 8pq + 4q^2 \\ &= 4(p^2 - 2pq + q^2) \\ &= 4(p-q)^2 \end{aligned}$$

62. $5a^2 - 10ab + 5b^2 = 5(a^2 - 2ab + b^2) = 5(a-b)^2$

63. $x^3 + 8 = x^3 + 2^3$
$= (x+2)(x^2 - 2x + 4)$

64. $y^3 - 64 = (y-4)(y^2 + 4y + 16)$

65. $m^3 - 1 = m^3 - 1^3$
$= (m-1)(m^2 + m + 1)$

66. $n^3 + 216 = (n+6)(n^2 - 6n + 36)$

67. $2y^3 - 128 = 2(y^3 - 64)$
$= 2(y^3 - 4^3)$
$= 2(y-4)(y^2 + 4y + 16)$

68. $8t^3 - 8 = 8(t^3 - 1) = 8(t-1)(t^2 + t + 1)$

69. $3a^5 - 24a^2 = 3a^2(a^3 - 8)$
$= 3a^2(a^3 - 2^3)$
$= 3a^2(a-2)(a^2 + 2a + 4)$

70. $250z^4 - 2z = 2z(125z^3 - 1)$
$= 2z(5z - 1)(25z^2 + 5z + 1)$

71. $t^6 + 1 = (t^2)^3 + 1^3$
$= (t^2 + 1)(t^4 - t^2 + 1)$

72. $27x^6 - 8 = (3x^2 - 2)(9x^4 + 6x^2 + 4)$

73. $18a^2b - 15ab^2 = 3ab \cdot 6a - 3ab \cdot 5b$
$= 3ab(6a - 5b)$

74. $4x^2y + 12xy^2 = 4xy(x + 3y)$

75. $x^3 - 4x^2 + 5x - 20 = x^2(x-4) + 5(x-4)$
$= (x-4)(x^2 + 5)$

76. $z^3 + 3z^2 - 3z - 9 = z^2(z+3) - 3(z+3)$
$= (z+3)(z^2 - 3)$

77. $8x^2 - 32 = 8(x^2 - 4)$
$= 8(x+2)(x-2)$

78. $6y^2 - 6 = 6(y^2 - 1) = 6(y+1)(y-1)$

79. $4y^2 - 5$

There are no common factors. We might try to factor this polynomial as a difference of squares, but there is no integer which yields 5 when squared. Thus, the polynomial is prime.

80. There are no common factors and there is no integer which yields 7 when squared, so $16x^2 - 7$ is prime.

81. $m^2 - 9n^2 = m^2 - (3n)^2$
$= (m + 3n)(m - 3n)$

82. $25t^2 - 16 = (5t + 4)(5t - 4)$

83. $x^2 + 9x + 20$

We look for two numbers with a product of 20 and a sum of 9. They are 4 and 5.

$x^2 + 9x + 20 = (x+4)(x+5)$

84. Note that $3(-2) = -6$ and $3 + (-2) = 1$. Then

$y^2 + y - 6 = (y+3)(y-2)$.

85. $y^2 - 6y + 5$

We look for two numbers with a product of 5 and a sum of -6. They are -5 and -1.

$y^2 - 6y + 5 = (y-5)(y-1)$

86. Note that $-7(3) = -21$ and $-7 + 3 = -4$.

$x^2 - 4x - 21 = (x-7)(x+3)$

87. $2a^2 + 9a + 4$

We use the FOIL method.

1. There is no common factor other than 1 or -1.
2. The factorization must be of the form $(2a+ \quad)(a+ \quad)$.
3. Factor the constant term, 4. The possibilities are $1 \cdot 4$, $-1(-4)$, and $2 \cdot 2$. The first two pairs of factors can be written in the opposite order as well: $4 \cdot 1$, $-4(-1)$.
4. Find a pair of factors for which the sum of the outside and the inside products is the middle term, $9a$. By trial, we determine that the factorization is $(2a+1)(a+4)$.

88. $3b^2 - b - 2 = (3b+2)(b-1)$

89. $6x^2 + 7x - 3$

We use the grouping method.

1. There is no common factor other than 1 or -1.
2. Multiply the leading coefficient and the constant: $6(-3) = -18$.
3. Try to factor -18 so that the sum of the factors is the coefficient of the middle term, 7. The factors we want are 9 and -2.
4. Split the middle term using the numbers found in step (3):

   $7x = 9x - 2x$

5. Factor by grouping.

   $6x^2 + 7x - 3 = 6x^2 + 9x - 2x - 3$
   $= 3x(2x+3) - (2x+3)$
   $= (2x+3)(3x-1)$

90. $8x^2 + 2x - 15 = (4x-5)(2x+3)$

91. $y^2 - 18y + 81 = y^2 - 2 \cdot y \cdot 9 + 9^2$
$= (y-9)^2$

92. $n^2 + 2n + 1 = (n+1)^2$

93. $9z^2 - 24z + 16 = (3z)^2 - 2 \cdot 3z \cdot 4 + 4^2$
$= (3z-4)^2$

94. $4z^2 + 20z + 25 = (2z+5)^2$

95. $x^2y^2 - 14xy + 49 = (xy)^2 - 2 \cdot xy \cdot 7 + 7^2$
$= (xy-7)^2$

96. $x^2y^2 - 16xy + 64 = (xy-8)^2$

97. $4ax^2 + 20ax - 56a = 4a(x^2 + 5x - 14)$
$= 4a(x+7)(x-2)$

98. $21x^2y + 2xy - 8y = y(21x^2 + 2x - 8)$
$= y(7x-4)(3x+2)$

99. $3z^3 - 24 = 3(z^3 - 8)$
$= 3(z^3 - 2^3)$
$= 3(z-2)(z^2 + 2z + 4)$

100. $4t^3 + 108 = 4(t^3 + 27)$
$= 4(t+3)(t^2 - 3t + 9)$

101. $16a^7b + 54ab^7$
$= 2ab(8a^6 + 27b^6)$
$= 2ab[(2a^2)^3 + (3b^2)^3]$
$= 2ab(2a^2 + 3b^2)(4a^4 - 6a^2b^2 + 9b^4)$

102. $24a^2x^4 - 375a^8x$
$= 3a^2x(8x^3 - 125a^6)$
$= 3a^2x(2x - 5a^2)(4x^2 + 10a^2x + 25a^4)$

103. $y^3 - 3y^2 - 4y + 12$
$= y^2(y-3) - 4(y-3)$
$= (y-3)(y^2-4)$
$= (y-3)(y+2)(y-2)$

104. $p^3 - 2p^2 - 9p + 18$
$= p^2(p-2) - 9(p-2)$
$= (p-2)(p^2-9)$
$= (p-2)(p+3)(p-3)$

105. $x^3 - x^2 + x - 1$
$= x^2(x-1) + (x-1)$
$= (x-1)(x^2+1)$

106. $x^3 - x^2 - x + 1$
$= x^2(x-1) - (x-1)$
$= (x-1)(x^2-1)$
$= (x-1)(x+1)(x-1)$, or
$(x-1)^2(x+1)$

107. $5m^4 - 20 = 5(m^4 - 4)$
$= 5(m^2+2)(m^2-2)$

108. $2x^2 - 288 = 2(x^2 - 144) = 2(x+12)(x-12)$

109. $2x^3 + 6x^2 - 8x - 24$
$= 2(x^3 + 3x^2 - 4x - 12)$
$= 2[x^2(x+3) - 4(x+3)]$
$= 2(x+3)(x^2-4)$
$= 2(x+3)(x+2)(x-2)$

110. $3x^3 + 6x^2 - 27x - 54$
$= 3(x^3 + 2x^2 - 9x - 18)$
$= 3[x^2(x+2) - 9(x+2)]$
$= 3(x+2)(x^2-9)$
$= 3(x+2)(x+3)(x-3)$

111. $4c^2 - 4cd - d^2 = (2c)^2 - 2 \cdot 2c \cdot d - d^2$
$= (2c-d)^2$

112. $9a^2 - 6ab + b^2 = (3a-b)^2$

113. $m^6 + 8m^3 - 20 = (m^3)^2 + 8m^3 - 20$

We look for two numbers with a product of -20 and a sum of 8. They are 10 and -2.

$$m^6 + 8m^3 - 20 = (m^3+10)(m^3-2)$$

114. $x^4 - 37x^2 + 36 = (x^2-1)(x^2-36)$
$= (x+1)(x-1)(x+6)(x-6)$

115. $p - 64p^4 = p(1 - 64p^3)$
$= p[1^3 - (4p)^3]$
$= p(1+4p)(1-4p+16p^2)$

116. $125a - 8a^4 = a(125 - 8a^3) = a(5-2a)(25+10a+4a^2)$

117. $A^2 + B^2$ can be factored when A and B have a common factor. For example, let $A = 2x$ and $B = 10$. Then $A^2 + B^2 = 4x^2 + 100 = 4(x^2+25)$.

118. $A^3 - B^3 = A^3 + (-B)^3$
$= (A + (-B))(A^2 - A(-B) + (-B)^2)$
$= (A-B)(A^2 + AB + B^2)$

119. $y^4 - 84 + 5y^2$
$= y^4 + 5y^2 - 84$
$= u^2 + 5u - 84$ Substituting u for y^2
$= (u+12)(u-7)$
$= (y^2+12)(y^2-7)$ Substituting y^2 for u

120. $11x^2 + x^4 - 80$
$= x^4 + 11x^2 - 80$
$= u^2 + 11u - 80$ Substituting u for x^2
$= (u+16)(u-5)$
$= (x^2+16)(x^2-5)$ Substituting x^2 for u

121. $y^2 - \frac{8}{49} + \frac{2}{7}y = y^2 + \frac{2}{7}y - \frac{8}{49}$
$= \left(y + \frac{4}{7}\right)\left(y - \frac{2}{7}\right)$

122. $t^2 - \frac{27}{100} + \frac{3}{5}t = t^2 + \frac{3}{5}t - \frac{27}{100} = \left(t + \frac{9}{10}\right)\left(t - \frac{3}{10}\right)$

123. $x^2 + 3x + \frac{9}{4} = x^2 + 2 \cdot x \cdot \frac{3}{2} + \left(\frac{3}{2}\right)^2$
$= \left(x + \frac{3}{2}\right)^2$

124. $x^2 - 5x + \frac{25}{4} = \left(x - \frac{5}{2}\right)^2$

125. $x^2 - x + \frac{1}{4} = x^2 - 2 \cdot x \cdot \frac{1}{2} + \left(\frac{1}{2}\right)^2$
$= \left(x - \frac{1}{2}\right)^2$

126. $x^2 - \frac{2}{3}x + \frac{1}{9} = \left(x - \frac{1}{3}\right)^2$

127. $(x + h)^3 - x^3$
$= [(x + h) - x][(x + h)^2 + x(x + h) + x^2]$
$= (x + h - x)(x^2 + 2xh + h^2 + x^2 + xh + x^2)$
$= h(3x^2 + 3xh + h^2)$

128. $(x + 0.01)^2 - x^2$
$= (x + 0.01 + x)(x + 0.01 - x)$
$= 0.01(2x + 0.01)$, or $0.02(x + 0.005)$

129. $(y - 4)^2 + 5(y - 4) - 24$
$= u^2 + 5u - 24$ Substituting u for $y - 4$
$= (u + 8)(u - 3)$
$= (y - 4 + 8)(y - 4 - 3)$ Substituting $y - 4$ for u
$= (y + 4)(y - 7)$

130. $6(2p + q)^2 - 5(2p + q) - 25$
$= 6u^2 - 5u - 25$ Substituting u for $2p + q$
$= (3u + 5)(2u - 5)$
$= [3(2p + q) + 5][2(2p + q) - 5]$ Substituting $2p + q$ for u
$= (6p + 3q + 5)(4p + 2q - 5)$

131. $x^{2n} + 5x^n - 24 = (x^n)^2 + 5x^n - 24$
$= (x^n + 8)(x^n - 3)$

132. $4x^{2n} - 4x^n - 3 = (2x^n - 3)(2x^n + 1)$

133. $x^2 + ax + bx + ab = x(x + a) + b(x + a)$
$= (x + a)(x + b)$

134. $bdy^2 + ady + bcy + ac$
$= dy(by + a) + c(by + a)$
$= (by + a)(dy + c)$

135. $25y^{2m} - (x^{2n} - 2x^n + 1)$
$= (5y^m)^2 - (x^n - 1)^2$
$= [5y^m + (x^n - 1)][5y^m - (x^n - 1)]$
$= (5y^m + x^n - 1)(5y^m - x^n + 1)$

136. $x^{6a} - t^{3b} = (x^{2a})^3 - (t^b)^3 =$
$(x^{2a} - t^b)(x^{4a} + x^{2a}t^b + t^{2b})$

137. $(y - 1)^4 - (y - 1)^2$
$= (y - 1)^2[(y - 1)^2 - 1]$
$= (y - 1)^2[y^2 - 2y + 1 - 1]$
$= (y - 1)^2(y^2 - 2y)$
$= y(y - 1)^2(y - 2)$

138. $x^6 - 2x^5 + x^4 - x^2 + 2x - 1$
$= x^4(x^2 - 2x + 1) - (x^2 - 2x + 1)$
$= (x^2 - 2x + 1)(x^4 - 1)$
$= (x - 1)^2(x^2 + 1)(x^2 - 1)$
$= (x - 1)^2(x^2 + 1)(x + 1)(x - 1)$
$= (x^2 + 1)(x + 1)(x - 1)^3$

Exercise Set R.5

1. Since $-\frac{3}{4}$ is defined for all real numbers, the domain is $\{x|x$ is a real number$\}$.

2. Since $8 - x = 0$ when $x = 8$, the domain is $\{x|x$ is a real number *and* $x \neq 8\}$.

3. $\frac{3x - 3}{x(x - 1)}$
The denominator is 0 when the factor $x = 0$ and also when $x - 1 = 0$, or $x = 1$. The domain is $\{x|x$ is a real number *and* $x \neq 0$ *and* $x \neq 1\}$.

4. $\frac{15x - 10}{2x(3x - 2)}$
Since $2x = 0$ when $x = 0$ and $3x - 2 = 0$ when $x = \frac{2}{3}$, the domain is
$\left\{x \middle| x \text{ is a real number } and\ x \neq 0\ and\ x \neq \frac{2}{3}\right\}$.

5. $\frac{x + 5}{x^2 + 4x - 5} = \frac{x + 5}{(x + 5)(x - 1)}$
We see that $x + 5 = 0$ when $x = -5$ and $x - 1 = 0$ when $x = 1$. Thus, the domain is $\{x|x$ is a real number *and* $x \neq -5$ *and* $x \neq 1\}$.

6. $\frac{(x^2 - 4)(x + 1)}{(x + 2)(x^2 - 1)} = \frac{(x^2 - 4)(x + 1)}{(x + 2)(x + 1)(x - 1)}$
$x + 2 = 0$ when $x = -2$; $x + 1 = 0$ when $x = -1$; $x - 1 = 0$ when $x = 1$. The domain is $\{x|x$ is a real number *and* $x \neq -2$ *and* $x \neq -1$ *and* $x \neq 1\}$.

7. We first factor the denominator completely.
$\frac{7x^2 - 28x + 28}{(x^2-4)(x^2+3x-10)} = \frac{7x^2 - 28x + 28}{(x+2)(x-2)(x+5)(x-2)}$
We see that $x + 2 = 0$ when $x = -2$, $x - 2 = 0$ when $x = 2$, and $x + 5 = 0$ when $x = -5$. Thus, the domain is $\{x|x$ is a real number *and* $x \neq -2$ *and* $x \neq 2$ *and* $x \neq -5\}$.

8. $\frac{7x^2+11x-6}{x(x^2-x-6)} = \frac{7x^2+11x-6}{x(x-3)(x+2)}$

The denominator is 0 when $x = 0$ or when $x - 3 = 0$ or when $x + 2 = 0$. Now $x - 3 = 0$ when $x = 3$ and $x + 2 = 0$ when $x = -2$. Thus, the domain is $\{x|x \text{ is a real number } and\ x \neq 0\ and\ x \neq 3\ and\ x \neq -2\}$.

9. $\frac{x^2-y^2}{(x-y)^2} \cdot \frac{1}{x+y}$

$= \frac{(x^2-y^2)\cdot 1}{(x-y)^2(x+y)}$

$= \frac{\cancel{(x+y)}\cancel{(x-y)}\cdot 1}{\cancel{(x-y)}(x-y)\cancel{(x+y)}}$

$= \frac{1}{x-y}$

10. $\frac{r-s}{r+s} \cdot \frac{r^2-s^2}{(r-s)^2} = \frac{(r-s)(r^2-s^2)}{(r+s)(r-s)^2}$

$= \frac{\cancel{(r-s)}\cancel{(r-s)}\cancel{(r+s)}\cdot 1}{\cancel{(r+s)}\cancel{(r-s)}\cancel{(r-s)}}$

$= 1$

11. $\frac{x^2-2x-35}{2x^3-3x^2} \cdot \frac{4x^3-9x}{7x-49}$

$= \frac{\cancel{(x-7)}(x+5)\cancel{(x)}(2x+3)\cancel{(2x-3)}}{\cancel{x}\cdot x\cancel{(2x-3)}(7)\cancel{(x-7)}}$

$= \frac{(x+5)(2x+3)}{7x}$

12. $\frac{x^2+2x-35}{3x^3-2x^2} \cdot \frac{9x^3-4x}{7x+49}$

$= \frac{\cancel{(x+7)}(x-5)\cancel{(x)}(3x+2)\cancel{(3x-2)}}{\cancel{x}\cdot x\cancel{(3x-2)}(7)\cancel{(x+7)}}$

$= \frac{(x-5)(3x+2)}{7x}$

13. $\frac{a^2-a-6}{a^2-7a+12} \cdot \frac{a^2-2a-8}{a^2-3a-10}$

$= \frac{\cancel{(a-3)}(a+2)\cancel{(a-4)}\cancel{(a+2)}}{\cancel{(a-4)}\cancel{(a-3)}(a-5)\cancel{(a+2)}}$

$= \frac{a+2}{a-5}$

14. $\frac{a^2-a-12}{a^2-6a+8} \cdot \frac{a^2+a-6}{a^2-2a-24}$

$= \frac{\cancel{(a-4)}(a+3)(a+3)\cancel{(a-2)}}{\cancel{(a-2)}\cancel{(a-4)}(a-6)(a+4)}$

$= \frac{(a+3)^2}{(a-6)(a+4)}$

15. $\frac{m^2-n^2}{r+s} \div \frac{m-n}{r+s}$

$= \frac{m^2-n^2}{r+s} \cdot \frac{r+s}{m-n}$

$= \frac{(m+n)\cancel{(m-n)}\cancel{(r+s)}}{\cancel{(r+s)}\cancel{(m-n)}}$

$= m+n$

16. $\frac{a^2-b^2}{x-y} \div \frac{a+b}{x-y}$

$= \frac{a^2-b^2}{x-y} \cdot \frac{x-y}{a+b}$

$= \frac{\cancel{(a+b)}(a-b)\cancel{(x-y)}}{\cancel{(x-y)}\cancel{(a+b)}\cdot 1}$

$= a-b$

17. $\frac{3x+12}{2x-8} \div \frac{(x+4)^2}{(x-4)^2}$

$= \frac{3x+12}{2x-8} \cdot \frac{(x-4)^2}{(x+4)^2}$

$= \frac{3\cancel{(x+4)}\cancel{(x-4)}(x-4)}{2\cancel{(x-4)}\cancel{(x+4)}(x+4)}$

$= \frac{3(x-4)}{2(x+4)}$

18. $\frac{a^2-a-2}{a^2-a-6} \div \frac{a^2-2a}{2a+a^2}$

$= \frac{a^2-a-2}{a^2-a-6} \cdot \frac{2a+a^2}{a^2-2a}$

$= \frac{\cancel{(a-2)}(a+1)\cancel{(a)}\cancel{(2+a)}}{(a-3)\cancel{(a+2)}\cancel{(a)}\cancel{(a-2)}}$

$= \frac{a+1}{a-3}$

19. $\frac{x^2-y^2}{x^3-y^3} \cdot \frac{x^2+xy+y^2}{x^2+2xy+y^2}$

$= \frac{(x+y)(x-y)(x^2+xy+y^2)}{(x-y)(x^2+xy+y^2)(x+y)(x+y)}$

$= \frac{1}{x+y} \cdot \frac{(x+y)(x-y)(x^2+xy+y^2)}{(x+y)(x-y)(x^2+xy+y^2)}$

$= \frac{1}{x+y} \cdot 1$ Removing a factor of 1

$= \frac{1}{x+y}$

20. $\frac{c^3+8}{c^2-4} \div \frac{c^2-2c+4}{c^2-4c+4}$

$= \frac{c^3+8}{c^2-4} \cdot \frac{c^2-4c+4}{c^2-2c+4}$

$= \frac{(c+2)(c^2-2c+4)(c-2)(c-2)}{(c+2)(c-2)(c^2-2c+4)}$

$= \frac{(c+2)(c^2-2c+4)(c-2)}{(c+2)(c^2-2c+4)(c-2)} \cdot \frac{c-2}{1}$

$= c-2$

21. $\dfrac{(x-y)^2-z^2}{(x+y)^2-z^2} \div \dfrac{x-y+z}{x+y-z}$

$= \dfrac{(x-y)^2-z^2}{(x+y)^2-z^2} \cdot \dfrac{x+y-z}{x-y+z}$

$= \dfrac{(x-y+z)(x-y-z)(x+y-z)}{(x+y+z)(x+y-z)(x-y+z)}$

$= \dfrac{(x-y+z)(x+y-z)}{(x-y+z)(x+y-z)} \cdot \dfrac{x-y-z}{x+y+z}$

$= 1 \cdot \dfrac{x-y-z}{x+y+z}$ Removing a factor of 1

$= \dfrac{x-y-z}{x+y+z}$

22. $\dfrac{(a+b)^2-9}{(a-b)^2-9} \cdot \dfrac{a-b-3}{a+b+3}$

$= \dfrac{(a+b+3)(a+b-3)(a-b-3)}{(a-b+3)(a-b-3)(a+b+3)}$

$= \dfrac{(a+b+3)(a-b-3)}{(a+b+3)(a-b-3)} \cdot \dfrac{a+b-3}{a-b+3}$

$= \dfrac{a+b-3}{a-b+3}$

23. $\dfrac{5}{2x} + \dfrac{1}{2x} = \dfrac{5+1}{2x}$

$= \dfrac{6}{2x}$

$= \dfrac{\not{2} \cdot 3}{\not{2} \cdot x}$

$= \dfrac{3}{x}$

24. $\dfrac{10}{9y} - \dfrac{4}{9y} = \dfrac{6}{9y}$

$= \dfrac{\not{3} \cdot 2}{\not{3} \cdot 3y}$

$= \dfrac{2}{3y}$

25. $\dfrac{3}{2a+3} + \dfrac{2a}{2a+3}$

$= \dfrac{3+2a}{2a+3}$

$= 1$

26. $\dfrac{a-3b}{a+b} + \dfrac{a+5b}{a+b} = \dfrac{2a+2b}{a+b}$

$= \dfrac{2(\cancel{a+b})}{1 \cdot (\cancel{a+b})}$

$= 2$

27. $\dfrac{5}{4z} - \dfrac{3}{8z}$, LCD is $8z$

$= \dfrac{5}{4z} \cdot \dfrac{2}{2} - \dfrac{3}{8z}$

$= \dfrac{10}{8z} - \dfrac{3}{8z}$

$= \dfrac{7}{8z}$

28. $\dfrac{12}{x^2y} + \dfrac{5}{xy^2}$, LCD is x^2y^2

$= \dfrac{12y}{x^2y^2} + \dfrac{5x}{x^2y^2}$

$= \dfrac{12y+5x}{x^2y^2}$

29. $\dfrac{3}{x+2} + \dfrac{2}{x^2-4}$

$= \dfrac{3}{x+2} + \dfrac{2}{(x+2)(x-2)}$, LCD is $(x+2)(x-2)$

$= \dfrac{3}{x+2} \cdot \dfrac{x-2}{x-2} + \dfrac{2}{(x+2)(x-2)}$

$= \dfrac{3x-6}{(x+2)(x-2)} + \dfrac{2}{(x+2)(x-2)}$

$= \dfrac{3x-4}{(x+2)(x-2)}$

30. $\dfrac{5}{a-3} - \dfrac{2}{a^2-9}$

$= \dfrac{5}{a-3} - \dfrac{2}{(a+3)(a-3)}$, LCD is $(a+3)(a-3)$

$= \dfrac{5(a+3)-2}{(a+3)(a-3)}$

$= \dfrac{5a+15-2}{(a+3)(a-3)}$

$= \dfrac{5a+13}{(a+3)(a-3)}$

31. $\dfrac{y}{y^2-y-20} - \dfrac{2}{y+4}$

$= \dfrac{y}{(y+4)(y-5)} - \dfrac{2}{y+4}$, LCD is $(y+4)(y-5)$

$= \dfrac{y}{(y+4)(y-5)} - \dfrac{2}{y+4} \cdot \dfrac{y-5}{y-5}$

$= \dfrac{y}{(y+4)(y-5)} - \dfrac{2y-10}{(y+4)(y-5)}$

$= \dfrac{y-(2y-10)}{(y+4)(y-5)}$

$= \dfrac{y-2y+10}{(y+4)(y-5)}$

$= \dfrac{-y+10}{(y+4)(y-5)}$

32. $\dfrac{6}{y^2+6y+9} - \dfrac{5}{y+3}$

$= \dfrac{6}{(y+3)^2} - \dfrac{5}{y+3}$, LCD is $(y+3)^2$

$= \dfrac{6-5(y+3)}{(y+3)^2}$

$= \dfrac{6-5y-15}{(y+3)^2}$

$= \dfrac{-5y-9}{(y+3)^2}$

33. $\dfrac{3}{x+y}+\dfrac{x-5y}{x^2-y^2}$

$=\dfrac{3}{x+y}+\dfrac{x-5y}{(x+y)(x-y)}$, LCD is $(x+y)(x-y)$

$=\dfrac{3}{x+y}\cdot\dfrac{x-y}{x-y}+\dfrac{x-5y}{(x+y)(x-y)}$

$=\dfrac{3x-3y}{(x+y)(x-y)}+\dfrac{x-5y}{(x+y)(x-y)}$

$=\dfrac{4x-8y}{(x+y)(x-y)}$

34. $\dfrac{a^2+1}{a^2-1}-\dfrac{a-1}{a+1}$

$=\dfrac{a^2+1}{(a+1)(a-1)}-\dfrac{a-1}{a+1}$, LCD is $(a+1)(a-1)$

$=\dfrac{a^2+1-(a-1)(a-1)}{(a+1)(a-1)}$

$=\dfrac{a^2+1-a^2+2a-1}{(a+1)(a-1)}$

$=\dfrac{2a}{(a+1)(a-1)}$

35. $\dfrac{y}{y-1}+\dfrac{2}{1-y}$

$=\dfrac{y}{y-1}+\dfrac{-1}{-1}\cdot\dfrac{2}{1-y}$

$=\dfrac{y}{y-1}+\dfrac{-2}{y-1}$

$=\dfrac{y-2}{y-1}$

36. $\dfrac{a}{a-b}+\dfrac{b}{b-a}=\dfrac{a}{a-b}+\dfrac{-b}{a-b}$

$=\dfrac{a-b}{a-b}$

$=1$

37. $\dfrac{x}{2x-3y}-\dfrac{y}{3y-2x}$

$=\dfrac{x}{2x-3y}-\dfrac{-1}{-1}\cdot\dfrac{y}{3y-2x}$

$=\dfrac{x}{2x-3y}-\dfrac{-y}{2x-3y}$

$=\dfrac{x+y}{2x-3y}$ $\quad[x-(-y)=x+y]$

38. $\dfrac{3a}{3a-2b}-\dfrac{2a}{2b-3a}=\dfrac{3a}{3a-2b}-\dfrac{-2a}{3a-2b}$

$=\dfrac{5a}{3a-2b}$

39. $\dfrac{9x+2}{3x^2-2x-8}+\dfrac{7}{3x^2+x-4}$

$=\dfrac{9x+2}{(3x+4)(x-2)}+\dfrac{7}{(3x+4)(x-1)}$,
LCD is $(3x+4)(x-2)(x-1)$

$=\dfrac{9x+2}{(3x+4)(x-2)}\cdot\dfrac{x-1}{x-1}+\dfrac{7}{(3x+4)(x-1)}\cdot\dfrac{x-2}{x-2}$

$=\dfrac{9x^2-7x-2}{(3x+4)(x-2)(x-1)}+\dfrac{7x-14}{(3x+4)(x-1)(x-2)}$

$=\dfrac{9x^2-16}{(3x+4)(x-2)(x-1)}$

$=\dfrac{\cancel{(3x+4)}(3x-4)}{\cancel{(3x+4)}(x-2)(x-1)}$

$=\dfrac{3x-4}{(x-2)(x-1)}$

40. $\dfrac{3y}{y^2-7y+10}-\dfrac{2y}{y^2-8y+15}$

$=\dfrac{3y}{(y-2)(y-5)}-\dfrac{2y}{(y-5)(y-3)}$,
LCD is $(y-2)(y-5)(y-3)$

$=\dfrac{3y(y-3)-2y(y-2)}{(y-2)(y-5)(y-3)}$

$=\dfrac{3y^2-9y-2y^2+4y}{(y-2)(y-5)(y-3)}$

$=\dfrac{y^2-5y}{(y-2)(y-5)(y-3)}$

$=\dfrac{y\cancel{(y-5)}}{(y-2)\cancel{(y-5)}(y-3)}$

$=\dfrac{y}{(y-2)(y-3)}$

41. $\dfrac{5a}{a-b}+\dfrac{ab}{a^2-b^2}+\dfrac{4b}{a+b}$

$=\dfrac{5a}{a-b}+\dfrac{ab}{(a+b)(a-b)}+\dfrac{4b}{a+b}$,
LCD is $(a+b)(a-b)$

$=\dfrac{5a}{a-b}\cdot\dfrac{a+b}{a+b}+\dfrac{ab}{(a+b)(a-b)}+\dfrac{4b}{a+b}\cdot\dfrac{a-b}{a-b}$

$=\dfrac{5a^2+5ab}{(a+b)(a-b)}+\dfrac{ab}{(a+b)(a-b)}+\dfrac{4ab-4b^2}{(a+b)(a-b)}$

$=\dfrac{5a^2+10ab-4b^2}{(a+b)(a-b)}$

42. $\dfrac{6a}{a-b}+\dfrac{3b}{b-a}+\dfrac{5}{a^2-b^2}$

$=\dfrac{6a}{a-b}+\dfrac{-3b}{a-b}+\dfrac{5}{(a+b)(a-b)}$,
LCD is $(a+b)(a-b)$

$=\dfrac{6a(a+b)+3b(a+b)+5}{(a+b)(a-b)}$

$=\dfrac{6a^2+6ab+3ab+3b^2+5}{(a+b)(a-b)}$

$=\dfrac{6a^2+9ab+3b^2+5}{(a+b)(a-b)}$

43.
$$\frac{7}{x+2} - \frac{x+8}{4-x^2} + \frac{3x-2}{4-4x+x^2}$$
$$= \frac{7}{x+2} - \frac{x+8}{(2+x)(2-x)} + \frac{3x-2}{(2-x)^2},$$
LCD is $(2+x)(2-x)^2$
$$= \frac{7}{2+x}\cdot\frac{(2-x)^2}{(2-x)^2} - \frac{x+8}{(2+x)(2-x)}\cdot\frac{2-x}{2-x} + \frac{3x-2}{(2-x)^2}\cdot\frac{2+x}{2+x}$$
$$= \frac{28-28x+7x^2-(16-6x-x^2)+3x^2+4x-4}{(2+x)(2-x)^2}$$
$$= \frac{28-28x+7x^2-16+6x+x^2+3x^2+4x-4}{(2+x)(2-x)^2}$$
$$= \frac{11x^2-18x+8}{(2+x)(2-x)^2}, \text{ or } \frac{11x^2-18x+8}{(x+2)(x-2)^2}$$

44.
$$\frac{6}{x+3} - \frac{x+4}{9-x^2} + \frac{2x-3}{9-6x+x^2}$$
$$= \frac{6}{x+3} - \frac{x+4}{(3+x)(3-x)} + \frac{2x-3}{(3-x)^2},$$
LCD is $(3+x)(3-x)^2$
$$= \frac{6(3-x)^2-(x+4)(3-x)+(2x-3)(3+x)}{(3+x)(3-x)^2}$$
$$= \frac{54-36x+6x^2+x^2+x-12+2x^2+3x-9}{(3+x)(3-x)^2}$$
$$= \frac{33-32x+9x^2}{(3+x)(3-x)^2}, \text{ or } \frac{9x^2-32x+33}{(x+3)(x-3)^2}$$

45.
$$\frac{1}{x+1} + \frac{x}{2-x} + \frac{x^2+2}{x^2-x-2}$$
$$= \frac{1}{x+1} + \frac{x}{2-x} + \frac{x^2+2}{(x+1)(x-2)}$$
$$= \frac{1}{x+1} + \frac{-1}{-1}\cdot\frac{x}{2-x} + \frac{x^2+2}{(x+1)(x-2)}$$
$$= \frac{1}{x+1} + \frac{-x}{x-2} + \frac{x^2+2}{(x+1)(x-2)},$$
LCD is $(x+1)(x-2)$
$$= \frac{1}{x+1}\cdot\frac{x-2}{x-2} + \frac{-x}{x-2}\cdot\frac{x+1}{x+1} + \frac{x^2+2}{(x+1)(x-2)}$$
$$= \frac{x-2}{(x+1)(x-2)} + \frac{-x^2-x}{(x+1)(x-2)} + \frac{x^2+2}{(x+1)(x-2)}$$
$$= \frac{x-2-x^2-x+x^2+2}{(x+1)(x-2)}$$
$$= \frac{0}{(x+1)(x-2)}$$
$$= 0$$

46.
$$\frac{x-1}{x-2} - \frac{x+1}{x+2} - \frac{x-6}{4-x^2}$$
$$= \frac{x-1}{x-2} - \frac{x+1}{x+2} - \frac{x-6}{(2+x)(2-x)}$$
$$= \frac{1-x}{2-x} - \frac{x+1}{x+2} - \frac{x-6}{(2+x)(2-x)},$$
LCD is $(2+x)(2-x)$
$$= \frac{(1-x)(2+x)-(x+1)(2-x)-(x-6)}{(2+x)(2-x)}$$
$$= \frac{2-x-x^2+x^2-x-2-x+6}{(2+x)(2-x)}$$
$$= \frac{6-3x}{(2+x)(2-x)}$$
$$= \frac{3\cancel{(2-x)}}{(2+x)\cancel{(2-x)}}$$
$$= \frac{3}{2+x}$$

47.
$$\frac{\frac{x^2-y^2}{xy}}{\frac{x-y}{y}} = \frac{x^2-y^2}{xy}\cdot\frac{y}{x-y}$$
$$= \frac{(x+y)\cancel{(x-y)}\cancel{y}}{x\cancel{y}\cancel{(x-y)}}$$
$$= \frac{x+y}{x}$$

48.
$$\frac{\frac{a-b}{b}}{\frac{a^2-b^2}{ab}} = \frac{a-b}{b}\cdot\frac{ab}{(a+b)(a-b)}$$
$$= \frac{a\cancel{b}\cancel{(a-b)}}{\cancel{b}(a+b)\cancel{(a-b)}}$$
$$= \frac{a}{a+b}$$

49.
$$\frac{\frac{x}{y}-\frac{y}{x}}{\frac{1}{y}+\frac{1}{x}} = \frac{\frac{x}{y}-\frac{y}{x}}{\frac{1}{y}+\frac{1}{x}}\cdot\frac{xy}{xy}, \text{ LCM is } xy$$
$$= \frac{\left(\frac{x}{y}-\frac{y}{x}\right)(xy)}{\left(\frac{1}{y}+\frac{1}{x}\right)(xy)}$$
$$= \frac{x^2-y^2}{x+y}$$
$$= \frac{\cancel{(x+y)}(x-y)}{\cancel{(x+y)}\cdot 1}$$
$$= x-y$$

50. $\dfrac{\dfrac{a}{b}-\dfrac{b}{a}}{\dfrac{1}{a}-\dfrac{1}{b}} = \dfrac{a^2-b^2}{b-a}$ Multiplying by $\dfrac{ab}{ab}$

$= \dfrac{(a+b)(a-b)}{b-a}$

$= \dfrac{(a+b)\cancel{(a-b)}}{-1\cdot\cancel{(a-b)}}$

$= -a-b$

51. $\dfrac{c+\dfrac{8}{c^2}}{1+\dfrac{2}{c}} = \dfrac{c\cdot\dfrac{c^2}{c^2}+\dfrac{8}{c^2}}{1\cdot\dfrac{c}{c}+\dfrac{2}{c}}$

$= \dfrac{\dfrac{c^3+8}{c^2}}{\dfrac{c+2}{c}}$

$= \dfrac{c^3+8}{c^2}\cdot\dfrac{c}{c+2}$

$= \dfrac{\cancel{(c+2)}(c^2-2c+4)\cancel{c}}{\cancel{c}\cdot c\cancel{(c+2)}}$

$= \dfrac{c^2-2c+4}{c}$

52. $\dfrac{a-\dfrac{a}{b}}{b-\dfrac{b}{a}} = \dfrac{\dfrac{ab-a}{b}}{\dfrac{ab-b}{a}}$

$= \dfrac{a(b-1)}{b}\cdot\dfrac{a}{b(a-1)}$

$= \dfrac{a^2(b-1)}{b^2(a-1)}$

53. $\dfrac{x^2+xy+y^2}{\dfrac{x^2}{y}-\dfrac{y^2}{x}} = \dfrac{x^2+xy+y^2}{\dfrac{x^2}{y}\cdot\dfrac{x}{x}-\dfrac{y^2}{x}\cdot\dfrac{y}{y}}$

$= \dfrac{x^2+xy+y^2}{\dfrac{x^3-y^3}{xy}}$

$= (x^2+xy+y^2)\cdot\dfrac{xy}{x^3-y^3}$

$= \dfrac{(x^2+xy+y^2)(xy)}{(x-y)(x^2+xy+y^2)}$

$= \dfrac{x^2+xy+y^2}{x^2+xy+y^2}\cdot\dfrac{xy}{x-y}$

$= 1\cdot\dfrac{xy}{x-y}$

$= \dfrac{xy}{x-y}$

54. $\dfrac{\dfrac{a^2}{b}+\dfrac{b^2}{a}}{a^2-ab+b^2} = \dfrac{\dfrac{a^3+b^3}{ab}}{a^2-ab+b^2}$

$= \dfrac{(a+b)(a^2-ab+b^2)}{ab}\cdot\dfrac{1}{a^2-ab+b^2}$

$= \dfrac{a+b}{ab}\cdot\dfrac{a^2-ab+b^2}{a^2-ab+b^2}$

$= \dfrac{a+b}{ab}$

55. $\dfrac{a-a^{-1}}{a+a^{-1}} = \dfrac{a-\dfrac{1}{a}}{a+\dfrac{1}{a}} = \dfrac{a\cdot\dfrac{a}{a}-\dfrac{1}{a}}{a\cdot\dfrac{a}{a}+\dfrac{1}{a}}$

$= \dfrac{\dfrac{a^2-1}{a}}{\dfrac{a^2+1}{a}}$

$= \dfrac{a^2-1}{a}\cdot\dfrac{a}{a^2+1}$

$= \dfrac{a^2-1}{a^2+1}$

56. $\dfrac{x^{-1}+y^{-1}}{x^{-3}+y^{-3}} = \dfrac{\dfrac{1}{x}+\dfrac{1}{y}}{\dfrac{1}{x^3}+\dfrac{1}{y3}}$

$= \dfrac{\left(\dfrac{1}{x}+\dfrac{1}{y}\right)(x^3y^3)}{\left(\dfrac{1}{x^3}+\dfrac{1}{y^3}\right)(x^3y^3)}$

$= \dfrac{x^2y^3+x^3y^2}{y^3+x^3}$

$= \dfrac{x^2y^2\cancel{(y+x)}}{\cancel{(y+x)}(y^2-yx+x^2)}$

$= \dfrac{x^2y^2}{y^2-yx+x^2}$

57. $$\frac{\frac{1}{x-3}+\frac{2}{x+3}}{\frac{3}{x-1}-\frac{4}{x+2}} = \frac{\frac{1}{x-3}\cdot\frac{x+3}{x+3}+\frac{2}{x+3}\cdot\frac{x-3}{x-3}}{\frac{3}{x-1}\cdot\frac{x+2}{x+2}-\frac{4}{x+2}\cdot\frac{x-1}{x-1}}$$

$$= \frac{\frac{x+3+2(x-3)}{(x-3)(x+3)}}{\frac{3(x+2)-4(x-1)}{(x-1)(x+2)}}$$

$$= \frac{\frac{x+3+2x-6}{(x-3)(x+3)}}{\frac{3x+6-4x+4}{(x-1)(x+2)}}$$

$$= \frac{\frac{3x-3}{(x-3)(x+3)}}{\frac{-x+10}{(x-1)(x+2)}}$$

$$= \frac{3x-3}{(x-3)(x+3)}\cdot\frac{(x-1)(x+2)}{-x+10}$$

$$= \frac{(3x-3)(x-1)(x+2)}{(x-3)(x+3)(-x+10)}, \text{ or}$$

$$\frac{3(x-1)^2(x+2)}{(x-3)(x+3)(-x+10)}$$

58. $$\frac{\frac{5}{x+1}-\frac{3}{x-2}}{\frac{1}{x-5}+\frac{2}{x+2}} = \frac{\frac{5(x-2)-3(x+1)}{(x+1)(x-2)}}{\frac{x+2+2(x-5)}{(x-5)(x+2)}}$$

$$= \frac{\frac{5x-10-3x-3}{(x+1)(x-2)}}{\frac{x+2+2x-10}{(x-5)(x+2)}}$$

$$= \frac{\frac{2x-13}{(x+1)(x-2)}}{\frac{3x-8}{(x-5)(x+2)}}$$

$$= \frac{2x-13}{(x+1)(x-2)}\cdot\frac{(x-5)(x+2)}{3x-8}$$

$$= \frac{(2x-13)(x-5)(x+2)}{(x+1)(x-2)(3x-8)}$$

59. $$\frac{\frac{a}{1-a}+\frac{1+a}{a}}{\frac{1-a}{a}+\frac{a}{1+a}} = \frac{\frac{a}{1-a}\cdot\frac{a}{a}+\frac{1+a}{a}\cdot\frac{1-a}{1-a}}{\frac{1-a}{a}\cdot\frac{1+a}{1+a}+\frac{a}{1+a}\cdot\frac{a}{a}}$$

$$= \frac{\frac{a^2+(1-a^2)}{a(1-a)}}{\frac{(1-a^2)+a^2}{a(1+a)}}$$

$$= \frac{1}{\not{a}(1-a)}\cdot\frac{\not{a}(1+a)}{1}$$

$$= \frac{1+a}{1-a}$$

60. $$\frac{\frac{1-x}{x}+\frac{x}{1+x}}{\frac{1+x}{x}+\frac{x}{1-x}} = \frac{\frac{1-x^2+x^2}{x(1+x)}}{\frac{1-x^2+x^2}{x(1-x)}}$$

$$= \frac{1}{x(1+x)}\cdot\frac{x(1-x)}{1}$$

$$= \frac{\not{x}(1-x)}{\not{x}(1+x)}$$

$$= \frac{1-x}{1+x}$$

61. $$\frac{\frac{1}{a^2}+\frac{2}{ab}+\frac{1}{b^2}}{\frac{1}{a^2}-\frac{1}{b^2}} = \frac{\frac{1}{a^2}+\frac{2}{ab}+\frac{1}{b^2}}{\frac{1}{a^2}-\frac{1}{b^2}}\cdot\frac{a^2b^2}{a^2b^2},$$

LCM is a^2b^2

$$= \frac{b^2+2ab+a^2}{b^2-a^2}$$

$$= \frac{\cancel{(b+a)}(b+a)}{\cancel{(b+a)}(b-a)}$$

$$= \frac{b+a}{b-a}$$

62. $$\frac{\frac{1}{x^2}-\frac{1}{y^2}}{\frac{1}{x^2}-\frac{2}{xy}+\frac{1}{y^2}} = \frac{y^2-x^2}{y^2-2xy+x^2}$$

Multiplying by $\frac{x^2y^2}{x^2y^2}$

$$= \frac{(y+x)\cancel{(y-x)}}{(y-x)\cancel{(y-x)}}$$

$$= \frac{y+x}{y-x}$$

63. When the least common denominator is used, the multiplication in the numerators is often simpler and there is usually less simplification required after the addition or subtraction is performed.

64. When there are three or more different binomial denominators, as in Exercise 53, Method 2 is usually preferable. Otherwise, some might prefer Method 1 while others will prefer Method 2.

65. $$\frac{(x+h)^2-x^2}{h} = \frac{x^2+2xh+h^2-x^2}{h}$$

$$= \frac{2xh+h^2}{h}$$

$$= \frac{\not{h}(2x+h)}{\not{h}\cdot 1}$$

$$= 2x+h$$

66. $$\frac{\frac{1}{x+h}-\frac{1}{x}}{h} = \frac{\frac{x-x-h}{x(x+h)}}{h} = \frac{-h}{x(x+h)}\cdot\frac{1}{h} = \frac{-1\cdot \cancel{h}}{x\cancel{h}(x+h)} = \frac{-1}{x(x+h)}$$

67. $$\frac{(x+h)^3-x^3}{h} = \frac{x^3+3x^2h+3xh^2+h^3-x^3}{h} = \frac{3x^2h+3xh^2+h^3}{h} = \frac{\cancel{h}(3x^2+3xh+h^2)}{\cancel{h}\cdot 1} = 3x^2+3xh+h^2$$

68. $$\frac{\frac{1}{(x+h)^2}-\frac{1}{x^2}}{h} = \frac{\frac{x^2-x^2-2xh-h^2}{x^2(x+h)^2}}{h} = \frac{-2xh-h^2}{x^2(x+h)^2}\cdot\frac{1}{h} = \frac{\cancel{h}(-2x-h)}{x^2\cancel{h}(x+h)^2} = \frac{-2x-h}{x^2(x+h)^2}$$

69. $$\left[\frac{\frac{x+1}{x-1}+1}{\frac{x+1}{x-1}-1}\right]^5 = \left[\frac{\frac{(x+1)+(x-1)}{x-1}}{\frac{(x+1)-(x-1)}{x-1}}\right]^5 = \left[\frac{2x}{x-1}\cdot\frac{x-1}{2}\right]^5 = \left[\frac{\cancel{2}x(\cancel{x-1})}{1\cdot\cancel{2}(\cancel{x-1})}\right]^5 = x^5$$

70. $$1+\cfrac{1}{1+\cfrac{1}{1+\cfrac{1}{1+\cfrac{1}{x}}}} = 1+\cfrac{1}{1+\cfrac{1}{1+\cfrac{1}{\frac{x+1}{x}}}} = 1+\cfrac{1}{1+\cfrac{1}{1+\frac{x}{x+1}}} = 1+\cfrac{1}{1+\cfrac{1}{\frac{2x+1}{x+1}}} = 1+\cfrac{1}{1+\frac{x+1}{2x+1}} = 1+\cfrac{1}{\frac{3x+2}{2x+1}} = 1+\frac{2x+1}{3x+2} = \frac{5x+3}{3x+2}$$

71. $$\frac{n(n+1)(n+2)}{2\cdot 3}+\frac{(n+1)(n+2)}{2}$$
$$= \frac{n(n+1)(n+2)}{2\cdot 3}+\frac{(n+1)(n+2)}{2}\cdot\frac{3}{3},$$ LCD is $2\cdot 3$
$$= \frac{n(n+1)(n+2)+3(n+1)(n+2)}{2\cdot 3}$$
$$= \frac{(n+1)(n+2)(n+3)}{2\cdot 3}$$ Factoring the numerator by grouping

72. $$\frac{n(n+1)(n+2)(n+3)}{2\cdot 3\cdot 4}+\frac{(n+1)(n+2)(n+3)}{2\cdot 3}$$
$$= \frac{n(n+1)(n+2)(n+3)+4(n+1)(n+2)(n+3)}{2\cdot 3\cdot 4},$$ LCD is $2\cdot 3\cdot 4$
$$= \frac{(n+1)(n+2)(n+3)(n+4)}{2\cdot 3\cdot 4}$$

73. $\frac{x^2-9}{x^3+27}\cdot\frac{5x^2-15x+45}{x^2-2x-3}+\frac{x^2+x}{4+2x}$

$=\frac{(x+3)(x-3)(5)(x^2-3x+9)}{(x+3)(x^2-3x+9)(x-3)(x+1)}+\frac{x^2+x}{4+2x}$

$=\frac{(x+3)(x-3)(x^2-3x+9)}{(x+3)(x-3)(x^2-3x+9)}\cdot\frac{5}{x+1}+\frac{x^2+x}{4+2x}$

$=1\cdot\frac{5}{x+1}+\frac{x^2+x}{4+2x}$

$=\frac{5}{x+1}+\frac{x^2+x}{2(2+x)}$

$=\frac{5\cdot 2(2+x)+(x^2+x)(x+1)}{2(x+1)(2+x)}$

$=\frac{20+10x+x^3+2x^2+x}{2(x+1)(2+x)}$

$=\frac{x^3+2x^2+11x+20}{2(x+1)(2+x)}$

74. $\frac{x^2+2x-3}{x^2-x-12}\div\frac{x^2-1}{x^2-16}-\frac{2x+1}{x^2+2x+1}$

$=\frac{x^2+2x-3}{x^2-x-12}\cdot\frac{x^2-16}{x^2-1}-\frac{2x+1}{x^2+2x+1}$

$=\frac{(x+3)(x-1)(x+4)(x-4)}{(x-4)(x+3)(x+1)(x-1)}-\frac{2x+1}{x^2+2x+1}$

$=\frac{x+4}{x+1}-\frac{2x+1}{(x+1)(x+1)}$

$=\frac{(x+4)(x+1)-(2x+1)}{(x+1)(x+1)}$

$=\frac{x^2+5x+4-2x-1}{(x+1)(x+1)}$

$=\frac{x^2+3x+3}{(x+1)^2}$

Exercise Set R.6

1. $\sqrt{(-11)^2}=|-11|=11$

2. $\sqrt{(-1)^2}=|-1|=1$

3. $\sqrt{16y^2}=\sqrt{(4y)^2}=|4y|=4|y|$

4. $\sqrt{36t^2}=|6t|=6|t|$

5. $\sqrt{(b+1)^2}=|b+1|$

6. $\sqrt{(2c-3)^2}=|2c-3|$

7. $\sqrt[3]{-27x^3}=\sqrt[3]{(-3x)^3}=-3x$

8. $\sqrt[3]{-8y^3}=-2y$

9. $\sqrt[4]{81x^8}=\sqrt[4]{(3x^2)^4}=|3x^2|=3x^2$

10. $\sqrt[4]{16z^{12}}=|2z^3|=2|z^3|=2z^2|z|$

11. $\sqrt[5]{32}=\sqrt[5]{2^5}=2$

12. $\sqrt[5]{-32}=-2$

13. $\sqrt{180}=\sqrt{36\cdot 5}=\sqrt{36}\cdot\sqrt{5}=6\sqrt{5}$

14. $\sqrt{48}=\sqrt{16\cdot 3}=4\sqrt{3}$

15. $\sqrt{72}=\sqrt{36\cdot 2}=\sqrt{36}\cdot\sqrt{2}=6\sqrt{2}$

16. $\sqrt{250}=\sqrt{25\cdot 10}=5\sqrt{10}$

17. $\sqrt[3]{54}=\sqrt[3]{27\cdot 2}=\sqrt[3]{27}\cdot\sqrt[3]{2}=3\sqrt[3]{2}$

18. $\sqrt[3]{135}=\sqrt[3]{27\cdot 5}=3\sqrt[3]{5}$

19. $\sqrt{128c^2d^4}=\sqrt{64c^2d^4\cdot 2}=|8cd^2|\sqrt{2}=8\sqrt{2}\,|c|d^2$

20. $\sqrt{162c^4d^6}=\sqrt{81c^4\cdot d^6\cdot 2}=9c^2|d^3|\sqrt{2}=$
$9\sqrt{2}c^2d^2|d|$

21. $\sqrt[4]{48x^6y^4}=\sqrt[4]{16x^4y^4\cdot 3x^2}=|2xy|\sqrt[4]{3x^2}=$
$2|x||y|\sqrt[4]{3x^2}$

22. $\sqrt[4]{243m^5n^{10}}=\sqrt[4]{81m^4n^8\cdot 3mn^2}=3|m|n^2\sqrt[4]{3mn^2}$

23. $\sqrt{x^2-4x+4}=\sqrt{(x-2)^2}=|x-2|$

24. $\sqrt{x^2+16x+64}=\sqrt{(x+8)^2}=|x+8|$

25. $\sqrt{10}\sqrt{30}=\sqrt{10\cdot 30}=\sqrt{300}=\sqrt{100\cdot 3}=$
$\sqrt{100}\cdot\sqrt{3}=10\sqrt{3}$

26. $\sqrt{28}\sqrt{14}=\sqrt{28\cdot 14}=\sqrt{14\cdot 2\cdot 14}=14\sqrt{2}$

27. $\sqrt{12}\sqrt{33}=\sqrt{12\cdot 33}=\sqrt{3\cdot 4\cdot 3\cdot 11}=\sqrt{3^2\cdot 4\cdot 11}=$
$3\cdot 2\sqrt{11}=6\sqrt{11}$

28. $\sqrt{15}\sqrt{35}=\sqrt{15\cdot 35}=\sqrt{3\cdot 5\cdot 5\cdot 7}=5\sqrt{21}$

29. $\sqrt{2x^3y}\sqrt{12xy}=\sqrt{24x^4y^2}=\sqrt{4x^4y^2\cdot 6}=2x^2y\sqrt{6}$

30. $\sqrt{3y^4z}\sqrt{20z}=\sqrt{60y^4z^2}=\sqrt{4y^4z^2\cdot 15}=2y^2z\sqrt{15}$

31. $\sqrt[3]{3x^2y}\sqrt[3]{36x}=\sqrt[3]{108x^3y}=\sqrt[3]{27x^3\cdot 4y}=3x\sqrt[3]{4y}$

32. $\sqrt[5]{8x^3y^4}\sqrt[5]{4x^4y}=\sqrt[5]{32x^7y^5}=2xy\sqrt[5]{x^2}$

33. $\sqrt[3]{2(x+4)}\sqrt[3]{4(x+4)^4}=\sqrt[3]{8(x+4)^5}$
$=\sqrt[3]{8(x+4)^3\cdot(x+4)^2}$
$=2(x+4)\sqrt[3]{(x+4)^2}$

34. $\sqrt[3]{4(x+1)^2}\sqrt[3]{18(x+1)^2}=\sqrt[3]{72(x+1)^4}$
$=\sqrt[3]{8(x+1)^3\cdot 9(x+1)}$
$=2(x+1)\sqrt[3]{9(x+1)}$

35. $\sqrt[6]{\frac{m^{12}n^{24}}{64}}=\sqrt[6]{\left(\frac{m^2n^4}{2}\right)^6}=\frac{m^2n^4}{2}$

36. $\sqrt[8]{\frac{m^{16}n^{24}}{2^8}}=\frac{m^2n^3}{2}$

37. $\dfrac{\sqrt[3]{40m}}{\sqrt[3]{5m}} = \sqrt[3]{\dfrac{40m}{5m}} = \sqrt[3]{8} = 2$

38. $\dfrac{\sqrt{40xy}}{\sqrt{8x}} = \sqrt{\dfrac{40xy}{8x}} = \sqrt{5y}$

39. $\dfrac{\sqrt[3]{3x^2}}{\sqrt[3]{24x^5}} = \sqrt[3]{\dfrac{3x^2}{24x^5}} = \sqrt[3]{\dfrac{1}{8x^3}} = \dfrac{1}{2x}$

40. $\dfrac{\sqrt{128a^2b^4}}{\sqrt{16ab}} = \sqrt{\dfrac{128a^2b^4}{16ab}} = \sqrt{8ab^3} =$
$\sqrt{4 \cdot 2 \cdot a \cdot b^2 \cdot b} = 2b\sqrt{2ab}$

41. $\sqrt[3]{\dfrac{64a^4}{27b^3}} = \sqrt[3]{\dfrac{64 \cdot a^3 \cdot a}{27 \cdot b^3}}$
$= \dfrac{\sqrt[3]{64a^3}\sqrt[3]{a}}{\sqrt[3]{27b^3}}$
$= \dfrac{4a\sqrt[3]{a}}{3b}$

42. $\sqrt{\dfrac{9x^7}{16y^8}} = \dfrac{\sqrt{9 \cdot x^6 \cdot x}}{\sqrt{16 \cdot y^8}} = \dfrac{3x^3\sqrt{x}}{4y^4}$

43. $\sqrt{\dfrac{7x^3}{36y^6}} = \sqrt{\dfrac{7 \cdot x^2 \cdot x}{36 \cdot y^6}}$
$= \dfrac{\sqrt{x^2}\sqrt{7x}}{\sqrt{36y^6}}$
$= \dfrac{x\sqrt{7x}}{6y^3}$

44. $\sqrt[3]{\dfrac{2yz}{250z^4}} = \sqrt[3]{\dfrac{y}{125z^3}} = \dfrac{\sqrt[3]{y}}{\sqrt[3]{125z^3}} = \dfrac{\sqrt[3]{y}}{5z}$

45. $9\sqrt{50} + 6\sqrt{2} = 9\sqrt{25 \cdot 2} + 6\sqrt{2}$
$= 9 \cdot 5\sqrt{2} + 6\sqrt{2}$
$= 45\sqrt{2} + 6\sqrt{2}$
$= (45 + 6)\sqrt{2}$
$= 51\sqrt{2}$

46. $11\sqrt{27} - 4\sqrt{3} = 11 \cdot 3\sqrt{3} - 4\sqrt{3} = 33\sqrt{3} - 4\sqrt{3} =$
$29\sqrt{3}$

47. $6\sqrt{20} - 4\sqrt{45} + \sqrt{80} = 6\sqrt{4 \cdot 5} - 4\sqrt{9 \cdot 5} + \sqrt{16 \cdot 5}$
$= 6 \cdot 2\sqrt{5} - 4 \cdot 3\sqrt{5} + 4\sqrt{5}$
$= 12\sqrt{5} - 12\sqrt{5} + 4\sqrt{5}$
$= (12 - 12 + 4)\sqrt{5}$
$= 4\sqrt{5}$

48. $2\sqrt{32} + 3\sqrt{8} - 4\sqrt{18} = 2 \cdot 4\sqrt{2} + 3 \cdot 2\sqrt{2} - 4 \cdot 3\sqrt{2} =$
$8\sqrt{2} + 6\sqrt{2} - 12\sqrt{2} = 2\sqrt{2}$

49. $8\sqrt{2x^2} - 6\sqrt{20x} - 5\sqrt{8x^2}$
$= 8x\sqrt{2} - 6\sqrt{4 \cdot 5x} - 5\sqrt{4x^2 \cdot 2}$
$= 8x\sqrt{2} - 6 \cdot 2\sqrt{5x} - 5 \cdot 2x\sqrt{2}$
$= 8x\sqrt{2} - 12\sqrt{5x} - 10x\sqrt{2}$
$= -2x\sqrt{2} - 12\sqrt{5x}$

50. $2\sqrt[3]{8x^2} + 5\sqrt[3]{27x^2} - 3\sqrt{x^3}$
$= 4\sqrt[3]{x^2} + 15\sqrt[3]{x^2} - 3x\sqrt{x}$
$= 19\sqrt[3]{x^2} - 3x\sqrt{x}$

51. $(\sqrt{3} - \sqrt{2})(\sqrt{3} + \sqrt{2})$
$= (\sqrt{3})^2 - (\sqrt{2})^2$
$= 3 - 2$
$= 1$

52. $(\sqrt{8} + 2\sqrt{5})(\sqrt{8} - 2\sqrt{5}) = 8 - 4 \cdot 5 = -12$

53. $(2\sqrt{3} + \sqrt{5})(\sqrt{3} - 3\sqrt{5})$
$= 2\sqrt{3} \cdot \sqrt{3} - 2\sqrt{3} \cdot 3\sqrt{5} + \sqrt{5} \cdot \sqrt{3} - \sqrt{5} \cdot 3\sqrt{5}$
$= 2 \cdot 3 - 6\sqrt{15} + \sqrt{15} - 3 \cdot 5$
$= 6 - 6\sqrt{15} + \sqrt{15} - 15$
$= -9 - 5\sqrt{15}$

54. $(\sqrt{6} - 4\sqrt{7})(3\sqrt{6} + 2\sqrt{7})$
$= 3 \cdot 6 + 2\sqrt{42} - 12\sqrt{42} - 8 \cdot 7$
$= 18 + 2\sqrt{42} - 12\sqrt{42} - 56$
$= -38 - 10\sqrt{42}$

55. $(1 + \sqrt{3})^2 = 1^2 + 2 \cdot 1 \cdot \sqrt{3} + (\sqrt{3})^2$
$= 1 + 2\sqrt{3} + 3$
$= 4 + 2\sqrt{3}$

56. $(\sqrt{2} - 5)^2 = 2 - 10\sqrt{2} + 25 = 27 - 10\sqrt{2}$

57. $(\sqrt{5} - \sqrt{6})^2 = (\sqrt{5})^2 - 2\sqrt{5} \cdot \sqrt{6} + (\sqrt{6})^2$
$= 5 - 2\sqrt{30} + 6$
$= 11 - 2\sqrt{30}$

58. $(\sqrt{3} + \sqrt{2})^2 = 3 + 2\sqrt{6} + 2 = 5 + 2\sqrt{6}$

59. We use the Pythagorean theorem to find b, the airplane's horizontal distance from the airport. We have $a = 3700$ and $c = 14,200$.

$$c^2 = a^2 + b^2$$
$$14,200^2 = 3700^2 + b^2$$
$$201,640,000 = 13,690,000 + b^2$$
$$187,950,000 = b^2$$
$$13,709.5 \approx b$$

The airplane is about 13,709.5 ft horizontally from the airport.

60.

c
a
b

2 mi $= 2 \cdot 5280$ ft $= 10,560$ ft

2 mi $+ 2$ ft $= 10,562$ ft

Then $b = \dfrac{1}{2} \cdot 10,560$ ft $= 5280$ ft and

$c = \dfrac{1}{2} \cdot 10,562$ ft $= 5281$ ft.

$$a^2 = c^2 - [illegible]$$
$$a^2 = (52[illegible] - (5280)^2$$
$$a^2 = 10,[illegible]$$
$$a \approx 102[illegible] \text{ ft}$$

61. a) $h^2 + \left(\frac{a}{2}\right)^2 = a^2$ Pythagorean theorem

$$h^2 + \frac{a^2}{4} = a^2$$
$$h^2 = \frac{3a^2}{4}$$
$$h = \sqrt{\frac{3a^2}{4}}$$
$$h = \frac{a}{2}\sqrt{3}$$

b) Using the result of part (a) we have

$$A = \frac{1}{2} \cdot \text{base} \cdot \text{height}$$
$$A = \frac{1}{2}a \cdot \frac{a}{2}\sqrt{3} \quad \left(\frac{a}{2} + \frac{a}{2} = a\right)$$
$$A = \frac{a^2}{4}\sqrt{3}$$

62. $c^2 = s^2 + s^2$

$$c^2 = 2s^2$$
$$c = s\sqrt{2} \quad \text{Length of third side}$$

63.

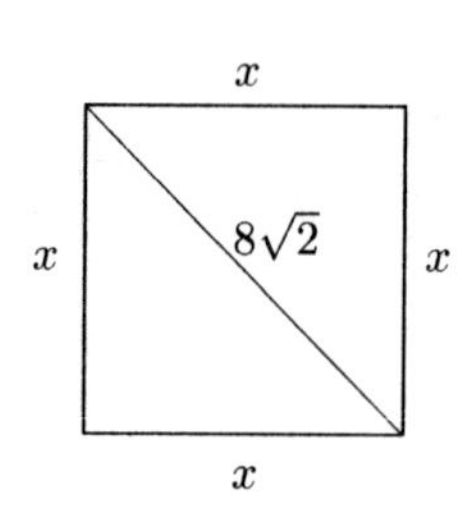

$$x^2 + x^2 = (8\sqrt{2})^2 \quad \text{Pythagorean theorem}$$
$$2x^2 = 128$$
$$x^2 = 64$$
$$x = 8$$

64.

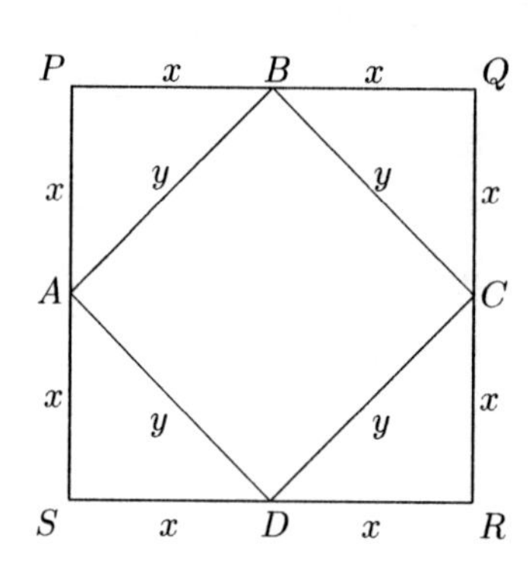

$$(2x)^2 = 100$$
$$4x^2 = 100$$
$$x^2 = 25$$
$$x = 5$$
$$A = y^2 = x^2 + x^2 = 5^2 + 5^2 = 25 + 25 = 50 \text{ ft}^2$$

65. $\sqrt{\frac{2}{3}} = \sqrt{\frac{2}{3} \cdot \frac{3}{3}} = \sqrt{\frac{6}{9}} = \frac{\sqrt{6}}{\sqrt{9}} = \frac{\sqrt{6}}{3}$

66. $\sqrt{\frac{3}{7}} = \sqrt{\frac{3}{7} \cdot \frac{7}{7}} = \sqrt{\frac{21}{49}} = \frac{\sqrt{21}}{7}$

67. $\frac{\sqrt[3]{5}}{\sqrt[3]{4}} = \frac{\sqrt[3]{5}}{\sqrt[3]{4}} \cdot \frac{\sqrt[3]{2}}{\sqrt[3]{2}} = \frac{\sqrt[3]{10}}{\sqrt[3]{8}} = \frac{\sqrt[3]{10}}{2}$

68. $\frac{\sqrt[3]{7}}{\sqrt[3]{25}} = \frac{\sqrt[3]{7}}{\sqrt[3]{25}} \cdot \frac{\sqrt[3]{5}}{\sqrt[3]{5}} = \frac{\sqrt[3]{35}}{\sqrt[3]{125}} = \frac{\sqrt[3]{35}}{5}$

69. $\sqrt[3]{\frac{16}{9}} = \sqrt[3]{\frac{16}{9} \cdot \frac{3}{3}} = \sqrt[3]{\frac{48}{27}} = \frac{\sqrt[3]{48}}{\sqrt[3]{27}} =$

$$\frac{\sqrt[3]{8 \cdot 6}}{3} = \frac{2\sqrt[3]{6}}{3}$$

70. $\sqrt[3]{\frac{3}{5}} = \sqrt[3]{\frac{3}{5} \cdot \frac{25}{25}} = \sqrt[3]{\frac{75}{125}} = \frac{\sqrt[3]{75}}{5}$

71. $\frac{6}{3+\sqrt{5}} = \frac{6}{3+\sqrt{5}} \cdot \frac{3-\sqrt{5}}{3-\sqrt{5}}$

$$= \frac{6(3-\sqrt{5})}{9-5}$$
$$= \frac{6(3-\sqrt{5})}{4}$$
$$= \frac{3(3-\sqrt{5})}{2} = \frac{9-3\sqrt{5}}{2}$$

72. $\frac{2}{\sqrt{3}-1} = \frac{2}{\sqrt{3}-1} \cdot \frac{\sqrt{3}+1}{\sqrt{3}+1}$

$$= \frac{\cancel{2}(\sqrt{3}+1)}{\cancel{2} \cdot 1}$$
$$= \sqrt{3}+1$$

73. $\frac{1-\sqrt{2}}{2\sqrt{3}-\sqrt{6}} = \frac{1-\sqrt{2}}{2\sqrt{3}-\sqrt{6}} \cdot \frac{2\sqrt{3}+\sqrt{6}}{2\sqrt{3}+\sqrt{6}}$
$= \frac{2\sqrt{3}+\sqrt{6}-2\sqrt{6}-\sqrt{12}}{4\cdot 3-6}$
$= \frac{2\sqrt{3}+\sqrt{6}-2\sqrt{6}-2\sqrt{3}}{12-6}$
$= \frac{-\sqrt{6}}{6}$, or $-\frac{\sqrt{6}}{6}$

74. $\frac{\sqrt{5}+4}{\sqrt{2}+3\sqrt{7}} = \frac{\sqrt{5}+4}{\sqrt{2}+3\sqrt{7}} \cdot \frac{\sqrt{2}-3\sqrt{7}}{\sqrt{2}-3\sqrt{7}}$
$= \frac{\sqrt{10}-3\sqrt{35}+4\sqrt{2}-12\sqrt{7}}{2-9\cdot 7}$
$= \frac{\sqrt{10}-3\sqrt{35}+4\sqrt{2}-12\sqrt{7}}{-61}$

75. $\frac{6}{\sqrt{m}-\sqrt{n}} = \frac{6}{\sqrt{m}-\sqrt{n}} \cdot \frac{\sqrt{m}+\sqrt{n}}{\sqrt{m}+\sqrt{n}}$
$= \frac{6(\sqrt{m}+\sqrt{n})}{(\sqrt{m})^2-(\sqrt{n})^2}$
$= \frac{6\sqrt{m}+6\sqrt{n}}{m-n}$

76. $\frac{3}{\sqrt{v}+\sqrt{w}} = \frac{3}{\sqrt{v}+\sqrt{w}} \cdot \frac{\sqrt{v}-\sqrt{w}}{\sqrt{v}-\sqrt{w}} = \frac{3\sqrt{v}-3\sqrt{w}}{v-w}$

77. $\frac{\sqrt{12}}{5} = \frac{\sqrt{12}}{5} \cdot \frac{\sqrt{3}}{\sqrt{3}} = \frac{\sqrt{36}}{5\sqrt{3}} = \frac{6}{5\sqrt{3}}$

78. $\frac{\sqrt{50}}{3} = \frac{\sqrt{50}}{3} \cdot \frac{\sqrt{2}}{\sqrt{2}} = \frac{\sqrt{100}}{3\sqrt{2}} = \frac{10}{3\sqrt{2}}$

79. $\sqrt[3]{\frac{7}{2}} = \sqrt[3]{\frac{7}{2}\cdot\frac{49}{49}} = \sqrt[3]{\frac{343}{98}} = \frac{\sqrt[3]{343}}{\sqrt[3]{98}} = \frac{7}{\sqrt[3]{98}}$

80. $\sqrt[3]{\frac{2}{5}} = \sqrt[3]{\frac{2}{5}\cdot\frac{4}{4}} = \sqrt[3]{\frac{8}{20}} = \frac{2}{\sqrt[3]{20}}$

81. $\frac{\sqrt{11}}{\sqrt{3}} = \frac{\sqrt{11}}{\sqrt{3}} \cdot \frac{\sqrt{11}}{\sqrt{11}} = \frac{\sqrt{121}}{\sqrt{33}} = \frac{11}{\sqrt{33}}$

82. $\frac{\sqrt{5}}{\sqrt{2}} = \frac{\sqrt{5}}{\sqrt{2}} \cdot \frac{\sqrt{5}}{\sqrt{5}} = \frac{\sqrt{25}}{\sqrt{10}} = \frac{5}{\sqrt{10}}$

83. $\frac{9-\sqrt{5}}{3-\sqrt{3}} = \frac{9-\sqrt{5}}{3-\sqrt{3}} \cdot \frac{9+\sqrt{5}}{9+\sqrt{5}}$
$= \frac{9^2-(\sqrt{5})^2}{27+3\sqrt{5}-9\sqrt{3}-\sqrt{15}}$
$= \frac{81-5}{27+3\sqrt{5}-9\sqrt{3}-\sqrt{15}}$
$= \frac{76}{27+3\sqrt{5}-9\sqrt{3}-\sqrt{15}}$

84. $\frac{8-\sqrt{6}}{5-\sqrt{2}} = \frac{8-\sqrt{6}}{5-\sqrt{2}} \cdot \frac{8+\sqrt{6}}{8+\sqrt{6}}$
$= \frac{64-6}{40+5\sqrt{6}-8\sqrt{2}-\sqrt{12}}$
$= \frac{58}{40+5\sqrt{6}-8\sqrt{2}-2\sqrt{3}}$

85. $\frac{\sqrt{a}+\sqrt{b}}{3a} = \frac{\sqrt{a}+\sqrt{b}}{3a} \cdot \frac{\sqrt{a}-\sqrt{b}}{\sqrt{a}-\sqrt{b}}$
$= \frac{(\sqrt{a})^2-(\sqrt{b})^2}{3a(\sqrt{a}-\sqrt{b})}$
$= \frac{a-b}{3a\sqrt{a}-3a\sqrt{b}}$

86. $\frac{\sqrt{p}-\sqrt{q}}{1+\sqrt{q}} = \frac{\sqrt{p}-\sqrt{q}}{1+\sqrt{q}} \cdot \frac{\sqrt{p}+\sqrt{q}}{\sqrt{p}+\sqrt{q}}$
$= \frac{p-q}{\sqrt{p}+\sqrt{q}+\sqrt{pq}+q}$

87. $x^{3/4} = \sqrt[4]{x^3}$

88. $y^{2/5} = \sqrt[5]{y^2}$

89. $16^{3/4} = (16^{1/4})^3 = (\sqrt[4]{16})^3 = 2^3 = 8$

90. $4^{7/2} = (\sqrt{4})^7 = 2^7 = 128$

91. $125^{-1/3} = \frac{1}{125^{1/3}} = \frac{1}{\sqrt[3]{125}} = \frac{1}{5}$

92. $32^{-4/5} = (\sqrt[5]{32})^{-4} = 2^{-4} = \frac{1}{16}$

93. $a^{5/4}b^{-3/4} = \frac{a^{5/4}}{b^{3/4}} = \frac{\sqrt[4]{a^5}}{\sqrt[4]{b^3}} = \frac{a\sqrt[4]{a}}{\sqrt[4]{b^3}}$, or $a\sqrt[4]{\frac{a}{b^3}}$

94. $x^{2/5}y^{-1/5} = \sqrt[5]{\frac{x^2}{y}}$

95. $m^{5/3}n^{7/3} = \sqrt[3]{m^5}\sqrt[3]{n^7} = \sqrt[3]{m^5n^7} = mn^2\sqrt[3]{m^2n}$

96. $p^{7/6}q^{11/6} = \sqrt[6]{p^7}\sqrt[6]{q^{11}} = \sqrt[6]{p^7q^{11}} = pq\sqrt[6]{pq^5}$

97. $(\sqrt[4]{13})^5 = \sqrt[4]{13^5} = 13^{5/4}$

98. $\sqrt[5]{17^3} = 17^{3/5}$

99. $\sqrt[3]{20^2} = 20^{2/3}$

100. $(\sqrt[5]{12})^4 = 12^{4/5}$

101. $\sqrt[3]{\sqrt{11}} = (\sqrt{11})^{1/3} = (11^{1/2})^{1/3} = 11^{1/6}$

102. $\sqrt[3]{\sqrt[4]{7}} = (7^{1/4})^{1/3} = 7^{1/12}$

103. $\sqrt{5}\sqrt[3]{5} = 5^{1/2}\cdot 5^{1/3} = 5^{1/2+1/3} = 5^{5/6}$

104. $\sqrt[3]{2}\sqrt{2} = 2^{1/3}\cdot 2^{1/2} = 2^{5/6}$

105. $\sqrt[5]{32^2} = 32^{2/5} = (32^{1/5})^2 = 2^2 = 4$

106. $\sqrt[3]{64^2} = 64^{2/3} = (64^{1/3})^2 = 4^2 = 16$

107. $(2a^{3/2})(4a^{1/2}) = 8a^{3/2+1/2} = 8a^2$

108. $(3a^{5/6})(8a^{2/3}) = 24a^{3/2} = 24a\sqrt{a}$

109. $\left(\frac{x^6}{9b^{-4}}\right)^{1/2} = \left(\frac{x^6}{3^2b^{-4}}\right)^{1/2} = \frac{x^3}{3b^{-2}}$, or $\frac{x^3b^2}{3}$

110. $\left(\frac{x^{2/3}}{4y^{-2}}\right)^{1/2} = \frac{x^{1/3}}{4^{1/2}y^{-1}} = \frac{\sqrt[3]{x}}{2y^{-1}}$, or $\frac{y\sqrt[3]{x}}{2}$

111. $\frac{x^{2/3}y^{5/6}}{x^{-1/3}y^{1/2}} = x^{2/3-(-1/3)}y^{5/6-1/2} = xy^{1/3} = x\sqrt[3]{y}$

112. $\frac{a^{1/2}b^{5/8}}{a^{1/4}b^{3/8}} = a^{1/4}b^{1/4} = \sqrt[4]{ab}$

113. $(m^{1/2}n^{5/2})^{2/3} = m^{\frac{1}{2}\cdot\frac{2}{3}}n^{\frac{5}{2}\cdot\frac{2}{3}} = m^{1/3}n^{5/3} =$
$\sqrt[3]{m}\sqrt[3]{n^5} = \sqrt[3]{mn^5} = n\sqrt[3]{mn^2}$

114. $(x^{5/3}y^{1/3}z^{2/3})^{3/5} = xy^{1/5}z^{2/5} = x\sqrt[5]{yz^2}$

115. $a^{3/4}(a^{2/3} + a^{4/3}) = a^{3/4+2/3} + a^{3/4+4/3} =$
$a^{17/12} + a^{25/12} = \sqrt[12]{a^{17}} + \sqrt[12]{a^{25}} =$
$a\sqrt[12]{a^5} + a^2\sqrt[12]{a}$

116. $m^{2/3}(m^{7/4} - m^{5/4}) = m^{29/12} - m^{23/12} =$
$\sqrt[12]{m^{29}} - \sqrt[12]{m^{23}} = m^2\sqrt[12]{m^5} - m\sqrt[12]{m^{11}}$

117. $\sqrt[3]{6}\sqrt{2} = 6^{1/3}2^{1/2} = 6^{2/6}2^{3/6}$
$= (6^2 2^3)^{1/6}$
$= \sqrt[6]{36\cdot 8}$
$= \sqrt[6]{288}$

118. $\sqrt{2}\sqrt[4]{8} = 2^{1/2}(2^3)^{1/4} = 2^{1/2}2^{3/4} = 2^{5/4} =$
$\sqrt[4]{2^5} = 2\sqrt[4]{2}$

119. $\sqrt[4]{xy}\sqrt[3]{x^2y} = (xy)^{1/4}(x^2y)^{1/3} = (xy)^{3/12}(x^2y)^{4/12}$
$= \left[(xy)^3(x^2y)^4\right]^{1/12}$
$= \left[x^3y^3x^8y^4\right]^{1/12}$
$= \sqrt[12]{x^{11}y^7}$

120. $\sqrt[3]{ab^2}\sqrt{ab} = (ab^2)^{1/3}(ab)^{1/2} = (ab^2)^{2/6}(ab)^{3/6} =$
$\sqrt[6]{(ab^2)^2(ab)^3} = \sqrt[6]{a^5b^7} = b\sqrt[6]{a^5b}$

121. $\sqrt[3]{a^4\sqrt{a^3}} = (a^4\sqrt{a^3})^{1/3} = (a^4a^{3/2})^{1/3}$
$= (a^{11/2})^{1/3}$
$= a^{11/6}$
$= \sqrt[6]{a^{11}}$
$= a\sqrt[6]{a^5}$

122. $\sqrt{a^3\sqrt[3]{a^2}} = (a^3\cdot a^{2/3})^{1/2} = (a^{11/3})^{1/2} = a^{11/6} =$
$\sqrt[6]{a^{11}} = a\sqrt[6]{a^5}$

123. $\frac{\sqrt{(a+x)^3}\sqrt[3]{(a+x)^2}}{\sqrt[4]{a+x}} = \frac{(a+x)^{3/2}(a+x)^{2/3}}{(a+x)^{1/4}}$
$= \frac{(a+x)^{26/12}}{(a+x)^{3/12}}$
$= (a+x)^{23/12}$
$= \sqrt[12]{(a+x)^{23}}$
$= (a+x)\sqrt[12]{(a+x)^{11}}$

124. $\frac{\sqrt[4]{(x+y)^2}\sqrt[3]{x+y}}{\sqrt{(x+y)^3}} = \frac{(x+y)^{2/4}(x+y)^{1/3}}{(x+y)^{3/2}} =$
$(x+y)^{-2/3} = \frac{1}{\sqrt[3]{(x+y)^2}}$

125. Choose specific values for a and b ($a \neq 0$, $b \neq 0$) and show that $\sqrt{a+b} \neq \sqrt{a} + \sqrt{b}$. For example let $a = 9$ and $b = 16$. Then $\sqrt{9+16} = \sqrt{25} = 5$, but $\sqrt{9} + \sqrt{16} = 3 + 4 = 7$.

126. Observe that $26 > 25$, so $\sqrt{26} > \sqrt{25}$, or $\sqrt{26} > 5$. Then $10\sqrt{26} - 50 > 10\cdot 5 - 50$, or $10\sqrt{26} - 50 > 0$.

127. $\sqrt{1+x^2} + \frac{1}{\sqrt{1+x^2}}$
$= \sqrt{1+x^2}\cdot\frac{1+x^2}{1+x^2} + \frac{1}{\sqrt{1+x^2}}\cdot\frac{\sqrt{1+x^2}}{\sqrt{1+x^2}}$
$= \frac{(1+x^2)\sqrt{1+x^2}}{1+x^2} + \frac{\sqrt{1+x^2}}{1+x^2}$
$= \frac{(2+x^2)\sqrt{1+x^2}}{1+x^2}$

128. $\sqrt{1-x^2} - \frac{x^2}{2\sqrt{1-x^2}}$
$= \sqrt{1-x^2} - \frac{x^2\sqrt{1-x^2}}{2(1-x^2)}$ Rationalizing the denominator
$= \frac{2(1-x^2)\sqrt{1-x^2} - x^2\sqrt{1-x^2}}{2(1-x^2)}$
$= \frac{(2-3x^2)\sqrt{1-x^2}}{2(1-x^2)}$

129. $\left(\sqrt{a\sqrt{a}}\right)^{\sqrt{a}} = \left(a^{\sqrt{a}/2}\right)^{\sqrt{a}} = a^{a/2}$

130. $\frac{(2a^3b^{5/4}c^{1/7})^4}{(54a^{-2}b^{2/3}c^{6/5})^{-1/3}} = \frac{16a^{12}b^5c^{4/7}}{54^{-1/3}a^{2/3}b^{-2/9}c^{-2/5}}$
$= 16\sqrt[3]{54}a^{34/3}b^{47/9}c^{34/35}$
$= 2^4\cdot 3\cdot 2^{1/3}a^{34/3}b^{47/9}c^{34/35}$
$= 3\cdot 2^{13/3}a^{34/3}b^{47/9}c^{34/35}$, or
$48\cdot 2^{1/3}a^{34/3}b^{47/9}c^{34/35}$

Chapter 1

Graphs, Functions, and Models

Exercise Set 1.1

1. To graph $(4, 0)$ we move from the origin 4 units to the right of the y-axis. Since the second coordinate is 0, we do not move up or down from the x-axis.

   To graph $(-3, -5)$ we move from the origin 3 units to the left of the y-axis. Then we move 5 units down from the x-axis.

   To graph $(-1, 4)$ we move from the origin 1 unit to the left of the y-axis. Then we move 4 units up from the x-axis.

   To graph $(0, 2)$ we do not move to the right or the left of the y-axis since the first coordinate is 0. From the origin we move 2 units up.

   To graph $(2, -2)$ we move from the origin 2 units to the right of the y-axis. Then we move 2 units down from the x-axis.

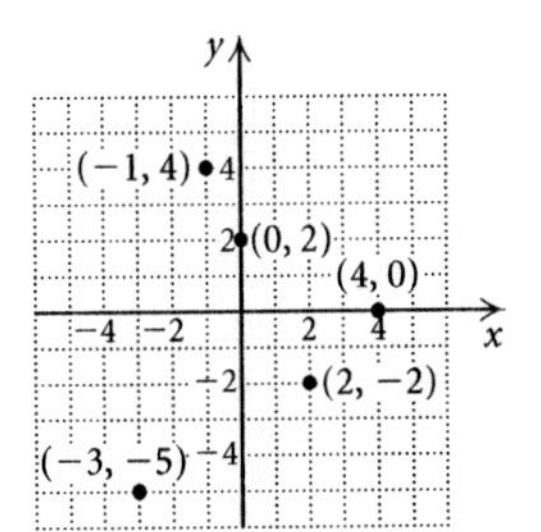

2.

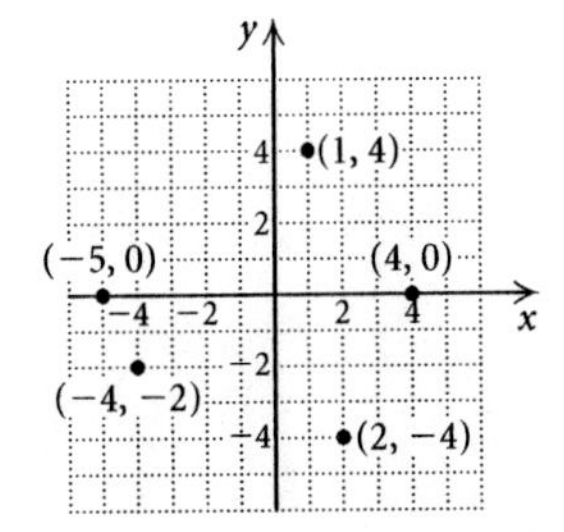

3. To graph $(-5, 1)$ we move from the origin 5 units to the left of the y-axis. Then we move 1 unit up from the x-axis.

   To graph $(5, 1)$ we move from the origin 5 units to the right of the y-axis. Then we move 1 unit up from the x-axis.

   To graph $(2, 3)$ we move from the origin 2 units to the right of the y-axis. Then we move 3 units up from the x-axis.

   To graph $(2, -1)$ we move from the origin 2 units to the right of the y-axis. Then we move 1 unit down from the x-axis.

   To graph $(0, 1)$ we do not move to the right or the left of the y-axis since the first coordinate is 0. From the origin we move 1 unit up.

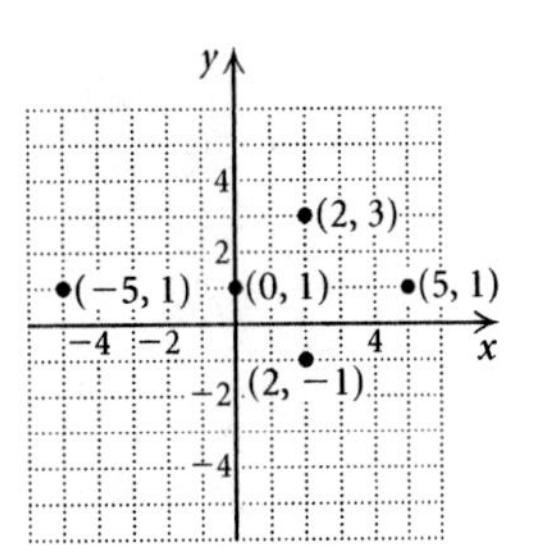

4.

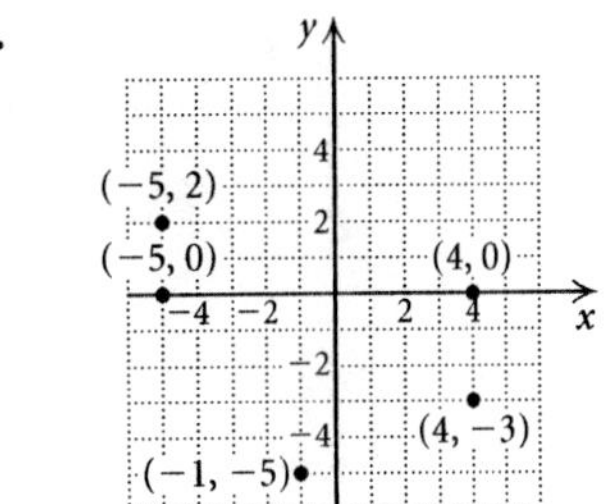

5. The first coordinate represents the year and the corresponding second coordinate represents the amount of sales. The ordered pairs are (1998, \$14 billion), (1999, \$19 billion), (2000, \$25 billion), (2001, \$32 billion), and (2002, \$38 billion).

6. (1996, 3.5 billion gallons), (1997, 3.8 billion gallons), (1998, 4.2 billion gallons), (1999, 4.6 billion gallons), (2000, 4.9 billion gallons), (2001, 5.5 billion gallons)

7. To determine whether $(1, -1)$ is a solution, substitute 1 for x and -1 for y.

$$\begin{array}{c|c} \multicolumn{2}{c}{y = 2x - 3} \\ \hline -1 \ ? & 2 \cdot 1 - 3 \\ & 2 - 3 \\ -1 & -1 \quad \text{TRUE} \end{array}$$

   The equation $-1 = -1$ is true, so $(1, -1)$ is a solution.

   To determine whether $(0, 3)$ is a solution, substitute 0 for x and -3 for y.

$$\begin{array}{c|c} \multicolumn{2}{c}{y = 2x - 3} \\ \hline 3 \ ? & 2 \cdot 0 - 3 \\ & 0 - 3 \\ 3 & -3 \quad \text{FALSE} \end{array}$$

   The equation $3 = -3$ is false, so $(0, 3)$ is not a solution.

8. For $(2, 5)$:

$$\begin{array}{c|c} \multicolumn{2}{c}{y = 3x - 1} \\ \hline 5 \ ? & 3 \cdot 2 - 1 \\ & 6 - 1 \\ 5 & 5 \quad \text{TRUE} \end{array}$$

   $(2, 5)$ is a solution.

For $(-2,-5)$:

$$\begin{array}{r|l} \multicolumn{2}{c}{y = 3x - 1} \\ \hline -5 & ?\ 3(-2) - 1 \\ & -6 - 1 \\ -5 & -7 \quad \text{FALSE} \end{array}$$

$(-2,-5)$ is not a solution.

9. To determine whether $\left(\frac{2}{3}, \frac{3}{4}\right)$ is a solution, substitute $\frac{2}{3}$ for x and $\frac{3}{4}$ for y.

$$\begin{array}{r|l} \multicolumn{2}{c}{6x - 4y = 1} \\ \hline 6\cdot\frac{2}{3} - 4\cdot\frac{3}{4} & ?\ 1 \\ 4 - 3 & \\ 1 & 1 \quad \text{TRUE} \end{array}$$

The equation $1 = 1$ is true, so $\left(\frac{2}{3}, \frac{3}{4}\right)$ is a solution.

To determine whether $\left(1, \frac{3}{2}\right)$ is a solution, substitute 1 for x and $\frac{3}{2}$ for y.

$$\begin{array}{r|l} \multicolumn{2}{c}{6x - 4y = 1} \\ \hline 6\cdot 1 - 4\cdot\frac{3}{2} & ?\ 1 \\ 6 - 6 & \\ 0 & 1 \quad \text{FALSE} \end{array}$$

The equation $0 = 1$ is false, so $\left(1, \frac{3}{2}\right)$ is not a solution.

10. For $(1.5, 2.6)$:

$$\begin{array}{r|l} \multicolumn{2}{c}{x^2 + y^2 = 9} \\ \hline (1.5)^2 + (2.6)^2 & ?\ 9 \\ 2.25 + 6.76 & \\ 9.01 & 9 \quad \text{FALSE} \end{array}$$

$(1.5, 2.6)$ is not a solution.

For $(-3, 0)$:

$$\begin{array}{r|l} \multicolumn{2}{c}{x^2 + y^2 = 9} \\ \hline (-3)^2 + 0^2 & ?\ 9 \\ 9 + 0 & \\ 9 & 9 \quad \text{TRUE} \end{array}$$

$(-3, 0)$ is a solution.

11. To determine whether $\left(-\frac{1}{2}, -\frac{4}{5}\right)$ is a solution, substitute $-\frac{1}{2}$ for a and $-\frac{4}{5}$ for b.

$$\begin{array}{r|l} \multicolumn{2}{c}{2a + 5b = 3} \\ \hline 2\left(-\frac{1}{2}\right) + 5\left(-\frac{4}{5}\right) & ?\ 3 \\ -1 - 4 & \\ -5 & 3 \quad \text{FALSE} \end{array}$$

The equation $-5 = 3$ is false, so $\left(-\frac{1}{2}, -\frac{4}{5}\right)$ is not a solution.

To determine whether $\left(0, \frac{3}{5}\right)$ is a solution, substitute 0 for a and $\frac{3}{5}$ for b.

$$\begin{array}{r|l} \multicolumn{2}{c}{2a + 5b = 3} \\ \hline 2\cdot 0 + 5\cdot\frac{3}{5} & ?\ 3 \\ 0 + 3 & \\ 3 & 3 \quad \text{TRUE} \end{array}$$

The equation $3 = 3$ is true, so $\left(0, \frac{3}{5}\right)$ is a solution.

12. For $\left(0, \frac{3}{2}\right)$:

$$\begin{array}{r|l} \multicolumn{2}{c}{3m + 4n = 6} \\ \hline 3\cdot 0 + 4\cdot\frac{3}{2} & ?\ 6 \\ 0 + 6 & \\ 6 & 6 \quad \text{TRUE} \end{array}$$

$\left(0, \frac{3}{2}\right)$ is a solution.

For $\left(\frac{2}{3}, 1\right)$:

$$\begin{array}{r|l} \multicolumn{2}{c}{3m + 4n = 6} \\ \hline 3\cdot\frac{2}{3} + 4\cdot 1 & ?\ 6 \\ 2 + 4 & \\ 6 & 6 \quad \text{TRUE} \end{array}$$

The equation $\left(\frac{2}{3}, 1\right)$ is true, so $\left(\frac{2}{3}, 1\right)$ is a solution.

13. To determine whether $(-0.75, 2.75)$ is a solution, substitute -0.75 for x and 2.75 for y.

$$\begin{array}{r|l} \multicolumn{2}{c}{x^2 - y^2 = 3} \\ \hline (-0.75)^2 - (2.75)^2 & ?\ 3 \\ 0.5625 - 7.5625 & \\ -7 & 3 \quad \text{FALSE} \end{array}$$

The equation $-7 = 3$ is false, so $(-0.75, 2.75)$ is not a solution.

To determine whether $(2, -1)$ is a solution, substitute 2 for x and -1 for y.

$$\begin{array}{r|l} \multicolumn{2}{c}{x^2 - y^2 = 3} \\ \hline 2^2 - (-1)^2 & ?\ 3 \\ 4 - 1 & \\ 3 & 3 \quad \text{TRUE} \end{array}$$

The equation $3 = 3$ is true, so $(2, -1)$ is a solution.

14. For $(2, -4)$:

$$\begin{array}{r|l} \multicolumn{2}{c}{5x + 2y^2 = 70} \\ \hline 5\cdot 2 + 2(-4)^2 & ?\ 70 \\ 10 + 2\cdot 16 & \\ 10 + 32 & \\ 42 & 70 \quad \text{FALSE} \end{array}$$

$(2, -4)$ is not a solution.

For $(4, -5)$:

$5x + 2y^2$	$= 70$
$5 \cdot 4 + 2(-5)^2$	? 70
$20 + 2 \cdot 25$	
$20 + 50$	
70	70 TRUE

$(4, -5)$ is a solution.

15. Graph $y = 3x + 5$.

We choose some values for x and find the corresponding y-values.

When $x = -3$, $y = 3x + 5 = 3(-3) + 5 = -9 + 5 = -4$.

When $x = -1$, $y = 3x + 5 = 3(-1) + 5 = -3 + 5 = 2$.

When $x = 0$, $y = 3x + 5 = 3 \cdot 0 + 5 = 0 + 5 = 5$

We list these points in a table, plot them, and draw the graph.

x	y	(x, y)
-3	-4	$(-3, -4)$
-1	2	$(-1, 2)$
0	5	$(0, 5)$

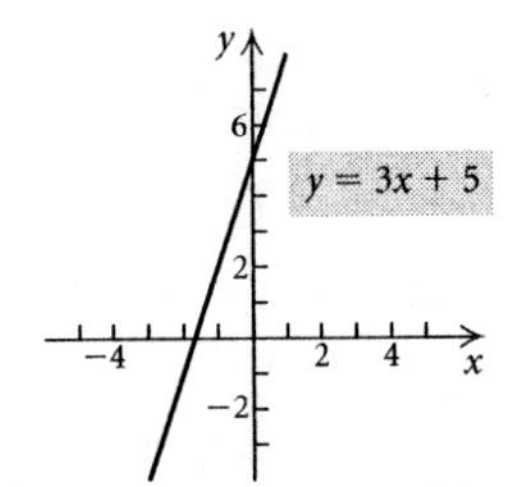

16.

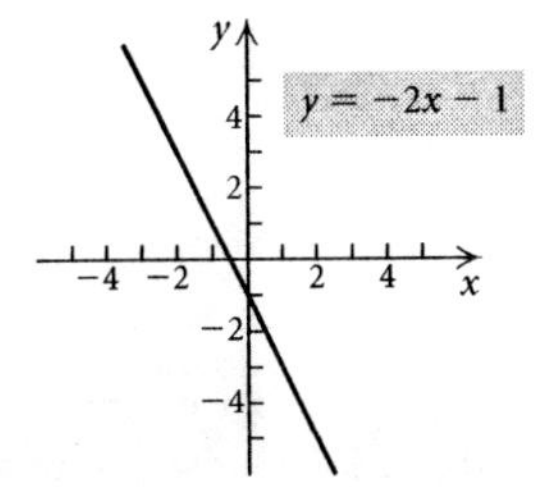

17. Graph $x - y = 3$.

Make a table of values, plot the points in the table, and draw the graph.

x	y	(x, y)
-2	-5	$(-2, -5)$
0	-3	$(0, -3)$
3	0	$(3, 0)$

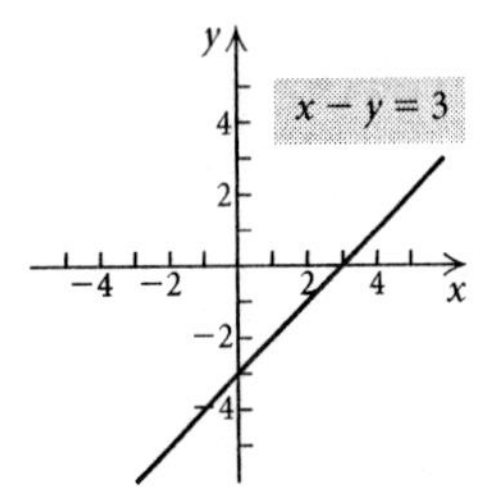

18.

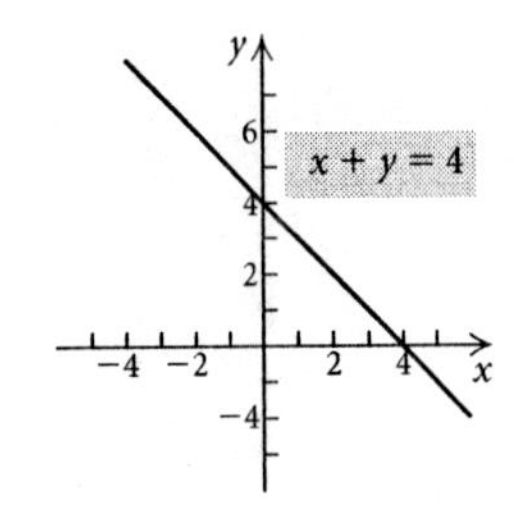

19. Graph $2x + y = 4$.

We could solve for y first.

$2x + y = 4$

$y = -2x + 4$ Subtracting $2x$ on both sides

Make a table of values, plot the points in the table, and draw the graph.

x	y	(x, y)
-1	6	$(-1, 6)$
0	4	$(0, 4)$
2	0	$(2, 0)$

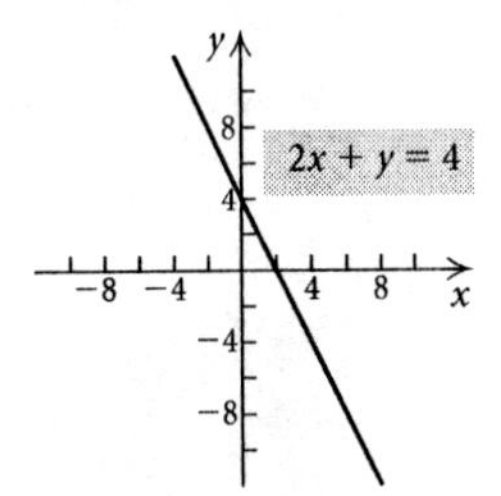

20.

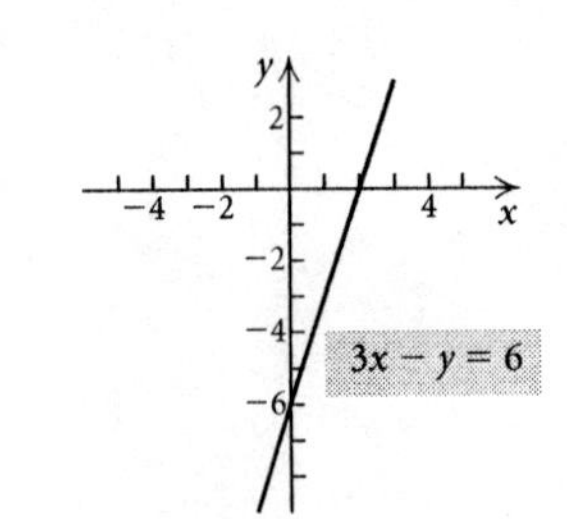

21. Graph $y = -\frac{3}{4}x + 3$.

By choosing multiples of 4 for x, we can avoid fraction values for y. Make a table of values, plot the points in the table, and draw the graph.

x	y	(x,y)
-4	6	$(-4,6)$
0	3	$(0,3)$
4	0	$(4,0)$

22.

23. Graph $5x - 2y = 8$.

We could solve for y first.

$5x - 2y = 8$

$-2y = -5x + 8$ Subtracting $5x$ on both sides

$y = \frac{5}{2}x - 4$ Multiplying by $-\frac{1}{2}$ on both sides

By choosing multiples of 2 for x we can avoid fraction values for y. Make a table of values, plot the points in the table, and draw the graph.

x	y	(x,y)
0	-4	$(0,-4)$
2	1	$(2,1)$
4	6	$(4,6)$

24.

25. Graph $x - 4y = 5$.

Make a table of values, plot the points in the table, and draw the graph.

x	y	(x,y)
-3	-2	$(-3,-2)$
1	-1	$(1,-1)$
5	0	$(5,0)$

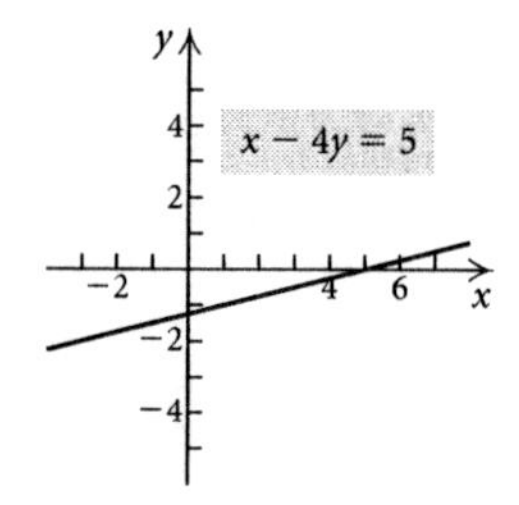

26.

27. Graph $3x - 4y = 12$.

In this case, it is convenient to find the intercepts along with a third point on the graph. Make a table of values, plot the points in the table, and draw the graph.

x	y	(x,y)
4	0	$(4,0)$
0	-3	$(0,-3)$
-4	-6	$(-4,-6)$

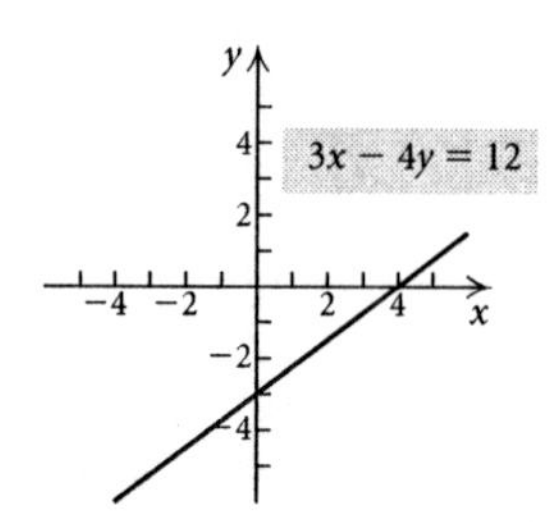

28.

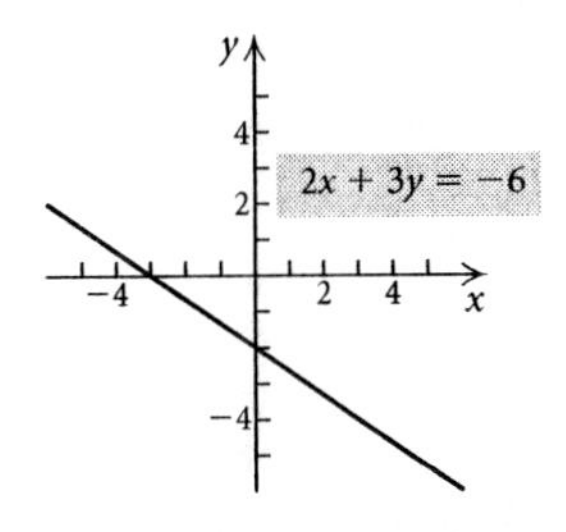

29. Graph $2x + 5y = -10$.

In this case, it is convenient to find the intercepts along with a third point on the graph. Make a table of values, plot the points in the table, and draw the graph.

x	y	(x, y)
-5	0	$(-5, 0)$
0	-2	$(0, -2)$
5	-4	$(5, -4)$

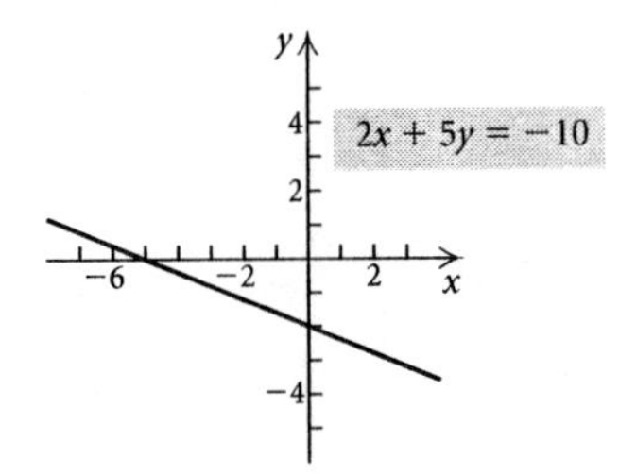

30.

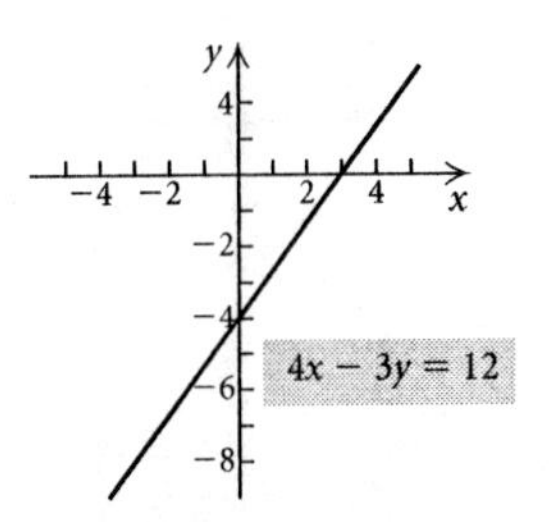

31. Graph $y = -x^2$.

Make a table of values, plot the points in the table, and draw the graph.

x	y	(x, y)
-2	-4	$(-2, -4)$
-1	-1	$(-1, -1)$
0	0	$(0, 0)$
1	-1	$(1, -1)$
2	-4	$(2, -4)$

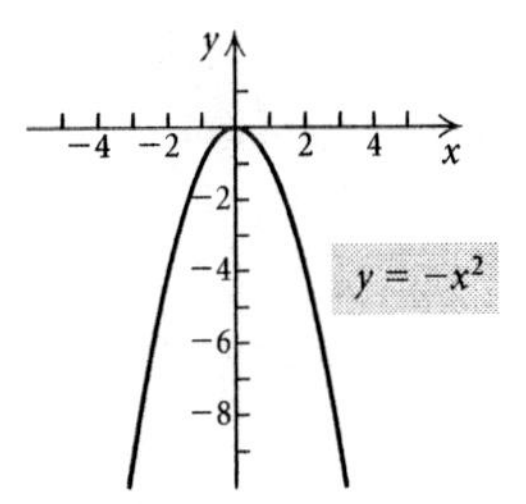

32.

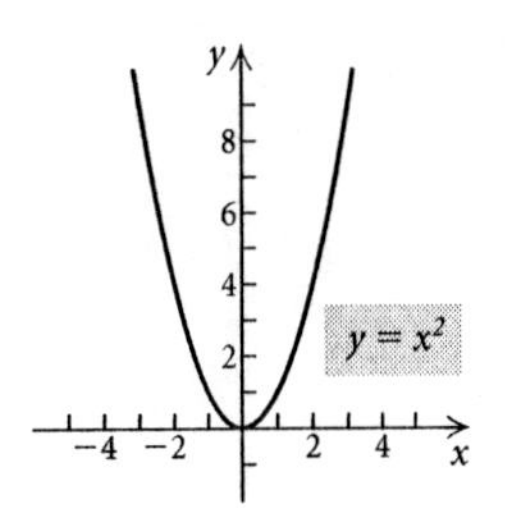

33. Graph $y = x^2 - 3$.

Make a table of values, plot the points in the table, and draw the graph.

x	y	(x, y)
-3	6	$(-3, 6)$
-1	-2	$(-1, -2)$
0	-3	$(0, -3)$
1	-2	$(1, -2)$
3	6	$(3, 6)$

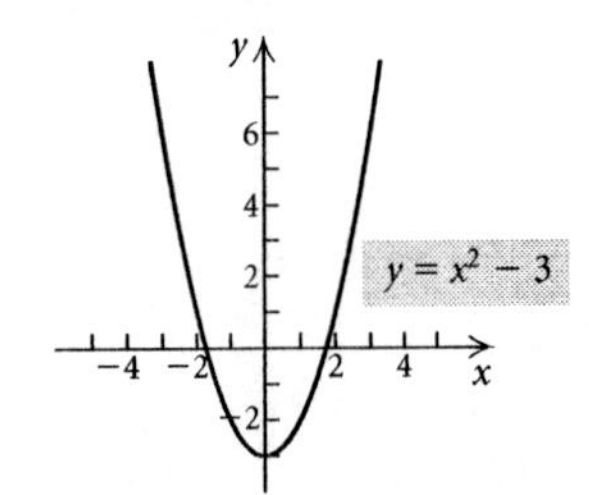

34.

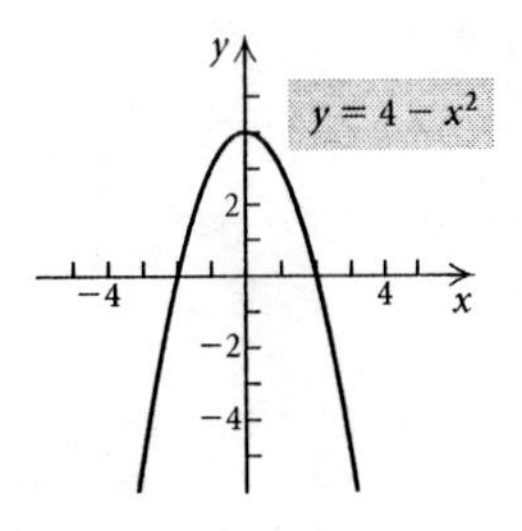

35. Graph $y = -x^2 + 2x + 3$.

Make a table of values, plot the points in the table, and draw the graph.

x	y	(x, y)
-2	-5	$(-2, -5)$
-1	0	$(-1, 0)$
0	3	$(0, 3)$
1	4	$(1, 4)$
2	3	$(2, 3)$
3	0	$(3, 0)$
4	-5	$(4, -5)$

36.

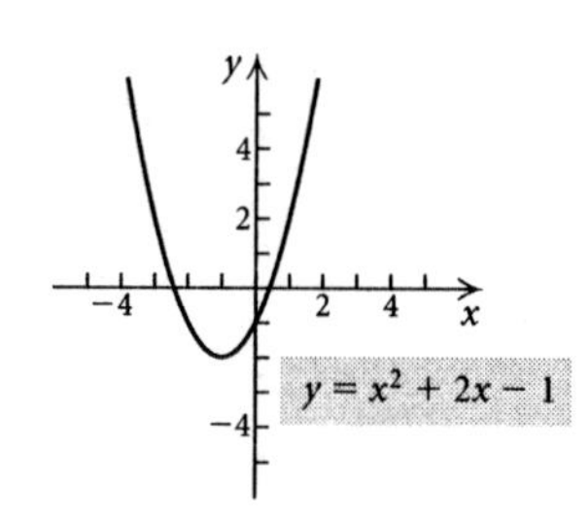

37. Either point can be considered as (x_1, y_1).

$$d = \sqrt{(4-5)^2 + (6-9)^2}$$
$$= \sqrt{(-1)^2 + (-3)^2} = \sqrt{10} \approx 3.162$$

38. $d = \sqrt{(-3-2)^2 + (7-11)^2} = \sqrt{41} \approx 6.403$

39. Either point can be considered as (x_1, y_1).

$$d = \sqrt{(6-9)^2 + (-1-5)^2}$$
$$= \sqrt{(-3)^2 + (-6)^2} = \sqrt{45} \approx 6.708$$

40. $d = \sqrt{(-4-(-1))^2 + (-7-3)^2} = \sqrt{109} \approx 10.440$

41. Either point can be considered as (x_1, y_1).

$$d = \sqrt{(-4.2-2.1)^2 + [3-(-6.4)]^2}$$
$$= \sqrt{(-6.3)^2 + (9.4)^2} = \sqrt{128.05} \approx 11.316$$

42. $d = \sqrt{\left[-\frac{3}{5} - \left(-\frac{3}{5}\right)\right]^2 + \left(-4 - \frac{2}{3}\right)^2} =$

$$\sqrt{\left(-\frac{14}{3}\right)^2} = \frac{14}{3}$$

43. Either point can be considered as (x_1, y_1).

$$d = \sqrt{\left(-\frac{1}{2} - \frac{5}{2}\right)^2 + (4-4)^2}$$
$$= \sqrt{(-3)^2 + 0^2} = \sqrt{9} = 3$$

44. $d = \sqrt{[0.6-(-8.1)]^2 + [-1.5-(-1.5)]^2} =$
$\sqrt{(8.7)^2} = 8.7$

45. Either point can be considered as (x_1, y_1).

$$d = \sqrt{(-\sqrt{6} - \sqrt{3})^2 + (0-(-\sqrt{5}))^2}$$
$$= \sqrt{6 + 2\sqrt{18} + 3 + 5} = \sqrt{14 + 2\sqrt{9 \cdot 2}}$$
$$= \sqrt{14 + 2 \cdot 3\sqrt{2}} = \sqrt{14 + 6\sqrt{2}} \approx 4.742$$

46. $d = \sqrt{(-\sqrt{2}-0)^2 + (1-\sqrt{7})^2} = \sqrt{2+1-2\sqrt{7}+7} =$
$\sqrt{10 - 2\sqrt{7}} \approx 2.170$

47. Either point can be considered as (x_1, y_1).
$d = \sqrt{(0-a)^2 + (0-b)^2} = \sqrt{a^2 + b^2}$

48. $d = \sqrt{[r-(-r)]^2 + [s-(-s)]^2} = \sqrt{4r^2 + 4s^2} =$
$2\sqrt{r^2 + s^2}$

49. First we find the length of the diameter:

$$d = \sqrt{(-3-9)^2 + (-1-4)^2}$$
$$= \sqrt{(-12)^2 + (-5)^2} = \sqrt{169} = 13$$

The length of the radius is one-half the length of the diameter, or $\frac{1}{2}(13)$, or 6.5.

50. Radius $= \sqrt{(-3-0)^2 + (5-1)^2} = \sqrt{25} = 5$

Diameter $= 2 \cdot 5 = 10$

51. First we find the distance between each pair of points.

For $(-4, 5)$ and $(6, 1)$:

$$d = \sqrt{(-4-6)^2 + (5-1)^2}$$
$$= \sqrt{(-10)^2 + 4^2} = \sqrt{116}$$

For $(-4, 5)$ and $(-8, -5)$:

$$d = \sqrt{(-4-(-8))^2 + (5-(-5))^2}$$
$$= \sqrt{4^2 + 10^2} = \sqrt{116}$$

For $(6, 1)$ and $(-8, -5)$:

$$d = \sqrt{(6-(-8))^2 + (1-(-5))^2}$$
$$= \sqrt{14^2 + 6^2} = \sqrt{232}$$

Since $(\sqrt{116})^2 + (\sqrt{116})^2 = (\sqrt{232})^2$, the points could be the vertices of a right triangle.

52. For $(-3, 1)$ and $(2, -1)$:

$$d = \sqrt{(-3-2)^2 + (1-(-1))^2} = \sqrt{29}$$

For $(-3, 1)$ and $(6, 9)$:

$$d = \sqrt{(-3-6)^2 + (1-9)^2} = \sqrt{145}$$

For $(2, -1)$ and $(6, 9)$:

$$d = \sqrt{(2-6)^2 + (-1-9)^2} = \sqrt{116}$$

Since $(\sqrt{29})^2 + (\sqrt{116})^2 = (\sqrt{145})^2$, the points could be the vertices of a right triangle.

53. First we find the distance between each pair of points.

For $(-4,3)$ and $(0,5)$:

$$d = \sqrt{(-4-0)^2+(3-5)^2} = \sqrt{(-4)^2+(-2)^2} = \sqrt{20}$$

For $(-4,3)$ and $(3,-4)$:

$$d = \sqrt{(-4-3)^2+[3-(-4)]^2} = \sqrt{(-7)^2+7^2} = \sqrt{98}$$

For $(0,5)$ and $(3,-4)$:

$$d = \sqrt{(0-3)^2+[5-(-4)]^2} = \sqrt{(-3)^2+9^2} = \sqrt{90}$$

The greatest distance is $\sqrt{98}$, so if the points are the vertices of a right triangle, then it is the hypotenuse. But $(\sqrt{20})^2+(\sqrt{90})^2 \neq (\sqrt{98})^2$, so the points are not the vertices of a right triangle.

54. See the graph of this rectangle in Exercise 65.

The segments with endpoints $(-3,4)$, $(2,-1)$ and $(5,2)$, $(0,7)$ are one pair of opposite sides. We find the length of each of these sides.

For $(-3,4)$, $(2,-1)$:

$$d = \sqrt{(-3-2)^2+(4-(-1))^2} = \sqrt{50}$$

For $(5,2)$, $(0,7)$:

$$d = \sqrt{(5-0)^2+(2-7)^2} = \sqrt{50}$$

The segments with endpoints $(2,-1)$, $(5,2)$ and $(0,7)$, $(-3,4)$ are the second pair of opposite sides. We find their lengths.

For $(2,-1)$, $(5,2)$:

$$d = \sqrt{(2-5)^2+(-1-2)^2} = \sqrt{18}$$

For $(0,7)$, $(-3,4)$:

$$d = \sqrt{(0-(-3))^2+(7-4)^2} = \sqrt{18}$$

The endpoints of the diagonals are $(-3,4)$, $(5,2)$ and $(2,-1)$, $(0,7)$. We find the length of each.

For $(-3,4)$, $(5,2)$:

$$d = \sqrt{(-3-5)^2+(4-2)^2} = \sqrt{68}$$

For $(2,-1)$, $(0,7)$:

$$d = \sqrt{(2-0)^2+(-1-7)^2} = \sqrt{68}$$

The opposite sides of the quadrilateral are the same length and the diagonals are the same length, so the quadrilateral is a rectangle.

55. We use the midpoint formula.

$$\left(\frac{4+(-12)}{2}, \frac{-9+(-3)}{2}\right) = \left(-\frac{8}{2}, -\frac{12}{2}\right) = (-4,-6)$$

56. $\left(\frac{7+9}{2}, \frac{-2+5}{2}\right) = \left(8, \frac{3}{2}\right)$

57. We use the midpoint formula.

$$\left(\frac{6.1+3.8}{2}, \frac{-3.8+(-6.1)}{2}\right) = \left(\frac{9.9}{2}, -\frac{9.9}{2}\right) =$$

$(4.95, -4.95)$

58. $\left(\frac{-0.5+4.8}{2}, \frac{-2.7+(-0.3)}{2}\right) = (2.15, -1.5)$

59. We use the midpoint formula.

$$\left(\frac{-6+(-6)}{2}, \frac{5+8}{2}\right) = \left(-\frac{12}{2}, \frac{13}{2}\right) = \left(-6, \frac{13}{2}\right)$$

60. $\left(\frac{1+(-1)}{2}, \frac{-2+2}{2}\right) = (0,0)$

61. We use the midpoint formula.

$$\left(\frac{\frac{1}{6}+\left(-\frac{2}{3}\right)}{2}, \frac{-\frac{3}{5}+\frac{5}{4}}{2}\right) = \left(\frac{-\frac{5}{6}}{2}, \frac{\frac{13}{20}}{2}\right) =$$

$\left(-\frac{5}{12}, \frac{13}{40}\right)$

62. $\left(\frac{\frac{2}{9}+\left(-\frac{2}{5}\right)}{2}, \frac{\frac{1}{3}+\frac{4}{5}}{2}\right) = \left(-\frac{4}{45}, \frac{17}{30}\right)$

63. We use the midpoint formula.

$$\left(\frac{\sqrt{3}+3\sqrt{3}}{2}, \frac{-1+4}{2}\right) = \left(\frac{4\sqrt{3}}{2}, \frac{3}{2}\right) = \left(2\sqrt{3}, \frac{3}{2}\right)$$

64. $\left(\frac{-\sqrt{5}+\sqrt{5}}{2}, \frac{2+\sqrt{7}}{2}\right) = \left(0, \frac{2+\sqrt{7}}{2}\right)$

65.

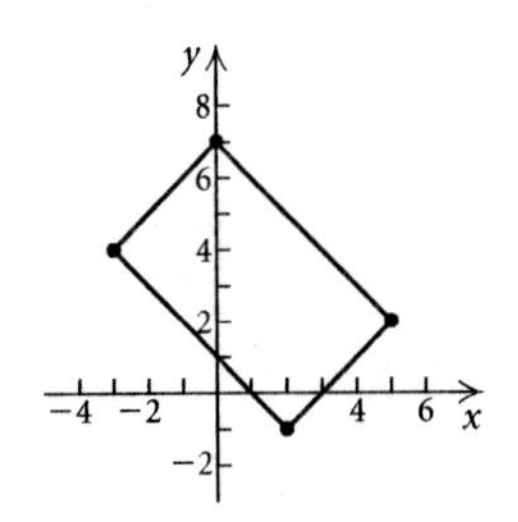

For the side with vertices $(-3,4)$ and $(2,-1)$:

$$\left(\frac{-3+2}{2}, \frac{4+(-1)}{2}\right) = \left(-\frac{1}{2}, \frac{3}{2}\right)$$

For the sides with vertices $(2,-1)$ and $(5,2)$:

$$\left(\frac{2+5}{2}, \frac{-1+2}{2}\right) = \left(\frac{7}{2}, \frac{1}{2}\right)$$

For the sides with vertices $(5,2)$ and $(0,7)$:

$$\left(\frac{5+0}{2}, \frac{2+7}{2}\right) = \left(\frac{5}{2}, \frac{9}{2}\right)$$

For the sides with vertices $(0,7)$ and $(-3,4)$:

$$\left(\frac{0+(-3)}{2}, \frac{7+4}{2}\right) = \left(-\frac{3}{2}, \frac{11}{2}\right)$$

For the quadrilateral whose vertices are the points found above, the diagonals have endpoints

$\left(-\frac{1}{2}, \frac{3}{2}\right)$, $\left(\frac{5}{2}, \frac{9}{2}\right)$ and $\left(\frac{7}{2}, \frac{1}{2}\right)$, $\left(-\frac{3}{2}, \frac{11}{2}\right)$.

We find the length of each of these diagonals.

For $\left(-\frac{1}{2},\frac{3}{2}\right)$, $\left(\frac{5}{2},\frac{9}{2}\right)$:

$$d=\sqrt{\left(-\frac{1}{2}-\frac{5}{2}\right)^2+\left(\frac{3}{2}-\frac{9}{2}\right)^2}$$
$$=\sqrt{(-3)^2+(-3)^2}=\sqrt{18}$$

For $\left(\frac{7}{2},\frac{1}{2}\right)$, $\left(-\frac{3}{2},\frac{11}{2}\right)$:

$$d=\sqrt{\left(\frac{7}{2}-\left(-\frac{3}{2}\right)\right)^2+\left(\frac{1}{2}-\frac{11}{2}\right)^2}$$
$$=\sqrt{5^2+(-5)^2}=\sqrt{50}$$

Since the diagonals do not have the same lengths, the midpoints are not vertices of a rectangle.

66.

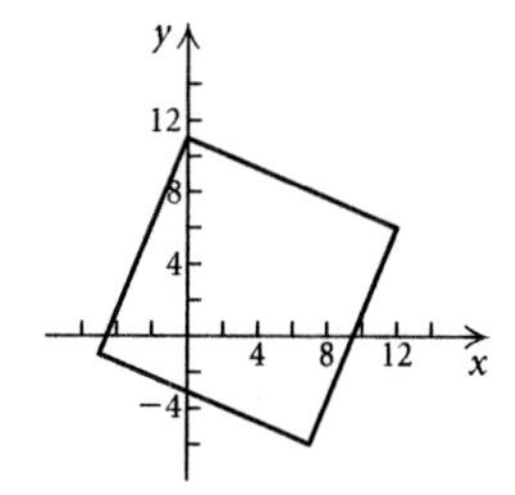

For the side with vertices $(-5,-1)$ and $(7,-6)$:

$$\left(\frac{-5+7}{2},\frac{-1+(-6)}{2}\right)=\left(1,-\frac{7}{2}\right)$$

For the side with vertices $(7,-6)$ and $(12,6)$:

$$\left(\frac{7+12}{2},\frac{-6+6}{2}\right)=\left(\frac{19}{2},0\right)$$

For the side with vertices $(12,6)$ and $(0,11)$:

$$\left(\frac{12+0}{2},\frac{6+11}{2}\right)=\left(6,\frac{17}{2}\right)$$

For the side with vertices $(0,11)$ and $(-5,-1)$:

$$\left(\frac{0+(-5)}{2},\frac{11+(-1)}{2}\right)=\left(-\frac{5}{2},5\right)$$

For the quadrilateral whose vertices are the points found above, one pair of opposite sides has endpoints $\left(1,-\frac{7}{2}\right)$, $\left(\frac{19}{2},0\right)$ and $\left(6,\frac{17}{2}\right)$, $\left(-\frac{5}{2},5\right)$. The length of each of these sides is $\frac{\sqrt{338}}{2}$. The other pair of opposite sides has endpoints $\left(\frac{19}{2},0\right)$, $\left(6,\frac{17}{2}\right)$ and $\left(-\frac{5}{2},5\right)$, $\left(1,-\frac{7}{2}\right)$. The length of each of these sides is also $\frac{\sqrt{338}}{2}$. The endpoints of the diagonals of the quadrilateral are $\left(1,-\frac{7}{2}\right)$, $\left(6,\frac{17}{2}\right)$ and $\left(\frac{19}{2},0\right)$, $\left(-\frac{5}{2},5\right)$. The length of each diagonal is 13. Since the four sides of the quadrilateral are the same length and the diagonals are the same length, the midpoints are vertices of a square.

67. We use the midpoint formula.

$$\left(\frac{\sqrt{7}+\sqrt{2}}{2},\frac{-4+3}{2}\right)=\left(\frac{\sqrt{7}+\sqrt{2}}{2},-\frac{1}{2}\right)$$

68. $\left(\frac{-3+1}{2},\frac{\sqrt{5}+\sqrt{2}}{2}\right)=\left(-1,\frac{\sqrt{5}+\sqrt{2}}{2}\right)$

69. $(x-h)^2+(y-k)^2=r^2$

$(x-2)^2+(y-3)^2=\left(\frac{5}{3}\right)^2$ Substituting

$(x-2)^2+(y-3)^2=\frac{25}{9}$

70. $(x-4)^2+(y-5)^2=(4.1)^2$

$(x-4)^2+(y-5)^2=16.81$

71. The length of a radius is the distance between $(-1,4)$ and $(3,7)$:

$$r=\sqrt{(-1-3)^2+(4-7)^2}$$
$$=\sqrt{(-4)^2+(-3)^2}=\sqrt{25}=5$$

$$(x-h)^2+(y-k)^2=r^2$$
$$[x-(-1)]^2+(y-4)^2=5^2$$
$$(x+1)^2+(y-4)^2=25$$

72. Find the length of a radius:

$$r=\sqrt{(6-1)^2+(-5-7)^2}=\sqrt{169}=13$$

$$(x-6)^2+[y-(-5)]^2=13^2$$
$$(x-6)^2+(y+5)^2=169$$

73. The center is the midpoint of the diameter:

$$\left(\frac{7+(-3)}{2},\frac{13+(-11)}{2}\right)=(2,1)$$

Use the center and either endpoint of the diameter to find the length of a radius. We use the point $(7,13)$:

$$r=\sqrt{(7-2)^2+(13-1)^2}$$
$$=\sqrt{5^2+12^2}=\sqrt{169}=13$$

$$(x-h)^2+(y-k)^2=r^2$$
$$(x-2)^2+(y-1)^2=13^2$$
$$(x-2)^2+(y-1)^2=169$$

74. The points $(-9,4)$ and $(-1,-2)$ are opposite vertices of the square and hence endpoints of a diameter of the circle. We use these points to find the center and radius.

Center: $\left(\frac{-9+(-1)}{2},\frac{4+(-2)}{2}\right)=(-5,1)$

Radius: $\frac{1}{2}\sqrt{(-9-(-1))^2+(4-(-2))^2}=\frac{1}{2}\cdot 10=5$

$$[x-(-5)]^2+(y-1)^2=5^2$$
$$(x+5)^2+(y-1)^2=25$$

75. Since the center is 2 units to the left of the y-axis and the circle is tangent to the y-axis, the length of a radius is 2.

$$(x-h)^2+(y-k)^2=r^2$$
$$[x-(-2)]^2+(y-3)^2=2^2$$
$$(x+2)^2+(y-3)^2=4$$

76. Since the center is 5 units below the x-axis and the circle is tangent to the x-axis, the length of a radius is 5.

$$(x-4)^2+[y-(-5)]^2=5^2$$
$$(x-4)^2+(y+5)^2=25$$

77.
$$x^2+y^2=4$$
$$(x-0)^2+(y-0)^2=2^2$$

Center: $(0,0)$; radius: 2

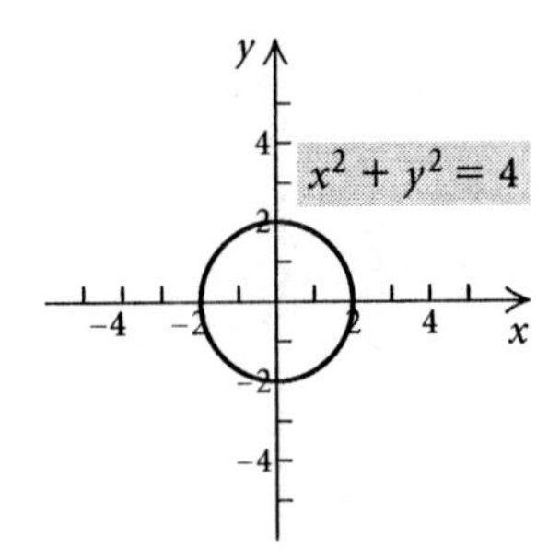

78.
$$x^2+y^2=81$$
$$(x-0)^2+(y-0)^2=9^2$$

Center: $(0,0)$; radius: 9

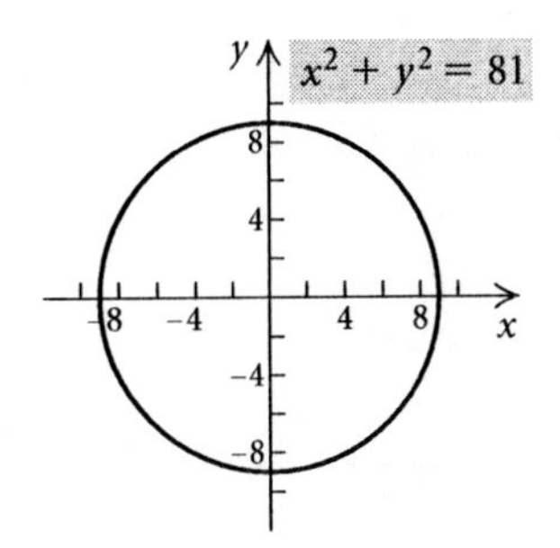

79.
$$x^2+(y-3)^2=16$$
$$(x-0)^2+(y-3)^2=4^2$$

Center: $(0,3)$; radius: 4

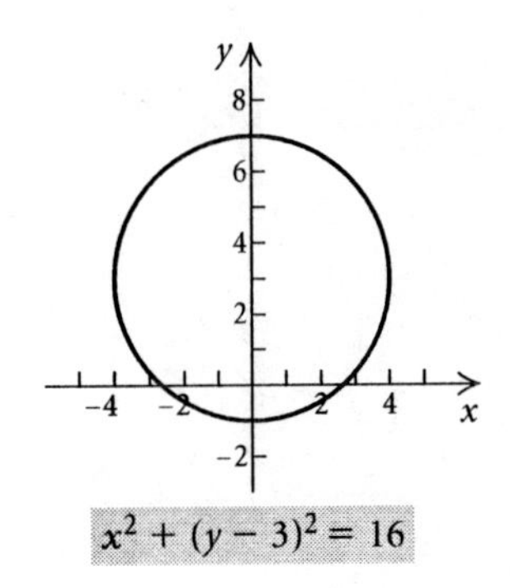

80.
$$(x+2)^2+y^2=100$$
$$[x-(-2)]^2+(y-0)^2=10^2$$

Center: $(-2,0)$; radius: 10

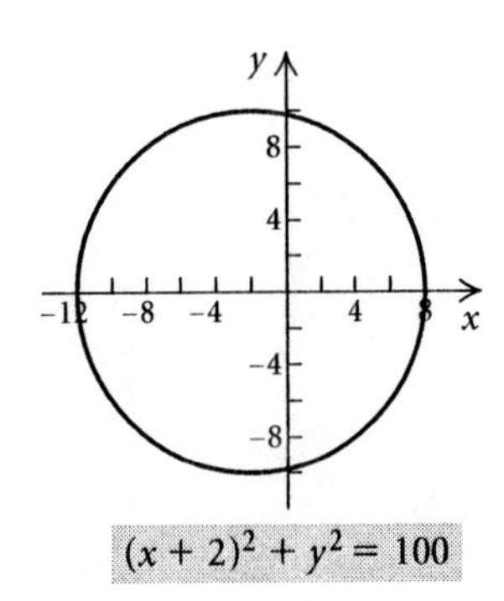

81.
$$(x-1)^2+(y-5)^2=36$$
$$(x-1)^2+(y-5)^2=6^2$$

Center: $(1,5)$; radius: 6

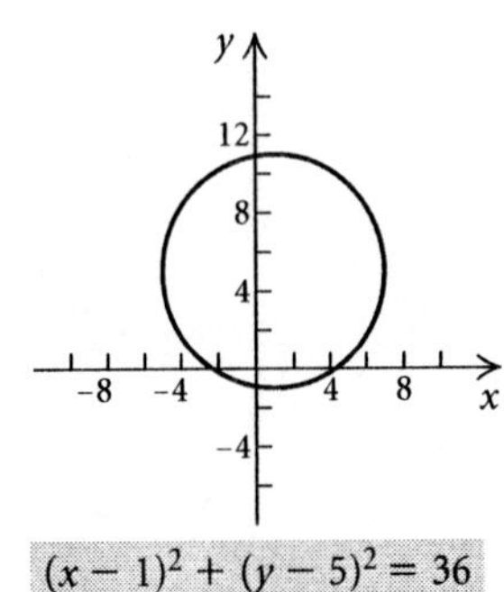

82.
$$(x-7)^2+(y+2)^2=25$$
$$(x-7)^2+[y-(-2)]^2=5^2$$

Center: $(7,-2)$; radius: 5

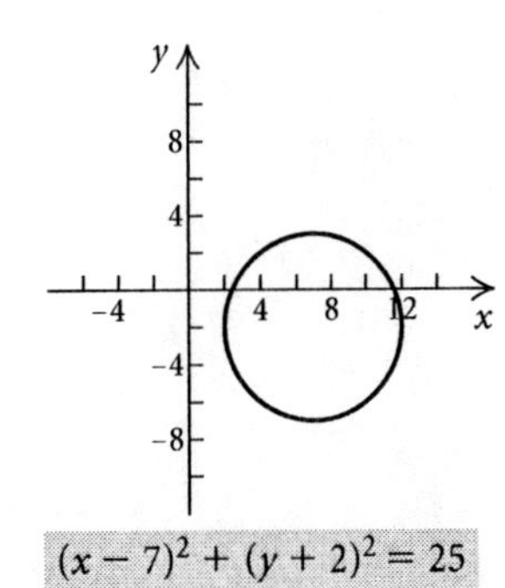

83.
$$(x+4)^2 + (y+5)^2 = 9$$
$$[x-(-4)]^2 + [y-(-5)]^2 = 3^2$$
Center: $(-4,-5)$; radius: 3

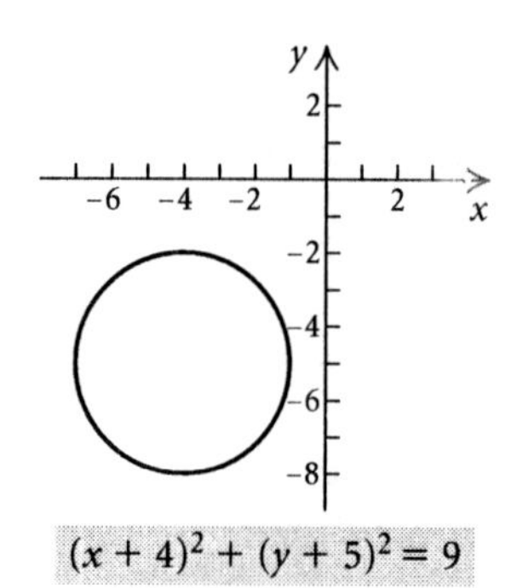

84.
$$(x+1)^2 + (y-2)^2 = 64$$
$$[x-(-1)]^2 + (y-2)^2 = 8^2$$
Center: $(-1,2)$; radius: 8

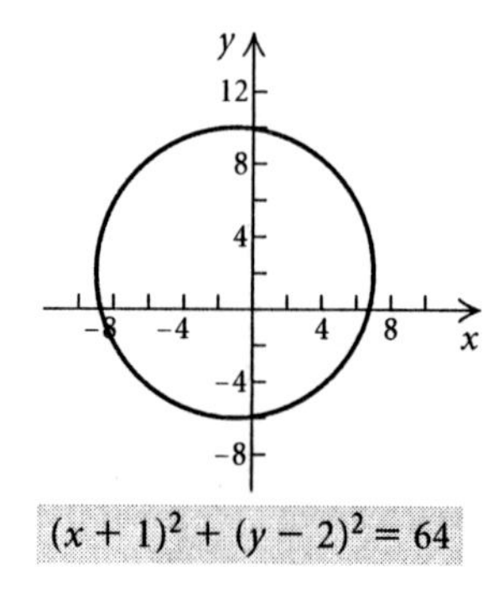

85. We regroup the terms and complete the square twice.
$$x^2 + y^2 - 6x - 2y - 6 = 0$$
$$(x^2 - 6x) + (y^2 - 2y) - 6 = 0$$
$$(x^2 - 6x + 9) + (y^2 - 2y + 1) - 9 - 1 - 6 = 0$$
$$\left[\tfrac{1}{2}(-6) = -3 \text{ and } (-3)^2 = 9;\ \tfrac{1}{2}(-2) = -1 \text{ and } (-1)^2 = 1\right]$$
$$(x-3)^2 + (y-1)^2 - 16 = 0$$
$$(x-3)^2 + (y-1)^2 = 16$$
$$(x-3)^2 + (y-1)^2 = 4^2$$
Center: $(3,1)$; radius: 4

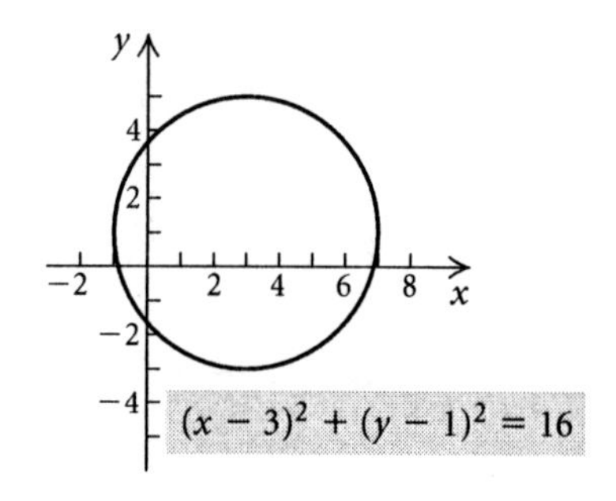

86.
$$x^2 + y^2 + 4x - 8y + 19 = 0$$
$$(x^2 + 4x) + (y^2 - 8y) + 19 = 0$$
$$(x^2 + 4x + 4) + (y^2 - 8y + 16) - 4 - 16 + 19 = 0$$
$$(x+2)^2 + (y-4)^2 - 1 = 0$$
$$(x+2)^2 + (y-4)^2 = 1$$
$$[x-(-2)]^2 + (y-4)^2 = 1^2$$
Center: $(-2,4)$; radius: 1

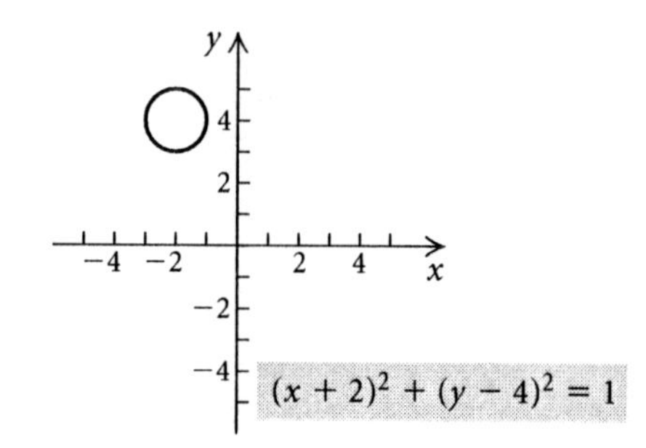

87. We regroup the terms and complete the square twice.
$$x^2 + y^2 + 2x + 2y = 7$$
$$(x^2 + 2x) + (y^2 + 2y) = 7$$
$$(x^2 + 2x + 1) + (y^2 + 2y + 1) - 1 - 1 = 7$$
$$\left[\tfrac{1}{2} \cdot 2 = 1 \text{ and } 1^2 = 1\right]$$
$$(x+1)^2 + (y+1)^2 - 2 = 7$$
$$(x+1)^2 + (y+1)^2 = 9$$
$$[(x-(-1)]^2 + [y-(-1)]^2 = 3^2$$
Center: $(-1,-1)$; radius: 3

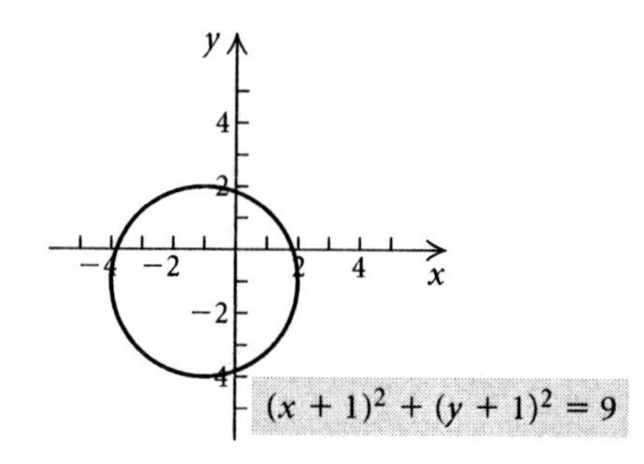

88.
$$x^2 + y^2 - 10x + 6y = -30$$
$$(x^2 - 10x) + (y^2 + 6y) = -30$$
$$(x^2 - 10x + 25) + (y^2 + 6y + 9) - 25 - 9 = -30$$
$$(x-5)^2 + (y+3)^2 - 34 = -30$$
$$(x-5)^2 + (y+3)^2 = 4$$
$$(x-5)^2 + [y-(-3)]^2 = 2^2$$
Center: $(5,-3)$; radius: 2

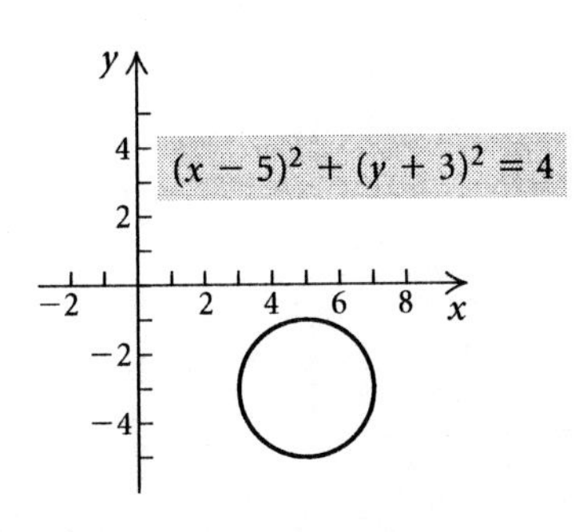

89. From the graph we see that the center of the circle is $(-2,1)$ and the radius is 3. The equation of the circle is $[x-(-2)]^2+(y-1)^2=3^2$, or $(x+2)^2+(y-1)^2=3^2$.

90. Center: $(3,-5)$, radius: 4

Equation: $(x-3)^2+[y-(-5)]^2=4^2$, or

$(x-3)^2+(y+5)^2=4^2$

91. From the graph we see that the center of the circle is $(5,-5)$ and the radius is 15. The equation of the circle is $(x-5)^2+[y-(-5)]^2=15^2$, or $(x-5)^2+(y+5)^2=15^2$.

92. Center: $(-8,2)$, radius: 4

Equation: $[x-(-8)]^2+(y-2)^2=4^2$, or

$(x+8)^2+(y-2)^2=4^2$

93. First solve the equation for y: $y=-4x+7$. Enter the equation in this form, select the standard window, and graph the equation.

$4x+y=7$

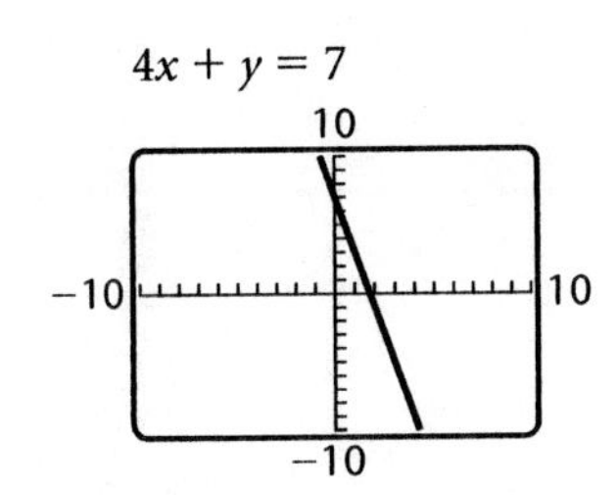

94. $5x+y=-8$

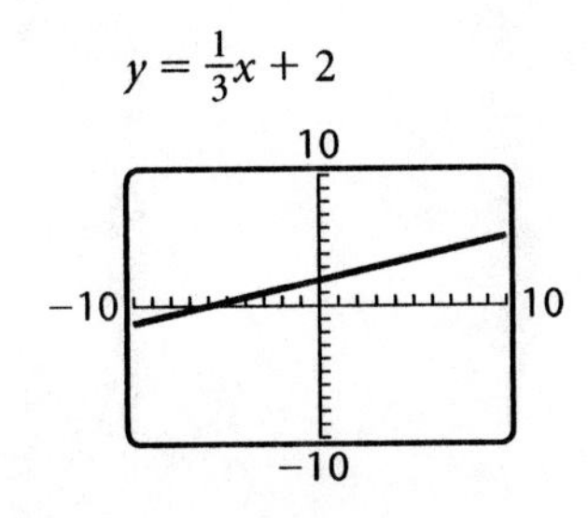

95. Enter the equation, select the standard window, and graph the equation.

$y=\frac{1}{3}x+2$

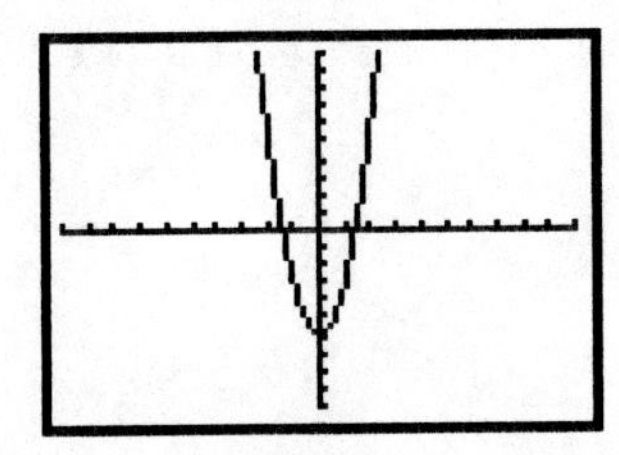

96. $y=\frac{3}{2}x-4$

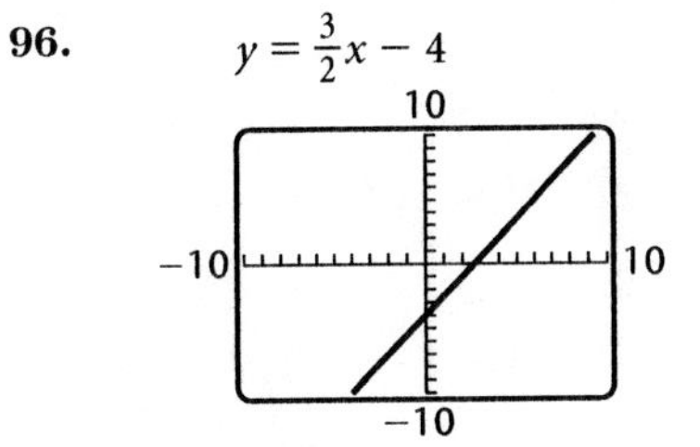

97. Enter the equation, select the standard window, and graph the equation.

$y=x^2+6$

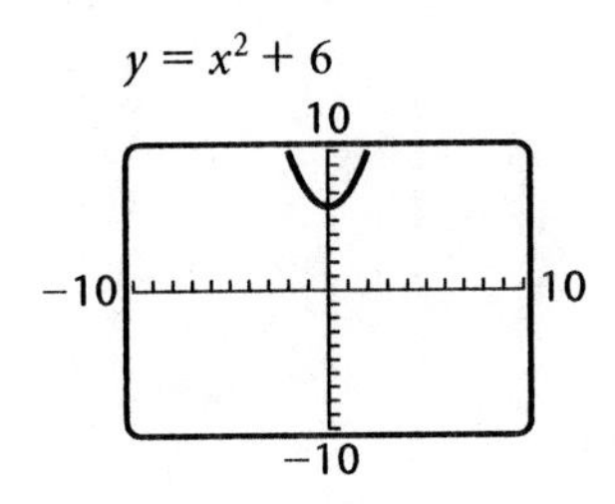

98. $y=5-x^2$

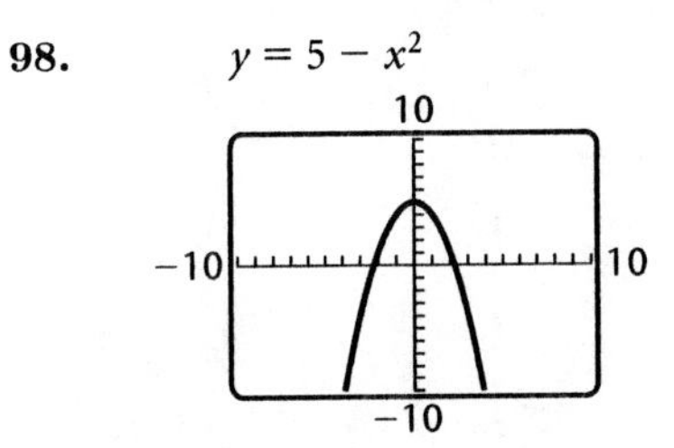

99. Enter the equation, select the standard window, and graph the equation.

$y=2-x^2$

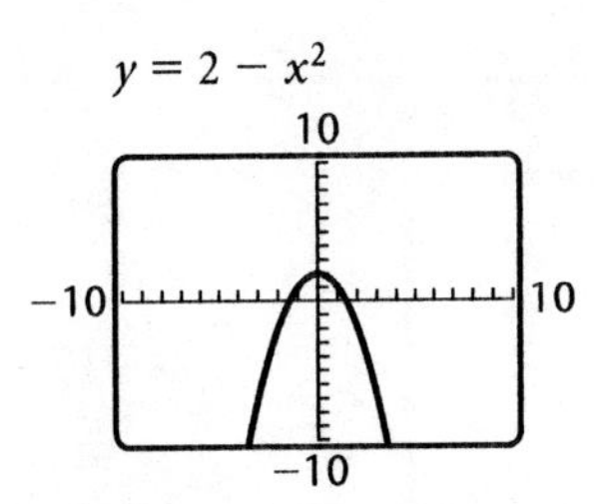

100. $y=x^2-5x+3$

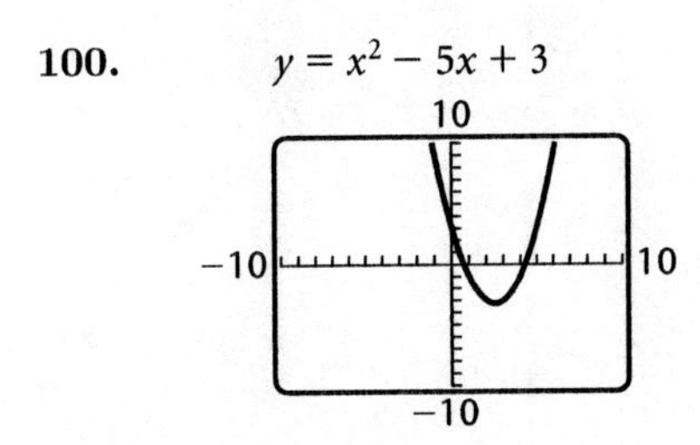

101. Standard window:

$[-4, 4, -4, 4]$

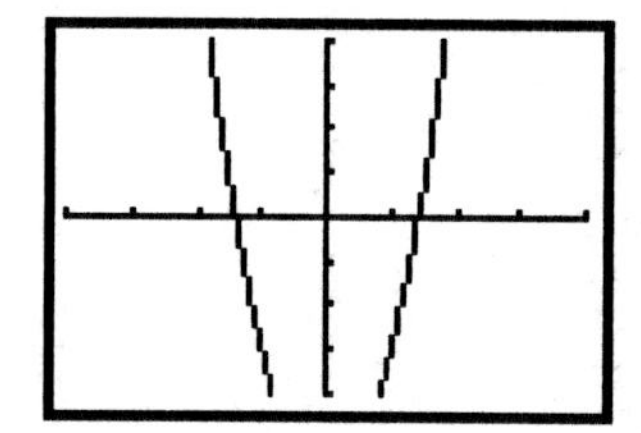

We see that the standard window is a better choice for this graph.

102. Standard window:

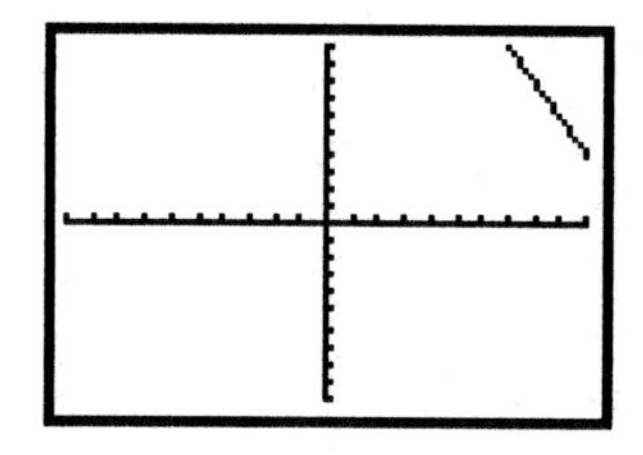

$[-15, 15, -10, 30]$, Xscl $= 3$, Yscl $= 5$

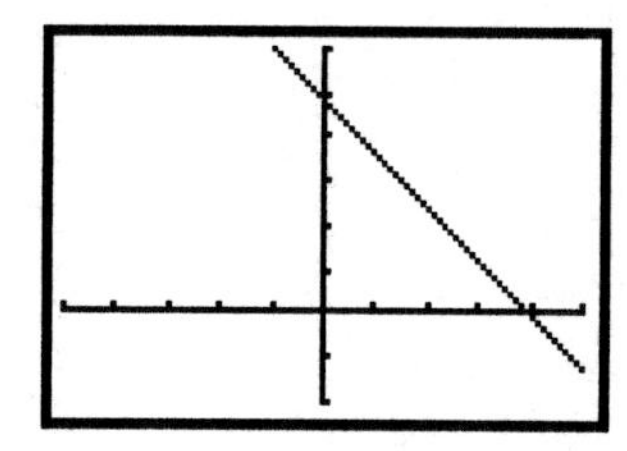

We see that $[-15, 15, -10, 30]$ is a better choice for this graph.

103. Standard window:

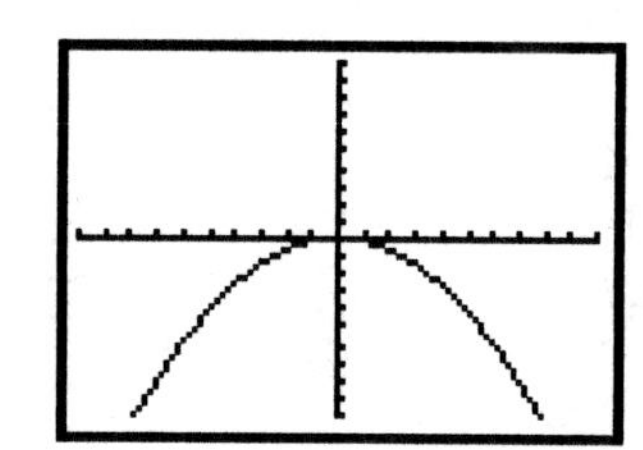

$[-1, 1, -0.3, 0.3]$, Xscl $= 0.1$, Yscl $= 0.1$

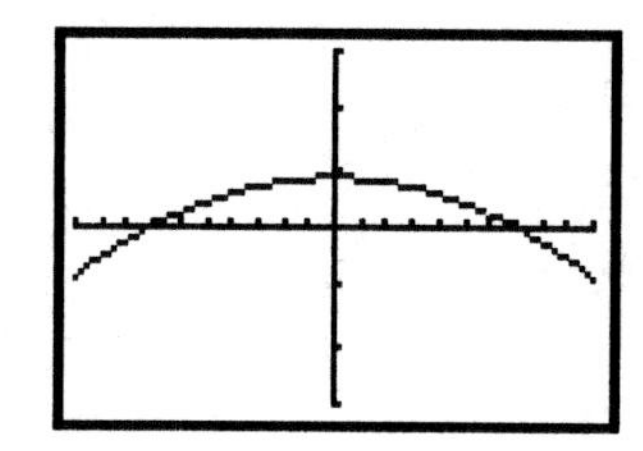

We see that $[-1, 1, -0.3, 0.3]$ is a better choice for this graph.

104. Standard window:

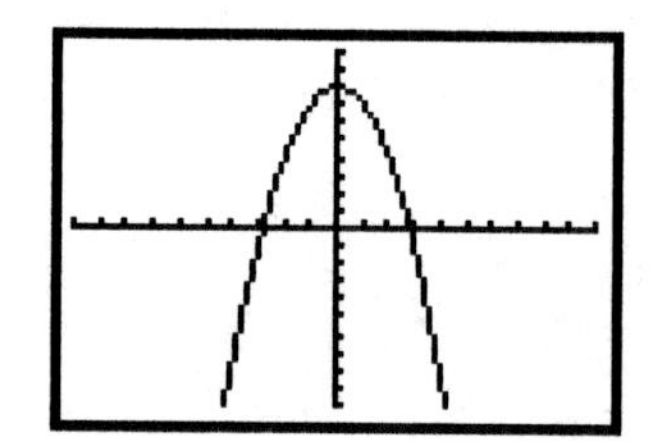

$[-3, 3, -3, 3]$

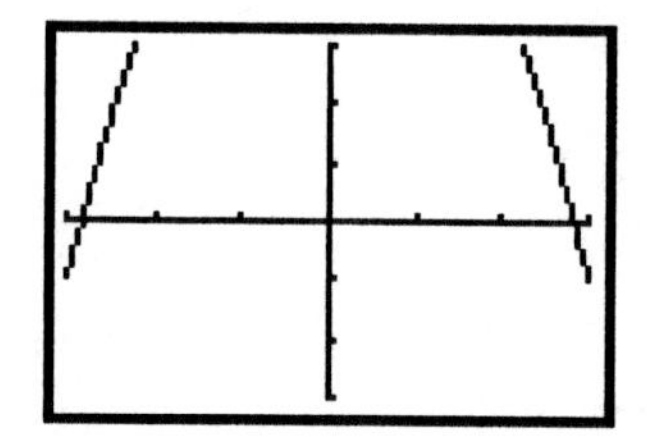

We see that the standard window is a better choice for this graph.

105. Square the viewing window. For the graph shown, one possibility is $[-12, 9, -4, 10]$.

106. Square the viewing window. For the window shown, one possibility is $[-10, 20, -15, 5]$.

107.

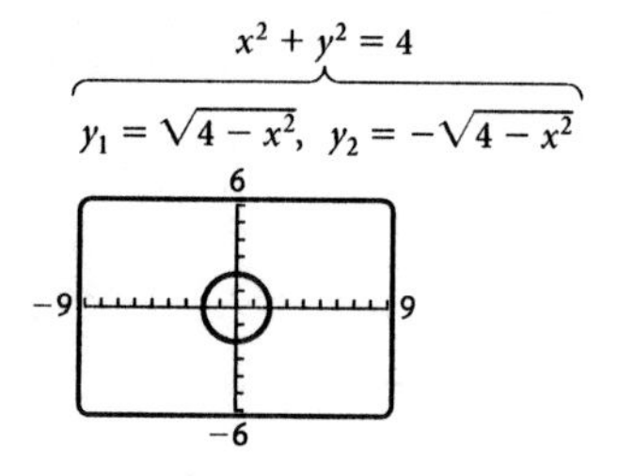

108.

$(x + 2)^2 + y^2 = 100$

10

-15 15

-10

109.

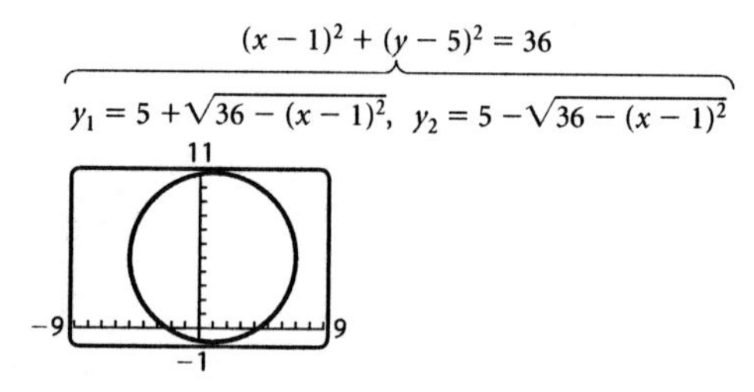

110.

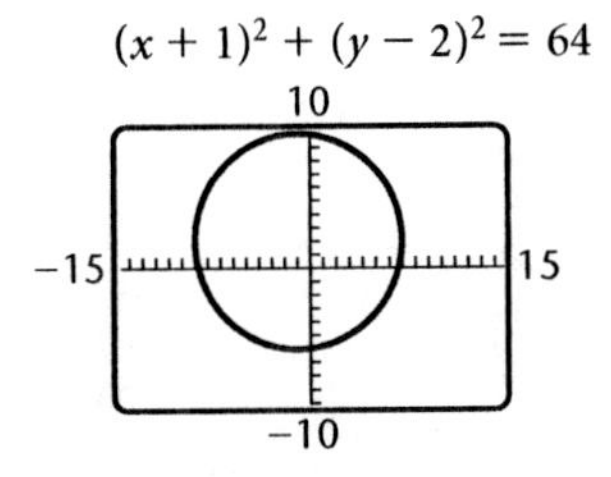

111. We graph $y_1 = 1 + \sqrt{16 - (x-3)^2}$ and $y_2 = 1 - \sqrt{16 - (x-3)^2}$ in the window $[-9, 9, -6, 6]$.

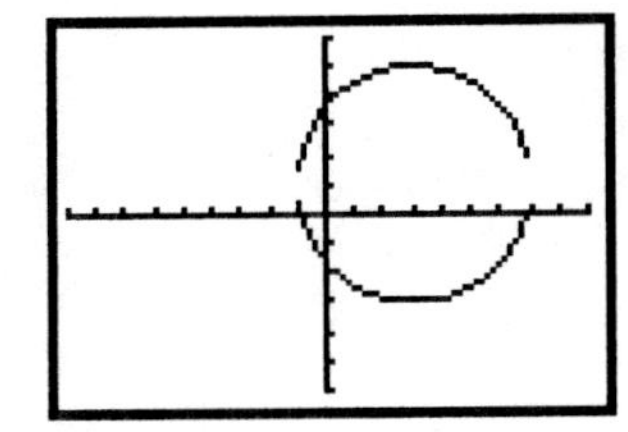

112. We graph $y_1 = -3 + \sqrt{4 - (x-5)^2}$ and $y_2 = -3 - \sqrt{4 - (x-5)^2}$ in the window $[-2, 10, -6, 2]$.

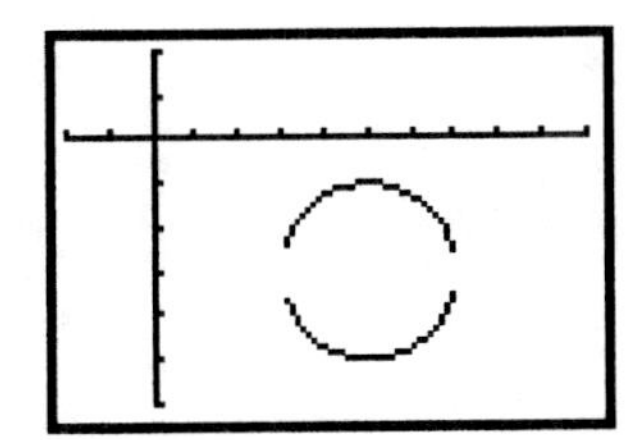

113. The Pythagorean theorem is used to derive the distance formula, and the distance formula is used to derive the equation of a circle in standard form.

114. Let $A = (a, b)$ and $B = (c, d)$. The coordinates of a point C one-half of the way from A to B are $\left(\frac{a+c}{2}, \frac{b+d}{2}\right)$. A point D that is one-half of the way from C to B is $\frac{1}{2} + \frac{1}{2} \cdot \frac{1}{2}$, or $\frac{3}{4}$ of the way from A to B. Its coordinates are $\left(\frac{\frac{a+c}{2} + c}{2}, \frac{\frac{b+d}{2} + d}{2}\right)$, or $\left(\frac{a+3c}{4}, \frac{b+3d}{4}\right)$. Then a point E that is one-half of the way from D to B is $\frac{3}{4} + \frac{1}{2} \cdot \frac{1}{4}$, or $\frac{7}{8}$ of the way from A to B. Its coordinates are $\left(\frac{\frac{a+3c}{4} + c}{2}, \frac{\frac{b+3d}{4} + d}{2}\right)$, or $\left(\frac{a+7c}{8}, \frac{b+7d}{8}\right)$.

115. If the point (p, q) is in the fourth quadrant, then $p > 0$ and $q < 0$. If $p > 0$, then $-p < 0$ so both coordinates of the point $(q, -p)$ are negative and $(q, -p)$ is in the third quadrant.

116. Use the distance formula:

$$d = \sqrt{(a+h-a)^2 + \left(\frac{1}{a+h} - \frac{1}{a}\right)^2} =$$

$$\sqrt{h^2 + \left(\frac{-h}{a(a+h)}\right)^2} = \sqrt{h^2 + \frac{h^2}{a^2(a+h)^2}} =$$

$$\sqrt{\frac{h^2a^2(a+h)^2 + h^2}{a^2(a+h)^2}} = \sqrt{\frac{h^2(a^2(a+h)^2 + 1)}{a^2(a+h)^2}} =$$

$$\left|\frac{h}{a(a+h)}\right| \sqrt{a^2(a+h)^2 + 1}$$

Find the midpoint:

$$\left(\frac{a+a+h}{2}, \frac{\frac{1}{a} + \frac{1}{a+h}}{2}\right) = \left(\frac{2a+h}{2}, \frac{2a+h}{2a(a+h)}\right)$$

117. Use the distance formula. Either point can be considered as (x_1, y_1).

$$\begin{aligned} d &= \sqrt{(a+h-a)^2 + (\sqrt{a+h} - \sqrt{a})^2} \\ &= \sqrt{h^2 + a + h - 2\sqrt{a^2+ah} + a} \\ &= \sqrt{h^2 + 2a + h - 2\sqrt{a^2+ah}} \end{aligned}$$

Next we use the midpoint formula.

$$\left(\frac{a+a+h}{2}, \frac{\sqrt{a}+\sqrt{a+h}}{2}\right) = \left(\frac{2a+h}{2}, \frac{\sqrt{a}+\sqrt{a+h}}{2}\right)$$

118.

$$\begin{aligned} C &= 2\pi r \\ 10\pi &= 2\pi r \\ 5 &= r \end{aligned}$$

Then $[x-(-5)]^2 + (y-8)^2 = 5^2$, or $(x+5)^2 + (y-8)^2 = 25$.

119. First use the formula for the area of a circle to find r^2:

$$\begin{aligned} A &= \pi r^2 \\ 36\pi &= \pi r^2 \\ 36 &= r^2 \end{aligned}$$

Then we have:

$$\begin{aligned} (x-h)^2 + (y-k)^2 &= r^2 \\ (x-2)^2 + [y-(-7)]^2 &= 36 \\ (x-2)^2 + (y+7)^2 &= 36 \end{aligned}$$

120. Let the point be $(x, 0)$. We set the distance from $(-4, -3)$ to $(x, 0)$ equal to the distance from $(-1, 5)$ to $(x, 0)$ and solve for x.

$$\begin{aligned} \sqrt{(-4-x)^2 + (-3-0)^2} &= \sqrt{(-1-x)^2 + (5-0)^2} \\ \sqrt{16 + 8x + x^2 + 9} &= \sqrt{1 + 2x + x^2 + 25} \\ \sqrt{x^2 + 8x + 25} &= \sqrt{x^2 + 2x + 26} \\ x^2 + 8x + 25 &= x^2 + 2x + 26 \quad \text{Squaring both sides} \\ 8x + 25 &= 2x + 26 \\ 6x &= 1 \\ x &= \frac{1}{6} \end{aligned}$$

The point is $\left(\frac{1}{6}, 0\right)$.

121. Let $(0, y)$ be the required point. We set the distance from $(-2, 0)$ to $(0, y)$ equal to the distance from $(4, 6)$ to $(0, y)$ and solve for y.

$$\sqrt{[0-(-2)]^2+(y-0)^2} = \sqrt{(0-4)^2+(y-6)^2}$$
$$\sqrt{4+y^2} = \sqrt{16+y^2-12y+36}$$
$$4+y^2 = 16+y^2-12y+36$$
Squaring both sides
$$-48 = -12y$$
$$4 = y$$

The point is $(0, 4)$.

122. We first find the distance between each pair of points.

For $(-1, -3)$ and $(-4, -9)$:

$$d_1 = \sqrt{[-1-(-4)]^2+[-3-(-9)]^2}$$
$$= \sqrt{3^2+6^2} = \sqrt{9+36}$$
$$= \sqrt{45} = 3\sqrt{5}$$

For $(-1, -3)$ and $(2, 3)$:

$$d_2 = \sqrt{(-1-2)^2+(-3-3)^2}$$
$$= \sqrt{(-3)^2+(-6)^2} = \sqrt{9+36}$$
$$= \sqrt{45} = 3\sqrt{5}$$

For $(-4, -9)$ and $(2, 3)$:

$$d_3 = \sqrt{(-4-2)^2+(-9-3)^2}$$
$$= \sqrt{(-6)^2+(-12)^2} = \sqrt{36+144}$$
$$= \sqrt{180} = 6\sqrt{5}$$

Since $d_1 + d_2 = d_3$, the points are collinear.

123. Label the drawing with additional information and lettering.

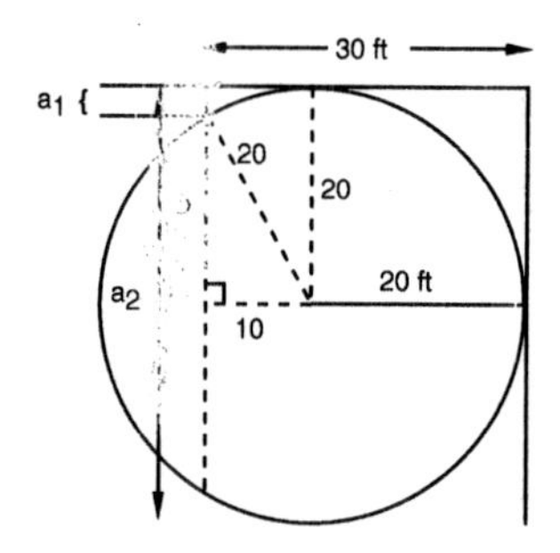

Find b using the Pythagorean theorem.

$$b^2+10^2 = 20^2$$
$$b^2+100 = 400$$
$$b^2 = 300$$
$$b = 10\sqrt{3}$$
$$b \approx 17.3$$

Find a_1:

$$a_1 = 20 - b \approx 20 - 17.3 \approx 2.7 \text{ ft}$$

Find a_2:

$$a_2 = 2b + a_1 \approx 2(17.3) + 2.7 \approx 37.3 \text{ ft}$$

124. a) When the circle is positioned on a coordinate system as shown in the text, the center lies on the y-axis and is equidistant from $(-4, 0)$ and $(0, 2)$.

Let $(0, y)$ be the coordinates of the center.

$$\sqrt{(-4-0)^2+(0-y)^2} = \sqrt{(0-0)^2+(2-y)^2}$$
$$4^2+y^2 = (2-y)^2$$
$$16+y^2 = 4-4y+y^2$$
$$12 = -4y$$
$$-3 = y$$

The center of the circle is $(0, -3)$.

b) Use the point $(-4, 0)$ and the center $(0, -3)$ to find the radius.

$$(-4-0)^2+[0-(-3)]^2 = r^2$$
$$25 = r^2$$
$$5 = r$$

The radius is 5 ft.

125.
$$x^2+y^2 = 1$$
$$\left(\frac{\sqrt{3}}{2}\right)^2+\left(-\frac{1}{2}\right)^2 \ ? \ 1$$
$$\frac{3}{4}+\frac{1}{4} \ \Big|$$
$$1 \ \Big| \ 1 \quad \text{TRUE}$$

$\left(\frac{\sqrt{3}}{2}, -\frac{1}{2}\right)$ lies on the unit circle.

126.
$$x^2+y^2 = 1$$
$$0^2+(-1)^2 \ ? \ 1$$
$$1 \ \Big| \ 1 \quad \text{TRUE}$$

$(0, -1)$ lies on the unit circle.

127.
$$x^2+y^2 = 1$$
$$\left(\frac{\sqrt{2}}{2}\right)^2+\left(\frac{\sqrt{2}}{2}\right)^2 \ ? \ 1$$
$$\frac{2}{4}+\frac{2}{4} \ \Big|$$
$$1 \ \Big| \ 1 \quad \text{TRUE}$$

$\left(\frac{\sqrt{2}}{2}, \frac{\sqrt{2}}{2}\right)$ lies on the unit circle.

128.
$$x^2+y^2 = 1$$
$$\left(\frac{1}{2}\right)^2+\left(-\frac{\sqrt{3}}{2}\right)^2 \ ? \ 1$$
$$\frac{1}{4}+\frac{3}{4} \ \Big|$$
$$1 \ \Big| \ 1 \quad \text{TRUE}$$

$\left(\frac{1}{2}, -\frac{\sqrt{3}}{2}\right)$ lies on the unit circle.

129. a), b) See the answer section in the text.

130. The coordinates of P are $\left(\frac{b}{2}, \frac{h}{2}\right)$ by the midpoint formula. By the distance formula, each of the distances from P to $(0, h)$, from P to $(0, 0)$, and from P to $(b, 0)$ is $\frac{\sqrt{b^2 + h^2}}{2}$.

Exercise Set 1.2

1. This correspondence is a function, because each member of the domain corresponds to exactly one member of the range.

2. This correspondence is a function, because each member of the domain corresponds to exactly one member of the range.

3. This correspondence is a function, because each member of the domain corresponds to exactly one member of the range.

4. This correspondence is not a function, because there is a member of the domain (1) that corresponds to more than one member of the range (4 and 6).

5. This correspondence is not a function, because there is a member of the domain (m) that corresponds to more than one member of the range (A and B).

6. This correspondence is a function, because each member of the domain corresponds to exactly one member of the range.

7. This correspondence is a function, because each member of the domain corresponds to exactly one member of the range.

8. This correspondence is a function, because each member of the domain corresponds to exactly one member of the range.

9. This correspondence is a function, because each car has exactly one license number.

10. This correspondence is not a function, because we can safely assume that at least one person uses more than one doctor.

11. This correspondence is a function, because each member of the family has exactly one eye color.

12. This correspondence is not a function, because we can safely assume that at least one band member plays more than one instrument.

13. This correspondence is not a function, because at least one student will have more than one neighboring seat occupied by another student.

14. This correspondence is a function, because each bag has exactly one weight.

15. The relation is a function, because no two ordered pairs have the same first coordinate and different second coordinates.

    The domain is the set of all first coordinates: $\{2, 3, 4\}$.

    The range is the set of all second coordinates: $\{10, 15, 20\}$.

16. The relation is a function, because no two ordered pairs have the same first coordinate and different second coordinates.

    Domain: $\{3, 5, 7\}$

    Range: $\{1\}$

17. The relation is not a function, because the ordered pairs $(-2, 1)$ and $(-2, 4)$ have the same first coordinate and different second coordinates.

    The domain is the set of all first coordinates: $\{-7, -2, 0\}$.

    The range is the set of all second coordinates: $\{3, 1, 4, 7\}$.

18. The relation is not a function, because of each of the ordered pairs has the same first coordinate and different second coordinates.

    Domain: $\{1\}$

    Range: $\{3, 5, 7, 9\}$

19. The relation is a function, because no two ordered pairs have the same first coordinate and different second coordinates.

    The domain is the set of all first coordinates: $\{-2, 0, 2, 4, -3\}$.

    The range is the set of all second coordinates: $\{1\}$.

20. The relation is not a function, because the ordered pairs $(5, 0)$ and $(5, -1)$ have the same first coordinates and different second coordinates. This is also true of the pairs $(3, -1)$ and $(3, -2)$.

    Domain: $\{5, 3, 0\}$

    Range: $\{0, -1, -2\}$

21. From the graph we see that, when the input is 1, the output is -2, so $h(1) = -2$. When the input is 3, the output is 2, so $h(3) = 2$. When the input is 4, the output is 1, so $h(4) = 1$.

22. $t(-4) = 3$; $t(0) = 3$; $t(3) = 3$

23. From the graph we see that, when the input is -4, the output is 3, so $s(-4) = 3$. When the input is -2, the output is 0, so $s(-2) = 0$. When the input is 0, the output is -3, so $s(0) = -3$.

24. $g(-4) = \frac{3}{2}$; $g(-1) = -3$; $g(0) = -\frac{5}{2}$

25. From the graph we see that, when the input is -1, the output is 2, so $f(-1) = 2$. When the input is 0, the output is 0, so $f(0) = 0$. When the input is 1, the output is -2, so $f(1) = -2$.

26. $g(-2) = 4$; $g(0) = -4$; $g(2.4) = -2.6176$

27. $g(x) = 3x^2 - 2x + 1$

    a) $g(0) = 3 \cdot 0^2 - 2 \cdot 0 + 1 = 1$

    b) $g(-1) = 3(-1)^2 - 2(-1) + 1 = 6$

    c) $g(3) = 3 \cdot 3^2 - 2 \cdot 3 + 1 = 22$

    d) $g(-x) = 3(-x)^2 - 2(-x) + 1 = 3x^2 + 2x + 1$

e) $g(1-t) = 3(1-t)^2 - 2(1-t) + 1 =$
$3(1-2t+t^2) - 2(1-t) + 1 = 3 - 6t + 3t^2 - 2 + 2t + 1 =$
$3t^2 - 4t + 2$

28. $f(x) = 5x^2 + 4x$

a) $f(0) = 5 \cdot 0^2 + 4 \cdot 0 = 0 + 0 = 0$

b) $f(-1) = 5(-1)^2 + 4(-1) = 5 - 4 = 1$

c) $f(3) = 5 \cdot 3^2 + 4 \cdot 3 = 45 + 12 = 57$

d) $f(t) = 5t^2 + 4t$

e) $f(t-1) = 5(t-1)^2 + 4(t-1) = 5t^2 - 6t + 1$

29. $g(x) = x^3$

a) $g(2) = 2^3 = 8$

b) $g(-2) = (-2)^3 = -8$

c) $g(-x) = (-x)^3 = -x^3$

d) $g(3y) = (3y)^3 = 27y^3$

e) $g(2+h) = (2+h)^3 = 8 + 12h + 6h^2 + h^3$

30. $f(x) = 2|x| + 3x$

a) $f(1) = 2|1| + 3 \cdot 1 = 2 + 3 = 5$

b) $f(-2) = 2|-2| + 3(-2) = 4 - 6 = -2$

c) $f(-x) = 2|-x| + 3(-x) = 2|x| - 3x$

d) $f(2y) = 2|2y| + 3 \cdot 2y = 4|y| + 6y$

e) $f(2-h) = 2|2-h| + 3(2-h) =$
$2|2-h| + 6 - 3h$

31. $g(x) = \dfrac{x-4}{x+3}$

a) $g(5) = \dfrac{5-4}{5+3} = \dfrac{1}{8}$

b) $g(4) = \dfrac{4-4}{4+7} = 0$

c) $g(-3) = \dfrac{-3-4}{-3+3} = \dfrac{-7}{0}$

Since division by 0 is not defined, $g(-3)$ does not exist.

d) $g(-16.25) = \dfrac{-16.25-4}{-16.25+3} = \dfrac{-20.25}{-13.25} = \dfrac{81}{53}$

e) $g(x+h) = \dfrac{x+h-4}{x+h+3}$

32. $f(x) = \dfrac{x}{2-x}$

a) $f(2) = \dfrac{2}{2-2} = \dfrac{2}{0}$

Since division by 0 is not defined, $f(2)$ does not exist.

b) $f(1) = \dfrac{1}{2-1} = 1$

c) $f(-16) = \dfrac{-16}{2-(-16)} = \dfrac{-16}{18} = -\dfrac{8}{9}$

d) $f(-x) = \dfrac{-x}{2-(-x)} = \dfrac{-x}{2+x}$

e) $f\left(-\dfrac{2}{3}\right) = \dfrac{-\dfrac{2}{3}}{2-\left(-\dfrac{2}{3}\right)} = \dfrac{-\dfrac{2}{3}}{\dfrac{8}{3}} = -\dfrac{1}{4}$

33. $g(x) = \dfrac{x}{\sqrt{1-x^2}}$

$g(0) = \dfrac{0}{\sqrt{1-0^2}} = \dfrac{0}{\sqrt{1}} = \dfrac{0}{1} = 0$

$g(-1) = \dfrac{-1}{\sqrt{1-(-1)^2}} = \dfrac{-1}{\sqrt{1-1}} = \dfrac{-1}{\sqrt{0}} = \dfrac{-1}{0}$

Since division by 0 is not defined, $g(-1)$ does not exist.

$g(5) = \dfrac{5}{\sqrt{1-5^2}} = \dfrac{5}{\sqrt{1-25}} = \dfrac{5}{\sqrt{-24}}$

Since $\sqrt{-24}$ is not defined as a real number, $g(5)$ does not exist as a real number.

$g\left(\dfrac{1}{2}\right) = \dfrac{\dfrac{1}{2}}{\sqrt{1-\left(\dfrac{1}{2}\right)^2}} = \dfrac{\dfrac{1}{2}}{\sqrt{1-\dfrac{1}{4}}} = \dfrac{\dfrac{1}{2}}{\sqrt{\dfrac{3}{4}}} =$

$\dfrac{\dfrac{1}{2}}{\dfrac{\sqrt{3}}{2}} = \dfrac{1}{2} \cdot \dfrac{2}{\sqrt{3}} = \dfrac{1 \cdot 2}{2\sqrt{3}} = \dfrac{1}{\sqrt{3}}$, or $\dfrac{\sqrt{3}}{3}$

34. $h(x) = x + \sqrt{x^2-1}$

$h(0) = 0 + \sqrt{0^2-1} = 0 + \sqrt{-1}$

Since $\sqrt{-1}$ is not defined as a real number, $h(0)$ does not exist as a real number.

$h(2) = 2 + \sqrt{2^2-1} = 2 + \sqrt{3}$

$h(-x) = -x + \sqrt{(-x)^2-1} = -x + \sqrt{x^2-1}$

35. We can substitute any real number for x. Thus, the domain is the set of all real numbers, or $(-\infty, \infty)$.

36. We can substitute any real number for x. Thus, the domain is the set of all real numbers, or $(-\infty, \infty)$.

37. The input 0 results in a denominator of 0. Thus, the domain is $\{x|x \neq 0\}$, or $(-\infty, 0) \cup (0, \infty)$.

38. The input 0 results in a denominator of 0. Thus, the domain is $\{x|x \neq 0\}$, or $(-\infty, 0) \cup (0, \infty)$.

39. We can substitute any real number in the numerator, but we must avoid inputs that make the denominator 0. We find these inputs.

$2 - x = 0$

$2 = x$

The domain is $\{x|x \neq 2\}$, or $(-\infty, 2) \cup (2, \infty)$.

40. We find the inputs that make the denominator 0:

$x + 4 = 0$

$x = -4$

The domain is $\{x|x \neq -4\}$, or $(-\infty, -4) \cup (-4, \infty)$.

41. We find the inputs that make the denominator 0:

$x^2 - 4x - 5 = 0$

$(x-5)(x+1) = 0$

$x - 5 = 0$ *or* $x + 1 = 0$

$x = 5$ *or* $x = -1$

The domain is $\{x|x \neq 5$ *and* $x \neq -1\}$, or $(-\infty, -1) \cup (-1, 5) \cup (5, \infty)$.

42. We can substitute any real number in the numerator. Find the inputs that make the denominator 0:

$$3x^2 - 10x - 8 = 0$$
$$(3x + 2)(x - 4) = 0$$
$$x = -\frac{2}{3} \text{ or } x = 4$$

Domain: $\left\{x \middle| x \neq -\frac{2}{3} \text{ and } x \neq 4\right\}$, or

$\left(-\infty, -\frac{2}{3}\right) \cup \left(\frac{2}{3}, 4\right) \cup (4, \infty)$

43. We can substitute any real number for which the radicand is nonnegative. We see that $8 - x \geq 0$ for $x \leq 8$, so the domain is $\{x|x \leq 8\}$, or $(-\infty, 8]$.

44. We can substitute any real number for x. Thus, the domain is the set of all real numbers, or $(-\infty, \infty)$.

45. We can substitute any real number for x. Thus, the domain is the set of all real numbers, or $(-\infty, \infty)$.

46. In the numerator we can substitute any real number for which the radicand is nonnegative. We see that $x + 1 \geq 0$ for $x \geq -1$. The denominator is 0 when $x = 0$, so 0 cannot be an input. Thus the domain is $\{x|x \geq -1 \text{ and } x \neq 0\}$, or $[-1, 0) \cup (0, \infty)$.

47.

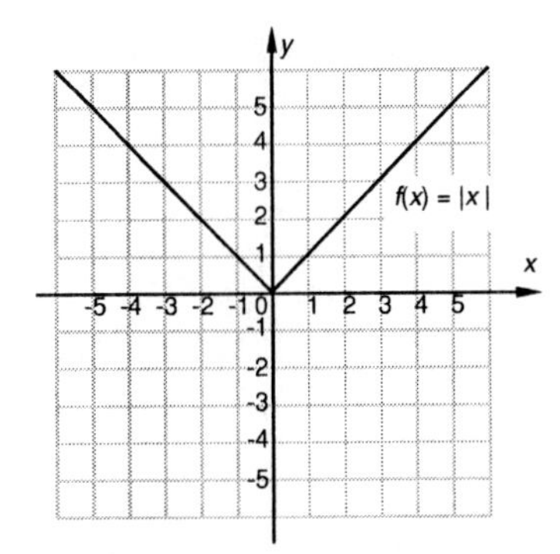

To find the domain we look for the inputs on the x-axis that correspond to a point on the graph. We see that each point on the x-axis corresponds to a point on the graph so the domain is the set of all real numbers, or $(-\infty, \infty)$.

To find the range we look for outputs on the y-axis. The number 0 is the smallest output, and every number greater than 0 is also an output. Thus, the range is $[0, \infty)$.

48.

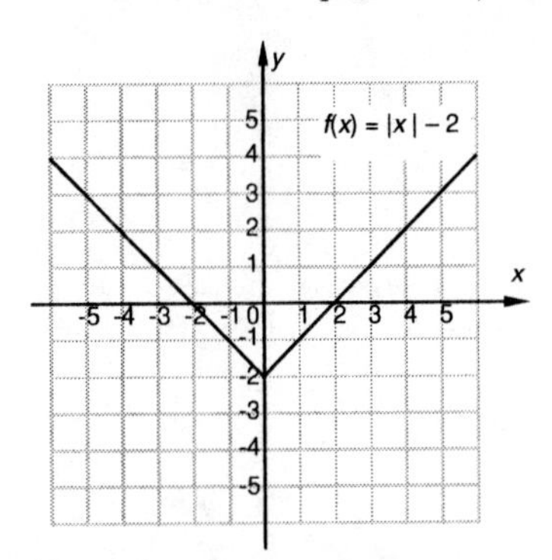

Domain: all real numbers, or $(-\infty, \infty)$

Range: $[-2, \infty)$

49.

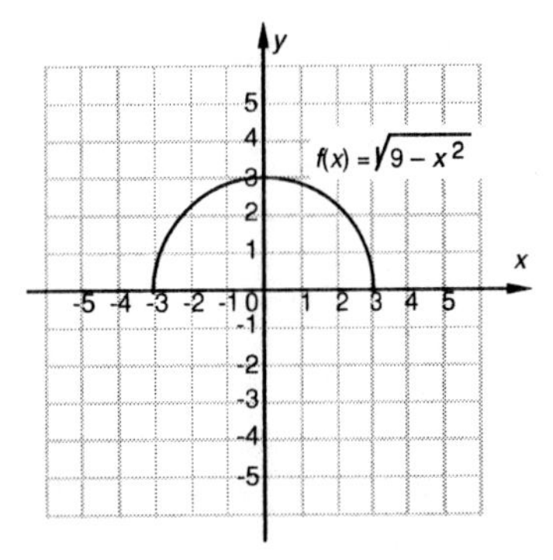

The inputs on the x-axis extend from -3 to 3, so the domain is $[-3, 3]$.

The outputs on the y-axis extend from 0 to 3, so the range is $[0, 3]$.

50.

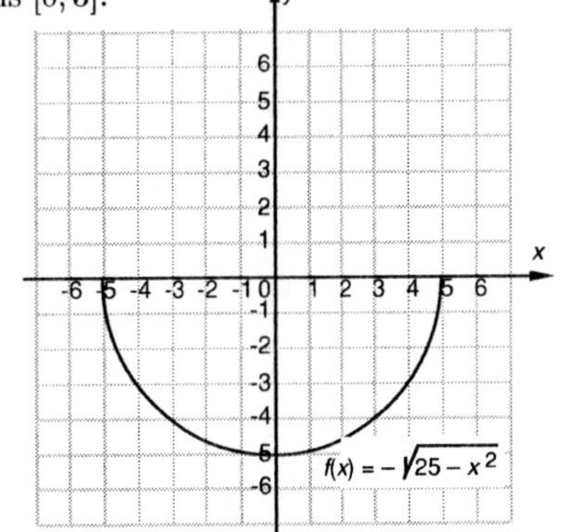

Domain: $[-5, 5]$

Range: $[-5, 0]$

51.

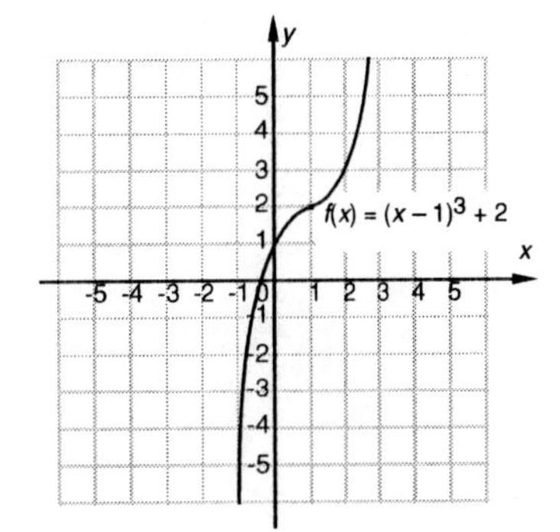

Each point on the x-axis corresponds to a point on the graph, so the domain is the set of all real numbers, or $(-\infty, \infty)$.

Each point on the y-axis also corresponds to a point on the graph, so the range is the set of all real numbers, $(-\infty, \infty)$.

52.

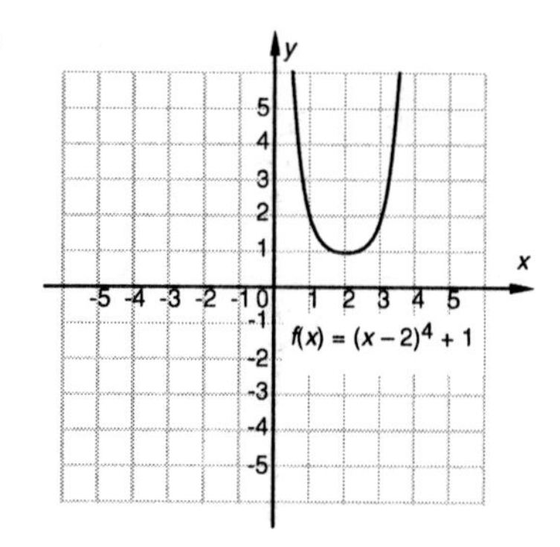

Domain: all real numbers, or $(-\infty, \infty)$

Range: $[1, \infty)$

53.

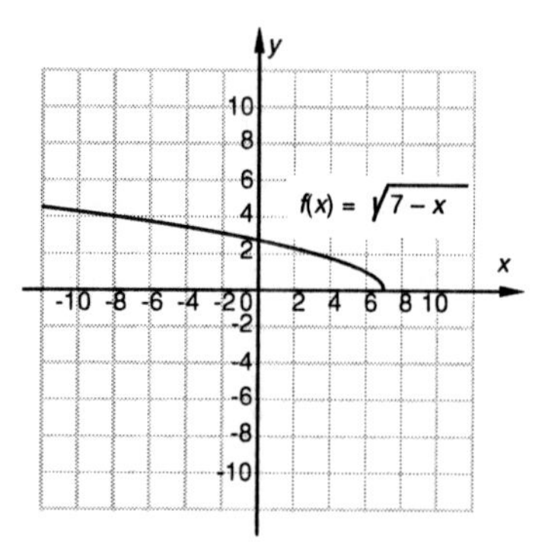

The largest input on the x-axis is 7 and every number less than 7 is also an input. Thus, the domain is $(-\infty, 7]$.

The number 0 is the smallest output, and every number greater than 0 is also an output. Thus, the range is $[0, \infty)$.

54.

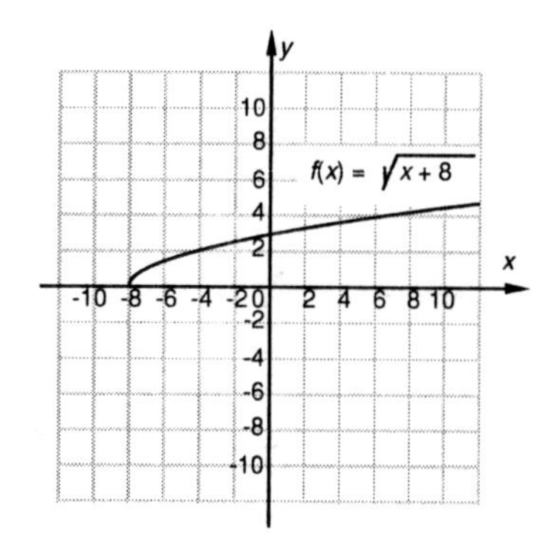

Domain: $[-8, \infty)$

Range: $[0, \infty)$

55.

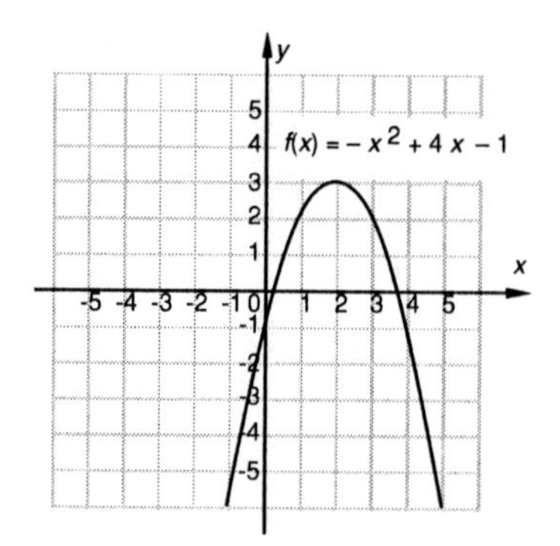

Each point on the x-axis corresponds to a point on the graph, so the domain is the set of all real numbers, or $(-\infty, \infty)$.

The largest output is 3 and every number less than 3 is also an output. Thus, the range is $(-\infty, 3]$.

56.

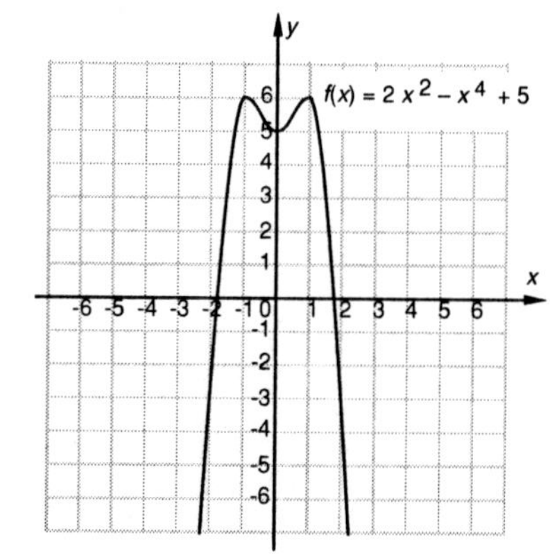

Domain: all real numbers, or $(-\infty, \infty)$

Range: $(-\infty, 6]$

57. This is not the graph of a function, because we can find a vertical line that crosses the graph more than once.

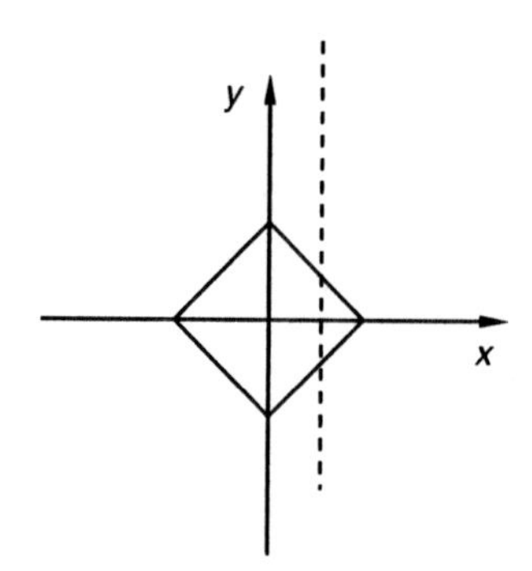

58. This is not the graph of a function, because we can find a vertical line that crosses the graph more than once.

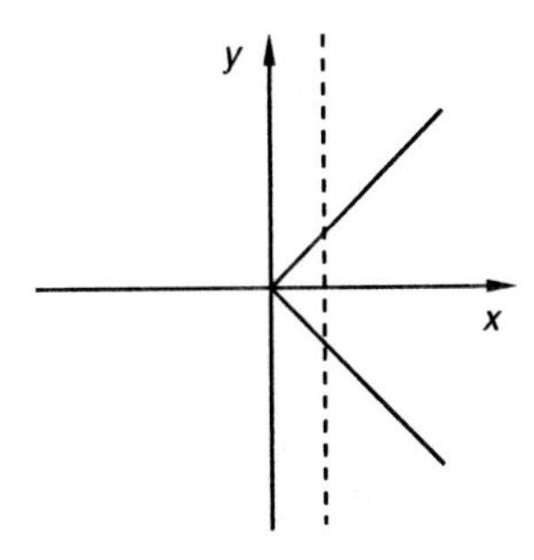

59. This is the graph of a function, because there is no vertical line that crosses the graph more than once.

60. This is the graph of a function, because there is no vertical line that crosses the graph more than once.

61. This is the graph of a function, because there is no vertical line that crosses the graph more than once.

62. This is the graph of a function, because there is no vertical line that crosses the graph more than once.

63. This is not the graph of a function, because we can find a vertical line that crosses the graph more than once.

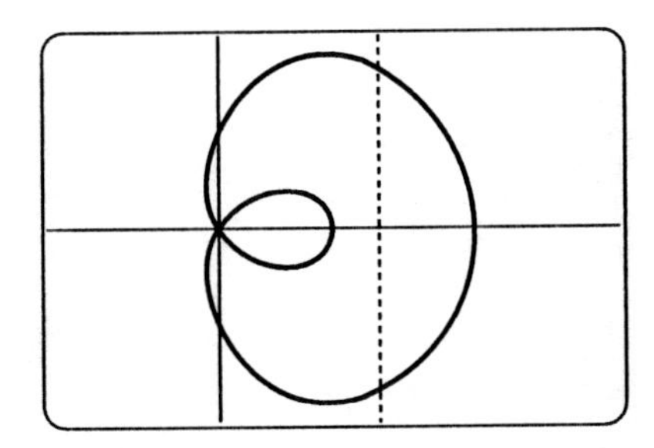

64. This is not the graph of a function, because we can find a vertical line that crosses the graph more than once.

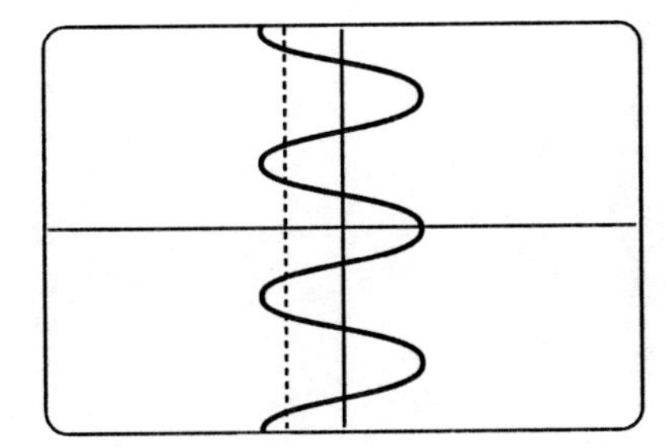

65. This is the graph of a function, because there is no vertical line that crosses the graph more than once.

    The inputs on the x-axis that correspond to points on the graph extend from 0 to 5, inclusive. Thus, the domain is $\{x|0 \le x \le 5\}$, or $[0, 5]$.

    The outputs on the y-axis extend from 0 to 3, inclusive. Thus, the range is $\{y|0 \le y \le 3\}$, or $[0, 3]$.

66. This is the graph of a function, because there is no vertical line that crosses the graph more than once.

    The inputs on the x-axis that correspond to points on the graph extend from -3 up to but not including 5. Thus, the domain is $\{x|-3 \le x < 5\}$, or $[-3, 5)$.

    The outputs on the y-axis extend from -4 up to but not including 1. Thus, the range is $\{y|-4 \le y < 1\}$, or $[-4, 1)$.

67. This is the graph of a function, because there is no vertical line that crosses the graph more than once.

    The inputs on the x-axis that correspond to points on the graph extend from -2π to 2π inclusive. Thus, the domain is $\{x|-2\pi \le x \le 2\pi\}$, or $[-2\pi, 2\pi]$.

    The outputs on the y-axis extend from -1 to 1, inclusive. Thus, the range is $\{y|-1 \le y \le 1\}$, or $[-1, 1]$.

68. This is the graph of a function, because there is no vertical line that crosses the graph more than once.

    The inputs on the x-axis that correspond to points on the graph extend from -2 to 1, inclusive. Thus, the domain is $\{x|-2 \le x \le 1\}$, or $[-2, 1]$.

    The outputs on the y-axis extend from -1 to 4, inclusive. Thus, the range is $\{y|-1 \le y \le 4\}$, or $[-1, 4]$.

69. $E(t) = 1000(100 - t) + 580(100 - t)^2$

    a) $E(99.5) = 1000(100-99.5)+580(100-99.5)^2$
    $= 1000(0.5) + 580(0.5)^2$
    $= 500 + 580(0.25) = 500 + 145$
    $= 645$ m above sea level

    b) $E(100) = 1000(100 - 100) + 580(100 - 100)^2$
    $= 1000 \cdot 0 + 580(0)^2 = 0 + 0$
    $= 0$ m above sea level, or at sea level

70. a) $P(2005) = 0.1522(2005) - 298.592 \approx \$6.57$
    $P(2010) = 0.1522(2010) - 298.592 = \$7.33$

    b) Solve: $8 = 0.1522x - 298.592$
    $x \approx 2014$

71. $T(0.5) = 0.5^{1.31} \approx 0.4$ acre
    $T(10) = 10^{1.31} \approx 20.4$ acres
    $T(20) = 20^{1.31} \approx 50.6$ acres
    $T(100) = 100^{1.31} \approx 416.9$ acres
    $T(200) = 200^{1.31} \approx 1033.6$ acres

72.

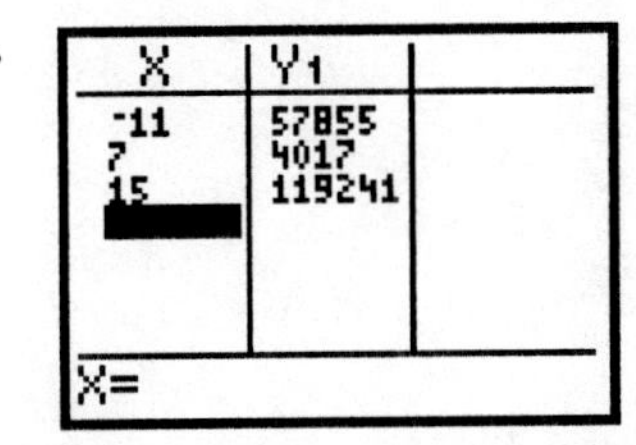

We see that $h(-11) = 57,885$, $h(7) = 4017$, and $h(15) = 119,241$.

73.

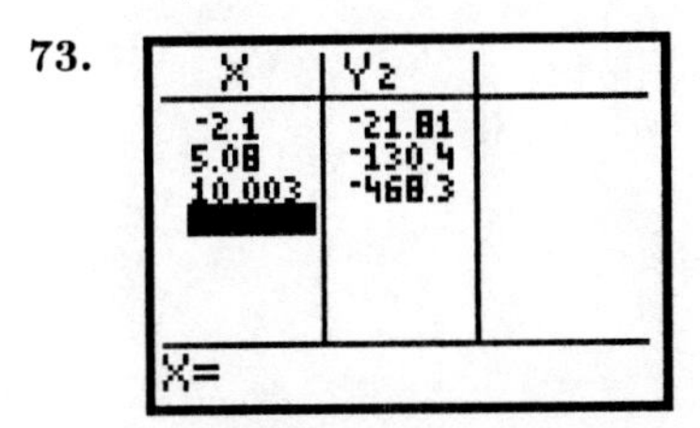

Rounding to the nearest tenth, we see that $g(-2.1) \approx -21.8$, $g(5.08) \approx -130.4$, and $g(10.003) \approx -468.3$.

74. a)

X	Y3
-4	-.25
-6	ERROR

X=

We see that $f(-4) = -0.25$. The ERROR message in the table indicates that $f(-6)$ does not exist.

b)

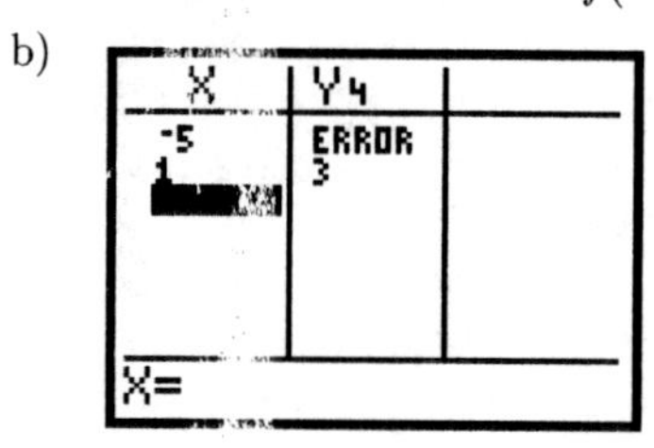

The ERROR message in the table indicates that $g(-5)$ does not exist. We see that $g(1) = 3$.

75. A function is a correspondence between two sets in which each member of the first set corresponds to exactly one member of the second set.

76. The domain of a function is the set of all inputs of the function. The range is the set of all outputs. The range depends on the domain.

77. To determine whether $(0, -7)$ is a solution, substitute 0 for x and -7 for y.

$$\begin{array}{c|l} y & = 0.5x + 7 \\ \hline -7 & ?\ 0.5(0) + 7 \\ & 0 + 7 \\ -7 & 7 \qquad \text{FALSE} \end{array}$$

The equation $-7 = 7$ is false, so $(0, -7)$ is not a solution.

To determine whether $(8, 11)$ is a solution, substitute 8 for x and 11 for y.

$$\begin{array}{c|l} y & = 0.5x + 7 \\ \hline 11 & ?\ 0.5(8) + 7 \\ & 4 + 7 \\ 11 & 11 \qquad \text{TRUE} \end{array}$$

The equation $11 = 11$ is true, so $(8, 11)$ is a solution.

78. For: $\left(\frac{4}{5}, -2\right)$:

$$\begin{array}{c|l} 15x - 10y & = 32 \\ \hline 15 \cdot \frac{4}{5} - 10(-2) & ?\ 32 \\ 12 + 20 & \\ 32 & 32 \qquad \text{TRUE} \end{array}$$

$\left(\frac{4}{5}, -2\right)$ is a solution.

For: $\left(\frac{11}{5}, \frac{1}{10}\right)$:

$$\begin{array}{c|l} 15x - 10y & = 32 \\ \hline 15 \cdot \frac{11}{5} - 10 \cdot \frac{1}{10} & ?\ 32 \\ 33 - 1 & \\ 32 & 32 \qquad \text{TRUE} \end{array}$$

$\left(\frac{11}{5}, \frac{1}{10}\right)$ is a solution.

79. Graph $y = (x - 1)^2$.

Make a table of values, plot the points in the table, and draw the graph.

x	y	(x, y)
-1	4	$(-1, 4)$
0	1	$(0, 1)$
1	0	$(1, 0)$
2	1	$(2, 1)$
3	4	$(3, 4)$

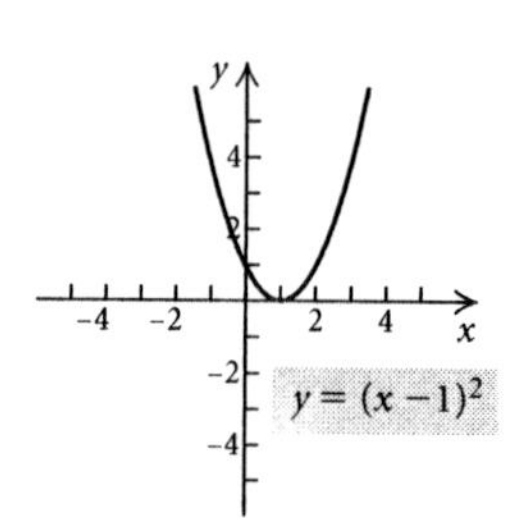

80.

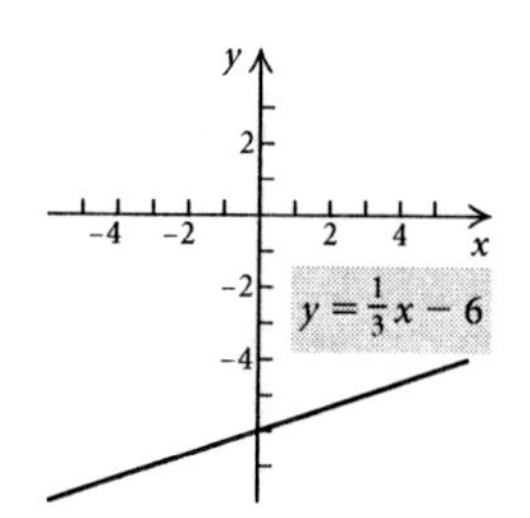

81. Graph $-2x - 5y = 10$.

Make a table of values, plot the points in the table, and draw the graph.

x	y	(x, y)
-5	0	$(-5, 0)$
0	-2	$(0, -2)$
5	-4	$(5, -4)$

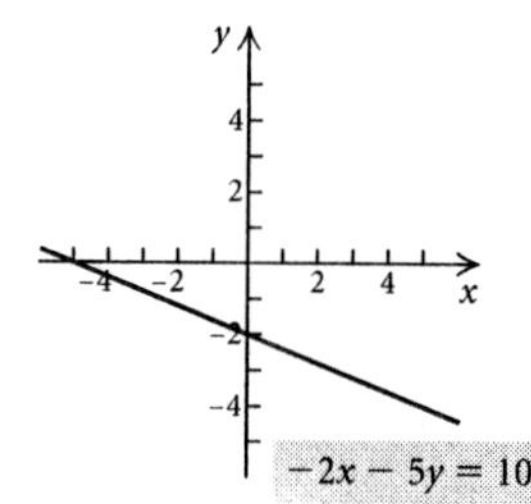

82.

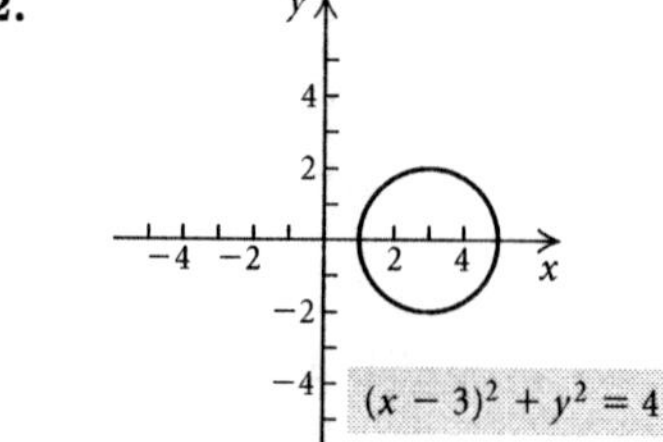

83. Answers may vary. Two possibilities are $f(x) = x$, $g(x) = x + 1$ and $f(x) = x^2$, $g(x) = x^2 - 4$.

84.

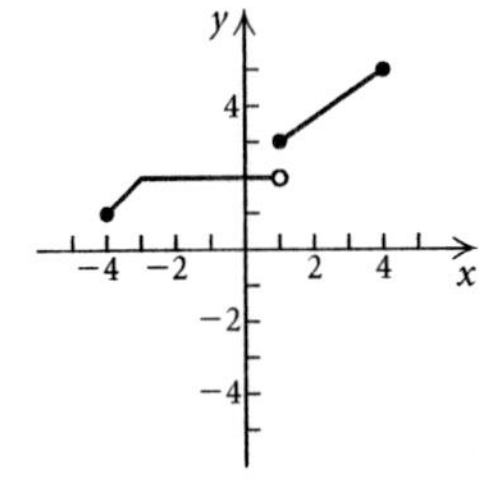

85.

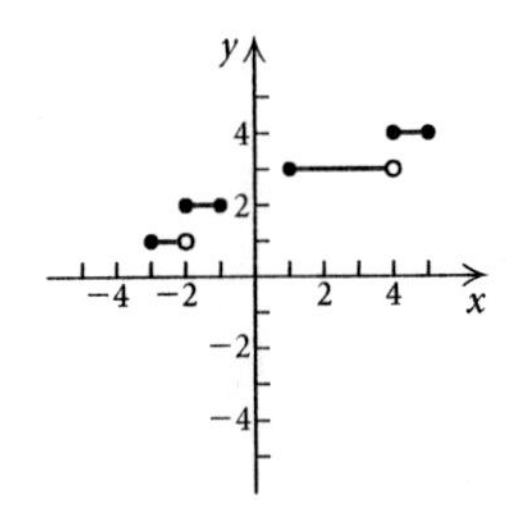

86. $f(x-1) = 5x$

$f(6) = f(7-1) = 5 \cdot 7 = 35$

87. First find the value of x for which $x+3 = -1$.

$$x + 3 = -1$$
$$x = -4$$

Then we have:

$g(x+3) = 2x+1$

$g(-1) = g(-4+3) = 2(-4)+1 = -8+1 = -7$

88. $f(x) = |x+3| - |x-4|$

a) If x is in the interval $(-\infty, -3)$, then $x+3 < 0$ and $x-4 < 0$. We have:

$$\begin{aligned} f(x) &= |x+3| - |x-4| \\ &= -(x+3) - [-(x-4)] \\ &= -(x+3) - (-x+4) \\ &= -x - 3 + x - 4 \\ &= -7 \end{aligned}$$

b) If x is in the interval $[-3, 4)$, then $x+3 \geq 0$ and $x-4 < 0$. We have:

$$\begin{aligned} f(x) &= |x+3| - |x-4| \\ &= x+3 - [-(x-4)] \\ &= x+3 - (-x+4) \\ &= x+3+x-4 \\ &= 2x-1 \end{aligned}$$

c) If x is in the interval $[4, \infty)$, then $x+3 > 0$ and $x-4 \geq 0$. We have:

$$\begin{aligned} f(x) &= |x+3| - |x-4| \\ &= x+3 - (x-4) \\ &= x+3-x+4 \\ &= 7 \end{aligned}$$

89. $f(x) = |x| + |x-1|$

a) If x is in the interval $(-\infty, 0)$, then $x < 0$ and $x-1 < 0$. We have:

$$\begin{aligned} f(x) &= |x| + |x-1| \\ &= -x - (x-1) \\ &= -x - x + 1 \\ &= -2x + 1 \end{aligned}$$

b) If x is in the interval $[0, 1)$, then $x \geq 0$ and $x-1 < 0$. We have:

$$\begin{aligned} f(x) &= |x| + |x-1| \\ &= x - (x-1) \\ &= x - x + 1 \\ &= 1 \end{aligned}$$

c) If x is in the interval $[1, \infty)$, then $x > 0$ and $x-1 \geq 0$. We have:

$$\begin{aligned} f(x) &= |x| + |x-1| \\ &= x + x - 1 \\ &= 2x + 1 \end{aligned}$$

Exercise Set 1.3

1. a) Yes. Each input is 1 more than the one that precedes it.

b) Yes. Each output is 3 more than the one that precedes it.

c) Yes. Constant changes in inputs result in constant changes in outputs.

2. a) Yes. Each input is 10 more than the one that precedes it.

b) No. The change in the outputs varies.

c) No. Constant changes in inputs do not result in constant changes in outputs.

3. a) Yes. Each input is 15 more than the one that precedes it.

b) No. The change in the outputs varies.

c) No. Constant changes in inputs do not result in constant changes in outputs.

4. a) Yes. Each input is 2 more than the one that precedes it.

b) Yes. Each output is 4 less than the one that precedes it.

c) Yes. Constant changes in inputs result in constant changes in outputs.

5. Two points on the line are $(0, 3)$ and $(5, 0)$.

$$m = \frac{y_2 - y_1}{x_2 - x_1} = \frac{0-3}{5-0} = \frac{-3}{5}, \text{ or } -\frac{3}{5}$$

6. $m = \dfrac{0-(-3)}{-2-(-2)} = \dfrac{3}{0}$

The slope is not defined.

7. $m = \dfrac{y_2 - y_1}{x_2 - x_1} = \dfrac{3-3}{3-0} = \dfrac{0}{3} = 0$

8. $m = \dfrac{1-(-4)}{5-(-3)} = \dfrac{5}{8}$

9. $m = \dfrac{y_2 - y_1}{x_2 - x_1} = \dfrac{2-4}{-1-9} = \dfrac{-2}{-10} = \dfrac{1}{5}$

10. $m = \dfrac{-1-7}{5-(-3)} = \dfrac{-8}{8} = -1$

11. $m = \dfrac{y_2 - y_1}{x_2 - x_1} = \dfrac{6-(-9)}{-5-4} = \dfrac{15}{-9} = -\dfrac{5}{3}$

12. $m = \dfrac{-13-(-1)}{2-(-6)} = \dfrac{-12}{8} = -\dfrac{3}{2}$

13. $m = \dfrac{y_2 - y_1}{x_2 - x_1} = \dfrac{-0.4-(-0.1)}{-0.3-0.7} = \dfrac{-0.3}{-1} = 0.3$

14. $m = \dfrac{-\frac{5}{7} - \left(-\frac{1}{4}\right)}{\frac{2}{7} - \left(-\frac{3}{4}\right)} = \dfrac{-\frac{20}{28} + \frac{7}{28}}{\frac{8}{28} + \frac{21}{28}} = \dfrac{-\frac{13}{28}}{\frac{29}{28}} =$

$-\dfrac{13}{28} \cdot \dfrac{28}{29} = -\dfrac{13}{29}$

15. $m = \dfrac{y_2 - y_1}{x_2 - x_1} = \dfrac{-2-(-2)}{4-2} = \dfrac{0}{2} = 0$

16. $m = \dfrac{-6-8}{7-(-9)} = \dfrac{-14}{16} = -\dfrac{7}{8}$

17. $m = \dfrac{y_2 - y_1}{x_2 - x_1} = \dfrac{\frac{3}{5} - \left(-\frac{3}{5}\right)}{-\frac{1}{2} - \frac{1}{2}} = \dfrac{\frac{6}{5}}{-1} = -\dfrac{6}{5}$

18. $m = \dfrac{-2.16 - 4.04}{3.14 - (-8.26)} = \dfrac{-6.2}{11.4} = -\dfrac{62}{114} = -\dfrac{31}{57}$

19. $m = \dfrac{y_2 - y_1}{x_2 - x_1} = \dfrac{-5-(-13)}{-8-16} = \dfrac{8}{-24} = -\dfrac{1}{3}$

20. $m = \dfrac{7-(-7)}{10-(-10)} = \dfrac{14}{0}$

The slope is not defined.

21. $m = \dfrac{y_2 - y_1}{x_2 - x_1} = \dfrac{2-(-3)}{\pi - \pi} = \dfrac{5}{0}$

Since division by 0 is not defined, the slope is not defined.

22. $m = \dfrac{-4-(-4)}{0.56-\sqrt{2}} = \dfrac{0}{0.56-\sqrt{2}} = 0$

23. $y = -\dfrac{1}{2}x + 3$

We find and plot two ordered pairs on the line and connect the points.

When $x = 0$, $y = -\dfrac{1}{2} \cdot 0 + 3 = 3$.

When $x = 2$, $y = -\dfrac{1}{2} \cdot 2 + 3 = -1 + 3 = 2$.

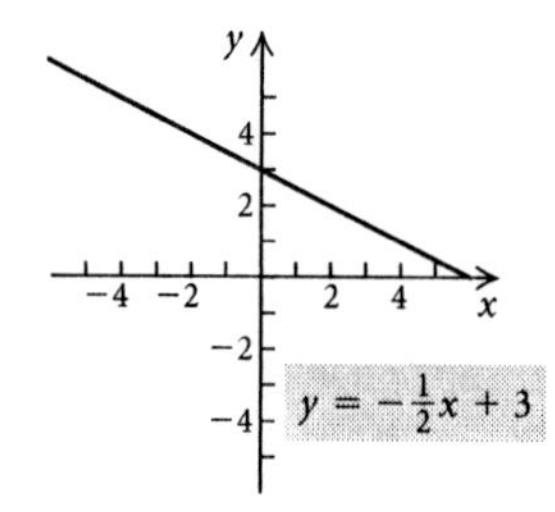

Since the equation is in the form $y = mx + b$ with $m = -\dfrac{1}{2}$, we know that the slope is $-\dfrac{1}{2}$.

24. 

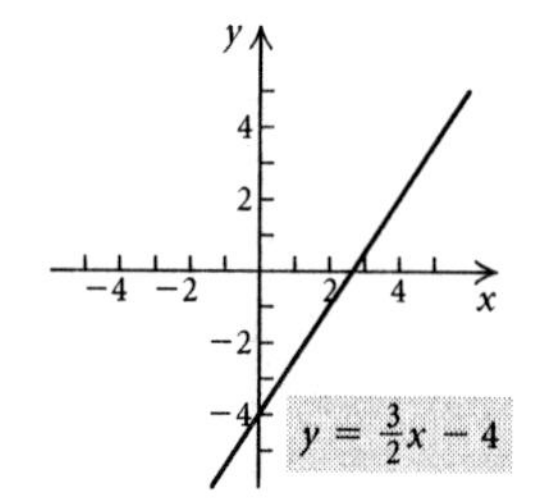

$m = \dfrac{3}{2}$

25. $2y - 3x = -6$

We find and plot two ordered pairs on the line and connect the points.

First let $y = 0$ and solve for x.

$$2 \cdot 0 - 3x = -6$$
$$-3x = -6$$
$$x = 2$$

The point $(2, 0)$ is on the graph.

Next let $x = 0$ and solve for y.

$$2y - 3 \cdot 0 = -6$$
$$2y = -6$$
$$y = -3$$

The point $(0, -3)$ is on the graph.

Plot the points $(2, 0)$ and $(0, -3)$ and connect them.

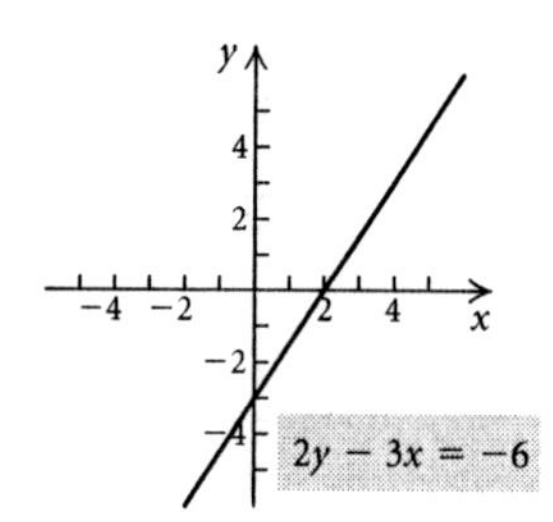

Now find the slope.

$m = \dfrac{y_2 - y_1}{x_2 - x_1} = \dfrac{-3-0}{0-2} = \dfrac{-3}{-2} = \dfrac{3}{2}$

26.

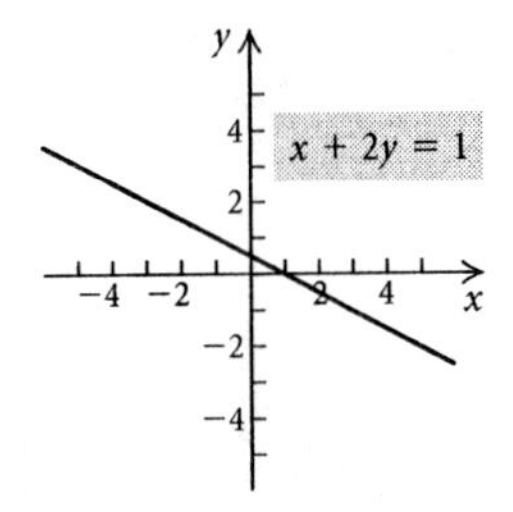

Two points on the graph are $(-3, 2)$ and $(1, 0)$.

$$m = \frac{0-2}{1-(-3)} = \frac{-2}{4} = -\frac{1}{2}$$

27. $5x + 2y = 10$

We find and plot two ordered pairs on the line and connect the points.

First let $y = 0$ and solve for x.

$$5x + 2 \cdot 0 = 10$$
$$5x = 10$$
$$x = 2$$

The point $(2, 0)$ is on the graph.

Next let $x = 0$ and solve for y.

$$5 \cdot 0 + 2y = 10$$
$$2y = 10$$
$$y = 5$$

The point $(0, 5)$ is on the graph.

Plot the points $(2, 0)$ and $(0, 5)$ and connect them.

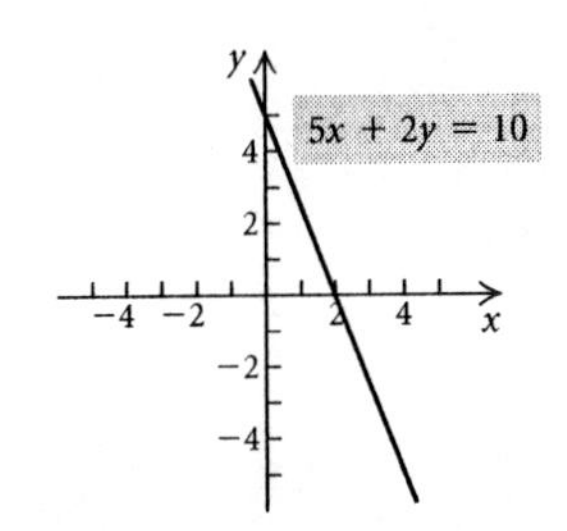

Now find the slope

$$m = \frac{5-0}{0-2} = \frac{5}{-2} = -\frac{5}{2}$$

28.

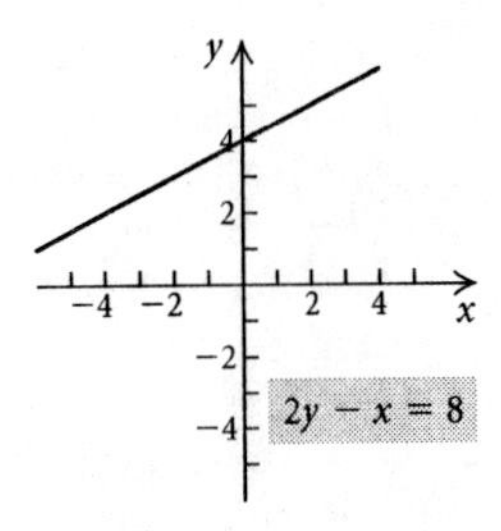

Two points on the graph are $(0, 4)$ and $(2, 5)$.

$$m = \frac{5-4}{2-0} = \frac{1}{2}$$

29. $y = -\dfrac{2}{3}$

This is the equation of a horizontal line whose graph is $\dfrac{2}{3}$ units below the x-axis.

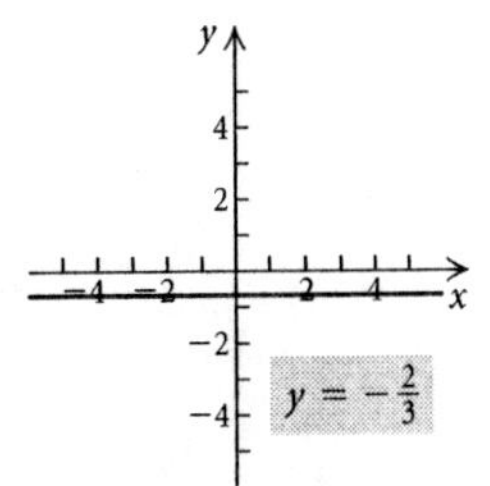

The slope of a horizontal line is 0.

30.

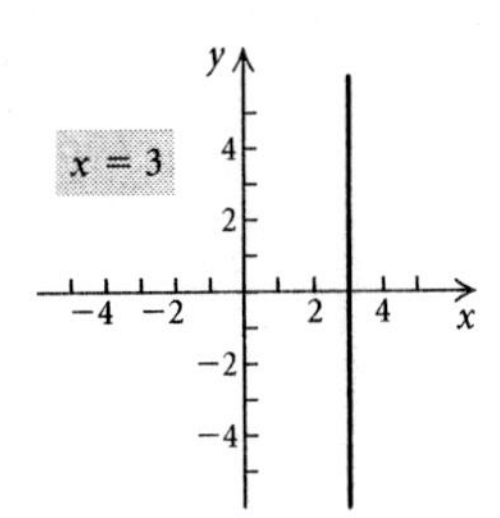

The slope of a vertical line is not defined.

31. We use the points $(1992, \$3275)$ and $(2002, \$8940)$ and find the average rate of change, or slope.

$$m = \frac{\$8940 - \$3275}{2002 - 1992} = \frac{\$5665}{10} = \$566.5$$

The average rate of change was \$566.50 per year.

32. $m = \dfrac{5.6 - 1.0}{2002 - 1999} = \dfrac{4.6}{3} \approx 1.5$

The average rate of change was about 1.5 billion e-mail messages per year.

33. We use the points $(1992, 6472)$ and $(2002, 20,099)$ and find the average rate of change, or slope.

$$m = \frac{20,099 - 6472}{2002 - 1992} = \frac{13,627}{10} = 1362.7 \approx 1363$$

The average rate of change was about 1363 visas per year.

34. $m = \dfrac{6062 - 210}{2002 - 1992} = \dfrac{5852}{10} = 585.2$

Rounding up, we find that the average rate of change was about 586 adoptions per year.

35. First express $1\frac{1}{2}$ hr as 90 min. Then we have the points $(50, 10)$ and $(50 + 90, 25)$, or $(50, 10)$ and $(140, 25)$.

$$\begin{aligned} \text{Speed} &= \text{average rate of change} \\ &= \frac{25-10}{140-50} \\ &= \frac{15}{90} \\ &= \frac{1}{6} \end{aligned}$$

The speed is $\dfrac{1}{6}$ km per minute.

36. Typing rate $= \dfrac{\frac{3}{4}-\frac{1}{6}}{6}$

$= \dfrac{\frac{7}{12}}{6}$

$= \dfrac{7}{72}$ of the paper per hour

37. a) $W(h) = 4h - 130$

b) $W(62) = 4 \cdot 62 - 130 = 248 - 130 = 118$ lb

c) Both the height and weight must be positive. Solving $h > 0$ *and* $4h - 130 > 0$, we find that the domain of the function is $\{h|h > 32.5\}$, or $(32.5, \infty)$.

38. $P(d) = \dfrac{1}{33}d + 1$

a) $P(0) = \dfrac{1}{33} \cdot 0 + 1 = 1$ atm

$P(5) = \dfrac{1}{33} \cdot 5 + 1 = 1\dfrac{5}{33}$ atm

$P(10) = \dfrac{1}{33} \cdot 10 + 1 = 1\dfrac{10}{33}$ atm

$P(33) = \dfrac{1}{33} \cdot 33 + 1 = 2$ atm

$P(200) = \dfrac{1}{33} \cdot 200 + 1 = \dfrac{233}{33}$ atm, or $7\dfrac{2}{33}$ atm

b) The depth must be nonnegative, so the domain is $\{d|d \geq 0\}$, or $[0, \infty)$.

39. $D(F) = 2F + 115$

a) $D(0) = 2 \cdot 0 + 115 = 115$ ft

$D(-20) = 2(-20) + 115 = -40 + 115 = 75$ ft

$D(10) = 2 \cdot 10 + 115 = 20 + 115 = 135$ ft

$D(32) = 2 \cdot 32 + 115 = 64 + 115 = 179$ ft

b) Below $-57.5°$, stopping distance is negative; above $32°$, ice doesn't form.

40. a) $M(x) = 2.89x + 70.64$

$M(26) = 2.89(26) + 70.64 = 145.78$ cm

b) The length of the humerus must be positive, so the domain is $\{x|x > 0\}$, or $(0, \infty)$. Realistically, however, we might expect the length of the humerus to be between 20 cm and 60 cm, so the domain could be $\{x|20 \leq x \leq 60\}$, or $[20, 60]$. Answers may vary.

41. a) $D(r) = \dfrac{11r + 5}{10} = \dfrac{11}{10}r + \dfrac{5}{10}$

The slope is $\dfrac{11}{10}$.

For each mph faster the car travels, it takes $\dfrac{11}{10}$ ft longer to stop.

b) $D(5) = \dfrac{11 \cdot 5 + 5}{10} = \dfrac{60}{10} = 6$ ft

$D(10) = \dfrac{11 \cdot 10 + 5}{10} = \dfrac{115}{10} = 11.5$ ft

$D(20) = \dfrac{11 \cdot 20 + 5}{10} = \dfrac{225}{10} = 22.5$ ft

$D(50) = \dfrac{11 \cdot 50 + 5}{10} = \dfrac{555}{10} = 55.5$ ft

$D(65) = \dfrac{11 \cdot 65 + 5}{10} = \dfrac{720}{10} = 72$ ft

c) The speed cannot be negative. $D(0) = \dfrac{1}{2}$ which says that a stopped car travels $\dfrac{1}{2}$ ft before stopping. Thus, 0 is not in the domain. The speed can be positive, so the domain is $\{r|r > 0\}$, or $(0, \infty)$.

42. $V(t) = \$5200 - \$512.50t$

a) $V(0) = \$5200 - \$512.50(0) =$
$\$5200 - \$0 = \$5200$

$V(1) = \$5200 - \$512.50(1) =$
$\$5200 - \$512.50 = \$4687.50$

$V(2) = \$5200 - \$512.50(2) =$
$\$5200 - \$1025 = \$4175$

$V(3) = \$5200 - \$512.50(3) =$
$\$5200 - \$1537.50 = \$3662.50$

$V(8) = \$5200 - \$512.50(8) =$
$\$5200 - \$4100 = \$1100$

b) Since the time must be nonnegative and not more than 8 years, the domain is $[0, 8]$. The value starts at \$5200 and declines to \$1100, so the range is $[1100, 5200]$.

43. $C(t) = 60 + 29t$

$C(6) = 60 + 29 \cdot 6 = \$234$

44. $C(t) = 65 + 80t$

$C(8) = 65 + 80 \cdot 8 = \$705$

45. Let $x =$ the number of shirts produced.

$C(x) = 800 + 3x$

$C(75) = 800 + 3 \cdot 75 = \$1025$

46. Let $x =$ the number of rackets restrung.

$C(x) = 950 + 18x$

$C(150) = 950 + 18 \cdot 150 = \$3650$

47. Left to the student.

48. The sign of the slope indicates the slant of a line. A line that slants up from left to right has positive slope because corresponding changes in x and y have the same sign. A line that slants down from left to right has negative slope, because corresponding changes in x and y have opposite signs. A horizontal line has zero slope, because there is no change in y for a given change in x. A vertical line has undefined slope, because there is no change in x for a given change in y and division by 0 is undefined. The larger the absolute value of slope, the steeper the line. This is because a larger absolute value corresponds to a greater change in y, compared to the change in x, than a smaller absolute value.

49. A vertical line ($x = a$) crosses the graph more than once.

50. $f(5) = 5^2 - 3 \cdot 5 = 10$

51. $f(x) = x^2 - 3x$

$f(-5) = (-5)^2 - 3(-5) = 25 + 15 = 40$

52. $f(x) = x^2 - 3x$

$f(-a) = (-a)^2 - 3(-a) = a^2 + 3a$

53. $f(x) = x^2 - 3x$

$f(a+h) = (a+h)^2 - 3(a+h) = a^2 + 2ah + h^2 - 3a - 3h$

54. We make a drawing and label it. Let h = the height of the triangle, in feet.

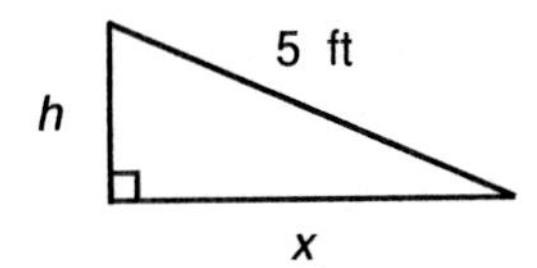

Using the Pythagorean theorem we have:

$$x^2 + h^2 = 25$$
$$x^2 = 25 - h^2$$
$$x = \sqrt{25 - h^2}$$

We know that the grade of the treadmill is 8%, or 0.08. Then we have

$$\frac{h}{x} = 0.08$$
$$\frac{h}{\sqrt{25 - h^2}} = 0.08 \quad \text{Substituting } \sqrt{25 - h^2} \text{ for } x$$
$$\frac{h^2}{25 - h^2} = 0.0064 \quad \text{Squaring both sides}$$
$$h^2 = 0.16 - 0.0064h^2$$
$$1.0064h^2 = 0.16$$
$$h^2 = \frac{0.16}{1.0064}$$
$$h \approx 0.4 \text{ ft}$$

55. $m = \dfrac{y_2 - y_1}{x_2 - x_1} = \dfrac{-2d - (-d)}{9c - (-c)} = \dfrac{-2d + d}{9c + c} = \dfrac{-d}{10c} = -\dfrac{d}{10c}$

56. $m = \dfrac{s - (s+t)}{r - r} = \dfrac{s - s - t}{0}$

The slope is not defined.

57. $m = \dfrac{y_2 - y_1}{x_2 - x_1} = \dfrac{z - z}{2 - q - (z + q)} = \dfrac{0}{z - q - z - q} =$

$\dfrac{0}{-2q} = 0$

58. $m = \dfrac{p - q - (p + q)}{a + b - (-a - b)} = \dfrac{p - q - p - q}{a + b + a + b} = \dfrac{-2q}{2a + 2b} =$

$-\dfrac{q}{a + b}$

59. $m = \dfrac{y_2 - y_1}{x_2 - x_1} = \dfrac{(a+h)^2 - a^2}{a + h - a} = \dfrac{a^2 + 2ah + h^2 - a^2}{h} =$

$\dfrac{2ah + h^2}{h} = \dfrac{h(2a + h)}{h} = 2a + h$

60. $m = \dfrac{3(a+h) + 1 - (3a + 1)}{a + h - a} =$

$\dfrac{3a + 3h + 1 - 3a - 1}{h} = \dfrac{3h}{h} = 3$

61. False. For example, let $f(x) = x + 1$. Then $f(cd) = cd + 1$, but $f(c)f(d) = (c+1)(d+1) = cd + c + d + 1 \neq cd + 1$ for $c \neq -d$.

62. False. For example, let $f(x) = x + 1$. Then $f(c + d) = c + d + 1$, but $f(c) + f(d) = c + 1 + d + 1 = c + d + 2$.

63. False. For example, let $f(x) = x + 1$. Then $f(c - d) = c - d + 1$, but $f(c) - f(d) = c + 1 - (d + 1) = c - d$.

64. False. For example, let $f(x) = x + 1$. Then $f(kx) = kx + 1$, but $kf(x) = k(x + 1) = kx + k \neq kx + 1$ for $k \neq 1$.

65.
$$f(x) = mx + b$$
$$f(x + 2) = f(x) + 2$$
$$m(x + 2) + b = mx + b + 2$$
$$mx + 2m + b = mx + b + 2$$
$$2m = 2$$
$$m = 1$$

Thus, $f(x) = 1 \cdot x + b$, or $f(x) = x + b$.

66.
$$3mx + b = 3(mx + b)$$
$$3mx + b = 3mx + 3b$$
$$b = 3b$$
$$0 = 2b$$
$$0 = b$$

Thus, $f(x) = mx + 0$, or $f(x) = mx$.

Exercise Set 1.4

1. $y = \dfrac{3}{5}x - 7$

The equation is in the form $y = mx + b$ where $m = \dfrac{3}{5}$ and $b = -7$. Thus, the slope is $\dfrac{3}{5}$, and the y-intercept is $(0, -7)$.

2. $f(x) = -2x + 3$

Slope: -2; y-intercept: $(0, 3)$

3. $x = -\dfrac{2}{5}$

This is the equation of a vertical line $\dfrac{2}{5}$ unit to the left of the y-axis. The slope is not defined, and there is no y-intercept.

4. $y = \dfrac{4}{7} = 0 \cdot x + \dfrac{4}{7}$

Slope: 0; y-intercept: $\left(0, \dfrac{4}{7}\right)$

5. $f(x) = 5 - \frac{1}{2}x$, or $f(x) = -\frac{1}{2}x + 5$

The second equation is in the form $y = mx + b$ where $m = -\frac{1}{2}$ and $b = 5$. Thus, the slope is $-\frac{1}{2}$ and the y-intercept is $(0, 5)$.

6. $y = 2 + \frac{3}{7}x$

Slope: $\frac{3}{7}$; y-intercept: $(0, 2)$

7. Solve the equation for y.

$$\begin{aligned} 3x + 2y &= 10 \\ 2y &= -3x + 10 \\ y &= -\frac{3}{2}x + 5 \end{aligned}$$

Slope: $-\frac{3}{2}$; y-intercept: $(0, 5)$

8. $$\begin{aligned} 2x - 3y &= 12 \\ -3y &= -2x + 12 \\ y &= \frac{2}{3}x - 4 \end{aligned}$$

Slope: $\frac{2}{3}$; y-intercept: $(0, -4)$

9. $y = -6 = 0 \cdot x - 6$

Slope: 0; y-intercept: $(0, -6)$

10. $x = 10$

This is the equation of a vertical line 10 units to the right of the y-axis. The slope is not defined, and there is no y-intercept.

11. Solve the equation for y.

$$\begin{aligned} 5y - 4x &= 8 \\ 5y &= 4x + 8 \\ y &= \frac{4}{5}x + \frac{8}{5} \end{aligned}$$

Slope: $\frac{4}{5}$; y-intercept: $\left(0, \frac{8}{5}\right)$

12. $$\begin{aligned} 5x - 2y + 9 &= 0 \\ -2y &= -5x - 9 \\ y &= \frac{5}{2}x + \frac{9}{2} \end{aligned}$$

Slope: $\frac{5}{2}$; y-intercept: $\left(0, \frac{9}{2}\right)$

13. We see that the y-intercept is $(0, -2)$. Another point on the graph is $(1, 2)$. Use these points to find the slope.

$$m = \frac{y_2 - y_1}{x_2 - x_1} = \frac{2 - (-2)}{1 - 0} = \frac{4}{1} = 4$$

We have $m = 4$ and $b = -2$, so the equation is $y = 4x - 2$.

14. We see that the y-intercept is $(0, 2)$. Another point on the graph is $(4, -1)$.

$$m = \frac{-1 - 2}{4 - 0} = -\frac{3}{4}$$

The equation is $y = -\frac{3}{4}x + 2$.

15. We see that the y-intercept is $(0, 0)$. Another point on the graph is $(3, -3)$. Use these points to find the slope.

$$m = \frac{y_2 - y_1}{x_2 - x_1} = \frac{-3 - 0}{3 - 0} = \frac{-3}{3} = -1$$

We have $m = -1$ and $b = 0$, so the equation is $y = -1 \cdot x + 0$, or $y = -x$.

16. We see that the y-intercept is $(0, -1)$. Another point on the graph is $(3, 1)$.

$$m = \frac{1 - (-1)}{3 - 0} = \frac{2}{3}$$

The equation is $y = \frac{2}{3}x - 1$.

17. We see that the y-intercept is $(0, -3)$. This is a horizontal line, so the slope is 0. We have $m = 0$ and $b = -3$, so the equation is $y = 0 \cdot x - 3$, or $y = -3$.

18. We see that the y-intercept is $(0, 0)$. Another point on the graph is $(3, 3)$.

$$m = \frac{3 - 0}{3 - 0} = \frac{3}{3} = 1$$

The equation is $y = 1 \cdot x + 0$, or $y = x$.

19. Graph $y = -\frac{1}{2}x - 3$.

Plot the y-intercept, $(0, -3)$. We can think of the slope as $\frac{-1}{2}$. Start at $(0, -3)$ and find another point by moving down 1 unit and right 2 units. We have the point $(2, -4)$.

We could also think of the slope as $\frac{1}{-2}$. Then we can start at $(0, -3)$ and get another point by moving up 1 unit and left 2 units. We have the point $(-2, -2)$. Connect the three points to draw the graph.

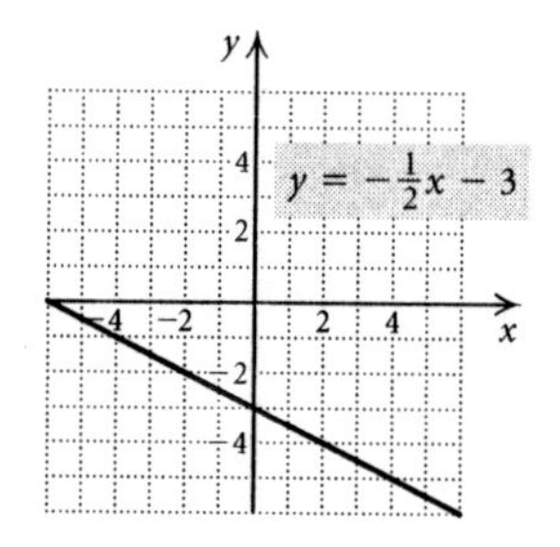

20.

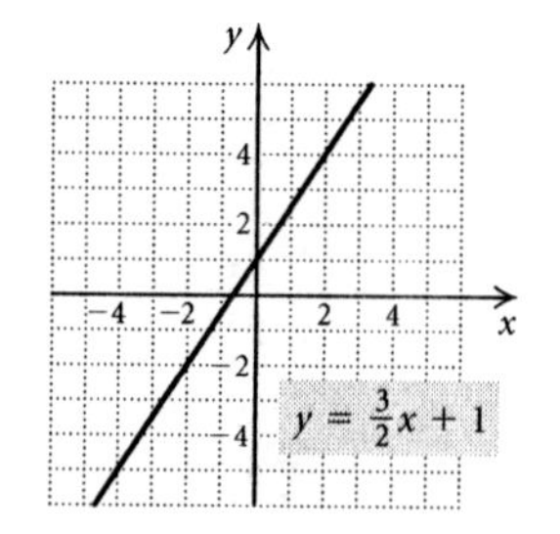

21. Graph $f(x) = 3x - 1$.

Plot the y-intercept, $(0, -1)$. We can think of the slope as $\frac{3}{1}$. Start at $(0, -1)$ and find another point by moving up 3 units and right 1 unit. We have the point $(1, 2)$. We can move from the point $(1,2)$ in a similar manner to get a third point, $(2,5)$. Connect the three points to draw the graph.

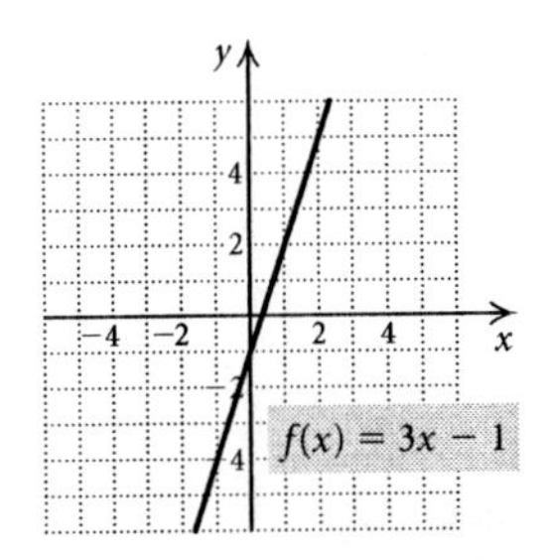

22.

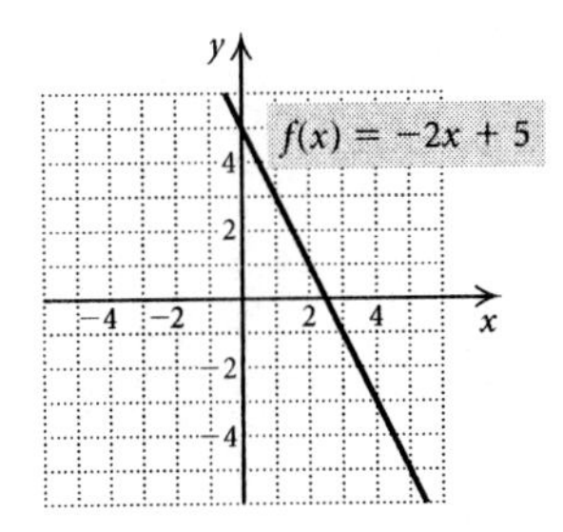

23. First solve the equation for y.

$$3x - 4y = 20$$
$$-4y = -3x + 20$$
$$y = \frac{3}{4}x - 5$$

Plot the y-intercept, $(0,-5)$. Then using the slope, $\frac{3}{4}$, start at $(0,-5)$ and find another point by moving up 3 units and right 4 units. We have the point $(4,-2)$. We can move from the point $(4,-2)$ in a similar manner to get a third point, $(8,1)$. Connect the three points to draw the graph.

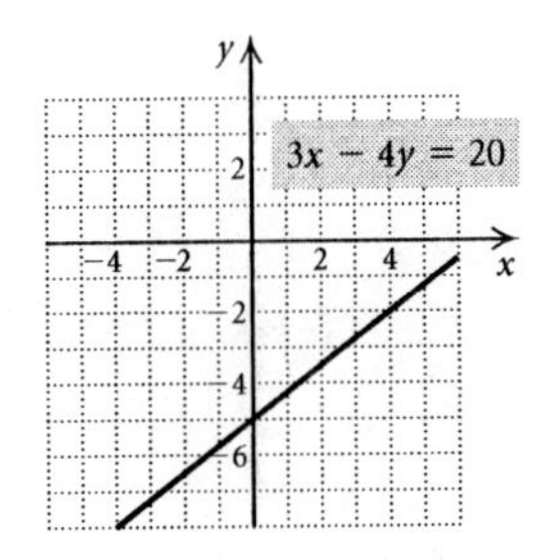

24.

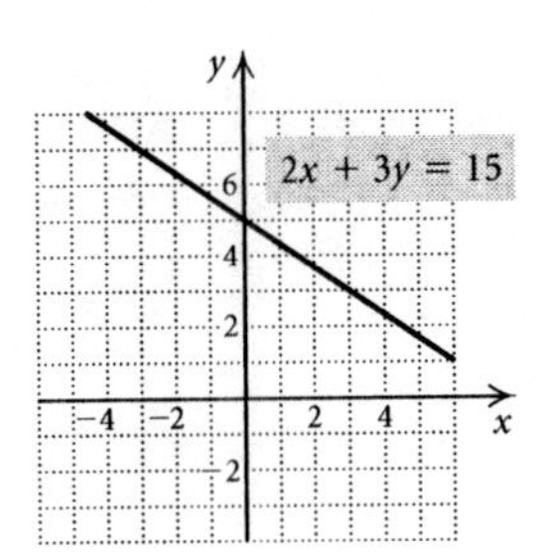

25. First solve the equation for y.

$$x + 3y = 18$$
$$3y = -x + 18$$
$$y = -\frac{1}{3}x + 6$$

Plot the y-intercept, $(0,6)$. We can think of the slope as $\frac{-1}{3}$. Start at $(0,6)$ and find another point by moving down 1 unit and right 3 units. We have the point $(3,5)$. We can move from the point $(3,5)$ in a similar manner to get a third point, $(6,4)$. Connect the three points and draw the graph.

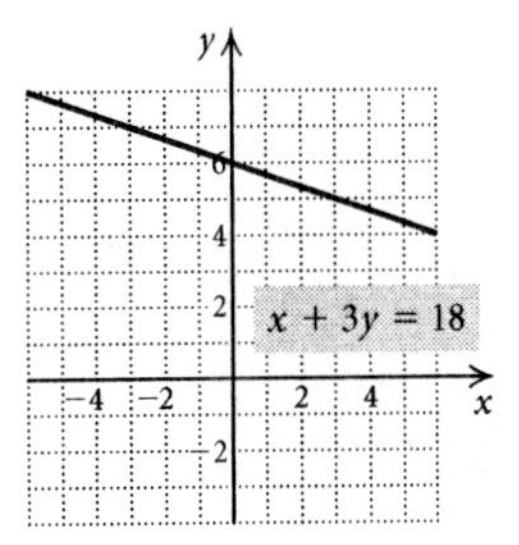

26.

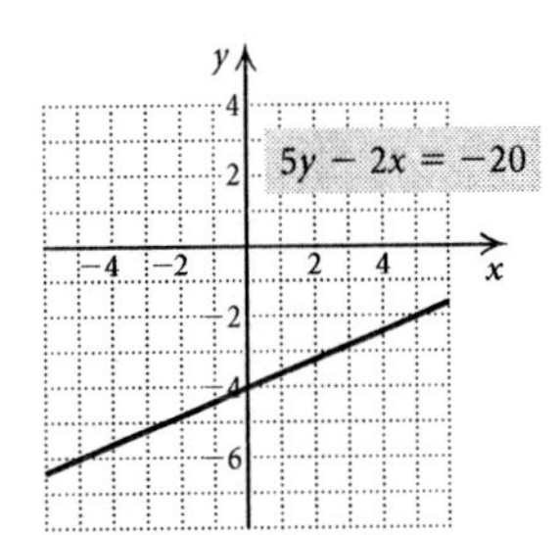

27. We substitute $\frac{2}{9}$ for m and 4 for b in the slope-intercept equation.

$$y = mx + b$$
$$y = \frac{2}{9}x + 4$$

28. $y = -\frac{3}{8}x + 5$

29. We substitute -4 for m and -7 for b in the slope-intercept equation.

$$y = mx + b$$
$$y = -4x - 7$$

30. $y = \frac{2}{7}x - 6$

31. We substitute -4.2 for m and $\frac{3}{4}$ for b in the slope-intercept equation.

$$y = mx + b$$
$$y = -4.2x + \frac{3}{4}$$

32. $y = -4x - \frac{3}{2}$

33. Using the point-slope equation:

$$y - y_1 = m(x - x_1)$$
$$y - 7 = \frac{2}{9}(x - 3) \quad \text{Substituting}$$
$$y - 7 = \frac{2}{9}x - \frac{2}{3}$$
$$y = \frac{2}{9}x + \frac{19}{3} \quad \text{Slope-intercept equation}$$

Using the slope-intercept equation:

Substitute $\frac{2}{9}$ for m, 3 for x, and 7 for y in the slope-intercept equation and solve for b.

$$y = mx + b$$
$$7 = \frac{2}{9} \cdot 3 + b$$
$$7 = \frac{2}{3} + b$$
$$\frac{19}{3} = b$$

Now substitute $\frac{2}{9}$ for m and $\frac{19}{3}$ for b in $y = mx + b$.

$$y = \frac{2}{9}x + \frac{19}{3}$$

34. Using the point-slope equation:

$$y - 6 = -\frac{3}{8}(x - 5)$$
$$y = -\frac{3}{8}x + \frac{63}{8}$$

Using the slope-intercept equation:

$$6 = -\frac{3}{8} \cdot 5 + b$$
$$\frac{63}{8} = b$$

We have $y = -\frac{3}{8}x + \frac{63}{8}$.

35. Using the point-slope equation:

$$y - y_1 = m(x - x_1)$$
$$y - (-2) = 3(x - 1)$$
$$y + 2 = 3x - 3$$
$$y = 3x - 5 \quad \text{Slope-intercept equation}$$

Using the slope-intercept equation:

$$y = mx + b$$
$$-2 = 3 \cdot 1 + b$$
$$-2 = 3 + b$$
$$-5 = b$$

Then we have $y = 3x - 5$.

36. Using the point-slope equation:

$$y - 1 = -2(x - (-5))$$
$$y = -2x - 9$$

Using the slope-intercept equation:

$$1 = -2(-5) + b$$
$$-9 = b$$

We have $y = -2x - 9$.

37. Using the point-slope equation:

$$y - y_1 = m(x - x_1)$$
$$y - (-1) = -\frac{3}{5}(x - (-4))$$
$$y + 1 = -\frac{3}{5}(x + 4)$$
$$y + 1 = -\frac{3}{5}x - \frac{12}{5}$$
$$y = -\frac{3}{5}x - \frac{17}{5} \quad \text{Slope-intercept equation}$$

Using the slope-intercept equation:

$$y = mx + b$$
$$-1 = -\frac{3}{5}(-4) + b$$
$$-1 = \frac{12}{5} + b$$
$$-\frac{17}{5} = b$$

Then we have $y = -\frac{3}{5}x - \frac{17}{5}$.

38. Using the point-slope equation:

$$y - (-5) = \frac{2}{3}(x - (-4))$$
$$y = \frac{2}{3}x - \frac{7}{3}$$

Using the slope-intercept equation:

$$-5 = \frac{2}{3}(-4) + b$$
$$-\frac{7}{3} = b$$

We have $y = \frac{2}{3}x - \frac{7}{3}$.

39. First we find the slope.

$$m = \frac{-4 - 5}{2 - (-1)} = \frac{-9}{3} = -3$$

Using the point-slope equation:

Using the point $(-1, 5)$, we get

$$y - 5 = -3(x - (-1)), \text{ or } y - 5 = -3(x + 1).$$

Using the point $(2, -4)$, we get

$$y - (-4) = -3(x - 2), \text{ or } y + 4 = -3(x - 2).$$

In either case, the slope-intercept equation is $y = -3x + 2$.

Using the slope-intercept equation and the point $(-1, 5)$:

$$y = mx + b$$
$$5 = -3(-1) + b$$
$$5 = 3 + b$$
$$2 = b$$

Then we have $y = -3x + 2$.

40. First we find the slope.

$$m = \frac{-11 - (-1)}{7 - 2} = \frac{-10}{5} = -2$$

Using the point-slope equation:

Using $(2,-1)$: $y-(-1)=-2(x-2)$, or

$y+1=-2(x-2)$

Using $(7,-11)$: $y-(-11)=-2(x-7)$, or

$y+11=-2(x-7)$

In either case, we have $y=-2x+3$.

Using the slope-intercept equation and the point $(2,-1)$:

$-1=-2\cdot 2+b$

$3=b$

We have $y=-2x+3$.

41. First we find the slope.

$$m=\frac{4-0}{-1-7}=\frac{4}{-8}=-\frac{1}{2}$$

Using the point-slope equation:

Using the point $(7,0)$, we get

$y-0=-\frac{1}{2}(x-7)$.

Using the point $(-1,4)$, we get

$y-4=-\frac{1}{2}(x-(-1))$, or

$y-4=-\frac{1}{2}(x+1)$.

In either case, the slope-intercept equation is $y=-\frac{1}{2}x+\frac{7}{2}$.

Using the slope-intercept equation and the point $(7,0)$:

$0=-\frac{1}{2}\cdot 7+b$

$\frac{7}{2}=b$

Then we have $y=-\frac{1}{2}x+\frac{7}{2}$.

42. First we find the slope.

$$m=\frac{-5-7}{-1-(-3)}=\frac{-12}{2}=-6$$

Using the point-slope equation:

Using $(-3,7)$: $y-7=-6(x-(-3))$, or

$y-7=-6(x+3)$

Using $(-1,-5)$: $y-(-5)=-6(x-(-1))$, or

$y+5=-6(x+1)$

In either case, we have $y=-6x-11$.

Using the slope-intercept equation and the point $(-1,-5)$:

$-5=-6(-1)+b$

$-11=b$

We have $y=-6x-11$.

43. First we find the slope.

$$m=\frac{-4-(-6)}{3-0}=\frac{2}{3}$$

We know the y-intercept is $(0,-6)$, so we substitute in the slope-intercept equation.

$y=mx+b$

$y=\frac{2}{3}x-6$

44. First we find the slope.

$$m=\frac{\frac{4}{5}-0}{0-(-5)}=\frac{\frac{4}{5}}{5}=\frac{4}{25}$$

We know the y-intercept is $\left(0,\frac{4}{5}\right)$, so we substitute in the slope-intercept equation.

$y=\frac{4}{25}x+\frac{4}{5}$

45. The equation of the horizontal line through $(0,-3)$ is of the form $y=b$ where b is -3. We have $y=-3$.

The equation of the vertical line through $(0,-3)$ is of the form $x=a$ where a is 0. We have $x=0$.

46. Horizontal line: $y=7$

Vertical line: $x=-\frac{1}{4}$

47. The equation of the horizontal line through $\left(\frac{2}{11},-1\right)$ is of the form $y=b$ where b is -1. We have $y=-1$.

The equation of the vertical line through $\left(\frac{2}{11},-1\right)$ is of the form $x=a$ where a is $\frac{2}{11}$. We have $x=\frac{2}{11}$.

48. Horizontal line: $y=0$

Vertical line: $x=0.03$

49. The slopes are $\frac{26}{3}$ and $-\frac{3}{26}$. Their product is -1, so the lines are perpendicular.

50. The slopes are -3 and $-\frac{1}{3}$. The slopes are not the same and their product is not -1, so the lines are neither parallel nor perpendicular.

51. The slopes are $\frac{2}{5}$ and $-\frac{2}{5}$. The slopes are not the same and their product is not -1, so the lines are neither parallel nor perpendicular.

52. The slopes are the same $\left(\frac{3}{2}=1.5\right)$ and the y-intercepts, -8 and 8, are different, so the lines are parallel.

53. We solve each equation for y.

$x+2y=5$ $\qquad$ $2x+4y=8$

$y=-\frac{1}{2}x+\frac{5}{2}$ $\qquad$ $y=-\frac{1}{2}x+2$

We see that $m_1=-\frac{1}{2}$ and $m_2=-\frac{1}{2}$. Since the slopes are the same and the y-intercepts, $\frac{5}{2}$ and 2, are different, the lines are parallel.

54. $2x-5y=-3$ $\qquad$ $2x+5y=4$

$y=\frac{2}{5}x+\frac{3}{5}$ $\qquad$ $y=-\frac{2}{5}x+\frac{4}{5}$

$m_1=\frac{2}{5}$, $m_2=-\frac{2}{5}$; $m_1\neq m_2$; $m_1m_2=-\frac{4}{25}\neq -1$

The lines are neither parallel nor perpendicular.

55. We solve each equation for y.

$y = 4x - 5 \qquad 4y = 8 - x$

$$y = -\frac{1}{4}x + 2$$

We see that $m_1 = 4$ and $m_2 = -\frac{1}{4}$. Since $m_1m_2 = 4\left(-\frac{1}{4}\right) = -1$, the lines are perpendicular.

56. $y = 7 - x$,

$y = x + 3$

$m_1 = -1$, $m_2 = 1$; $m_1m_2 = -1 \cdot 1 = -1$

The lines are perpendicular.

57. $y = \frac{2}{7}x + 1$; $m = \frac{2}{7}$

The line parallel to the given line will have slope $\frac{2}{7}$. We use the point-slope equation for a line with slope $\frac{2}{7}$ and containing the point $(3, 5)$:

$$y - y_1 = m(x - x_1)$$
$$y - 5 = \frac{2}{7}(x - 3)$$
$$y - 5 = \frac{2}{7}x - \frac{6}{7}$$
$$y = \frac{2}{7}x + \frac{29}{7} \quad \text{Slope-intercept form}$$

The slope of the line perpendicular to the given line is the opposite of the reciprocal of $\frac{2}{7}$, or $-\frac{7}{2}$. We use the point-slope equation for a line with slope $-\frac{7}{2}$ and containing the point $(3, 5)$:

$$y - y_1 = m(x - x_1)$$
$$y - 5 = -\frac{7}{2}(x - 3)$$
$$y - 5 = -\frac{7}{2}x + \frac{21}{2}$$
$$y = -\frac{7}{2}x + \frac{31}{2} \quad \text{Slope-intercept form}$$

58. $f(x) = 2x + 9$

$m = 2$, $-\frac{1}{m} = -\frac{1}{2}$

Parallel line: $y - 6 = 2(x - (-1))$

$y = 2x + 8$

Perpendicular line: $y - 6 = -\frac{1}{2}(x - (-1))$

$$y = -\frac{1}{2}x + \frac{11}{2}$$

59. $y = -0.3x + 4.3$; $m = -0.3$

The line parallel to the given line will have slope -0.3. We use the point-slope equation for a line with slope -0.3 and containing the point $(-7, 0)$:

$$y - y_1 = m(x - x_1)$$
$$y - 0 = -0.3(x - (-7))$$
$$y = -0.3x - 2.1 \quad \text{Slope-intercept form}$$

The slope of the line perpendicular to the given line is the opposite of the reciprocal of -0.3, or $\frac{1}{0.3} = \frac{10}{3}$.

We use the point-slope equation for a line with slope $\frac{10}{3}$ and containing the point $(-7, 0)$:

$$y - y_1 = m(x - x_1)$$
$$y - 0 = \frac{10}{3}(x - (-7))$$
$$y = \frac{10}{3}x + \frac{70}{3} \quad \text{Slope-intercept form}$$

60. $2x + y = -4$

$y = -2x - 4$

$m = -2$, $-\frac{1}{m} = \frac{1}{2}$

Parallel line: $y - (-5) = -2(x - (-4))$

$y = -2x - 13$

Perpendicular line: $y - (-5) = \frac{1}{2}(x - (-4))$

$$y = \frac{1}{2}x - 3$$

61. $3x + 4y = 5$

$$4y = -3x + 5$$
$$y = -\frac{3}{4}x + \frac{5}{4};\ m = -\frac{3}{4}$$

The line parallel to the given line will have slope $-\frac{3}{4}$. We use the point-slope equation for a line with slope $-\frac{3}{4}$ and containing the point $(3, -2)$:

$$y - y_1 = m(x - x_1)$$
$$y - (-2) = -\frac{3}{4}(x - 3)$$
$$y + 2 = -\frac{3}{4}x + \frac{9}{4}$$
$$y = -\frac{3}{4}x + \frac{1}{4} \quad \text{Slope-intercept form}$$

The slope of the line perpendicular to the given line is the opposite of the reciprocal of $-\frac{3}{4}$, or $\frac{4}{3}$. We use the point-slope equation for a line with slope $\frac{4}{3}$ and containing the point $(3, -2)$:

$$y - y_1 = m(x - x_1)$$
$$y - (-2) = \frac{4}{3}(x - 3)$$
$$y + 2 = \frac{4}{3}x - 4$$
$$y = \frac{4}{3}x - 6 \quad \text{Slope-intercept form}$$

62. $y = 4.2(x - 3) + 1$

$y = 4.2x - 11.6$

$m = 4.2$; $-\frac{1}{m} = -\frac{1}{4.2} = -\frac{5}{21}$

Parallel line: $y-(-2)=4.2(x-8)$

$$y=4.2x-35.6$$

Perpendicular line: $y-(-2)=-\frac{5}{21}(x-8)$

$$y=-\frac{5}{21}x-\frac{2}{21}$$

63. $x=-1$ is the equation of a vertical line. The line parallel to the given line is a vertical line containing the point $(3,-3)$, or $x=3$.

The line perpendicular to the given line is a horizontal line containing the point $(3,-3)$, or $y=-3$.

64. $y=-1$ is a horizontal line.

Parallel line: $y=-5$

Perpendicular line: $x=4$

65. a) For model 1 we will use the data points $(0,7.8)$ and $(20,6.4)$. First we find the slope of the line.

$$m=\frac{6.4-7.8}{20-0}=\frac{-1.4}{20}=-0.07$$

We know that the y-intercept is $(0,7.8)$, so we substitute in the slope-intercept equation.

$$y=mx+b$$
$$y=-0.07x+7.8$$

For model 2 we will use the data points $(10,7.3)$ and $(31,4.9)$. First we find the slope of the line.

$$m=\frac{4.9-7.3}{31-10}=\frac{-2.4}{21}=-\frac{2.4}{21}\cdot\frac{10}{10}=-\frac{24}{210}=-\frac{4}{35}$$

Now we use the slope-intercept equation and the point $(10,7.3)$.

$$y=mx+b$$
$$7.3=-\frac{4}{35}(10)+b$$
$$\frac{73}{10}=-\frac{8}{7}+b$$
$$\frac{591}{70}=b$$

Then we have $y=-\frac{4}{35}x+\frac{591}{70}$.

b) In 2005, $x=2005-1970=35$.

With model 1: $y=-0.07x+7.8$

$$y=-0.07(35)+7.8$$
$$y=5.35$$

Using model 1, we estimate that in 2005 the average length of a hospital stay will be about 5.4 days.

With model 2: $y=-\frac{4}{35}x+\frac{591}{70}$

$$y=-\frac{4}{35}\cdot 35+\frac{591}{70}$$
$$y=-4+\frac{591}{70}$$
$$y=\frac{311}{70}\approx 4.4$$

Using model 2, we estimate that in 2005 the average length of a hospital stay will be about 4.4 days.

c) The estimate found using model 2 is more in keeping with the decreases shown in the graph in the text than is the estimate found using model 1. Thus, model 2 appears to fit the data more closely.

66. a) For model 1 we will use the data points $(0,432)$ and $(2,710)$.

$$m=\frac{710-432}{20-0}=139$$
$$y=139x+432$$

For model 2 we will use the data points $(1,546)$ and $(3,850)$.

$$m=\frac{850-546}{3-1}=152$$

Using the slope-intercept equation and the point $(1,546)$, we have:

$$546=152.1+b$$
$$394=b$$

Then we have $y=152x+394$.

b) In 2006, $t=2006-1998=8$.

Using model 1: $y=139\cdot 8+432=1544$ thousand, or 1,544,000 motorcycles

Using model 2: $y=152\cdot 8+394=1610$ thousand, or 1,610,000 motorcycles.

c) The estimate found using model 1 appears to be more in keeping with the size of the increases shown in the graph in the text, so it appears to fit the data better.

67. Answers will vary depending on the data points used. We will use $(1,\ 10,424)$ and $(3,\ 11,717)$.

$$m=\frac{11,717-10,424}{3-1}=\frac{1293}{2}=646.5$$

Use the point-slope equation.

$$y-10,424=646.5(x-1)$$
$$y-10,424=646.5x-646.5$$
$$y=646.5x+9777.5$$

In 2004-2005, $y=646.5(7)+9777.5=\$14,303$.

In 2006-2007, $y=646.5(9)+9777.5=\$15,596$.

In 2010-2011, $y=646.5(13)+9777.5=\$18,182$.

68. Answers will vary depending on the data points used. We will use $(1,4.33)$ and $(5,4.85)$.

$$m=\frac{4.85-4.33}{5-1}=0.1325\approx 0.13$$
$$y-4.33=0.13(x-1)$$
$$y=0.13x+4.2$$

In 2005, $x=2005-1994=11$, and $y=0.13(11)+4.2=$ 5.63 million foreign travelers.

In 2008, $x=2008-1994=14$, and $y=0.13(14)+4.2=$ 6.02 million foreign travelers.

69. Answers will vary depending on the data points used. We will use $(3,77.5)$ and $(5,82.7)$.

$$m=\frac{82.7-77.5}{5-3}=\frac{5.2}{2}=2.6$$

Use the sl·pe intercept equation and the point $(3, 77.5)$.

$$y = mx + b$$
$$77.5 = 2.6(3) + b$$
$$77.5 = 7.8 + b$$
$$69.7 = b$$

We have $y = 2.6x + 69.7$.

In 2004, $x = 2004 - 1994 = 10$, and $y = 2.6(10) + 69.7 = 95.7$ billion eggs.

In 2008, $x = 2008 - 1994 = 14$, and $y = 2.6(14) + 69.7 = 106.1$ billion eggs.

In 2012, $x = 2012 - 1994 = 18$, and $y = 2.6(18) + 69.7 = 116.5$ billion eggs.

70. Answers will vary depending on the data points used. We will use $(5, 11.4)$ and $(15, 13)$.

$$m = \frac{13 - 11.4}{15 - 5} = \frac{1.6}{10} = 0.16$$

Use the point-slope equation and the point $(15, 13)$.

$$y - 13 = 0.16(x - 15)$$
$$y = 0.16x + 10.6$$

In 2010, $x = 2010 - 1986 = 24$, and $y = 0.16(24) + 10.6 = 14.44 \approx 14.4\%$.

In 2014, $x = 2014 - 1986 = 28$, and $y = 0.16(28) + 10.6 = 15.08 \approx 15.1\%$.

71. a) Using the linear regression feature on a graphing calculator, we get $M = 0.2H + 156$.

b) For $H = 40$: $M = 0.2(40) + 156 = 164$ beats per minute

For $H = 65$: $M = 0.2(65) + 156 = 169$ beats per minute

For $H = 76$: $M = 0.2(76) + 156 \approx 171$ beats per minute

For $H = 84$: $M = 0.2(84) + 156 \approx 173$ beats per minute

c) $r = 1$; all the data points are on the regression line so it should be a good predictor.

72. a) $y = 0.072050673x + 81.99920823$

b) For $x = 24$:
$y = 0.072050673(24) + 81.99920823 \approx 84\%$

For $x = 6$:
$y = 0.072050673(6) + 81.99920823 \approx 82\%$

For $x = 18$:
$y = 0.072050673(18) + 81.99920823 \approx 83\%$

c) $r = 0.0636$; since there is a very low correlation, the regression line is not a good predictor.

73. a) Using the linear regression feature on a graphing calculator, we get $y = 0.02x + 1.77$, where x is the number of years after 1999 and y is in billions of dollars.

b) In 2005, $x = 2005 - 1999 = 6$, and
$y = 0.02(6) + 1.77 = \$1.89$ billion.

In 2010, $x = 2010 - 1999 = 11$,and
$y = 0.02(11) + 1.77 = \$1.99$ billion.

c) $r \approx 0.3162$; since there is a low correlation, the regression line does not fit the data closely.

74. a) $y = 131.4x + 432.2$, where x is the number of years after 1998 and y is in thousands.

b) In 2006, $y = 131.4(8) + 432.2 = 1483.4$ thousand, or 1,483,400 motorcycles. This value is 60,600 less than the value found using model 1.

c) $r \approx 0.9956$; this value is close to 1, so the line fits the data well.

75. a) Using the linear regression feature on a graphing calculator, we get $y = -0.093641074x + 8.028026378$, where x is the number of years after 1970.

b) In 2005, $x = 2005 - 1970 = 35$,
so $y = -0.093641074(35) + 8.028026378 \approx 4.8$ days. The value is 0.4 days more than the value found using model 2.

c) $r \approx 0.9786$; the value is close to 1 so the line fits the data well.

76. A linear function cannot be used to express the yearly salary as a function of the number of years the employee has worked because the increases in salary are not a constant amount. Each increase is larger than the preceding one because it is calculated on a larger base amount than the preceding one.

77. $\left|-\frac{3}{4}\right| > \left|\frac{2}{5}\right|$, so the line with slope $-\frac{3}{4}$ is steeper.

78. $m = \dfrac{-7-7}{5-5} = \dfrac{-14}{0}$

The slope is not defined.

79.
$$m = \frac{y_2 - y_1}{x_2 - x_1} = \frac{-1-(-8)}{-5-2} = \frac{-1+8}{-7} = \frac{7}{-7} = -1$$

80. $r = \dfrac{d}{2} = \dfrac{5}{2}$

$$(x-0)^2 + (y-3)^2 = \left(\frac{5}{2}\right)^2$$
$$x^2 + (y-3)^2 = \frac{25}{4}, \text{ or}$$
$$x^2 + (y-3)^2 = 6.25$$

81.
$$(x-h)^2 + (y-k)^2 = r^2$$
$$[x-(-7)]^2 + [y-(-1)]^2 = \left(\frac{9}{5}\right)^2$$
$$(x+7)^2 + (y+1)^2 = \frac{81}{25}$$

82. $m = \dfrac{920.58}{13,740} = 0.067$

The road grade is 6.7%.

We find an equation of the line with slope 0.067 and containing the point $(13,740, 920.58)$:

$$y - 920.58 = 0.067(x - 13,740)$$
$$y - 920.58 = 0.067x - 920.58$$
$$y = 0.067x$$

83. The slope of the line containing $(-3, k)$ and $(4, 8)$ is
$$\frac{8-k}{4-(-3)} = \frac{8-k}{7}.$$
The slope of the line containing $(5, 3)$ and $(1, -6)$ is
$$\frac{-6-3}{1-5} = \frac{-9}{-4} = \frac{9}{4}.$$
The slopes must be equal in order for the lines to be parallel:
$$\frac{8-k}{7} = \frac{9}{4}$$
$$32 - 4k = 63 \quad \text{Multiplying by 28}$$
$$-4k = 31$$
$$k = -\frac{31}{4}, \text{ or } -7.75$$

84. The slope of the line containing $(-1, 3)$ and $(2, 9)$ is
$$\frac{9-3}{2-(-1)} = \frac{6}{3} = 2.$$
Then the slope of the desired line is $-\frac{1}{2}$. We find the equation of that line:
$$y - 5 = -\frac{1}{2}(x - 4)$$
$$y - 5 = -\frac{1}{2}x + 2$$
$$y = -\frac{1}{2}x + 7$$

Exercise Set 1.5

1. a) For x-values from -5 to 1, the y-values increase from -3 to 3. Thus the function is increasing on the interval $(-5, 1)$.

b) For x-values from 3 to 5, the y-values decrease from 3 to 1. Thus the function is decreasing on the interval $(3, 5)$.

c) For x-values from 1 to 3, y is 3. Thus the function is constant on $(1, 3)$.

2. a) For x-values from 1 to 3, the y-values increase from 1 to 2. Thus, the function is increasing on the interval $(1, 3)$.

b) For x-values from -5 to 1, the y-values decrease from 4 to 1. Thus the function is decreasing on the interval $(-5, 1)$.

c) For x-values from 3 to 5, y is 2. Thus the function is constant on $(3, 5)$.

3. a) For x-values from -3 to -1, the y-values increase from -4 to 4. Also, for x-values from 3 to 5, the y-values increase from 2 to 6. Thus the function is increasing on $(-3, -1)$ and on $(3, 5)$.

b) For x-values from 1 to 3, the y-values decrease from 3 to 2. Thus the function is decreasing on the interval $(1, 3)$.

c) For x-values from -5 to -3, y is 1. Thus the function is constant on $(-5, -3)$.

4. a) For x-values from 1 to 2, the y-values increase from 1 to 2. Thus the function is increasing on the interval $(1, 2)$.

b) For x-values from -5 to -2, the y-values decrease from 3 to 1. For x-values from -2 to 1, the y-values decrease from 3 to 1. And for x-values from 3 to 5, the y-values decrease from 2 to 1. Thus the function is decreasing on $(-5, -2)$, on $(-2, 1)$, and on $(3, 5)$.

c) For x-values from 2 to 3, y is 2. Thus the function is constant on $(2, 3)$.

5. a) For x-values from $-\infty$ to -8, the y-values increase from $-\infty$ to 2. Also, for x-values from -3 to -2, the y-values increase from -2 to 3. Thus the function is increasing on $(-\infty, -8)$ and on $(-3, -2)$.

b) For x-values from -8 to -6, the y-values decrease from 2 to -2. Thus the function is decreasing on the interval $(-8, -6)$.

c) For x-values from -6 to -3, y is -2. Also, for x-values from -2 to ∞, y is 3. Thus the function is constant on $(-6, -3)$ and on $(-2, \infty)$.

6. a) For x-values from 1 to 4, the y-values increase from 2 to 11. Thus the function is increasing on the interval $(1, 4)$.

b) For x-values from -1 to 1, the y-values decrease from 6 to 2. Also, for x-values from 4 to ∞, the y-values decrease from 11 to $-\infty$. Thus the function is decreasing on $(-1, 1)$ and on $(4, \infty)$.

c) For x-values from $-\infty$ to -1, y is 3. Thus the function is constant on $(-\infty, -1)$.

7. The x-values extend from -5 to 5, so the domain is $[-5, 5]$.

The y-values extend from -3 to 3, so the range is $[-3, 3]$.

8. Domain: $[-5, 5]$; range: $[1, 4]$

9. The x-values extend from -5 to -1 and from 1 to 5, so the domain is $[-5, -1] \cup [1, 5]$.

The y-values extend from -4 to 6, so the range is $[-4, 6]$.

10. Domain: $[-5, 5]$; range: $[1, 3]$

11. The x-values extend from $-\infty$ to ∞, so the domain is $(-\infty, \infty)$.

The y-values extend from $-\infty$ to 3, so the range is $(-\infty, 3]$.

12. Domain: $(-\infty, \infty)$; range: $(-\infty, 11]$

13. From the graph we see that a relative maximum value of the function is 3.25. It occurs at $x = 2.5$. There is no relative minimum value.

The graph starts rising, or increasing, from the left and stops increasing at the relative maximum. From this point, the graph decreases. Thus the function is increasing on $(-\infty, 2.5)$ and is decreasing on $(2.5, \infty)$.

14. From the graph we see that a relative minimum value of 2 occurs at $x = 1$. There is no relative maximum value.

The graph starts falling, or decreasing, from the left and stops decreasing at the relative minimum. From this point, the graph increases. Thus the function is increasing on $(1, \infty)$ and is decreasing on $(-\infty, 1)$.

15. From the graph we see that a relative maximum value of the function is 2.370. It occurs at $x = -0.667$. We also see that a relative minimum value of 0 occurs at $x = 2$.

The graph starts rising, or increasing, from the left and stops increasing at the relative maximum. From this point it decreases to the relative minimum and then increases again. Thus the function is increasing on $(-\infty, -0.667)$ and on $(2, \infty)$. It is decreasing on $(-0.667, 2)$.

16. From the graph we see that a relative maximum value of 2.921 occurs at $x = 3.601$. A relative minimum value of 0.995 occurs at $x = 0.103$.

The graph starts decreasing from the left and stops decreasing at the relative minimum. From this point it increases to the relative maximum and then decreases again. Thus the function is increasing on $(0.103, 3.601)$ and is decreasing on $(-\infty, 0.103)$ and on $(3.601, \infty)$.

17.

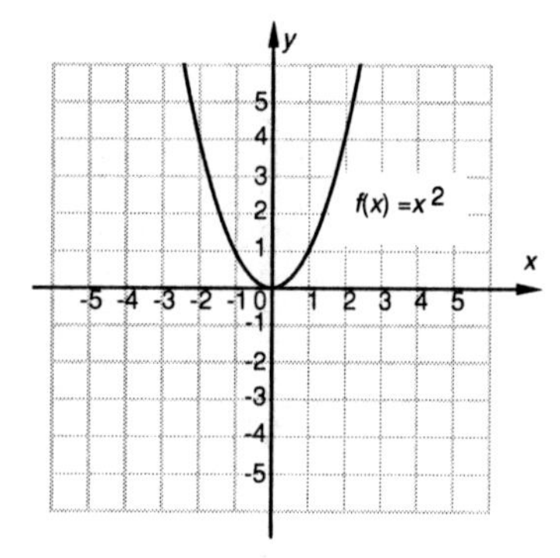

The function is increasing on $(0, \infty)$ and decreasing on $(-\infty, 0)$. We estimate that the minimum is 0 at $x = 0$. There are no maxima.

18.

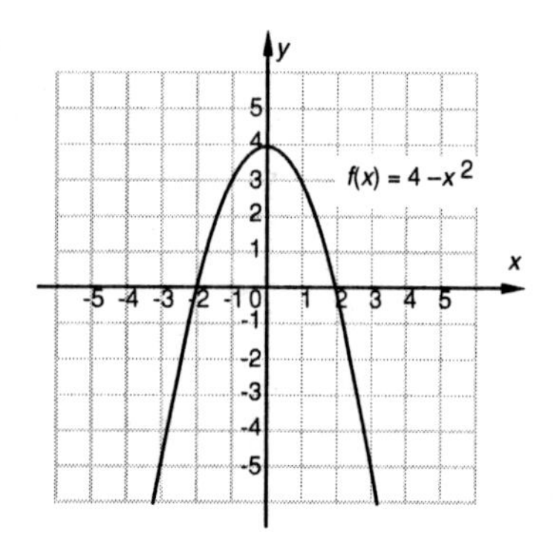

Increasing: $(-\infty, 0)$

Decreasing: $(0, \infty)$

Maximum: 4 at $x = 0$

Minima: none

19.

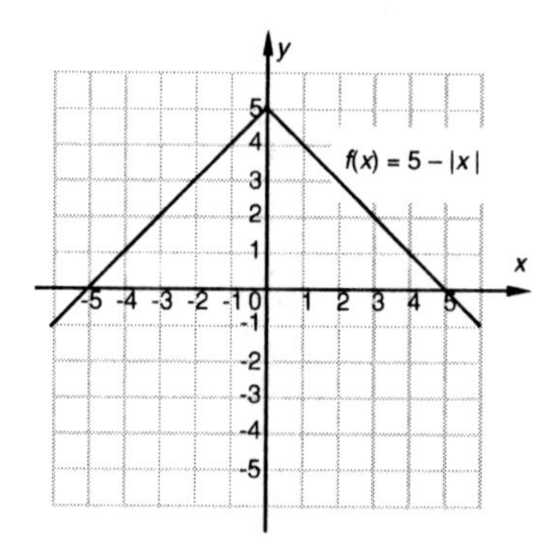

The function is increasing on $(-\infty, 0)$ and decreasing on $(0, \infty)$. We estimate that the maximum is 5 at $x = 0$. There are no minima.

20.

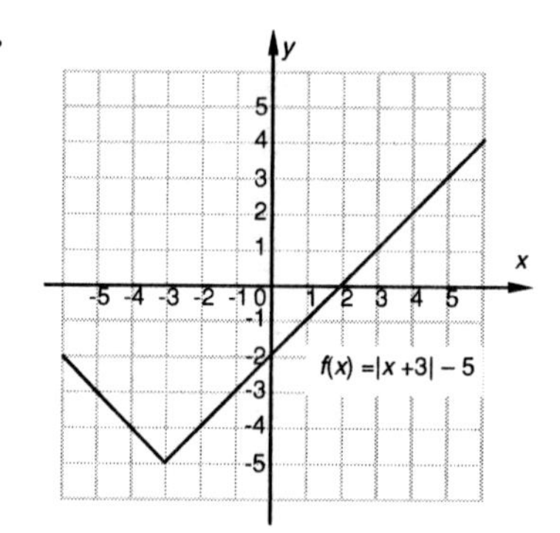

Increasing: $(-3, \infty)$

Decreasing: $(-\infty, -3)$

Maxima: none

Minimum: -5 at $x = -3$

21.

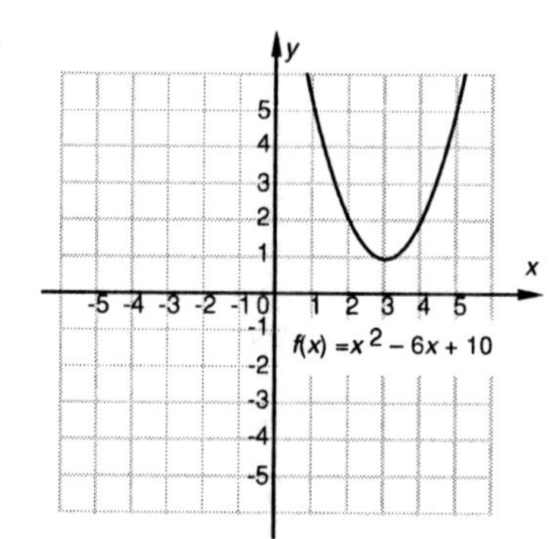

The function is decreasing on $(-\infty, 3)$ and increasing on $(3, \infty)$. We estimate that the minimum is 1 at $x = 3$. There are no maxima.

22.

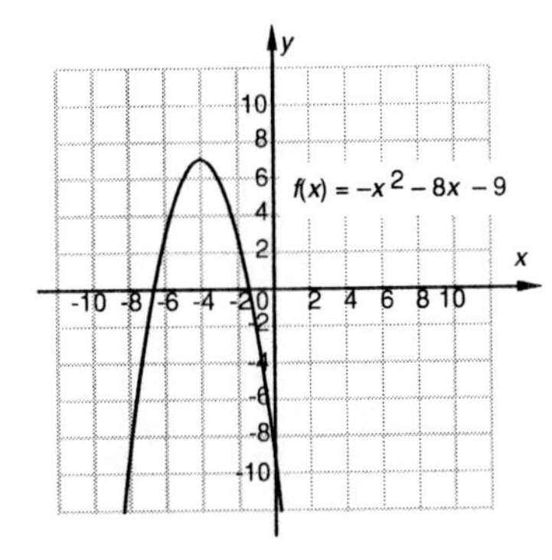

Increasing: $(-\infty, -4)$

Decreasing: $(-4, \infty)$

Maximum: 7 at $x = -4$

Minima: none

23. If $x =$ the length of the rectangle, in meters, then the width is $\dfrac{60-2x}{2}$, or $30 - x$. We use the formula Area = length × width:

$$A(x) = x(30 - x)$$
$$A(x) = 30x - x^2$$

24. Let $h =$ the height of the flag, in inches. Then the length of the base $= 2h - 7$.

$$A(h) = \frac{1}{2}(2h-7)(h)$$
$$A(h) = h^2 - \frac{7}{2}h$$

25. After t minutes, the balloon has risen $120t$ ft. We use the Pythagorean theorem.

$$[d(t)]^2 = (120t)^2 + (400)^2$$
$$d(t) = \sqrt{(120t)^2 + (400)^2}$$

We only considered the positive square root since distance must be nonnegative.

26. Use the Pythagorean theorem.

$$[h(d)]^2 + (3700)^2 = d^2$$
$$[h(d)]^2 = d^2 - (3700)^2$$
$$h(d) = \sqrt{d^2 - (3700)^2} \quad \text{Taking the positive square root}$$

27. Let $w =$ the width of the rectangle. Then the length $= \dfrac{40-2w}{2}$, or $20 - w$. Divide the rectangle into quadrants as shown below.

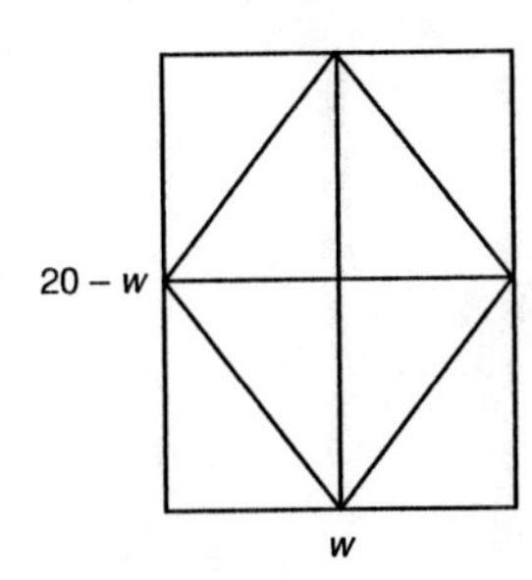

In each quadrant there are two congruent triangles. One triangle is part of the rhombus and both are part of the rectangle. Thus, in each quadrant the area of the rhombus is one-half the area of the rectangle. Then, in total, the area of the rhombus is one-half the area of the rectangle.

$$A(w) = \frac{1}{2}(20-w)(w)$$
$$A(w) = 10w - \frac{w^2}{2}$$

28. Let $w =$ the width, in feet. Then the length $= \dfrac{16-2w}{2}$, or $8 - w$.

$$A(w) = (8-w)w$$
$$A(w) = 8w - w^2$$

29. We will use similar triangles, expressing all distances in feet. $\left(6 \text{ in.} = \frac{1}{2} \text{ ft}, s \text{ in.} = \frac{s}{12} \text{ ft, and } d \text{ yd} = 3d \text{ ft}\right)$ We have

$$\frac{3d}{7} = \frac{\frac{1}{2}}{\frac{s}{12}}$$
$$\frac{s}{12} \cdot 3d = 7 \cdot \frac{1}{2}$$
$$\frac{sd}{4} = \frac{7}{2}$$
$$d = \frac{4}{s} \cdot \frac{7}{2}, \text{ so}$$
$$d(s) = \frac{14}{s}.$$

30. The volume of the tank is the sum of the volume of a sphere with radius r and a right circular cylinder with radius r and height 6 ft.

$$V(r) = \frac{4}{3}\pi r^3 + 6\pi r^2$$

31. a) If the length $= x$ feet, then the width $= 30 - x$ feet.

$$A(x) = x(30 - x)$$
$$A(x) = 30x - x^2$$

b) The length of the rectangle must be positive and less than 30 ft, so the domain of the function is $\{x|0 < x < 30\}$, or $(0, 30)$.

c) We see from the graph that the maximum value of the area function on the interval $(0, 30)$ appears to be 225 when $x = 15$. Then the dimensions that yield the maximum area are length = 15 ft and width = 30 − 15, or 15 ft.

32. a) $A(x) = x(360 - 3x)$, or $360x - 3x^2$

b) The domain is $\left\{x \middle| 0 < x < \frac{360}{3}\right\}$, or $\{x|0 < x < 120\}$, or $(0, 120)$.

c) The maximum value occurs when $x = 60$ so the width of each corral should be 60 yd and the total length of the two corrals should be $360 - 3 \cdot 60$, or 180 yd.

33. a) When a square with sides of length x are cut from each corner, the length of each of the remaining sides of the piece of cardboard is $12 - 2x$. Then the dimensions of the box are x by $12-2x$ by $12-2x$. We use the formula Volume = length × width × height to find the volume of the box:

$$V(x) = (12-2x)(12-2x)(x)$$
$$V(x) = (144 - 48x + 4x^2)(x)$$
$$V(x) = 144x - 48x^2 + 4x^3$$

This can also be expressed as $V(x) = 4x(x-6)^2$, or $V(x) = 4x(6-x)^2$.

b) The length of the sides of the square corners that are cut out must be positive and less than half the length of a side of the piece of cardboard. Thus, the domain of the function is $\{x|0 < x < 6\}$, or $(0, 6)$.

c) We see from the graph that the maximum value of the area function on the interval $(0, 6)$ appears to be 128 when $x = 2$. When $x = 2$, then $12 - 2x = 12 - 2 \cdot 2 = 8$, so the dimensions that yield the maximum volume are 8 cm by 8 cm by 2 cm.

34. a) $V(x) = 8x(14 - 2x)$, or $112x - 16x^2$

b) The domain is $\left\{x \middle| 0 < x < \frac{14}{2}\right\}$, or $\{x|0 < x < 7\}$, or $(0, 7)$.

c) The maximum occurs when $x = 3.5$, so the file should be 3.5 in. tall.

35. $g(x) = \begin{cases} x+4, & \text{for } x \le 1, \\ 8-x, & \text{for } x > 1 \end{cases}$

Since $-4 \le 1$, $g(-4) = -4 + 4 = 0$.

Since $0 \le 1$, $g(0) = 0 + 4 = 4$.

Since $1 \le 1$, $g(1) = 1 + 4 = 5$.

Since $3 > 1$, $g(3) = 8 - 3 = 5$.

36. $f(x) = \begin{cases} 3, & \text{for } x \le -2, \\ \frac{1}{2}x + 6, & \text{for } x > -2 \end{cases}$

$f(-5) = 3$

$f(-2) = 3$

$f(0) = \frac{1}{2} \cdot 0 + 6 = 6$

$f(2) = \frac{1}{2} \cdot 2 + 6 = 7$

37. $h(x) = \begin{cases} -3x - 18, & \text{for } x < -5, \\ 1, & \text{for } -5 \le x < 1, \\ x + 2, & \text{for } x \ge 1 \end{cases}$

Since -5 is in the interval $[-5, 1)$, $h(-5) = 1$.

Since 0 is in the interval $[-5, 1)$, $h(0) = 1$.

Since $1 \ge 1$, $h(1) = 1 + 2 = 3$.

Since $4 \ge 1$, $h(4) = 4 + 2 = 6$.

38. $f(x) = \begin{cases} -5x - 8, & \text{for } x < -2, \\ \frac{1}{2}x + 5, & \text{for } -2 \le x \le 4, \\ 10 - 2x, & \text{for } x > 4 \end{cases}$

Since $-4 < -2$, $f(-4) = -5(-4) - 8 = 12$.

Since -2 is in the interval $[-2, 4]$, $f(-2) = \frac{1}{2}(-2) + 5 = 4$.

Since 4 is in the interval $[-2, 4]$, $f(4) = \frac{1}{2} \cdot 4 + 5 = 7$.

Since $6 > 4$, $f(6) = 10 - 2 \cdot 6 = -2$.

39. $f(x) = \begin{cases} \frac{1}{2}x, & \text{for } x < 0, \\ x + 3, & \text{for } x \ge 0 \end{cases}$

We create the graph in two parts. Graph $f(x) = \frac{1}{2}x$ for inputs x less than 0. The graph $f(x) = x + 3$ for inputs x greater than or equal to 0.

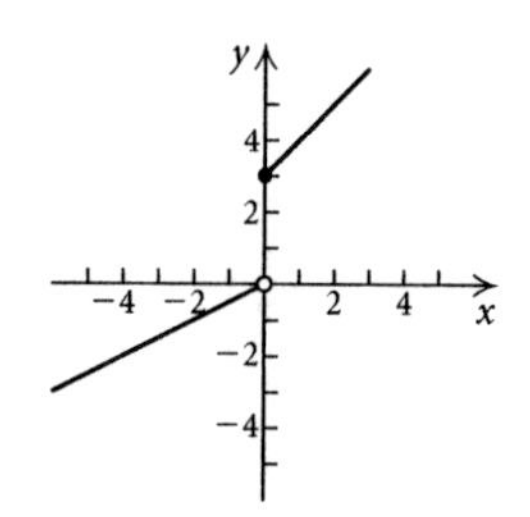

40. $f(x) = \begin{cases} -\frac{1}{3}x + 2, & \text{for } x \le 0, \\ x - 5, & \text{for } x > 0 \end{cases}$

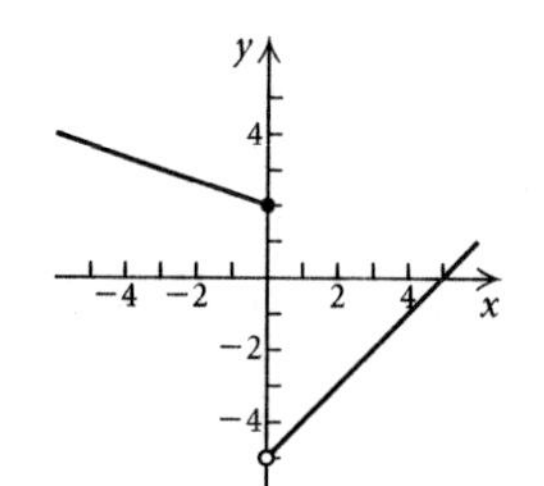

41. $f(x) = \begin{cases} -\frac{3}{4}x + 2, & \text{for } x < 4, \\ -1, & \text{for } x \ge 4 \end{cases}$

We create the graph in two parts. Graph $f(x) = -\frac{3}{4}x + 2$ for inputs x less than 4. The graph $f(x) = -1$ for inputs x greater than or equal to 4.

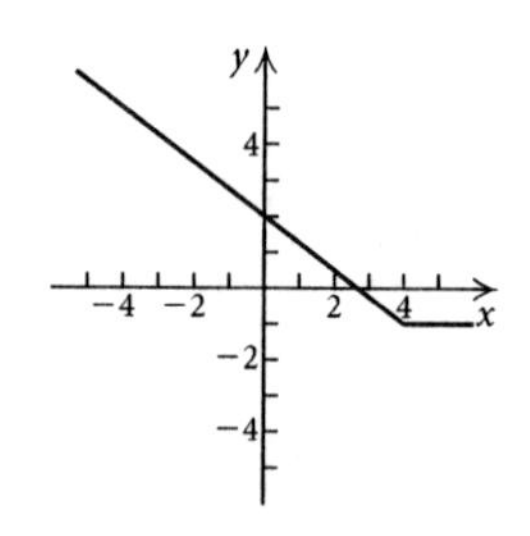

42. $f(x) = \begin{cases} 4, & \text{for } x \le -2, \\ x+1, & \text{for } -2 < x < 3 \\ -x, & \text{for } x \ge 3 \end{cases}$

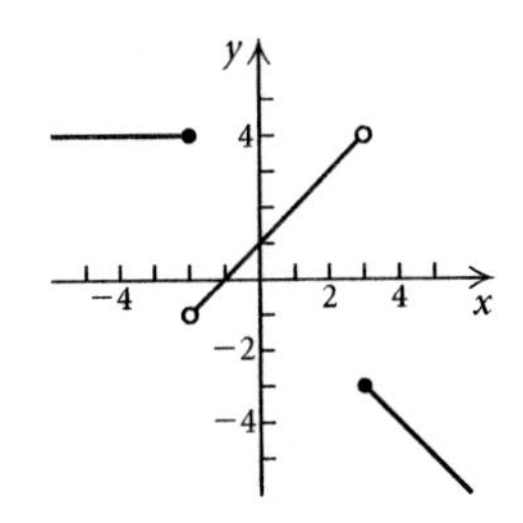

43. $f(x) = \begin{cases} x+1, & \text{for } x \le -3, \\ -1, & \text{for } -3 < x < 4 \\ \frac{1}{2}x, & \text{for } x \ge 4 \end{cases}$

We create the graph in three parts. Graph $f(x) = x + 1$ for inputs x less than or equal to -3. Graph $f(x) = -1$ for inputs greater than -3 and less than 4. Then graph $f(x) = \frac{1}{2}x$ for inputs greater than or equal to 4.

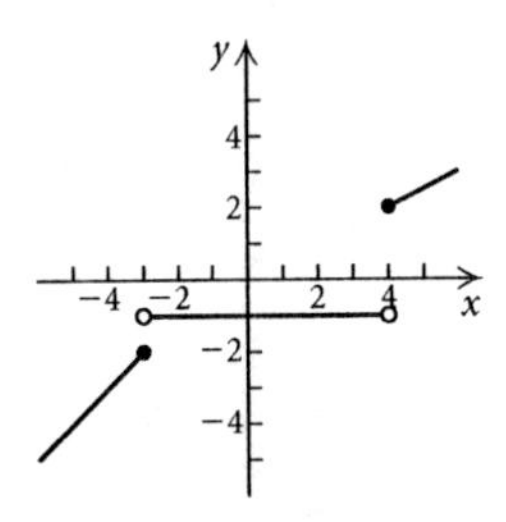

44. $f(x) = \begin{cases} \dfrac{x^2-9}{x+3}, & \text{for } x \ne -3, \\ 5, & \text{for } x = -3 \end{cases}$

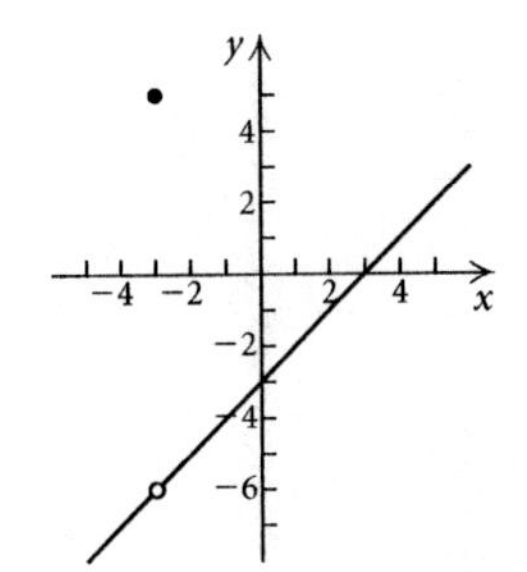

45. $f(x) = \begin{cases} 2, & \text{for } x = 5, \\ \dfrac{x^2-25}{x-5}, & \text{for } x \ne 5 \end{cases}$

When $x \ne 5$, the denominator of $(x^2 - 25)/(x - 5)$ is nonzero so we can simplify:

$$\frac{x^2-25}{x-5} = \frac{(x+5)(x-5)}{x-5} = x+5.$$

Thus, $f(x) = x + 5$, for $x \ne 5$.

The graph of this part of the function consists of a line with a "hole" at the point $(5, 10)$, indicated by an open dot. At $x = 5$, we have $f(5) = 2$, so the point $(5, 2)$ is plotted below the open dot.

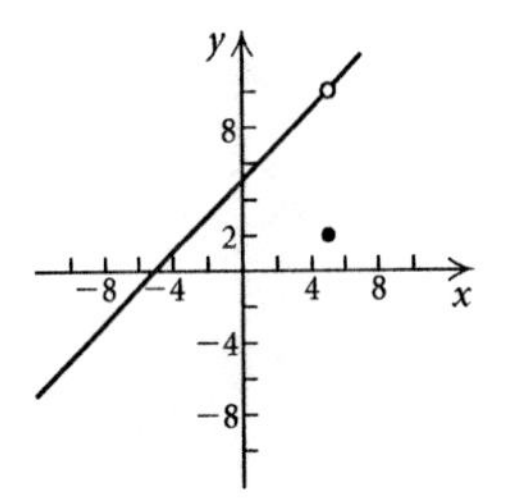

46. $f(x) = \begin{cases} \dfrac{x^2+3x+2}{x+1}, & \text{for } x \ne -1, \\ 7, & \text{for } x = -1 \end{cases}$

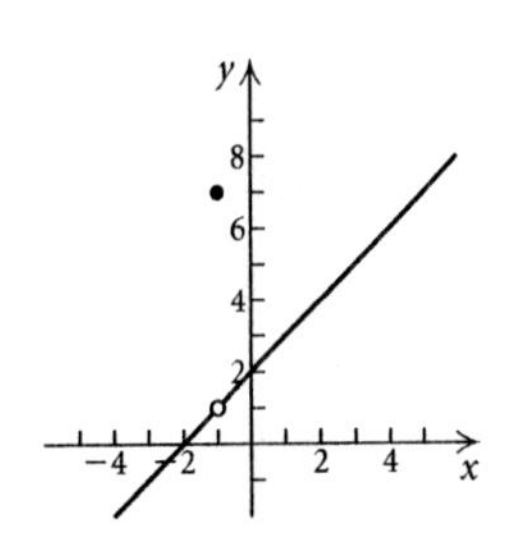

47. $f(x) = \text{int}(x)$

See Example 7.

48. $f(x) = 2\,\text{int}(x)$

This function can be defined by a piecewise function with an infinite number of statements:

$$f(x) = \begin{cases} \vdots \\ -4, & \text{for } -2 \le x < -1, \\ -2, & \text{for } -1 \le x < -0, \\ 0, & \text{for } 0 \le x < 1, \\ 2, & \text{for } 1 \le x < 2, \\ \vdots \end{cases}$$

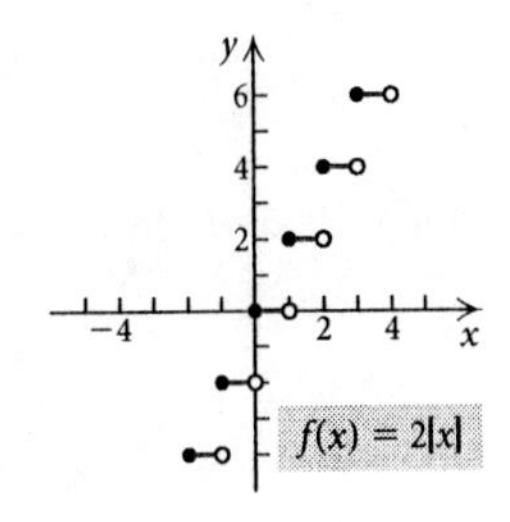

49. $f(x) = 1 + \text{int}(x)$

This function can be defined by a piecewise function with an infinite number of statements:

$$f(x) = \begin{cases} \cdot \\ \cdot \\ \cdot \\ -1, & \text{for } -2 \le x < -1, \\ 0, & \text{for } -1 \le x < -0, \\ 1, & \text{for } 0 \le x < 1, \\ 2, & \text{for } 1 \le x < 2, \\ \cdot \\ \cdot \\ \cdot \end{cases}$$

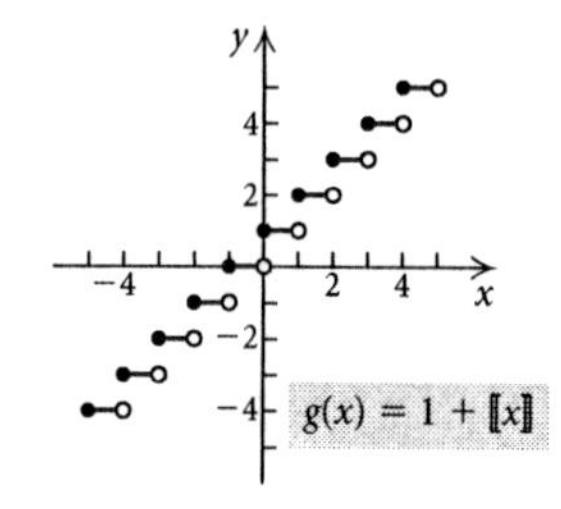

50. $f(x) = \frac{1}{2}\text{int}(x) - 2$

This function can be defined by a piecewise function with an infinite number of statements:

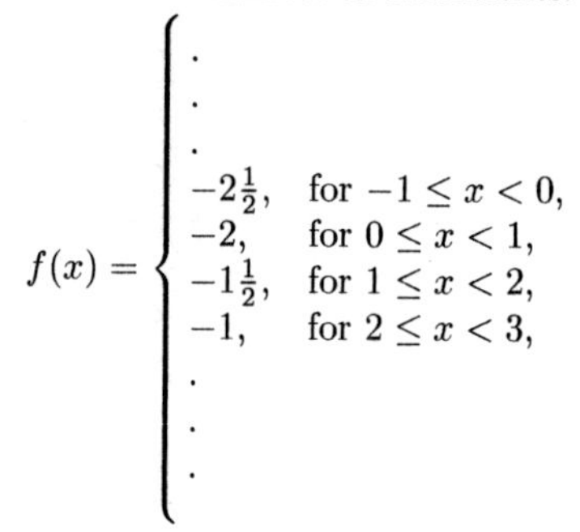

$$f(x) = \begin{cases} \cdot \\ \cdot \\ \cdot \\ -2\frac{1}{2}, & \text{for } -1 \le x < 0, \\ -2, & \text{for } 0 \le x < 1, \\ -1\frac{1}{2}, & \text{for } 1 \le x < 2, \\ -1, & \text{for } 2 \le x < 3, \\ \cdot \\ \cdot \\ \cdot \end{cases}$$

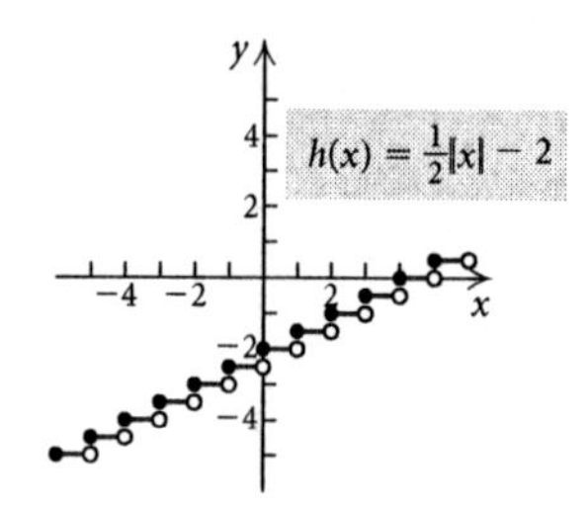

51. From the graph we see that the domain is $(-\infty, \infty)$ and the range is $(-\infty, 0) \cup [3, \infty)$.

52. Domain: $(-\infty, \infty)$; range: $(-5, \infty)$

53. From the graph we see that the domain is $(-\infty, \infty)$ and the range is $[-1, \infty)$.

54. Domain: (∞, ∞); range: $(-\infty, -3] \cup (-1, 4]$

55. From the graph we see that the domain is $(-\infty, \infty)$ and the range is $(-\infty, -2] \cup \{-1\} \cup [2, \infty)$, or $\{y | y \le -2 \text{ or } y = -1 \text{ or } y \ge 2\}$.

56. Domain: $(-\infty, \infty)$; range: $(-\infty, -6) \cup (-6, \infty)$

57. From the graph we see that the domain is $(-\infty, \infty)$ and the range is $\{-5, -2, 4\}$. An equation for the function is:

$$f(x) = \begin{cases} -2, & \text{for } x < 2, \\ -5, & \text{for } x = 2, \\ 4, & \text{for } x > 2 \end{cases}$$

58. Domain: $(-\infty, \infty)$; range: $\{y | y = -3 \text{ or } y \ge 0\}$

$$g(x) = \begin{cases} -3, & \text{for } x < 0, \\ x, & \text{for } x \ge 0 \end{cases}$$

59. From the graph we see that the domain is $(-\infty, \infty)$ and the range is $(-\infty, -1] \cup [2, \infty)$. Finding the slope of each segment and using the slope-intercept or point-slope formula, we find that an equation for the function is:

$$g(x) = \begin{cases} x, & \text{for } x \le -1, \\ 2, & \text{for } -1 < x \le 2, \\ x, & \text{for } x > 2 \end{cases}$$

This can also be expressed as follows:

$$g(x) = \begin{cases} x, & \text{for } x \le -1, \\ 2, & \text{for } -1 < x < 2, \\ x, & \text{for } x \ge 2 \end{cases}$$

60. Domain: $(-\infty, \infty)$; range: $\{-2\} \cup [0, \infty)$, or $\{y | y = -2 \text{ or } y \ge 0\}$. An equation for the function is:

$$h(x) = \begin{cases} |x|, & \text{for } x < 3, \\ -2, & \text{for } x \ge 3 \end{cases}$$

This can also be expressed as follows:

$$h(x) = \begin{cases} -x, & \text{for } x \le 0, \\ x, & \text{for } 0 < x < 3, \\ -2, & \text{for } x \ge 3 \end{cases}$$

It can also be expressed as follows:

$$h(x) = \begin{cases} -x, & \text{for } x < 0, \\ x, & \text{for } 0 \le x < 3, \\ -2, & \text{for } x \ge 3 \end{cases}$$

61. From the graph we see that the domain is $[-5, 3]$ and the range is $(-3, 5)$. Finding the slope of each segment and using the slope-intercept or point-slope formula, we find that an equation for the function is:

$$h(x) = \begin{cases} x + 8, & \text{for } -5 \le x < -3, \\ 3, & \text{for } -3 \le x \le 1, \\ 3x - 6, & \text{for } 1 < x \le 3 \end{cases}$$

62. Domain: $[-4, \infty)$; range: $[-2, 4]$

$$f(x) = \begin{cases} -2x - 4, & \text{for } -4 \le x \le -1, \\ x - 1, & \text{for } -1 < x < 2, \\ 2, & \text{for } x \ge 2 \end{cases}$$

This can also be expressed as:

$$f(x) = \begin{cases} -2x - 4, & \text{for } -4 \le x \le -1, \\ x - 1, & \text{for } -1 \le x < 2, \\ 2, & \text{for } x \ge 2 \end{cases}$$

63. a) $y = -0.1x^2 + 1.2x + 98.6$

110

0 12

90

b) Using the MAXIMUM feature we find that the relative maximum is 102.2 at $t = 6$. Thus, we know that the patient's temperature was the highest at $t = 6$, or 6 days after the onset of the illness and that the highest temperature was 102.2°F.

64. a)

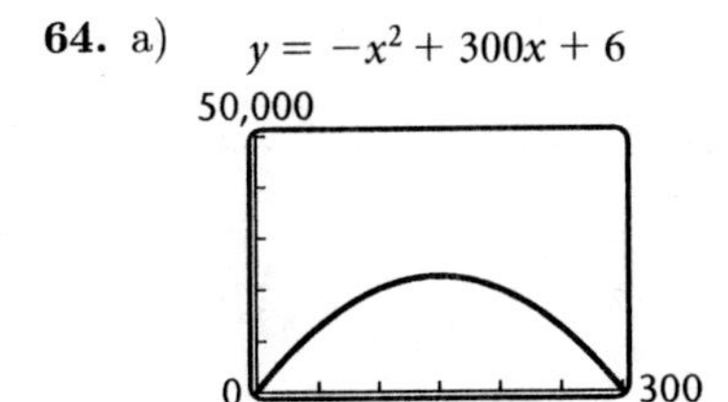

b) $22,506$ at $a = 150$

c) The greatest number of games will be sold when \$150 thousand is spent on advertising. For that amount, 22,506 games will be sold.

65. Graph $y = \dfrac{8x}{x^2+1}$.

Increasing: $(-1, 1)$

Decreasing: $(-\infty, -1)$, $(1, \infty)$

66. Graph $y = \dfrac{-4}{x^2+1}$.

Increasing: $(0, \infty)$

Decreasing: $(-\infty, 0)$

67. Graph $y = x\sqrt{4 - x^2}$, for $-2 \le x \le 2$.

Increasing: $(-1.414, 1.414)$

Decreasing: $(-2, -1.414)$, $(1.414, 2)$

68. Graph $y = -0.8x\sqrt{9 - x^2}$, for $-3 \le x \le 3$.

Increasing: $(-3, -2.121)$, $(2.121, 3)$

Decreasing: $(-2.121, 2.121)$

69. a) The length of a diameter of the circle (and a diagonal of the rectangle) is $2 \cdot 8$, or 16 ft. Let $l =$ the length of the rectangle. Use the Pythagorean theorem to write l as a function of x.

$$x^2 + l^2 = 16^2$$
$$x^2 + l^2 = 256$$
$$l^2 = 256 - x^2$$
$$l = \sqrt{256 - x^2}$$

Since the length must be positive, we considered only the positive square root.

Use the formula Area = length × width to find the area of the rectangle:

$$A(x) = x\sqrt{256 - x^2}$$

b) The width of the rectangle must be positive and less than the diameter of the circle. Thus, the domain of the function is $\{x \mid 0 < x < 16\}$, or $(0, 16)$.

c)

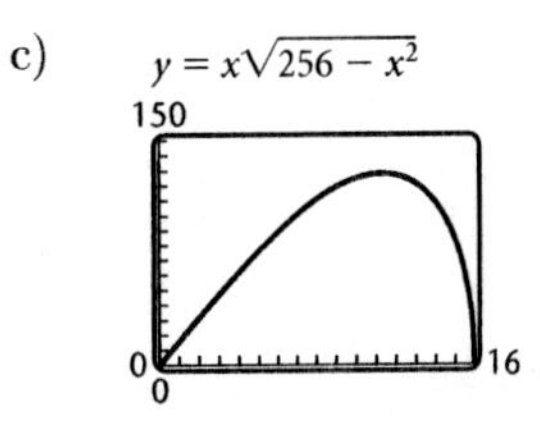

d) Using the MAXIMUM feature, we find that the maximum area occurs when x is about 11.314. When $x \approx 11.314$, $\sqrt{256 - x^2} \approx \sqrt{256 - (11.314)^2} \approx 11.313$. Thus, the dimensions that maximize the area are about 11.314 ft by 11.313 ft. (Answers may vary slightly due to rounding differences.)

70. a) Let $h(x) =$ the height of the box.

$$320 = x \cdot x \cdot h(x)$$
$$\frac{320}{x^2} = h(x)$$

Area of the bottom: x^2

Area of each side: $x\left(\dfrac{320}{x^2}\right)$, or $\dfrac{320}{x}$

Area of the top: x^2

$$C(x) = 1.5x^2 + +4(2.5)\left(\frac{320}{x}\right) + 1 \cdot x^2$$
$$C(x) = 2.5x^2 + \frac{3200}{x}$$

b) The length of the base must be positive, so the domain of the function is $\{x \mid x > 0\}$, or $(0, \infty)$.

c)

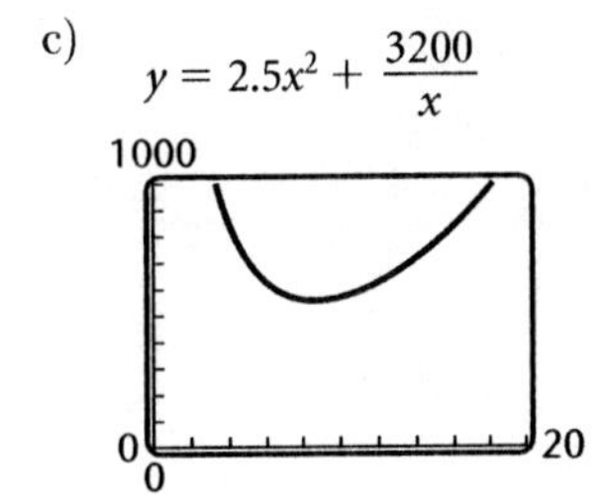

d) Using the MIMIMUM feature, we find that the minimum cost occurs when $x \approx 8.618$. Thus, the dimensions that minimize the cost are about 8.618 ft by 8.618 ft by $\dfrac{320}{(8.618)^2}$, or about 4.309 ft.

71. Some possibilities are outdoor temperature during a 24 hour period, sales of a new product, and temperature during an illness.

72. For continuous functions, relative extrema occur at points for which the function changes from increasing to decreasing or vice versa.

73. Function; domain; range; domain; exactly one; range

74. Midpoint formula

75. x-intercept

76. Constant; identity

77. If $[\text{int}(x)]^2 = 25$, then $\text{int}(x) = -5$ or $\text{int}(x) = 5$. For $-5 \leq x < -4$, $\text{int}(x) = -5$. For $5 \leq x < 6$, $\text{int}(x) = 5$. Thus, the possible inputs for x are $\{x| -5 \leq x < -4 \text{ or } 5 \leq x < 6\}$.

78. If $\text{int}(x+2) = -3$, then $-3 \leq x+2 < -2$, or $-5 \leq x < -4$. The possible inputs for x are $\{x| -5 \leq x < -4\}$.

79. a) We add labels to the drawing in the text.

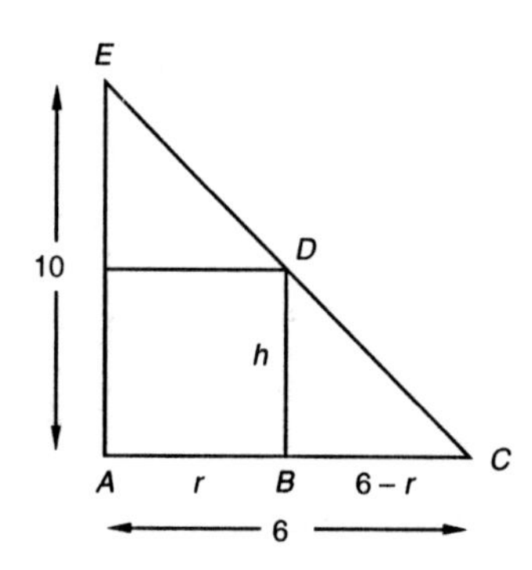

We write a proportion involving the lengths of the sides of the similar triangles BCD and ACE. Then we solve it for h.

$$\frac{h}{6-r} = \frac{10}{6}$$

$$h = \frac{10}{6}(6-r) = \frac{5}{3}(6-r)$$

$$h = \frac{30-5r}{3}$$

Thus, $h(r) = \dfrac{30-5r}{3}$.

b) $$V = \pi r^2 h$$

$$V(r) = \pi r^2\left(\frac{30-5r}{3}\right) \quad \text{Substituting for } h$$

c) We first express r in terms of h.

$$h = \frac{30-5r}{3}$$

$$3h = 30-5r$$

$$5r = 30-3h$$

$$r = \frac{30-3h}{5}$$

$$V = \pi r^2 h$$

$$V(h) = \pi\left(\frac{30-3h}{5}\right)^2 h$$

Substituting for r

We can also write $V(h) = \pi h\left(\dfrac{30-3h}{5}\right)^2$.

80. a) The distance from A to S is $4-x$.

Using the Pythagorean theorem, we find that the distance from S to C is $\sqrt{1+x^2}$.

Then $C(x) = 3000(4-x)+5000\sqrt{1+x^2}$, or $12,000-3000x+5000\sqrt{1+x^2}$.

b) Graph $y = 12,000-3000x+5000\sqrt{1+x^2}$ in a window such as $[0, 5, 10,000, 20,000]$, Xscl = 1, Yscl = 1000. Using the MINIMUM feature, we find that cost is minimized when $x = 0.75$, so the line should come to shore 0.75 mi from B.

Exercise Set 1.6

1. $$\begin{aligned}(f+g)(5) &= f(5)+g(5)\\ &= (5^2-3)+(2\cdot 5+1)\\ &= 25-3+10+1\\ &= 33\end{aligned}$$

2. $$\begin{aligned}(fg)(0) &= f(0)\cdot g(0)\\ &= (0^2-3)(2\cdot 0+1)\\ &= -3(1) = -3\end{aligned}$$

3. $$\begin{aligned}(f-g)(-1) &= f(-1)-g(-1)\\ &= ((-1)^2-3)-(2(-1)+1)\\ &= -2-(-1) = -2+1\\ &= -1\end{aligned}$$

4. $$\begin{aligned}(fg)(2) &= f(2)\cdot g(2)\\ &= (2^2-3)(2\cdot 2+1)\\ &= 1\cdot 5 = 5\end{aligned}$$

5. $$\begin{aligned}(f/g)\left(-\frac{1}{2}\right) &= \frac{f\left(-\frac{1}{2}\right)}{g\left(-\frac{1}{2}\right)}\\ &= \frac{\left(-\frac{1}{2}\right)^2-3}{2\left(-\frac{1}{2}\right)+1}\\ &= \frac{\frac{1}{4}-3}{-1+1}\\ &= \frac{-\frac{11}{4}}{0}\end{aligned}$$

Since division by 0 is not defined, $(f/g)\left(-\dfrac{1}{2}\right)$ does not exist.

6. $$\begin{aligned}(f-g)(0) &= f(0)-g(0)\\ &= (0^2-3)-(2\cdot 0+1)\\ &= -3-1 = -4\end{aligned}$$

7. $$\begin{aligned}(fg)\left(-\frac{1}{2}\right) &= f\left(-\frac{1}{2}\right)\cdot g\left(-\frac{1}{2}\right)\\ &= \left[\left(-\frac{1}{2}\right)^2-3\right]\left[2\left(-\frac{1}{2}\right)+1\right]\\ &= -\frac{11}{4}\cdot 0 = 0\end{aligned}$$

8. $(f/g)(-\sqrt{3}) = \dfrac{f(-\sqrt{3})}{g(-\sqrt{3})}$
$= \dfrac{(-\sqrt{3})^2-3}{2(-\sqrt{3})+1}$
$= \dfrac{0}{-2\sqrt{3}+1} = 0$

9. $(g-f)(-1) = g(-1) - f(-1)$
$= [2(-1)+1] - [(-1)^2 - 3]$
$= (-2+1) - (1-3)$
$= -1 - (-2)$
$= -1 + 2$
$= 1$

10. $(g/f)\left(-\dfrac{1}{2}\right) = \dfrac{g\left(-\frac{1}{2}\right)}{f\left(-\frac{1}{2}\right)}$
$= \dfrac{2\left(-\frac{1}{2}\right)+1}{\left(-\frac{1}{2}\right)^2-3}$
$= \dfrac{0}{-\frac{11}{4}}$
$= 0$

11. $(h-g)(-4) = h(-4) - g(-4)$
$= (-4+4) - \sqrt{-4-1}$
$= 0 - \sqrt{-5}$

Since $\sqrt{-5}$ is not a real number, $(h-g)(-4)$ does not exist.

12. $(gh)(10) = g(10) \cdot h(10)$
$= \sqrt{10-1}(10+4)$
$= \sqrt{9}(14)$
$= 3 \cdot 14 = 42$

13. $(g/h)(1) = \dfrac{g(1)}{h(1)}$
$= \dfrac{\sqrt{1-1}}{1+4}$
$= \dfrac{\sqrt{0}}{5}$
$= \dfrac{0}{5} = 0$

14. $(h/g)(1) = \dfrac{h(1)}{g(1)}$
$= \dfrac{1+4}{\sqrt{1-1}}$
$= \dfrac{5}{0}$

Since division by 0 is not defined, $(h/g)(1)$ does not exist.

15. $(g+h)(1) = g(1) + h(1)$
$= \sqrt{1-1} + (1+4)$
$= \sqrt{0} + 5$
$= 0 + 5 = 5$

16. $(hg)(3) = h(3) \cdot g(3)$
$= (3+4)\sqrt{3-1}$
$= 7\sqrt{2}$

17. $f(x) = 2x+3$, $g(x) = 3-5x$

a) The domain of f and of g is the set of all real numbers, or $(-\infty, \infty)$. Then the domain of $f+g$, $f-g$, ff, and fg is also $(-\infty, \infty)$. For f/g we must exclude $\frac{3}{5}$ since $g\left(\frac{3}{5}\right) = 0$. Then the domain of f/g is $\left(-\infty, \frac{3}{5}\right) \cup \left(\frac{3}{5}, \infty\right)$. For g/f we must exclude $-\frac{3}{2}$ since $f\left(-\frac{3}{2}\right) = 0$. The domain of g/f is $\left(-\infty, -\frac{3}{2}\right) \cup \left(-\frac{3}{2}, \infty\right)$.

b) $(f+g)(x) = f(x) + g(x) = (2x+3) + (3-5x) = -3x+6$

$(f-g)(x) = f(x) - g(x) = (2x+3) - (3-5x) = 2x+3-3+5x = 7x$

$(fg)(x) = f(x) \cdot g(x) = (2x+3)(3-5x) = 6x - 10x^2 + 9 - 15x = -10x^2 - 9x + 9$

$(ff)(x) = f(x) \cdot f(x) = (2x+3)(2x+3) = 4x^2 + 12x + 9$

$(f/g)(x) = \dfrac{f(x)}{g(x)} = \dfrac{2x+3}{3-5x}$

$(g/f)(x) = \dfrac{g(x)}{f(x)} = \dfrac{3-5x}{2x+3}$

18. $f(x) = -x+1$, $g(x) = 4x-2$

a) The domain of f, g, $f+g$, $f-g$, fg, and ff is $(-\infty, \infty)$. Since $g\left(\frac{1}{2}\right) = 0$, the domain of f/g is $\left(-\infty, \frac{1}{2}\right) \cup \left(\frac{1}{2}, \infty\right)$. Since $f(1) = 0$, the domain of g/f is $(-\infty, 1) \cup (1, \infty)$.

b) $(f+g)(x) = (-x+1) + (4x-2) = 3x-1$

$(f-g)(x) = (-x+1) - (4x-2) = -x+1-4x+2 = -5x+3$

$(fg)(x) = (-x+1)(4x-2) = -4x^2 + 6x - 2$

$(ff)(x) = (-x+1)(-x+1) = x^2 - 2x + 1$

$(f/g)(x) = \dfrac{-x+1}{4x-2}$

$(g/f)(x) = \dfrac{4x-2}{-x+1}$

19. $f(x) = x-3$, $g(x) = \sqrt{x+4}$

a) Any number can be an input in f, so the domain of f is the set of all real numbers, or $(-\infty, \infty)$.

The inputs of g must be nonnegative, so we have $x+4 \geq 0$, or $x \geq -4$. Thus, the domain of g is $[-4, \infty)$.

The domain of $f+g$, $f-g$, and fg is the set of all numbers in the domains of both f and g. This is $[-4, \infty)$.

The domain of ff is the domain of f, or $(-\infty, \infty)$.

The domain of f/g is the set of all numbers in the domains of f and g, excluding those for which $g(x) = 0$. Since $g(-4) = 0$, the domain of f/g is $(-4, \infty)$.

The domain of g/f is the set of all numbers in the domains of g and f, excluding those for which $f(x) = 0$. Since $f(3) = 0$, the domain of g/f is $[-4, 3) \cup (3, \infty)$.

b) $(f+g)(x) = f(x) + g(x) = x - 3 + \sqrt{x+4}$
$(f-g)(x) = f(x) - g(x) = x - 3 - \sqrt{x+4}$
$(fg)(x) = f(x) \cdot g(x) = (x-3)\sqrt{x+4}$
$(ff)(x) = \left[f(x)\right]^2 = (x-3)^2 = x^2 - 6x + 9$
$(f/g)(x) = \dfrac{f(x)}{g(x)} = \dfrac{x-3}{\sqrt{x+4}}$
$(g/f)(x) = \dfrac{g(x)}{f(x)} = \dfrac{\sqrt{x+4}}{x-3}$

20. $f(x) = x + 2$, $g(x) = \sqrt{x-1}$

a) The domain of f is $(-\infty, \infty)$. The domain of g consists of all the values of x for which $x - 1$ is nonnegative, or $[1, \infty)$. Then the domain of $f + g$, $f - g$, and fg is $[1, \infty)$. The domain of ff is $(-\infty, \infty)$. Since $g(1) = 0$, the domain of f/g is $(1, \infty)$. Since $f(-2) = 0$ and -2 is not in the domain of g, the domain of g/f is $[1, \infty)$.

b) $(f+g)(x) = x + 2 + \sqrt{x-1}$
$(f-g)(x) = x + 2 - \sqrt{x+1}$
$(fg)(x) = (x+2)\sqrt{x-1}$
$(ff)(x) = (x+2)(x+2) = x^2 + 4x + 4$
$(f/g)(x) = \dfrac{x+2}{\sqrt{x-1}}$
$(g/f)(x) = \dfrac{\sqrt{x-1}}{x+2}$

21. $f(x) = 2x - 1$, $g(x) = -2x^2$

a) The domain of f and of g is $(-\infty, \infty)$. Then the domain of $f + g$, $f - g$, fg, and ff is $(-\infty, \infty)$. For f/g, we must exclude 0 since $g(0) = 0$. The domain of f/g is $(-\infty, 0) \cup (0, \infty)$. For g/f, we must exclude $\dfrac{1}{2}$ since $f\left(\dfrac{1}{2}\right) = 0$. The domain of g/f is $\left(-\infty, \dfrac{1}{2}\right) \cup \left(\dfrac{1}{2}, \infty\right)$.

b) $(f+g)(x) = f(x) + g(x) = (2x-1) + (-2x^2) = -2x^2 + 2x - 1$
$(f-g)(x) = f(x) - g(x) = (2x-1) - (-2x^2) = 2x^2 + 2x - 1$
$(fg)(x) = f(x) \cdot g(x) = (2x-1)(-2x^2) = -4x^3 + 2x^2$
$(ff)(x) = f(x) \cdot f(x) = (2x-1)(2x-1) = 4x^2 - 4x + 1$
$(f/g)(x) = \dfrac{f(x)}{g(x)} = \dfrac{2x-1}{-2x^2}$
$(g/f)(x) = \dfrac{g(x)}{f(x)} = \dfrac{-2x^2}{2x-1}$

22. $f(x) = x^2 - 1$, $g(x) = 2x + 5$

a) The domain of f and of g is the set of all real numbers, or $(-\infty, \infty)$. Then the domain of $f+g$, $f-g$, fg and ff is $(-\infty, \infty)$. Since $g\left(-\dfrac{5}{2}\right) = 0$, the domain of f/g is $\left(-\infty, -\dfrac{5}{2}\right) \cup \left(-\dfrac{5}{2}, \infty\right)$. Since $f(1) = 0$ and $f(-1) = 0$, the domain of g/f is $(-\infty, -1) \cup (-1, 1) \cup (1, \infty)$.

b) $(f+g)(x) = x^2 - 1 + 2x + 5 = x^2 + 2x + 4$
$(f-g)(x) = x^2 - 1 - (2x+5) = x^2 - 2x - 6$
$(fg)(x) = (x^2-1)(2x+5) = 2x^3 + 5x^2 - 2x - 5$
$(ff)(x) = (x^2-1)^2 = x^4 - 2x^2 + 1$
$(f/g)(x) = \dfrac{x^2-1}{2x+5}$
$(g/f)(x) = \dfrac{2x+5}{x^2-1}$

23. $f(x) = \sqrt{x-3}$, $g(x) = \sqrt{x+3}$

a) Since $f(x)$ is nonnegative for values of x in $[3, \infty)$, this is the domain of f. Since $g(x)$ is nonnegative for values of x in $[-3, \infty)$, this is the domain of g. The domain of $f+g$, $f-g$, and fg is the intersection of the domains of f and g, or $[3, \infty)$. The domain of ff is the same as the domain of f, or $[3, \infty)$. For f/g, we must exclude -3 since $g(-3) = 0$. This is not in $[3, \infty)$, so the domain of f/g is $[3, \infty)$. For g/f, we must exclude 3 since $f(3) = 0$. The domain of g/f is $(3, \infty)$.

b) $(f+g)(x) = f(x) + g(x) = \sqrt{x-3} + \sqrt{x+3}$
$(f-g)(x) = f(x) - g(x) = \sqrt{x-3} - \sqrt{x+3}$
$(fg)(x) = f(x) \cdot g(x) = \sqrt{x-3} \cdot \sqrt{x+3} = \sqrt{x^2-9}$
$(ff)(x) = f(x) \cdot f(x) = \sqrt{x-3} \cdot \sqrt{x-3} = |x-3|$
$(f/g)(x) = \dfrac{\sqrt{x-3}}{\sqrt{x+3}}$
$(g/f)(x) = \dfrac{\sqrt{x+3}}{\sqrt{x-3}}$

24. $f(x) = \sqrt{x}$, $g(x) = \sqrt{2-x}$

a) The domain of f is $[0, \infty)$. The domain of g is $(-\infty, 2]$. Then the domain of $f+g$, $f-g$, and fg is $[0, 2]$. The domain of ff is the same as the domain of f, $[0, \infty)$. Since $g(2) = 0$, the domain of f/g is $[0, 2)$. Since $f(0) = 0$, the domain of g/f is $(0, 2]$.

b) $(f+g)(x) = \sqrt{x} + \sqrt{2-x}$
$(f-g)(x) = \sqrt{x} - \sqrt{2-x}$
$(fg)(x) = \sqrt{x} \cdot \sqrt{2-x} = \sqrt{2x - x^2}$
$(ff)(x) = \sqrt{x} \cdot \sqrt{x} = \sqrt{x^2} = |x|$
$(f/g)(x) = \dfrac{\sqrt{x}}{\sqrt{2-x}}$
$(g/f)(x) = \dfrac{\sqrt{2-x}}{\sqrt{x}}$

25. $f(x) = x + 1$, $g(x) = |x|$

a) The domain of f and of g is $(-\infty, \infty)$. Then the domain of $f + g$, $f - g$, fg, and ff is $(-\infty, \infty)$. For f/g, we must exclude 0 since $g(0) = 0$. The domain of f/g is $(-\infty, 0) \cup (0, \infty)$. For g/f, we must exclude -1 since $f(-1) = 0$. The domain of g/f is $(-\infty, -1) \cup (-1, \infty)$.

b) $(f + g)(x) = f(x) + g(x) = x + 1 + |x|$

$(f - g)(x) = f(x) - g(x) = x + 1 - |x|$

$(fg)(x) = f(x) \cdot g(x) = (x + 1)|x|$

$(ff)(x) = f(x) \cdot f(x) = (x+1)(x+1) = x^2 + 2x + 1$

$(f/g)(x) = \dfrac{x + 1}{|x|}$

$(g/f)(x) = \dfrac{|x|}{x + 1}$

26. $f(x) = 4|x|$, $g(x) = 1 - x$

a) The domain of f and of g is $(-\infty, \infty)$. Then the domain of $f+g$, $f-g$, fg, and ff is $(-\infty, \infty)$. Since $g(1) = 0$, the domain of f/g is $(-\infty, 1) \cup (1, \infty)$. Since $f(0) = 0$, the domain of g/f is $(-\infty, 0) \cup (0, \infty)$.

b) $(f + g)(x) = 4|x| + 1 - x$

$(f - g)(x) = 4|x| - (1 - x) = 4|x| - 1 + x$

$(fg)(x) = 4|x|(1 - x) = 4|x| - 4x|x|$

$(ff)(x) = 4|x| \cdot 4|x| = 16x^2$

$(f/g)(x) = \dfrac{4|x|}{1 - x}$

$(g/f)(x) = \dfrac{1 - x}{4|x|}$

27. $f(x) = x^3$, $g(x) = 2x^2 + 5x - 3$

a) Since any number can be an input for either f or g, the domain of f, g, $f + g$, $f - g$, fg, and ff is the set of all real numbers, or $(-\infty, \infty)$.

Since $g(-3) = 0$ and $g\left(\dfrac{1}{2}\right) = 0$, the domain of f/g is $(-\infty, -3) \cup \left(-3, \dfrac{1}{2}\right) \cup \left(\dfrac{1}{2}, \infty\right)$.

Since $f(0) = 0$, the domain of g/f is $(-\infty, 0) \cup (0, \infty)$.

b) $(f + g)(x) = f(x) + g(x) = x^3 + 2x^2 + 5x - 3$

$(f - g)(x) = f(x) - g(x) = x^3 - (2x^2 + 5x - 3) = x^3 - 2x^2 - 5x + 3$

$(fg)(x) = f(x) \cdot g(x) = x^3(2x^2 + 5x - 3) = 2x^5 + 5x^4 - 3x^3$

$(ff)(x) = f(x) \cdot f(x) = x^3 \cdot x^3 = x^6$

$(f/g)(x) = \dfrac{f(x)}{g(x)} = \dfrac{x^3}{2x^2 + 5x - 3}$

$(g/f)(x) = \dfrac{g(x)}{f(x)} = \dfrac{2x^2 + 5x - 3}{x^3}$

28. $f(x) = x^2 - 4$, $g(x) = x^3$

a) The domain of f and of g is $(-\infty, \infty)$. Then the domain of $f+g$, $f-g$, fg, and ff is $(-\infty, \infty)$. Since $g(0) = 0$, the domain of f/g is $(-\infty, 0) \cup (0, \infty)$. Since $f(-2) = 0$ and $f(2) = 0$, the domain of g/f is $(-\infty, -2) \cup (-2, 2) \cup (2, \infty)$.

b) $(f + g)(x) = x^2 - 4 + x^3$, or $x^3 + x^2 - 4$

$(f - g)(x) = x^2 - 4 - x^3$, or $-x^3 + x^2 - 4$

$(fg)(x) = (x^2 - 4)(x^3) = x^5 - 4x^3$

$(ff)(x) = (x^2 - 4)(x^2 - 4) = x^4 - 8x^2 + 16$

$(f/g)(x) = \dfrac{x^2 - 4}{x^3}$

$(g/f)(x) = \dfrac{x^3}{x^2 - 4}$

29. $f(x) = \dfrac{4}{x + 1}$, $g(x) = \dfrac{1}{6 - x}$

a) Since $x + 1 = 0$ when $x = -1$, we must exclude -1 from the domain of f. It is $(-\infty, -1) \cup (-1, \infty)$. Since $6 - x = 0$ when $x = 6$, we must exclude 6 from the domain of g. It is $(-\infty, 6) \cup (6, \infty)$. The domain of $f + g$, $f - g$, and fg is the intersection of the domains of f and g, or $(-\infty, -1) \cup (-1, 6) \cup (6, \infty)$. The domain of ff is the same as the domain of f, or $(-\infty, -1) \cup (-1, \infty)$. Since there are no values of x for which $g(x) = 0$ or $f(x) = 0$, the domain of f/g and g/f is $(-\infty, -1) \cup (-1, 6) \cup (6, \infty)$.

b) $(f + g)(x) = f(x) + g(x) = \dfrac{4}{x + 1} + \dfrac{1}{6 - x}$

$(f - g)(x) = f(x) - g(x) = \dfrac{4}{x + 1} - \dfrac{1}{6 - x}$

$(fg)(x) = f(x) \cdot g(x) = \dfrac{4}{x+1} \cdot \dfrac{1}{6-x} = \dfrac{4}{(x+1)(6-x)}$

$(ff)(x) = f(x) \cdot f(x) = \dfrac{4}{x + 1} \cdot \dfrac{4}{x + 1} = \dfrac{16}{(x + 1)^2}$, or $\dfrac{16}{x^2 + 2x + 1}$

$(f/g)(x) = \dfrac{\frac{4}{x + 1}}{\frac{1}{6 - x}} = \dfrac{4}{x + 1} \cdot \dfrac{6 - x}{1} = \dfrac{4(6 - x)}{x + 1}$

$(g/f)(x) = \dfrac{\frac{1}{6 - x}}{\frac{4}{x + 1}} = \dfrac{1}{6 - x} \cdot \dfrac{x + 1}{4} = \dfrac{x + 1}{4(6 - x)}$

30. $f(x) = 2x^2$, $g(x) = \dfrac{2}{x - 5}$

a) The domain of f is $(-\infty, \infty)$. Since $x - 5 = 0$ when $x = 5$, the domain of g is $(-\infty, 5) \cup (5, \infty)$. Then the domain of $f + g$, $f - g$, and fg is $(-\infty, 5) \cup (5, \infty)$. The domain of ff is $(-\infty, \infty)$. Since there are no values of x for which $g(x) = 0$, the domain of f/g is $(-\infty, 5) \cup (5, \infty)$. Since $f(0) = 0$, the domain of g/f is $(-\infty, 0) \cup (0, 5) \cup (5, \infty)$.

b) $(f+g)(x) = 2x^2 + \dfrac{2}{x-5}$

$(f-g)(x) = 2x^2 - \dfrac{2}{x-5}$

$(fg)(x) = 2x^2 \cdot \dfrac{2}{x-5} = \dfrac{4x^2}{x-5}$

$(ff)(x) = 2x^2 \cdot 2x^2 = 4x^4$

$(f/g)(x) = \dfrac{2x^2}{\frac{2}{x-5}} = 2x^2 \cdot \dfrac{x-5}{2} = x^2(x-5) = x^3 - 5x^2$

$(g/f)(x) = \dfrac{\frac{2}{x-5}}{2x^2} = \dfrac{2}{x-5} \cdot \dfrac{1}{2x^2} = \dfrac{1}{x^2(x-5)} = \dfrac{1}{x^3-5x^2}$

31. $f(x) = \dfrac{1}{x}$, $g(x) = x - 3$

a) Since $f(0)$ is not defined, the domain of f is $(-\infty, 0) \cup (0, \infty)$. The domain of g is $(-\infty, \infty)$. Then the domain of $f+g$, $f-g$, fg, and ff is $(-\infty, 0) \cup (0, \infty)$. Since $g(3) = 0$, the domain of f/g is $(-\infty, 0) \cup (0, 3) \cup (3, \infty)$. There are no values of x for which $f(x) = 0$, so the domain of g/f is $(-\infty, 0) \cup (0, \infty)$.

b) $(f+g)(x) = f(x) + g(x) = \dfrac{1}{x} + x - 3$

$(f-g)(x) = f(x) - g(x) = \dfrac{1}{x} - (x-3) = \dfrac{1}{x} - x + 3$

$(fg)(x) = f(x) \cdot g(x) = \dfrac{1}{x} \cdot (x-3) = \dfrac{x-3}{x}$, or $1 - \dfrac{3}{x}$

$(ff)(x) = f(x) \cdot f(x) = \dfrac{1}{x} \cdot \dfrac{1}{x} = \dfrac{1}{x^2}$

$(f/g)(x) = \dfrac{f(x)}{g(x)} = \dfrac{\frac{1}{x}}{x-3} = \dfrac{1}{x} \cdot \dfrac{1}{x-3} = \dfrac{1}{x(x-3)}$

$(g/f)(x) = \dfrac{g(x)}{f(x)} = \dfrac{x-3}{\frac{1}{x}} = (x-3) \cdot \dfrac{x}{1} = x(x-3)$, or $x^2 - 3x$

32. $f(x) = \sqrt{x+6}$, $g(x) = \dfrac{1}{x}$

a) The domain of $f(x)$ is $[-6, \infty)$. The domain of $g(x)$ is $(-\infty, 0) \cup (0, \infty)$. Then the domain of $f+g$, $f-g$, and fg is $[-6, 0) \cup (0, \infty)$. The domain of ff is $[-6, \infty)$. Since there are no values of x for which $g(x) = 0$, the domain of f/g is $[-6, 0) \cup (0, \infty)$. Since $f(-6) = 0$, the domain of g/f is $(-6, 0) \cup (0, \infty)$.

b) $(f+g)(x) = \sqrt{x+6} + \dfrac{1}{x}$

$(f-g)(x) = \sqrt{x+6} - \dfrac{1}{x}$

$(fg)(x) = \sqrt{x+6} \cdot \dfrac{1}{x} = \dfrac{\sqrt{x+6}}{x}$

$(ff)(x) = \sqrt{x+6} \cdot \sqrt{x+6} = |x+6|$

$(f/g)(x) = \dfrac{\sqrt{x+6}}{\frac{1}{x}} = \sqrt{x+6} \cdot \dfrac{x}{1} = x\sqrt{x+6}$

$(g/f)(x) = \dfrac{\frac{1}{x}}{\sqrt{x+6}} = \dfrac{1}{x} \cdot \dfrac{1}{\sqrt{x+6}} = \dfrac{1}{x\sqrt{x+6}}$

33. From the graph, we see that the domain of F is $[0, 9]$ and the domain of G is $[3, 10]$. The domain of $F+G$ is the set of numbers in the domains of both F and G. This is $[3, 9]$.

34. The domain of $F-G$ and FG is the set of numbers in the domains of both F and G. (See Exercise 33.) This is $[3, 9]$.

The domain of F/G is the set of numbers in the domains of both F and G, excluding those for which $G = 0$. Since $G > 0$ for all values of x in its domain, the domain of F/G is $[3, 9]$.

35. The domain of G/F is the set of numbers in the domains of both F and G (See Exercise 33.), excluding those for which $F = 0$. Since $F(6) = 0$ and $F(8) = 0$, the domain of G/F is $[3, 6) \cup (6, 8) \cup (8, 9]$.

36. $(F+G)(x) = F(x) + G(x)$

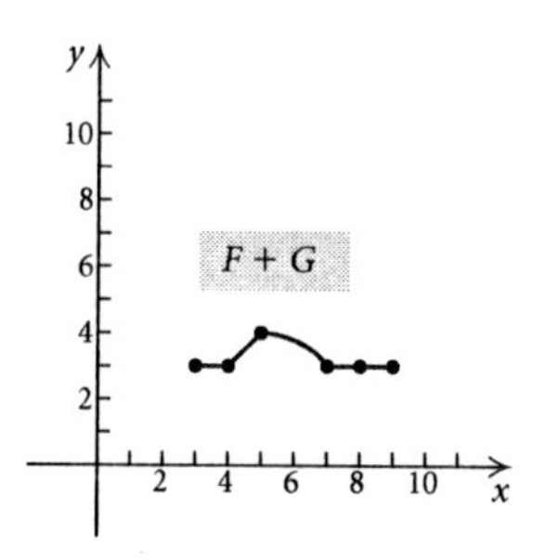

37.

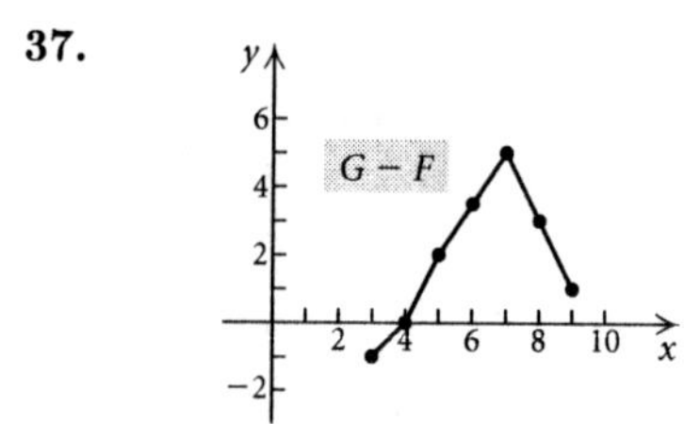

38.

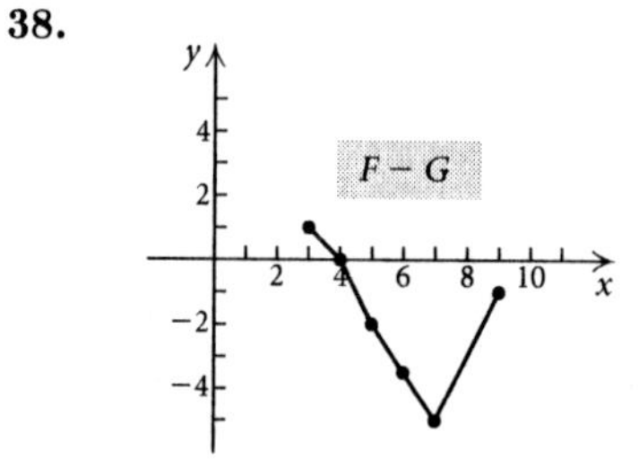

39. a) $P(x) = R(x) - C(x) = 60x - 0.4x^2 - (3x + 13) =$
$60x - 0.4x^2 - 3x - 13 = -0.4x^2 + 57x - 13$

b) $R(100) = 60 \cdot 100 - 0.4(100)^2 = 6000 - 0.4(10,000) =$
$6000 - 4000 = 2000$

$C(100) = 3 \cdot 100 + 13 = 300 + 13 = 313$

$P(100) = R(100) - C(100) = 2000 - 313 = 1687$

40. a) $P(x) = 200x - x^2 - (5000 + 8x) =$
$200x - x^2 - 5000 - 8x = -x^2 + 192x - 5000$

b) $R(175) = 200(175) - 175^2 = 4375$

$C(175) = 5000 + 8 \cdot 175 = 6400$

$P(175) = R(175) - C(175) = 4375 - 6400 = -2025$

(We could also use the function found in part (a) to find $P(175)$.)

41. $f(x) = x^2 + 1$

$f(x+h) = (x+h)^2 + 1 = x^2 + 2xh + h^2 + 1$

$$\frac{f(x+h)-f(x)}{h} = \frac{x^2+2xh+h^2+1-(x^2+1)}{h} = \frac{x^2+2xh+h^2+1-x^2-1}{h} = \frac{2xh+h^2}{h} = \frac{h(2x+h)}{h} = \frac{h}{h}\cdot\frac{2x+h}{1} = 2x+h$$

42. $f(x) = 2 - x^2$

$$\frac{f(x+h)-f(x)}{h} = \frac{2-(x+h)^2-(2-x^2)}{h} = \frac{2-x^2-2xh-h^2-2+x^2}{h} = \frac{-2xh-h^2}{h} = \frac{h(-2x-h)}{h} = -2x-h$$

43. $f(x) = 3x - 5$

$f(x+h) = 3(x+h) - 5 = 3x + 3h - 5$

$$\frac{f(x+h)-f(x)}{h} = \frac{3x+3h-5-(3x-5)}{h} = \frac{3x+3h-5-3x+5}{h} = \frac{3h}{h} = 3$$

44. $f(x) = -\frac{1}{2}x + 7$

$$\frac{f(x+h)-f(x)}{h} = \frac{-\frac{1}{2}(x+h)+7-\left(-\frac{1}{2}+7\right)}{h} = \frac{-\frac{1}{2}x-\frac{1}{2}h+7+\frac{1}{2}-7}{h} = \frac{-\frac{1}{2}h}{h} = -\frac{1}{2}$$

45. $f(x) = 3x^2 - 2x + 1$

$f(x+h) = 3(x+h)^2 - 2(x+h) + 1 =$

$3(x^2 + 2xh + h^2) - 2(x+h) + 1 =$

$3x^2 + 6xh + 3h^2 - 2x - 2h + 1$

$f(x) = 3x^2 - 2x + 1$

$$\frac{f(x+h)-f(x)}{h} = \frac{(3x^2+6xh+3h^2-2x-2h+1)-(3x^2-2x+1)}{h} = \frac{3x^2+6xh+3h^2-2x-2h+1-3x^2+2x-1}{h} = \frac{6xh+3h^2-2h}{h} = \frac{h(6x+3h-2)}{h\cdot 1} = \frac{h}{h}\cdot\frac{6x+3h-2}{1} = 6x+3h-2$$

46. $f(x) = 5x^2 + 4x$

$$\frac{f(x+h)-f(x)}{h} = \frac{(5x^2+10xh+5h^2+4x+4h)-(5x^2+4x)}{h} = \frac{10xh+5h^2+4h}{h} = 10x+5h+4$$

47. $f(x) = 4 + 5|x|$

$f(x+h) = 4 + 5|x+h|$

$$\frac{f(x+h)-f(x)}{h} = \frac{4+5|x+h|-(4+5|x|)}{h} = \frac{4+5|x+h|-4-5|x|}{h} = \frac{5|x+h|-5|x|}{h}$$

48. $f(x) = 2|x| + 3x$

$$\frac{f(x+h)-f(x)}{h} = \frac{(2|x+h|+3x+3h)-(2|x|+3x)}{h} = \frac{2|x+h|-2|x|+3h}{h}$$

49. $f(x) = x^3$

$f(x+h) = (x+h)^3 = x^3 + 3x^2h + 3xh^2 + h^3$

$f(x) = x^3$

$$\frac{f(x+h)-f(x)}{h} = \frac{x^3+3x^2h+3xh^2+h^3-x^3}{h} = \frac{3x^2h+3xh^2+h^3}{h} = \frac{h(3x^2+3xh+h^2)}{h\cdot 1} = \frac{h}{h}\cdot\frac{3x^2+3xh+h^2}{1} = 3x^2+3xh+h^2$$

50. $f(x) = x^3 - 2x$

$$\frac{f(x+h)-f(x)}{h} = \frac{(x+h)^3-2(x+h)-(x^3-2x)}{h} = \frac{x^3+3x^2h+3xh^2+h^3-2x-2h-x^3+2x}{h} = \frac{3x^2h+3xh^2+h^3-2h}{h} = \frac{h(3x^2+3xh+h^2-2)}{h} = 3x^2+3xh+h^2-2$$

51. $f(x) = \frac{x-4}{x+3}$

$$\frac{f(x+h)-f(x)}{h} = \frac{\frac{x+h-4}{x+h+3}-\frac{x-4}{x+3}}{h} = \frac{\frac{x+h-4}{x+h+3}-\frac{x-4}{x+3}}{h}\cdot\frac{(x+h+3)(x+3)}{(x+h+3)(x+3)} = \frac{(x+h-4)(x+3)-(x-4)(x+h+3)}{h(x+h+3)(x+3)} =$$

$\dfrac{x^2+hx-4x+3x+3h-12-(x^2+hx+3x-4x-4h-12)}{h(x+h+3)(x+3)} =$

$\dfrac{x^2+hx-x+3h-12-x^2-hx+x+4h+12}{h(x+h+3)(x+3)} =$

$\dfrac{7h}{h(x+h+3)(x+3)} = \dfrac{h}{h} \cdot \dfrac{7}{(x+h+3)(x+3)} =$

$\dfrac{7}{(x+h+3)(x+3)}$

52. $f(x) = \dfrac{x}{2-x}$

$\dfrac{f(x+h)-f(x)}{h} = \dfrac{\dfrac{x+h}{2-(x+h)} - \dfrac{x}{2-x}}{h} =$

$\dfrac{\dfrac{(x+h)(2-x)-x(2-x-h)}{(2-x-h)(2-x)}}{h} =$

$\dfrac{\dfrac{2x-x^2+2h-hx-2x+x^2+hx}{(2-x-h)(2-x)}}{h} =$

$\dfrac{\dfrac{2h}{(2-x-h)(2-x)}}{h} =$

$\dfrac{2h}{(2-x-h)(2-x)} \cdot \dfrac{1}{h} = \dfrac{2}{(2-x-h)(2-x)}$

53. $(f \circ g)(-1) = f(g(-1)) = f((-1)^2 - 2(-1) - 6) =$
$f(1+2-6) = f(-3) = 3(-3)+1 = -9+1 = -8$

54. $(g \circ f)(-2) = g(f(-2)) = g(3(-2)+1) = g(-5) =$
$(-5)^2 - 2(-5) - 6 = 25+10-6 = 29$

55. $(h \circ f)(1) = h(f(1)) = h(3 \cdot 1 + 1) = h(3+1) =$
$h(4) = 4^3 = 64$

56. $(g \circ h)\left(\frac{1}{2}\right) = g\left(h\left(\frac{1}{2}\right)\right) = g\left(\left(\frac{1}{2}\right)^3\right) = g\left(\frac{1}{8}\right) =$
$\left(\frac{1}{8}\right)^2 - 2\left(\frac{1}{8}\right) - 6 = \frac{1}{64} - \frac{1}{4} - 6 = -\frac{399}{64}$, or $-6\frac{15}{64}$

57. $(g \circ f)(5) = g(f(5)) = g(3 \cdot 5 + 1) = g(15+1) =$
$g(16) = 16^2 - 2 \cdot 16 - 6 = 218$

58. $(f \circ g)\left(\frac{1}{3}\right) = f\left(g\left(\frac{1}{3}\right)\right) = f\left(\left(\frac{1}{3}\right)^2 - 2\left(\frac{1}{3}\right) - 6\right) =$
$f\left(\frac{1}{9} - \frac{2}{3} - 6\right) = f\left(-\frac{59}{9}\right) = 3\left(-\frac{59}{9}\right) + 1 = -\frac{56}{3}$

59. $(f \circ h)(-3) = f(h(-3)) = f((-3)^3) = f(-27) =$
$3(-27)+1 = -81+1 = -80$

60. $(h \circ g)(3) = h(g(3)) = h(3^2 - 2 \cdot 3 - 6) =$
$h(9-6-6) = h(-3) = (-3)^3 = -27$

61. $(f \circ g)(x) = f(g(x)) = f(x-3) = x-3+3 = x$
$(g \circ f)(x) = g(f(x)) = g(x+3) = x+3-3 = x$
The domain of f and of g is $(-\infty, \infty)$, so the domain of $f \circ g$ and of $g \circ f$ is $(-\infty, \infty)$.

62. $(f \circ g)(x) = f\left(\frac{5}{4}x\right) = \frac{4}{5} \cdot \frac{5}{4}x = x$
$(g \circ f)(x) = g\left(\frac{4}{5}x\right) = \frac{5}{4} \cdot \frac{4}{5}x = x$
The domain of f and of g is $(-\infty, \infty)$, so the domain of $f \circ g$ and of $g \circ f$ is $(-\infty, \infty)$.

63. $(f \circ g)(x) = f(g(x)) = f\left(\frac{1}{x}\right) = \dfrac{4}{1 - 5 \cdot \frac{1}{x}} = \dfrac{4}{1 - \frac{5}{x}} =$
$\dfrac{4}{\frac{x-5}{x}} = 4 \cdot \dfrac{x}{x-5} = \dfrac{4x}{x-5}$

$(g \circ f)(x) = g(f(x)) = g\left(\dfrac{4}{1-5x}\right) = \dfrac{1}{\frac{4}{1-5x}} =$
$1 \cdot \dfrac{1-5x}{4} = \dfrac{1-5x}{4}$

The domain of f is $\left\{x \middle| x \neq \frac{1}{5}\right\}$ and the domain of g is $\{x|x \neq 0\}$. Consider the domain of $f \circ g$. Since 0 is not in the domain of g, 0 is not in the domain of $f \circ g$. Since $\frac{1}{5}$ is not in the domain of f, we know that $g(x)$ cannot be $\frac{1}{5}$. We find the value(s) of x for which $g(x) = \frac{1}{5}$.

$\frac{1}{x} = \frac{1}{5}$

$5 = x$ Multiplying by $5x$

Thus 5 is also not in the domain of $f \circ g$. Then the domain of $f \circ g$ is $\{x|x \neq 0$ *and* $x \neq 5\}$, or $(-\infty, 0) \cup (0, 5) \cup (5, \infty)$.

Now consider the domain of $g \circ f$. Recall that $\frac{1}{5}$ is not in the domain of f, so it is not in the domain of $g \circ f$. Now 0 is not in the domain of g but $f(x)$ is never 0, so the domain of $g \circ f$ is $\left\{x \middle| x \neq \frac{1}{5}\right\}$, or $\left(-\infty, \frac{1}{5}\right) \cup \left(\frac{1}{5}, \infty\right)$.

64. $(f \circ g)(x) = f\left(\dfrac{1}{2x+1}\right) = \dfrac{6}{\frac{1}{2x+1}} = 6 \cdot \dfrac{2x+1}{1} =$
$6(2x+1)$, or $12x+6$

$(g \circ f)(x) = g\left(\frac{6}{x}\right) = \dfrac{1}{2 \cdot \frac{6}{x} + 1} = \dfrac{1}{\frac{12}{x} + 1} = \dfrac{1}{\frac{12+x}{x}} =$
$1 \cdot \dfrac{x}{12+x} = \dfrac{x}{12+x}$

The domain of f is $\{x|x \neq 0\}$ and the domain of g is $\left\{x \middle| x = -\frac{1}{2}\right\}$. Consider the domain of $f \circ g$. Since $-\frac{1}{2}$ is not in the domain of g, $-\frac{1}{2}$ is not in the domain of $f \circ g$. Now 0 is not in the domain of f but $g(x)$ is never 0, so the domain of $f \circ g$ is $\left\{x \middle| x \neq -\frac{1}{2}\right\}$, or $\left(-\infty, -\frac{1}{2}\right) \cup \left(-\frac{1}{2}, \infty\right)$.

Now consider the domain of $g \circ f$. Since 0 is not in the domain of f, we find the value(s) of x for which $f(x) = -\frac{1}{2}$.

$$\frac{6}{x} = -\frac{1}{2}$$
$$-12 = x$$

Then the domain of $g \circ f$ is $\left\{x \middle| x \neq -12 \text{ and } x \neq -\frac{1}{2}\right\}$, or $(-\infty, -12) \cup \left(-12, -\frac{1}{2}\right) \cup \left(-\frac{1}{2}, \infty\right)$.

65. $(f \circ g)(x) = f(g(x)) = f\left(\frac{x+7}{3}\right) =$

$3\left(\frac{x+7}{3}\right) - 7 = x + 7 - 7 = x$

$(g \circ f)(x) = g(f(x)) = g(3x - 7) = \frac{(3x-7)+7}{3} =$

$\frac{3x}{3} = x$

The domain of f and of g is $(-\infty, \infty)$, so the domain of $f \circ g$ and of $g \circ f$ is $(-\infty, \infty)$.

66. $(f \circ g)(x) = f(1.5x + 1.2) = \frac{2}{3}(1.5x + 1.2) - \frac{4}{5}$

$x + 0.8 - \frac{4}{5} = x$

$(g \circ f)(x) = g\left(\frac{2}{3}x - \frac{4}{5}\right) = 1.5\left(\frac{2}{3}x - \frac{4}{5}\right) + 1.2 =$

$x - 1.2 + 1.2 = x$

The domain of f and of g is $(-\infty, \infty)$, so the domain of $f \circ g$ and of $g \circ f$ is $(-\infty, \infty)$.

67. $(f \circ g)(x) = f(g(x)) = f(\sqrt{x}) = 2\sqrt{x} + 1$

$(g \circ f)(x) = g(f(x)) = g(2x + 1) = \sqrt{2x + 1}$

The domain of f is $(-\infty, \infty)$ and the domain of g is $\{x|x \geq 0\}$. Thus the domain of $f \circ g$ is $\{x|x \geq 0\}$, or $[0, \infty)$.

Now consider the domain of $g \circ f$. There are no restrictions on the domain of f, but the domain of g is $\{x|x \geq 0\}$. Since $f(x) \geq 0$ for $x \geq -\frac{1}{2}$, the domain of $g \circ f$ is $\left\{x \middle| x \geq -\frac{1}{2}\right\}$, or $\left[-\frac{1}{2}, \infty\right)$.

68. $(f \circ g)(x) = f\left(\frac{2}{x}\right) = \sqrt{\frac{2}{x} - 4}$

$(g \circ f)(x) = g(\sqrt{x - 4}) = \frac{2}{\sqrt{x - 4}}$

The domain of f is $\{x|x \geq 4\}$ and the domain of g is $\{x|x \neq 0\}$, so 0 is not in the domain of $f \circ g$. Since $g(x) \geq 4$ when $0 < x \leq \frac{1}{2}$, the domain of $f \circ g$ is $\left(0, \frac{1}{2}\right]$.

Now consider the domain of $g \circ f$. Recall that the domain of f is $\{x|x \geq 4\}$. Since 0 is not in the domain of g and $f(x) = 0$ when $x = 4$, the domain of $g \circ f$ is $(4, \infty)$.

69. $(f \circ g)(x) = f(g(x)) = f(0.05) = 20$

$(g \circ f)(x) = g(f(x)) = g(20) = 0.05$

The domain of f and of g is $(-\infty, \infty)$, so the domain of $f \circ g$ and of $g \circ f$ is $(-\infty, \infty)$.

70. $(f \circ g)(x) = (\sqrt[4]{x})^4 = x$

$(g \circ f)(x) = \sqrt[4]{x^4} = |x|$

The domain of f is $(-\infty, \infty)$ and the domain of g is $\{x|x \geq 0\}$, so the domain of $f \circ g$ is $\{x|x \geq 0\}$, or $[0, \infty)$.

Now consider the domain of $g \circ f$. There are no restrictions on the domain of f and $f(x) \geq 0$ for all values of x, so the domain is $(-\infty, \infty)$.

71. $(f \circ g)(x) = f(g(x)) = f(x^2 - 5) =$

$\sqrt{x^2 - 5 + 5} = \sqrt{x^2} = |x|$

$(g \circ f)(x) = g(f(x)) = g(\sqrt{x + 5}) =$

$(\sqrt{x + 5})^2 - 5 = x + 5 - 5 = x$

The domain of f is $\{x|x \geq -5\}$ and the domain of g is $(-\infty, \infty)$. Since $x^2 \geq 0$ for all values of x, then $x^2 - 5 \geq -5$ for all values of x and the domain of $g \circ f$ is $(-\infty, \infty)$.

Now consider the domain of $f \circ g$. There are no restrictions on the domain of g, so the domain of $f \circ g$ is the same as the domain of f, $\{x|x \geq -5\}$, or $[-5, \infty)$.

72. $(f \circ g)(x) = (\sqrt[5]{x + 2})^5 - 2 = x + 2 - 2 = x$

$(g \circ f)(x) = \sqrt[5]{x^5 - 2 + 2} = \sqrt[5]{x^5} = x$

The domain of f and of g is $(-\infty, \infty)$, so the domain of $f \circ g$ and of $g \circ f$ is $(-\infty, \infty)$.

73. $(f \circ g)(x) = f(g(x)) = f(\sqrt{3 - x}) = (\sqrt{3 - x})^2 + 2 =$

$3 - x + 2 = 5 - x$

$(g \circ f)(x) = g(f(x)) = g(x^2 + 2) = \sqrt{3 - (x^2 + 2)} =$

$\sqrt{3 - x^2 - 2} = \sqrt{1 - x^2}$

The domain of f is $(-\infty, \infty)$ and the domain of g is $\{x|x \leq 3\}$, so the domain of $f \circ g$ is $\{x|x \leq 3\}$, or $(-\infty, 3]$.

Now consider the domain of $g \circ f$. There are no restrictions on the domain of f and the domain of g is $\{x|x \leq 3\}$, so we find the values of x for which $f(x) \leq 3$. We see that $x^2 + 2 \leq 3$ for $-1 \leq x \leq 1$, so the domain of $g \circ f$ is $\{x| -1 \leq x \leq 1\}$, or $[-1, 1]$.

74. $(f \circ g)(x) = f(\sqrt{x^2 - 25}) = 1 - (\sqrt{x^2 - 25})^2 =$

$1 - (x^2 - 25) = 1 - x^2 + 25 = 26 - x^2$

$(g \circ f)(x) = g(1 - x^2) = \sqrt{(1 - x^2)^2 - 25} =$

$\sqrt{1 - 2x^2 + x^4 - 25} = \sqrt{x^4 - 2x^2 - 24}$

The domain of f is $(-\infty, \infty)$ and the domain of g is $\{x|x \leq -5 \text{ or } x \geq 5\}$, so the domain of $f \circ g$ is $\{x|x \leq -5 \text{ or } x \geq 5\}$, or $(-\infty, -5] \cup [5, \infty)$.

Now consider the domain of $g \circ f$. There are no restrictions on the domain of f and the domain of g is $\{x|x \leq -5 \text{ or } x \geq 5\}$, so we find the values of x for which $f(x) \leq -5$ or $f(x) \geq 5$. We see that $1 - x^2 \leq -5$ when $x \leq -\sqrt{6}$ or $x \geq \sqrt{6}$ and $1 - x^2 \geq 5$ has no solution, so the domain of $g \circ f$ is $\{x|x \leq -\sqrt{6} \text{ or } x \geq \sqrt{6}\}$, or $(-\infty, -\sqrt{6}] \cup [\sqrt{6}, \infty)$.

75. $(f \circ g)(x) = f(g(x)) = f\left(\frac{1}{1+x}\right) =$

$$\frac{1-\left(\frac{1}{1+x}\right)}{\frac{1}{1+x}} = \frac{\frac{1+x-1}{1+x}}{\frac{1}{1+x}} =$$

$$\frac{x}{1+x} \cdot \frac{1+x}{1} = x$$

$(g \circ f)(x) = g(f(x)) = g\left(\frac{1-x}{x}\right) =$

$$\frac{1}{1+\left(\frac{1-x}{x}\right)} = \frac{1}{\frac{x+1-x}{x}} =$$

$$\frac{1}{\frac{1}{x}} = 1 \cdot \frac{x}{1} = x$$

The domain of f is $\{x|x \neq 0\}$ and the domain of g is $\{x|x \neq -1\}$, so we know that -1 is not in the domain of $f \circ g$. Since 0 is not in the domain of f, values of x for which $g(x) = 0$ are not in the domain of $f \circ g$. But $g(x)$ is never 0, so the domain of $f \circ g$ is $\{x|x \neq -1\}$, or $(-\infty, -1) \cup (-1, \infty)$.

Now consider the domain of $g \circ f$. Recall that 0 is not in the domain of f. Since -1 is not in the domain of g, we know that $g(x)$ cannot be -1. We find the value(s) of x for which $f(x) = -1$.

$$\frac{1-x}{x} = -1$$

$1 - x = -x$ Multiplying by x

$1 = 0$ False equation

We see that there are no values of x for which $f(x) = -1$, so the domain of $g \circ f$ is $\{x|x \neq 0\}$, or $(-\infty, 0) \cup (0, \infty)$.

76. $(f \circ g)(x) = \dfrac{\left(\frac{3x-4}{5x-2}\right)^2 - 1}{\left(\frac{3x-4}{5x-2}\right)^2 + 1} = \dfrac{-8x^2 - 2x + 6}{17x^2 - 22x + 10}$

$(g \circ f)(x) = \dfrac{3\left(\frac{x^2-1}{x^2+1}\right) - 4}{5\left(\frac{x^2-1}{x^2+1}\right) - 2} = \dfrac{-x^2 - 7}{3x^2 - 7}$

The domain of f is $(-\infty, \infty)$ and the domain of g is $\left\{x\middle|x \neq \frac{2}{5}\right\}$, so the domain of $f \circ g$ is $\left\{x\middle|x \neq \frac{2}{5}\right\}$, or $\left(-\infty, \frac{2}{5}\right) \cup \left(\frac{2}{5}, \infty\right)$.

Now consider the domain of $g \circ f$. There are no restrictions on the domain of f, but the domain of g is $\left\{x\middle|x \neq \frac{2}{5}\right\}$. We find the values of x for which $f(x) = \frac{2}{5}$.

$$\frac{x^2-1}{x^2+1} = \frac{2}{5}$$

$$5x^2 - 5 = 2x^2 + 2$$

$$3x^2 = 7$$

$$x^2 = \frac{7}{3}$$

$$x = \pm\sqrt{\frac{7}{3}}, \text{ or } \pm\frac{\sqrt{21}}{3}$$

Then the domain of $g \circ f$ is $\left\{x\middle|x \neq -\frac{\sqrt{21}}{3} \text{ and } x \neq \frac{\sqrt{21}}{3}\right\}$, or $\left(-\infty, -\frac{\sqrt{21}}{3}\right) \cup \left(-\frac{\sqrt{21}}{3}, \frac{\sqrt{21}}{3}\right) \cup \left(\frac{\sqrt{21}}{3}, \infty\right)$.

77. $(f \circ g)(x) = f(g(x)) = f(x+1) =$
$(x+1)^3 - 5(x+1)^2 + 3(x+1) + 7 =$
$x^3 + 3x^2 + 3x + 1 - 5x^2 - 10x - 5 + 3x + 3 + 7 =$
$x^3 - 2x^2 - 4x + 6$

$(g \circ f)(x) = g(f(x)) = g(x^3 - 5x^2 + 3x + 7) =$
$x^3 - 5x^2 + 3x + 7 + 1 = x^3 - 5x^2 + 3x + 8$

The domain of f and of g is $(-\infty, \infty)$, so the domain of $f \circ g$ and of $g \circ f$ is $(-\infty, \infty)$.

78. $(g \circ f)(x) = x^3 + 2x^2 - 3x - 9 - 1 =$
$x^3 + 2x^2 - 3x - 10$
$(g \circ f)(x) = (x-1)^3 + 2(x-1)^2 - 3(x-1) - 9 =$
$x^3 - 3x^2 + 3x - 1 + 2x^2 - 4x + 2 - 3x + 3 - 9 =$
$x^3 - x^2 - 4x - 5$

The domain of f and of g is $(-\infty, \infty)$, so the domain of $f \circ g$ and of $g \circ f$ is $(-\infty, \infty)$.

79. $h(x) = (4 + 3x)^5$

This is $4 + 3x$ to the 5th power. The most obvious answer is $f(x) = x^5$ and $g(x) = 4 + 3x$.

80. $f(x) = \sqrt[3]{x}$, $g(x) = x^2 - 8$

81. $h(x) = \dfrac{1}{(x-2)^4}$

This is 1 divided by $(x-2)$ to the 4th power. One obvious answer is $f(x) = \frac{1}{x^4}$ and $g(x) = x - 2$. Another possibility is $f(x) = \frac{1}{x}$ and $g(x) = (x-2)^4$.

82. $f(x) = \dfrac{1}{\sqrt{x}}$, $g(x) = 3x + 7$

83. $f(x) = \dfrac{x-1}{x+1}$, $g(x) = x^3$

84. $f(x) = |x|$, $g(x) = 9x^2 - 4$

85. $f(x) = x^6$, $g(x) = \dfrac{2+x^3}{2-x^3}$

86. $f(x) = x^4$, $g(x) = \sqrt{x} - 3$

87. $f(x) = \sqrt{x}$, $g(x) = \dfrac{x-5}{x+2}$

88. $f(x) = \sqrt{1+x}$, $g(x) = \sqrt{1+x}$

89. $f(x) = x^3 - 5x^2 + 3x - 1$, $g(x) = x + 2$

90. $f(x) = 2x^{5/3} + 5x^{2/3}$, $g(x) = x - 1$, or
$f(x) = 2x^5 + 5x^2$, $g(x) = (x - 1)^{1/3}$

91. $f(x) = y(x) \circ s(x) = y(s(x))$
$f(x) = 2(x - 32 + 12) = 2(x - 20)$

92. a) Use the distance formula, distance = rate × time. Substitute 3 for the rate and t for time.

$r(t) = 3t$

b) Use the formula for the area of a circle.

$A(r) = \pi r^2$

c) $(A \circ r)(t) = A(r(t)) = A(3t) = \pi(3t)^2 = 9\pi t^2$

This function gives the area of the ripple in terms of time t.

93.

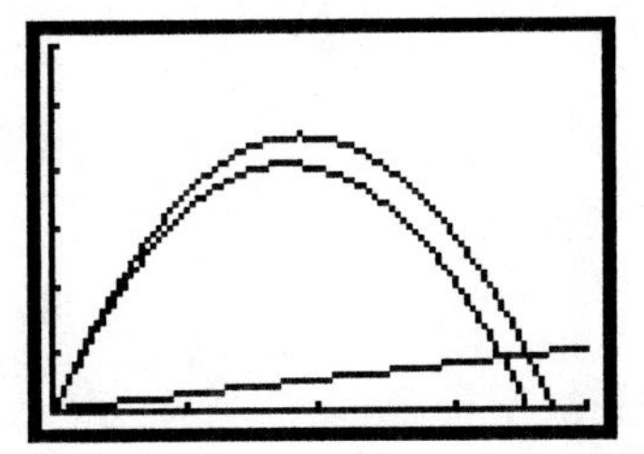

94.

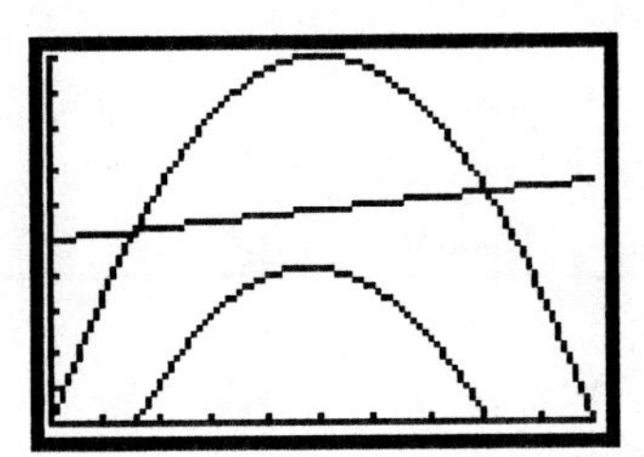

95. The graph of $y = (h - g)(x)$ will be the same as the graph of $y = h(x)$ moved down b units.

96. The values of x that must be excluded from the domain of $(f \circ g)(x)$ are the value of x that are not in the domain of g as well as the values of x for which $g(x)$ takes values that are not in the domain of f.

The values of x that must be excluded from the domain of $(g \circ f)(x)$ are the values of x that are not in the domain of f as well as the values of x for which $f(x)$ takes values that are not in the domain of g.

97. Equations $(a) - (f)$ are in the form $y = mx + b$, so we can read the y-intercepts directly from the equations. Equations (g) and (h) can be written in this form as $y = \frac{2}{3}x - 2$ and $y = -2x + 3$, respectively. We see that only equation (c) has y-intercept $(0, 1)$.

98. None (See Exercise 97.)

99. If a line slopes down from left to right, its slope is negative. The equations $y = mx + b$ for which m is negative are (b), (d), (f), and (h). (See Exercise 97.)

100. The equation for which $|m|$ is greatest is the equation with the steepest slant. This is equation (b). (See Exercise 97.)

101. The only equation that has $(0, 0)$ as a solution is (a).

102. Equations (c) and (g) have the same slope. (See Exercise 97.)

103. Only equations (c) and (g) have the same slope and different y-intercepts. They represent parallel lines.

104. The only equations for which the product of the slopes is -1 are (a) and (f).

105. Answers may vary. One example is $f(x) = 2x + 5$ and $g(x) = \dfrac{x-5}{2}$. Other examples are found in Exercises 61, 62, 65, 66, 72, and 75.

106. Answers may vary; $f(x) = \dfrac{1}{x+7}$, $g(x) = \dfrac{1}{x-3}$

107. The domain of $h(x)$ is $\left\{x \middle| x \neq \frac{7}{3}\right\}$, and the domain of $g(x)$ is $\{x|x \neq 3\}$, so $\frac{7}{3}$ and 3 are not in the domain of $(h/g)(x)$. We must also exclude the value of x for which $g(x) = 0$.

$$\frac{x^4 - 1}{5x - 15} = 0$$
$$x^4 - 1 = 0 \quad \text{Multiplying by } 5x - 15$$
$$x^4 = 1$$
$$x = \pm 1$$

Then the domain of $(h/g)(x)$ is
$\left\{x \middle| x \neq \frac{7}{3} \text{ and } x \neq 3 \text{ and } x \neq -1 \text{ and } x \neq 1\right\}$, or
$(-\infty, -1) \cup (-1, 1) \cup \left(1, \frac{7}{3}\right) \cup \left(\frac{7}{3}, 3\right) \cup (3, \infty)$.

108. The domain of $h+f$, $h-f$, and hf consists of all numbers that are in the domain of both h and f, or $\{-4, 0, 3\}$.

The domain of h/f consists of all numbers that are in the domain of both h and f, excluding any for which the value of f is 0, or $\{-4, 0\}$.

Exercise Set 1.7

1. If the graph were folded on the x-axis, the parts above and below the x-axis would not coincide, so the graph is not symmetric with respect to the x-axis.

If the graph were folded on the y-axis, the parts to the left and right of the y-axis would coincide, so the graph is symmetric with respect to the y-axis.

If the graph were rotated 180°, the resulting graph would not coincide with the original graph, so it is not symmetric with respect to the origin.

2. If the graph were folded on the x-axis, the parts above and below the x-axis would not coincide, so the graph is not symmetric with respect to the x-axis.

If the graph were folded on the y-axis, the parts to the left and right of the y-axis would coincide, so the graph is symmetric with respect to the y-axis.

If the graph were rotated 180°, the resulting graph would not coincide with the original graph, so it is not symmetric with respect to the origin.

3. If the graph were folded on the x-axis, the parts above and below the x-axis would coincide, so the graph is symmetric with respect to the x-axis.

   If the graph were folded on the y-axis, the parts to the left and right of the y-axis would not coincide, so the graph is not symmetric with respect to the y-axis.

   If the graph were rotated 180°, the resulting graph would not coincide with the original graph, so it is not symmetric with respect to the origin.

4. If the graph were folded on the x-axis, the parts above and below the x-axis would not coincide, so the graph is not symmetric with respect to the x-axis.

   If the graph were folded on the y-axis, the parts to the left and right of the y-axis would not coincide, so the graph is not symmetric with respect to the y-axis.

   If the graph were rotated 180°, the resulting graph would coincide with the original graph, so it is symmetric with respect to the origin.

5. If the graph were folded on the x-axis, the parts above and below the x-axis would not coincide, so the graph is not symmetric with respect to the x-axis.

   If the graph were folded on the y-axis, the parts to the left and right of the y-axis would not coincide, so the graph is not symmetric with respect to the y-axis.

   If the graph were rotated 180°, the resulting graph would coincide with the original graph, so it is symmetric with respect to the origin.

6. If the graph were folded on the x-axis, the parts above and below the x-axis would coincide, so the graph is symmetric with respect to the x-axis.

   If the graph were folded on the y-axis, the parts to the left and right of the y-axis would coincide, so the graph is symmetric with respect to the y-axis.

   If the graph were rotated 180°, the resulting graph would coincide with the original graph, so it is symmetric with respect to the origin.

7. Test for symmetry with respect to the x-axis:

$$\begin{aligned} 5x-5y&=0 && \text{Original equation}\\ 5x-5(-y)&=0 && \text{Replacing } y \text{ by } -y\\ 5x+5y&=0 && \text{Simplifying} \end{aligned}$$

   The last equation is not equivalent to the original equation, so the graph is not symmetric with respect to the x-axis.

   Test for symmetry with respect to the y-axis:

$$\begin{aligned} 5x-5y&=0 && \text{Original equation}\\ 5(-x)-5y&=0 && \text{Replacing } x \text{ by } -x\\ -5x-5y&=0 && \text{Simplifying}\\ 5x+5y&=0 \end{aligned}$$

   The last equation is not equivalent to the original equation, so the graph is not symmetric with respect to the y-axis.

   Test for symmetry with respect to the origin:

$$\begin{aligned} 5x-5y&=0 && \text{Original equation}\\ 5(-x)-5(-y)&=0 && \text{Replacing } x \text{ by } -x \text{ and } y \text{ by } -y\\ -5x+5y&=0 && \text{Simplifying}\\ 5x-5y&=0 \end{aligned}$$

   The last equation is equivalent to the original equation, so the graph is symmetric with respect to the origin.

8. Test for symmetry with respect to the x-axis:

$$\begin{aligned} 6x+7y&=0 && \text{Original equation}\\ 6x+7(-y)&=0 && \text{Replacing } y \text{ by } -y\\ 6x-7y&=0 && \text{Simplifying} \end{aligned}$$

   The last equation is not equivalent to the original equation, so the graph is not symmetric with respect to the x-axis.

   Test for symmetry with respect to the y-axis:

$$\begin{aligned} 6x+7y&=0 && \text{Original equation}\\ 6(-x)+7y&=0 && \text{Replacing } x \text{ by } -x\\ 6x-7y&=0 && \text{Simplifying} \end{aligned}$$

   The last equation is not equivalent to the original equation, so the graph is not symmetric with respect to the y-axis.

   Test for symmetry with respect to the origin:

$$\begin{aligned} 6x+7y&=0 && \text{Original equation}\\ 6(-x)+7(-y)&=0 && \text{Replacing } x \text{ by } -x \text{ and } y \text{ by } -y\\ 6x+7y&=0 && \text{Simplifying} \end{aligned}$$

   The last equation is equivalent to the original equation, so the graph is symmetric with respect to the origin.

9. Test for symmetry with respect to the x-axis:

$$\begin{aligned} 3x^2-2y^2&=3 && \text{Original equation}\\ 3x^2-2(-y)^2&=3 && \text{Replacing } y \text{ by } -y\\ 3x^2-2y^2&=3 && \text{Simplifying} \end{aligned}$$

   The last equation is equivalent to the original equation, so the graph is symmetric with respect to the x-axis.

   Test for symmetry with respect to the y-axis:

$$\begin{aligned} 3x^2-2y^2&=3 && \text{Original equation}\\ 3(-x)^2-2y^2&=3 && \text{Replacing } x \text{ by } -x\\ 3x^2-2y^2&=3 && \text{Simplifying} \end{aligned}$$

   The last equation is equivalent to the original equation, so the graph is symmetric with respect to the y-axis.

   Test for symmetry with respect to the origin:

$$\begin{aligned} 3x^2-2y^2&=3 && \text{Original equation}\\ 3(-x)^2-2(-y)^2&=3 && \text{Replacing } x \text{ by } -x \text{ and } y \text{ by } -y\\ 3x^2-2y^2&=3 && \text{Simplifying} \end{aligned}$$

   The last equation is equivalent to the original equation, so the graph is symmetric with respect to the origin.

10. Test for symmetry with respect to the x-axis:

$$\begin{aligned} 5y&=7x^2-2x && \text{Original equation}\\ 5(-y)&=7x^2-2x && \text{Replacing } y \text{ by } -y\\ 5y&=-7x^2+2x && \text{Simplifying} \end{aligned}$$

The last equation is not equivalent to the original equation, so the graph is not symmetric with respect to the x-axis.

Test for symmetry with respect to the y-axis:

$5y = 7x^2 - 2x$ Original equation

$5y = 7(-x)^2 - 2(-x)$ Replacing x by $-x$

$5y = 7x^2 + 2x$ Simplifying

The last equation is not equivalent to the original equation, so the graph is not symmetric with respect to the y-axis.

Test for symmetry with respect to the origin:

$5y = 7x^2 - 2x$ Original equation

$5(-y) = 7(-x)^2 - 2(-x)$ Replacing x by $-x$ and y by $-y$

$-5y = 7x^2 + 2x$ Simplifying

$5y = -7x^2 - 2x$

The last equation is not equivalent to the original equation, so the graph is not symmetric with respect to the origin.

11. Test for symmetry with respect to the x-axis:

$y = |2x|$ Original equation

$-y = |2x|$ Replacing y by $-y$

$y = -|2x|$ Simplifying

The last equation is not equivalent to the original equation, so the graph is not symmetric with respect to the x-axis.

Test for symmetry with respect to the y-axis:

$y = |2x|$ Original equation

$y = |2(-x)|$ Replacing x by $-x$

$y = |-2x|$ Simplifying

$y = |2x|$

The last equation is equivalent to the original equation, so the graph is symmetric with respect to the y-axis.

Test for symmetry with respect to the origin:

$y = |2x|$ Original equation

$-y = |2(-x)|$ Replacing x by $-x$ and y by $-y$

$-y = |-2x|$ Simplifying

$-y = |2x|$

$y = -|2x|$

The last equation is not equivalent to the original equation, so the graph is not symmetric with respect to the origin.

12. Test for symmetry with respect to the x-axis:

$y^3 = 2x^2$ Original equation

$(-y)^3 = 2x^2$ Replacing y by $-y$

$-y^3 = 2x^2$ Simplifying

$y^3 = -2x^2$

The last equation is not equivalent to the original equation, so the graph is not symmetric with respect to the x-axis.

Test for symmetry with respect to the y-axis:

$y^3 = 2x^2$ Original equation

$y^3 = 2(-x)^2$ Replacing x by $-x$

$y^3 = 2x^2$ Simplifying

The last equation is equivalent to the original equation, so the graph is symmetric with respect to the y-axis.

Test for symmetry with respect to the origin:

$y^3 = 2x^2$ Original equation

$(-y)^3 = 2(-x)^2$ Replacing x by $-x$ and y by $-y$

$-y^3 = 2x^2$ Simplifying

$y^3 = -2x^2$

The last equation is not equivalent to the original equation, so the graph is not symmetric with respect to the origin.

13. Test for symmetry with respect to the x-axis:

$2x^4 + 3 = y^2$ Original equation

$2x^4 + 3 = (-y)^2$ Replacing y by $-y$

$2x^4 + 3 = y^2$ Simplifying

The last equation is equivalent to the original equation, so the graph is symmetric with respect to the x-axis.

Test for symmetry with respect to the y-axis:

$2x^4 + 3 = y^2$ Original equation

$2(-x)^4 + 3 = y^2$ Replacing x by $-x$

$2x^4 + 3 = y^2$ Simplifying

The last equation is equivalent to the original equation, so the graph is symmetric with respect to the y-axis.

Test for symmetry with respect to the origin:

$2x^4 + 3 = y^2$ Original equation

$2(-x)^4 + 3 = (-y)^2$ Replacing x by $-x$ and y by $-y$

$2x^4 + 3 = y^2$ Simplifying

The last equation is equivalent to the original equation, so the graph is symmetric with respect to the origin.

14. Test for symmetry with respect to the x-axis:

$2y^2 = 5x^2 + 12$ Original equation

$2(-y)^2 = 5x^2 + 12$ Replacing y by $-y$

$2y^2 = 5x^2 + 12$ Simplifying

The last equation is equivalent to the original equation, so the graph is symmetric with respect to the x-axis.

Test for symmetry with respect to the y-axis:

$2y^2 = 5x^2 + 12$ Original equation

$2y^2 = 5(-x)^2 + 12$ Replacing x by $-x$

$2y^2 = 5x^2 + 12$ Simplifying

The last equation is equivalent to the original equation, so the graph is symmetric with respect to the y-axis.

Test for symmetry with respect to the origin:

$2y^2 = 5x^2 + 12$ Original equation

$2(-y)^2 = 5(-x)^2 + 12$ Replacing x by $-x$ and y by $-y$

$2y^2 = 5x^2 + 12$ Simplifying

The last equation is equivalent to the original equation, so the graph is symmetric with respect to the origin.

15. Test for symmetry with respect to the x-axis:

$3y^3 = 4x^3 + 2$ Original equation
$3(-y)^3 = 4x^3 + 2$ Replacing y by $-y$
$-3y^3 = 4x^3 + 2$ Simplifying
$3y^3 = -4x^3 - 2$

The last equation is not equivalent to the original equation, so the graph is not symmetric with respect to the x-axis.

Test for symmetry with respect to the y-axis:

$3y^3 = 4x^3 + 2$ Original equation
$3y^3 = 4(-x)^3 + 2$ Replacing x by $-x$
$3y^3 = -4x^3 + 2$ Simplifying

The last equation is not equivalent to the original equation, so the graph is not symmetric with respect to the y-axis.

Test for symmetry with respect to the origin:

$3y^3 = 4x^3 + 2$ Original equation
$3(-y)^3 = 4(-x)^3 + 2$ Replacing x by $-x$ and y by $-y$
$-3y^3 = -4x^3 + 2$ Simplifying
$3y^3 = 4x^3 - 2$

The last equation is not equivalent to the original equation, so the graph is not symmetric with respect to the origin.

16. Test for symmetry with respect to the x-axis:

$3x = |y|$ Original equation
$3x = |-y|$ Replacing y by $-y$
$3x = |y|$ Simplifying

The last equation is equivalent to the original equation, so the graph is symmetric with respect to the x-axis.

Test for symmetry with respect to the y-axis:

$3x = |y|$ Original equation
$3(-x) = |y|$ Replacing x by $-x$
$-3x = |y|$ Simplifying

The last equation is not equivalent to the original equation, so the graph is not symmetric with respect to the y-axis.

Test for symmetry with respect to the origin:

$3x = |y|$ Original equation
$3(-x) = |-y|$ Replacing x by $-x$ and y by $-y$
$-3x = |y|$ Simplifying

The last equation is not equivalent to the original equation, so the graph is not symmetric with respect to the origin.

17. Test for symmetry with respect to the x-axis:

$xy = 12$ Original equation
$x(-y) = 12$ Replacing y by $-y$
$-xy = 12$ Simplifying
$xy = -12$

The last equation is not equivalent to the original equation, so the graph is not symmetric with respect to the x-axis.

Test for symmetry with respect to the y-axis:

$xy = 12$ Original equation
$-xy = 12$ Replacing x by $-x$
$xy = -12$ Simplifying

The last equation is not equivalent to the original equation, so the graph is not symmetric with respect to the y-axis.

Test for symmetry with respect to the origin:

$xy = 12$ Original equation
$-x(-y) = 12$ Replacing x by $-x$ and y by $-y$
$xy = 12$ Simplifying

The last equation is equivalent to the original equation, so the graph is symmetric with respect to the origin.

18. Test for symmetry with respect to the x-axis:

$xy - x^2 = 3$ Original equation
$x(-y) - x^2 = 3$ Replacing y by $-y$
$xy + x^2 = -3$ Simplifying

The last equation is not equivalent to the original equation, so the graph is not symmetric with respect to the x-axis.

Test for symmetry with respect to the y-axis:

$xy - x^2 = 3$ Original equation
$-xy - (-x)^2 = 3$ Replacing x by $-x$
$xy + x^2 = -3$ Simplifying

The last equation is not equivalent to the original equation, so the graph is not symmetric with respect to the y-axis.

Test for symmetry with respect to the origin:

$xy - x^2 = 3$ Original equation
$-x(-y) - (-x)^2 = 3$ Replacing x by $-x$ and y by $-y$
$xy - x^2 = 3$ Simplifying

The last equation is equivalent to the original equation, so the graph is symmetric with respect to the origin.

19. x-axis: Replace y with $-y$; $(-5, -6)$
y-axis: Replace x with $-x$; $(5, 6)$
Origin: Replace x with $-x$ and y with $-y$; $(5, -6)$

20. x-axis: Replace y with $-y$; $\left(\frac{7}{2}, 0\right)$
y-axis: Replace x with $-x$; $\left(-\frac{7}{2}, 0\right)$
Origin: Replace x with $-x$ and y with $-y$; $\left(-\frac{7}{2}, 0\right)$

21. x-axis: Replace y with $-y$; $(-10, 7)$
y-axis: Replace x with $-x$; $(10, -7)$
Origin: Replace x with $-x$ and y with $-y$; $(10, 7)$

22. x-axis: Replace y with $-y$; $\left(1, -\frac{3}{8}\right)$
y-axis: Replace x with $-x$; $\left(-1, \frac{3}{8}\right)$
Origin: Replace x with $-x$ and y with $-y$; $\left(-1, -\frac{3}{8}\right)$

23. x-axis: Replace y with $-y$; $(0, 4)$
y-axis: Replace x with $-x$; $(0, -4)$
Origin: Replace x with $-x$ and y with $-y$; $(0, 4)$

24. x-axis: Replace y with $-y$; $(8, 3)$
y-axis: Replace x with $-x$; $(-8, -3)$
Origin: Replace x with $-x$ and y with $-y$; $(-8, 3)$

25. The graph is symmetric with respect to the y-axis, so the function is even.

26. The graph is symmetric with respect to the y-axis, so the function is even.

27. The graph is symmetric with respect to the origin, so the function is odd.

28. The graph is not symmetric with respect to either the y-axis or the origin, so the function is neither even nor odd.

29. The graph is not symmetric with respect to either the y-axis or the origin, so the function is neither even nor odd.

30. The graph is not symmetric with respect to either the y-axis or the origin, so the function is neither even nor odd.

31. $f(x) = -3x^3 + 2x$
$f(-x) = -3(-x)^3 + 2(-x) = 3x^3 - 2x$
$-f(x) = -(-3x^3 + 2x) = 3x^3 - 2x$
$f(-x) = -f(x)$, so f is odd.

32. $f(x) = 7x^3 + 4x - 2$
$f(-x) = 7(-x)^3 + 4(-x) - 2 = -7x^3 - 4x - 2$
$-f(x) = -(7x^3 + 4x - 2) = -7x^3 - 4x + 2$
$f(x) \neq f(-x)$, so f is not even.
$f(-x) \neq -f(x)$, so f is not odd.
Thus, $f(x) = 7x^3 + 4x - 2$ is neither even nor odd.

33. $f(x) = 5x^2 + 2x^4 - 1$
$f(-x) = 5(-x)^2 + 2(-x)^4 - 1 = 5x^2 + 2x^4 - 1$
$f(x) = f(-x)$, so f is even.

34. $f(x) = x + \frac{1}{x}$
$f(-x) = -x + \frac{1}{-x} = -x - \frac{1}{x}$
$-f(x) = -\left(x + \frac{1}{x}\right) = -x - \frac{1}{x}$
$f(-x) = -f(x)$, so f is odd.

35. $f(x) = x^{17}$
$f(-x) = (-x)^{17} = -x^{17}$
$-f(x) = -x^{17}$
$f(-x) = -f(x)$, so f is odd.

36. $f(x) = \sqrt[3]{x}$
$f(-x) = \sqrt[3]{-x} = -\sqrt[3]{x}$
$-f(x) = -\sqrt[3]{x}$
$f(-x) = -f(x)$, so f is odd.

37. $f(x) = \frac{1}{x^2}$
$f(-x) = \frac{1}{(-x)^2} = \frac{1}{x^2}$
$f(x) = f(-x)$, so f is even.

38. $f(x) = x - |x|$
$f(-x) = (-x) - |(-x)| = -x - |x|$
$-f(x) = -(x - |x|) = -x + |x|$
$f(x) \neq f(-x)$, so f is not even.
$f(-x) \neq -f(x)$, so f is not odd.
Thus, $f(x) = x - |x|$ is neither even nor odd.

39. $f(x) = 8$
$f(-x) = 8$
$f(x) = f(-x)$, so f is even.

40. $f(x) = \sqrt{x^2 + 1}$
$f(-x) = \sqrt{(-x)^2 + 1} = \sqrt{x^2 + 1}$
$f(x) = f(-x)$, so f is even.

41. Shift the graph of $f(x) = x^2$ right 3 units.

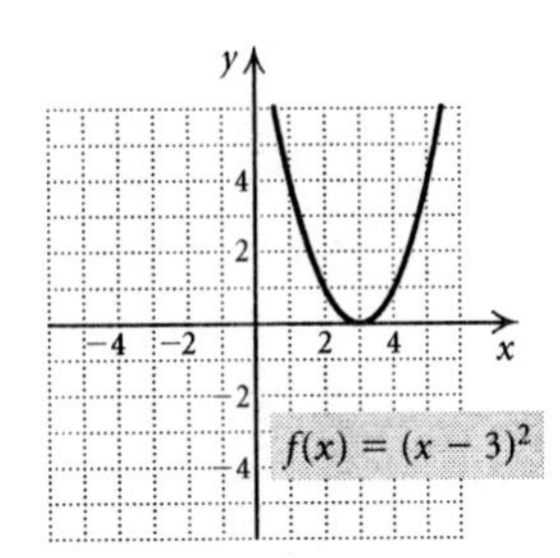

42. Shift the graph of $g(x) = x^2$ up $\frac{1}{2}$ unit.

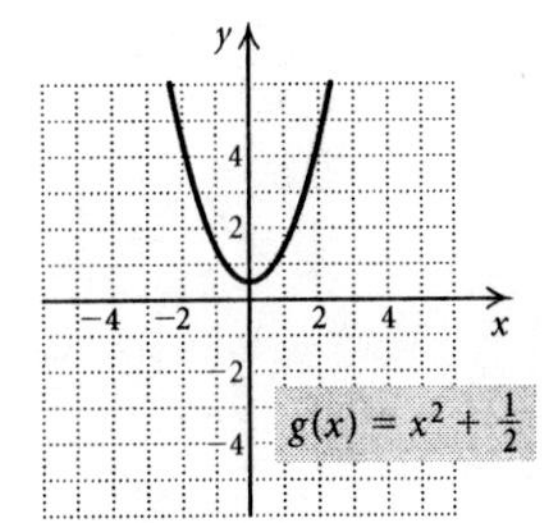

43. Shift the graph of $g(x) = x$ down 3 units.

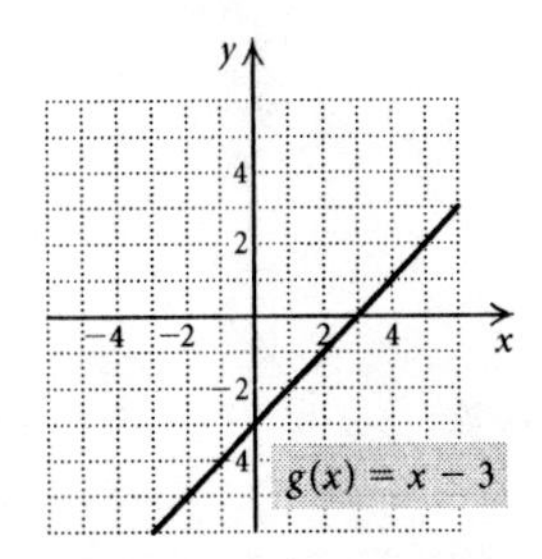

44. Reflect the graph of $g(x) = x$ across the x-axis and then shift it down 2 units.

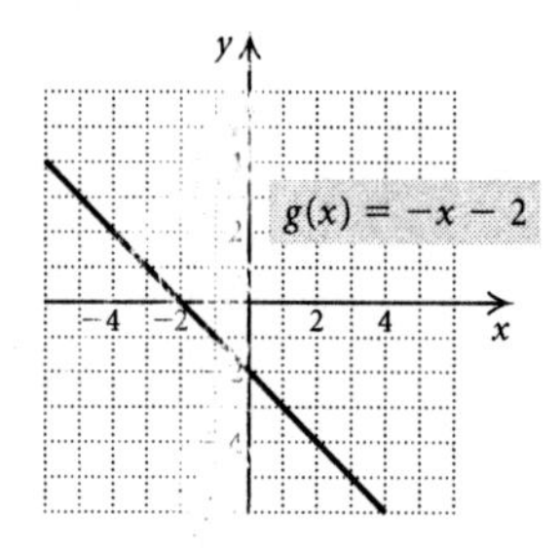

45. Reflect the graph of $h(x) = \sqrt{x}$ across the x-axis.

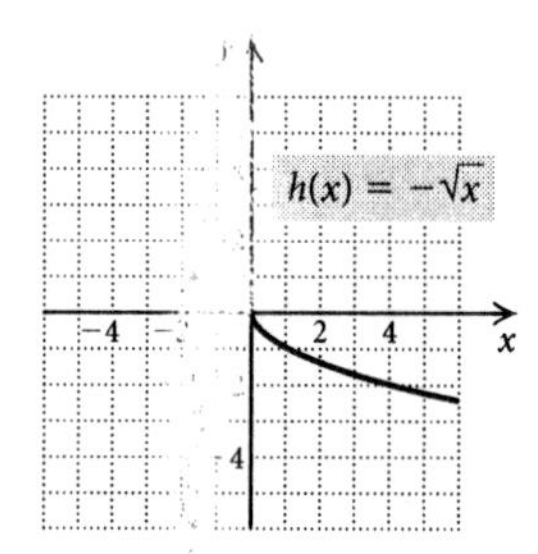

46. Shift the graph of $g(x) = \sqrt{x}$ right 1 unit.

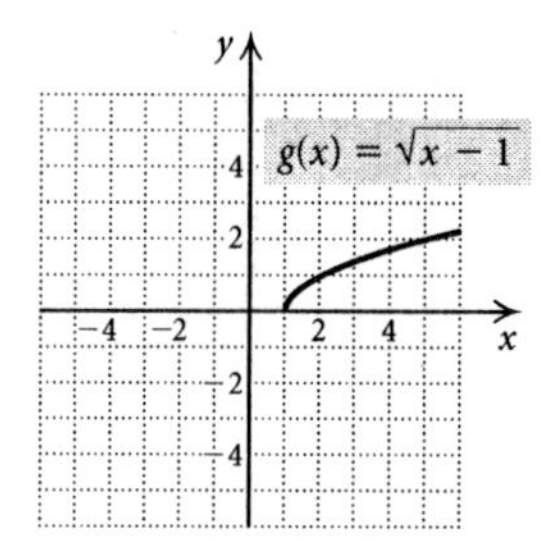

47. Shift the graph of $h(x) = \frac{1}{x}$ up 4 units.

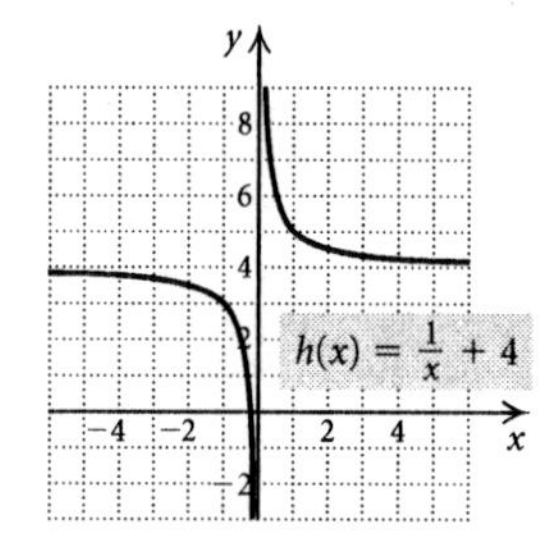

48. Shift the graph of $g(x) = \frac{1}{x}$ right 2 units.

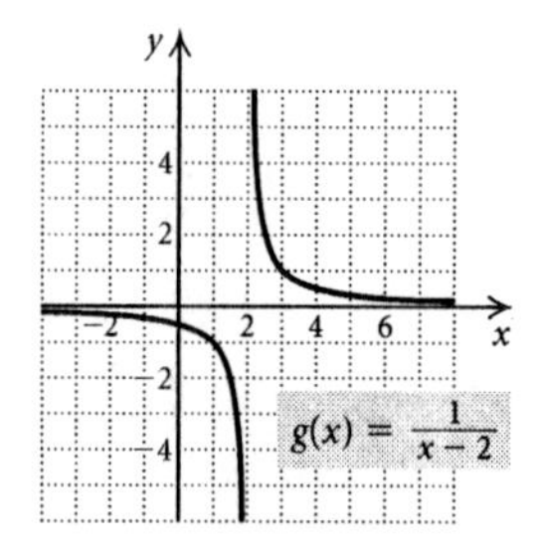

49. First stretch the graph of $h(x) = x$ vertically by multiplying each y-coordinate by 3. Then reflect it across the x-axis and shift it up 3 units.

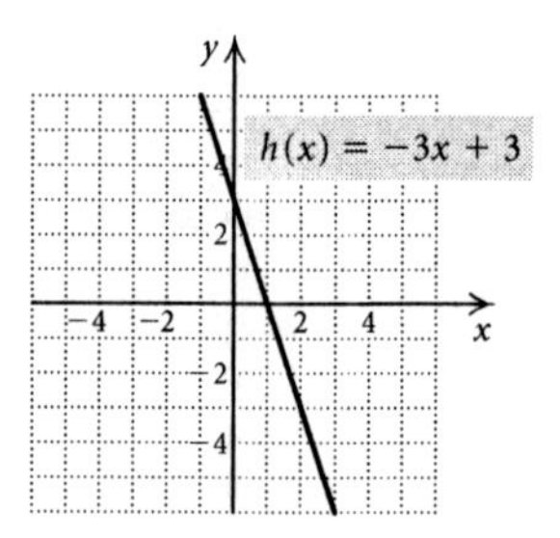

50. First stretch the graph of $f(x) = x$ vertically by multiplying each y-coordinate by 2. Then shift it up 1 unit.

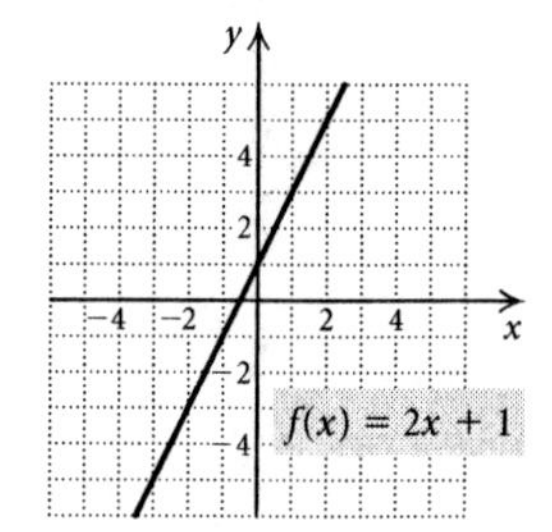

51. First shrink the graph of $h(x) = |x|$ vertically by multiplying each y-coordinate by $\frac{1}{2}$. Then shift it down 2 units.

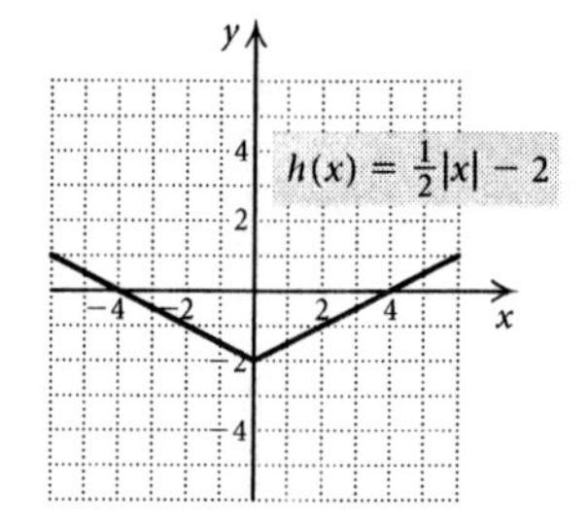

52. Reflect the graph of $g(x) = |x|$ across the x-axis and shift it up 2 units.

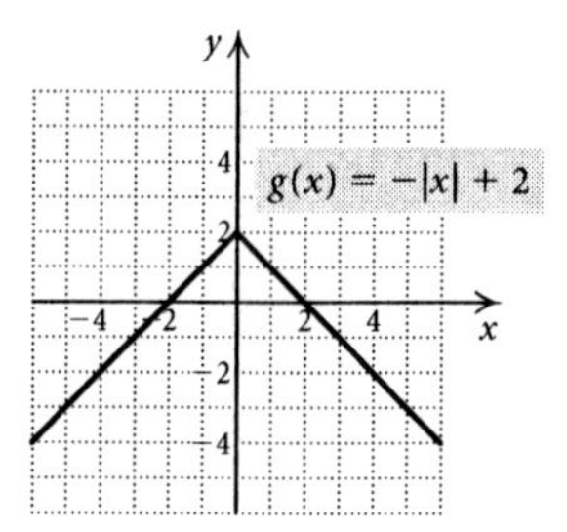

53. Shift the graph of $g(x) = x^3$ right 2 units and reflect it across the x-axis.

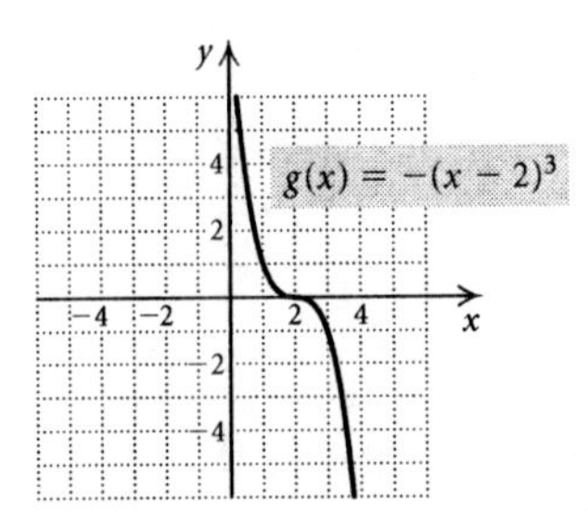

54. Shift the graph of $f(x) = x^3$ left 1 unit.

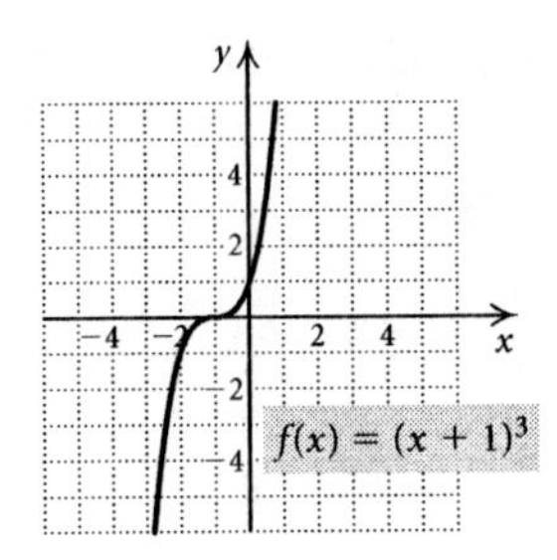

55. Shift the graph of $g(x) = x^2$ left 1 unit and down 1 unit.

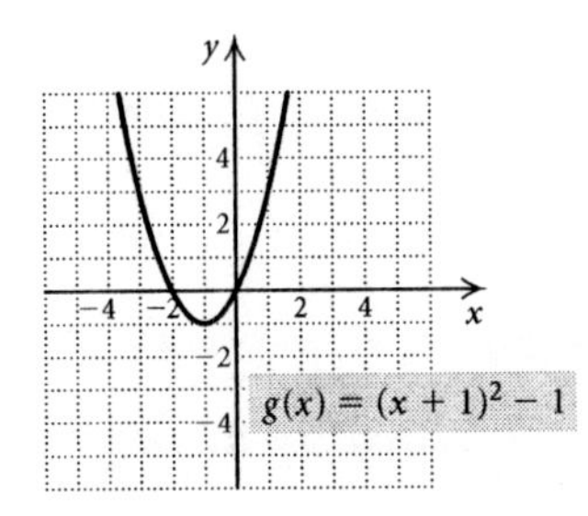

56. Reflect the graph of $h(x) = x^2$ across the x-axis and down 4 units.

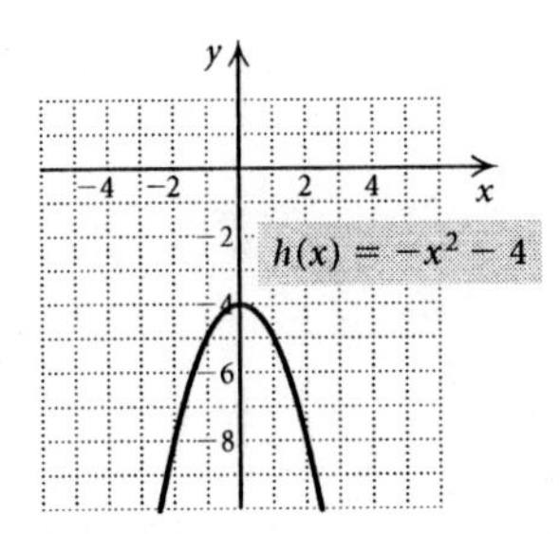

57. First shrink the graph of $g(x) = x^3$ vertically by multiplying each y-coordinate by $\frac{1}{3}$. Then shift it up 2 units.

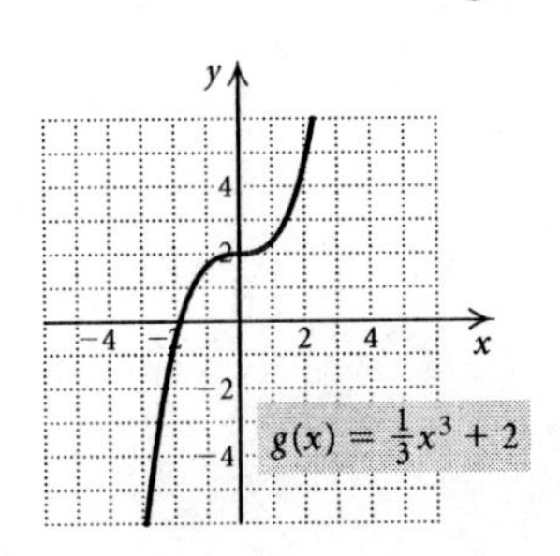

58. Reflect the graph of $h(x) = x^3$ across the y-axis.

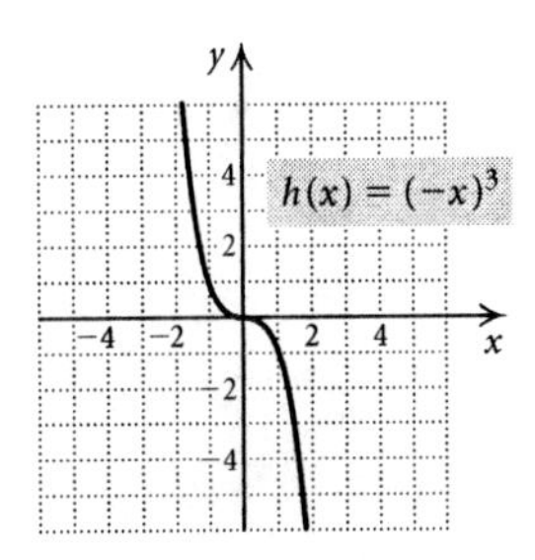

59. Shift the graph of $f(x) = \sqrt{x}$ left 2 units.

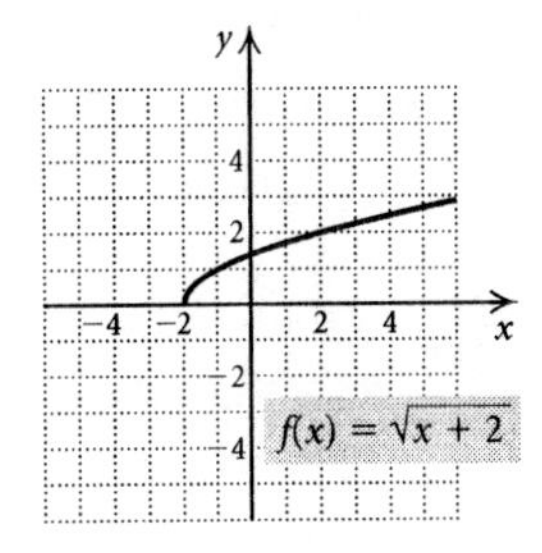

60. First shift the graph of $f(x) = \sqrt{x}$ right 1 unit. Shrink it vertically by multiplying each y-coordinate by $\frac{1}{2}$ and then reflect it across the x-axis.

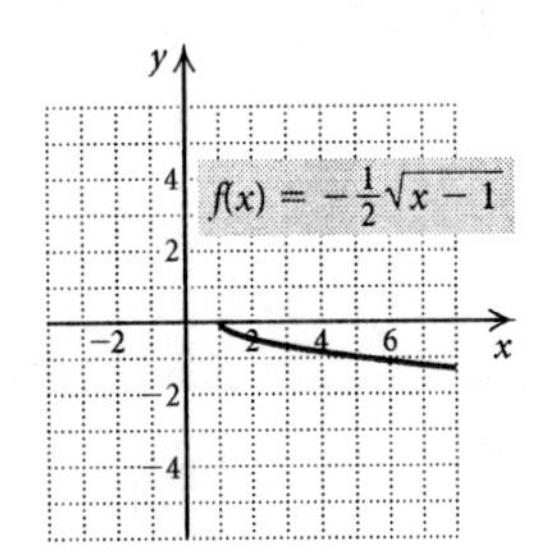

61. Shift the graph of $f(x) = \sqrt[3]{x}$ down 2 units.

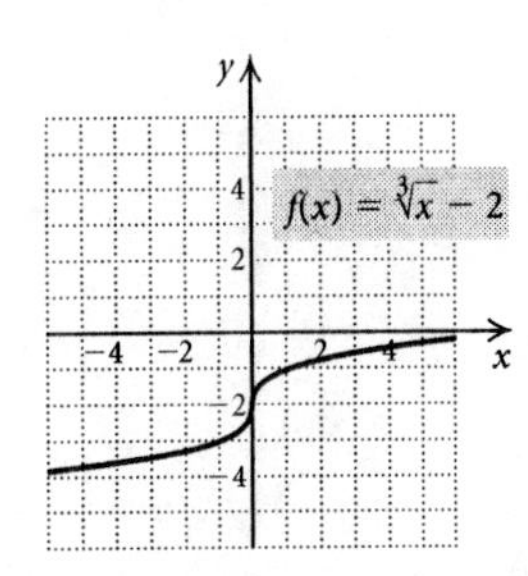

62. Shift the graph of $h(x) = \sqrt[3]{x}$ left 1 unit.

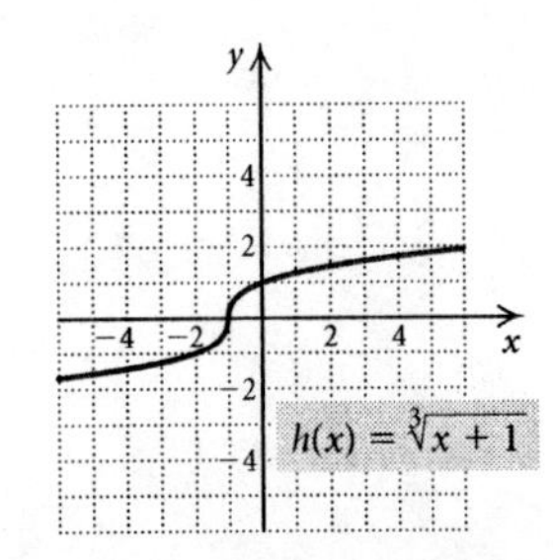

63. Think of the graph of $f(x) = |x|$. Since $g(x) = f(3x)$, the graph of $g(x) = |3x|$ is the graph of $f(x) = |x|$ shrunk horizontally by dividing each x-coordinate by 3 $\left(\text{or multiplying each } x\text{-coordinate by } \frac{1}{3}\right)$.

64. Think of the graph of $g(x) = \sqrt[3]{x}$. Since $f(x) = \frac{1}{2}g(x)$, the graph of $f(x) = \frac{1}{2}\sqrt[3]{x}$ is the graph of $g(x) = \sqrt[3]{x}$ shrunk vertically by multiplying each y-coordinate by $\frac{1}{2}$.

65. Think of the graph of $f(x) = \frac{1}{x}$. Since $h(x) = 2f(x)$, the graph of $h(x) = \frac{2}{x}$ is the graph of $f(x) = \frac{1}{x}$ stretched vertically by multiplying each y-coordinate by 2.

66. Think of the graph of $g(x) = |x|$. Since $f(x) = g(x-3)-4$, the graph of $f(x) = |x-3| - 4$ is the graph of $g(x) = |x|$ shifted right 3 units and down 4 units.

67. Think of the graph of $g(x) = \sqrt{x}$. Since $f(x) = 3g(x) - 5$, the graph of $f(x) = 3\sqrt{x} - 5$ is the graph of $g(x) = \sqrt{x}$ stretched vertically by multiplying each y-coordinate by 3 and then shifted down 5 units.

68. Think of the graph of $g(x) = \frac{1}{x}$. Since $f(x) = 5 - g(x)$, or $f(x) = -g(x) + 5$, the graph of $f(x) = 5 - \frac{1}{x}$ is the graph of $g(x) = \frac{1}{x}$ reflected across the x-axis and then shifted up 5 units.

69. Think of the graph of $f(x) = |x|$. Since $g(x) = f\left(\frac{1}{3}x\right) - 4$, the graph of $g(x) = \left|\frac{1}{3}x\right| - 4$ is the graph of $f(x) = |x|$ stretched horizontally by multiplying each x-coordinate by 3 and then shifted down 4 units.

70. Think of the graph of $g(x) = x^3$. Since $f(x) = \frac{2}{3}g(x) - 4$, the graph of $f(x) = \frac{2}{3}x^3 - 4$ is the graph of $g(x) = x^3$ shrunk vertically by multiplying each y-coordinate by $\frac{2}{3}$ and then shifted down 4 units.

71. Think of the graph of $g(x) = x^2$. Since $f(x) = -\frac{1}{4}g(x-5)$, the graph of $f(x) = -\frac{1}{4}(x-5)^2$ is the graph of $g(x) = x^2$ shifted right 5 units, shrunk vertically by multiplying each y-coordinate by $\frac{1}{4}$, and reflected across the x-axis.

72. Think of the graph of $g(x) = x^3$. Since $f(x) = g(-x) - 5$, the graph of $f(x) = (-x)^3 - 5$ is the graph of $g(x) = x^3$ reflected across the y-axis and shifted down 5 units.

73. Think of the graph of $g(x) = \frac{1}{x}$. Since $f(x) = g(x+3) + 2$, the graph of $f(x) = \frac{1}{x+3} + 2$ is the graph of $g(x) = \frac{1}{x}$ shifted left 3 units and up 2 units.

74. Think of the graph of $f(x) = \sqrt{x}$. Since $g(x) = f(-x)+5$, the graph of $g(x) = \sqrt{-x} + 5$ is the graph of $f(x) = \sqrt{x}$ reflected across the y-axis and shifted up 5 units.

75. Think of the graph of $f(x) = x^2$. Since $h(x) = -f(x-3)+5$, the graph of $h(x) = -(x-3)^2+5$ is the graph of $f(x) = x^2$ shifted right 3 units, reflected across the x-axis, and shifted up 5 units.

76. Think of the graph of $g(x) = x^2$. Since $f(x) = 3g(x+4)-3$, the graph of $f(x) = 3(x+4)^2-3$ is the graph of $g(x) = x^2$ shifted left 4 units, stretched vertically by multiplying each y-coordinate by 3, and then shifted down 3 units.

77. The graph of $y = g(x)$ is the graph of $y = f(x)$ shrunk vertically by a factor of $\frac{1}{2}$. Multiply the y-coordinate by $\frac{1}{2}$: $(-12, 2)$.

78. The graph of $y = g(x)$ is the graph of $y = f(x)$ shifted right 2 units. Add 2 to the x-coordinate: $(-10, 4)$.

79. The graph of $y = g(x)$ is the graph of $y = f(x)$ reflected across the y-axis, so we reflect the point across the y-axis: $(12, 4)$.

80. The graph of $y = g(x)$ is the graph of $y = f(x)$ shrunk horizontally. The x-coordinates of $y = g(x)$ are $\frac{1}{4}$ the corresponding x-coordinates of $y = f(x)$, so we divide the x-coordinate by 4 $\left(\text{or multiply it by } \frac{1}{4}\right)$: $(-3, 4)$.

81. The graph of $y = g(x)$ is the graph of $y = f(x)$ shifted down 2 units. Subtract 2 from the y-coordinate: $(-12, 2)$.

82. The graph of $y = g(x)$ is the graph of $y = f(x)$ stretched horizontally. The x-coordinates of $y = g(x)$ are twice the corresponding x-coordinates of $y = f(x)$, so we multiply the x-coordinate by 2 $\left(\text{or divide it by } \frac{1}{2}\right)$: $(-24, 4)$.

83. The graph of $y = g(x)$ is the graph of $y = f(x)$ stretched vertically by a factor of 4. Multiply the y-coordinate by 4: $(-12, 16)$.

84. The graph of $y = g(x)$ is the graph $y = f(x)$ reflected across the x-axis. Reflect the point across the x-axis: $(-12, -4)$.

85. Shape: $h(x) = x^2$

Turn $h(x)$ upside-down (that is, reflect it across the x-axis): $g(x) = -h(x) = -x^2$

Shift $g(x)$ right 8 units: $f(x) = g(x-8) = -(x-8)^2$

86. Shape: $h(x) = \sqrt{x}$

Shift $h(x)$ left 6 units: $g(x) = h(x+6) = \sqrt{x+6}$

Shift $g(x)$ down 5 units: $f(x) = g(x) - 5 = \sqrt{x+6} - 5$

87. Shape: $h(x) = |x|$

Shift $h(x)$ left 7 units: $g(x) = h(x+7) = |x+7|$

Shift $g(x)$ up 2 units: $f(x) = g(x) + 2 = |x+7| + 2$

88. Shape: $h(x) = x^3$

Turn $h(x)$ upside-down (that is, reflect it across the x-axis): $g(x) = -h(x) = -x^3$

Shift $g(x)$ right 5 units: $f(x) = g(x-5) = -(x-5)^3$

89. Shape: $h(x) = \frac{1}{x}$

Shrink $h(x)$ vertically by a factor of $\frac{1}{2}$ $\left(\text{that is, multiply each function value by } \frac{1}{2}\right)$:

$g(x) = \frac{1}{2}h(x) = \frac{1}{2} \cdot \frac{1}{x}$, or $\frac{1}{2x}$

Shift $g(x)$ down 3 units: $f(x) = g(x) - 3 = \frac{1}{2x} - 3$

90. Shape: $h(x) = x^2$

Shift $h(x)$ right 6 units: $g(x) = h(x-6) = (x-6)^2$

Shift $g(x)$ up 2 units: $f(x) = g(x) + 2 = (x-6)^2 + 2$

91. Shape: $m(x) = x^2$

Turn $m(x)$ upside-down (that is, reflect it across the x-axis): $h(x) = -m(x) = -x^2$

Shift $h(x)$ right 3 units: $g(x) = h(x-3) = -(x-3)^2$

Shift $g(x)$ up 4 units: $f(x) = g(x) + 4 = -(x-3)^2 + 4$

92. Shape: $h(x) = |x|$

Stretch $h(x)$ horizontally by a factor of 2 $\left(\text{that is, multiply each } x\text{-value by } \frac{1}{2}\right)$: $g(x) = h\left(\frac{1}{2}x\right) = \left|\frac{1}{2}x\right|$

Shift $g(x)$ down 5 units: $f(x) = g(x) - 5 = \left|\frac{1}{2}x\right| - 5$

93. Shape: $m(x) = \sqrt{x}$

Reflect $m(x)$ across the y-axis: $h(x) = m(-x) = \sqrt{-x}$

Shift $h(x)$ left 2 units: $g(x) = h(x+2) = \sqrt{-(x+2)}$

Shift $g(x)$ down 1 unit: $f(x) = g(x) - 1 = \sqrt{-(x+2)} - 1$

94. Shape: $h(x) = \frac{1}{x}$

Reflect $h(x)$ across the x-axis: $g(x) = -h(x) = -\frac{1}{x}$

Shift $g(x)$ up 1 unit: $f(x) = g(x) + 1 = -\frac{1}{x} + 1$

95. Each y-coordinate is multiplied by -2. We plot and connect $(-4,0)$, $(-3,4)$, $(-1,4)$, $(2,-6)$, and $(5,0)$.

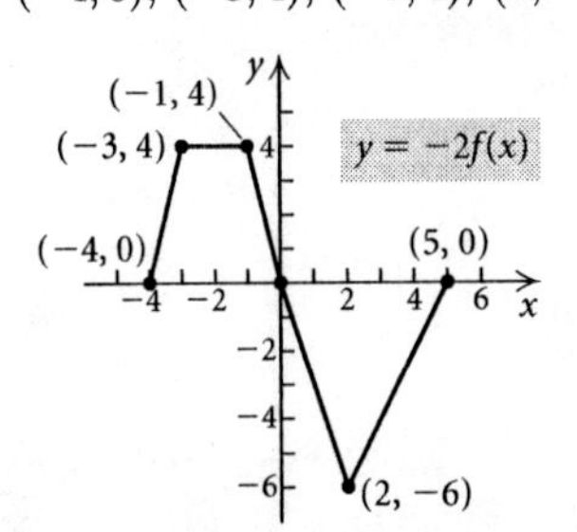

96. Each y-coordinate is multiplied by $\frac{1}{2}$. We plot and connect $(-4,0)$, $(-3,-1)$, $(-1,-1)$, $(2,1.5)$, and $(5,0)$.

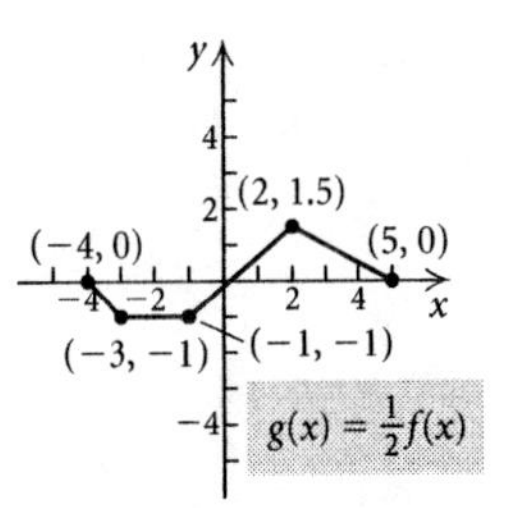

97. The graph is reflected across the y-axis and stretched horizontally by a factor of 2. That is, each x-coordinate is multiplied by -2 $\left(\text{or divided by } -\frac{1}{2}\right)$. We plot and connect $(8,0)$, $(6,-2)$, $(2,-2)$, $(-4,3)$, and $(-10,0)$.

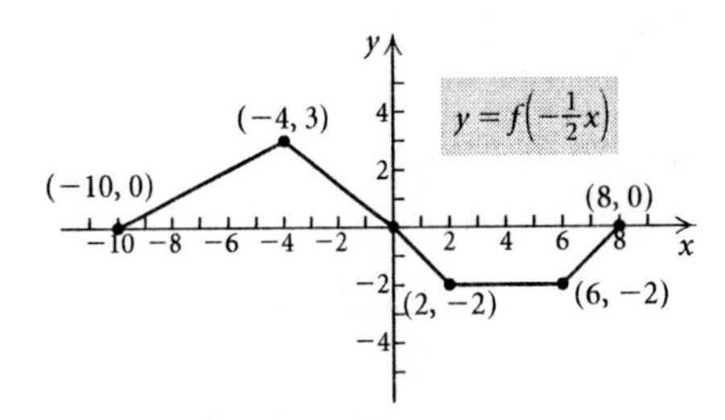

98. The graph is shrunk horizontally by a factor of 2. That is, each x-coordinate is divided by 2 $\left(\text{or multiplied by } \frac{1}{2}\right)$. We plot and connect $(-2,0)$, $(-1.5,-2)$, $(-0.5,-2)$, $(1,3)$, and $(2.5,0)$.

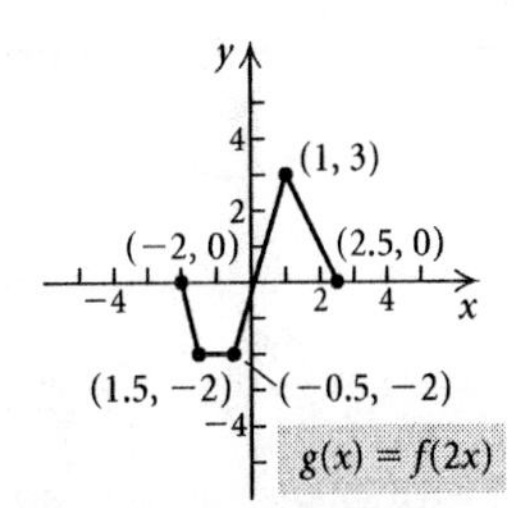

99. The graph is shifted right 1 unit so each x-coordinate is increased by 1. The graph is also reflected across the x-axis, shrunk vertically by a factor of 2, and shifted up 3 units. Thus, each y-coordinate is multiplied by $-\frac{1}{2}$ and then increased by 3. We plot and connect $(-3,3)$, $(-2,4)$, $(0,4)$, $(3,1.5)$, and $(6,3)$.

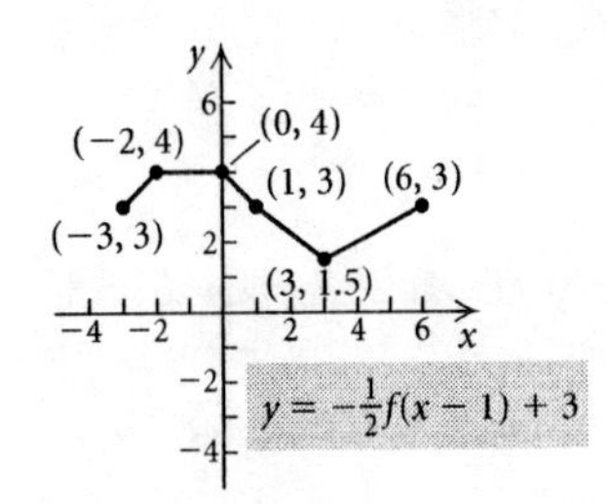

100. The graph is shifted left 1 unit so each x-coordinate is decreased by 1. The graph is also reflected across the x-axis, stretched vertically by a factor of 3, and shifted down 4 units. Thus, each y-coordinate is multiplied by -3 and then decreased by 4. We plot and connect $(-5,-4)$, $(-4,2)$, $(-2,2)$, $(1,-13)$, and $(4,-4)$.

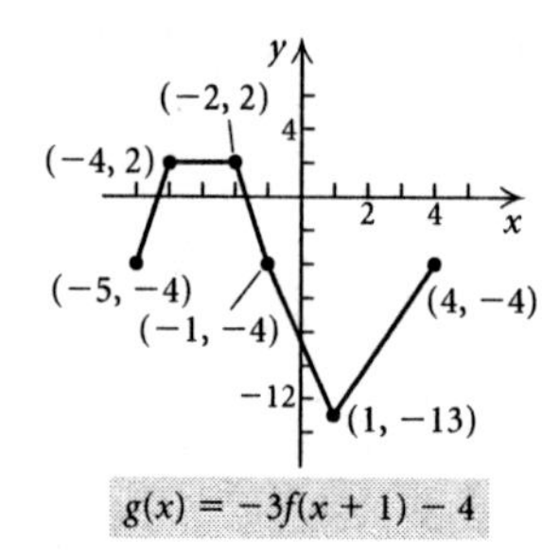

101. The graph is reflected across the y-axis so each x-coordinate is replaced by its opposite.

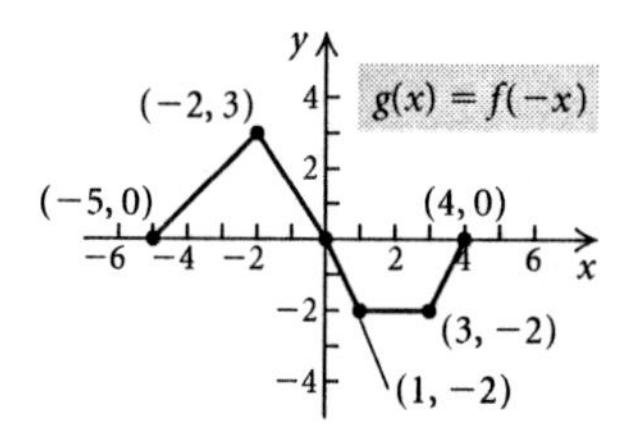

102. The graph is reflected across the x-axis so each y-coordinate is replaced by its opposite.

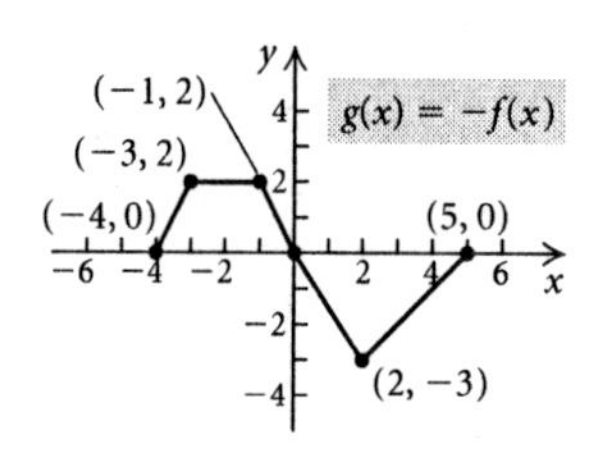

103. The graph is shifted left 2 units so each x-coordinate is decreased by 2. It is also reflected across the x-axis so each y-coordinate is replaced with its opposite. In addition, the graph is shifted up 1 unit, so each y-coordinate is then increased by 1.

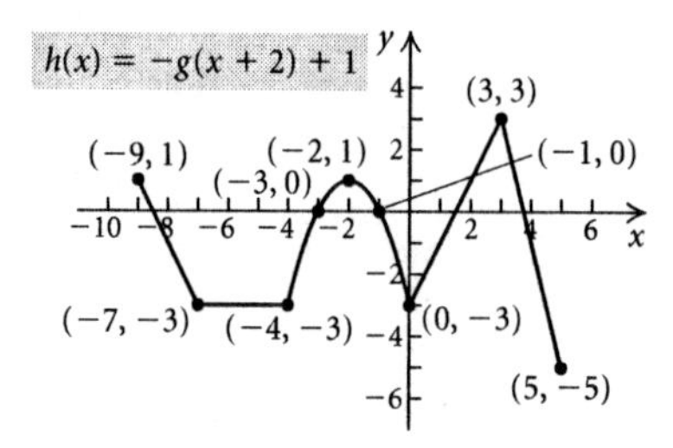

104. The graph is reflected across the y-axis so each x-coordinate is replaced with its opposite. It is also shrunk vertically by a factor of $\frac{1}{2}$, so each y-coordinate is multiplied by $\frac{1}{2}$ (or divided by 2).

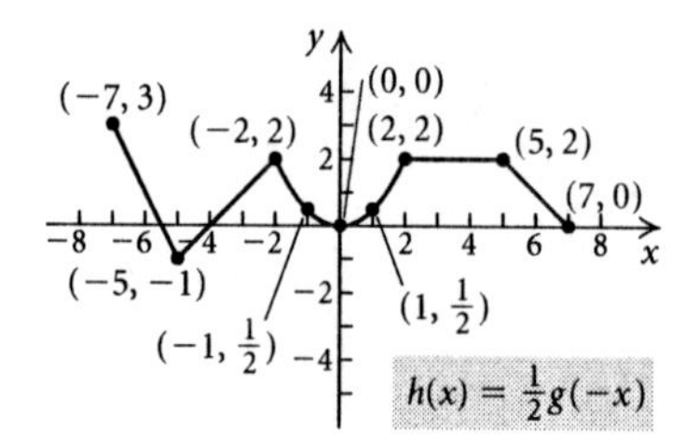

105. The graph is shrunk horizontally. The x-coordinates of $y = h(x)$ are one-half the corresponding x-coordinates of $y = g(x)$.

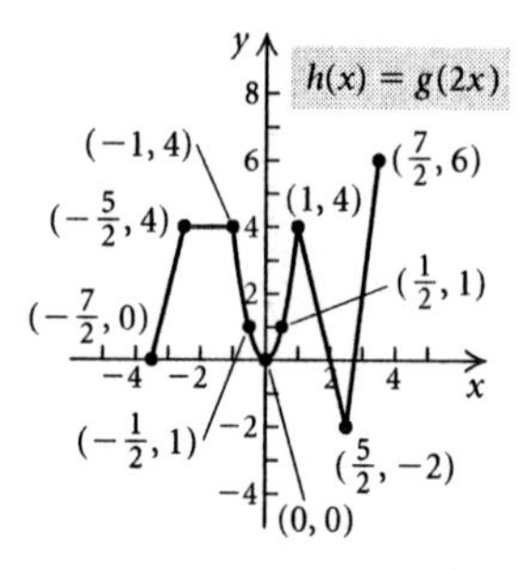

106. The graph is shifted right 1 unit, so each x-coordinate is increased by 1. It is also stretched vertically by a factor of 2, so each y-coordinate is multiplied by 2 $\left(\text{or divided by } \frac{1}{2}\right)$. In addition, the graph is shifted down 3 units, so each y-coordinate is decreased by 3.

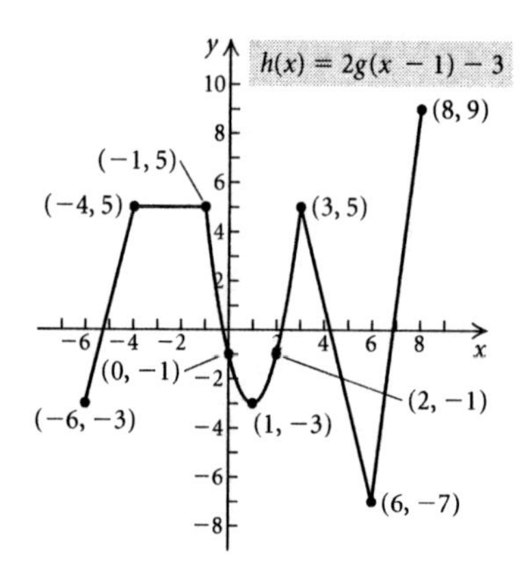

107. $g(x) = f(-x) + 3$

The graph of $g(x)$ is the graph of $f(x)$ reflected across the y-axis and shifted up 3 units. This is graph (f).

108. $g(x) = f(x) + 3$

The graph of $g(x)$ is the graph of $f(x)$ shifted up 3 units. This is graph (h).

109. $g(x) = -f(x) + 3$

The graph of $g(x)$ is the graph of $f(x)$ reflected across the x-axis and shifted up 3 units. This is graph (f).

110. $g(x) = -f(-x)$

The graph of $g(x)$ is the graph of $f(x)$ reflected across the x-axis and the y-axis. This is graph (a).

111. $g(x) = \frac{1}{3}f(x-2)$

The graph of $g(x)$ is the graph of $f(x)$ shrunk vertically by a factor of 3 $\left(\text{that is, each } y\text{-coordinate is multiplied by } \frac{1}{3}\right)$ and then shifted right 2 units. This is graph (d).

112. $g(x) = \frac{1}{3}f(x) - 3$

The graph of $g(x)$ is the graph of $f(x)$ shrunk vertically by a factor of 3 $\left(\text{that is, each } y\text{-coordinate is multiplied by } \frac{1}{3}\right)$ and then shifted down 3 units. This is graph (e).

113. $g(x) = \frac{1}{3}f(x+2)$

The graph of $g(x)$ is the graph of $f(x)$ shrunk vertically by a factor of 3 $\left(\text{that is, each } y\text{-coordinate is multiplied by } \frac{1}{3}\right)$ and then shifted left 2 units. This is graph (c).

114. $g(x) = -f(x+2)$

The graph of $g(x)$ is the graph $f(x)$ reflected across the x-axis and shifted left 2 units. This is graph (b).

115. $f(-x) = 2(-x)^4 - 35(-x)^3 + 3(-x) - 5 = 2x^4 + 35x^3 - 3x - 5 = g(x)$

116. $f(-x) = \frac{1}{4}(-x)^4 + \frac{1}{5}(-x)^3 - 81(-x)^2 - 17 = \frac{1}{4}x^4 - \frac{1}{5}x^3 - 81x^2 - 17 \neq g(x)$

117. The graph of $f(x) = x^3 - 3x^2$ is shifted up 2 units. A formula for the transformed function is $g(x) = f(x) + 2$, or $g(x) = x^3 - 3x^2 + 2$.

118. Each y-coordinate of the graph of $f(x) = x^3 - 3x^2$ is multiplied by $\frac{1}{2}$. A formula for the transformed function is $h(x) = \frac{1}{2}f(x)$, or $h(x) = \frac{1}{2}(x^3 - 3x^2)$.

119. The graph of $f(x) = x^3 - 3x^2$ is shifted left 1 unit. A formula for the transformed function is $k(x) = f(x+1)$, or $k(x) = (x+1)^3 - 3(x+1)^2$.

120. The graph of $f(x) = x^3 - 3x^2$ is shifted right 2 units and up 1 unit. A formula for the transformed function is $t(x) = f(x-2) + 1$, or $t(x) = (x-2)^3 - 3(x-2)^2 + 1$.

121. The graph of $f(x) = 0$ is symmetric with respect to the x-axis, the y-axis, and the origin. This function is both even and odd.

122. If all of the exponents are even numbers, then $f(x)$ is an even function. If $a_0 = 0$ and all of the exponents are odd numbers, then $f(x)$ is an odd function.

123. For every point (x, y) on the graph of $y = f(x)$, its reflection across the y-axis $(-x, y)$ is on the graph of $y = f(-x)$.

124. The graph of $f(x) = |x^2 - 9|$ looks like the graph of $g(x) = x^2 - 9$ with the points with negative y-coordinates reflected across the x-axis.

125. $f(x) = 5x^2 - 7$

a) $f(-3) = 5(-3)^2 - 7 = 5 \cdot 9 - 7 = 45 - 7 = 38$

b) $f(3) = 5 \cdot 3^2 - 7 = 5 \cdot 9 - 7 = 45 - 7 = 38$

c) $f(a) = 5a^2 - 7$

d) $f(-a) = 5(-a)^2 - 7 = 5a^2 - 7$

126. $f(x) = 4x^3 - 5x$

a) $f(2) = 4 \cdot 2^3 - 5 \cdot 2 = 4 \cdot 8 - 5 \cdot 2 = 32 - 10 = 22$

b) $f(-2) = 4(-2)^3 - 5(-2) = 4(-8) - 5(-2) = -32 + 10 = -22$

c) $f(a) = 4a^3 - 5a$

d) $f(-a) = 4(-a)^3 - 5(-a) = 4(-a^3) - 5(-a) = -4a^3 + 5a$

127. First find the slope of the given line.

$$8x - y = 10$$
$$8x = y + 10$$
$$8x - 10 = y$$

The slope of the given line is 8. The slope of a line perpendicular to this line is the opposite of the reciprocal of 8, or $-\frac{1}{8}$.

$$y - y_1 = m(x - x_1)$$
$$y - 1 = -\frac{1}{8}[x - (-1)]$$
$$y - 1 = -\frac{1}{8}(x + 1)$$
$$y - 1 = -\frac{1}{8}x - \frac{1}{8}$$
$$y = -\frac{1}{8}x + \frac{7}{8}$$

128. $2x - 9y + 1 = 0$

$$2x + 1 = 9y$$
$$\frac{2}{9}x + \frac{1}{9} = y$$

Slope: $\frac{2}{9}$; y-intercept: $\left(0, \frac{1}{9}\right)$

129. Each point for which $f(x) < 0$ is reflected across the x-axis.

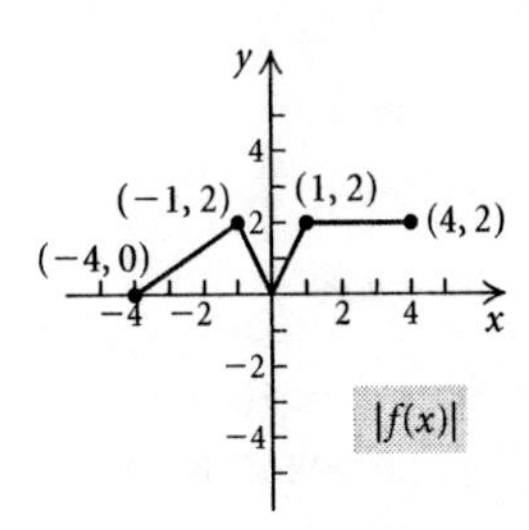

130. The graph of $y = f(|x|)$ consists of the points of $y = f(x)$ for which $x \geq 0$ along with their reflections across the y-axis.

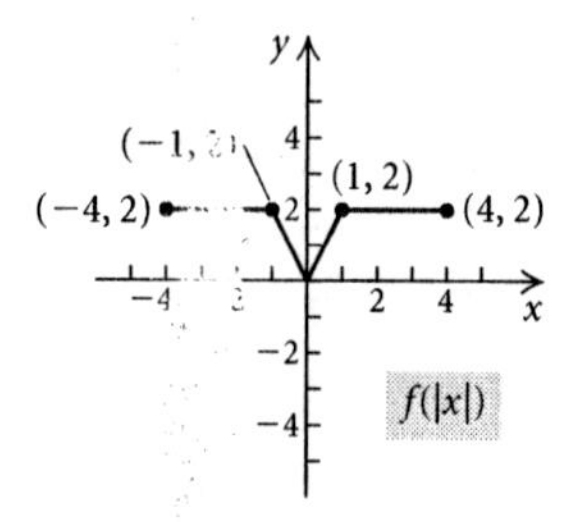

131. Each point for which $g(x) < 0$ is reflected across the x-axis.

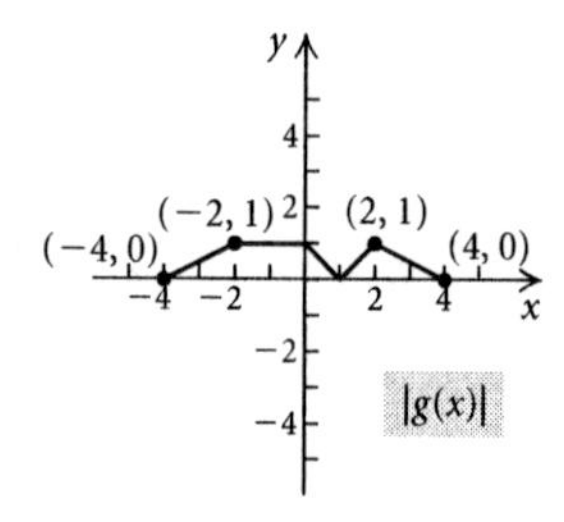

132. The graph of $y = g(|x|)$ consists of the points of $y = g(x)$ for which $x \geq 0$ along with their reflections across the y-axis.

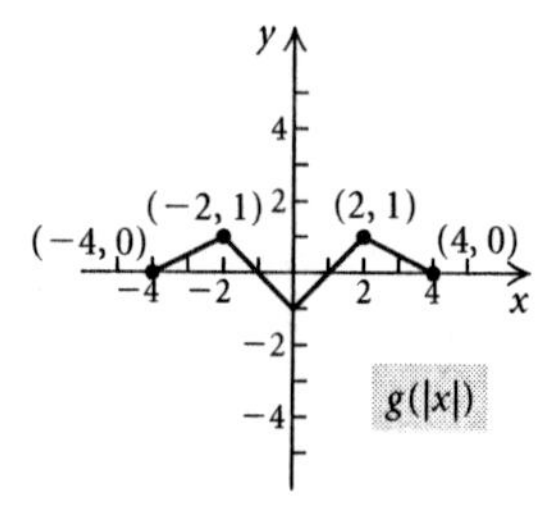

133. $f(x) = x\sqrt{10 - x^2}$

$f(-x) = -x\sqrt{10 - (-x)^2} = -x\sqrt{10 - x^2}$

$-f(x) = -x\sqrt{10 - x^2}$

Since $f(-x) = -f(x)$, f is odd.

134. $f(x) = \dfrac{x^2 + 1}{x^3 + 1}$

$f(-x) = \dfrac{(-x)^2 + 1}{(-x)^3 + 1} = \dfrac{x^2 + 1}{-x^3 + 1}$

$-f(x) = -\dfrac{x^2 + 1}{x^3 + 1}$

Since $f(x) \neq f(-x)$, f is not even.

Since $f(-x) \neq -f(x)$, f is not odd.

Thus, $f(x) = \dfrac{x^2 + 1}{x^3 + 1}$ is neither even nor odd.

135. If the graph were folded on the x-axis, the parts above and below the x-axis would coincide, so the graph is symmetric with respect to the x-axis.

If the graph were folded on the y-axis, the parts to the left and right of the y-axis would not coincide, so the graph is not symmetric with respect to the y-axis.

If the graph were rotated 180°, the resulting graph would not coincide with the original graph, so it is not symmetric with respect to the origin.

136. If the graph were folded on the x-axis, the parts above and below the x-axis would not coincide, so the graph is not symmetric with respect to the x-axis.

If the graph were folded on the y-axis, the parts to the left and right of the y-axis would not coincide, so the graph is not symmetric with respect to the y-axis.

If the graph were rotated 180°, the resulting graph would coincide with the original graph, so it is symmetric with respect to the origin.

137. If the graph were folded on the x-axis, the parts above and below the x-axis would coincide, so the graph is symmetric with respect to the x-axis.

If the graph were folded on the y-axis, the parts to the left and right of the y-axis would not coincide, so the graph is not symmetric with respect to the y-axis.

If the graph were rotated 180°, the resulting graph would not coincide with the original graph, so it is not symmetric with respect to the origin.

138. Call the transformed function $g(x)$.

Then $g(5) = 4 - f(-3) = 4 - f(5 - 8)$,

$g(8) = 4 - f(0) = 4 - f(8 - 8)$,

and $g(11) = 4 - f(3) = 4 - f(11 - 8)$.

Thus $g(x) = 4 - f(x - 8)$, or $g(x) = 4 - |x - 8|$.

139. $f(2 - 3) = f(-1) = 5$, so $b = 5$.

(The graph of $y = f(x - 3)$ is the graph of $y = f(x)$ shifted right 3 units, so the point $(-1, 5)$ on $y = f(x)$ is transformed to the point $(-1 + 3, 5)$, or $(2, 5)$ on $y = f(x - 3)$.)

140. Let $f(x) = g(x) = x$. Now f and g are odd functions, but $(fg)(x) = x^2 = (fg)(-x)$. Thus, the product is even, so the statement is false.

141. Let $f(x)$ and $g(x)$ be even functions. Then by definition, $f(x) = f(-x)$ and $g(x) = g(-x)$. Thus, $(f + g)(x) = f(x) + g(x) = f(-x) + g(-x) = (f + g)(-x)$ and $f + g$ is even. The statement is true.

142. Let $f(x)$ be an even function, and let $g(x)$ be an odd function. By definition $f(x) = f(-x)$ and $g(-x) = -g(x)$, or $g(x) = -g(-x)$. Then $fg(x) = f(x) \cdot g(x) = f(-x) \cdot [-g(-x)] = -f(-x) \cdot g(-x) = -fg(-x)$, and fg is odd. The statement is true.

143. See the answer section in the text.

144. $O(-x) = \dfrac{f(-x) - f(-(-x))}{2} = \dfrac{f(-x) - f(x)}{2}$,

$-O(x) = -\dfrac{f(x) - f(-x)}{2} = \dfrac{f(-x) - f(x)}{2}$. Thus,

$O(-x) = -O(x)$ and O is odd.

145. a), b) See the answer section in the text.

Chapter 2

Functions, Equations, and Inequalities

Exercise Set 2.1

1. $4x + 5 = 21$

$4x = 16$ Subtracting 5 on both sides

$x = 4$ Dividing by 4 on both sides

The solution is 4.

2. $2y - 1 = 3$

$2y = 4$

$y = 2$

The solution is 2.

3. $4x + 3 = 0$

$4x = -3$ Subtracting 3 on both sides

$x = -\frac{3}{4}$ Dividing by 4 on both sides

The solution is $-\frac{3}{4}$.

4. $3x - 16 = 0$

$3x = 16$

$x = \frac{16}{3}$

The solution is $\frac{16}{3}$.

5. $3 - x = 12$

$-x = 9$ Subtracting 3 on both sides

$x = -9$ Multiplying (or dividing) by -1 on both sides

The solution is -9.

6. $4 - x = -5$

$-x = -9$

$x = 9$

The solution is 9.

7. $8 = 5x - 3$

$11 = 5x$ Adding 3 on both sides

$\frac{11}{5} = x$ Dividing by 5 on both sides

The solution is $\frac{11}{5}$.

8. $9 = 4x - 8$

$17 = 4x$

$\frac{17}{4} = x$

The solution is $\frac{17}{4}$.

9. $y + 1 = 2y - 7$

$1 = y - 7$ Subtracting y on both sides

$8 = y$ Adding 7 on both sides

The solution is 8.

10. $5 - 4x = x - 13$

$18 = 5x$

$\frac{18}{5} = x$, or

$3.6 = x$

The solution is $\frac{18}{5}$, or 3.6.

11. $2x + 7 = x + 3$

$x + 7 = 3$ Subtracting x on both sides

$x = -4$ Subtracting 7 on both sides

The solution is -4.

12. $5x - 4 = 2x + 5$

$3x - 4 = 5$

$3x = 9$

$x = 3$

The solution is 3.

13. $3x - 5 = 2x + 1$

$x - 5 = 1$ Subtracting $2x$ on both sides

$x = 6$ Adding 5 on both sides

The solution is 6.

14. $4x + 3 = 2x - 7$

$2x = -10$

$x = -5$

The solution is -5.

15. $4x - 5 = 7x - 2$

$-5 = 3x - 2$ Subtracting $4x$ on both sides

$-3 = 3x$ Adding 2 on both sides

$-1 = x$ Dividing by 3 on both sides

The solution is -1.

16. $5x + 1 = 9x - 7$

$8 = 4x$

$2 = x$

The solution is 2.

17. $5x - 2 + 3x = 2x + 6 - 4x$

$8x - 2 = 6 - 2x$ Collecting like terms

$8x + 2x = 6 + 2$ Adding $2x$ and 2 on both sides

$10x = 8$ Collecting like terms

$x = \frac{8}{10}$ Dividing by 10 on both sides

$x = \frac{4}{5}$, or 0.8 Simplifying

The solution is $\frac{4}{5}$, or 0.8.

18. $5x - 17 - 2x = 6x - 1 - x$

$3x - 17 = 5x - 1$

$-2x = 16$

$x = -8$

The solution is -8.

19. $7(3x + 6) = 11 - (x + 2)$

$21x + 42 = 11 - x - 2$ Using the distributive property

$21x + 42 = 9 - x$ Collecting like terms

$21x + x = 9 - 42$ Adding x and subtracting 42 on both sides

$22x = -33$ Collecting like terms

$x = -\frac{33}{22}$ Dividing by 22 on both sides

$x = -\frac{3}{2}$, or -1.5 Simplifying

The solution is $-\frac{3}{2}$, or -1.5.

20. $4(5y + 3) = 3(2y - 5)$

$20y + 12 = 6y - 15$

$14y = -27$

$y = -\frac{27}{14}$

The solution is $-\frac{27}{14}$.

21. $3(x + 1) = 5 - 2(3x + 4)$

$3x + 3 = 5 - 6x - 8$ Removing parentheses

$3x + 3 = -6x - 3$ Collecting like terms

$9x + 3 = -3$ Adding $6x$

$9x = -6$ Subtracting 3

$x = -\frac{2}{3}$ Dividing by 9

The solution is $-\frac{2}{3}$.

22. $4(3x + 2) - 7 = 3(x - 2)$

$12x + 8 - 7 = 3x - 6$

$12x + 1 = 3x - 6$

$9x + 1 = -6$

$9x = -7$

$x = -\frac{7}{9}$

The solution is $-\frac{7}{9}$.

23. $2(x - 4) = 3 - 5(2x + 1)$

$2x - 8 = 3 - 10x - 5$ Using the distributive property

$2x - 8 = -10x - 2$ Collecting like terms

$12x = 6$ Adding $10x$ and 8 on both sides

$x = \frac{1}{2}$ Dividing by 12 on both sides

The solution is $\frac{1}{2}$.

24. $3(2x - 5) + 4 = 2(4x + 3)$

$6x - 15 + 4 = 8x + 6$

$6x - 11 = 8x + 6$

$-2x = 17$

$x = -\frac{17}{2}$

The solution is $-\frac{17}{2}$.

25. ***Familiarize.*** Let $t =$ the average tuition at a public college or university in 2001. We know that 9.6% of this amount is \$356.

Translate.

\$356 is 9.6% of the average tuition in 2001.

$$356 = 0.096 \cdot t$$

Carry out. We solve the equation.

$356 = 0.096 \cdot t$

$3708 \approx t$ Dividing by 0.096

Check. Since 9.6% of \$3708, or 0.096(\$3708), is approximately \$356, the answer checks.

State. In 2001 the average tuition at a public college or university was about \$3708.

26. Let $s =$ the amount of a student's expenditure for books that goes to the college store.

Solve: $s = 0.232(501)$

$s \approx \$116.23$

27. ***Familiarize.*** Let $d =$ the percent of spending on films in 2002 that went for DVDs.

Translate.

\$8.1 billion is what percent of \$12.4 billion?

$$8.1 = d \cdot 12.4$$

Carry out. We solve the equation.

$$8.1 = d \cdot 12.4$$

$$0.653 \approx d \quad \text{Dividing by 12.4}$$

$$65.3\% \approx d \quad \text{Writing percent notation}$$

Check. Since 65.3% of \$12.4 billion, or 0.653(\$12.4 billion), is approximately \$8.1 billion, the answer checks.

State. In 2002, about 65.3% of the spending on copies of films for home viewing went for DVDs.

28. Let $t =$ the amount *Training Day* earned in its initial theatrical release, in millions of dollars.

Solve: $86.1 = t + 9.8$

$t = \$76.3$ million

29. ***Familiarize.*** Let $h =$ the number of U.S. households with broadband Internet access in 2003, in millions.

Translate.

Number of households in 2006 is 25.6 million more than number of households in 2003.

$$54.8 = 25.6 + h$$

Carry out. We solve the equation.

$$54.8 = 25.6 + h$$

$$29.2 = h \quad \text{Subtracting 25.6}$$

Check. Since 29.2 million more than 25.6 million or 25.6 million + 29.2 million is 54.8 million, the answer checks.

State. In 2003 29.2 million households had broadband Internet access.

30. Let $m =$ the number of calories in a Big Mac.

Solve: $m + (m + 20) = 1200$

$m = 590$, so a Big Mac has 590 calories and an order of Super-Size fries has $590 + 20$, or 610 calories.

31. ***Familiarize.*** Let $c =$ the average daily calorie requirement for many adults.

Translate.

1560 calories is $\frac{3}{4}$ of the average dailyrequirement for many adults.

$$1560 = \frac{3}{4} \cdot c$$

Carry out. We solve the equation.

$$1560 = \frac{3}{4} \cdot c$$

$$\frac{4}{3} \cdot 1560 = c \quad \text{Multiplying by } \frac{4}{3}$$

$$2080 = c$$

Check. $\frac{3}{4} \cdot 2080 = 1560$, so the answer checks.

State. The average daily calorie requirement for many adults is 2080 calories.

32. Let $h =$ the number of households represented by a Nielsen rating of 9.4.

Solve: $h = 9.4(1,067,000)$

$h = 10,029,800$ households

33. ***Familiarize.*** Let $P =$ the amount Tamisha borrowed. We will use the formula $I = Prt$ to find the interest owed. For $r = 5\%$, or 0.05, and $t = 1$, we have $I = P(0.05)(1)$, or $0.05P$.

Translate.

Amount borrowed plus interest is \$1365.

$$P + 0.05P = 1365$$

Carry out. We solve the equation.

$$P + 0.05P = 1365$$

$$1.05P = 1365 \quad \text{Adding}$$

$$P = 1300 \quad \text{Dividing by 1.05}$$

Check. The interest due on a loan of \$1300 for 1 year at a rate of 5% is \$1300(0.05)(1), or \$65, and \$1300 + \$65 = \$1365. The answer checks.

State. Tamisha borrowed \$1300.

34. Let $P =$ the amount invested.

Solve: $P + 0.04P = \$1560$

$P = \$1500$

35. ***Familiarize.*** Let $s =$ Ryan's sales for the month. Then his commission is 8% of s, or $0.08s$.

Translate.

Base salary plus commission is total pay.

$$1500 + 0.08s = 2284$$

Carry out. We solve the equation.

$$1500 + 0.08s = 2284$$

$$0.08s = 784 \quad \text{Subtracting 1500}$$

$$s = 9800$$

Check. 8% of \$9800, or 0.08(\$9800), is \$784 and \$1500 + \$784 = \$2284. The answer checks.

State. Ryan's sales for the month were \$9800.

36. Let $s =$ the amount of sales for which the two choices will be equal.

Solve: $1800 = 1600 + 0.04s$

$s = \$5000$

37. ***Familiarize.*** Let $d =$ the number of miles Diego traveled in the cab.

Translate.

Pickup fee plus cost per mile times number of miles traveled is \$19.75.

$$1.75 + 1.50 \cdot d = 19.75$$

Carry out. We solve the equation.

$$1.75 + 1.50 \cdot d = 19.75$$
$$1.5d = 18 \quad \text{Subtracting 1.75}$$
$$d = 12 \quad \text{Dividing by 1.5}$$

Check. If Diego travels 12 mi, his fare is \$1.75+\$1.50·12, or \$1.75 + \$18, or \$19.75. The answer checks.

State. Diego traveled 12 mi in the cab.

38. Let w = Soledad's regular hourly wage. She worked 48 − 40, or 8 hr, of overtime.

Solve: $40w + 8(1.5w) = 442$

$w = \$8.50$

39. ***Familiarize.*** We make a drawing.

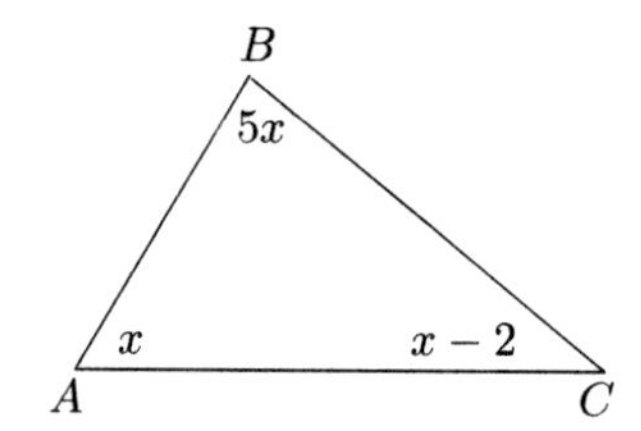

We let x = the measure of angle A. Then $5x$ = the measure of angle B, and $x - 2$ = the measure of angle C. The sum of the angle measures is 180°.

Translate.

Measure of angle A + Measure of angle B + Measure of angle C = 180

$$x + 5x + x - 2 = 180$$

Carry out. We solve the equation.

$$x + 5x + x - 2 = 180$$
$$7x - 2 = 180$$
$$7x = 182$$
$$x = 26$$

If $x = 26$, then $5x = 5 \cdot 26$, or 130, and $x - 2 = 26 - 2$, or 24.

Check. The measure of angle B, 130°, is five times the measure of angle A, 26°. The measure of angle C, 24°, is 2° less than the measure of angle A, 26°. The sum of the angle measures is 26° + 130° + 24°, or 180°. The answer checks.

State. The measure of angles A, B, and C are 26°, 130°, and 24°, respectively.

40. Let x = the measure of angle A.

Solve: $x + 2x + x + 20 = 180$

$x = 40°$, so the measure of angle A is 40°; the measure of angle B is 2 · 40°, or 80°; and the measure of angle C is 40° + 20°, or 60°.

41. ***Familiarize.*** Using the labels on the drawing in the text, we let w = the width of the test plot and $w + 25$ = the length, in meters. Recall that for a rectangle, Perimeter = 2 · length + 2 · width.

Translate.

Perimeter = 2 · length + 2 · width

$$322 = 2(w + 25) + 2 \cdot w$$

Carry out. We solve the equation.

$$322 = 2(w + 25) + 2 \cdot w$$
$$322 = 2w + 50 + 2w$$
$$322 = 4w + 50$$
$$272 = 4w$$
$$68 = w$$

When $w = 68$, then $w + 25 = 68 + 25 = 93$.

Check. The length is 25 m more than the width: $93 = 68 + 25$. The perimeter is $2 \cdot 93 + 2 \cdot 68$, or $186 + 136$, or 322 m. The answer checks.

State. The length is 93 m; the width is 68 m.

42. Let w = the width of the garden.

Solve: $2 \cdot 2w + 2 \cdot w = 39$

$w = 6.5$, so the width is 6.5 m, and the length is 2(6.5), or 13 m.

43. ***Familiarize.*** Let l = the length of the soccer field and $l - 35$ = the width, in yards.

Translate. We use the formula for the perimeter of a rectangle. We substitute 330 for P and $l - 35$ for w.

$$P = 2l + 2w$$
$$330 = 2l + 2(l - 35)$$

Carry out. We solve the equation.

$$330 = 2l + 2(l - 35)$$
$$330 = 2l + 2l - 70$$
$$330 = 4l - 70$$
$$400 = 4l$$
$$100 = l$$

If $l = 100$, then $l - 35 = 100 - 35 = 65$.

Check. The width, 65 yd, is 35 yd less than the length, 100 yd. Also, the perimeter is

$$2 \cdot 100 \text{ yd} + 2 \cdot 65 \text{ yd} = 200 \text{ yd} + 130 \text{ yd} = 330 \text{ yd}.$$

The answer checks.

State. The length of the field is 100 yd, and the width is 65 yd.

44. Let h = the height of the poster and $\frac{2}{3}h$ = the width, in inches.

Solve: $100 = 2 \cdot h + 2 \cdot \frac{2}{3}h$

$h = 30$, so the height is 30 in. and the width is $\frac{2}{3} \cdot 30$, or 20 in.

45. ***Familiarize.*** Let w = the number of pounds of Jocelyn's body weight that is water.

Translate.

50% of body weight is water.

$$0.5 \times 135 = w$$

Carry out. We solve the equation.

$0.5 \times 135 = w$

$67.5 = w$

Check. Since 50% of 138 is 67.5, the answer checks.

State. 67.5 lb of Jocelyn's body weight is water.

46. Let w = the number of pounds of Reggie's body weight that is water.

Solve: $0.6 \times 186 = w$

$w = 111.6$ lb

47. ***Familiarize.*** We make a drawing. Let r = the speed of the Central Railway freight train. Then $r+14$ = the speed of the Amtrak passenger train. Also let t = the time the train travels.

Passenger train
r mph t hr   330 mi

Passenger train
$r + 14$ mph t hr   400 mi

We can also organize the information in a table.

$d = r \cdot t$

	Distance	Rate	Time
Freight train	330	r	t
Passenger train	400	$r+14$	t

Translate. Using the formula $d = rt$ in each row of the table, we get two equations.

$330 = rt$ and $400 = (r+14)t$

Solve each equation for t.

$\frac{330}{r} = t$ and $\frac{400}{r+14} = t$

Thus, we have the equation

$$\frac{330}{r} = \frac{400}{r+14}.$$

Carry out. We solve the equation.

$$\frac{330}{r} = \frac{400}{r+14}, \text{ LCD is } r(r+14)$$

$$r(r+14) \cdot \frac{330}{r} = r(r+14) \cdot \frac{400}{r+14}$$

$$330(r+14) = 400r$$

$$330r + 4620 = 400r$$

$$4620 = 70r$$

$$66 = r$$

When $r = 66$, $r + 14 = 66 + 14 = 80$.

Check. The freight train travels 330 mi in 330/66, or 5 hr and the passenger train travels 400 mi in 400/80, or 5 hr. Since the times are the same the answer checks.

State. The speed of the Central Railway freight train is 66 mph, and the speed of the Amtrak passenger train is 80 mph.

48. Let t = the time the private airplane travels.

	Distance	Rate	Time
Private airplane	d	180	t
Jet	d	900	$t-2$

From the table we have the following equations:

$d = 180t$ and $d = 900(t-2)$

Solve: $180t = 900(t-2)$

$t = 2.5$

In 2.5 hr the private airplane travels 180(2.5), or 450 km. This is the distance from the airport at which it is overtaken by the jet.

49. ***Familiarize.*** Let t = the number of hours it takes the paddleboat to travel 36 mi upstream. The boat travels upstream at a rate of $12 - 4$, or 8 mph.

Translate. We use the formula $d = rt$.

$36 = 8 \cdot t$

Carry out. We solve the equation.

$36 = 8 \cdot t$

$4.5 = t$

Check. At a rate of 8 mph, in 4.5 hr the boat travels 8(4.5), or 36 mi. The answer checks.

State. It takes the paddleboat 4.5 hr to travel 36 mi upstream.

50. Let t = the number of hours it will take the plane to travel 1050 mi into the wind. The speed into the headwind is $450 - 30$, or 420 mph.

Solve: $1050 = 420 \cdot t$

$t = 2.5$ hr

51. ***Familiarize.*** Let t = the number of hours it will take Angelo to travel 20 km downstream. The kayak travels downstream at a rate of $14 + 2$, or 16 km/h.

Translate. We use the formula $d = rt$.

$20 = 16t$

Carry out. We solve the equation.

$20 = 16t$

$1.25 = t$

Check. At a rate of 16 km/h, in 1.25 hr the kayak will travel 16(1.25), or 20 km. The answer checks.

State. It will take Angelo 1.25 hr to travel 20 km downstream.

52. Let t = the number of hours it will take the plane to travel 700 mi with the wind. The speed with the wind is $375+25$, or 400 mph.

Solve: $700 = 400t$

$t = 1.75$ hr

53. ***Familiarize.*** Let x = the amount invested at 3% interest. Then $5000 - x$ = the amount invested at 4%. We organize the information in a table, keeping in mind the simple interest formula, $I = Prt$.

	Amount invested	Interest rate	Time	Amount of interest
3% investment	x	3%, or 0.03	1 yr	$x(0.03)(1)$, or $0.03x$
4% investment	$5000-x$	4%, or 0.04	1 yr	$(5000-x)(0.04)(1)$, or $0.04(5000-x)$
Total	5000			176

Translate.

$$\underbrace{\text{Interest on 3\% investment}}_{0.03x} \ \underbrace{\text{plus}}_{+} \ \underbrace{\text{interest on 4\% investment}}_{0.04(5000-x)} \ \underbrace{\text{is}}_{=} \ \underbrace{\$176.}_{176}$$

Carry out. We solve the equation.

$$0.03x + 0.04(5000 - x) = 176$$
$$0.03x + 200 - 0.04x = 176$$
$$-0.01x + 200 = 176$$
$$-0.01x = -24$$
$$x = 2400$$

If $x = 2400$, then $5000 - x = 5000 - 2400 = 2600$.

Check. The interest on \$2400 at 3% for 1 yr is $\$2400(0.03)(1) = \$72$. The interest on \$2600 at 4% for 1 yr is $\$2600(0.04)(1) = \$104$. Since $\$72 + \$104 = \$176$, the answer checks.

State. \$2400 was invested at 3%, and \$2600 was invested at 4%.

54. Let x = the amount borrowed at 5%. Then $9000 - x$ = the amount invested at 6%.

Solve: $0.05x + 0.06(9000 - x) = 492$

$x = 4800$, so \$4800 was borrowed at 5% and \$9000 \$4800 = \$4200 was borrowed at 6%.

55. ***Familiarize.*** Let w = the number of violations reqported by Division II schools. Then $1989 - v$ = the number of violations reported by Division I schools.

Translate.

$$\underbrace{\text{Number of Division I violations}}_{1989 - v} \ \underbrace{\text{is}}_{=} \ \underbrace{6.5}_{6.5} \ \underbrace{\text{times}}_{\times} \ \underbrace{\text{number of Division II violations.}}_{v}$$

Carry out. We solve the equation.

$$1989 - v = 6.5v$$
$$1989 = 7.5v$$
$$265 \approx v$$

If $v = 265$, then $1989 - v = 1989 - 265 = 1724$.

Check. $6.5(265) = 1722.5 \approx 1724$, so the answer checks.

State. Division I schools reported 1724 violations, and Division II schools reported 265 violations.

56. Let p = the number of working pharmacists in the United States in 1975.

Solve: $224,500 = 1.84p$

$p \approx 122,011$ pharamcists

57. ***Familiarize.*** Let m = the number of MSM users, in millions. Then $m + 23.1$ = the number of AOL users, in millions.

Translate.

$$\underbrace{\text{Number of AOL users}}_{m + 23.1} \ \underbrace{\text{plus}}_{+} \ \underbrace{\text{Number of MSN users}}_{m} \ \underbrace{\text{is}}_{=} \ \underbrace{\text{81.9 million.}}_{81.9}$$

Carry out. We solve the equation.

$$m + 23.1 + m = 81.9$$
$$2m + 23.1 = 81.9$$
$$2m = 58.8$$
$$m = 29.4$$

If $m = 29.4$, then $m + 23.1 = 29.4 + 23.1 = 52.5$.

Check. The number of AOL users, 52.5 million, is 23.1 million more than 29.4 million, the number of MSN users. Also, 52.5 million + 29.4 million = 81.9 million, the total number of users. The answer checks.

State. AOL's Instant Messenger Service had 52.5 million users, and MSN's Messenger Service had 29.4 million users.

58. Let p = the number of license plates sold in Florida.

Solve: $1,017,866 = 0.056p$

$p \approx 18,176,179$ plates

59. ***Familiarize.*** Let x = the number of years it will take Horseshoe Falls to migrate one-fourth mile upstream. We will express one-fourth mile in feet:

$$\frac{1}{4}\text{ mi} \times \frac{5280\text{ ft}}{1\text{ mi}} = 1320\text{ ft}$$

Translate.

$$\underbrace{\text{Rate per year}}_{2} \ \underbrace{\text{times}}_{\cdot} \ \underbrace{\text{number of years}}_{x} \ \underbrace{\text{is}}_{=} \ \underbrace{\text{1320 ft.}}_{1320}$$

Carry out. We solve the equation.

$$2x = 1320$$
$$x = 660$$

Check. At a rate of 2 ft per year, in 660 yr the falls will migrate $2 \cdot 660$, or 1320 ft. The answer checks.

State. It will take Horseshoe Falls 660 yr to migrate one-fourth mile upstream.

60. Express one-half mile in inches:

$$\frac{1}{2}\text{ mi} \times \frac{5280\text{ ft}}{1\text{ mi}} \times \frac{12\text{ in.}}{1\text{ ft}} = 31,680\text{ in.}$$

Let n = the number of inches the volcano rises in a year.

Solve: $50,000n = 31,680$

$n = 0.6336$ in.

61. $x+5=0$ Setting $f(x)=0$

$x+5-5=0-5$ Subtracting 5 on both sides

$x=-5$

The zero of the function is -5.

62. $5x+20=0$

$5x=-20$

$x=-4$

63. $-x+18=0$ Setting $f(x)=0$

$-x+18+x=0+x$ Adding x on both sides

$18=x$

The zero of the function is 18.

64. $8+x=0$

$x=-8$

65. $16-x=0$ Setting $f(x)=0$

$16-x+x=0+x$ Adding x on both sides

$16=x$

The zero of the function is 16.

66. $-2x+7=0$

$-2x=-7$

$x=\frac{7}{2}$, or 3.5

67. $x+12=0$ Setting $f(x)=0$

$x+12-12=0-12$ Subtracting 12 on both sides

$x=-12$

The zero of the function is -12.

68. $8x+2=0$

$8x=-2$

$x=-\frac{1}{4}$, or -0.25

69. $-x+6=0$ Setting $f(x)=0$

$-x+6+x=0+x$ Adding x on both sides

$6=x$

The zero of the function is 6.

70. $4+x=0$

$x=-4$

71. $20-x=0$ Setting $f(x)=0$

$20-x+x=0+x$ Adding x on both sides

$20=x$

The zero of the function is 20.

72. $-3x+13=0$

$-3x=-13$

$x=\frac{13}{3}$, or $4.\overline{3}$

73. $x-6=0$ Setting $f(x)=0$

$x=6$ Adding 6 on both sides

The zero of the function is 6.

74. $3x-9=0$

$3x=9$

$x=3$

75. $-x+15=0$ Setting $f(x)=0$

$15=x$ Adding x on both sides

The zero of the function is 15.

76. $4-x=0$

$4=x$

77. a) The graph crosses the x-axis at $(4,0)$. This is the x-intercept.

b) The zero of the function is the first coordinate of the x-intercept. It is 4.

78. a) $(5,0)$

b) 5

79. a) The graph crosses the x-axis at $(-2,0)$. This is the x-intercept.

b) The zero of the function is the first coordinate of the x-intercept. It is -2.

80. a) $(2,0)$

b) 2

81. a) The graph crosses the x-axis at $(-4,0)$. This is the x-intercept.

b) The zero of the function is the first coordinate of the x-intercept. It is -4.

82. a) $(-2,0)$

b) -2

83. $A=\frac{1}{2}bh$

$2A=bh$ Multiplying by 2 on both sides

$\frac{2A}{h}=b$ Dividing by h on both sides

84. $A=\pi r^2$

$\frac{A}{r^2}=\pi$

85. $P=2l+2w$

$P-2l=2w$ Subtracting $2l$ on both sides

$\frac{P-2l}{2}=w$ Dividing by 2 on both sides

86. $A=P+Prt$

$A-P=Prt$

$\frac{A-P}{Pt}=r$

87. $A = \frac{1}{2}h(b_1 + b_2)$

$2A = h(b_1 + b_2)$ Multiplying by 2 on both sides

$\frac{2A}{b_1 + b_2} = h$ Dividing by $b_1 + b_2$ on both sides

88. $A = \frac{1}{2}h(b_1 + b_2)$

$\frac{2A}{h} = b_1 + b_2$

$\frac{2A}{h} - b_1 = b_2$, or

$\frac{2A - b_1 h}{h} = b_2$

89. $V = \frac{4}{3}\pi r^3$

$3V = 4\pi r^3$ Multiplying by 3 on both sides

$\frac{3V}{4r^3} = \pi$ Dividing by $4r^3$ on both sides

90. $V = \frac{4}{3}\pi r^3$

$\frac{3V}{4\pi} = r^3$

91. $F = \frac{9}{5}C + 32$

$F - 32 = \frac{9}{5}C$ Subtracting 32 on both sides

$\frac{5}{9}(F - 32) = C$ Multiplying by $\frac{5}{9}$ on both sides

92. $Ax + By = C$

$By = C - Ax$

$y = \frac{C - Ax}{B}$

93. $Ax + By = C$

$Ax = C - By$ Subtracting By on both sides

$A = \frac{C - By}{x}$ Dividing by x on both sides

94. $2w + 2h + l = p$

$2w = p - 2h - l$

$w = \frac{p - 2h - l}{2}$

95. $2w + 2h + l = p$

$2h = p - 2w - l$ Subtracting $2w$ and l

$h = \frac{p - 2w - l}{2}$ Dividing by 2

96. $3x + 4y = 12$

$4y = 12 - 3x$

$y = \frac{12 - 3x}{4}$

97. $2x - 3y = 6$

$-3y = 6 - 2x$ Subtracting $2x$

$y = \frac{6 - 2x}{-3}$, or Dividing by -3

$\frac{2x - 6}{3}$

98. $T = \frac{3}{10}(I - 12,000)$

$\frac{10}{3}T = I - 12,000$

$\frac{10}{3}T + 12,000 = I$, or

$\frac{10T + 36,000}{3} = I$

99. $a = b + bcd$

$a = b(1 + cd)$ Factoring

$\frac{a}{1 + cd} = b$ Dividing by $1 + cd$

100. $q = p - np$

$q = p(1 - n)$

$\frac{q}{1 - n} = p$

101. $z = xy - xy^2$

$z = x(y - y^2)$ Factoring

$\frac{z}{y - y^2} = x$ Dividing by $y - y^2$

102. $st = t - 4$

$st - t = -4$

$t(s - 1) = -4$

$t = \frac{-4}{s - 1}$, or $\frac{4}{1 - s}$

103. Left to the student

104. Left to the student

105. The graph of $f(x) = mx + b$, $m \neq 0$, is a straight line that is not horizontal. The graph of such a line intersects the x-axis exactly once. Thus, the function has exactly one zero.

106. If a person wanted to convert several Fahrenheit temperatures to Celsius, it would be useful to solve the formula for C and then use the formula in that form.

107. First find the slope of the given line.

$3x + 4y = 7$

$4y = -3x + 7$

$y = -\frac{3}{4}x + \frac{7}{4}$

The slope is $-\frac{3}{4}$. Now write a slope-intersect equation of the line containing $(-1, 4)$ with slope $-\frac{3}{4}$.

$$y - 4 = -\frac{3}{4}[x - (-1)]$$
$$y - 4 = -\frac{3}{4}(x + 1)$$
$$y - 4 = -\frac{3}{4}x - \frac{3}{4}$$
$$y = -\frac{3}{4}x + \frac{13}{4}$$

108. $m = \frac{4-(-2)}{-5-3} = \frac{6}{-8} = -\frac{3}{4}$

$$y - 4 = -\frac{3}{4}(x - (-5))$$
$$y - 4 = -\frac{3}{4}x - \frac{15}{4}$$
$$y = -\frac{3}{4}x + \frac{1}{4}$$

109. The domain of f is the set of all real numbers as is the domain of g, so the domain of $f + g$ is the set of all real numbers, or $(-\infty, \infty)$.

110. The domain of f is the set of all real numbers as is the domain of g. When $x = -2$, $g(x) = 0$, so the domain of f/g is $(-\infty, -2) \cup (-2, \infty)$.

111. $(f - g)(x) = f(x) - g(x) = 2x - 1 - (3x + 6) =$
$2x - 1 - 3x - 6 = -x - 7$

112. $fg(-1) = f(-1) \cdot g(-1) = [2(-1) - 1][3(-1) + 6] =$
$-3 \cdot 3 = -9$

113. $f(x) = 7 - \frac{3}{2}x = -\frac{3}{2}x + 7$

The function can be written in the form $y = mx + b$, so it is a linear function.

114. $f(x) = \frac{3}{2x} + 5$ cannot be written in the form $f(x) = mx + b$, so it is not a linear function.

115. $f(x) = x^2 + 1$ cannot be written in the form $f(x) = mx + b$, so it is not a linear function.

116. $f(x) = \frac{3}{4}x - (2.4)^2$ is in the form $f(x) = mx + b$, so it is a linear function.

117.
$$2x - \{x - [3x - (6x + 5)]\} = 4x - 1$$
$$2x - \{x - [3x - 6x - 5]\} = 4x - 1$$
$$2x - \{x - [-3x - 5]\} = 4x - 1$$
$$2x - \{x + 3x + 5\} = 4x - 1$$
$$2x - \{4x + 5\} = 4x - 1$$
$$2x - 4x - 5 = 4x - 1$$
$$-2x - 5 = 4x - 1$$
$$-6x - 5 = -1$$
$$-6x = 4$$
$$x = -\frac{2}{3}$$

The solution is $-\frac{2}{3}$.

118.
$$14 - 2[3 + 5(x - 1)] = 3\{x - 4[1 + 6(2 - x)]\}$$
$$14 - 2[3 + 5x - 5] = 3\{x - 4[1 + 12 - 6x]\}$$
$$14 - 2[5x - 2] = 3\{x - 4[13 - 6x]\}$$
$$14 - 10x + 4 = 3\{x - 52 + 24x\}$$
$$18 - 10x = 3\{25x - 52\}$$
$$18 - 10x = 75x - 156$$
$$174 = 85x$$
$$\frac{174}{85} = x$$

119. The size of the cup was reduced 8 oz − 6 oz, or 2 oz, and $\frac{2 \text{ oz}}{8 \text{ oz}} = 0.25$, so the size was reduced 25%. The price per ounce of the 8 oz cup was $\frac{89¢}{8 \text{ oz}}$, or 11.25¢/oz. The price per ounce of the 6 oz cup is $\frac{71¢}{6 \text{ oz}}$, or $11.8\overline{3}$¢/oz. Since the price per ounce was not reduced, it is clear that the price per ounce was not reduced by the same percent as the size of the cup. The price was increased by $11.8\overline{3} - 11.125$¢, or $0.708\overline{3}$¢ per ounce. This is an increase of $\frac{0.708\overline{3}¢}{11.8\overline{3}¢} \approx 0.064$, or about 6.4% per ounce.

120. The size of the container was reduced 100 oz − 80 oz, or 20 oz, and $\frac{20 \text{ oz}}{100 \text{ oz}} = 0.2$, so the size of the container was reduced 20%. The price per ounce of the 100-oz container was $\frac{\$6.99}{100 \text{ oz}}$, or \$0.0699/oz. The price per ounce of the 80-oz container is $\frac{\$5.75}{80 \text{ oz}}$, or \$0.071875. Since the price per ounce was not reduced, it is clear that the price per ounce was not reduced by the same percent as the size of the container. The price increased by \$0.071875 − \$0.0699, or \$0.001975. This is an increase of $\frac{\$0.001975}{\$0.0699} \approx 0.028$, or about 2.8% per ounce.

121. We use a proportion to determine the number of calories c burned running for 75 minutes, or 1.25 hr.

$$\frac{720}{1} = \frac{c}{1.25}$$
$$720(1.25) = c$$
$$900 = c$$

Next we use a proportion to determine how long the person would have to walk to use 900 calories. Let t represent this time, in hours. We express 90 min as 1.5 hr.

$$\frac{1.5}{480} = \frac{t}{900}$$
$$\frac{900(1.5)}{480} = t$$
$$2.8125 = t$$

Then, at a rate of 4 mph, the person would have to walk 4(2.8125), or 11.25 mi.

122. Let x = the number of copies of *Dr. Atkins' New Diet Revolution* that were sold. Then $4075 - x$ = the number of copies of *The King of Torts* that were sold.

Solve: $\frac{x}{4075 - x} = \frac{10}{6.3}$

$x = 2500$, so 2500 copies of *Dr. Atkins' New Diet Revolution* were sold and $4075 - x = 4075 - 2500 = 1575$ copies of *The King of Torts* were sold.

Exercise Set 2.2

1.
$$\begin{aligned}&(-5+3i)+(7+8i)\\&=(-5+7)+(3i+8i) \quad \text{Collecting the real parts and the imaginary parts}\\&=2+(3+8)i\\&=2+11i\end{aligned}$$

2. $(-6-5i)+(9+2i)=(-6+9)+(-5i+2i)=3-3i$

3.
$$\begin{aligned}&(4-9i)+(1-3i)\\&=(4+1)+(-9i-3i) \quad \text{Collecting the real parts and the imaginary parts}\\&=5+(-9-3)i\\&=5-12i\end{aligned}$$

4. $(7-2i)+(4-5i)=(7+4)+(-2i-5i)=11-7i$

5.
$$\begin{aligned}&(12+3i)+(-8+5i)\\&=(12-8)+(3i+5i)\\&=4+8i\end{aligned}$$

6. $(-11+4i)+(6+8i)=(-11+6)+(4i+8i)=-5+12i$

7.
$$\begin{aligned}&(-1-i)+(-3-i)\\&=(-1-3)+(-i-i)\\&=-4-2i\end{aligned}$$

8. $(-5-i)+(6+2i)=(-5+6)+(-i+2i)=1+i$

9.
$$\begin{aligned}(3+\sqrt{-16})+(2+\sqrt{-25})&=(3+4i)+(2+5i)\\&=(3+2)+(4i+5i)\\&=5+9i\end{aligned}$$

10. $(7-\sqrt{-36})+(2+\sqrt{-9})=(7-6i)+(2+3i)=(7+2)+(-6i+3i)=9-3i$

11.
$$\begin{aligned}&(10+7i)-(5+3i)\\&=(10-5)+(7i-3i) \quad \text{The 5 and the } 3i \text{ are both being subtracted.}\\&=5+4i\end{aligned}$$

12. $(-3-4i)-(8-i)=(-3-8)+[-4i-(-i)]=-11-3i$

13.
$$\begin{aligned}&(13+9i)-(8+2i)\\&=(13-8)+(9i-2i) \quad \text{The 8 and the } 2i \text{ are both being subtracted.}\\&=5+7i\end{aligned}$$

14. $(-7+12i)-(3-6i)=(-7-3)+[12i-(-6i)]=-10+18i$

15.
$$\begin{aligned}&(6-4i)-(-5+i)\\&=[6-(-5)]+(-4i-i)\\&=(6+5)+(-4i-i)\\&=11-5i\end{aligned}$$

16. $(8-3i)-(9-i)=(8-9)+[-3i-(-i)]=-1-2i$

17.
$$\begin{aligned}&(-5+2i)-(-4-3i)\\&=[-5-(-4)]+[2i-(-3i)]\\&=(-5+4)+(2i+3i)\\&=-1+5i\end{aligned}$$

18. $(-6+7i)-(-5-2i)=[-6-(-5)]+[7i-(-2i)]=-1+9i$

19.
$$\begin{aligned}&(4-9i)-(2+3i)\\&=(4-2)+(-9i-3i)\\&=2-12i\end{aligned}$$

20. $(10-4i)-(8+2i)=(10-8)+(-4i-2i)=2-6i$

21.
$$\begin{aligned}&7i(2-5i)\\&=14i-35i^2 \quad \text{Using the distributive law}\\&=14i+35 \quad i^2=-1\\&=35+14i \quad \text{Writing in the form } a+bi\end{aligned}$$

22. $3i(6+4i)=18i+12i^2=18i-12=-12+18i$

23.
$$\begin{aligned}&-2i(-8+3i)\\&=16i-6i^2 \quad \text{Using the distributive law}\\&=16i+6 \quad i^2=-1\\&=6+16i \quad \text{Writing in the form } a+bi\end{aligned}$$

24. $-6i(-5+i)=30i-6i^2=30i+6=6+30i$

25.
$$\begin{aligned}&(1+3i)(1-4i)\\&=1-4i+3i-12i^2 \quad \text{Using FOIL}\\&=1-4i+3i-12(-1) \quad i^2=-1\\&=1-i+12\\&=13-i\end{aligned}$$

26. $(1-2i)(1+3i)=1+3i-2i-6i^2=1+i+6=7+i$

27.
$$\begin{aligned}&(2+3i)(2+5i)\\&=4+10i+6i+15i^2 \quad \text{Using FOIL}\\&=4+10i+6i-15 \quad i^2=-1\\&=-11+16i\end{aligned}$$

28. $(3-5i)(8-2i)=24-6i-40i+10i^2=24-6i-40i-10=14-46i$

29.
$$\begin{aligned}&(-4+i)(3-2i)\\&=-12+8i+3i-2i^2 \quad \text{Using FOIL}\\&=-12+8i+3i+2 \quad i^2=-1\\&=-10+11i\end{aligned}$$

30. $(5-2i)(-1+i) = -5+5i+2i-2i^2 =$
$-5+5i+2i+2 = -3+7i$

31. $(8-3i)(-2-5i)$
$= -16-40i+6i+15i^2$
$= -16-40i+6i-15 \qquad i^2=-1$
$= -31-34i$

32. $(7-4i)(-3-3i) = -21-21i+12i+12i^2 =$
$-21-21i+12i-12 = -33-9i$

33. $(3+\sqrt{-16})(2+\sqrt{-25})$
$= (3+4i)(2+5i)$
$= 6+15i+8i+20i^2$
$= 6+15i+8i-20 \qquad i^2=-1$
$= -14+23i$

34. $(7-\sqrt{-16})(2+\sqrt{-9}) = (7-4i)(2+3i) =$
$14+21i-8i-12i^2 = 14+21i-8i+12 =$
$26+13i$

35. $(5-4i)(5+4i) = 5^2-(4i)^2$
$= 25-16i^2$
$= 25+16 \qquad i^2=-1$
$= 41$

36. $(5+9i)(5-9i) = 25-81i^2 = 25+81 = 106$

37. $(3+2i)(3-2i)$
$= 9-6i+6i-4i^2$
$= 9-6i+6i+4 \qquad i^2=-1$
$= 13$

38. $(8+i)(8-i) = 64-8i+8i-i^2 =$
$64-8i+8i+1 = 65$

39. $(7-5i)(7+5i)$
$= 49+35i-35i-25i^2$
$= 49+35i-35i+25 \qquad i^2=-1$
$= 74$

40. $(6-8i)(6+8i) = 36+48i-48i-64i^2 =$
$36+48i-48i+64 = 100$

41. $(4+2i)^2$
$= 16+2\cdot 4\cdot 2i+(2i)^2$ Recall $(A+B)^2 = A^2+2AB+B^2$
$= 16+16i+4i^2$
$= 16+16i-4 \qquad i^2=-1$
$= 12+16i$

42. $(5-4i)^2 = 25-40i+16i^2 = 25-40i-16 =$
$9-40i$

43. $(-2+7i)^2$
$= (-2)^2+2(-2)(7i)+(7i)^2$ Recall $(A+B)^2 = A^2+2AB+B^2$
$= 4-28i+49i^2$
$= 4-28i-49 \qquad i^2=-1$
$= -45-28i$

44. $(-3+2i)^2 = 9-12i+4i^2 = 9-12i-4 = 5-12i$

45. $(1-3i)^2$
$= 1^2-2\cdot 1\cdot(3i)+(3i)^2$
$= 1-6i+9i^2$
$= 1-6i-9 \qquad i^2=-1$
$= -8-6i$

46. $(2-5i)^2 = 4-20i+25i^2 = 4-20i-25 =$
$-21-20i$

47. $(-1-i)^2$
$= (-1)^2-2(-1)(i)+i^2$
$= 1+2i+i^2$
$= 1+2i-1 \qquad i^2=-1$
$= 2i$

48. $(-4-2i)^2 = 16+16i+4i^2 = 16+16i-4 =$
$12+16i$

49. $(3+4i)^2$
$= 9+2\cdot 3\cdot 4i+(4i)^2$
$= 9+24i+16i^2$
$= 9+24i-16 \qquad i^2=-1$
$= -7+24i$

50. $(6+5i)^2 = 36+60i+25i^2 = 36+60i-25 =$
$11+60i$

51. $\frac{3}{5-11i}$
$= \frac{3}{5-11i}\cdot\frac{5+11i}{5+11i}$ $5-11i$ is the conjugate of $5+11i$.
$= \frac{3(5+11i)}{(5-11i)(5+11i)}$
$= \frac{15+33i}{25-121i^2}$
$= \frac{15+33i}{25+121} \qquad i^2=-1$
$= \frac{15+33i}{146}$
$= \frac{15}{146}+\frac{33}{146}i$ Writing in the form $a+bi$

52. $\dfrac{i}{2+i} = \dfrac{i}{2+i} \cdot \dfrac{2-i}{2-i}$

$= \dfrac{2i - i^2}{4 - i^2}$

$= \dfrac{2i+1}{4+1}$

$= \dfrac{1}{5} + \dfrac{2}{5}i$

53. $\dfrac{5}{2+3i}$

$= \dfrac{5}{2+3i} \cdot \dfrac{2-3i}{2-3i}$ $2-3i$ is the conjugate of $2+3i$.

$= \dfrac{5(2-3i)}{(2+3i)(2-3i)}$

$= \dfrac{10-15i}{4-9i^2}$

$= \dfrac{10-15i}{4+9}$ $i^2 = -1$

$= \dfrac{10-15i}{13}$

$= \dfrac{10}{13} - \dfrac{15}{13}i$ Writing in the form $a+bi$

54. $\dfrac{-3}{4-5i} = \dfrac{-3}{4-5i} \cdot \dfrac{4+5i}{4+5i}$

$= \dfrac{-12-15i}{16-25i^2}$

$= \dfrac{-12-15i}{16+25}$

$= -\dfrac{12}{41} - \dfrac{15}{41}i$

55. $\dfrac{4+i}{-3-2i}$

$= \dfrac{4+i}{-3-2i} \cdot \dfrac{-3+2i}{-3+2i}$ $-3+2i$ is the conjugate of the divisor.

$= \dfrac{(4+i)(-3+2i)}{(-3-2i)(-3+2i)}$

$= \dfrac{-12+5i+2i^2}{9-4i^2}$

$= \dfrac{-12+5i-2}{9+4}$ $i^2 = -1$

$= \dfrac{-14+5i}{13}$

$= -\dfrac{14}{13} + \dfrac{5}{13}i$ Writing in the form $a+bi$

56. $\dfrac{5-i}{-7+2i} = \dfrac{5-i}{-7+2i} \cdot \dfrac{-7-2i}{-7-2i}$

$= \dfrac{-35-3i+2i^2}{49-4i^2}$

$= \dfrac{-35-3i-2}{49+4}$

$= -\dfrac{37}{53} - \dfrac{3}{53}i$

57. $\dfrac{5-3i}{4+3i}$

$= \dfrac{5-3i}{4+3i} \cdot \dfrac{4-3i}{4-3i}$ $4-3i$ is the conjugate of $4+3i$.

$= \dfrac{(5-3i)(4-3i)}{(4+3i)(4-3i)}$

$= \dfrac{20-27i+9i^2}{16-9i^2}$

$= \dfrac{20-27i-9}{16+9}$ $i^2 = -1$

$= \dfrac{11-27i}{25}$

$= \dfrac{11}{25} - \dfrac{27}{25}i$ Writing in the form $a+bi$

58. $\dfrac{6+5i}{3-4i} = \dfrac{6+5i}{3-4i} \cdot \dfrac{3+4i}{3+4i}$

$= \dfrac{18+39i+20i^2}{9-16i^2}$

$= \dfrac{18+39i-20}{9+16}$

$= -\dfrac{2}{25} + \dfrac{39}{25}i$

59. $\dfrac{2+\sqrt{3}i}{5-4i}$

$= \dfrac{2+\sqrt{3}i}{5-4i} \cdot \dfrac{5+4i}{5+4i}$ $5+4i$ is the conjugate of the divisor.

$= \dfrac{(2+\sqrt{3}i)(5+4i)}{(5-4i)(5+4i)}$

$= \dfrac{10+8i+5\sqrt{3}i+4\sqrt{3}i^2}{25-16i^2}$

$= \dfrac{10+8i+5\sqrt{3}i-4\sqrt{3}}{25+16}$ $i^2 = -1$

$= \dfrac{10-4\sqrt{3}+(8+5\sqrt{3})i}{41}$

$= \dfrac{10-4\sqrt{3}}{41} + \dfrac{8+5\sqrt{3}}{41}i$ Writing in the form $a+bi$

60. $\dfrac{\sqrt{5}+3i}{1-i} = \dfrac{\sqrt{5}+3i}{1-i} \cdot \dfrac{1+i}{1+i}$

$= \dfrac{\sqrt{5}+\sqrt{5}i+3i+3i^2}{1-i^2}$

$= \dfrac{\sqrt{5}+\sqrt{5}i+3i-3}{1+1}$

$= \dfrac{\sqrt{5}-3}{2} + \dfrac{\sqrt{5}+3}{2}i$

61. $\dfrac{1+i}{(1-i)^2}$

$= \dfrac{1+i}{1-2i+i^2}$

$= \dfrac{1+i}{1-2i-1}$ $\quad i^2=-1$

$= \dfrac{1+i}{-2i}$

$= \dfrac{1+i}{-2i}\cdot\dfrac{2i}{2i}$ $\quad 2i$ is the conjugate of $-2i$.

$= \dfrac{(1+i)(2i)}{(-2i)(2i)}$

$= \dfrac{2i+2i^2}{-4i^2}$

$= \dfrac{2i-2}{4}$ $\quad i^2=-1$

$= -\dfrac{2}{4}+\dfrac{2}{4}i$

$= -\dfrac{1}{2}+\dfrac{1}{2}i$

62. $\dfrac{1-i}{(1+i)^2} = \dfrac{1-i}{1+2i+i^2}$

$= \dfrac{1-i}{1+2i-1}$

$= \dfrac{1-i}{2i}$

$= \dfrac{1-i}{2i}\cdot\dfrac{-2i}{-2i}$

$= \dfrac{-2i+2i^2}{-4i^2}$

$= \dfrac{-2i-2}{4}$

$= -\dfrac{1}{2}-\dfrac{1}{2}i$

63. $\dfrac{4-2i}{1+i}+\dfrac{2-5i}{1+i}$

$= \dfrac{6-7i}{1+i}$ $\quad$ Adding

$= \dfrac{6-7i}{1+i}\cdot\dfrac{1-i}{1-i}$ $\quad 1-i$ is the conjugate of $1+i$.

$= \dfrac{(6-7i)(1-i)}{(1+i)(1-i)}$

$= \dfrac{6-13i+7i^2}{1-i^2}$

$= \dfrac{6-13i-7}{1+1}$ $\quad i^2=-1$

$= \dfrac{-1-13i}{2}$

$= -\dfrac{1}{2}-\dfrac{13}{2}i$

64. $\dfrac{3+2i}{1-i}+\dfrac{6+2i}{1-i} = \dfrac{9+4i}{1-i}$

$= \dfrac{9+4i}{1-i}\cdot\dfrac{1+i}{1+i}$

$= \dfrac{9+13i+4i^2}{1-i^2}$

$= \dfrac{9+13i-4}{1+1}$

$= \dfrac{5}{2}+\dfrac{13}{2}i$

65. $i^{11} = i^{10}\cdot i = (i^2)^5\cdot i = (-1)^5\cdot i = -1\cdot i = -i$

66. $i^7 = i^6\cdot i = (i^2)^3\cdot i = (-1)^3\cdot i = -1\cdot i = -i$

67. $i^{35} = i^{34}\cdot i = (i^2)^{17}\cdot i = (-1)^{17}\cdot i = -1\cdot i = -i$

68. $i^{24} = (i^2)^{12} = (-1)^{12} = 1$

69. $i^{64} = (i^2)^{32} = (-1)^{32} = 1$

70. $i^{42} = (i^2)^{21} = (-1)^{21} = -1$

71. $(-i)^{71} = (-1\cdot i)^{71} = (-1)^{71}\cdot i^{71} = -i^{70}\cdot i = -(i^2)^{35}\cdot i = -(-1)^{35}\cdot i = -(-1)i = i$

72. $(-i)^6 = i^6 = (i^2)^3 = (-1)^3 = -1$

73. $(5i)^4 = 5^4\cdot i^4 = 625(i^2)^2 = 625(-1)^2 = 625\cdot 1 = 625$

74. $(2i)^5 = 32i^5 = 32\cdot i^4\cdot i = 32(i^2)^2\cdot i = 32(-1)^2\cdot i = 32\cdot 1\cdot i = 32i$

75. Left to the student

76. Left to the student

77. The sum of two imaginary numbers is not always an imaginary number. For example, $(2+i)+(3-i)=5$, a real number.

78. The product of two imaginary numbers is not always an imaginary number. For example, $i\cdot i = i^2 = -1$, a real number.

79. First find the slope of the given line.

$$3x-6y=7$$
$$-6y=-3x+7$$
$$y=\frac{1}{2}x-\frac{7}{6}$$

The slope is $\frac{1}{2}$. The slope of the desired line is the opposite of the reciprocal of $\frac{1}{2}$, or -2. Write a slope-intercept equation of the line containing $(3,-5)$ with slope -2.

$$y-(-5)=-2(x-3)$$
$$y+5=-2x+6$$
$$y=-2x+1$$

80. The domain of f is the set of all real numbers as is the domain of g. Then the domain of $(f-g)(x)$ is the set of all real numbers, or $(-\infty,\infty)$.

81. The domain of f is the set of all real numbers as is the domain of g. When $x = -\frac{5}{3}$, $g(x) = 0$, so the domain of f/g is $\left(-\infty, -\frac{5}{3}\right) \cup \left(-\frac{5}{3}, \infty\right)$.

82. $(f-g)(x) = f(x) - g(x) = x^2 + 4 - (3x+5) = x^2 - 3x - 1$

83. $(f/g)(2) = \frac{f(2)}{g(2)} = \frac{2^2+4}{3\cdot 2+5} = \frac{4+4}{6+5} = \frac{8}{11}$

84.
$$\frac{f(x+h) - f(x)}{h}$$
$$= \frac{(x+h)^2 - 3(x+h) + 4 - (x^2 - 3x + 4)}{h}$$
$$= \frac{x^2 + 2xh + h^2 - 3x - 3h + 4 - x^2 + 3x - 4}{h}$$
$$= \frac{2xh + h^2 - 3h}{h}$$
$$= \frac{h(2x + h - 3)}{h}$$
$$= 2x + h - 3$$

85. $(a+bi)+(a-bi) = 2a$, a real number. Thus, the statement is true.

86. $(a+bi)+(c+di) = (a+c)+(b+d)i$. The conjugate of this sum is $(a+c)-(b+d)i = a+c-bi-di = (a-bi)+(c-di)$, the sum of the conjugates of the individual complex numbers. Thus, the statement is true.

87. $(a+bi)(c+di) = (ac-bd) + (ad+bc)i$. The conjugate of the product is $(ac-bd) - (ad+bc)i =$ $(a-bi)(c-di)$, the product of the conjugates of the individual complex numbers. Thus, the statement is true.

88. $\frac{1}{z} = \frac{1}{a+bi} \cdot \frac{a-bi}{a-bi} = \frac{a}{a^2+b^2} + \frac{-b}{a^2+b^2}i$

89. $z\bar{z} = (a+bi)(a-bi) = a^2 - b^2i^2 = a^2 + b^2$

90.
$$z + 6\bar{z} = 7$$
$$a + bi + 6(a - bi) = 7$$
$$a + bi + 6a - 6bi = 7$$
$$7a - 5bi = 7$$

Then $7a = 7$, so $a = 1$, and $-5b = 0$, so $b = 0$. Thus, $z = 1$.

Exercise Set 2.3

1. $(2x-3)(3x-2) = 0$

$2x - 3 = 0$ *or* $3x - 2 = 0$ Using the principle of zero products

$2x = 3$ *or* $3x = 2$

$x = \frac{3}{2}$ *or* $x = \frac{2}{3}$

The solutions are $\frac{3}{2}$ and $\frac{2}{3}$.

2. $(5x-2)(2x+3) = 0$

$x = \frac{2}{5}$ *or* $x = -\frac{3}{2}$

The solutions are $\frac{2}{5}$ and $-\frac{3}{2}$.

3. $x^2 - 8x - 20 = 0$

$(x-10)(x+2) = 0$ Factoring

$x - 10 = 0$ *or* $x + 2 = 0$ Using the principle of zero products

$x = 10$ *or* $x = -2$

The solutions are 10 and -2.

4. $x^2 + 6x + 8 = 0$

$(x+2)(x+4) = 0$

$x = -2$ *or* $x = -4$

The solutions are -2 and -4.

5. $3x^2 + x - 2 = 0$

$(3x-2)(x+1) = 0$ Factoring

$3x - 2 = 0$ *or* $x + 1 = 0$ Using the principle of zero products

$x = \frac{2}{3}$ *or* $x = -1$

The solutions are $\frac{2}{3}$ and -1.

6. $10x^2 - 16x + 6 = 0$

$2(5x-3)(x-1) = 0$

$x = \frac{3}{5}$ *or* $x = 1$

The solutions are $\frac{3}{5}$ and 1.

7. $4x^2 - 12 = 0$

$4x^2 = 12$

$x^2 = 3$

$x = \sqrt{3}$ *or* $x = -\sqrt{3}$ Using the principle of square roots

The solutions are $\sqrt{3}$ and $-\sqrt{3}$.

8. $6x^2 = 36$

$x^2 = 6$

$x = \sqrt{6}$ *or* $x = -\sqrt{6}$

The solutions are $\sqrt{6}$ and $-\sqrt{6}$.

9. $3x^2 = 21$

$x^2 = 7$

$x = \sqrt{7}$ *or* $x = -\sqrt{7}$ Using the principle of square roots

The solutions are $\sqrt{7}$ and $-\sqrt{7}$.

10. $2x^2 - 10 = 0$

$2x^2 = 10$

$x^2 = 5$

$x = \sqrt{5}$ *or* $x = -\sqrt{5}$

The solutions are $\sqrt{5}$ and $-\sqrt{5}$.

11. $5x^2 + 10 = 0$

$$5x^2 = -10$$
$$x^2 = -2$$
$$x = \sqrt{2}i \text{ or } x = -\sqrt{2}i$$

The solutions are $\sqrt{2}i$ and $-\sqrt{2}i$.

12. $4x^2 + 12 = 0$

$$4x^2 = -12$$
$$x^2 = -3$$
$$x = \sqrt{3}i \text{ or } x = -\sqrt{3}i$$

The solutions are $\sqrt{3}i$ and $-\sqrt{3}i$.

13. $2x^2 - 34 = 0$

$$2x^2 = 34$$
$$x^2 = 17$$
$$x = \sqrt{17} \text{ or } x = -\sqrt{17}$$

The solutions are $\sqrt{17}$ and $-\sqrt{17}$.

14. $3x^3 = 33$

$$x^2 = 11$$
$$x = \sqrt{11} \text{ or } x = -\sqrt{11}$$

The solutions are $\sqrt{11}$ and $-\sqrt{11}$.

15. $2x^2 = 6x$

$$2x^2 - 6x = 0 \quad \text{Subtracting } 6x \text{ on both sides}$$
$$2x(x - 3) = 0$$
$$2x = 0 \text{ or } x - 3 = 0$$
$$x = 0 \text{ or } x = 3$$

The solutions are 0 and 3.

16. $18x + 9x^2 = 0$

$$9x(2 + x) = 0$$
$$x = 0 \text{ or } x = -2$$

The solutions are -2 and 0.

17. $3y^3 - 5y^2 - 2y = 0$

$$y(3y^2 - 5y - 2) = 0$$
$$y(3y + 1)(y - 2) = 0$$
$$y = 0 \text{ or } 3y + 1 = 0 \text{ or } y - 2 = 0$$
$$y = 0 \text{ or } y = -\frac{1}{3} \text{ or } y = 2$$

The solutions are $-\frac{1}{3}$, 0 and 2.

18. $3t^3 + 2t = 5t^2$

$$3t^3 - 5t^2 + 2t = 0$$
$$t(t - 1)(3t - 2) = 0$$
$$t = 0 \text{ or } t = 1 \text{ or } t = \frac{2}{3}$$

The solutions are 0, $\frac{2}{3}$, and 1.

19. $7x^3 + x^2 - 7x - 1 = 0$

$$x^2(7x + 1) - (7x + 1) = 0$$
$$(x^2 - 1)(7x + 1) = 0$$
$$(x + 1)(x - 1)(7x + 1) = 0$$
$$x + 1 = 0 \text{ or } x - 1 = 0 \text{ or } 7x + 1 = 0$$
$$x = -1 \text{ or } x = 1 \text{ or } x = -\frac{1}{7}$$

The solutions are -1, $-\frac{1}{7}$, and 1.

20. $3x^3 + x^2 - 12x - 4 = 0$

$$x^2(3x + 1) - 4(3x + 1) = 0$$
$$(3x + 1)(x^2 - 4) = 0$$
$$(3x + 1)(x + 2)(x - 2) = 0$$
$$x = -\frac{1}{3} \text{ or } x = -2 \text{ or } x = 2$$

The solutions are -2, $-\frac{1}{3}$, and 2.

21. a) The graph crosses the x-axis at $(-4, 0)$ and at $(2, 0)$. These are the x-intercepts.

b) The zeros of the function are the first coordinates of the x-intercepts of the graph. They are -4 and 2.

22. a) $(-5, 0)$, $(3, 0)$

b) -5, 3

23. a) The graph crosses the x-axis at $(-1, 0)$ and at $(3, 0)$. These are the x-intercepts.

b) The zeros of the function are the first coordinates of the x-intercepts of the graph. They are -1 and 3.

24. a) $(-3, 0)$, $(1, 0)$

b) -3, 1

25. a) The graph crosses the x-axis at $(-2, 0)$ and at $(2, 0)$. These are the x-intercepts.

b) The zeros of the function are the first coordinates of the x-intercepts of the graph. They are -2 and 2.

26. a) $(-1, 0)$, $(1, 0)$

b) -1, 1

27. $x^2 + 6x = 7$

$$x^2 + 6x + 9 = 7 + 9 \quad \text{Completing the square: } \tfrac{1}{2} \cdot 6 = 3 \text{ and } 3^2 = 9$$
$$(x + 3)^2 = 16 \quad \text{Factoring}$$
$$x + 3 = \pm 4 \quad \text{Using the principle of square roots}$$
$$x = -3 \pm 4$$
$$x = -3 - 4 \text{ or } x = -3 + 4$$
$$x = -7 \text{ or } x = 1$$

The solutions are -7 and 1.

28.
$$x^2+8x=-15$$
$$x^2+8x+16=-15+16 \quad (\tfrac{1}{2}\cdot 8=4 \text{ and } 4^2=16)$$
$$(x+4)^2=1$$
$$x+4=\pm 1$$
$$x=-4\pm 1$$
$$x=-4-1 \quad or \quad x=-4+1$$
$$x=-5 \quad or \quad x=-3$$
The solutions are -5 and -3.

29.
$$x^2=8x-9$$
$$x^2-8x=-9 \quad \text{Subtracting } 8x$$
$$x^2-8x+16=-9+16 \quad \text{Completing the square: } \tfrac{1}{2}(-8)=-4 \text{ and } (-4)^2=16$$
$$(x-4)^2=7 \quad \text{Factoring}$$
$$x-4=\pm\sqrt{7} \quad \text{Using the principle of square roots}$$
$$x=4\pm\sqrt{7}$$
The solutions are $4-\sqrt{7}$ and $4+\sqrt{7}$, or $4\pm\sqrt{7}$.

30.
$$x^2=22+10x$$
$$x^2-10x=22$$
$$x^2-10x+25=22+25 \quad (\tfrac{1}{2}(-10)=-5 \text{ and } (-5)^2=25)$$
$$(x-5)^2=47$$
$$x-5=\pm\sqrt{47}$$
$$x=5\pm\sqrt{47}$$
The solutions are $5-\sqrt{47}$ and $5+\sqrt{47}$, or $5\pm\sqrt{47}$.

31. $x^2+8x+25=0$
$$x^2+8x=-25 \quad \text{Subtracting 25}$$
$$x^2+8x+16=-25+16 \quad \text{Completing the square: } \tfrac{1}{2}\cdot 8=4 \text{ and } 4^2=16$$
$$(x+4)^2=-9 \quad \text{Factoring}$$
$$x+4=\pm 3i \quad \text{Using the principle of square roots}$$
$$x=-4\pm 3i$$
The solutions are $-4-3i$ and $-4+3i$, or $-4\pm 3i$.

32. $x^2+6x+13=0$
$$x^2+6x=-13$$
$$x^2+6x+9=-13+9 \quad (\tfrac{1}{2}\cdot 6=3 \text{ and } 3^2=9)$$
$$(x+3)^2=-4$$
$$x+3=\pm 2i$$
$$x=-3\pm 2i$$
The solution are $-3-2i$ and $-3+2i$, or $-3\pm 2i$.

33. $3x^2+5x-2=0$
$$3x^2+5x=2 \quad \text{Adding 2}$$
$$x^2+\frac{5}{3}x=\frac{2}{3} \quad \text{Dividing by 3}$$
$$x^2+\frac{5}{3}x+\frac{25}{36}=\frac{2}{3}+\frac{25}{36} \quad \text{Completing the square: } \tfrac{1}{2}\cdot\tfrac{5}{3}=\tfrac{5}{6} \text{ and } (\tfrac{5}{6})^2=\tfrac{25}{36}$$
$$\left(x+\frac{5}{6}\right)^2=\frac{49}{36} \quad \text{Factoring and simplifying}$$
$$x+\frac{5}{6}=\pm\frac{7}{6} \quad \text{Using the principle of square roots}$$
$$x=-\frac{5}{6}\pm\frac{7}{6}$$
$$x=-\frac{5}{6}-\frac{7}{6} \quad or \quad x=-\frac{5}{6}+\frac{7}{6}$$
$$x=-\frac{12}{6} \quad or \quad x=\frac{2}{6}$$
$$x=-2 \quad or \quad x=\frac{1}{3}$$
The solutions are -2 and $\frac{1}{3}$.

34. $2x^2-5x-3=0$
$$2x^2-5x=3$$
$$x^2-\frac{5}{2}x=\frac{3}{2}$$
$$x^2-\frac{5}{2}x+\frac{25}{16}=\frac{3}{2}+\frac{25}{16} \quad (\tfrac{1}{2}(-\tfrac{5}{2})=-\tfrac{5}{4} \text{ and } (-\tfrac{5}{4})^2=\tfrac{25}{16})$$
$$\left(x-\frac{5}{4}\right)^2=\frac{49}{16}$$
$$x-\frac{5}{4}=\pm\frac{7}{4}$$
$$x=\frac{5}{4}\pm\frac{7}{4}$$
$$x=\frac{5}{4}-\frac{7}{4} \quad or \quad x=\frac{5}{4}+\frac{7}{4}$$
$$x=-\frac{1}{2} \quad or \quad x=3$$
The solutions are $-\frac{1}{2}$ and 3.

35.
$$x^2-2x=15$$
$$x^2-2x-15=0$$
$$(x-5)(x+3)=0 \quad \text{Factoring}$$
$$x-5=0 \quad or \quad x+3=0$$
$$x=5 \quad or \quad x=-3$$
The solutions are 5 and -3.

36.
$$x^2+4x=5$$
$$x^2+4x-5=0$$
$$(x+5)(x-1)=0$$

$x+5=0 \quad or \quad x-1=0$

$x=-5 \quad or \quad x=1$

The solutions are -5 and 1.

37. $5m^2+3m=2$

$5m^2+3m-2=0$

$(5m-2)(m+1)=0$ Factoring

$5m-2=0 \quad or \quad m+1=0$

$m=\frac{2}{5} \quad or \quad m=-1$

The solutions are $\frac{2}{5}$ and -1.

38. $2y^2-3y-2=0$

$(2y+1)(y-2)=0$

$2y+1=0 \quad or \quad y-2=0$

$y=-\frac{1}{2} \quad or \quad y=2$

The solutions are $-\frac{1}{2}$ and 2.

39. $3x^2+6=10x$

$3x^2-10x+6=0$

We use the quadratic formula. Here $a=3$, $b=-10$, and $c=6$.

$$x=\frac{-b\pm\sqrt{b^2-4ac}}{2a}$$

$$=\frac{-(-10)\pm\sqrt{(-10)^2-4\cdot3\cdot6}}{2\cdot3} \quad \text{Substituting}$$

$$=\frac{10\pm\sqrt{28}}{6}=\frac{10\pm2\sqrt{7}}{6}$$

$$=\frac{2(5\pm\sqrt{7})}{2\cdot3}=\frac{5\pm\sqrt{7}}{3}$$

The solutions are $\frac{5-\sqrt{7}}{3}$ and $\frac{5+\sqrt{7}}{3}$, or $\frac{5\pm\sqrt{7}}{3}$.

40. $3t^2+8t+3=0$

$$t=\frac{-8\pm\sqrt{8^2-4\cdot3\cdot3}}{2\cdot3}$$

$$=\frac{-8\pm\sqrt{28}}{6}=\frac{-8\pm2\sqrt{7}}{6}$$

$$=\frac{2(-4\pm\sqrt{7})}{2\cdot3}=\frac{-4\pm\sqrt{7}}{3}$$

The solutions are $\frac{-4-\sqrt{7}}{3}$ and $\frac{-4+\sqrt{7}}{3}$, or $\frac{-4\pm\sqrt{7}}{3}$.

41. $x^2+x+2=0$

We use the quadratic formula. Here $a=1$, $b=1$, and $c=2$.

$$x=\frac{-b\pm\sqrt{b^2-4ac}}{2a}$$

$$=\frac{-1\pm\sqrt{1^2-4\cdot1\cdot2}}{2\cdot1} \quad \text{Substituting}$$

$$=\frac{-1\pm\sqrt{-7}}{2}$$

$$=\frac{-1\pm\sqrt{7}i}{2}=-\frac{1}{2}\pm\frac{\sqrt{7}}{2}i$$

The solutions are $-\frac{1}{2}-\frac{\sqrt{7}}{2}i$ and $-\frac{1}{2}+\frac{\sqrt{7}}{2}i$, or $-\frac{1}{2}\pm\frac{\sqrt{7}}{2}i$.

42. $x^2+1=x$

$x^2-x+1=0$

$$x=\frac{-(-1)\pm\sqrt{(-1)^2-4\cdot1\cdot1}}{2\cdot1}$$

$$=\frac{1\pm\sqrt{-3}}{2}=\frac{1\pm\sqrt{3}i}{2}$$

$$=\frac{1}{2}\pm\frac{\sqrt{3}}{2}i$$

The solutions are $\frac{1}{2}-\frac{\sqrt{3}}{2}i$ and $\frac{1}{2}+\frac{\sqrt{3}}{2}i$, or $\frac{1}{2}\pm\frac{\sqrt{3}}{2}i$.

43. $5t^2-8t=3$

$5t^2-8t-3=0$

We use the quadratic formula. Here $a=5$, $b=-8$, and $c=-3$.

$$t=\frac{-b\pm\sqrt{b^2-4ac}}{2a}$$

$$=\frac{-(-8)\pm\sqrt{(-8)^2-4\cdot5(-3)}}{2\cdot5}$$

$$=\frac{8\pm\sqrt{124}}{10}=\frac{8\pm2\sqrt{31}}{10}$$

$$=\frac{2(4\pm\sqrt{31})}{2\cdot5}=\frac{4\pm\sqrt{31}}{5}$$

The solutions are $\frac{4-\sqrt{31}}{5}$ and $\frac{4+\sqrt{31}}{5}$, or $\frac{4\pm\sqrt{31}}{5}$.

44. $5x^2+2=x$

$5x^2-x+2=0$

$$x=\frac{-(-1)\pm\sqrt{(-1)^2-4\cdot5\cdot2}}{2\cdot5}$$

$$=\frac{1\pm\sqrt{-39}}{10}=\frac{1\pm\sqrt{39}i}{10}$$

$$=\frac{1}{10}\pm\frac{\sqrt{39}}{10}i$$

The solutions are $\frac{1}{10}-\frac{\sqrt{39}}{10}i$ and $\frac{1}{10}+\frac{\sqrt{39}}{10}i$, or $\frac{1}{10}\pm\frac{\sqrt{39}}{10}i$.

45. $3x^2 + 4 = 5x$

$3x^2 - 5x + 4 = 0$

We use the quadratic formula. Here $a = 3$, $b = -5$, and $c = 4$.

$$x = \frac{-b \pm \sqrt{b^2 - 4ac}}{2a} = \frac{-(-5) \pm \sqrt{(-5)^2 - 4 \cdot 3 \cdot 4}}{2 \cdot 3} = \frac{5 \pm \sqrt{-23}}{6} = \frac{5 \pm \sqrt{23}i}{6} = \frac{5}{6} \pm \frac{\sqrt{23}}{6}i$$

The solutions are $\frac{5}{6} - \frac{\sqrt{23}}{6}i$ and $\frac{5}{6} + \frac{\sqrt{23}}{6}i$, or $\frac{5}{6} \pm \frac{\sqrt{23}}{6}i$.

46. $2t^2 - 5t = 1$

$2t^2 - 5t - 1 = 0$

$$t = \frac{-(-5) \pm \sqrt{(-5)^2 - 4 \cdot 2(-1)}}{2 \cdot 2} = \frac{5 \pm \sqrt{33}}{4}$$

The solutions are $\frac{5 - \sqrt{33}}{4}$ and $\frac{5 + \sqrt{33}}{4}$, or $\frac{5 \pm \sqrt{33}}{4}$.

47. $x^2 - 8x + 5 = 0$

We use the quadratic formula. Here $a = 1$, $b = -8$, and $c = 5$.

$$x = \frac{-b \pm \sqrt{b^2 - 4ac}}{2a} = \frac{-(-8) \pm \sqrt{(-8)^2 - 4 \cdot 1 \cdot 5}}{2 \cdot 1} = \frac{8 \pm \sqrt{44}}{2} = \frac{8 \pm 2\sqrt{11}}{2} = \frac{2(4 \pm \sqrt{11})}{2} = 4 \pm \sqrt{11}$$

The solutions are $4 - \sqrt{11}$ and $4 + \sqrt{11}$, or $4 \pm \sqrt{11}$.

48. $x^2 - 6x + 3 = 0$

$$x = \frac{-(-6) \pm \sqrt{(-6)^2 - 4 \cdot 1 \cdot 3}}{2 \cdot 1} = \frac{6 \pm \sqrt{24}}{2} = \frac{6 \pm 2\sqrt{6}}{2} = \frac{2(3 \pm \sqrt{6})}{2} = 3 \pm \sqrt{6}$$

The solutions are $3 - \sqrt{6}$ and $3 + \sqrt{6}$, or $3 \pm \sqrt{6}$.

49. $3x^2 + x = 5$

$3x^2 + x - 5 = 0$

We use the quadratic formula. We have $a = 3$, $b = 1$, and $c = -5$.

$$x = \frac{-b \pm \sqrt{b^2 - 4ac}}{2a} = \frac{-1 \pm \sqrt{1^2 - 4 \cdot 3 \cdot (-5)}}{2 \cdot 3} = \frac{-1 \pm \sqrt{61}}{6}$$

The solutions are $\frac{-1 - \sqrt{61}}{6}$ and $\frac{-1 + \sqrt{61}}{6}$, or $\frac{-1 \pm \sqrt{61}}{6}$.

50. $5x^2 + 3x = 1$

$5x^2 + 3x - 1 = 0$

$$x = \frac{-3 \pm \sqrt{3^2 - 4 \cdot 5 \cdot (-1)}}{2 \cdot 5} = \frac{-3 \pm \sqrt{29}}{10}$$

The solutions are $\frac{-3 - \sqrt{29}}{10}$ and $\frac{-3 + \sqrt{29}}{10}$, or $\frac{-3 \pm \sqrt{29}}{10}$.

51. $2x^2 + 1 = 5x$

$2x^2 - 5x + 1 = 0$

We use the quadratic formula. We have $a = 2$, $b = -5$, and $c = 1$.

$$x = \frac{-b \pm \sqrt{b^2 - 4ac}}{2a} = \frac{-(-5) \pm \sqrt{(-5)^2 - 4 \cdot 2 \cdot 1}}{2 \cdot 2} = \frac{5 \pm \sqrt{17}}{4}$$

The solutions are $\frac{5 - \sqrt{17}}{4}$ and $\frac{5 + \sqrt{17}}{4}$, or $\frac{5 \pm \sqrt{17}}{4}$.

52. $4x^2 + 3 = x$

$4x^2 - x + 3 = 0$

$$x = \frac{-(-1) \pm \sqrt{(-1)^2 - 4 \cdot 4 \cdot 3}}{2 \cdot 4} = \frac{1 \pm \sqrt{-47}}{8} = \frac{1 \pm \sqrt{47}i}{8} = \frac{1}{8} \pm \frac{\sqrt{47}}{8}i$$

The solutions are $\frac{1}{8} - \frac{\sqrt{47}}{8}i$ and $\frac{1}{8} + \frac{\sqrt{47}}{8}i$, or $\frac{1}{8} \pm \frac{\sqrt{47}}{8}i$.

53. $5x^2 + 2x = -2$

$5x^2 + 2x + 2 = 0$

We use the quadratic formula. We have $a = 5$, $b = 2$, and $c = 2$.

$$x = \frac{-b \pm \sqrt{b^2-4ac}}{2a}$$
$$= \frac{-2 \pm \sqrt{2^2-4\cdot 5\cdot 2}}{2\cdot 5}$$
$$= \frac{-2 \pm \sqrt{-36}}{10} = \frac{-2 \pm 6i}{10}$$
$$= \frac{2(-1 \pm 3i)}{2\cdot 5} = \frac{-1 \pm 3i}{5}$$
$$= -\frac{1}{5} \pm \frac{3}{5}i$$

The solutions are $-\frac{1}{5} - \frac{3}{5}i$ and $-\frac{1}{5} + \frac{3}{5}i$, or $-\frac{1}{5} \pm \frac{3}{5}i$.

54. $3x^2 + 3x = -4$

$3x^2 + 3x + 4 = 0$

$$x = \frac{-3 \pm \sqrt{3^2 - 4\cdot 3\cdot 4}}{2\cdot 3}$$
$$= \frac{-3 \pm \sqrt{-39}}{6} = \frac{-3 \pm \sqrt{39}i}{6}$$
$$= -\frac{1}{2} \pm \frac{\sqrt{39}}{6}i$$

The solutions are $-\frac{1}{2} - \frac{\sqrt{39}}{6}i$ and $-\frac{1}{2} + \frac{\sqrt{39}}{6}i$ or $-\frac{1}{2} \pm \frac{\sqrt{39}}{6}i$.

55. $4x^2 = 8x + 5$

$4x^2 - 8x - 5 = 0$

$a = 4,\ b = -8,\ c = -5$

$b^2 - 4ac = (-8)^2 - 4\cdot 4(-5) = 144$

Since $b^2 - 4ac > 0$, there are two different real-number solutions.

56. $4x^2 - 12x + 9 = 0$

$b^2 - 4ac = (-12)^2 - 4\cdot 4\cdot 9 = 0$

There is one real-number solution.

57. $x^2 + 3x + 4 = 0$

$a = 1,\ b = 3,\ c = 4$

$b^2 - 4ac = 3^2 - 4\cdot 1\cdot 4 = -7$

Since $b^2 - 4ac < 0$, there are two different imaginary-number solutions.

58. $x^2 - 2x + 4 = 0$

$b^2 - 4ac = (-2)^2 - 4\cdot 1\cdot 4 = -12 < 0$

There are two different imaginary-number solutions.

59. $5t^2 - 7t = 0$

$a = 5,\ b = -7,\ c = 0$

$b^2 - 4ac = (-7)^2 - 4\cdot 5\cdot 0 = 49$

Since $b^2 - 4ac > 0$, there are two different real-number solutions.

60. $5t^2 - 4t = 11$

$5t^2 - 4t - 11 = 0$

$b^2 - 4ac = (-4)^2 - 4\cdot 5(-11) = 236 > 0$

There are two different real-number solutions.

61. $x^2 + 6x + 5 = 0$ Setting $f(x) = 0$

$(x+5)(x+1) = 0$ Factoring

$x + 5 = 0$ *or* $x + 1 = 0$

$x = -5$ *or* $x = -1$

The zeros of the function are -5 and -1.

62. $x^2 - x - 2 = 0$

$(x+1)(x-2) = 0$

$x + 1 = 0$ *or* $x - 2 = 0$

$x = -1$ *or* $x = 2$

The zeros of the function are -1 and 2.

63. $x^2 - 3x - 3 = 0$

$a = 1,\ b = -3,\ c = -3$

$$x = \frac{-b \pm \sqrt{b^2-4ac}}{2a}$$
$$= \frac{-(-3) \pm \sqrt{(-3)^2 - 4\cdot 1\cdot (-3)}}{2\cdot 1}$$
$$= \frac{3 \pm \sqrt{9+12}}{2}$$
$$= \frac{3 \pm \sqrt{21}}{2}$$

The zeros of the function are $\frac{3-\sqrt{21}}{2}$ and $\frac{3+\sqrt{21}}{2}$, or $\frac{3\pm\sqrt{21}}{2}$.

64. $3x^2 + 8x + 2 = 0$

$$x = \frac{-8 \pm \sqrt{8^2 - 4\cdot 3\cdot 2}}{2\cdot 3}$$
$$= \frac{-8 \pm \sqrt{40}}{6} = \frac{-8 \pm 2\sqrt{10}}{6}$$
$$= \frac{-4 \pm \sqrt{10}}{3}$$

The zeros of the function are $\frac{-4-\sqrt{10}}{3}$ and $\frac{-4+\sqrt{10}}{3}$, or $\frac{-4\pm\sqrt{10}}{3}$.

65. $x^2 - 5x + 1 = 0$

$a = 1,\ b = -5,\ c = 1$

$$x = \frac{-b \pm \sqrt{b^2 - 4ac}}{2a}$$
$$= \frac{-(-5) \pm \sqrt{(-5)^2 - 4 \cdot 1 \cdot 1}}{2 \cdot 1}$$
$$= \frac{5 \pm \sqrt{25 - 4}}{2}$$
$$= \frac{5 \pm \sqrt{21}}{2}$$

The zeros of the function are $\frac{5 - \sqrt{21}}{2}$ and $\frac{5 + \sqrt{21}}{2}$, or $\frac{5 \pm \sqrt{21}}{2}$.

66. $x^2 - 3x - 7 = 0$

$$x = \frac{-(-3) \pm \sqrt{(-3)^2 - 4 \cdot 1 \cdot (-7)}}{2 \cdot 1}$$
$$= \frac{3 \pm \sqrt{37}}{2}$$

The zeros of the function are $\frac{3 - \sqrt{37}}{2}$ and $\frac{3 + \sqrt{37}}{2}$, or $\frac{3 \pm \sqrt{37}}{2}$.

67. $x^2 + 2x - 5 = 0$

$a = 1,\ b = 2,\ c = -5$

$$x = \frac{-b \pm \sqrt{b^2 - 4ac}}{2a}$$
$$= \frac{-2 \pm \sqrt{2^2 - 4 \cdot 1 \cdot (-5)}}{2 \cdot 1}$$
$$= \frac{-2 \pm \sqrt{4 + 20}}{2} = \frac{-2 \pm \sqrt{24}}{2}$$
$$= \frac{-2 \pm 2\sqrt{6}}{2} = -1 \pm \sqrt{6}$$

The zeros of the function are $-1 + \sqrt{6}$ and $-1 - \sqrt{6}$, or $-1 \pm \sqrt{6}$.

68. $x^2 - x - 4 = 0$

$$x = \frac{-(-1) \pm \sqrt{(-1)^2 - 4 \cdot 1 \cdot (-4)}}{2 \cdot 1}$$
$$= \frac{1 \pm \sqrt{17}}{2}$$

The zeros of the function are $\frac{1 + \sqrt{17}}{2}$ or $\frac{1 - \sqrt{17}}{2}$, or $\frac{1 \pm \sqrt{17}}{2}$.

69. $2x^2 - x + 4 = 0$

$a = 2,\ b = -1,\ c = 4$

$$x = \frac{-b \pm \sqrt{b^2 - 4ac}}{2a}$$
$$= \frac{-(-1) \pm \sqrt{(-1)^2 - 4 \cdot 2 \cdot 4}}{2 \cdot 2}$$
$$= \frac{1 \pm \sqrt{-31}}{4} = \frac{1 \pm \sqrt{31}i}{4}$$
$$= \frac{1}{4} \pm \frac{\sqrt{31}}{4}i$$

The zeros of the function are $\frac{1}{4} - \frac{\sqrt{31}}{4}i$ and $\frac{1}{4} + \frac{\sqrt{31}}{4}i$, or $\frac{1}{4} \pm \frac{\sqrt{31}}{4}i$.

70. $2x^2 + 3x + 2 = 0$

$$x = \frac{-3 \pm \sqrt{3^2 - 4 \cdot 2 \cdot 2}}{2 \cdot 2}$$
$$= \frac{-3 \pm \sqrt{-7}}{4} = \frac{-3 \pm \sqrt{7}i}{4}$$
$$= -\frac{3}{4} \pm \frac{\sqrt{7}}{4}i$$

The zeros of the function are $-\frac{3}{4} - \frac{\sqrt{7}}{4}i$ and $-\frac{3}{4} + \frac{\sqrt{7}}{4}i$, or $-\frac{3}{4} \pm \frac{\sqrt{7}}{4}i$.

71. $3x^2 - x - 1 = 0$

$a = 3,\ b = -1,\ c = -1$

$$x = \frac{-b \pm \sqrt{b^2 - 4ac}}{2a}$$
$$= \frac{-(-1) \pm \sqrt{(-1)^2 - 4 \cdot 3 \cdot (-1)}}{2 \cdot 3}$$
$$= \frac{1 \pm \sqrt{13}}{6}$$

The zeros of the function are $\frac{1 - \sqrt{13}}{6}$ and $\frac{1 + \sqrt{13}}{6}$, or $\frac{1 \pm \sqrt{13}}{6}$.

72. $3x^2 + 5x + 1 = 0$

$$x = \frac{-5 \pm \sqrt{5^2 - 4 \cdot 3 \cdot 1}}{2 \cdot 3}$$
$$= \frac{-5 \pm \sqrt{13}}{6}$$

The zeros of the function are $\frac{-5 - \sqrt{13}}{6}$ and $\frac{-5 + \sqrt{13}}{6}$, or $\frac{-5 \pm \sqrt{13}}{6}$.

73. $5x^2 - 2x - 1 = 0$

$a = 5,\ b = -2,\ c = -1$

$$x = \frac{-b \pm \sqrt{b^2 - 4ac}}{2a}$$
$$= \frac{-(-2) \pm \sqrt{(-2)^2 - 4 \cdot 5 \cdot (-1)}}{2 \cdot 5}$$
$$= \frac{2 \pm \sqrt{24}}{10} = \frac{2 \pm 2\sqrt{6}}{10}$$
$$= \frac{2(1 \pm \sqrt{6})}{2 \cdot 5} = \frac{1 \pm \sqrt{6}}{5}$$

The zeros of the function are $\frac{1 - \sqrt{6}}{5}$ and $\frac{1 + \sqrt{6}}{5}$, or $\frac{1 \pm \sqrt{6}}{5}$.

74. $4x^2 - 4x - 5 = 0$

$$x = \frac{-(-4) \pm \sqrt{(-4)^2 - 4 \cdot 4 \cdot (-5)}}{2 \cdot 4}$$
$$= \frac{4 \pm \sqrt{96}}{8} = \frac{4 \pm 4\sqrt{6}}{8}$$
$$= \frac{4(1 \pm \sqrt{6})}{4 \cdot 2} = \frac{1 \pm \sqrt{6}}{2}$$

The zeros of the function are $\frac{1 - \sqrt{6}}{2}$ and $\frac{1 + \sqrt{6}}{2}$, or $\frac{1 \pm \sqrt{6}}{2}$.

75. $4x^2 + 3x - 3 = 0$

$a = 4,\ b = 3,\ c = -3$

$$x = \frac{-b \pm \sqrt{b^2 - 4ac}}{2a}$$
$$= \frac{-3 \pm \sqrt{3^2 - 4 \cdot 4 \cdot (-3)}}{2 \cdot 4}$$
$$= \frac{-3 \pm \sqrt{57}}{8}$$

The zeros of the function are $\frac{-3 - \sqrt{57}}{8}$ and $\frac{-3 + \sqrt{57}}{8}$, or $\frac{-3 \pm \sqrt{57}}{8}$.

76. $x^2 + 6x - 3 = 0$

$$x = \frac{-6 \pm \sqrt{6^2 - 4 \cdot 1 \cdot (-3)}}{2 \cdot 1}$$
$$= \frac{-6 \pm \sqrt{48}}{2} = \frac{-6 \pm 4\sqrt{3}}{2}$$
$$= \frac{2(-3 \pm 2\sqrt{3})}{2} = -3 \pm 2\sqrt{3}$$

The zeros of the function are $-3 - 2\sqrt{3}$ and $-3 + 2\sqrt{3}$, or $-3 \pm 2\sqrt{3}$.

77. $x^4 - 3x^2 + 2 = 0$

Let $u = x^2$.

$u^2 - 3u + 2 = 0$ Substituting u for x^2

$(u - 1)(u - 2) = 0$

$u - 1 = 0$ *or* $u - 2 = 0$

$u = 1$ *or* $u = 2$

Now substitute x^2 for u and solve for x.

$x^2 = 1$ *or* $x^2 = 2$

$x = \pm 1$ *or* $x = \pm\sqrt{2}$

The solutions are -1, 1, $-\sqrt{2}$, and $\sqrt{2}$.

78. $x^4 + 3 = 4x^2$

$x^4 - 4x^2 + 3 = 0$

Let $u = x^2$.

$u^2 - 4u + 3 = 0$ Substituting u for x^2

$(u - 1)(u - 3) = 0$

$u - 1 = 0$ *or* $u - 3 = 0$

$u = 1$ *or* $u = 3$

Substitute x^2 for u and solve for x.

$x^2 = 1$ *or* $x^2 = 3$

$x = \pm 1$ *or* $x = \pm\sqrt{3}$

The solutions are -1, 1, $-\sqrt{3}$, and $\sqrt{3}$.

79. $x^4 + 3x^2 = 10$

$x^4 + 3x^2 - 10 = 0$

Let $u = x^2$.

$u^2 + 3u - 10 = 0$ Substituting u for x^2

$(u + 5)(u - 2) = 0$

$u + 5 = 0$ *or* $u - 2 = 0$

$u = -5$ *or* $u = 2$

Now substitute x^2 for u and solve for x.

$x^2 = -5$ *or* $x^2 = 2$

$x = \pm\sqrt{5}i$ *or* $x = \pm\sqrt{2}$

The solutions are $-\sqrt{5}i$, $\sqrt{5}i$, $-\sqrt{2}$, and $\sqrt{2}$.

80. $x^4 - 8x^2 = 9$

$x^4 - 8x^2 - 9 = 0$

Let $u = x^2$.

$u^2 - 8u - 9 = 0$ Substituting u for x^2

$(u - 9)(u + 1) = 0$

$u - 9 = 0$ *or* $u + 1 = 0$

$u = 9$ *or* $u = -1$

Now substitute x^2 for u and solve for x.

$x^2 = 9$ *or* $x^2 = -1$

$x = \pm 3$ *or* $x = \pm i$

The solutions are -3, 3, i, and $-i$.

81. $y^6 - 9y^3 + 8 = 0$

Let $u = y^3$.

$u^2 - 9u + 8 = 0$ Substituting u for y^3

$(u - 1)(u - 8) = 0$

$u - 1 = 0$ *or* $u - 8 = 0$

$u = 1$ *or* $u = 8$

Now substitute y^3 for u and solve for y.

$y^3 = 1$ *or* $y^3 = 8$

$y = 1$ *or* $y = 2$

The solutions are 1 and 2.

82. $y^6 - 26y^3 - 27 = 0$

Let $u = y^3$.

$u^2 - 26u - 27 = 0$ Substituting u for y^3

$(u+1)(u-27) = 0$

$u + 1 = 0$ *or* $u - 27 = 0$

$u = -1$ *or* $u = 27$

Now substitute y^3 for u and solve for y.

$y^3 = -1$ *or* $y^3 = 27$

$y = -1$ *or* $y = 3$

The solutions are -1 and 3.

83. $x - 3\sqrt{x} - 4 = 0$

Let $u = \sqrt{x}$.

$u^2 - 3u - 4 = 0$ Substituting u for $\sqrt{x}$

$(u+1)(u-4) = 0$

$u + 1 = 0$ *or* $u - 4 = 0$

$u = -1$ *or* $u = 4$

Now substitute $\sqrt{x}$ for u and solve for x.

$\sqrt{x} = -1$ *or* $\sqrt{x} = 4$

No solution $\quad x = 16$

Note that $\sqrt{x}$ must be nonnegative, so $\sqrt{x} = -1$ has no solution. The number 16 checks and is the solution. The solution is 16.

84. $2x - 9\sqrt{x} + 4 = 0$

Let $u = \sqrt{x}$.

$2u^2 - 9u + 4 = 0$ Substituting u for $\sqrt{x}$

$(2u-1)(u-4) = 0$

$2u - 1 = 0$ *or* $u - 4 = 0$

$u = \frac{1}{2}$ *or* $u = 4$

Substitute $\sqrt{x}$ for u and solve for u.

$\sqrt{x} = \frac{1}{2}$ *or* $\sqrt{x} = 4$

$x = \frac{1}{4}$ *or* $x = 16$

Both numbers check. The solutions are $\frac{1}{4}$ and 16.

85. $m^{2/3} - 2m^{1/3} - 8 = 0$

Let $u = m^{1/3}$.

$u^2 - 2u - 8 = 0$ Substituting u for $m^{1/3}$

$(u+2)(u-4) = 0$

$u + 2 = 0$ *or* $u - 4 = 0$

$u = -2$ *or* $u = 4$

Now substitute $m^{1/3}$ for u and solve for m.

$m^{1/3} = -2$ *or* $m^{1/3} = 4$

$(m^{1/3})^3 = (-2)^3$ *or* $(m^{1/3})^3 = 4^3$ Using the principle of powers

$m = -8$ *or* $m = 64$

The solutions are -8 and 64.

86. $t^{2/3} + t^{1/3} - 6 = 0$

Let $u = t^{1/3}$.

$u^2 + u - 6 = 0$

$(u+3)(u-2) = 0$

$u + 3 = 0$ *or* $u - 2 = 0$

$u = -3$ *or* $u = 2$

Substitute $t^{1/3}$ for u and solve for t.

$t^{1/3} = -3$ *or* $t^{1/3} = 2$

$t = -27$ *or* $t = 8$

The solutions are -27 and 8.

87. $x^{1/2} - 3x^{1/4} + 2 = 0$

Let $u = x^{1/4}$.

$u^2 - 3u + 2 = 0$ Substituting u for $x^{1/4}$

$(u-1)(u-2) = 0$

$u - 1 = 0$ *or* $u - 2 = 0$

$u = 1$ *or* $u = 2$

Now substitute $x^{1/4}$ for u and solve for x.

$x^{1/4} = 1$ *or* $x^{1/4} = 2$

$(x^{1/4})^4 = 1^4$ *or* $(x^{1/4})^4 = 2^4$

$x = 1$ *or* $x = 16$

The solutions are 1 and 16.

88. $x^{1/2} - 4x^{1/4} = -3$

$x^{1/2} - 4x^{1/4} + 3 = 0$

Let $u = x^{1/4}$.

$u^2 - 4u + 3 = 0$

$(u-1)(u-3) = 0$

$u - 1 = 0$ *or* $u - 3 = 0$

$u = 1$ *or* $u = 3$

Substitute $x^{1/4}$ for u and solve for x.

$x^{1/4} = 1$ *or* $x^{1/4} = 3$

$x = 1$ *or* $x = 81$

The solutions are 1 and 81.

89. $(2x-3)^2 - 5(2x-3) + 6 = 0$

Let $u = 2x - 3$.

$u^2 - 5u + 6 = 0$ Substituting u for $2x - 3$

$(u-2)(u-3) = 0$

$u - 2 = 0$ *or* $u - 3 = 0$

$u = 2$ *or* $u = 3$

Now substitute $2x - 3$ for u and solve for x.

$2x-3=2$ *or* $2x-3=3$

$2x=5$ *or* $2x=6$

$x=\frac{5}{2}$ *or* $x=3$

The solutions are $\frac{5}{2}$ and 3.

90. $(3x+2)^2+7(3x+2)-8=0$

Let $u=3x+2$.

$u^2+7u-8=0$ Substituting u for $3x+2$

$(u+8)(u-1)=0$

$u+8=0$ *or* $u-1=0$

$u=-8$ *or* $u=1$

Substitute $3x+2$ for u and solve for x.

$3x+2=-8$ *or* $3x+2=1$

$3x=-10$ *or* $3x=-1$

$x=-\frac{10}{3}$ *or* $x=-\frac{1}{3}$

The solutions are $-\frac{10}{3}$ and $-\frac{1}{3}$.

91. $(2t^2+t)^2-4(2t^2+t)+3=0$

Let $u=2t^2+t$.

$u^2-4u+3=0$ Substituting u for $2t^2+t$

$(u-1)(u-3)=0$

$u-1=0$ *or* $u-3=0$

$u=1$ *or* $u=3$

Now substitute $2t^2+t$ for u and solve for t.

$2t^2+t=1$ *or* $2t^2+t=3$

$2t^2+t-1=0$ *or* $2t^2+t-3=0$

$(2t-1)(t+1)=0$ *or* $(2t+3)(t-1)=0$

$2t-1=0$ *or* $t+1=0$ *or* $2t+3=0$ *or* $t-1=0$

$t=\frac{1}{2}$ *or* $t=-1$ *or* $t=-\frac{3}{2}$ *or* $t=1$

The solutions are $\frac{1}{2}$, -1, $-\frac{3}{2}$ and 1.

92. $12=(m^2-5m)^2+(m^2-5m)$

$0=(m^2-5m)^2+(m^2-5m)-12$

Let $u=m^2-5m$.

$0=u^2+u-12$ Substituting u for m^2-5m

$0=(u+4)(u-3)$

$u+4=0$ *or* $u-3=0$

$u=-4$ *or* $u=3$

Substitute m^2-5m for u and solve for m.

$m^2-5m=-4$ *or* $m^2-5m=3$

$m^2-5m+4=0$ *or* $m^2-5m-3=0$

$(m-1)(m-4)=0$ *or*

$$m=\frac{-(-5)\pm\sqrt{(-5)^2-4\cdot 1(-3)}}{2\cdot 1}$$

$m=1$ *or* $m=4$ *or* $m=\frac{5\pm\sqrt{37}}{2}$

The solutions are 1, 4, $\frac{5-\sqrt{37}}{2}$, and $\frac{5+\sqrt{37}}{2}$, or 1, 4, and $\frac{5\pm\sqrt{37}}{2}$.

93. ***Familiarize and Translate.*** We will use the formula $s=16t^2$, substituting 2120 for s.

$2120=16t^2$

Carry out. We solve the equation.

$2120=16t^2$

$132.5=t^2$ Dividing by 16 on both sides

$11.5\approx t$ Taking the square root on both sides

Check. When $t=11.5$, $s=16(11.5)^2=2116\approx 2120$. The answer checks.

State. It would take an object about 11.5 sec to reach the ground.

94. Solve: $2063=16t^2$

$t\approx 11.4$ sec

95. Substitute 9.7 for $w(x)$ and solve for x.

$-0.01x^2+0.27x+8.60=9.7$

$-0.01x^2+0.27x-1.1=0$

$a=-0.01,\ b=0.27,\ c=-1.1$

$$x=\frac{-b\pm\sqrt{b^2-4ac}}{2a}$$

$$=\frac{-0.27\pm\sqrt{(0.27)^2-4(-0.01)(-1.1)}}{2(-0.01)}$$

$$=\frac{-0.27\pm\sqrt{0.0289}}{-0.02}$$

$x\approx 5$ *or* $x\approx 22$

There were 9.7 million self-employed workers in the United States 5 years after 1980 and also 22 years after 1980, or in 1985 and in 2002.

96. Solve: $-0.01x^2+0.27x+8.60=9.1$

$x\approx 2$ or $x\approx 25$, so there were or will be 9.1 self-employed workers in the United States 2 years after 1980, or in 1982, and 25 years after 1980, or in 2005.

97. ***Familiarize.*** Let $w=$ the width of the rug. Then $w+1=$ the length.

Translate. We use the Pythagorean equation.

$w^2+(w+1)^2=5^2$

Carry out. We solve the equation.

$w^2+(w+1)^2=5^2$

$w^2+w^2+2w+1=25$

$2w^2+2w+1=25$

$2w^2+2w-24=0$

$2(w+4)(w-3)=0$

$w+4=0$ *or* $w-3=0$

$w=-4$ *or* $w=3$

Since the width cannot be negative, we consider only 3. When $w=3$, $w+1=3+1=4$.

Check. The length, 4 ft, is 1 ft more than the width, 3 ft. The length of a diagonal of a rectangle with width 3 ft and length 4 ft is $\sqrt{3^2+4^2} = \sqrt{9+16} = \sqrt{25} = 5$. The answer checks.

State. The length is 4 ft, and the width is 3 ft.

98. Let x = the length of the longer leg.

Solve: $x^2 + (x-7)^2 = 13^2$

$x = -5$ *or* $x = 12$

Only 12 has meaning in the original problem. The length of one leg is 12 cm, and the length of the other leg is $12-7$, or 5 cm.

99. ***Familiarize.*** Let n = the smaller number. Then $n+5$ = the larger number.

Translate.

$$\underbrace{\text{The product of the numbers}}_{\downarrow\; n(n+5)} \underset{=}{\text{is}} \underset{36}{36.}$$

Carry out.

$$n(n+5) = 36$$
$$n^2 + 5n = 36$$
$$n^2 + 5n - 36 = 0$$
$$(n+9)(n-4) = 0$$
$$n+9 = 0 \quad or \quad n-4 = 0$$
$$n = -9 \quad or \quad n = 4$$

If $n = -9$, then $n+5 = -9+5 = -4$. If $n = 4$, then $n+5 = 4+5 = 9$.

Check. The number -4 is 5 more than -9 and $(-4)(-9) = 36$, so the pair -9 and -4 check. The number 9 is 5 more than 4 and $9 \cdot 4 = 36$, so the pair 4 and 9 also check.

State. The numbers are -9 and -4 or 4 and 9.

100. Let n = the larger number.

Solve: $n(n-6) = 72$

$n = -6$ *or* $n = 12$

When $n = -6$, then $n - 6 = -6 - 6 = -12$, so one pair of numbers is -6 and -12. When $n = 12$, then $n - 6 = 12 - 6 = 6$, so the other pair of numbers is 6 and 12.

101. ***Familiarize.*** We add labels to the drawing in the text.

We let x represent the length of a side of the square in each corner. Then the length and width of the resulting base are represented by $20-2x$ and $10-2x$, respectively. Recall that for a rectangle, Area = length × width.

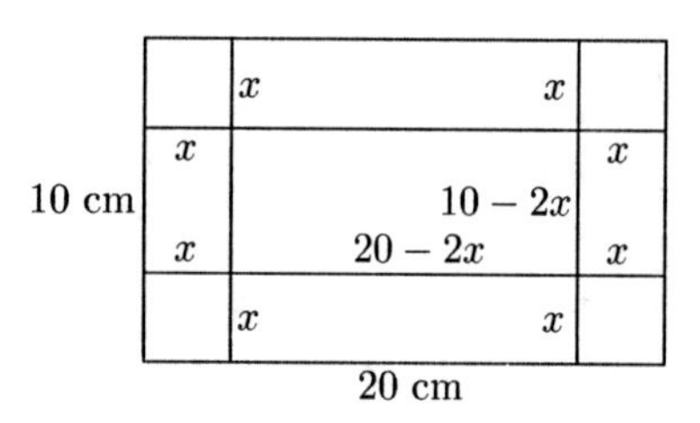

Translate.

$$\underbrace{\text{The area of the base}}_{(20-2x)(10-2x)} \underset{=}{\text{is}} \underbrace{96\text{ cm}^2.}_{96}$$

Carry out. We solve the equation.

$$200 - 60x + 4x^2 = 96$$
$$4x^2 - 60x + 104 = 0$$
$$x^2 - 15x + 26 = 0$$
$$(x-13)(x-2) = 0$$
$$x - 13 = 0 \quad or \quad x - 2 = 0$$
$$x = 13 \quad or \quad x = 2$$

Check. When $x = 13$, both $20 - 2x$ and $10 - 2x$ are negative numbers, so we only consider $x = 2$. When $x = 2$, then $20 - 2x = 20 - 2 \cdot 2 = 16$ and $10 - 2x = 10 - 2 \cdot 2 = 6$, and the area of the base is $16 \cdot 6$, or 96 cm^2. The answer checks.

State. The length of the sides of the squares is 2 cm.

102. Let w = the width of the frame.

Solve: $(32 - 2w)(28 - 2w) = 192$

$w = 8$ or $w = 22$

Only 8 has meaning in the original problem. The width of the frame is 8 cm.

103. ***Familiarize.*** We have $P = 2l + 2w$, or $28 = 2l + 2w$. Solving for w, we have

$$28 = 2l + 2w$$
$$14 = l + w \quad \text{Dividing by 2}$$
$$14 - l = w.$$

Then we have l = the length of the rug and $14 - l$ = the width, in feet. Recall that the area of a rectangle is the product of the length and the width.

Translate.

$$\underbrace{\text{The area}}_{\downarrow\; l(14-l)} \underset{=}{\text{is}} \underbrace{48\text{ ft}^2.}_{\downarrow\; 48}$$

Carry out. We solve the equation.

$$l(14-l) = 48$$
$$14l - l^2 = 48$$
$$0 = l^2 - 14l + 48$$
$$0 = (l-6)(l-8)$$
$$l - 6 = 0 \quad or \quad l - 8 = 0$$
$$l = 6 \quad or \quad l = 8$$

If $l = 6$, then $14 - l = 14 - 6 = 8$.

If $l - 8$, then $14 - l = 14 - 8 = 6$.

In either case, the dimensions are 8 ft by 6 ft. Since we usually consider the length to be greater than the width, we let 8 ft = the length and 6 ft = the width.

Check. The perimeter is $2 \cdot 8$ ft $+ 2 \cdot 6$ ft $= 16$ ft $+ 12$ ft $=$ 28 ft. The answer checks.

State. The length of the rug is 8 ft, and the width is 6 ft.

104. We have $170 = 2l + 2w$, so $w = 85 - l$.

Solve: $l(85 - l) = 1750$

$l = 35$ *or* $l = 50$

Choosing the larger number to be the length, we find that the length of the petting area is 50 m, and the width is 35 m.

105. $f(x) = 4 - 5x = -5x + 4$

The function can be written in the form $y = mx + b$, so it is a linear function.

106. $f(x) = 4 - 5x^2 = -5x^2 + 4$

The function can be written in the form $f(x) = ax^2+bx+c$, $a \neq 0$, so it is a quadratic function.

107. $f(x) = 7x^2$

The function is in the form $f(x) = ax^2 + bx + c$, $a \neq 0$, so it is a quadratic function.

108. $f(x) = 23x + 6$

The function is in the form $f(x) = mx + b$, so it is a linear function.

109. $f(x) = 1.2x - (3.6)^2$

The function is in the form $f(x) = mx + b$, so it is a linear function.

110. $f(x) = 2 - x - x^2 = -x^2 - x + 2$

The function can be written in the form $f(x) = ax^2+bx+c$, $a \neq 0$, so it is a quadratic function.

111. Graph $y = x^2 - 8x + 12$ and use the Zero feature twice.

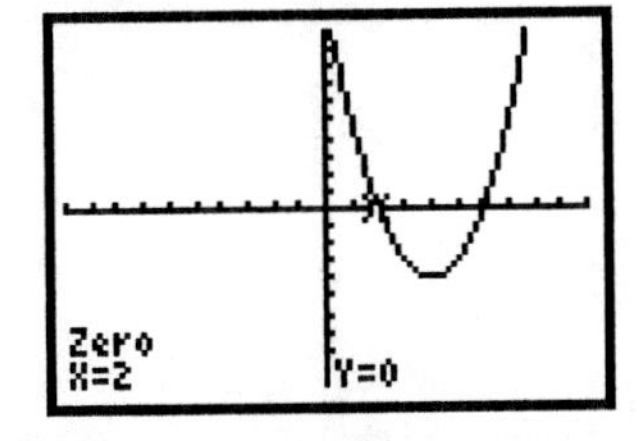

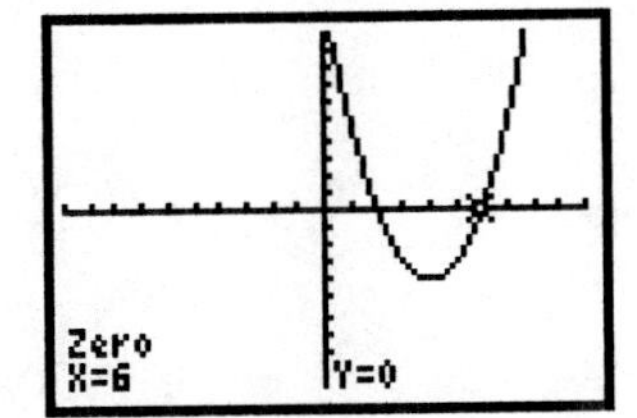

The solutions are 2 and 6.

112. Graph $y = 5x^2 + 42x + 16$ and use the Zero feature twice.

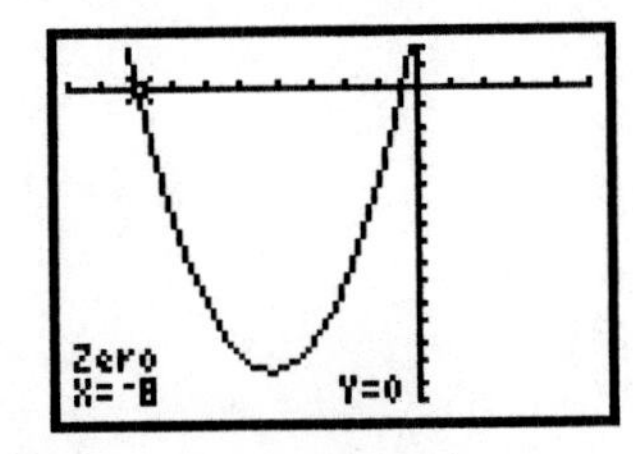

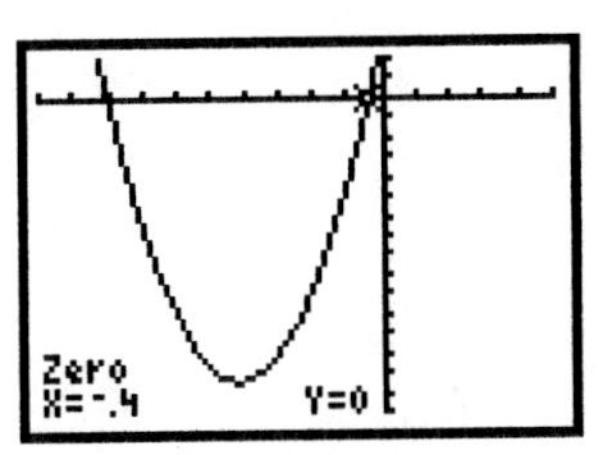

The solutions are -8 and -0.4.

113. Graph $y = 7x^2 - 43x + 6$ and use the Zero feature twice.

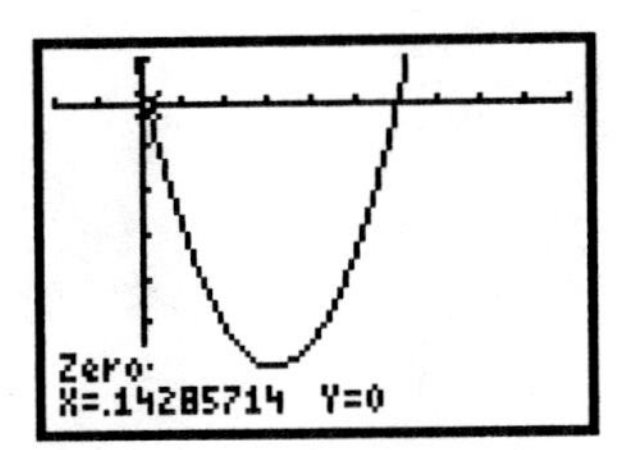

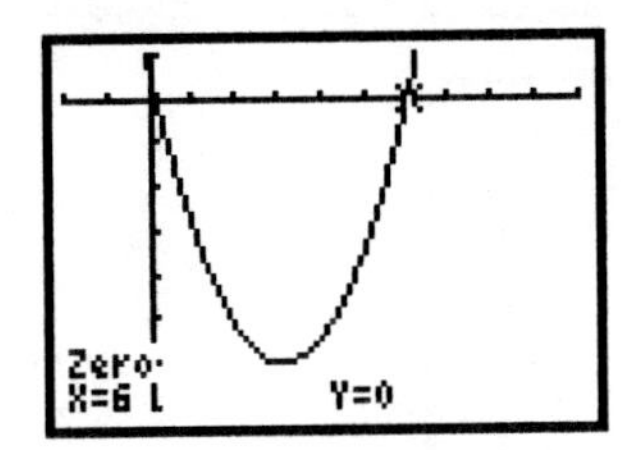

One solution is approximately 0.143 and the other is 6.

114. Graph $y = 10x^2 - 23x + 12$ and use the Zero feature twice.

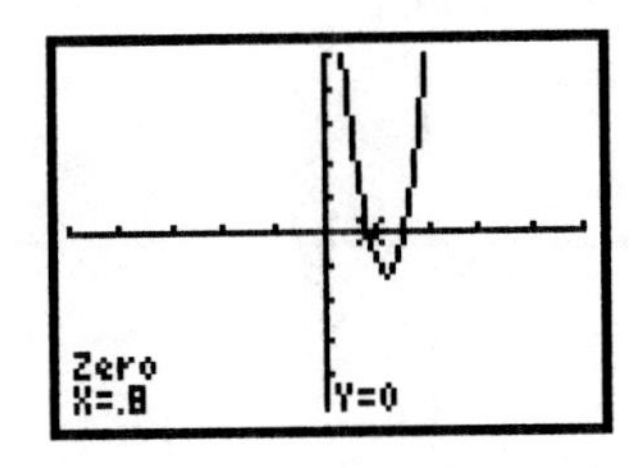

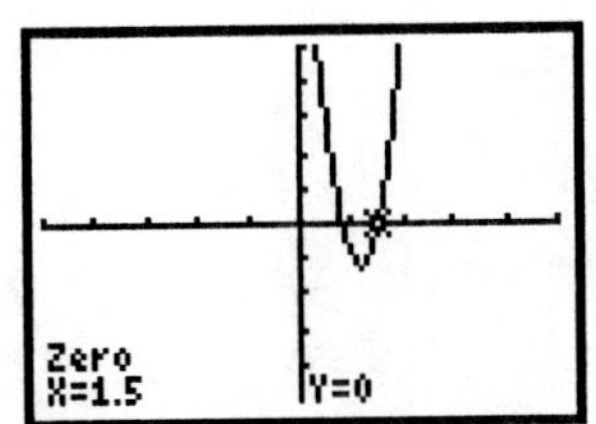

The solutions are 0.8 and 1.5.

115. Graph $y_1 = 6x + 1$ and $y_2 = 4x^2$ and use the Intersect feature twice.

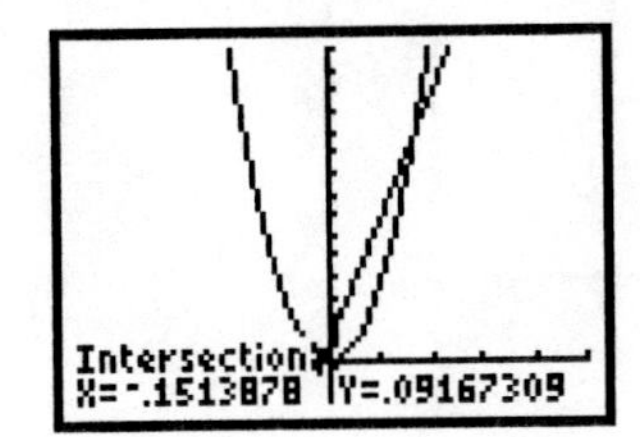

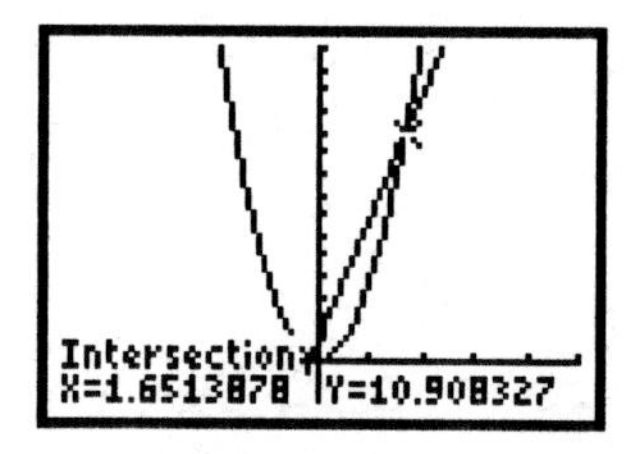

The solutions are approximately -0.151 and 1.651.

116. Graph $y_1 = 3x^2 + 5x$ and $y_2 = 3$ and use the Intersect feature twice.

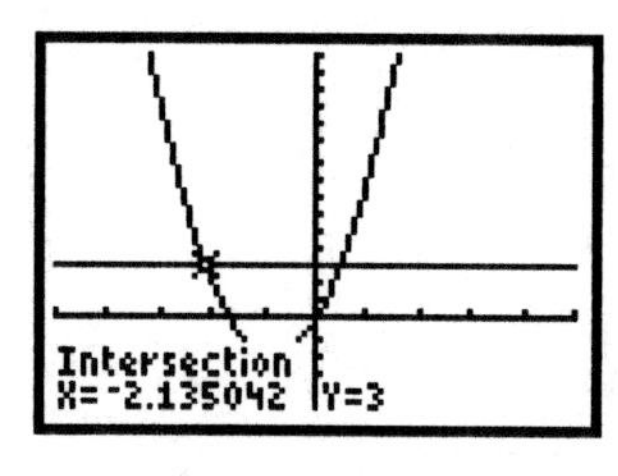

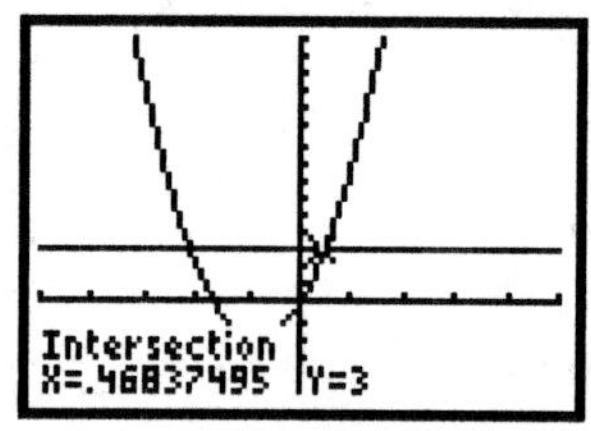

The solutions are approximately -2.135 and 0.468.

117. Graph $y = 2x^2 - 5x - 4$ and use the Zero feature twice.

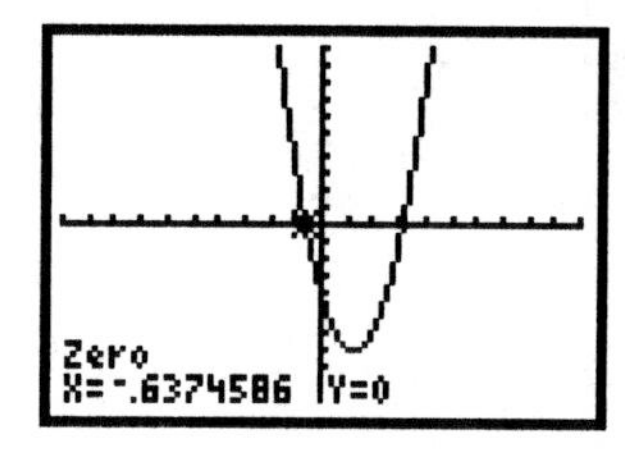

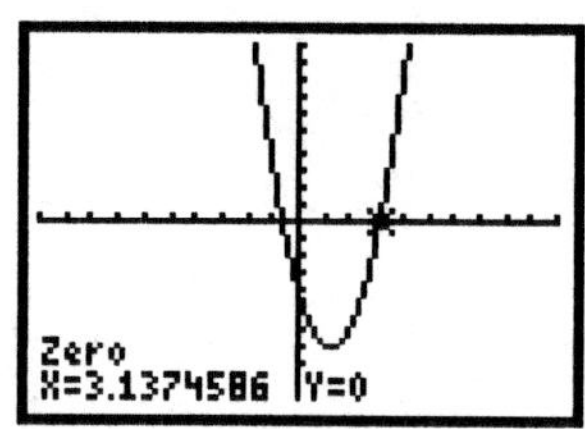

The zeros are approximately -0.637 and 3.137.

118. Graph $y = 4x^2 - 3x - 2$ and use the Zero feature twice.

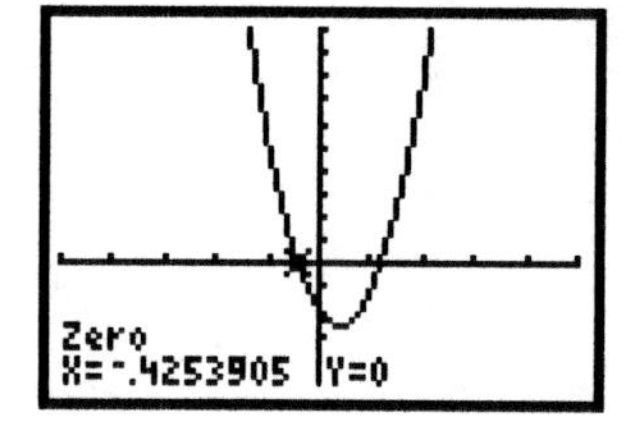

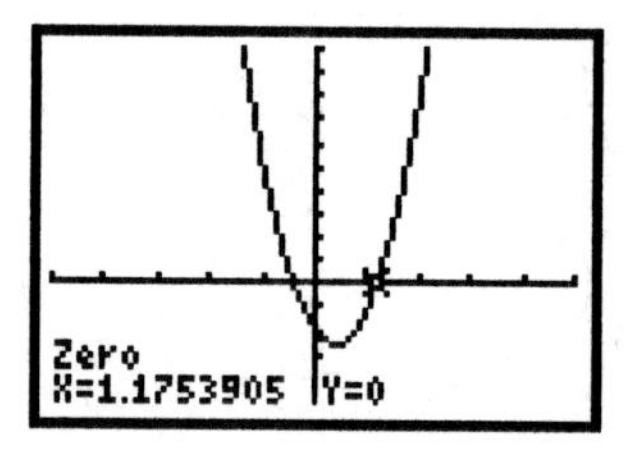

The zeros are approximately -0.425 and 1.175.

119. Graph $y = 3x^2 + 2x - 4$ and use the Zero feature twice.

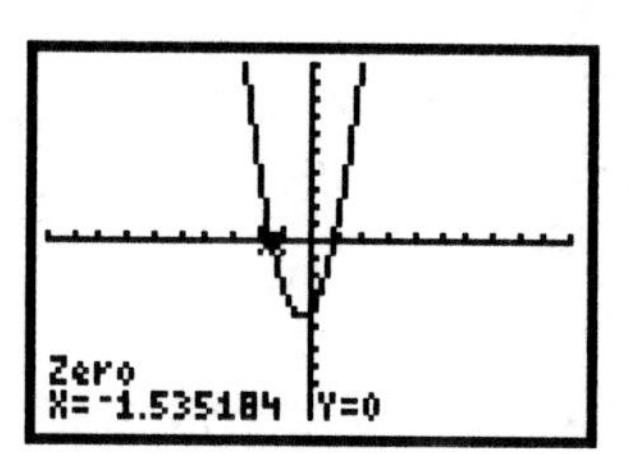

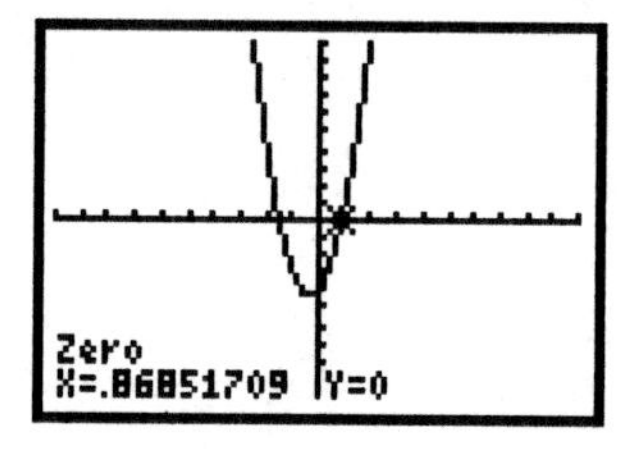

The zeros are approximately -1.535 and 0.869.

120. Graph $y = 9x^2 - 8x - 7$ and use the Zero feature twice.

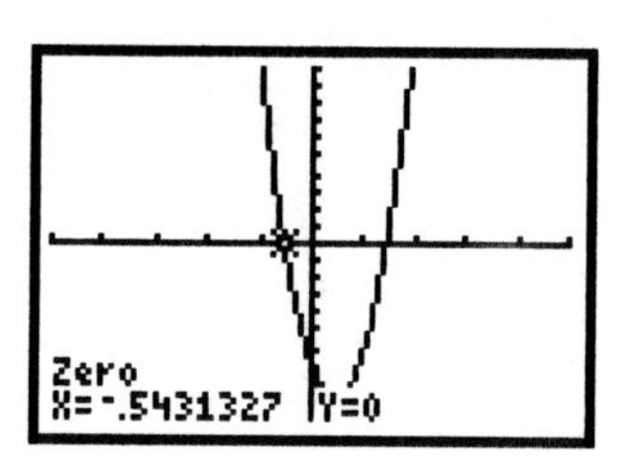

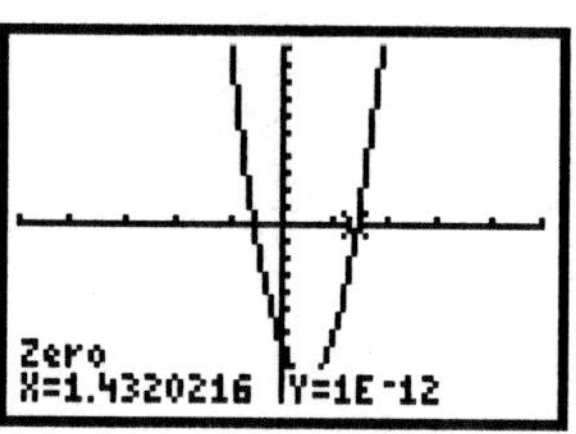

The zeros are approximately -0.543 and 1.432.

121. Graph $y = 5.02x^2 - 4.19x - 2.057$ and use the Zero feature twice.

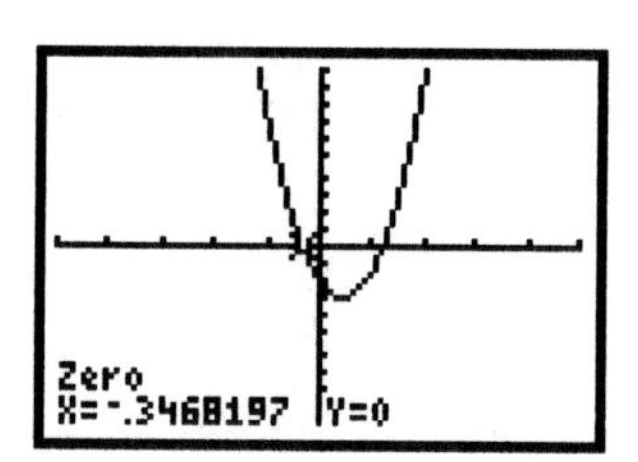

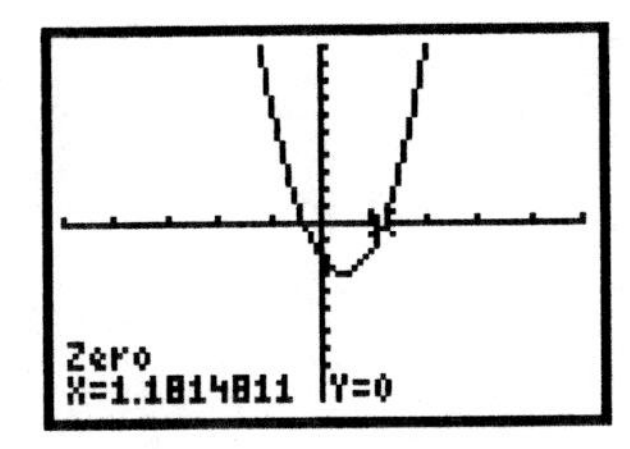

The zeros are approximately -0.347 and 1.181.

122. Graph $y = 1.21x^2 - 2.34x - 5.63$ and use the Zero feature twice.

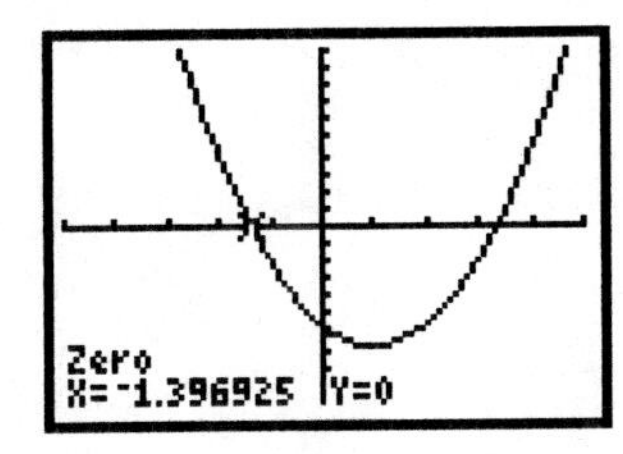

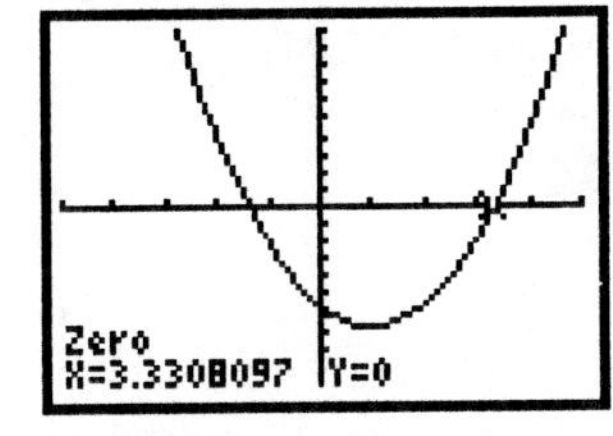

The zeros are approximately -1.397 and 3.331.

123. No; consider the quadratic formula $x = \dfrac{-b \pm \sqrt{b^2 - 4ac}}{2a}$. If $b^2 - 4ac = 0$, then $x = \dfrac{-b}{2a}$, so there is one real zero. If $b^2 - 4ac > 0$, then $\sqrt{b^2 - 4ac}$ is a real number and there are two real zeros. If $b^2 - 4ac < 0$, then $\sqrt{b^2 - 4ac}$ is an imaginary number and there are two imaginary zeros. Thus, a quadratic function cannot have one real zero and one imaginary zero.

124. Use the discriminant. If $b^2 - 4ac < 0$, there are no x-intercepts. If $b^2 - 4ac = 0$, there is one x-intercept. If $b^2 - 4ac > 0$, there are two x-intercepts.

125. $1998 - 1980 = 18$, so we substitute 18 for x.
$a(18) = 9096(18) + 387,725 = 551,453$ associate's degrees

126. $2010 - 1980 = 30$

$a(30) = 9096(30) + 387,725 = 660,605$ associate's degrees

127. Test for symmetry with respect to the x-axis:

$$\begin{aligned} 3x^2 + 4y^2 &= 5 && \text{Original equation} \\ 3x^2 + 4(-y)^2 &= 5 && \text{Replacing } y \text{ by } -y \\ 3x^2 + 4y^2 &= 5 && \text{Simplifying} \end{aligned}$$

The last equation is equivalent to the original equation, so the graph is symmetric with respect to the x-axis.

Test for symmetry with respect to the y-axis:

$$\begin{aligned} 3x^2 + 4y^2 &= 5 && \text{Original equation} \\ 3(-x)^2 + 4y^2 &= 5 && \text{Replacing } x \text{ by } -x \\ 3x^2 + 4y^2 &= 5 && \text{Simplifying} \end{aligned}$$

The last equation is equivalent to the original equation, so the equation is symmetric with respect to the y-axis.

Test for symmetry with respect to the origin:

$$\begin{aligned} 3x^2 + 4y^2 &= 5 && \text{Original equation} \\ 3(-x)^2 + 4(-y)^2 &= 5 && \text{Replacing } x \text{ by } -x \\ & && \text{and } y \text{ by } -y \\ 3x^2 + 4y^2 &= 5 && \text{Simplifying} \end{aligned}$$

The last equation is equivalent to the original equation, so the equation is symmetric with respect to the origin.

128. Test for symmetry with respect to the x-axis:

$$\begin{aligned} y^3 &= 6x^2 && \text{Original equation} \\ (-y)^3 &= 6x^2 && \text{Replacing } y \text{ by } -y \\ -y^3 &= 6x && \text{Simplifying} \end{aligned}$$

The last equation is not equivalent to the original equation, so the graph is not symmetric with respect to the x-axis.

Test for symmetry with respect to the y-axis:

$$\begin{aligned} y^3 &= 6x^2 && \text{Original equation} \\ y^3 &= 6(-x)^2 && \text{Replacing } x \text{ by } -x \\ y^3 &= 6x^2 && \text{Simplifying} \end{aligned}$$

The last equation is equivalent to the original equation, so the equation is symmetric with respect to the y-axis.

Test for symmetry with respect to the origin:

$$\begin{aligned} y^3 &= 6x^2 && \text{Original equation} \\ (-y)^3 &= 6(-x)^2 && \text{Replacing } x \text{ by } -x \\ & && y \text{ by } -y \\ -y^3 &= 6x^2 && \text{Simplifying} \end{aligned}$$

The last equation is not equivalent to the original equation, so the graph is not symmetric with respect to the origin.

129.
$$\begin{aligned} f(x) &= 2x^3 - x \\ f(-x) &= 2(-x)^3 - (-x) = -2x^3 + x \\ -f(x) &= -2x^3 + x \end{aligned}$$
$f(x) \neq f(-x)$ so f is not even
$f(-x) = -f(x)$, so f is odd.

130.
$$\begin{aligned} f(x) &= 4x^2 + 2x - 3 \\ f(-x) &= 4(-x)^2 + 2(-x) - 3 = 4x^2 - 2x - 3 \\ -f(x) &= -4x^2 - 2x + 3 \end{aligned}$$
$f(x) \neq f(-x)$ so f is not even
$f(-x) \neq -f(x)$, so f is not odd.
Thus $f(x) = 4x^2 + 2x - 3$ is neither even nor odd.

131. a)
$$\begin{aligned} kx^2 - 17x + 33 &= 0 \\ k(3)^2 - 17(3) + 33 &= 0 && \text{Substituting 3 for } x \\ 9k - 51 + 33 &= 0 \\ 9k &= 18 \\ k &= 2 \end{aligned}$$

b) $2x^2 - 17x + 33 = 0$ Substituting 2 for k

$(2x - 11)(x - 3) = 0$

$2x - 11 = 0$ *or* $x - 3 = 0$

$x = \frac{11}{2}$ *or* $x = 3$

The other solution is $\frac{11}{2}$.

132. a) $kx^2 - 2x + k = 0$

$k(-3)^2 - 2(-3) + k = 0$ Substituting -3 for x

$9k + 6 + k = 0$

$10k = -6$

$k = -\frac{3}{5}$

b) $-\frac{3}{5}x^2 - 2x - \frac{3}{5} = 0$ Substituting $-\frac{3}{5}$ for k

$3x^2 + 10x + 3 = 0$ Multiplying by -5

$(3x + 1)(x + 3) = 0$

$3x + 1 = 0$ *or* $x + 3 = 0$

$3x = -1$ *or* $x = -3$

$x = -\frac{1}{3}$ *or* $x = -3$

The other solution is $-\frac{1}{3}$.

133. a) $(1 + i)^2 - k(1 + i) + 2 = 0$ Substituting $1 + i$ for x

$1 + 2i - 1 - k - ki + 2 = 0$

$2 + 2i = k + ki$

$2(1 + i) = k(1 + i)$

$2 = k$

b) $x^2 - 2x + 2 = 0$ Substituting 2 for k

$$x = \frac{-(-2) \pm \sqrt{(-2)^2 - 4 \cdot 1 \cdot 2}}{2 \cdot 1}$$

$$= \frac{2 \pm \sqrt{-4}}{2}$$

$$= \frac{2 \pm 2i}{2} = 1 \pm i$$

The other solution is $1 - i$.

134. a) $x^2 - (6 + 3i)x + k = 0$

$3^2 - (6 + 3i) \cdot 3 + k = 0$ Substituting 3 for x

$9 - 18 - 9i + k = 0$

$k = 9 + 9i$

b) $x^2 - (6 + 3i)x + 9 + 9i = 0$

$$x = \frac{-[-(6+3i)] \pm \sqrt{[-(6+3i)]^2 - 4(1)(9+9i)}}{2 \cdot 1}$$

$$x = \frac{6 + 3i \pm \sqrt{36 + 36i - 9 - 36 - 36i}}{2}$$

$$x = \frac{6 + 3i \pm \sqrt{-9}}{2} = \frac{6 + 3i \pm 3i}{2}$$

$x = \frac{6 + 3i + 3i}{2}$ *or* $x = \frac{6 + 3i - 3i}{2}$

$x = \frac{6 + 6i}{2}$ *or* $x = \frac{6}{2}$

$x = 3 + 3i$ *or* $x = 3$

The other solution is $3 + 3i$.

135. $(x - 2)^3 = x^3 - 2$

$x^3 - 6x^2 + 12x - 8 = x^3 - 2$

$0 = 6x^2 - 12x + 6$

$0 = 6(x^2 - 2x + 1)$

$0 = 6(x - 1)(x - 1)$

$x - 1 = 0$ *or* $x - 1 = 0$

$x = 1$ *or* $x = 1$

The solution is 1.

136. $(x + 1)^3 = (x - 1)^3 + 26$

$x^3 + 3x^2 + 3x + 1 = x^3 - 3x^2 + 3x - 1 + 26$

$x^3 + 3x^2 + 3x + 1 = x^3 - 3x^2 + 3x + 25$

$6x^2 - 24 = 0$

$6(x^2 - 4) = 0$

$6(x + 2)(x - 2) = 0$

$x + 2 = 0$ *or* $x - 2 = 0$

$x = -2$ *or* $x = 2$

The solutions are -2 and 2.

137. $(6x^3 + 7x^2 - 3x)(x^2 - 7) = 0$

$x(6x^2 + 7x - 3)(x^2 - 7) = 0$

$x(3x - 1)(2x + 3)(x^2 - 7) = 0$

$x=0$ *or* $3x - 1=0$ *or* $2x + 3=0$ *or* $x^2 - 7 = 0$

$x=0$ *or* $x=\frac{1}{3}$ *or* $x=-\frac{3}{2}$ *or* $x = \sqrt{7}$ *or* $x = -\sqrt{7}$

The exact solutions are $-\sqrt{7}$, $-\frac{3}{2}$, 0, $\frac{1}{3}$, and $\sqrt{7}$.

138. $\left(x - \frac{1}{5}\right)\left(x^2 - \frac{1}{4}\right) + \left(x - \frac{1}{5}\right)\left(x^2 + \frac{1}{8}\right) = 0$

$\left(x - \frac{1}{5}\right)\left(2x^2 - \frac{1}{8}\right) = 0$

$\left(x - \frac{1}{5}\right)(2)\left(x + \frac{1}{4}\right)\left(x - \frac{1}{4}\right) = 0$

$x = \frac{1}{5}$ *or* $x = -\frac{1}{4}$ *or* $x = \frac{1}{4}$

The solutions are $-\frac{1}{4}$, $\frac{1}{5}$, and $\frac{1}{4}$.

139. $x^2 + x - \sqrt{2} = 0$

$$x = \frac{-b \pm \sqrt{b^2 - 4ac}}{2a}$$

$$= \frac{-1 \pm \sqrt{1^2 - 4 \cdot 1(-\sqrt{2})}}{2 \cdot 1} = \frac{-1 \pm \sqrt{1 + 4\sqrt{2}}}{2}$$

The solutions are $\frac{-1 \pm \sqrt{1 + 4\sqrt{2}}}{2}$.

140. $x^2 + \sqrt{5}x - \sqrt{3} = 0$

Use the quadratic formula. Here $a = 1$, $b = \sqrt{5}$, and $c = -\sqrt{3}$.

$$x = \frac{-b \pm \sqrt{b^2 - 4ac}}{2a}$$

$$= \frac{-\sqrt{5} \pm \sqrt{(\sqrt{5})^2 - 4 \cdot 1(-\sqrt{3})}}{2 \cdot 1}$$

$$= \frac{-\sqrt{5} \pm \sqrt{5 + 4\sqrt{3}}}{2}$$

The solutions are $\frac{-\sqrt{5} \pm \sqrt{5 + 4\sqrt{3}}}{2}$.

141. $2t^2 + (t - 4)^2 = 5t(t - 4) + 24$

$2t^2 + t^2 - 8t + 16 = 5t^2 - 20t + 24$

$0 = 2t^2 - 12t + 8$

$0 = t^2 - 6t + 4$ Dividing by 2

Use the quadratic formula.

$$t = \frac{-b \pm \sqrt{b^2 - 4ac}}{2a}$$

$$= \frac{-(-6) \pm \sqrt{(-6)^2 - 4 \cdot 1 \cdot 4}}{2 \cdot 1}$$

$$= \frac{6 \pm \sqrt{20}}{2} = \frac{6 \pm 2\sqrt{5}}{2}$$

$$= \frac{2(3 \pm \sqrt{5})}{2} = 3 \pm \sqrt{5}$$

The solutions are $3 \pm \sqrt{5}$.

142. $9t(t + 2) - 3t(t - 2) = 2(t + 4)(t + 6)$

$9t^2 + 18t - 3t^2 + 6t = 2t^2 + 20t + 48$

$4t^2 + 4t - 48 = 0$

$4(t + 4)(t - 3) = 0$

$t + 4 = 0$ *or* $t - 3 = 0$

$t = -4$ *or* $t = 3$

The solutions are -4 and 3.

143. $\sqrt{x - 3} - \sqrt[4]{x - 3} = 2$

Substitute u for $\sqrt[4]{x - 3}$.

$u^2 - u - 2 = 0$

$(u - 2)(u + 1) = 0$

$u - 2 = 0$ *or* $u + 1 = 0$

$u = 2$ *or* $u = -1$

Substitute $\sqrt[4]{x - 3}$ for u and solve for x.

$\sqrt[4]{x - 3} = 2$ *or* $\sqrt[4]{x - 3} = 1$

$x - 3 = 16$ No solution

$x = 19$

The value checks. The solution is 19.

144. $x^6 - 28x^3 + 27 = 0$

Substitute u for x^3.

$u^2 - 28u + 27 = 0$

$(u - 27)(u - 1) = 0$

$u = 27$ or $u = 1$

Substitute x^3 for u and solve for x.

$x^3 = 27$ *or* $x^3 = 1$

$x^3 - 27 = 0$ *or* $x^3 - 1 = 0$

$(x-3)(x^2+3x+9) = 0$ *or* $(x-1)(x^2+x+1) = 0$

Using the principle of zero products and, where necessary, the quadratic formula, we find that the solutions are $3, -\frac{3}{2} \pm \frac{3\sqrt{3}}{2}i, 1$, and $-\frac{1}{2} \pm \frac{\sqrt{3}}{2}i$.

145. $\left(y + \frac{2}{y}\right)^2 + 3y + \frac{6}{y} = 4$

$\left(y + \frac{2}{y}\right)^2 + 3\left(y + \frac{2}{y}\right) - 4 = 0$

Substitute u for $y + \frac{2}{y}$.

$u^2 + 3u - 4 = 0$

$(u + 4)(u - 1) = 0$

$u = -4$ *or* $u = 1$

Substitute $y + \frac{2}{y}$ for u and solve for y.

$y + \frac{2}{y} = -4$ *or* $y + \frac{2}{y} = 1$

$y^2 + 2 = -4y$ *or* $y^2 + 2 = y$

$y^2 + 4y + 2 = 0$ *or* $y^2 - y + 2 = 0$

$y = \frac{-4 \pm \sqrt{4^2 - 4 \cdot 1 \cdot 2}}{2 \cdot 1}$ *or*

$y = \frac{-(-1) \pm \sqrt{(-1)^2 - 4 \cdot 1 \cdot 2}}{2 \cdot 1}$

$y = \frac{-4 \pm \sqrt{8}}{2}$ *or* $y = \frac{1 \pm \sqrt{-7}}{2}$

$y = \frac{-4 \pm 2\sqrt{2}}{2}$ *or* $y = \frac{1 \pm \sqrt{7}i}{2}$

$y = -2 \pm \sqrt{2}$ *or* $y = \frac{1}{2} \pm \frac{\sqrt{7}}{2}i$

The solutions are $-2 \pm \sqrt{2}$ and $\frac{1}{2} \pm \frac{\sqrt{7}}{2}i$.

146. $x^2+3x+1-\sqrt{x^2+3x+1}=8$

$x^2+3x+1-\sqrt{x^2+3x+1}-8=0$

$u^2-u-8=0$

$u=\frac{1+\sqrt{33}}{2}$ *or* $u=\frac{1-\sqrt{33}}{2}$

$\sqrt{x^2+3x+1}=\frac{1+\sqrt{33}}{2}$ *or*

$\sqrt{x^2+3x+1}=\frac{1-\sqrt{33}}{2}$

$x^2+3x+1=\frac{34+2\sqrt{33}}{4}$ *or*

$x^2+3x+1=\frac{34-2\sqrt{33}}{4}$

$x^2+3x+\frac{-15-\sqrt{33}}{2}=0$ *or*

$x^2+3x+\frac{-15+\sqrt{33}}{2}=0$

$x=\frac{-3\pm\sqrt{39+2\sqrt{33}}}{2}$ *or*

$x=\frac{-3\pm\sqrt{39-2\sqrt{33}}}{2}$

Only $\frac{-3\pm\sqrt{39+2\sqrt{33}}}{2}$ checks. The solutions are $\frac{-3\pm\sqrt{39+2\sqrt{33}}}{2}$.

147. $\frac{1}{2}at+v_0t+x_0=0$

Use the quadratic formula. Here $a=\frac{1}{2}a$, $b=v_0$, and $c=x_0$.

$$t=\frac{-v_0\pm\sqrt{(v_0)^2-4\cdot\frac{1}{2}a\cdot x_0}}{2\cdot\frac{1}{2}a}$$

$$t=\frac{-v_0\pm\sqrt{v_0^2-2ax_0}}{a}$$

Exercise Set 2.4

1. a) The minimum function value occurs at the vertex, so the vertex is $\left(-\frac{1}{2},-\frac{9}{4}\right)$.

b) The line of symmetry is a vertical line through the vertex. It is $x=-\frac{1}{2}$.

c) The minimum value of the function is $-\frac{9}{4}$.

2. a) $\left(-\frac{1}{2},\frac{25}{4}\right)$

b) $x=-\frac{1}{2}$

c) Maximum: $\frac{25}{4}$

3. $f(x)=x^2-8x+12$ — 16 completes the square for x^2-8x.

$=x^2-8x+16-16+12$ — Adding $16-16$ on the right side

$=(x^2-8x+16)-16+12$

$=(x-4)^2-4$ — Factoring and simplifying

$=(x-4)^2+(-4)$ — Writing in the form $f(x)=a(x-h)^2+k$

a) Vertex: $(4,-4)$

b) Line of symmetry: $x=4$

c) Minimum value: -4

d) We plot the vertex and find several points on either side of it. Then we plot these points and connect them with a smooth curve.

x	$f(x)$
4	−4
2	0
1	5
5	−3
6	0

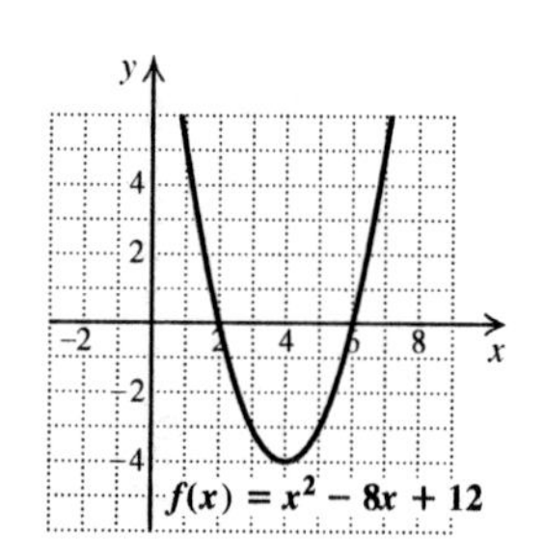

4. $g(x)=x^2+7x-8$

$=x^2+7x+\frac{49}{4}-\frac{49}{4}-8$ $\left(\frac{1}{2}\cdot7=\frac{7}{2}\text{ and }\left(\frac{7}{2}\right)^2=\frac{49}{4}\right)$

$=\left(x+\frac{7}{2}\right)^2-\frac{81}{4}$

$=\left[x-\left(-\frac{7}{2}\right)\right]^2+\left(-\frac{81}{4}\right)$

a) Vertex: $\left(-\frac{7}{2},-\frac{81}{4}\right)$

b) Line of symmetry: $x=-\frac{7}{2}$

c) Minimum value: $-\frac{81}{4}$

d)

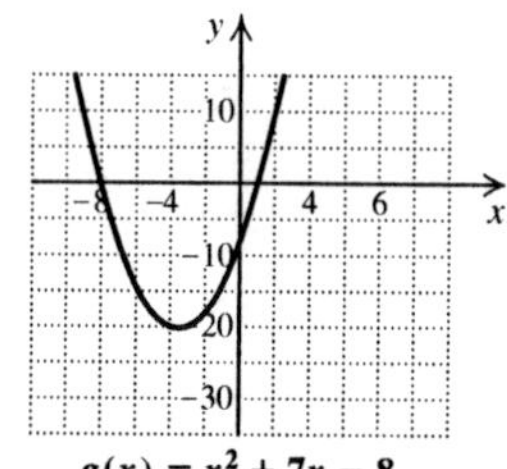

5. $f(x) = x^2 - 7x + 12$ $\qquad \frac{49}{4}$ completes the square for $x^2 - 7x$.

$= x^2 - 7x + \frac{49}{4} - \frac{49}{4} + 12$ Adding $\frac{49}{4} - \frac{49}{4}$ on the right side

$= \left(x^2 - 7x + \frac{49}{4}\right) - \frac{49}{4} + 12$

$= \left(x - \frac{7}{2}\right)^2 - \frac{1}{4}$ Factoring and simplifying

$= \left(x - \frac{7}{2}\right)^2 + \left(-\frac{1}{4}\right)$ Writing in the form $f(x) = a(x-h)^2 + k$

a) Vertex: $\left(\frac{7}{2}, -\frac{1}{4}\right)$

b) Line of symmetry: $x = \frac{7}{2}$

c) Minimum value: $-\frac{1}{4}$

d) We plot the vertex and find several points on either side of it. Then we plot these points and connect them with a smooth curve.

x	$f(x)$
$\frac{7}{2}$	$-\frac{1}{4}$
4	0
5	2
3	0
1	6

$f(x) = x^2 - 7x + 12$

6. $g(x) = x^2 - 5x + 6$

$= x^2 - 5x + \frac{25}{4} - \frac{25}{4} + 6$ $\quad \left(\frac{1}{2}(-5) = -\frac{5}{2}\right.$ and $\left.\left(-\frac{5}{2}\right)^2 = \frac{25}{4}\right)$

$= \left(x - \frac{5}{2}\right)^2 - \frac{1}{4}$

$= \left(x - \frac{5}{2}\right)^2 + \left(-\frac{1}{4}\right)$

a) Vertex: $\left(\frac{5}{2}, -\frac{1}{4}\right)$

b) Line of symmetry: $x = \frac{5}{2}$

c) Minimum value: $-\frac{1}{4}$

d)

$g(x) = x^2 - 5x + 6$

7. $f(x) = x^2 + 4x + 5$ $\qquad$ 4 completes the square for $x^2 + 4x$

$= x^2 + 4x + 4 - 4 + 5$ Adding $4 - 4$ on the right side

$= (x + 2)^2 + 1$ Factoring and simplifying

$= [x - (-2)]^2 + 1$ Writing in the form $f(x) = a(x-h)^2 + k$

a) Vertex: $(-2, 1)$

b) Line of symmetry: $x = -2$

c) Minimum value: 1

d) We plot the vertex and find several points on either side of it. Then we plot these points and connect them with a smooth curve.

x	$f(x)$
-2	1
-1	2
0	5
-3	2
-4	5

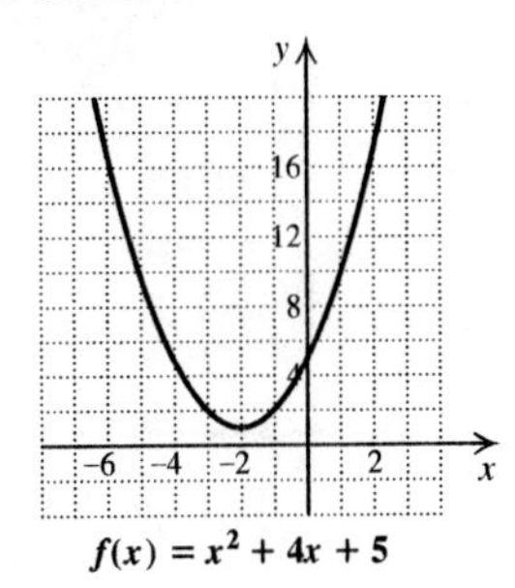

$f(x) = x^2 + 4x + 5$

8. $f(x) = x^2 + 2x + 6$

$= x^2 + 2x + 1 - 1 + 6$ $\quad \left(\frac{1}{2} \cdot 2 = 1 \textit{ and } 1^2 = 1\right)$

$= (x + 1)^2 + 5$

$= [x - (-1)]^2 + 5$

a) Vertex: $(-1, 5)$

b) Line of symmetry: $x = -1$

c) Minimum value: 5

d)

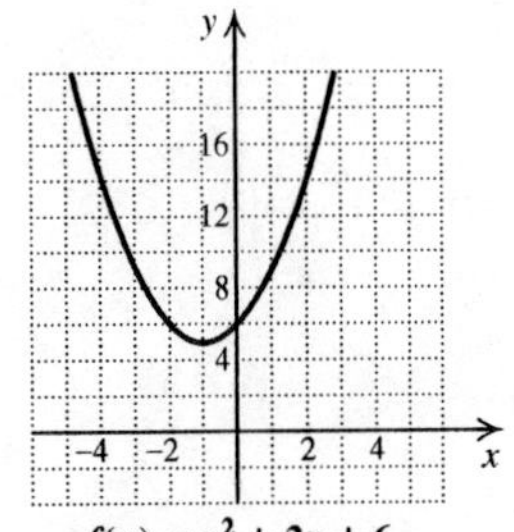

$f(x) = x^2 + 2x + 6$

9. $f(x) = -x^2 - 6x + 3$

$= -(x^2 + 6x) + 3$ 9 completes the square for $x^2 + 6x$.

$= -(x^2 + 6x + 9 - 9) + 3$

$= -(x + 3)^2 - (-9) + 3$ Removing -9 from the parentheses

$= -(x + 3)^2 + 9 + 3$

$= -[x - (-3)]^2 + 12$

a) Vertex: $(-3, 12)$

b) Line of symmetry: $x = -3$

c) Maximum value: 12

d) We plot the vertex and find several points on either side of it. Then we plot these points and connect them with a smooth curve.

x	$f(x)$
-3	12
0	3
1	-4
-6	3
-7	-4

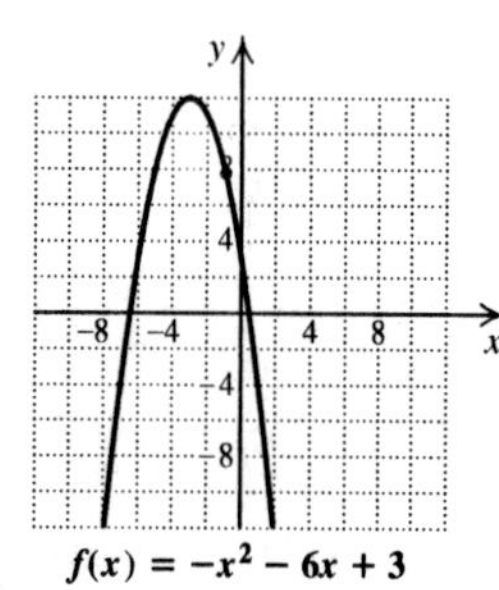

$f(x) = -x^2 - 6x + 3$

10. $f(x) = -x^2 - 8x + 5$

$= -(x^2 + 8x) + 5$

$= -(x^2 + 8x + 16 - 16) + 5$

$\left(\frac{1}{2} \cdot 8 = 4 \text{ and } 4^2 = 16\right)$

$= -(x^2 + 8x + 16) - (-16) + 5$

$= -(x^2 + 8x + 16) + 21$

$= -[x - (-4)]^2 + 21$

a) Vertex: $(-4, 21)$

b) Line of symmetry: $x = -4$

c) Maximum value: 21

d)

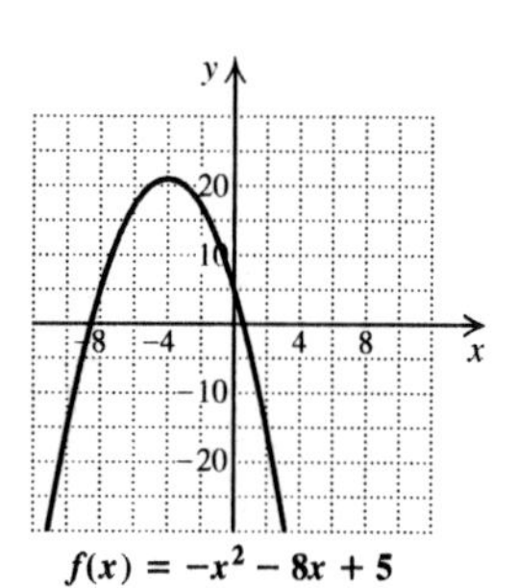

$f(x) = -x^2 - 8x + 5$

11. $g(x) = 2x^2 + 6x + 8$

$= 2(x^2 + 3x) + 8$ Factoring 2 out of the first two terms

$= 2\left(x^2 + 3x + \frac{9}{4} - \frac{9}{4}\right) + 8$ Adding $\frac{9}{4} - \frac{9}{4}$ inside the parentheses

$= 2\left(x^2 + 3x + \frac{9}{4}\right) - 2 \cdot \frac{9}{4} + 8$ Removing $-\frac{9}{4}$ from within the parentheses

$= 2\left(x + \frac{3}{2}\right)^2 + \frac{7}{2}$ Factoring and simplifying

$= 2\left[x - \left(-\frac{3}{2}\right)\right]^2 + \frac{7}{2}$

a) Vertex: $\left(-\frac{3}{2}, \frac{7}{2}\right)$

b) Line of symmetry: $x = -\frac{3}{2}$

c) Minimum value: $\frac{7}{2}$

d) We plot the vertex and find several points on either side of it. Then we plot these points and connect them with a smooth curve.

x	$f(x)$
$-\frac{3}{2}$	$\frac{7}{2}$
-1	4
0	8
-2	4
-3	8

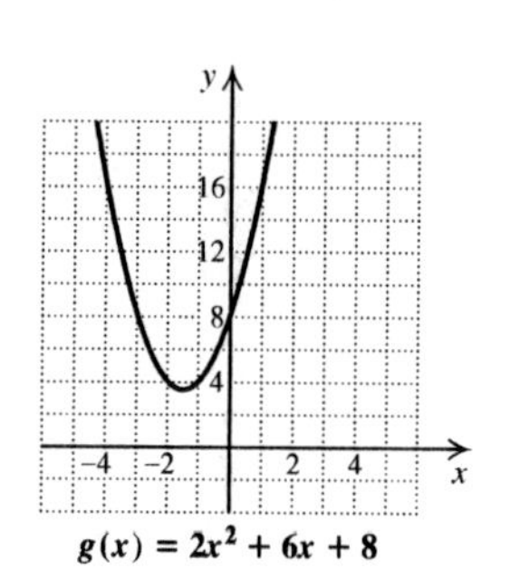

$g(x) = 2x^2 + 6x + 8$

12. $f(x) = 2x^2 - 10x + 14$

$= 2(x^2 - 5x) + 14$

$= 2\left(x^2 - 5x + \frac{25}{4} - \frac{25}{4}\right) + 14$

$= 2\left(x^2 - 5x + \frac{25}{4}\right) - 2 \cdot \frac{25}{4} + 14$

$= 2\left(x - \frac{5}{2}\right)^2 + \frac{3}{2}$

a) Vertex: $\left(\frac{5}{2}, \frac{3}{2}\right)$

b) Line of symmetry: $x = \frac{5}{2}$

c) Minimum value: $\frac{3}{2}$

d)

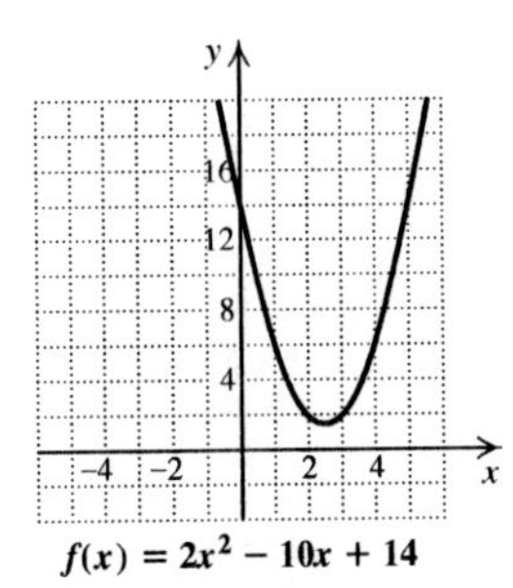

$f(x) = 2x^2 - 10x + 14$

13. $g(x) = -2x^2 + 2x + 1$

$= -2(x^2 - x) + 1$ Factoring -2 out of the first two terms

$= -2\left(x^2 - x + \frac{1}{4} - \frac{1}{4}\right) + 1$ Adding $\frac{1}{4} - \frac{1}{4}$ inside the parentheses

$= -2\left(x^2 - x + \frac{1}{4}\right) - 2\left(-\frac{1}{4}\right) + 1$

Removing $-\frac{1}{4}$ from within the parentheses

$= -2\left(x - \frac{1}{2}\right)^2 + \frac{3}{2}$

a) Vertex: $\left(\frac{1}{2}, \frac{3}{2}\right)$

b) Line of symmetry: $x = \frac{1}{2}$

c) Maximum value: $\frac{3}{2}$

d) We plot the vertex and find several points on either side of it. Then we plot these points and connect them with a smooth curve.

x	$f(x)$
$\frac{1}{2}$	$\frac{3}{2}$
1	1
2	−3
0	1
−1	−3

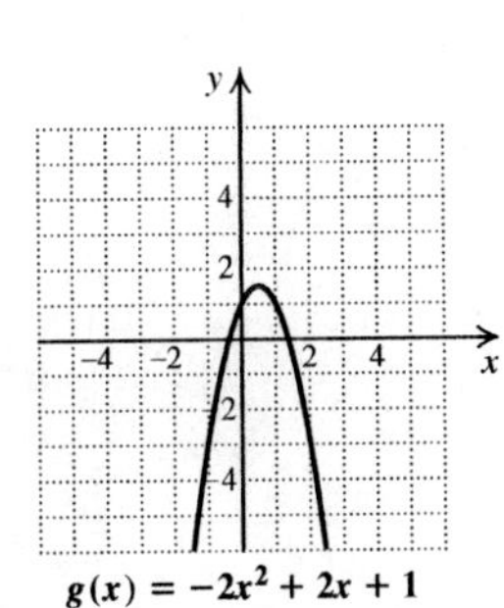

$g(x) = -2x^2 + 2x + 1$

14. $f(x) = -3x^2 - 3x + 1$

$= -3(x^2 + x) + 1$

$= -3\left(x^2 + x + \frac{1}{4} - \frac{1}{4}\right) + 1$

$= -3\left(x^2 + x + \frac{1}{4}\right) - 3\left(-\frac{1}{4}\right) + 1$

$= -3\left(x + \frac{1}{2}\right)^2 + \frac{7}{4}$

$= -3\left[x - \left(-\frac{1}{2}\right)\right]^2 + \frac{7}{4}$

a) Vertex: $\left(-\frac{1}{2}, \frac{7}{4}\right)$

b) Line of symmetry: $x = -\frac{1}{2}$

c) Maximum value: $\frac{7}{4}$

d)

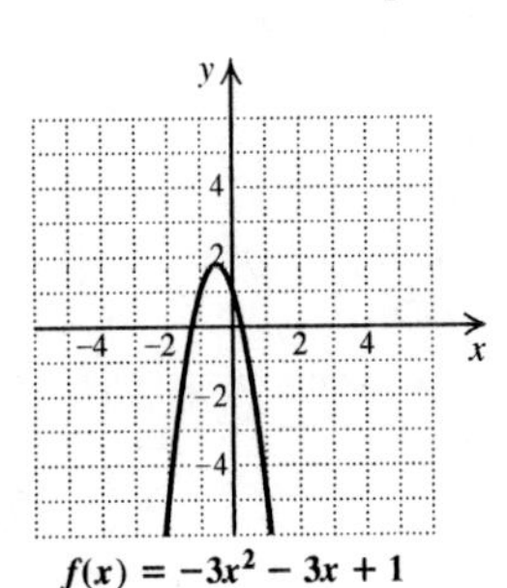

$f(x) = -3x^2 - 3x + 1$

15. The graph of $y = (x + 3)^2$ has vertex $(-3, 0)$ and opens up. It is graph (f).

16. The graph of $y = -(x - 4)^2 + 3$ has vertex $(4, 3)$ and opens down. It is graph (e).

17. The graph of $y = 2(x - 4)^2 - 1$ has vertex $(4, -1)$ and opens up. It is graph (b).

18. The graph of $y = x^2 - 3$ has vertex $(0, -3)$ and opens up. It is graph (g).

19. The graph of $y = -\frac{1}{2}(x + 3)^2 + 4$ has vertex $(-3, 4)$ and opens down. It is graph (h).

20. The graph of $y = (x - 3)^2$ has vertex $(3, 0)$ and opens up. It is graph (a).

21. The graph of $y = -(x + 3)^2 + 4$ has vertex $(-3, 4)$ and opens down. It is graph (c).

22. The graph of $y = 2(x - 1)^2 - 4$ has vertex $(1, -4)$ and opens up. It is graph (d).

23. $f(x) = x^2 - 6x + 5$

a) The x-coordinate of the vertex is

$-\frac{b}{2a} = -\frac{-6}{2 \cdot 1} = 3.$

Since $f(3) = 3^2 - 6 \cdot 3 + 5 = -4$, the vertex is $(3, -4)$.

b) Since $a = 1 > 0$, the graph opens up so the second coordinate of the vertex, -4, is the minimum value of the function.

c) The range is $[-4, \infty)$.

d) Since the graph opens up, function values decrease to the left of the vertex and increase to the right of the vertex. Thus, $f(x)$ is increasing on $(3, \infty)$ and decreasing on $(-\infty, 3)$.

24. $f(x) = x^2 + 4x - 5$

a) $-\frac{b}{2a} = -\frac{4}{2 \cdot 1} = -2$

$f(-2) = (-2)^2 + 4(-2) - 5 = -9$

The vertex is $(-2, -9)$.

b) Since $a = 1 > 0$, the graph opens up. The minimum value of $f(x)$ is -9.

c) Range: $[-9, \infty)$

d) Increasing: $(-2, \infty)$; decreasing: $(-\infty, -2)$

25. $f(x) = 2x^2 + 4x - 16$

a) The x-coordinate of the vertex is

$$-\frac{b}{2a} = -\frac{4}{2 \cdot 2} = -1.$$

Since $f(-1) = 2(-1)^2 + 4(-1) - 16 = -18$, the vertex is $(-1, -18)$.

b) Since $a = 2 > 0$, the graph opens up so the second coordinate of the vertex, -18, is the minimum value of the function.

c) The range is $[-18, \infty)$.

d) Since the graph opens up, function values decrease to the left of the vertex and increase to the right of the vertex. Thus, $f(x)$ is increasing on $(-1, \infty)$ and decreasing on $(-\infty, -1)$.

26. $f(x) = \frac{1}{2}x^2 - 3x + \frac{5}{2}$

a) $-\frac{b}{2a} = -\frac{-3}{2 \cdot \frac{1}{2}} = 3$

$f(3) = \frac{1}{2} \cdot 3^2 - 3 \cdot 3 + \frac{5}{2} = -2$

The vertex is $(3, -2)$.

b) Since $a = \frac{1}{2} > 0$, the graph opens up. The minimum value of $f(x)$ is -2.

c) Range: $[-2, \infty)$

d) Increasing: $(3, \infty)$; decreasing: $(-\infty, 3)$

27. $f(x) = -\frac{1}{2}x^2 + 5x - 8$

a) The x-coordinate of the vertex is

$$-\frac{b}{2a} = -\frac{5}{2\left(-\frac{1}{2}\right)} = 5.$$

Since $f(5) = -\frac{1}{2} \cdot 5^2 + 5 \cdot 5 - 8 = \frac{9}{2}$, the vertex is $\left(5, \frac{9}{2}\right)$.

b) Since $a = -\frac{1}{2} < 0$, the graph opens down so the second coordinate of the vertex, $\frac{9}{2}$, is the maximum value of the function.

c) The range is $\left(-\infty, \frac{9}{2}\right]$.

d) Since the graph opens down, function values increase to the left of the vertex and decreases to the right of the vertex. Thus, $f(x)$ is increasing on $(-\infty, 5)$ and decreasing on $(5, \infty,)$.

28. $f(x) = -2x^2 - 24x - 64$

a) $-\frac{b}{2a} = -\frac{-24}{2(-2)} = -6.$

$f(-6) = -2(-6)^2 - 24(-6) - 64 = 8$

The vertex is $(-6, 8)$.

b) Since $a = -2 < 0$, the graph opens down. The maximum value of $f(x)$ is 8.

c) Range: $(-\infty, 8]$

d) Increasing: $(-\infty, -6)$; decreasing: $(-6, \infty)$

29. $f(x) = 3x^2 + 6x + 5$

a) The x-coordinate of the vertex is

$$-\frac{b}{2a} = -\frac{6}{2 \cdot 3} = -1.$$

Since $f(-1) = 3(-1)^2 + 6(-1) + 5 = 2$, the vertex is $(-1, 2)$.

b) Since $a = 3 > 0$, the graph opens up so the second coordinate of the vertex, 2, is the minimum value of the function.

c) The range is $[2, \infty)$.

d) Since the graph opens up, function values decrease to the left of the vertex and increase to the right of the vertex. Thus, $f(x)$ is increasing on $(-1, \infty)$ and decreasing on $(-\infty, -1)$.

30. $f(x) = -3x^2 + 24x - 49$

a) $-\frac{b}{2a} = -\frac{24}{2(-3)} = 4.$

$f(4) = -3 \cdot 4^2 + 24 \cdot 4 - 49 = -1$

The vertex is $(4, -1)$.

b) Since $a = -3 < 0$, the graph opens down. The maximum value of $f(x)$ is -1.

c) Range: $(-\infty, -1]$

d) Increasing: $(-\infty, 4)$; decreasing: $(4, \infty)$

31. $g(x) = -4x^2 - 12x + 9$

a) The x-coordinate of the vertex is

$$-\frac{b}{2a} = -\frac{-12}{2(-4)} = -\frac{3}{2}.$$

Since $g\left(-\frac{3}{2}\right) = -4\left(-\frac{3}{2}\right)^2 - 12\left(-\frac{3}{2}\right) + 9 = 18$, the vertex is $\left(-\frac{3}{2}, 18\right)$.

b) Since $a = -4 < 0$, the graph opens down so the second coordinate of the vertex, 18, is the maximum value of the function.

c) The range is $(-\infty, 18]$.

d) Since the graph opens down, function values increase to the left of the vertex and decrease to the right of the vertex. Thus, $g(x)$ is increasing on $\left(-\infty, -\frac{3}{2}\right)$ and decreasing on $\left(-\frac{3}{2}, \infty\right)$.

32. $g(x) = 2x^2 - 6x + 5$

a) $-\frac{b}{2a} = -\frac{-6}{2 \cdot 2} = \frac{3}{2}$

$g\left(\frac{3}{2}\right) = 2\left(\frac{3}{2}\right)^2 - 6\left(\frac{3}{2}\right) + 5 = \frac{1}{2}$

The vertex is $\left(\frac{3}{2}, \frac{1}{2}\right)$.

b) Since $a = 2 > 0$, the graph opens up. The minimum value of $g(x)$ is $\frac{1}{2}$.

c) Range: $\left[\frac{1}{2}, \infty\right)$

d) Increasing: $\left(\frac{3}{2}, \infty\right)$; decreasing: $\left(-\infty, \frac{3}{2}\right)$

33. ***Familiarize and Translate***. The function $s(t) = -16t^2 + 20t + 6$ is given in the statement of the problem.

Carry out. The function $s(t)$ is quadratic and the coefficient of t^2 is negative, so $s(t)$ has a maximum value. It occurs at the vertex of the graph of the function. We find the first coordinate of the vertex. This is the time at which the ball reaches its maximum height.

$$t = -\frac{b}{2a} = -\frac{20}{2(-16)} = 0.625$$

The second coordinate of the vertex gives the maximum height.

$$s(0.625) = -16(0.625)^2 + 20(0.625) + 6 = 12.25$$

Check. Completing the square, we write the function in the form $s(t) = -16(t - 0.625)^2 + 12.25$. We see that the coordinates of the vertex are $(0.625, 12.25)$, so the answer checks.

State. The ball reaches its maximum height after 0.625 seconds. The maximum height is 12.25 ft.

34. Find the first coordinate of the vertex:

$$t = -\frac{60}{2(-16)} = 1.875$$

Then $s(1.875) = -16(1.875)^2 + 60(1.875) + 30 = 86.25$. Thus the maximum height is reached after 1.875 sec. The maximum height is 86.25 ft.

35. ***Familiarize and Translate***. The function $s(t) = -16t^2 + 120t + 80$ is given in the statement of the problem.

Carry out. The function $s(t)$ is quadratic and the coefficient of t^2 is negative, so $s(t)$ has a maximum value. It occurs at the vertex of the graph of the function. We find the first coordinate of the vertex. This is the time at which the rocket reaches its maximum height.

$$t = -\frac{b}{2a} = -\frac{120}{2(-16)} = 3.75$$

The second coordinate of the vertex gives the maximum height.

$$s(3.75) = -16(3.75)^2 + 120(3.75) + 80 = 305$$

Check. Completing the square, we write the function in the form $s(t) = -16(t - 3.75)^2 + 305$. We see that the coordinates of the vertex are $(3.75, 305)$, so the answer checks.

State. The rocket reaches its maximum height after 3.75 seconds. The maximum height is 305 ft.

36. Find the first coordinate of the vertex:

$$t = -\frac{150}{2(-16)} = 4.6875$$

Then $s(4.6875) = -16(4.6875)^2 + 150(4.6875) + 40 = 391.5625$. Thus the maximum height is reached after 4.6875 sec. The maximum height is 391.5625 ft.

37. ***Familiarize***. Using the label in the text, we let x = the height of the file. Then the length = 10 and the width = $18 - 2x$.

Translate. Since the volume of a rectangular solid is length × width × height we have

$$V(x) = 10(18 - 2x)x, \text{ or } -20x^2 + 180x.$$

Carry out. Since $V(x)$ is a quadratic function with $a = -20 < 0$, the maximum function value occurs at the vertex of the graph of the function. The first coordinate of the vertex is

$$-\frac{b}{2a} = -\frac{180}{2(-20)} = 4.5.$$

Check. When $x = 4.5$, then $18 - 2x = 9$ and $V(x) = 10 \cdot 9(4.5)$, or 405. As a partial check, we can find $V(x)$ for a value of x less than 4.5 and for a value of x greater than 4.5. For instance, $V(4.4) = 404.8$ and $V(4.6) = 404.8$. Since both of these values are less than 405, our result appears to be correct.

State. The file should be 4.5 in. tall in order to maximize the volume.

38. Let w = the width of the garden. Then the length = $32 - 2w$ and the area is given by $A(w) = (32 - 2w)w$, or $-2w^2 + 32w$. The maximum function value occurs at the vertex of the graph of $A(w)$. The first coordinate of the vertex is

$$-\frac{b}{2a} = -\frac{32}{2(-2)} = 8.$$

When $w = 8$, then $32 - 2w = 16$ and the area is $16 \cdot 8$, or 128 ft^2. A garden with dimensions 8 ft by 16 ft yields this area.

39. ***Familiarize***. Let b = the length of the base of the triangle. Then the height = $20 - b$.

Translate. Since the area of a triangle is $\frac{1}{2} \times$ base × height, we have

$$A(b) = \frac{1}{2}b(20 - b), \text{ or } -\frac{1}{2}b^2 + 10b.$$

Carry out. Since $A(b)$ is a quadratic function with $a = -\frac{1}{2} < 0$, the maximum function value occurs at the vertex of the graph of the function. The first coordinate of the vertex is

$$-\frac{b}{2a} = -\frac{10}{2\left(-\frac{1}{2}\right)} = 10.$$

When $b = 10$, then $20 - b = 20 - 10 = 10$, and the area is $\frac{1}{2} \cdot 10 \cdot 10 = 50$ cm^2.

Check. As a partial check, we can find $A(b)$ for a value of b less than 10 and for a value of b greater than 10. For instance, $V(9.9) = 49.995$ and $V(10.1) = 49.995$. Since both of these values are less than 50, our result appears to be correct.

State. The area is a maximum when the base and the height are both 10 cm.

40. Let b = the length of the base. Then $69 - b$ = the height and $A(b) = b(69-b)$, or $-b^2+69b$. The maximum function value occurs at the vertex of the graph of $A(b)$. The first coordinate of the vertex is

$$-\frac{b}{2a} = -\frac{69}{2(-1)} = 34.5.$$

When $b = 34.5$, then $69-b = 34.5$. The area is a maximum when the base and height are both 34.5 cm.

41. $C(x) = 0.1x^2 - 0.7x + 2.425$

Since $C(x)$ is a quadratic function with $a = 0.1 > 0$, a minimum function value occurs at the vertex of the graph of $C(x)$. The first coordinate of the vertex is

$$-\frac{b}{2a} = -\frac{-0.7}{2(0.1)} = 3.5.$$

Thus, 3.5 hundred, or 350 bicycles should be built to minimize the average cost per bicycle.

42. $P(x) = R(x) - C(x)$

$P(x) = 5x - (0.001x^2 + 1.2x + 60)$

$P(x) = -0.001x^2 + 3.8x - 60$

Since $P(x)$ is a quadratic function with $a = -0.001 < 0$, a maximum function value occurs at the vertex of the graph of the function. The first coordinate of the vertex is

$$-\frac{b}{2a} = -\frac{3.8}{2(-0.001)} = 1900.$$

$P(1900) = -0.001(1900)^2 + 3.8(1900) - 60 = 3550$

Thus, the maximum profit is \$3550. It occurs when 1900 units are sold.

43. $P(x) = R(x) - C(x)$

$P(x) = (50x - 0.5x^2) - (10x + 3)$

$P(x) = -0.5x^2 + 40x - 3$

Since $P(x)$ is a quadratic function with $a = -0.5 < 0$, a maximum function value occurs at the vertex of the graph of the function. The first coordinate of the vertex is

$$-\frac{b}{2a} = -\frac{40}{2(-0.5)} = 40.$$

$P(40) = -0.5(40)^2 + 40 \cdot 40 - 3 = 797$

Thus, the maximum profit is \$797. It occurs when 40 units are sold.

44. $P(x) = R(x) - C(x)$

$P(x) = 20x - 0.1x^2 - (4x + 2)$

$P(x) = -0.1x^2 + 16x - 2$

Since $P(x)$ is a quadratic function with $a = -0.1 < 0$, a maximum function value occurs at the vertex of the graph of the function. The first coordinate of the vertex is

$$-\frac{b}{2a} = -\frac{16}{2(-0.1)} = 80.$$

$P(80) = -0.1(80)^2 + 16(80) - 2 = 638$

Thus, the maximum profit is \$638. It occurs when 80 units are sold.

45. ***Familiarize.*** We let s = the height of the elevator shaft, t_1 = the time it takes the screwdriver to reach the bottom of the shaft, and t_2 = the time it takes the sound to reach the top of the shaft.

Translate. We know that $t_1 + t_2 = 5$. Using the information in Example 4 we also know that

$$s = 16t_1^2, \quad \textit{or} \quad t_1 = \frac{\sqrt{x}}{4} \text{ and}$$

$$s = 1100t_2, \quad \textit{or} \quad t_2 = \frac{s}{1100}.$$

Then $\dfrac{\sqrt{s}}{4} + \dfrac{s}{1100} = 5.$

Carry out. We solve the last equation above.

$$\frac{\sqrt{s}}{4} + \frac{s}{1100} = 5$$

$$275\sqrt{s} + s = 5500 \quad \text{Multiplying by 1100}$$

$$2 + 275\sqrt{s} - 5500 = 0$$

Let $u = \sqrt{s}$ and substitute.

$u^2 + 275u - 5500 = 0$

$$u = \frac{-b + \sqrt{b^2 - 4ac}}{2a} \quad \text{We only want the positive solution.}$$

$$= \frac{-275 + \sqrt{275^2 - 4 \cdot 1(-5500)}}{2 \cdot 1}$$

$$= \frac{-275 + \sqrt{97,625}}{2} \approx 18.725$$

Since $u \approx 18.725$, we have $\sqrt{s} = 18.725$, so $s \approx 350.6$.

Check. If $s \approx 350.6$, then $t_1 = \dfrac{\sqrt{s}}{4} = \dfrac{\sqrt{350.6}}{4} \approx 4.68$ and $t_2 = \dfrac{s}{1100} = \dfrac{350.6}{1100} \approx 0.32$, so $t_1 + t_2 = 4.68 + 0.32 = 5$.

The result checks.

State. The elevator shaft is about 350.6 ft tall.

46. Let s = the height of the cliff, t_1 = the time it takes the balloon to hit the ground, and t_2 = the time it takes for the sound to reach the top of the cliff. Then we have

$t_1 + t_2 = 3,$

$s = 16t_1^2, \quad or \quad t_1 = \frac{\sqrt{s}}{4}$, and

$s = 1100t_2, \quad or \quad t_2 = \frac{s}{1100}$, so

$\frac{\sqrt{s}}{4} + \frac{s}{1100} = 3.$

Solving the last equation, we find that $s \approx 132.7$ ft.

47. ***Familiarize.*** Using the labels on the drawing in the text, we let $x =$ the width of each corral and $240 - 3x =$ the total length of the corrals.

Translate. Since the area of a rectangle is length $\times$ width, we have

$$A(x) = (240 - 3x)x = -3x^2 + 240x.$$

Carry out. Since $A(x)$ is a quadratic function with $a = -3 < 0$, the maximum function value occurs at the vertex of the graph of $A(x)$. The first coordinate of the vertex is

$$-\frac{b}{2a} = -\frac{240}{2(-3)} = 40.$$

$$A(40) = -3(40)^2 + 240(40) = 4800$$

Check. As a partial check we can find $A(x)$ for a value of x less than 40 and for a value of x greater than 40. For instance, $A(39.9) = 4799.97$ and $A(40.1) = 4799.97$. Since both of these values are less than 4800, our result appears to be correct.

State. The largest total area that can be enclosed is 4800 yd^2.

48. $\frac{1}{2} \cdot 2\pi x + 2x + 2y = 24$, so $y = 12 - \frac{\pi x}{2} - x$.

$$A(x) = \frac{1}{2} \cdot \pi x^2 + 2x\left(12 - \frac{\pi x}{2} - x\right)$$

$$A(x) = \frac{\pi x^2}{2} + 24x - \pi x^2 - 2x^2$$

$$A(x) = 24x - \frac{\pi x^2}{2} - 2x^2, \text{ or } 24x - \left(\frac{\pi}{2} + 2\right)x^2$$

Since $A(x)$ is a quadratic function with $a = -\left(\frac{\pi}{2} + 2\right) < 0$, the maximum function value occurs at the vertex of the graph of $A(x)$. The first coordinate of the vertex is

$$\frac{-b}{2a} = -\frac{24}{2\left[-\left(\frac{\pi}{2} + 2\right)\right]} = \frac{24}{\pi + 4}.$$

When $x = \frac{24}{\pi + 4}$, then $y = \frac{24}{\pi + 4}$. Thus, the maximum amount of light will enter when the dimensions of the rectangular part of the window are $2x$ by y, or $\frac{48}{\pi + 4}$ ft by $\frac{24}{\pi + 4}$ ft, or approximately 6.72 ft by 3.36 ft.

49. Left to the student

50. Left to the student

51. Left to the student

52. Left to the student

53. Answers will vary. The problem could be similar to Examples 5 and 6 or Exercises 33 through 48.

54. Completing the square was used in Section 2.3 to solve quadratic equations. It was used again in this section to write quadratic functions in the form $f(x) = a(x - h)^2 + k$.

55. The x-intercepts of $g(x)$ are also $(x_1, 0)$ and $(x_2, 0)$. This is true because $f(x)$ and $g(x)$ have the same zeros. Consider $g(x) = 0$, or $-ax^2 - bx - c = 0$. Multiplying by -1 on both sides, we get an equivalent equation $ax^2 + bx + c = 0$, or $f(x) = 0$.

56.
$$\frac{f(x+h) - f(x)}{h} = \frac{3(x+h) - 7 - (3x - 7)}{h} = \frac{3x + 3h - 7 - 3x + 7}{h} = \frac{3h}{h} = 3$$

57. $f(x) = 2x^2 - x + 4$

$$f(x+h) = 2(x+h)^2 - (x+h) + 4 = 2(x^2 + 2xh + h^2) - (x+h) + 4 = 2x^2 + 4xh + 2h^2 - x - h - 4$$

$$\frac{f(x+h) - f(x)}{h}$$
$$= \frac{2x^2 + 4xh + 2h^2 - x - h - 4 - (2x^2 - x + 4)}{h}$$
$$= \frac{2x^2 + 4xh + 2h^2 - x - h - 4 - 2x^2 + x - 4}{h}$$
$$= \frac{4xh + 2h^2 - h}{h} = \frac{h(4x + 2h - 1)}{h}$$
$$= 4x + 2h - 1$$

58.

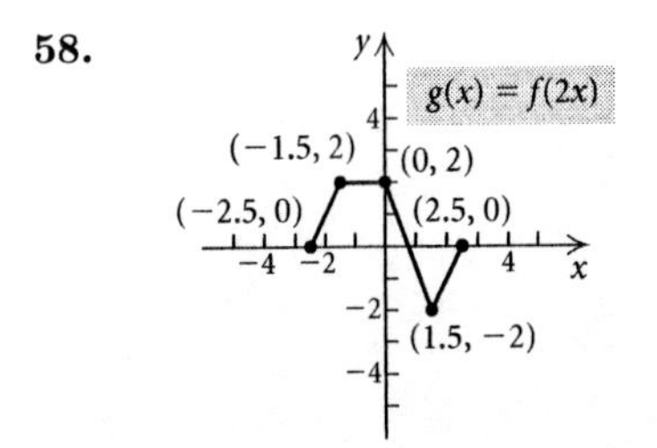

59. The graph of $f(x)$ is stretched vertically and reflected across the x-axis.

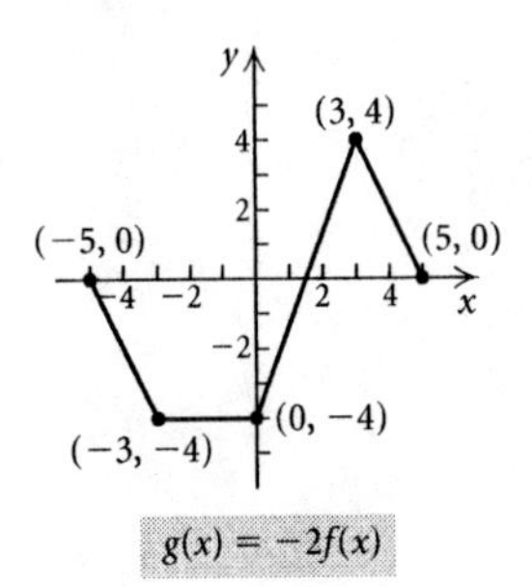

60. $f(x) = -4x^2 + bx + 3$

The x-coordinate of the vertex of $f(x)$ is $-\dfrac{b}{2(-4)}$, or $\dfrac{b}{8}$.

Now we find b such that $f\left(\dfrac{b}{8}\right) = 50$.

$$-4\left(\frac{b}{8}\right)^2 + b\cdot\frac{b}{8} + 3 = 50$$
$$-\frac{b^2}{16} + \frac{b^2}{8} + 3 = 50$$
$$\frac{b^2}{16} = 47$$
$$b^2 = 16\cdot 47$$
$$b = \pm\sqrt{16\cdot 47}$$
$$b = \pm 4\sqrt{47}$$

61. $f(x) = -0.2x^2 - 3x + c$

The x-coordinate of the vertex of $f(x)$ is $-\dfrac{b}{2a} = -\dfrac{-3}{2(-0.2)} = -7.5$. Now we find c such that $f(-7.5) = -225$.

$$-0.2(-7.5)^2 - 3(-7.5) + c = -225$$
$$-11.25 + 22.5 + c = -225$$
$$c = -236.25$$

62. $f(x) = a(x-h)^2 + k$

$1 = a(-3-4)^2 - 5$, so $a = \dfrac{6}{49}$. Then $f(x) = \dfrac{6}{49}(x-4)^2 - 5$.

63.

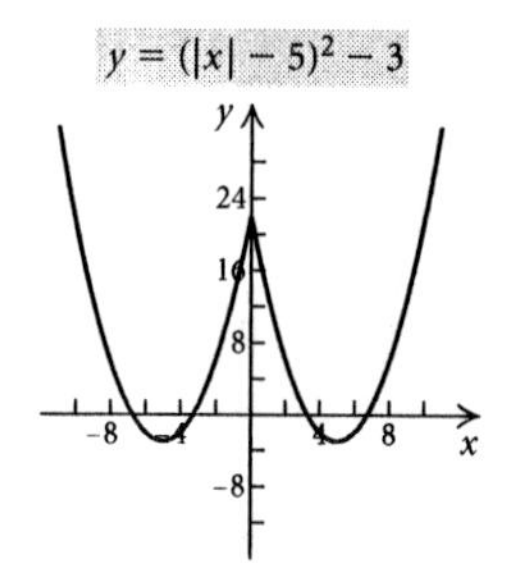

64. First we find the radius r of a circle with circumference x:

$$2\pi r = x$$
$$r = \frac{x}{2\pi}$$

Then we find the length s of a side of a square with perimeter $24 - x$:

$$4s = 24 - x$$
$$s = \frac{24-x}{4}$$

Then S = area of circle + area of square

$$S = \pi r^2 + s^2$$
$$S(x) = \pi\left(\frac{x}{2\pi}\right)^2 + \left(\frac{24-x}{4}\right)^2$$
$$S(x) = \left(\frac{1}{4\pi} + \frac{1}{16}\right)x^2 - 3x + 36$$

Since $S(x)$ is a quadratic function with $a = \dfrac{1}{4\pi} + \dfrac{1}{16} > 0$, the minimum function value occurs at the vertex of the graph of $S(x)$. The first coordinate of the vertex is

$$-\frac{b}{2a} = -\frac{-3}{2\left(\frac{1}{4\pi} + \frac{1}{16}\right)} = \frac{24\pi}{4+\pi}.$$

Then the string should be cut so that one piece is $\dfrac{24\pi}{4+\pi}$ in., or about 10.56 in. The other piece will be $24 - \dfrac{24\pi}{4+\pi}$, or $\dfrac{96}{4+\pi}$ in., or about 13.44 in.

Exercise Set 2.5

1. $\dfrac{1}{4} + \dfrac{1}{5} = \dfrac{1}{t}$, LCD is $20t$

$$20t\left(\frac{1}{4} + \frac{1}{5}\right) = 20t\cdot\frac{1}{t}$$
$$20t\cdot\frac{1}{4} + 20t\cdot\frac{1}{5} = 20t\cdot\frac{1}{t}$$
$$5t + 4t = 20$$
$$9t = 20$$
$$t = \frac{20}{9}$$

Check:

$\frac{1}{4} + \frac{1}{5} = \frac{1}{t}$		
$\frac{1}{4} + \frac{1}{5}$?	$\frac{1}{\frac{20}{9}}$	
$\frac{5}{20} + \frac{4}{20}$	$1\cdot\frac{9}{20}$	
$\frac{9}{20}$	$\frac{9}{20}$	TRUE

The solution is $\dfrac{20}{9}$.

2. $\dfrac{1}{3} - \dfrac{5}{6} = \dfrac{1}{x}$, LCD is $6x$

$2x - 5x = 6$ Multiplying by $6x$

$$-3x = 6$$
$$x = -2$$

-2 checks. The solution is -2.

3. $\frac{x+2}{4} - \frac{x-1}{5} = 15$, LCD is 20

$$20\left(\frac{x+2}{4} - \frac{x-1}{5}\right) = 20 \cdot 15$$
$$5(x+2) - 4(x-1) = 300$$
$$5x + 10 - 4x + 4 = 300$$
$$x + 14 = 300$$
$$x = 286$$

The solution is 286.

4. $\frac{t+1}{3} - \frac{t-1}{2} = 1$, LCD is 6

$2t + 2 - 3t + 3 = 6$ Multiplying by 6
$$-t = 1$$
$$t = -1$$

The solution is -1.

5. $\frac{1}{2} + \frac{2}{x} = \frac{1}{3} + \frac{3}{x}$, LCD is $6x$

$$6x\left(\frac{1}{2} + \frac{2}{x}\right) = 6x\left(\frac{1}{3} + \frac{3}{x}\right)$$
$$3x + 12 = 2x + 18$$
$$3x - 2x = 18 - 12$$
$$x = 6$$

Check:

$\frac{1}{2} + \frac{2}{x} = \frac{1}{3} + \frac{3}{x}$		
$\frac{1}{2} + \frac{2}{6}$?	$\frac{1}{3} + \frac{3}{6}$	
$\frac{1}{2} + \frac{1}{3}$	$\frac{1}{3} + \frac{1}{2}$	TRUE

The solution is 6.

6. $\frac{1}{t} + \frac{1}{2t} + \frac{1}{3t} = 5$, LCD is $6t$

$6 + 3 + 2 = 30t$ Multiplying by $6t$
$$11 = 30t$$
$$\frac{11}{30} = t$$

$\frac{11}{30}$ checks. The solution is $\frac{11}{30}$.

7. $\frac{3x}{x+2} + \frac{6}{x} = \frac{12}{x^2+2x}$

$\frac{3x}{x+2} + \frac{6}{x} = \frac{12}{x(x+2)}$, LCD is $x(x+2)$

$$x(x+2)\left(\frac{3x}{x+2} + \frac{6}{x}\right) = x(x+2) \cdot \frac{12}{x(x+2)}$$
$$3x \cdot x + 6(x+2) = 12$$
$$3x^2 + 6x + 12 = 12$$
$$3x^2 + 6x = 0$$
$$3x(x+2) = 0$$
$$3x = 0 \text{ or } x + 2 = 0$$
$$x = 0 \text{ or } x = -2$$

Neither 0 nor -2 checks, so the equation has no solution.

8. $\frac{5x}{x-4} - \frac{20}{x} = \frac{80}{x^2-4x}$

$\frac{5x}{x-4} - \frac{20}{x} = \frac{80}{x(x-4)}$, LCD is $x(x-4)$

$5x^2 - 20x + 80 = 80$ Multiplying by $x(x-4)$
$$5x^2 - 20x = 0$$
$$5x(x-4) = 0$$
$$x = 0 \text{ or } x = 4$$

Neither 0 nor 4 checks. There is no solution.

9. $\frac{4}{x^2-1} - \frac{2}{x-1} = \frac{3}{x+1}$,
LCD is $(x+1)(x-1)$

$$(x+1)(x-1)\left(\frac{4}{(x+1)(x-1)} - \frac{2}{x-1}\right) = (x+1)(x-1) \cdot \frac{3}{x+1}$$
$$4 - 2(x+1) = 3(x-1)$$
$$4 - 2x - 2 = 3x - 3$$
$$2 - 2x = 3x - 3$$
$$2 + 3 = 3x + 2x$$
$$5 = 5x$$
$$1 = x$$

Check:

$\frac{4}{x^2-1} - \frac{2}{x-1} = \frac{3}{x+1}$	
$\frac{4}{1^2-1} - \frac{2}{1-1}$?	$\frac{3}{1+1}$
$\frac{4}{0} - \frac{2}{0}$	$\frac{3}{2}$

Division by zero is undefined.

There is no solution.

10. $\frac{3y+5}{y^2+5y} + \frac{y+4}{y+5} = \frac{y+1}{y}$

$\frac{3y+5}{y(y+5)} + \frac{y+4}{y+5} = \frac{y+1}{y}$, LCD is $y(y+5)$

$3y + 5 + y^2 + 4y = y^2 + 6y + 5$ Multiplying by $y(y+5)$
$$y = 0$$

0 does not check. There is no solution.

11.
$$\frac{490}{x^2-49} = \frac{5x}{x-7} - \frac{35}{x+7}$$
$$\frac{490}{(x+7)(x-7)} = \frac{5x}{x-7} - \frac{35}{x+7},$$
LCD is $(x+7)(x-7)$
$$(x+7)(x-7)\left(\frac{490}{(x+7)(x-7)}\right) = (x+7)(x-7)\left(\frac{5x}{x-7} - \frac{35}{x+7}\right)$$
$$490 = 5x(x+7) - 35(x-7)$$
$$490 = 5x^2 + 35x - 35x + 245$$
$$0 = 5x^2 - 245$$
$$0 = 5(x+7)(x-7)$$
$x+7=0$ or $x-7=0$

$x=-7$ or $x=7$

Neither -7 nor 7 checks, so the equation has no solution.

12.
$$\frac{3}{m+2} + \frac{2}{m} = \frac{4m-4}{m^2-4}$$
$$\frac{3}{m+2} + \frac{2}{m} = \frac{4m-4}{(m+2)(m-2)},$$
LCD is $m(m+2)(m-2)$
$$3m^2 - 6m + 2m^2 - 8 = 4m^2 - 4m$$
Multiplying by $m(m+2)(m-2)$
$$m^2 - 2m - 8 = 0$$
$$(m-4)(m+2) = 0$$
$m=4$ or $m=-2$

Only 4 checks. The solution is 4.

13.
$$\frac{1}{x-6} - \frac{1}{x} = \frac{6}{x^2-6x}$$
$$\frac{1}{x-6} - \frac{1}{x} = \frac{6}{x(x-6)},$$ LCD is $x(x-6)$
$$x(x-6)\left(\frac{1}{x-6} - \frac{1}{x}\right) = x(x-6)\cdot\frac{6}{x(x-6)}$$
$$x - (x-6) = 6$$
$$x - x + 6 = 6$$
$$6 = 6$$
We get an equation that is true for all real numbers. Note, however, that when $x = 6$ or $x = 0$, division by 0 occurs in the original equation. Thus, the solution set is $\{x|x$ is a real number *and* $x \neq 6$ *and* $x \neq 0\}$, or $(-\infty, 0) \cup (0, 6) \cup (6, \infty)$.

14.
$$\frac{8}{x^2-4} = \frac{x}{x-2} - \frac{2}{x+2}$$
$$\frac{8}{(x+2)(x-2)} = \frac{x}{x-2} - \frac{2}{x+2},$$
LCD is $(x+2)(x-2)$
$$8 = x^2 + 2x - 2x + 4$$
Multiplying by $(x+2)(x-2)$
$$0 = x^2 - 4$$
$$0 = (x+2)(x-2)$$
$x=-2$ or $x=2$

Neither -2 nor 2 checks. There is no solution.

15.
$$\frac{x}{x-4} - \frac{4}{x+4} = \frac{32}{x^2-16}$$
$$\frac{x}{x-4} - \frac{4}{x+4} = \frac{32}{(x+4)(x-4)},$$
LCD is $(x+4)(x-4)$
$$(x+4)(x-4)\left(\frac{x}{x-4} - \frac{4}{x+4}\right) = (x+4)(x-4)\cdot\frac{32}{(x+4)(x-4)}$$
$$x(x+4) - 4(x-4) = 32$$
$$x^2 + 4x - 4x + 16 = 32$$
$$x^2 + 16 = 32$$
$$x^2 = 16$$
$$x = \pm 4$$
Neither 4 nor -4 checks, so the equation has no solution.

16.
$$\frac{x}{x-1} - \frac{1}{x+1} = \frac{2}{x^2-1}$$
$$\frac{x}{x-1} - \frac{1}{x+1} = \frac{2}{(x+1)(x-1)},$$
LCD is $(x+1)(x-1)$
$$x^2 + x - x + 1 = 2$$
$$x^2 = 1$$
$$x = \pm 1$$
Neither 1 nor -1 checks. There is no solution.

17.
$$\frac{x}{x+3} + \frac{3}{x-3} = \frac{18}{x^2-9}$$
$$\frac{x}{x+3} + \frac{3}{x-3} = \frac{18}{(x+3)(x-3)},$$
LCD is $(x+3)(x-3)$
$$(x+3)(x-3)\left(\frac{x}{x+3} + \frac{3}{x-3}\right) = (x+3)(x-3)\cdot\frac{18}{(x+3)(x-3)}$$
$$x(x-3) + 3(x+3) = 18$$
$$x^2 - 3x + 3x + 9 = 18$$
$$x^2 + 9 = 18$$
$$x^2 = 9$$
$$x = \pm 3$$
Neither 3 nor -3 checks, so the equation has no solution.

18.
$$\frac{x}{x-5} - \frac{5}{x+5} = \frac{50}{x^2-25}$$
$$\frac{x}{x-5} - \frac{5}{x+5} = \frac{50}{(x+5)(x-5)},$$
LCD is $(x+5)(x-5)$
$$x^2 + 5x - 5x + 25 = 50$$
$$x^2 = 25$$
$$x = \pm 5$$
Neither 5 nor -5 checks. There is no solution.

19. $$\frac{1}{5x+20}-\frac{1}{x^2-16}=\frac{3}{x-4}$$

$$\frac{1}{5(x+4)}-\frac{1}{(x+4)(x-4)}=\frac{3}{x-4},$$

LCD is $5(x+4)(x-4)$

$$5(x+4)(x-4)\left(\frac{1}{5(x+4)}-\frac{1}{(x+4)(x-4)}\right)=5(x+4)(x-4)\cdot\frac{3}{x-4}$$

$$x-4-5=15(x+4)$$
$$x-9=15x+60$$
$$-14x-9=60$$
$$-14x=69$$
$$x=-\frac{69}{14}$$

$-\frac{69}{14}$ checks, so the solution is $-\frac{69}{14}$.

20. $$\frac{1}{4x+12}-\frac{1}{x^2-9}=\frac{5}{x-3}$$

$$\frac{1}{4(x+3)}-\frac{1}{(x+3)(x-3)}=\frac{5}{x-3},$$

LCD is $4(x+3)(x-3)$

$$x-3-4=20x+60$$
$$-19x=67$$
$$x=-\frac{67}{19}$$

$-\frac{67}{19}$ checks. The solution is $-\frac{67}{19}$.

21. $$\frac{2}{5x+5}-\frac{3}{x^2-1}=\frac{4}{x-1}$$

$$\frac{2}{5(x+1)}-\frac{3}{(x+1)(x-1)}=\frac{4}{x-1},$$

LCD is $5(x+1)(x-1)$

$$5(x+1)(x-1)\left(\frac{2}{5(x+1)}-\frac{3}{(x+1)(x-1)}\right)=5(x+1)(x-1)\cdot\frac{4}{x-1}$$

$$2(x-1)-5\cdot 3=20(x+1)$$
$$2x-2-15=20x+20$$
$$2x-17=20x+20$$
$$-18x-17=20$$
$$-18x=37$$
$$x=-\frac{37}{18}$$

$-\frac{37}{18}$ checks, so the solution is $-\frac{37}{18}$.

22. $$\frac{1}{3x+6}-\frac{1}{x^2-4}=\frac{3}{x-2}$$

$$\frac{1}{3(x+2)}-\frac{1}{(x+2)(x-2)}=\frac{3}{x-2},$$

LCD is $3(x+2)(x-2)$

$$x-2-3=9x+18$$
$$x-5=9x+18$$
$$-8x=23$$
$$x=-\frac{23}{8}$$

$-\frac{23}{8}$ checks. The solution is $-\frac{23}{8}$.

23. $$\frac{8}{x^2-2x+4}=\frac{x}{x+2}+\frac{24}{x^3+8},$$

LCD is $(x+2)(x^2-2x+4)$

$$(x+2)(x^2-2x+4)\cdot\frac{8}{x^2-2x+4}=$$
$$(x+2)(x^2-2x+4)\left(\frac{x}{x+2}+\frac{24}{(x+2)(x^2-2x+4)}\right)$$

$$8(x+2)=x(x^2-2x+4)+24$$
$$8x+16=x^3-2x^2+4x+24$$
$$0=x^3-2x^2-4x+8$$
$$0=x^2(x-2)-4(x-2)$$
$$0=(x-2)(x^2-4)$$
$$0=(x-2)(x+2)(x-2)$$

$x-2=0$ or $x+2=0$ or $x-2=0$

$x=2$ or $x=-2$ or $x=2$

Only 2 checks. The solution is 2.

24. $$\frac{18}{x^2-3x+9}-\frac{x}{x+3}=\frac{81}{x^3+27},$$

LCD is $(x+3)(x^2-3x+9)$

$18x+54-x^3+3x^2-9x=81$ Multiplying by $(x+3)(x^2-3x+9)$

$$-x^3+3x^2+9x-27=0$$
$$-x^2(x-3)+9(x-3)=0$$
$$(x-3)(9-x^2)=0$$
$$(x-3)(3+x)(3-x)=0$$

$x=3$ or $x=-3$

Only 3 checks. The solution is 3.

25. $$\sqrt{3x-4}=1$$
$$(\sqrt{3x-4})^2=1^2$$
$$3x-4=1$$
$$3x=5$$
$$x=\frac{5}{3}$$

Check:

$\sqrt{3x-4}=1$	
$\sqrt{3\cdot\frac{5}{3}-4}$? 1	
$\sqrt{5-4}$	
$\sqrt{1}$	
1	1 TRUE

The solution is $\frac{5}{3}$.

26. $$\sqrt{4x+1}=3$$
$$4x+1=9$$
$$4x=8$$
$$x=2$$

The answer checks. The solution is 2.

27.
$$\sqrt{2x-5} = 2$$
$$(\sqrt{2x-5})^2 = 2^2$$
$$2x-5 = 4$$
$$2x = 9$$
$$x = \frac{9}{2}$$

Check:

$\sqrt{2x-5} = 2$	
$\sqrt{2\cdot\frac{9}{2}-5}$?	2
$\sqrt{9-5}$	
$\sqrt{4}$	
2	2 TRUE

The solution is $\frac{9}{2}$.

28.
$$\sqrt{3x+2} = 6$$
$$3x+2 = 36$$
$$3x = 34$$
$$x = \frac{34}{3}$$

The answer checks. The solution is $\frac{34}{3}$.

29.
$$\sqrt{7-x} = 2$$
$$(\sqrt{7-x})^2 = 2^2$$
$$7-x = 4$$
$$-x = -3$$
$$x = 3$$

Check:

$\sqrt{7-x} = 2$	
$\sqrt{7-3}$?	2
$\sqrt{4}$	
2	2 TRUE

The solution is 3.

30.
$$\sqrt{5-x} = 1$$
$$5-x = 1$$
$$4 = x$$

The answer checks. The solution is 4.

31.
$$\sqrt{1-2x} = 3$$
$$(\sqrt{1-2x})^2 = 3^2$$
$$1-2x = 9$$
$$-2x = 8$$
$$x = -4$$

Check:

$\sqrt{1-2x} = 3$	
$\sqrt{1-2(-4)}$?	3
$\sqrt{1+8}$	
$\sqrt{9}$	
3	3 TRUE

The solution is -4.

32.
$$\sqrt{2-7x} = 2$$
$$2-7x = 4$$
$$-7x = 2$$
$$x = -\frac{2}{7}$$

The answer checks. The solution is $-\frac{2}{7}$.

33.
$$\sqrt[3]{5x-2} = -3$$
$$(\sqrt[3]{5x-2})^3 = (-3)^3$$
$$5x-2 = -27$$
$$5x = -25$$
$$x = -5$$

Check:

$\sqrt[3]{5x-2} = -3$	
$\sqrt[3]{5(-5)-2}$?	-3
$\sqrt[3]{-25-2}$	
$\sqrt[3]{-27}$	
-3	-3 TRUE

The solution is -5.

34.
$$\sqrt[3]{2x+1} = -5$$
$$2x+1 = -125$$
$$2x = -126$$
$$x = -63$$

The answer checks. The solution is -63.

35.
$$\sqrt[4]{x^2-1} = 1$$
$$(\sqrt[4]{x^2-1})^4 = 1^4$$
$$x^2-1 = 1$$
$$x^2 = 2$$
$$x = \pm\sqrt{2}$$

Check:

$\sqrt[4]{x^2-1} = 1$	
$\sqrt[4]{(\pm\sqrt{2})^2-1}$?	1
$\sqrt[4]{2-1}$	
$\sqrt[4]{1}$	
1	1 TRUE

The solutions are $\pm\sqrt{2}$.

36. $\sqrt[5]{3x+4} = 2$

$3x + 4 = 32$

$3x = 28$

$x = \frac{28}{3}$

The answer checks. The solution is $\frac{28}{3}$.

37. $\sqrt{y-1} + 4 = 0$

$\sqrt{y-1} = -4$

The principal square root is never negative. Thus, there is no solution.

If we do not observe the above fact, we can continue and reach the same answer.

$(\sqrt{y-1})^2 = (-4)^2$

$y - 1 = 16$

$y = 17$

Check:

$\sqrt{y-1}+4 = 0$	
$\sqrt{17-1}+4$?	0
$\sqrt{16}+4$	
$4+4$	
8	0 FALSE

Since 17 does not check, there is no solution.

38. $\sqrt{m+1} - 5 = 8$

$\sqrt{m+1} = 13$

$m + 1 = 169$

$m = 168$

The answer checks. The solution is 168.

39. $\sqrt{b+3} - 2 = 1$

$\sqrt{b+3} = 3$

$(\sqrt{b+3})^2 = 3^2$

$b + 3 = 9$

$b = 6$

Check:

$\sqrt{b+3}-2 = 1$	
$\sqrt{6+3}-2$?	1
$\sqrt{9}-2$	
$3-2$	
1	1 TRUE

The solution is 6.

40. $\sqrt{x-4} + 1 = 5$

$\sqrt{x-4} = 4$

$x - 4 = 16$

$x = 20$

The answer checks. The solution is 20.

41. $\sqrt{z+2} + 3 = 4$

$\sqrt{z+2} = 1$

$(\sqrt{z+2})^2 = 1^2$

$z + 2 = 1$

$z = -1$

Check:

$\sqrt{z+2}+3 = 4$	
$\sqrt{-1+2}+3$?	4
$\sqrt{1}+3$	
$1+3$	
4	4 TRUE

The solution is -1.

42. $\sqrt{y-5} - 2 = 3$

$\sqrt{y-5} = 5$

$y - 5 = 25$

$y = 30$

The answer checks. The solution is 30.

43. $\sqrt{2x+1} - 3 = 3$

$\sqrt{2x+1} = 6$

$(\sqrt{2x+1})^2 = 6^2$

$2x + 1 = 36$

$2x = 35$

$x = \frac{35}{2}$

Check:

$\sqrt{2x+1}-3 = 3$	
$\sqrt{2\cdot\frac{35}{2}+1}-3$?	3
$\sqrt{35+1}-3$	
$\sqrt{36}-3$	
$6-3$	
3	3 TRUE

The solution is $\frac{35}{2}$.

44. $\sqrt{3x-1} + 2 = 7$

$\sqrt{3x-1} = 5$

$3x - 1 = 25$

$3x = 26$

$x = \frac{26}{3}$

The answer checks. The solution is $\frac{26}{3}$.

45. $\sqrt{2-x} - 4 = 6$

$\sqrt{2-x} = 10$

$(\sqrt{2-x})^2 = 10^2$

$2 - x = 100$

$-x = 98$

$x = -98$

Check:

$$\begin{array}{c|c} \sqrt{2-x}-4 & 6 \\ \hline \sqrt{2-(-98)}-4 & ?\ 6 \\ \sqrt{100}-4 & \\ 10-4 & \\ 6 & 6 \end{array} \quad \text{TRUE}$$

The solution is -98.

46. $\sqrt{5-x}+2=8$

$$\begin{aligned} \sqrt{5-x} &= 6 \\ 5-x &= 36 \\ -x &= 31 \\ x &= -31 \end{aligned}$$

The answer checks. The solution is -31.

47. $\sqrt[3]{6x+9}+8=5$

$$\begin{aligned} \sqrt[3]{6x+9} &= -3 \\ (\sqrt[3]{6x+9})^3 &= (-3)^3 \\ 6x+9 &= -27 \\ 6x &= -36 \\ x &= -6 \end{aligned}$$

Check:

$$\begin{array}{c|c} \sqrt[3]{6x+9}+8 & 5 \\ \hline \sqrt[3]{6(-6)+9}+8 & ?\ 5 \\ \sqrt[3]{-27}+8 & \\ -3+8 & \\ 5 & 5 \end{array} \quad \text{TRUE}$$

The solution is -6.

48. $\sqrt[5]{2x-3}-1=1$

$$\begin{aligned} \sqrt[5]{2x-3} &= 2 \\ 2x-3 &= 32 \\ 2x &= 35 \\ x &= \frac{35}{2} \end{aligned}$$

The answer checks. The solution is $\frac{35}{2}$.

49. $\sqrt{x+4}+2=x$

$$\begin{aligned} \sqrt{x+4} &= x-2 \\ (\sqrt{x+4})^2 &= (x-2)^2 \\ x+4 &= x^2-4x+4 \\ 0 &= x^2-5x \\ 0 &= x(x-5) \end{aligned}$$

$x=0$ *or* $x-5=0$

$x=0$ *or* $x=5$

Check:

For 0:

$$\begin{array}{c|c} \sqrt{x+4}+2 & x \\ \hline \sqrt{0+4}+2 & ?\ 0 \\ 2+2 & \\ 4 & 0 \end{array} \quad \text{FALSE}$$

For 5:

$$\begin{array}{c|c} \sqrt{x+4}+2 & x \\ \hline \sqrt{5+4}+2 & ?\ 5 \\ \sqrt{9}+2 & \\ 3+2 & \\ 5 & 5 \end{array} \quad \text{TRUE}$$

The number 5 checks but 0 does not. The solution is 5.

50. $\sqrt{x+1}+1=x$

$$\begin{aligned} \sqrt{x+1} &= x-1 \\ x+1 &= x^2-2x+1 \\ 0 &= x^2-3x \\ 0 &= x(x-3) \end{aligned}$$

$x=0$ *or* $x=3$

Only 3 checks. The solution is 3.

51. $\sqrt{x-3}+5=x$

$$\begin{aligned} \sqrt{x-3} &= x-5 \\ (\sqrt{x-3})^2 &= (x-5)^2 \\ x-3 &= x^2-10x+25 \\ 0 &= x^2-11x+28 \\ 0 &= (x-4)(x-7) \end{aligned}$$

$x-4=0$ *or* $x-7=0$

$x=4$ *or* $x=7$

Check:

For 4:

$$\begin{array}{c|c} \sqrt{x-3}+5 & x \\ \hline \sqrt{4-3}+5 & ?\ 4 \\ \sqrt{1}+5 & \\ 1+5 & \\ 6 & 4 \end{array} \quad \text{FALSE}$$

For 7:

$$\begin{array}{c|c} \sqrt{x-3}+5 & x \\ \hline \sqrt{7-3}+5 & ?\ 7 \\ \sqrt{4}+5 & \\ 2+5 & \\ 7 & 7 \end{array} \quad \text{TRUE}$$

The number 7 checks but 4 does not. The solution is 7.

52. $\sqrt{x+3}-1=x$

$$\begin{aligned} \sqrt{x+3} &= x+1 \\ x+3 &= x^2+2x+1 \\ 0 &= x^2+x-2 \\ 0 &= (x+2)(x-1) \end{aligned}$$

$x=-2$ *or* $x=1$

Only 1 checks. The solution is 1.

53. $\sqrt{x+7} = x+1$

$(\sqrt{x+7})^2 = (x+1)^2$

$x+7 = x^2+2x+1$

$0 = x^2+x-6$

$0 = (x+3)(x-2)$

$x+3=0$ *or* $x-2=0$

$x=-3$ *or* $x=2$

Check:

For -3:

$\sqrt{x+7}$	$=$	$x+1$	
$\sqrt{-3+7}$	?	$-3+1$	
$\sqrt{4}$		-2	
2		-2	FALSE

For 2:

$\sqrt{x+7}$	$=$	$x+1$	
$\sqrt{2+7}$	?	$2+1$	
$\sqrt{9}$		3	
3		3	TRUE

The number 2 checks but -3 does not. The solution is 2.

54. $\sqrt{6x+7} = x+2$

$6x+7 = x^2+4x+4$

$0 = x^2-2x-3$

$0 = (x-3)(x+1)$

$x=3$ *or* $x=-1$

Both values check. The solutions are 3 and -1.

55. $\sqrt{3x+3} = x+1$

$(\sqrt{3x+3})^2 = (x+1)^2$

$3x+3 = x^2+2x+1$

$0 = x^2-x-2$

$0 = (x-2)(x+1)$

$x-2=0$ *or* $x+1=0$

$x=2$ *or* $x=-1$

Check:

For 2:

$\sqrt{3x+3}$	$=$	$x+1$	
$\sqrt{3\cdot 2+3}$	?	$2+1$	
$\sqrt{9}$		3	
3		3	TRUE

For -1:

$\sqrt{3x+3}$	$=$	$x+1$	
$\sqrt{3(-1)+3}$	?	$-1+1$	
$\sqrt{0}$		0	
0		0	TRUE

Both numbers check. The solutions are 2 and -1.

56. $\sqrt{2x+5} = x-5$

$2x+5 = x^2-10x+25$

$0 = x^2-12x+20$

$0 = (x-2)(x-10)$

$x=2$ *or* $x=10$

Only 10 checks. The solution is 10.

57. $\sqrt{5x+1} = x-1$

$(\sqrt{5x+1})^2 = (x-1)^2$

$5x+1 = x^2-2x+1$

$0 = x^2-7x$

$0 = x(x-7)$

$x=0$ *or* $x-7=0$

$x=0$ *or* $x=7$

Check:

For 0:

$\sqrt{5x+1}$	$=$	$x-1$	
$\sqrt{5\cdot 0+1}$	?	$0-1$	
$\sqrt{1}$		-1	
1		-1	FALSE

For 7:

$\sqrt{5x+1}$	$=$	$x-1$	
$\sqrt{5\cdot 7+1}$	?	$7-1$	
$\sqrt{36}$		6	
6		6	TRUE

The number 7 checks but 0 does not. The solution is 7.

58. $\sqrt{7x+4} = x+2$

$7x+4 = x^2+4x+4$

$0 = x^2-3x$

$0 = x(x-3)$

$x=0$ *or* $x=3$

Both numbers check. The solutions are 0 and 3.

59. $\sqrt{x-3}+\sqrt{x+2} = 5$

$\sqrt{x+2} = 5-\sqrt{x-3}$

$(\sqrt{x+2})^2 = (5-\sqrt{x-3})^2$

$x+2 = 25-10\sqrt{x-3}+(x-3)$

$x+2 = 22-10\sqrt{x-3}+x$

$10\sqrt{x-3} = 20$

$\sqrt{x-3} = 2$

$(\sqrt{x-3})^2 = 2^2$

$x-3 = 4$

$x = 7$

Check:

$\sqrt{x-3}+\sqrt{x+2}=5$		
$\sqrt{7-3}+\sqrt{7+2}$	? 5	
$\sqrt{4}+\sqrt{9}$		
$2+3$		
5	5	TRUE

The solution is 7.

60. $\sqrt{x}-\sqrt{x-5}=1$

$$\begin{aligned}\sqrt{x} &= \sqrt{x-5}+1\\ x &= x-5+2\sqrt{x-5}+1\\ 4 &= 2\sqrt{x-5}\\ 2 &= \sqrt{x-5}\\ 4 &= x-5\\ 9 &= x\end{aligned}$$

The answer checks. The solution is 9.

61. $\sqrt{3x-5}+\sqrt{2x+3}+1=0$

$$\sqrt{3x-5}+\sqrt{2x+3}=-1$$

The principal square root is never negative. Thus the sum of two principal square roots cannot equal -1. There is no solution.

62.

$$\begin{aligned}\sqrt{2m-3} &= \sqrt{m+7}-2\\ 2m-3 &= m+7-4\sqrt{m+7}+4\\ m-14 &= -4\sqrt{m+7}\\ m^2-28m+196 &= 16m+112\\ m^2-44m+84 &= 0\\ (m-2)(m-42) &= 0\end{aligned}$$

$m=2$ *or* $m=42$

Only 2 checks. The solution is 2.

63. $\sqrt{x}-\sqrt{3x-3}=1$

$$\begin{aligned}\sqrt{x} &= \sqrt{3x-3}+1\\ (\sqrt{x})^2 &= (\sqrt{3x-3}+1)^2\\ x &= (3x-3)+2\sqrt{3x-3}+1\\ 2-2x &= 2\sqrt{3x-3}\\ 1-x &= \sqrt{3x-3}\\ (1-x)^2 &= (\sqrt{3x-3})^2\\ 1-2x+x^2 &= 3x-3\\ x^2-5x+4 &= 0\\ (x-4)(x-1) &= 0\end{aligned}$$

$x=4$ *or* $x=1$

The number 4 does not check, but 1 does. The solution is 1.

64. $\sqrt{2x+1}-\sqrt{x}=1$

$$\begin{aligned}\sqrt{2x+1} &= \sqrt{x}+1\\ 2x+1 &= x+2\sqrt{x}+1\\ x &= 2\sqrt{x}\\ x^2 &= 4x\\ x^2-4x &= 0\\ x(x-4) &= 0\end{aligned}$$

$x=0$ *or* $x=4$

Both values check. The solutions are 0 and 4.

65. $\sqrt{2y-5}-\sqrt{y-3}=1$

$$\begin{aligned}\sqrt{2y-5} &= \sqrt{y-3}+1\\ (\sqrt{2y-5})^2 &= (\sqrt{y-3}+1)^2\\ 2y-5 &= (y-3)+2\sqrt{y-3}+1\\ y-3 &= 2\sqrt{y-3}\\ (y-3)^2 &= (2\sqrt{y-3})^2\\ y^2-6y+9 &= 4(y-3)\\ y^2-6y+9 &= 4y-12\\ y^2-10y+21 &= 0\\ (y-7)(y-3) &= 0\end{aligned}$$

$y=7$ *or* $y=3$

Both numbers check. The solutions are 7 and 3.

66. $\sqrt{4p+5}+\sqrt{p+5}=3$

$$\begin{aligned}\sqrt{4p+5} &= 3-\sqrt{p+5}\\ 4p+5 &= 9-6\sqrt{p+5}+p+5\\ 3p-9 &= -6\sqrt{p+5}\\ p-3 &= -2\sqrt{p+5}\\ p^2-6p+9 &= 4p+20\\ p^2-10p-11 &= 0\\ (p-11)(p+1) &= 0\end{aligned}$$

$p=11$ *or* $p=-1$

Only -1 checks. The solution is -1.

67. $\sqrt{y+4}-\sqrt{y-1}=1$

$$\begin{aligned}\sqrt{y+4} &= \sqrt{y-1}+1\\ (\sqrt{y+4})^2 &= (\sqrt{y-1}+1)^2\\ y+4 &= y-1+2\sqrt{y-1}+1\\ 4 &= 2\sqrt{y-1}\\ 2 &= \sqrt{y-1} \quad \text{Dividing by 2}\\ 2^2 &= (\sqrt{y-1})^2\\ 4 &= y-1\\ 5 &= y\end{aligned}$$

The answer checks. The solution is 5.

68. $\sqrt{y+7}+\sqrt{y+16}=9$

$$\begin{aligned}\sqrt{y+7} &= 9-\sqrt{y+16}\\ y+7 &= 81-18\sqrt{y+16}+y+16\\ -90 &= -18\sqrt{y+16}\\ 5 &= \sqrt{y+16}\\ 25 &= y+16\\ 9 &= y\end{aligned}$$

The answer checks. The solution is 9.

69. $\sqrt{x+5}+\sqrt{x+2}=3$

$$\begin{aligned}\sqrt{x+5} &= 3-\sqrt{x+2}\\ (\sqrt{x+5})^2 &= (3-\sqrt{x+2})^2\\ x+5 &= 9-6\sqrt{x+2}+x+2\\ -6 &= -6\sqrt{x+2}\\ 1 &= \sqrt{x+2} \qquad \text{Dividing by } -6\\ 1^2 &= (\sqrt{x+2})^2\\ 1 &= x+2\\ -1 &= x\end{aligned}$$

The answer checks. The solution is -1.

70. $\sqrt{6x+6}=5+\sqrt{21-4x}$

$$\begin{aligned}6x+6 &= 25+10\sqrt{21-4x}+21-4x\\ 10x-40 &= 10\sqrt{21-4x}\\ x-4 &= \sqrt{21-4x}\\ x^2-8x+16 &= 21-4x\\ x^2-4x-5 &= 0\\ (x-5)(x+1) &= 0\end{aligned}$$

$x=5$ *or* $x=-1$

Only 5 checks. The solution is 5.

71. $x^{1/3}=-2$

$(x^{1/3})^3=(-2)^3 \qquad (x^{1/3}=\sqrt[3]{x})$

$x=-8$

The value checks. The solution is -8.

72. $t^{1/5}=2$

$t=32$

The value checks. The solution is 32.

73. $t^{1/4}=3$

$(t^{1/4})^4=3^4 \qquad (t^{1/4}=\sqrt[4]{t})$

$t=81$

The value checks. The solution is 81.

74. $m^{1/2}=-7$

The principal square root is never negative. There is no solution.

75. $|x|=7$

The solutions are those numbers whose distance from 0 on a number line is 7. They are -7 and 7. That is,

$x=-7$ *or* $x=7$.

The solutions are -7 and 7.

76. $|x|=4.5$

$x=-4.5$ *or* $x=4.5$

The solutions are -4.5 and 4.5.

77. $|x|=-10.7$

The absolute value of a number is nonnegative. Thus, the equation has no solution.

78. $|x|=-\frac{3}{5}$

The absolute value of a number is nonnegative. Thus, there is no solution.

79. $|x-1|=4$

$x-1=-4$ *or* $x-1=4$

$x=-3$ *or* $x=5$

The solutions are -3 and 5.

80. $|x-7|=5$

$x-7=-5$ *or* $x-7=5$

$x=2$ *or* $x=12$

The solutions are 2 and 12.

81. $|3x|=1$

$3x=-1$ *or* $3x=1$

$x=-\frac{1}{3}$ *or* $x=\frac{1}{3}$

The solutions are $-\frac{1}{3}$ and $\frac{1}{3}$.

82. $|5x|=4$

$5x=-4$ *or* $5x=4$

$x=-\frac{4}{5}$ *or* $x=\frac{4}{5}$

The solutions are $-\frac{4}{5}$ and $\frac{4}{5}$.

83. $|x|=0$

The distance of 0 from 0 on a number line is 0. That is,

$x=0$.

The solution is 0.

84. $|6x|=0$

$6x=0$

$x=0$

The solution is 0.

85. $|3x+2|=1$

$3x+2=-1$ *or* $3x+2=1$

$3x=-3$ *or* $3x=-1$

$x=-1$ *or* $x=-\frac{1}{3}$

The solutions are -1 and $-\frac{1}{3}$.

86. $|7x-4|=8$

$7x-4=-8$ *or* $7x-4=8$

$7x=-4$ *or* $7x=12$

$x=-\frac{4}{7}$ *or* $x=\frac{12}{7}$

The solutions are $-\frac{4}{7}$ and $\frac{12}{7}$.

87. $\left|\frac{1}{2}x-5\right|=17$

$\frac{1}{2}x-5=-17$ *or* $\frac{1}{2}x-5=17$

$\frac{1}{2}x=-12$ *or* $\frac{1}{2}x=22$

$x=-24$ *or* $x=44$

The solutions are -24 and 44.

88. $\left|\frac{1}{3}x-4\right|=13$

$\frac{1}{3}x-4=-13$ *or* $\frac{1}{3}x-4=13$

$\frac{1}{3}x=-9$ *or* $\frac{1}{3}x=17$

$x=-27$ *or* $x=51$

The solutions are -27 and 51.

89. $|x-1|+3=6$

$|x-1|=3$

$x-1=-3$ *or* $x-1=3$

$x=-2$ *or* $x=4$

The solutions are -2 and 4.

90. $|x+2|-5=9$

$|x+2|=14$

$x+2=-14$ *or* $x+2=14$

$x=-16$ *or* $x=12$

The solutions are -16 and 12.

91. $|x+3|-2=8$

$|x+3|=10$

$x+3=-10$ *or* $x+3=10$

$x=-13$ *or* $x=7$

The solutions are -13 and 7.

92. $|x-4|+3=9$

$|x-4|=6$

$x-4=-6$ *or* $x-4=6$

$x=-2$ *or* $x=10$

The solutions are -2 and 10.

93. $|3x+1|-4=-1$

$|3x+1|=3$

$3x+1=-3$ *or* $3x+1=3$

$3x=-4$ *or* $3x=2$

$x=-\frac{4}{3}$ *or* $x=\frac{2}{3}$

The solutions are $-\frac{4}{3}$ and $\frac{2}{3}$.

94. $|2x-1|-5=-3$

$|2x-1|=2$

$2x-1=-2$ *or* $2x-1=2$

$2x=-1$ *or* $2x=3$

$x=-\frac{1}{2}$ *or* $x=\frac{3}{2}$

The solutions are $-\frac{1}{2}$ and $\frac{3}{2}$.

95. $|4x-3|+1=7$

$|4x-3|=6$

$4x-3=-6$ *or* $4x-3=6$

$4x=-3$ *or* $4x=9$

$x=-\frac{3}{4}$ *or* $x=\frac{9}{4}$

The solutions are $-\frac{3}{4}$ and $\frac{9}{4}$.

96. $|5x+4|+2=5$

$|5x+4|=3$

$5x+4=-3$ *or* $5x+4=3$

$5x=-7$ *or* $5x=-1$

$x=-\frac{7}{5}$ *or* $x=-\frac{1}{5}$

The solutions are $-\frac{7}{5}$ and $-\frac{1}{5}$.

97. $12-|x+6|=5$

$-|x+6|=-7$

$|x+6|=7$ Multiplying by -1

$x+6=-7$ *or* $x+6=7$

$x=-13$ *or* $x=1$

The solutions are -13 and 1.

98. $9-|x-2|=7$

$2=|x-2|$

$x-2=-2$ *or* $x-2=2$

$x=0$ *or* $x=4$

The solutions are 0 and 4.

99. $\frac{P_1V_1}{T_1}=\frac{P_2V_2}{T_2}$

$P_1V_1T_2=P_2V_2T_1$ Multiplying by T_1T_2 on both sides

$\frac{P_1V_1T_2}{P_2V_2}=T_1$ Dividing by P_2V_2 on both sides

100. $$\frac{1}{F}=\frac{1}{m}+\frac{1}{p}$$
$$mp=Fp+Fm$$
$$mp=F(p+m)$$
$$\frac{mp}{p+m}=F$$

101. $$\frac{1}{R}=\frac{1}{R_1}+\frac{1}{R_2}$$
$$RR_1R_2\cdot\frac{1}{R}=RR_1R_2\left(\frac{1}{R_1}+\frac{1}{R_2}\right)$$
Multiplying by RR_1R_2 on both sides
$$R_1R_2=RR_2+RR_1$$
$$R_1R_2-RR_2=RR_1$$ Subtracting RR_2 on both sides
$$R_2(R_1-R)=RR_1$$ Factoring
$$R_2=\frac{RR_1}{R_1-R}$$ Dividing by R_1-R on both sides

102. $$A=P(1+i)^2$$
$$\frac{A}{P}=(1+i)^2$$
$$\sqrt{\frac{A}{P}}=1+i$$
$$\sqrt{\frac{A}{P}}-1=i$$

103. $$\frac{1}{F}=\frac{1}{m}+\frac{1}{p}$$
$$Fmp\cdot\frac{1}{F}=Fmp\left(\frac{1}{m}+\frac{1}{p}\right)$$ Multiplying by Fmp on both sides
$$mp=Fp+Fm$$
$$mp-Fp=Fm$$ Subtracting Fp on both sides
$$p(m-F)=Fm$$ Factoring
$$p=\frac{Fm}{m-F}$$ Dividing by $m-F$ on both sides

104. Left to the student

105. Left to the student

106. When both sides of an equation are multiplied by the LCD, the resulting equation might not be equivalent to the original equation. One or more of the possible solutions of the resulting equation might make a denominator of the original equation 0.

107. When both sides of an equation are raised to an even power, the resulting equation might not be the equivalent to the original equation. For example, the solution set of $x=-2$ is $\{-2\}$, but the solution set of $x^2=(-2)^2$, or $x^2=4$, is $\{-2,2\}$.

108. $$-3x+9=0$$
$$-3x=-9$$
$$x=3$$
The zero of the function is 3.

109. $$15-2x=0$$ Setting $f(x)=0$
$$15=2x$$
$$\frac{15}{2}=x,\text{ or}$$
$$7.5=x$$
The zero of the function is $\frac{15}{2}$, or 7.5.

110. The amount of the increase is $28.4-24.4$, or 4.0 lb. Let $p=$ the percent of increase.

Solve: $p=\dfrac{4.0}{24.4}$

$p\approx 0.164$, or 16.4%

111. ***Familiarize.*** Let $a=$ the number of adults who passed the GED test in 1990.

Translate.

Number of adults who passed the test in 2000	was	82,000	more than	number who passed the test in 1990.
↓	↓	↓	↓	↓
501,000	=	82,000	+	a

Carry out. We solve the equation.
$$501,000=82,000+a$$
$$419,000=a$$ Subtracting 82,000

Check. 82,000 more than 419,000 is $82,000+419,000$, or 501,000. The answer checks.

State. In 1990, 419,000 adults passed the GED test.

112. $$\frac{x+3}{x+2}-\frac{x+4}{x+3}=\frac{x+5}{x+4}-\frac{x+6}{x+5},$$
LCD is $(x+2)(x+3)(x+4)(x+5)$
$$x^4+15x^3+83x^2+201x+180-x^4-15x^3-$$
$$82x^2-192x-160=x^4+15x^3+81x^2+185x+$$
$$150-x^4-15x^3-80x^2-180x-144$$
$$x^2+9x+20=x^2+5x+6$$
$$4x=-14$$
$$x=-\frac{7}{2}$$
The number $-\frac{7}{2}$ checks. The solution is $-\frac{7}{2}$.

113. $$(x-3)^{2/3}=2$$
$$[(x-3)^{2/3}]^3=2^3$$
$$(x-3)^2=8$$
$$x^2-6x+9=8$$
$$x^2-6x+1=0$$
$$a=1,\ b=-6,\ c=1$$

$$
\begin{aligned}
x &= \frac{-b \pm \sqrt{b^2 - 4ac}}{2a} \\
&= \frac{-(-6) \pm \sqrt{(-6)^2 - 4 \cdot 1 \cdot 1}}{2 \cdot 1} \\
&= \frac{6 \pm \sqrt{32}}{2} = \frac{6 \pm 4\sqrt{2}}{2} \\
&= \frac{2(3 \pm 2\sqrt{2})}{2} = 3 \pm 2\sqrt{2}
\end{aligned}
$$

Both values check. The solutions are $3 \pm 2\sqrt{2}$.

114.

$$
\begin{aligned}
\sqrt{15 + \sqrt{2x + 80}} &= 5 \\
\left(\sqrt{15 + \sqrt{2x + 80}}\right)^2 &= 5^2 \\
15 + \sqrt{2x + 80} &= 25 \\
\sqrt{2x + 80} &= 10 \\
(\sqrt{2x + 80})^2 &= 10^2 \\
2x + 80 &= 100 \\
2x &= 20 \\
x &= 10
\end{aligned}
$$

This number checks. The solution is 10.

115.

$$
\begin{aligned}
\sqrt{x + 5} + 1 &= \frac{6}{\sqrt{x + 5}}, \quad \text{LCD is } \sqrt{x + 5} \\
x + 5 + \sqrt{x + 5} &= 6 \quad \text{Multiplying by } \sqrt{x + 5} \\
\sqrt{x + 5} &= 1 - x \\
x + 5 &= 1 - 2x + x^2 \\
0 &= x^2 - 3x - 4 \\
0 &= (x - 4)(x + 1)
\end{aligned}
$$

$x = 4$ *or* $x = -1$

Only -1 checks. The solution set is -1.

116.

$$
\begin{aligned}
x^{2/3} &= x \\
(x^{2/3})^3 &= x^3 \\
x^2 &= x^3 \\
0 &= x^3 - x^2 \\
0 &= x^2(x - 1)
\end{aligned}
$$

$x^2 = 0$ *or* $x - 1 = 0$

$x = 0$ *or* $x = 1$

Both numbers check. The solutions are 0 and 1.

Exercise Set 2.6

1. $x + 6 < 5x - 6$

$6 + 6 < 5x - x$ Subtracting x and adding 6 on both sides

$12 < 4x$

$\frac{12}{4} < x$ Dividing by 4 on both sides

$3 < x$

This inequality could also be solved as follows:

$x + 6 < 5x - 6$

$x - 5x < -6 - 6$ Subtracting $5x$ and 6 on both sides

$-4x < -12$

$x > \frac{-12}{-4}$ Dividing by -4 on both sides and reversing the inequality symbol

$x > 3$

The solution set is $\{x|x > 3\}$, or $(3, \infty)$. The graph is shown below.

0 3

2. $3 - x < 4x + 7$

$-5x < 4$

$x > -\frac{4}{5}$

The solution set is $\left\{x \middle| x > -\frac{4}{5}\right\}$, or $\left(-\frac{4}{5}, \infty\right)$. The graph is shown below.

$-\frac{4}{5}$ 0

3. $3x - 3 + 2x \geq 1 - 7x - 9$

$5x - 3 \geq -7x - 8$ Collecting like terms

$5x + 7x \geq -8 + 3$ Adding $7x$ and 3 on both sides

$12x \geq -5$

$x \geq -\frac{5}{12}$ Dividing by 12 on both sides

The solution set is $\left\{x \middle| x \geq -\frac{5}{12}\right\}$, or $\left[-\frac{5}{12}, \infty\right)$. The graph is shown below.

$-\frac{5}{12}$ 0

4. $5y - 5 + y \leq 2 - 6y - 8$

$6y - 5 \leq -6y - 6$

$12y \leq -1$

$y \leq -\frac{1}{12}$

The solution set is $\left\{y \middle| y \leq -\frac{1}{12}\right\}$, or $\left(-\infty, -\frac{1}{12}\right]$. The graph is shown below.

$-\frac{1}{12}$ 0

5. $14 - 5y \leq 8y - 8$

$14 + 8 \leq 8y + 5y$

$22 \leq 13y$

$\frac{22}{13} \leq y$

This inequality could also be solved as follows:

$$14 - 5y \le 8y - 8$$
$$-5y - 8y \le -8 - 14$$
$$-13y \le -22$$
$$y \ge \frac{22}{13}$$ Dividing by -13 on both sides and reversing the inequality symbol

The solution set is $\left\{y\middle|y \ge \frac{22}{13}\right\}$, or $\left[\frac{22}{13}, \infty\right)$. The graph is shown below.

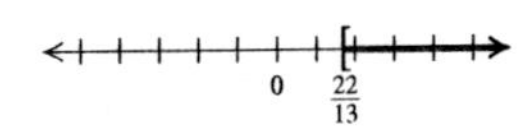

6. $8x - 7 < 6x + 3$

$$2x < 10$$
$$x < 5$$

The solution set is $\{x|x < 5\}$, or $(-\infty, 5)$. The graph is shown below.

7. $-\frac{3}{4}x \ge -\frac{5}{8} + \frac{2}{3}x$

$$\frac{5}{8} \ge \frac{3}{4}x + \frac{2}{3}x$$
$$\frac{5}{8} \ge \frac{9}{12}x + \frac{8}{12}x$$
$$\frac{5}{8} \ge \frac{17}{12}x$$
$$\frac{12}{17} \cdot \frac{5}{8} \ge \frac{12}{17} \cdot \frac{17}{12}x$$
$$\frac{15}{34} \ge x$$

The solution set is $\left\{x\middle|x \le \frac{15}{34}\right\}$, or $\left(-\infty, \frac{15}{34}\right]$. The graph is shown below.

8. $-\frac{5}{6}x \le \frac{3}{4} + \frac{8}{3}x$

$$-\frac{21}{6}x \le \frac{3}{4}$$
$$x \ge -\frac{3}{14}$$

The solution set is $\left\{x\middle|x \ge -\frac{3}{14}\right\}$, or $\left[-\frac{3}{14}, \infty\right)$. The graph is shown below.

9. $4x(x-2) < 2(2x-1)(x-3)$

$$4x(x-2) < 2(2x^2 - 7x + 3)$$
$$4x^2 - 8x < 4x^2 - 14x + 6$$
$$-8x < -14x + 6$$
$$-8x + 14x < 6$$
$$6x < 6$$
$$x < \frac{6}{6}$$
$$x < 1$$

The solution set is $\{x|x < 1\}$, or $(-\infty, 1)$. The graph is shown below.

10. $(x+1)(x+2) > x(x+1)$

$$x^2 + 3x + 2 > x^2 + x$$
$$2x > -2$$
$$x > -1$$

The solution set is $\{x|x > -1\}$, or $(-1, \infty)$. The graph is shown below.

11. $-2 \le x + 1 < 4$

$-3 \le x < 3$ Subtracting 1

The solution set is $[-3, 3)$. The graph is shown below.

12. $-3 < x + 2 \le 5$

$$-5 < x \le 3$$

$(-5, 3]$

13. $5 \le x - 3 \le 7$

$8 \le x \le 10$ Adding 3

The solution set is $[8, 10]$. The graph is shown below.

14. $-1 < x - 4 < 7$

$$3 < x < 11$$

$(3, 11)$

15. $-3 \le x + 4 \le 3$

$-7 \le x \le -1$ Subtracting 4

The solution set is $[-7, -1]$. The graph is shown below.

16. $-5 < x + 2 < 15$

$-7 < x < 13$

$(-7, 13)$

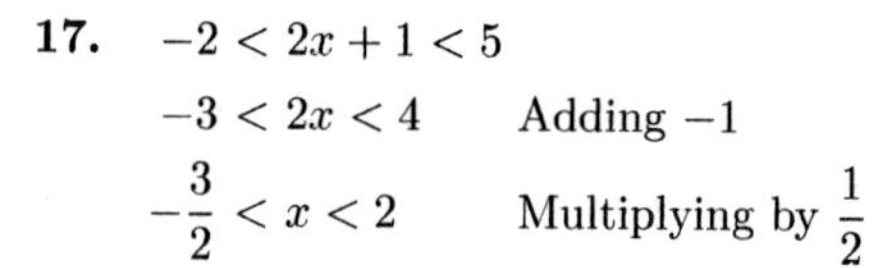

17. $-2 < 2x + 1 < 5$

$-3 < 2x < 4$ Adding -1

$-\frac{3}{2} < x < 2$ Multiplying by $\frac{1}{2}$

The solution set is $\left(-\frac{3}{2}, 2\right)$. The graph is shown below.

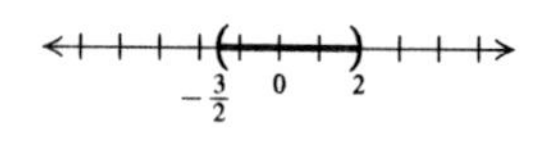

18. $-3 \le 5x + 1 \le 3$

$-4 \le 5x \le 2$

$-\frac{4}{5} \le x \le \frac{2}{5}$

$\left[-\frac{4}{5}, \frac{2}{5}\right]$

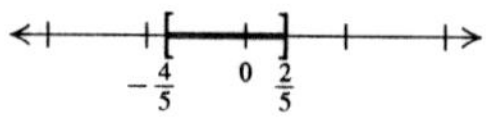

19. $-4 \le 6 - 2x < 4$

$-10 \le -2x < -2$ Adding -6

$5 \ge x > 1$ Multiplying by $-\frac{1}{2}$

or $1 < x \le 5$

The solution set is $(1, 5]$. The graph is shown below.

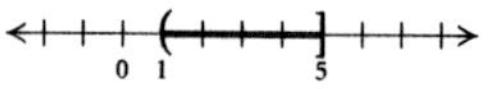

20. $-3 < 1 - 2x \le 3$

$-4 < -2x \le 2$

$2 > x \ge -1$

$[-1, 2)$

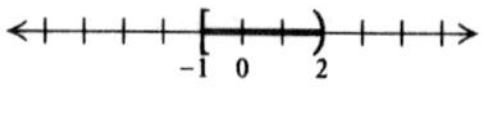

21. $-5 < \frac{1}{2}(3x + 1) < 7$

$-10 < 3x + 1 < 14$ Multiplying by 2

$-11 < 3x < 13$ Adding -1

$-\frac{11}{3} < x < \frac{13}{3}$ Multiplying by $\frac{1}{3}$

The solution set is $\left(-\frac{11}{3}, \frac{13}{3}\right)$. The graph is shown below.

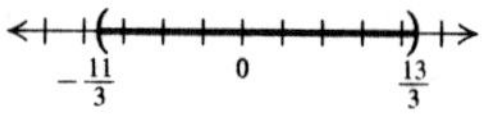

22. $\frac{2}{3} \le -\frac{4}{5}(x - 3) < 1$

$-\frac{5}{6} \ge x - 3 > -\frac{5}{4}$

$\frac{13}{6} \ge x > \frac{7}{4}$

$\left(\frac{7}{4}, \frac{13}{6}\right]$

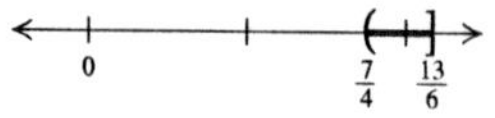

23. $3x \le -6$ *or* $x - 1 > 0$

$x \le -2$ *or* $x > 1$

The solution set is $(-\infty, -2] \cup (1, \infty)$. The graph is shown below.

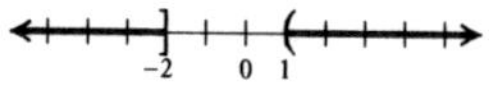

24. $2x < 8$ *or* $x + 3 \ge 10$

$x < 4$ *or* $x \ge 7$

$(-\infty, 4) \cup [7, \infty)$

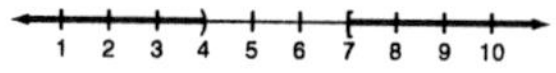

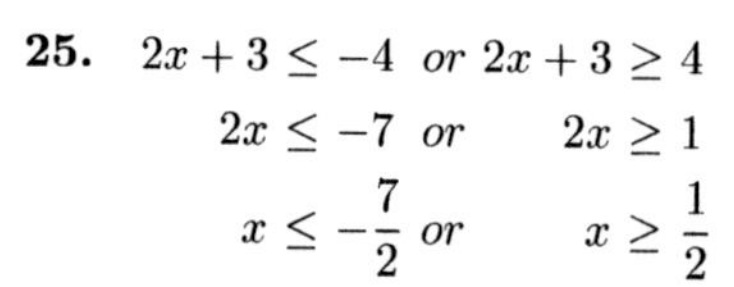

25. $2x + 3 \le -4$ *or* $2x + 3 \ge 4$

$2x \le -7$ *or* $2x \ge 1$

$x \le -\frac{7}{2}$ *or* $x \ge \frac{1}{2}$

The solution set is $\left(-\infty, -\frac{7}{2}\right] \cup \left[\frac{1}{2}, \infty\right)$. The graph is shown below.

26. $3x - 1 < -5$ *or* $3x - 1 > 5$

$3x < -4$ *or* $3x > 6$

$x < -\frac{4}{3}$ *or* $x > 2$

$\left(-\infty, -\frac{4}{3}\right) \cup (2, \infty)$

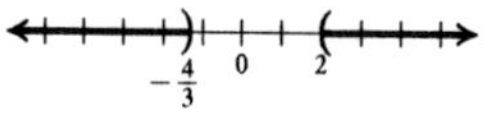

27. $2x - 20 < -0.8$ *or* $2x - 20 > 0.8$

$2x < 19.2$ *or* $2x > 20.8$

$x < 9.6$ *or* $x > 10.4$

The solution set is $(-\infty, 9.6) \cup (10.4, \infty)$. The graph is shown below.

28. $5x + 11 \leq -4$ *or* $5x + 11 \geq 4$

$5x \leq -15$ *or* $5x \geq -7$

$x \leq -3$ *or* $x \geq -\frac{7}{5}$

$(-\infty, -3] \cup \left[-\frac{7}{5}, \infty\right)$

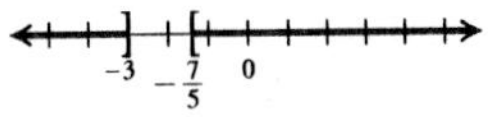

29. $x + 14 \leq -\frac{1}{4}$ *or* $x + 14 \geq \frac{1}{4}$

$x \leq -\frac{57}{4}$ *or* $x \geq -\frac{55}{4}$

The solution set is $\left(-\infty, -\frac{57}{4}\right] \cup \left[-\frac{55}{4}, \infty\right)$. The graph is shown below.

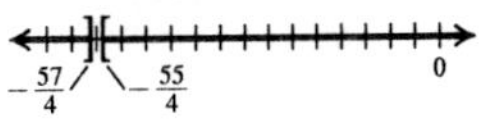

30. $x - 9 < -\frac{1}{2}$ *or* $x - 9 > \frac{1}{2}$

$x < \frac{17}{2}$ *or* $x > \frac{19}{2}$

$\left(-\infty, \frac{17}{2}\right) \cup \left(\frac{19}{2}, \infty\right)$

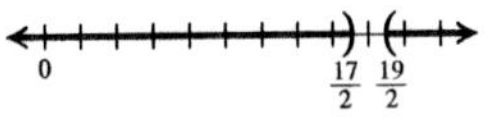

31. $|x| < 7$

To solve we look for all numbers x whose distance from 0 is less than 7. These are the numbers between -7 and 7. That is, $-7 < x < 7$. The solution set and its graph are as follows:

$(-7, 7)$

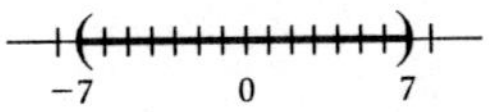

32. $|x| \leq 4.5$

$-4.5 \leq x \leq 4.5$

The solution set is $[-4.5, 4.5]$.

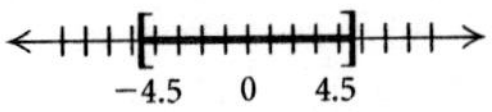

33. $|x| \geq 4.5$

To solve we look for all numbers x whose distance from 0 is greater than or equal to 4.5. That is, $x \leq -4.5$ or $x \geq 4.5$. The solution set and its graph are as follows.

$\{x | x \leq -4.5 \text{ or } x \geq 4.5\}$, or $(-\infty, -4.5] \cup [4.5, \infty)$

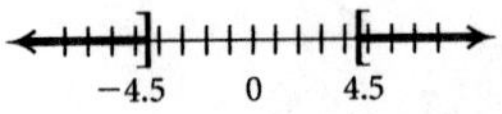

34. $|x| > 7$

$x < -7$ or $x > 7$

The solution set is $(-\infty, -7) \cup (7, \infty)$.

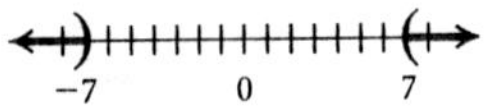

35. $|x + 8| < 9$

$-9 < x + 8 < 9$

$-17 < x < 1$ Subtracting 8

The solution set is $(-17, 1)$. The graph is shown below.

36. $|x + 6| < 10$

$-10 \leq x + 6 \leq 10$

$-16 \leq x \leq 4$

The solution set is $[-16, 4]$.

37. $|x + 8| \geq 9$

$x + 8 \leq -9$ *or* $x + 8 \geq 9$

$x \leq -17$ *or* $x \geq 1$ Subtracting 8

The solution set is $(-\infty, -17] \cup [1, \infty)$. The graph is shown below.

38. $|x + 6| > 10$

$x + 6 < -10$ *or* $x + 6 > 10$

$x < -16$ *or* $x > 4$

The solution set is $(-\infty, -16) \cup (4, \infty)$.

39. $\left|x - \frac{1}{4}\right| < \frac{1}{2}$

$-\frac{1}{2} < x - \frac{1}{4} < \frac{1}{2}$

$-\frac{1}{4} < x < \frac{3}{4}$ Adding $\frac{1}{4}$

The solution set is $\left(-\frac{1}{4}, \frac{3}{4}\right)$. The graph is shown below.

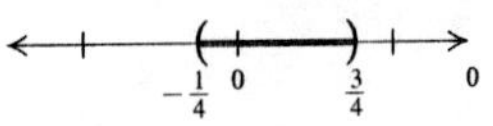

40. $|x - 0.5| \leq 0.2$

$-0.2 \leq x - 0.5 \leq 0.2$

$0.3 \leq x \leq 0.7$

The solution set is $[0.3, 0.7]$.

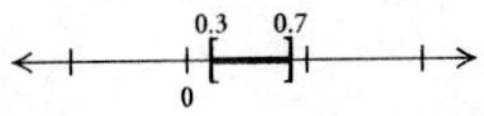

41. $|3x| < 1$

$-1 < 3x < 1$

$-\frac{1}{3} < x < \frac{1}{3}$ Dividing by 3

The solution set is $\left(-\frac{1}{3}, \frac{1}{3}\right)$. The graph is shown below.

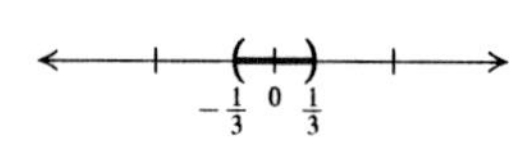

42. $|5x| \leq 4$

$-4 \leq 5x \leq 4$

$-\frac{4}{5} \leq x \leq \frac{4}{5}$

The solution set is $\left[-\frac{4}{5}, \frac{4}{5}\right]$.

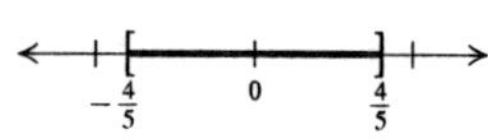

43. $|2x + 3| \leq 9$

$-9 \leq 2x + 3 \leq 9$

$-12 \leq 2x \leq 6$ Subtracting 3

$-6 \leq x \leq 3$ Dividing by 2

The solution set is $[-6, 3]$. The graph is shown below.

44. $|3x + 4| < 13$

$-13 < 3x + 4 < 13$

$-17 < 3x < 9$

$-\frac{17}{3} < x < 3$

The solution set is $\left(-\frac{17}{3}, 3\right)$.

45. $|x - 5| > 0.1$

$x - 5 < -0.1$ *or* $x - 5 > 0.1$

$x < 4.9$ *or* $x > 5.1$ Adding 5

The solution set is $(-\infty, 4.9) \cup (5.1, \infty)$. The graph is shown below.

46. $|x - 7| \geq 0.4$

$x - 7 \leq -0.4$ *or* $x - 7 \geq 0.4$

$x \leq 6.6$ *or* $x \geq 7.4$

The solution set is $(-\infty, 6.6] \cup [7.4, \infty)$.

47. $|6 - 4x| \leq 8$

$-8 \leq 6 - 4x \leq 8$

$-14 \leq -4x \leq 2$ Subtracting 6

$\frac{14}{4} \geq x \geq -\frac{2}{4}$ Dividing by -4 and reversing the inequality symbols

$\frac{7}{2} \geq x \geq -\frac{1}{2}$ Simplifying

The solution set is $\left[-\frac{1}{2}, \frac{7}{2}\right]$. The graph is shown below.

48. $|5 - 2x| > 10$

$5 - 2x < -10$ *or* $5 - 2x > 10$

$-2x < -15$ *or* $-2x > 5$

$x > \frac{15}{2}$ *or* $x < -\frac{5}{2}$

The solution set is $\left(-\infty, -\frac{5}{2}\right) \cup \left(\frac{15}{2}, \infty\right)$.

49. $\left|x + \frac{2}{3}\right| \leq \frac{5}{3}$

$-\frac{5}{3} \leq x + \frac{2}{3} \leq \frac{5}{3}$

$-\frac{7}{3} \leq x \leq 1$ Subtracting $\frac{2}{3}$

The solution set is $\left[-\frac{7}{3}, 1\right]$. The graph is shown below.

50. $\left|x + \frac{3}{4}\right| < \frac{1}{4}$

$-\frac{1}{4} < x + \frac{3}{4} < \frac{1}{4}$

$-1 < x < -\frac{1}{2}$

The solution set is $\left(-1, -\frac{1}{2}\right)$.

51. $\left|\frac{2x + 1}{3}\right| > 5$

$\frac{2x + 1}{3} < -5$ *or* $\frac{2x + 1}{3} > 5$

$2x + 1 < -15$ *or* $2x + 1 > 15$ Multiplying by 3

$2x < -16$ *or* $2x > 14$ Subtracting 1

$x < -8$ *or* $x > 7$ Dividing by 2

The solution set is $\{x|x < -8 \text{ or } x > 7\}$, or $(-\infty, -8) \cup (7, \infty)$. The graph is shown below.

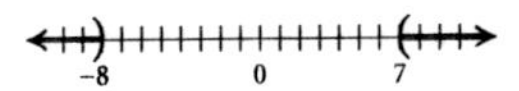

52. 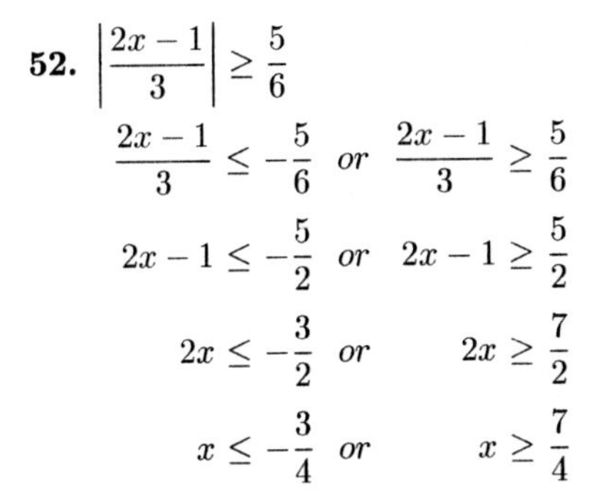

$$\left|\frac{2x-1}{3}\right| \geq \frac{5}{6}$$

$$\frac{2x-1}{3} \leq -\frac{5}{6} \quad or \quad \frac{2x-1}{3} \geq \frac{5}{6}$$

$$2x-1 \leq -\frac{5}{2} \quad or \quad 2x-1 \geq \frac{5}{2}$$

$$2x \leq -\frac{3}{2} \quad or \quad 2x \geq \frac{7}{2}$$

$$x \leq -\frac{3}{4} \quad or \quad x \geq \frac{7}{4}$$

The solution set is $\left(-\infty, -\frac{3}{4}\right] \cup \left[\frac{7}{4}, \infty\right)$.

53. $|2x-4| < -5$

Since $|2x-4| \geq 0$ for all x, there is no x such that $|2x-4|$ would be less than -5. There is no solution.

54. $|3x+5| < 0$

$|3x+5| \geq 0$ for all x, so there is no solution.

55. ***Familiarize and Translate.*** Spending is given by the equation $y = 12.7x + 15.2$. We want to know when the spending will be more than \$66 billion, so we have

$12.7x + 15.2 > 66.$

Carry out. We solve the inequality.

$$12.7x + 15.2 > 66$$
$$12.7x > 50.8$$
$$x > 4$$

Check. When $x = 4$, the spending is $12.7(4) + 15.2 = 66$. As a partial check, we could try a value of x less than 4 and one greater than 4. When $x = 3.9$, we have $y = 12.7(3.9) + 15.2 = 64.73 < 66$; when $x = 4.1$, we have $y = 12.7(4.1) + 15.2 = 67.27 > 66$. Since $y = 66$ when $x = 4$ and $y > 66$ when $x = 4.1 > 4$, the answer is probably correct.

State. The spending will be more than \$66 billion more than 4 yr after 2002.

56. Solve: $5x + 5 \geq 20$

$x \geq 3$, so 3 or more y after 2002, or in 2005 and later, there will be at least 20 million homes with devices installed that receive and manage broadband TV and Internet content.

57. ***Familiarize.*** Let t = the number of hours worked. Then Acme Movers charge $100 + 30t$ and Hank's Movers charge $55t$.

Translate.

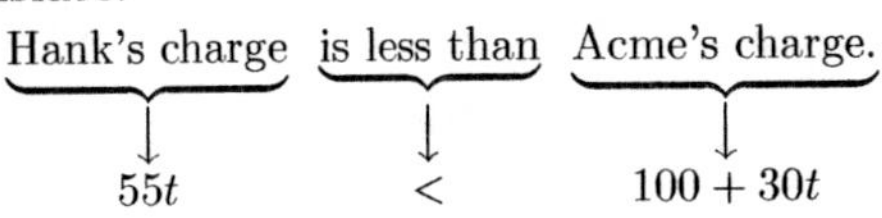

Carry out. We solve the inequality.

$$55t < 100 + 30t$$
$$25t < 100$$
$$t < 4$$

Check. When $t = 4$, Hank's Movers charge $55 \cdot 4$, or \$220 and Acme Movers charge $100 + 30 \cdot 4 = 100 + 120 = \$220$, so the charges are the same. As a partial check, we find the charges for a value of $t < 4$. When $t = 3.5$, Hank's Movers charge $55(3.5) = \$192.50$ and Acme Movers charge $100 + 30(3.5) = 100 + 105 = \$205$. Since Hank's charge is less than Acme's, the answer is probably correct.

State. For times less than 4 hr it costs less to hire Hank's Movers.

58. Let x = the amount invested at 4%. Then $12,000 - x$ = the amount invested at 6%.

Solve: $0.04x + 0.06(12,000 - x) \geq 650$

$x \leq 3500$, so at most \$3500 can be invested at 4%.

59. ***Familiarize.*** Let x = the amount invested at 4%. Then $7500 - x$ = the amount invested at 5%. Using the simple-interest formula, $I = Prt$, we see that in one year the 4% investment earns $0.04x$ and the 5% investment earns $0.05(7500 - x)$.

Translate.

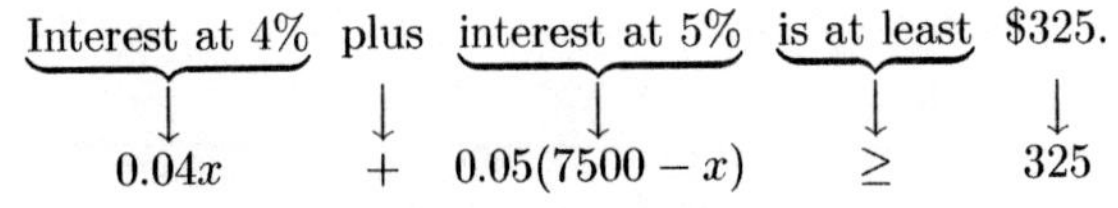

Carry out. We solve the inequality.

$$0.04x + 0.05(7500 - x) \geq 325$$
$$0.04x + 375 - 0.05x \geq 325$$
$$-0.01x + 375 \geq 325$$
$$-0.01x \geq -50$$
$$x \leq 5000$$

Check. When \$5000 is invested at 4%, then \$7500−\$5000, or \$2500, is invested at 5%. In one year the 4% investment earns 0.04(\$5000), or \$200, in simple interest and the 5% investment earns 0.05(\$2500), or \$125, so the total interest is \$200 + \$125, or \$325. As a partial check, we determine the total interest when an amount greater than \$5000 is invested at 4%. Suppose \$5001 is invested at 4%. Then \$2499 is invested at 5%, and the total interest is 0.04(\$5001) + 0.05(\$2499), or \$324.99. Since this amount is less than \$325, the answer is probably correct.

State. The most that can be invested at 4% is \$5000.

60. Let c = the number of check written per month.

Solve: $0.20c < 6 + 0.05c$

$c < 40$, so the Smart Checking plan will cost less than the Consumer Checking plan when fewer than 40 checks are written per month.

61. ***Familiarize.*** Let $c =$ the number of checks written per month. Then the No Frills plan costs $0.35c$ per month and the Simple Checking plan costs $5 + 0.10c$ per month.

Translate.

Simple Checking cost | is less than | No Frills cost.

$5 + 0.10c \quad < \quad 0.35c$

Carry out We solve the inequality.

$$5 + 0.10c < 0.35c$$
$$5 < 0.25c$$
$$20 < c$$

Check. When 20 checks are written the No Frills plan costs 0.35(20), or \$7 per month and the Simple Checking plan costs 5 + 0.10(20), or \$7, so the costs are the same. As a partial check, we compare the cost for some number of checks greater than 20. When 21 checks are written, the No Frills plan costs 0.35(21), or \$7.35 and the Simple Checking plan costs 5+0.10(21), or \$7.10. Since the Simple Checking plan costs less than the No Frills plan, the answer is probably correct.

State. The Simple Checking plan costs less when more than 20 checks are written per month.

62. Let $s =$ the monthly sales.

Solve: $750 + 0.1s > 1000 + 0.08(s - 2000)$

$s > 4500$, so Plan A is better for monthly sales greater than \$4500.

63. ***Familiarize.*** Let $s =$ the monthly sales. Then the amount of sales in excess of \$8000 is $s - 8000$.

Translate.

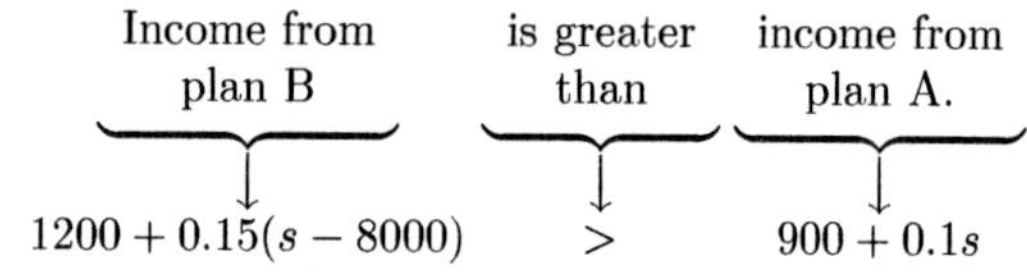

Carry out. We solve the inequality.

$$1200 + 0.15(s - 8000) > 900 + 0.1s$$
$$1200 + 0.15s - 1200 > 900 + 0.1s$$
$$0.15s > 900 + 0.1s$$
$$0.05s > 900$$
$$s > 18,000$$

Check. For sales of \$18,000 the income from plan A is \$900+0.1(\$18,000), or \$2700, and the income from plan B is 1200+0.15(18,000 − 8000), or \$2700 so the incomes are the same. As a partial check we can compare the incomes for an amount of sales greater than \$18,000. For sales of \$18,001, for example, the income from plan A is \$900 + 0.1(\$18,001), or \$2700.10, and the income from plan B is \$1200 + 0.15(\$18,001 − \$8000), or \$2700.15. Since plan B is better than plan A in this case, the answer is probably correct.

State. Plan B is better than plan A for monthly sales greater than \$18,000.

64. Solve: $200 + 12n > 20n$

$n < 25$

65. Left to the student

66. Left to the student

67. Absolute value is nonnegative.

68. $|x| \geq 0 > p$ for any real number x.

69. y-intercept

70. distance formula

71. relation

72. function

73. horizontal line

74. parallel

75. decreasing

76. Symmetric with respect to the y-axis

77. $2x \leq 5 - 7x < 7 + x$

$$2x \leq 5 - 7x \quad and \quad 5 - 7x < 7 + x$$
$$9x \leq 5 \quad and \quad -8x < 2$$
$$x \leq \frac{5}{9} \quad and \quad x > -\frac{1}{4}$$

The solution set is $\left(-\frac{1}{4}, \frac{5}{9}\right]$.

78. $x \leq 3x - 2 \leq 2 - x$

$$x \leq 3x - 2 \quad and \quad 3x - 2 \leq 2 - x$$
$$-2x \leq -2 \quad and \quad 4x \leq 4$$
$$x \geq 1 \quad and \quad x \leq 1$$

The solution is 1.

79. $|3x - 1| > 5x - 2$

$$3x - 1 < -(5x - 2) \quad or \quad 3x - 1 > 5x - 2$$
$$3x - 1 < -5x + 2 \quad or \quad 1 > 2x$$
$$8x < 3 \quad or \quad \frac{1}{2} > x$$
$$x < \frac{3}{8} \quad or \quad \frac{1}{2} > x$$

The solution set is $\left(-\infty, \frac{3}{8}\right) \cup \left(-\infty, \frac{1}{2}\right)$. This is equivalent to $\left(-\infty, \frac{1}{2}\right)$.

80. $|x + 2| \leq |x - 5|$

Divide the set of real numbers into three intervals:

$(-\infty, -2)$, $[-2, 5)$, and $[5, \infty,)$.

Find the solution set of $|x + 2| \leq |x - 5|$ in each interval. Then find the union of the three solution sets.

If $x < -2$, then $|x+2| = -(x+2)$ and $|x-5| = -(x-5)$.

Solve: $x < -2$ *and* $-(x+2) \leq -(x-5)$

$x < -2$ *and* $-x-2 \leq -x+5$

$x < -2$ *and* $-2 \leq 5$

The solution set for this interval is $(-\infty, -2)$.

If $-2 \leq x < 5$, then $|x+2| = x+2$ and $|x-5| = -(x-5)$.

Solve: $-2 \leq x < 5$ *and* $x+2 \leq -(x-5)$

$-2 \leq x < 5$ *and* $x+2 \leq -x+5$

$-2 \leq x < 5$ *and* $2x \leq 3$

$-2 \leq x < 5$ *and* $x \leq \frac{3}{2}$

The solution set for this interval is $\left[-2, \frac{3}{2}\right]$.

If $x \geq 5$, then $|x+2| = x+2$ and $|x-5| = x-5$.

Solve: $x \geq 5$ *and* $x+2 \leq x-5$

$x \geq 5$ *and* $2 \leq -5$

The solution set for this interval is $\emptyset$.

The union of the above three solution set is $\left(-\infty, \frac{3}{2}\right]$. This is the solution set of $|x+2| \leq |x-5|$.

81. $|p-4| + |p+4| < 8$

If $p < -4$, then $|p-4| = -(p-4)$ and $|p+4| = -(p+4)$.

Solve: $-(p-4) + [-(p+4)] < 8$

$-p+4-p-4 < 8$

$-2p < 8$

$p > -4$

Since this is false for all values of p in the interval $(-\infty, -4)$ there is no solution in this interval.

If $p \geq -4$, then $|p+4| = p+4$.

Solve: $|p-4| + p + 4 < 8$

$|p-4| < 4-p$

$p-4 > -(4-p)$ *and* $p-4 < 4-p$

$p-4 > p-4$ *and* $2p < 8$

$-4 > -4$ *and* $p < 4$

Since $-4 > -4$ is false for all values of p, there is no solution in the interval $[-4, \infty)$.

Thus, $|p-4| + |p+4| < 8$ has no solution.

82. $|x| + |x+1| < 10$

If $x < -1$, then $|x| = -x$ and $|x+1| = -(x+1)$ and we have:

$x < -1$ *and* $-x + [-(x+1)] < 10$

$x < -1$ *and* $-x-x-1 < 10$

$x < -1$ *and* $-2x-1 < 10$

$x < -1$ *and* $-2x < 10$

$x < -1$ *and* $x > -\frac{11}{2}$

The solution set for this interval is $\left(-\frac{11}{2}, -1\right)$.

If $-1 \leq x < 0$, then $|x| = -x$ and $|x+1| = x+1$ and we have:

$-1 \leq x$ *and* $-x+x+1 < 10$

$-1 \leq x$ *and* $1 < 10$

The solution set for this interval is $[-1, 0]$.

If $x \geq 0$, then $|x| = x$ and $|x+1| = x+1$ and we have:

$x \geq 0$ *and* $x+x+1 < 10$

$x \geq 0$ *and* $2x+1 < 10$

$x \geq 0$ *and* $2x < 9$

$x \geq 0$ *and* $x < \frac{9}{2}$

The solution set for this interval is $\left[0, \frac{9}{2}\right)$.

The union of the three solution sets above is $\left(-\frac{11}{2}, \frac{9}{2}\right)$. This is the solution set of $|x| + |x+1| < 10$.

83. $|x-3| + |2x+5| > 6$

Divide the set of real numbers into three intervals: $\left(-\infty, -\frac{5}{2}\right)$, $\left[-\frac{5}{2}, 3\right)$, and $[3, \infty)$.

Find the solution set of $|x-3| + |2x+5| > 6$ in each interval. Then find the union of the three solution sets.

If $x < -\frac{5}{2}$, then $|x-3| = -(x-3)$ and $|2x+5| = -(2x+5)$.

Solve: $x < -\frac{5}{2}$ *and* $-(x-3) + [-(2x+5)] > 6$

$x < -\frac{5}{2}$ *and* $-x+3-2x-5 > 6$

$x < -\frac{5}{2}$ *and* $-3x > 8$

$x < -\frac{5}{2}$ *and* $x < -\frac{8}{3}$

The solution set in this interval is $\left(-\infty, -\frac{8}{3}\right)$.

If $-\frac{5}{2} \leq x < 3$, then $|x-3| = -(x-3)$ and $|2x+5| = 2x+5$.

Solve: $-\frac{5}{2} \leq x < 3$ *and* $-(x-3) + 2x + 5 > 6$

$-\frac{5}{2} \leq x < 3$ *and* $-x+3+2x+5 > 6$

$-\frac{5}{2} \leq x < 3$ *and* $x > -2$

The solution set in this interval is $(-2, 3)$.

If $x \geq 3$, then $|x-3| = x-3$ and $|2x+5| = 2x+5$.

Solve: $x \geq 3$ *and* $x-3+2x+5 > 6$

$x \geq 3$ *and* $3x > 4$

$x \geq 3$ *and* $x > \frac{4}{3}$

The solution set in this interval is $[3, \infty)$.

The union of the above solution sets is $\left(-\infty, -\frac{8}{3}\right) \cup (-2, \infty)$. This is the solution set of

$|x-3| + |2x+5| > 6$.

Chapter 3

Polynomial and Rational Functions

Exercise Set 3.1

1. $g(x) = \frac{1}{2}x^3 - 10x + 8$

   The degree of the polynomial is 3, so the polynomial is cubic. The leading term is $\frac{1}{2}x^3$ and the leading coefficient is $\frac{1}{2}$.

2. $f(x) = 15x^2 - 10 + 0.11x^4 - 7x^3 =$
   $0.11x^4 - 7x^3 + 15x^2 - 10$

   The degree of the polynomial is 4, so the polynomial is quartic. The leading term is $0.11x^4$ and the leading coefficient is 0.11.

3. $h(x) = 0.9x - 0.13$

   The degree of the polynomial is 1, so the polynomial is linear. The leading term is $0.9x$ and the leading coefficient is 0.9.

4. $f(x) = -6 = -6x^0$

   The degree of the polynomial is 0, so the polynomial is constant. The leading term and leading coefficient are both -6.

5. $g(x) = 305x^4 + 4021$

   The degree of the polynomial is 4, so the polynomial is quartic. The leading term is $305x^4$ and the leading coefficient is 305.

6. $h(x) = 2.4x^3 + 5x^2 - x + \frac{7}{8}$

   The degree of the polynomial is 3, so the polynomial is cubic. The leading term is $2.4x^3$ and the leading coefficient is 2.4.

7. $h(x) = -5x^2 + 7x^3 + x^4 = x^4 + 7x^3 - 5x^2$

   The degree of the polynomial is 4, so the polynomial is quartic. The leading term is x^4 and the leading coefficient is 1 ($x^4 = 1 \cdot x^4$).

8. $f(x) = 2 - x^2 = -x^2 + 2$

   The degree of the polynomial is 2, so the polynomial is quadratic. The leading term is $-x^2$ and the leading coefficient is -1.

9. $g(x) = 4x^3 - \frac{1}{2}x^2 + 8$

   The degree of the polynomial is 3, so the polynomial is cubic. The leading term is $4x^3$ and the leading coefficient is 4.

10. $f(x) = 12 + x = x + 12$

    The degree of the polynomial is 1, so the polynomial is linear. The leading term is x and the leading coefficient is 1 ($x = 1 \cdot x$).

11. $f(x) = -3x^3 - x + 4$

    The leading term is $-3x^3$. The degree, 3, is odd and the leading coefficient, -3, is negative. Thus the end behavior of the graph is like that of (d).

12. $f(x) = \frac{1}{4}x^4 + \frac{1}{2}x^3 - 6x^2 + x - 5$

    The leading term is $\frac{1}{4}x^4$. The degree, 4, is even and the leading coefficient, $\frac{1}{4}$, is positive. Thus the end behavior of the graph is like that of (a).

13. $f(x) = -x^6 + \frac{3}{4}x^4$

    The leading term is $-x^6$. The degree, 6, is even and the leading coefficient, -1, is negative. Thus the end behavior of the graph is like that of (b).

14. $f(x) = \frac{2}{5}x^5 - 2x^4 + x^3 - \frac{1}{2}x + 3$

    The leading term is $\frac{2}{5}x^5$. The degree, 5, is odd and the leading coefficient, $\frac{2}{5}$, is positive. Thus the end behavior of the graph is like that of (c).

15. $f(x) = -3.5x^4 + x^6 + 0.1x^7 = 0.1x^7 + x^6 - 3.5x^4$

    The leading term is $0.1x^7$. The degree, 7, is odd and the leading coefficient, 0.1, is positive. Thus the end behavior of the graph is like that of (c).

16. $f(x) = -x^3 + x^5 - 0.5x^6 = -0.5x^6 + x^5 - x^3$

    The leading term is $-0.5x^6$. The degree, 6, is even and the leading coefficient, -0.5, is negative. Thus the end behavior of the graph is like that of (b).

17. $f(x) = 10 + \frac{1}{10}x^4 - \frac{2}{5}x^3 = \frac{1}{10}x^4 - \frac{2}{5}x^3 + 10$

    The leading term is $\frac{1}{10}x^4$. The degree, 4, is even and the leading coefficient, $\frac{1}{10}$, is positive. Thus the end behavior of the graph is like that of (a).

18. $f(x) = 2x + x^3 - x^5 = -x^5 + x^3 + 2x$

    The leading term is $-x^5$. The degree, 5, is odd and the leading coefficient, -1, is negative. Thus the end behavior of the graph is like that of (d).

19. $f(x) = \frac{1}{4}x^2 - 5$

The leading term is $\frac{1}{4}x^2$. The sign of the leading coefficient, $\frac{1}{4}$, is positive and the degree, 2, is even, so we would choose either graph (b) or graph (d). Note also that $f(0) = -5$, so the y-intercept is $(0, -5)$. Thus, graph (d) is the graph of this function.

20. $f(x) = -0.5x^6 - x^5 + 4x^4 - 5x^3 - 7x^2 + x - 3$

The leading term is $-0.5x^6$. The sign of the leading coefficient, -0.5, is negative and the degree, 6, is even. Thus, graph (a) is the graph of this function.

21. $f(x) = x^5 - x^4 + x^2 + 4$

The leading term is x^5. The sign of the leading coefficient, 1, is positive and the degree, 5, is odd. Thus, graph (f) is the graph of this function.

22. $f(x) = -\frac{1}{3}x^3 - 4x^2 + 6x + 42$

The leading term is $-\frac{1}{3}x^3$. The sign of the leading coefficient, $-\frac{1}{3}$, is negative and the degree, 3, is odd, so we would choose either graph (c) or graph (e). Note also that $f(0) = 42$, so the y-intercept is $(0, 42)$. Thus, graph (c) is the graph of this function.

23. $f(x) = x^4 - 2x^3 + 12x^2 + x - 20$

The leading term is x^4. The sign of the leading coefficient, 1, is positive and the degree, 4, is even, so we would choose either graph (b) or graph (d). Note also that $f(0) = -20$, so the y-intercept is $(0, -20)$. Thus, graph (b) is the graph of this function.

24. $f(x) = -0.3x^7 + 0.11x^6 - 0.25x^5 + x^4 + x^3 - 6x - 5$

The leading term is $-0.3x^7$. The sign of the leading coefficient, -0.3, is negative and the degree, 7, is odd, so we would choose either graph (c) or graph (e). Note also that $f(0) = -5$, so the y-intercept is $(0, -5)$. Thus, graph (e) is the graph of this function.

25. $f(x) = x^3 - 9x^2 + 14x + 24$

$f(4) = 4^3 - 9 \cdot 4^2 + 14 \cdot 4 + 24 = 0$

Since $f(4) = 0$, 4 is a zero of $f(x)$.

$f(5) = 5^3 - 9 \cdot 5^2 + 14 \cdot 5 + 24 = -6$

Since $f(5) \neq 0$, 5 is not a zero of $f(x)$.

$f(-2) = (-2)^3 - 9(-2)^2 + 14(-2) + 24 = -48$

Since $f(-2) \neq 0$, -2 is not a zero of $f(x)$.

26. $f(x) = 2x^3 - 3x^2 + x + 6$

$f(2) = 2 \cdot 2^3 - 3 \cdot 2^2 + 2 + 6 = 12$

$f(2) \neq 0$, so 2 is not a zero of $f(x)$.

$f(3) = 2 \cdot 3^3 - 3 \cdot 3^2 + 3 + 6 = 36$

$f(3) \neq 0$, so 3 is not a zero of $f(x)$.

$f(-1) = 2(-1)^3 - 3(-1)^2 + (-1) + 6 = 0$

$f(-1) = 0$, so -1 is a zero of $f(x)$.

27. $g(x) = x^4 - 6x^3 + 8x^2 + 6x - 9$

$g(2) = 2^4 - 6 \cdot 2^3 + 8 \cdot 2^2 + 6 \cdot 2 - 9 = 3$

Since $g(2) \neq 0$, 2 is not a zero of $g(x)$.

$g(3) = 3^4 - 6 \cdot 3^3 + 8 \cdot 3^2 + 6 \cdot 3 - 9 = 0$

Since $g(3) = 0$, 3 is a zero of $g(x)$.

$g(-1) = (-1)^4 - 6(-1)^3 + 8(-1)^2 + 6(-1) - 9 = 0$

Since $g(-1) = 0$, -1 is a zero of $g(x)$.

28. $g(x) = x^4 - x^3 - 3x^2 + 5x - 2$

$g(1) = 1^4 - 1^3 - 3 \cdot 1^2 + 5 \cdot 1 - 2 = 0$

Since $g(1) = 0$, 1 is a zero of $g(x)$.

$g(-2) = (-2)^4 - (-2)^3 - 3(-2)^2 + 5(-2) - 2 = 0$

Since $g(-2) = 0$, -2 is a zero of $g(x)$.

$g(3) = 3^4 - 3^3 - 3 \cdot 3^2 + 5 \cdot 3 - 2 = 40$

Since $g(3) \neq 0$, 3 is not a zero of $g(x)$.

29. $f(x) = (x+3)^2(x-1) = (x+3)(x+3)(x-1)$

To solve $f(x) = 0$ we use the principle of zero products, solving $x + 3 = 0$ and $x - 1 = 0$. The zeros of $f(x)$ are -3 and 1.

The factor $x + 3$ occurs twice. Thus the zero -3 has a multiplicity of two.

The factor $x - 1$ occurs only one time. Thus the zero 1 has a multiplicity of one.

30. $f(x) = (x+5)^3(x-4)(x+1)^2$

-5, multiplicity 3; 4, multiplicity 1;

-1, multiplicity 2

31. $f(x) = -2(x-4)(x-4)(x-4)(x+6) = 2(x-4)^3(x+6)$

To solve $f(x) = 0$ we use the principle of zero products, solving $x - 4 = 0$ and $x + 6 = 0$. The zeros of $f(x)$ are 4 and -6.

The factor $x - 4$ occurs three times. Thus the zero 4 has a multiplicity of 3.

The factor $x + 6$ occurs only one time. Thus the zero -6 has a multiplicity of 1.

32. $f(x) = \left(x+\frac{1}{2}\right)(x+7)(x+7)(x+5) = \left(x+\frac{1}{2}\right)(x+7)^2(x+5)$

$-\frac{1}{2}$, multiplicity 1; -7, multiplicity 2;

-5, multiplicity 1

33. $f(x) = (x^2 - 9)^3 = [(x+3)(x-3)]^3 = (x+3)^3(x-3)^3$

To solve $f(x) = 0$ we use the principle of zero products, solving $x + 3 = 0$ and $x - 3 = 0$. The zeros of $f(x)$ are -3 and 3.

The factors $x + 3$ and $x - 3$ each occur three times so each zero has a multiplicity of 3.

34. $f(x) = (x^2 - 4)^2 = [(x+2)(x-2)]^2 = (x+2)^2(x-2)^2$

-2, multiplicity 2; 2, multiplicity 2

35. $f(x) = x^3(x-1)^2(x+4)$

To solve $f(x) = 0$ we use the principle of zero products, solving $x = 0$, $x - 1 = 0$, and $x + 4 = 0$. The zeros of $f(x)$ are 0, 1, and -4.

The factor x occurs three times. Thus the zero 0 has a multiplicity of three.

The factor $x - 1$ occurs twice. Thus the zero 1 has a multiplicity of two.

The factor $x + 4$ occurs only one time. Thus the zero -4 has a multiplicity of one.

36. $f(x) = x^2(x+3)^2(x-4)(x+1)^4$

0, multiplicity 2; -3, multiplicity 2;

4, multiplicity 1; -1, multiplicity 4

37. $f(x) = -8(x-3)^2(x+4)^3x^4$

To solve $f(x) = 0$ we use the principle of zero products, solving $x - 3 = 0$, $x + 4 = 0$, and $x = 0$. The zeros of $f(x)$ are 3, -4, and 0.

The factor $x - 3$ occurs twice. Thus the zero 3 has a multiplicity of 2.

The factor $x + 4$ occurs three times. Thus the zero -4 has a multiplicity of 3.

The factor x occurs four times. Thus the zero 0 has a multiplicity of 4.

38.
$$\begin{aligned} f(x) &= (x^2 - 5x + 6)^2 \\ &= [(x-3)(x-2)]^2 \\ &= (x-3)^2(x-2)^2 \end{aligned}$$

3, multiplicity 2; 2, multiplicity 2

39. $f(x) = x^4 - 4x^2 + 3$

We factor as follows:
$$\begin{aligned} f(x) &= (x^2 - 3)(x^2 - 1) \\ &= (x - \sqrt{3})(x + \sqrt{3})(x-1)(x+1) \end{aligned}$$

The zeros of the function are $\sqrt{3}$, $-\sqrt{3}$, 1, and -1. Each has a multiplicity of 1.

40.
$$\begin{aligned} f(x) &= x^4 - 10x^2 + 9 \\ &= (x^2 - 9)(x^2 - 1) \\ &= (x+3)(x-3)(x+1)(x-1) \end{aligned}$$

± 3, ± 1; each has a multiplicity of 1.

41. $f(x) = x^3 + 3x^2 - x - 3$

We factor by grouping:
$$\begin{aligned} f(x) &= x^2(x+3) - (x+3) \\ &= (x^2 - 1)(x+3) \\ &= (x-1)(x+1)(x+3) \end{aligned}$$

The zeros of the function are 1, -1, and -3. Each has a multiplicity of 1.

42.
$$\begin{aligned} f(x) &= x^3 - x^2 - 2x + 2 \\ &= x^2(x-1) - 2(x-1) \\ &= (x^2 - 2)(x-1) \\ &= (x - \sqrt{2})(x + \sqrt{2})(x-1) \end{aligned}$$

$\sqrt{2}$, $-\sqrt{2}$, 1; each has a multiplicity of 1.

43.
$$\begin{aligned} f(x) &= 2x^3 - x^2 - 8x + 4 \\ &= x^2(2x-1) - 4(2x-1) \\ &= (2x-1)(x^2 - 4) \\ &= (2x-1)(x+2)(x-2) \end{aligned}$$

The zeros of the function are $\frac{1}{2}$, -2, and 2. Each has a multiplicity of 1.

44.
$$\begin{aligned} f(x) &= 3x^3 + x^2 - 48x - 16 \\ &= x^2(3x+1) - 16(3x+1) \\ &= (3x+1)(x^2 - 16) \\ &= (3x+1)(x+4)(x-4) \end{aligned}$$

$-\frac{1}{3}$, -4, 4; each has a multiplicity of 1

45. $f(x) = -x^3 - 2x^2$

1. The leading term is $-x^3$. The degree, 3, is odd and the leading coefficient, -1, is negative so as $|x| \to \infty$, $f(x) \to -\infty$ and as $|x| \to -\infty$, $f(x) \to \infty$.

2. We solve $f(x) = 0$.

$$\begin{aligned} -x^3 - 2x^2 &= 0 \\ -x^2(x+2) &= 0 \\ -x^2 = 0 \quad &or \quad x + 2 = 0 \\ x^2 = 0 \quad &or \quad x = -2 \\ x = 0 \quad &or \quad x = -2 \end{aligned}$$

The zeros of the function are 0 and -2, so the x-intercepts of the graph are $(0, 0)$ and $(-2, 0)$.

3. The zeros divide the x-axis into 3 intervals, $(-\infty, -2)$, $(-2, 0)$, and $(0, \infty)$. We choose a value for x from each interval and find $f(x)$. This tells us the sign of $f(x)$ for all values of x in that interval.

In $(-\infty, -2)$, test -3:

$f(-3) = -(-3)^3 - 2(-3)^2 = 9 > 0$

In $(-2, 0)$, test -1:

$f(-1) = -(-1)^3 - 2(-1)^2 = -1 < 0$

In $(0, \infty)$, test 1:

$f(1) = -1^3 - 2 \cdot 1^2 = -3 < 0$

Thus the graph lies above the x-axis on $(-\infty, -2)$ and below the x-axis on $(-2, 0)$ and $(0, \infty)$. We also know the points $(-3, 9)$, $(-1, -1)$, and $(1, -3)$ are on the graph.

4. From Step 2 we see that the y-intercept is $(0, 0)$.

5. We find additional points on the graph and then draw the graph.

x	$f(x)$
-2.5	3.125
-1.5	-1.125
1.5	-7.875

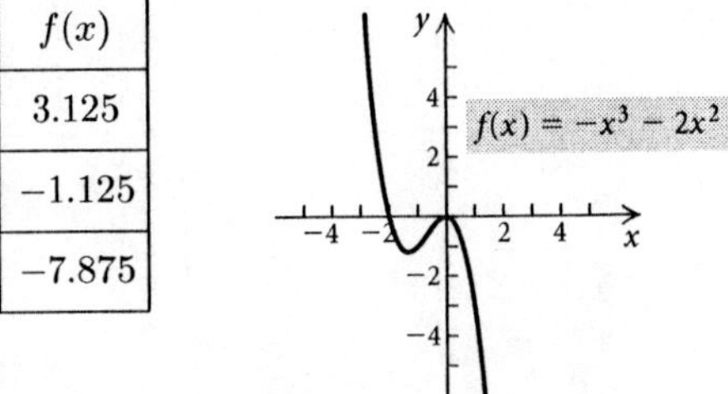

6. Checking the graph as described on page 254 in the text, we see that it appears to be correct.

46.

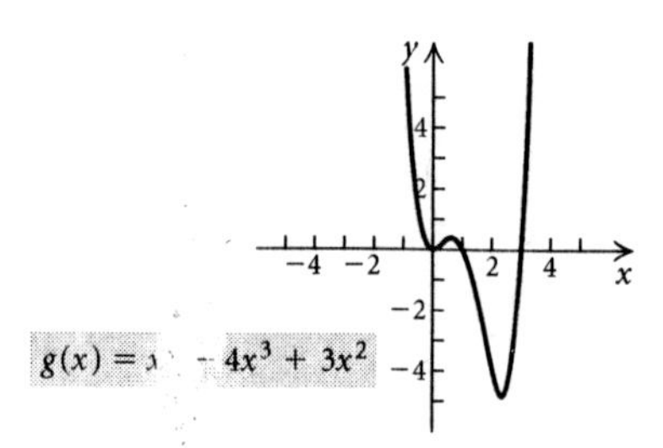

$g(x) = x^4 - 4x^3 + 3x^2$

47. $h(x) = x^5 - 4x^3$

1. The leading term is x^5. The degree, 5, is odd and the leading coefficient, 1, is positive so as $|x| \to \infty$, $h(x) \to \infty$ and as $|x| \to -\infty$, $h(x) \to -\infty$.
2. We solve $h(x) = 0$.

$$x^5 - 4x^3 = 0$$
$$x^3(x^2 - 4) = 0$$
$$x^3(x + 2)(x - 2) = 0$$
$$x^3 = 0 \quad or \quad x + 2 = 0 \quad or \quad x - 2 = 0$$
$$x = 0 \quad or \quad x = -2 \quad or \quad x = 2$$

The zeros of the function are 0, -2, and 2 so the x-intercepts of the graph are $(0, 0)$, $(-2, 0)$, and $(2, 0)$.

3. The zeros divide the x-axis into 4 intervals, $(-\infty, -2)$, $(-2, 0)$, $(0, 2)$, and $(2, \infty)$. We choose a value for x from each interval and find $h(x)$. This tells us the sign of $h(x)$ for all values of x in that interval.

In $(-\infty, -2)$, test -3:

$h(-3) = (-3)^5 - 4(-3)^3 = -135 < 0$

In $(-2, 0)$, test -1:

$h(-1) = (-1)^5 - 4(-1)^3 = 3 > 0$

In $(0, 2)$, test 1:

$h(1) = 1^5 - 4 \cdot 1^3 = -3 < 0$

In $(2, \infty)$, test 3:

$h(3) = 3^5 - 4 \cdot 3^3 = 135 > 0$

Thus the graph lies below the x-axis on $(-\infty, -2)$ and on $(0, 2)$. It lies above the x-axis on $(-2, 0)$ and on $(2, \infty)$. We also know the points $(-3, -135)$, $(-1, 3)$, $(1, -3)$, and $(3, 135)$ are on the graph.

4. From Step 2 we see that the y-intercept is $(0, 0)$.
5. We find additional points on the graph and then draw the graph.

x	$h(x)$
-2.5	-35.2
-1.5	5.9
1.5	-5.9
2.5	35.2

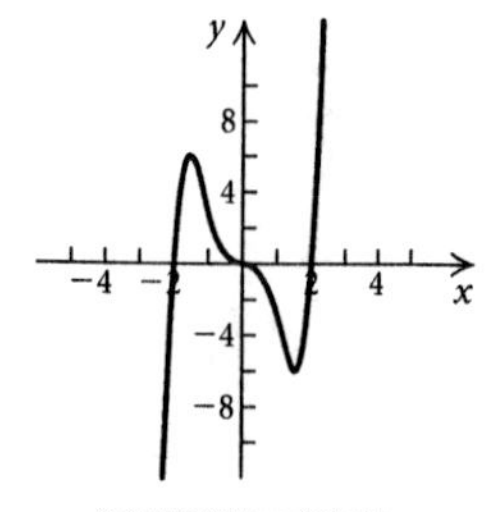

$h(x) = x^5 - 4x^3$

6. Checking the graph as described on page 254 in the text, we see that it appears to be correct.

48.

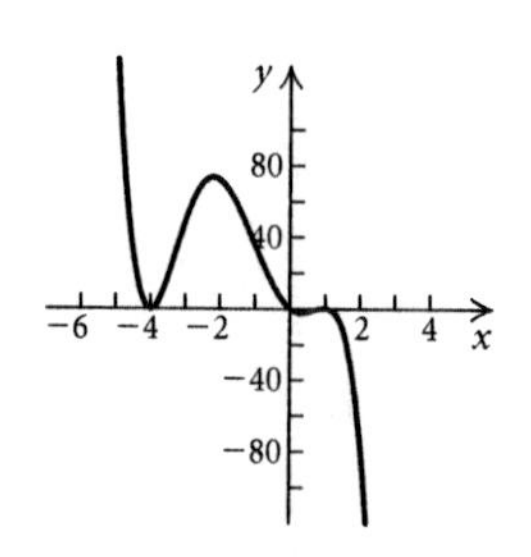

$g(x) = -x(x - 1)^2(x + 4)^2$

49. $h(x) = x(x - 4)(x + 1)(x - 2)$

1. The leading term is $x \cdot x \cdot x \cdot x$, or x^4. The degree, 4, is even and the leading coefficient, 1, is positive so as $|x| \to \infty$, $h(x) \to \infty$ and as $|x| \to -\infty$, $h(x) \to \infty$.
2. We see that the zeros of the function are 0, 4, -1, and 2 so the x-intercepts of the graph are $(0, 0)$, $(4, 0)$, $(-1, 0)$, and $(2, 0)$.
3. The zeros divide the x-axis into 5 intervals, $(-\infty, -1)$, $(-1, 0)$, $(0, 2)$, $(2, 4)$, and $(4, \infty)$. We choose a value for x from each interval and find $h(x)$. This tells us the sign of $h(x)$ for all values of x in that interval.

In $(-\infty, -1)$, test -2:

$h(-2) = -2(-2 - 4)(-2 + 1)(-2 - 2) = 48 > 0$

In $(-1, 0)$, test -0.5:

$h(-0.5) = (-0.5)(-0.5 - 4)(-0.5 + 1)(-0.5 - 2) = -2.8125 < 0$

In $(0, 2)$, test 1:

$h(1) = 1(1 - 4)(1 + 1)(1 - 2) = 6 > 0$

In $(2, 4)$, test 3:

$h(3) = 3(3 - 4)(3 + 1)(3 - 2) = -12 < 0$

In $(4, \infty)$, test 5:

$h(5) = 5(5 - 4)(5 + 1)(5 - 2) = 90 > 0$

Thus the graph lies above the x-axis on $(-\infty, -1)$, $(0, 2)$, and $(4, \infty)$. It lies below the x-axis on $(-1, 0)$ and on $(2, 4)$. We also know the points $(-2, 48)$, $(-0.5, -2.8125)$, $(1, 6)$, $(3, -12)$, and $(5, 90)$ are on the graph.

4. From Step 2 we see that the y-intercept is $(0, 0)$.
5. We find additional points on the graph and then draw the graph.

x	$h(x)$
-1.5	14.4
1.5	4.7
2.5	-6.6
4.5	30.9

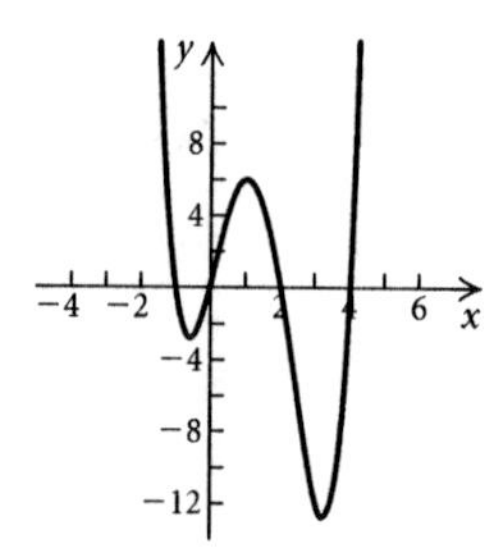

$h(x) = x(x - 4)(x + 1)(x - 2)$

6. Checking the graph as described on page 254 in the text, we see that it appears to be correct.

50.

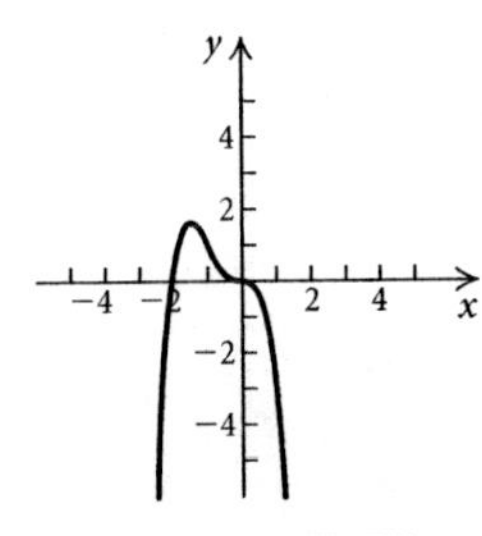

51. $f(x) = \frac{1}{2}x^3 + \frac{5}{2}x^2$

1. The leading term is $\frac{1}{2}x^3$. The degree, 3, is odd and the leading coefficient, $\frac{1}{2}$, is positive so as $|x| \to \infty$, $f(x) \to \infty$ and as $|x| \to -\infty$, $f(x) \to -\infty$.
2. We solve $f(x) = 0$.

$$\frac{1}{2}x^3 + \frac{5}{2}x^2 = 0$$
$$\frac{1}{2}x^2(x+5) = 0$$
$$\frac{1}{2}x^2 = 0 \quad or \quad x+5 = 0$$
$$x^2 = 0 \quad or \quad x = -5$$
$$x = 0 \quad or \quad x = -5$$

The zeros of the function are 0 and -5, so the x-intercepts of the graph are $(0,0)$ and $(-5,0)$.

3. The zeros divide the x-axis into 3 intervals, $(-\infty,-5)$, $(-5,0)$, and $(0,\infty)$. We choose a value for x from each interval and find $f(x)$. This tells us the sign of $f(x)$ for all values of x in that interval.

In $(-\infty,-5)$, test -6:

$f(-6) = \frac{1}{2}(-6)^3 + \frac{5}{2}(-6)^2 = -18 < 0$

In $(-5,0)$, test -1:

$f(-1) = \frac{1}{2}(-1)^3 + \frac{5}{2}(-1)^2 = 2 > 0$

In $(0,\infty)$, test 1:

$f(1) = \frac{1}{2}\cdot 1^3 + \frac{5}{2}\cdot 1^2 = 3 > 0$

Thus the graph lies below the x-axis on $(-\infty,-5)$ and above the x-axis on $(-5,0)$ and $(0,\infty)$. We also know the points $(-6,-18)$, $(-1,2)$, and $(1,3)$ are on the graph.

4. From Step 2 we see that the y-intercept is $(0,0)$.
5. We find additional points on the graph and then draw the graph.

x	$f(x)$
-5.5	-7.6
-3	9
-2	6
2	14

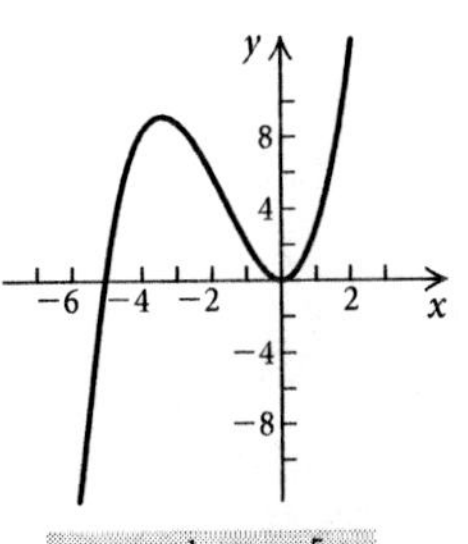

6. Checking the graph as described on page 254 in the text, we see that it appears to be correct.

52.

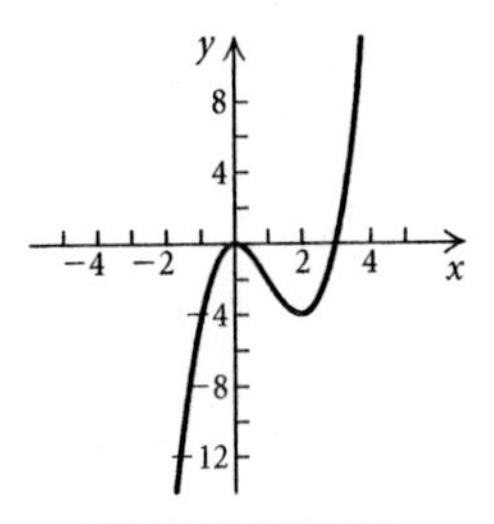

53. $g(x) = (x-2)^3(x+3)$

1. The leading term is $x\cdot x\cdot x\cdot x$, or x^4. The degree, 4, is even and the leading coefficient, 1, is positive so as $|x| \to \infty$, $g(x) \to \infty$ and as $|x| \to -\infty$, $g(x) \to \infty$.
2. We see that the zeros of the function are 2 and -3 so the x-intercepts of the graph are $(2,0)$ and $(-3,0)$.
3. The zeros divide the x-axis into 3 intervals, $(-\infty,-3)$, $(-3,2)$, and $(2,\infty)$. We choose a value for x from each interval and find $g(x)$. This tells us the sign of $g(x)$ for all values of x in that interval.

In $(-\infty,-3)$, test -4:

$g(-4) = (-4-2)^3(-4+3) = 216 > 0$

In $(-3,2)$, test 0:

$g(0) = (0-2)^3(0+3) = -24 < 0$

In $(2,\infty)$, test 3:

$g(3) = (3-2)^3(3+3) = 6 > 0$

Thus the graph lies above the x-axis on $(-\infty,-3)$ and on $(2,\infty)$. It lies below the x-axis on $(-3,2)$. We also know the points $(-4,216)$, $(0,-24)$, and $(3,6)$ are on the graph.

4. From Step 3 we know that $g(0) = -24$ so the y-intercept is $(0,-24)$.

5. We find additional points on the graph and then draw the graph.

x	$g(x)$
-3.5	83.2
-2	-64
-1	-54
3.5	23.9

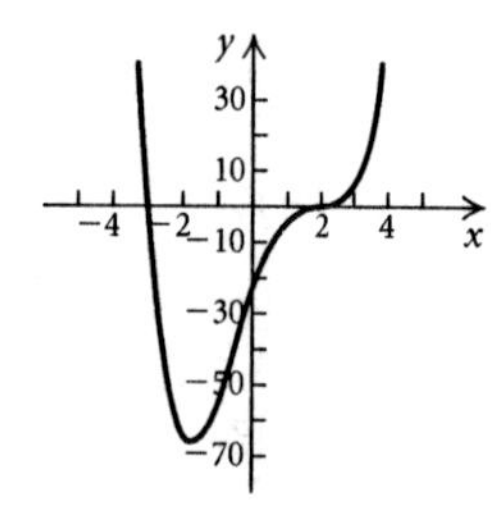

$g(x) = (x-2)^3(x+3)$

6. Checking the graph as described on page 254 in the text, we see that it appears to be correct.

54.

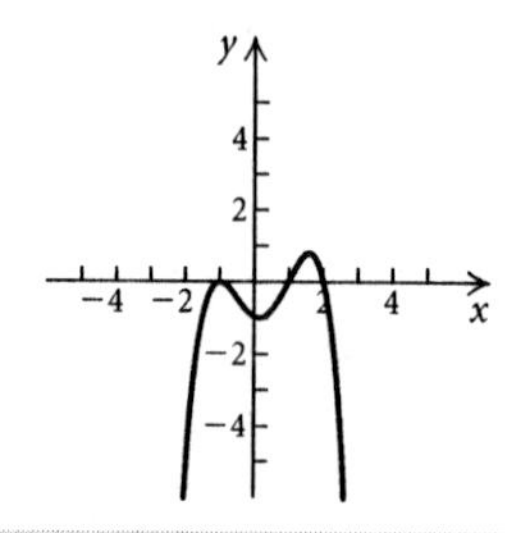

$f(x) = -\frac{1}{2}(x-2)(x+1)^2(x-1)$

55. $f(x) = x^3 - x$

1. The leading term is x^3. The degree, 3, is odd and the leading coefficient, 1, is positive so as $|x| \to \infty$, $f(x) \to \infty$ and as $|x| \to -\infty$, $f(x) \to -\infty$.

2. We solve $f(x) = 0$.

$$x^3 - x = 0$$
$$x(x^2-1) = 0$$
$$x(x+1)(x-1) = 0$$
$$x = 0 \quad or \quad x+1 = 0 \quad or \quad x-1 = 0$$
$$x = 0 \quad or \quad x = -1 \quad or \quad x = 1$$

The zeros of the function are 0, -1, and 1, so the x-intercepts of the graph are $(0,0)$, $(-1,0)$, and $(1,0)$.

3. The zeros divide the x-axis into 4 intervals, $(-\infty,-1)$, $(-1,0)$, $(0,1)$, and $(1,\infty)$. We choose a value for x from each interval and find $f(x)$. This tells us the sign of $f(x)$ for all values of x in that interval.

In $(-\infty,-1)$, test -2:

$f(-2) = (-2)^3 - (-2) = -6 < 0$

In $(-1,0)$, test -0.5:

$f(-0.5) = (-0.5)^3 - (-0.5) = 0.375 > 0$

In $(0,1)$, test 0.5:

$f(0.5) = (0.5)^3 - 0.5 = -0.375 < 0$

In $(1,\infty)$, test 2:

$f(2) = 2^3 - 2 = 6 > 0$

Thus the graph lies below the x-axis on $(-\infty,-1)$ and on $(0,1)$. It lies above the x-axis on $(-1,0)$ and on $(1,\infty)$. We also know the points $(-2,-6)$, $(-0.5,0.375)$, $(0.5,-0.375)$, and $(2,6)$ are on the graph.

4. From Step 2 we see that the y-intercept is $(0,0)$.

5. We find additional points on the graph and then draw the graph.

x	$f(x)$
-3	-24
-0.75	0.3
0.25	-0.2
3	24

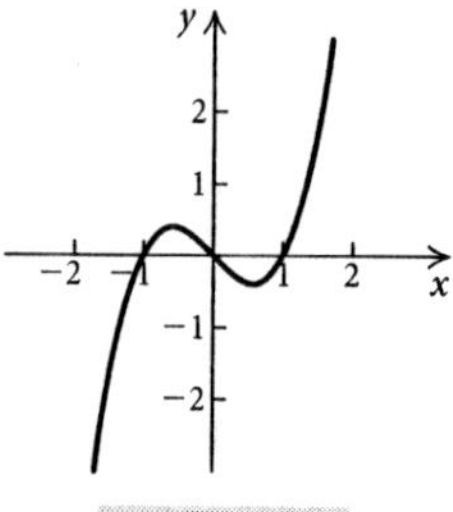

$f(x) = x^3 - x$

6. Checking the graph as described on page 254 in the text, we see that it appears to be correct.

56.

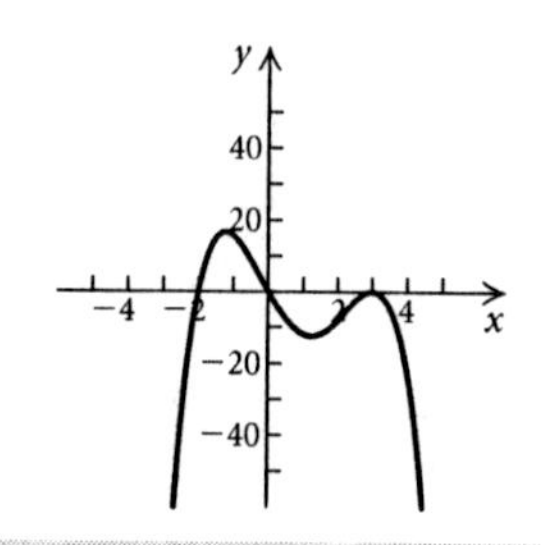

$h(x) = -x(x-3)(x-3)(x+2)$

57. $f(x) = (x-2)^2(x+1)^4$

1. The leading term is $x \cdot x \cdot x \cdot x \cdot x \cdot x$, or x^6. The degree, 6, is even and the leading coefficient, 1, is positive so as $|x| \to \infty$, $f(x) \to \infty$ and as $|x| \to -\infty$, $f(x) \to \infty$.

2. We see that the zeros of the function are 2 and -1 so the x-intercepts of the graph are $(2,0)$ and $(-1,0)$.

3. The zeros divide the x-axis into 3 intervals, $(-\infty,-1)$, $(-1,2)$, and $(2,\infty)$. We choose a value for x from each interval and find $f(x)$. This tells us the sign of $f(x)$ for all values of x in that interval.

In $(-\infty,-1)$, test -2:

$f(-2) = (-2-2)^2(-2+1)^4 = 16 > 0$

In $(-1,2)$, test 0:

$f(0) = (0-2)^2(0+1)^4 = 4 > 0$

In $(2,\infty)$, test 3:

$f(3) = (3-2)^2(3+1)^4 = 256 > 0$

Thus the graph lies above the x-axis on all 3 intervals. We also know the points $(-2,16)$, $(0,4)$, and $(3,256)$ are on the graph.

4. From Step 3 we know that $f(0) = 4$ so the y-intercept is $(0,4)$.

5. We find additional points on the graph and then draw the graph.

x	$f(x)$
-1.5	0.8
-0.5	0.4
1	16
1.5	9.8

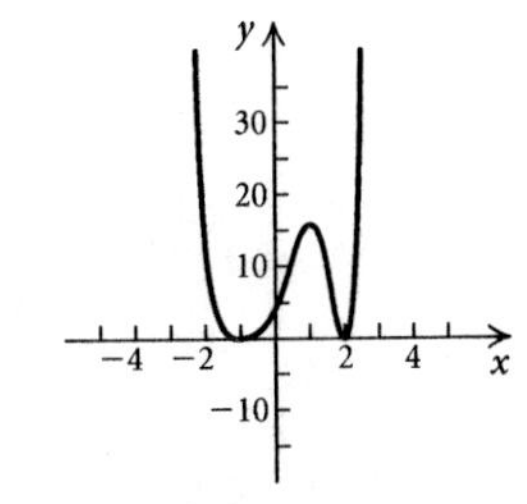

$f(x) = (x-2)^2(x+1)^4$

6. Checking the graph as described on page 254 in the text, we see that it appears to be correct.

58.

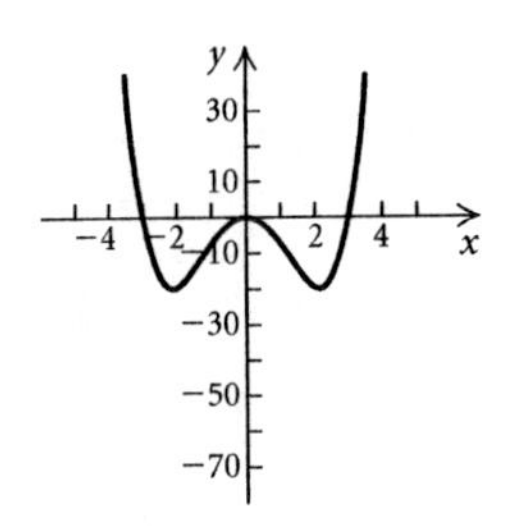

$g(x) = x^4 - 9x^2$

59. $g(x) = -(x-1)^4$

1. The leading term is $-1 \cdot x \cdot x \cdot x \cdot x$, or $-x^4$. The degree, 4, is even and the leading coefficient, -1, is negative so as $|x| \to \infty$, $g(x) \to -\infty$ and as $|x| \to -\infty$, $g(x) \to -\infty$.
2. We see that the zero of the function is 1, so the x-intercept is $(1, 0)$.
3. The zero divides the x-axis into 2 intervals, $(-\infty, 1)$ and $(1, \infty)$. We choose a value for x from each interval and find $g(x)$. This tells us the sign of $g(x)$ for all values of x in that interval.
   In $(-\infty, 1)$, test 0:
   $g(0) = -(0-1)^4 = -1 < 0$
   In $(1, \infty)$, test 2:
   $g(2) = -(2-1)^4 = -1 < 0$
   Thus the graph lies below the x-axis on both intervals. We also know the points $(0, -1)$ and $(2, -1)$ are on the graph.
4. From Step 3 we know that $g(0) = -1$ so the y-intercept is $(0, -1)$.
5. We find additional points on the graph and then draw the graph.

x	$g(x)$
-1	-16
-0.5	-5.1
1.5	0.1
3	-16

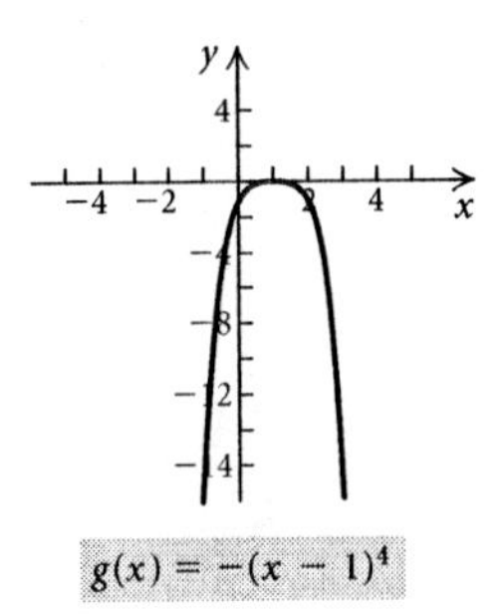

$g(x) = -(x-1)^4$

6. Checking the graph as described on page 254 in the text, we see that it appears to be correct.

60.

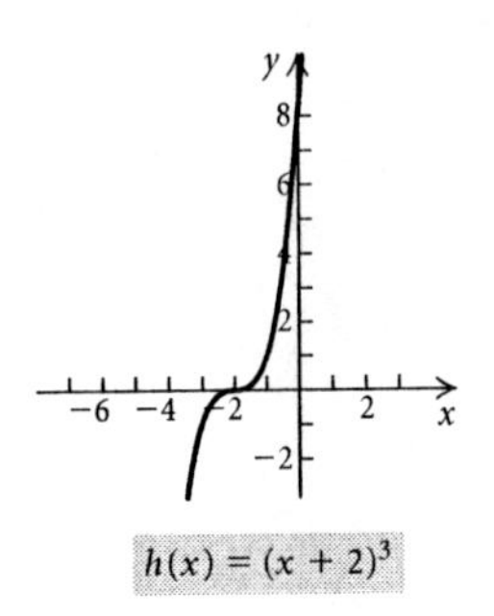

$h(x) = (x+2)^3$

61. $h(x) = x^3 + 3x^2 - x - 3$

1. The leading term is x^3. The degree, 3, is odd and the leading coefficient, 1, is positive so as $|x| \to \infty$, $h(x) \to \infty$ and as $|x| \to -\infty$, $h(x) \to -\infty$.
2. We solve $h(x) = 0$.
   $$x^3 + 3x^2 - x - 3 = 0$$
   $$x^2(x+3) - (x+3) = 0$$
   $$(x+3)(x^2-1) = 0$$
   $$(x+3)(x+1)(x-1) = 0$$
   $$x+3 = 0 \quad or \quad x+1 = 0 \quad or \quad x-1 = 0$$
   $$x = -3 \quad or \quad x = -1 \quad or \quad x = 1$$
   The zeros of the function are -3, -1, and 1 so the x-intercepts of the graph are $(-3, 0)$, $(-1, 0)$, and $(1, 0)$.
3. The zeros divide the x-axis into 4 intervals, $(-\infty, -3)$, $(-3, -1)$, $(-1, 1)$, and $(1, \infty)$. We choose a value for x from each interval and find $h(x)$. This tells us the sign of $h(x)$ for all values of x in that interval.
   In $(-\infty, -3)$, test -4:
   $h(-4) = (-4)^3 + 3(-4)^2 - (-4) - 3 = -15 < 0$
   In $(-3, -1)$, test -2:
   $h(-2) = (-2)^3 + 3(-2)^2 - (-2) - 3 = 3 > 0$
   In $(-1, 1)$, test 0:
   $h(0) = 0^3 + 3 \cdot 0^2 - 0 - 3 = -3 < 0$
   In $(1, \infty)$, test 2:
   $h(2) = 2^3 + 3 \cdot 2^2 - 2 - 3 = 15 > 0$
   Thus the graph lies below the x-axis on $(-\infty, -3)$ and on $(-1, 1)$ and above the x-axis on $(-3, -1)$ and on $(1, \infty)$. We also know the points $(-4, -15)$, $(-2, 3)$, $(0, -3)$, and $(2, 15)$ are on the graph.

4. From Step 3 we know that $h(0) = -3$ so the y-intercept is $(0, -3)$.

5. We find additional points on the graph and then draw the graph.

x	$h(x)$
-4.5	-28.9
-2.5	2.6
0.5	-2.6
2.5	28.9

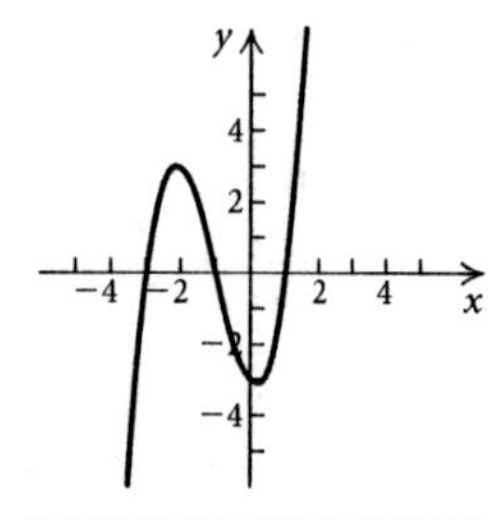

$h(x) = x^3 + 3x^2 - x - 3$

6. Checking the graph as described on page 254 in the text, we see that it appears to be correct.

62.

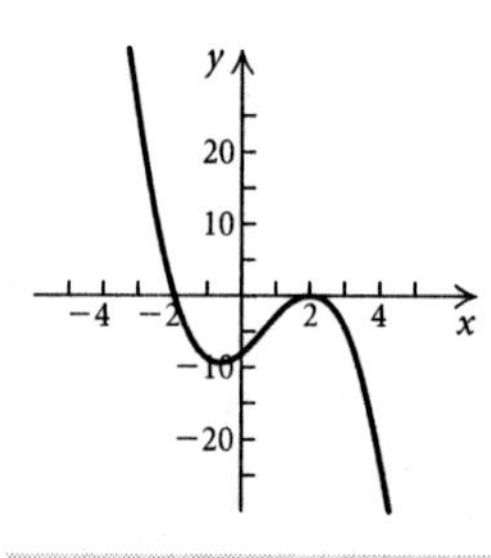

$g(x) = -x^3 + 2x^2 + 4x - 8$

63. $f(x) = 6x^3 - 8x^2 - 54x + 72$

1. The leading term is $6x^3$. The degree, 3, is odd and the leading coefficient, 6, is positive so as $|x| \to \infty$, $f(x) \to \infty$ and as $|x| \to -\infty$, $f(x) \to -\infty$.

2. We solve $f(x) = 0$.

$$6x^3 - 8x^2 - 54x + 72 = 0$$
$$2(3x^3 - 4x^2 - 27x + 36) = 0$$
$$2[x^2(3x - 4) - 9(3x - 4)] = 0$$
$$2(3x - 4)(x^2 - 9) = 0$$
$$2(3x - 4)(x + 3)(x - 3) = 0$$
$$3x - 4 = 0 \quad or \quad x + 3 = 0 \quad or \quad x - 3 = 0$$
$$x = \frac{4}{3} \quad or \quad x = -3 \quad or \quad x = 3$$

The zeros of the function are $\frac{4}{3}$, -3, and 3, so the x-intercepts of the graph are $\left(\frac{4}{3}, 0\right)$, $(-3, 0)$, and $(3, 0)$.

3. The zeros divide the x-axis into 4 intervals, $(-\infty, -3)$, $\left(-3, \frac{4}{3}\right)$, $\left(\frac{4}{3}, 3\right)$, and $(3, \infty)$. We choose a value for x from each interval and find $f(x)$. This tells us the sign of $f(x)$ for all values of x in that interval.

In $(-\infty, -3)$, test -4:

$f(-4) = 6(-4)^3 - 8(-4)^2 - 54(-4) + 72 = -224 < 0$

In $\left(-3, \frac{4}{3}\right)$, test 0:

$f(0) = 6 \cdot 0^3 - 8 \cdot 0^2 - 54 \cdot 0 + 72 = 72 > 0$

In $\left(\frac{4}{3}, 3\right)$, test 2:

$f(2) = 6 \cdot 2^3 - 8 \cdot 2^2 - 54 \cdot 2 + 72 = -20 < 0$

In $(3, \infty)$, test 4:

$f(4) = 6 \cdot 4^3 - 8 \cdot 4^2 - 54 \cdot 4 + 72 = 112 > 0$

Thus the graph lies below the x-axis on $(-\infty, -3)$ and on $\left(\frac{4}{3}, 3\right)$ and above the x-axis on $\left(-3, \frac{4}{3}\right)$ and on $(3, \infty)$. We also know the points $(-4, -224)$, $(0, 72)$, $(2, -20)$, and $(4, 112)$ are on the graph.

4. From Step 3 we know that $f(0) = 72$ so the y-intercept is $(0, 72)$.

5. We find additional points on the graph and then draw the graph.

x	$f(x)$
-1	112
1	16
3.5	42.25

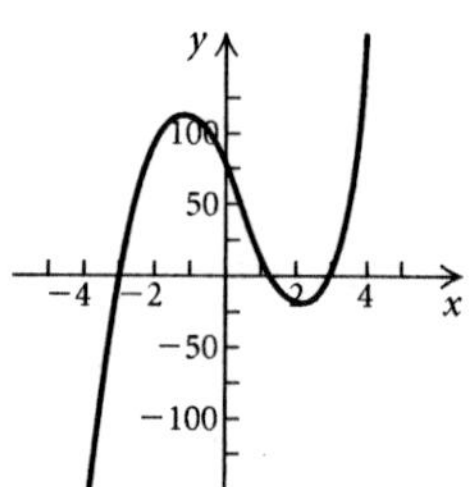

$f(x) = 6x^3 - 8x^2 - 54x + 72$

6. Checking the graph as described on page 254 in the text, we see that it appears to be correct.

64.

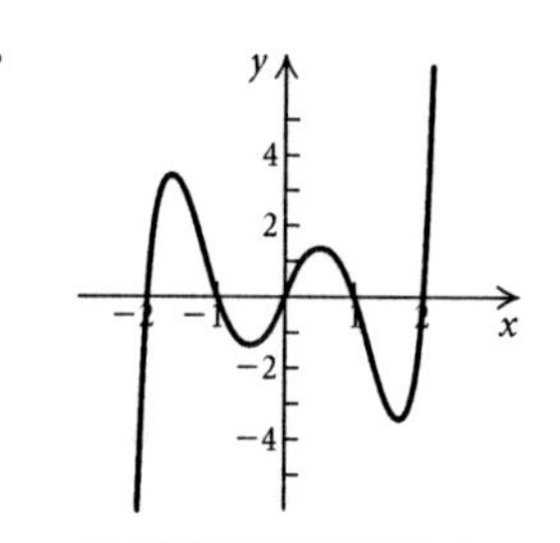

$h(x) = x^5 - 5x^3 + 4x$

65. $f(-5) = (-5)^3 + 3(-5)^2 - 9(-5) - 13 = -18$

$f(-4) = (-4)^3 + 3(-4)^2 - 9(-4) - 13 = 7$

By the intermediate value theorem, since $f(-5)$ and $f(-4)$ have opposite signs then $f(x)$ has a zero between -5 and -4.

66. $f(1) = 1^3 + 3 \cdot 1^2 - 9 \cdot 1 - 13 = -18$

$f(2) = 2^3 + 3 \cdot 2^2 - 9 \cdot 2 - 13 = -11$

Since both $f(1)$ and $f(2)$ are negative, we cannot use the intermediate value theorem to determine if there is a zero between 1 and 2.

67. $f(-3) = 3(-3)^2 - 2(-3) - 11 = 22$

$f(-2) = 3(-2)^2 - 2(-2) - 11 = 5$

Since both $f(-3)$ and $f(-2)$ are positive, we cannot use the intermediate value theorem to determine if there is a zero between -3 and -2.

68. $f(2) = 3 \cdot 2^2 - 2 \cdot 2 - 11 = -3$

$f(3) = 3 \cdot 3^2 - 2 \cdot 3 - 11 = 10$

By the intermediate value theorem, since $f(2)$ and $f(3)$ have opposite signs then $f(x)$ has a zero between 2 and 3.

69. $f(2) = 2^4 - 2 \cdot 2^2 - 6 = 2$

$f(3) = 3^4 - 2 \cdot 3^2 - 6 = 57$

Since both $f(2)$ and $f(3)$ are positive, we cannot use the intermediate value theorem to determine if there is a zero between 2 and 3.

70. $f(1) = 2 \cdot 1^5 - 7 \cdot 1 + 1 = -4$

$f(2) = 2 \cdot 2^5 - 7 \cdot 2 + 1 = 51$

By the intermediate value theorem, since $f(1)$ and $f(2)$ have opposite signs then $f(x)$ has a zero between 1 and 2.

71. $f(4) = 4^3 - 5 \cdot 4^2 + 4 = -12$

$f(5) = 5^3 - 5 \cdot 5^2 + 4 = 4$

By the intermediate value theorem, since $f(4)$ and $f(5)$ have opposite signs then $f(x)$ has a zero between 4 and 5.

72. $f(-3) = (-3)^4 - 3(-3)^2 + (-3) - 1 = 50$

$f(-2) = (-2)^4 - 3(-2)^2 + (-2) - 1 = 1$

Since both $f(-3)$ and $f(-2)$ are positive, we cannot use the intermediate value theorem to determine if there is a zero between -3 and -2.

73. For 1995, $x = 1995 - 1993 = 2$.

$f(2) = 0.854(2)^4 - 16.115(2)^3 + 86.769(2)^2-$
$139.235(2) + 735.664 \approx 689$

In 1995 about 689,000 acres of tobacco were harvested.

For 2000, $x = 2000 - 1993 = 7$.

$f(7) = 0.854(7)^4 - 16.115(7)^3 + 86.769(7)^2-$
$139.235(7) + 735.664 \approx 536$

In 2000 about 536,000 acres of tobacco were harvested.

74. Solve $294 = 4.9t^2 + 34.3t$.

$t = 5$ sec

75. $V = 48T^2$

$36 = 48T^2$ Substituting

$0.75 = T^2$

$0.866 \approx T$ We only want the positive solution.

Anfernee Hardaway's hang time is 0.866 sec.

76. For 1993: $f(1) \approx \$86$ billion

For 1996: $f(4) \approx \$74$ billion

For 2000: $f(8) \approx \$48$ billion

For 2002: $f(10) \approx \$91$ billion

For 2003: $f(11) \approx \$138$ billion

77. First find the number of games played.

$N(x) = x^2 - x$

$N(9) = 9^2 - 9 = 72$

Now multiply the number of games by the cost per game to find the total cost.

$72 \cdot 45 = 3240$

It will cost \$3240 to play the entire schedule.

78. a) $P(15) = 0.015(15)^3 = 50.625$ watts per hour

b) Solve $120 = 0.015v^3$.

$v = 20$ mph

79. For 1997, $x = 1997 - 1996 = 1$.

$f(1) = -3.394(1)^4 + 35.838(1)^3 - 98.955(1)^2+$
$41.930(1) + 174.974 \approx 150$ bald eagles

For 1999, $x = 1999 - 1996 = 3$.

$f(3) = -3.394(3)^4 + 35.838(3)^3 - 98.955(3)^2+$
$41.930(3) + 174.974 \approx 103$ bald eagles

For 2001, $x = 2001 - 1996 = 5$.

$f(5) = -3.394(5)^4 + 35.838(5)^3 - 98.955(5)^2+$
$41.930(5) + 174.974 \approx 269$ bald eagles

For 2002, $x = 2002 - 1996 = 6$.

$f(6) = -3.394(6)^4 + 35.838(6)^3 - 98.955(6)^2+$
$41.930(6) + 174.974 \approx 207$ bald eagles

80. For 1970: $f(0) \approx 360$ thousand

For 1980: $f(10) \approx 254$ thousand

For 1990: $f(20) \approx 268$ thousand

For 1998: $f(28) \approx 220$ thousand

For 2000: $f(30) \approx 233$ thousand

81. $A = P(1+i)^t$

a) $4368.10 = 4000(1+i)^2$ Substituting

$\dfrac{4368.10}{4000} = (1+i)^2$

$\pm 1.045 = 1 + i$ Taking the square root on both sides

$-1 \pm 1.045 = i$

$-1 - 1.045 = i$ *or* $-1 + 1.045 = i$

$-2.045 = i$ *or* $0.045 = i$

Only the positive result has meaning in this application. The interest rate is 0.045, or 4.5%.

b) $13,310 = 10,000(1+i)^3$ Substituting

$\dfrac{13,310}{10,000} = (1+i)^3$

$1.10 \approx 1 + i$ Taking the cube root on both sides

$0.10 \approx i$

The interest rate is 0.1, or 10%.

82. 5 ft, 7 in. = 67 in.

$W(67) = \left(\dfrac{67}{12.3}\right)^3 \approx 162$ lb

83. $g(x) = x^3 - 1.2x + 1$

Graph the function and use the Zero, Maximum, and Minimum features.

Relative maximum: 1.506 when $x \approx -0.632$

Relative minimum: 0.494 when $x \approx 0.632$

Range: $(-\infty, \infty)$

84. $h(x) = -\frac{1}{2}x^4 + 3x^3 - 5x^2 + 3x + 6$

Graph the function and use the Zero, Maximum, and Minimum features.

Relative maxima: 6.59375 when $x = 0.5$, 10.5 when $x = 3$

Relative minimum: 6.5 when $x = 1$

Range: $(-\infty, 10.5]$

85. $f(x) = x^6 - 3.8$

Graph the function and use the Zero and Minimum features.

There is no relative maximum.

Relative minimum: -3.8 when $x = 0$

Range: $[-3.8, \infty)$

86. $h(x) = 2x^3 - x^4 + 20$

Graph the function and use the Zero and Maximum features.

Relative maximum: 21.688 when $x = 1.5$

There is no relative minimum.

Range: $(-\infty, 21.688]$

87. $f(x) = x^2 + 10x - x^5$

Graph the function and use the Zero, Maximum, and Minimum features.

Relative maximum: 11.012 when $x \approx 1.258$

Relative minimum: -8.183 when $x \approx -1.116$

Range: $(-\infty, \infty)$

88. $f(x) = 2x^4 - 5.6x^2 + 10$

Graph the function and use the Zero, Maximum, and Minimum features.

Relative maximum: 10 when $x = 0$

Relative minima: 6.08 when $x \approx -1.183$ and when $x \approx 1.183$

Range: $[6.08, \infty)$

89. a) For each function, x represents the number of years after 1900.

Linear: $y = 3.730909091x + 83.72727273$

Quadratic: $y = 0.025011655x^2 + 1.22974359x + 121.2447552$

Cubic: $y = -0.0011822067x^3 + 0.2023426573x^2 - 5.532478632x + 163.8041958$

Quartic: $y = -0.00002634033x^4 + 0.0040858586x^3 - 0.1269114219x^2 + 1.052602953x + 144.8391608$

b) In 2005, $x = 2005 - 1900 = 105$; in 2010, $x = 2010 - 1900 = 110$.

For the linear function: $f(105) \approx 475$ and $f(110) \approx 494$.

For the quadratic function: $f(105) \approx 526$ and $f(110) \approx 559$.

For the cubic function: $f(105) \approx 445$ and $f(110) \approx 430$.

For the power function: $f(105) \approx 384$ and $f(110) \approx 307$.

It is reasonable to assume that the average acreage will continue to decrease as it did from 1990 to 2000. For this reason we rule out the estimates made using the linear and quadratic functions. The rate of decrease appears to be more like that estimated by the cubic function than the quartic function. Thus we say that the esimates of 445 acres and 430 acres given by cubic function are the most realistic. Answers may vary.

90. a) For each function, x represents the number of years after 1980.

Linear: $y = 170.4126917x + 1891.215503$

Quadratic: $y = -2.567331153x^2 + 223.3644434x + 1751.197548$

Cubic: $y = -0.3576190681x^3 + 9.016361061x^2 + 129.212158x + 1831.569976$

Quartic: $y = -0.0512181196x^4 + 1.833287973x^3 - 19.88479957x^2 + 243.3857579x + 1810.850991$

b) For the linear function: $f(25) \approx \$6152$ and $f(30) \approx \$7004$.

For the quadratic function: $f(25) \approx \$5731$ and $f(30) \approx \$6142$.

For the cubic function: $f(25) \approx \$5109$ and $f(30) \approx \$4167$

For the quartic function: $f(25) \approx \$4106$ and $f(30) \approx -\$772$.

The quartic function does not give a realistic estimate for the cost in 2010. It is reasonable to assume that the costs will continue to increase, so the cubic function does not give realistic estimates either. The rate of increase in the cost appears to be more like that estimated by the quadratic function than the linear function. Thus, we say that the estimates given by the quadratic function are the most realistic. Answers may vary. Some might argue that the linear function gives the most realistic estimates.

91. a) For each function, x represents the number of years after 1992.

Linear: $y = 378.7575758x + 3701.090909$

Quadratic: $y = 43.10227273x^2 - 9.162878788x + 4218.318182$

b) In 2005, $x = 2005 - 1992 = 13$.

For the linear function: $f(13) \approx \$8625$ billion.

For the quadratic function: $f(13) \approx \$11,383$ billion.

The estimate given by the quadratic function seems more consistent with the rate of increase in the table. Answers may vary.

92. a) Linear: $y = 4.550243651x + 10.30607732$

Cubic: $y = 0.0102548814x^3 - 0.3401188859x^2 +$

$7.397498939x + 6.618361098$

Since R^2 is higher for the cubic function, it has the better fit.

b) $f(5) \approx 36$ yr; $f(10) \approx 57$ yr; $f(15) \approx 76$ yr

93. The range of a polynomial function with an odd degree is $(-\infty, \infty)$. The range of a polynomial function with an even degree is $[s, \infty)$ for some real number s if $a_n > 0$ and is $(-\infty, s]$ for some real number s if $a_n < 0$.

94. In terms of the domain, "continuous" means that all real numbers are in the domain. In terms of the graphs, "continuous" means that there are no holes or breaks and no sharp corners.

95.
$$\begin{aligned} d &= \sqrt{(x_2 - x_1)^2 + (y_2 - y_1)^2} \\ &= \sqrt{[-1 - (-5)]^2 + (0 - 3)^2} \\ &= \sqrt{4^2 + (-3)^2} = \sqrt{16 + 9} \\ &= \sqrt{25} = 5 \end{aligned}$$

96. $d = \sqrt{(-2 - 4)^2 + (-4 - 2)^2} = \sqrt{36 + 36} = \sqrt{72} = 6\sqrt{2}$

97.
$$\begin{aligned} (x - 3)^2 + (y + 5)^2 &= 49 \\ (x - 3)^2 + [y - (-5)]^2 &= 7^2 \end{aligned}$$

Center: $(3, -5)$; radius: 7

98. The center of the circle is the midpoint of a segment that is a diameter:

$$\left(\frac{-6 + (-2)}{2}, \frac{5 + 1}{2}\right) = (-4, 3).$$

The length of a radius is the distance from the center to one of the endpoints of the diameter:

$$\begin{aligned} r &= \sqrt{[-6 - (-4)]^2 + (5 - 3)^2} \\ &= \sqrt{4 + 4} = \sqrt{8} \\ &= 2\sqrt{2} \end{aligned}$$

99.
$$\begin{aligned} 2y - 3 &\geq 1 - y + 5 \\ 2y - 3 &\geq 6 - y && \text{Collecting like terms} \\ 3y - 3 &\geq 6 && \text{Adding } y \\ 3y &\geq 9 && \text{Adding 3} \\ y &\geq 3 && \text{Dividing by 3} \end{aligned}$$

The solution set is $\{y | y \geq 3\}$, or $[3, \infty)$.

100.
$$\begin{aligned} (x - 2)(x + 5) &> x(x - 3) \\ x^2 + 3x - 10 &> x^2 - 3x \\ 6x - 10 &> 0 \\ 6x &> 10 \\ x &> \frac{5}{3} \end{aligned}$$

The solution set is $\left\{x \middle| x > \frac{5}{3}\right\}$, or $\left(\frac{5}{3}, \infty\right)$.

101.
$$\begin{aligned} |x + 6| &\geq 7 \\ x + 6 \leq -7 \quad &or \quad x + 6 \geq 7 \\ x \leq -13 \quad &or \quad x \geq 1 \end{aligned}$$

The solution set is $\{x | x \leq -13 \text{ or } x \geq 1\}$, or $(-\infty, -13] \cup [1, \infty)$.

102.
$$\begin{aligned} \left|x + \frac{1}{4}\right| &\leq \frac{2}{3} \\ -\frac{2}{3} \leq x &+ \frac{1}{4} \leq \frac{2}{3} \\ -\frac{11}{12} \leq &x \leq \frac{5}{12} \end{aligned}$$

The solution set is $\left\{x \middle| -\frac{11}{12} \leq x \leq \frac{5}{12}\right\}$, or $\left[-\frac{11}{12}, \frac{5}{12}\right]$.

103. ***Familiarize***. We will use the compound interest formula. The \$2000 deposit is invested for two years and grows to an amount A_1 given by $A_1 = 2000(1 + i)^2$. The \$1200 deposit is invested for one year and grows to an amount A_2 given by $A_2 = 1200(1 + i)$.

Translate. There is a total of \$3573.80 in both accounts at the end of the second year, so we have

$A_1 + A_2 = 3573.80$, or

$2000(1 + i)^2 + 1200(1 + i) = 3573.80.$

Carry out. We solve the equation.

$2000(1 + i)^2 + 1200(1 + i) - 3573.8 = 0$

Substitute u for $1 + i$.

$2000u^2 + 1200u - 3573.8 = 0$

Using the quadratic formula we find that $u = 1.07$ or $u = -3.34$. Only the positive value has meaning in this application. Then since $u = 1.07$, we have $1 + i = 1.07$, or $i = 0.07$.

Check. At an interest rate of 0.07, or 7%, in two years \$2000 would grow to $2000(1 + 0.07)^2$, or \$2289.80. In one year \$1200 would grow to $1200(1 + 0.07)$, or \$1284. Now \$2289.80 + \$1284 = \$3573.80, so the result checks.

State. The interest rate is 7%.

Exercise Set 3.2

1. a)

$$
\begin{array}{r}
x^3 - 7x^2 + 8x + 16 \\
x+1 \overline{)\, x^4 - 6x^3 + x^2 + 24x - 20} \\
\underline{x^4 + x^3} \qquad\qquad\qquad\quad \\
-7x^3 + x^2 \qquad\qquad \\
\underline{-7x^3 - 7x^2} \qquad\qquad \\
8x^2 + 24x \quad\;\; \\
\underline{8x^2 + 8x} \quad\;\; \\
16x - 20 \\
\underline{16x + 16} \\
-4
\end{array}
$$

Since the remainder is not 0, $x+1$ is not a factor of $f(x)$.

b)

$$
\begin{array}{r}
x^3 - 4x^2 - 7x + 10 \\
x-2 \overline{)\, x^4 - 6x^3 + x^2 + 24x - 20} \\
\underline{x^4 - 2x^3} \qquad\qquad\qquad\quad \\
-4x^3 + x^2 \qquad\qquad \\
\underline{-4x^3 + 8x^2} \qquad\qquad \\
-7x^2 + 24x \quad\;\; \\
\underline{-7x^2 + 14x} \quad\;\; \\
10x - 20 \\
\underline{10x - 20} \\
0
\end{array}
$$

Since the remainder is 0, $x-2$ is a factor of $f(x)$.

c)

$$
\begin{array}{r}
x^3 - 11x^2 + 56x - 256 \\
x+5 \overline{)\, x^4 - 6x^3 + x^2 + 24x - 20} \\
\underline{x^4 + 5x^3} \qquad\qquad\qquad\quad \\
-11x^3 + x^2 \qquad\qquad \\
\underline{-11x^3 - 55x^2} \qquad\qquad \\
56x^2 + 24x \quad\;\; \\
\underline{56x^2 + 280x} \quad\;\; \\
-256x - 20 \\
\underline{-256x - 1280} \\
1260
\end{array}
$$

Since the remainder is not 0, $x+5$ is not a factor of $f(x)$.

2. a)

$$
\begin{array}{r}
x^2 - 6x + 13 \\
x+5 \overline{)\, x^3 - x^2 - 17x - 15} \\
\underline{x^3 + 5x^2} \qquad\qquad\quad \\
-6x^2 - 17x \qquad \\
\underline{-6x^2 - 30x} \qquad \\
13x - 15 \\
\underline{13x + 65} \\
-80
\end{array}
$$

Since the remainder is not 0, $x+5$ is not a factor of $h(x)$.

b)

$$
\begin{array}{r}
x^2 - 2x - 15 \\
x+1 \overline{)\, x^3 - x^2 - 17x - 15} \\
\underline{x^3 + x^2} \qquad\qquad\quad \\
-2x^2 - 17x \qquad \\
\underline{-2x^2 - 2x} \qquad \\
-15x - 15 \\
\underline{-15x - 15} \\
0
\end{array}
$$

Since the remainder is 0, $x+1$ is a factor of $h(x)$.

c)

$$
\begin{array}{r}
x^2 - 4x - 5 \\
x+3 \overline{)\, x^3 - x^2 - 17x - 15} \\
\underline{x^3 + 3x^2} \qquad\qquad\quad \\
-4x^2 - 17x \qquad \\
\underline{-4x^2 - 12x} \qquad \\
-5x - 15 \\
\underline{-5x - 15} \\
0
\end{array}
$$

Since the remainder is 0, $x+3$ is a factor of $h(x)$.

3. a)

$$
\begin{array}{r}
x^2 + 2x - 3 \\
x-4 \overline{)\, x^3 - 2x^2 - 11x + 12} \\
\underline{x^3 - 4x^2} \qquad\qquad\quad \\
2x^2 - 11x \qquad \\
\underline{2x^2 - 8x} \qquad \\
-3x + 12 \\
\underline{-3x + 12} \\
0
\end{array}
$$

Since the remainder is 0, $x-4$ is a factor of $g(x)$.

b)

$$
\begin{array}{r}
x^2 + x - 8 \\
x-3 \overline{)\, x^3 - 2x^2 - 11x + 12} \\
\underline{x^3 - 3x^2} \qquad\qquad\quad \\
x^2 - 11x \qquad \\
\underline{x^2 - 3x} \qquad \\
-8x + 12 \\
\underline{-8x + 24} \\
-12
\end{array}
$$

Since the remainder is not 0, $x-3$ is not a factor of $g(x)$.

c)

$$
\begin{array}{r}
x^2 - x - 12 \\
x-1 \overline{)\, x^3 - 2x^2 - 11x + 12} \\
\underline{x^3 - x^2} \qquad\qquad\quad \\
-x^2 - 11x \qquad \\
\underline{-x^2 + x} \qquad \\
-12x + 12 \\
\underline{-12x + 12} \\
0
\end{array}
$$

Since the remainder is 0, $x-1$ is a factor of $g(x)$.

4. a)

$$
\begin{array}{r}
x^3 + 2x^2 - 7x + 4 \\
x+6 \overline{)\, x^4 + 8x^3 + 5x^2 - 38x + 24} \\
\underline{x^4 + 6x^3} \qquad\qquad\qquad\quad \\
2x^3 + 5x^2 \qquad\qquad \\
\underline{2x^3 + 12x^2} \qquad\qquad \\
-7x^2 - 38x \quad\;\; \\
\underline{-7x^2 - 42x} \quad\;\; \\
4x + 24 \\
\underline{4x + 24} \\
0
\end{array}
$$

Since the remainder is 0, $x+6$ is a factor of $f(x)$.

b)
$$\begin{array}{r}
x^3+7x^2-2x-36 \\
x+1\overline{)\,x^4+8x^3+5x^2-38x+24} \\
\underline{x^4+x^3} \qquad\qquad\qquad \\
7x^3+5x^2 \qquad\qquad \\
\underline{7x^3+7x^2} \qquad\qquad \\
-2x^2-38x \qquad \\
\underline{-2x^2-2x} \qquad \\
-36x+24 \\
\underline{-36x-36} \\
60
\end{array}$$

Since the remainder is not 0, $x+1$ is not a factor of $f(x)$.

c)
$$\begin{array}{r}
x^3+12x^2+53x+174 \\
x-4\overline{)\,x^4+8x^3+5x^2-38x+24} \\
\underline{x^4-4x^3} \qquad\qquad\qquad \\
12x^3+5x^2 \qquad\qquad \\
\underline{12x^3-48x^2} \qquad\qquad \\
53x^2-38x \qquad \\
\underline{53x^2-212x} \qquad \\
174x+24 \\
\underline{174x-696} \\
720
\end{array}$$

Since the remainder is not 0, $x-4$ is not a factor of $f(x)$.

5.
$$\begin{array}{r}
x^2-2x+4 \\
x+2\overline{)\,x^3+0x^2+0x-8} \\
\underline{x^3+2x^2} \qquad\qquad \\
-2x^2+0x \qquad \\
\underline{-2x^2-4x} \qquad \\
4x-8 \\
\underline{4x+8} \\
-16
\end{array}$$

$x^3-8=(x+2)(x^2-2x+4)-16$

6.
$$\begin{array}{r}
2x^2+3x+10 \\
x-3\overline{)\,2x^3-3x^2+x-1} \\
\underline{2x^3-6x^2} \qquad\qquad \\
3x^2+x \qquad \\
\underline{3x^2-9x} \qquad \\
10x-1 \\
\underline{10x-30} \\
29
\end{array}$$

$2x^3-3x^2+x-1=(x-3)(2x^2+3x+10)+29$

7.
$$\begin{array}{r}
x^2-3x+2 \\
x+9\overline{)\,x^3+6x^2-25x+18} \\
\underline{x^3+9x^2} \qquad\qquad \\
-3x^2-25x \qquad \\
\underline{-3x^2-27x} \qquad \\
2x+18 \\
\underline{2x+18} \\
0
\end{array}$$

$x^3+6x^2-25x+18=(x+9)(x^2-3x+2)+0$

8.
$$\begin{array}{r}
x^2-4x-5 \\
x-5\overline{)\,x^3-9x^2+15x+25} \\
\underline{x^3-5x^2} \qquad\qquad \\
-4x^2+15x \qquad \\
\underline{-4x^2+20x} \qquad \\
-5x+25 \\
\underline{-5x+25} \\
0
\end{array}$$

$x^3-9x^2+15x+25=(x-5)(x^2-4x-5)+0$

9.
$$\begin{array}{r}
x^3-2x^2+2x-4 \\
x+2\overline{)\,x^4+0x^3-2x^2+0x+3} \\
\underline{x^4+2x^3} \qquad\qquad\qquad \\
-2x^3-2x^2 \qquad\qquad \\
\underline{-2x^3-4x^2} \qquad\qquad \\
2x^2+0x \qquad \\
\underline{2x^2+4x} \qquad \\
-4x+3 \\
\underline{-4x-8} \\
11
\end{array}$$

$x^4-2x^2+3=(x+2)(x^3-2x^2+2x-4)+11$

10.
$$\begin{array}{r}
x^3+7x^2+7x+7 \\
x-1\overline{)\,x^4+6x^3+0x^2+0x+0} \\
\underline{x^4-x^3} \qquad\qquad\qquad \\
7x^3+0x^2 \qquad\qquad \\
\underline{7x^3-7x^2} \qquad\qquad \\
7x^2+0x \qquad \\
\underline{7x^2-7x} \qquad \\
7x+0 \\
\underline{7x-7} \\
7
\end{array}$$

$x^4+6x^3=(x-1)(x^3+7x^2+7x+7)+7$

11. $(2x^4+7x^3+x-12)\div(x+3)$

$=(2x^4+7x^3+0x^2+x-12)\div[x-(-3)]$

$$\begin{array}{r|rrrrr}
-3 & 2 & 7 & 0 & 1 & -12 \\
 & & -6 & -3 & 9 & -30 \\
\hline
 & 2 & 1 & -3 & 10 & -42
\end{array}$$

The quotient is $2x^3+x^2-3x+10$. The remainder is -42.

12.
$$\begin{array}{r|rrrr}
2 & 1 & -7 & 13 & 3 \\
 & & 2 & -10 & 6 \\
\hline
 & 1 & -5 & 3 & 9
\end{array}$$

$Q(x)=x^2-5x+3,\ R(x)=9$

13. $(x^3-2x^2-8)\div(x+2)$

$=(x^3-2x^2+0x-8)\div[x-(-2)]$

$$\begin{array}{r|rrrr}
-2 & 1 & -2 & 0 & -8 \\
 & & -2 & 8 & -16 \\
\hline
 & 1 & -4 & 8 & -24
\end{array}$$

The quotient is x^2-4x+8. The remainder is -24.

14.
$$\begin{array}{r|rrrr}
2 & 1 & 0 & -3 & 10 \\
 & & 2 & 4 & 2 \\
\hline
 & 1 & 2 & 1 & 12
\end{array}$$

$Q(x)=x^2+2x+1,\ R(x)=12$

15. $(3x^3 - x^2 + 4x - 10) \div (x + 1)$

$= (3x^3 - x^2 + 4x - 10) \div [x - (-1)]$

$$\begin{array}{r|rrrr} -1 & 3 & -1 & 4 & -10 \\ & & -3 & 4 & -8 \\ \hline & 3 & -4 & 8 & -18 \end{array}$$

The quotient is $3x^2 - 4x + 8$. The remainder is -18.

16.

$$\begin{array}{r|rrrrr} -3 & 4 & 0 & 0 & -2 & 5 \\ & & -12 & 36 & -108 & 330 \\ \hline & 4 & -12 & 36 & -110 & 335 \end{array}$$

$Q(x) = 4x^3 - 12x^2 + 36x - 110$, $R(x) = 335$

17. $(x^5 + x^3 - x) \div (x - 3)$

$= (x^5 + 0x^4 + x^3 + 0x^2 - x + 0) \div (x - 3)$

$$\begin{array}{r|rrrrrr} 3 & 1 & 0 & 1 & 0 & -1 & 0 \\ & & 3 & 9 & 30 & 90 & 267 \\ \hline & 1 & 3 & 10 & 30 & 89 & 267 \end{array}$$

The quotient is $x^4 + 3x^3 + 10x^2 + 30x + 89$.

The remainder is 267.

18.

$$\begin{array}{r|rrrrrrrr} -1 & 1 & -1 & 1 & -1 & 0 & 0 & 0 & 2 \\ & & -1 & 2 & -3 & 4 & -4 & 4 & -4 \\ \hline & 1 & -2 & 3 & -4 & 4 & -4 & 4 & -2 \end{array}$$

$Q(x) = x^6 - 2x^5 + 3x^4 - 4x^3 + 4x^2 - 4x + 4$, $R(x) = -2$

19. $(x^4 - 1) \div (x - 1)$

$= (x^4 + 0x^3 + 0x^2 + 0x - 1) \div (x - 1)$

$$\begin{array}{r|rrrrr} 1 & 1 & 0 & 0 & 0 & -1 \\ & & 1 & 1 & 1 & 1 \\ \hline & 1 & 1 & 1 & 1 & 0 \end{array}$$

The quotient is $x^3 + x^2 + x + 1$. The remainder is 0.

20.

$$\begin{array}{r|rrrrrr} -2 & 1 & 0 & 0 & 0 & 0 & 32 \\ & & -2 & 4 & -8 & 16 & -32 \\ \hline & 1 & -2 & 4 & -8 & 16 & 0 \end{array}$$

$Q(x) = x^4 - 2x^3 + 4x^2 - 8x + 16$, $R(x) = 0$

21. $(2x^4 + 3x^2 - 1) \div \left(x - \frac{1}{2}\right)$

$(2x^4 + 0x^3 + 3x^2 + 0x - 1) \div \left(x - \frac{1}{2}\right)$

$$\begin{array}{r|rrrrr} \frac{1}{2} & 2 & 0 & 3 & 0 & -1 \\ & & 1 & \frac{1}{2} & \frac{7}{4} & \frac{7}{8} \\ \hline & 2 & 1 & \frac{7}{2} & \frac{7}{4} & -\frac{1}{8} \end{array}$$

The quotient is $2x^3 + x^2 + \frac{7}{2}x + \frac{7}{4}$. The remainder is $-\frac{1}{8}$.

22.

$$\begin{array}{r|rrrrr} \frac{1}{4} & 3 & 0 & -2 & 0 & 2 \\ & & \frac{3}{4} & \frac{3}{16} & -\frac{29}{64} & -\frac{29}{256} \\ \hline & 3 & \frac{3}{4} & -\frac{29}{16} & -\frac{29}{64} & \frac{483}{256} \end{array}$$

$Q(x) = 3x^3 + \frac{3}{4}x^2 - \frac{29}{16}x - \frac{29}{64}$, $R(x) = \frac{483}{256}$

23. $f(x) = x^3 - 6x^2 + 11x - 6$

Find $f(1)$.

$$\begin{array}{r|rrrr} 1 & 1 & -6 & 11 & -6 \\ & & 1 & -5 & 6 \\ \hline & 1 & -5 & 6 & 0 \end{array}$$

$f(1) = 0$

Find $f(-2)$.

$$\begin{array}{r|rrrr} -2 & 1 & -6 & 11 & -6 \\ & & -2 & 16 & -54 \\ \hline & 1 & -8 & 27 & -60 \end{array}$$

$f(-2) = -60$

Find $f(3)$.

$$\begin{array}{r|rrrr} 3 & 1 & -6 & 11 & -6 \\ & & 3 & -9 & 6 \\ \hline & 1 & -3 & 2 & 0 \end{array}$$

$f(3) = 0$

24.

$$\begin{array}{r|rrrr} -3 & 1 & 7 & -12 & -3 \\ & & -3 & -12 & 72 \\ \hline & 1 & 4 & -24 & 69 \end{array}$$

$f(-3) = 69$

$$\begin{array}{r|rrrr} -2 & 1 & 7 & -12 & -3 \\ & & -2 & -10 & 44 \\ \hline & 1 & 5 & -22 & 41 \end{array}$$

$f(-2) = 41$

$$\begin{array}{r|rrrr} 1 & 1 & 7 & -12 & -3 \\ & & 1 & 8 & -4 \\ \hline & 1 & 8 & -4 & -7 \end{array}$$

$f(1) = -7$

25. $f(x) = x^4 - 3x^3 + 2x + 8$

Find $f(-1)$.

$$\begin{array}{r|rrrrr} -1 & 1 & -3 & 0 & 2 & 8 \\ & & -1 & 4 & -4 & 2 \\ \hline & 1 & -4 & 4 & -2 & 10 \end{array}$$

$f(-1) = 10$

Find $f(4)$.

$$\begin{array}{r|rrrrr} 4 & 1 & -3 & 0 & 2 & 8 \\ & & 4 & 4 & 16 & 72 \\ \hline & 1 & 1 & 4 & 18 & 80 \end{array}$$

$f(4) = 80$

Find $f(-5)$.

$$\begin{array}{r|rrrrr} -5 & 1 & -3 & 0 & 2 & 8 \\ & & -5 & 40 & -200 & 990 \\ \hline & 1 & -8 & 40 & -198 & 998 \end{array}$$

$f(-5) = 998$

26.

$$\begin{array}{r|rrrrr} -10 & 2 & 0 & 1 & -10 & 1 \\ & & -20 & 200 & -2010 & 20,200 \\ \hline & 2 & -20 & 201 & -2020 & 20,201 \end{array}$$

$f(-10) = 20,201$

$$\begin{array}{r|rrrrr} 2 & 2 & 0 & 1 & -10 & 1 \\ & & 4 & 8 & 18 & 16 \\ \hline & 2 & 4 & 9 & 8 & \boxed{17} \end{array}$$

$f(2) = 17$

$$\begin{array}{r|rrrrr} 3 & 2 & 0 & 1 & -10 & 1 \\ & & 6 & 18 & 57 & 141 \\ \hline & 2 & 6 & 19 & 47 & \boxed{142} \end{array}$$

$f(3) = 142$

27. $f(x) = 2x^5 - 3x^4 + 2x^3 - x + 8$

Find $f(20)$.

$$\begin{array}{r|rrrrrr} 20 & 2 & -3 & 2 & 0 & -1 & 8 \\ & & 40 & 740 & 14,840 & 296,800 & 5,935,980 \\ \hline & 2 & 37 & 742 & 14,840 & 296,799 & \boxed{5,935,988} \end{array}$$

$f(20) = 5,935,988$

Find $f(-3)$.

$$\begin{array}{r|rrrrrr} -3 & 2 & -3 & 2 & 0 & -1 & 8 \\ & & -6 & 27 & -87 & 261 & -780 \\ \hline & 2 & -9 & 29 & -87 & 260 & \boxed{-772} \end{array}$$

$f(-3) = -772$

28.

$$\begin{array}{r|rrrrrr} -10 & 1 & -10 & 20 & 0 & -5 & -100 \\ & & -10 & 200 & -2200 & 22,000 & -219,950 \\ \hline & 1 & -20 & 220 & -2200 & 21,995 & \boxed{-220,050} \end{array}$$

$f(-10) = -220,050$

$$\begin{array}{r|rrrrrr} 5 & 1 & -10 & 20 & 0 & -5 & -100 \\ & & 5 & -25 & -25 & -125 & -650 \\ \hline & 1 & -5 & -5 & -25 & -130 & \boxed{-750} \end{array}$$

$f(5) = -750$

29. $f(x) = x^4 - 16$

Find $f(2)$.

$$\begin{array}{r|rrrrr} 2 & 1 & 0 & 0 & 0 & -16 \\ & & 2 & 4 & 8 & 16 \\ \hline & 1 & 2 & 4 & 8 & \boxed{0} \end{array}$$

$f(2) = 0$

Find $f(-2)$.

$$\begin{array}{r|rrrrr} -2 & 1 & 0 & 0 & 0 & -16 \\ & & -2 & 4 & -8 & 16 \\ \hline & 1 & -2 & 4 & -8 & \boxed{0} \end{array}$$

$f(-2) = 0$

Find $f(3)$.

$$\begin{array}{r|rrrrr} 3 & 1 & 0 & 0 & 0 & -16 \\ & & 3 & 9 & 27 & 81 \\ \hline & 1 & 3 & 9 & 27 & \boxed{65} \end{array}$$

$f(3) = 65$

Find $f(1 - \sqrt{2})$.

$$\begin{array}{r|rrrrr} 1-\sqrt{2} & 1 & 0 & 0 & 0 & -16 \\ & & 1-\sqrt{2} & 3-2\sqrt{2} & 7-5\sqrt{2} & 17-12\sqrt{2} \\ \hline & 1 & 1-\sqrt{2} & 3-2\sqrt{2} & 7-5\sqrt{2} & \boxed{1-12\sqrt{2}} \end{array}$$

$f(1 - \sqrt{2}) = 1 - 12\sqrt{2}$

30.

$$\begin{array}{r|rrrrrr} 2 & 1 & 0 & 0 & 0 & 0 & 32 \\ & & 2 & 4 & 8 & 16 & 32 \\ \hline & 1 & 2 & 4 & 8 & 16 & \boxed{64} \end{array}$$

$f(2) = 64$

$f(-2) = 0$ (See Exercise 20.)

$$\begin{array}{r|rrrrrr} 3 & 1 & 0 & 0 & 0 & 0 & 32 \\ & & 3 & 9 & 27 & 81 & 243 \\ \hline & 1 & 3 & 9 & 27 & 81 & \boxed{275} \end{array}$$

$f(3) = 275$

$$\begin{array}{l} \underline{2+3i\,|} \\ \begin{array}{rrrrrr} 1 & 0 & 0 & 0 & 0 & 32 \\ & 2+3i & -5+12i & -46+9i & -119-120i & 122-597i \\ \hline 1 & 2+3i & -5+12i & -46+9i & -119-120i & \boxed{154-597i} \end{array} \end{array}$$

$f(2 + 3i) = 154 - 597i$

31. $f(x) = 3x^3 + 5x^2 - 6x + 18$

If -3 is a zero of $f(x)$, then $f(-3) = 0$. Find $f(-3)$ using synthetic division.

$$\begin{array}{r|rrrr} -3 & 3 & 5 & -6 & 18 \\ & & -9 & 12 & -18 \\ \hline & 3 & -4 & 6 & \boxed{0} \end{array}$$

Since $f(-3) = 0$, -3 is a zero of $f(x)$.

If 2 is a zero of $f(x)$, then $f(2) = 0$. Find $f(2)$ using synthetic division.

$$\begin{array}{r|rrrr} 2 & 3 & 5 & -6 & 18 \\ & & 6 & 22 & 32 \\ \hline & 3 & 11 & 16 & \boxed{50} \end{array}$$

Since $f(2) \neq 0$, 2 is not a zero of $f(x)$.

32.

$$\begin{array}{r|rrrr} -4 & 3 & 11 & -2 & 8 \\ & & -12 & 4 & -8 \\ \hline & 3 & -1 & 2 & \boxed{0} \end{array}$$

$f(-4) = 0$, so -4 is a zero of $f(x)$.

$$\begin{array}{r|rrrr} 2 & 3 & 11 & -2 & 8 \\ & & 6 & 34 & 64 \\ \hline & 3 & 17 & 32 & \boxed{72} \end{array}$$

$f(2) \neq 0$, so 2 is not a zero of $f(x)$.

33. $h(x) = x^4 + 4x^3 + 2x^2 - 4x - 3$

If -3 is a zero of $h(x)$, then $h(-3) = 0$. Find $h(-3)$ using synthetic division.

$$\begin{array}{r|rrrrr} -3 & 1 & 4 & 2 & -4 & -3 \\ & & -3 & -3 & 3 & 3 \\ \hline & 1 & 1 & -1 & -1 & \boxed{0} \end{array}$$

Since $h(-3) = 0$, -3 is a zero of $h(x)$.

If 1 is a zero of $h(x)$, then $h(1) = 0$. Find $h(1)$ using synthetic division.

$$\begin{array}{r|rrrrr} 1 & 1 & 4 & 2 & -4 & -3 \\ & & 1 & 5 & 7 & 3 \\ \hline & 1 & 5 & 7 & 3 & \boxed{0} \end{array}$$

Since $h(1) = 0$, 1 is a zero of $h(x)$.

34. $$\begin{array}{r|rrrrr} 2 & 1 & -6 & 1 & 24 & -20 \\ & & 2 & -8 & -14 & 20 \\ \hline & 1 & -4 & -7 & 10 & 0 \end{array}$$

$g(2) = 0$, so 2 is a zero of $g(x)$.

$$\begin{array}{r|rrrrr} -1 & 1 & -6 & 1 & 24 & -20 \\ & & -1 & 7 & -8 & -16 \\ \hline & 1 & -7 & 8 & 16 & -36 \end{array}$$

$g(-1) \neq 0$, so -1 is not a zero of $g(x)$.

35. $g(x) = x^3 - 4x^2 + 4x - 16$

If i is a zero of $g(x)$, then $g(i) = 0$. Find $g(i)$ using synthetic division. Keep in mind that $i^2 = -1$.

$$\begin{array}{r|rrrr} i & 1 & -4 & 4 & -16 \\ & & i & -4i-1 & 3i+4 \\ \hline & 1 & -4+i & 3-4i & -12+3i \end{array}$$

Since $g(i) \neq 0$, i is not a zero of $g(x)$.

If $-2i$ is a zero of $g(x)$, then $g(-2i) = 0$. Find $g(-2i)$ using synthetic division. Keep in mind that $i^2 = -1$.

$$\begin{array}{r|rrrr} -2i & 1 & -4 & 4 & -16 \\ & & -2i & 8i-4 & 16 \\ \hline & 1 & -4-2i & 8i & 0 \end{array}$$

Since $g(-2i) = 0$, $-2i$ is a zero of $g(x)$.

36. $$\begin{array}{r|rrrr} \frac{1}{3} & 1 & -1 & -\frac{1}{9} & \frac{1}{9} \\ & & \frac{1}{3} & -\frac{2}{9} & -\frac{1}{9} \\ \hline & 1 & -\frac{2}{3} & -\frac{1}{3} & 0 \end{array}$$

$h\left(\dfrac{1}{3}\right) = 0$, so $\dfrac{1}{3}$ is a zero of $h(x)$.

$$\begin{array}{r|rrrr} 2 & 1 & -1 & -\frac{1}{9} & \frac{1}{9} \\ & & 2 & 2 & \frac{34}{9} \\ \hline & 1 & 1 & \frac{17}{9} & \frac{35}{9} \end{array}$$

$h(2) \neq 0$, so 2 is not a zero of $h(x)$.

37. $f(x) = x^3 - \dfrac{7}{2}x^2 + x - \dfrac{3}{2}$

If -3 is a zero of $f(x)$, then $f(-3) = 0$. Find $f(-3)$ using synthetic division.

$$\begin{array}{r|rrrr} -3 & 1 & -\frac{7}{2} & 1 & -\frac{3}{2} \\ & & -3 & \frac{39}{2} & -\frac{123}{2} \\ \hline & 1 & -\frac{13}{2} & \frac{41}{2} & -63 \end{array}$$

Since $f(-3) \neq 0$, -3 is not a zero of $f(x)$.

If $\dfrac{1}{2}$ is a zero of $f(x)$, then $f\left(\dfrac{1}{2}\right) = 0$.

Find $f\left(\dfrac{1}{2}\right)$ using synthetic division.

$$\begin{array}{r|rrrr} \frac{1}{2} & 1 & -\frac{7}{2} & 1 & -\frac{3}{2} \\ & & \frac{1}{2} & -\frac{3}{2} & -\frac{1}{4} \\ \hline & 1 & -3 & -\frac{1}{2} & -\frac{7}{4} \end{array}$$

Since $f\left(\dfrac{1}{2}\right) \neq 0$, $\dfrac{1}{2}$ is not a zero of $f(x)$.

38. $$\begin{array}{r|rrrr} i & 1 & 2 & 1 & 2 \\ & & i & -1+2i & -2 \\ \hline & 1 & 2+i & 2i & 0 \end{array}$$

$f(i) = 0$, so i is a zero of $f(x)$.

$$\begin{array}{r|rrrr} -i & 1 & 2 & 1 & 2 \\ & & -i & -1-2i & -2 \\ \hline & 1 & 2-i & -2i & 0 \end{array}$$

$f(-i) = 0$, so $-i$ is a zero of $f(x)$.

$$\begin{array}{r|rrrr} -2 & 1 & 2 & 1 & 2 \\ & & -2 & 0 & -2 \\ \hline & 1 & 0 & 1 & 0 \end{array}$$

$f(-2) = 0$, so -2 is a zero of $f(x)$.

39. $f(x) = x^3 + 4x^2 + x - 6$

Try $x - 1$. Use synthetic division to see whether $f(1) = 0$.

$$\begin{array}{r|rrrr} 1 & 1 & 4 & 1 & -6 \\ & & 1 & 5 & 6 \\ \hline & 1 & 5 & 6 & 0 \end{array}$$

Since $f(1) = 0$, $x - 1$ is a factor of $f(x)$. Thus $f(x) = (x-1)(x^2+5x+6)$.

Factoring the trinomial we get

$f(x) = (x-1)(x+2)(x+3)$.

To solve the equation $f(x) = 0$, use the principle of zero products.

$(x-1)(x+2)(x+3) = 0$

$x - 1 = 0$ *or* $x + 2 = 0$ *or* $x + 3 = 0$

$x = 1$ *or* $x = -2$ *or* $x = -3$

The solutions are 1, -2, and -3.

40. $$\begin{array}{r|rrrr} 2 & 1 & 5 & -2 & -24 \\ & & 2 & 14 & 24 \\ \hline & 1 & 7 & 12 & 0 \end{array}$$

$$\begin{aligned} f(x) &= (x-2)(x^2+7x+12) \\ &= (x-2)(x+3)(x+4) \end{aligned}$$

The solutions of $f(x) = 0$ are 2, -3, and -4.

41. $f(x) = x^3 - 6x^2 + 3x + 10$

Try $x - 1$. Use synthetic division to see whether $f(1) = 0$.

$$\begin{array}{r|rrrr} 1 & 1 & -6 & 3 & 10 \\ & & 1 & -5 & -2 \\ \hline & 1 & -5 & -2 & 8 \end{array}$$

Since $f(1) \neq 0$, $x - 1$ is not a factor of $P(x)$.

Try $x+1$. Use synthetic division to see whether $f(-1) = 0$.

$$\begin{array}{r|rrrr} -1 & 1 & -6 & 3 & 10 \\ & & -1 & 7 & -10 \\ \hline & 1 & -7 & 10 & 0 \end{array}$$

Since $f(-1) = 0$, $x + 1$ is a factor of $f(x)$.

Thus $f(x) = (x+1)(x^2 - 7x + 10)$.

Factoring the trinomial we get

$f(x) = (x+1)(x-2)(x-5)$.

To solve the equation $f(x) = 0$, use the principle of zero products.

$(x+1)(x-2)(x-5)=0$

$x+1=0$ *or* $x-2=0$ *or* $x-5=0$

$x=-1$ *or* $x=2$ *or* $x=5$

The solutions are -1, 2, and 5.

42.
$$\begin{array}{r|rrrr} 1 & 1 & 2 & -13 & 10 \\ & & 1 & 3 & -10 \\ \hline & 1 & 3 & -10 & 0 \end{array}$$

$f(x)=(x-1)(x^2+3x-10)$

$=(x-1)(x-2)(x+5)$

The solutions of $f(x)=0$ are 1, 2, and -5.

43. $f(x)=x^3-x^2-14x+24$

Try $x+1$, $x-1$, and $x+2$. Using synthetic division we find that $f(-1)\neq 0$, $f(1)\neq 0$ and $f(-2)\neq 0$. Thus $x+1$, $x-1$, and $x+2$, are not factors of $f(x)$.

Try $x-2$. Use synthetic division to see whether $f(2)=0$.

$$\begin{array}{r|rrrr} 2 & 1 & -1 & -14 & 24 \\ & & 2 & 2 & -24 \\ \hline & 1 & 1 & -12 & 0 \end{array}$$

Since $f(2)=0$, $x-2$ is a factor of $f(x)$. Thus $f(x)=(x-2)(x^2+x-12)$.

Factoring the trinomial we get

$f(x)=(x-2)(x+4)(x-3)$

To solve the equation $f(x)=0$, use the principle of zero products.

$(x-2)(x+4)(x-3)=0$

$x-2=0$ *or* $x+4=0$ *or* $x-3=0$

$x=2$ *or* $x=-4$ *or* $x=3$

The solutions are 2, -4, and 3.

44.
$$\begin{array}{r|rrrr} 2 & 1 & -3 & -10 & 24 \\ & & 2 & -2 & -24 \\ \hline & 1 & -1 & -12 & 0 \end{array}$$

$f(x)=(x-2)(x^2-x-12)$

$=(x-2)(x-4)(x+3)$

The solutions of $f(x)=0$ are 2, 4, and -3.

45. $f(x)=x^4-7x^3+9x^2+27x-54$

Try $x+1$ and $x-1$. Using synthetic division we find that $f(-1)\neq 0$ and $f(1)\neq 0$. Thus $x+1$ and $x-1$ are not factors of $f(x)$. Try $x+2$. Use synthetic division to see whether $f(-2)=0$.

$$\begin{array}{r|rrrrr} -2 & 1 & -7 & 9 & 27 & -54 \\ & & -2 & 18 & -54 & 54 \\ \hline & 1 & -9 & 27 & -27 & 0 \end{array}$$

Since $f(-2)=0$, $x+2$ is a factor of $f(x)$. Thus $f(x)=(x+2)(x^3-9x^2+27x-27)$.

We continue to use synthetic division to factor $g(x)=x^3-9x^2+27x-27$. Trying $x+2$ again and $x-2$ we find that $g(-2)\neq 0$ and $g(2)\neq 0$. Thus $x+2$ and $x-2$ are not factors of $g(x)$. Try $x-3$.

$$\begin{array}{r|rrrr} 3 & 1 & -9 & 27 & -27 \\ & & 3 & -18 & 27 \\ \hline & 1 & -6 & 9 & 0 \end{array}$$

Since $g(3)=0$, $x-3$ is a factor of $x^3-9x^2+27x-27$.

Thus $f(x)=(x+2)(x-3)(x^2-6x+9)$.

Factoring the trinomial we get

$f(x)=(x+2)(x-3)(x-3)^2$, or $f(x)=(x+2)(x-3)^3$.

To solve the equation $f(x)=0$, use the principle of zero products.

$(x+2)(x-3)(x-3)(x-3)=0$

$x+2=0$ *or* $x-3=0$ *or* $x-3=0$ *or* $x-3=0$

$x=-2$ *or* $x=3$ *or* $x=3$ *or* $x=3$

The solutions are -2 and 3.

46.
$$\begin{array}{r|rrrrr} 1 & 1 & -4 & -7 & 34 & -24 \\ & & 1 & -3 & -10 & 24 \\ \hline & 1 & -3 & -10 & 24 & 0 \end{array}$$

$$\begin{array}{r|rrrr} 2 & 1 & -3 & -10 & 24 \\ & & 2 & -2 & -24 \\ \hline & 1 & -1 & -12 & 0 \end{array}$$

$f(x)=(x-1)(x-2)(x^2-x-12)$

$=(x-1)(x-2)(x-4)(x+3)$

The solutions of $f(x)=0$ are 1, 2, 4, and -3.

47. $f(x)=x^4-x^3-19x^2+49x-30$

Try $x-1$. Use synthetic division to see whether $f(1)=0$.

$$\begin{array}{r|rrrrr} 1 & 1 & -1 & -19 & 49 & -30 \\ & & 1 & 0 & -19 & 30 \\ \hline & 1 & 0 & -19 & 30 & 0 \end{array}$$

Since $f(1)=0$, $x-1$ is a factor of $f(x)$. Thus $f(x)=(x-1)(x^3-19x+30)$.

We continue to use synthetic division to factor $g(x)=x^3-19x+30$. Trying $x-1$, $x+1$, and $x+2$ we find that $g(1)\neq 0$, $g(-1)\neq 0$, and $g(-2)\neq 0$. Thus $x-1$, $x+1$, and $x+2$ are not factors of $x^3-19x+30$. Try $x-2$.

$$\begin{array}{r|rrrr} 2 & 1 & 0 & -19 & 30 \\ & & 2 & 4 & -30 \\ \hline & 1 & 2 & -15 & 0 \end{array}$$

Since $g(2)=0$, $x-2$ is a factor of $x^3-19x+30$.

Thus $f(x)=(x-1)(x-2)(x^2+2x-15)$.

Factoring the trinomial we get

$f(x)=(x-1)(x-2)(x-3)(x+5)$.

To solve the equation $f(x)=0$, use the principle of zero products.

$(x-1)(x-2)(x-3)(x+5)=0$

$x-1=0$ *or* $x-2=0$ *or* $x-3=0$ *or* $x+5=0$

$x=1$ *or* $x=2$ *or* $x=3$ *or* $x=-5$

The solutions are 1, 2, 3, and -5.

48.
$$\begin{array}{r|rrrrr} -1 & 1 & 11 & 41 & 61 & 30 \\ & & -1 & -10 & -31 & -30 \\ \hline & 1 & 10 & 31 & 30 & 0 \end{array}$$

$$\begin{array}{r|rrrr} -2 & 1 & 10 & 31 & 30 \\ & & -2 & -16 & -30 \\ \hline & 1 & 8 & 15 & 0 \end{array}$$

$$f(x) = (x+1)(x+2)(x^2+8x+15)$$
$$= (x+1)(x+2)(x+3)(x+5)$$

The solutions of $f(x) = 0$ are -1, -2, -3, and -5.

49. $f(x) = x^4 - x^3 - 7x^2 + x + 6$

1. The leading term is x^4. The degree, 4, is even and the leading coefficient, 1, is positive so as $x \to \infty$, $f(x) \to \infty$ and as $x \to -\infty$, $f(x) \to \infty$.

2. Find the zeros of the function. We first use synthetic division to determine if $f(1) = 0$.

$$\begin{array}{r|rrrrr} 1 & 1 & -1 & -7 & 1 & 6 \\ & & 1 & 0 & -7 & -6 \\ \hline & 1 & 0 & -7 & -6 & 0 \end{array}$$

   1 is a zero of the function and we have $f(x) = (x-1)(x^3 - 7x - 6)$.

   Synthetic division shows that -1 is a zero of $g(x) = x^3 - 7x - 6$.

$$\begin{array}{r|rrrr} -1 & 1 & 0 & -7 & -6 \\ & & -1 & 1 & 6 \\ \hline & 1 & -1 & -6 & 0 \end{array}$$

   Then we have $f(x) = (x-1)(x+1)(x^2 - x - 6)$.

   To find the other zeros we solve the following equation:

$$x^2 - x - 6 = 0$$
$$(x-3)(x+2) = 0$$
$$x - 3 = 0 \quad or \quad x + 2 = 0$$
$$x = 3 \quad or \quad x = -2$$

   The zeros of the function are 1, -1, 3, and -2 so the x-intercepts of the graph are $(1, 0)$, $(-1, 0)$, $(3, 0)$, and $(-2, 0)$.

3. The zeros divide the x-axis into five intervals, $(-\infty, -2)$, $(-2, -1)$, $(-1, 1)$, $(1, 3)$, and $(3, \infty)$. We choose a value for x from each interval and find $f(x)$. This tells us the sign of $f(x)$ for all values of x in the interval.

   In $(-\infty, -2)$, test -3:

   $f(-3) = (-3)^4 - (-3)^3 - 7(-3)^2 + (-3) + 6 = 48 > 0$

   In $(-2, -1)$, test -1.5:

   $f(-1.5) = (-1.5)^4 - (-1.5)^3 - 7(-1.5)^2 + (-1.5) + 6 = -2.8125 < 0$

   In $(-1, 1)$, test 0:

   $f(0) = 0^4 - 0^3 - 7 \cdot 0^2 + 0 + 6 = 6 > 0$

   In $(1, 3)$, test 2:

   $f(2) = 2^4 - 2^3 - 7 \cdot 2^2 + 2 + 6 = -12 < 0$

   In $(3, \infty)$, test 4:

   $f(4) = 4^4 - 4^3 - 7 \cdot 4^2 + 4 + 6 = 90 > 0$

   Thus the graph lies above the x-axis on $(-\infty, -2)$, on $(-1, 1)$, and on $(3, \infty)$. It lies below the x-axis on $(-2, -1)$ and on $(1, 3)$. We also know the points $(-3, 48)$, $(-1.5, -2.8125)$, $(0, 6)$, $(2, -12)$, and $(4, 90)$ are on the graph.

4. From Step 3 we see that $f(0) = 6$ so the y-intercept is $(0, 6)$.

5. We find additional points on the graph and draw the graph.

x	$f(x)$
-2.5	14.3
-0.5	3.9
0.5	4.7
2.5	-11.8

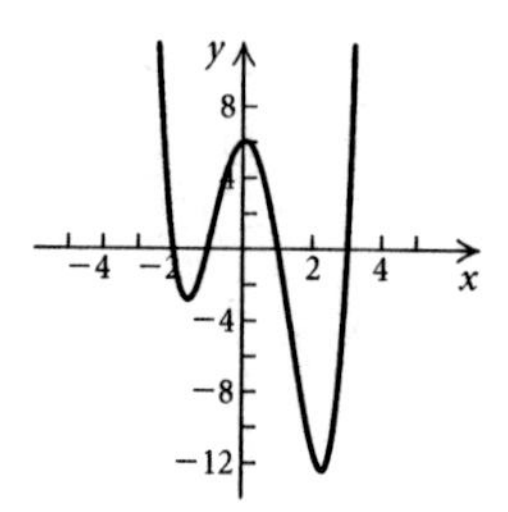

$f(x) = x^4 - x^3 - 7x^2 + x + 6$

6. Checking the graph as described on page 254 in the text, we see that it appears to be correct.

50.

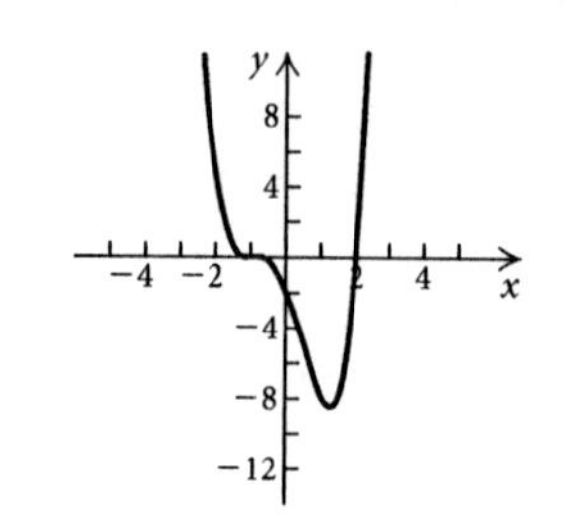

$f(x) = x^4 + x^3 - 3x^2 - 5x - 2$

51. $f(x) = x^3 - 7x + 6$

1. The leading term is x^3. The degree, 3, is odd and the leading coefficient, 1, is positive so as $x \to \infty$, $f(x) \to \infty$ and as $x \to -\infty$, $f(x) \to -\infty$.

2. Find the zeros of the function. We first use synthetic division to determine if $f(1) = 0$.

$$\begin{array}{r|rrrr} 1 & 1 & 0 & -7 & 6 \\ & & 1 & 1 & -6 \\ \hline & 1 & 1 & -6 & 0 \end{array}$$

   1 is a zero of the function and we have $f(x) = (x-1)(x^2 + x - 6)$. To find the other zeros we solve the following equation.

$$x^2 + x - 6 = 0$$
$$(x+3)(x-2) = 0$$
$$x + 3 = 0 \quad or \quad x - 2 = 0$$
$$x = -3 \quad or \quad x = 2$$

   The zeros of the function are 1, -3, and 2 so the x-intercepts of the graph are $(1, 0)$, $(-3, 0)$, and $(2, 0)$.

3. The zeros divide the x-axis into four intervals, $(-\infty, -3)$, $(-3, 1)$, $(1, 2)$, and $(2, \infty)$. We choose a value for x from each interval and find $f(x)$. This tells us the sign of $f(x)$ for all values of x in the interval.

   In $(-\infty, -3)$, test -4:

   $f(-4) = (-4)^3 - 7(-4) + 6 = -30 < 0$

   In $(-3, 1)$, test 0:

   $f(0) = 0^3 - 7 \cdot 0 + 6 = 6 > 0$

   In $(1, 2)$, test 1.5:

   $f(1.5) = (1.5)^3 - 7(1.5) + 6 = -1.125 < 0$

In $(2, \infty)$, test 3:

$f(3) = 3^3 - 7 \cdot 3 + 6 = 12 > 0$

Thus the graph lies below the x-axis on $(-\infty, -3)$ and on $(1, 2)$. It lies above the x-axis on $(-3, 1)$ and on $(2, \infty)$. We also know the points $(-4, -30)$, $(0, 6)$, $(1.5, -1.125)$, and $(3, 12)$ are on the graph.

4. From Step 3 we see that $f(0) = 6$ so the y-intercept is $(0, 6)$.

5. We find additional points on the graph and draw the graph.

x	$f(x)$
-3.5	-12.4
-2	12
2.5	4.1
4	42

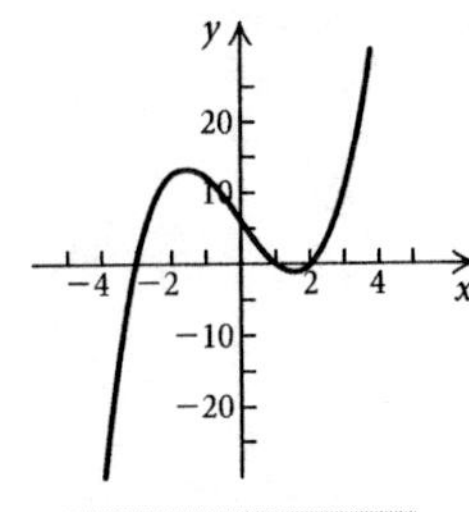

$f(x) = x^3 - 7x + 6$

6. Checking the graph as described on page 254 in the text, we see that it appears to be correct.

52.

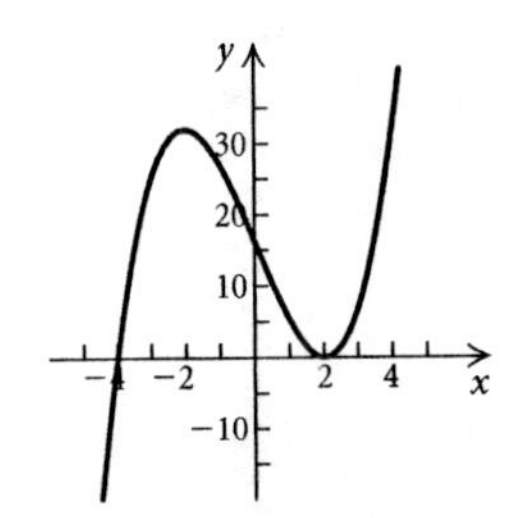

$f(x) = x^3 - 12x + 16$

53. $f(x) = -x^3 + 3x^2 + 6x - 8$

1. The leading term is $-x^3$. The degree, 3, is odd and the leading coefficient, -1, is negative so as $x \to \infty$, $f(x) \to -\infty$ and as $x \to -\infty$, $f(x) \to \infty$.

2. Find the zeros of the function. We first use synthetic division to determine if $f(1) = 0$.

$$\begin{array}{r|rrrr} 1 & -1 & 3 & 6 & -8 \\ & & -1 & 2 & 8 \\ \hline & -1 & 2 & 8 & 0 \end{array}$$

1 is a zero of the function and we have $f(x) = (x-1)(-x^2+2x+8)$. To find the other zeros we solve the following equation.

$$-x^2 + 2x + 8 = 0$$
$$x^2 - 2x - 8 = 0$$
$$(x-4)(x+2) = 0$$
$$x - 4 = 0 \quad or \quad x + 2 = 0$$
$$x = 4 \quad or \quad x = -2$$

The zeros of the function are 1, 4, and -2 so the x-intercepts of the graph are $(1, 0)$, $(4, 0)$, and $(-2, 0)$.

3. The zeros divide the x-axis into four intervals, $(-\infty, -2)$, $(-2, 1)$, $(1, 4)$, and $(4, \infty)$. We choose a value for x from each interval and find $f(x)$. This tells us the sign of $f(x)$ for all values of x in the interval.

In $(-\infty, -2)$, test -3:

$f(-3) = (-3)^3 + 3(-3)^2 + 6(-3) - 8 = 28 > 0$

In $(-2, 1)$, test 0:

$f(0) = -0^3 + 3 \cdot 0^2 + 6 \cdot 0 - 8 = -8 < 0$

In $(1, 4)$, test 2:

$f(2) = -2^3 + 3 \cdot 2^2 + 6 \cdot 2 - 8 = 8 > 0$

In $(4, \infty)$, test 5:

$f(5) = -5^3 + 3 \cdot 5^2 + 6 \cdot 5 - 8 = -28 < 0$

Thus the graph lies above the x-axis on $(-\infty, -2)$ and on $(1, 4)$. It lies below the x-axis on $(-2, 1)$ and on $(4, \infty)$. We also know the points $(-3, 28)$, $(0, -8)$, $(2, 8)$, and $(5, -28)$ are on the graph.

4. From Step 3 we see that $f(0) = -8$ so the y-intercept is $(0, -8)$.

5. We find additional points on the graph and draw the graph.

x	$f(x)$
-2.5	11.4
-1	-10
3	10
4.5	-11.4

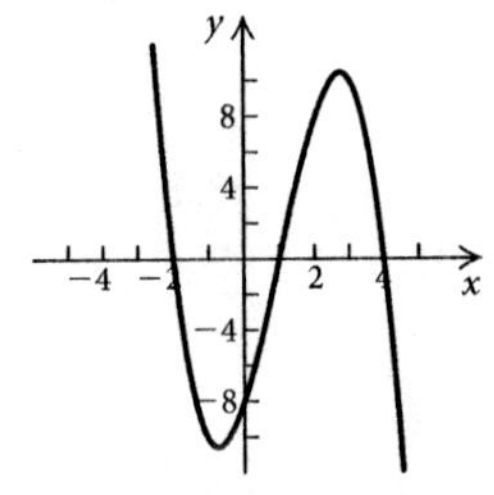

$f(x) = -x^3 + 3x^2 + 6x - 8$

6. Checking the graph as described on page 254 in the text, we see that it appears to be correct.

54.

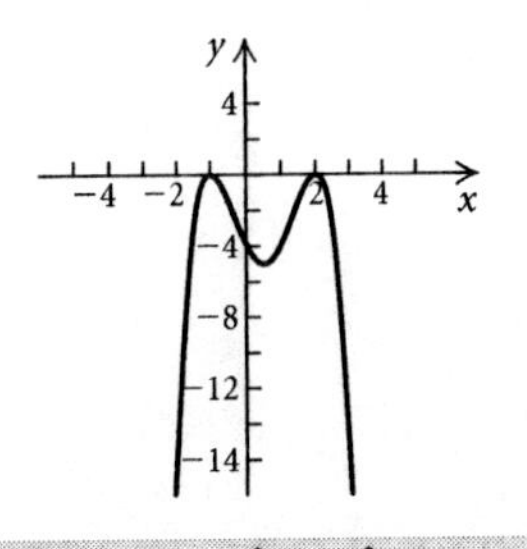

$f(x) = -x^4 + 2x^3 + 3x^2 - 4x - 4$

55. Left to the student

56. No; the polynomial cannot have more than n linear factors.

57. We can write $P(x) = p(x) \cdot q(x)$, where $q(x)$ is also a factor of $P(x)$. If $p(2) = 0$, then $P(2) = p(2) \cdot q(2) = 0 \cdot q(2) = 0$, so $P(2) = 0$.

If $P(2) = 0$ it does not necessarily follow that $p(2) = 0$. Consider the function $P(x) = p(x) \cdot q(x)$ where $p(x) = x - 1$ and $q(x) = x - 2$. Then $P(2) =$ $p(2) \cdot q(2) = (2-1)(2-2) = 1 \cdot 0 = 0$, but $p(2) = 1 \neq 0$.

58. $2x - 7 = 5x + 8$

$-7 = 3x + 8$ Subtracting $2x$ on both sides

$-15 = 3x$ Subtracting 8 on both sides

$-5 = x$ Dividing by 3 on both sides

The solution is -5.

59. $2x^2 + 12 = 5x$

$2x^2 - 5x + 12 = 0$

$a = 2,\ b = -5,\ c = 12$

$$x = \frac{-b \pm \sqrt{b^2 - 4ac}}{2a}$$

$$x = \frac{-(-5) \pm \sqrt{(-5)^2 - 4 \cdot 2 \cdot 12}}{2 \cdot 2}$$

$$= \frac{5 \pm \sqrt{-71}}{4}$$

$$= \frac{5 \pm i\sqrt{71}}{4}$$

The solutions are $\frac{5 + i\sqrt{71}}{4}$ and $\frac{5 - i\sqrt{71}}{4}$, or $\frac{5 \pm i\sqrt{71}}{4}$.

60. $7x^2 + 4x = 3$

$7x^2 + 4x - 3 = 0$

$(7x - 3)(x + 1) = 0$

$7x - 3 = 0$ *or* $x + 1 = 0$

$7x = 3$ *or* $x = -1$

$x = \frac{3}{7}$ *or* $x = -1$

The solutions are $\frac{3}{7}$ and -1.

61. We substitute -14 for $g(x)$ and solve for x.

$-14 = x^2 + 5x - 14$

$0 = x^2 + 5x$

$0 = x(x + 5)$

$x = 0$ *or* $x + 5 = 0$

$x = 0$ *or* $x = -5$

When the output is -14, the input is 0 or -5.

62. $g(3) = 3^2 + 5 \cdot 3 - 14 = 10$

63. We substitute -20 for $g(x)$ and solve for x.

$-20 = x^2 + 5x - 14$

$0 = x^2 + 5x + 6$

$0 = (x + 3)(x + 2)$

$x + 3 = 0$ *or* $x + 2 = 0$

$x = -3$ *or* $x = -2$

When the output is -20, the input is -3 or -2.

64. We use the points $(0, 27.2)$ and $(12, 72)$ to find m.

$$m = \frac{72 - 27.2}{12 - 0} = \frac{44.8}{12} = \frac{448}{120} = \frac{56}{15}$$

Since we have the point $(0, 27.2)$, we know that $b = 27.2$, so we have $f(x) = \frac{56}{15}x + 27.2$, where x is the number of years since 1989 and $f(x)$ is in billions of dollars.

For 2000, $f(11) = \frac{56}{15} \cdot 11 + 27.2 \approx \$68.3$ billion

For 2005, $f(16) = \frac{56}{15} \cdot 16 + 27.2 \approx \$86.9$ billion

For 2010, $f(21) = \frac{56}{15} \cdot 21 + 27.2 = \$105.6$ billion

65. Let b and h represent the length of the base and the height of the triangle, respectively.

$b + h = 30$, so $b = 30 - h$.

$$A = \frac{1}{2}bh = \frac{1}{2}(30 - h)h = -\frac{1}{2}h^2 + 15h$$

Find the value of h for which A is a maximum:

$$h = \frac{-15}{2(-1/2)} = 15$$

When $h = 15$, $b = 30 - 15 = 15$.

The area is a maximum when the base and the height are each 15 in.

66. a) -5, -3, 4, 6, and 7 are zeros of the function, so $x + 5$, $x + 3$, $x - 4$, $x - 6$, and $x - 7$ are factors.

b) We first write the product of the factors:

$F(x) = (x + 5)(x + 3)(x - 4)(x - 6)(x - 7)$

Note that $F(0) = 5 \cdot 3(-4)(-6)(-7) < 0$, but that the y-intercept of the graph is positive. Thus we must reflect $F(x)$ across the x-axis to obtain a function $P(x)$ with the given graph. We have:

$P(x) = -F(x)$

$P(x) = -(x + 5)(x + 3)(x - 4)(x - 6)(x - 7)$

c) Yes; two examples are $f(x) = c \cdot P(x)$, for any nonzero constant c, and $g(x) = (x - a)P(x)$.

d) No; only the function in part (b) has the given graph.

67. a) -4, -3, 2, and 5 are zeros of the function, so $x + 4$, $x + 3$, $x - 2$, and $x - 5$ are factors.

b) We first write the product of the factors:

$P(x) = (x + 4)(x + 3)(x - 2)(x - 5)$

Note that $P(0) = 4 \cdot 3(-2)(-5) > 0$ and the graph shows a positive y-intercept, so this function is a correct one.

c) Yes; two examples are $f(x) = c \cdot P(x)$ for any nonzero constant c and $g(x) = (x - a)P(x)$.

d) No; only the function in part (b) has the given graph.

68. Divide $x^2 + kx + 4$ by $x - 1$.

$$\begin{array}{r|ccc} 1 & 1 & k & 4 \\ & & 1 & k+1 \\ \hline & 1 & k+1 & k+5 \end{array}$$

The remainder is $k + 5$.

Divide $x^2 + kx + 4$ by $x + 1$.

$$\begin{array}{r|ccc} -1 & 1 & k & 4 \\ & & -1 & -k+1 \\ \hline & 1 & k-1 & -k+5 \end{array}$$

The remainder is $-k + 5$.

Let $k+5=-k+5$ and solve for k.

$k+5=-k+5$

$2k=0$

$k=0$

69. Divide $x^3-kx^2+3x+7k$ by $x+2$.

$$\begin{array}{r|rrrr} -2 & 1 & -k & 3 & 7k \\ & & -2 & 2k+4 & -4k-14 \\ \hline & 1 & -k-2 & 2k+7 & 3k-14 \end{array}$$

Thus $P(-2)=3k-14$.

We know that if $x+2$ is a factor of $f(x)$, then $f(-2)=0$.

We solve $0=3k-14$ for k.

$0=3k-14$

$\frac{14}{3}=k$

70. $y=\frac{1}{13}x^3-\frac{1}{14}x$

$y=x\left(\frac{1}{13}x^2-\frac{1}{14}\right)$

We use the principle of zero products to find the zeros of the polynomial.

$x=0$ *or* $\frac{1}{13}x^2-\frac{1}{14}=0$

$x=0$ *or* $\frac{1}{13}x^2=\frac{1}{14}$

$x=0$ *or* $x^2=\frac{13}{14}$

$x=0$ *or* $x=\pm\sqrt{\frac{13}{14}}$

$x=0$ *or* $x\approx\pm 0.9636$

Only 0 and 0.9636 are in the interval $[0,2]$.

71. $\frac{2x^2}{x^2-1}+\frac{4}{x+3}=\frac{32}{3x^2-x-3}$,

LCM is $(x+1)(x-1)(x+3)$

$$(x+1)(x-1)(x+3)\left[\frac{2x^2}{(x+1)(x-1)}+\frac{4}{x+3}\right]=$$

$$(x+1)(x-1)(x+3)\cdot\frac{32}{(x+1)(x-1)(x+3)}$$

$2x^2(x+3)+4(x+1)(x-1)=32$

$2x^3+6x^2+4x^2-4=32$

$2x^3+10x^2-36=0$

$x^3+5x^2-18=0$

Using synthetic division and several trials, we find that -3 is a factor of $f(x)=x^3+5x^2-18$:

$$\begin{array}{r|rrrr} -3 & 1 & 5 & 0 & -18 \\ & & -3 & -6 & 18 \\ \hline & 1 & 2 & -6 & 0 \end{array}$$

Then we have:

$(x+3)(x^2+2x-6)=0$

$x+3=0$ *or* $x^2+2x-6=0$

$x=-3$ *or* $x=-1\pm\sqrt{7}$

Only $-1\pm\sqrt{7}$ check.

72. $\frac{6x^2}{x^2+11}+\frac{60}{x^3-7x^2+11x-77}=\frac{1}{x-7}$,

LCM is $(x^2+11)(x-7)$

$$(x^2+11)(x-7)\left[\frac{6x^2}{x^2+11}+\frac{60}{(x^2+11)(x-7)}\right]=$$

$$(x^2+11)(x-7)\cdot\frac{1}{x-7}$$

$6x^2(x-7)+60=x^2+11$

$6x^3-42x^2+60=x^2+11$

$6x^3-43x^2+49=0$

Use synthetic division to find factors of $f(x)=6x^3-43x^2+49$.

$$\begin{array}{r|rrrr} -1 & 6 & -43 & 0 & 49 \\ & & -6 & 49 & -49 \\ \hline & 6 & -49 & 49 & 0 \end{array}$$

Then we have:

$(x+1)(6x^2-49x+49)=0$

$(x+1)(6x-7)(x-7)=0$

$x=-1$ or $x=\frac{7}{6}$ or $x=7$

Only -1 and $\frac{7}{6}$ check.

73. Answers may vary. One possibility is $P(x)=x^{15}-x^{14}$.

74.

$$\begin{array}{r|rrr} 3+2i & 1 & -4 & -2 \\ & & 3+2i & -7+4i \\ \hline & 1 & -1+2i & -9+4i \end{array}$$

The answer is $x-1+2i$, R $-9+4i$.

75.

$$\begin{array}{r|rrr} i & 1 & -3 & 7 \\ & & i & -3i-1 \\ \hline & 1 & -3+i & 6-3i \end{array} \qquad (i^2=-1)$$

The answer is $x-3+i$, R $6-3i$.

Exercise Set 3.3

1. Find a polynomial function of degree 3 with -2, 3, and 5 as zeros.

Such a function has factors $x+2$, $x-3$, and $x-5$, so we have $f(x)=a_n(x+2)(x-3)(x-5)$.

The number a_n can be any nonzero number. The simplest polynomial will be obtained if we let it be 1. Multiplying the factors, we obtain

$f(x)=(x+2)(x-3)(x-5)$

$=(x^2-x-6)(x-5)$

$=x^3-6x^2-x+30.$

2. $f(x)=(x+1)(x)(x-4)$

$=(x^2+x)(x-4)$

$=x^3-3x^2-4x$

3. Find a polynomial function of degree 3 with -3, $2i$, and $-2i$ as zeros.

Such a function has factors $x+3$, $x-2i$, and $x+2i$, so we have $f(x) = a_n(x+3)(x-2i)(x+2i)$.

The number a_n can be any nonzero number. The simplest polynomial will be obtained if we let it be 1. Multiplying the factors, we obtain

$$\begin{aligned} f(x) &= (x+3)(x-2i)(x+2i) \\ &= (x+3)(x^2+4) \\ &= x^3+3x^2+4x+12. \end{aligned}$$

4. $$\begin{aligned} f(x) &= (x-2)(x-i)(x+i) \\ &= (x-2)(x^2+1) \\ &= x^3-2x^2+x-2 \end{aligned}$$

5. Find a polynomial function of degree 3 with $\sqrt{2}$, $-\sqrt{2}$, and 3 as zeros.

Such a function has factors $x-\sqrt{2}$, $x+\sqrt{2}$, and $x-3$, so we have $f(x) = a_n(x-\sqrt{2})(x+\sqrt{2})(x-3)$.

The number a_n can be any nonzero number. The simplest polynomial will be obtained if we let it be 1. Multiplying the factors, we obtain

$$\begin{aligned} f(x) &= (x-\sqrt{2})(x+\sqrt{2})(x-3) \\ &= (x^2-2)(x-3) \\ &= x^3-3x^2-2x+6. \end{aligned}$$

6. $$\begin{aligned} f(x) &= (x+5)(x-\sqrt{3})(x+\sqrt{3}) \\ &= (x+5)(x^2-3) \\ &= x^3+5x^2-3x-15 \end{aligned}$$

7. Find a polynomial function of degree 3 with $1-\sqrt{3}$, $1+\sqrt{3}$, and -2 as zeros.

Such a function has factors $x-(1-\sqrt{3})$, $x-(1+\sqrt{3})$, and $x+2$, so we have

$f(x) = a_n[x-(1-\sqrt{3})][x-(1+\sqrt{3})](x+2)$.

The number a_n can be any nonzero number. The simplest polynomial will be obtained if we let it be 1. Multiplying the factors, we obtain

$$\begin{aligned} f(x) &= [x-(1-\sqrt{3})][x-(1+\sqrt{3})](x+2) \\ &= [(x-1)+\sqrt{3}][(x-1)-\sqrt{3}](x+2) \\ &= [(x-1)^2-(\sqrt{3})^2](x+2) \\ &= (x^2-2x+1-3)(x+2) \\ &= (x^2-2x-2)(x+2) \\ &= x^3-2x^2-2x+2x^2-4x-4 \\ &= x^3-6x-4. \end{aligned}$$

8. $$\begin{aligned} f(x) &= (x+4)[x-(1-\sqrt{5})][x-(1+\sqrt{5})] \\ &= (x+4)[(x-1)+\sqrt{5}][(x-1)-\sqrt{5}] \\ &= (x+4)(x^2-2x+1-5) \\ &= (x+4)(x^2-2x-4) \\ &= x^3-2x^2-4x+4x^2-8x-16 \\ &= x^3+2x^2-12x-16 \end{aligned}$$

9. Find a polynomial function of degree 3 with $1+6i$, $1-6i$, and -4 as zeros.

Such a function has factors $x-(1+6i)$, $x-(1-6i)$, and $x+4$, so we have

$f(x) = a_n[x-(1+6i)][x-(1-6i)](x+4)$.

The number a_n can be any nonzero number. The simplest polynomial will be obtained if we let it be 1. Multiplying the factors, we obtain

$$\begin{aligned} f(x) &= [x-(1+6i)][x-(1-6i)](x+4) \\ &= [(x-1)-6i][(x-1)+6i](x+4) \\ &= [(x-1)^2-(6i)^2](x+4) \\ &= (x^2-2x+1+36)(x+4) \\ &= (x^2-2x+37)(x+4) \\ &= x^3-2x^2+37x+4x^2-8x+148 \\ &= x^3+2x^2+29x+148. \end{aligned}$$

10. $$\begin{aligned} f(x) &= [x-(1+4i)][x-(1-4i)](x+1) \\ &= (x^2-2x+17)(x+1) \\ &= x^3-x^2+15x+17 \end{aligned}$$

11. Find a polynomial function of degree 3 with $-\frac{1}{3}$, 0, and 2 as zeros.

Such a function has factors $x+\frac{1}{3}$, $x-0$ (or x), and $x-2$ so we have

$f(x) = a_n\left(x+\frac{1}{3}\right)(x)(x-2)$.

The number a_n can be any nonzero number. The simplest polynomial will be obtained if we let it be 1. Multiplying the factors, we obtain

$$\begin{aligned} f(x) &= \left(x+\frac{1}{3}\right)(x)(x-2) \\ &= \left(x^2+\frac{1}{3}x\right)(x-2) \\ &= x^3-\frac{5}{3}x^2-\frac{2}{3}x. \end{aligned}$$

12. $$\begin{aligned} f(x) &= (x+3)(x)\left(x-\frac{1}{2}\right) \\ &= (x^2+3x)(\left(x-\frac{1}{2}\right) \\ &= x^3+\frac{5}{2}x^2-\frac{3}{2}x \end{aligned}$$

13. A polynomial function of degree 5 has at most 5 zeros. Since 5 zeros are given, these are all of the zeros of the desired function. We proceed as in Exercises 1-11, letting $a_n = 1$.

$$\begin{aligned} f(x) &= (x+1)^3(x-0)(x-1) \\ &= (x^3+3x^2+3x+1)(x^2-x) \\ &= x^5+2x^4-2x^2-x \end{aligned}$$

14. $$\begin{aligned} f(x) &= (x+2)(x-3)^2(x+1) \\ &= x^4-3x^3-7x^2+15x+18 \end{aligned}$$

15. A polynomial function of degree 4 has at most 4 zeros. Since 4 zeros are given, these are all of the zeros of the desired function. We proceed as in Exercises 1-11, letting $a_n = 1$.

$$\begin{aligned} f(x) &= (x+1)^3(x-0) \\ &= (x^3+3x^2+3x+1)(x) \\ &= x^4+3x^3+3x^2+x \end{aligned}$$

16. $$\begin{aligned} f(x) &= \left(x+\frac{1}{2}\right)^2(x-0)(x-1)^2 \\ &= x^5 - x^4 - \frac{3}{4}x^3 + \frac{1}{2}x^2 + \frac{1}{4}x \end{aligned}$$

17. A polynomial function of degree 4 can have at most 4 zeros. Since $f(x)$ has rational coefficients, in addition to the three zeros given, the other zero is the conjugate of $\sqrt{3}$, or $-\sqrt{3}$.

18. A polynomial function of degree 4 can have at most 4 zeros. Since $f(x)$ has rational coefficients, in addition to the three zeros given, the other zero is the conjugate of $-\sqrt{2}$, or $\sqrt{2}$.

19. A polynomial function of degree 4 can have at most 4 zeros. Since $f(x)$ has rational coefficients, the other zeros are the conjugates of the given zeros. They are i and $2+\sqrt{5}$.

20. A polynomial function of degree 4 can have at most 4 zeros. Since $f(x)$ has rational coefficients, the other zeros are the conjugates of the given zeros. They are $-i$ and $-3-\sqrt{3}$.

21. A polynomial function of degree 4 can have at most 4 zeros. Since $f(x)$ has rational coefficients, in addition to the three zeros given, the other zero is the conjugate of $3i$, or $-3i$.

22. A polynomial function of degree 4 can have at most 4 zeros. Since $f(x)$ has rational coefficients, in addition to the three zeros given, the other zero is the conjugate of $-2i$, or $2i$.

23. A polynomial function of degree 4 can have at most 4 zeros. Since $f(x)$ has rational coefficients, the other zeros are the conjugates of the given zeros. They are $-4+3i$ and $2+\sqrt{3}$.

24. A polynomial function of degree 4 can have at most 4 zeros. Since $f(x)$ has rational coefficients, the other zeros are the conjugates of the given zeros. They are $6+5i$ and $-1-\sqrt{7}$.

25. A polynomial function $f(x)$ of degree 5 has at most 5 zeros. Since $f(x)$ has rational coefficients, in addition to the 3 given zeros, the other zeros are the conjugates of $\sqrt{5}$ and $-4i$, or $-\sqrt{5}$ and $4i$.

26. A polynomial function $f(x)$ of degree 5 has at most 5 zeros. Since $f(x)$ has rational coefficients, in addition to the 3 given zeros, the other zeros are the conjugates of $-\sqrt{3}$ and $2i$, or $\sqrt{3}$ and $-2i$.

27. A polynomial function $f(x)$ of degree 5 has at most 5 zeros. Since $f(x)$ has rational coefficients, the other zero is the conjugate of $2-i$, or $2+i$.

28. A polynomial function $f(x)$ of degree 5 has at most 5 zeros. Since $f(x)$ has rational coefficients, the other zero is the conjugate of $1-i$, or $1+i$.

29. A polynomial function $f(x)$ of degree 5 has at most 5 zeros. Since $f(x)$ has rational coefficients, in addition to the 3 given zeros, the other zeros are the conjugates of $-3+4i$ and $4-\sqrt{5}$, or $-3-4i$ and $4+\sqrt{5}$.

30. A polynomial function $f(x)$ of degree 5 has at most 5 zeros. Since $f(x)$ has rational coefficients, in addition to the 3 given zeros, the other zeros are the conjugates of $-3-3i$ and $2+\sqrt{13}$, or $-3+3i$ and $2-\sqrt{13}$.

31. A polynomial function $f(x)$ of degree 5 has at most 5 zeros. Since $f(x)$ has rational coefficients, the other zero is the conjugate of $4-i$, or $4+i$.

32. A polynomial function $f(x)$ of degree 5 has at most 5 zeros. Since $f(x)$ has rational coefficients, the other zero is the conjugate of $-3+\sqrt{2}$, or $-3-\sqrt{2}$.

33. Find a polynomial function of lowest degree with rational coefficients that has $1+i$ and 2 as some of its zeros. $1-i$ is also a zero.

Thus the polynomial function is

$f(x) = a_n(x-2)[x-(1+i)][x-(1-i)]$.

If we let $a_n = 1$, we obtain

$$\begin{aligned} f(x) &= (x-2)[(x-1)-i][(x-1)+i] \\ &= (x-2)[(x-1)^2-i^2] \\ &= (x-2)(x^2-2x+1+1) \\ &= (x-2)(x^2-2x+2) \\ &= x^3-4x^2+6x-4. \end{aligned}$$

34. $$\begin{aligned} f(x) &= [x-(2-i)][x-(2+i)](x+1) \\ &= x^3-3x^2+x+5 \end{aligned}$$

35. Find a polynomial function of lowest degree with rational coefficients that has $4i$ as one of its zeros. $-4i$ is also a zero.

Thus the polynomial function is

$f(x) = a_n(x-4i)(x+4i)$.

If we let $a_n = 1$, we obtain

$f(x) = (x-4i)(x+4i) = x^2+16$.

36. $f(x) = (x+5i)(x-5i) = x^2+25$

37. Find a polynomial function of lowest degree with rational coefficients that has $-4i$ and 5 as some of its zeros.

$4i$ is also a zero.

Thus the polynomial function is

$f(x) = a_n(x-5)(x+4i)(x-4i)$.

If we let $a_n = 1$, we obtain

$$\begin{aligned} f(x) &= (x-5)[x^2-(4i)^2] \\ &= (x-5)(x^2+16) \\ &= x^3-5x^2+16x-80 \end{aligned}$$

38. $f(x) = (x-3)(x+i)(x-i) = x^3-3x^2+x-3$

39. Find a polynomial function of lowest degree with rational coefficients that has $1 - i$ and $-\sqrt{5}$ as some of its zeros. $1 + i$ and $\sqrt{5}$ are also zeros.

Thus the polynomial function is

$f(x) = a_n[x - (1 - i)][x - (1 + i)](x + \sqrt{5})(x - \sqrt{5})$.

If we let $a_n = 1$, we obtain

$$\begin{aligned} f(x) &= [x - (1 - i)][x - (1 + i)](x + \sqrt{5})(x - \sqrt{5}) \\ &= [(x - 1) + i][(x - 1) - i](x + \sqrt{5})(x - \sqrt{5}) \\ &= (x^2 - 2x + 1 + 1)(x^2 - 5) \\ &= (x^2 - 2x + 2)(x^2 - 5) \\ &= x^4 - 2x^3 + 2x^2 - 5x^2 + 10x - 10 \\ &= x^4 - 2x^3 - 3x^2 + 10x - 10 \end{aligned}$$

40. $$\begin{aligned} f(x) &= [x-(2-\sqrt{3})][x-(2+\sqrt{3})][x-(1+i)][x-(1-i)] \\ &= x^4 - 6x^3 + 11x^2 - 10x + 2 \end{aligned}$$

41. Find a polynomial function of lowest degree with rational coefficients that has $\sqrt{5}$ and $-3i$ as some of its zeros.

$-\sqrt{5}$ and $3i$ are also zeros.

Thus the polynomial function is

$f(x) = a_n(x - \sqrt{5})(x + \sqrt{5})(x + 3i)(x - 3i)$.

If we let $a_n = 1$, we obtain

$$\begin{aligned} f(x) &= (x^2 - 5)(x^2 + 9) \\ &= x^4 + 4x^2 - 45 \end{aligned}$$

42. $$\begin{aligned} f(x) &= (x + \sqrt{2})(x - \sqrt{2})(x - 4i)(x + 4i) \\ &= x^4 + 14x^2 - 32 \end{aligned}$$

43. $f(x) = x^3 + 5x^2 - 2x - 10$

Since -5 is a zero of $f(x)$, we have $f(x) = (x + 5) \cdot Q(x)$. We use synthetic division to find $Q(x)$.

$$\begin{array}{r|rrrr} -5 & 1 & 5 & -2 & -10 \\ & & -5 & 0 & 10 \\ \hline & 1 & 0 & -2 & 0 \end{array}$$

Then $f(x) = (x + 5)(x^2 - 2)$. To find the other zeros we solve $x^2 - 2 = 0$.

$$\begin{aligned} x^2 - 2 &= 0 \\ x^2 &= 2 \\ x &= \pm\sqrt{2} \end{aligned}$$

The other zeros are $-\sqrt{2}$ and $\sqrt{2}$.

44. $$\begin{array}{r|rrrr} 1 & 1 & -1 & 1 & -1 \\ & & 1 & 0 & 1 \\ \hline & 1 & 0 & 1 & 0 \end{array}$$

$f(x) = (x - 1)(x^2 + 1)$

Solving $x^2 + 1 = 0$, we find that the other zeros are $-i$ and i.

45. If $-i$ is a zero of $f(x) = x^4 - 5x^3 + 7x^2 - 5x + 6$, i is also a zero. Thus $x + i$ and $x - i$ are factors of the polynomial. Since $(x + i)(x - i) = x^2 + 1$, we know that $f(x) = (x^2 + 1) \cdot Q(x)$. Divide $x^4 - 5x^3 + 7x^2 - 5x + 6$ by $x^2 + 1$.

$$\begin{array}{r|l} & x^2 - 5x + 6 \\ \hline x^2 + 1 & x^4 - 5x^3 + 7x^2 - 5x + 6 \\ & \underline{x^4 \qquad\quad + x^2} \\ & \quad -5x^3 + 6x^2 - 5x \\ & \quad \underline{-5x^3 \qquad\quad - 5x} \\ & \qquad\qquad 6x^2 \qquad + 6 \\ & \qquad\qquad \underline{6x^2 \qquad + 6} \\ & \qquad\qquad\qquad\qquad 0 \end{array}$$

Thus

$$\begin{aligned} x^4 - 5x^3 + 7x^2 - 5x + 6 &= (x+i)(x-i)(x^2 - 5x + 6) \\ &= (x+i)(x-i)(x-2)(x-3) \end{aligned}$$

Using the principle of zero products we find the other zeros to be i, 2, and 3.

46. $(x - 2i)$ and $(x + 2i)$ are both factors of $P(x) = x^4 - 16$.

$(x - 2i)(x + 2i) = x^2 + 4$

$$\begin{array}{r|l} & x^2 - 4 \\ \hline x^2 + 4 & x^4 + 0x^2 - 16 \\ & \underline{x^4 + 4x^2} \\ & \quad -4x^2 - 16 \\ & \quad \underline{-4x^2 - 16} \\ & \qquad\qquad 0 \end{array}$$

$$\begin{aligned} (x - 2i)(x + 2i)(x^2 - 4) &= 0 \\ (x - 2i)(x + 2i)(x + 2)(x - 2) &= 0 \end{aligned}$$

The other zeros are $-2i$, -2, and 2.

47. $x^3 - 6x^2 + 13x - 20 = 0$

If 4 is a zero, then $x - 4$ is a factor. Use synthetic division to find another factor.

$$\begin{array}{r|rrrr} 4 & 1 & -6 & 13 & -20 \\ & & 4 & -8 & 20 \\ \hline & 1 & -2 & 5 & 0 \end{array}$$

$(x - 4)(x^2 - 2x + 5) = 0$

$x - 4 = 0$ *or* $x^2 - 2x + 5 = 0$ Principle of zero products

$x = 4$ *or* $x = \dfrac{2 \pm \sqrt{4 - 20}}{2}$ Quadratic formula

$x = 4$ *or* $x = \dfrac{2 \pm 4i}{2} = 1 \pm 2i$

The other zeros are $1 + 2i$ and $1 - 2i$.

48. $$\begin{array}{r|rrrr} 2 & 1 & 0 & 0 & -8 \\ & & 2 & 4 & 8 \\ \hline & 1 & 2 & 4 & 0 \end{array}$$

$(x - 2)(x^2 + 2x + 4) = 0$

$x = 2$ *or* $x = -1 \pm \sqrt{3}i$

The other zeros are $-1 + \sqrt{3}i$ and $-1 - \sqrt{3}i$.

49. $f(x) = x^5 - 3x^2 + 1$

According to the rational zeros theorem, any rational zero of f must be of the form p/q, where p is a factor of the constant term, 1, and q is a factor of the coefficient of x^5, 1.

$\dfrac{\text{Possibilities for } p}{\text{Possibilities for } q} : \dfrac{\pm 1}{\pm 1}$

Possibilities for p/q: 1, -1

50. $f(x) = x^7 + 37x^5 - 6x^2 + 12$

$\dfrac{\text{Possibilities for } p}{\text{Possibilities for } q} : \dfrac{\pm 1, \pm 2, \pm 3, \pm 4, \pm 6, \pm 12}{\pm 1}$

Possibilities for p/q: $1, -1, 2, -2, 3, -3, 4, -4, 6, -6, 12, -12$

51. $f(x) = 2x^4 - 3x^3 - x + 8$

According to the rational zeros theorem, any rational zero of f must be of the form p/q, where p is a factor of the constant term, 8, and q is a factor of the coefficient of x^4, 2.

$\dfrac{\text{Possibilities for } p}{\text{Possibilities for } q} : \dfrac{\pm 1, \pm 2, \pm 4, \pm 8}{\pm 1, \pm 2}$

Possibilities for p/q: $1, -1, 2, -2, 4, -4, 8, -8, \frac{1}{2}, -\frac{1}{2}$

52. $f(x) = 3x^3 - x^2 + 6x - 9$

$\dfrac{\text{Possibilities for } p}{\text{Possibilities for } q} : \dfrac{\pm 1, \pm 3, \pm 9}{\pm 1, \pm 3}$

Possibilities for p/q: $1, -1, 3, -3, 9, -9, \frac{1}{3}, -\frac{1}{3}$

53. $f(x) = 15x^6 + 47x^2 + 2$

According to the rational zeros theorem, any rational zero of f must be of the form p/q, where p is a factor of 2 and q is a factor of 15.

$\dfrac{\text{Possibilities for } p}{\text{Possibilities for } q} : \dfrac{\pm 1, \pm 2}{\pm 1, \pm 3, \pm 5, \pm 15}$

Possibilities for p/q: $1, -1, 2, -2, \frac{1}{3}, -\frac{1}{3}, \frac{2}{3}, -\frac{2}{3}, \frac{1}{5}, -\frac{1}{5}, \frac{2}{5}, -\frac{2}{5}, \frac{1}{15}, -\frac{1}{15}, \frac{2}{15}, -\frac{2}{15}$

54. $f(x) = 10x^{25} + 3x^{17} - 35x + 6$

$\dfrac{\text{Possibilities for } p}{\text{Possibilities for } q} : \dfrac{\pm 1, \pm 2, \pm 3, \pm 6}{\pm 1, \pm 2, \pm 5, \pm 10}$

Possibilities for p/q: $1, -1, 2, -2, 3, -3, 6, -6, \frac{1}{2}, -\frac{1}{2}, \frac{3}{2}, -\frac{3}{2}, \frac{1}{5}, -\frac{1}{5}, \frac{2}{5}, -\frac{2}{5}, \frac{3}{5}, -\frac{3}{5}, \frac{6}{5}, -\frac{6}{5}, \frac{1}{10}, -\frac{1}{10}, \frac{3}{10}, -\frac{3}{10}$

55. $f(x) = x^3 + 3x^2 - 2x - 6$

a) $\dfrac{\text{Possibilities for } p}{\text{Possibilities for } q} : \dfrac{\pm 1, \pm 2, \pm 3, \pm 6}{\pm 1}$

Possibilities for p/q: $1, -1, 2, -2, 3, -3, 6, -6$

From the graph of $y = x^3 + 3x^2 - 2x - 6$, we see that, of the possibilities above, only -3 might be a zero. We use synthetic division to determine whether -3 is indeed a zero.

$$\begin{array}{r|rrrr} -3 & 1 & 3 & -2 & -6 \\ & & -3 & 0 & 6 \\ \hline & 1 & 0 & -2 & 0 \end{array}$$

Then we have $f(x) = (x + 3)(x^2 - 2)$.

We find the other zeros:

$$x^2 - 2 = 0$$
$$x^2 = 2$$
$$x = \pm\sqrt{2}.$$

There is only one rational zero, -3. The other zeros are $\pm\sqrt{2}$. (Note that we could have used factoring by grouping to find this result.)

b) $f(x) = (x + 3)(x - \sqrt{2})(x + \sqrt{2})$

56. $f(x) = x^3 - x^2 - 3x + 3$

a) $\dfrac{\text{Possibilities for } p}{\text{Possibilities for } q} : \dfrac{\pm 1, \pm 3}{\pm 1}$

Possibilities for p/q: $1, -1, 3, -3$

From the graph of $y = x^3 - x^2 - 3x + 3$, we see that, of the possibilities above, only 1 might be a zero.

$$\begin{array}{r|rrrr} 1 & 1 & -1 & -3 & 3 \\ & & 1 & 0 & -3 \\ \hline & 1 & 0 & -3 & 0 \end{array}$$

$f(x) = (x - 1)(x^2 - 3)$

Now $x^2 - 3 = 0$ for $x = \pm\sqrt{3}$. Thus, there is only one rational zero, 1. The other zeros are $\pm\sqrt{3}$. (Note that we would have used factoring by grouping to find this result.)

b) $f(x) = (x - 1)(x - \sqrt{3})(x + \sqrt{3})$

57. $f(x) = x^3 - 3x + 2$

a) $\dfrac{\text{Possibilities for } p}{\text{Possibilities for } q} : \dfrac{\pm 1, \pm 2}{\pm 1}$

Possibilities for p/q: $1, -1, 2, -2$

From the graph of $y = x^3 - 3x + 2$, we see that, of the possibilities above, -2 and 1 might be a zeros. We use synthetic division to determine whether -2 is a zero.

$$\begin{array}{r|rrrr} -2 & 1 & 0 & -3 & 2 \\ & & -2 & 4 & -2 \\ \hline & 1 & -2 & 1 & 0 \end{array}$$

Then we have $f(x) = (x + 2)(x^2 - 2x + 1) = (x + 2)(x - 1)^2$.

Now $(x - 1)^2 = 0$ for $x = 1$. Thus, the rational zeros are -2 and 1. (The zero 1 has a multiplicity of 2.) These are the only zeros.

b) $f(x) = (x + 2)(x - 1)^2$

58. $f(x) = x^3 - 2x + 4$

a) $\dfrac{\text{Possibilities for } p}{\text{Possibilities for } q} : \dfrac{\pm 1, \pm 2, \pm 4}{\pm 1}$

Possibilities for p/q: $1, -1, 2, -2, 4, -4$

From the graph of $y = x^3 - 2x + 4$, we see that, of the possibilities above, only -2 might be a zero.

$$\begin{array}{r|rrrr} -2 & 1 & 0 & -2 & 4 \\ & & -2 & 4 & -4 \\ \hline & 1 & -2 & 2 & 0 \end{array}$$

$f(x) = (x+2)(x^2 - 2x + 2)$

Using the quadratic formula, we find that the other zeros are $1 \pm i$. The only rational zero is -2. The other zeros are $1 \pm i$.

b) $f(x) = (x+2)[x - (1+i)][x - (1-i)]$
$\quad = (x+2)(x-1-i)(x-1+i)$

59. $f(x) = x^3 - 5x^2 + 11x + 17$

a) $\dfrac{\text{Possibilities for } p}{\text{Possibilities for } q} : \dfrac{\pm 1, \pm 17}{\pm 1}$

Possibilities for p/q: $1, -1, 17, -17$

From the graph of $y = x^3 - 5x^2 + 11x + 17$, we see that, of the possibilities above, we see that only -1 might be a zero. We use synthetic division to determine whether -1 is indeed a zero.

$$\begin{array}{r|rrrr} -1 & 1 & -5 & 11 & 17 \\ & & -1 & 6 & -17 \\ \hline & 1 & -6 & 17 & 0 \end{array}$$

Then we have $f(x) = (x+1)(x^2 - 6x + 17)$. We use the quadratic formula to find the other zeros.

$x^2 - 6x + 17 = 0$

$$x = \frac{-(-6) \pm \sqrt{(-6)^2 - 4 \cdot 1 \cdot 17}}{2 \cdot 1}$$
$$= \frac{6 \pm \sqrt{-32}}{2} = \frac{6 \pm 4\sqrt{2}i}{2}$$
$$= 3 \pm 2\sqrt{2}i$$

The only rational zero is -1. The other zeros are $3 \pm 2\sqrt{2}i$.

b) $f(x) = (x+1)[x - (3 + 2\sqrt{2}i)][x - (3 - 2\sqrt{2}i)]$
$\quad = (x+1)(x - 3 - 2\sqrt{2}i)(x - 3 + 2\sqrt{2}i)$

60. $f(x) = 2x^3 + 7x^2 + 2x - 8$

a) $\dfrac{\text{Possibilities for } p}{\text{Possibilities for } q} : \dfrac{\pm 1, \pm 2, \pm 4, \pm 8}{\pm 1, \pm 2}$

Possibilities for p/q: $1, -1, 2, -2, 4, -4, 8, -8, \frac{1}{2}, -\frac{1}{2}$

From the graph of $y = 2x^3 + 7x^2 + 2x - 8$, we see that, of the possibilities above, only -2 and 1 might be a zeros.

$$\begin{array}{r|rrrr} -2 & 2 & 7 & 2 & -8 \\ & & -4 & -6 & 8 \\ \hline & 2 & 3 & -4 & 0 \end{array}$$

$f(x) = (x+2)(2x^2 + 3x - 4)$

Using the quadratic formula, we find that the other zeros are $\dfrac{-3 \pm \sqrt{41}}{4}$. The only rational zero is -2. The other zeros are $\dfrac{-3 \pm \sqrt{41}}{4}$.

b) $f(x) = (x+2)\left(x - \dfrac{-3+\sqrt{41}}{4}\right)\left(x - \dfrac{-3-\sqrt{41}}{4}\right)$

61. $f(x) = 5x^4 - 4x^3 + 19x^2 - 16x - 4$

a) $\dfrac{\text{Possibilities for } p}{\text{Possibilities for } q} : \dfrac{\pm 1, \pm 2, \pm 4}{\pm 1, \pm 5}$

Possibilities for p/q: $1, -1, 2, -2, 4, -4, \frac{1}{5}, -\frac{1}{5}, \frac{2}{5}, -\frac{2}{5}, \frac{4}{5}, -\frac{4}{5}$

From the graph of $y = 5x^4 - 4x^3 + 19x^2 - 16x - 4$, we see that, of the possibilities above, only $-\frac{2}{5}$, $-\frac{1}{5}$ and 1 might be zeros. We use synthetic division to determine whether 1 is a zero.

$$\begin{array}{r|rrrrr} 1 & 5 & -4 & 19 & -16 & -4 \\ & & 5 & 1 & 20 & 4 \\ \hline & 5 & 1 & 20 & 4 & 0 \end{array}$$

Then we have

$f(x) = (x-1)(5x^3 + x^2 + 20x + 4)$
$\quad = (x-1)[x^2(5x+1) + 4(5x+1)]$
$\quad = (x-1)(5x+1)(x^2+4)$.

We find the other zeros:

$5x + 1 = 0 \quad \textit{or} \quad x^2 + 4 = 0$
$5x = -1 \quad \textit{or} \quad x^2 = -4$
$x = -\frac{1}{5} \quad \textit{or} \quad x = \pm 2i$

The rational zeros are $-\frac{1}{5}$ and 1. The other zeros are $\pm 2i$.

b) From part (a) we see that

$f(x) = (5x+1)(x-1)(x+2i)(x-2i)$.

62. $f(x) = 3x^4 - 4x^3 + x^2 + 6x - 2$

a) $\dfrac{\text{Possibilities for } p}{\text{Possibilities for } q} : \dfrac{\pm 1, \pm 2}{\pm 1, \pm 3}$

Possibilities for p/q: $1, -1, 2, -2, \frac{1}{3}, -\frac{1}{3}, \frac{2}{3}, -\frac{2}{3}$

From the graph of $y = 3x^4 - 4x^3 + x^2 + 6x - 2$, we see that, of the possibilities above, only -1 and $\frac{1}{3}$ might be zeros.

$$\begin{array}{r|rrrrr} -1 & 3 & -4 & 1 & 6 & -2 \\ & & -3 & 7 & -8 & 2 \\ \hline & 3 & -7 & 8 & -2 & 0 \end{array}$$

$$\begin{array}{r|rrrr} \frac{1}{3} & 3 & -7 & 8 & -2 \\ & & 1 & -2 & 2 \\ \hline & 3 & -6 & 6 & 0 \end{array}$$

$f(x) = (x+1)\left(x - \frac{1}{3}\right)(3x^2 - 6x + 6)$
$\quad = (x+1)\left(x - \frac{1}{3}\right)(3)(x^2 - 2x + 2)$

Using the quadratic formula, we find that the other zeros are $1 \pm i$.

The rational zeros are -1 and $\frac{1}{3}$. The other zeros are $1 \pm i$.

b) $f(x) = 3(x+1)\left(x-\frac{1}{3}\right)[x-(1+i)][x-(1-i)]$

$= (x+1)(3x-1)(x-1-i)(x-1+i)$

63. $f(x) = x^4 - 3x^3 - 20x^2 - 24x - 8$

a) $\dfrac{\text{Possibilities for } p}{\text{Possibilities for } q} : \dfrac{\pm 1, \pm 2, \pm 4, \pm 8}{\pm 1}$

Possibilities for p/q: $1, -1, 2, -2, 4, -4, 8, -8$

From the graph of $y = x^4 - 3x^3 - 20x^2 - 24x - 8$, we see that, of the possibilities above, only -2 and -1 might be zeros. We use synthetic division to determine if -2 is a zero.

$$\begin{array}{r|rrrrr} -2 & 1 & -3 & -20 & -24 & -8 \\ & & -2 & 10 & 20 & 8 \\ \hline & 1 & -5 & -10 & -4 & 0 \end{array}$$

We see that -2 is a zero. Now we determine whether -1 is a zero.

$$\begin{array}{r|rrrr} -1 & 1 & -5 & -10 & -4 \\ & & -1 & 6 & 4 \\ \hline & 1 & -6 & -4 & 0 \end{array}$$

Then we have $f(x) = (x+2)(x+1)(x^2 - 6x - 4)$. Use the quadratic formula to find the other zeros.

$x^2 - 6x - 4 = 0$

$$x = \frac{-(-6) \pm \sqrt{(-6)^2 - 4 \cdot 1 \cdot (-4)}}{2 \cdot 1}$$

$$= \frac{6 \pm \sqrt{52}}{2} = \frac{6 \pm 2\sqrt{13}}{2}$$

$$= 3 \pm \sqrt{13}$$

The rational zeros are -2 and -1. The other zeros are $3 \pm \sqrt{13}$.

b) $f(x) = (x+2)(x+1)[x-(3+\sqrt{13})][x-(3-\sqrt{13})]$

$= (x+2)(x+1)(x-3-\sqrt{13})(x-3+\sqrt{13})$

64. $f(x) = x^4 + 5x^3 - 27x^2 + 31x - 10$

a) $\dfrac{\text{Possibilities for } p}{\text{Possibilities for } q} : \dfrac{\pm 1, \pm 2, \pm 5, \pm 10}{\pm 1}$

Possibilities for p/q: $1, -1, 2, -2, 5, -5, 10, -10$

From the graph of $y = x^4 + 5x^3 - 27x^2 + 31x - 10$, we see that, of the possibilities above, only 1 and 2 might be zeros.

$$\begin{array}{r|rrrrr} 1 & 1 & 5 & -27 & 31 & -10 \\ & & 1 & 6 & -21 & 10 \\ \hline & 1 & 6 & -21 & 10 & 0 \end{array}$$

$$\begin{array}{r|rrrr} 2 & 1 & 6 & -21 & 10 \\ & & 2 & 16 & -10 \\ \hline & 1 & 8 & -5 & 0 \end{array}$$

$f(x) = (x-1)(x-2)(x^2 + 8x - 5)$

Using the quadratic formula, we find that the other zeros are $-4 \pm \sqrt{21}$.

The rational zeros are 1 and 2. The other zeros are $-4 \pm \sqrt{21}$.

b) $f(x) = (x-1)(x-2)[x-(-4+\sqrt{21})][x-(-4-\sqrt{21})]$

$= (x-1)(x-2)(x+4-\sqrt{21})(x+4+\sqrt{21})$

65. $f(x) = x^3 - 4x^2 + 2x + 4$

a) $\dfrac{\text{Possibilities for } p}{\text{Possibilities for } q} : \dfrac{\pm 1, \pm 2, \pm 4}{\pm 1}$

Possibilities for p/q: $1, -1, 2, -2, 4, -4$

From the graph of $y = x^3 - 4x^2 + 2x + 4$, we see that, of the possibilities above, only -1, 1, and 2 might be zeros. Synthetic division shows that neither -1 nor 1 is a zero. Try 2.

$$\begin{array}{r|rrrr} 2 & 1 & -4 & 2 & 4 \\ & & 2 & -4 & -4 \\ \hline & 1 & -2 & -2 & 0 \end{array}$$

Then we have $f(x) = (x-2)(x^2 - 2x - 2)$. Use the quadratic formula to find the other zeros.

$x^2 - 2x - 2 = 0$

$$x = \frac{-(-2) \pm \sqrt{(-2)^2 - 4 \cdot 1 \cdot (-2)}}{2 \cdot 1}$$

$$= \frac{2 \pm \sqrt{12}}{2} = \frac{2 \pm 2\sqrt{3}}{2}$$

$$= 1 \pm \sqrt{3}$$

The only rational zero is 2. The other zeros are $1 \pm \sqrt{3}$.

b) $f(x) = (x-2)[x-(1+\sqrt{3})][x-(1-\sqrt{3})]$

$= (x-2)(x-1-\sqrt{3})(x-1+\sqrt{3})$

66. $f(x) = x^3 - 8x^2 + 17x - 4$

a) $\dfrac{\text{Possibilities for } p}{\text{Possibilities for } q} : \dfrac{\pm 1, \pm 2, \pm 4}{\pm 1}$

Possibilities for p/q: $1, -1, 2, -2, 4, -4$

From the graph of $y = x^3 - 8x^2 + 17x - 4$, we see that, of the possibilities above, only 4 might be a zero.

$$\begin{array}{r|rrrr} 4 & 1 & -8 & 17 & -4 \\ & & 4 & -16 & 4 \\ \hline & 1 & -4 & 1 & 0 \end{array}$$

$f(x) = (x-4)(x^2 - 4x + 1)$

Using the quadratic formula, we find that the other zeros are $2 \pm \sqrt{3}$.

The only rational zero is 4. The other zeros are $2 \pm \sqrt{3}$.

b) $f(x) = (x-4)[x-(2+\sqrt{3})][x-(2-\sqrt{3})]$

$= (x-4)(x-2-\sqrt{3})(x-2+\sqrt{3})$

67. $f(x) = x^3 + 8$

a) $\dfrac{\text{Possibilities for } p}{\text{Possibilities for } q} : \dfrac{\pm 1, \pm 2, \pm 4, \pm 8}{\pm 1}$

Possibilities for p/q: $1, -1, 2, -2, 4, -4, 8, -8$

From the graph of $y = x^3 + 8$, we see that, of the possibilities above, only -2 might be a zero. We use synthetic division to see if it is.

$$\begin{array}{r|rrrr} -2 & 1 & 0 & 0 & 8 \\ & & -2 & 4 & -8 \\ \hline & 1 & -2 & 4 & 0 \end{array}$$

We have $f(x) = (x+2)(x^2-2x+4)$. Use the quadratic formula to find the other zeros.

$x^2 - 2x + 4 = 0$

$$x = \frac{-(-2) \pm \sqrt{(-2)^2 - 4\cdot 1\cdot 4}}{2\cdot 1}$$
$$= \frac{2 \pm \sqrt{-12}}{2} = \frac{2 \pm 2\sqrt{3}i}{2}$$
$$= 1 \pm \sqrt{3}i$$

The only rational zero is -2. The other zeros are $1 \pm \sqrt{3}i$.

b) $f(x) = (x+2)[x-(1+\sqrt{3}i)][x-(1-\sqrt{3}i)]$
$= (x+2)(x-1-\sqrt{3}i)(x-1+\sqrt{3}i)$

68. $f(x) = x^3 - 8$

a) As in Exercise 43, the possibilities for p/q are 1, -1, 2, -2, 4, -4, 8, and -8.

From the graph of $y = x^3 - 8$, we see that, of the possibilities above, only 2 might be a zero.

$$\begin{array}{r|rrrr} 2 & 1 & 0 & 0 & -8 \\ & & 2 & 4 & 8 \\ \hline & 1 & 2 & 4 & 0 \end{array}$$

$f(x) = (x-2)(x^2+2x+4)$

Using the quadratic formula, we find that the other zeros are $-1 \pm \sqrt{3}i$.

The only rational zero is 2. The other zeros are $-1 \pm \sqrt{3}i$.

b) $f(x) = (x-2)[x-(-1+\sqrt{3}i)][x-(-1-\sqrt{3}i)]$
$= (x-2)(x+1-\sqrt{3}i)(x+1+\sqrt{3}i)$

69. $f(x) = \frac{1}{3}x^3 - \frac{1}{2}x^2 - \frac{1}{6}x + \frac{1}{6}$
$= \frac{1}{6}(2x^3 - 3x^2 - x + 1)$

a) The second form of the equation is equivalent to the first and has the advantage of having integer coefficients. Thus, we can use the rational zeros theorem for $g(x) = 2x^3 - 3x^2 - x + 1$. The zeros of $g(x)$ are the same as the zeros of $f(x)$. We find the zeros of $g(x)$.

$$\frac{\text{Possibilities for } p}{\text{Possibilities for } q} : \frac{\pm 1}{\pm 1, \pm 2}$$

Possibilities for p/q: $1, -1, \frac{1}{2}, -\frac{1}{2}$

From the graph of $y = 2x^3 - 3x^2 - x + 1$, we see that, of the possibilities above, only $-\frac{1}{2}$ and $\frac{1}{2}$ might be zeros. Synthetic division shows that $-\frac{1}{2}$ is not a zero. Try $\frac{1}{2}$.

$$\begin{array}{r|rrrr} \frac{1}{2} & 2 & -3 & -1 & 1 \\ & & 1 & -1 & -1 \\ \hline & 2 & -2 & -2 & 0 \end{array}$$

We have $g(x) = \left(x - \frac{1}{2}\right)(2x^2 - 2x - 2) = \left(x - \frac{1}{2}\right)(2)(x^2 - x - 1)$. Use the quadratic formula to find the other zeros.

$x^2 - x - 1 = 0$

$$x = \frac{-(-1) \pm \sqrt{(-1)^2 - 4\cdot 1\cdot(-1)}}{2\cdot 1}$$
$$= \frac{1 \pm \sqrt{5}}{2}$$

The only rational zero is $\frac{1}{2}$. The other zeros are $\frac{1 \pm \sqrt{5}}{2}$.

b) $f(x) = \frac{1}{6}g(x)$
$$= \frac{1}{6}\left(x - \frac{1}{2}\right)(2)\left[x - \frac{1+\sqrt{5}}{2}\right]\left[x - \frac{1-\sqrt{5}}{2}\right]$$
$$= \frac{1}{3}\left(x - \frac{1}{2}\right)\left(x - \frac{1+\sqrt{5}}{2}\right)\left(x - \frac{1-\sqrt{5}}{2}\right)$$

70. $f(x) = \frac{2}{3}x^3 - \frac{1}{2}x^2 + \frac{2}{3}x - \frac{1}{2}$
$= \frac{1}{6}(4x^3 - 3x^2 + 4x - 3)$

a) Find the zeros of $g(x) = 4x^3 - 3x^2 + 4x - 3$.

$$\frac{\text{Possibilities for } p}{\text{Possibilities for } q} : \frac{\pm 1, \pm 3}{\pm 1, \pm 2, \pm 4}$$

Possibilities for p/q: $1, -1, 3, -3, \frac{1}{2}, -\frac{1}{2}, \frac{3}{2}, -\frac{3}{2}, \frac{1}{4}, -\frac{1}{4}, \frac{3}{4}, -\frac{3}{4}$

From the graph of $y = 4x^3 - 3x^2 + 4x - 3$, we see that, of the possibilities above, only $\frac{1}{2}$, $\frac{3}{4}$, and 1 might be zeros. Synthetic division shows that $\frac{1}{2}$ is not a zero. Try $\frac{3}{4}$.

$$\begin{array}{r|rrrr} \frac{3}{4} & 4 & -3 & 4 & -3 \\ & & 3 & 0 & 3 \\ \hline & 4 & 0 & 4 & 0 \end{array}$$

$g(x) = \left(x - \frac{3}{4}\right)(4x^2 + 4) = \left(x - \frac{3}{4}\right)(4)(x^2+1)$

Now $x^2 + 1 = 0$ when $x = \pm i$. Thus, the only rational zero is $\frac{3}{4}$. The other zeros are $\pm i$. (Note that we could have used factoring by grouping to find this result.)

b) $f(x) = \frac{1}{6}g(x)$
$$= \frac{1}{6}\left(x - \frac{3}{4}\right)(4)(x+i)(x-i)$$
$$= \frac{2}{3}\left(x - \frac{3}{4}\right)(x+i)(x-i)$$

71. $f(x) = x^4 + 32$

According to the rational zeros theorem, the possible rational zeros are $\pm 1, \pm 2, \pm 4, \pm 8, \pm 16$, and ± 32. The graph of $y = x^4 + 32$ has no x-intercepts, so $f(x)$ has no real-number zeros and hence no rational zeros.

72. $f(x) = x^6 + 8$

Possible rational zeros: $\pm 1, \pm 2, \pm 4, \pm 8$

The graph of $y = x^6 + 8$ has no x-intercepts, so $f(x)$ has no real-number zeros and hence no rational zeros.

73. $f(x) = x^3 - x^2 - 4x + 3$

According to the rational zeros theorem, the possible rational zeros are ± 1 and ± 3. The graph of $y = x^3 - x^2 - 4x + 3$ shows that none of these is a zero. Thus, there are no rational zeros.

74. $f(x) = 2x^3 + 3x^2 + 2x + 3$

Possible rational zeros: $\pm 1, \pm 3, \pm \frac{1}{2}, \pm \frac{3}{2}$

The graph of $y = 2x^3 + 3x^2 + 2x + 3$ shows that, of these possibilities, only $-\frac{3}{2}$ might be a zero.

$$\begin{array}{r|rrrr} -\frac{3}{2} & 2 & 3 & 2 & 3 \\ & & -3 & 0 & -3 \\ \hline & 2 & 0 & 2 & 0 \end{array}$$

$$f(x) = \left(x + \frac{3}{2}\right)(2x^2 + 2) = \left(x + \frac{3}{2}\right)(2)(x^2 + 1)$$

Since $g(x) = x^2 + 1$ has no real-number zeros, the only rational zero is $-\frac{3}{2}$. (We could have used factoring by grouping to find this result.)

75. $f(x) = x^4 + 2x^3 + 2x^2 - 4x - 8$

According to the rational zeros theorem, the possible rational zeros are ± 1, ± 2, ± 4, and ± 8. The graph of $y = x^4 + 2x^3 + 2x^2 - 4x - 8$ shows that none of the possibilities is a zero. Thus, there are no rational zeros.

76. $f(x) = x^4 + 6x^3 + 17x^2 + 36x + 66$

Possible rational zeros: $\pm 1, \pm 2, \pm 3, \pm 6, \pm 11,$
$\pm 22, \pm 33, \pm 66$

The graph of $y = x^4 + 6x^3 + 17x^2 + 36x + 66$ has no x-intercepts, so $f(x)$ has no real-number zeros and hence no rational zeros.

77. $f(x) = x^5 - 5x^4 + 5x^3 + 15x^2 - 36x + 20$

According to the rational zeros theorem, the possible rational zeros are ± 1, ± 2, ± 4, ± 5, ± 10, and ± 20. The graph of $y = x^5 - 5x^4 + 5x^3 + 15x^2 - 36x + 20$ shows that, of these possibilities, only -2, 1 and 2 might be zeros. We try -2.

$$\begin{array}{r|rrrrrr} -2 & 1 & -5 & 5 & 15 & -36 & 20 \\ & & -2 & 14 & -38 & 46 & -20 \\ \hline & 1 & -7 & 19 & -23 & 10 & 0 \end{array}$$

Thus, -2 is a zero. Now try 1.

$$\begin{array}{r|rrrrr} 1 & 1 & -7 & 19 & -23 & 10 \\ & & 1 & -6 & 13 & 10 \\ \hline & 1 & -6 & 13 & -10 & 0 \end{array}$$

1 is also a zero. Try 2.

$$\begin{array}{r|rrrr} 2 & 1 & -6 & 13 & -10 \\ & & 2 & -8 & 10 \\ \hline & 1 & -4 & 5 & 0 \end{array}$$

2 is also a zero.

We have $f(x) = (x + 2)(x - 1)(x - 2)(x^2 - 4x + 5)$. The discriminant of $x^2 - 4x + 5$ is $(-4)^2 - 4 \cdot 1 \cdot 5$, or $4 < 0$, so $x^2 - 4x + 5$ has two nonreal zeros. Thus, the rational zeros are -2, 1, and 2.

78. $f(x) = x^5 - 3x^4 - 3x^3 + 9x^2 - 4x + 12$

Possible rational zeros: ± 1, ± 2, ± 3, ± 4, ± 6, ± 12. The graph of $y = x^5 - 3x^4 - 3x^3 + 9x^2 - 4x + 12$ shows that, of these possibilities, only -2, 2 and 3 might be zeros.

$$\begin{array}{r|rrrrrr} -2 & 1 & -3 & -3 & 9 & -4 & 12 \\ & & -2 & 10 & -14 & 10 & -12 \\ \hline & 1 & -5 & 7 & -5 & 6 & 0 \end{array}$$

-2 is a zero.

$$\begin{array}{r|rrrrr} 2 & 1 & -5 & 7 & -5 & 6 \\ & & 2 & -6 & 2 & -6 \\ \hline & 1 & -3 & 1 & -3 & 0 \end{array}$$

2 is a zero.

$$\begin{array}{r|rrrr} 3 & 1 & -3 & 1 & -3 \\ & & 3 & 0 & 3 \\ \hline & 1 & 0 & 1 & 0 \end{array}$$

3 is also a zero.

$f(x) = (x + 2)(x - 2)(x - 3)(x^2 + 1)$

Since $g(x) = x^2 + 1$ has no real-number zeros, the rational zeros are -2, 2, and 3.

79. $f(x) = 3x^5 - 2x^2 + x - 1$

The number of variations in sign in $f(x)$ is 3. Then the number of positive real zeros is either 3 or less than 3 by 2, 4, 6, and so on. Thus, the number of positive real zeros is 3 or 1.

$$\begin{aligned} f(-x) &= 3(-x)^5 - 2(-x)^2 + (-x) - 1 \\ &= -3x^5 - 2x^2 - x - 1 \end{aligned}$$

There are no variations in sign in $f(-x)$, so there are 0 negative real zeros.

80. $g(x) = 5x^6 - 3x^3 + x^2 - x$

The number of variations in sign in $g(x)$ is 3. Then the number of positive real zeros is either 3 or less than 3 by 2, 4, 6, and so on. Thus, the number of positive real zeros is 3 or 1.

$$\begin{aligned} g(-x) &= 5(-x)^6 - 3(-x)^3 + (-x)^2 - (-x) \\ &= 5x^6 + 3x^3 + x^2 + x \end{aligned}$$

There are no variations in sign in $g(-x)$, so there are 0 negative real zeros.

81. $h(x) = 6x^7 + 2x^2 + 5x + 4$

There are no variations in sign in $h(x)$, so there are 0 positive real zeros.

$$h(-x) = 6(-x)^7 + 2(-x)^2 + 5(-x) + 4$$
$$= -6x^7 + 2x^2 - 5x + 4$$

The number of variations in sign in $h(-x)$ is 3. Thus, there are 3 or 1 negative real zeros.

82. $P(x) = -3x^5 - 7x^3 - 4x - 5$

There are no variations in sign in $P(x)$, so there are 0 positive real zeros.

$$P(-x) = -3(-x)^5 - 7(-x)^3 - 4(-x) - 5$$
$$= 3x^5 + 7x^3 + 4x - 5$$

There is 1 variation in sign in $P(-x)$, so there is 1 negative real zero.

83. $F(p) = 3p^{18} + 2p^4 - 5p^2 + p + 3$

There are 2 variations in sign in $F(p)$, so there are 2 or 0 positive real zeros.

$$F(-p) = 3(-p)^{18} + 2(-p)^4 - 5(-p)^2 + (-p) + 3$$
$$= 3p^{18} + 2p^4 - 5p^2 - p + 3$$

There are 2 variations in sign in $F(-p)$, so there are 2 or 0 negative real zeros.

84. $H(t) = 5t^{12} - 7t^4 + 3t^2 + t + 1$

There are 2 variations in sign in $H(t)$, so there are 2 or 0 positive real zeros.

$$H(-t) = 5(-t)^{12} - 7(-t)^4 + 3(-t)^2 + (-t) + 1$$
$$= 5t^{12} - 7t^4 + 3t^2 - t + 1$$

There are 4 variations in sign in $H(-t)$, so there are 4, 2, or 0 negative real zeros.

85. $C(x) = 7x^6 + 3x^4 - x - 10$

There is 1 variation in sign in $C(x)$, so there is 1 positive real zero.

$$C(-x) = 7(-x)^6 + 3(-x)^4 - (-x) - 10$$
$$= 7x^6 + 3x^4 + x - 10$$

There is 1 variation in sign in $C(-x)$, so there is 1 negative real zero.

86. $g(z) = -z^{10} + 8z^7 + z^3 + 6z - 1$

There are 2 variations in sign in $g(z)$, so there are 2 or 0 positive real zeros.

$$g(-z) = -(-z)^{10} + 8(-z)^7 + (-z)^3 + 6(-z) - 1$$
$$= -z^{10} - 8z^7 - z^3 - 6z - 1$$

There are no variations in sign in $g(-z)$, so there are 0 negative real zeros.

87. $h(t) = -4t^5 - t^3 + 2t^2 + 1$

There is 1 variation in sign in $h(t)$, so there is 1 positive real zero.

$$h(-t) = -4(-t)^5 - (-t)^3 + 2(-t)^2 + 1$$
$$= 4t^5 + t^3 + 2t^2 + 1$$

There are no variations in sign in $h(-t)$, so there are 0 negative real zeros.

88. $P(x) = x^6 + 2x^4 - 9x^3 - 4$

There is 1 variation in sign in $P(x)$, so there is 1 positive real zero.

$$P(-x) = (-x)^6 + 2(-x)^4 - 9(-x)^3 - 4$$
$$= x^6 + +2x^4 + 9x^3 - 4$$

There is 1 variation in sign in $P(-x)$, so there is 1 negative real zero.

89. $f(y) = y^4 + 13y^3 - y + 5$

There are 2 variations in sign in $f(y)$, so there are 2 or 0 positive real zeros.

$$f(-y) = (-y)^4 + 13(-y)^3 - (-y) + 5$$
$$= y^4 - 13y^3 + y + 5$$

There are 2 variations in sign in $f(-y)$, so there are 2 or 0 negative real zeros.

90. $Q(x) = x^4 - 2x^2 + 12x - 8$

There are 3 variations in sign in $Q(x)$, so there are 3 or 1 positive real zeros.

$$Q(-x) = (-x)^4 - 2(-x)^2 + 12(-x) - 8$$
$$= x^4 - 2x^2 - 12x - 8$$

There is 1 variation in sign in $Q(-x)$, so there is 1 negative real zero.

91. $r(x) = x^4 - 6x^2 + 20x - 24$

There are 3 variations in sign in $r(x)$, so there are 3 or 1 positive real zeros.

$$r(-x) = (-x)^4 - 6(-x)^2 + 20(-x) - 24$$
$$= x^4 - 6x^2 - 20x - 24$$

There is 1 variation in sign in $r(-x)$, so there is 1 negative real zero.

92. $f(x) = x^5 - 2x^3 - 8x$

There is 1 variation in sign in $f(x)$, so there is 1 positive real zero.

$$f(-x) = (-x)^5 - 2(-x)^3 - 8(-x)$$
$$= -x^5 + 2x^3 + 8x$$

There is 1 variation in sign in $f(-x)$, so there is 1 negative real zero.

93. $R(x) = 3x^5 - 5x^3 - 4x$

There is 1 variation in sign in $R(x)$, so there is 1 positive real zero.

$$R(-x) = 3(-x)^5 - 5(-x)^3 - 4(-x)$$
$$= -3x^5 + 5x^3 + 4x$$

There is 1 variation in sign in $R(-x)$, so there is 1 negative real zero.

94. $f(x) = x^4 - 9x^2 - 6x + 4$

There are 2 variations in sign in $f(x)$, so there are 2 or 0 positive real zeros.

$$f(-x) = (-x)^4 - 9(-x)^2 - 6(-x) + 4$$
$$= x^4 - 9x^2 + 6x + 4$$

There are 2 variations in sign in $f(-x)$, so there are 2 or 0 negative real zeros.

95. $f(x) = 4x^3 + x^2 - 8x - 2$

1. The leasing term is $4x^3$. The degree, 3, is odd and the leading coefficient, 4, is positive so as $x \to \infty$, $f(x) \to \infty$ and $x \to -\infty$, $f(x) \to -\infty$.

2. We find the rational zeros p/q of $f(x)$.

   $\frac{\text{Possibilities for } p}{\text{Possibilities for } q} : \frac{\pm 1, \pm 2}{\pm 1, \pm 2, \pm 4}$

   Possibilities for p/q: $1, -1, 2, -2, \frac{1}{2}, -\frac{1}{2}, \frac{1}{4}, -\frac{1}{4}$

   Synthetic division shows that $-\frac{1}{4}$ is a zero.

   $$\begin{array}{r|rrrr} -\frac{1}{4} & 4 & 1 & -8 & -2 \\ & & -1 & 0 & 2 \\ \hline & 4 & 0 & -8 & 0 \end{array}$$

   We have $f(x) = \left(x + \frac{1}{4}\right)(4x^2 - 8) = 4\left(x + \frac{1}{4}\right)(x^2 - 2)$. Solving $x^2 - 2 = 0$ we get $x = \pm\sqrt{2}$. Thus the zeros of the function are $-\frac{1}{4}$, $-\sqrt{2}$, and $\sqrt{2}$ so the x-intercepts of the graph are $\left(-\frac{1}{4}, 0\right)$, $(-\sqrt{2}, 0)$, and $(\sqrt{2}, 0)$.

3. The zeros divide the x-axis into 4 intervals, $(-\infty, -\sqrt{2})$, $\left(-\sqrt{2}, -\frac{1}{4}\right)$, $\left(-\frac{1}{4}, \sqrt{2}\right)$, and $(\sqrt{2}, \infty)$. We choose a value for x from each interval and find $f(x)$. This tells us the sign of $f(x)$ for all values of x in that interval.

   In $(-\infty, -\sqrt{2})$, test -2:

   $f(-2) = 4(-2)^3 + (-2)^2 - 8(-2) - 2 = -14 < 0$

   In $\left(-\sqrt{2}, -\frac{1}{4}\right)$, test -1:

   $f(-1) = 4(-1)^3 + (-1)^2 - 8(-1) - 2 = 3 > 0$

   In $\left(-\frac{1}{4}, \sqrt{2}\right)$, test 0:

   $f(0) = 4 \cdot 0^3 + 0^2 - 8 \cdot 0 - 2 = -2 < 0$

   In $(\sqrt{2}, \infty)$, test 2:

   $f(2) = 4 \cdot 2^3 + 2^2 - 8 \cdot 2 - 2 = 18 > 0$

   Thus the graph lies below the x-axis on $(-\infty, -\sqrt{2})$ and on $\left(-\frac{1}{4}, \sqrt{2}\right)$. It lies above the x-axis on $\left(-\sqrt{2}, -\frac{1}{4}\right)$ and on $(\sqrt{2}, \infty)$. We also know the points $(-2, -14)$, $(-1, 3)$, $(0, -2)$, and $(2, 18)$ are on the graph.

4. From Step 3 we see that $f(0) = -2$ so the y-intercept is $(0, -2)$.

5. We find additional points on the graph and then draw the graph.

x	$f(x)$
-1.5	-1.25
-0.5	1.75
1	-5
1.5	1.75

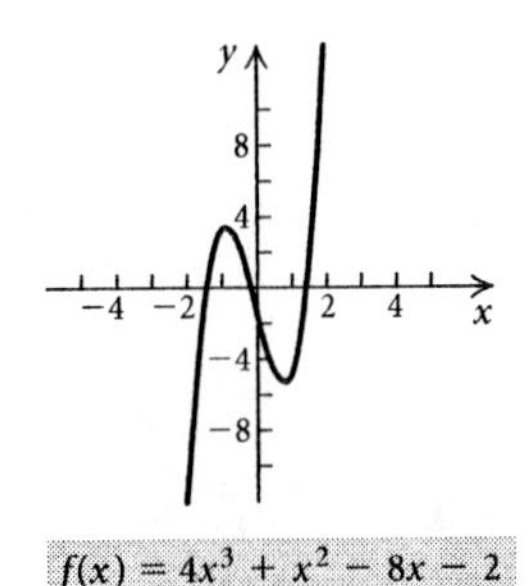

$f(x) = 4x^3 + x^2 - 8x - 2$

6. Checking the graph as described on page 254 in the text, we see that it appears to be correct.

96.

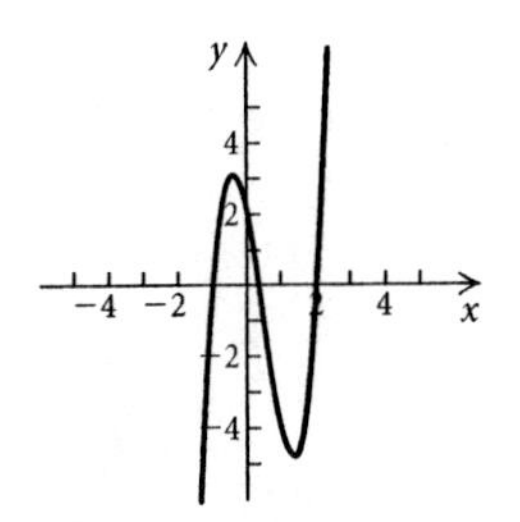

$f(x) = 3x^3 - 4x^2 - 5x + 2$

97. $f(x) = 2x^4 - 3x^3 - 2x^2 + 3x$

1. The leading term is $2x^4$. The degree, 4, is even and the leading coefficient, 2, is positive so as $x \to \infty$, $f(x) \to \infty$ and as $x \to -\infty$, $f(x) \to \infty$.

2. We find the rational zeros p/q of $f(x)$. First note that $f(x) = x(2x^3 - 3x^2 - 2x + 3)$, so x is a zero. Now consider $g(x) = 2x^3 - 3x^2 - 2x + 3$.

   $\frac{\text{Possibilities for } p}{\text{Possibilities for } q} : \frac{\pm 1, \pm 3}{\pm 1, \pm 2}$

   Possibilities for p/q: $1, -1, 3, -3, \frac{1}{2}, -\frac{1}{2}, \frac{3}{2}, -\frac{3}{2}$

   We try 1.

   $$\begin{array}{r|rrrr} 1 & 2 & -3 & -2 & 3 \\ & & 2 & -1 & -3 \\ \hline & 2 & -1 & -3 & 0 \end{array}$$

   Then $f(x) = x(x-1)(2x^2 - x - 3)$. Using the principle of zero products to solve $2x^2 - x - 3 = 0$, we get $x = \frac{3}{2}$ or $x = -1$.

   Thus the zeros of the function are 0, 1, $\frac{3}{2}$, and -1 so the x-intercepts of the graph are $(0, 0)$, $(1, 0)$, $\left(\frac{3}{2}, 0\right)$, and $(-1, 0)$.

3. The zeros divide the x-axis into 5 intervals, $(-\infty, -1)$, $(-1, 0)$, $(0, 1)$, $\left(1, \frac{3}{2}\right)$, and $\left(\frac{3}{2}, \infty\right)$. We choose a value for x from each interval and find $f(x)$. This tells us the sign of $f(x)$ for all values of x in that interval.

In $(-\infty,-1)$, test -2:
$f(-2)=2(-2)^4-3(-2)^3-2(-2)^2+3(2)=42>0$
In $(-1,0)$, test -0.5:
$f(-0.5)=2(-0.5)^4-3(-0.5)^3-2(-0.5)^2+3(-0.5)=$
$-1.5<0$
In $(0,1)$, test 0.5:
$f(0.5)=2(0.5)^4-3(0.5)^3-2(0.5)^2+3(0.5)=0.75>0$
In $\left(1,\frac{3}{2}\right)$, test 1.25:
$f(1.25)=2(1.25)^4-3(1.25)^3-2(1.25)^2+3(1.25)=$
$-0.3515625<0$
In $\left(\frac{3}{2},\infty\right)$, test 2:
$f(2)=2\cdot 2^4-3\cdot 2^3-2\cdot 2^2+3\cdot 2=6>0$
Thus the graph lies above the x-axis on $(-\infty,-1)$, on $(0,1)$, and on $\left(\frac{3}{2},\infty\right)$. It lies below the x-axis on $(-1,0)$ and on $\left(1,\frac{3}{2}\right)$. We also know the points $(-2,42)$, $(-0.5,-1.5)$, $(0.5,0.75)$, $(1.25,-0.3515625)$, and $(2,6)$ are on the graph.

4. From Step 2 we know that $f(0)=0$ so the y-intercept is $(0,0)$.
5. We find additional points on the graph and then draw the graph.

x	$f(x)$
-1.5	11.25
2.5	26.25
3	72

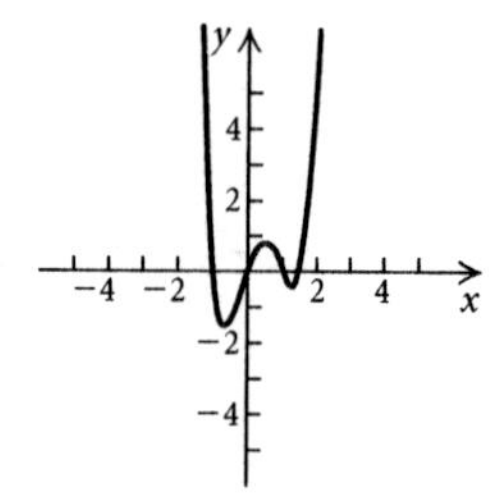

$f(x)=2x^4-3x^3-2x^2+3x$

6. Checking the graph as described on page 254 in the text, we see that it appears to be correct.

98.

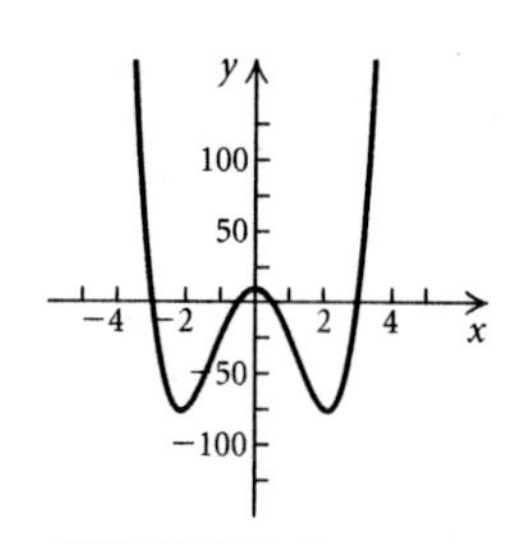

$f(x)=4x^4-37x^2+9$

99. Left to the student

100. Yes; let c be a zero of $P(x)$. Then $P(c)=0$, so $-P(c)=Q(c)=0$. Thus, every zero of $P(x)$ is a zero of $Q(x)$. Now let r be a zero of $Q(x)$. Then $Q(r)=0$, so $-P(r)=0$ and $P(r)=0$. Every zero of $Q(x)$ is also a zero of $P(x)$. Thus, $P(x)$ and $Q(x)$ have the same zeros.

101. No; since imaginary zeros of polynomials with rational coefficients occur in conjugate pairs, a third-degree polynomial with rational coefficients can have at most two imaginary zeros. Thus, there must be at least one real zero.

102. $f(x)=3x^2-6x-1$

a) $-\frac{b}{2a}=-\frac{-6}{2\cdot 3}=1$
$f(1)=3\cdot 1^2-6\cdot 1-1=-4$
The vertex is $(1,-4)$.

b) $x=1$

c) Minimum: -4 at $x=1$

103. $f(x)=x^2-8x+10$

a) $-\frac{b}{2a}=-\frac{-8}{2\cdot 1}=-(-4)=4$
$f(4)=4^2-8\cdot 4+10=-6$
The vertex is $(4,-6)$.

b) The axis of symmetry is $x=4$.

c) Since the coefficient of x^2 is positive, there is a minimum function value. It is the second coordinate of the vertex, -6. It occurs when $x=4$.

104. $x^2-8x-33=0$
$(x-11)(x+3)=0$
$x=11$ *or* $x=-3$
The zeros are -3 and 11.

105. $-\frac{4}{5}x+8=0$
$-\frac{4}{5}x=-8$ Subtracting 8
$-\frac{5}{4}\left(-\frac{4}{5}x\right)=-\frac{5}{4}(-8)$ Multiplying by $-\frac{5}{4}$
$x=10$
The zero is 10.

106. $f(x)=-x^2-3x+6$
The degree is 2, so the function is quadratic.
Leading term: $-x^2$; leading coefficient: -1
Since the degree is even and the leading coefficient is negative, as $x\to\infty$, $f(x)\to-\infty$, and as $x\to-\infty$, $f(x)\to-\infty$.

107. $g(x)=-x^3-2x^2$
The degree is 3, so the function is cubic.
Leading term: $-x^3$; leading coefficient: -1
Since the degree is odd and the leading coefficient is negative, as $x\to\infty$, $g(x)\to-\infty$ and as $x\to-\infty$, $g(x)\to\infty$.

108. $h(x)=x-2$
The degree is 1, so the function is linear.
Leading term: x; leading coefficient: 1
Since the degree is odd and the leading coefficient is positive, as $x\to\infty$, $h(x)\to\infty$ and as $x\to-\infty$, $h(x)\to-\infty$.

109. $f(x) = -\dfrac{4}{9}$

The degree is 0, so this is a constant function.

Leading term: $-\dfrac{4}{9}$; leading coefficient: $-\dfrac{4}{9}$;

for all x, $f(x) = -\dfrac{4}{9}$

110. $h(x) = x^3 + \dfrac{1}{2}x^2 - 4x - 3$

The degree is 3, so the function is cubic.

Leading term: x^3; leading coefficient: 1

Since the degree is odd and the leading coefficient is positive, as $x \to \infty$, $h(x) \to \infty$ and as $x \to -\infty$, $h(x) \to -\infty$.

111. $g(x) = x^4 - 2x^3 + x^2 - x + 2$

The degree is 4, so the function is quartic.

Leading term: x^4; leading coefficient: 1

Since the degree is even and the leading coefficient is positive, as $x \to \infty$, $g(x) \to \infty$ and as $x \to -\infty$, $g(x) \to \infty$.

112.
$$\begin{array}{r|rrrr} -i & 1 & 3i & -4i & -2 \\ & & -i & 2 & -4-2i \\ \hline & 1 & 2i & 2-4i & -6-2i \end{array}$$

$Q(x) = x^2 + 2ix + (2 - 4i)$, $R(x) = -6 - 2i$

113.
$$\begin{aligned} &(x^4 - y^4) \div (x - y) \\ &= (x^4 + 0x^3 + 0x^2 + 0x - y^4) \div (x - y) \end{aligned}$$

$$\begin{array}{r|rrrrr} y & 1 & 0 & 0 & 0 & -y^4 \\ & & y & y^2 & y^3 & y^4 \\ \hline & 1 & y & y^2 & y^3 & 0 \end{array}$$

The quotient is $x^3 + x^2y + xy^2 + y^3$. The remainder is 0.

114. By the rational zeros theorem, only ± 1, ± 2, ± 3, ± 4, ± 6, and ± 12 can be rational solutions of $x^4 - 12 = 0$. Since none of them is a solution, the equation has no rational solutions. But $\sqrt[4]{12}$ is a solution of the equation, so $\sqrt[4]{12}$ must be irrational.

115. $f(x) = 2x^3 - 5x^2 - 4x + 3$

a) $2x^3 - 5x^2 - 4x + 3 = 0$

$\dfrac{\text{Possibilities for } p}{\text{Possibilities for } q} : \dfrac{\pm 1, \pm 3}{\pm 1, \pm 2}$

Possibilities for p/q: $1, -1, 3, -3, \dfrac{1}{2}, -\dfrac{1}{2}, \dfrac{3}{2}, -\dfrac{3}{2}$

The first possibility that is a solution of $f(x) = 0$ is -1:

$$\begin{array}{r|rrrr} -1 & 2 & -5 & -4 & 3 \\ & & -2 & 7 & -3 \\ \hline & 2 & -7 & 3 & 0 \end{array}$$

Thus, -1 is a solution.

Then we have:

$(x + 1)(2x^2 - 7x + 3) = 0$

$(x + 1)(2x - 1)(x - 3) = 0$

The other solutions are $\dfrac{1}{2}$ and 3.

b) The graph of $y = f(x - 1)$ is the graph of $y = f(x)$ shifted 1 unit right. Thus, we add 1 to each solution of $f(x) = 0$ to find the solutions of $f(x - 1) = 0$. The solutions are $-1 + 1$, or 0; $\dfrac{1}{2} + 1$, or $\dfrac{3}{2}$; and $3 + 1$, or 4.

c) The graph of $y = f(x + 2)$ is the graph of $y = f(x)$ shifted 2 units left. Thus, we subtract 2 from each solution of $f(x) = 0$ to find the solutions of $f(x + 2) = 0$. The solutions are $-1 - 2$, or -3; $\dfrac{1}{2} - 2$, or $-\dfrac{3}{2}$; and $3 - 2$, or 1.

d) The graph of $y = f(2x)$ is a horizontal shrinking of the graph of $y = f(x)$ by a factor of 2. We divide each solution of $f(x) = 0$ by 2 to find the solutions of $f(2x) = 0$. The solutions are $\dfrac{-1}{2}$ or $-\dfrac{1}{2}$; $\dfrac{1/2}{2}$, or $\dfrac{1}{4}$; and $\dfrac{3}{2}$.

116. $P(x) = x^6 - x^5 - 72x^4 - 81x^2 + 486x + 5832$

a) $x^6 - 6x^5 - 72x^4 - 81x^2 + 486x + 5832 = 0$

Synthetic division shows that we can factor as follows:

$$\begin{aligned} P(x) &= (x-3)(x+3)(x+6)(x^3-12x^2+9x-108) \\ &= (x-3)(x+3)(x+6)[x^2(x-12) + 9(x-12)] \\ &= (x-3)(x+3)(x+6)(x-12)(x^2+9) \end{aligned}$$

The rational zeros are 3, -3, -6, and 12.

117. $P(x) = 2x^5 - 33x^4 - 84x^3 + 2203x^2 - 3348x - 10{,}080$

a) $2x^5 - 33x^4 - 84x^3 + 2203x^2 - 3348x - 10{,}080 = 0$

Trying some of the many possibilities for p/q, we find that 4 is a zero.

$$\begin{array}{r|rrrrrr} 4 & 2 & -33 & -84 & 2203 & -3348 & -10{,}080 \\ & & 8 & -100 & -736 & 5868 & 10{,}080 \\ \hline & 2 & -25 & -184 & 1467 & 2520 & 0 \end{array}$$

Then we have:

$(x - 4)(2x^4 - 25x^3 - 184x^2 + 1467x + 2520) = 0$

We now use the fourth degree polynomial above to find another zero. Synthetic division shows that 4 is not a double zero, but 7 is a zero.

$$\begin{array}{r|rrrrr} 7 & 2 & -25 & -184 & 1467 & 2520 \\ & & 14 & -77 & -1827 & -2520 \\ \hline & 2 & -11 & -261 & -360 & 0 \end{array}$$

Now we have:

$(x - 4)(x - 7)(2x^3 - 11x^2 - 261x - 360) = 0$

Use the third degree polynomial above to find a third zero. Synthetic division shows that 7 is not a double zero, but 15 is a zero.

$$\begin{array}{r|rrrr} 15 & 2 & -11 & -261 & -360 \\ & & 30 & 285 & 360 \\ \hline & 2 & 19 & 24 & 0 \end{array}$$

We have:

$$P(x) = (x-4)(x-7)(x-15)(2x^2+19x+24)$$
$$= (x-4)(x-7)(x-15)(2x+3)(x+8)$$

The rational zeros are 4, 7, 15, $-\frac{3}{2}$, and -8.

Exercise Set 3.4

1. Graph (d) is the graph of $f(x) = \dfrac{8}{x^2-4}$.

   $x^2-4=0$ when $x=\pm 2$, so $x=-2$ and $x=2$ are vertical asymptotes.

   The x-axis, $y=0$, is the horizontal asymptote because the degree of the numerator is less than the degree of the denominator.

   There is no oblique asymptote.

2. Graph (f) is the graph of $f(x) = \dfrac{8}{x^2+4}$.

   $x^2+4=0$ has no real solutions, so there is no vertical asymptote.

   The x-axis, $y=0$, is the horizontal asymptote because the degree of the numerator is less than the degree of the denominator.

   There is no oblique asymptote.

3. Graph (e) is the graph of $f(x) = \dfrac{8x}{x^2-4}$.

   As in Exercise 1, $x=-2$ and $x=2$ are vertical asymptotes.

   The x-axis, $y=0$, is the horizontal asymptote because the degree of the numerator is less than the degree of the denominator.

   There is no oblique asymptote.

4. Graph (a) is the graph of $f(x) = \dfrac{8x^2}{x^2-4}$.

   As in Exercise 1, $x=2$ and $x=-2$ are vertical asymptotes.

   The numerator and denominator have the same degree, so $y=8/1$, or $y=8$, is the horizontal asymptote.

   There is no oblique asymptote.

5. Graph (c) is the graph of $f(x) = \dfrac{8x^3}{x^2-4}$.

   As in Exercise 1, $x=-2$ and $x=2$ are vertical asymptotes.

   The degree of the numerator is greater than the degree of the denominator, so there is no horizontal asymptote but there is a vertical asymptote. To find it we first divide to find an equivalent expression.

$$\begin{array}{r} 8x \\ x^2-4\overline{)\,8x^3} \\ \underline{8x^3-32x} \\ 32x \end{array}$$

$$\frac{8x^3}{x^2-4} = 8x + \frac{32x}{x^2-4}$$

   Now we multiply by 1, using $(1/x^2)/(1/x^2)$.

$$\frac{32x}{x^2-4}\cdot\frac{\frac{1}{x^2}}{\frac{1}{x^2}} = \frac{\frac{32}{x}}{1-\frac{4}{x^2}}$$

   As $|x|$ becomes very large, each expression with x in the denominator tends toward zero.

   Then, as $|x|\to\infty$, we have

$$\frac{\frac{32}{x}}{1-\frac{4}{x^2}} \to \frac{0}{1-0},\text{ or } 0.$$

   Thus, as $|x|$ becomes very large, the graph of $f(x)$ gets very close to the graph of $y=8x$, so $y=8x$ is the oblique asymptote.

6. Graph (b) is the graph of $f(x) = \dfrac{8x^3}{x^2+4}$.

   As in Exercise 2, there is no vertical asymptote.

   The degree of the numerator is greater than the degree of the denominator, so there is no horizontal asymptote but there is a vertical asymptote. To find it we first divide to find an equivalent expression.

$$\frac{8x^3}{x^2+4} = 8x - \frac{32x}{x^2+4}$$

   Now $\dfrac{32x}{x^2+4} = \dfrac{\frac{32}{x}}{1+\frac{4}{x^2}}$ and, as $|x|\to\infty$,

$$\frac{\frac{32}{x}}{1+\frac{4}{x^2}} \to \frac{0}{1+0},\text{ or } 0.$$

   Thus, as $y=8x$ is the oblique asymptote.

7. $g(x) = \dfrac{1}{x^2}$

   The zero of the denominator is 0, so the vertical asymptote is $x=0$.

8. $f(x) = \dfrac{4}{x+10}$

   $x+10=0$ when $x=-10$, so the vertical asymptote is $x=-10$.

9. $h(x) = \dfrac{x+7}{2-x}$

   $2-x=0$ when $x=2$, so the vertical asymptote is $x=2$.

10. $g(x) = \dfrac{x^4+2}{x}$

    The zero of the denominator is 0, so the vertical asymptote is $x=0$.

11. $f(x) = \dfrac{3-x}{(x-4)(x+6)}$

    The zeros of the denominator are 4 and -6, so the vertical asymptotes are $x=4$ and $x=-6$.

12. $h(x) = \dfrac{x^2+4}{x(x+5)(x-2)}$

    The zeros of the denominator are 0, -5, and 2, so the vertical asymptotes are $x=0$, $x=-5$, and $x=2$.

13. $g(x) = \dfrac{x^2}{2x^2 - x - 3} = \dfrac{x^2}{(2x-3)(x+1)}$

The zeros of the denominator are $\dfrac{3}{2}$ and -1, so the vertical asymptotes are $x = \dfrac{3}{2}$ and $x = -1$.

14. $f(x) = \dfrac{x+5}{x^2+4x-32} = \dfrac{x+5}{(x+8)(x-4)}$

The zeros of the denominator are -8 and 4, so the vertical asymptotes are $x = -8$ and $x = 4$.

15. $f(x) = \dfrac{3x^2+5}{4x^2-3}$

The numerator and denominator have the same degree and the ratio of the leading coefficients is $\dfrac{3}{4}$, so $y = \dfrac{3}{4}$ is the horizontal asymptote.

16. $g(x) = \dfrac{x+6}{x^3+2x^2}$

The degree of the numerator is less than the degree of the denominator, so $y = 0$ is the horizontal asymptote.

17. $h(x) = \dfrac{x^2-4}{2x^4+3}$

The degree of the numerator is less than the degree of the denominator, so $y = 0$ is the horizontal asymptote.

18. $f(x) = \dfrac{x^5}{x^5+x}$

The numerator and denominator have the same degree and the ratio of the leading coefficients is 1, so $y = 1$ is the horizontal asymptote.

19. $g(x) = \dfrac{x^3-2x^2+x-1}{x^2-16}$

The degree of the numerator is greater than the degree of the denominator, so there is no horizontal asymptote.

20. $h(x) = \dfrac{8x^4+x-2}{2x^4-10}$

The numerator and denominator have the same degree and the ratio of the leading coefficients is 4, so $y = 4$ is the horizontal asymptote.

21. $g(x) = \dfrac{x^2+4x-1}{x+3}$

$$\begin{array}{r} x+1 \\ x+3\,\overline{)\,x^2+4x-1} \\ \underline{x^2+3x} \\ x-1 \\ \underline{x+3} \\ -4 \end{array}$$

Then $g(x) = x + 1 + \dfrac{-4}{x+3}$. The oblique asymptote is $y = x + 1$.

22. $f(x) = \dfrac{x^2-6x}{x-5}$

$$\begin{array}{r} x-1 \\ x-5\,\overline{)\,x^2-6x+0} \\ \underline{x^2-5x} \\ -x+0 \\ \underline{-x+5} \\ -5 \end{array}$$

Then $f(x) = x - 1 + \dfrac{-5}{x-5}$. The oblique asymptote is $y = x - 1$.

23. $h(x) = \dfrac{x^4-2}{x^3+1}$

$$\begin{array}{r} x \\ x^3+1\,\overline{)\,x^4+0x^3+0x^2+0x} \\ \underline{x^4 \qquad\qquad\quad +x} \\ -x \end{array}$$

Then $h(x) = x + \dfrac{-x}{x^3+1}$. The oblique asymptote is $y = x$.

24. $g(x) = \dfrac{12x^3-x}{6x^2+4}$

$$\begin{array}{r} 2x \\ 6x^2+4\,\overline{)\,12x^3-x} \\ \underline{12x^3+8x} \\ -9x \end{array}$$

Then $g(x) = 2x + \dfrac{-9x}{6x^2+4}$. The oblique asymptote is $y = 2x$.

25. $f(x) = \dfrac{x^3-x^2+x-4}{x^2+2x-1}$

$$\begin{array}{r} x-3 \\ x^2+2x-1\,\overline{)\,x^3-x^2+x-4} \\ \underline{x^3+2x^2-x} \\ -3x^2+2x-4 \\ \underline{-3x^2-6x+3} \\ 8x-7 \end{array}$$

Then $f(x) = x - 3 + \dfrac{8x-7}{x^2+2x-1}$. The oblique asymptote is $y = x - 3$.

26. $h(x) = \dfrac{5x^3-x^2+x-1}{x^2-x+2}$

$$\begin{array}{r} 5x+4 \\ x^2-x+2\,\overline{)\,5x^3-x^2+x-1} \\ \underline{5x^3-5x^2+10x} \\ 4x^2-9x-1 \\ \underline{4x^2-4x+8} \\ -5x-9 \end{array}$$

Then $h(x) = 5x + 4 + \dfrac{-5x-9}{x^2-x+2}$. The oblique asymptote is $y = 5x + 4$.

27. $f(x) = \dfrac{1}{x}$

1. 0 is the zero of the denominator, so the domain excludes 0. It is $(-\infty, 0) \cup (0, \infty)$. The line $x = 0$, or the y-axis, is the vertical asymptote.

2. Because the degree of the numerator is less than the degree of the denominator, the x-axis, or $y = 0$, is the horizontal asymptote. There are no oblique asymptotes.

3. The numerator has no zeros, so there is no x-intercept.

4. Since 0 is not in the domain of the function, there is no y-intercept.

5. Find other function values to determine the shape of the graph and then draw it.

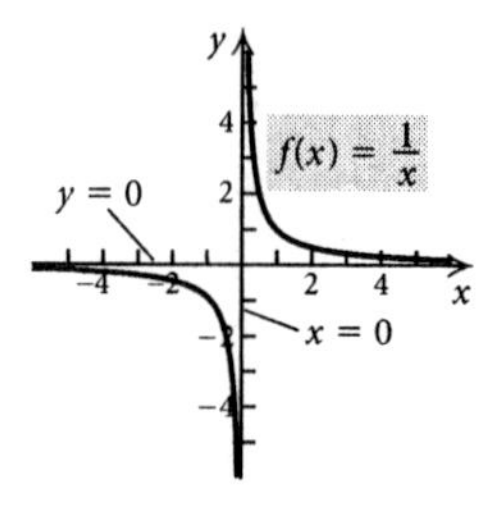

28. $g(x) = \frac{1}{x^2}$

1. 0 is the zero of the denominator, so the domain excludes 0. It is $(-\infty, 0) \cup (0, \infty)$. The line $x = 0$, or the y-axis, is the vertical asymptote.

2. Because the degree of the numerator is less than the degree of the denominator, the x-axis, or $y = 0$, is the horizontal asymptote. There is no oblique asymptote.

3. The numerator has no zeros, so there is no x-intercept.

4. Since 0 is not in the domain of the function, there is no y-intercept.

5. Find other function values to determine the shape of the graph and then draw it.

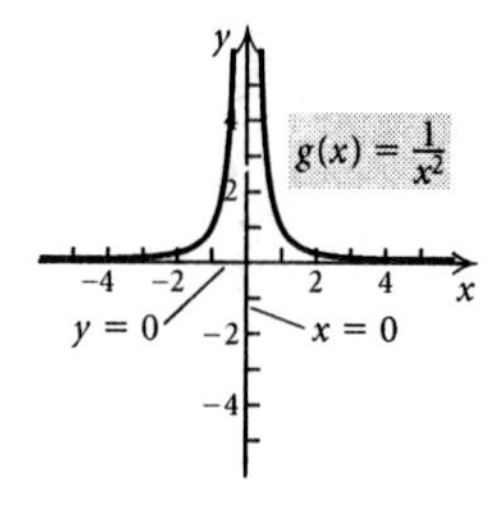

29. $h(x) = -\frac{4}{x^2}$

1. 0 is the zero of the denominator, so the domain excludes 0. It is $(-\infty, 0) \cup (0, \infty)$. The line $x = 0$, or the y-axis, is the vertical asymptote.

2. Because the degree of the numerator is less than the degree of the denominator, the x-axis, or $y = 0$, is the horizontal asymptote. There is no oblique asymptote.

3. The numerator has no zeros, so there is no x-intercept.

4. Since 0 is not in the domain of the function, there is no y-intercept.

5. Find other function values to determine the shape of the graph and then draw it.

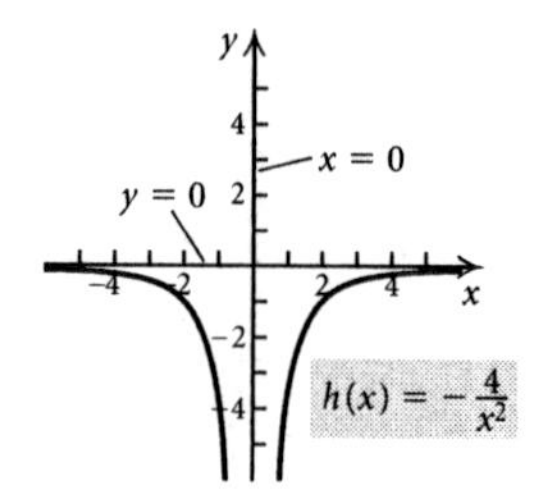

30. $f(x) = -\frac{6}{x}$

1. 0 is the zero of the denominator, so the domain excludes 0. It is $(-\infty, 0) \cup (0, \infty)$. The line $x = 0$, or the y-axis, is the vertical asymptote.

2. Because the degree of the numerator is less than the degree of the denominator, the x-axis, or $y = 0$, is the horizontal asymptote. There is no oblique asymptote.

3. The numerator has no zeros, so there is no x-intercept.

4. Since 0 is not in the domain of the function, there is no y-intercept.

5. Find other function values to determine the shape of the graph and then draw it.

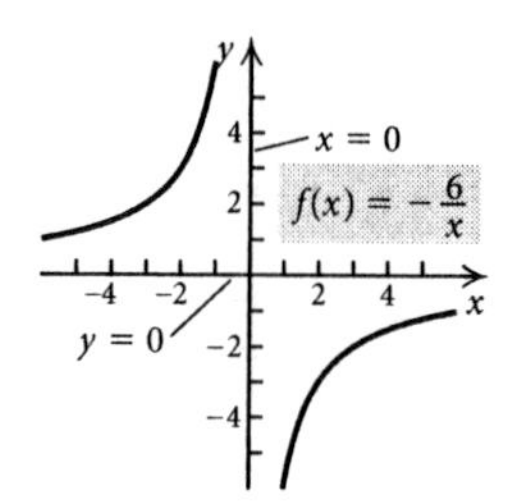

31. $g(x) = \frac{x^2 - 4x + 3}{x + 1}$

1. The denominator, $x + 1$, is 0 when $x = -1$, so the domain excludes -1. It is $(-\infty, -1) \cup (-1, \infty)$. The line $x = -1$ is the vertical asymptote.

2. The degree of the numerator is 1 greater than the degree of the denominator, so we divide to find the oblique asymptote.

$$\begin{array}{r} x - 5 \\ x + 1 \overline{) x^2 - 4x + 3} \\ \underline{x^2 + x } \\ -5x + 3 \\ \underline{-5x - 5} \\ 8 \end{array}$$

The oblique asymptote is $y = x - 5$. There is no horizontal asymptote.

3. $x^2 - 4x + 3 = (x - 1)(x - 3)$, so the zeros of the numerator are 1 and 3. Thus the x-intercepts are $(1, 0)$ and $(3, 0)$.

4. $g(0) = \dfrac{0^2 - 4 \cdot 0 + 3}{0 + 1} = 3$, so the y-intercept is $(0, 3)$.
5. Find other function values to determine the shape of the graph and then draw it.

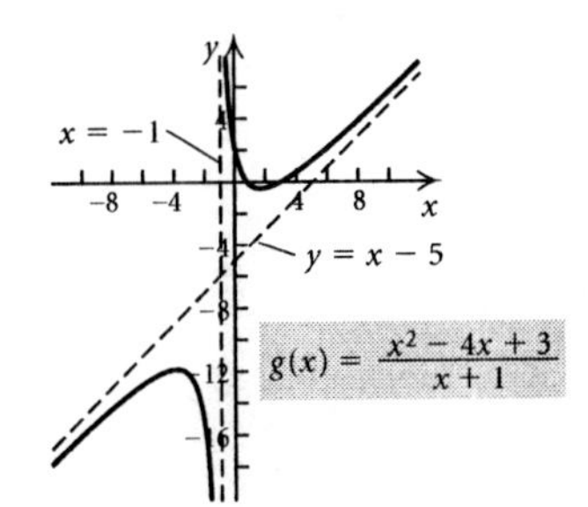

32. $h(x) = \dfrac{2x^2 - x - 3}{x - 1}$

1. The denominator is 0 when $x = 1$, so the domain excludes 1. It is $(-\infty, 1) \cup (1, \infty)$. The line $x = 1$ is the vertical asymptote.
2. The degree of the numerator is 1 greater than the degree of the denominator, so we divide to find the oblique asymptote.

$$\begin{array}{r|l} & 2x + 1 \\ x - 1 & 2x^2 - x - 3 \\ & \underline{2x^2 - 2x} \\ & \quad\; x - 3 \\ & \quad\; \underline{x - 1} \\ & \quad\;\; -2 \end{array}$$

The oblique asymptote is $y = 2x + 1$. There is no horizontal asymptote.

3. $2x^2 - x - 3 = (2x - 3)(x + 1)$, so the zeros of the numerator are $\frac{3}{2}$ and -1. Thus the x-intercepts are $\left(\frac{3}{2}, 0\right)$ and $(-1, 0)$.
4. $h(0) = \dfrac{2 \cdot 0^2 - 0 - 3}{0 - 1} = 3$, so the y-intercept is $(0, 3)$.
5. Find other function values to determine the shape of the graph and then draw it.

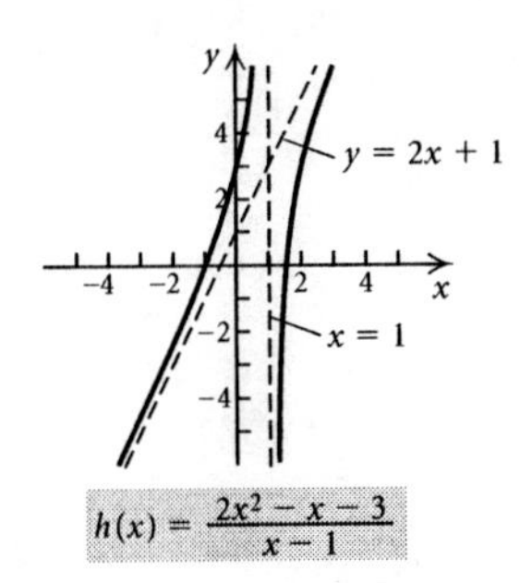

33. $f(x) = \dfrac{1}{x + 3}$

1. -3 is the zero of the denominator, so the domain excludes -3. It is $(-\infty, -3) \cup (-3, \infty)$. The line $x = -3$ is the vertical asymptote.
2. Because the degree of the numerator is less than the degree of the denominator, the x-axis, or $y = 0$, is the horizontal asymptote. There is no oblique asymptote.
3. The numerator has no zeros, so there is no x-intercept.
4. $f(0) = \dfrac{1}{0 + 3} = \dfrac{1}{3}$, so $\left(0, \dfrac{1}{3}\right)$ is the y-intercept.
5. Find other function values to determine the shape of the graph and then draw it.

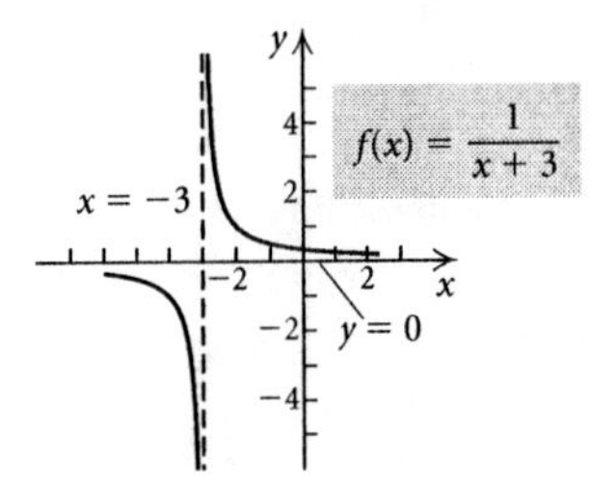

34. $f(x) = \dfrac{1}{x - 5}$

1. 5 is the zero of the denominator, so the domain is $(-\infty, 5) \cup (5, \infty)$ and $x = 5$ is the vertical asymptote.
2. Because the degree of the numerator is less than the degree of the denominator, the x-axis, or $y = 0$, is the horizontal asymptote. There is no oblique asymptote.
3. The numerator has no zeros, so there is no x-intercept.
4. $f(0) = \dfrac{1}{0 - 5} = -\dfrac{1}{5}$, so $\left(0, -\dfrac{1}{5}\right)$ is the y-intercept.
5. Find other function values to determine the shape of the graph and then draw it.

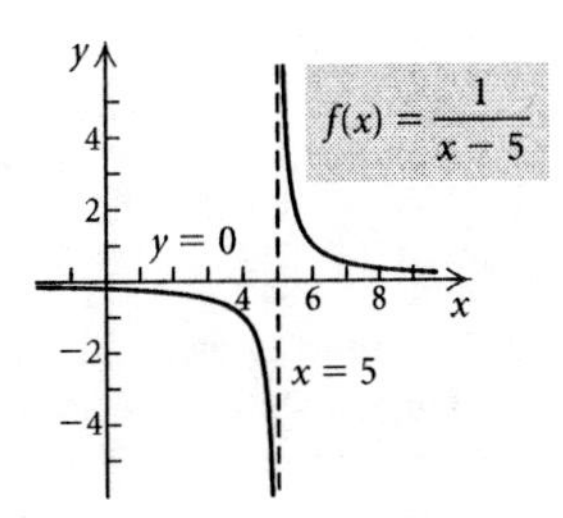

35. $f(x) = \dfrac{-2}{x - 5}$

1. 5 is the zero of the denominator, so the domain excludes 5. It is $(-\infty, 5) \cup (5, \infty)$. The line $x = 5$ is the vertical asymptote.
2. Because the degree of the numerator is less than the degree of the denominator, the x-axis, or $y = 0$, is the horizontal asymptote. There is no oblique asymptote.
3. The numerator has no zeros, so there is no x-intercept.
4. $f(0) = \dfrac{-2}{0 - 5} = \dfrac{2}{5}$, so $\left(0, \dfrac{2}{5}\right)$ is the y-intercept.

5. Find other function values to determine the shape of the graph and then draw it.

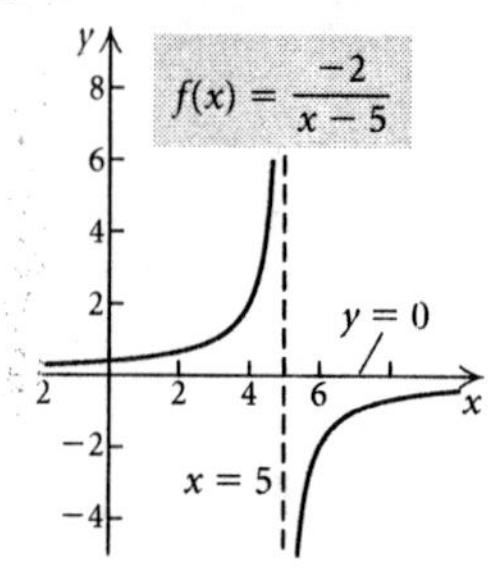

36. $f(x) = \dfrac{3}{3-x}$

1. 3 is the zero of the denominator, so the domain is $(-\infty, 3) \cup (3, \infty)$ and $x = 3$ is the vertical asymptote.
2. Because the degree of the numerator is less than the degree of the denominator, the x-axis, or $y = 0$, is the horizontal asymptote. There is no oblique asymptote.
3. The numerator has no zeros, so there is no x-intercept.
4. $f(0) = \dfrac{3}{3-0} = 1$, so $(0, 1)$ is the y-intercept.
5. Find other function values to determine the shape of the graph and then draw it.

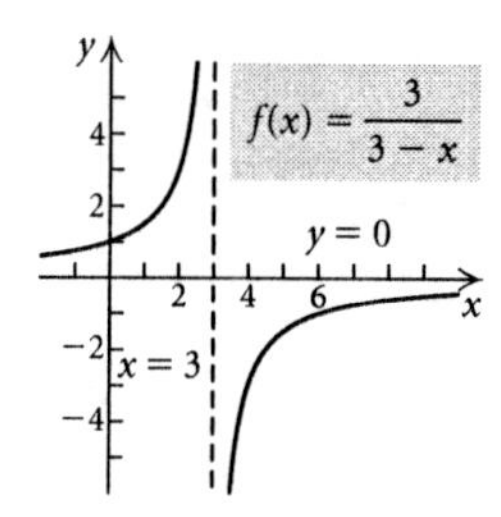

37. $f(x) = \dfrac{2x+1}{x}$

1. 0 is the zero of the denominator, so the domain excludes 0. It is $(-\infty, 0) \cup (0, \infty)$. The line $x = 0$, or the y-axis, is the vertical asymptote.
2. The numerator and denominator have the same degree, so the horizontal asymptote is determined by the ratio of the leading coefficients, $2/1$, or 2. Thus, $y = 2$ is the horizontal asymptote. There is no oblique asymptote.
3. The zero of the numerator is the solution of $2x+1 = 0$, or $-\dfrac{1}{2}$. The x-intercept is $\left(-\dfrac{1}{2}, 0\right)$.
4. Since 0 is not in the domain of the function, there is no y-intercept.
5. Find other function values to determine the shape of the graph and then draw it.

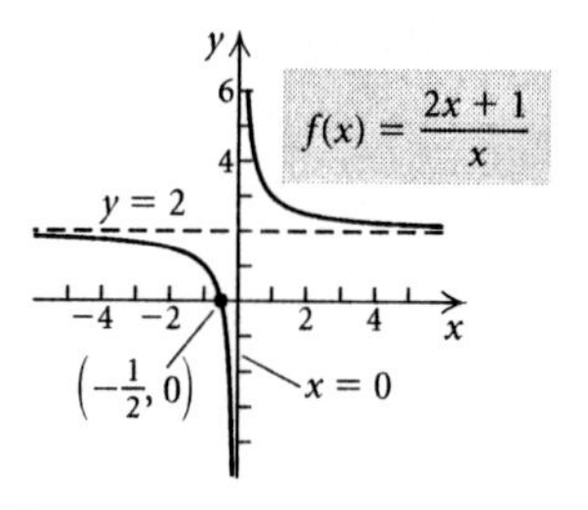

38. $f(x) = \dfrac{3x-1}{x}$

1. 0 is the zero of the denominator, so the domain is $(-\infty, 0) \cup (0, \infty)$. The line $x = 0$, or the y-axis, is the vertical asymptote.
2. The numerator and denominator have the same degree, so the horizontal asymptote is determined by the ratio of the leading coefficients, $3/1$, or 3. Thus, $y = 3$ is the horizontal asymptote. There is no oblique asymptote.
3. The zero of the numerator is $\dfrac{1}{3}$, so the x-intercept is $\left(\dfrac{1}{3}, 0\right)$.
4. Since 0 is not in the domain of the function, there is no y-intercept.
5. Find other function values to determine the shape of the graph and then draw it.

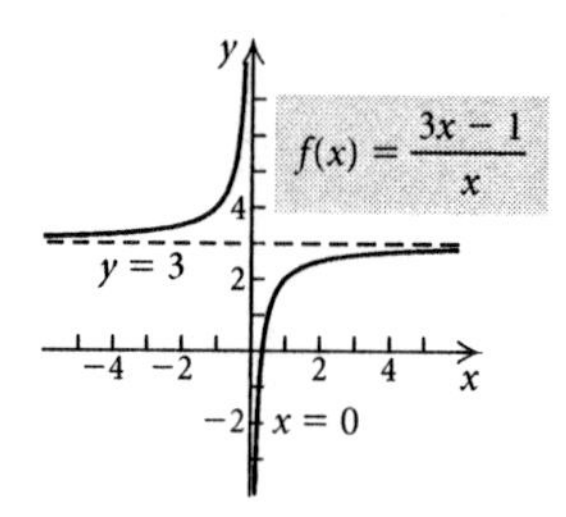

39. $f(x) = \dfrac{1}{(x-2)^2}$

1. 2 is the zero of the denominator, so the domain excludes 2. It is $(-\infty, 2) \cup (2, \infty)$. The line $x = 2$ is the vertical asymptote.
2. Because the degree of the numerator is less than the degree of the denominator, the x-axis, or $y = 0$, is the horizontal asymptote. There is no oblique asymptote.
3. The numerator has no zeros, so there is no x-intercept.
4. $f(0) = \dfrac{1}{(0-2)^2} = \dfrac{1}{4}$, so $\left(0, \dfrac{1}{4}\right)$ is the y-intercept.
5. Find other function values to determine the shape of the graph and then draw it.

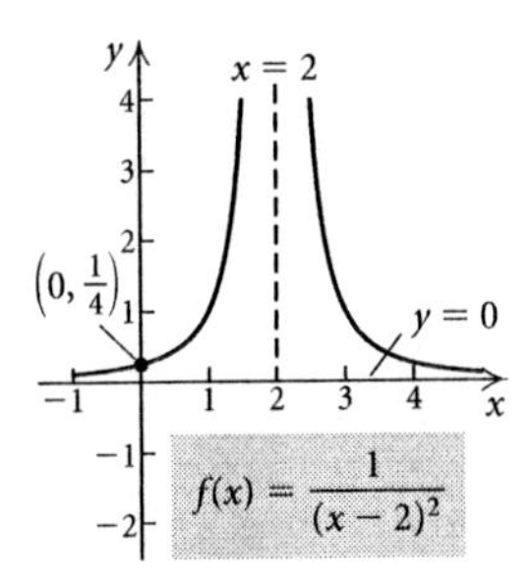

40. $f(x) = \dfrac{-2}{(x-3)^2}$

1. 3 is the zero of the denominator, so the domain is $(-\infty, 3) \cup (3, \infty)$ and $x = 3$ is the vertical asymptote.
2. Because the degree of the numerator is less than the degree of the denominator, the x-axis, or $y = 0$, is the horizontal asymptote. There is no oblique asymptote.
3. The numerator has no zeros, so there is no x-intercept.
4. $f(0) = \dfrac{-2}{(0-3)^2} = -\dfrac{2}{9}$, so $\left(0, -\dfrac{2}{9}\right)$ is the y-intercept.
5. Find other function values to determine the shape of the graph and then draw it.

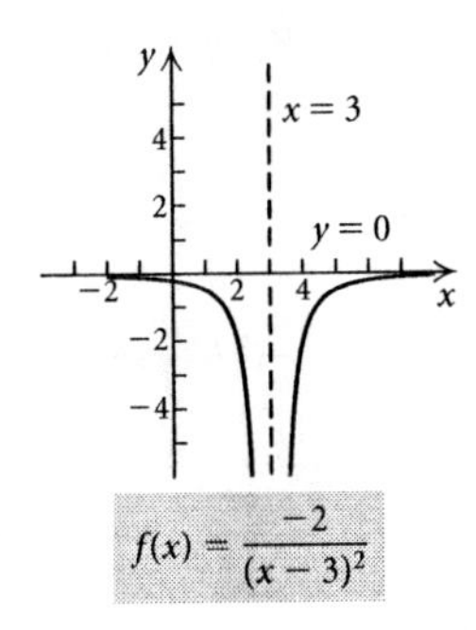

41. $f(x) = -\dfrac{1}{x^2}$

1. 0 is the zero of the denominator, so the domain excludes 0. It is $(-\infty, 0) \cup (0, \infty)$. The line $x = 0$, or the y-axis, is the vertical asymptote.
2. Because the degree of the numerator is less than the degree of the denominator, the x-axis, or $y = 0$, is the horizontal asymptote. There is no oblique asymptote.
3. The numerator has no zeros, so there is no x-intercept.
4. Since 0 is not in the domain of the function, there is no y-intercept.
5. Find other function values to determine the shape of the graph and then draw it.

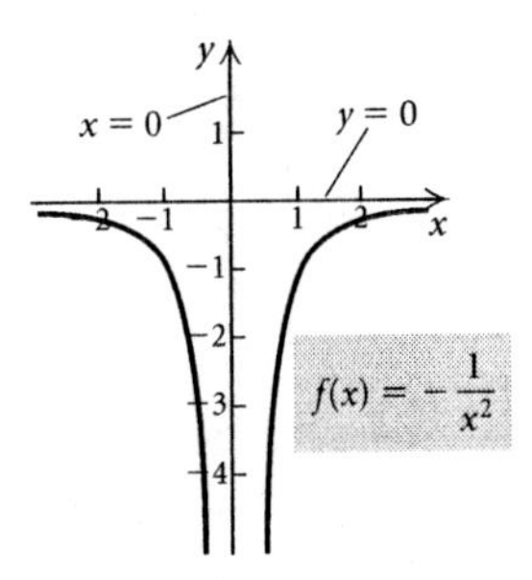

42. $f(x) = \dfrac{1}{3x^2}$

1. 0 is the zero of the denominator, so the domain is $(-\infty, 0) \cup (0, \infty)$. The line $x = 0$, or the y-axis, is the vertical asymptote.
2. Because the degree of the numerator is less than the degree of the denominator, the x-axis, or $y = 0$, is the horizontal asymptote. There is no oblique asymptote.
3. The numerator has no zeros, so there is no x-intercept.
4. Since 0 is not in the domain of the function, there is no y-intercept.
5. Find other function values to determine the shape of the graph and then draw it.

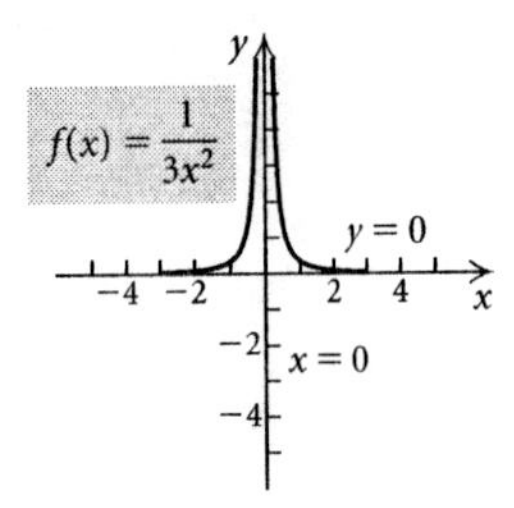

43. $f(x) = \dfrac{1}{x^2 + 3}$

1. The denominator has no real-number zeros, so the domain is $(-\infty, \infty)$ and there is no vertical asymptote.
2. Because the degree of the numerator is less than the degree of the denominator, the x-axis, or $y = 0$, is the horizontal asymptote. There is no oblique asymptote.
3. The numerator has no zeros, so there is no x-intercept.
4. $f(0) = \dfrac{1}{0^2 + 3} = \dfrac{1}{3}$, so $\left(0, \dfrac{1}{3}\right)$ is the y-intercept.
5. Find other function values to determine the shape of the graph and then draw it.

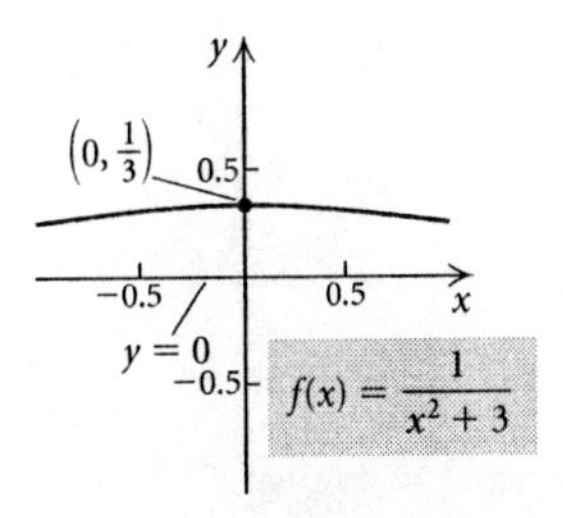

44. $f(x) = \dfrac{-1}{x^2+2}$

1. The denominator has no real-number zeros, so the domain is $(-\infty, \infty)$ and there is no vertical asymptote.
2. Because the degree of the numerator is less than the degree of the denominator, the x-axis, or $y = 0$, is the horizontal asymptote. There is no oblique asymptote.
3. The numerator has no zeros, so there is no x-intercept.
4. $f(0) = \dfrac{-1}{0^2+2} = -\dfrac{1}{2}$, so $\left(0, -\dfrac{1}{2}\right)$ is the y-intercept.
5. Find other function values to determine the shape of the graph and then draw it.

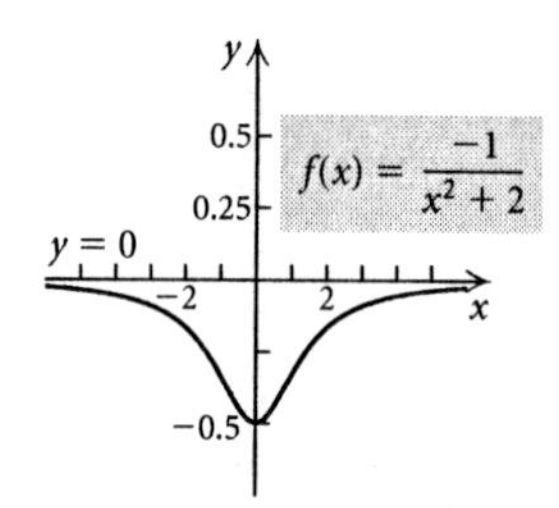

45. $f(x) = \dfrac{x^2-4}{x-2} = \dfrac{(x+2)(x-2)}{x-2} = x+2,\ x \neq 2$

The graph is the same as the graph of $f(x) = x+2$ except at $x = 2$, where there is a hole. The zero of $f(x) = x+2$ is -2, so the x-intercept is $(-2, 0)$; $f(0) = 2$, so the y-intercept is $(0, 2)$.

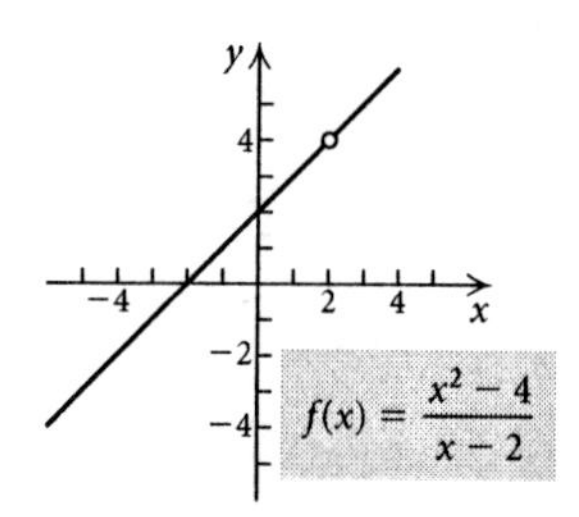

46. $f(x) = \dfrac{x^2-9}{x+3} = \dfrac{(x+3)(x-3)}{x+3} = x-3,\ x \neq -3$

The zero of $f(x) = x-3$ is 3, so the x-intercept is $(3, 0)$; $f(0) = -3$, so the y-intercept is $(0, -3)$.

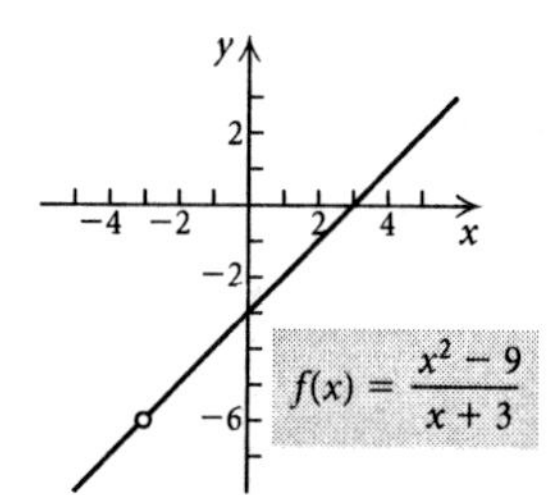

47. $f(x) = \dfrac{x-1}{x+2}$

1. -2 is the zero of the denominator, so the domain excludes -2. It is $(-\infty, -2) \cup (-2, \infty)$. The line $x = -2$ is the vertical asymptote.
2. The numerator and denominator have the same degree, so the horizontal asymptote is determined by the ratio of the leading coefficients, $1/1$, or 1. Thus, $y = 1$ is the horizontal asymptote. There is no oblique asymptote.
3. The zero of the numerator is 1, so the x-intercept is $(1, 0)$.
4. $f(0) = \dfrac{0-1}{0+2} = -\dfrac{1}{2}$, so $\left(0, -\dfrac{1}{2}\right)$ is the y-intercept.
5. Find other function values to determine the shape of the graph and then draw it.

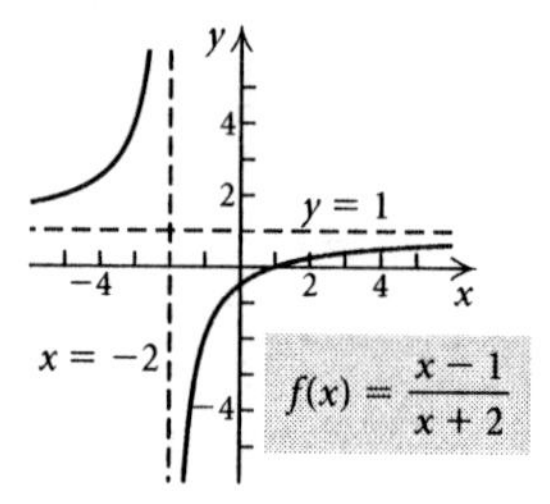

48. $f(x) = \dfrac{x-2}{x+1}$

1. -1 is the zero of the denominator, so the domain is $(-\infty, -1) \cup (-1, \infty)$ and $x = -1$ is the vertical asymptote.
2. The numerator and denominator have the same degree, so the horizontal asymptote is determined by the ratio of the leading coefficients, $1/1$, or 1. Thus, $y = 1$ is the horizontal asymptote. There is no oblique asymptote.
3. The zero of the numerator is 2, so the x-intercept is $(2, 0)$.
4. $f(0) = \dfrac{0-2}{0+1} = -2$, so $(0, -2)$ is the y-intercept.
5. Find other function values to determine the shape of the graph and then draw it.

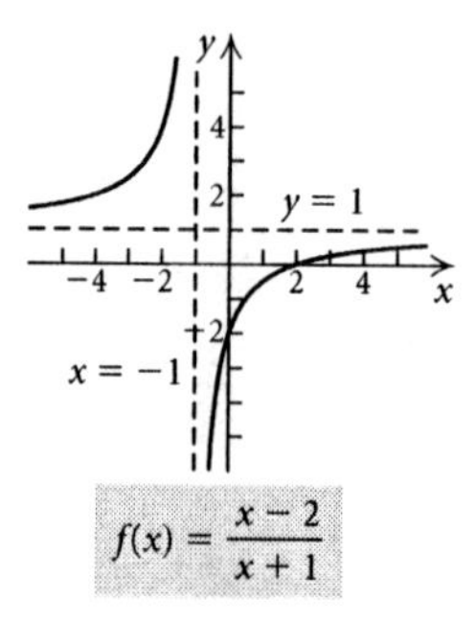

49. $f(x) = \dfrac{x+3}{2x^2-5x-3}$

1. The zeros of the denominator are the solutions of $2x^2-5x-3 = 0$. Since $2x^2-5x-3 = (2x+1)(x-3)$, the zeros are $-\dfrac{1}{2}$ and 3. Thus, the domain is $\left(-\infty, -\dfrac{1}{2}\right) \cup \left(-\dfrac{1}{2}, 3\right) \cup (3, \infty)$ and the lines $x = -\dfrac{1}{2}$ and $x = 3$ are vertical asymptotes.

2. Because the degree of the numerator is less than the degree of the denominator, the x-axis, or $y = 0$, is the horizontal asymptote. There is no oblique asymptote.

3. -3 is the zero of the numerator, so $(-3, 0)$ is the x-intercept.

4. $f(0) = \dfrac{0+3}{2 \cdot 0^2 - 5 \cdot 0 - 3} = -1$, so $(0, -1)$ is the y-intercept.

5. Find other function values to determine the shape of the graph and then draw it.

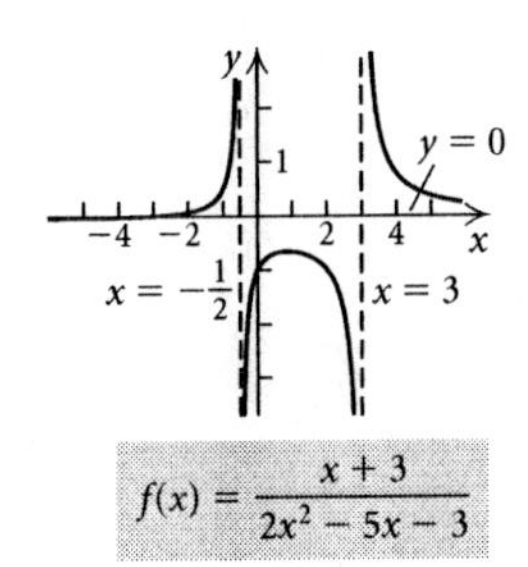

50. $f(x) = \dfrac{3x}{x^2 + 5x + 4}$

1. $x^2 + 5x + 4 = (x + 4)(x + 1)$, so the domain excludes -4 and -1. It is $(-\infty, -4) \cup (-4, -1) \cup (-1, \infty)$ and the lines $x = -4$ and $x = -1$ are vertical asymptotes.

2. Because the degree of the numerator is less than the degree of the denominator, the x-axis, or $y = 0$, is the horizontal asymptote. There is no oblique asymptote.

3. 0 is the zero of the numerator, so $(0, 0)$ is the x-intercept.

4. From part (3) we see that $(0, 0)$ is the y-intercept.

5. Find other function values to determine the shape of the graph and then draw it.

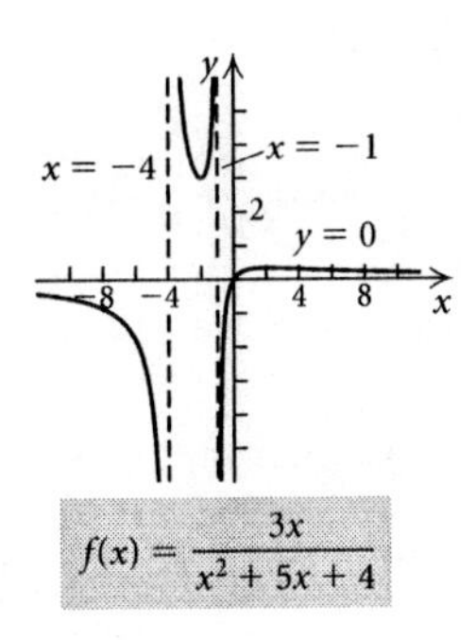

51. $f(x) = \dfrac{x^2 - 9}{x + 1}$

1. -1 is the zero of the denominator, so the domain excludes -1. It is $(-\infty, -1) \cup (-1, \infty)$. The line $x = -1$ is the vertical asymptote.

2. Because the degree of the numerator is one greater than the degree of the denominator, there is an oblique asymptote. Using division, we find that $\dfrac{x^2 - 9}{x + 1} = x - 1 + \dfrac{-8}{x + 1}$. As $|x|$ becomes very large, the graph of $f(x)$ gets close to the graph of $y = x - 1$. Thus, the line $y = x - 1$ is the oblique asymptote.

3. Since $x^2 - 9 = (x + 3)(x - 3)$, the zeros of the numerator are -3 and 3. Thus, the x-intercepts are $(-3, 0)$ and $(3, 0)$.

4. $f(0) = \dfrac{0^2 - 9}{0 + 1} = -9$, so $(0, -9)$ is the y-intercept.

5. Find other function values to determine the shape of the graph and then draw it.

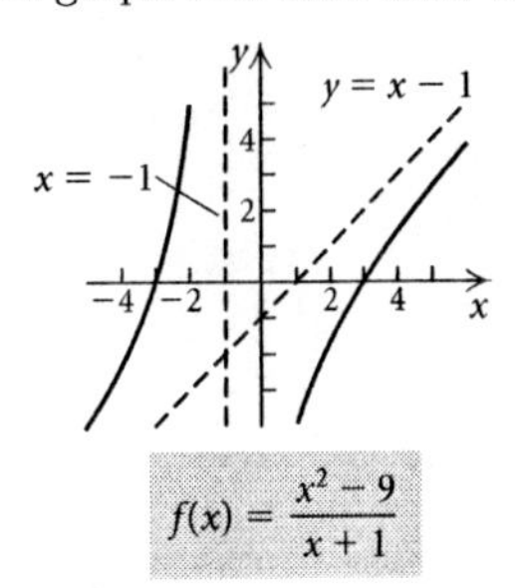

52. $f(x) = \dfrac{x^2 - 4}{x - 1}$

1. 1 is the zero of the denominator, so the domain is $(-\infty, 1) \cup (1, \infty)$ and $x = 1$ is the vertical asymptote.

2. $\dfrac{x^2 - 4}{x - 1} = x + 1 + \dfrac{-3}{x - 1}$, so $y = x + 1$ is the oblique asymptote.

3. $x^2 - 4 = (x + 2)(x - 2)$, so the x-intercepts are $(-2, 0)$ and $(2, 0)$.

4. $f(0) = \dfrac{0^2 - 4}{0 - 1} = 4$, so $(0, 4)$ is the y-intercept.

5. Find other function values to determine the shape of the graph and then draw it.

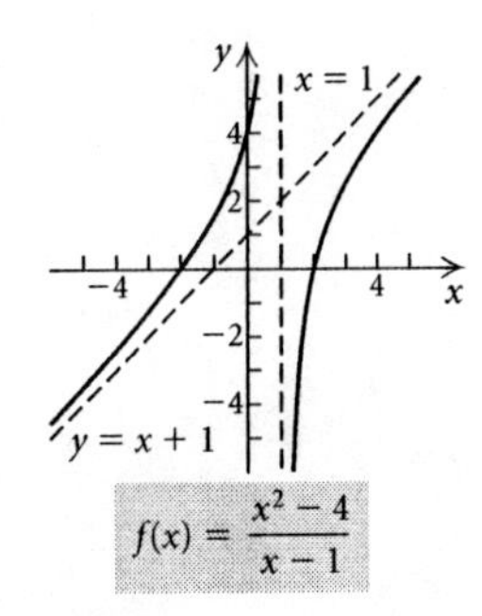

53. $f(x) = \dfrac{x^2 + x - 2}{2x^2 + 1}$

1. The denominator has no real-number zeros, so the domain is $(-\infty, \infty)$ and there is no vertical asymptote.

2. The numerator and denominator have the same degree, so the horizontal asymptote is determined by the ratio of the leading coefficients, 1/2. Thus, $y = 1/2$ is the horizontal asymptote. There is no oblique asymptote.

3. Since $x^2 + x - 2 = (x+2)(x-1)$, the zeros of the numerator are -2 and 1. Thus, the x-intercepts are $(-2, 0)$ and $(1, 0)$.

4. $f(0) = \dfrac{0^2 + 0 - 2}{2 \cdot 0^2 + 1} = -2$, so $(0, -2)$ is the y-intercept.

5. Find other function values to determine the shape of the graph and then draw it.

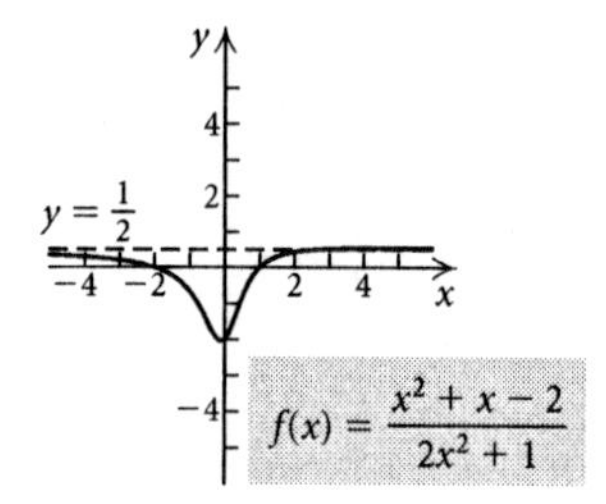

54. $f(x) = \dfrac{x^2 - 2x - 3}{3x^2 + 2}$

1. The denominator has no real-number zeros, so the domain is $(-\infty, \infty)$ and there is no vertical asymptote.

2. The numerator and denominator have the same degree, so the horizontal asymptote is determined by the ratio of the leading coefficients, 1/3. Thus, $y = 1/3$ is the horizontal asymptote. There is no oblique asymptote.

3. $x^2 - 2x - 3 = (x+1)(x-3)$, so the x-intercepts are $(-1, 0)$ and $(3, 0)$.

4. $f(0) = \dfrac{0^2 - 2 \cdot 0 - 3}{3 \cdot 0^2 + 2} = -\dfrac{3}{2}$, so $\left(0, -\dfrac{3}{2}\right)$ is the y-intercept.

5. Find other function values to determine the shape of the graph and then draw it.

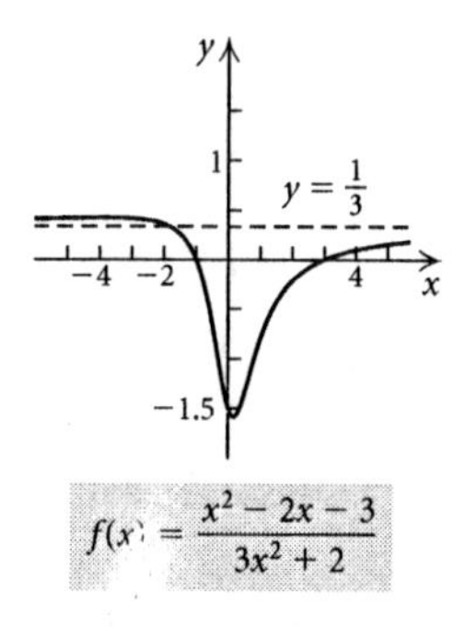

55. $g(x) = \dfrac{3x^2 - x - 2}{x - 1} = \dfrac{(3x+2)(x-1)}{x-1} = 3x + 2$, $x \neq 1$

The graph is the same as the graph of $g(x) = 3x + 2$ except at $x = 1$, where there is a hole.

The zero of $g(x) = 3x + 2$ is $-\dfrac{2}{3}$, so the x-intercept is $\left(-\dfrac{2}{3}, 0\right)$; $g(0) = 2$, so the y-intercept is $(0, 2)$.

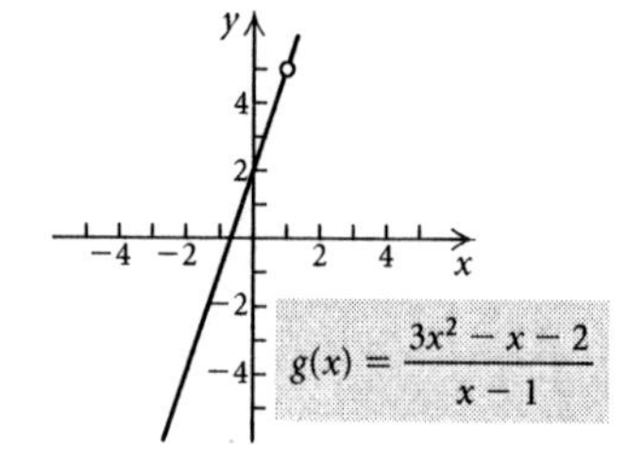

56. $f(x) = \dfrac{2x + 1}{2x^2 - 5x - 3} = \dfrac{2x+1}{(2x+1)(x-3)} = \dfrac{1}{x-3}$, $x \neq -\dfrac{1}{2}$

$f(x) = \dfrac{1}{x-3}$ has no zero, so there is no x-intercept; $f(0) = -\dfrac{1}{3}$, so the y-intercept is $\left(0, -\dfrac{1}{3}\right)$.

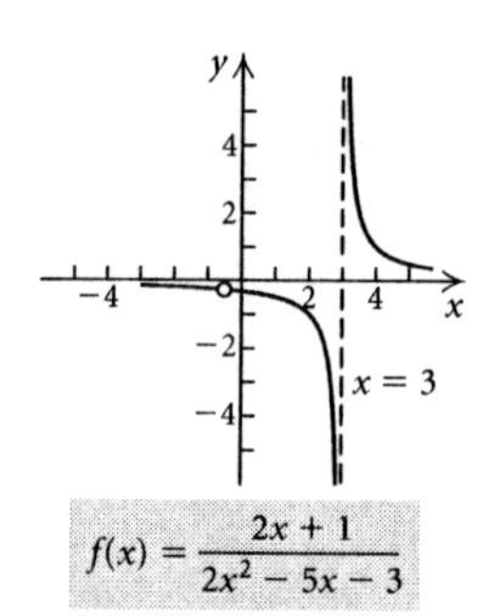

57. $f(x) = \dfrac{x - 1}{x^2 - 2x - 3}$

1. The zeros of the denominator are the solutions of $x^2 - 2x - 3 = 0$. Since $x^2 - 2x - 3 = (x+1)(x-3)$, the zeros are -1 and 3. Thus, the domain is $(-\infty, -1) \cup (-1, 3) \cup (3, \infty)$ and the lines $x = -1$ and $y = 3$ are the vertical asymptotes.

2. Because the degree of the numerator is less than the degree of the denominator, the x-axis, or $y = 0$, is the horizontal asymptote. There is no oblique asymptote.

3. 1 is the zero of the numerator, so $(1, 0)$ is the x-intercept.

4. $f(0) = \dfrac{0 - 1}{0^2 - 2 \cdot 0 - 3} = \dfrac{1}{3}$, so $\left(0, \dfrac{1}{3}\right)$ is the y-intercept.

5. Find other function values to determine the shape of the graph and then draw it.

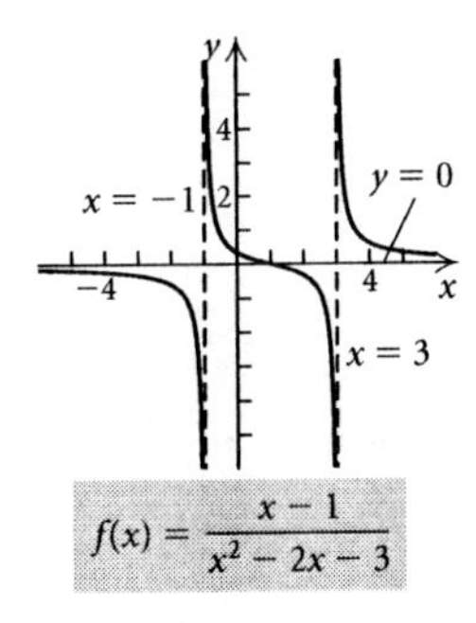

58. $f(x) = \dfrac{x+2}{x^2+2x-15}$

1. $x^2+2x-15 = (x+5)(x-3)$, so the domain excludes -5 and 3. It is $(-\infty,-5)\cup(-5,3)\cup(3,\infty)$ and the lines $x=-5$ and $x=3$ are vertical asymptotes.
2. Because the degree of the numerator is less than the degree of the denominator, the x-axis, or $y=0$, is the horizontal asymptote. There is no oblique asymptote.
3. -2 is the zero of the numerator, so $(-2,0)$ is the x-intercept.
4. $f(0) = \dfrac{0+2}{0^2+2\cdot 0-15} = -\dfrac{2}{15}$, so $\left(0,-\dfrac{2}{15}\right)$ is the y-intercept.
5. Find other function values to determine the shape of the graph and then draw it.

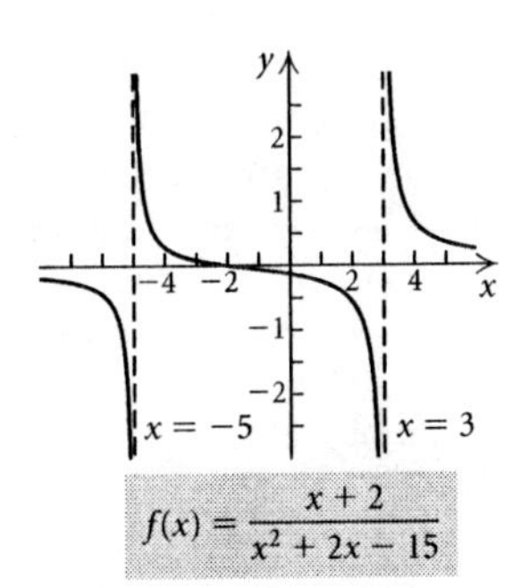

59. $f(x) = \dfrac{x-3}{(x+1)^3}$

1. -1 is the zero of the denominator, so the domain excludes -1. It is $(-\infty,-1)\cup(-1,\infty)$. The line $x=-1$ is the vertical asymptote.
2. Because the degree of the numerator is less than the degree of the denominator, the x-axis, or $y=0$, is the horizontal asymptote. There is no oblique asymptote.
3. 3 is the zero of the numerator, so $(3,0)$ is the x-intercept.
4. $f(0) = \dfrac{0-3}{(0+1)^3} = -3$, so $(0,-3)$ is the y-intercept.
5. Find other function values to determine the shape of the graph and then draw it.

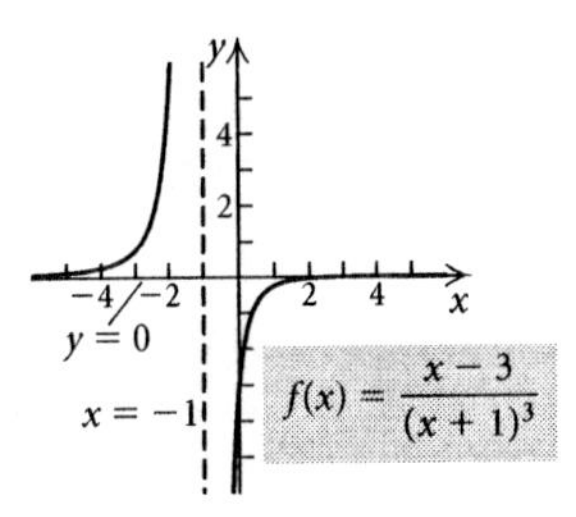

60. $f(x) = \dfrac{x+2}{(x-1)^3}$

1. 1 is the zero of the denominator, so the domain is $(-\infty,1)\cup(1,\infty)$ and $x=1$ is the vertical asymptote.
2. Because the degree of the numerator is less than the degree of the denominator, the x-axis, or $y=0$, is the horizontal asymptote. There is no oblique asymptote.
3. -2 is the zero of the numerator, so $(-2,0)$ is the x-intercept.
4. $f(0) = \dfrac{0+2}{(0-1)^3} = -2$, so $(0,-2)$, is the y-intercept.
5. Find other function values to determine the shape of the graph and then draw it.

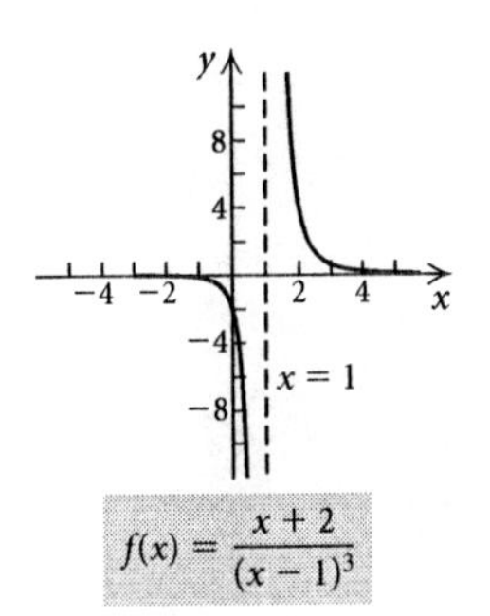

61. $f(x) = \dfrac{x^3+1}{x}$

1. 0 is the zero of the denominator, so the domain excludes 0. It is $(-\infty,0)\cup(0,\infty)$. The line $x=0$, or the y-axis, is the vertical asymptote.
2. Because the degree of the numerator is more than one greater than the degree of the denominator, there is no horizontal or oblique asymptote.
3. The real-number zero of the numerator is -1, so the x-intercept is $(-1,0)$.
4. Since 0 is not in the domain of the function, there is no y-intercept.
5. Find other function values to determine the shape of the graph and then draw it.

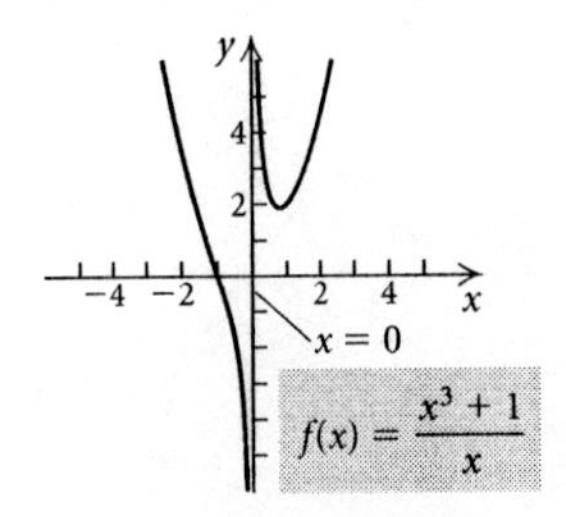

62. $f(x) = \dfrac{x^3 - 1}{x}$

1. 0 is the zero of the denominator, so the domain excludes 0. It is $(-\infty, 0) \cup (0, \infty)$. The line $x = 0$, or the y-axis, is the vertical asymptote.
2. Because the degree of the numerator is more than one greater than the degree of the denominator, there is no horizontal or oblique asymptote.
3. The real-number zero of the numerator is 1, so the x-intercept is $(1, 0)$.
4. Since 0 is not in the domain of the function, there is no y-intercept.
5. Find other function values to determine the shape of the graph and then draw it.

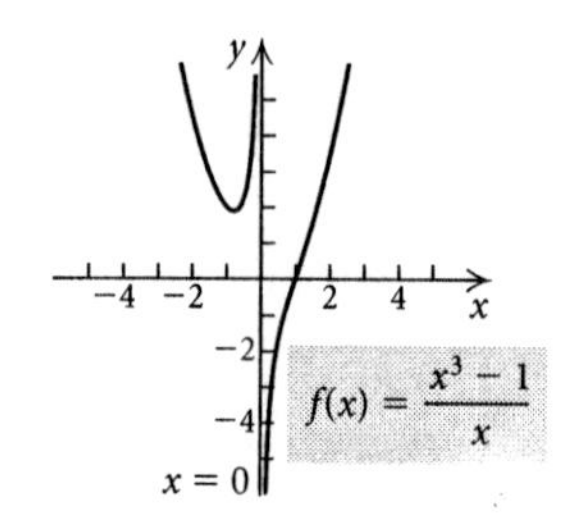

63. $f(x) = \dfrac{x^3 + 2x^2 - 15x}{x^2 - 5x - 14}$

1. The zeros of the denominator are the solutions of $x^2 - 5x - 14 = 0$. Since $x^2 - 5x - 14 = (x+2)(x-7)$, the zeros are -2 and 7. Thus, the domain is $(-\infty, -2) \cup (-2, 7) \cup (7, \infty)$ and the lines $x = -2$ and $x = 7$ are the vertical asymptotes.
2. Because the degree of the numerator is one greater than the degree of the denominator, there is an oblique asymptote. Using division, we find that $\dfrac{x^3 + 2x^2 - 15x}{x^2 - 5x - 14} = x + 7 + \dfrac{34x + 98}{x^2 - 5x - 14}$. As $|x|$ becomes very large, the graph of $f(x)$ gets close to the graph of $y = x + 7$. Thus, the line $y = x + 7$ is the oblique asymptote.
3. The zeros of the numerator are the solutions of $x^3 + 2x^2 - 15x = 0$. Since $x^3 + 2x^2 - 15x = x(x + 5)(x - 3)$, the zeros are 0, -5, and 3. Thus, the x-intercepts are $(-5, 0)$, $(0, 0)$, and $(3, 0)$.
4. From part (3) we see that $(0, 0)$ is the y-intercept.
5. Find other function values to determine the shape of the graph and then draw it.

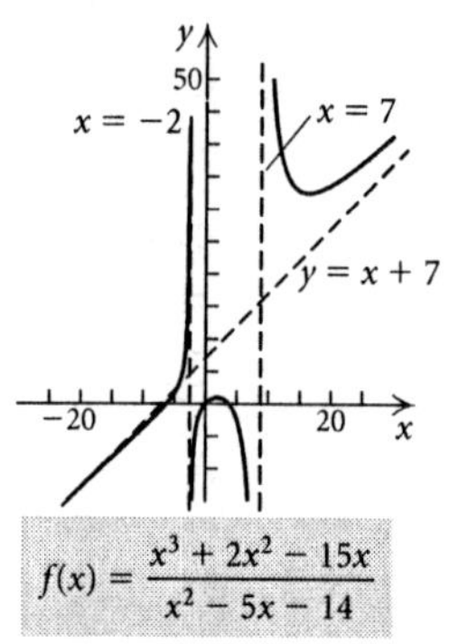

64. $f(x) = \dfrac{x^3 + 2x^2 - 3x}{x^2 - 25}$

1. The zeros of the denominator are -5 and 5, so the domain excludes -5 and 5. It is $(-\infty, -5) \cup (-5, 5) \cup (5, \infty)$ and the lines $x = -5$ and $x = 5$ are the vertical asymptotes.
2. $\dfrac{x^3 + 2x^2 - 3x}{x^2 - 25} = x + 2 + \dfrac{-28x + 50}{x^2 - 25}$, so $y = x + 2$ is the oblique asymptote.
3. $x^3 + 2x^2 - 3x = x(x+3)(x-1)$, so the x-intercepts are $(0, 0)$, $(-3, 0)$, and $(1, 0)$.
4. From part (3) we see that $(0, 0)$ is the y-intercept.
5. Find other function values to determine the shape of the graph and then draw it.

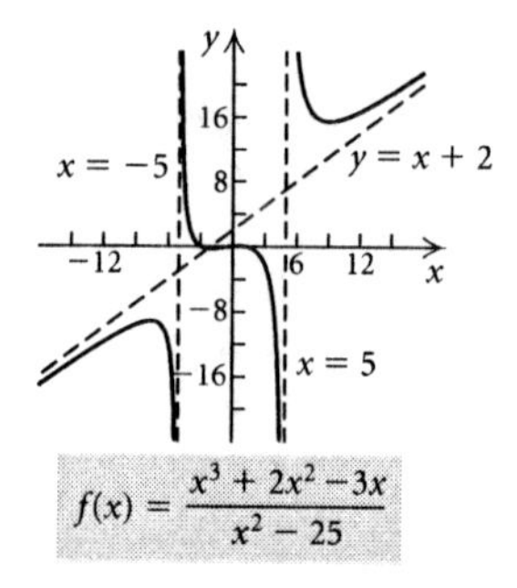

65. $f(x) = \dfrac{5x^4}{x^4 + 1}$

1. The denominator has no real-number zeros, so the domain is $(-\infty, \infty)$ and there is no vertical asymptote.
2. The numerator and denominator have the same degree, so the horizontal asymptote is determined by the ratio of the leading coefficients, $5/1$, or 5. Thus, $y = 5$ is the horizontal asymptote. There is no oblique asymptote.
3. The zero of the numerator is 0, so $(0, 0)$ is the x-intercept.
4. From part (3) we see that $(0, 0)$ is the y-intercept.
5. Find other function values to determine the shape of the graph and then draw it.

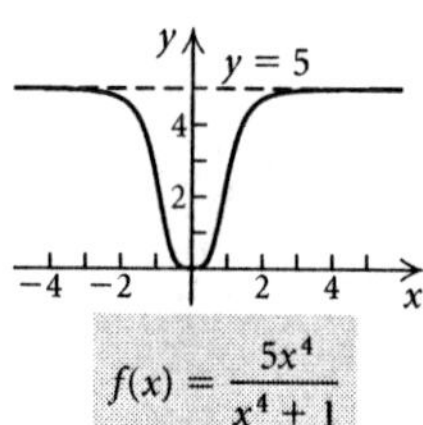

66. $f(x) = \dfrac{x + 1}{x^2 + x - 6}$

1. $x^2 + x - 6 = (x + 3)(x - 2)$, so the domain excludes -3 and 2. It is $(-\infty, -3) \cup (-3, 2) \cup (2, \infty)$ and the lines $x = -3$ and $x = 2$ are vertical asymptotes.
2. Because the degree of the numerator is less than the degree of the denominator, the x-axis, or $y = 0$, is the horizontal asymptote. There is no oblique asymptote.

3. The zero of the numerator is -1, so the x-intercept is $(-1, 0)$.

4. $f(0) = \dfrac{0+1}{0^2+0-6} = -\dfrac{1}{6}$, so $\left(0, -\dfrac{1}{6}\right)$ is the y-intercept.

5. Find other function values to determine the shape of the graph and then draw it.

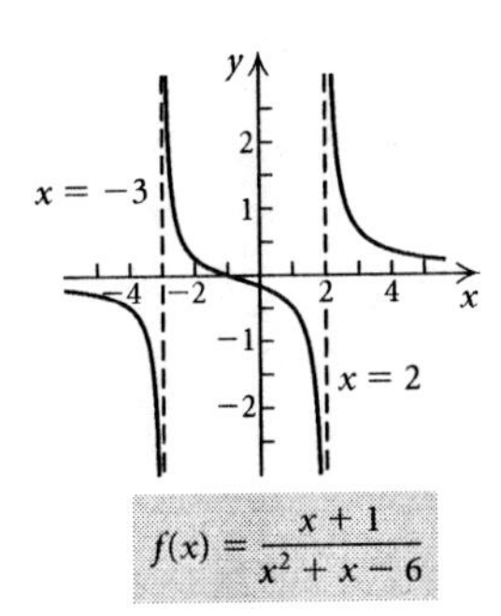

67. $f(x) = \dfrac{x^2}{x^2-x-2}$

1. The zeros of the denominator are the solutions of $x^2-x-2=0$. Since $x^2-x-2=(x+1)(x-2)$, the zeros are -1 and 2. Thus, the domain is $(-\infty, -1) \cup (-1, 2) \cup (2, \infty)$ and the lines $x=-1$ and $x=2$ are the vertical asymptotes.

2. The numerator and denominator have the same degree, so the horizontal asymptote is determined by the ratio of the leading coefficients, $1/1$, or 1. Thus, $y=1$ is the horizontal asymptote. There is no oblique asymptote.

3. The zero of the numerator is 0, so the x-intercept is $(0, 0)$.

4. From part (3) we see that $(0, 0)$ is the y-intercept.

5. Find other function values to determine the shape of the graph and then draw it.

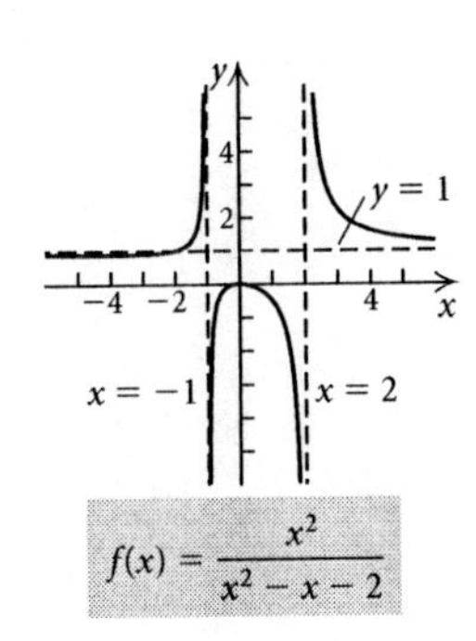

68. $f(x) = \dfrac{x^2-x-2}{x+2}$

1. -2 is the zero of the denominator, so the domain is $(-\infty, -2) \cup (-2, \infty)$ and the line $x=-2$ is the vertical asymptote.

2. $\dfrac{x^2-x-2}{x+2} = x-3+\dfrac{4}{x+2}$, so $y=x-3$ is the oblique asymptote.

3. $x^2-x-2=(x+1)(x-2)$, so the x-intercepts are $(-1, 0)$ and $(2, 0)$.

4. $f(0) = \dfrac{0^2-0-2}{0+2} = -1$, so $(0, -1)$ is the y-intercept.

5. Find other function values to determine the shape of the graph and then draw it.

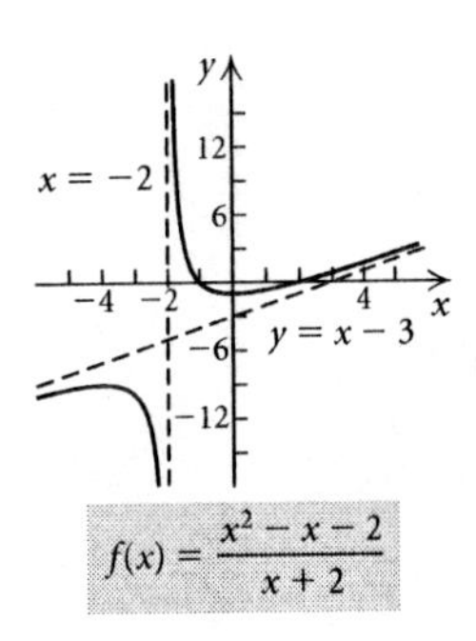

69. Answers may vary. The numbers -4 and 5 must be zeros of the denominator. A function that satisfies these conditions is

$$f(x) = \frac{1}{(x+4)(x-5)}, \text{ or } f(x) = \frac{1}{x^2-x-20}.$$

70. Answers may vary. The numbers -4 and 5 must be zeros of the denominator and -2 must be a zero of the numerator.

$$f(x) = \frac{x+2}{(x+4)(x-5)}, \text{ or } f(x) = \frac{x+2}{x^2-x-20}.$$

71. Answers may vary. The numbers -4 and 5 must be zeros of the denominator and -2 must be a zero of the numerator. In addition, the numerator and denominator must have the same degree and the ratio of their leading coefficients must be $3/2$. A function that satisfies these conditions is

$$f(x) = \frac{3x(x+2)}{2(x+4)(x-5)}, \text{ or } f(x) = \frac{3x^2+6x}{2x^2-2x-40}.$$

Another function that satisfies these conditions is

$$g(x) = \frac{3(x+2)^2}{2(x+4)(x-5)}, \text{ or } g(x) = \frac{3x^2+12x+12}{2x^2-2x-40}.$$

72. Answers may vary. The degree of the numerator must be 1 greater than the degree of the denominator and the quotient, when long division is performed, must be $x-1$. If we let the remainder be 1, a function that satisfies these conditions is $f(x) = x-1+\dfrac{1}{x}$, or $f(x) = \dfrac{x^2-x+1}{x}$.

73. a) The horizontal asymptote of $N(t)$ is the ratio of the leading coefficients of the numerator and denominator, $0.8/5$, or 0.16. Thus, $N(t) \to 0.16$ as $t \to \infty$.

b) The medication never completely disappears from the body; a trace amount remains.

74. a) $A(x) \to 2/1$, or 2 as $x \to \infty$.

b) As more videotapes are produced, the average cost approaches \$2.

75. a) $P(0) = 0$; $P(1) = 45.455$ thousand, or $45,455$;

$P(3) = 55.556$ thousand, or $55,556$;

$P(8) = 29.197$ thousand, or $29,197$

b) The degree of the numerator is less than the degree of the denominator, so the x-axis is the horizontal asymptote. Thus, $P(t) \to 0$ as $t \to \infty$.

c) Eventually, no one will live in Lordsburg.

76. a) $S =$ area of base $+ 4 \cdot$ area of a side

$= x^2 + 4xy$

Now we express y in terms of x:

Volume $= 108 = x^2y$

$$\frac{108}{x^2} = y$$

Thus, $S = x^2 + 4x\left(\frac{108}{x^2}\right)$, or

$$S = x^2 + \frac{432}{x}.$$

b) 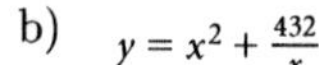

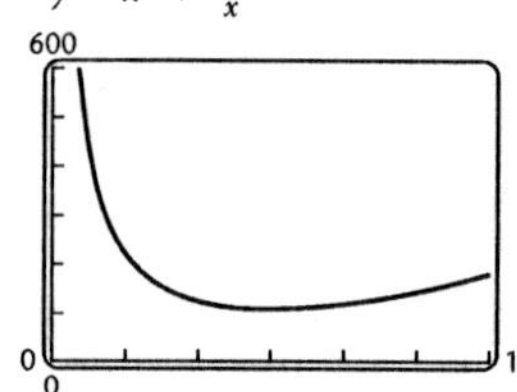

c) Using the Minimum feature on a graphing calculator, we estimate that the minimum surface area is 108 cm^2. This occurs when $x = 6$ cm.

77. Using the Maximum feature on a graphing calculator, we find that the maximum population is about 58.926 thousand, or 58,926. This occurs when $t \approx 2.12$ months.

78. $y_1 = \dfrac{x^3+4}{x} = x^2 + \dfrac{4}{x}$

As $|x| \to \infty$, $\dfrac{4}{x} \to 0$ and the value of $y_1 \to x^2$. Thus, the parabola $y_2 = x^2$ can be thought of as a nonlinear asymptote for y_1. The graph confirms this.

79. The denominator has no real-number zeros.

80. A horizontal asymptote occurs when the degree of the numerator of a rational function is less than or equal to the degree of the denominator. An oblique asymptote occurs when the degree of the numerator is 1 greater than the degree of the denominator. Thus, a rational function cannot have both a horizontal asymptote and an oblique asymptote.

81. slope

82. slope-intercept equation

83. point-slope equation

84. domain, range, domain, range

85. $f(-x) = -f(x)$

86. x-intercept

87. midpoint formula

88. vertical lines

89. $f(x) = \dfrac{x^5 + 2x^3 + 4x^2}{x^2+2} = x^3 + 4 + \dfrac{-8}{x^2+2}$

As $|x| \to \infty$, $\dfrac{-8}{x^2+2} \to 0$ and the value of $f(x) \to x^3 + 4$.
Thus, the nonlinear asymptote is $y = x^3 + 4$.

90. $f(x) = \dfrac{x^4 + 3x^2}{x^2+1} = x^2 + 2 + \dfrac{-2}{x^2+1}$

As $|x| \to \infty$, $\dfrac{-2}{x^2+1} \to 0$ and the value of $f(x) \to x^2 + 2$.
Thus, the nonlinear asymptote is $y = x^2 + 2$.

91. 

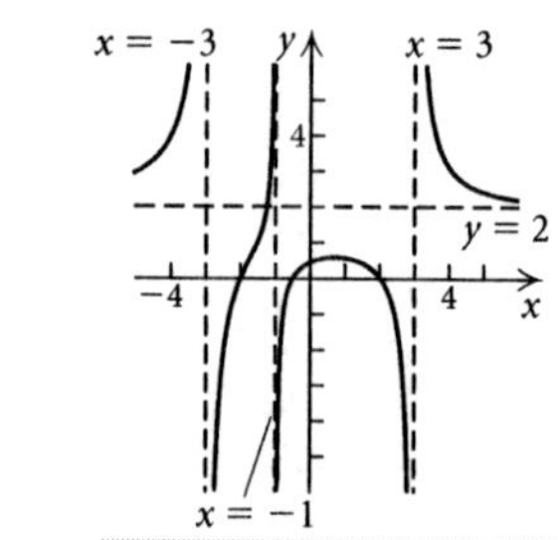

$f(x) = \dfrac{2x^3 + x^2 - 8x - 4}{x^3 + x^2 - 9x - 9}$

92.

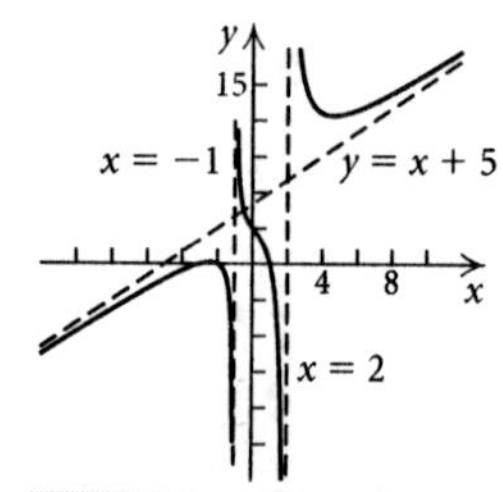

$f(x) = \dfrac{x^3 + 4x^2 + x - 6}{x^2 - x - 2}$

93. $f(x) = \sqrt{\dfrac{72}{x^2 - 4x - 21}}$

The radicand must be nonnegative and the denominator must be nonzero. Thus, the values of x for which $x^2 - 4x - 21 > 0$ comprise the domain. By inspecting the graph of $y = x^2 - 4x - 21$ we see that the domain is $\{x | x < -3 \text{ or } x > 7\}$, or $(-\infty, -3) \cup (7, \infty)$.

94. $f(x) = \sqrt{x^2 - 4x - 21}$

The radicand must be nonnegative. By inspecting the graph of $y = x^2 - 4x - 21$ we see that the domain is $\{x | x \le -3 \text{ or } x \ge 7\}$, or $(-\infty, -3] \cup [7, \infty)$.

Exercise Set 3.5

1. $x^2 + 2x - 15 = 0$

$(x+5)(x-3) = 0$

$x + 5 = 0 \quad or \quad x - 3 = 0$

$x = -5 \quad or \quad x = 3$

The solution set is $\{-5, 3\}$.

2. Solve $x^2 + 2x - 15 < 0$.

From Exercise 1 we know the solutions of the related equation are -5 and 3. These numbers divide the x-axis into the intervals $(-\infty, -5)$, $(-5, 3)$, and $(3, \infty)$. We test a value in each interval.

$(-\infty, -5)$: $f(-6) = 9 > 0$

$(-5, 3)$: $f(0) = -15 < 0$

$(3, \infty)$: $f(4) = 9 > 0$

Function values are negative only in the interval $(-5, 3)$. The solution set is $(-5, 3)$.

3. Solve $x^2 + 2x - 15 \le 0$.

From Exercise 2 we know the solution set of $x^2+2x-15 < 0$ is $(-5, 3)$. The solution set of $x^2+2x-15 \le 0$ includes the endpoints of this interval. Thus the solution set is $[-5, 3]$.

4. Solve $x^2 + 2x - 15 > 0$.

From our work in Exercise 2 we see that the solution set is $(-\infty, -5) \cup (3, \infty)$.

5. Solve $x^2 + 2x - 15 \ge 0$.

From Exercise 4 we know the solution set of $x^2+2x-15 > 0$ is $(-\infty, -5) \cup (3, \infty)$. The solution set of $x^2 + 2x - 15 \ge 0$ includes the endpoints -5 and 3. Thus the solution set is $(-\infty, -5] \cup [3, \infty)$.

6.
$$x^5 - 9x^3 = 0$$
$$x^3(x^2 - 9) = 0$$
$$x^3(x + 3)(x - 3) = 0$$

The solution set is $\{0, -3, 3\}$.

7. Solve $x^5 - 9x^3 < 0$.

From Exercise 6 we know the solutions of the related equation are -3, 0, and 3. These numbers divide the x-axis into the intervals $(-\infty, -3)$, $(-3, 0)$, $(0, 3)$, and $(3, \infty)$. We test a value in each interval.

$(-\infty, -3)$: $g(-4) = -448 < 0$

$(-3, 0)$: $g(-1) = 8 > 0$

$(0, 3)$: $g(1) = -8 < 0$

$(3, \infty)$: $g(4) = 448 > 0$

Function values are negative on $(-\infty, -3)$ and on $(0, 3)$. The solution set is $(-\infty, -3) \cup (0, 3)$.

8. Solve $x^5 - 9x^3 \le 0$.

From Exercise 7 we know the solution set of $x^5 - 9x^3 < 0$ is $(-\infty, -3) \cup (0, 3)$. The solution set of $x^5 - 9x^3 \le 0$ includes the endpoints -3, 0, and 3. Thus the solution set is $(-\infty, -3] \cup [0, 3]$.

9. Solve $x^5 - 9x^3 > 0$.

From our work in Exercise 7 we see that the solution set is $(-3, 0) \cup (3, \infty)$.

10. Solve $x^5 - 9x^3 \ge 0$.

From Exercise 9 we know the solution set of $x^5 - 9x^3 > 0$ is $(-3, 0) \cup (3, \infty)$. The solution set of $x^5 - 9x^3 \ge 0$ includes the endpoints -3, 0, and 3. Thus the solution set is $[-3, 0] \cup [3, \infty)$.

11. First we find an equivalent inequality with 0 on one side.

$$x^3 + 6x^2 < x + 30$$
$$x^3 + 6x^2 - x - 30 < 30$$

From the graph we see that the x-intercepts of the related function occur at $x = -5$, $x = -3$, and $x = 2$. They divide the x-axis into the intervals $(-\infty, -5)$, $(-5, -3)$, $(-3, 2)$, and $(2, \infty)$. From the graph we see that the function has negative values only on $(-\infty, -5)$ and $(-3, 2)$. Thus, the solution set is $(-\infty, -5) \cup (-3, 2)$.

12. From the graph we see that the x-intercepts of the related function occur at $x = -4$, $x = -3$, $x = 2$, and $x = 5$. They divide the x-axis into the intervals $(-\infty, -4)$, $(-4, -3)$, $(-3, 2)$, $(2, 5)$, and $(5, \infty)$. The function has positive values only on $(-\infty, -4)$, $(-3, 2)$, and $(5, \infty)$. Since the inequality symbol is $\ge$, the endpoints of the intervals are included in the solution set. It is $(-\infty, -4] \cup [-3, 2] \cup [5, \infty)$.

13. By observing the graph or the denominator of the function, we see that the function is not defined for $x = -2$ or $x = 2$. We also see that 0 is a zero of the function. These numbers divide the x-axis into the intervals $(-\infty, -2)$, $(-2, 0)$, $(0, 2)$, and $(2, \infty)$. From the graph we see that the function has positive values only on $(-2, 0)$ and $(2, \infty)$. Since the inequality symbol is $\ge$, 0 must be included in the solution set. It is $(-2, 0] \cup (2, \infty)$.

14. By observing the graph or the denominator of the function, we see that the function is not defined for $x = -2$ or $x = 2$. We also see that the function has no zeros. Thus, the number line is divided into the intervals $(-\infty, -2)$, $(-2, 2)$, and $(2, \infty)$. From the graph we see that the function has negative values only on $(-2, 2)$. Thus, the solution set is $(-2, 2)$.

15. $(x - 1)(x + 4) < 0$

The related equation is $(x - 1)(x + 4) = 0$. Using the principle of zero products, we find that the solutions of the related equation are 1 and -4. These numbers divide the x-axis into the intervals $(-\infty, -4)$, $(-4, 1)$, and $(1, \infty)$. We let $f(x) = (x - 1)(x + 4)$ and test a value in each interval.

$(-\infty, -4)$: $f(-5) = 6 > 0$

$(-4, 1)$: $f(0) = -4 < 0$

$(1, \infty)$: $f(2) = 6 > 0$

Function values are negative only in the interval $(-4, 1)$. The solution set is $(-4, 1)$.

16. $(x + 3)(x - 5) < 0$

The related equation is $(x + 3)(x - 5) = 0$. Its solutions are -3 and 5. These numbers divide the x-axis into the intervals $(-\infty, -3)$, $(-3, 5)$, and $(5, \infty)$. Let $f(x) = (x + 3)(x - 5)$ and test a value in each interval.

$(-\infty, -3)$: $f(-4) = 9 > 0$

$(-3, 5)$: $f(0) = -15 < 0$

$(5, \infty)$: $f(6) = 9 > 0$

Function values are negative only in the interval $(-3, 5)$. The solution set is $(-3, 5)$.

17. $(x-4)(x+2) \geq 0$

The related equation is $(x-4)(x+2)=0$. Using the principle of zero products, we find that the solutions of the related equation are 4 and -2. These numbers divide the x-axis into the intervals $(-\infty,-2)$, $(-2,4)$, and $(4,\infty)$. We let $f(x)=(x-4)(x+2)$ and test a value in each interval.

$(-\infty,-2)$: $f(-3)=7>0$

$(-2,4)$: $f(0)=-8<0$

$(4,\infty)$: $f(5)=7>0$

Function values are positive on $(-\infty,-2)$ and $(4,\infty)$. Since the inequality symbol is $\geq$, the endpoints of the intervals must be included in the solution set. It is $(-\infty,-2]\cup[4,\infty)$.

18. $(x-2)(x+1) \geq 0$

The related equation is $(x-2)(x+1)=0$. Its solutions are 2 and -1. These numbers divide the x-axis into the intervals $(-\infty,-1)$, $(-1,2)$, and $(2,\infty)$. We let $f(x)=(x-2)(x+1)$ and test a value in each interval.

$(-\infty,-1)$: $f(-2)=4>0$

$(-1,2)$: $f(0)=-2<0$

$(2,\infty)$: $f(3)=4>0$

Function values are positive on $(-\infty,-1)$ and $(2,\infty)$. Since the inequality symbol is $\geq$, the endpoints of the intervals must be included in the solution set. It is $(-\infty,-1]\cup[2,\infty)$.

19.

$x^2+x-2>0$ Polynomial inequality

$x^2+x-2=0$ Related equation

$(x+2)(x-1)=0$ Factoring

Using the principle of zero products, we find that the solutions of the related equation are -2 and 1. These numbers divide the x-axis into the intervals $(-\infty,-2)$, $(-2,1)$, and $(1,\infty)$. We let $f(x)=x^2+x-2$ and test a value in each interval.

$(-\infty,-2)$: $f(-3)=4>0$

$(-2,1)$: $f(0)=-2<0$

$(1,\infty)$: $f(2)=4>0$

Function values are positive on $(-\infty,-2)$ and $(1,\infty)$. The solution set is $(-\infty,-2)\cup(1,\infty)$.

20.

$x^2-x-6>0$ Polynomial inequality

$x^2-x-6=0$ Related equation

$(x-3)(x+2)=0$ Factoring

The solutions of the related equation are 3 and -2. These numbers divide the x-axis into the intervals $(-\infty,-2)$, $(-2,3)$, and $(3,\infty)$. We let $f(x)=x^2-x-6$ and test a value in each interval.

$(-\infty,-2)$: $f(-3)=6>0$

$(-2,3)$: $f(0)=-6<0$

$(3,\infty)$: $f(4)=6>0$

Function values are positive on $(-\infty,-2)$ and $(3,\infty)$. The solution set is $(-\infty,-2)\cup(3,\infty)$.

21.

$x^2>25$ Polynomial inequality

$x^2-25>0$ Equivalent inequality with 0 on one side

$x^2-25=0$ Related equation

$(x+5)(x-5)=0$ Factoring

Using the principle of zero products, we find that the solutions of the related equation are -5 and 5. These numbers divide the x-axis into the intervals $(-\infty,-5)$, $(-5,5)$, and $(5,\infty)$. We let $f(x)=x^2-25$ and test a value in each interval.

$(-\infty,-5)$: $f(-6)=11>0$

$(-5,5)$: $f(0)=-25<0$

$(5,\infty)$: $f(6)=11>0$

Function values are positive on $(-\infty,-5)$ and $(5,\infty)$. The solution set is $(-\infty,-5)\cup(5,\infty)$.

22.

$x^2\leq 1$ Polynomial inequality

$x^2-1\leq 0$ Equivalent inequality with 0 on one side

$x^2-1=0$ Related equation

$(x+1)(x-1)=0$ Factoring

The solutions of the related equation are -1 and 1. These numbers divide the x-axis into the intervals $(-\infty,-1)$, $(-1,1)$, and $(1,\infty)$. We let $f(x)=x^2-1$ and test a value in each interval.

$(-\infty,-1)$: $f(-2)=3>0$

$(-1,1)$: $f(0)=-1<0$

$(1,\infty)$: $f(2)=3>0$

Function values are negative only on $(-1,1)$. Since the inequality symbol is $\leq$, the endpoints of the interval must be included in the solution set. It is $[-1,1]$.

23.

$4-x^2\leq 0$ Polynomial inequality

$4-x^2=0$ Related equation

$(2+x)(2-x)=0$ Factoring

Using the principle of zero products, we find that the solutions of the related equation are -2 and 2. These numbers divide the x-axis into the intervals $(-\infty,-2)$, $(-2,2)$, and $(2,\infty)$. We let $f(x)=4-x^2$ and test a value in each interval.

$(-\infty,-2)$: $f(-3)=-5<0$

$(-2,2)$: $f(0)=4>0$

$(2,\infty)$: $f(3)=-5<0$

Function values are negative on $(-\infty,-2)$ and $(2,\infty)$. Since the inequality symbol is $\leq$, the endpoints of the intervals must be included in the solution set. It is $(-\infty,-2]\cup[2,\infty)$.

24.

$11-x^2\geq 0$ Polynomial inequality

$11-x^2=0$ Related equation

The solutions of the related equation are $\pm\sqrt{11}$. These numbers divide the x-axis into the intervals $(-\infty,-\sqrt{11})$, $(-\sqrt{11},\sqrt{11})$, and $(\sqrt{11},\infty)$. We let $f(x)=11-x^2$ and test a value in each interval.

$(-\infty, -\sqrt{11})$: $f(-4) = -5 < 0$

$(-\sqrt{11}, \sqrt{11})$: $f(0) = 11 > 0$

$(\sqrt{11}, \infty)$: $f(4) = -5 < 0$

Function values are positive only on $(-\sqrt{11}, \sqrt{11})$. Since the inequality symbol is $\geq$, the endpoints of the interval must be included in the solution set. It is $[-\sqrt{11}, \sqrt{11}]$.

25.
$6x - 9 - x^2 < 0$ Polynomial inequality

$6x - 9 - x^2 = 0$ Related equation

$-(x^2 - 6x + 9) = 0$ Factoring out -1 and rearranging

$-(x-3)(x-3) = 0$ Factoring

Using the principle of zero products, we find that the solution of the related equation is 3. This number divides the x-axis into the intervals $(-\infty, 3)$ and $(3, \infty)$. We let $f(x) = 6x - 9 - x^2$ and test a value in each interval.

$(-\infty, 3)$: $f(-4) = -49 < 0$

$(3, \infty)$: $f(4) = -49 < 0$

Function values are negative on both intervals. The solution set is $(-\infty, 3) \cup (3, \infty)$.

26.
$x^2 + 2x + 1 \leq 0$ Polynomial inequality

$x^2 + 2x + 1 = 0$ Related equation

$(x+1)(x+1) = 0$ Factoring

The solution of the related equation is -1. This number divides the x-axis into the intervals $(-\infty, -1)$, and $(-1, \infty)$. We let $f(x) = x^2 + 2x + 1$ and test a value in each interval.

$(-\infty, -1)$: $f(-2) = 1 > 0$

$(-1, \infty)$: $f(0) = 1 > 0$

Function values are negative in neither interval. The function is equal to 0 when $x = -1$. Thus, the solution set is $\{-1\}$.

27.
$x^2 + 12 < 4x$ Polynomial inequality

$x^2 - 4x + 12 < 0$ Equivalent inequality with 0 on one side

$x^2 - 4x + 12 = 0$ Related equation

Using the quadratic formula, we find that the related equation has no real-number solutions. The graph lies entirely above the x-axis, so the inequality has no solution. We could determine this algebraically by letting $f(x) = x^2 - 4x + 12$ and testing any real number (since there are no real-number solutions of $f(x) = 0$ to divide the x-axis into intervals). For example, $f(0) = 12 > 0$, so we see algebraically that the inequality has no solution. The solution set is $\emptyset$.

28.
$x^2 - 8 > 6x$ Polynomial inequality

$x^2 - 6x - 8 > 0$ Equivalent inequality with 0 on one side

$x^2 - 6x - 8 = 0$ Related equation

Using the quadratic formula, we find that the solutions of the related equation are $3 - \sqrt{17}$ and $3 + \sqrt{17}$. These numbers divide the x-axis into the intervals $(-\infty, 3 - \sqrt{17})$, $(3 - \sqrt{17}, 3 + \sqrt{17})$, and $(3 + \sqrt{17}, \infty)$. We let $f(x) = x^2 - 6x - 8$ and test a value in each interval.

$(-\infty, 3 - \sqrt{17})$: $f(-2) = 8 > 0$

$(3 - \sqrt{17}, 3 + \sqrt{17})$: $f(0) = -8 < 0$

$(3 + \sqrt{17}, \infty)$: $f(8) = 8 > 0$

Function values are positive on $(-\infty, 3 - \sqrt{17})$ and $(3 + \sqrt{17}, \infty)$. The solution set is $(-\infty, 3 - \sqrt{17}) \cup (3 + \sqrt{17}, \infty)$ or approximately $(-\infty, -1.123) \cup (7.123, \infty)$.

29.
$4x^3 - 7x^2 \leq 15x$ Polynomial inequality

$4x^3 - 7x^2 - 15x \leq 0$ Equivalent inequality with 0 on one side

$4x^3 - 7x^2 - 15x = 0$ Related equation

$x(4x+5)(x-3) = 0$ Factoring

Using the principle of zero products, we find that the solutions of the related equation are 0, $-\frac{5}{4}$, and 3. These numbers divide the x-axis into the intervals $\left(-\infty, -\frac{5}{4}\right)$, $\left(-\frac{5}{4}, 0\right)$, $(0, 3)$, and $(3, \infty)$. We let $f(x) = 4x^3 - 7x^2 - 15x$ and test a value in each interval.

$\left(-\infty, -\frac{5}{4}\right)$: $f(-2) = -30 < 0$

$\left(-\frac{5}{4}, 0\right)$: $f(-1) = 4 > 0$

$(0, 3)$ $f(1) = -18 < 0$

$(3, \infty)$: $f(4) = 84 > 0$

Function values are negative on $\left(-\infty, -\frac{5}{4}\right)$ and $(0, 3)$. Since the inequality symbol is $\leq$, the endpoints of the intervals must be included in the solution set. It is $\left(-\infty, -\frac{5}{4}\right] \cup [0, 3]$.

30.
$2x^3 - x^2 < 5x$ Polynomial inequality

$2x^3 - x^2 - 5x < 0$ Equivalent inequality with 0 on one side

$2x^3 - x^2 - 5x = 0$ Related equation

$x(2x^2 - x - 5) = 0$

$x = 0$ *or* $2x^2 - x - 5 = 0$

$x = 0$ *or* $x = \frac{1 \pm \sqrt{41}}{4}$

The solutions of the related equation divide the x-axis into four intervals. We let $f(x) = 2x^3 - x^2 - 5x$ and test a value in each interval. Note that $\frac{1 - \sqrt{41}}{4} \approx -1.3508$ and $\frac{1 + \sqrt{41}}{4} \approx 1.8508$.

$\left(-\infty, \frac{1 - \sqrt{41}}{4}\right)$: $f(-2) = -10 < 0$

$\left(\frac{1 - \sqrt{41}}{4}, 0\right)$: $f(-1) = 2 > 0$

$\left(0, \frac{1 + \sqrt{41}}{4}\right)$: $f(1) = -4 < 0$

$\left(\frac{1+\sqrt{41}}{4}, \infty\right)$: $f(2) = 2 > 0$

Function values are negative on $\left(-\infty, \frac{1-\sqrt{41}}{4}\right)$ and $\left(0, \frac{1+\sqrt{41}}{4}\right)$. The solution set is $\left(-\infty, \frac{1-\sqrt{41}}{4}\right) \cup \left(0, \frac{1+\sqrt{41}}{4}\right)$, or approximately $(-\infty, -1.3508) \cup (0, 1.8508)$.

31.

$x^3 + 3x^2 - x - 3 \geq 0$	Polynomial inequality
$x^3 + 3x^2 - x - 3 = 0$	Related equation
$x^2(x+3) - (x+3) = 0$	Factoring
$(x^2 - 1)(x+3) = 0$	
$(x+1)(x-1)(x+3) = 0$	

Using the principle of zero products, we find that the solutions of the related equation are -1, 1, and -3. These numbers divide the x-axis into the intervals $(-\infty, -3)$, $(-3, -1)$, $(-1, 1)$, and $(1, \infty)$. We let $f(x) = x^3 + 3x^2 - x - 3$ and test a value in each interval.

$(-\infty, -3)$: $f(-4) = -15 < 0$

$(-3, -1)$: $f(-2) = 3 > 0$

$(-1, 1)$: $f(0) = -3 < 0$

$(1, \infty)$: $f(2) = 15 > 0$

Function values are positive on $(-3, -1)$ and $(1, \infty)$. Since the inequality symbol is $\geq$, the endpoints of the intervals must be included in the solution set. It is $[-3, -1] \cup [1, \infty)$.

32.

$x^3 + x^2 - 4x - 4 \geq 0$	Polynomial inequality
$x^3 + x^2 - 4x - 4 = 0$	Related equation
$(x+2)(x-2)(x+1) = 0$	Factoring

The solutions of the related equation are -2, 2, and -1. These numbers divide the x-axis into the intervals $(-\infty, -2)$, $(-2, -1)$, $(-1, 2)$, and $(2, \infty)$. We let $f(x) = x^3 + x^2 - 4x - 4$ and test a value in each interval.

$(-\infty, -2)$: $f(-3) = -10 < 0$

$(-2, -1)$: $f(-1.5) = 0.875 > 0$

$(-1, 2)$: $f(0) = -4 < 0$

$(2, \infty)$: $f(3) = 20 > 0$

Function values are positive only on $(-2, -1)$ and $(2, \infty)$. Since the inequality symbol is $\geq$, the endpoints of the interval must be included in the solution set. It is $[-2, -1] \cup [2, \infty)$.

33.

$x^3 - 2x^2 < 5x - 6$	Polynomial inequality
$x^3 - 2x^2 - 5x + 6 < 0$	Equivalent inequality with 0 on one side
$x^3 - 2x^2 - 5x + 6 = 0$	Related equation

Using the techniques of Section 3.3, we find that the solutions of the related equation are -2, 1, and 3. They divide the x-axis into the intervals $(-\infty, -2)$, $(-2, 1)$, $(1, 3)$, and $(3, \infty)$. Let $f(x) = x^3 - 2x^2 - 5x + 6$ and test a value in each interval.

$(-\infty, -2)$: $f(-3) = -24 < 0$

$(-2, 1)$: $f(0) = 6 > 0$

$(1, 3)$: $f(2) = -4 < 0$

$(3, \infty)$: $f(4) = 18 > 0$

Function values are negative on $(-\infty, -2)$ and $(1, 3)$. The solution set is $(-\infty, -2) \cup (1, 3)$.

34.

$x^3 + x \leq 6 - 4x^2$	Polynomial inequality
$x^3 + 4x^2 + x - 6 \leq 0$	Equivalent inequality with 0 on one side
$x^3 + 4x^2 + x - 6 = 0$	Related equation

Using the techniques of Section 3.3, we find that the solutions of the related equation are -3, -2, and 1. They divide the x-axis into the intervals $(-\infty, -3)$, $(-3, -2)$, $(-2, 1)$, and $(1, \infty)$. Let $f(x) = x^3 + 4x^2 + x - 6$ and test a value in each interval.

$(-\infty, -3)$: $f(-4) = -10 < 0$

$(-3, -2)$: $f(-2.5) = 0.875 > 0$

$(-2, 1)$: $f(0) = -6 < 0$

$(1, \infty)$: $f(2) = 20 > 0$

Function values are negative on $(-\infty, -3)$ and $(-2, 1)$. Since the inequality symbol is $\leq$, the endpoints of the intervals must be included in the solution set. It is $(-\infty, -3] \cup [-2, 1]$.

35.

$x^5 + x^2 \geq 2x^3 + 2$	Polynomial inequality
$x^5 - 2x^3 + x^2 - 2 \geq 0$	Related inequality with 0 on one side
$x^5 - 2x^3 + x^2 - 2 = 0$	Related equation
$x^3(x^2 - 2) + x^2 - 2 = 0$	Factoring
$(x^3 + 1)(x^2 - 2) = 0$	

Using the principle of zero products, we find that the real-number solutions of the related equation are -1, $-\sqrt{2}$, and $\sqrt{2}$. These numbers divide the x-axis into the intervals $(-\infty, -\sqrt{2})$, $(-\sqrt{2}, -1)$, $(-1, \sqrt{2})$, and $(\sqrt{2}, \infty)$. We let $f(x) = x^5 - 2x^3 + x^2 - 2$ and test a value in each interval.

$(-\infty, -\sqrt{2})$: $f(-2) = -14 < 0$

$(-\sqrt{2}, -1)$: $f(-1.3) \approx 0.37107 > 0$

$(-1, \sqrt{2})$: $f(0) = -2 < 0$

$(\sqrt{2}, \infty)$: $f(2) = 18 > 0$

Function values are positive on $(-\sqrt{2}, -1)$ and $(\sqrt{2}, \infty)$. Since the inequality symbol is $\geq$, the endpoints of the intervals must be included in the solution set. It is $[-\sqrt{2}, -1] \cup [\sqrt{2}, \infty)$.

36.

$$x^5 + 24 > 3x^3 + 8x^2 \quad \text{Polynomial inequality}$$

$$x^5 - 3x^3 - 8x^2 + 24 > 0 \quad \text{Equivalent inequality with 0 on one side}$$

$$x^5 - 3x^3 - 8x^2 + 24 = 0 \quad \text{Related equation}$$

$$x^3(x^2 - 3) - 8(x^2 - 3) = 0 \quad \text{Factoring}$$

$$(x^3 - 8)(x^2 - 3) = 0$$

Using the principle of zero products, we find that the real-number solutions of the related equation are 2, $-\sqrt{3}$, and $\sqrt{3}$. These numbers divide the x-axis into the intervals $(-\infty, -\sqrt{3})$, $(-\sqrt{3}, \sqrt{3})$, $(\sqrt{3}, 2)$, and $(2, \infty)$. We let $f(x) = x^5 - 3x^3 - 8x^2 + 24$ and test a value in each interval.

$(-\infty, -\sqrt{3})$: $f(-2) = -16 < 0$

$(-\sqrt{3}, \sqrt{3})$: $f(0) = 24 > 0$

$(\sqrt{3}, 2)$: $f(1.8) \approx -0.5203 < 0$

$(2, \infty)$: $f(3) = 114 > 0$

Function values are positive on $(-\sqrt{3}, \sqrt{3})$ and $(2, \infty)$. The solution set is $(-\sqrt{3}, \sqrt{3}) \cup (2, \infty)$.

37.

$$2x^3 + 6 \le 5x^2 + x \quad \text{Polynomial inequality}$$

$$2x^3 - 5x^2 - x + 6 \le 0 \quad \text{Equivalent inequality with 0 on one side}$$

$$2x^3 - 5x^2 - x + 6 = 0 \quad \text{Related equation}$$

Using the techniques of Section 3.3, we find that the solutions of the related equation are -1, $\frac{3}{2}$, and 2. We can also use the graph of $y = 2x^3 - 5x^2 - x + 6$ to find these solutions. They divide the x-axis into the intervals $(-\infty, -1)$, $\left(-1, \frac{3}{2}\right)$, $\left(\frac{3}{2}, 2\right)$, and $(2, \infty)$. Let $f(x) = 2x^3 - 5x^2 - x + 6$ and test a value in each interval.

$(-\infty, -1)$: $f(-2) = -28 < 0$

$\left(-1, \frac{3}{2}\right)$: $f(0) = 6 > 0$

$\left(\frac{3}{2}, 2\right)$: $f(1.6) = -0.208 < 0$

$(2, \infty)$: $f(3) = 12 > 0$

Function values are negative in $(-\infty, -1)$ and $\left(\frac{3}{2}, 2\right)$. Since the inequality symbol is $\le$, the endpoints of the intervals must be included in the solution set. The solution set is $\left(-\infty, -1\right] \cup \left[\frac{3}{2}, 2\right]$.

38.

$$2x^3 + x^2 < 10 + 11x \quad \text{Polynomial inequality}$$

$$2x^3 + x^2 - 11x - 10 < 0 \quad \text{Equivalent inequality with 0 on one side}$$

$$2x^3 + x^2 - 11x - 10 = 0 \quad \text{Related equation}$$

Using the techniques of Section 3.3, we find that the real-number solutions of the related equation are -2, -1, and $\frac{5}{2}$. These numbers divide the x-axis into the intervals $(-\infty, -2)$, $(-2, -1)$, $\left(-1, \frac{5}{2}\right)$, and $\left(\frac{5}{2}, \infty\right)$. We let $f(x) = 2x^3 + x^2 - 11x - 10$ and test a value in each interval.

$(-\infty, -2)$: $f(-3) = -22 < 0$

$(-2, -1)$: $f(-1.5) = 2 > 0$

$\left(-1, \frac{5}{2}\right)$: $f(0) = -10 < 0$

$\left(\frac{5}{2}, \infty\right)$: $f(3) = 20 > 0$

Function values are negative on $(-\infty, -2)$ and $\left(-1, \frac{5}{2}\right)$. The solution set is $(-\infty, -2) \cup \left(-1, \frac{5}{2}\right)$.

39.

$$x^3 + 5x^2 - 25x \le 125 \quad \text{Polynomial inequality}$$

$$x^3 + 5x^2 - 25x - 125 \le 0 \quad \text{Equivalent inequality with 0 on one side}$$

$$x^3 + 5x^2 - 25x - 125 = 0 \quad \text{Related equation}$$

$$x^2(x + 5) - 25(x + 5) = 0 \quad \text{Factoring}$$

$$(x^2 - 25)(x + 5) = 0$$

$$(x + 5)(x - 5)(x + 5) = 0$$

Using the principle of zero products, we find that the solutions of the related equation are -5 and 5. These numbers divide the x-axis into the intervals $(-\infty, -5)$, $(-5, 5)$, and $(5, \infty)$. We let $f(x) = x^3 + 5x^2 - 25x - 125$ and test a value in each interval.

$(-\infty, -5)$: $f(-6) = -11 < 0$

$(-5, 5)$: $f(0) = -125 < 0$

$(5, \infty)$: $f(6) = 121 > 0$

Function values are negative on $(-\infty, -5)$ and $(-5, 5)$. Since the inequality symbol is $\le$, the endpoints of the intervals must be included in the solution set. It is $(-\infty, -5] \cup [-5, 5]$ or $(-\infty, 5]$.

40.

$$x^3 - 9x + 27 \ge 3x^2 \quad \text{Polynomial inequality}$$

$$x^3 - 3x^2 - 9x + 27 \ge 0 \quad \text{Equivalent inequality with 0 on one side}$$

$$x^3 - 3x^2 - 9x + 27 = 0 \quad \text{Related equation}$$

$$x^2(x - 3) - 9(x - 3) = 0 \quad \text{Factoring}$$

$$(x^2 - 9)(x - 3) = 0$$

$$(x + 3)(x - 3)(x - 3) = 0$$

The solutions of the related equation are -3 and 3. These numbers divide the x-axis into the intervals $(-\infty, -3)$, $(-3, 3)$, and $(3, \infty)$. We let $f(x) = x^3 - 3x^2 - 9x + 27$ and test a value in each interval.

$(-\infty, -3)$: $f(-4) = -49 < 0$

$(-3, 3)$: $f(0) = 27 > 0$

$(3, \infty)$: $f(4) = 7 > 0$

Function values are positive only on $(-3, 3)$ and $(3, \infty)$. Since the inequality symbol is $\ge$, the endpoints of the intervals must be included in the solution set. It is $[-3,3]\cup[3,\infty)$, or $[-3,\infty)$.

41. $\dfrac{1}{x+4} > 0$ Rational inequality

$\dfrac{1}{x+4} = 0$ Related equation

The denominator of $f(x) = \dfrac{1}{x+4}$ is 0 when $x=-4$, so the function is not defined for $x=-4$. The related equation has no solution. Thus, the only critical value is -4. It divides the x-axis into the intervals $(-\infty,-4)$ and $(-4,\infty)$. We test a value in each interval.

$(-\infty,-4)$: $f(-5) = -1 < 0$

$(-4,\infty)$: $f(0) = \dfrac{1}{4} > 0$

Function values are positive on $(-4,\infty)$. This can also be determined from the graph of $y = \dfrac{1}{x+4}$. The solution set is $(-4,\infty)$.

42. $\dfrac{1}{x-3} \le 0$ Rational inequality

$\dfrac{1}{x-3} = 0$ Related equation

The denominator of $f(x) = \dfrac{1}{x-3}$ is 0 when $x=3$, so the function is not defined for $x=3$. The related equation has no solution. Thus, the only critical value is 3. It divides the x-axis into the intervals $(-\infty,3)$ and $(3,\infty)$. We test a value in each interval.

$(-\infty,3)$: $f(0) = -\dfrac{1}{3} < 0$

$(3,\infty)$: $f(4) = 1 > 0$

Function values are negative on $(-\infty,3)$. Note that since 3 is not in the domain of $f(x)$, it cannot be included in the solution set. It is $(-\infty,3)$.

43. $\dfrac{-4}{2x+5} < 0$ Rational inequality

$\dfrac{-4}{2x+5} = 0$ Related equation

The denominator of $f(x) = \dfrac{-4}{2x+5}$ is 0 when $x = -\dfrac{5}{2}$, so the function is not defined for $x = -\dfrac{5}{2}$. The related equation has no solution. Thus, the only critical value is $-\dfrac{5}{2}$. It divides the x-axis into the intervals $\left(-\infty,-\dfrac{5}{2}\right)$ and $\left(-\dfrac{5}{2},\infty\right)$. We test a value in each interval.

$\left(-\infty,-\dfrac{5}{2}\right)$: $f(-3) = 4 > 0$

$\left(-\dfrac{5}{2},\infty\right)$: $f(0) = -\dfrac{4}{5} < 0$

Function values are negative on $\left(-\dfrac{5}{2},\infty\right)$. The solution set is $\left(-\dfrac{5}{2},\infty\right)$.

44. $\dfrac{-2}{5-x} \ge 0$ Rational inequality

$\dfrac{-2}{5-x} = 0$ Related equation

The denominator of $f(x) = \dfrac{-2}{5-x}$ is 0 when $x=5$, so the function is not defined for $x=5$. The related equation has no solution. Thus, the only critical value is 5. It divides the x-axis into the intervals $(-\infty,5)$ and $(5,\infty)$. We test a value in each interval.

$(-\infty,5)$: $f(0) = -\dfrac{2}{5} < 0$

$(5,\infty)$: $f(6) = 2 > 0$

Function values are positive on $(5,\infty)$. Note that since 5 is not in the domain of $f(x)$, it cannot be included in the solution set. It is $(5,\infty)$.

45. $\dfrac{x-4}{x+3} - \dfrac{x+2}{x-1} \le 0$

The denominator of $f(x) = \dfrac{x-4}{x+3} - \dfrac{x+2}{x-1}$ is 0 when $x = -3$ or $x=1$, so the function is not defined for these values of x. We solve the related equation $f(x)=0$.

$$\frac{x-4}{x+3} - \frac{x+2}{x-1} = 0$$
$$(x+3)(x-1)\left(\frac{x-4}{x+3} - \frac{x+2}{x-1}\right) = (x+3)(x-1)\cdot 0$$
$$(x-1)(x-4) - (x+3)(x+2) = 0$$
$$x^2-5x+4-(x^2+5x+6) = 0$$
$$-10x-2 = 0$$
$$-10x = 2$$
$$x = -\frac{1}{5}$$

The critical values are -3, $-\dfrac{1}{5}$, and 1. They divide the x-axis into the intervals $(-\infty,-3)$, $\left(3,-\dfrac{1}{5}\right)$, $\left(-\dfrac{1}{5},1\right)$, and $(1,\infty)$. We test a value in each interval.

$(-\infty,-3)$: $f(-4) = 7.6 > 0$

$\left(-3,-\dfrac{1}{5}\right)$: $f(-1) = -2 < 0$

$\left(-\dfrac{1}{5},1\right)$: $f(0) = \dfrac{2}{3} > 0$

$(1,\infty)$: $f(2) = -4.4 < 0$

Function values are negative on $\left(-3,-\dfrac{1}{5}\right)$ and $(1,\infty)$. Note that since the inequality symbol is $\le$ and $f\left(-\dfrac{1}{5}\right) = 0$, then $-\dfrac{1}{5}$ must be included in the solution set. Note also that since neither -3 nor 1 is in the domain of $f(x)$, they are not included in the solution set. It is $\left(-3,-\dfrac{1}{5}\right]\cup(1,\infty)$.

46. $\dfrac{x+1}{x-2} + \dfrac{x-3}{x-1} < 0$

The denominator of $f(x) = \dfrac{x+1}{x-2} + \dfrac{x-3}{x-1}$ is 0 when $x=2$

or $x = 1$, so the function is not defined for these values of x. We solve the related equation $f(x) = 0$.

$$\frac{x+1}{x-2} + \frac{x-3}{x-1} = 0$$

$$(x-1)(x+1) + (x-2)(x-3) = 0$$

Multiplying by $(x-2)(x-1)$

$$x^2 - 1 + x^2 - 5x + 6 = 0$$

$$2x^2 - 5x + 5 = 0$$

This equation has no real-number solutions, so the critical values are 1 and 2. They divide the x-axis into the intervals $(-\infty, 1)$, $(1, 2)$, and $(2, \infty)$. We test a value in each interval.

$(-\infty, 1)$: $f(0) = 2.5 > 0$

$(1, 2)$: $f(1.5) = -8 < 0$

$(2, \infty)$: $f(3) = 4 > 0$

Function values are negative on $(1, 2)$. The solution set is $(1, 2)$.

47. $\dfrac{2x-1}{x+3} \geq \dfrac{x+1}{3x+1}$ Rational inequality

$\dfrac{2x-1}{x+3} - \dfrac{x+1}{3x+1} \geq 0$ Equivalent inequality with 0 on one side

The denominator of $f(x) = \dfrac{2x-1}{x+3} - \dfrac{x+1}{3x+1}$ is 0 when $x = -3$ or $x = -\dfrac{1}{3}$, so the function is not defined for these values of x. We solve the related equation $f(x) = 0$.

$$\frac{2x-1}{x+3} - \frac{x+1}{3x+1} = 0$$

$$(x+3)(3x+1)\left(\frac{2x-1}{x+3} - \frac{x+1}{3x+1}\right) = (x+3)(3x+1) \cdot 0$$

$$(3x+1)(2x-1) - (x+3)(x+1) = 0$$

$$6x^2 - x - 1 - (x^2 + 4x + 3) = 0$$

$$5x^2 - 5x - 4 = 0$$

Using the quadratic formula we find that $x = \dfrac{5 \pm \sqrt{105}}{10}$. Then the critical values are -3, $\dfrac{5-\sqrt{105}}{10}$, $-\dfrac{1}{3}$, and $\dfrac{5+\sqrt{105}}{10}$. They divide the x-axis into the intervals $(-\infty, -3)$, $\left(-3, \dfrac{5-\sqrt{105}}{10}\right)$, $\left(\dfrac{5-\sqrt{105}}{10}, -\dfrac{1}{3}\right)$, $\left(-\dfrac{1}{3}, \dfrac{5+\sqrt{105}}{10}\right)$, and $\left(\dfrac{5+\sqrt{105}}{10}, \infty\right)$. We test a value in each interval.

$(-\infty, -3)$: $f(-4) \approx 8.7273 > 0$

$\left(-3, \dfrac{5-\sqrt{105}}{10}\right)$: $f(-2) = -5.2 < 0$

$\left(\dfrac{5-\sqrt{105}}{10}, -\dfrac{1}{3}\right)$: $f(-0.4) \approx 2.3077 > 0$

$\left(-\dfrac{1}{3}, \dfrac{5+\sqrt{105}}{10}\right)$: $f(0) \approx -1.333 < 0$

$\left(\dfrac{5+\sqrt{105}}{10}, \infty\right)$: $f(2) \approx 0.1714 > 0$

Function values are positive on $(-\infty, -3)$, $\left(\dfrac{5-\sqrt{105}}{10}, -\dfrac{1}{3}\right)$ and $\left(\dfrac{5+\sqrt{105}}{10}, \infty\right)$. Note that since the inequality symbol is $\geq$ and $f\left(\dfrac{5 \pm \sqrt{105}}{10}\right) = 0$, then $\dfrac{5 \pm \sqrt{105}}{10}$ must be included in the solution set. Note also that since neither -3 nor $-\dfrac{1}{3}$ is in the domain of $f(x)$, they are not included in the solution set. It is

$(-\infty, -3) \cup \left[\dfrac{5-\sqrt{105}}{10}, -\dfrac{1}{3}\right) \cup \left[\dfrac{5+\sqrt{105}}{10}, \infty\right)$.

48. $\dfrac{x+5}{x-4} > \dfrac{3x+2}{2x+1}$ Rational inequality

$\dfrac{x+5}{x-4} - \dfrac{3x+2}{2x+1} > 0$ Equivalent inequality with 0 on one side

The denominator of $f(x) = \dfrac{x+5}{x-4} - \dfrac{3x+2}{2x+1}$ is 0 when $x = 4$ or $x = -\dfrac{1}{2}$, so the function is not defined for these values of x. We solve the related equation $f(x) = 0$.

$$\frac{x+5}{x-4} - \frac{3x+2}{2x+1} = 0$$

$$(2x+1)(x+5) - (x-4)(3x+2) = 0$$

Multiplying by $(x-4)(2x+1)$

$$2x^2 + 11x + 5 - (3x^2 - 10x - 8) = 0$$

$$-x^2 + 21x + 13 = 0$$

Using the quadratic formula we find that $x = \dfrac{21 \pm \sqrt{493}}{2}$. Then the critical values are $\dfrac{21-\sqrt{493}}{2}$, $-\dfrac{1}{2}$, 4, and $\dfrac{21+\sqrt{493}}{2}$. They divide the x-axis into the intervals $\left(-\infty, \dfrac{21-\sqrt{493}}{2}\right)$, $\left(\dfrac{21-\sqrt{493}}{2}, -\dfrac{1}{2}\right)$, $\left(-\dfrac{1}{2}, 4\right)$, $\left(4, \dfrac{21+\sqrt{493}}{2}\right)$, and $\left(\dfrac{21+\sqrt{493}}{2}, \infty\right)$. We test a value in each interval.

$\left(-\infty, \dfrac{21-\sqrt{493}}{2}\right)$: $f(-1) = -1.8 < 0$

$\left(\dfrac{21-\sqrt{493}}{2}, -\dfrac{1}{2}\right)$: $f(-0.55) = 2.522 > 0$

$\left(-\dfrac{1}{2}, 4\right)$: $f(0) = -3.25 < 0$

$\left(4, \dfrac{21+\sqrt{493}}{2}\right)$: $f(5) \approx 8.4545 > 0$

$\left(\dfrac{21+\sqrt{493}}{2}, \infty\right)$: $f(22) \approx -0.0111 < 0$

Function values are positive on $\left(\dfrac{21-\sqrt{493}}{2}, -\dfrac{1}{2}\right)$ and

$\left(4, \dfrac{21+\sqrt{493}}{2}\right)$. The solution set is

$\left(\dfrac{21-\sqrt{493}}{2}, -\dfrac{1}{2}\right) \cup \left(4, \dfrac{21+\sqrt{493}}{2}\right)$.

49. $\dfrac{x+1}{x-2} \geq 3$ Rational inequality

$\dfrac{x+1}{x-2} - 3 \geq 0$ Equivalent inequality with 0 on one side

The denominator of $f(x) = \dfrac{x+1}{x-2} - 3$ is 0 when $x = 2$, so the function is not defined for this value of x. We solve the related equation $f(x) = 0$.

$$\begin{aligned}
\frac{x+1}{x-2} - 3 &= 0 \\
(x-2)\left(\frac{x+1}{x-2} - 3\right) &= (x-2)\cdot 0 \\
x+1-3(x-2) &= 0 \\
x+1-3x+6 &= 0 \\
-2x+7 &= 0 \\
-2x &= -7 \\
x &= \frac{7}{2}
\end{aligned}$$

The critical values are 2 and $\dfrac{7}{2}$. They divide the x-axis into the intervals $(-\infty, 2)$, $\left(2, \dfrac{7}{2}\right)$, and $\left(\dfrac{7}{2}, \infty\right)$. We test a value in each interval.

$(-\infty, 2)$: $f(0) = -3.5 < 0$

$\left(2, \dfrac{7}{2}\right)$: $f(3) = 1 > 0$

$\left(\dfrac{7}{2}, \infty\right)$: $f(4) = -0.5 < 0$

Function values are positive on $\left(2, \dfrac{7}{2}\right)$. Note that since the inequality symbol is $\geq$ and $f\left(\dfrac{7}{2}\right) = 0$, then $\dfrac{7}{2}$ must be included in the solution set. Note also that since 2 is not in the domain of $f(x)$, it is not included in the solution set. It is $\left(2, \dfrac{7}{2}\right]$.

50. $\dfrac{x}{x-5} < 2$ Rational inequality

$\dfrac{x}{x-5} - 2 < 0$ Equivalent inequality with 0 on one side

The denominator of $f(x) = \dfrac{x}{x-5} - 2$ is 0 when $x = 5$, so the function is not defined for this value of x. We solve the related equation $f(x) = 0$.

$$\begin{aligned}
\frac{x}{x-5} - 2 &= 0 \\
x - 2(x-5) &= 0 \quad \text{Multiplying by } x-5 \\
x - 2x + 10 &= 0 \\
-x + 10 &= 0 \\
x &= 10
\end{aligned}$$

The critical values are 5 and 10. They divide the x-axis into the intervals $(-\infty, 5)$, $(5, 10)$, and $(10, \infty)$. We test a value in each interval.

$(-\infty, 5)$: $f(0) = -2 < 0$

$(5, 10)$: $f(6) = 4 > 0$

$(10, \infty)$: $f(11) = -\dfrac{1}{6} < 0$

Function values are negative on $(-\infty, 5)$ and $(10, \infty)$. The solution set is $(-\infty, 5) \cup (10, \infty)$.

51. $x - 2 > \dfrac{1}{x}$ Rational inequality

$x - 2 - \dfrac{1}{x} > 0$ Equivalent inequality with 0 on one side

The denominator of $f(x) = x - 2 - \dfrac{1}{x}$ is 0 when $x = 0$, so the function is not defined for this value of x. We solve the related equation $f(x) = 0$.

$$\begin{aligned}
x - 2 - \frac{1}{x} &= 0 \\
x\left(x - 2 - \frac{1}{x}\right) &= x \cdot 0 \\
x^2 - 2x - x \cdot \frac{1}{x} &= 0 \\
x^2 - 2x - 1 &= 0
\end{aligned}$$

Using the quadratic formula we find that $x = 1 \pm \sqrt{2}$. The critical values are $1 - \sqrt{2}$, 0, and $1 + \sqrt{2}$. They divide the x-axis into the intervals $(-\infty, 1-\sqrt{2})$, $(1-\sqrt{2}, 0)$, $(0, 1+\sqrt{2})$, and $(1+\sqrt{2}, \infty)$. We test a value in each interval.

$(-\infty, 1-\sqrt{2})$: $f(-1) = -2 < 0$

$(1-\sqrt{2}, 0)$: $f(-0.1) = 7.9 > 0$

$(0, 1+\sqrt{2})$: $f(1) = -2 < 0$

$(1+\sqrt{2}, \infty)$: $f(3) = \dfrac{2}{3} > 0$

Function values are positive on $(1-\sqrt{2}, 0)$ and $(1+\sqrt{2}, \infty)$. The solution set is $(1-\sqrt{2}, 0) \cup (1+\sqrt{2}, \infty)$.

52. $4 \geq \dfrac{4}{x} + x$ Rational inequality

$4 - \dfrac{4}{x} - x \geq 0$ Equivalent inequality with 0 on one side

The denominator of $f(x) = 4 - \dfrac{4}{x} - x$ is 0 when $x = 0$, so the function is not defined for this value of x. We solve the related equation $f(x) = 0$.

$$\begin{aligned}
4 - \frac{4}{x} - x &= 0 \\
4x - 4 - x^2 &= 0 \quad \text{Multiplying by } x \\
-(x-2)^2 &= 0 \\
x &= 2
\end{aligned}$$

The critical values are 0 and 2. They divide the x-axis into the intervals $(-\infty, 0)$, $(0, 2)$, and $(2, \infty)$. We test a value in each interval.

$(-\infty, 0)$: $f(-1) = 9 > 0$

$(0, 2)$: $f(1) = -1 < 0$

$(2, \infty)$: $f(3) = -\frac{1}{3} < 0$

Function values are positive on $(-\infty, 0)$. Note that since the inequality symbol is $\geq$ and $f(2) = 0$, then 2 must be included in the solution set. Note also that since 0 is not in the domain of $f(x)$, it is not included in the solution set. It is $(-\infty, 0) \cup \{2\}$.

53.
$$\frac{2}{x^2 - 4x + 3} \leq \frac{5}{x^2 - 9}$$
$$\frac{2}{x^2 - 4x + 3} - \frac{5}{x^2 - 9} \leq 0$$
$$\frac{2}{(x-1)(x-3)} - \frac{5}{(x+3)(x-3)} \leq 0$$

The denominator of $f(x) = \frac{2}{(x-1)(x-3)} - \frac{5}{(x+3)(x-3)}$ is 0 when $x = 1$, 3, or -3, so the function is not defined for these values of x. We solve the related equation $f(x) = 0$.

$$\frac{2}{(x-1)(x-3)} - \frac{5}{(x+3)(x-3)} = 0$$
$$(x-1)(x-3)(x+3)\left(\frac{2}{(x-1)(x-3)} - \frac{5}{(x+3)(x-3)}\right) = (x-1)(x-3)(x+3) \cdot 0$$
$$2(x+3) - 5(x-1) = 0$$
$$2x + 6 - 5x + 5 = 0$$
$$-3x + 11 = 0$$
$$-3x = -11$$
$$x = \frac{11}{3}$$

The critical values are -3, 1, 3, and $\frac{11}{3}$. They divide the x-axis into the intervals $(-\infty, -3)$, $(-3, 1)$, $(1, 3)$, $\left(3, \frac{11}{3}\right)$, and $\left(\frac{11}{3}, \infty\right)$. We test a value in each interval.

$(-\infty, -3)$: $f(-4) \approx -0.6571 < 0$

$(-3, 1)$: $f(0) \approx 1.2222 > 0$

$(1, 3)$: $f(2) = -1 < 0$

$\left(3, \frac{11}{3}\right)$: $f(3.5) \approx 0.6154 > 0$

$\left(\frac{11}{3}, \infty\right)$: $f(4) \approx -0.0476 < 0$

Function values are negative on $(-\infty, -3)$, $(1, 3)$, and $\left(\frac{11}{3}, \infty\right)$. Note that since the inequality symbol is $\leq$ and $f\left(\frac{11}{3}\right) = 0$, then $\frac{11}{3}$ must be included in the solution set. Note also that since -3, 1, and 3 are not in the domain of $f(x)$, they are not included in the solution set. It is $(-\infty, -3) \cup (1, 3) \cup \left[\frac{11}{3}, \infty\right)$.

54.
$$\frac{3}{x^2 - 4} \leq \frac{5}{x^2 + 7x + 10}$$
$$\frac{3}{(x+2)(x-2)} - \frac{5}{(x+2)(x+5)} \leq 0$$

The denominator of $f(x) = \frac{3}{(x+2)(x-2)} - \frac{5}{(x+2)(x+5)}$ is 0 when $x = -2$, 2, or -5, so the function is not defined for these values of x. We solve the related equation $f(x) = 0$.

$$\frac{3}{(x+2)(x-2)} - \frac{5}{(x+2)(x+5)} = 0$$
$$3(x+5) - 5(x-2) = 0 \quad \text{Multiplying by } (x+2)(x-2)(x+5)$$
$$3x + 15 - 5x + 10 = 0$$
$$-2x + 25 = 0$$
$$x = \frac{25}{2}$$

The critical values are -5, -2, 2, and $\frac{25}{2}$. They divide the x-axis into the intervals $(-\infty, -5)$, $(-5, -2)$, $(-2, 2)$, $\left(2, \frac{25}{2}\right)$, and $\left(\frac{25}{2}, \infty\right)$. We test a value in each interval.

$(-\infty, -5)$: $f(-6) \approx -1.156 < 0$

$(-5, -2)$: $f(-3) = 3.1 > 0$

$(-2, 2)$: $f(0) = -1.25 < 0$

$\left(2, \frac{25}{2}\right)$: $f(3) = 0.475 > 0$

$\left(\frac{25}{2}, \infty\right)$: $f(13) \approx -0.0003 < 0$

Function values are negative on $(-\infty, -5)$, $(-2, 2)$, and $\left(\frac{25}{2}, \infty\right)$. Note that since the inequality symbol is $\leq$ and $f\left(\frac{25}{2}\right) = 0$, then $\frac{25}{2}$ must be included in the solution set. Note also that since -5, -2, and 2 are not in the domain of $f(x)$, they are not included in the solution set. It is $(-\infty, -5) \cup (-2, 2) \cup \left[\frac{25}{2}, \infty\right)$.

55.
$$\frac{3}{x^2 + 1} \geq \frac{6}{5x^2 + 2}$$
$$\frac{3}{x^2 + 1} - \frac{6}{5x^2 + 2} \geq 0$$

The denominator of $f(x) = \frac{3}{x^2 + 1} - \frac{6}{5x^2 + 2}$ has no real-number zeros. We solve the related equation $f(x) = 0$.

$$\frac{3}{x^2+1}-\frac{6}{5x^2+2}=0$$

$$(x^2+1)(5x^2+2)\left(\frac{3}{x^2+1}-\frac{6}{5x^2+2}\right)=(x^2+1)(5x^2+2)\cdot 0$$

$$3(5x^2+2)-6(x^2+1)=0$$

$$15x^2+6-6x^2-6=0$$

$$9x^2=0$$

$$x^2=0$$

$$x=0$$

The only critical value is 0. It divides the x-axis into the intervals $(-\infty,0)$ and $(0,\infty)$. We test a value in each interval.

$(-\infty,0)$: $f(-1)\approx 0.64286>0$

$(0,0)$: $f(1)\approx 0.64286>0$

Function values are positive on both intervals. Note that since the inequality symbol is $\geq$ and $f(0)=0$, then 0 must be included in the solution set. It is $(-\infty,0]\cup[0,\infty)$, or $(-\infty,\infty)$.

56.

$$\frac{4}{x^2-9}<\frac{3}{x^2-25}$$

$$\frac{4}{x^2-9}-\frac{3}{x^2-25}<0$$

$$\frac{4}{(x+3)(x-3)}-\frac{3}{(x+5)(x-5)}<0$$

The denominator of $f(x)=\frac{4}{(x+3)(x-3)}-\frac{3}{(x+5)(x-5)}$ is 0 when $x=-3$, 3, -5, or 5, so the function is not defined for these values of x. We solve the related equation $f(x)=0$.

$$\frac{4}{(x+3)(x-3)}-\frac{3}{(x+5)(x-5)}=0$$

$$4(x+5)(x-5)-3(x+3)(x-3)=0$$

Multiplying by $(x+3)(x-3)(x+5)(x-5)$

$$4x^2-100-3x^2+27=0$$

$$x^2-73=0$$

$$x=\pm\sqrt{73}$$

The critical values are $-\sqrt{73}$, -5, -3, 3, 5, and $\sqrt{73}$. They divide the x-axis into the intervals $(-\infty,-\sqrt{73})$, $(-\sqrt{73},-5)$, $(-5,-3)$, $(-3,3)$, $(3,5)$, $(5,\sqrt{73})$, and $(\sqrt{73},\infty)$. We test a value in each interval.

$(-\infty,-\sqrt{73})$: $f(-9)\approx 0.00198>0$

$(-\sqrt{73},-5)$: $f(-6)\approx -0.1246<0$

$(-5,-3)$: $f(-4)\approx 0.90476>0$

$(-3,3)$: $f(0)\approx -0.3244<0$

$(3,5)$: $f(4)\approx 0.90476>0$

$(5,\sqrt{73})$: $f(6)\approx -0.1246<0$

$(\sqrt{73},\infty)$: $f(9)\approx 0.00198>0$

Function values are negative on $(-\sqrt{73},-5)$, $(-3,3)$, and $(5,\sqrt{73})$. The solution set is $(-\sqrt{73},-5)\cup(-3,3)\cup(5,\sqrt{73})$.

57.

$$\frac{5}{x^2+3x}<\frac{3}{2x+1}$$

$$\frac{5}{x^2+3x}-\frac{3}{2x+1}<0$$

$$\frac{5}{x(x+3)}-\frac{3}{2x+1}<0$$

The denominator of $f(x)=\frac{5}{x(x+3)}-\frac{3}{2x+1}$ is 0 when $x=0$, -3, or $-\frac{1}{2}$, so the function is not defined for these values of x. We solve the related equation $f(x)=0$.

$$\frac{5}{x(x+3)}-\frac{3}{2x+1}=0$$

$$x(x+3)(2x+1)\left(\frac{5}{x(x+3)}-\frac{3}{2x+1}\right)=x(x+3)(2x+1)\cdot 0$$

$$5(2x+1)-3x(x+3)=0$$

$$10x+5-3x^2-9x=0$$

$$-3x^2+x+5=0$$

Using the quadratic formula we find that $x=\frac{1\pm\sqrt{61}}{6}$. The critical values are -3, $\frac{1-\sqrt{61}}{6}$, $-\frac{1}{2}$, 0, and $\frac{1+\sqrt{61}}{6}$. They divide the x-axis into the intervals $(-\infty,-3)$, $\left(-3,\frac{1-\sqrt{61}}{6}\right)$, $\left(\frac{1-\sqrt{61}}{6},-\frac{1}{2}\right)$, $\left(-\frac{1}{2},0\right)$, $\left(0,\frac{1+\sqrt{61}}{6}\right)$, and $\left(\frac{1+\sqrt{61}}{6},\infty\right)$.

We test a value in each interval.

$(-\infty,-3)$: $f(-4)\approx 1.6786>0$

$\left(-3,\frac{1-\sqrt{61}}{6}\right)$: $f(-2)=-1.5<0$

$\left(\frac{1-\sqrt{61}}{6},-\frac{1}{2}\right)$: $f(-1)=0.5>0$

$\left(-\frac{1}{2},0\right)$: $f(-0.1)\approx -20.99<0$

$\left(0,\frac{1+\sqrt{61}}{6}\right)$: $f(1)=0.25>0$

$\left(\frac{1+\sqrt{61}}{6},\infty\right)$: $f(2)=-0.1<0$

Function values are negative on $\left(-3,\frac{1-\sqrt{61}}{6}\right)$, $\left(-\frac{1}{2},0\right)$ and $\left(\frac{1+\sqrt{61}}{6},\infty\right)$. The solution set is $\left(-3,\frac{1-\sqrt{61}}{6}\right)\cup\left(-\frac{1}{2},0\right)\cup\left(\frac{1+\sqrt{61}}{6},\infty\right)$.

58. $\dfrac{2}{x^2+3} > \dfrac{3}{5+4x^2}$

$\dfrac{2}{x^2+3} - \dfrac{3}{5+4x^2} > 0$

The denominator of $f(x) = \dfrac{2}{x^2+3} - \dfrac{3}{5+4x^2}$ has no real-number zeros. We solve the related equation $f(x) = 0$.

$\dfrac{2}{x^2+3} - \dfrac{3}{5+4x^2} = 0$

$2(5+4x^2) - 3(x^2+3) = 0$ Multiplying by $(x^2+3)(5+4x^2)$

$10 + 8x^2 - 3x^2 - 9 = 0$

$5x^2 + 1 = 0$

This equation has no real-number solutions. Thus, there are no critical values. We test a value in $(-\infty, \infty)$: $f(0) = \dfrac{1}{15} > 0$. The function is positive on $(-\infty, \infty)$. This is the solution set.

59. $\dfrac{5x}{7x-2} > \dfrac{x}{x+1}$

$\dfrac{5x}{7x-2} - \dfrac{x}{x+1} > 0$

The denominator of $f(x) = \dfrac{5x}{7x-2} - \dfrac{x}{x+1}$ is 0 when $x = \dfrac{2}{7}$ or $x = -1$, so the function is not defined for these values of x. We solve the related equation $f(x) = 0$.

$\dfrac{5x}{7x-2} - \dfrac{x}{x+1} = 0$

$(7x-2)(x+1)\left(\dfrac{5x}{7x-2} - \dfrac{x}{x+1}\right) = (7x-2)(x+1) \cdot 0$

$5x(x+1) - x(7x-2) = 0$

$5x^2 + 5x - 7x^2 + 2x = 0$

$-2x^2 + 7x = 0$

$-x(2x-7) = 0$

$x = 0$ *or* $x = \dfrac{7}{2}$

The critical values are -1, 0, $\dfrac{2}{7}$, and $\dfrac{7}{2}$. They divide the x-axis into the intervals $(-\infty, -1)$, $(-1, 0)$, $\left(0, \dfrac{2}{7}\right)$, $\left(\dfrac{2}{7}, \dfrac{7}{2}\right)$, and $\left(\dfrac{7}{2}, \infty\right)$. We test a value in each interval.

$(-\infty, -1)$: $f(-2) = -1.375 < 0$

$(-1, 0)$: $f(-0.5) \approx 1.4545 > 0$

$\left(0, \dfrac{2}{7}\right)$: $f(0.1) \approx -0.4755 < 0$

$\left(\dfrac{2}{7}, \dfrac{7}{2}\right)$: $f(1) = 0.5 > 0$

$\left(\dfrac{7}{2}, \infty\right)$: $f(4) \approx -0.0308 < 0$

Function values are positive on $(-1, 0)$ and $\left(\dfrac{2}{7}, \dfrac{7}{2}\right)$. The solution set is $(-1, 0) \cup \left(\dfrac{2}{7}, \dfrac{7}{2}\right)$.

60. $\dfrac{x^2-x-2}{x^2+5x+6} < 0$

$\dfrac{x^2-x-2}{(x+3)(x+2)} < 0$

The denominator of $f(x) = \dfrac{x^2-x-2}{(x+3)(x+2)}$ is 0 when $x = -3$ or $x = -2$, so the function is not defined for these values of x. We solve the related equation $f(x) = 0$.

$\dfrac{x^2-x-2}{(x+3)(x+2)} = 0$

$(x-2)(x+1) = 0$

$x = 2$ *or* $x = -1$

The critical values are -3, -2, -1, and 2. They divide the x-axis into the intervals $(-\infty, -3)$, $(-3, -2)$, $(-2, -1)$, $(-1, 2)$, and $(2, \infty)$. We test a value in each interval.

$(-\infty, -3)$: $f(-4) = 9 > 0$

$(-3, -2)$: $f(-2.5) = -27 < 0$

$(-2, -1)$: $f(-1.5) \approx 2.3333 > 0$

$(-1, 2)$: $f(0) \approx -0.3333 < 0$

$(2, \infty)$: $f(3) \approx 0.13333 > 0$

Function values are negative on $(-3, -2)$ and $(-1, 2)$. The solution set is $(-3, -2) \cup (-1, 2)$.

61. $\dfrac{x}{x^2+4x-5} + \dfrac{3}{x^2-25} \le \dfrac{2x}{x^2-6x+5}$

$\dfrac{x}{x^2+4x-5} + \dfrac{3}{x^2-25} - \dfrac{2x}{x^2-6x+5} \le 0$

$\dfrac{x}{(x+5)(x-1)} + \dfrac{3}{(x+5)(x-5)} - \dfrac{2x}{(x-5)(x-1)} \le 0$

The denominator of

$f(x) = \dfrac{x}{(x+5)(x-1)} + \dfrac{3}{(x+5)(x-5)} - \dfrac{2x}{(x-5)(x-1)}$

is 0 when $x = -5$, 1, or 5, so the function is not defined for these values of x. We solve the related equation $f(x) = 0$.

$\dfrac{x}{(x+5)(x-1)} + \dfrac{3}{(x+5)(x-5)} - \dfrac{2x}{(x-5)(x-1)} = 0$

$x(x-5) + 3(x-1) - 2x(x+5) = 0$

Multiplying by $(x+5)(x-1)(x-5)$

$x^2 - 5x + 3x - 3 - 2x^2 - 10x = 0$

$-x^2 - 12x - 3 = 0$

$x^2 + 12x + 3 = 0$

Using the quadratic formula, we find that $x = -6 \pm \sqrt{33}$. The critical values are $-6-\sqrt{33}$, -5, $-6+\sqrt{33}$, 1, and 5. They divide the x-axis into the intervals $(-\infty, -6-\sqrt{33})$, $(-6-\sqrt{33}, -5)$, $(-5, -6+\sqrt{33})$, $(-6+\sqrt{33}, 1)$, $(1, 5)$, and $(5, \infty)$. We test a value in each interval.

$(-\infty, -6-\sqrt{33})$: $f(-12) \approx 0.00194 > 0$

$(-6-\sqrt{33}, -5)$: $f(-6) \approx -0.4286 < 0$

$(-5, -6+\sqrt{33})$: $f(-1) \approx 0.16667 > 0$

$(-6+\sqrt{33},1)$: $f(0)=-0.12<0$

$(1,5)$: $f(2)\approx 1.4762>0$

$(5,\infty)$: $f(6)\approx -2.018<0$

Function values are negative on $(-6-\sqrt{33},-5)$, $(-6+\sqrt{33},1)$, and $(5,\infty)$. Note that since the inequality symbol is $\le$ and $f(-6\pm\sqrt{33})=0$, then $-6-\sqrt{33}$ and $-6+\sqrt{33}$ must be included in the solution set. Note also that since -5, 1, and 5 are not in the domain of $f(x)$, they are not included in the solution set. It is $[-6-\sqrt{33},-5)\cup[-6+\sqrt{33},1)\cup(5,\infty)$.

62. $$\frac{2x}{x^2-9}+\frac{x}{x^2+x-12}\ge\frac{3x}{x^2+7x+12}$$

$$\frac{2x}{(x+3)(x-3)}+\frac{x}{(x+4)(x-3)}-\frac{3x}{(x+4)(x+3)}\ge 0$$

The denominator of $f(x)=\dfrac{2x}{(x+3)(x-3)}+\dfrac{x}{(x+4)(x-3)}-\dfrac{3x}{(x+4)(x+3)}$ is 0 when $x=-3$, 3, or -4, so the function is not defined for these values of x. We solve the related equation $f(x)=0$.

$$\frac{2x}{(x+3)(x-3)}+\frac{x}{(x+4)(x-3)}-\frac{3x}{(x+4)(x+3)}=0$$
$$2x(x+4)+x(x+3)-3x(x-3)=0$$
$$2x^2+8x+x^2+3x-3x^2+9x=0$$
$$20x=0$$
$$x=0$$

The critical values are -4, -3, 0, and 3. They divide the x-axis into the intervals $(-\infty,-4)$, $(-4,-3)$, $(-3,0)$, $(0,3)$, and $(3,\infty)$. We test a value in each interval.

$(-\infty,-4)$: $f(-5)=6.25>0$

$(-4,-3)$: $f(-3.5)\approx -43.08<0$

$(-3,0)$: $f(-1)\approx 0.83333>0$

$(0,3)$: $f(1)=-0.5<0$

$(3,\infty)$: $f(4)\approx 1.4286>0$

Function values are positive on $(-\infty,-4)$, $(-3,0)$, and $(3,\infty)$. Note that since the inequality symbol is $\ge$ and $f(0)=0$, then 0 must be included in the solution set. Note also that since -4, -3, and 3 are not in the domain of $f(x)$, they are not included in the solution set. It is $(-\infty,-4)\cup(-3,0]\cup(3,\infty)$.

63. We write and solve a rational inequality.

$$\frac{4t}{t^2+1}+98.6>100$$
$$\frac{4t}{t^2+1}-1.4>0$$

The denominator of $f(t)=\dfrac{4t}{t^2+1}-1.4$ has no real-number zeros. We solve the related equation $f(t)=0$.

$$\frac{4t}{t^2+1}-1.4=0$$
$$4t-1.4(t^2+1)=0\quad\text{Multiplying by } t^2+1$$
$$4t-1.4t^2-1.4=0$$

Using the quadratic formula, we find that $t=\dfrac{4\pm\sqrt{8.16}}{2.8}$; that is, $t\approx 0.408$ or $t\approx 2.449$. These numbers divide the t-axis into the intervals $(-\infty,0.408)$, $(0.408,2.449)$, and $(2.449,\infty)$. We test a value in each interval.

$(-\infty,0.408)$: $f(0)=-1.4<0$

$(0.408,2.449)$: $f(1)=0.6>0$

$(2.449,\infty)$: $f(3)=-0.2<0$

Function values are positive on $(0.408,2.449)$. The solution set is $(0.408,2.449)$.

64. We write and solve a rational inequality.

$$\frac{500t}{2t^2+9}\ge 40$$
$$\frac{500t}{2t^2+9}-40\ge 0$$

The denominator of $f(t)=\dfrac{500t}{2t^2+9}-40$ has no real-number zeros. We solve the related equation $f(t)=0$.

$$\frac{500t}{2t^2+9}-40=0$$
$$500t-80t^2-360=0\quad\text{Multiplying by } 2t^2+9$$

Using the quadratic formula, we find that $t=\dfrac{25\pm\sqrt{337}}{8}$; that is, $t\approx 0.830$ or $t\approx 5.420$. These numbers divide the t-axis into the intervals $(-\infty,0.830)$, $(0.830,5.420)$, and $(5.420,\infty)$. We test a value in each interval.

$(-\infty,0.830)$: $f(0)=-40<0$

$(0.830,5.420)$: $f(1)\approx 5.4545>0$

$(5.420,\infty)$: $f(6)\approx -2.963<0$

Function values are positive on $(0.830,5.420)$. The solution set is $[0.830,5.420]$.

65. a) We write and solve a polynomial inequality.

$$-3x^2+630x-6000>0\quad (x\ge 0)$$

We first solve the related equation.

$$-3x^2+630x-6000=0$$
$$x^2-210x+2000=0\quad\text{Dividing by } -3$$
$$(x-10)(x-200)=0\quad\text{Factoring}$$

Using the principle of zero products or by observing the graph of $y=-3x^2+630-6000$, we see that the solutions of the related equation are 10 and 200. These numbers divide the x-axis into the intervals $(-\infty,10)$, $(10,200)$, and $(200,\infty)$. Since we are restricting our discussion to nonnegative values of x, we consider the intervals $[0,10)$, $(10,200)$, and $(200,\infty)$.

We let $f(x)=-3x^2+630x-6000$ and test a value in each interval.

$[0,10)$: $f(0)=-6000<0$

$(10,200)$: $f(11)=567>0$

$(200,\infty)$: $f(201)=-573<0$

Function values are positive only on $(10, 200)$. The solution set is $\{x|10 < x < 200\}$, or $(10, 200)$.

b) From part (a), we see that function values are negative on $[0, 10)$ and $(200, \infty)$. Thus, the solution set is $\{x|0 < x < 10 \text{ or } x > 200\}$, or $(0, 10) \cup (200, \infty)$.

66. a) We write and solve a polynomial inequality.

$$-16t^2 + 32t + 1920 > 1920$$
$$-16t^2 + 32t > 0$$
$$-16t^2 + 32t = 0 \quad \text{Related equation}$$
$$-16t(t - 2) = 0 \quad \text{Factoring}$$

The solutions of the related equation are 0 and 2. This could also be determined from the graph of $y = -16x^2 + 32x$. These numbers divide the t-axis into the intervals $(-\infty, 0)$, $(0, 2)$, and $(2, \infty)$. Since only nonnegative values of t have meaning in this application, we restrict our discussion to the intervals $(0, 2)$ and $(2, \infty)$. We let $f(x) = -16t^2 + 32t$ and test a value in each interval.

$(0, 2)$: $f(1) = 16 > 0$

$(2, \infty)$: $f(3) = -48 < 0$

Function values are positive on $(0, 2)$. The solution set is $\{t|0 < t < 2\}$, or $(0, 2)$.

b) We write and solve a polynomial inequality.

$$-16t^2 + 32t + 1920 < 640$$
$$-16t^2 + 32t + 1280 < 0$$
$$-16t^2 + 32t^2 + 1280 = 0 \quad \text{Related equation}$$
$$t^2 - 2t - 80 = 0$$
$$(t - 10)(t + 8) = 0$$

The solutions of the related equation are 10 and -8. These numbers divide the t-axis into the intervals $(-\infty, -8)$, $(-8, 10)$, and $(10, \infty)$. As in part (a), we will not consider negative values of t. In addition, note that the nonnegative solution of $S(t) = 0$ is 12. This means that the object reaches the ground in 12 sec. Thus, we also restrict our discussion to values of t such that $t \leq 12$. We consider the intervals $[0, 10)$ and $(10, 12]$. We let $f(x) = -16t^2 + 32t + 1280$ and test a value in each interval.

$[0, 10)$: $f(0) = 1280 > 0$

$(10, 12]$: $f(11) = -304 < 0$

Function values are negative on $(10, 12]$. The solution set is $\{x|10 < x \leq 12\}$, or $(10, 12]$.

67. We write an inequality.

$$27 \leq \frac{n(n-3)}{2} \leq 230$$
$$54 \leq n(n - 3) \leq 460 \quad \text{Multiplying by 2}$$
$$54 \leq n^2 - 3n \leq 460$$

We write this as two inequalities.

$$54 \leq n^2 - 3n \quad \text{and} \quad n^2 - 3n \leq 460$$

Solve each inequality.

$$n^2 - 3n \geq 54$$
$$n^2 - 3n - 54 \geq 0$$
$$n^2 - 3n - 54 = 0 \quad \text{Related equation}$$
$$(n + 6)(n - 9) = 0$$
$$n = -6 \text{ or } n = 9$$

Since only positive values of n have meaning in this application, we consider the intervals $(0, 9)$ and $(9, \infty)$. Let $f(n) = n^2 - 3n - 54$ and test a value in each interval.

$(0, 9)$: $f(1) = -56 < 0$

$(9, \infty)$: $f(10) = 16 > 0$

Function values are positive on $(9, \infty)$. Since the inequality symbol is $\geq$, 9 must also be included in the solution set for this portion of the inequality. It is $\{n|n \geq 9\}$.

Now solve the second inequality.

$$n^2 - 3n \leq 460$$
$$n^2 - 3n - 460 \leq 0$$
$$n^2 - 3n - 460 = 0 \quad \text{Related equation}$$
$$(n + 20)(n - 23) = 0$$
$$n = -20 \text{ or } n = 23$$

We consider only positive values of n as above. Thus, we consider the intervals $(0, 23)$ and $(23, \infty)$. Let $f(n) = n^2 - 3n - 460$ and test a value in each interval.

$(0, 23)$: $f(1) = -462 < 0$

$(23, \infty)$: $f(24) = 44 > 0$

Function values are negative on $(0, 23)$. Since the inequality symbol is $\leq$, 23 must also be included in the solution set for this portion of the inequality. It is $\{n|0 < n \leq 23\}$.

The solution set of the original inequality is $\{n|n \geq 9 \text{ and } 0 < n \leq 23\}$, or $\{n|9 \leq n \leq 23\}$.

68. We write an inequality.

$$66 \leq \frac{n(n-1)}{2} \leq 300$$
$$132 \leq n^2 - n \leq 600$$

We write this as two inequalities.

$$132 \leq n^2 - n \text{ and } n^2 - n \leq 600$$

Solve each inequality.

$$n^2 - n \geq 132$$
$$n^2 - n - 132 \geq 0$$
$$n^2 - n - 132 = 0 \quad \text{Related equation}$$
$$(n + 11)(n - 12) = 0$$
$$n = -11 \text{ or } n = 12$$

Since only positive values of n have meaning in this application, we consider the intervals $(0, 12)$ and $(12, \infty)$. Let $f(n) = n^2 - n - 132$ and test a value in each interval.

$(0, 12)$: $f(1) = -132 < 0$

$(12, \infty)$: $f(13) = 24 > 0$

Function values are positive on $(12, \infty)$. Since the inequality symbol is $\geq$, 12 must also be included in the solution set for this portion of the inequality. It is $\{n|n \geq 12\}$.

Now solve the second inequality.

$$n^2 - n \le 600$$
$$n^2 - n - 600 \le 0$$
$$n^2 - n - 600 = 0 \quad \text{Related equation}$$
$$(n + 24)(n - 25) = 0$$
$$n = -24 \ \text{or} \ n = 25$$

We consider only positive values of n as above. Thus, we consider the intervals $(0, 25)$ and $(25, \infty)$. Let $f(n) = n^2 - n - 600$ and test a value in each interval.

$(0, 25)$: $f(1) = -600 < 0$

$(25, \infty)$: $f(26) = 50 > 0$

Function values are negative on $(0, 25)$. Since the inequality symbol is $\le$, 25 must also be included in the solution set for this portion of the inequality. It is $\{n|0 < n \le 25\}$.

The solution set of the original inequality is $\{n|n \ge 12 \text{ and } 0 < n \le 25\}$, or $\{n|12 \le n \le 25\}$.

69. Left to the student

70. We need to know these values because they cannot be included in the solution set. The sign of the function often changes at these values as well.

71. A quadratic inequality $ax^2+bx+c \le 0$, $a > 0$, or $ax^2+bx+c \ge 0$, $a < 0$ has a solution set that is a closed interval.

72. $r = \dfrac{7/2}{2} = \dfrac{7}{4}$

$$(x - 0)^2 + [y - (-3)]^2 = \left(\frac{7}{4}\right)^2$$
$$x^2 + (y + 3)^2 = \frac{49}{16}$$

73.
$$(x - h)^2 + (y - k)^2 = r^2$$
$$[x - (-2)]^2 + (y - 4)^2 = 3^2$$
$$(x + 2)^2 + (y - 4)^2 = 9$$

74. $g(x) = x^2 - 10x + 2$

a) $-\dfrac{b}{2a} = -\dfrac{-10}{2 \cdot 1} = 5$

$g(5) = 5^2 - 10 \cdot 5 + 2 = -23$

The vertex is $(5, -23)$.

b) Minimum: -23 at $x = 5$

c) $[-23, \infty)$

75. $h(x) = -2x^2 + 3x - 8$

a) $-\dfrac{b}{2a} = -\dfrac{3}{2(-2)} = \dfrac{3}{4}$

$$h\left(\frac{3}{4}\right) = -2\left(\frac{3}{4}\right)^2 + 3 \cdot \frac{3}{4} - 8 = -\frac{55}{8}$$

The vertex is $\left(\dfrac{3}{4}, -\dfrac{55}{8}\right)$.

b) The coefficient of x^2 is negative, so there is a maximum value. It is the second coordinate of the vertex, $-\dfrac{55}{8}$. It occurs at $x = \dfrac{3}{4}$.

c) The range is $\left(-\infty, -\dfrac{55}{8}\right]$.

76.
$$x^4 - 6x^2 + 5 > 0$$
$$x^4 - 6x^2 + 5 = 0 \quad \text{Related equation}$$
$$(x^2 - 1)(x^2 - 5) = 0$$
$$x = \pm 1 \ \text{or} \ x = \pm\sqrt{5}$$

Let $f(x) = x^4 - 6x^2 + 5$ and test a value in each of the intervals determined by the solutions of the related equation.

$(-\infty, -\sqrt{5})$: $f(-3) = 32 > 0$

$(-\sqrt{5}, -1)$: $f(-2) = -3 < 0$

$(-1, 1)$: $f(0) = 5 > 0$

$(1, \sqrt{5})$: $f(2) = -3 < 0$

$(\sqrt{5}, \infty)$: $f(3) = 32 > 0$

The solution set is $(-\infty, -\sqrt{5}) \cup (-1, 1) \cup (\sqrt{5}, \infty)$.

77.
$$x^4 + 3x^2 > 4x - 15$$
$$x^4 + 3x^2 - 4x + 15 > 0$$

The graph of $y = x^4 + 3x^2 - 4x + 15$ lies entirely above the x-axis. Thus, the solution set is the set of all real numbers, or $(-\infty, \infty)$.

78.
$$\left|\frac{x+3}{x-4}\right| < 2$$
$$-2 < \frac{x+3}{x-4} < 2$$
$$-2 < \frac{x+3}{x-4} \ \text{and} \ \frac{x+3}{x-4} < 2$$

First solve $-2 < \dfrac{x+3}{x-4}$.

$$\frac{x+3}{x-4} + 2 > 0$$

The denominator of $f(x) = \dfrac{x+3}{x-4} + 2$ is 0 when $x = 4$, so the function is not defined for this value of x. Now solve the related equation.

$$\frac{x+3}{x-4} + 2 = 0$$
$$x + 3 + 2(x - 4) = 0 \quad \text{Multiplying by } x - 4$$
$$x + 3 + 2x - 8 = 0$$
$$3x - 5 = 0$$
$$x = \frac{5}{3}$$

The critical values are $\dfrac{5}{3}$ and 4. Test a value in each of the intervals determined by them.

$\left(-\infty, \dfrac{5}{3}\right)$: $f(0) = 1.25 > 0$

$\left(\dfrac{5}{3}, 4\right)$: $f(2) = -0.5 < 0$

$(4, \infty)$: $f(5) = 10 > 0$

The solution set for this portion of the inequality is $\left(-\infty, \dfrac{5}{3}\right) \cup (4, \infty)$.

Next solve $\dfrac{x+3}{x-4} < 2$, or $\dfrac{x+3}{x-4} - 2 < 0$.

The denominator of $f(x) = \dfrac{x+3}{x-4} - 2$ is 0 when $x = 4$, so the function is not defined for this value of x. Now solve the related equation.

$$\frac{x+3}{x-4} - 2 = 0$$
$$x + 3 - 2(x-4) = 0 \quad \text{Multiplying by } x-4$$
$$x + 3 - 2x + 8 = 0$$
$$-x + 11 = 0$$
$$x = 11$$

The critical values are 4 and 11. Test a value in each of the intervals determined by them.

$(-\infty, 4)$: $f(0) = -2.75 < 0$

$(4, 11)$: $f(5) = 6 > 0$

$(11, \infty)$: $f(12) = -0.125 < 0$

The solution set for this portion of the inequality is $(-\infty, 4) \cup (11, \infty)$.

The solution set of the original inequality is

$\left(\left(-\infty, \frac{5}{3}\right) \cup (4, \infty)\right)$ and $((-\infty, 4) \cup (11, \infty))$, or

$\left(-\infty, \frac{5}{3}\right) \cup (11, \infty)$.

79. $|x^2 - 5| = |5 - x^2| = 5 - x^2$ when $5 - x^2 \geq 0$. Thus we solve $5 - x^2 \geq 0$.

$$5 - x^2 \geq 0$$
$$5 - x^2 = 0 \quad \text{Related equation}$$
$$5 = x^2$$
$$\pm\sqrt{5} = x$$

Let $f(x) = 5 - x^2$ and test a value in each of the intervals determined by the solutions of the related equation.

$(-\infty, -\sqrt{5})$: $f(-3) = -4 < 0$

$(-\sqrt{5}, \sqrt{5})$: $f(0) = 5 > 0$

$(\sqrt{5}, \infty)$: $f(3) = -4 < 0$

Function values are positive on $(-\sqrt{5}, \sqrt{5})$. Since the inequality symbol is $\geq$, the endpoints of the interval must be included in the solution set. It is $[-\sqrt{5}, \sqrt{5}]$.

80. $(7-x)^{-2} < 0$

$$\frac{1}{(7-x)^2} < 0$$

Since $(7-x)^2 \geq 0$ for all real numbers x, then $\dfrac{1}{(7-x)^2} > 0$ for all values of x in the domain of $f(x) = \dfrac{1}{(7-x)^2}$. Thus, the solution set is $\emptyset$.

81.
$$2|x|^2 - |x| + 2 \leq 5$$
$$2|x|^2 - |x| - 3 \leq 0$$
$$2|x|^2 - |x| - 3 = 0 \quad \text{Related equation}$$
$$(2|x| - 3)(|x| + 1) = 0 \quad \text{Factoring}$$
$$2|x| - 3 = 0 \quad \text{or} \quad |x| + 1 = 0$$
$$|x| = \frac{3}{2} \quad \text{or} \quad |x| = -1$$

The solution of the first equation is $x = -\frac{3}{2}$ or $x = \frac{3}{2}$. The second equation has no solution. Let $f(x) = 2|x|^2 - |x| - 3$ and test a value in each interval determined by the solutions of the related equation.

$\left(-\infty, -\frac{3}{2}\right)$: $f(-2) = 3 > 0$

$\left(-\frac{3}{2}, \frac{3}{2}\right)$: $f(0) = -3 < 0$

$\left(\frac{3}{2}, \infty\right)$: $f(2) = 3 > 0$

Function values are negative on $\left(-\frac{3}{2}, \frac{3}{2}\right)$. Since the inequality symbol is $\leq$, the endpoints of the interval must also be included in the solution set. It is $\left[-\frac{3}{2}, \frac{3}{2}\right]$.

82. $\left|2 - \dfrac{1}{x}\right| \leq 2 + \left|\dfrac{1}{x}\right|$

Note that $\frac{1}{x}$ is not defined when $x = 0$. Thus, $x \neq 0$. Also note that $2 - \frac{1}{x} = 0$ when $x = \frac{1}{2}$. The numbers 0 and $\frac{1}{2}$ divide the x-axis into the intervals $(-\infty, 0)$, $\left(0, \frac{1}{2}\right)$, and $\left(\frac{1}{2}, \infty\right)$. Find the solution set of the inequality for each interval. Then find the union of the three solution sets.

If $x < 0$, then $2 - \frac{1}{x} > 0$ and $\frac{1}{x} < 0$, so $\left|2 - \frac{1}{x}\right| = 2 - \frac{1}{x}$ and $\left|\frac{1}{x}\right| = -\frac{1}{x}$.

$$\text{Solve: } x < 0 \ \textit{and} \ 2 - \frac{1}{x} \leq 2 - \frac{1}{x}$$
$$x < 0 \ \textit{and} \quad 2 \leq 2$$

The solution set for this interval is $(-\infty, 0)$.

If $0 < x < \frac{1}{2}$, then $2 - \frac{1}{x} < 0$ and $\frac{1}{x} > 0$, so $\left|2 - \frac{1}{x}\right| = -\left(2 - \frac{1}{x}\right) = -2 + \frac{1}{x}$ and $\left|\frac{1}{x}\right| = \frac{1}{x}$.

$$\text{Solve: } 0 < x < \frac{1}{2} \ \textit{and} \ -2 + \frac{1}{x} \leq 2 + \frac{1}{x}$$
$$0 < x < \frac{1}{2} \ \textit{and} \quad -2 \leq 2$$

The solution set for this interval is $\left(0, \frac{1}{2}\right)$.

If $x \geq \frac{1}{2}$, then $2 - \frac{1}{x} > 0$ and $\frac{1}{x} > 0$, so $\left|2 - \frac{1}{x}\right| = 2 - \frac{1}{x}$ and $\left|\frac{1}{x}\right| = \frac{1}{x}$.

$$\text{Solve: } x \geq \frac{1}{2} \ \textit{and} \ 2 - \frac{1}{x} \leq 2 + \frac{1}{x}$$
$$x \geq \frac{1}{2} \ \textit{and} \quad -\frac{1}{x} \leq \frac{1}{x}$$
$$x \geq \frac{1}{2} \ \textit{and} \quad -1 \leq -1$$

Then the solution set for this interval is $\left[\frac{1}{2},\infty\right)$.

Then the solution set of the original inequality is

$(-\infty,0)\cup\left(0,\frac{1}{2}\right)\cup\left[\frac{1}{2},\infty\right)$, or $(-\infty,0)\cup(0,\infty)$.

83. $\left|1+\frac{1}{x}\right|<3$

$-3<1+\frac{1}{x}<3$

$-3<1+\frac{1}{x}$ *and* $1+\frac{1}{x}<3$

First solve $-3<1+\frac{1}{x}$.

$0<4+\frac{1}{x}$, or $\frac{1}{x}+4>0$

The denominator of $f(x)=\frac{1}{x}+4$ is 0 when $x=0$, so the function is not defined for this value of x. Now solve the related equation.

$\frac{1}{x}+4=0$

$1+4x=0$ Multiplying by x

$x=-\frac{1}{4}$

The critical values are $-\frac{1}{4}$ and 0. Test a value in each of the intervals determined by them.

$\left(\infty,-\frac{1}{4}\right)$: $f(-1)=3>0$

$\left(-\frac{1}{4},0\right)$: $f(-0.1)=-6<0$

$(0,\infty)$: $f(1)=5>0$

The solution set for this portion of the inequality is $\left(-\infty,-\frac{1}{4}\right)\cup(0,\infty)$.

Next solve $1+\frac{1}{x}<3$, or $\frac{1}{x}-2<0$. The denominator of $f(x)=\frac{1}{x}-2$ is 0 when $x=0$, so the function is not defined for this value of x. Now solve the related equation.

$\frac{1}{x}-2=0$

$1-2x=0$ Multiplying by x

$x=\frac{1}{2}$

The critical values are 0 and $\frac{1}{2}$. Test a value in each of the intervals determined by them.

$(-\infty,0)$: $f(-1)=-3<0$

$\left(0,\frac{1}{2}\right)$: $f(0.1)=8>0$

$\left(\frac{1}{2},\infty\right)$: $f(1)=-1<0$

The solution set for this portion of the inequality is $(-\infty,0)\cup\left(\frac{1}{2},\infty\right)$.

The solution set of the original inequality is

$\left(\left(-\infty,-\frac{1}{4}\right)\cup(0,\infty)\right)$*and*$\left((-\infty,0)\cup\left(\frac{1}{2},\infty\right)\right)$,

or $\left(-\infty,-\frac{1}{4}\right)\cup\left(\frac{1}{2},\infty\right)$.

84. $|1+5x-x^2|\geq 5$

$1+5x-x^2\leq -5$ *or* $1+5x-x^2\geq 5$

First solve $1+5x-x^2\leq -5$.

$1+5x-x^2\leq -5$

$6+5x-x^2\leq 0$

$6+5x-x^2=0$ Related equation

$(6-x)(1+x)=0$

$x=6$ *or* $x=-1$

Let $f(x)=6+5x-x^2$ and test a value in each of the intervals determined by the solution of the related equation.

$(-\infty,-1)$: $f(-2)=-8<0$

$(-1,6)$: $f(0)=6>0$

$(6,\infty)$: $f(7)=-8<0$

The solution set for this portion of the inequality is $(-\infty,-1]\cup[6,\infty)$.

Next solve $1+5x-x^2\geq 5$.

$1+5x-x^2\geq 5$

$-4+5x-x^2\geq 0$

$x^2-5x+4\leq 0$

$x^2-5x+4=0$ Related equation

$(x-1)(x-4)=0$

$x=1$ *or* $x=4$

Let $f(x)=x^2-5x+4$ and test a value in each of the intervals determined by the solution of the related equation.

$(-\infty,1)$: $f(0)=4>0$

$(1,4)$: $f(2)=-2<0$

$(4,\infty)$: $f(5)=4>0$

The solution set for this portion of the inequality is $[1,4]$.

The solution set of the original inequality is $((-\infty,-1]\cup[6,\infty))$ *or* $[1,4]$, or $(-\infty,-1]\cup[1,4]\cup[6,\infty)$.

85. $|x^2+3x-1|<3$

$-3<x^2+3x-1<3$

$-3<x^2+3x-1$ *and* $x^2+3x-1<3$

First solve $-3<x^2+3x-1$, or $x^2+3x-1>-3$.

$x^2+3x-1>-3$

$x^2+3x+2>0$

$x^2+3x+2=0$ Related equation

$(x+2)(x+1)=0$

$x=-2$ *or* $x=-1$

Let $f(x)=x^2+3x+2$ and test a value in each of the intervals determined by the solution of the related equation.

$(-\infty, -2)$: $f(-3) = 2 > 0$

$(-2, -1)$: $f(-1.5) = -0.25 < 0$

$(-1, \infty)$: $f(0) = 2 > 0$

The solution set for this portion of the inequality is $(-\infty, -2) \cup (-1, \infty)$.

Next solve $x^2 + 3x - 1 < 3$.

$$x^2 + 3x - 1 < 3$$
$$x^2 + 3x - 4 < 0$$
$$x^2 + 3x - 4 = 0 \quad \text{Related equation}$$
$$(x + 4)(x - 1) = 0$$
$$x = -4 \quad or \quad x = 1$$

Let $f(x) = x^2 + 3x - 4$ and test a value in each of the intervals determined by the solution of the related equation.

$(-\infty, -4)$: $f(-5) = 6 > 0$

$(-4, 1)$: $f(0) = -4 < 0$

$(1, \infty)$: $f(2) = 6 > 0$

The solution set for this portion of the inequality is $(-4, 1)$.

The solution set of the original inequality is $((-\infty, -2) \cup (-1, \infty))$ *and* $(-4, 1)$, or $(-4, -2) \cup (-1, 1)$.

86. First find the polynomial with solutions -4, 3, and 7.

$$(x + 4)(x - 3)(x - 7) = 0$$
$$x^3 - 6x^2 - 19x + 84 = 0$$

Test a point in each of the four intervals determined by -4, 3, and 7.

$(-\infty, -4)$: $(-5 + 4)(-5 - 3)(-5 - 7) = -96 < 0$

$(-4, 3)$: $(0 + 4)(0 - 3)(0 - 7) = 84 > 0$

$(3, 7)$: $(4 + 4)(4 - 3)(4 - 7) = -24 < 0$

$(7, \infty)$: $(8 + 4)(8 - 3)(8 - 7) = 60 > 0$

Then a polynomial inequality for which the solution set is $[-4, 3] \cup [7, \infty)$ is $x^3 - 6x^2 - 19x + 84 \geq 0$. Answers may vary.

87. First find a quadratic equation with solutions -4 and 3.

$$(x + 4)(x - 3) = 0$$
$$x^2 + x - 12 = 0$$

Test a point in each of the three intervals determined by -4 and 3.

$(-\infty, -4)$: $(-5 + 4)(-5 - 3) = 8 > 0$

$(-4, 3)$: $(0 + 4)(0 - 3) = -12 < 0$

$(3, \infty)$: $(4 + 4)(4 - 3) = 8 > 0$

Then a quadratic inequality for which the solution set is $(-4, 3)$ is $x^2 + x - 12 < 0$. Answers may vary.

Exercise Set 3.6

1.
$$y = kx$$
$$54 = k \cdot 12$$
$$\frac{54}{12} = k, \text{ or } k = \frac{9}{2}$$

The variation constant is $\frac{9}{2}$, or 4.5. The equation of variation is $y = \frac{9}{2}x$, or $y = 4.5x$.

2.
$$y = kx$$
$$0.1 = k(0.2)$$
$$\frac{1}{2} = k \quad \text{Variation constant}$$

Equation of variation: $y = \frac{1}{2}x$, or $y = 0.5x$.

3.
$$y = \frac{k}{x}$$
$$3 = \frac{k}{12}$$
$$36 = k$$

The variation constant is 36. The equation of variation is $y = \frac{36}{x}$.

4.
$$y = \frac{k}{x}$$
$$12 = \frac{k}{5}$$
$$60 = k \quad \text{Variation constant}$$

Equation of variation: $y = \frac{60}{x}$

5.
$$y = kx$$
$$1 = k \cdot \frac{1}{4}$$
$$4 = k$$

The variation constant is 4. The equation of variation is $y = 4x$.

6.
$$y = \frac{k}{x}$$
$$0.1 = \frac{k}{0.5}$$
$$0.05 = k \quad \text{Variation constant}$$

Equation of variation: $y = \frac{0.05}{x}$

7.
$$y = \frac{k}{x}$$
$$32 = \frac{k}{\frac{1}{8}}$$
$$\frac{1}{8} \cdot 32 = k$$
$$4 = k$$

The variation constant is 4. The equation of variation is $y = \frac{4}{x}$.

8. $y = kx$

$3 = k \cdot 33$

$\frac{1}{11} = k$ Variation constant

Equation of variation: $y = \frac{1}{11}x$

9. $y = kx$

$\frac{3}{4} = k \cdot 2$

$\frac{1}{2} \cdot \frac{3}{4} = k$

$\frac{3}{8} = k$

The variation constant is $\frac{3}{8}$. The equation of variation is $y = \frac{3}{8}x$.

10. $y = \frac{k}{x}$

$\frac{1}{5} = \frac{k}{35}$

$7 = k$ Variation constant

Equation of variation: $y = \frac{7}{x}$

11. $y = \frac{k}{x}$

$1.8 = \frac{k}{0.3}$

$0.54 = k$

The variation constant is 0.54. The equation of variation is $y = \frac{0.54}{x}$.

12. $y = kx$

$0.9 = k(0.4)$

$\frac{9}{4} = k$ Variation constant

Equation of variation: $y = \frac{9}{4}x$, or $y = 2.25x$

13. $T = \frac{k}{P}$ T varies inversely as P.

$5 = \frac{k}{7}$ Substituting

$35 = k$ Variation constant

$T = \frac{35}{P}$ Equation of variation

$T = \frac{35}{10}$ Substituting

$T = 3.5$

It will take 10 bricklayers 3.5 hr to complete the job.

14. $A = kG$

$9.66 = k \cdot 9$

$\frac{161}{150} = k$

$A = \frac{161}{150}G$

$A = \frac{161}{150} \cdot 4$

$A \approx \$4.29$

15. Let F = the number of grams of fat and w = the weight.

$F = kw$ F varies directly as w.

$60 = k \cdot 120$ Substituting

$\frac{60}{120} = k$, or Solving for k

$\frac{1}{2} = k$ Variation constant

$F = \frac{1}{2}w$ Equation of variation

$F = \frac{1}{2} \cdot 180$ Substituting

$F = 90$

The maximum daily fat intake for a person weighing 180 lb is 90 g.

16. $t = \frac{k}{r}$

$5 = \frac{k}{80}$

$400 = r$

$t = \frac{400}{r}$

$t = \frac{400}{70}$

$t = \frac{40}{7}$, or $5\frac{5}{7}$ hr

17. $W = \frac{k}{L}$ W varies inversely as L.

$1200 = \frac{k}{8}$ Substituting

$9600 = k$ Variation constant

$W = \frac{9600}{L}$ Equation of variation

$W = \frac{9600}{14}$ Substituting

$W = 685\frac{5}{7}$

A 14-m beam can support $685\frac{5}{7}$ kg, or about 686 kg.

18. $N = kP$

$29 = k \cdot 19{,}011{,}000$ Substituting

$\frac{29}{19{,}011{,}000} = k$ Variation constant

$N = \frac{29}{19{,}011{,}000}P$

$N = \frac{29}{19{,}011{,}000} \cdot 4{,}418{,}000$ Substituting

$N \approx 7$

Colorado has 7 representatives.

19. $M = kE$ $\quad$ M varies directly as E.

$38 = k \cdot 95$ $\quad$ Substituting

$\frac{2}{5} = k$ $\quad$ Variation constant

$M = \frac{2}{5}E$ $\quad$ Equation of variation

$M = \frac{2}{5} \cdot 100$ $\quad$ Substituting

$M = 40$

A 100-lb person would weigh 40 lb on Mars.

20.
$$t = \frac{k}{r}$$
$$45 = \frac{k}{600}$$
$$27,000 = k$$
$$t = \frac{27,000}{r}$$
$$t = \frac{27,000}{1000}$$
$$t = 27 \text{ min}$$

21. $d = km$ $\quad$ d varies directly as m.

$40 = k \cdot 3$ $\quad$ Substituting

$\frac{40}{3} = k$ $\quad$ Variation constant

$d = \frac{40}{3}m$ $\quad$ Equation of variation

$d = \frac{40}{3} \cdot 5 = \frac{200}{3}$ $\quad$ Substituting

$d = 66\frac{2}{3}$

A 5-kg mass will stretch the spring $66\frac{2}{3}$ cm.

22.
$$f = kF$$
$$6.3 = k \cdot 150$$
$$0.042 = k$$
$$f = 0.042F$$
$$f = 0.042(80)$$
$$f = 3.36$$

23. $P = \frac{k}{W}$ $\quad$ P varies inversely as W.

$330 = \frac{k}{3.2}$ $\quad$ Substituting

$1056 = k$ $\quad$ Variation constant

$P = \frac{1056}{W}$ $\quad$ Equation of variation

$550 = \frac{1056}{W}$ $\quad$ Substituting

$550W = 1056$ $\quad$ Multiplying by W

$W = \frac{1056}{550}$ $\quad$ Dividing by 550

$W = 1.92$ $\quad$ Simplifying

A tone with a pitch of 550 vibrations per second has a wavelength of 1.92 ft.

24.
$$B = kP$$
$$4445 = k \cdot 9880$$
$$\frac{889}{1976} = k$$
$$B = \frac{889}{1976}P$$
$$B = \frac{889}{1976} \cdot 74,650$$
$$B \approx 33,585 \text{ lb}$$

25.
$$y = \frac{k}{x^2}$$

$0.15 = \frac{k}{(0.1)^2}$ $\quad$ Substituting

$$0.15 = \frac{k}{0.01}$$
$$0.15(0.01) = k$$
$$0.0015 = k$$

The equation of variation is $y = \frac{0.0015}{x^2}$.

26.
$$y = \frac{k}{x^2}$$
$$6 = \frac{k}{3^2}$$
$$54 = k$$
$$y = \frac{54}{x^2}$$

27. $y = kx^2$

$0.15 = k(0.1)^2$ $\quad$ Substituting

$$0.15 = 0.01k$$
$$\frac{0.15}{0.01} = k$$
$$15 = k$$

The equation of variation is $y = 15x^2$.

28.
$$y = kx^2$$
$$6 = k \cdot 3^2$$
$$\frac{2}{3} = k$$
$$y = \frac{2}{3}x^2$$

29. $y = kxz$

$56 = k \cdot 7 \cdot 8$ Substituting

$56 = 56k$

$1 = k$

The equation of variation is $y = xz$.

30. $y = \frac{kx}{z}$

$4 = \frac{k \cdot 12}{15}$

$5 = k$

$y = \frac{5x}{z}$

31. $y = kxz^2$

$105 = k \cdot 14 \cdot 5^2$ Substituting

$105 = 350k$

$\frac{105}{350} = k$

$\frac{3}{10} = k$

The equation of variation is $y = \frac{3}{10}xz^2$.

32. $y = k \cdot \frac{xz}{w}$

$\frac{3}{2} = k \cdot \frac{2 \cdot 3}{4}$

$1 = k$

$y = \frac{xz}{w}$

33. $y = k\frac{xz}{wp}$

$\frac{3}{28} = k\frac{3 \cdot 10}{7 \cdot 8}$ Substituting

$\frac{3}{28} = k \cdot \frac{30}{56}$

$\frac{3}{28} \cdot \frac{56}{30} = k$

$\frac{1}{5} = k$

The equation of variation is $y = \frac{xz}{5wp}$.

34. $y = k \cdot \frac{xz}{w^2}$

$\frac{12}{5} = k \cdot \frac{16 \cdot 3}{5^2}$

$\frac{5}{4} = k$

$y = \frac{5xz}{4w^2}$

35. $I = \frac{k}{d^2}$

$90 = \frac{k}{5^2}$ Substituting

$90 = \frac{k}{25}$

$2250 = k$

The equation of variation is $I = \frac{2250}{d^2}$.

Substitute 40 for I and find d.

$40 = \frac{2250}{d^2}$

$40d^2 = 2250$

$d^2 = 56.25$

$d = 7.5$

The distance from 5 m to 7.5 m is $7.5 - 5$, or 2.5 m, so it is 2.5 m further to a point where the intensity is 40 W/m^2.

36. $D = kAv$

$222 = k \cdot 37.8 \cdot 40$

$\frac{37}{252} = k$

$D = \frac{37}{252}Av$

$430 = \frac{37}{252} \cdot 51v$

$v \approx 57.4$ mph

37. $d = kr^2$

$200 = k \cdot 60^2$ Substituting

$200 = 3600k$

$\frac{200}{3600} = k$

$\frac{1}{18} = k$

The equation of variation is $d = \frac{1}{18}r^2$.

Substitute 72 for d and find r.

$72 = \frac{1}{18}r^2$

$1296 = r^2$

$36 = r$

A car can travel 36 mph and still stop in 72 ft.

38. $W = \frac{k}{d^2}$

$220 = \frac{k}{(3978)^2}$

$3,481,386,480 = k$

$W = \frac{3,481,386,480}{d^2}$

$W = \frac{3,481,386,480}{(3978 + 200)^2}$

$W \approx 199$ lb

39. $E = \frac{kR}{I}$

We first find k.

$2.24 = \frac{k \cdot 28}{112\frac{1}{3}}$ Substituting

$2.24\left(\frac{112\frac{1}{3}}{28}\right) = k$ Multiplying by $\frac{112\frac{1}{3}}{28}$

$9 \approx k$

The equation of variation is $E = \dfrac{9R}{I}$.

Substitute 2.24 for E and 300 for I and solve R.

$$2.24 = \frac{9R}{300}$$

$$2.24\left(\frac{300}{9}\right) = R \quad \text{Multiplying by } \frac{300}{9}$$

$$75 \approx R$$

Kevin Brown would have given up 75 earned runs if he had pitched 300 innings.

40. $V = \dfrac{kT}{P}$

$231 = \dfrac{k \cdot 42}{20}$

$110 = k$

$V = \dfrac{110T}{P}$

$V = \dfrac{110 \cdot 30}{15}$

$V = 220 \text{ cm}^3$

41. Let $y(x) = kx^2$. Then $y(2x) = k(2x)^2 = k \cdot 4x^2 = 4 \cdot kx^2 = 4 \cdot y(x)$. Thus, doubling x causes y to be quadrupled.

42. Let $y = k_1x$ and $x = \dfrac{k_2}{z}$. Then $y = k_1 \cdot \dfrac{k_2}{z}$, or $y = \dfrac{k_1k_2}{z}$, so y varies inversely as z.

43.

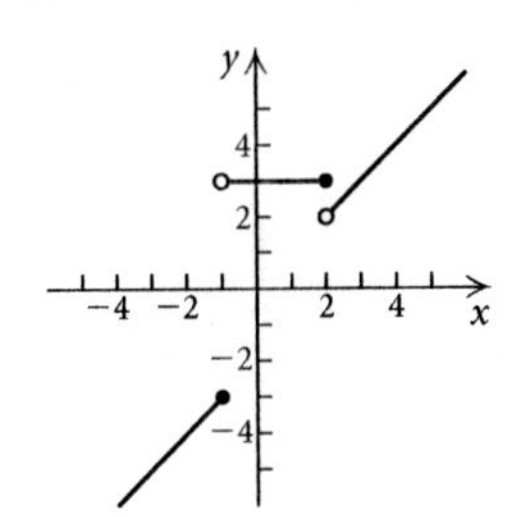

44. Test for symmetry with respect to the x-axis.

$y = 3x^4 - 3$ Original equation

$-y = 3x^4 - 3$ Replacing y by $-y$

$y = -3x^4 + 3$ Simplifying

The last equation is not equivalent to the original equation, so the graph is not symmetric with respect to the x-axis.

Test for symmetry with respect to the y-axis.

$y = 3x^4 - 3$ Original equation

$y = 3(-x)^4 - 3$ Replacing x by $-x$

$y = 3x^4 - 3$ Simplifying

The last equation is equivalent to the original equation, so the graph is symmetric with respect to the y-axis.

Test for symmetry with respect to the origin:

$y = 3x^4 - 3$

$-y = 3(-x)^4 - 3$ Replacing x by $-x$ and y by $-y$

$-y = 3x^4 - 3$

$y = -3x^4 + 3$ Simplifying

The last equation is not equivalent to the original equation, so the graph is not symmetric with respect to the origin.

45. Test for symmetry with respect to the x-axis.

$y^2 = x$ Original equation

$(-y)^2 = x$ Replacing y by $-y$

$y^2 = x$ Simplifying

The last equation is equivalent to the original equation, so the graph is symmetric with respect to the x-axis.

Test for symmetry with respect to the y-axis:

$y^2 = x$ Original equation

$y^2 = -x$ Replacing x by $-x$

The last equation is not equivalent to the original equation, so the graph is not symmetric with respect to the y-axis.

Test for symmetry with respect to the origin:

$y^2 = x$ Original equation

$(-y)^2 = -x$ Replacing x by $-x$ and y by $-y$

$y^2 = -x$ Simplifying

The last equation is not equivalent to the original equation, so the graph is not symmetric with respect to the origin.

46. Test for symmetry with respect to the x-axis:

$2x - 5y = 0$ Original equation

$2x - 5(-y) = 0$ Replacing y by $-y$

$2x + 5y = 0$ Simplifying

The last equation is not equivalent to the original equation, so the graph is not symmetric with respect to the x-axis.

Test for symmetry with respect to the y-axis:

$2x - 5y = 0$ Original equation

$2(-x) - 5y = 0$ Replacing x by $-x$

$-2x - 5y = 0$ Simplifying

The last equation is not equivalent to the original equation, so the graph is not symmetric with respect to the y-axis.

Test for symmetry with respect to the origin:

$2x - 5y = 0$ Original equation

$2(-x) - 5(-y) = 0$ Replacing x by $-x$ and y by $-y$

$-2x + 5y = 0$

$2x - 5y = 0$ Simplifying

The last equation is equivalent to the original equation, so the graph is symmetric with respect to the origin.

47. Let V represent the volume and p represent the price of a jar of peanut butter.

$V = kp$ V varies directly as p.

$\pi\left(\frac{3}{2}\right)^2(5) = k(1.8)$ Substituting

$6.25\pi = k$ Variation constant

$V = 6.25\pi p$ Equation of variation

$\pi(1.625)^2(5.5) = 6.25\pi p$ Substituting

$2.32 \approx p$

If cost is directly proportional to volume, the larger jar should cost \$2.32.

Now let W represent the weight and p represent the price of a jar of peanut butter.

$W = kp$

$18 = k(1.8)$ Substituting

$10 = k$ Variation constant

$W = 10p$ Equation of variation

$28 = 10p$ Substituting

$2.8 = p$

If cost is directly proportional to weight, the larger jar should cost \$2.80.

48. a) $7xy = 14$

$y = \frac{2}{x}$

Inversely

b) $x - 2y = 12$

$y = \frac{x}{2} - 6$

Neither

c) $-2x + y = 0$

$y = 2x$

Directly

d) $x = \frac{3}{4}y$

$y = \frac{4}{3}x$

Directly

e) $\frac{x}{y} = 2$

$y = \frac{1}{2}x$

Directly

49. We are told $A = kd^2$, and we know $A = \pi r^2$ so we have:

$kd^2 = \pi r^2$

$kd^2 = \pi\left(\frac{d}{2}\right)^2$ $\quad r = \frac{d}{2}$

$kd^2 = \frac{\pi d^2}{4}$

$k = \frac{\pi}{4}$ Variation constant

50. $Q = \frac{kp^2}{q^3}$

Q varies directly as the square of p and inversely as the cube of q.

Chapter 4

Exponential and Logarithmic Functions

Exercise Set 4.1

1. We interchange the first and second coordinates of each ordered pair to find the inverse of the relation. It is

   $\{(8,7), (8,-2), (-4,3), (-8,8)\}$.

2. $\{(1,0), (6,5), (-4,-2)\}$

3. We interchange the first and second coordinates of each ordered pair to find the inverse of the relation. It is

   $\{(-1,-1), (4,-3)\}$.

4. $\{(3,-1), (5,2), (5,-3), (0,2)\}$

5. Interchange x and y.

   $y = 4x - 5$
   $\downarrow \quad \downarrow$
   $x = 4y - 5$

6. $2y^2 + 5x^2 = 4$

7. Interchange x and y.

   $x^3 y = -5$
   $\downarrow\downarrow$
   $y^3 x = -5$

8. $x = 3y^2 - 5y + 9$

9. Interchange x and y.

   $x = y^2 - 2y$
   $\downarrow \quad \downarrow \downarrow$
   $y = x^2 - 2x$

10. $y = \dfrac{1}{2}x + 4$

11. Graph $x = y^2 - 3$. Some points on the graph are $(-3,0)$, $(-2,-1)$, $(-2,1)$, $(1,-2)$, and $(1,2)$. Plot these points and draw the curve. Then reflect the graph across the line $y = x$.

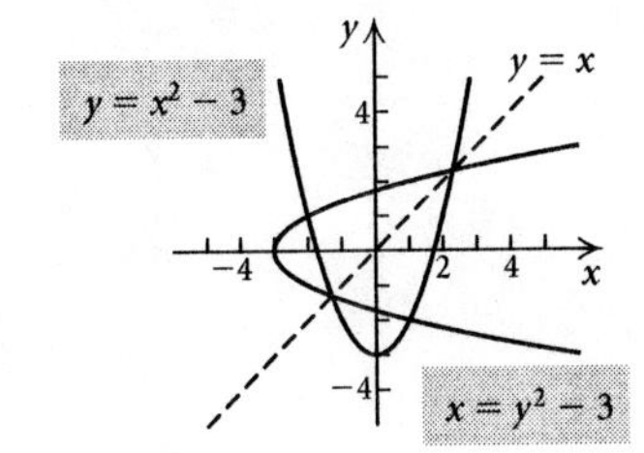

12.

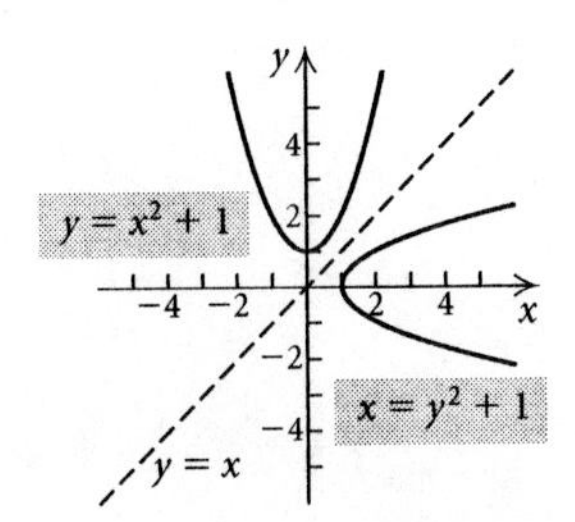

13. Graph $y = 3x - 2$. The intercepts are $(0,-2)$ and $\left(\dfrac{2}{3}, 0\right)$. Plot these points and draw the line. Then reflect the graph across the line $y = x$.

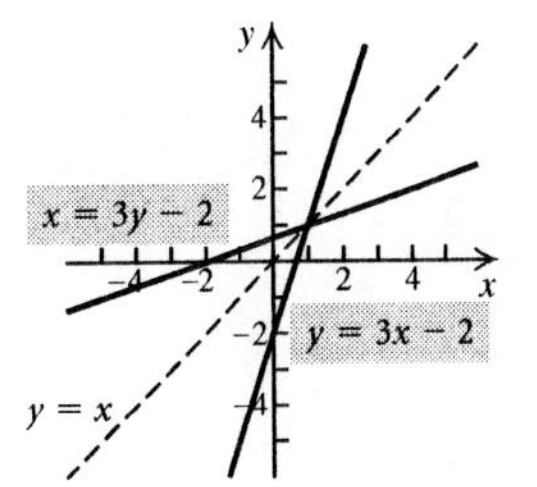

14.

15. Graph $y = |x|$. Some points on the graph are $(0,0)$, $(-2,2)$, $(2,2)$, $(-5,5)$, and $(5,5)$. Plot these points and draw the graph. Then reflect the graph across the line $y = x$.

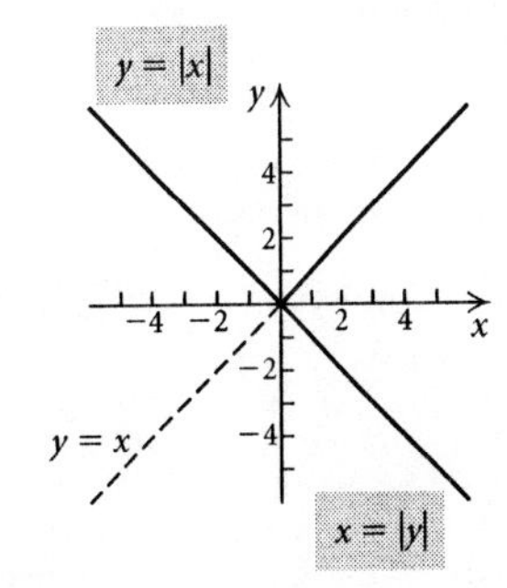

16.

17. We show that if $f(a) = f(b)$, then $a = b$.

$$\frac{1}{3}a - 6 = \frac{1}{3}b - 6$$

$$\frac{1}{3}a = \frac{1}{3}b \quad \text{Adding 6}$$

$$a = b \quad \text{Multiplying by 3}$$

Thus f is one-to-one.

18. Assume $f(a) = f(b)$.

$$4 - 2a = 4 - 2b$$

$$-2a = -2b$$

$$a = b$$

Then f is one-to-one.

19. We show that if $f(a) = f(b)$, then $a = b$.

$$a^3 + \frac{1}{2} = b^3 + \frac{1}{2}$$

$$a^3 = b^3 \quad \text{Subtracting } \frac{1}{2}$$

$$a = b \quad \text{Taking cube roots}$$

Thus f is one-to-one.

20. Assume $f(a) = f(b)$.

$$\sqrt[3]{a} = \sqrt[3]{b}$$

$$a = b \quad \text{Using the principle of powers}$$

Then f is one-to-one.

21. $g(-1) = 1 - (-1)^2 = 1 - 1 = 0$ and $g(1) = 1 - 1^2 = 1 - 1 = 0$, so $g(-1) = g(1)$ but $-1 \neq 1$. Thus the function is not one-to-one.

22. $g(-1) = 4$ and $g(1) = 4$, so $g(-1) = g(1)$ but $-1 \neq 1$. Thus the function is not one-to-one.

23. $f(-2) = (-2)^4 - (-2)^2 = 16 - 4 = 12$ and $f(2) = 2^4 - 2^2 = 16 - 4 = 12$, so $f(-2) = f(2)$ but $-2 \neq 2$. Thus the function is not one-to-one.

24. $g(-1) = 1$ and $g(1) = 1$ so $g(-1) = g(1)$ but $-1 \neq 1$. Thus the function is not one-to-one.

25. The function is one-to-one, because no horizontal line crosses the graph more than once.

26. The function is one-to-one, because no horizontal line crosses the graph more than once.

27. The function is not one-to-one, because there are many horizontal lines that cross the graph more than once.

28. The function is not one-to-one, because there are many horizontal lines that cross the graph more than once.

29. The function is not one-to-one, because there are many horizontal lines that cross the graph more than once.

30. The function is one-to-one, because no horizontal line crosses the graph more than once.

31. The function is one-to-one, because no horizontal line crosses the graph more than once.

32. The function is one-to-one, because no horizontal line crosses the graph more than once.

33. The graph of $f(x) = 5x - 8$ is shown below.

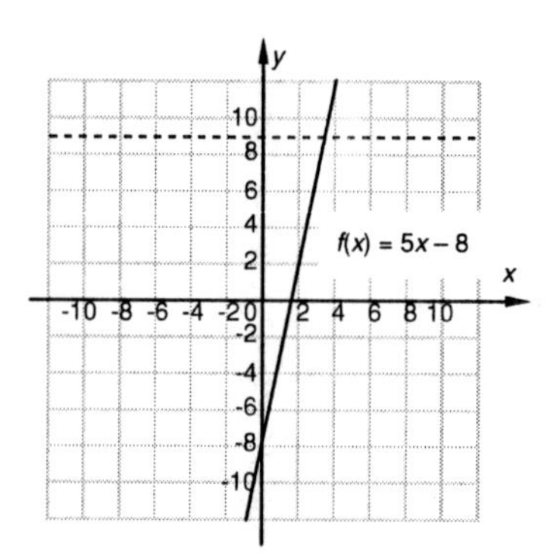

Since there is no horizontal line that crosses the graph more than once, the function is one-to-one.

34. The graph of $f(x) = 3 + 4x$ is shown below.

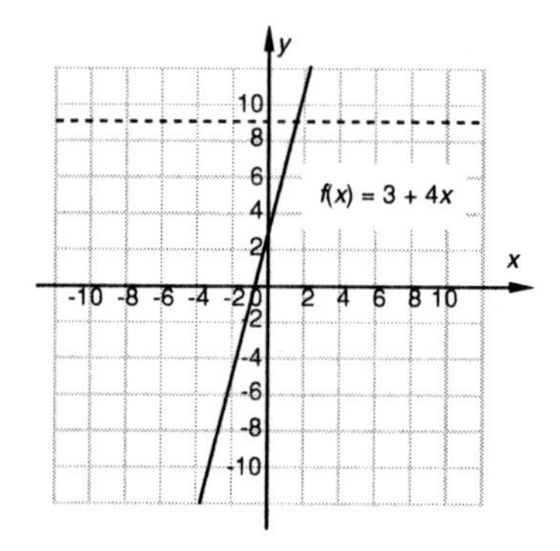

Since there is no horizontal line that crosses the graph more than once, the function is one-to-one.

35. The graph of $f(x) = 1 - x^2$ is shown below.

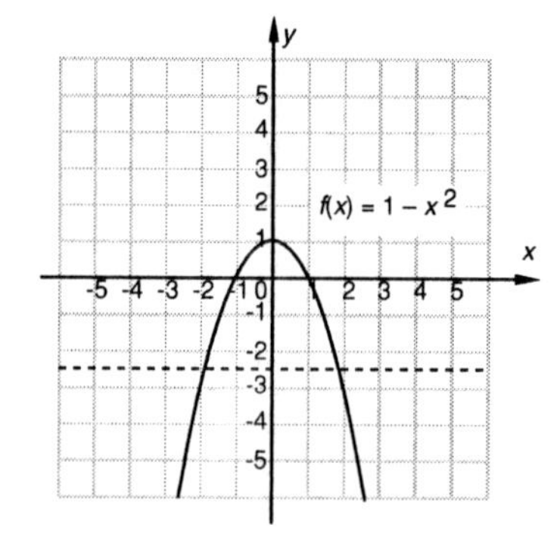

Since there are many horizontal lines that cross the graph more than once, the function is not one-to-one.

36. The graph of $f(x) = |x| - 2$ is shown below.

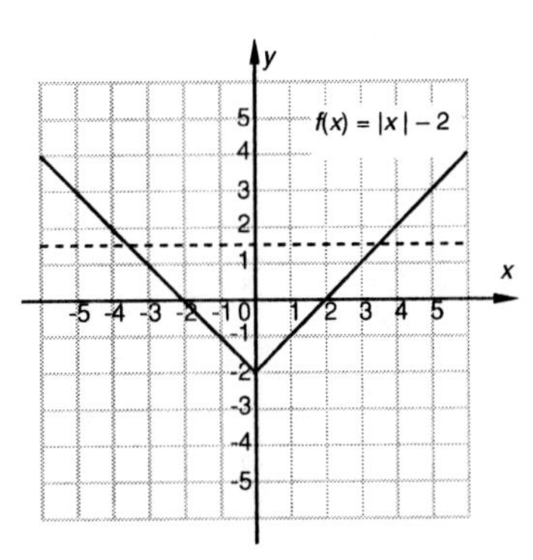

Since there are many horizontal lines that cross the graph more than once, the function is not one-to-one.

37. The graph of $f(x) = |x + 2|$ is shown below.

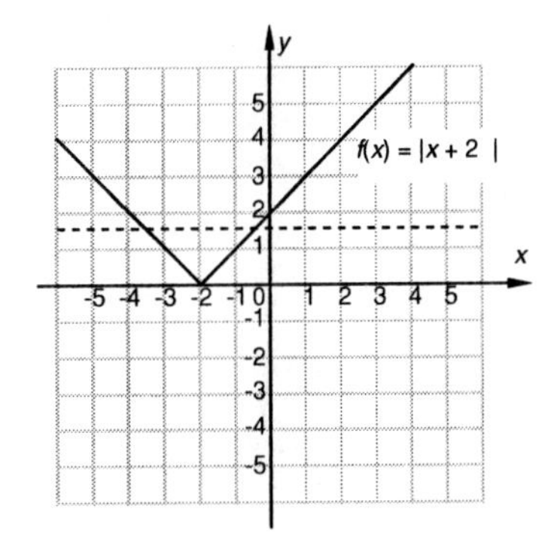

Since there are many horizontal lines that cross the graph more than once, the function is not one-to-one.

38. The graph of $f(x) = -0.8$ is shown below.

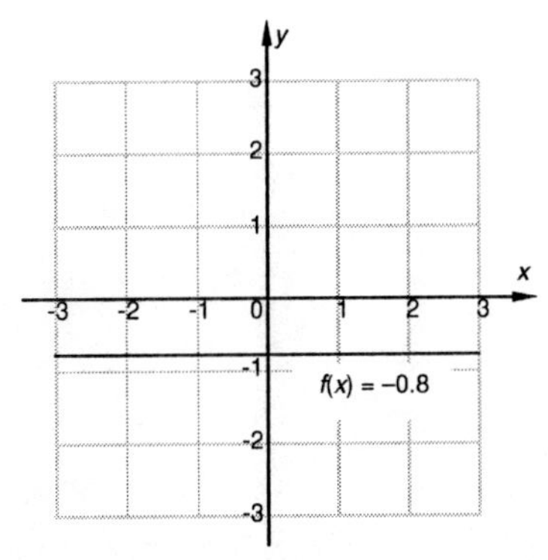

Since the horizontal line $y = -0.8$ crosses the graph more than once, the function is not one-to-one.

39. The graph of $f(x) = -\frac{4}{x}$ is shown below.

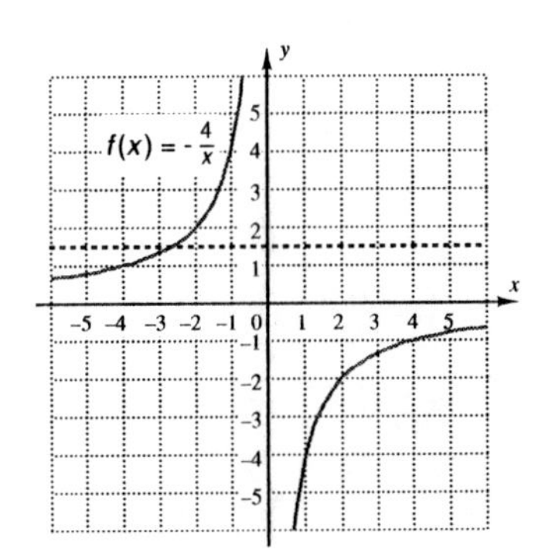

Since there is no horizontal line that crosses the graph more than once, the function is one-to-one.

40. The graph of $f(x) = \frac{2}{x+3}$ is shown below.

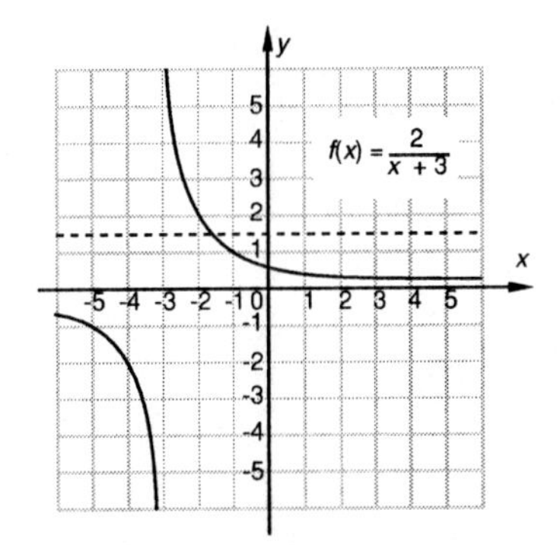

Since there is no horizontal line that crosses the graph more than once, the function is one-to-one.

41. The graph of $f(x) = \frac{2}{3}$ is shown below.

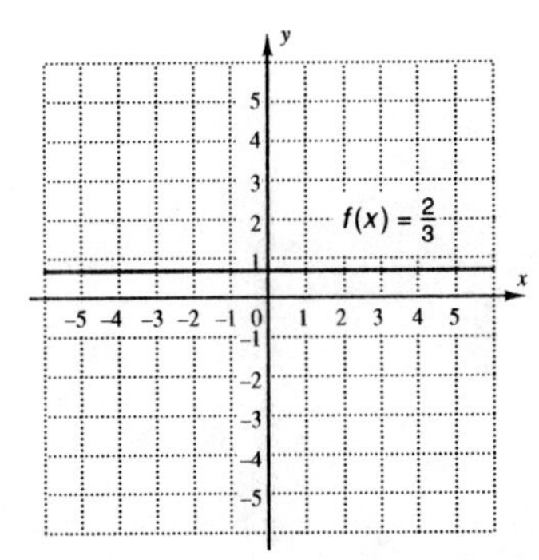

Since the horizontal line $y = \frac{2}{3}$ crosses the graph more than once, the function is not one-to-one.

42. The graph of $f(x) = \frac{1}{2}x^2 + 3$ is shown below.

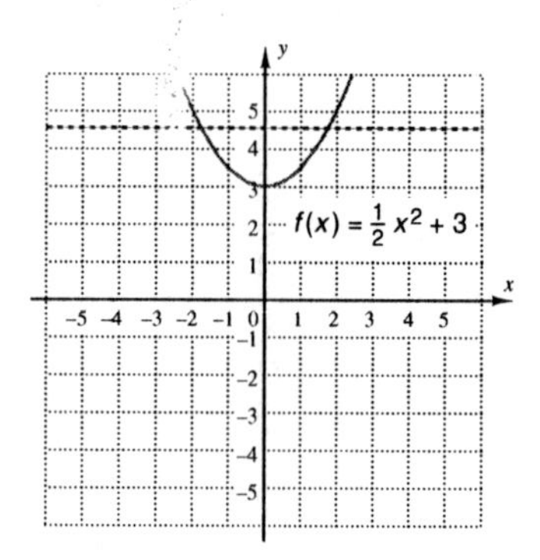

Since there are many horizontal lines that cross the graph more than once, the function is not one-to-one.

43. The graph of $f(x) = \sqrt{25 - x^2}$ is shown below.

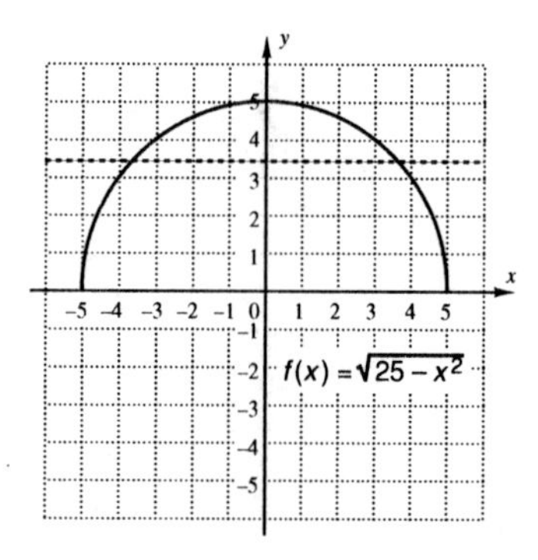

Since there are many horizontal lines that cross the graph more than once, the function is not one-to-one.

44. The graph of $f(x) = -x^3 + 2$ is shown below.

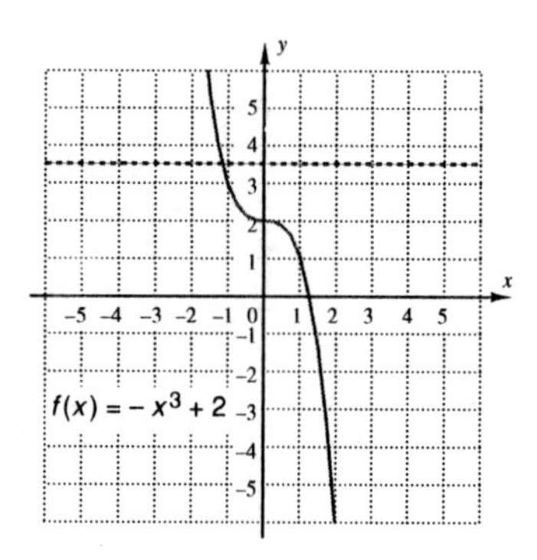

Since there is no horizontal line that crosses the graph more than once, the function is one-to-one.

45. a) The graph of $f(x) = x + 4$ is shown below. It passes the horizontal line test, so it is one-to-one.

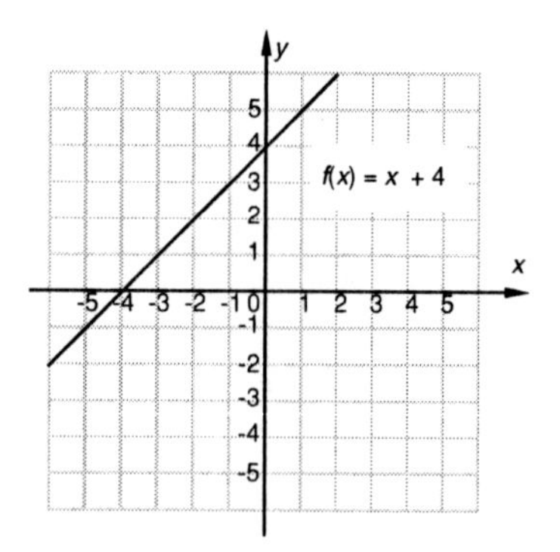

b) Replace $f(x)$ with y: $y = x + 4$

Interchange x and y: $x = y + 4$

Solve for y: $x - 4 = y$

Replace y with $f^{-1}(x)$: $f^{-1}(x) = x - 4$

46. a) The graph of $f(x) = 7 - x$ is shown below. It passes the horizontal line test, so it is one-to-one.

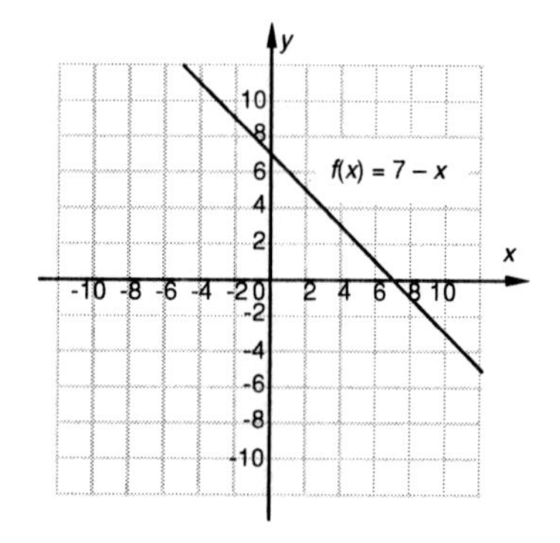

b) Replace $f(x)$ with y: $y = 7 - x$

Interchange x and y: $x = 7 - y$

Solve for y: $y = 7 - x$

Replace y with $f^{-1}(x)$: $f^{-1}(x) = 7 - x$

47. a) The graph of $f(x) = 2x - 1$ is shown below. It passes the horizontal line test, so it is one-to-one.

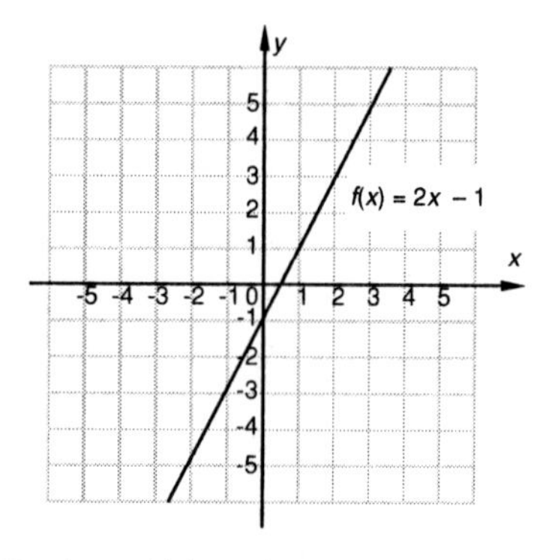

b) Replace $f(x)$ with y: $y = 2x - 1$

Interchange x and y: $x = 2y - 1$

Solve for y: $\frac{x + 1}{2} = y$

Replace y with $f^{-1}(x)$: $f^{-1}(x) = \frac{x + 1}{2}$

48. a) The graph of $f(x) = 5x + 8$ is shown below. It passes the horizontal line test, so it is one-to-one.

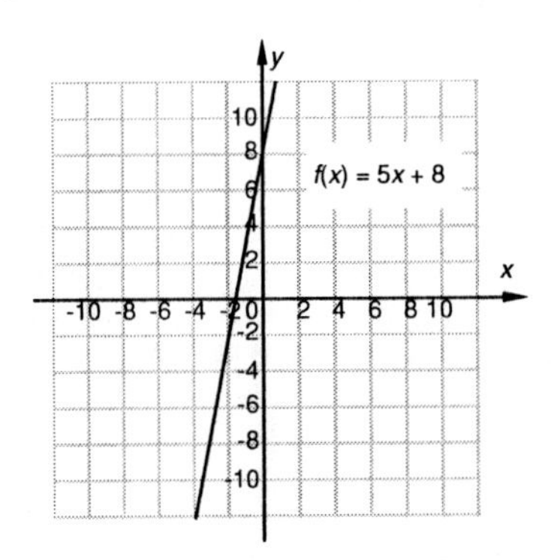

b) Replace $f(x)$ with y: $y = 5x + 8$

Interchange x and y: $x = 5y + 8$

Solve for y: $\frac{x-8}{5} = y$

Replace y with $f^{-1}(x)$: $f^{-1}(x) = \frac{x-8}{5}$

49. a) The graph of $f(x) = \frac{4}{x+7}$ is shown below. It passes the horizontal line test, so the function is one-to-one.

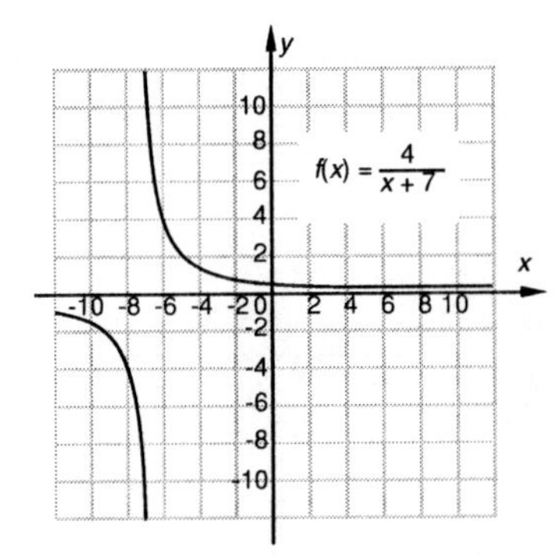

b) Replace $f(x)$ with y: $y = \frac{4}{x+7}$

Interchange x and y: $x = \frac{4}{y+7}$

Solve for y: $x(y+7) = 4$

$$y + 7 = \frac{4}{x}$$

$$y = \frac{4}{x} - 7$$

Replace y with $f^{-1}(x)$: $f^{-1}(x) = \frac{4}{x} - 7$

50. a) The graph of $f(x) = -\frac{3}{x}$ is shown below. It passes the horizontal line test, so it is one-to-one.

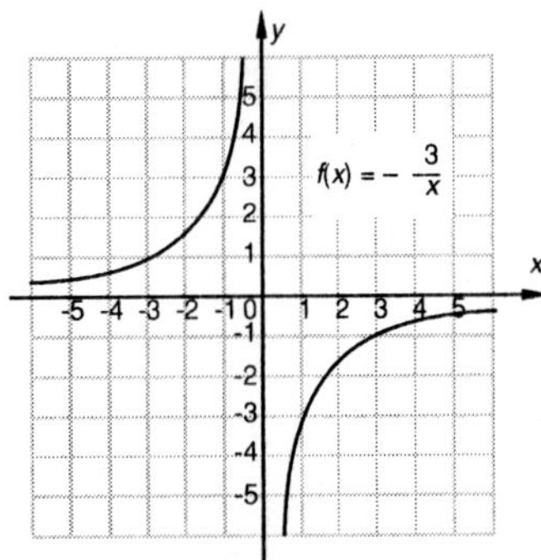

b) Replace $f(x)$ with y: $y = -\frac{3}{x}$

Interchange x and y: $x = -\frac{3}{y}$

Solve for y: $y = -\frac{3}{x}$

Replace y with $f^{-1}(x)$: $f^{-1}(x) = -\frac{3}{x}$

51. a) The graph of $f(x) = \frac{x+4}{x-3}$ is shown below. It passes the horizontal line test, so the function is one-to-one.

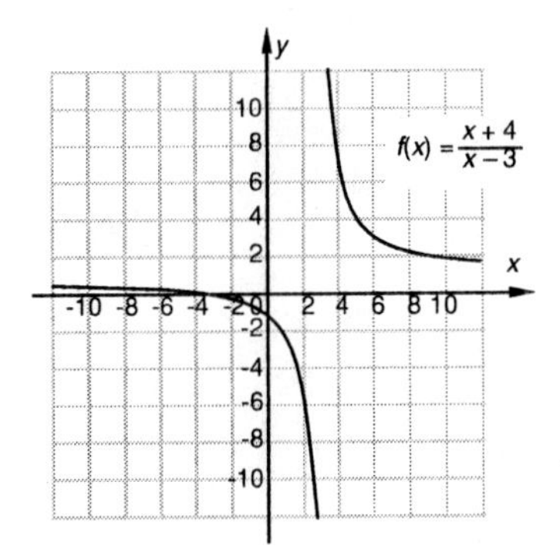

b) Replace $f(x)$ with y: $y = \frac{x+4}{x-3}$

Interchange x and y: $x = \frac{y+4}{y-3}$

Solve for y: $(y-3)x = y + 4$

$$xy - 3x = y + 4$$

$$xy - y = 3x + 4$$

$$y(x-1) = 3x + 4$$

$$y = \frac{3x+4}{x-1}$$

Replace y with $f^{-1}(x)$: $f^{-1}(x) = \frac{3x+4}{x-1}$

52. a) The graph of $f(x) = \frac{5x-3}{2x+1}$ is shown below. It passes the horizontal line test, so it is one-to-one.

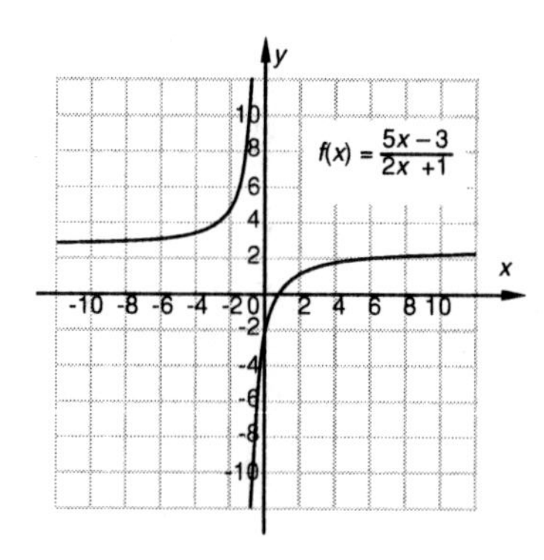

b) Replace $f(x)$ with y: $y = \dfrac{5x-3}{2x+1}$

Interchange x and y: $x = \dfrac{5y-3}{2y+1}$

Solve for y: $\dfrac{x+3}{5-2x} = y$

Replace y with $f^{-1}(x)$: $f^{-1}(x) = \dfrac{x+3}{5-2x}$

53. a) The graph of $f(x) = x^3 - 1$ is shown below. It passes the horizontal line test, so the function is one-to-one.

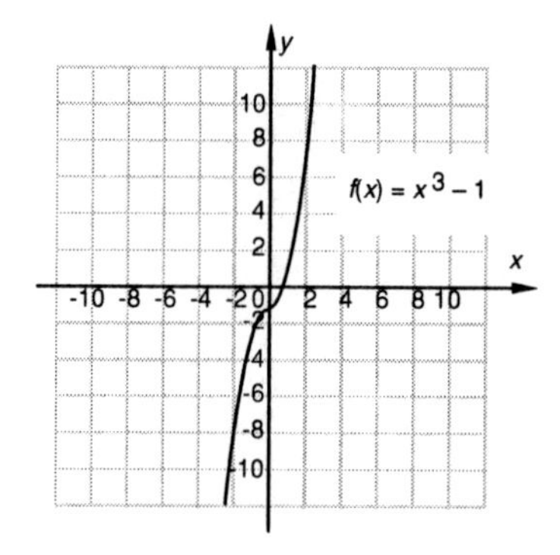

b) Replace $f(x)$ with y: $y = x^3 - 1$

Interchange x and y: $x = y^3 - 1$

Solve for y: $x + 1 = y^3$

$\sqrt[3]{x+1} = y$

Replace y with $f^{-1}(x)$: $f^{-1}(x) = \sqrt[3]{x+1}$

54. a) The graph of $f(x) = (x+5)^3$ is shown below. It passes the horizontal line test, so it is one-to-one.

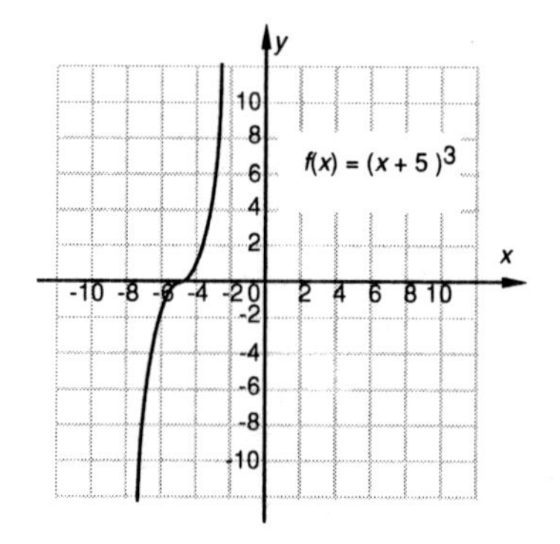

b) Replace $f(x)$ with y: $y = (x+5)^3$

Interchange x and y: $x = (y+5)^3$

Solve for y: $\sqrt[3]{x} - 5 = y$

Replace y with $f^{-1}(x)$: $f^{-1}(x) = \sqrt[3]{x} - 5$

55. a) The graph of $f(x) = x\sqrt{4-x^2}$ is shown below. Since there are many horizontal lines that cross the graph more than once, the function is not one-to-one and thus does not have an inverse that is a function.

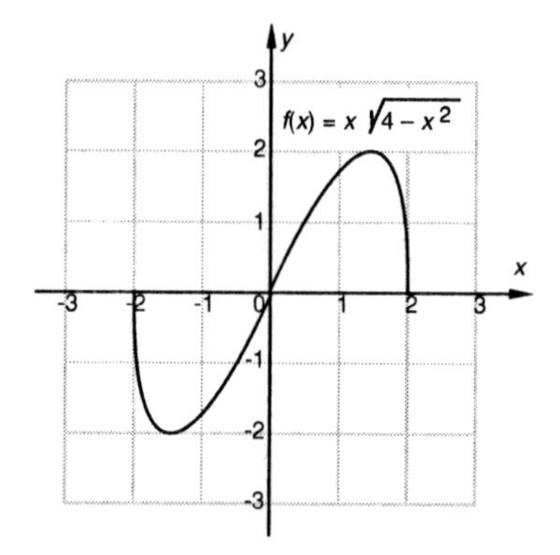

56. a) The graph of $f(x) = 2x^2 - x - 1$ is shown below. Since there are many horizontal lines that cross the graph more than once, the function is not one-to-one and thus does not have an inverse that is a function.

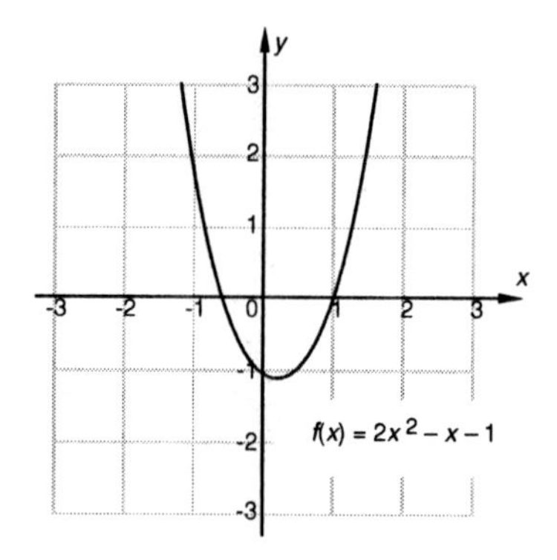

57. a) The graph of $f(x) = 5x^2 - 2$, $x \geq 0$ is shown below. It passes the horizontal line test, so it is one-to-one.

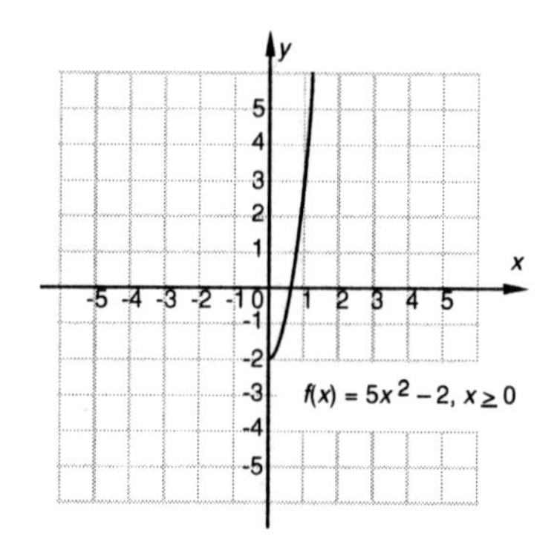

b) Replace $f(x)$ with y: $y = 5x^2 - 2$

Interchange x and y: $x = 5y^2 - 2$

Solve for y: $x + 2 = 5y^2$

$$\frac{x+2}{5} = y^2$$

$$\sqrt{\frac{x+2}{5}} = y$$

(We take the principal square root, because $x \geq 0$ in the original equation.)

Replace y with $f^{-1}(x)$: $f^{-1}(x) = \sqrt{\frac{x+2}{5}}$ for all x in the range of $f(x)$, or $f^{-1}(x) = \sqrt{\frac{x+2}{5}}$, $x \geq -2$

58. a) The graph of $f(x) = 4x^2 + 3$, $x \geq 0$ is shown below. It passes the horizontal line test, so the function is one-to-one.

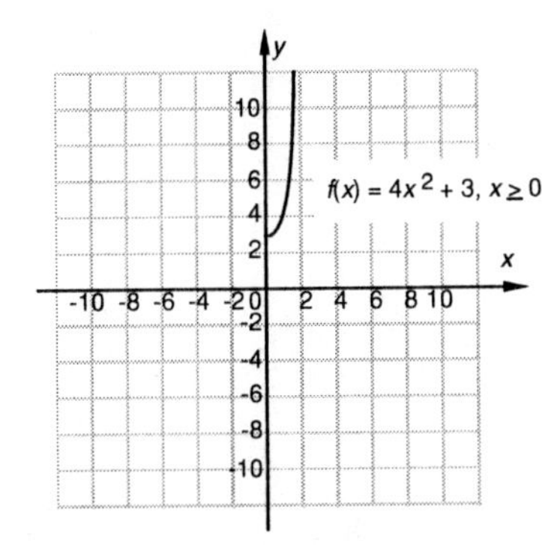

b) Replace $f(x)$ with y: $y = 4x^2 + 3$

Interchange x and y: $x = 4y^2 + 3$

Solve for y: $x - 3 = 4y^2$

$$\frac{x-3}{4} = y^2$$

$$\frac{\sqrt{x-3}}{2} = y$$

(We take the principal square root since $x \geq 0$ in the original function.)

Replace y with $f^{-1}(x)$: $f^{-1}(x) = \frac{\sqrt{x-3}}{2}$ for all x in the range of $f(x)$, or $f^{-1}(x) = \frac{\sqrt{x-3}}{2}$, $x \geq 3$

59. a) The graph of $f(x) = \sqrt{x+1}$ is shown below. It passes the horizontal line test, so the function is one-to-one.

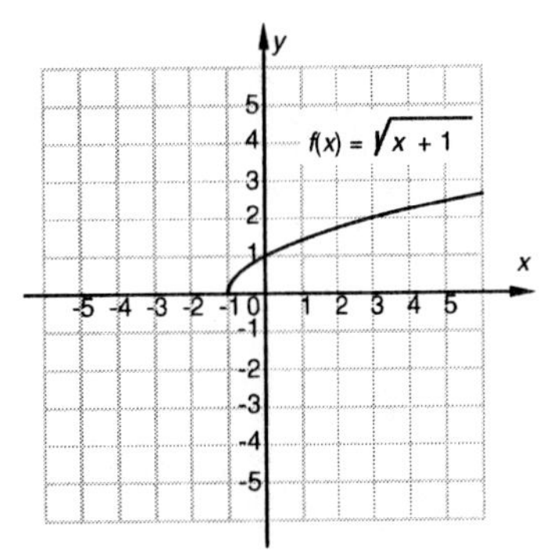

b) Replace $f(x)$ with y: $y = \sqrt{x+1}$

Interchange x and y: $x = \sqrt{y+1}$

Solve for y: $x^2 = y + 1$

$$x^2 - 1 = y$$

Replace y with $f^{-1}(x)$: $f^{-1}(x) = x^2 - 1$ for all x in the range of $f(x)$, or $f^{-1}(x) = x^2 - 1$, $x \geq 0$.

60. a) The graph of $f(x) = \sqrt[3]{x-8}$ is shown below. It passes the horizontal line test, so the function is one-to-one.

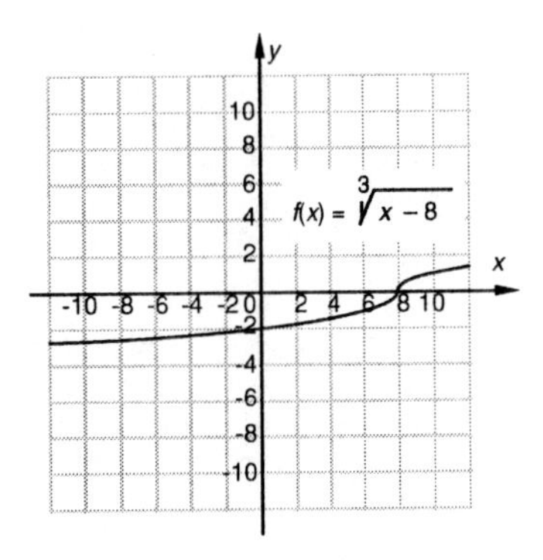

b) Replace $f(x)$ with y: $y = \sqrt[3]{x-8}$

Interchange x and y: $x = \sqrt[3]{y-8}$

Solve for y: $x^3 + 8 = y$

Replace y with $f^{-1}(x)$: $f^{-1}(x) = x^3 + 8$

61. $f(x) = 3x$

The function f multiplies an input by 3. Then to reverse this procedure, f^{-1} would divide each of its inputs by 3. Thus, $f^{-1}(x) = \frac{x}{3}$, or $f^{-1}(x) = \frac{1}{3}x$.

62. $f(x) = \frac{1}{4}x + 7$

The function f multiplies an input by $\frac{1}{4}$ and then adds 7. To reverse this procedure, f^{-1} would subtract 7 from each of its inputs and then multiply by 4. Thus, $f^{-1}(x) = 4(x - 7)$.

63. $f(x) = -x$

The outputs of f are the opposites, or additive inverses, of the inputs. Then the outputs of f^{-1} are the opposites of its inputs. Thus, $f^{-1}(x) = -x$.

64. $f(x) = \sqrt[3]{x} - 5$

The function f takes the cube root of an input and then subtracts 5. To reverse this procedure, f^{-1} would add 5 to each of its inputs and then raise the result to the third power. Thus, $f^{-1}(x) = (x+5)^3$.

65. $f(x) = \sqrt[3]{x-5}$

The function f subtracts 5 from each input and then takes the cube root of the result. To reverse this procedure, f^{-1} would raise each input to the third power and then add 5 to the result. Thus, $f^{-1}(x) = x^3 + 5$.

66. $f(x) = x^{-1}$

The outputs of f are the reciprocals of the inputs. Then the outputs of f^{-1} are the reciprocals of its inputs. Thus, $f^{-1}(x) = x^{-1}$.

67. We reflect the graph of f across the line $y = x$. The reflections of the labeled points are $(-5,-5)$, $(-3,0)$, $(1,2)$, and $(3,5)$.

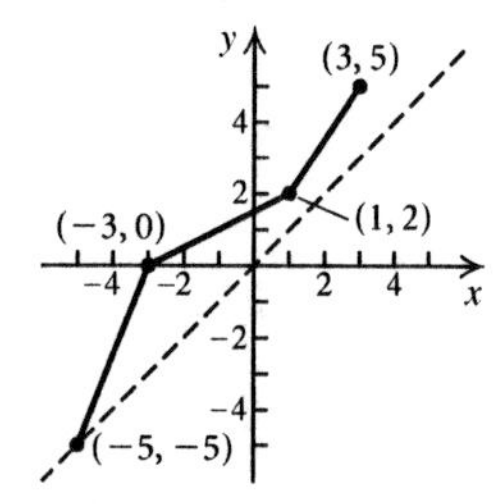

68. We reflect the graph of f across the line $y = x$. The reflections of the labeled points are $(-3,-6)$, $(2,-4)$, $(3,1)$, and $(5,2)$.

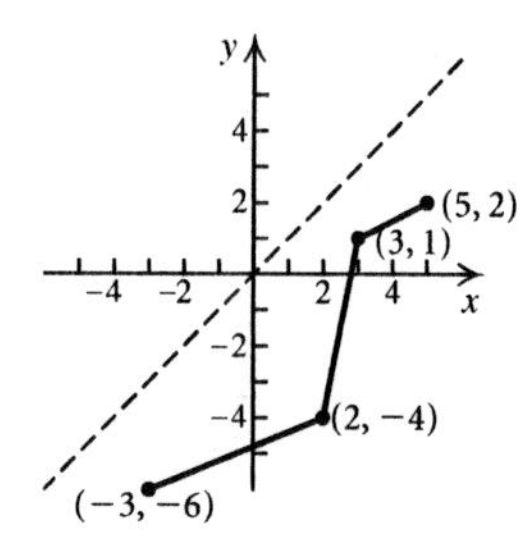

69. We reflect the graph of f across the line $y = x$. The reflections of the labeled points are $(-6,-2)$, $(1,-1)$, $(2,0)$, and $(5.375, 1.5)$.

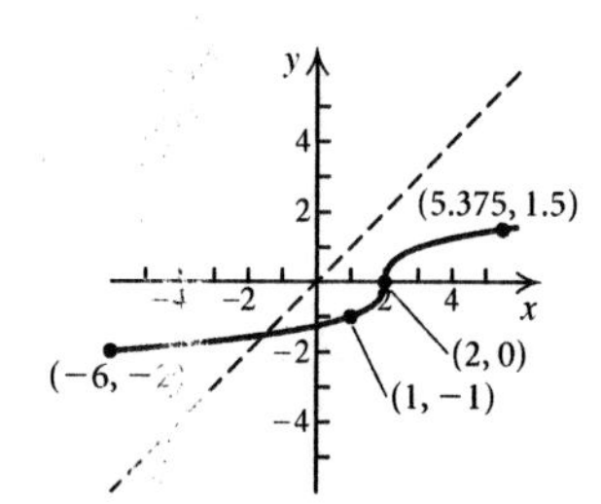

70. We reflect the graph of f across the line $y = x$. The reflections of the labeled points are $(-3,-5)$, $(-2,0)$, $(-1,3)$, and $(0,4)$.

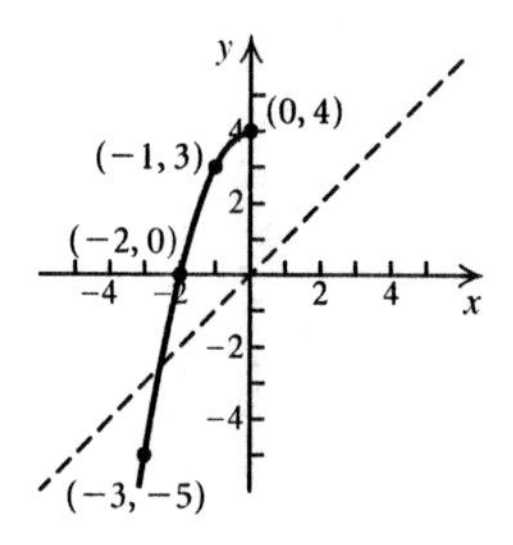

71. We reflect the graph of f across the line $y = x$.

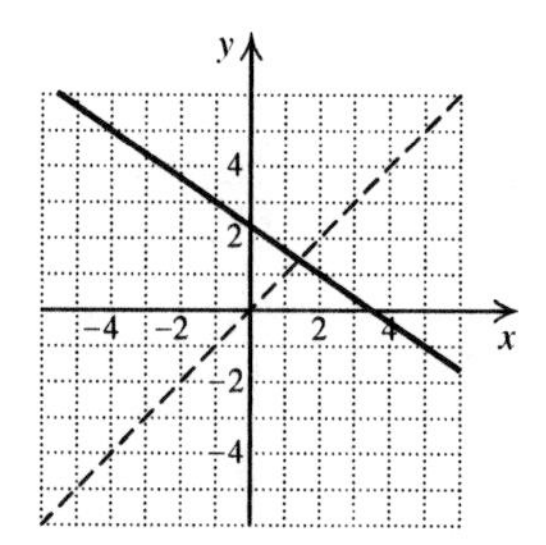

72. We reflect the graph of f across the line $y = x$.

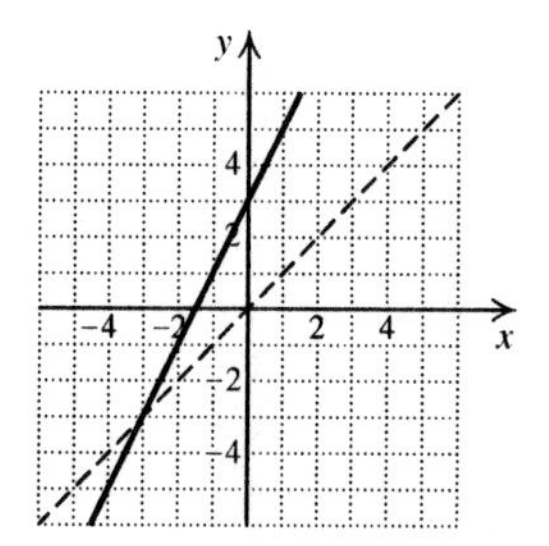

73. We find $(f^{-1} \circ f)(x)$ and $(f \circ f^{-1})(x)$ and check to see that each is x.

$(f^{-1} \circ f)(x) = f^{-1}(f(x)) = f^{-1}\left(\frac{7}{8}x\right) =$

$\frac{8}{7}\left(\frac{7}{8}x\right) = x$

$(f \circ f^{-1})(x) = f(f^{-1}(x)) = f\left(\frac{8}{7}x\right) = \frac{7}{8}\left(\frac{8}{7}x\right) = x$

74. $(f^{-1} \circ f)(x) = 4\left(\frac{x+5}{4}\right) - 5 = x + 5 - 5 = x$

$(f \circ f^{-1})(x) = \frac{4x - 5 + 5}{4} = \frac{4x}{4} = x$

75. We find $(f^{-1} \circ f)(x)$ and $(f \circ f^{-1})(x)$ and check to see that each is x.

$(f^{-1} \circ f)(x) = f^{-1}(f(x)) = f^{-1}\left(\frac{1-x}{x}\right) =$

$\frac{1}{\frac{1-x}{x} + 1} = \frac{1}{\frac{1-x+x}{x}} = \frac{1}{\frac{1}{x}} = x$

$(f \circ f^{-1})(x) = f(f^{-1}(x)) = f\left(\frac{1}{x+1}\right) =$

$\frac{1 - \frac{1}{x+1}}{\frac{1}{x+1}} = \frac{\frac{x+1-1}{x+1}}{\frac{1}{x+1}} = \frac{\frac{x}{x+1}}{\frac{1}{x+1}} = x$

76. $(f^{-1} \circ f)(x) = (\sqrt[3]{x+4})^3 - 4 = x + 4 - 4 = x$

$(f \circ f^{-1})(x) = \sqrt[3]{x^3 - 4 + 4} = \sqrt[3]{x^3} = x$

77. $(f^{-1} \circ f)(x) = f^{-1}(f(x)) = \dfrac{5\left(\frac{2}{5}x + 1\right) - 5}{2} =$

$\dfrac{2x + 5 - 5}{2} = \dfrac{2x}{2} = x$

$(f \circ f^{-1})(x) = f(f^{-1}(x) = \dfrac{2}{5}\left(\dfrac{5x-5}{2}\right) + 1 =$

$x - 1 + 1 = x$

78. $(f^{-1} \circ f)(x) = \dfrac{4\left(\frac{x+6}{3x-4}\right) + 6}{3\left(\frac{x+6}{3x-4}\right) - 1} = \dfrac{\frac{4x+24}{3x-4} + 6}{\frac{3x+18}{3x-4} - 1} =$

$\dfrac{\frac{4x + 24 + 18x - 24}{3x-4}}{\frac{3x + 18 - 3x + 4}{3x-4}} = \dfrac{22x}{3x-4} \cdot \dfrac{3x-4}{22} = \dfrac{22x}{22} = x$

$(f \circ f^{-1})(x) = \dfrac{\frac{4x+6}{3x-1} + 6}{3\left(\frac{4x+6}{3x-1}\right) - 4} = \dfrac{\frac{4x + 6 + 18x - 6}{3x-1}}{\frac{12x+18}{3x-1} - 4} =$

$\dfrac{\frac{22x}{3x-1}}{\frac{12x + 18 - 12x + 4}{3x-1}} = \dfrac{22x}{3x-1} \cdot \dfrac{3x-1}{22} = \dfrac{22x}{22} = x$

79. Replace $f(x)$ with y: $y = 5x - 3$

Interchange x and y: $x = 5y - 3$

Solve for y: $x + 3 = 5y$

$\dfrac{x+3}{5} = y$

Replace y with $f^{-1}(x)$: $f^{-1}(x) = \dfrac{x+3}{5}$, or $\dfrac{1}{5}x + \dfrac{3}{5}$

The domain and range of f are $(-\infty, \infty)$, so the domain and range of f^{-1} are also $(-\infty, \infty)$.

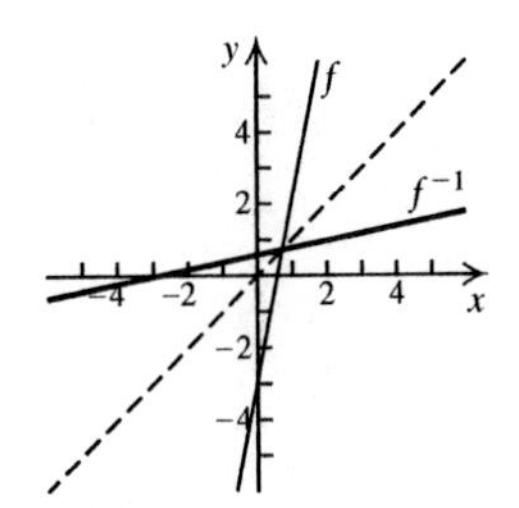

80. Replace $f(x)$ with y: $y = 2 - x$

Interchange x and y: $x = 2 - y$

Solve for y: $y = 2 - x$

Replace y with $f^{-1}(x)$: $f^{-1}(x) = 2 - x$

The domain and range of f are $(-\infty, \infty)$, so the domain and range of f^{-1} are also $(-\infty, \infty)$.

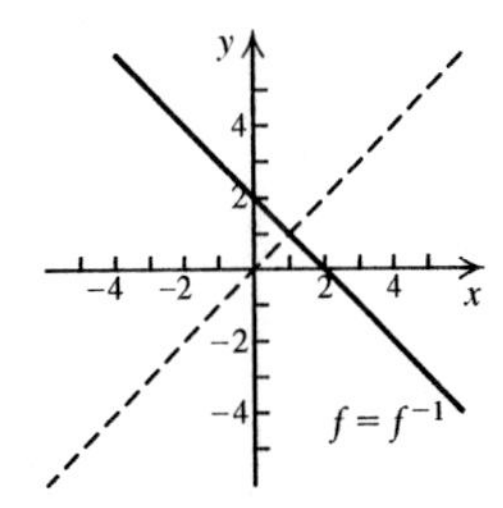

81. Replace $f(x)$ with y: $y = \dfrac{2}{x}$

Interchange x and y: $x = \dfrac{2}{y}$

Solve for y: $xy = 2$

$y = \dfrac{2}{x}$

Replace y with $f^{-1}(x)$: $f^{-1}(x) = \dfrac{2}{x}$

The domain and range of f are $(-\infty, 0) \cup (0, \infty)$, so the domain and range of f^{-1} are also $(-\infty, 0) \cup (0, \infty)$.

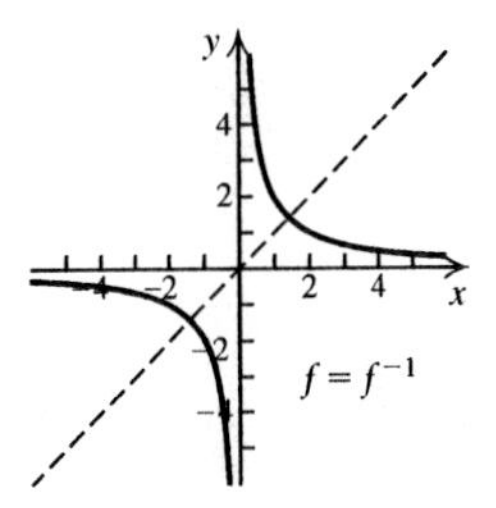

82. Replace $f(x)$ with y: $y = -\dfrac{3}{x+1}$

Interchange x and y: $x = -\dfrac{3}{y+1}$

Solve for y: $xy + x = -3$

$xy = -3 - x$

$y = \dfrac{-3-x}{x}$, *or* $-\dfrac{3}{x} - 1$

Replace y with $f^{-1}(x)$: $f^{-1}(x) = -\dfrac{3}{x} - 1$

The domain of f is $(-\infty, -1) \cup (-1, \infty)$ and the range of f is $(-\infty, 0) \cup (0, \infty)$. Thus the domain of f^{-1} is $(-\infty, 0) \cup (0, \infty)$ and the range of f^{-1} is $(-\infty, -1) \cup (-1, \infty)$.

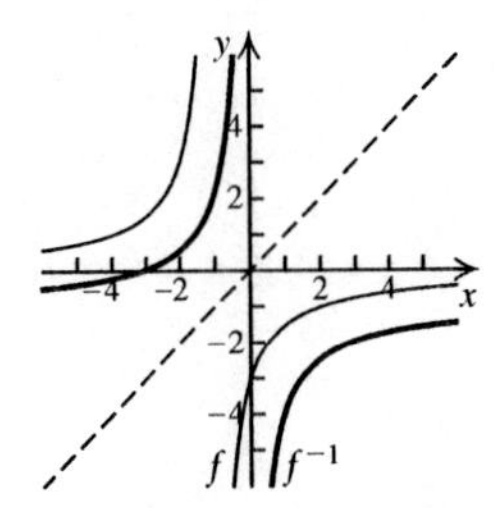

83. Replace $f(x)$ with y: $y = \dfrac{1}{3}x^3 - 2$

Interchange x and y: $x = \dfrac{1}{3}y^3 - 2$

Solve for y: $x+2=\frac{1}{3}y^3$

$3x+6=y^3$

$\sqrt[3]{3x+6}=y$

Replace y with $f^{-1}(x)$: $f^{-1}(x)=\sqrt[3]{3x+6}$

The domain and range of f are $(-\infty,\infty)$, so the domain and range of f^{-1} are also $(-\infty,\infty)$.

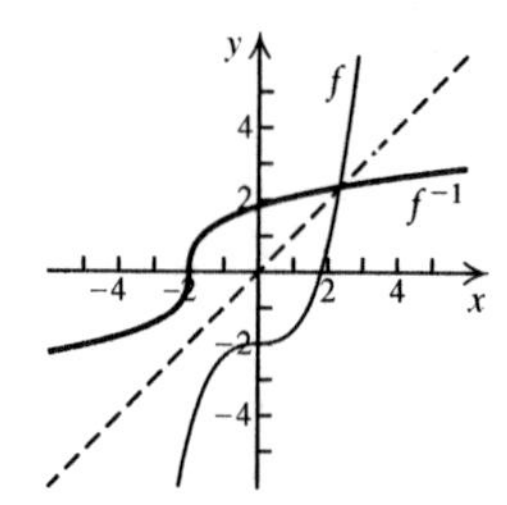

84. Replace $f(x)$ with y: $y=\sqrt[3]{x}-1$

Interchange x and y: $x=\sqrt[3]{y}-1$

Solve for y: $x+1=\sqrt[3]{y}$

$(x+1)^3=y$

Replace y with $f^{-1}(x)$: $f^{-1}(x)=(x+1)^3$

The domain and range of f are $(-\infty,\infty)$, so the domain and range of f^{-1} are also $(-\infty,\infty)$.

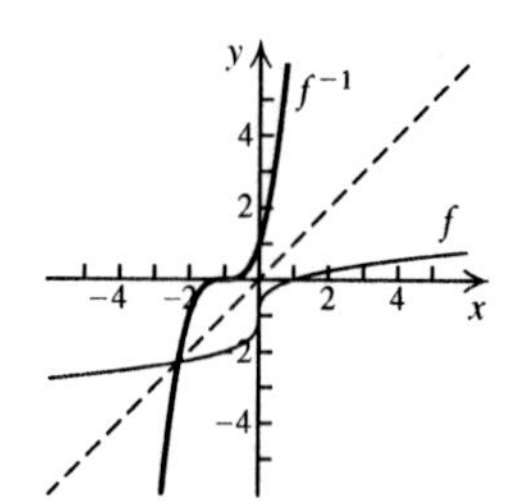

85. Replace $f(x)$ with y: $y=\frac{x+1}{x-3}$

Interchange x and y: $x=\frac{y+1}{y-3}$

Solve for y: $xy-3x=y+1$

$xy-y=3x+1$

$y(x-1)=3x+1$

$y=\frac{3x+1}{x-1}$

Replace y with $f^{-1}(x)$: $f^{-1}(x)=\frac{3x+1}{x-1}$

The domain of f is $(-\infty,3)\cup(3,\infty)$ and the range of f is $(-\infty,1)\cup(1,\infty)$. Thus the domain of f^{-1} is $(-\infty,1)\cup(1,\infty)$ and the range of f^{-1} is $(-\infty,3)\cup(3,\infty)$.

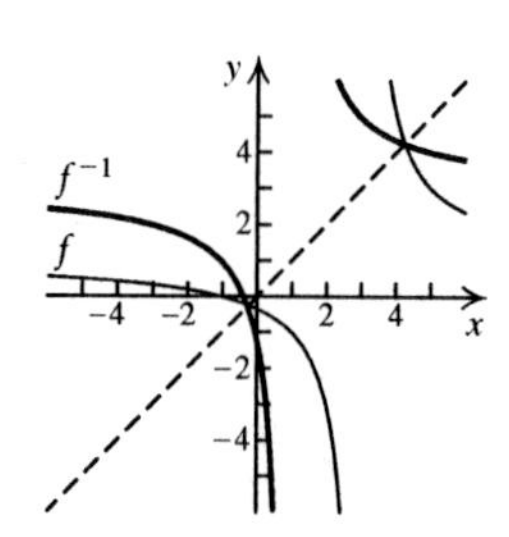

86. Replace $f(x)$ with y: $y=\frac{x-1}{x+2}$

Interchange x and y: $x=\frac{y-1}{y+2}$

Solve for y: $xy+2x=y-1$

$2x+1=y-xy$

$2x+1=y(1-x)$

$\frac{2x+1}{1-x}=y$

Replace y with $f^{-1}(x)$: $f^{-1}(x)=\frac{2x+1}{1-x}$

The domain of f is $(-\infty,-2)\cup(-2,\infty)$ and the range of f is $(-\infty,1)\cup(1,\infty)$. Thus the domain of f^{-1} is $(-\infty,1)\cup(1,\infty)$ and the range of f^{-1} is $(-\infty,-2)\cup(-2,\infty)$.

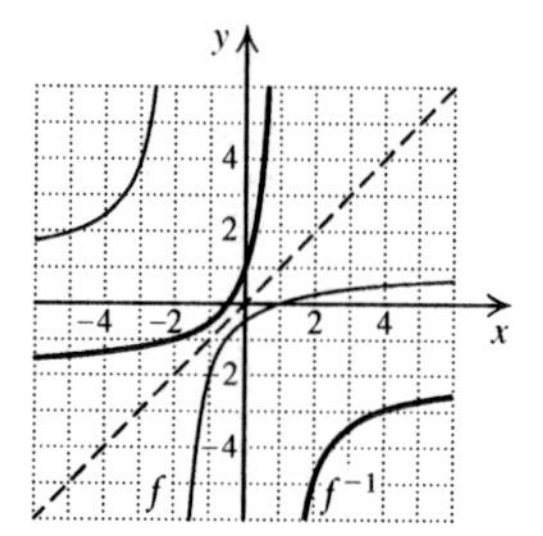

87. Since $f(f^{-1}(x))=f^{-1}(f(x))=x$, then $f(f^{-1}(5))=5$ and $f^{-1}(f(a))=a$.

88. Since $f^{-1}(f(x))=f(f^{-1}(x))=x$, then $f^{-1}(f(p))=p$ and $f(f^{-1}(1253))=1253$.

89. a) $g(x)=2(x+12)$

$g(6)=2(6+12)=2\cdot 18=36$

$g(8)=2(8+12)=2\cdot 20=40$

$g(10)=2(10+12)=2\cdot 22=44$

$g(14)=2(14+12)=2\cdot 26=52$

$g(18)=2(18+12)=2\cdot 30=60$

b) The graph passes the horizontal line test and thus has an inverse that is a function.

Replace $g(x)$ with y: $y=2(x+12)$

Interchange x and y: $x=2(y+12)$

Solve for y: $\frac{x-24}{2}=y$

Replace y with $g^{-1}(x)$: $g^{-1}(x)=\frac{x-24}{2}$, or $\frac{x}{2}-12$

c) $g^{-1}(36)=\frac{36-24}{2}=\frac{12}{2}=6$

$g^{-1}(40)=\frac{40-24}{2}=\frac{16}{2}=8$

$g^{-1}(44)=\frac{44-24}{2}=\frac{20}{2}=10$

$g^{-1}(52)=\frac{52-24}{2}=\frac{28}{2}=14$

$g^{-1}(60)=\frac{60-24}{2}=\frac{36}{2}=18$

90. $C(x) = \dfrac{100 + 5x}{x}$

Replace $C(x)$ with y: $y = \dfrac{100 + 5x}{x}$

Interchange x and y: $x = \dfrac{100 + 5y}{y}$

Solve for y: $y = \dfrac{100}{x - 5}$

Replace y with $C^{-1}(x)$: $C^{-1}(x) = \dfrac{100}{x - 5}$

$C^{-1}(x)$ gives the number of people in the group, where x is the cost per person, in dollars.

91. $y_1 = 0.8x + 1.7,$
$y_2 = \dfrac{x - 1.7}{0.8}$

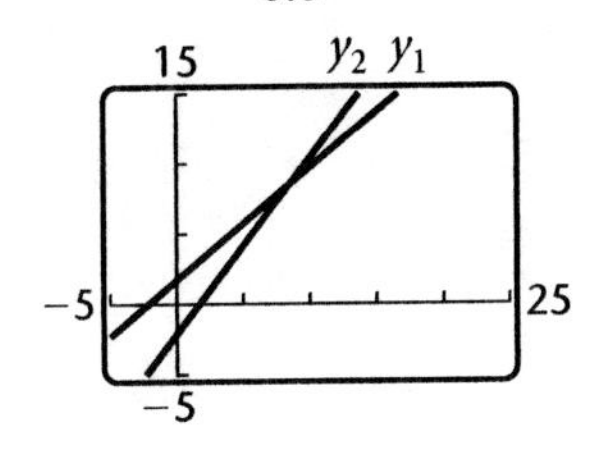

Both the domain and the range of f are the set of all real numbers. Then both the domain and the range of f^{-1} are also the set of all real numbers.

92. $y_1 = 2.7 - 1.08x,$
$y_2 = \dfrac{2.7 - x}{1.08}$

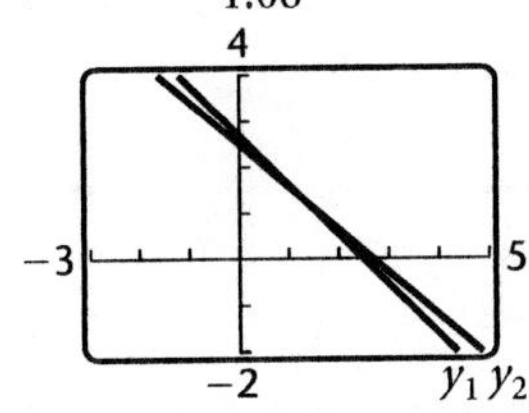

Both the domain and the range of f are the set of all real numbers. Then both the domain and the range of f^{-1} are also the set of all real numbers.

93. $y_1 = \frac{1}{2}x - 4,$
$y_2 = 2x + 8$

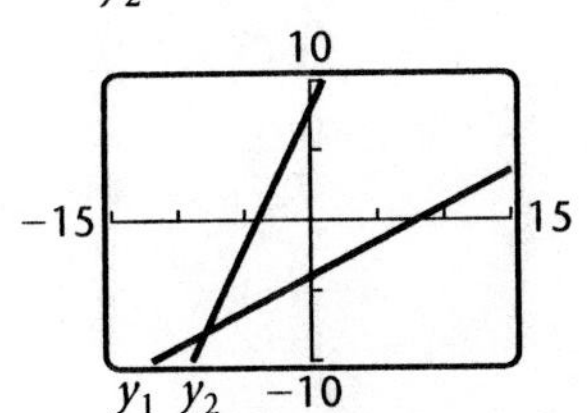

Both the domain and the range of f are the set of all real numbers. Then both the domain and the range of f^{-1} are also the set of all real numbers.

94. $y_1 = x^3 - 1,$
$y_2 = \sqrt[3]{x + 1}$

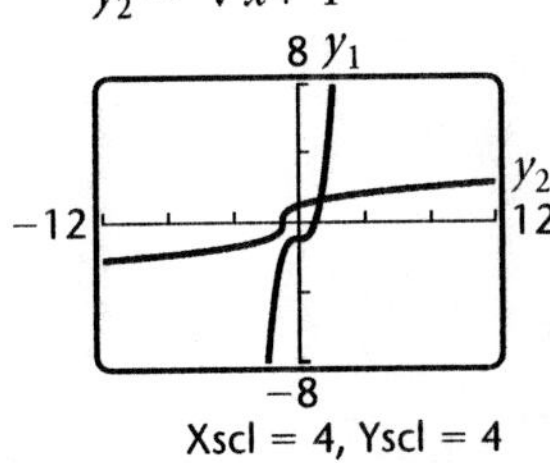

Both the domain and the range of f are the set of all real numbers. Then both the domain and the range of f^{-1} are also the set of all real numbers.

95. $y_1 = \sqrt{x - 3},$
$y_2 = x^2 + 3, x \geq 0$

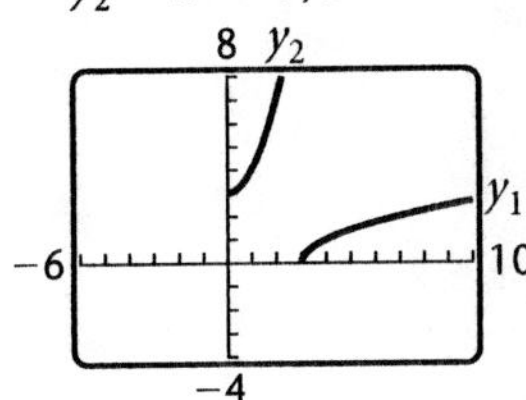

The domain of f is $[3, \infty)$ and the range of f is $[0, \infty)$. Then the domain of f^{-1} is $[0, \infty)$ and the range of f^{-1} is $[3, \infty)$.

96. $y_1 = -\dfrac{2}{x}, \; y_2 = -\dfrac{2}{x}$

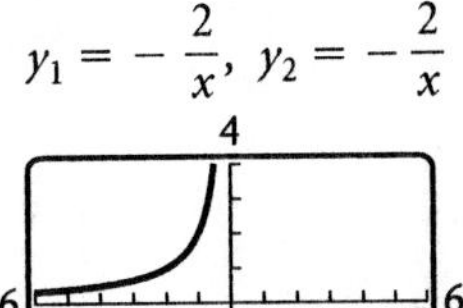

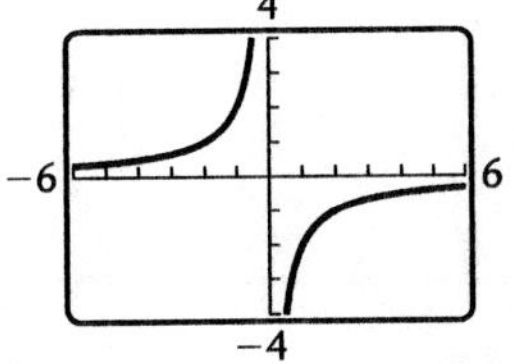

Both the domain and the range of f are $(-\infty, 0) \cup (0, \infty)$. Since $f^{-1} = f$, f^{-1} has the same domain and range.

97. $y_1 = x^2 - 4, x \geq 0; \; y_2 = \sqrt{4 + x}$

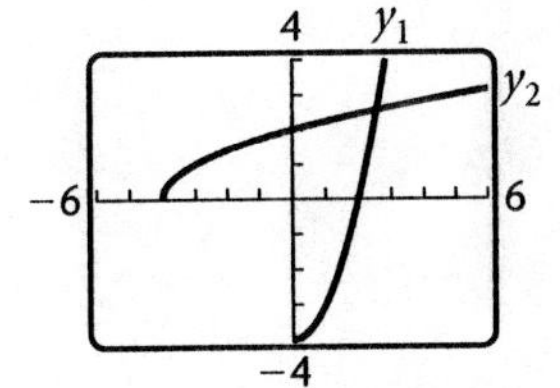

Since it is specified that $x \geq 0$, the domain of f is $[0, \infty)$. The range of f is $[-4, \infty)$. Then the domain of f^{-1} is $[-4, \infty)$ and the range of f^{-1} is $[0, \infty)$.

98. $y_1 = 3 - x^2, x \geq 0;$
$y_2 = \sqrt{3 - x}$

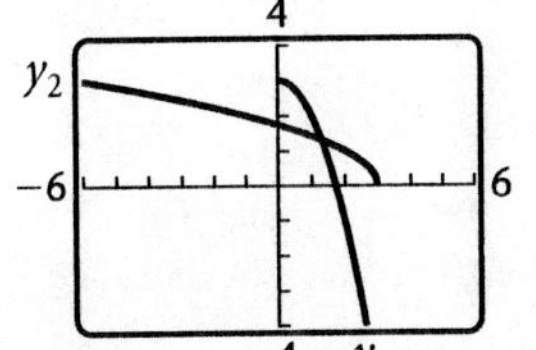

Since it is specified that $x \geq 0$, the domain of f is $[0, \infty)$. The range of f is $(-\infty, 3]$. Then the domain of f^{-1} is $(-\infty, 3]$ and the range of f^{-1} is $[0, \infty)$.

99. 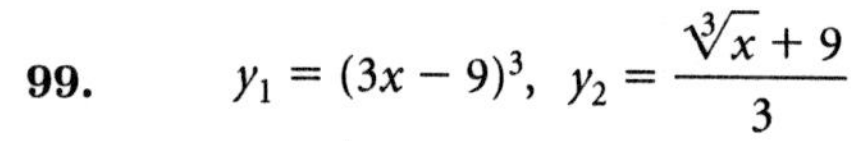

$$y_1 = (3x-9)^3, \quad y_2 = \frac{\sqrt[3]{x}+9}{3}$$

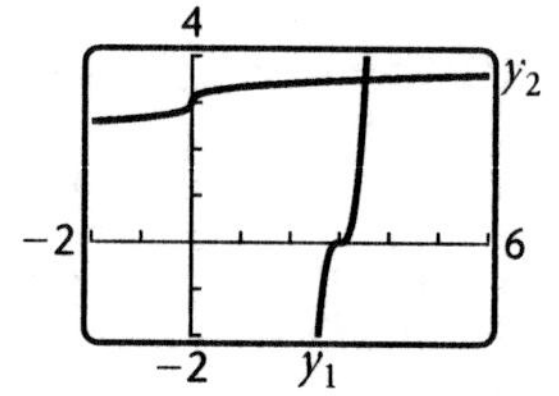

Both the domain and the range of f are the set of all real numbers. Then both the domain and the range of f^{-1} are also the set of all real numbers.

100.

$$y_1 = \sqrt[3]{\frac{x-3.2}{1.4}},$$
$$y_2 = 1.4x^3 + 3.2$$

Both the domain and the range of f are the set of all real numbers. Then both the domain and the range of f^{-1} are also the set of all real numbers.

101. a) $D(r) = \dfrac{11r+5}{10}$

$D(0) = \dfrac{11 \cdot 0 + 5}{10} = \dfrac{5}{10} = 0.5$

$D(10) = \dfrac{11 \cdot 10 + 5}{10} = \dfrac{115}{10} = 11.5$

$D(20) = \dfrac{11 \cdot 20 + 5}{10} = \dfrac{225}{10} = 22.5$

$D(50) = \dfrac{11 \cdot 50 + 5}{10} = \dfrac{555}{10} = 55.5$

$D(65) = \dfrac{11 \cdot 65 + 5}{10} = \dfrac{720}{10} = 72$

b) Replace $D(r)$ with y: $y = \dfrac{11r+5}{10}$

Interchange r and y: $r = \dfrac{11y+5}{10}$

Solve for y: $10r = 11y + 5$

$10r - 5 = 11y$

$\dfrac{10r-5}{11} = y$

Replace y with $D^{-1}(r)$: $D^{-1}(r) = \dfrac{10r-5}{11}$

$D^{-1}(r)$ represents the speed, in miles per hour, that the car is traveling when the reaction distance is r feet.

c) $y_1 = \dfrac{11x+5}{10}, \quad y_2 = \dfrac{10x-5}{11}$

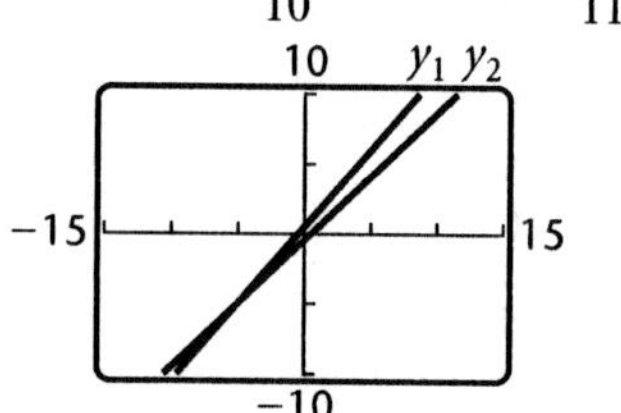

102. a) In 2005, $t = 2005 - 1995$, or 10.

$N(10) = 0.6514(10) + 53.1599 = 59.6739$

In 2010, $t = 2010 - 1995$, or 15.

$N(15) = 0.6514(15) + 53.1599 = 62.9309$

b) $y_1 = 0.6514x + 53.1599,$

$y_2 = \dfrac{x - 53.1599}{0.6514}$

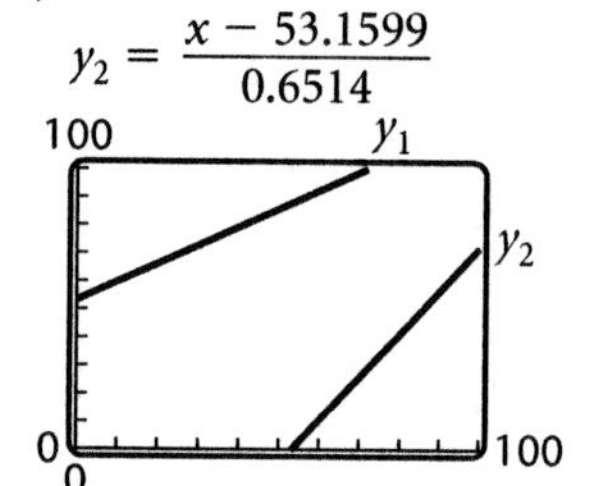

c) $N^{-1}(t)$ represents the number of years after 1995 when t loaves of bread are consumed per person per year.

103. For an even function f, $f(x) = f(-x)$ so we have $f(x) = f(-x)$ but $x \neq -x$ (for $x \neq 0$). Thus f is not one-to-one and hence it does not have an inverse.

104. C and F are inverses.

105. The functions for which the coefficient of x^2 is negative have a maximum value. These are (b), (d), (f), and (h).

106. The graphs of the functions for which the coefficient of x^2 is positive open up. These are (a), (c), (e), and (g).

107. Since $|2| > 1$ the graph of $f(x) = 2x^2$ can be obtained by stretching the graph of $f(x) = x^2$ vertically. Since $0 < \left|\frac{1}{4}\right| < 1$, the graph of $f(x) = \frac{1}{4}x^2$ can be obtained by shrinking the graph of $y = x^2$ vertically. Thus the graph of $f(x) = 2x^2$, or (a) is narrower.

108. Since $|-5| > 0$ and $0 < \left|\frac{2}{3}\right| < 1$, the graph of (d) is narrower.

109. We can write (f) as $f(x) = -2[x - (-3)]^2 + 1$. Thus the graph of (f) has vertex $(-3, 1)$.

110. For the functions that can be written in the form $f(x) = a(x-0)^2 + k$, or $f(x) = ax^2 + k$, the line of symmetry is $x = 0$. These are (a), (b), (c), and (d).

111. The graph of $f(x) = x^2 - 3$ is a parabola with vertex $(0, -3)$. If we consider x-values such that $x \geq 0$, then the graph is the right-hand side of the parabola and it passes the horizontal line test. We find the inverse of $f(x) = x^2 - 3$, $x \geq 0$.

Replace $f(x)$ with y: $y = x^2 - 3$

Interchange x and y: $x = y^2 - 3$

Solve for y: $x + 3 = y^2$

$\sqrt{x+3} = y$

(We take the principal square root, because $x \geq 0$ in the original equation.)

Replace y with $f^{-1}(x)$: $f^{-1}(x) = \sqrt{x+3}$ for all x in the range of $f(x)$, or $f^{-1}(x) = \sqrt{x+3}$, $x \geq -3$.

Answers may vary. There are other restrictions that also make $f(x)$ one-to-one.

112. No; the graph of f does not pass the horizontal line test.

113. Answers may vary. $f(x) = \dfrac{3}{x}$, $f(x) = 1 - x$, $f(x) = x$.

114. First find $f^{-1}(x)$.

Replace $f^{-1}(x)$ with y: $y = ax + b$

Interchange x and y: $x = ay + b$

Solve for y: $x - b = ay$

$\dfrac{x-b}{a} = y$

Replace y with $f^{-1}(x)$: $f^{-1}(x) = \dfrac{x-b}{a} = \dfrac{1}{a}x - \dfrac{b}{a}$

Now we find the values of a and b for which $ax + b = \dfrac{1}{a}x - \dfrac{b}{a}$. We see that $a = \dfrac{1}{a}$ for $a = \pm 1$. If $a = 1$, we have $x + b = x - b$, so $b = 0$. If $a = -1$, we have $-x + b = -x + b$, so b can be any real number.

115. If the graph $y = f(x)$ (or the graph of $y = f^{-1}(x)$) is symmetric with respect to the line $y = x$, then $f^{-1}(x) = f(x)$.

Exercise Set 4.2

1. $e^4 \approx 54.5982$

2. $e^{10} \approx 22{,}026.4658$

3. $e^{-2.458} \approx 0.0856$

4. $\left(\dfrac{1}{e^3}\right)^2 \approx 0.0025$

5. $f(x) = -2^x - 1$

$f(0) = -2^0 - 1 = -1 - 1 = -2$

The only graph with y-intercept $(0, -2)$ is (f).

6. $f(x) = -\left(\dfrac{1}{2}\right)^x$

$f(0) = -\left(\dfrac{1}{2}\right)^0 = -1$

Since the y-intercept is $(0, -1)$, the correct graph is (a) or (c). Check another point on the graph.

$f(-1) = -\left(\dfrac{1}{2}\right)^{-1} = -2$, so the point $(-1, -2)$ is on the graph. Thus (c) is the correct choice.

7. $f(x) = e^x + 3$

This is the graph of $f(x) = e^x$ shifted up 3 units. Then (e) is the correct choice.

8. $f(x) = e^{x+1}$

This is the graph of $f(x) = e^x$ shifted left 1 unit. Then (b) is the correct choice.

9. $f(x) = 3^{-x} - 2$

$f(0) = 3^{-0} - 2 = 1 - 2 = -1$

Since the y-intercept is $(0, -1)$, the correct graph is (a) or (c). Check another point on the graph. $f(-1) = 3^{-(-1)} - 2 = 3 - 2 = 1$, so $(-1, 1)$ is on the graph. Thus (a) is the correct choice.

10. $f(x) = 1 - e^x$

$f(0) = 1 - e^0 = 1 - 1 = 0$

The only graph with y-intercept $(0, 0)$ is (d).

11. Graph $f(x) = 3^x$.

Compute some function values, plot the corresponding points, and connect them with a smooth curve.

x	$y = f(x)$	(x, y)
-3	$\frac{1}{27}$	$\left(-3, \frac{1}{27}\right)$
-2	$\frac{1}{9}$	$\left(-2, \frac{1}{9}\right)$
-1	$\frac{1}{3}$	$\left(-1, \frac{1}{3}\right)$
0	1	$(0, 1)$
1	3	$(1, 3)$
2	9	$(2, 9)$
3	27	$(3, 27)$

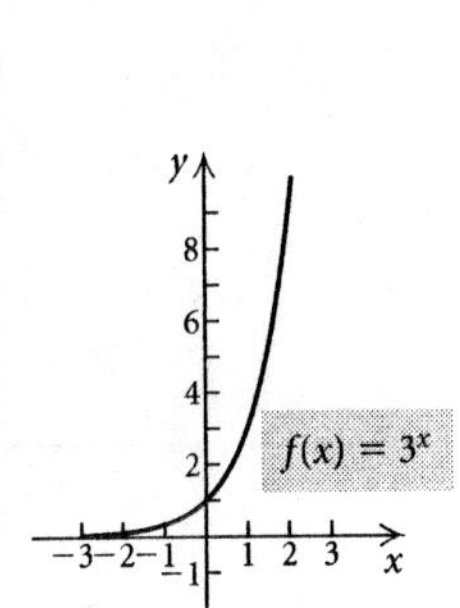

12. Graph $f(x) = 5^x$.

x	$y = f(x)$	(x, y)
-3	$\frac{1}{125}$	$\left(-3, \frac{1}{125}\right)$
-2	$\frac{1}{25}$	$\left(-2, \frac{1}{25}\right)$
-1	$\frac{1}{5}$	$\left(-1, \frac{1}{5}\right)$
0	1	$(0, 1)$
1	5	$(1, 5)$
2	25	$(2, 25)$
3	125	$(3, 125)$

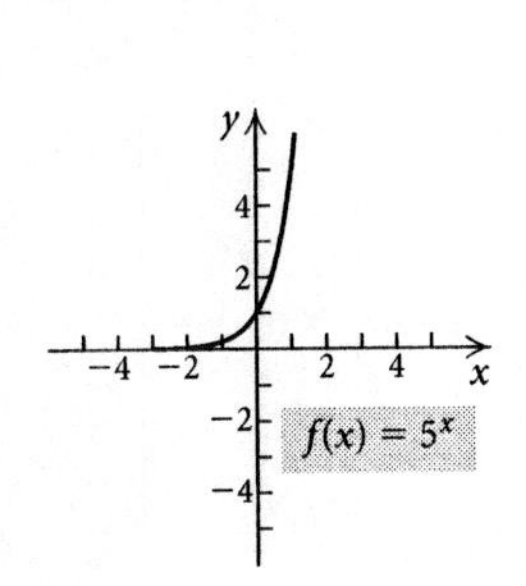

13. Graph $f(x) = 6^x$.

Compute some function values, plot the corresponding points, and connect them with a smooth curve.

x	$y = f(x)$	(x, y)
-3	$\frac{1}{216}$	$\left(-3, \frac{1}{216}\right)$
-2	$\frac{1}{36}$	$\left(-2, \frac{1}{36}\right)$
-1	$\frac{1}{6}$	$\left(-1, \frac{1}{6}\right)$
0	1	$(0, 1)$
1	6	$(1, 6)$
2	36	$(2, 36)$
3	216	$(3, 216)$

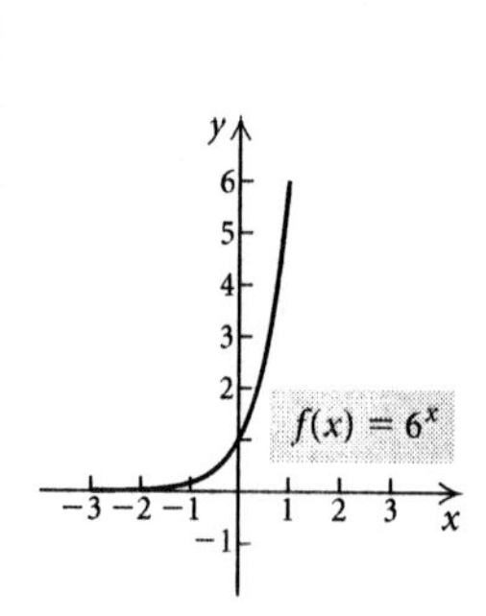

14. Graph $f(x) = 3^{-x}$.

x	$y = f(x)$	(x, y)
-3	27	$(-3, 27)$
-2	9	$(-2, 9)$
-1	3	$(-1, 3)$
0	1	$(0, 1)$
1	$\frac{1}{3}$	$\left(1, \frac{1}{3}\right)$
2	$\frac{1}{9}$	$\left(2, \frac{1}{9}\right)$
3	$\frac{1}{27}$	$\left(3, \frac{1}{27}\right)$

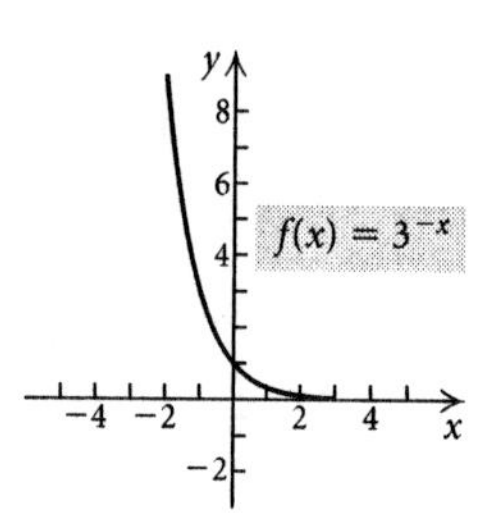

15. Graph $f(x) = \left(\frac{1}{4}\right)^x$.

Compute some function values, plot the corresponding points, and connect them with a smooth curve.

x	$y = f(x)$	(x, y)
-3	64	$(-3, 64)$
-2	16	$(-2, 16)$
-1	4	$(-1, 4)$
0	1	$(0, 1)$
1	$\frac{1}{4}$	$\left(1, \frac{1}{4}\right)$
2	[illegible]	$\left(2, \frac{1}{16}\right)$
3	[illegible]	$\left(3, \frac{1}{64}\right)$

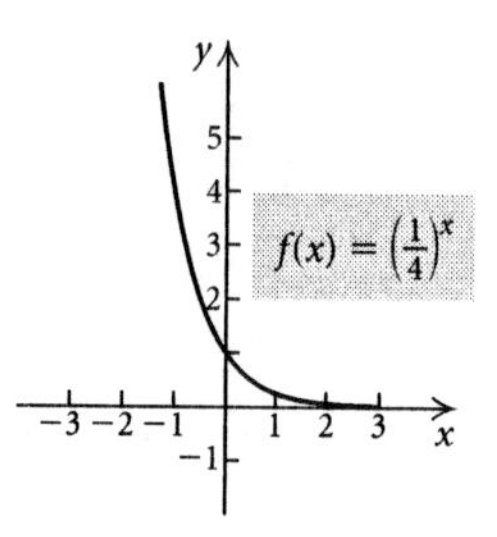

16. Graph $f(x) = \left(\frac{2}{3}\right)^x$.

x	$y = f(x)$	(x, y)
-3	$\frac{27}{8}$	$\left(-3, \frac{27}{8}\right)$
-2	$\frac{9}{4}$	$\left(-2, \frac{9}{4}\right)$
-1	$\frac{3}{2}$	$\left(-1, \frac{3}{2}\right)$
0	1	$(0, 1)$
1	$\frac{2}{3}$	$\left(1, \frac{2}{3}\right)$
2	$\frac{4}{9}$	$\left(2, \frac{4}{9}\right)$
3	$\frac{8}{27}$	$\left(3, \frac{8}{27}\right)$

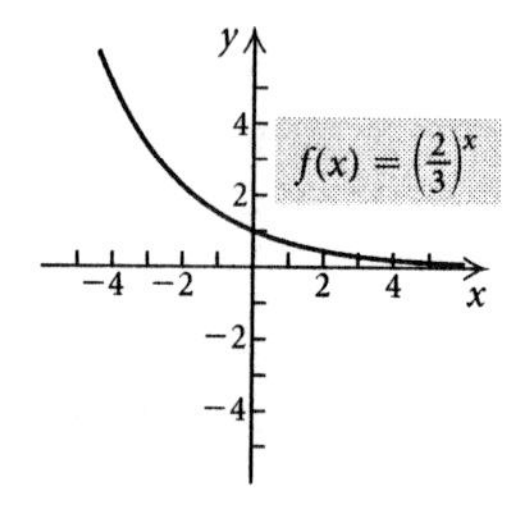

17. Graph $y = -2^x$.

x	y	(x, y)
-3	$-\frac{1}{8}$	$\left(-3, -\frac{1}{8}\right)$
-2	$-\frac{1}{4}$	$\left(-2, -\frac{1}{4}\right)$
-1	$-\frac{1}{2}$	$\left(-1, -\frac{1}{2}\right)$
0	-1	$(0, -1)$
1	-2	$(1, -2)$
2	-4	$(2, -4)$
3	-8	$(3, -8)$

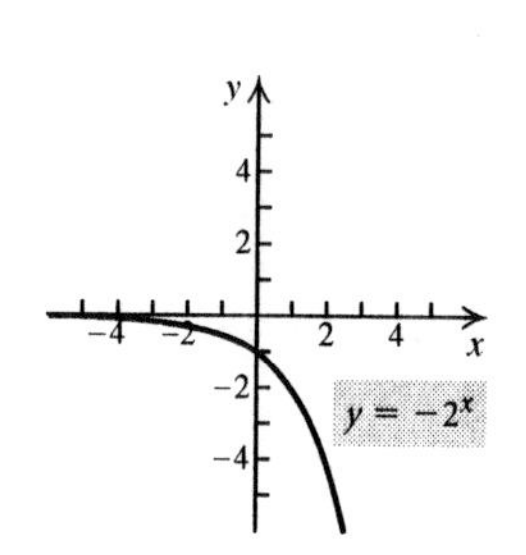

18. Graph $y = 3 - 3^x$.

x	y	(x, y)
-3	$\frac{80}{27}$	$\left(-3, \frac{80}{27}\right)$
-2	$\frac{26}{9}$	$\left(-2, \frac{26}{9}\right)$
-1	$\frac{8}{3}$	$\left(-1, \frac{8}{3}\right)$
0	2	$(0, 2)$
1	0	$(1, 0)$
2	-6	$(2, -6)$
3	-24	$(3, -24)$

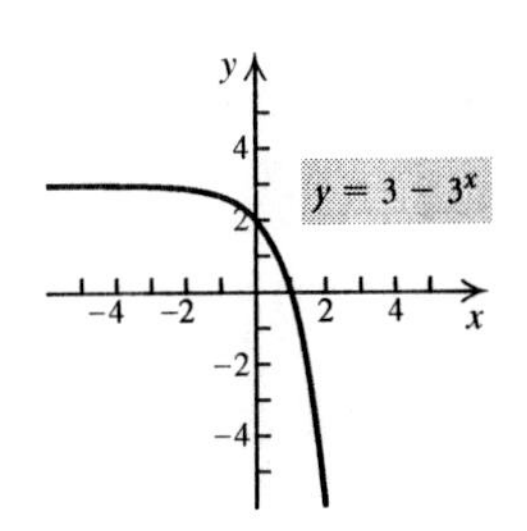

19. Graph $f(x) = -0.25^x + 4$.

x	$y = f(x)$	(x, y)
-3	-60	$(-3, -60)$
-2	-12	$(-2, -12)$
-1	0	$(-1, 0)$
0	3	$(0, 3)$
1	3.75	$(1, 3.75)$
2	3.94	$(2, 3.94)$
3	3.98	$(3, 3.98)$

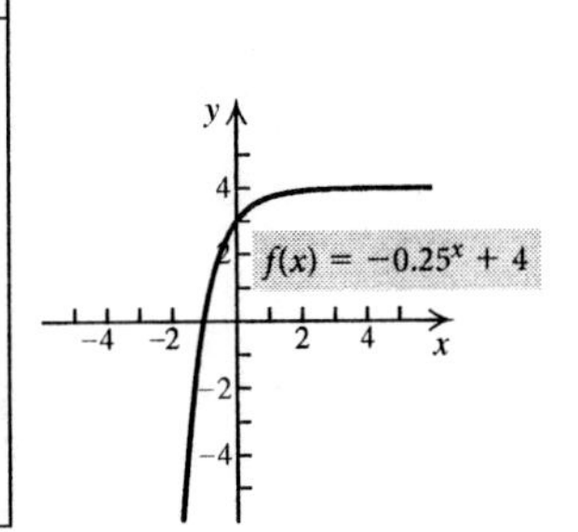

20. Graph $f(x) = 0.6^x - 3$.

x	$y = f(x)$	(x, y)
-3	1.63	$(-3, 1.63)$
-2	-0.22	$(-2, -0.22)$
-1	-1.33	$(-1, -1.33)$
0	-2	$(0, -2)$
1	-2.4	$(1, -2.4)$
2	-2.64	$(2, -2.64)$
3	-2.78	$(3, -2.78)$

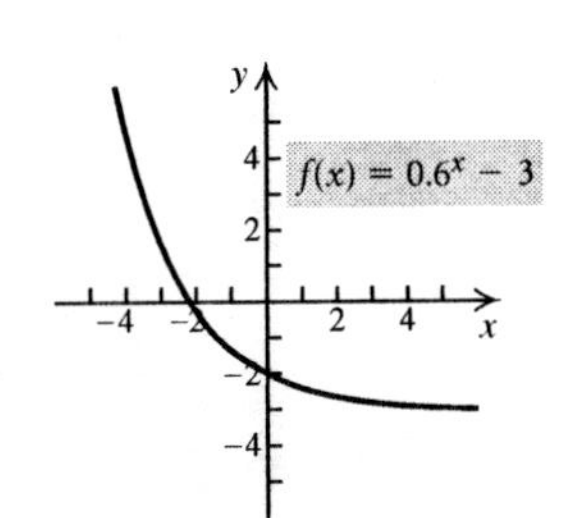

21. Graph $f(x) = 1 + e^{-x}$.

x	$y = f(x)$	(x, y)
-3	21.1	$(-3, 21.1)$
-2	8.4	$(-2, 8.4)$
-1	3.7	$(-1, 3.7)$
0	2	$(0, 2)$
1	1.4	$(1, 1.4)$
2	1.1	$(2, 1.1)$
3	1.0	$(3, 1.0)$

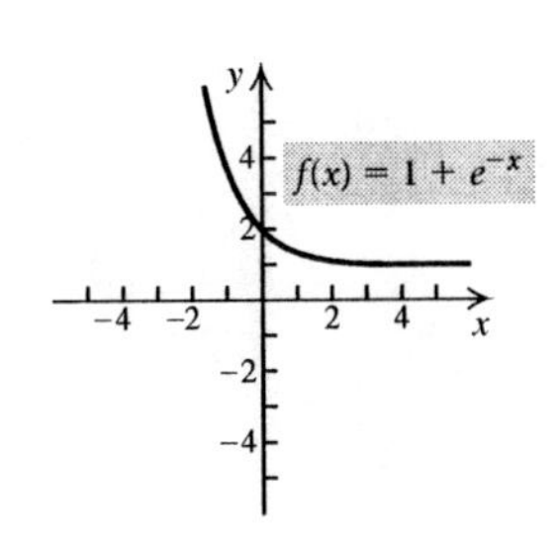

22. Graph $f(x) = 2 - e^{-x}$.

x	$y = f(x)$	(x, y)
-3	-18.1	$(-3, -18.1)$
-2	-5.4	$(-2, -5.4)$
-1	-0.8	$(-1, -0.8)$
0	1	$(0, 1)$
1	1.6	$(1, 1.6)$
2	1.9	$(2, 1.9)$
3	2.0	$(3, 2.0)$

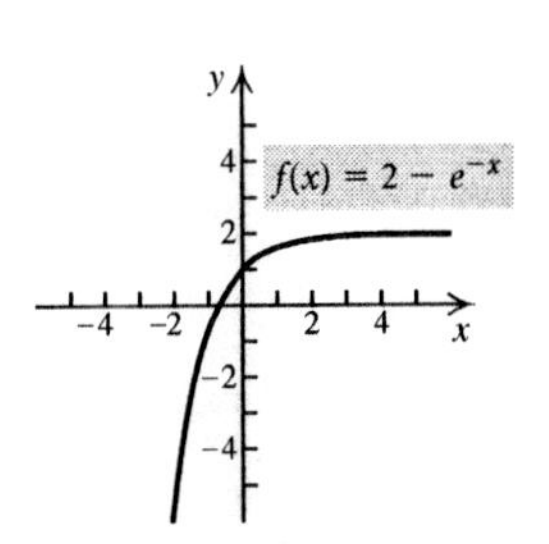

23. Graph $y = \frac{1}{4}e^x$.

Choose values for x and compute the corresponding y-values. Plot the points (x, y) and connect them with a smooth curve.

x	y	(x, y)
-3	0.0124	$(-3, 0.0124)$
-2	0.0338	$(-2, 0.0338)$
-1	0.0920	$(-1, 0.0920)$
0	0.25	$(0, 0.25)$
1	0.6796	$(1, 0.6796)$
2	1.8473	$(2, 1.8473)$
3	5.0214	$(3, 5.0214)$

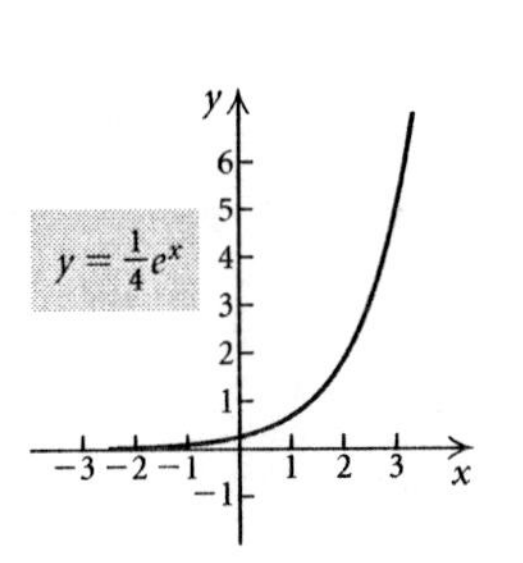

24. Graph $y = 2e^{-x}$.

x	y	(x, y)
-3	40.1711	$(-3, 40.1711)$
-2	14.7781	$(-2, 14.7781)$
-1	5.4366	$(-1, 5.4366)$
0	2	$(0, 2)$
1	0.7358	$(1, 0.7358)$
2	0.2707	$(2, 0.2707)$
3	0.0996	$(3, 0.0996)$

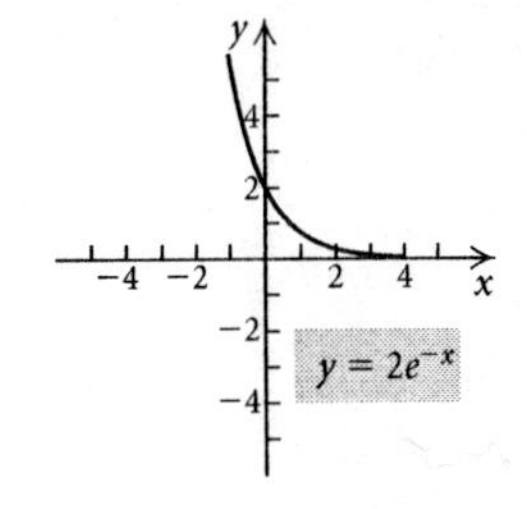

25. Graph $f(x) = 1 - e^{-x}$.

Compute some function values, plot the corresponding points, and connect them with a smooth curve.

x	y	(x, y)
-3	-19.0855	$(-3, -19.0855)$
-2	-6.3891	$(-2, -6.3891)$
-1	-1.7183	$(-1, -1.7183)$
0	0	$(0, 0)$
1	0.6321	$(1, 0.6321)$
2	0.8647	$(2, 0.8647)$
3	0.9502	$(3, 0.9502)$

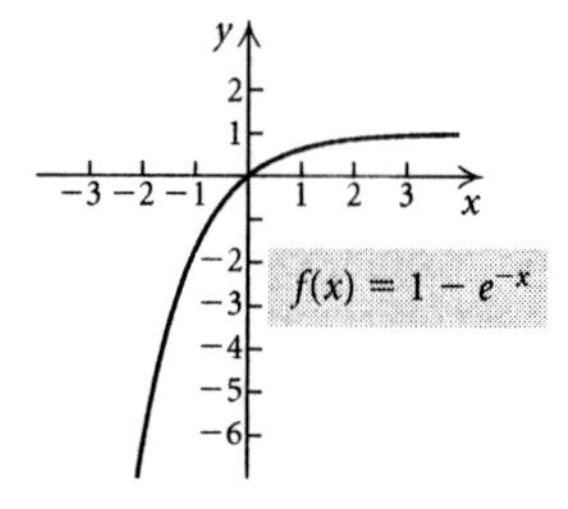

26. Graph $f(x) = e^x - 2$.

x	y	(x, y)
-3	-1.9502	$(-3, -1.9502)$
-2	-1.8647	$(-2, -1.8647)$
-1	-1.6321	$(-1, -1.6321)$
0	-1	$(0, -1)$
1	0.7183	$(1, 0.7183)$
2	5.3891	$(2, 5.3891)$
3	18.0855	$(3, 18.0855)$

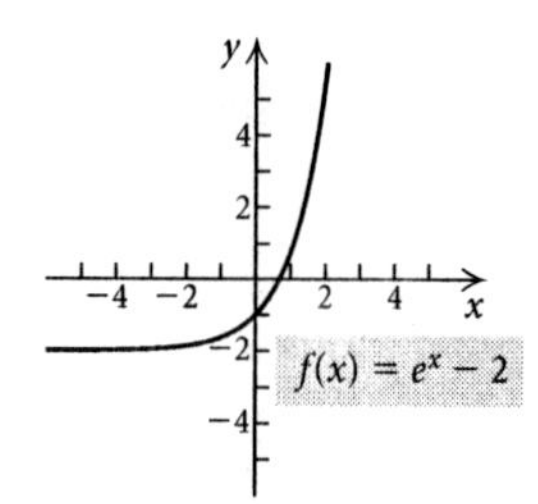

27. Shift the graph of $y = 2^x$ left 1 unit.

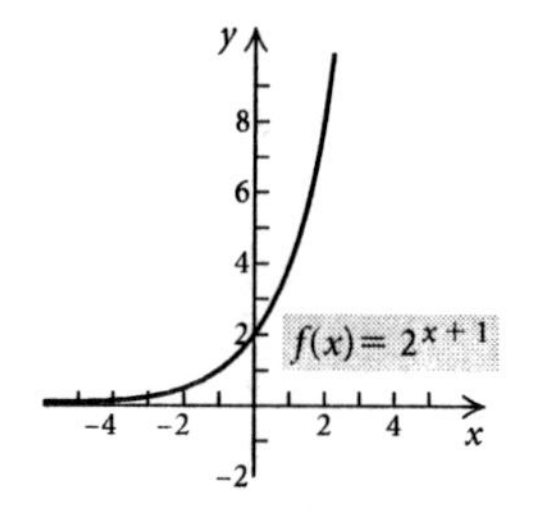

28. Shift the graph of $y = 2^x$ right 1 unit.

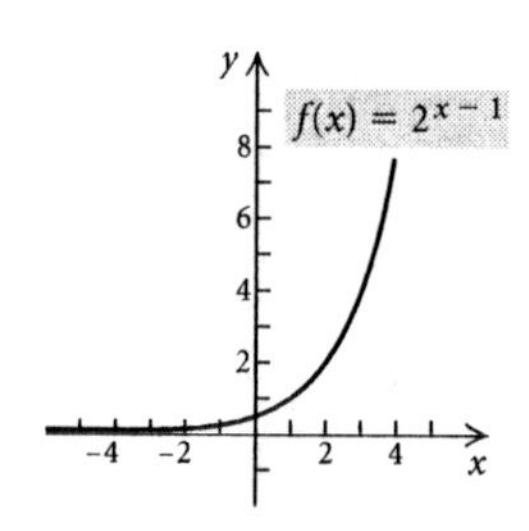

29. Shift the graph of $y = 2^x$ down 3 units.

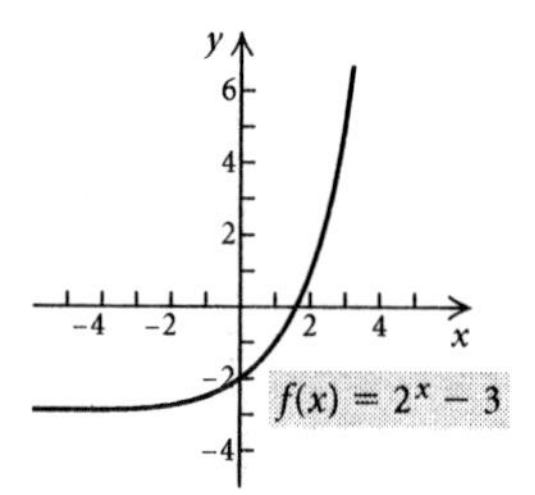

30. Shift the graph of $y = 2^x$ up 1 unit.

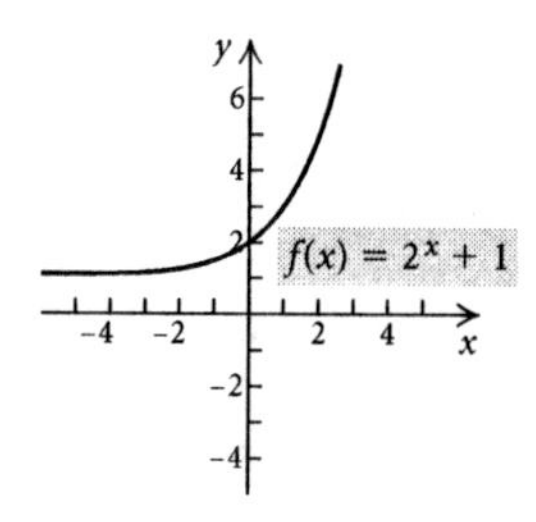

31. Reflect the graph of $y = 3^x$ across the y-axis, then across the x-axis, and then shift it up 4 units.

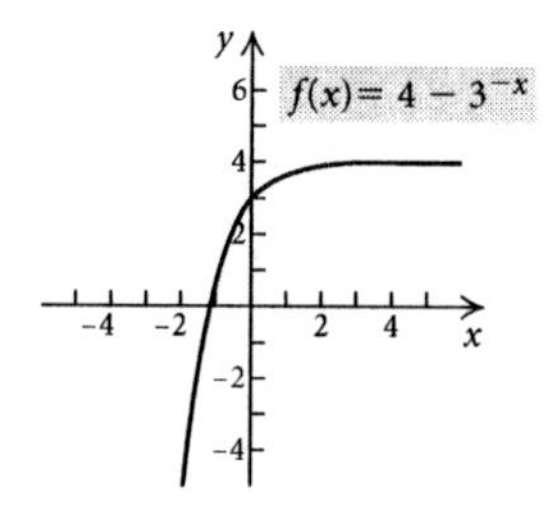

32. Shift the graph of $y = 2^x$ right 1 unit and down 3 units.

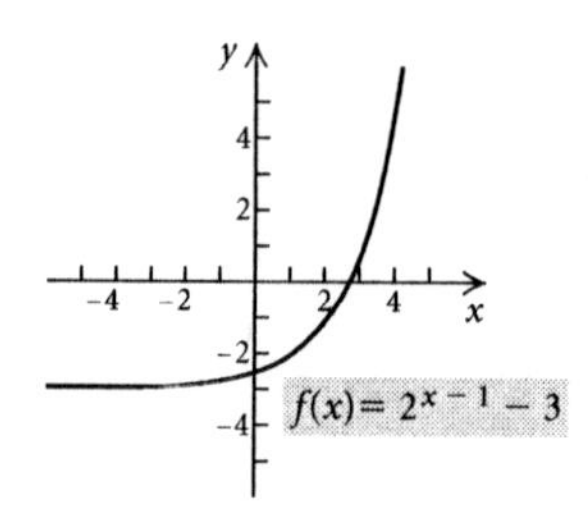

33. Shift the graph of $y = \left(\frac{3}{2}\right)^x$ right 1 unit.

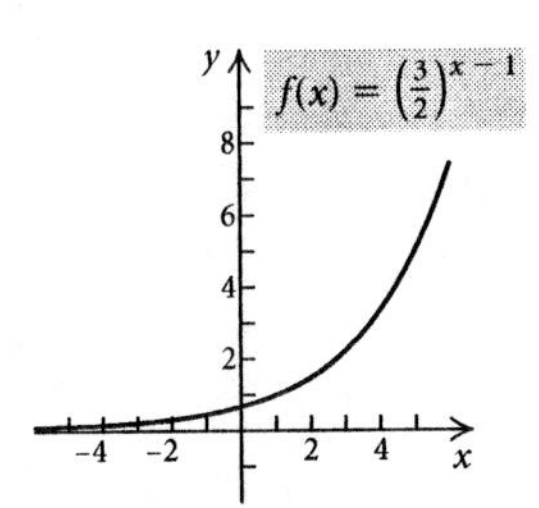

34. Reflect the graph of $y = 3^x$ across the y-axis and then shift it right 4 units.

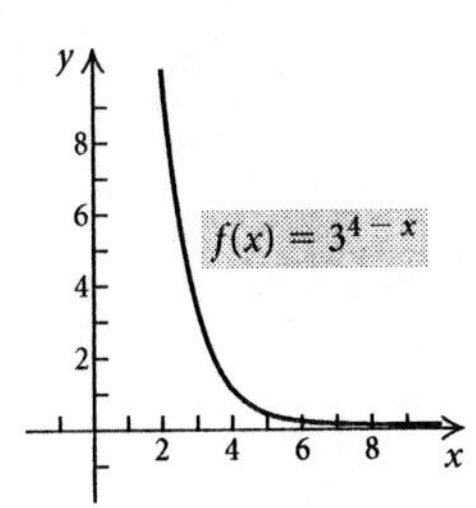

35. Shift the graph of $y = 2^x$ left 3 units and down 5 units.

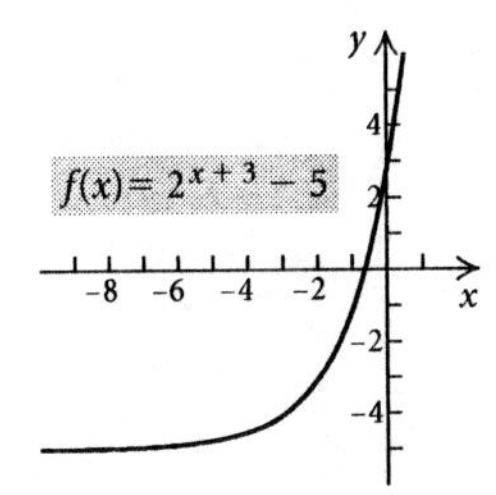

36. Shift the graph of $y = 3^x$ right 2 units and reflect it across the x-axis.

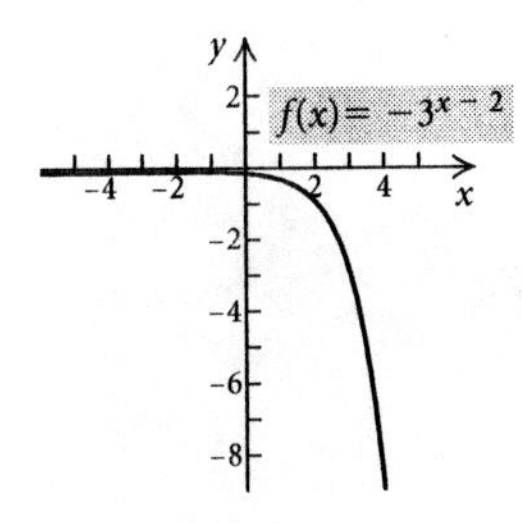

37. Shrink the graph of $y = e^x$ horizontally.

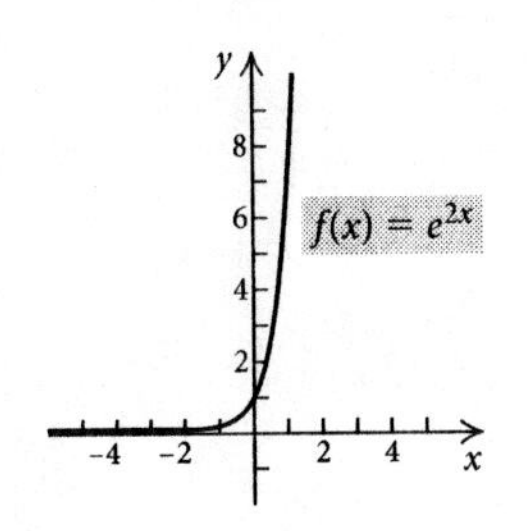

38. Stretch the graph of $y = e^x$ horizontally and reflect it across the y-axis.

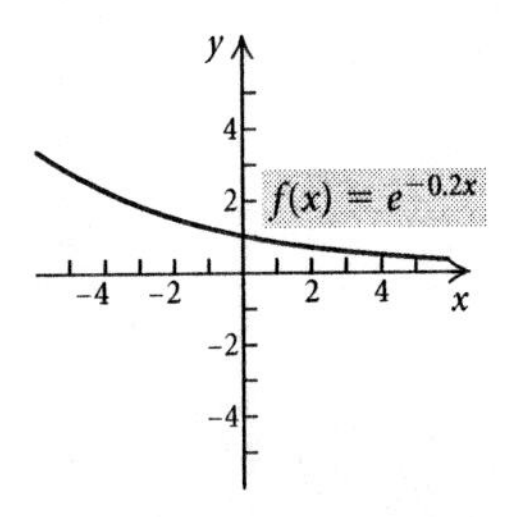

39. Shift the graph of $y = e^x$ left 1 unit and reflect it across the y-axis.

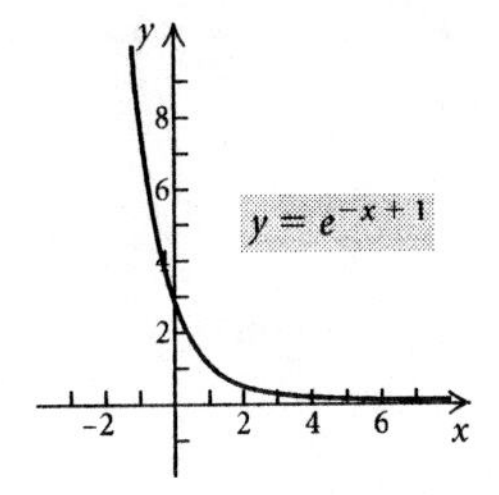

40. Shrink the graph of $y = e^x$ horizontally and shift it up 1 unit.

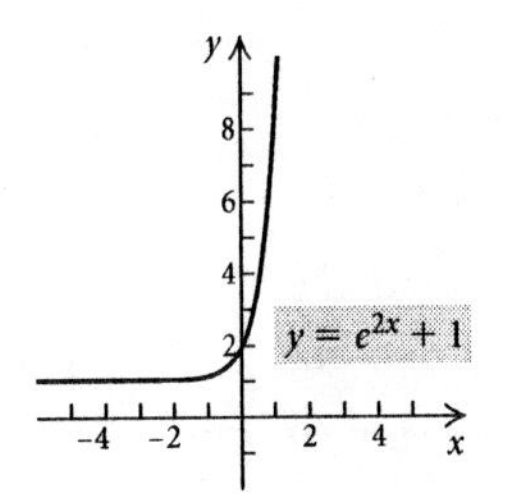

41. Reflect the graph of $y = e^x$ across the y-axis and then across the x-axis; shift it up 1 unit and then stretch it vertically.

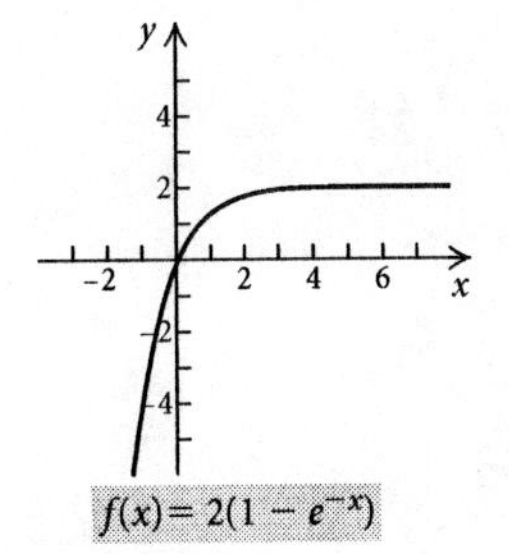

42. Stretch the graph of $y = e^x$ horizontally and reflect it across the y-axis; then reflect it down across the x-axis and shift it up 1 unit.

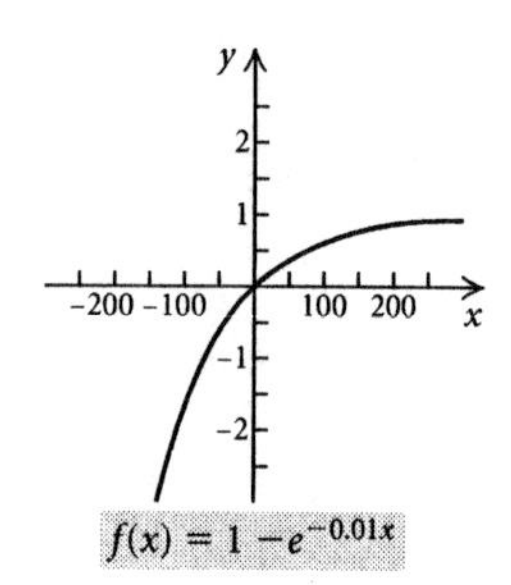

43. a) We use the formula $A = P\left(1+\frac{r}{n}\right)^{nt}$ and substitute 82,000 for P, 0.045 for r, and 4 for n.

$$A(t) = 82,000\left(1+\frac{0.045}{4}\right)^{4t} = 82,000(1.01125)^{4t}$$

b) $A(0) = 82,000(1.01125)^{4\cdot 0} = \$82,000$

$A(2) = 82,000(1.01125)^{4\cdot 2} \approx \$89,677.22$

$A(5) = 82,000(1.01125)^{4\cdot 5} \approx \$102,561.54$

$A(10) = 82,000(1.01125)^{4\cdot 10} \approx \$128,278.90$

44. a) $A(t) = 750\left(1+\frac{0.07}{2}\right)^{2t} = 750(1.035)^{2t}$

b) $A(1) = 750(1.035)^{2\cdot 1} \approx \$803.42$

$A(6) = 750(1.035)^{2\cdot 6} \approx \$1133.30$

$A(10) = 750(1.035)^{2\cdot 10} \approx \$1492.34$

$A(15) = 750(1.035)^{2\cdot 15} \approx \$2105.10$

$A(25) = 750(1.035)^{2\cdot 25} \approx \$4188.70$

45. We use the formula $A = P\left(1+\frac{r}{n}\right)^{nt}$ and substitute 3000 for P, 0.05 for r, and 4 for n.

$$A(t) = 3000\left(1+\frac{0.05}{4}\right)^{4t} = 3000(1.0125)^{4t}$$

On Jacob's sixteenth birthday, $t = 16 - 6 = 10$.

$A(10) = 3000(1.0125)^{4\cdot 10} = 4930.86$

When the CD matures \$4930.86 will be available.

46. a) $A(t) = 10,000\left(1+\frac{0.064}{2}\right)^{2t} = 10,000(1.032)^{2t}$

b) $A(0) = 10,000(1.032)^{2\cdot 0} = \$10,000$

$A(4) = 10,000(1.032)^{2\cdot 4} \approx \$12,865.82$

$A(8) = 10,000(1.032)^{2\cdot 8} \approx \$16,552.94$

$A(10) = 10,000(1.032)^{2\cdot 10} \approx \$18,775.61$

$A(18) = 10,000(1.032)^{2\cdot 18} \approx \$31,079.15$

47. We use the formula $A = P\left(1+\frac{r}{n}\right)^{nt}$ and substitute 3000 for P, 0.04 for r, 2 for n, and 2 for t.

$$A = 3000\left(1+\frac{0.04}{2}\right)^{2\cdot 2} \approx \$3247.30$$

48. $A = 12,500\left(1+\frac{0.03}{4}\right)^{4\cdot 3} \approx \$13,672.59$

49. We use the formula $A = P\left(1+\frac{r}{n}\right)^{nt}$ and substitute 120,000 for P, 0.025 for r, 1 for n, and 10 for t.

$$A = 120,000\left(1+\frac{0.025}{1}\right)^{1\cdot 10} \approx \$153,610.15$$

50. $A = 120,000\left(1+\frac{0.025}{4}\right)^{4\cdot 10} \approx \$153,963.22$

51. We use the formula $A = P\left(1+\frac{r}{n}\right)^{nt}$ and substitute 53,500 for P, 0.055 for r, 4 for n, and 6.5 for t.

$$A = 53,500\left(1+\frac{0.055}{4}\right)^{4(6.5)} \approx \$76,305.59$$

52. $A = 6250\left(1+\frac{0.0675}{2}\right)^{2(4.5)} \approx \$8425.97$

53. We use the formula $A = P\left(1+\frac{r}{n}\right)^{nt}$ and substitute 17,400 for P, 0.081 for r, 365 for n, and 5 for t.

$$A = 17,400\left(1+\frac{0.081}{365}\right)^{365\cdot 5} \approx \$26,086.69$$

54. $A = 900\left(1+\frac{0.073}{365}\right)^{365(7.25)} \approx \$1527.81$

55. a) $N(0) = 350,000\left(\frac{2}{3}\right)^0 = 350,000$

$N(1) = 350,000\left(\frac{2}{3}\right)^1 \approx 233,333$

$N(4) = 350,000\left(\frac{2}{3}\right)^4 \approx 69,136$

$N(10) = 350,000\left(\frac{2}{3}\right)^{10} \approx 6070$

b)

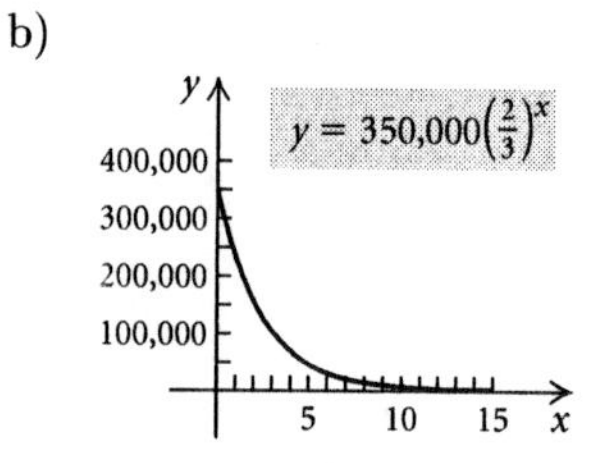

56. a) $N(10) = 3000(2)^{10/20} \approx 4243$;

$N(20) = 3000(2)^{20/20} = 6000$;

$N(30) = 3000(2)^{30/20} \approx 8485$;

$N(40) = 3000(2)^{40/20} = 12,000$;

$N(60) = 3000(2)^{60/20} = 24,000$

b)

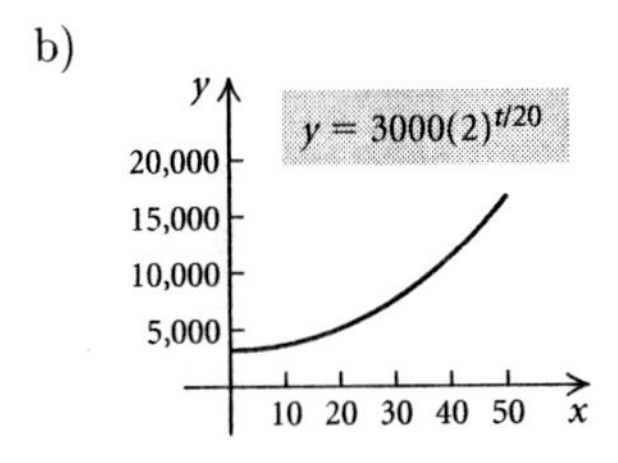

57. In 2003, $t = 2003 - 1998 = 5$.
$B(5) = 5.402(1.340)^5 \approx 23.3\%$
In 2004, $t = 2004 - 1998 = 6$.
$B(6) = 5.402(1.340)^6 \approx 31.3\%$
In 2005, $t = 2005 - 1998 = 7$.
$B(7) = 5.402(1.340)^7 \approx 41.9\%$

58. In 2005, $t = 2005 - 1997 = 8$.
$I(8) = 85,163.85(1.95)^8 \approx 17,804,598$
In 2008, $t = 2008 - 1997 = 11$.
$I(11) = 85,163.85(1.95)^{11} \approx 132,018,870$

59. In 2003, $t = 2003 - 1997 = 6$.
$E(6) = 256.43(1.22)^6 \approx \$845.5$ million
$I(6) = 464.38(1.42)^6 \approx \$3807.2$ million
In 2005, $t = 2005 - 1997 = 8$.
$E(8) = 256.43(1.22)^8 \approx \$1258.5$ million
$I(8) = 256.43(1.42)^8 \approx \$7676.8$ million
In 2008, $t = 2008 - 1997 = 11$.
$E(11) = 256.43(1.22)^{11} \approx \$2285.2$ million
$I(11) = 464.38(1.42)^{11} \approx \$21,980.9$ million

60. In 2005, $t = 2005 - 1964 = 41$.
$K(41) = 35.37(0.99)^{41} \approx 23\%$
In 2010, $t = 2010 - 1964 = 46$.
$K(46) = 35.37(0.99)^{46} \approx 22\%$
In 2015, $t = 2015 - 1964 = 51$.
$K(51) = 35.37(0.99)^{51} \approx 21\%$

61. $V(0) = 1800(0.8)^0 = \$1800$
$V(1) = 1800(0.8)^1 = \$1440$
$V(2) = 1800(0.8)^2 = \$1152$
$V(5) = 1800(0.8)^5 \approx \$589.82$
$V(10) = 1800(0.8)^{10} \approx \$193.27$

62. In 1992, $t = 1992 - 1985 = 7$.
$C(7) = 0.362(1.468)^7 \approx 5.3$ million
In 1999, $t = 1999 - 1985 = 14$.
$C(14) = 0.362(1.468)^{14} \approx 78.1$ million
In 2004, $t = 2004 - 1985 = 19$.
$C(19) = 0.362(1.468)^{19} \approx 532.7$ million

63. In 2005, $t = 2005 - 1981 = 24$.
$N(24) = 46.6(1.018)^{24} \approx 71.5$ billion cubic feet
In 2010, $t = 2010 - 1981 = 29$.
$N(29) = 46.6(1.018)^{29} \approx 78.2$ billion cubic feet

64. $S(10) = 200[1 - (0.86)^{10}] \approx 155.7$ words per minute
$S(20) = 200[1 - (0.86)^{20}] \approx 190.2$ words per minute
$S(40) = 200[1 - (0.86)^{40}] \approx 199.5$ words per minute
$S(100) \approx 200[1 - (0.86)^{100}] \approx 199.9999$ words per minute

65. $f(25) = 100(1 - e^{-0.04(25)}) \approx 63\%$

66. $V(1) = \$58(1 - e^{-1.1(1)}) + \$20 \approx \$58.69$
$V(2) = \$58(1 - e^{-1.1(2)}) + \$20 \approx \$71.57$
$V(4) = \$58(1 - e^{-1.1(4)}) + \$20 \approx \$77.29$
$V(6) = \$58(1 - e^{-1.1(6)}) + \$20 \approx \$77.92$
$V(12) = \$58(1 - e^{-1.1(12)}) + \$20 \approx \$78.00$

67. Graph (c) is the graph of $y = 3^x - 3^{-x}$.

68. Graph (j) is the graph of $y = 3^{-(x+1)^2}$.

69. Graph (a) is the graph of $f(x) = -2.3^x$.

70. Graph (d) is the graph of $f(x) = 30,000(1.4)^x$.

71. Graph (l) is the graph of $y = 2^{-|x|}$.

72. Graph (n) is the graph of $y = 2^{-(x-1)}$.

73. Graph (g) is the graph of $f(x) = (0.58)^x - 1$.

74. Graph (b) is the graph of $y = 2^x + 2^{-x}$.

75. Graph (i) is the graph of $g(x) = e^{|x|}$.

76. Graph (h) is the graph of $f(x) = |2^x - 1|$.

77. Graph (k) is the graph of $y = 2^{-x^2}$.

78. Graph (e) is the graph of $y = |2^{x^2} - 8|$.

79. Graph (m) is the graph of $g(x) = \dfrac{e^x - e^{-x}}{2}$.

80. Graph (f) is the graph of $f(x) = \dfrac{e^x + e^{-x}}{2}$.

81. Some differences are as follows: The range of f is $(-\infty, \infty)$ whereas the range of g is $(0, \infty)$; f has no asymptotes but g has a horizontal asymptote, the x-axis; the y-intercept of f is $(0, 0)$ and the y-intercept of g is $(0, 1)$.

82. The most interest will be earned the eighth year, because the principle is greatest during that year.

83. They are inverses.

84. They are inverses.

85.
$$\begin{aligned}(1 - 4i)(7 + 6i) &= 7 + 6i - 28i - 24i^2 \\ &= 7 + 6i - 28i + 24 \\ &= 31 - 22i\end{aligned}$$

86. $$\frac{2-i}{3+1} = \frac{2-i}{3+i} \cdot \frac{3-i}{3-i}$$
$$= \frac{6-5i+i^2}{9-i^2}$$
$$= \frac{6-5i-1}{9+1}$$
$$= \frac{5-5i}{10}$$
$$= \frac{1}{2} - \frac{1}{2}i$$

87. $2x^2 - 13x - 7 = 0$ Setting $f(x) = 0$

$(2x+1)(x-7) = 0$

$2x + 1 = 0$ *or* $x - 7 = 0$

$2x = -1$ *or* $x = 7$

$x = -\frac{1}{2}$ *or* $x = 7$

The zeros of the function are $-\frac{1}{2}$ and 7, and the x-intercepts are $\left(-\frac{1}{2}, 0\right)$ and $(7, 0)$.

88. $h(x) = x^3 - 3x^2 + 3x - 1$

The possible real-number solutions are of the form p/q where $p = \pm 1$ and $q = \pm 1$. Then the possibilities for p/q are 1 and -1. We try 1.

$$\begin{array}{r|rrrr} 1 & 1 & -3 & 3 & -1 \\ & & 1 & -2 & 1 \\ \hline & 1 & -2 & 1 & 0 \end{array}$$

$h(x) = (x-1)(x^2 - 2x + 1)$

Now find the zeros of $h(x)$.

$(x-1)(x^2 - 2x + 1) = 0$

$(x-1)(x-1)(x-1) = 0$

$x - 1 = 0$ *or* $x - 1 = 0$ *or* $x - 1 = 0$

$x = 1$ *or* $x = 1$ *or* $x = 1$

The zero of the function is 1 and the x-intercept is $(1, 0)$.

89. $x^4 - x^2 = 0$ Setting $h(x) = 0$

$x^2(x^2 - 1) = 0$

$x^2(x+1)(x-1) = 0$

$x^2 = 0$ *or* $x + 1 = 0$ *or* $x - 1 = 0$

$x = 0$ *or* $x = -1$ *or* $x = 1$

The zeros of the function are 0, -1, and 1, and the x-intercepts are $(0, 0)$, $(-1, 0)$, and $(1, 0)$.

90. $x^3 + x^2 - 12x = 0$

$x(x^2 + x - 12) = 0$

$x(x+4)(x-3) = 0$

$x = 0$ *or* $x + 4 = 0$ *or* $x - 3 = 0$

$x = 0$ *or* $x = -4$ *or* $x = 3$

The zeros of the function are 0, -4, and 3, and the x-intercepts are $(0, 0)$, $(-4, 0)$, and $(3, 0)$.

91. $x^3 + 6x^2 - 16x = 0$

$x(x^2 + 6x - 16) = 0$

$x(x+8)(x-2) = 0$

$x = 0$ *or* $x + 8 = 0$ *or* $x - 2 = 0$

$x = 0$ *or* $x = -8$ *or* $x = 2$

The solutions are 0, -8, and 2.

92. $3x^2 - 6 = 5x$

$3x^2 - 5x - 6 = 0$

$$x = \frac{-(-5) \pm \sqrt{(-5)^2 - 4 \cdot 3 \cdot (-6)}}{2 \cdot 3}$$
$$= \frac{5 \pm \sqrt{97}}{6}$$

93. $7^\pi \approx 451.8078726$ and $\pi^7 \approx 3020.293228$, so π^7 is larger.

$70^{80} \approx 4.054 \times 10^{147}$ and $80^{70} \approx 1.646 \times 10^{133}$, so 70^{80} is larger.

Exercise Set 4.3

1. Graph $x = 3^y$.

Choose values for y and compute the corresponding x-values. Plot the points (x, y) and connect them with a smooth curve.

x	y	(x, y)
$\frac{1}{27}$	-3	$\left(\frac{1}{27}, -3\right)$
$\frac{1}{9}$	-2	$\left(\frac{1}{9}, -2\right)$
$\frac{1}{3}$	-1	$\left(\frac{1}{3}, -1\right)$
1	0	$(1, 0)$
3	1	$(3, 1)$
9	2	$(9, 2)$
27	3	$(27, 3)$

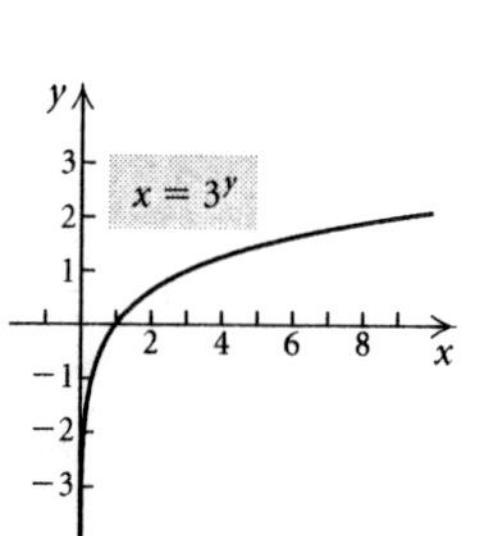

2. Graph $x = 4^y$.

x	y	(x, y)
$\frac{1}{64}$	-3	$\left(\frac{1}{64}, -3\right)$
$\frac{1}{16}$	-2	$\left(\frac{1}{16}, -2\right)$
$\frac{1}{4}$	-1	$\left(\frac{1}{4}, -1\right)$
1	0	$(1, 0)$
4	1	$(4, 1)$
16	2	$(16, 2)$
64	3	$(64, 3)$

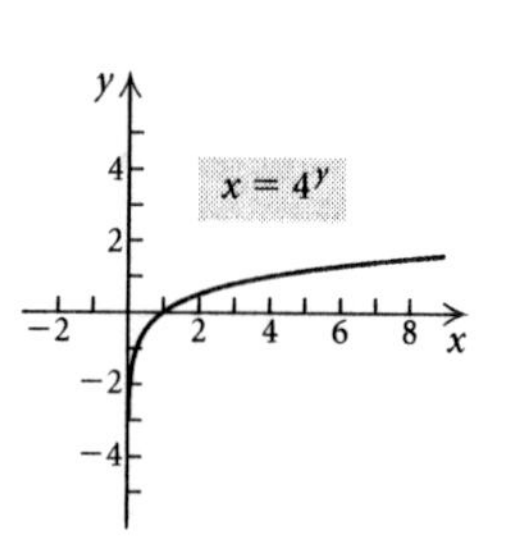

3. Graph $x = \left(\frac{1}{2}\right)^y$.

Choose values for y and compute the corresponding x-values. Plot the points (x, y) and connect them with a smooth curve.

x	y	(x, y)
8	-3	$(8, -3)$
4	-2	$(4, -2)$
2	-1	$(2, -1)$
1	0	$(1, 0)$
$\frac{1}{2}$	1	$\left(\frac{1}{2}, 1\right)$
$\frac{1}{4}$	2	$\left(\frac{1}{4}, 2\right)$
$\frac{1}{8}$	3	$\left(\frac{1}{8}, 3\right)$

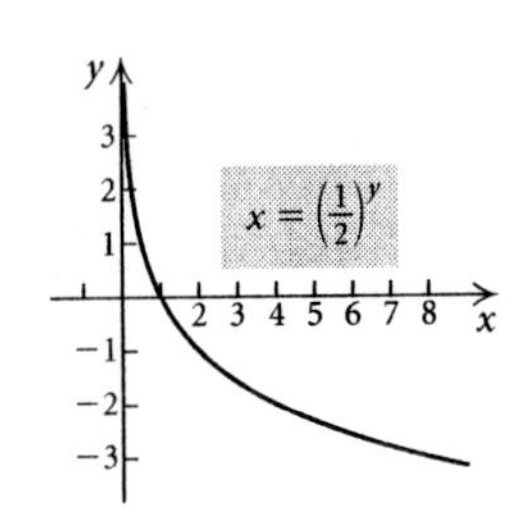

4. Graph $x = \left(\frac{4}{3}\right)^y$.

x	y	(x, y)
$\frac{27}{64}$	-3	$\left(\frac{27}{64}, -3\right)$
$\frac{9}{16}$	-2	$\left(\frac{9}{16}, -2\right)$
$\frac{3}{4}$	-1	$\left(\frac{3}{4}, -1\right)$
1	0	$(1, 0)$
$\frac{4}{3}$	1	$\left(\frac{4}{3}, 1\right)$
$\frac{16}{9}$	2	$\left(\frac{16}{9}, 2\right)$
$\frac{64}{27}$	3	$\left(\frac{64}{27}, 3\right)$

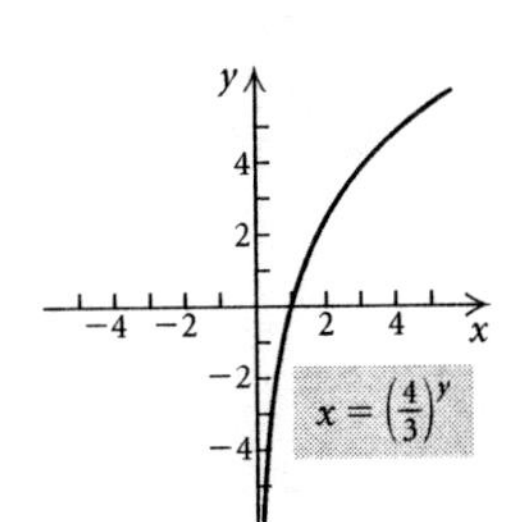

5. Graph $y = \log_3 x$.

The equation $y = \log_3 x$ is equivalent to $x = 3^y$. We can find ordered pairs that are solutions by choosing values for y and computing the corresponding x-values.

For $y = -2$, $x = 3^{-2} = \frac{1}{9}$.

For $y = -1$, $x = 3^{-1} = \frac{1}{3}$.

For $y = 0$, $x = 3^0 = 1$.

For $y = 1$, $x = 3^1 = 3$.

For $y = 2$, $x = 3^2 = 9$.

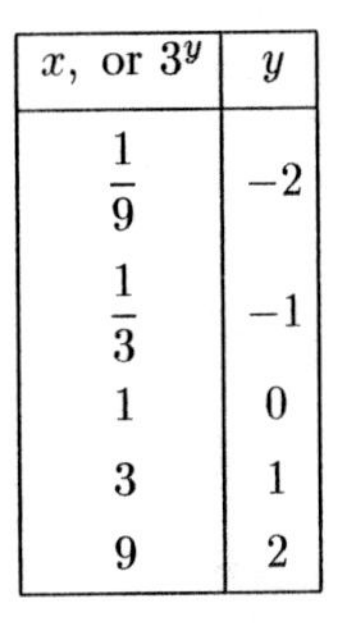

x, or 3^y	y
$\frac{1}{9}$	-2
$\frac{1}{3}$	-1
1	0
3	1
9	2

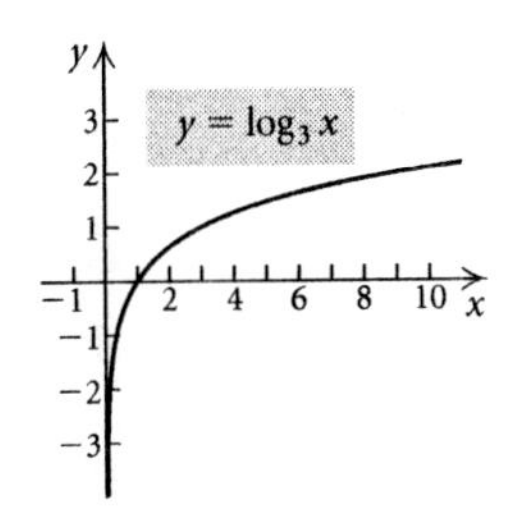

6. $y = \log_4 x$ is equivalent to $x = 4^y$.

x, or 4^y	y
$\frac{1}{16}$	-2
$\frac{1}{4}$	-1
1	0
4	1
16	2

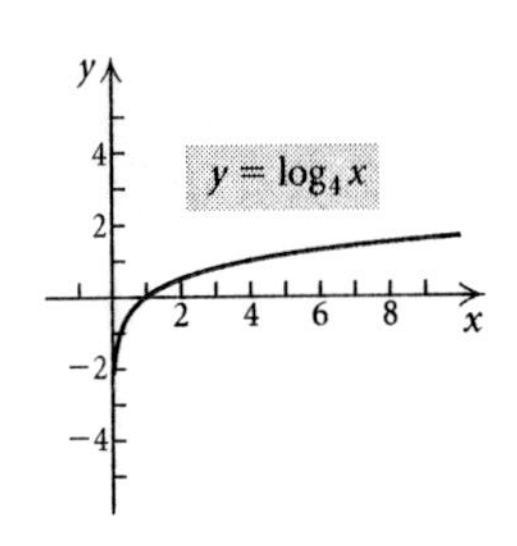

7. Graph $f(x) = \log x$.

Think of $f(x)$ as y. The equation $y = \log x$ is equivalent to $x = 10^y$. We can find ordered pairs that are solutions by choosing values for y and computing the corresponding x-values.

For $y = -2$, $x = 10^{-2} = 0.01$.

For $y = -1$, $x = 10^{-1} = 0.1$.

For $y = 0$, $x = 10^0 = 1$.

For $y = 1$, $x = 10^1 = 10$.

For $y = 2$, $x = 10^2 = 100$.

x, or 10^y	y
0.01	-2
0.1	-1
1	0
10	1
100	2

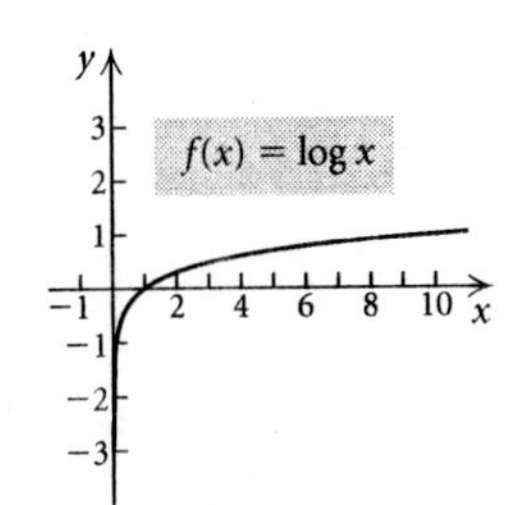

8. See Example 10.

9. $\log_2 16 = 4$ because the exponent to which we raise 2 to get 16 is 4.

10. $\log_3 9 = 2$, because the exponent to which we raise 3 to get 9 is 2.

11. $\log_5 125 = 3$, because the exponent to which we raise 5 to get 125 is 3.

12. $\log_2 64 = 6$, because the exponent to which we raise 2 to get 64 is 6.

13. $\log 0.001 = -3$, because the exponent to which we raise 10 to get 0.001 is -3.

14. $\log 100 = 2$, because the exponent to which we raise 10 to get 100 is 2.

15. $\log_2 \frac{1}{4} = -2$, because the exponent to which we raise 2 to get $\frac{1}{4}$ is -2.

16. $\log_8 2 = \frac{1}{3}$, because the exponent to which we raise 8 to get 2 is $\frac{1}{3}$.

17. $\ln 1 = 0$, because the exponent to which we raise e to get 1 is 0.

18. $\ln e = 1$, because the exponent to which we raise e to get e is 1.

19. $\log 10 = 1$, because the exponent to which we raise 10 to get 10 is 1.

20. $\log 1 = 0$, because the exponent to which we raise 10 to get 1 is 0.

21. $\log_5 5^4 = 4$, because the exponent to which we raise 5 to get 5^4 is 4.

22. $\log \sqrt{10} = \log 10^{1/2} = \frac{1}{2}$, because the exponent to which we raise 10 to get $10^{1/2}$ is $\frac{1}{2}$.

23. $\log_3 \sqrt[4]{3} = \log_3 3^{1/4} = \frac{1}{4}$, because the exponent to which we raise 3 to get $3^{1/4}$ is $\frac{1}{4}$.

24. $\log 10^{8/5} = \frac{8}{5}$, because the exponent to which we raise 10 to get $10^{8/5}$ is $\frac{8}{5}$.

25. $\log 10^{-7} = -7$, because the exponent to which we raise 10 to get 10^{-7} is -7.

26. $\log_5 1 = 0$, because the exponent to which we raise 5 to get 1 is 0.

27. $\log_{49} 7 = \frac{1}{2}$, because the exponent to which we raise 49 to get 7 is $\frac{1}{2}$. $(49^{1/2} = \sqrt{49} = 7)$

28. $\log_3 3^{-2} = -2$, because the exponent to which we raise 3 to get 3^{-2} is -2.

29. $\ln e^{3/4} = \frac{3}{4}$, because the exponent to which we raise e to get $e^{3/4}$ is $\frac{3}{4}$.

30. $\log_2 \sqrt{2} = \log_2 2^{1/2} = \frac{1}{2}$, because the exponent to which we raise 2 to get $2^{1/2}$ is $\frac{1}{2}$.

31. $\log_4 1 = 0$, because the exponent to which we raise 4 to get 1 is 0.

32. $\ln e^{-5} = -5$, because the exponent to which we raise e to get e^{-5} is -5.

33. $\ln \sqrt{e} = \ln e^{1/2} = \frac{1}{2}$, because the exponent to which we raise e to get $e^{1/2}$ is $\frac{1}{2}$.

34. $\log_{64} 4 = \frac{1}{3}$, because the exponent to which we raise 64 to get 4 is $\frac{1}{3}$. $(64^{1/3} = \sqrt[3]{64} = 4)$

35.

The exponent is the logarithm.

$10^3 = 1000 \Rightarrow 3 = \log_{10} 1000$

The base remains the same.

36. $5^{-3} = \frac{1}{125} \Rightarrow \log_5 \frac{1}{125} = -3$

37.

The exponent is the logarithm.

$8^{1/3} = 2 \Rightarrow \log_8 2 = \frac{1}{3}$

The base remains the same.

38. $10^{0.3010} = 2 \Rightarrow \log_{10} 2 = 0.0310$

39. $e^3 = t \Rightarrow \log_e t = 3$

40. $Q^t = x \Rightarrow \log_Q x = t$

41. $e^2 = 7.3891 \Rightarrow \log_e 7.3891 = 2$

42. $e^{-1} = 0.3679 \Rightarrow \log_e 0.3679 = -1$

43. $p^k = 3 \Rightarrow \log_p 3 = k$

44. $e^{-t} = 4000 \Rightarrow \log_e 4000 = -t$

45.

The logarithm is the exponent.

$\log_5 5 = 1 \Rightarrow 5^1 = 5$

The base remains the same.

46. $t = \log_4 7 \Rightarrow 7 = 4^t$

47. $\log 0.01 = -2$ is equivalent to $\log_{10} 0.01 = -2$.

The logarithm is the exponent.

$\log_{10} 0.01 = -2 \Rightarrow 10^{-2} = 0.01$

The base remains the same.

48. $\log 7 = 0.845 \Rightarrow 10^{0.845} = 7$

49. $\ln 30 = 3.4012 \Rightarrow e^{3.4012} = 30$

50. $\ln 0.38 = -0.9676 \Rightarrow e^{-0.9676} = 0.38$

51. $\log_a M = -x \Rightarrow a^{-x} = M$

52. $\log_t Q = k \Rightarrow t^k = Q$

53. $\log_a T^3 = x \Rightarrow a^x = T^3$

54. $\ln W^5 = t \Rightarrow e^t = W^5$

55. $\log 3 \approx 0.4771$

56. $\log 8 \approx 0.9031$

57. $\log 532 \approx 2.7259$

58. $\log 93,100 \approx 4.9689$

59. $\log 0.57 \approx -0.2441$

60. $\log 0.082 \approx -1.0862$

61. $\log(-2)$ does not exist. (The calculator gives an error message.)

62. $\ln 50 \approx 3.9120$

63. $\ln 2 \approx 0.6931$

64. $\ln(-4)$ does not exist. (The calculator gives an error message.)

65. $\ln 809.3 \approx 6.6962$

66. $\ln 0.00037 \approx -7.9020$

67. $\ln(-1.32)$ does not exist. (The calculator gives an error message.)

68. $\ln 0$ does not exist. (The calculator gives an error message.)

69. Let $a = 10$, $b = 4$, and $M = 100$ and substitute in the change-of-base formula.

$$\log_4 100 = \frac{\log_{10} 100}{\log_{10} 4} \approx 3.3219$$

70. $\log_3 20 = \dfrac{\log 20}{\log 3} \approx 2.7268$

71. Let $a = 10$, $b = 100$, and $M = 0.3$ and substitute in the change-of-base formula.

$$\log_{100} 0.3 = \frac{\log_{10} 0.3}{\log_{10} 100} \approx -0.2614$$

72. $\log_\pi 100 = \dfrac{\log 100}{\log \pi} \approx 4.0229$

73. Let $a = 10$, $b = 200$, and $M = 50$ and substitute in the change-of-base formula.

$$\log_{200} 50 = \frac{\log_{10} 50}{\log_{10} 200} \approx 0.7384$$

74. $\log_{5.3} 1700 = \dfrac{\log 1700}{\log 5.3} \approx 4.4602$

75. Let $a = e$, $b = 3$, and $M = 12$ and substitute in the change-of-base formula.

$$\log_3 12 = \frac{\ln 12}{\ln 3} \approx 2.2619$$

76. $\log_4 25 = \dfrac{\ln 25}{\ln 4} \approx 2.3219$

77. Let $a = e$, $b = 100$, and $M = 15$ and substitute in the change-of-base formula.

$$\log_{100} 15 = \frac{\ln 15}{\ln 100} \approx 0.5880$$

78. $\log_9 100 = \dfrac{\ln 100}{\ln 9} \approx 2.0959$

79. Graph $y = 3^x$ and then reflect this graph across the line $y = x$ to get the graph of $y = \log_3 x$.

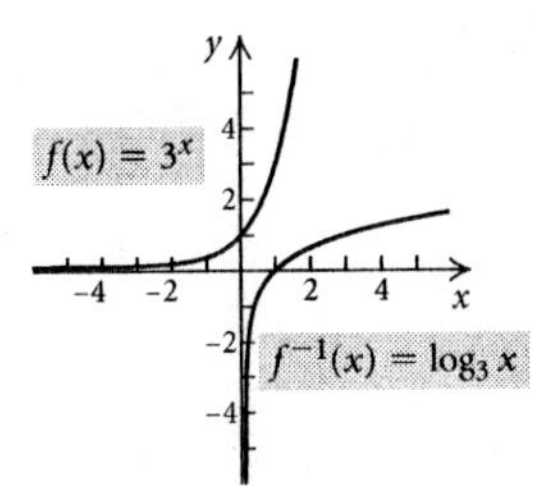

80. Graph $y = \log_4 x$ and then reflect this graph across the line $y = x$ to get the graph of $y = 4^x$.

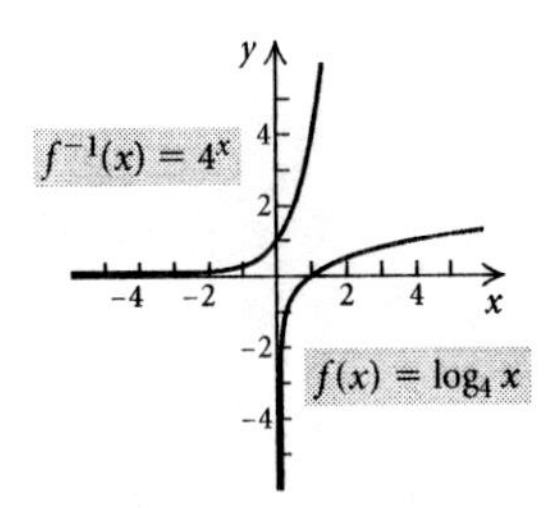

81. Graph $y = \log x$ and then reflect this graph across the line $y = x$ to get the graph of $y = 10^x$.

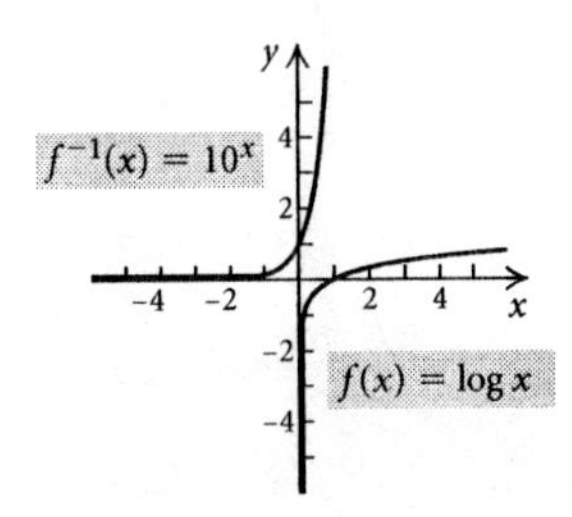

82. Graph $y = e^x$ and then reflect this graph across the line $y = x$ to get the graph of $y = \ln x$.

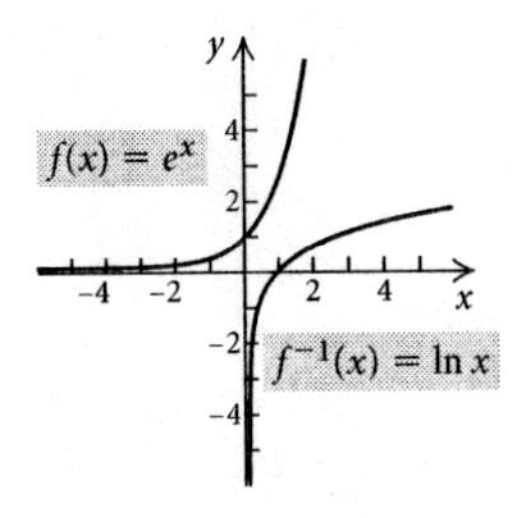

83. Shift the graph of $y = \log_2 x$ left 3 units.

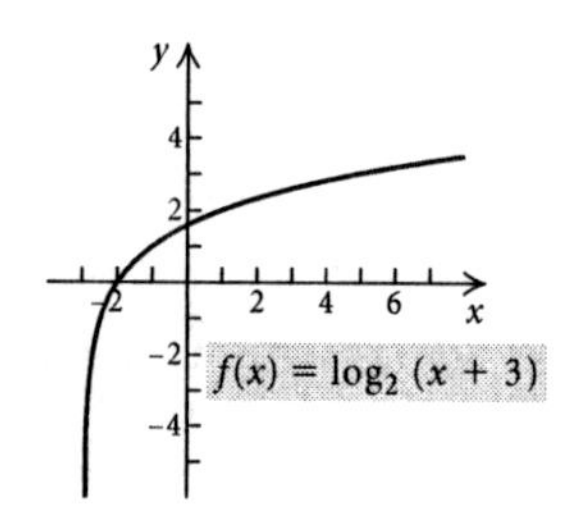

Domain: $(-3, \infty)$

Vertical asymptote: $x = -3$

84. Shift the graph of $y = \log_3 x$ right 2 units.

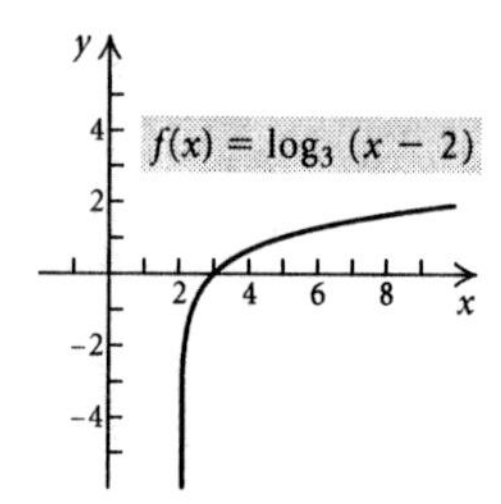

Domain: $(2, \infty)$

Vertical asymptote: $x = 2$

85. Shift the graph of $y = \log_3 x$ down 1 unit.

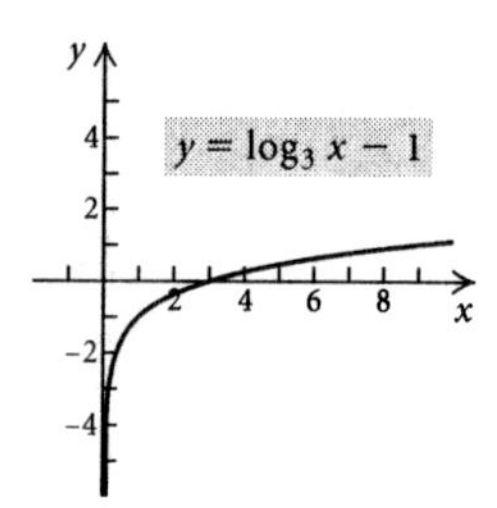

Domain: $(0, \infty)$

Vertical asymptote: $x = 0$

86. Shift the graph of $y = \log_2 x$ up 3 units.

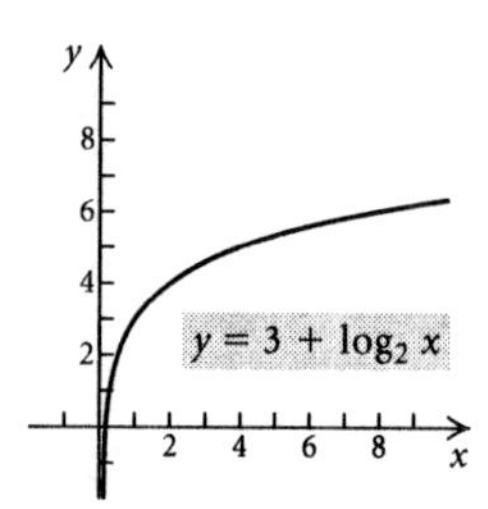

Domain: $(0, \infty)$

Vertical asymptote: $x = 0$

87. Stretch the graph of $y = \ln x$ vertically.

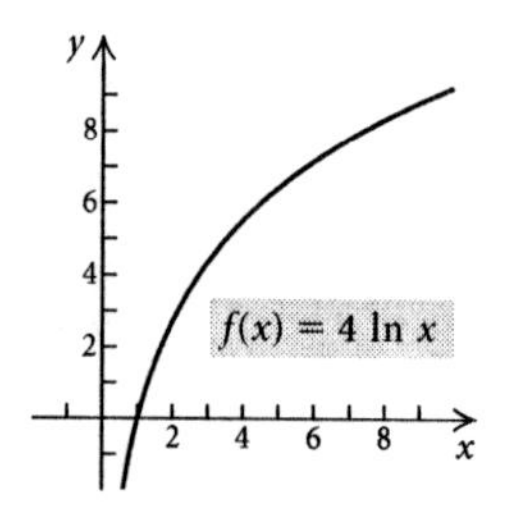

Domain: $(0, \infty)$

Vertical asymptote: $x = 0$

88. Shrink the graph of $y = \ln x$ vertically.

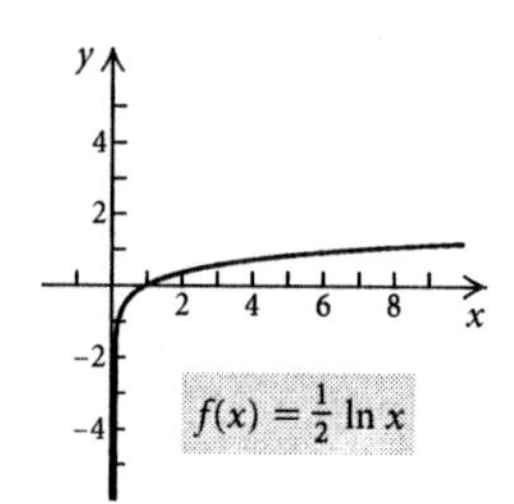

Domain: $(0, \infty)$

Vertical asymptote: $x = 0$

89. Reflect the graph of $y = \ln x$ across the x-axis and then shift it up 2 units.

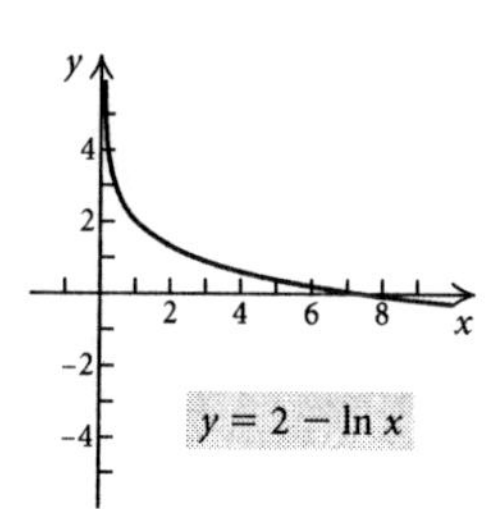

Domain: $(0, \infty)$

Vertical asymptote: $x = 0$

90. Shift the graph of $y = \ln x$ left 1 unit.

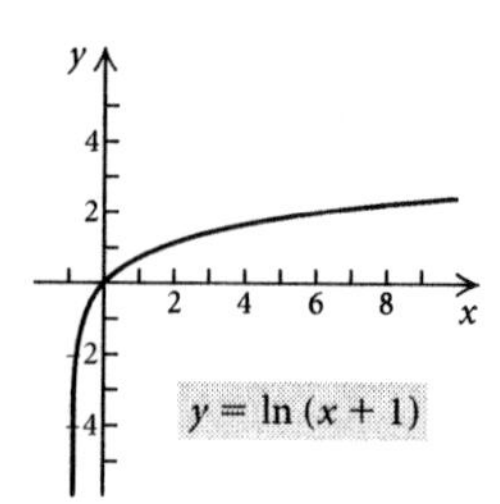

Domain: $(-1, \infty)$

Vertical asymptote: $x = -1$

91. a) We substitute 711.5 for P, since P is in thousands.

$$w(711.5) = 0.37 \ln 711.5 + 0.05$$
$$\approx 2.5 \text{ ft/sec}$$

b) We substitute 478.4 for P, since P is in thousands.

$w(478.4) = 0.37 \ln 478.4 + 0.05$

≈ 2.3 ft/sec

c) We substitute 277 for P, since P is in thousands.

$w(277) = 0.37 \ln 277 + 0.05$

≈ 2.1 ft/sec

d) We substitute 113 for P, since P is in thousands.

$w(113) = 0.37 \ln 113 + 0.05$

≈ 1.8 ft/sec

e) We substitute 416.5 for P, since P is in thousands.

$w(416.5) = 0.37 \ln 416.5 + 0.05$

≈ 2.3 ft/sec

f) We substitute 328 for P, since P is in thousands.

$w(328) = 0.37 \ln 328 + 0.05$

≈ 2.2 ft/sec

g) We substitute 8008.3 for P, since P is in thousands.

$w(8008.3) = 0.37 \ln 8008.3 + 0.05$

≈ 3.4 ft/sec

h) We substitute 2896 for P, since P is in thousands.

$w(2896) = 0.37 \ln 2896 + 0.05$

≈ 3.0 ft/sec

92. a) $R = \log \dfrac{10^{7.85} \cdot I_0}{I_0} = \log 10^{7.85} = 7.85$

b) $R = \log \dfrac{10^{8.25} \cdot I_0}{I_0} = \log 10^{8.25} = 8.25$

c) $R = \log \dfrac{10^{9.6} \cdot I_0}{I_0} = \log 10^{9.6} = 9.6$

d) $R = \log \dfrac{10^{7.85} \cdot I_0}{I_0} = \log 10^{7.85} = 7.85$

e) $R = \log \dfrac{10^{6.9} \cdot I_0}{I_0} = \log 10^{6.9} = 6.9$

93. a) $S(0) = 78 - 15\log(0+1)$

$= 78 - 15\log 1$

$= 78 - 15 \cdot 0$

$= 78\%$

b) $S(4) = 78 - 15\log(4+1)$

$= 78 - 15\log 5$

$\approx 78 - 15(0.698970)$

$\approx 67.5\%$

$S(24) = 78 - 15\log(24+1)$

$= 78 - 15\log 25$

$\approx 78 - 15(1.397940)$

$\approx 57\%$

94. a) $pH = -\log[1.6 \times 10^{-4}] \approx -(-3.8) \approx 3.8$

b) $pH = -\log[0.0013] \approx -(-2.9) \approx 2.9$

c) $pH = -\log[6.3 \times 10^{-7}] \approx -(-6.2) \approx 6.2$

d) $pH = -\log[1.6 \times 10^{-8}] \approx -(-7.8) \approx 7.8$

e) $pH = -\log[6.3 \times 10^{-5}] \approx -(-4.2) \approx 4.2$

95. a) $7 = -\log[H^+]$

$-7 = \log[H^+]$

$H^+ = 10^{-7}$ Using the definition of logarithm

b) $5.4 = -\log[H^+]$

$-5.4 = \log[H^+]$

$H^+ = 10^{-5.4}$ Using the definition of logarithm

$H^+ \approx 4.0 \times 10^{-6}$

c) $3.2 = -\log[H^+]$

$-3.2 = \log[H^+]$

$H^+ = 10^{-3.2}$ Using the definition of logarithm

$H^+ \approx 6.3 \times 10^{-4}$

d) $4.8 = -\log[H^+]$

$-4.8 = \log[H^+]$

$H^+ = 10^{-4.8}$ Using the definition of logarithm

$H^+ \approx 1.6 \times 10^{-5}$

96. a) $N(1) = 1000 + 200\ln 1 = 1000$

b) $N(5) = 1000 + 200 \ln 5 \approx 1332$

97. a) $L = 10\log \dfrac{2510 \cdot I_0}{I_0}$

$= 10\log 2510$

≈ 34 decibels

b) $L = 10\log \dfrac{2,500,000 \cdot I_0}{I_0}$

$= 10\log 2,500,000$

≈ 64 decibels

c) $L = 10\log \dfrac{10^6 \cdot I_0}{I_0}$

$= 10\log 10^6$

$= 60$ decibels

d) $L = 10\log \dfrac{10^9 \cdot I_0}{I_0}$

$= 10\log 10^9$

$= 90$ decibels

98. Left to the student

99. Reflect the graph of $f(x) = e^x$ across the line $y = x$ to obtain the graph of $h(x) = \ln x$. Then shift this graph up 3 units to obtain the graph of $g(x) = 3 + \ln x$.

100. Reflect the graph of $f(x) = \ln x$ across the line $y = x$ to obtain the graph of $h(x) = e^x$. Then shift this graph 2 units right to obtain the graph of $g(x) = e^{x-2}$.

101. $y = 6 = 0 \cdot x + 6$

Slope: 0; y-intercept $(0, 6)$

102.
$$\begin{aligned} 3x - 10y &= 14 \\ 3x - 14 &= 10y \\ \frac{3}{10}x - \frac{7}{5} &= y \end{aligned}$$

Slope: $\frac{3}{10}$; y-intercept: $\left(0, -\frac{7}{5}\right)$

103. $y = 2x - \frac{3}{13}$

Slope: 2, y-intercept: $\left(0, -\frac{3}{13}\right)$

104. $x = -4$

Slope: not defined; y-intercept: none

105.
$$\begin{array}{r|rrrr} -5 & 1 & -6 & 3 & 10 \\ & & -5 & 55 & -290 \\ \hline & 1 & -11 & 58 & -280 \end{array}$$

The remainder is -280, so $f(-5) = -280$.

106.
$$\begin{array}{r|rrrrr} -1 & 1 & -2 & 0 & 1 & -6 \\ & & -1 & 3 & -3 & 2 \\ \hline & 1 & -3 & 3 & -2 & -4 \end{array}$$

$f(-1) = -4$

107.
$$\begin{aligned} f(x) &= (x - \sqrt{7})(x + \sqrt{7})(x - 0) \\ &= (x^2 - 7)(x) \\ &= x^3 - 7x \end{aligned}$$

108.
$$\begin{aligned} f(x) &= (x - 4i)(x + 4i)(x - 1) \\ &= (x^2 + 16)(x - 1) \\ &= x^3 - x^2 + 16x - 16 \end{aligned}$$

109. Using the change-of-base formula, we get

$$\frac{\log_5 8}{\log_5 2} = \log_2 8 = 3.$$

110. Using the change-of-base formula, we get

$$\frac{\log_3 64}{\log_3 16} = \log_{16} 64.$$

Let $\log_{16} 64 = x$. Then we have

$$\begin{aligned} 16^x &= 64 \quad \text{Using the definition of logarithm} \\ (2^4)^x &= 2^6 \\ 2^{4x} &= 2^6, \text{ so} \\ 4x &= 6 \\ x &= \frac{6}{4} = \frac{3}{2} \end{aligned}$$

Thus, $\frac{\log_3 64}{\log_3 16} = \frac{3}{2}$.

111. $f(x) = \log_5 x^3$

x^3 must be positive. Since $x^3 > 0$ for $x > 0$, the domain is $(0, \infty)$.

112. $f(x) = \log_4 x^2$

x^2 must be positive, so the domain is $(-\infty, 0) \cup (0, \infty)$.

113. $f(x) = \ln |x|$

$|x|$ must be positive. Since $|x| > 0$ for $x \neq 0$, the domain is $(-\infty, 0) \cup (0, \infty)$.

114. $f(x) = \log(3x - 4)$

$3x - 4$ must be positive. We have

$$\begin{aligned} 3x - 4 &> 0 \\ x &> \frac{4}{3} \end{aligned}$$

The domain is $\left(\frac{4}{3}, \infty\right)$.

115. Graph $y = \log_2(2x + 5) = \frac{\log(2x + 5)}{\log 2}$. Observe that outputs are negative for inputs between $-\frac{5}{2}$ and -2. Thus, the solution set is $\left(-\frac{5}{2}, -2\right)$.

116. Graph $y_1 = \log_2(x - 3) = \frac{\log(x - 3)}{\log 2}$ and $y_2 = 4$. Observe that the graph of y_1 lies on or above the graph of y_2 for all inputs greater than or equal to 19. Thus, the solution set is $[19, \infty)$.

117. Graph (d) is the graph of $f(x) = \ln |x|$.

118. Graph (c) is the graph of $f(x) = |\ln x|$.

119. Graph (b) is the graph of $f(x) = \ln x^2$.

120. Graph (a) is the graph of $g(x) = |\ln(x - 1)|$.

Exercise Set 4.4

1. Use the product rule.

$\log_3(81 \cdot 27) = \log_3 81 + \log_3 27$

2. $\log_2(8 \cdot 64) = \log_2 8 + \log_2 64$

3. Use the product rule.

$\log_5(5 \cdot 125) = \log_5 5 + \log_5 125$

4. $\log_4(64 \cdot 32) = \log_4 64 + \log_4 32$

5. Use the product rule.

$\log_t 8Y = \log_t 8 + \log_t Y$

6. $\log 0.2x = \log 0.2 + \log x$

7. Use the product rule.

$\ln xy = \ln x + \ln y$

8. $\ln ab = \ln a + \ln b$

9. Use the power rule.

$\log_b t^3 = 3 \log_b t$

10. $\log_a x^4 = 4 \log_a x$

11. Use the power rule.

$\log y^8 = 8 \log y$

12. $\ln y^5 = 5 \ln y$

13. Use the power rule.

$\log_c K^{-6} = -6 \log_c K$

14. $\log_b Q^{-8} = -8 \log_b Q$

15. Use the power rule.

$\ln \sqrt[3]{4} = \ln 4^{1/3} = \frac{1}{3} \ln 4$

16. $\ln \sqrt{a} = \ln a^{1/2} = \frac{1}{2} \ln a$

17. Use the quotient rule.

$\log_t \frac{M}{8} = \log_t M - \log_t 8$

18. $\log_a \frac{76}{13} = \log_a 76 - \log_a 13$

19. Use the quotient rule.

$\log \frac{x}{y} = \log x - \log y$

20. $\ln \frac{a}{b} = \ln a - \ln b$

21. Use the quotient rule.

$\ln \frac{r}{s} = \ln r - \ln s$

22. $\log_b \frac{3}{w} = \log_b 3 - \log_b w$

23. $\log_a 6xy^5z^4$

$= \log_a 6 + \log_a x + \log_a y^5 + \log_a z^4$ Product rule

$= \log_a 6 + \log_a x + 5\log_a y + 4\log_a z$ Power rule

24. $\log_a x^3y^2z$

$= \log_a x^3 + \log_a y^2 + \log_a z$

$= 3\log_a x + 2\log_a y + \log_a z$

25. $\log_b \frac{p^2q^5}{m^4b^9}$

$= \log_b p^2q^5 - \log_b m^4b^9$ Quotient rule

$= \log_b p^2 + \log_b q^5 - (\log_b m^4 + \log_b b^9)$ Product rule

$= \log_b p^2 + \log_b q^5 - \log_b m^4 - \log_b b^9$

$= \log_b p^2 + \log_b q^5 - \log_b m^4 - 9$ $(\log_b b^9 = 9)$

$= 2\log_b p + 5\log_b q - 4\log_b m - 9$ Power rule

26. $\log_b \frac{x^2y}{b^3} = \log_b x^2y - \log_b b^3$

$= \log_b x^2 + \log_b y - \log_b b^3$

$= \log_b x^2 + \log_b y - 3$

$= 2\log_b x + \log_b y - 3$

27. $\ln \frac{2}{3x^3y}$

$= \ln 2 - \ln 3x^3y$ Quotient rule

$= \ln 2 - (\ln 3 + \ln x^3 + \ln y)$ Product rule

$= \ln 2 - \ln 3 - \ln x^3 - \ln y$

$= \ln 2 - \ln 3 - 3\ln x - \ln y$ Power rule

28. $\log \frac{5a}{4b^2} = \log 5a - \log 4b^2$

$= \log 5 + \log a - (\log 4 + \log b^2)$

$= \log 5 + \log a - \log 4 - \log b^2$

$= \log 5 + \log a - \log 4 - 2\log b$

29. $\log \sqrt{r^3t}$

$= \log(r^3t)^{1/2}$

$= \frac{1}{2}\log r^3t$ Power rule

$= \frac{1}{2}(\log r^3 + \log t)$ Product rule

$= \frac{1}{2}(3\log r + \log t)$ Power rule

$= \frac{3}{2}\log r + \frac{1}{2}\log t$

30. $\ln \sqrt[3]{5x^5} = \ln(5x^5)^{1/3}$

$= \frac{1}{3}\ln 5x^5$

$= \frac{1}{3}(\ln 5 + \ln x^5)$

$= \frac{1}{3}(\ln 5 + 5\ln x)$

$= \frac{1}{3}\ln 5 + \frac{5}{3}\ln x$

31. $\frac{1}{2}\log_a \frac{x^6}{p^5q^8}$

$= \frac{1}{2}[\log_a x^6 - \log_a(p^5q^8)]$ Quotient rule

$= \frac{1}{2}[\log_a x^6 - (\log_a p^5 + \log_a q^8)]$ Product rule

$= \frac{1}{2}(\log_a x^6 - \log_a p^5 - \log_a q^8)$

$= \frac{1}{2}(6\log_a x - 5\log_a p - 8\log_a q)$ Power rule

$= 3\log_a x - \frac{5}{2}\log_a p - 4\log_a q$

32. $\log_c \sqrt[3]{\frac{y^3z^2}{x^4}}$

$= \frac{1}{3}\log_c \frac{y^3z^2}{x^4}$

$= \frac{1}{3}(\log_c y^3z^2 - \log_c x^4)$

$= \frac{1}{3}(\log_c y^3 + \log_c z^2 - \log_c x^4)$

$= \frac{1}{3}(3\log_c y + 2\log_c z - 4\log_c x)$

$= \log_c y + \frac{2}{3}\log_c z - \frac{4}{3}\log_c x$

33. $\log_a \sqrt[4]{\dfrac{m^8n^{12}}{a^3b^5}}$

$= \frac{1}{4}\log_a \frac{m^8n^{12}}{a^3b^5}$ Power rule

$= \frac{1}{4}(\log_a m^8n^{12} - \log_a a^3b^5)$ Quotient rule

$= \frac{1}{4}[\log_a m^8 + \log_a n^{12} - (\log_a a^3 + \log_a b^5)]$ Product rule

$= \frac{1}{4}(\log_a m^8 + \log_a n^{12} - \log_a a^3 - \log_a b^5)$

$= \frac{1}{4}(\log_a m^8 + \log_a n^{12} - 3 - \log_a b^5)$ $(\log_a a^3 = 3)$

$= \frac{1}{4}(8\log_a m + 12\log_a n - 3 - 5\log_a b)$ Power rule

$= 2\log_a m + 3\log_a n - \frac{3}{4} - \frac{5}{4}\log_a b$

34. $\log_a \sqrt{\dfrac{a^6b^8}{a^2b^5}} = \log_a \sqrt{a^4b^3}$

$= \frac{1}{2}(\log_a a^4 + \log_a b^3)$

$= \frac{1}{2}(4 + 3\log_a b)$

$= 2 + \frac{3}{2}\log_a b$

35. $\log_a 75 + \log_a 2$

$= \log_a(75 \cdot 2)$ Product rule

$= \log_a 150$

36. $\log 0.01 + \log 1000 = \log(0.01 \cdot 1000) = \log 10 = 1$

37. $\log 10,000 - \log 100$

$= \log \frac{10,000}{100}$ Quotient rule

$= \log 100$

$= 2$

38. $\ln 54 - \ln 6 = \ln \frac{54}{6} = \ln 9$

39. $\frac{1}{2}\log n + 3\log m$

$= \log n^{1/2} + \log m^3$ Power rule

$= \log n^{1/2}m^3$, or Product rule

$\log m^3\sqrt{n}$ $\quad n^{1/2} = \sqrt{n}$

40. $\frac{1}{2}\log a - \log 2 = \log a^{1/2} - \log 2$

$= \log \frac{a^{1/2}}{2}$, or $\log \frac{\sqrt{a}}{2}$

41. $\frac{1}{2}\log_a x + 4\log_a y - 3\log_a x$

$= \log_a x^{1/2} + \log_a y^4 - \log_a x^3$ Power rule

$= \log_a x^{1/2}y^4 - \log_a x^3$ Product rule

$= \log_a \frac{x^{1/2}y^4}{x^3}$ Quotient rule

$= \log_a x^{-5/2}y^4$, or $\log_a \frac{y^4}{x^{5/2}}$ Simplifying

42. $\frac{2}{5}\log_a x - \frac{1}{3}\log_a y = \log_a x^{2/5} - \log_a y^{1/3} =$

$\log_a \frac{x^{2/5}}{y^{1/3}}$

43. $\ln x^2 - 2\ln\sqrt{x}$

$= \ln x^2 - \ln(\sqrt{x})^2$ Power rule

$= \ln x^2 - \ln x$ $\quad [(\sqrt{x})^2 = x]$

$= \ln \frac{x^2}{x}$ Quotient rule

$= \ln x$

44. $\ln 2x + 3(\ln x - \ln y) = \ln 2x + 3\ln \frac{x}{y}$

$= \ln 2x + \ln\left(\frac{x}{y}\right)^3$

$= \ln 2x\left(\frac{x}{y}\right)^3$

$= \ln \frac{2x^4}{y^3}$

45. $\ln(x^2 - 4) - \ln(x + 2)$

$= \ln \frac{x^2 - 4}{x + 2}$ Quotient rule

$= \ln \frac{(x+2)(x-2)}{x+2}$ Factoring

$= \ln(x - 2)$ Removing a factor of 1

46. $\log(x^3 - 8) - \log(x - 2)$

$= \log \frac{x^3 - 8}{x - 2}$

$= \log \frac{(x-2)(x^2+2x+4)}{x-2}$

$= \log(x^2 + 2x + 4)$

47. $\log(x^2 - 5x - 14) - \log(x^2 - 4)$

$= \log \frac{x^2 - 5x - 14}{x^2 - 4}$ Quotient rule

$= \log \frac{(x+2)(x-7)}{(x+2)(x-2)}$ Factoring

$= \log \frac{x-7}{x-2}$ Removing a factor of 1

48. $\log_a \frac{a}{\sqrt{x}} - \log_a \sqrt{ax} = \log_a \frac{a}{\sqrt{x}\sqrt{ax}}$

$= \log_a \frac{\sqrt{a}}{x}$

$= \log_a \sqrt{a} - \log_a x$

$= \frac{1}{2}\log_a a - \log_a x$

$= \frac{1}{2} - \log_a x$

49. $\ln x - 3[\ln(x-5) + \ln(x+5)]$

$= \ln x - 3\ln[(x-5)(x+5)]$ Product rule

$= \ln x - 3\ln(x^2-25)$

$= \ln x - \ln(x^2-25)^3$ Power rule

$= \ln \frac{x}{(x^2-25)^3}$ Quotient rule

50. $\frac{2}{3}[\ln(x^2-9) - \ln(x+3)] + \ln(x+y)$

$= \frac{2}{3}\ln\frac{x^2-9}{x+3} + \ln(x+y)$

$= \frac{2}{3}\ln\frac{(x+3)(x-3)}{x+3} + \ln(x+y)$

$= \frac{2}{3}\ln(x-3) + \ln(x+y)$

$= \ln(x-3)^{2/3} + \ln(x+y)$

$= \ln[(x-3)^{2/3}(x+y)]$

51. $\frac{3}{2}\ln 4x^6 - \frac{4}{5}\ln 2y^{10}$

$= \frac{3}{2}\ln 2^2x^6 - \frac{4}{5}\ln 2y^{10}$ Writing 4 as 2^2

$= \ln(2^2x^6)^{3/2} - \ln(2y^{10})^{4/5}$ Power rule

$= \ln(2^3x^9) - \ln(2^{4/5}y^8)$

$= \ln\frac{2^3x^9}{2^{4/5}y^8}$ Quotient rule

$= \ln\frac{2^{11/5}x^9}{y^8}$

52. $120(\ln\sqrt[5]{x^3} + \ln\sqrt[3]{y^2} - \ln\sqrt[4]{16z^5})$

$= 120\left(\ln\frac{\sqrt[5]{x^3}\sqrt[3]{y^2}}{\sqrt[4]{16z^5}}\right)$

$= 120\left(\frac{x^{3/5}y^{2/3}}{2z^{5/4}}\right)$

$= \ln\left(\frac{x^{3/5}y^{2/3}}{2z^{5/4}}\right)^{120}$

$= \ln\frac{x^{72}y^{80}}{2^{120}z^{150}}$

53. $\log_a \frac{2}{11} = \log_a 2 - \log_a 11$ Quotient rule

$\approx 0.301 - 1.041$

≈ -0.74

54. $\log_a 14 = \log_a(2\cdot 7)$

$= \log_a 2 + \log_a 7$

$\approx 0.301 + 0.845$

≈ 1.146

55. $\log_a 98 = \log_a(7^2\cdot 2)$

$= \log_a 7^2 + \log_a 2$ Product rule

$= 2\log_a 7 + \log_a 2$ Power rule

$\approx 2(0.845) + 0.301$

≈ 1.991

56. $\log_a \frac{1}{7} = \log_a 1 - \log_a 7$

$\approx 0 - 0.845$

≈ -0.845

57. $\frac{\log_a 2}{\log_a 7} \approx \frac{0.301}{0.845} \approx 0.356$

58. $\log_a 9$ cannot be found using the given information.

59. $\log_b 125 = \log_b 5^3$

$= 3\log_b 5$ Power rule

$\approx 3(1.609)$

≈ 4.827

60. $\log_b \frac{5}{3} = \log_b 5 - \log_b 3$

$\approx 1.609 - 1.099$

≈ 0.51

61. $\log_b \frac{1}{6} = \log_b 1 - \log_b 6$ Quotient rule

$= \log_b 1 - \log_b(2\cdot 3)$

$= \log_b 1 - (\log_b 2 + \log_b 3)$ Product rule

$= \log_b 1 - \log_b 2 - \log_b 3$

$\approx 0 - 0.693 - 1.099$

≈ -1.792

62. $\log_b 30 = \log_b(2\cdot 3\cdot 5)$

$= \log_b 2 + \log_b 3 + \log_b 5$

$\approx 0.693 + 1.099 + 1.609$

≈ 3.401

63. $\log_b \frac{3}{b} = \log_b 3 - \log_b b$ Quotient rule

$\approx 1.099 - 1$

≈ 0.099

64. $\log_b 15b = \log_b(3\cdot 5\cdot b)$

$= \log_b 3 + \log_b 5 + \log_b b$

$\approx 1.099 + 1.609 + 1$

≈ 3.708

65. $\log_p p^3 = 3$ $(\log_a a^x = x)$

66. $\log_t t^{2713} = 2713$

67. $\log_e e^{|x-4|} = |x-4|$ $(\log_a a^x = x)$

68. $\log_q q^{\sqrt{3}} = \sqrt{3}$

69. $3^{\log_3 4x} = 4x \qquad (a^{\log_a x} = x)$

70. $5^{\log_5 (4x-3)} = 4x - 3$

71. $10^{\log w} = w \qquad (a^{\log_a x} = x)$

72. $e^{\ln x^3} = x^3$

73. $\ln e^{8t} = 8t \qquad (\log_a a^x = x)$

74. $\log 10^{-k} = -k$

75.
$$\begin{aligned}\log_b \sqrt{b} &= \log_b b^{1/2}\\ &= \frac{1}{2}\log_b b \quad \text{Power rule}\\ &= \frac{1}{2}\cdot 1 \quad (\log_b b = 1)\\ &= \frac{1}{2}\end{aligned}$$

76.
$$\begin{aligned}\log_b \sqrt{b^3} &= \log_b b^{3/2}\\ &= \frac{3}{2}\log_b b\\ &= \frac{3}{2}\cdot 1\\ &= \frac{3}{2}\end{aligned}$$

77. $f(x) = a^x$, $g(x) = \log_a x$

Since f and g are inverses, we know that $(f \circ g)(x) = x$ and $(g \circ f)(x) = x$. Now $(f \circ g)(x) = f(g(x)) = f(\log_a x) = a^{\log_a x}$, so we know that $a^{\log_a x} = x$. Also $(g \circ f)(x) = g(f(x)) = g(a^x) = \log_a a^x$, so we know that $\log_a a^x = x$. These results are alternate proofs of the Logarithm of a Base to a Power property and the Base to a Logarithmic Power property.

78. $\log_a ab^3 \neq (\log_a a)(\log_a b^3)$. If the first step had been correct, then so would the second step. The correct procedure follows.

$\log_a ab^3 = \log_a a + \log_a b^3 = 1 + 3\log_a b$

79. The degree of $f(x) = 5 - x^2 + x^4$ is 4, so the function is quartic.

80. The variable in $f(x) = 2^x$ is in the exponent, so $f(x)$ is an exponential function.

81. $f(x) = -\frac{3}{4}$ is of the form $f(x) = mx + b$ $\left(\text{with } m = 0 \text{ and } b = -\frac{3}{4}\right)$, so it is a linear function. In fact, it is a constant function.

82. The variable in $f(x) = 4^x - 8$ is in the exponent, so $f(x)$ is an exponential function.

83. $f(x) = -\frac{3}{x}$ is of the form $f(x) = \frac{p(x)}{q(x)}$ where $p(x)$ and $q(x)$ are polynomials and $q(x)$ is not the zero polynomial, so $f(x)$ is a rational function.

84. $f(x) = \log x + 6$ is a logarithmic function.

85. The degree of $f(x) = -\frac{1}{3}x^3 - 4x^2 + 6x + 42$ is 3, so the function is cubic.

86. $f(x) = \frac{x^2 - 1}{x^2 + x - 6}$ is of the form $f(x) = \frac{p(x)}{q(x)}$ where $p(x)$ and $q(x)$ are polynomials and $q(x)$ is not the zero polynomial, so $f(x)$ is a rational function.

87. $f(x) = \frac{1}{2}x + 3$ is of the form $f(x) = mx + b$, so it is a linear function.

88. The degree of $f(x) = 2x^2 - 6x + 3$ is 2, so the function is quadratic.

89.
$$\begin{aligned}5^{\log_5 8} &= 2x\\ 8 &= 2x \quad (a^{\log_a x} = x)\\ 4 &= x\end{aligned}$$
The solution is 4.

90.
$$\begin{aligned}\ln e^{3x-5} &= -8\\ 3x - 5 &= -8\\ 3x &= -3\\ x &= -1\end{aligned}$$
The solution is -1.

91.
$$\begin{aligned}&\log_a(x^2 + xy + y^2) + \log_a(x - y)\\ &= \log_a[(x^2 + xy + y^2)(x - y)] \quad \text{Product rule}\\ &= \log_a(x^3 - y^3) \quad \text{Multiplying}\end{aligned}$$

92.
$$\begin{aligned}&\log_a(a^{10} - b^{10}) - \log_a(a + b)\\ &= \log_a \frac{a^{10} - b^{10}}{a + b}, \text{ or}\\ &\log_a(a^9 - a^8b + a^7b^2 - a^6b^3 + a^5b^4 - a^4b^5 +\\ &a^3b^6 - a^2b^7 + ab^8 - b^9)\end{aligned}$$

93.
$$\begin{aligned}&\log_a \frac{x - y}{\sqrt{x^2 - y^2}}\\ &= \log_a \frac{x - y}{(x^2 - y^2)^{1/2}}\\ &= \log_a(x - y) - \log_a(x^2 - y^2)^{1/2} \quad \text{Quotient rule}\\ &= \log_a(x - y) - \frac{1}{2}\log_a(x^2 - y^2) \quad \text{Power rule}\\ &= \log_a(x - y) - \frac{1}{2}\log_a[(x + y)(x - y)]\\ &= \log_a(x - y) - \frac{1}{2}[\log_a(x + y) + \log_a(x - y)]\\ &\qquad \text{Product rule}\\ &= \log_a(x - y) - \frac{1}{2}\log_a(x + y) - \frac{1}{2}\log_a(x - y)\\ &= \frac{1}{2}\log_a(x - y) - \frac{1}{2}\log_a(x + y)\end{aligned}$$

94. $\log_a \sqrt{9-x^2}$

$= \log_a (9-x^2)^{1/2}$

$= \frac{1}{2}\log_a (9-x^2)$

$= \frac{1}{2}\log_a [(3+x)(3-x)]$

$= \frac{1}{2}[\log_a (3+x) + \log_a (3-x)]$

$= \frac{1}{2}\log_a (3+x) + \frac{1}{2}\log_a (3-x)$

95. $\log_a \frac{\sqrt[4]{y^2z^5}}{\sqrt[4]{x^3z^{-2}}}$

$= \log_a \sqrt[4]{\frac{y^2z^5}{x^3z^{-2}}}$

$= \log_a \sqrt[4]{\frac{y^2z^7}{x^3}}$

$= \log_a \left(\frac{y^2z^7}{x^3}\right)^{1/4}$

$= \frac{1}{4}\log_a \left(\frac{y^2z^7}{x^3}\right)$ Power rule

$= \frac{1}{4}(\log_a y^2z^7 - \log_a x^3)$ Quotient rule

$= \frac{1}{4}(\log_a y^2 + \log_a z^7 - \log_a x^3)$ Product rule

$= \frac{1}{4}(2\log_a y + 7\log_a z - 3\log_a x)$ Power rule

$= \frac{1}{4}(2\cdot 3 + 7\cdot 4 - 3\cdot 2)$

$= \frac{1}{4}\cdot 28$

$= 7$

96. $\log_a M + \log_a N = \log_a (M+N)$

Let $a = 10$, $M = 1$, and $N = 10$. Then $\log_{10} 1 + \log_{10} 10 = 0 + 1 = 1$, but $\log_{10}(1+10) = \log_{10} 11 \approx 1.0414$. Thus, the statement is false.

97. $\log_a M - \log_a N = \log_a \frac{M}{N}$

This is the quotient rule, so it is true.

98. $\frac{\log_a M}{\log_a N} = \log_a M - \log_a N$

Let $M = a^2$ and $N = a$. Then $\frac{\log_a a^2}{\log_a a} = \frac{2}{1} = 2$, but $\log_a a^2 - \log_a a = 2 - 1 = 1$. Thus, the statement is false.

99. $\frac{\log_a M}{x} = \frac{1}{x}\log_a M = \log_a M^{1/x}$. The statement is true by the power rule.

100. $\log_a x^3 = 3\log_a x$ is true by the power rule.

101. $\log_a 8x = \log_a 8 + \log_a x = \log_a x + \log_a 8$. The statement is true by the product rule and the commutative property of addition.

102. $\log_N (M\cdot N)^x = x\log_N (M\cdot N)$

$= x(\log_N M + \log_N N)$

$= x(\log_N M + 1)$

$= x\log_N M + x$

The statement is true.

103. $\log_a \left(\frac{1}{x}\right) = \log_a x^{-1} = -1\cdot\log_a x = -1\cdot 2 = -2$

104. $\log_a x = 2$

$a^2 = x$

Let $\log_{1/a} x = n$ and solve for n.

$\log_{1/a} a^2 = n$ Substituting a^2 for x

$\left(\frac{1}{a}\right)^n = a^2$

$(a^{-1})^n = a^2$

$a^{-n} = a^2$

$-n = 2$

$n = -2$

Thus, $\log_{1/a} x = -2$ when $\log_a x = 2$.

105. We use the change-of-base formula.

$\log_{10} 11\cdot\log_{11} 12\cdot\log_{12} 13\cdots \log_{998} 999\cdot\log_{999} 1000$

$= \log_{10} 11\cdot\frac{\log_{10} 12}{\log_{10} 11}\cdot\frac{\log_{10} 13}{\log_{10} 12}\cdots \frac{\log_{10} 999}{\log_{10} 998}\cdot\frac{\log_{10} 1000}{\log_{10} 999}$

$= \frac{\log_{10} 11}{\log_{10} 11}\cdot\frac{\log_{10} 12}{\log_{10} 12}\cdots\frac{\log_{10} 999}{\log_{10} 999}\cdot\log_{10} 1000$

$= \log_{10} 1000$

$= 3$

106. $\log_a x + \log_a y - mz = 0$

$\log_a x + \log_a y = mz$

$\log_a xy = mz$

$a^{mz} = xy$

107. $\ln a - \ln b + xy = 0$

$\ln a + \ln b = -xy$

$\ln ab = -xy$

Then, using the definition of a logarithm, we have $e^{-xy} = ab$.

108. $\log_a \frac{1}{x} = \log_a 1 - \log_a x = -\log_a x$.

Let $-\log_a x = y$. Then $\log_a x = -y$ and $x = a^{-y} = a^{-1\cdot y} = \left(\frac{1}{a}\right)^y$, so $\log_{1/a} x = y$. Thus, $\log_a \left(\frac{1}{x}\right) = -\log_a x = \log_{1/a} x$.

109. $\log_a\left(\frac{x+\sqrt{x^2-5}}{5}\right)$

$= \log_a\left(\frac{x+\sqrt{x^2-5}}{5}\cdot\frac{x-\sqrt{x^2-5}}{x-\sqrt{x^2-5}}\right)$

$= \log_a\left(\frac{5}{5(x-\sqrt{x^2-5})}\right) = \log_a\left(\frac{1}{x-\sqrt{x^2-5}}\right)$

$= \log_a 1 - \log_a(x-\sqrt{x^2-5})$

$= -\log_a(x-\sqrt{x^2-5})$

Exercise Set 4.5

1. $3^x = 81$

$3^x = 3^4$

$x = 4$ The exponents are the same.

The solution is 4.

2. $2^x = 32$

$2^x = 2^5$

$x = 5$

The solution is 5.

3. $2^{2x} = 8$

$2^{2x} = 2^3$

$2x = 3$ The exponents are the same.

$x = \frac{3}{2}$

The solution is $\frac{3}{2}$.

4. $3^{7x} = 27$

$3^{7x} = 3^3$

$7x = 3$

$x = \frac{3}{7}$

The solution is $\frac{3}{7}$.

5. $2^x = 33$

$\log 2^x = \log 33$ Taking the common logarithm on both sides

$x\log 2 = \log 33$ Power rule

$x = \frac{\log 33}{\log 2}$

$x \approx \frac{1.5185}{0.3010}$

$x \approx 5.044$

The solution is 5.044.

6. $2^x = 40$

$\log 2^x = \log 40$

$x\log 2 = \log 40$

$x = \frac{\log 40}{\log 2}$

$x \approx \frac{1.6021}{0.3010}$

$x \approx 5.322$

The solution is 5.322.

7. $5^{4x-7} = 125$

$5^{4x-7} = 5^3$

$4x - 7 = 3$

$4x = 10$

$x = \frac{10}{4} = \frac{5}{2}$

The solution is $\frac{5}{2}$.

8. $4^{3x-5} = 16$

$4^{3x-5} = 4^2$

$3x - 5 = 2$

$3x = 7$

$x = \frac{7}{3}$

The solution is $\frac{7}{3}$.

9. $27 = 3^{5x}\cdot 9^{x^2}$

$3^3 = 3^{5x}\cdot(3^2)^{x^2}$

$3^3 = 3^{5x}\cdot 3^{2x^2}$

$3^3 = 3^{5x+2x^2}$

$3 = 5x + 2x^2$

$0 = 2x^2 + 5x - 3$

$0 = (2x-1)(x+3)$

$x = \frac{1}{2}$ *or* $x = -3$

The solutions are -3 and $\frac{1}{2}$.

10. $3^{x^2+4x} = \frac{1}{27}$

$3^{x^2+4x} = 3^{-3}$

$x^2 + 4x = -3$

$x^2 + 4x + 3 = 0$

$(x+3)(x+1) = 0$

$x = -3$ *or* $x = -1$

The solutions are -3 and -1.

11. $84^x = 70$

$\log 84^x = \log 70$

$x\log 84 = \log 70$

$x = \frac{\log 70}{\log 84}$

$x \approx \frac{1.8451}{1.9243}$

$x \approx 0.959$

The solution is 0.959.

12.
$$\begin{aligned} 28^x &= 10^{-3x} \\ \log 28^x &= \log 10^{-3x} \\ x\log 28 &= -3x \\ x\log 28 + 3x &= 0 \\ x(\log 28 + 3) &= 0 \\ x &= 0 \end{aligned}$$

The solution is 0.

13.
$$\begin{aligned} e^{-c} &= 5^{2c} \\ \ln e^{-c} &= \ln 5^{2c} \\ -c &= 2c \ln 5 \\ 0 &= c + 2c \ln 5 \\ 0 &= c(1 + 2 \ln 5) \\ 0 &= c \quad \text{Dividing by } 1 + 2 \ln 5 \end{aligned}$$

The solution is 0.

14.
$$\begin{aligned} 15^x &= 30 \\ \log 15^x &= \log 30 \\ x\log 15 &= \log 30 \\ x &= \frac{\log 30}{\log 15} \\ x &\approx 1.256 \end{aligned}$$

The solution is 1.256.

15.
$$\begin{aligned} e^t &= 1000 \\ \ln e^t &= \ln 1000 \\ t &= \ln 1000 \quad \text{Using } \log_a a^x = x \\ t &\approx 6.908 \end{aligned}$$

The solution is 6.908.

16.
$$\begin{aligned} e^{-t} &= 0.04 \\ \ln e^{-t} &= \ln 0.04 \\ -t &= \ln 0.04 \\ t &= -\ln 0.04 \approx 3.219 \end{aligned}$$

The solution is 3.219.

17.
$$\begin{aligned} e^{-0.03t} &= 0.08 \\ \ln e^{-0.03t} &= \ln 0.08 \\ -0.03t &= \ln 0.08 \\ t &= \frac{\ln 0.08}{-0.03} \\ t &\approx \frac{-2.5257}{-0.03} \\ t &\approx 84.191 \end{aligned}$$

The solution is 84.191.

18.
$$\begin{aligned} 1000e^{0.09t} &= 5000 \\ e^{0.09t} &= 5 \\ \ln e^{0.09t} &= \ln 5 \\ 0.09t &= \ln 5 \\ t &= \frac{\ln 5}{0.09} \\ t &\approx 17.883 \end{aligned}$$

The solution is 17.883.

19.
$$\begin{aligned} 3^x &= 2^{x-1} \\ \ln 3^x &= \ln 2^{x-1} \\ x\ln 3 &= (x-1)\ln 2 \\ x\ln 3 &= x\ln 2 - \ln 2 \\ \ln 2 &= x\ln 2 - x\ln 3 \\ \ln 2 &= x(\ln 2 - \ln 3) \\ \frac{\ln 2}{\ln 2 - \ln 3} &= x \\ \frac{0.6931}{0.6931 - 1.0986} &\approx x \\ -1.710 &\approx x \end{aligned}$$

The solution is -1.710.

20.
$$\begin{aligned} 5^{x+2} &= 4^{1-x} \\ \log 5^{x+2} &= \log 4^{1-x} \\ (x+2)\log 5 &= (1-x)\log 4 \\ x\log 5 + 2\log 5 &= \log 4 - x\log 4 \\ x\log 5 + x\log 4 &= \log 4 - 2\log 5 \\ x(\log 5 + \log 4) &= \log 4 - 2\log 5 \\ x &= \frac{\log 4 - 2\log 5}{\log 5 + \log 4} \\ x &\approx -0.612 \end{aligned}$$

The solution is -0.612.

21.
$$\begin{aligned} (3.9)^x &= 48 \\ \log(3.9)^x &= \log 48 \\ x\log 3.9 &= \log 48 \\ x &= \frac{\log 48}{\log 3.9} \\ x &\approx \frac{1.6812}{0.5911} \\ x &\approx 2.844 \end{aligned}$$

The solution is 2.844.

22.
$$\begin{aligned} 250 - (1.87)^x &= 0 \\ 250 &= (1.87)^x \\ \log 250 &= \log(1.87)^x \\ \log 250 &= x\log 1.87 \\ \frac{\log 250}{\log 1.87} &= x \\ 8.821 &\approx x \end{aligned}$$

The solution is 8.821.

23.
$$\begin{aligned} e^x + e^{-x} &= 5 \\ e^{2x} + 1 &= 5e^x \quad \text{Multiplying by } e^x \\ e^{2x} - 5e^x + 1 &= 0 \quad \text{This equation is quadratic in } e^x. \\ e^x &= \frac{5 \pm \sqrt{21}}{2} \\ x &= \ln\left(\frac{5 \pm \sqrt{21}}{2}\right) \approx \pm 1.567 \end{aligned}$$

The solutions are -1.567 and 1.567.

24. $e^x - 6e^{-x} = 1$

$e^{2x} - 6 = e^x$

$e^{2x} - e^x - 6 = 0$

$(e^x - 3)(e^x + 2) = 0$

$e^x = 3$ *or* $e^x = -2$

$\ln e^x = \ln 3$ No solution

$x = \ln 3$

$x \approx 1.099$

The solution is 1.099.

25. $\dfrac{e^x + e^{-x}}{e^x - e^{-x}} = 3$

$e^x + e^{-x} = 3e^x - 3e^{-x}$ Multiplying by $e^x - e^{-x}$

$4e^{-x} = 2e^x$ Subtracting e^x and adding e^{-x}

$2e^{-x} = e^x$

$2 = e^{2x}$ Multiplying by e^x

$\ln 2 = \ln e^{2x}$

$\ln 2 = 2x$

$\dfrac{\ln 2}{2} = x$

$0.347 \approx x$

The solution is 0.347.

26. $\dfrac{5^x - 5^{-x}}{5^x + 5^{-x}} = 8$

$5^{-x} - 5^{-x} = 8 \cdot 5^x + 8 \cdot 5^{-x}$

$-9 \cdot 5^{-x} = 7 \cdot 5^x$

$-9 = 7 \cdot 5^{2x}$ Multiplying by 5^x

$-\dfrac{9}{7} = 5^{2x}$

The number 5 raised to any power is non-negative. Thus, the equation has no solution.

27. $\log_5 x = 4$

$x = 5^4$ Writing an equivalent exponential equation

$x = 625$

The solution is 625.

28. $\log_2 x = -3$

$x = 2^{-3}$

$x = \dfrac{1}{8}$

The solution is $\dfrac{1}{8}$.

29. $\log x = -4$ The base is 10.

$x = 10^{-4}$, or 0.0001

The solution is 0.0001.

30. $\log x = 1$

$x = 10^1 = 10$

The solution is 10.

31. $\ln x = 1$ The base is e.

$x = e^1 = e$

The solution is e.

32. $\ln x = -2$

$x = e^{-2}$, or $\dfrac{1}{e^2}$

The solution is e^{-2}, or $\dfrac{1}{e^2}$.

33. $\log_2(10 + 3x) = 5$

$2^5 = 10 + 3x$

$32 = 10 + 3x$

$22 = 3x$

$\dfrac{22}{3} = x$

The answer checks. The solution is $\dfrac{22}{3}$.

34. $\log_5(8 - 7x) = 3$

$5^3 = 8 - 7x$

$125 = 8 - 7x$

$117 = -7x$

$-\dfrac{117}{7} = x$

The answer checks. The solution is $-\dfrac{117}{7}$.

35. $\log x + \log(x - 9) = 1$ The base is 10.

$\log_{10}[x(x - 9)] = 1$

$x(x - 9) = 10^1$

$x^2 - 9x = 10$

$x^2 - 9x - 10 = 0$

$(x - 10)(x + 1) = 0$

$x = 10$ or $x = -1$

Check: For 10:

$\log x + \log(x - 9) = 1$	
$\log 10 + \log(10 - 9)$	? 1
$\log 10 + \log 1$	
$1 + 0$	
1	1 TRUE

For -1:

$\log x + \log(x - 9) = 1$	
$\log(-1) + \log(-1 - 9)$	? 1

The number -1 does not check, because negative numbers do not have logarithms. The solution is 10.

36. $\log_2(x + 1) + \log_2(x - 1) = 3$

$\log_2[(x + 1)(x - 1)] = 3$

$(x + 1)(x - 1) = 2^3$

$x^2 - 1 = 8$

$x^2 = 9$

$x = \pm 3$

The number 3 checks, but -3 does not. The solution is 3.

37. $\log_2(x+20) - \log_2(x+2) = \log_2 x$

$$\log_2 \frac{x+20}{x+2} = \log_2 x$$

$$\frac{x+20}{x+2} = x \quad \text{Using the property of logarithmic equality}$$

$$x+20 = x^2+2x \quad \text{Multiplying by } x+2$$

$$0 = x^2+x-20$$

$$0 = (x+5)(x-4)$$

$x+5=0$ *or* $x-4=0$

$x=-5$ *or* $x=4$

Check: For -5:

$$\log_2(x+20) - \log_2(x+2) = \log_2 x$$

$$\log_2(-5+20) - \log_2(-5+2) \ ? \ \log_2(-5)$$

The number -5 does not check, because negative numbers do not have logarithms.

For 4:

$$\log_2(x+20) - \log_2(x+2) = \log_2 x$$

$$\log_2(4+20) - \log_2(4+2) \ ? \ \log_2 4$$

$$\log_2 24 - \log_2 6$$

$$\log_2 \frac{24}{6}$$

$$\log_2 4 \quad | \quad \log_2 4 \quad \text{TRUE}$$

The solution is 4.

38. $\log(x+5) - \log(x-3) = \log 2$

$$\log \frac{x+5}{x-3} = \log 2$$

$$\frac{x+5}{x-3} = 2$$

$$x+5 = 2x-6$$

$$11 = x$$

The answer checks. The solution is 11.

39. $\log_8(x+1) - \log_8 x = 2$

$$\log_8\left(\frac{x+1}{x}\right) = 2 \quad \text{Quotient rule}$$

$$\frac{x+1}{x} = 8^2$$

$$\frac{x+1}{x} = 64$$

$$x+1 = 64x$$

$$1 = 63x$$

$$\frac{1}{63} = x$$

The answer checks. The solution is $\frac{1}{63}$.

40. $\log x - \log(x+3) = -1$

$$\log_{10} \frac{x}{x+3} = -1$$

$$\frac{x}{x+3} = 10^{-1}$$

$$\frac{x}{x+3} = \frac{1}{10}$$

$$10x = x+3$$

$$9x = 3$$

$$x = \frac{1}{3}$$

The answer checks. The solution is $\frac{1}{3}$.

41. $\log x + \log(x+4) = \log 12$

$$\log x(x+4) = \log 12$$

$$x(x+4) = 12 \quad \text{Using the property of logarithmic equality}$$

$$x^2+4x = 12$$

$$x^2+4x-12 = 0$$

$$(x+6)(x-2) = 0$$

$x+6=0$ *or* $x-2=0$

$x=-6$ *or* $x=2$

Check: For -6:

$$\log x + \log(x+4) = \log 12$$

$$\log(-6) + \log(-6+4) \ ? \ \log 12$$

The number -6 does not check, because negative numbers do not have logarithms.

For 2:

$$\log x + \log(x+4) = \log 12$$

$$\log 2 + \log(2+4) \ ? \ \log 12$$

$$\log 2 + \log 6$$

$$\log(2 \cdot 6)$$

$$\log 12 \quad | \quad \log 12 \quad \text{TRUE}$$

The solution is 2.

42. $\ln x - \ln(x-4) = \ln 3$

$$\ln \frac{x}{x-4} = \ln 3$$

$$\frac{x}{x-4} = 3$$

$$x = 3x-12$$

$$12 = 2x$$

$$6 = x$$

The answer checks. The solution is 6.

43. $\log_4(x+3) + \log_4(x-3) = 2$

$$\log_4[(x+3)(x-3)] = 2 \quad \text{Product rule}$$

$$(x+3)(x-3) = 4^2$$

$$x^2-9 = 16$$

$$x^2 = 25$$

$$x = \pm 5$$

The number 5 checks, but -5 does not. The solution is 5.

44. $\ln(x+1) - \ln x = \ln 4$

$$\ln \frac{x+1}{x} = \ln 4$$
$$\frac{x+1}{x} = 4$$
$$x+1 = 4x$$
$$1 = 3x$$
$$\frac{1}{3} = x$$

The answer checks. The solution is $\frac{1}{3}$.

45. $\log(2x+1) - \log(x-2) = 1$

$$\log\left(\frac{2x+1}{x-2}\right) = 1 \quad \text{Quotient rule}$$
$$\frac{2x+1}{x-2} = 10^1 = 10$$
$$2x+1 = 10x - 20$$

Multiplying by $x-2$

$$21 = 8x$$
$$\frac{21}{8} = x$$

The answer checks. The solution is $\frac{21}{8}$.

46. $\log_5(x+4) + \log_5(x-4) = 2$

$$\log_5[(x+4)(x-4)] = 2$$
$$x^2 - 16 = 25$$
$$x^2 = 41$$
$$x = \pm\sqrt{41}$$

Only $\sqrt{41}$ checks. The solution is $\sqrt{41}$.

47. $\ln(x+8) + \ln(x-1) = 2\ln x$

$$\ln(x+8)(x-1) = \ln x^2$$
$$(x+8)(x-1) = x^2 \quad \text{Using the property of logarithmic equality}$$
$$x^2 + 7x - 8 = x^2$$
$$7x - 8 = 0$$
$$7x = 8$$
$$x = \frac{8}{7}$$

The answer checks. The solution is $\frac{8}{7}$.

48. $\log_3 x + \log_3(x+1) = \log_3 2 + \log_3(x+3)$

$$\log_3 x(x+1) = \log_3 2(x+3)$$
$$x(x+1) = 2(x+3)$$
$$x^2 + x = 2x + 6$$
$$x^2 - x - 6 = 0$$
$$(x-3)(x+2) = 0$$
$$x = 3 \quad or \quad x = -2$$

The number 3 checks, but -2 does not. The solution is 3.

49. $e^{7.2x} = 14.009$

Graph $y_1 = e^{7.2x}$ and $y_2 = 14.009$ and find the first coordinate of the point of intersection using the Intersect feature. The solution is 0.367.

50. $0.082e^{0.05x} = 0.034$

Graph $y_1 = 0.082e^{0.05x}$ and $y_2 = 0.034$ and find the first coordinate of the point of intersection using the Intersect feature. The solution is -17.607.

51. $xe^{3x} - 1 = 3$

Graph $y_1 = xe^{3x} - 1$ and $y_2 = 3$ and find the first coordinate of the point of intersection using the Intersect feature. The solution is 0.621.

52. $5e^{5x} + 10 = 3x + 40$

Graph $y_1 = 5e^{5x} + 10$ and $y_2 = 3x + 40$ and find the first coordinates of the points of intersection using the Intersect feature. The solutions are -10 and 0.366.

53. $4\ln(x+3.4) = 2.5$

Graph $y_1 = 4\ln(x+3.4)$ and $y_2 = 2.5$ and find the first coordinate of the point of intersection using the Intersect feature. The solution is -1.532.

54. $\ln x^2 = -x^2$

Graph $y_1 = \ln x^2$ and $y_2 = -x^2$ and find the first coordinates of the points of intersection using the Intersect feature. The solutions are -0.753 and 0.753.

55. $\log_8 x + \log_8(x+2) = 2$

Graph $y_1 = \frac{\log x}{\log 8} + \frac{\log(x+2)}{\log 8}$ and $y_2 = 2$ and find the first coordinate of the point of intersection using the intersect feature. The solution is 7.062.

56. $\log_3 x + 7 = 4 - \log_5 x$

Graph $y_1 = \frac{\log x}{\log 3} + 7$ and $y_2 = 4 - \frac{\log x}{\log 5}$ and find the first coordinate of the point of intersection using the Intersect feature. The solution is 0.141.

57. $\log_5(x+7) - \log_5(2x-3) = 1$

Graph $y_1 = \frac{\log(x+7)}{\log 5} - \frac{\log(2x-3)}{\log 5}$ and $y_2 = 1$ and find the first coordinate of the point of intersection using the Intersect feature. The solution is 2.444.

58. Graph $y_1 = \ln 3x$ and $y_2 = 3x - 8$ and use the Intersect feature to find the points of intersection. They are $(0.0001, -7.9997)$ and $(3.445, 2.336)$.

59. Solving the first equation for y, we get $y = \frac{12.4 - 2.3x}{3.8}$. Graph $y_1 = \frac{12.4 - 2.3x}{3.8}$ and $y_2 = 1.1\ln(x - 2.05)$ and use the Intersect feature to find the point of intersection. It is $(4.093, 0.786)$.

60. Graph $y_1 = 2.3\ln(x+10.7)$ and $y_2 = 10e^{-0.07x^2}$ and use the Intersect feature to find the points of intersection. They are $(-9.694, 0.014)$, $(-3.334, 4.593)$, and $(2.714, 5.971)$.

61. Graph $y_1 = 2.3\ln(x+10.7)$ and $y_2 = 10e^{-0.007x^2}$ and use the Intersect feature to find the point of intersection. It is $(7.586, 6.684)$.

62. The final result would have been the same, but to find t we would have computed $\dfrac{\log 2500}{0.08 \log e}$.

It seems best to take the natural logarithm on both sides since the final computation for t is simpler.

63. Use the graph of $y = \ln x$ to estimate the x-value that corresponds to the value on the right-hand side of the equation.

64. $f(x) = -x^2 + 6x - 8$

a) $-\dfrac{b}{2a} = -\dfrac{6}{2(-1)} = 3$

$f(3) = -3^2 + 6\cdot 3 - 8 = 1$

The vertex is $(3, 1)$.

b) $x = 3$

c) Maximum: 1 at $x = 3$

65. $g(x) = x^2 - 6$

a) $-\dfrac{b}{2a} = -\dfrac{0}{2\cdot 1} = 0$

$g(0) = 0^2 - 6 = -6$

The vertex is $(0, -6)$.

b) The axis of symmetry is $x = 0$.

c) Since the coefficient of the x^2-term is positive, the function has a minimum value. It is the second coordinate of the vertex, -6, and it occurs when $x = 0$.

66. $H(x) = 3x^2 - 12x + 16$

a) $-\dfrac{b}{2a} = -\dfrac{-12}{2\cdot 3} = 2$

$H(2) = 3\cdot 2^2 - 12\cdot 2 + 16 = 4$

The vertex is $(2, 4)$.

b) $x = 2$

c) Minimum: 4 at $x = 2$

67. $G(x) = -2x^2 - 4x - 7$

a) $-\dfrac{b}{2a} = -\dfrac{-4}{2(-2)} = -1$

$G(-1) = -2(-1)^2 - 4(-1) - 7 = -5$

The vertex is $(-1, -5)$.

b) The axis of symmetry is $x = -1$.

c) Since the coefficient of the x^2-term is negative, the function has a maximum value. It is the second coordinate of the vertex, -5, and it occurs when $x = -1$.

68.
$$\begin{aligned} \ln(\ln x) &= 2 \\ \ln x &= e^2 \\ x &= e^{e^2} \approx 1618.178 \end{aligned}$$

The answer checks. The solution is e^{e^2}, or 1618.178.

69.
$$\begin{aligned} \ln(\log x) &= 0 \\ \log x &= e^0 \\ \log x &= 1 \\ x &= 10^1 = 10 \end{aligned}$$

The answer checks. The solution is 10.

70.
$$\begin{aligned} \ln \sqrt[4]{x} &= \sqrt{\ln x} \\ \frac{1}{4}\ln x &= \sqrt{\ln x} \\ \frac{1}{16}(\ln x)^2 &= \ln x \quad \text{Squaring both sides} \\ \frac{1}{16}(\ln x)^2 - \ln x &= 0 \end{aligned}$$

Let $u = \ln x$ and substitute.

$$\begin{aligned} &\frac{1}{16}u^2 - u = 0 \\ &u\left(\frac{1}{16}u - 1\right) = 0 \\ &u = 0 \quad \text{or} \quad \frac{1}{16}u - 1 = 0 \\ &u = 0 \quad \text{or} \quad u = 16 \\ &\ln x = 0 \quad \text{or} \quad \ln x = 16 \\ &x = e^0 \quad \text{or} \quad x = e^{16} \\ &x = 1 \quad \text{or} \quad x = e^{16} \approx 8,886,110.521 \end{aligned}$$

Both answers check. The solutions are 1 and e^{16}, or 1 and 8,886,110.521.

71.
$$\begin{aligned} \sqrt{\ln x} &= \ln \sqrt{x} \\ \sqrt{\ln x} &= \frac{1}{2}\ln x \quad \text{Power rule} \\ \ln x &= \frac{1}{4}(\ln x)^2 \quad \text{Squaring both sides} \\ 0 &= \frac{1}{4}(\ln x)^2 - \ln x \end{aligned}$$

Let $u = \ln x$ and substitute.

$$\begin{aligned} &\frac{1}{4}u^2 - u = 0 \\ &u\left(\frac{1}{4}u - 1\right) = 0 \\ &u = 0 \quad \text{or} \quad \frac{1}{4}u - 1 = 0 \\ &u = 0 \quad \text{or} \quad \frac{1}{4}u = 1 \\ &u = 0 \quad \text{or} \quad u = 4 \\ &\ln x = 0 \quad \text{or} \quad \ln x = 4 \\ &x = e^0 = 1 \quad \text{or} \quad x = e^4 \approx 54.598 \end{aligned}$$

Both answers check. The solutions are 1 and e^4, or 1 and 54.598.

72. $\log_3(\log_4 x) = 0$

$\log_4 x = 3^0$

$\log_4 x = 1$

$x = 4^1$

$x = 4$

The answer checks. The solution is 4.

73. $(\log_3 x)^2 - \log_3 x^2 = 3$

$(\log_3 x)^2 - 2\log_3 x - 3 = 0$

Let $u = \log_3 x$ and substitute:

$u^2 - 2u - 3 = 0$

$(u-3)(u+1) = 0$

$u = 3$ *or* $u = -1$

$\log_3 x = 3$ *or* $\log_3 x = -1$

$x = 3^3$ *or* $x = 3^{-1}$

$x = 27$ *or* $x = \frac{1}{3}$

Both answers check. The solutions are $\frac{1}{3}$ and 27.

74. $(\log x)^2 - \log x^2 = 3$

$(\log x)^2 - 2\log x - 3 = 0$

Let $u = \log x$ and substitute.

$u^2 - 2u - 3 = 0$

$(u+1)(u-3) = 0$

$u = -1$ *or* $u = 3$

$\log x = -1$ *or* $\log x = 3$

$x = \frac{1}{10}$ *or* $x = 1000$

Both answers check. The solutions are $\frac{1}{10}$ and 1000.

75. $\ln x^2 = (\ln x)^2$

$2\ln x = (\ln x)^2$

$0 = (\ln x)^2 - 2\ln x$

Let $u = \ln x$ and substitute.

$0 = u^2 - 2u$

$0 = u(u-2)$

$u = 0$ *or* $u = 2$

$\ln x = 0$ *or* $\ln x = 2$

$x = 1$ *or* $x = e^2 \approx 7.389$

Both answers check. The solutions are 1 and e^2, or 1 and 7.389.

76. $e^{2x} - 9 \cdot e^x + 14 = 0$

$(e^x - 2)(e^x - 7) = 0$

$e^x = 2$ *or* $e^x = 7$

$\ln e^x = \ln 2$ *or* $\ln e^x = \ln 7$

$x = \ln 2$ *or* $x = \ln 7$

$x \approx 0.693$ *or* $x \approx 1.946$

The solutions are 0.693 and 1.946.

77. $5^{2x} - 3 \cdot 5^x + 2 = 0$

$(5^x - 1)(5^x - 2) = 0$ This equation is quadratic in 5^x.

$5^x = 1$ *or* $5^x = 2$

$\log 5^x = \log 1$ *or* $\log 5^x = \log 2$

$x \log 5 = 0$ *or* $x \log 5 = \log 2$

$x = 0$ *or* $x = \frac{\log 2}{\log 5} \approx 0.431$

The solutions are 0 and 0.431.

78. $x\left(\ln \frac{1}{6}\right) = \ln 6$

$x(\ln 1 - \ln 6) = \ln 6$

$-x \ln 6 = \ln 6 \quad (\ln 1 = 0)$

$x = -1$

The solution is -1.

79. $\log_3 |x| = 2$

$|x| = 3^2$

$|x| = 9$

$x = -9$ *or* $x = 9$

Both answers check. The solutions are -9 and 9.

80. $x^{\log x} = \frac{x^3}{100}$

$\log x^{\log x} = \log \frac{x^3}{100}$

$\log x \cdot \log x = \log x^3 - \log 100$

$(\log x)^2 = 3\log x - 2$

$(\log x)^2 - 3\log x + 2 = 0$

Let $u = \log x$ and substitute.

$u^2 - 3u + 2 = 0$

$(u-1)(u-2) = 0$

$u = 1$ *or* $u = 2$

$\log x = 1$ *or* $\log x = 2$

$x = 10$ *or* $x = 10^2 = 100$

Both answers check. The solutions are 10 and 100.

81. $\ln x^{\ln x} = 4$

$\ln x \cdot \ln x = 4$

$(\ln x)^2 = 4$

$\ln x = \pm 2$

$\ln x = -2$ *or* $\ln x = 2$

$x = e^{-2}$ *or* $x = e^2$

$x \approx 0.135$ *or* $x \approx 7.389$

Both answers check. The solutions are e^{-2} and e^2, or 0.135 and 7.389.

82. $\dfrac{(e^{3x+1})^2}{e^4} = e^{10x}$

$\dfrac{e^{6x+2}}{e^4} = e^{10x}$

$e^{6x-2} = e^{10x}$

$6x - 2 = 10x$

$-2 = 4x$

$-\dfrac{1}{2} = x$

The solution is $-\dfrac{1}{2}$.

83. $\dfrac{\sqrt{(e^{2x} \cdot e^{-5x})^{-4}}}{e^x \div e^{-x}} = e^7$

$\dfrac{\sqrt{e^{12x}}}{e^{x-(-x)}} = e^7$

$\dfrac{e^{6x}}{e^{2x}} = e^7$

$e^{4x} = e^7$

$4x = 7$

$x = \dfrac{7}{4}$

The solution is $\dfrac{7}{4}$.

84. $e^x < \dfrac{4}{5}$

$\ln e^x < \ln 0.8$

$x < -0.223$

The solution set is $(-\infty, -0.223)$.

85. $|\log_5 x| + 3\log_5 |x| = 4$

Note that we must have $x > 0$. First consider the case when $0 < x < 1$. When $0 < x < 1$, then $\log_5 x < 0$, so $|\log_5 x| = -\log_5 x$ and $|x| = x$. Thus we have:

$-\log_5 x + 3\log_5 x = 4$

$2\log_5 x = 4$

$\log_5 x^2 = 4$

$x^2 = 5^4$

$x = 5^2$

$x = 25$ (Recall that $x > 0$.)

25 cannot be a solution since we assumed $0 < x < 1$.

Now consider the case when $x > 1$. In this case $\log_5 x > 0$, so $|\log_5 x| = \log_5 x$ and $|x| = x$. Thus we have:

$\log_5 x + 3\log_5 x = 4$

$4\log_5 x = 4$

$\log_5 x = 1$

$x = 5$

This answer checks. The solution is 5.

86. $|2^{x^2} - 8| = 3$

$2^{x^2} - 8 = -3$ *or* $2^{x^2} - 8 = 3$

$2^{x^2} = 5$ *or* $2^{x^2} = 11$

$\log 2^{x^2} = \log 5$ *or* $\log 2^{x^2} = \log 11$

$x^2 \log 2 = \log 5$ *or* $x^2 \log 2 = \log 11$

$x^2 = \dfrac{\log 5}{\log 2}$ *or* $x^2 = \dfrac{\log 11}{\log 2}$

$x = \pm 1.524$ *or* $x = \pm 1.860$

The solutions are -1.860, -1.524, 1.524, and 1.860.

87. $a = \log_8 225$, so $8^a = 225 = 15^2$.

$b = \log_2 15$, so $2^b = 15$.

Then $8^a = (2^b)^2$

$(2^3)^a = 2^{2b}$

$2^{3a} = 2^{2b}$

$3a = 2b$

$a = \dfrac{2}{3}b$.

88. $\log_5 125 = 3$ and $\log_{125} 5 = \dfrac{1}{3}$, so $a = (\log_{125} 5)^{\log_5 125}$ is equivalent to $a = \left(\dfrac{1}{3}\right)^3 = \dfrac{1}{27}$. Then $\log_3 a = \log_3 \dfrac{1}{27} = -3$.

89. $\log_2[\log_3(\log_4 x)] = 0$ yields $x = 64$.

$\log_3[\log_2(\log_4 y)] = 0$ yields $y = 16$.

$\log_4[\log_3(\log_2 z)] = 0$ yields $z = 8$.

Then $x + y + z = 64 + 16 + 8 = 88$.

90. $f(x) = e^x - e^{-x}$

Replace $f(x)$ with y: $y = e^x - e^{-x}$

Interchange x and y: $x = e^y - e^{-y}$

Solve for y: $xe^y = e^{2y} - 1$ Multiplying by e^y

$0 = e^{2y} - xe^y - 1$

Using the quadratic formula with $a = 1$, $b = -x$, and $c = -1$ and taking the positive square root (since $e^y > 0$), we get $e^y = \dfrac{x + \sqrt{x^2+4}}{2}$. Then we have

$\ln e^y = \ln\left(\dfrac{x + \sqrt{x^2+4}}{2}\right)$

$y = \ln\left(\dfrac{x + \sqrt{x^2+4}}{2}\right)$

Replace y with $f^{-1}(x)$:

$f^{-1}(x) = \ln\left(\dfrac{x + \sqrt{x^2+4}}{2}\right)$.

Exercise Set 4.6

1. a) Substitute 6.2 for P_0 and 0.012 for k in $P(t) = P_0e^{kt}$. We have:

   $P(t) = 6.2e^{0.012t}$, where $P(t)$ is in billions and t is the number of years after 2001.

   b) In 2005, $t = 2005 - 2001 = 4$.

   $P(4) = 6.2e^{0.012(4)} \approx 6.5$ billion

   In 2010, $t = 2010 - 2001 = 9$.

   $P(9) = 6.2e^{0.012(9)} \approx 6.9$ billion

   c) Substitute 8 for $P(t)$ and solve for t.

   $$8 = 6.2e^{0.012t}$$
   $$\frac{8}{6.2} = e^{0.012t}$$
   $$\ln\frac{8}{6.2} = \ln e^{0.012t}$$
   $$\ln\frac{8}{6.2} = 0.012t$$
   $$\frac{\ln\frac{8}{6.2}}{0.012} = t$$
   $$21.2 \approx t$$

   The world population will be 8 billion about 21.2 yr after 2001.

   d) $T = \frac{\ln 2}{0.012} \approx 57.8$ yr

2. a) $P(t) = 100e^{0.117t}$

   b) $P(7) = 100e^{0.117(7)} \approx 227$

   Note that 2 weeks $= 2 \cdot 7$ days $= 14$ days.

   $P(14) = 100e^{0.117(14)} \approx 514$

   c) $t = \frac{\ln 2}{0.117} \approx 5.9$ days

3. a) $T = \frac{\ln 2}{0.025} \approx 27.7$ yr

   b) $k = \frac{\ln 2}{693} \approx 0.1\%$ per yr

   c) $T = \frac{\ln 2}{0.033} \approx 21.0$ yr

   d) $T = \frac{\ln 2}{0.002} \approx 346.6$ yr

   e) $k = \frac{\ln 2}{46.2} \approx 1.5\%$ per yr

   f) $k = \frac{\ln 2}{36.5} \approx 1.9\%$ per yr

   g) $k = \frac{\ln 2}{0.009} \approx 77.0$ yr

   h) $k = \frac{\ln 2}{99.0} \approx 0.7\%$ per yr

   i) $k = \frac{\ln 2}{0.011} \approx 63.0$ yr

   j) $k = \frac{\ln 2}{0.003} \approx 231.0$ yr

4. a) In 2000, $t = 2000 - 1950 = 50$.

   $$813,869 = 219,997e^{50k}$$
   $$\frac{813,869}{219,997} = e^{50k}$$
   $$\ln\left(\frac{813,869}{219,997}\right) = \ln e^{50k}$$
   $$\ln\left(\frac{813,869}{219,997}\right) = 50k$$
   $$\frac{\ln\left(\frac{813,869}{219,997}\right)}{50} = k$$
   $$0.026 \approx k$$

   $P(t) = 219,997e^{0.026t}$, where t is the number of years after 1950.

   b) In 2003, $t = 2003 - 1950 = 53$.

   $P(53) = 219,997e^{0.026(53)} \approx 872,719$

   In 2006, $t = 2006 - 1950 = 56$.

   $P(56) = 219,997e^{0.026(56)} \approx 943,517$

   In 2012, $t = 2012 - 1950 = 62$.

   $P(62) = 219,997e^{0.026(62)} \approx 1,102,807$

5.
   $$P(t) = P_0e^{kt}$$
   $$24,313,062,400 = 5,938,000e^{0.013t}$$
   $$\frac{24,313,062,400}{5,938,000} = e^{0.013t}$$
   $$\ln\left(\frac{24,313,062,400}{5,938,000}\right) = \ln e^{0.013t}$$
   $$\ln\left(\frac{24,313,062,400}{5,938,000}\right) = 0.013t$$
   $$\frac{\ln\left(\frac{24,313,062,400}{5,938,000}\right)}{0.013} = t$$
   $$640 \approx t$$

   There will be one person for every square yard of land about 640 yr after 2001.

6. In 2005, $t = 2005 - 1626 = 379$.

   $P(379) = 24e^{0.08(379)} \approx \$353,199,380,400,000$, or

   about $350,000,000,000,000

7. a) Substitute 10,000 for P_0 and 5.4%, or 0.054 for k.

   $P(t) = 10,000e^{0.054t}$

   b) $P(1) = 10,000e^{0.054(1)} \approx \$10,555$

   $P(2) = 10,000e^{0.054(2)} \approx \$11,140$

   $P(5) = 10,000e^{0.054(5)} \approx \$13,100$

   $P(10) = 10,000e^{0.054(10)} \approx \$17,160$

   c) $T = \frac{\ln 2}{0.054} \approx 12.8$ yr

8. a) $T = \frac{\ln 2}{0.062} \approx 11.2$ yr

   $P(5) = 35,000e^{0.062(5)} \approx \$47,719.88$

b) $$\begin{aligned} 7130.90 &= 5000e^{5k} \\ 1.4618 &= e^{5k} \\ \ln 1.4618 &= \ln e^{5k} \\ \ln 1.4618 &= 5k \\ \frac{\ln 1.4618}{5} &= k \\ 0.071 &\approx k \\ 7.1\% &\approx k \end{aligned}$$

$$T = \frac{\ln 2}{0.071} \approx 9.8 \text{ yr}$$

c) $$\begin{aligned} 11,414.71 &= P_0 e^{0.084(5)} \\ \frac{11,414.71}{e^{0.084(5)}} &= P_0 \\ \$7500 &\approx P_0 \end{aligned}$$

$$T = \frac{\ln 2}{0.084} \approx 8.3 \text{ yr}$$

d) $k = \dfrac{\ln 2}{11} \approx 0.063$, or 6.3%

$$\begin{aligned} 17,539.32 &= P_0 e^{0.063(5)} \\ \frac{17,539.32}{e^{0.063(5)}} &= P_0 \\ \$12,800 &\approx P_0 \end{aligned}$$

9. We use the function found in Example 5. If the mummy has lost 46% of its carbon-14 from an initial amount P_0, then $54\%P_0$, or $0.54P_0$ remains. We substitute in the function.

$$\begin{aligned} 0.54P_0 &= P_0 e^{-0.00012t} \\ 0.54 &= e^{-0.00012t} \\ \ln 0.54 &= \ln e^{-0.00012t} \\ \ln 0.54 &= -0.00012t \\ \frac{\ln 0.54}{-0.00012} &= t \\ 5135 &\approx t \end{aligned}$$

The mummy is about 5135 years old.

10. $35\%P_0$ of the carbon-14 has been lost, so $65\%P_0$, $0.65P_0$ remains.

$$\begin{aligned} 0.65P_0 &= P_0 e^{-0.00012t} \\ 0.65 &= e^{-0.00012t} \\ \ln 0.65 &= \ln e^{-0.00012t} \\ \ln 0.65 &= -0.00012t \\ \frac{\ln 0.65}{-0.00012} &= t \\ 3590 &\approx t \end{aligned}$$

The statue is about 3590 years old.

11. a) $K = \dfrac{\ln 2}{3} \approx 0.231$, or 23.1% per min

b) $k = \dfrac{\ln 2}{22} \approx 0.0315$, or 3.15% per yr

c) $T = \dfrac{\ln 2}{0.096} \approx 7.2$ days

d) $T = \dfrac{\ln 2}{0.063} \approx 11$ yr

e) $k = \dfrac{\ln 2}{25} \approx 0.028$, or 2.8% per yr

f) $k = \dfrac{\ln 2}{4560} \approx 0.00015$, or 0.015% per yr

g) $k = \dfrac{\ln 2}{23,105} \approx 0.00003$, or 0.003% per yr

12. a) $t = 2001 - 1950 = 51$

$$\begin{aligned} N(t) &= N_0 e^{-kt} \\ 2,158,000 &= 5,388,000e^{-k(51)} \\ \frac{2,158,000}{5,388,000} &= e^{-51k} \\ \ln\left(\frac{2,158,000}{5,388,000}\right) &= \ln e^{-51k} \\ \ln\left(\frac{2,158,000}{5,388,000}\right) &= -51k \\ \frac{\ln\left(\frac{2,158,000}{5,388,000}\right)}{-51} &= k \\ 0.018 &\approx k \end{aligned}$$

$N(t) = 5,388,000e^{-0.018t}$

b) In 2005, $t = 2005 - 1950 = 55$.

$N(55) = 5,388,000e^{-0.018(55)} \approx 2,002,055$

In 2010, $t = 2010 - 1950 = 60$.

$N(60) = 5,388,000e^{-0.018(60)} \approx 1,829,741$

c) $$\begin{aligned} 100,000 &= 5,388,000e^{-0.018t} \\ \frac{100,000}{5,388,000} &= e^{-0.018t} \\ \ln\left(\frac{100,000}{5,388,000}\right) &= \ln e^{-0.018t} \\ \ln\left(\frac{100,000}{5,388,000}\right) &= -0.018t \\ \frac{\ln\left(\frac{100,000}{5,388,000}\right)}{-0.018} &= t \\ 221 &\approx t \end{aligned}$$

Only 100,000 farms will remain about 221 years after 1950, or in 2171.

13. a) Substitute 2002 − 1984, or 18, for t; 3395 for P_0; and 145 for $P(18)$ in $P(t) = P_0 e^{-kt}$ and solve for k.

$$\begin{aligned} 145 &= 3395e^{-18k} \\ \frac{145}{3395} &= e^{-18k} \\ \ln\left(\frac{145}{3395}\right) &= \ln e^{-18k} \\ \ln\left(\frac{145}{3395}\right) &= -18k \\ \frac{\ln\left(\frac{145}{3395}\right)}{-18} &= k \\ 0.175 &\approx k \end{aligned}$$

The desired function is $P(t) = 3395e^{-0.175t}$, where $P(t)$ is in dollars and t is the number of years after 1984.

b) In 2004, $t = 2004 - 1984 = 20$.

$P(20) = 3395e^{-0.175(20)} \approx \$103$

In 2008, $t = 2008 - 1984 = 24$.

$P(24) = 3395e^{-0.175(24)} \approx \$51$

c)
$$39 = 3395e^{-0.175t}$$
$$\frac{39}{3395} = e^{-0.175t}$$
$$\ln\left(\frac{39}{3395}\right) = \ln e^{-0.175t}$$
$$\ln\left(\frac{39}{3395}\right) = -0.175t$$
$$\frac{\ln\left(\frac{39}{3395}\right)}{-0.175} = t$$
$$26 \approx t$$

At this decay rate, the price of a cellphone will be \$39 about 26 yr after 1984, or in 2010.

14. a) $t = 2003 - 1987 = 16$
$$50 = 8e^{k(16)}$$
$$6.25 = e^{16k}$$
$$\ln 6.25 = \ln e^{16k}$$
$$\ln 6.25 = 16k$$
$$\frac{\ln 6.25}{16} = k$$
$$0.1145 \approx k$$

$V(t) = 8e^{0.1145t}$, where t is the number of years after 1987.

b) In 2008, $t = 2008 - 1987 = 21$.

$V(21) = 8e^{0.1145(21)} \approx \$89$

c) $T = \dfrac{\ln 2}{0.1145} \approx 6.1$ yr

d)
$$2000 = 8e^{0.1145t}$$
$$250 = e^{0.1145t}$$
$$\ln 250 = \ln e^{0.1145t}$$
$$\ln 250 = 0.1145t$$
$$\frac{\ln 250}{0.1145} = t$$
$$48 \approx t$$

The value of the card will be \$2000 about 48 yr after 1987, or in 2035.

15. a) $N(0) = \dfrac{3500}{1 + 19.9e^{-0.6(0)}} \approx 167$

b) $N(2) = \dfrac{3500}{1 + 19.9e^{-0.6(2)}} \approx 500$

$N(5) = \dfrac{3500}{1 + 19.9e^{-0.6(5)}} \approx 1758$

$N(8) = \dfrac{3500}{1 + 19.9e^{-0.6(8)}} \approx 3007$

$N(12) = \dfrac{3500}{1 + 19.9e^{-0.6(12)}} \approx 3449$

$N(16) = \dfrac{3500}{1 + 19.9e^{-0.6(16)}} \approx 3495$

c) As $t \to \infty$, $N(t) \to 3500$; the number of people infected approaches 3500 but never actually reaches it.

16. $P(0) = \dfrac{2500}{1 + 5.25e^{-0.32(0)}} = 400$

$P(1) = \dfrac{2500}{1 + 5.25e^{-0.32(1)}} \approx 520$

$P(5) = \dfrac{2500}{1 + 5.25e^{-0.32(5)}} \approx 1214$

$P(10) = \dfrac{2500}{1 + 5.25e^{-0.32(10)}} \approx 2059$

$P(15) = \dfrac{2500}{1 + 5.25e^{-0.32(15)}} \approx 2396$

$P(20) = \dfrac{2500}{1 + 5.25e^{-0.32(20)}} \approx 2478$

17. To find k we substitute 105 for T_1, 0 for T_0, 5 for t, and 70 for $T(t)$ and solve for k.
$$70 = 0 + (105 - 0)e^{-5k}$$
$$70 = 105e^{-5k}$$
$$\frac{70}{105} = e^{-5k}$$
$$\ln\frac{70}{105} = \ln e^{-5k}$$
$$\ln\frac{70}{105} = -5k$$
$$\frac{\ln\frac{70}{105}}{-5} = k$$
$$0.081 \approx k$$

The function is $T(t) = 105e^{-0.081t}$.

Now we find $T(10)$.

$T(10) = 105e^{-0.081(10)} \approx 46.7$ °F

18. To find k we substitute 375 for T_1, 72 for T_0, 6 for t, and 325 for $T(t)$.

$$325 = 72 + (375 - 72)e^{-6k}$$
$$253 = 303e^{-6k}$$
$$\frac{253}{303} = e^{-6k}$$
$$\ln\frac{253}{303} = \ln e^{-6k}$$
$$\ln\frac{253}{303} = -6k$$
$$\frac{\ln\frac{253}{303}}{-6} = k$$
$$0.03 \approx k$$

The function is $T(t) = 72 + 303e^{-0.03t}$.

$T(15) = 72 + 303e^{-0.03(15)} \approx 265°\text{F}$

19. To find k we substitute 43 for T_1, 68 for T_0, 12 for t, and 55 for $T(t)$ and solve for k.

$$55 = 68 + (43 - 68)e^{-12k}$$
$$-13 = -25e^{-12k}$$
$$0.52 = e^{-12k}$$
$$\ln 0.52 = \ln e^{-12k}$$
$$\ln 0.52 = -12k$$
$$0.0545 \approx k$$

The function is $T(t) = 68 - 25e^{-0.0545t}$.

Now we find $T(20)$.

$T(20) = 68 - 25e^{-0.0545(20)} \approx 59.6°\text{F}$

20. To find k we substitute 94.6 for T_1, 70 for T_0, 60 for t (1 hr = 60 min), and 93.4 for $T(t)$.

$$93.4 = 70 + |94.6 - 70|e^{-k(60)}$$
$$23.4 = 24.6e^{-60k}$$
$$\frac{23.4}{24.6} = e^{-60k}$$
$$\ln\frac{23.4}{24.6} = \ln e^{-60k}$$
$$\ln\frac{23.4}{24.6} = -60k$$
$$k = \frac{\ln\frac{23.4}{24.6}}{-60} \approx 0.0008$$

The function is $T(t) = 70 + 24.6e^{-0.0008t}$.

We substitute 98.6 for $T(t)$ and solve for t.

$$98.6 = 70 + 24.6e^{-0.0008t}$$
$$28.6 = 24.6e^{-0.0008t}$$
$$\frac{28.6}{24.6} = e^{-0.0008t}$$
$$\ln\frac{28.6}{24.6} = \ln e^{-0.0008t}$$
$$\ln\frac{28.6}{24.6} = -0.0008t$$
$$t = \frac{\ln\frac{28.6}{24.6}}{-0.0008} \approx -188$$

The murder was committed at approximately 188 minutes, or about 3 hours, before 12:00 PM, or at about 9:00 AM. (Answers may vary slightly due to rounding differences.)

21. a) $y = 4.195491964(1.025306189)^x$

We can convert this equation to an equation with base e, if desired.

$$y = 4.195491964e^{x(\ln 1.025306189)}$$
$$= 4.195491964e^{0.024991289x}$$

In each case x is the number of years after 2000 and y is in millions.

The coefficient of correlation r is approximately 0.9954. Since this is close to 1, the function is a good fit.

b)

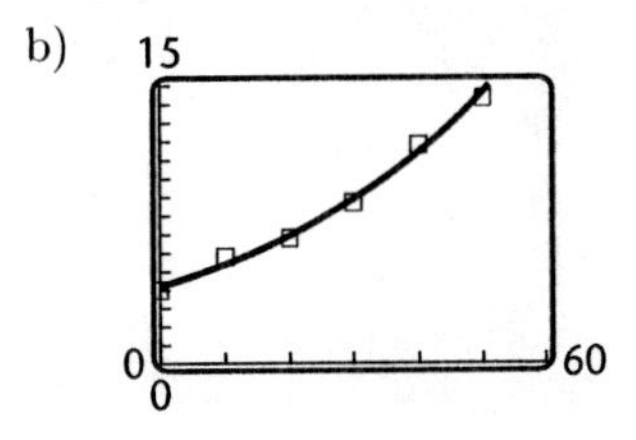

c) In 2005, $t = 2005 - 2000 = 5$.

$y = 4.195491964(1.025306189)^5 \approx 4.8$ million

In 2025, $t = 2025 - 2000 = 25$.

$y = 4.195491964(1.025306189)^{25} \approx 7.8$ million

In 2100, $t = 2100 - 2000 = 100$.

$y = 4.195491964(1.025306189)^{100} \approx 51.1$ million

22. a) Using the logarithmic regression feature on a graphing calculator, we get $y = 84.94353992 - 0.5412834098 \ln x$.

b) For $x = 8$, $y \approx 83.8\%$.

For $x = 10$, $y \approx 83.7\%$.

For $x = 24$, $y \approx 83.2\%$.

For $x = 36$, $y \approx 83.0\%$.

c) $82 = 84.94353992 - 0.5412834098 \ln x$

Graph $y_1 = 84.94353992 - 0.5412834098 \ln x$ and $y_2 = 82$ and find the first coordinate of the point of intersection of the graphs. It is approximately 230, so test scores will fall below 82% after about 230 months.

23. a)

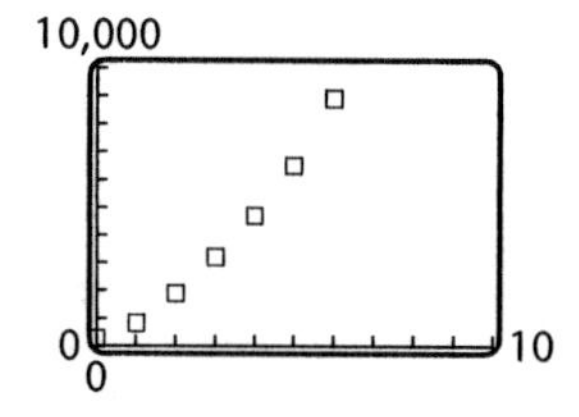

b) Linear: $y = 1429.214286x - 530.0714286$; $r^2 \approx 0.9641$

Quadratic: $y = 158.0952381x^2 + 480.6428571x + 260.4047619$; $R^2 \approx 0.9995$

Exponential: $y = 445.8787388(1.736315606)^x$; $r^2 \approx 0.9361$

The value of R^2 is highest for the quadratic function, so we determine that this function fits the data best.

c)

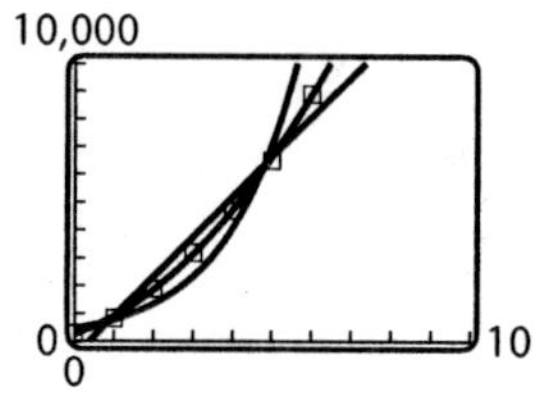

d) In 2010, $x = 2010 - 1996 = 14$.

For the linear function, when $x = 14$, $y \approx \$19,479$ million, or \$19.479 billion.

For the quadratic function, when $x = 14$, $y \approx \$37,976$ million, or \$37.976 billion.

For the exponential function, when $x = 14$, $y \approx \$1,009,295$ million, or \$1009.295 billion.

Given the rate at which the revenues in the table are increasing, it appears that the quadratic function provides the most realistic prediction. Answers may vary.

24. a) Using the exponential regression feature on a graphing calculator, we get $y=22.28866067(1.092917808)^x$, where x is the number of years after 1980 and y is in millions of dollars.

b) In 2007, $x = 2007 - 1980 = 27$.

$y = 22.28866067(1.092917808)^{27} \approx \$245.4$ million

In 2010, $x = 2010 - 1980 = 30$.

$y = 22.28866067(1.092917808)^{30} \approx \$320.4$ million

c) Graph $y_1 = 22.28866067(1.092917808)^x$ and $y_2 = 400$ and find the first coordinate of the point of intersection of the graphs. It is approximately 32, so the total cost will first exceed \$400 million for the first convention year that is more than 32 yr after 1980, or after 2012. The first convention year after 2012 is 2016 so this is the answer.

25. a) Using the exponential regression feature on a graphing calculator, we get $y=21.13877717(1.038467095)^x$, where x is the number of years after 1980 and y is a percent.

b) In 2005, $x = 2005 - 1980 = 25$.

$y = 21.13877717(1.038467095)^{25} \approx 54.3\%$

In 2010, $x = 2010 - 1980 = 30$.

$y = 21.13877717(1.038467095)^{30} \approx 65.6\%$

c) Graph $y_1 = 21.13877717(1.038467095)^x$ and $y_2 = 70$ and find the first coordinate of the point of intersection of the graphs. It is approximately 31.7, so the percentage will first exceed 70% 32 yr after 1980, or in 2012.

26. a) Using the logistic regression feature on a graphing calculator, we get

$$y = \frac{99.98884912}{1 + 489.2438401e^{-0.1299899024x}}.$$

b) For $x = 55$, $y \approx 72.2\%$.

For $x = 100$, $y \approx 99.9\%$.

c) $y = 100$ is the horizontal asymptote; as more and more ads are run, the percent of people who bought the product approaches 100%.

27. Answers will vary.

28. Measure the atmospheric pressure P at the top of the building. Substitute that value in the equation $P = 14.7e^{-0.00005a}$, and solve for the height, or altitude, a, of the top of the building. Also measure the atmospheric pressure at the base of the building and solve for the altitude of the base. Then subtract to find the height of the building.

29. Multiplication principle for inequalities

30. Product rule

31. Principle of zero products

32. Principle of square roots

33. Power rule

34. Multiplication principle for equations

35.
$$\begin{aligned} P(t) &= P_0e^{kt} \\ 50,000 &= P_0e^{0.07(18)} \\ \frac{50,000}{e^{0.07(18)}} &= P_0 \\ \$14,182.70 &\approx P_0 \end{aligned}$$

36. a)
$$\begin{aligned} P &= P_0e^{kt} \\ \frac{P}{e^{kt}} &= P_0, \text{ or} \\ Pe^{-kt} &= P_0 \end{aligned}$$

b) $P_0 = 50,000e^{-0.064(18)} \approx \$15,800.21$

37. $$\begin{aligned} 480e^{-0.003p} &= 150e^{0.004p} \\ \frac{480}{150} &= \frac{e^{0.004p}}{e^{-0.003p}} \\ 3.2 &= e^{0.007p} \\ \ln 3.2 &= \ln e^{0.007p} \\ \ln 3.2 &= 0.007p \\ \frac{\ln 3.2}{0.007} &= p \\ \$166.16 &\approx p \end{aligned}$$

38. $$\begin{aligned} P(4000) &= P_0 e^{-0.00012(4000)} \\ &= 0.619P_0, \text{ or } 61.9\%P_0 \end{aligned}$$

Thus, about 61.9% of the carbon-14 remains, so about 38.1% has been lost.

39. $$\begin{aligned} i &= \frac{V}{R}\left[1 - e^{-(R/L)t}\right] \\ \frac{iR}{V} &= 1 - e^{-(R/L)t} \\ e^{-(R/L)t} &= 1 - \frac{iR}{V} \\ \ln e^{-(R/L)t} &= \ln\left(1 - \frac{iR}{V}\right) \\ -\frac{R}{L}t &= \ln\left(1 - \frac{iR}{V}\right) \\ t &= -\frac{L}{R}\left[\ln\left(1 - \frac{iR}{V}\right)\right] \end{aligned}$$

40. a) At 1 m: $I = I_0 e^{-1.4(1)} \approx 0.247I_0$

24.7% of I_0 remains.

At 3 m: $I = I_0 e^{-1.4(3)} \approx 0.015I_0$

1.5% of I_0 remains.

At 5 m: $I = I_0 e^{-1.4(5)} \approx 0.0009I_0$

0.09% of I_0 remains.

At 50 m: $I = I_0 e^{-1.4(50)} \approx (3.98 \times 10^{-31})I_0$

Now, $3.98 \times 10^{-31} = (3.98 \times 10^{-29}) \times 10^{-2}$, so

$(3.98 \times 10^{-29})\%$ remains.

b) $I = I_0 e^{-1.4(10)} \approx 0.0000008I_0$

Thus, 0.00008% remains.

41. $$\begin{aligned} y &= ae^x \\ \ln y &= \ln(ae^x) \\ \ln y &= \ln a + \ln e^x \\ \ln y &= \ln a + x \\ Y &= x + \ln a \end{aligned}$$

This function is of the form $y = mx + b$, so it is linear.

42. $$\begin{aligned} y &= ax^b \\ \ln y &= \ln(ax^b) \\ \ln y &= \ln a + b \ln x \\ Y &= \ln a + bX \end{aligned}$$

This function is of the form $y = mx + b$, so it is linear.

Chapter 5

The Trigonometric Functions

Exercise Set 5.1

1. We use the definitions.

$$\sin\phi = \frac{\text{opp}}{\text{hyp}} = \frac{15}{17}$$

$$\cos\phi = \frac{\text{adj}}{\text{hyp}} = \frac{8}{17}$$

$$\tan\phi = \frac{\text{opp}}{\text{adj}} = \frac{15}{8}$$

$$\csc\phi = \frac{\text{hyp}}{\text{opp}} = \frac{17}{15}$$

$$\sec\phi = \frac{\text{hyp}}{\text{adj}} = \frac{17}{8}$$

$$\cot\phi = \frac{\text{adj}}{\text{opp}} = \frac{8}{15}$$

2. $\sin\beta = \frac{\text{opp}}{\text{hyp}} = \frac{0.3}{0.5} = \frac{3}{5}$

$$\cos\beta = \frac{\text{adj}}{\text{hyp}} = \frac{0.4}{0.5} = \frac{4}{5}$$

$$\tan\beta = \frac{\text{opp}}{\text{adj}} = \frac{0.3}{0.4} = \frac{3}{4}$$

$$\csc\beta = \frac{\text{hyp}}{\text{opp}} = \frac{0.5}{0.3} = \frac{5}{3}$$

$$\sec\beta = \frac{\text{hyp}}{\text{adj}} = \frac{0.5}{0.4} = \frac{5}{4}$$

$$\cot\beta = \frac{\text{adj}}{\text{opp}} = \frac{0.4}{0.3} = \frac{4}{3}$$

3. We use the definitions.

$$\sin\alpha = \frac{\text{opp}}{\text{hyp}} = \frac{3\sqrt{3}}{6} = \frac{\sqrt{3}}{2}$$

$$\cos\alpha = \frac{\text{adj}}{\text{hyp}} = \frac{3}{6} = \frac{1}{2}$$

$$\tan\alpha = \frac{\text{opp}}{\text{adj}} = \frac{3\sqrt{3}}{3} = \sqrt{3}$$

$$\csc\alpha = \frac{\text{hyp}}{\text{opp}} = \frac{6}{3\sqrt{3}} = \frac{2}{\sqrt{3}}, \text{ or } \frac{2\sqrt{3}}{3}$$

$$\sec\alpha = \frac{\text{hyp}}{\text{adj}} = \frac{6}{3} = 2$$

$$\cot\alpha = \frac{\text{adj}}{\text{opp}} = \frac{3}{3\sqrt{3}} = \frac{1}{\sqrt{3}}, \text{ or } \frac{\sqrt{3}}{3}$$

4. $\sin\theta = \frac{\text{opp}}{\text{hyp}} = \frac{6}{\frac{13}{2}} = \frac{12}{13}$

$$\cos\theta = \frac{\text{adj}}{\text{hyp}} = \frac{\frac{5}{2}}{\frac{13}{2}} = \frac{5}{13}$$

$$\tan\theta = \frac{\text{opp}}{\text{adj}} = \frac{6}{\frac{5}{2}} = \frac{12}{5}$$

$$\csc\theta = \frac{\text{hyp}}{\text{opp}} = \frac{\frac{13}{2}}{6} = \frac{13}{12}$$

$$\sec\theta = \frac{\text{hyp}}{\text{adj}} = \frac{\frac{13}{2}}{\frac{5}{2}} = \frac{13}{5}$$

$$\cot\theta = \frac{\text{adj}}{\text{opp}} = \frac{\frac{5}{2}}{6} = \frac{5}{12}$$

5. First we use the Pythagorean theorem to find the length of the hypotenuse, c.

$$a^2 + b^2 = c^2$$
$$4^2 + 7^2 = c^2$$
$$65 = c^2$$
$$\sqrt{65} = c$$

Then we use the definitions to find the trigonometric function values of ϕ.

$$\sin\phi = \frac{\text{opp}}{\text{hyp}} = \frac{7}{\sqrt{65}}, \text{ or } \frac{7\sqrt{65}}{65}$$

$$\cos\phi = \frac{\text{adj}}{\text{hyp}} = \frac{4}{\sqrt{65}}, \text{ or } \frac{4\sqrt{65}}{65}$$

$$\tan\phi = \frac{\text{opp}}{\text{adj}} = \frac{7}{4}$$

$$\csc\phi = \frac{\text{hyp}}{\text{opp}} = \frac{\sqrt{65}}{7}$$

$$\sec\phi = \frac{\text{hyp}}{\text{adj}} = \frac{\sqrt{65}}{4}$$

$$\cot\phi = \frac{\text{adj}}{\text{opp}} = \frac{4}{7}$$

6. First we use the Pythagorean theorem to find the length of the side opposite θ, a.

$$a^2 + b^2 = c^2$$
$$a^2 + (8.2)^2 = 9^2$$
$$a^2 = 13.76$$
$$a = \sqrt{13.76} = 4\sqrt{0.86} \approx 3.7$$

Now we find the trigonometric function values of θ.

$$\sin\theta = \frac{\text{opp}}{\text{hyp}} = \frac{\sqrt{13.76}}{9} \approx \frac{3.7}{9} \approx 0.4111$$

$$\cos\theta = \frac{\text{adj}}{\text{hyp}} = \frac{8.2}{9} \approx 0.9111$$

$$\tan\theta = \frac{\text{opp}}{\text{adj}} = \frac{\sqrt{13.76}}{8.2} \approx \frac{3.7}{8.2} \approx 0.4512$$

$$\csc\theta = \frac{\text{hyp}}{\text{opp}} = \frac{9}{\sqrt{13.76}} \approx \frac{9}{3.7} \approx 2.4324$$

$$\sec\theta = \frac{\text{hyp}}{\text{adj}} = \frac{9}{8.2} \approx 1.0976$$

$$\cot\theta = \frac{\text{adj}}{\text{opp}} = \frac{8.2}{\sqrt{13.76}} \approx \frac{8.2}{3.7} \approx 2.2162$$

7. $\csc\alpha = \dfrac{1}{\sin\alpha} = \dfrac{1}{\frac{\sqrt{5}}{3}} = \dfrac{3}{\sqrt{5}}$, or $\dfrac{3}{\sqrt{5}} \cdot \dfrac{\sqrt{5}}{\sqrt{5}} = \dfrac{3\sqrt{5}}{5}$

$$\sec\alpha = \frac{1}{\cos\alpha} = \frac{1}{\frac{2}{3}} = \frac{3}{2}$$

$$\cot\alpha = \frac{1}{\tan\alpha} = \frac{1}{\frac{\sqrt{5}}{2}} = \frac{2}{\sqrt{5}}, \text{ or } \frac{2}{\sqrt{5}} \cdot \frac{\sqrt{5}}{\sqrt{5}} = \frac{2\sqrt{5}}{5}$$

8. $\csc\beta = \dfrac{1}{\sin\beta} = \dfrac{1}{\frac{2\sqrt{2}}{3}} = \dfrac{3}{2\sqrt{2}}$, or $\dfrac{3}{2\sqrt{2}} \cdot \dfrac{\sqrt{2}}{\sqrt{2}} = \dfrac{3\sqrt{2}}{4}$

$$\sec\beta = \frac{1}{\cos\beta} = \frac{1}{\frac{1}{3}} = 3$$

$$\cot\beta = \frac{1}{\tan\beta} = \frac{1}{2\sqrt{2}}, \text{ or } \frac{1}{2\sqrt{2}} \cdot \frac{\sqrt{2}}{\sqrt{2}} = \frac{\sqrt{2}}{4}$$

9. We know from the definition of the sine function that the ratio $\frac{24}{25}$ is $\frac{\text{opp}}{\text{hyp}}$. Let's consider a right triangle in which the hypotenuse has length 25 and the side opposite θ has length 24.

Use the Pythagorean theorem to find the length of the side adjacent to θ.

$$a^2 + b^2 = c^2$$
$$a^2 + 24^2 = 25^2$$
$$a^2 = 625 - 576 = 49$$
$$a = 7$$

Use the lengths of the three sides to find the other five ratios.

$\cos\theta = \dfrac{7}{25}$, $\tan\theta = \dfrac{24}{7}$, $\csc\theta = \dfrac{25}{24}$, $\sec\theta = \dfrac{25}{7}$,

$\cot\theta = \dfrac{7}{24}$

10. $\cos\sigma = 0.7 = \dfrac{7}{10}$

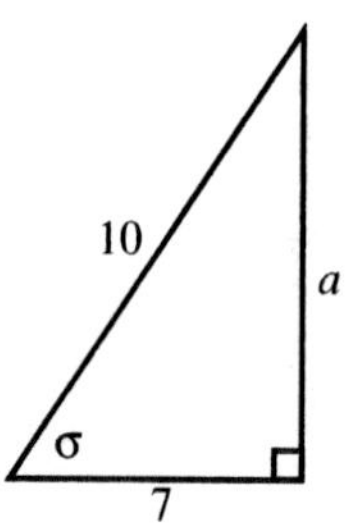

$$a^2 = 10^2 - 7^2 = 51$$
$$a = \sqrt{51}$$

$\sin\sigma = \dfrac{\sqrt{51}}{10}$; $\tan\sigma = \dfrac{\sqrt{51}}{7}$,

$\csc\sigma = \dfrac{10}{\sqrt{51}}$, or $\dfrac{10\sqrt{51}}{51}$,

$\sec\sigma = \dfrac{10}{7}$; $\cot\sigma = \dfrac{7}{\sqrt{51}}$, or $\dfrac{7\sqrt{51}}{51}$

11. We know from the definition of the tangent function that 2, or the ratio $\frac{2}{1}$ is $\frac{\text{opp}}{\text{adj}}$. Let's consider a right triangle in which the side opposite ϕ has length 2 and the side adjacent to ϕ has length 1.

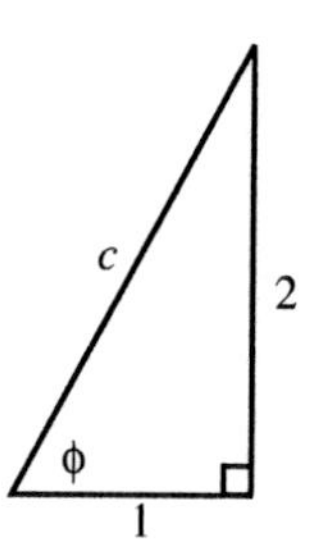

Use the Pythagorean theorem to find the length of the hypotenuse.

$$a^2 + b^2 = c^2$$
$$1^2 + 2^2 = c^2$$
$$1 + 4 = c^2$$
$$5 = c^2$$
$$\sqrt{5} = c$$

Use the lengths of the three sides to find the other five ratios.

$\sin\phi = \dfrac{2}{\sqrt{5}}$, or $\dfrac{2\sqrt{5}}{5}$; $\cos\phi = \dfrac{1}{\sqrt{5}}$, or $\dfrac{\sqrt{5}}{5}$;

$\csc\phi = \dfrac{\sqrt{5}}{2}$; $\sec\phi = \sqrt{5}$; $\cot\phi = \dfrac{1}{2}$

12.

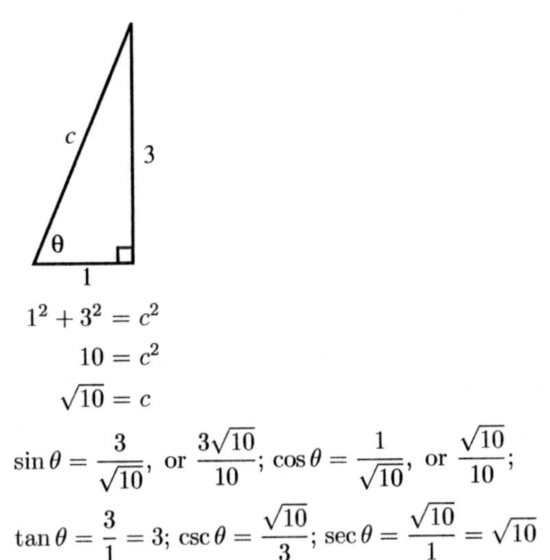

$$1^2 + 3^2 = c^2$$
$$10 = c^2$$
$$\sqrt{10} = c$$

$\sin\theta = \frac{3}{\sqrt{10}}$, or $\frac{3\sqrt{10}}{10}$; $\cos\theta = \frac{1}{\sqrt{10}}$, or $\frac{\sqrt{10}}{10}$;

$\tan\theta = \frac{3}{1} = 3$; $\csc\theta = \frac{\sqrt{10}}{3}$; $\sec\theta = \frac{\sqrt{10}}{1} = \sqrt{10}$

13. $\csc\theta = 1.5 = \frac{1.5}{1} = \frac{15}{10} = \frac{3}{2}$

We know from the definition of the cosecant function that the ratio $\frac{3}{2}$ is $\frac{\text{hyp}}{\text{opp}}$. Let's consider a right triangle in which the hypotenuse has length 3 and the side opposite θ has length 2.

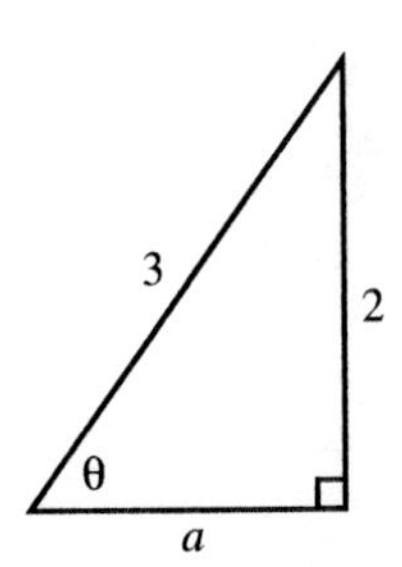

Use the Pythagorean theorem to find the length of the side adjacent to θ.

$$a^2 + b^2 = c^2$$
$$a^2 + 2^2 = 3^2$$
$$a^2 = 9 - 4 = 5$$
$$a = \sqrt{5}$$

Use the lengths of the three sides to find the other five ratios.

$\sin\theta = \frac{2}{3}$; $\cos\theta = \frac{\sqrt{5}}{3}$; $\tan\theta = \frac{2}{\sqrt{5}}$, or $\frac{2\sqrt{5}}{5}$;

$\sec\theta = \frac{3}{\sqrt{5}}$, or $\frac{3\sqrt{5}}{5}$; $\cot\theta = \frac{\sqrt{5}}{2}$

14.

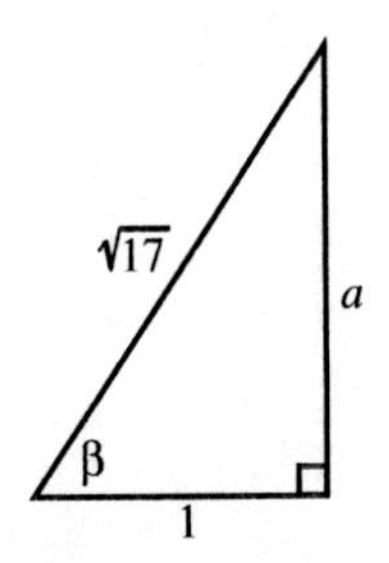

$$a^2 = (\sqrt{17})^2 - 1^2 = 16$$
$$a = 4$$

$\sin\beta = \frac{4}{\sqrt{17}}$, or $\frac{4\sqrt{17}}{17}$; $\cos\beta = \frac{1}{\sqrt{17}}$, or $\frac{\sqrt{17}}{17}$;

$\tan\beta = 4$; $\csc\beta = \frac{\sqrt{17}}{4}$; $\cot\beta = \frac{1}{4}$

15. We know from the definition of the cosine function that the ratio of $\frac{\sqrt{5}}{5}$ is $\frac{\text{adj}}{\text{hyp}}$. Let's consider a right triangle in which the side adjacent to β has length $\sqrt{5}$ and the hypotenuse has length 5.

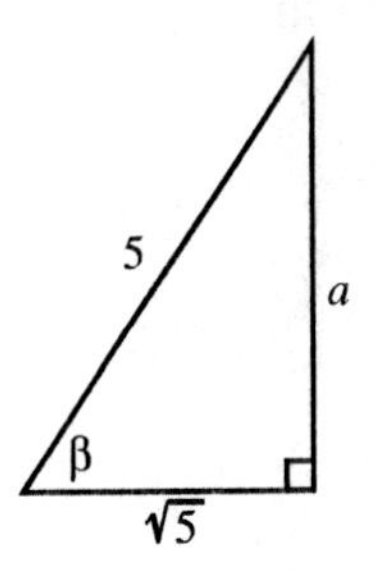

Use the Pythagorean theorem to find the length of the side opposite β.

$$a^2 + b^2 = c^2$$
$$a^2 + (\sqrt{5})^2 = 5^2$$
$$a^2 + 5 = 25$$
$$a^2 = 25 - 5 = 20$$
$$a = \sqrt{20} = 2\sqrt{5}$$

Use the lengths of the three sides to find the other five ratios.

$\sin\beta = \frac{2\sqrt{5}}{5}$; $\tan\beta = \frac{2\sqrt{5}}{\sqrt{5}} = 2$; $\csc\beta = \frac{5}{2\sqrt{5}}$, or $\frac{\sqrt{5}}{2}$;

$\sec\beta = \frac{5}{\sqrt{5}}$, or $\sqrt{5}$; $\cot\beta = \frac{\sqrt{5}}{2\sqrt{5}} = \frac{1}{2}$

16.

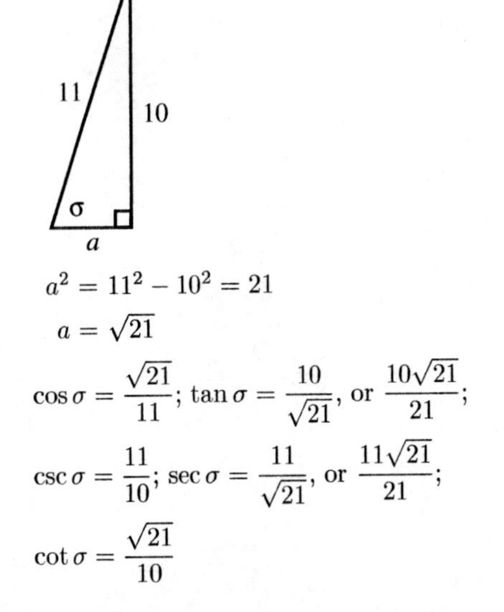

$$a^2 = 11^2 - 10^2 = 21$$
$$a = \sqrt{21}$$

$\cos\sigma = \frac{\sqrt{21}}{11}$; $\tan\sigma = \frac{10}{\sqrt{21}}$, or $\frac{10\sqrt{21}}{21}$;

$\csc\sigma = \frac{11}{10}$; $\sec\sigma = \frac{11}{\sqrt{21}}$, or $\frac{11\sqrt{21}}{21}$;

$\cot\sigma = \frac{\sqrt{21}}{10}$

17.

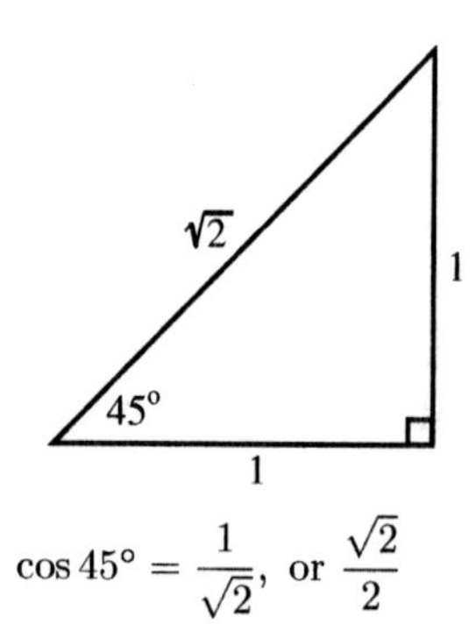

$\cos 45° = \dfrac{1}{\sqrt{2}}$, or $\dfrac{\sqrt{2}}{2}$

18. See the triangle in Exercise 19.

$\tan 30° = \dfrac{1}{\sqrt{3}}$, or $\dfrac{\sqrt{3}}{3}$

19.

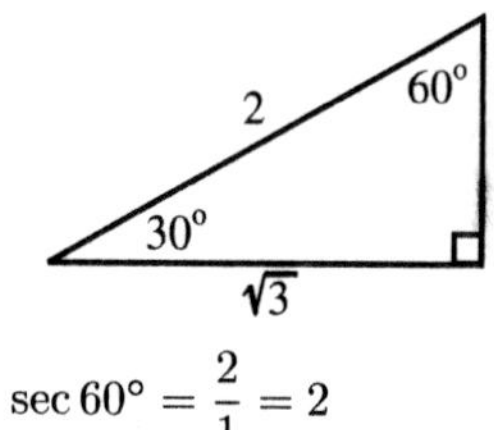

$\sec 60° = \dfrac{2}{1} = 2$

20. See the triangle in Exercise 17.

$\sin 45° = \dfrac{1}{\sqrt{2}}$, or $\dfrac{\sqrt{2}}{2}$

21. See the triangle in Exercise 19.

$\cot 60° = \dfrac{1}{\sqrt{3}}$, or $\dfrac{\sqrt{3}}{3}$

22. See the triangle in Exercise 17.

$\csc 45° = \dfrac{\sqrt{2}}{1} = \sqrt{2}$

23. See the triangle in Exercise 19.

$\sin 30° = \dfrac{1}{2}$

24. See the triangle in Exercise 19.

$\cos 60° = \dfrac{1}{2}$

25. See the triangle in Exercise 17.

$\tan 45° = 1$

26. See the triangle in Exercise 19.

$\sec 30° = \dfrac{2}{\sqrt{3}}$, or $\dfrac{2\sqrt{3}}{3}$

27. See the triangle in Exercise 19.

$\csc 30° = 2$

28. See the triangle in Exercise 19.

$\cot 60° = \dfrac{1}{\sqrt{3}}$, or $\dfrac{\sqrt{3}}{3}$

29. We know the measure of an acute angle of a right triangle and the length of the side opposite the angle, and we want to find the length of the adjacent side. We can use the tangent ratio or the cotangent ratio. Here we use the cotangent:

$$\begin{aligned}\cot 30° &= \frac{a}{36}\\ 36\cot 30° &= a\\ 36\sqrt{3} &= a\\ 62.4 \text{ m} &\approx a\end{aligned}$$

30. Since we know the measure of an acute angle of a right triangle and the lengths of the sides opposite and adjacent to the angle, we can use the sine, cosine, cosecant, or secant ratio to determine the length of the hypotenuse. We will use the cosecant function.

$$\begin{aligned}\csc 45° &= \frac{\text{hyp}}{\text{opp}} = \frac{h}{90}\\ 90\csc 45° &= h\\ 90\sqrt{2} &= h\\ 127.3 &\approx h\end{aligned}$$

The distance from third base to first base is about 127.3 ft.

31. Using a calculator, enter $9° 43'$. The result is $9° 43' \approx 9.72°$.

32. $52° 15' = 52.25°$

33. Using a calculator, enter $35° 50''$. The result is $35° 50'' \approx 35.01°$.

34. $64° 53' \approx 64.88°$

35. Using a calculator, enter $3° 2'$. The result is $3° 2' \approx 3.03°$.

36. $19° 47' 23'' \approx 19.79°$

37. Using a calculator, enter $49° 38' 46''$. The result is $49° 38' 46'' \approx 49.65°$.

38. $76° 11' 34'' \approx 76.19°$

39. Using a calculator, enter $0° 15' 5''$. The result is $15' 5'' \approx 0.25°$.

40. Using a calculator, enter $68° 0' 2''$. The result is $68° 2'' \approx 68.00°$.

41. Using a calculator, enter $5° 0' 53''$. The result is $5° 53'' \approx 5.01°$.

42. Using a calculator, enter $0° 44' 10''$. The result is $44' 10'' \approx 0.74°$.

43. Enter 17.6° on a calculator and use the DMS feature:

$17.6° = 17° 36'$

44. $20.14° = 20° 8' 24''$

45. Enter 83.025° on a calculator and use the DMS feature:

$83.025° = 83° 1' 30''$

46. $67.84° = 67° 50' 24''$

47. Enter 11.75° on a calculator and use the DMS feature:

$11.75° = 11° \, 45'$

48. $29.8° = 29° \, 48'$

49. Enter 47.8268° on a calculator and use the DMS feature:

$47.8268° \approx 47° \, 49' \, 36''$

50. $0.253° \approx 0° \, 15' \, 11''$, or $15' \, 11''$

51. Enter 0.9° on a calculator and use the DMS feature:

$0.9° = 0° \, 54' \, 0''$, or $54'$

52. $30.2505° \approx 30° \, 15' \, 2''$

53. Enter 39.45° on a calculator and use the DMS feature:

$39.45° = 39° \, 27'$

54. $2.4° = 2°24'$

55. Use a calculator set in degree mode.

$\cos 51° \approx 0.6293$

56. $\cot 17° \approx 3.2709$

57. Use a calculator set in degree mode.

$\tan 4°13' \approx 0.0737$

58. $\sin 26.1° \approx 0.4399$

59. Use a calculator set in degree mode. We find the reciprocal of cos 38.43° by entering $\dfrac{1}{\cos 38.43}$ or $(\cos 38.43)^{-1}$.

$\sec 38.43° \approx 1.2765$

60. $\cos 74°10' \, 40'' \approx 0.2727$

61. Use a calculator set in degree mode.

$\cos 40.35° \approx 0.7621$

62. $\csc 45.2° = \dfrac{1}{\sin 45.2°} \approx 1.4093$

63. Use a calculator set in degree mode.

$\sin 69° \approx 0.9336$

64. $\tan 63°48' \approx 2.0323$

65. Use a calculator set in degree mode.

$\tan 85.4° \approx 12.4288$

66. $\cos 4° \approx 0.9976$

67. Use a calculator set in degree mode. We find the reciprocal of sin 89.5° by entering $\dfrac{1}{\sin 89.5}$ or $(\sin 89.5)^{-1}$.

$\csc 89.5° \approx 1.0000$

68. $\sec 35.28° = \dfrac{1}{\cos 35.28°} \approx 1.2250$

69. Use a calculator set in degree mode. We find the reciprocal of $\tan 30°25' \, 6''$ by entering $\dfrac{1}{\tan 30' \, 25' \, 6'}$ or $(\tan 30' \, 25' \, 6')^{-1}$.

$\cot 30°25' \, 6'' \approx 1.7032$

70. $\sin 59.2° \approx 0.8590$

71. On a graphing calculator, press [2nd] [SIN] .5125 [ENTER].

$\theta = 30.8°$

72. $\tan \theta = 2.032$, so $\theta \approx 63.8°$.

73. $\tan \theta = 0.2226$

On a graphing calculator, press [2nd] [TAN] .2226 [ENTER].

$\theta = 12.5°$

74. $\cos \theta = 0.3842$, so $\theta \approx 67.4°$.

75. $\sin \theta = 0.9022$

On a graphing calculator, press [2nd] [SIN] .9022 [ENTER].

$\theta = 64.4°$

76. $\tan \theta = 3.056$, so $\theta \approx 71.9°$.

77. $\cos \theta = 0.6879$

On a graphing calculator, press [2nd] [COS] .6879 [ENTER].

$\theta = 46.5°$

78. $\sin \theta = 0.4005$, so $\theta \approx 23.6°$.

79. $\cot \theta = \dfrac{1}{\tan \theta} = 2.127$

Thus, $\tan \theta = \dfrac{1}{2.127}$.

On a graphing calculator, press [2nd] [TAN] $(1 \div 2.127)$ [ENTER].

$\theta \approx 25.2°$

80. $\csc \theta = \dfrac{1}{\sin \theta} = 1.147$

Thus, $\sin \theta = \dfrac{1}{1.147}$, so $\theta \approx 60.7°$.

81. $\sec \theta = \dfrac{1}{\cos \theta} = 1.279$

Thus, $\cos \theta = \dfrac{1}{1.279}$.

On a graphing calculator, press [2nd] [COS] $(1 \div 1.279)$ [ENTER].

$\theta \approx 38.6°$

82. $\cot \theta = \dfrac{1}{\tan \theta} = 1.351$

Thus, $\tan \theta = \dfrac{1}{1.351}$, so $\theta \approx 36.5°$.

83.

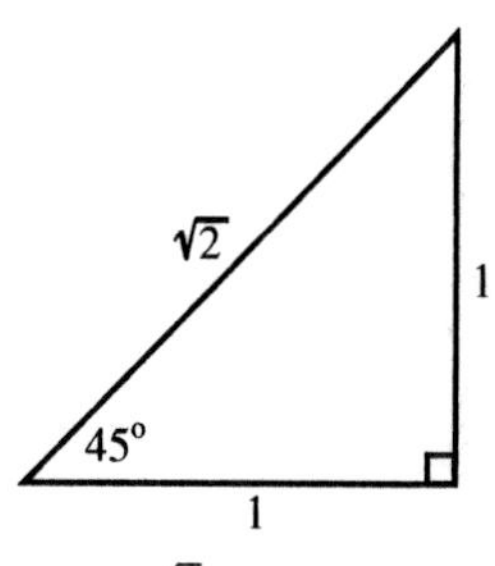

$\sin\theta = \dfrac{\sqrt{2}}{2}$, so $\theta = 45°$.

84. See the triangle in Exercise 85.

$\cot\theta = \dfrac{\sqrt{3}}{3}$, so $\theta = 60°$.

85.

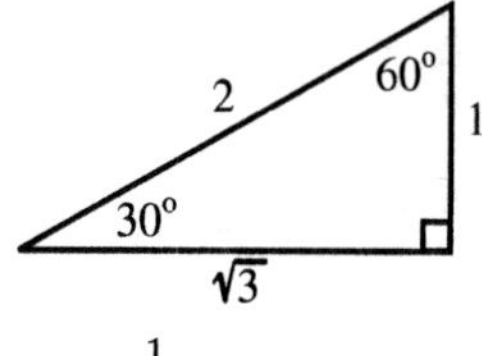

$\cos\theta = \dfrac{1}{2}$, so $\theta = 60°$.

86. See the triangle in Exercise 85.

$\sin\theta = \dfrac{1}{2}$, so $\theta = 30°$.

87. See the triangle in Exercise 83.

$\tan\theta = 1$, so $\theta = 45°$.

88. See the triangle in Exercise 85.

$\cos\theta = \dfrac{\sqrt{3}}{2}$, so $\theta = 30°$.

89. See the triangle in Exercise 85.

$\csc\theta = \dfrac{2\sqrt{3}}{3}$, or $\dfrac{2}{\sqrt{3}}$, so $\theta = 60°$.

90. See the triangle in Exercise 85.

$\tan\theta = \sqrt{3}$, or $\dfrac{\sqrt{3}}{1}$, so $\theta = 60°$.

91. See the triangle in Exercise 85.

$\cot\theta = \sqrt{3}$, or $\dfrac{\sqrt{3}}{1}$, so $\theta = 30°$.

92. See the triangle in Exercise 83.

$\sec\theta = \sqrt{2}$, or $\dfrac{\sqrt{2}}{1}$, so $\theta = 45°$.

93. The cosine and sine functions are cofunctions. The cosine and secant functions are reciprocals.

$\cos 20° = \sin 70° = \dfrac{1}{\sec 20°}$

94. $\sin 64° = \cos 26° = \dfrac{1}{\csc 64°}$

95. The tangent and cotangent functions are cofunctions and reciprocals.

$\tan 52° = \cot 38° = \dfrac{1}{\cot 52°}$

96. $\sec 13° = \csc 77° = \dfrac{1}{\cos 13°}$

97. Since 25° and 65° are complementary angles, we have

$$\begin{aligned}
\sin 25° &= \cos 65° \approx 0.4226,\\
\cos 25° &= \sin 65° \approx 0.9063,\\
\tan 25° &= \cot 65° \approx 0.4663,\\
\csc 25° &= \sec 65° \approx 2.3662,\\
\sec 25° &= \csc 65° \approx 1.1034,\\
\cot 25° &= \tan 65° \approx 2.1445.
\end{aligned}$$

98. Since 82° and 8° are complementary angles, we have

$$\begin{aligned}
\sin 82° &= \cos 8° \approx 0.9903,\\
\cos 82° &= \sin 8° \approx 0.1392,\\
\tan 82° &= \cot 8° \approx 7.1154,\\
\csc 82° &= \sec 8° \approx 1.0098,\\
\sec 82° &= \csc 8° \approx 7.1853,\\
\cot 82° &= \tan 8° \approx 0.1405.
\end{aligned}$$

99. Since $18°\,49'\,55''$ and $71°\,10'\,5''$ are complementary angles, we have

$$\begin{aligned}
\sin 18°\,49'\,55'' &= \cos 71°\,10'\,5'' \approx 0.3228,\\
\cos 18°\,49'\,55'' &= \sin 71°\,10'\,5'' \approx 0.9465,\\
\tan 18°\,49'\,55'' &= \cot 71°\,10'\,5'' = \frac{1}{\tan 71°\,10'\,5''} \approx \\
&\frac{1}{2.9321} \approx 0.3411,\\
\csc 18°\,49'\,55'' &= \sec 71°\,10'\,5'' = \frac{1}{\cos 71°\,10'\,5''} \approx \\
&\frac{1}{0.3228} \approx 3.0979,\\
\sec 18°\,49'\,55'' &= \csc 71°\,10'\,5'' = \frac{1}{\sin 71°\,10'\,5''} \approx \\
&\frac{1}{0.9465} \approx 1.0565,\\
\cot 18°\,49'\,55'' &= \tan 71°\,10'\,5'' \approx 2.9321.
\end{aligned}$$

100. Since 38.7° and 51.3° are complementary angles, we have

$$\begin{aligned}
\sin 51.3° &= \cos 38.7° \approx 0.7804,\\
\cos 51.3° &= \sin 38.7° \approx 0.6252,\\
\tan 51.3° &= \cot 38.7° \approx \frac{1}{0.8012} \approx 1.2481,\\
\csc 51.3° &= \sec 38.7° \approx \frac{1}{0.7804} \approx 1.2814,\\
\sec 51.3° &= \csc 38.7° \approx \frac{1}{0.6252} \approx 1.5995,\\
\cot 51.3° &= \tan 38.7° \approx 0.8012.
\end{aligned}$$

101. Since 82°and 8°are complementary angles, we have

$$\sin 8^\circ = \cos 82^\circ = q,$$
$$\cos 8^\circ = \sin 82^\circ = p,$$
$$\tan 8^\circ = \cot 82^\circ = \frac{1}{\tan 82^\circ} = \frac{1}{r},$$
$$\csc 8^\circ = \sec 82^\circ = \frac{1}{\cos 82^\circ} = \frac{1}{q},$$
$$\sec 8^\circ = \csc 82^\circ = \frac{1}{\sin 82^\circ} = \frac{1}{p},$$
$$\cot 8^\circ = \tan 82^\circ = r.$$

102. Left to the student.

103. Since 30° and 60° are complementary angles, once the function values for one angle are memorized the cofunction identities can be used to find the function values for the other angle.

104. If f and g are reciprocal functions, then $f(\theta) = \dfrac{1}{g(\theta)}$. If f and g are cofunctions, then $f(\theta) = g(90^\circ - \theta)$.

105.

x	$f(x)$
-4	0.1353
-2	0.3679
0	1
2	2.7183
4	7.3891

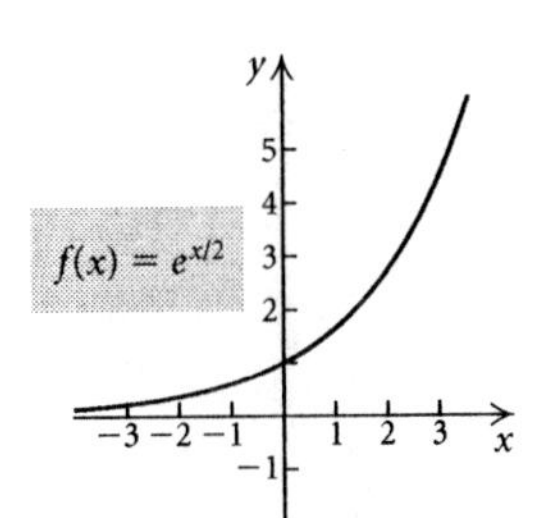

106.

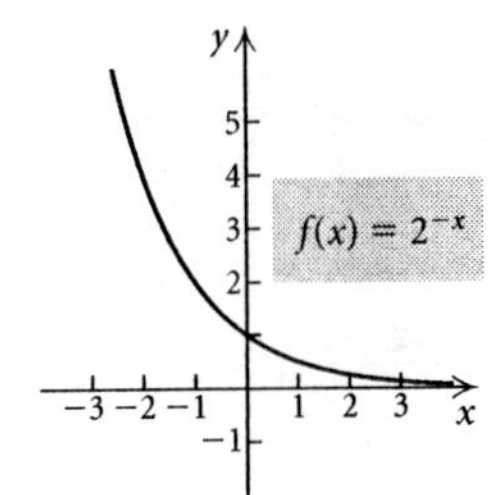

107.

x	$h(x)$
0.5	-0.6931
1	0
2	0.6932
4	1.3863
5	1.6094

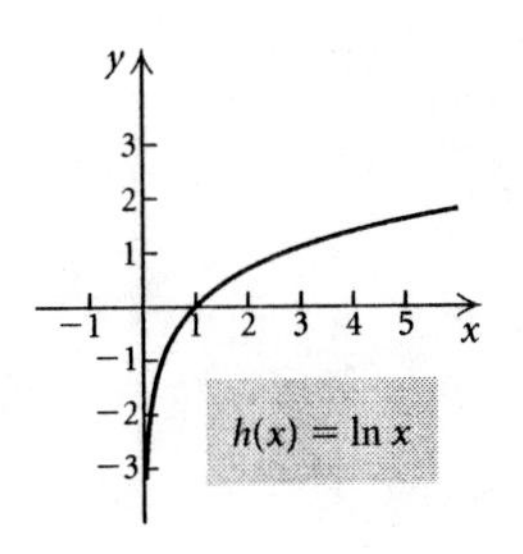

108.

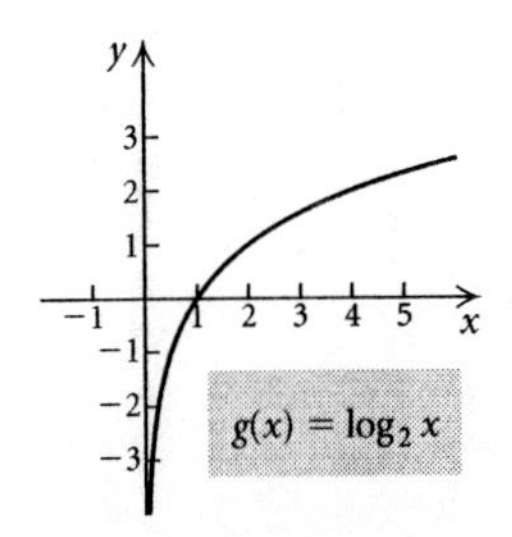

109.
$$5^x = 625$$
$$5^x = 5^4$$
$$x = 4 \quad \text{Equating exponents}$$
The solution is 4.

110.
$$e^t = 10,000$$
$$\ln e^t = \ln 10,000$$
$$t = \ln 10,000 \approx 9.21$$

111.
$$\log_7 x = 3$$
$$x = 7^3 = 343$$
The solution is 343.

112.
$$\log(3x+1) - \log(x-1) = 2$$
$$\log \frac{3x+1}{x-1} = 2$$
$$\frac{3x+1}{x-1} = 10^2$$
$$\frac{3x+1}{x-1} = 100$$
$$3x+1 = 100x - 100$$
$$101 = 97x$$
$$\frac{101}{97} = x$$

113. Since 49.2° and 40.8° are complementary angles, we have $\sin 40.8^\circ = \cos 49.2^\circ = \dfrac{1}{\sec 49.2^\circ} = \dfrac{1}{1.5304} \approx 0.6534.$

114. First find the length of the third side, a.
$$a^2 + b^2 = c^2$$
$$a^2 + \left(\frac{1}{q}\right)^2 = q^2$$
$$a^2 + \frac{1}{q^2} = q^2$$
$$a^2 = q^2 - \frac{1}{q^2}$$
$$a^2 = \frac{q^4 - 1}{q^2}$$
$$a = \frac{\sqrt{q^4-1}}{q}$$

$$\sin \alpha = \frac{\frac{1}{q}}{q} = \frac{1}{q^2}$$
$$\cos \alpha = \frac{\frac{\sqrt{q^4-1}}{q}}{q} = \frac{\sqrt{q^4-1}}{q^2}$$
$$\tan \alpha = \frac{\frac{1}{q}}{\frac{\sqrt{q^4-1}}{q}} = \frac{1}{\sqrt{q^4-1}}$$
$$\csc \alpha = \frac{q}{\frac{1}{q}} = q^2$$

$$\sec\alpha = \frac{q}{\frac{\sqrt{q^4-1}}{q}} = \frac{q^2}{\sqrt{q^4-1}}$$

$$\cot\alpha = \frac{\frac{\sqrt{q^4-1}}{q}}{\frac{1}{q}} = \sqrt{q^4-1}$$

115. Area $= \frac{1}{2}\times\text{base}\times\text{height}$

$= \frac{1}{2}ab$

$= \frac{1}{2}c\ \sin A\cdot b$

$\left(\sin A = \frac{a}{c},\ \text{so } c\sin A = a\right)$

$= \frac{1}{2}bc\ \sin A$

116. Let h = the height of the triangle. Then Area $= \frac{1}{2}bh$ where $\sin\theta = \frac{h}{a}$, or $h = a\ \sin\theta$, so Area $= \frac{1}{2}ab\ \sin\theta$.

Exercise Set 5.2

1.

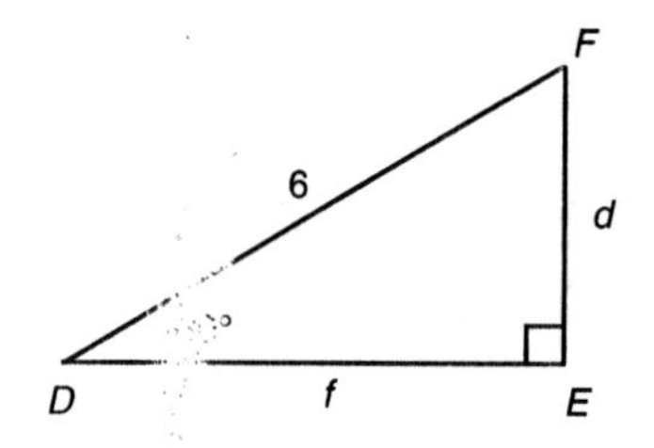

To solve this triangle find F, d, and f.

$F = 90° - 30° = 60°$

$\frac{d}{6} = \sin 30°$

$d = 6\sin 30°$

$d = 3$

$\frac{f}{6} = \cos 30°$

$f = 6\cos 30°$

$f = 3\sqrt{3} \approx 5.2$

2. $A = 90° - 45° = 45°$

$\frac{a}{10} = \cos 45°$

$a = 5\sqrt{2} \approx 7.1$

$\frac{b}{10} = \sin 45°$

$b = 5\sqrt{2} \approx 7.1$

3.

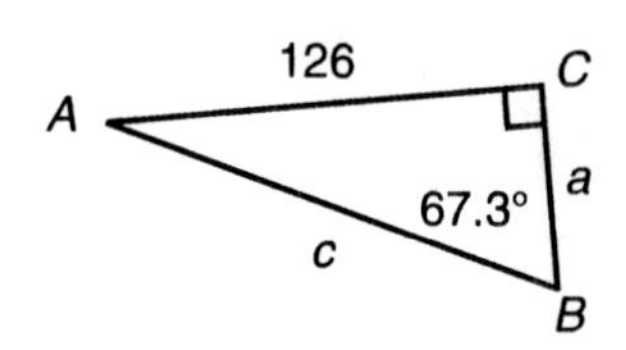

To solve this triangle, find A, a, and c.

$A = 90° - 67.3° = 22.7°$

$\cot 67.3° = \frac{a}{126}$

$a = 126\cot 67.3°$

$a \approx 52.7$

$\csc 67.3° = \frac{c}{126}$

$c = 126\csc 67.3°$

$c \approx 136.6$

4. $T = 90° - 26.7° = 63.3°$

$\csc 26.7° = \frac{s}{0.17}$

$s \approx 0.38$

$\cot 26.7° = \frac{t}{0.17}$

$t \approx 0.34$

5.

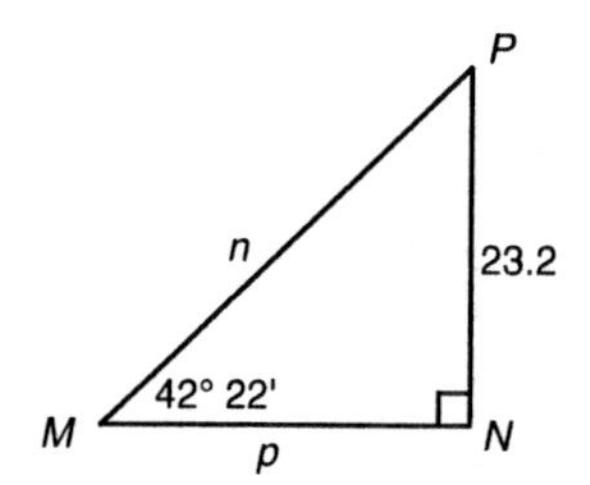

To solve this triangle, find P, p, and n.

$P = 90° - 42°\,22' = 47°38'$

$\frac{n}{23.2} = \csc 42°\,22'$

$n = 23.2\csc 42°\,22'$

$n \approx 34.4$

$\frac{p}{23.2} = \cot 42°\,22'$

$p = 23.2\cot 42°\,22'$

$p \approx 25.4$

6. $H = 90° - 28°\,34' = 61°\,26'$

$\csc 28°\,34' = \frac{f}{17.3}$

$f \approx 36.2$

$\cot 28°\,34' = \frac{h}{17.3}$

$h \approx 31.8$

7.

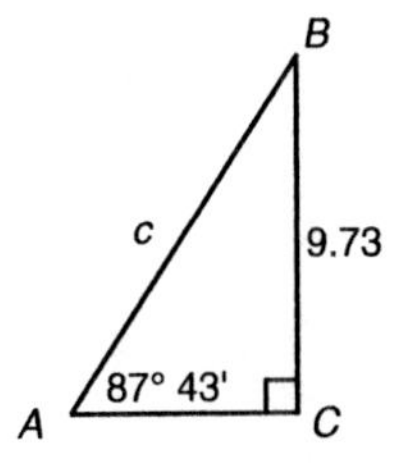

To solve this triangle, find B, b, and c.

$B = 90° - 87°\,43' = 2°17'$

$$\cot 87°\,43' = \frac{b}{9.73}$$
$$b = 9.73 \cot 87°\,43'$$
$$b \approx 0.39$$

$$\csc 87°\,43' = \frac{c}{9.73}$$
$$c = 9.73 \csc 87°\,43'$$
$$c \approx 9.74$$

8. $\tan A = \dfrac{12.5}{18.3}$

$$A \approx 34.3°$$
$$B \approx 90° - 34.3° \approx 55.7°$$
$$\csc 34.3° = \frac{c}{12.5}$$
$$c \approx 22.2$$

9.

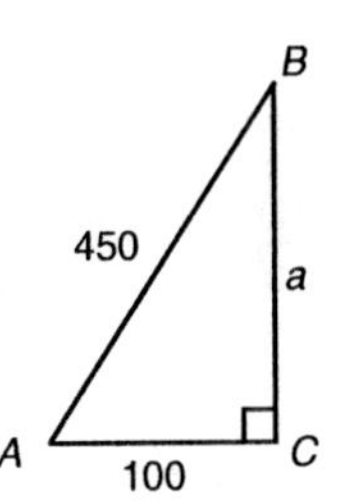

To solve this triangle, find A, B, and a.

$$\cos A = \frac{100}{450}$$
$$A \approx 77.2°$$

$$B = 90° - A \approx 90° - 77.2° \approx 12.8°$$
$$\sin A = \frac{a}{450}$$
$$\sin 77.2° \approx \frac{a}{450}$$
$$a \approx 450 \sin 77.2° \approx 439$$

10. $A = 90° - 56.5° = 33.5°$

$$\cos 56.5° = \frac{a}{0.0447}$$
$$a \approx 0.0247$$

$$\sin 56.5° = \frac{b}{0.0447}$$
$$b \approx 0.0373$$

11.

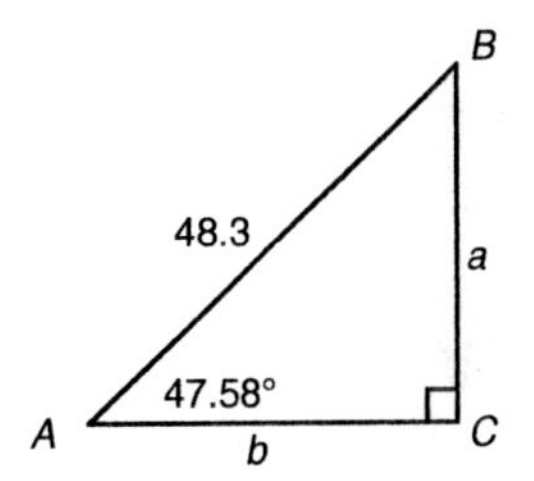

To solve this triangle, find B, a, and b.

$B = 90° - 47.58° = 42.42°$

$$\frac{a}{48.3} = \sin 47.58°$$
$$a = 48.3 \sin 47.58°$$
$$a \approx 35.7$$

$$\frac{b}{48.3} = \cos 47.58°$$
$$b = 48.3 \cos 47.58°$$
$$b \approx 32.6$$

12. $A = 90° - 20.6° = 69.4°$

$$\cot 69.4° = \frac{b}{7.5}$$
$$b \approx 2.8$$

$$\csc 69.4° = \frac{c}{7.5}$$
$$c \approx 8.0$$

13.

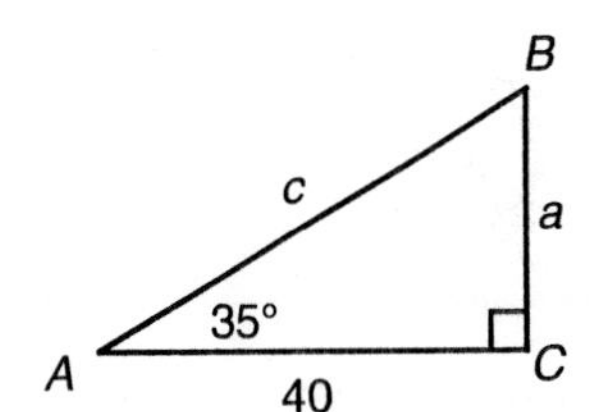

To solve this triangle find B, a, and c.

$B = 90° - A = 90° - 35° = 55°$

$$\tan A = \frac{a}{40}$$
$$\tan 35° = \frac{a}{40}$$
$$a = 40 \tan 35°$$
$$a \approx 28.0$$

$$\sec A = \frac{c}{40}$$
$$\sec 35° = \frac{c}{40}$$
$$c = 40 \sec 35°$$
$$c \approx 48.8$$

14. $A = 90° - 69.3° = 20.7°$

$$\cot 69.3° = \frac{a}{93.4}$$
$$a \approx 35.3$$

$$\csc 69.3° = \frac{c}{93.4}$$
$$c \approx 99.8$$

15.

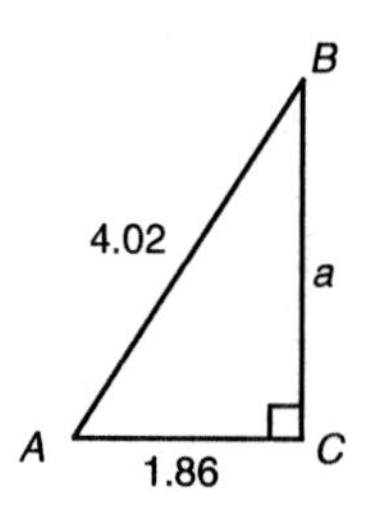

To solve this triangle find a, A, and B.

$$\cos A = \frac{1.86}{4.02}$$
$$A \approx 62.4°$$
$$B \approx 90° - 62.4° \approx 27.6°$$
$$\sin A = \frac{a}{4.02}$$
$$\sin 62.4° \approx \frac{a}{4.02}$$
$$a \approx 4.02 \sin 62.4°$$
$$a \approx 3.56$$

16. $\sin A = \dfrac{10.2}{20.4} = \dfrac{1}{2}$

$$A = 30°$$
$$B = 90° - 30° = 60°$$
$$\cos 30° = \frac{b}{20.4}$$
$$b \approx 17.7$$

17. First we find the distance from point B to the raft. We know the length of the side opposite the 50°angle and want to find the length of the hypotenuse, c. We use the sine ratio.

$$\sin 50° = \frac{40 \text{ ft}}{c}$$
$$c = \frac{40 \text{ ft}}{\sin 50°} \approx 52.2 \text{ ft}$$

Allowing 5 ft of rope at each end, Bryan needs about 52.2 ft $+ 2 \cdot 5$ ft, or about 62.2 ft of rope.

18. Let $s =$ the length of the new side.

$$\cos 53° = \frac{14.5}{s}$$
$$s = \frac{14.5}{\cos 53°} \approx 24.1 \text{ ft}$$

19. We know the length of the side adjacent to the 23°angle and want to find the length of the opposite side. We use the tangent ratio.

$$\tan 23° = \frac{d}{6 \text{ ft}}$$
$$d = 6 \text{ ft} \cdot \tan 23° \approx 2.5 \text{ ft}$$

The front legs should be about 2.5 ft from the wall.

20. Let $t =$ the height of the tree.

$$\tan 70° = \frac{t}{40 \text{ ft}}$$
$$t = 40 \text{ ft} \cdot \tan 70° \approx 110 \text{ ft}$$

21.

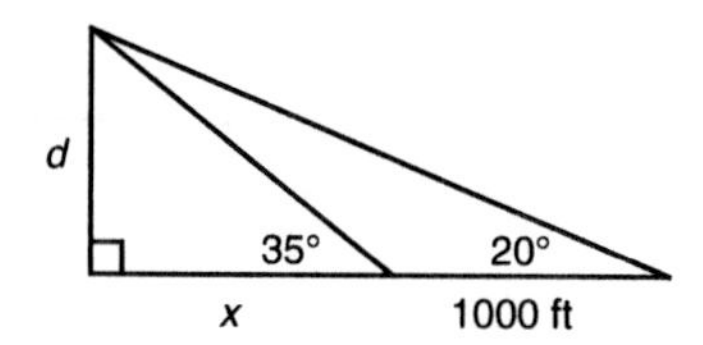

We first find the distance x, using the tangent function with 35°and 20°.

$$\tan 35° = \frac{d}{x} \quad \text{and} \quad \tan 20° = \frac{d}{x + 800}$$
$$d = x \tan 35° \quad \text{and} \quad d = (x+800) \tan 20°$$

Substitute $x \tan 35°$ for d in the second equation and solve for x.

$$x \tan 35° = (x + 800) \tan 20°$$
$$x \tan 35° = x \tan 20° + 800 \tan 20°$$
$$x(\tan 35° - \tan 20°) = 800 \tan 20°$$
$$x = \frac{800 \tan 20°}{\tan 35° - \tan 20°}$$
$$x \approx 866$$

Then we find d using the tangent function.

$$\tan 35° \approx \frac{d}{866}$$
$$d \approx 866 \tan 35° \approx 606$$

The height of the sand dune is about 606 ft.

22.

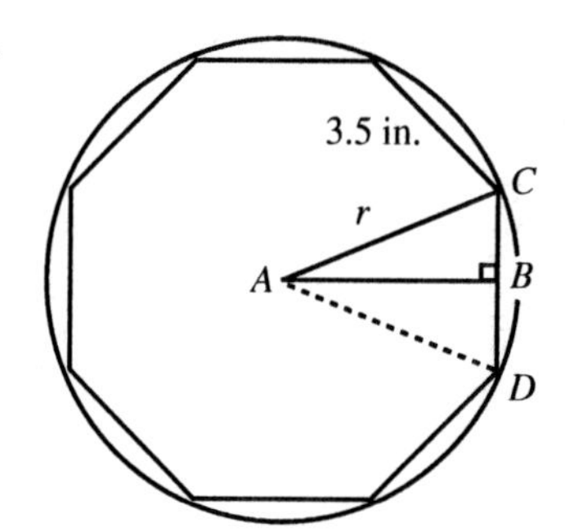

The measure of $\angle ADC$ is $\frac{1}{8} \cdot 360°$, or 45°. Then the measure of $\angle ABC$ is $\frac{1}{2} \cdot 45°$, or 22.5°. The length of the segment CD is 3.5 in., so the length of BC is $\frac{1}{2}$(3.5 in.), or 1.75 in. In $\angle ABC$, we know the length of the side opposite the 22.5° angle and we want to find the length of the hypotenuse. We use the sine ratio.

$$\sin 22.5° = \frac{1.75 \text{ in.}}{r}$$
$$r = \frac{1.75 \text{ in.}}{\sin 22.5°} \approx 4.6 \text{ in.}$$

23.

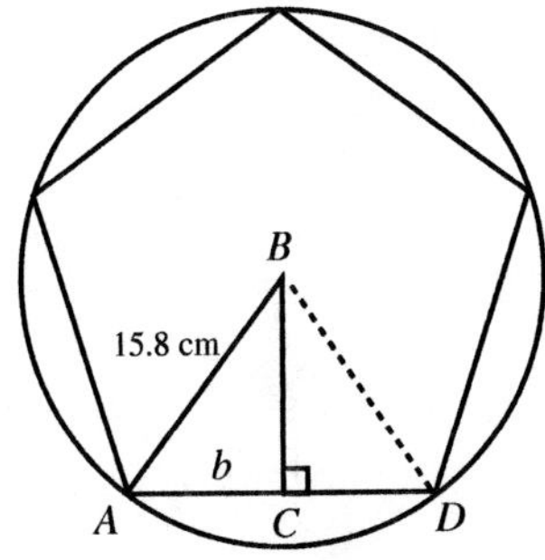

The measure of $\angle ABD$ is $\frac{1}{5} \cdot 360°$, or $72°$. Then the measure of $\angle ABC$ is $\frac{1}{2} \cdot 72°$, or $36°$. We know the length of the hypotenuse of ΔABC and want to find the length of the side opposite the 36°angle. We use the sine ratio.

$$\sin 36° = \frac{b}{15.8 \text{ cm}}$$
$$b = 15.8 \text{ cm} \cdot \sin 36° \approx 9.29 \text{ cm}$$

Then the length of each side of the pentagon is about 2(9.29 cm), or 18.58 cm, and the perimeter of the pentagon is about 5(18.58 cm), or 92.9 cm.

24.

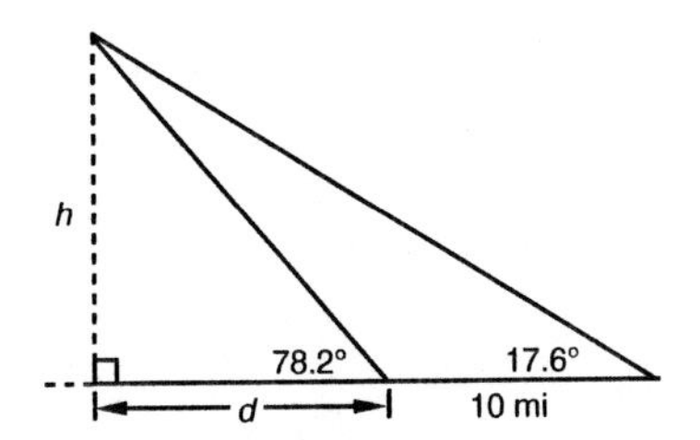

First find d. Then solve for h.

$$\frac{h}{d} = \tan 78.2° \quad \text{and} \quad \frac{h}{d+10} = \tan 17.6°$$
$$h = d \tan 78.2° \quad \text{and} \quad h = (d+10)\tan 17.6°$$

$$d \tan 78.2° = (d+10)\tan 17.6°$$
$$d \tan 78.2° = d \tan 17.6° + 10 \tan 17.6°$$
$$d(\tan 78.2° - \tan 17.6°) = 10 \tan 17.6°$$
$$d = \frac{10 \tan 17.6°}{\tan 78.2° - \tan 17.6°}$$
$$d \approx 0.7097$$

$$\tan 78.2° \approx \frac{h}{0.7097}$$
$$h \approx 0.7097(\tan 78.2°) \approx 3.4 \text{ mi}$$

25.

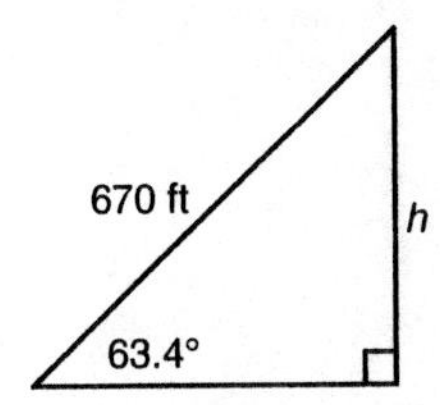

We know the length of the hypotenuse and want to find the length of the side opposite the 63.4°angle. We use the sine ratio.

$$\sin 63.4° = \frac{h}{670 \text{ ft}}$$
$$h \approx 670 \text{ ft} \cdot \sin 63.4° \approx 599 \text{ ft}$$

The kite was about 599 ft high.

26.

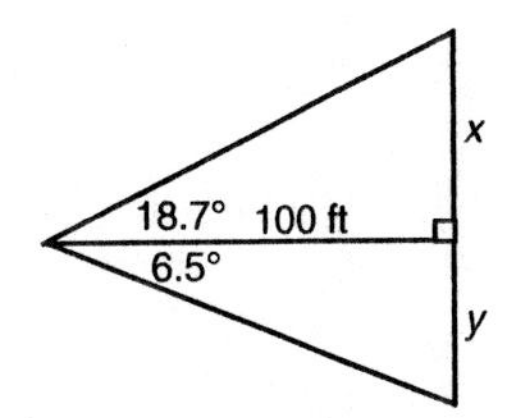

$$\tan 18.7° = \frac{x}{100 \text{ ft}}$$
$$x = 100 \text{ ft} \cdot \tan 18.7° \approx 33.8 \text{ ft}$$

$$\tan 6.5° = \frac{y}{100 \text{ ft}}$$
$$y = 100 \text{ ft} \cdot \tan 6.5° \approx 11.4 \text{ ft}$$

The height of the building is $x + y$:

$x + y \approx 33.8 \text{ ft} + 11.4 \text{ ft} \approx 45 \text{ ft}$

27.

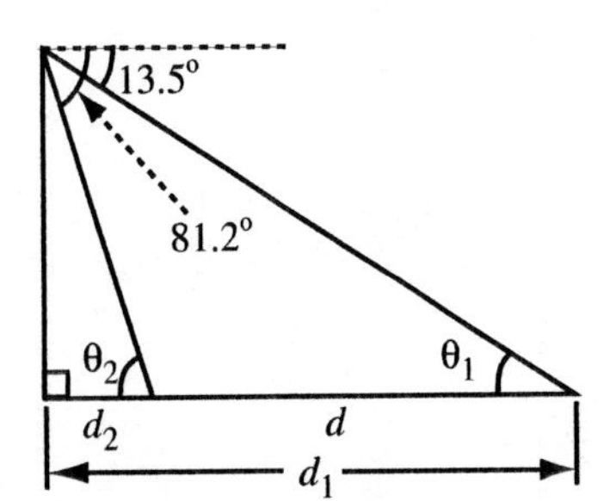

The distance to be found is d, which is $d_1 - d_2$.

From geometry we know that $\theta_1 = 13.5°$and $\theta_2 = 81.2°$.

$$\frac{d_1}{2} = \cot \theta_1 \quad \text{and} \quad \frac{d_2}{2} = \cot \theta_2$$
$$d_1 = 2 \cot 13.5° \quad \text{and} \quad d_2 = 2 \cot 81.2°$$
$$d = d_1 - d_2 = 2 \cot 13.5° - 2 \cot 81.2° \approx 8$$

The towns are about 8 km apart.

28.

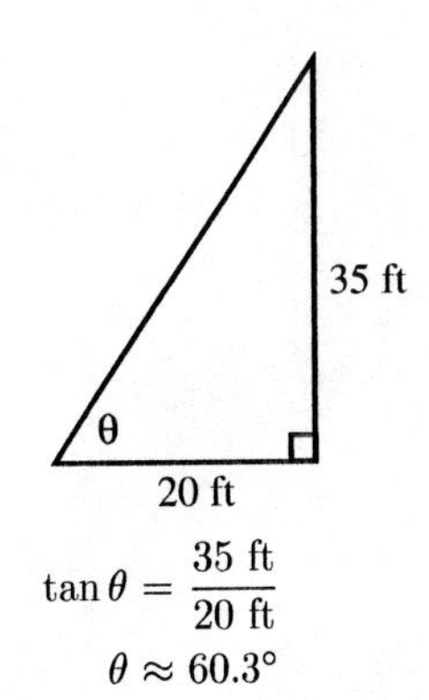

$$\tan \theta = \frac{35 \text{ ft}}{20 \text{ ft}}$$
$$\theta \approx 60.3°$$

29.

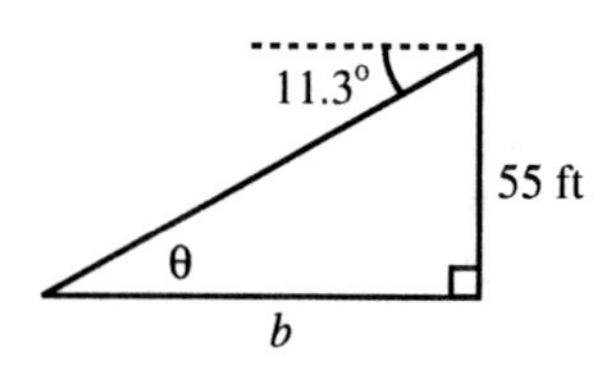

From geometry we know that $\theta = 11.3°$. We know the length of the side opposite θ and want to find the length of the side adjacent to θ. We use the tangent ratio.

$$\tan 11.3° = \frac{55 \text{ ft}}{b}$$

$$b = \frac{55 \text{ ft}}{\tan 11.3°} \approx 275 \text{ ft}$$

The boat is about 275 ft from the foot of the lighthouse.

30.

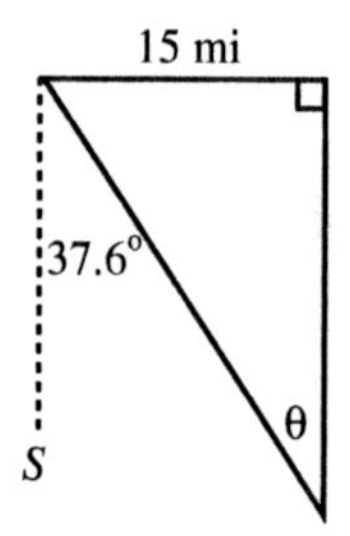

From geometry we know that $\theta = 37.6°$.

$$\tan 37.6° = \frac{15 \text{ mi}}{d}$$

$$d = \frac{15 \text{ mi}}{\tan 37.6°} \approx 19.5 \text{ mi}$$

31.

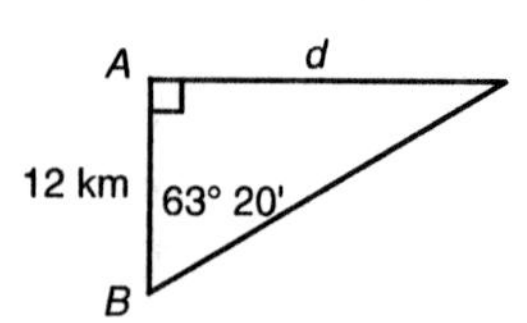

We know the length of the side adjacent to the $63° \, 20'$ angle and want to find the length of the opposite side. We use the tangent ratio.

$$\tan 63° \, 20' = \frac{d}{12 \text{ km}}$$

$$d = 12 \text{ km} \cdot \tan 63° \, 20' \approx 24 \text{ km}$$

The lobster boat is about 24 km from the lighthouse.

32.

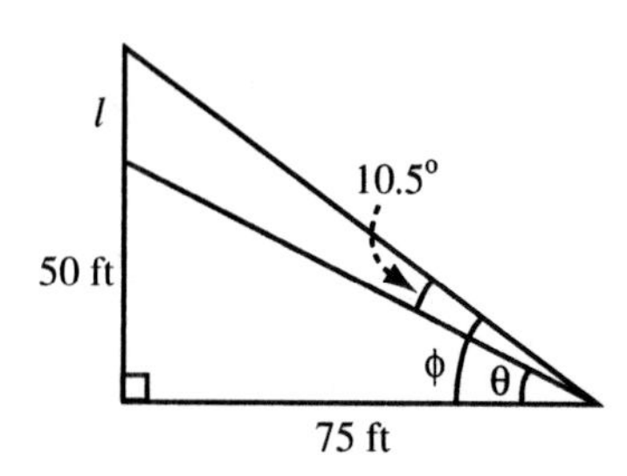

$$\tan\theta = \frac{50}{75}$$

$$\theta \approx 33.7°$$

$$\phi = \theta + 10.5° \approx 33.7° + 10.5° \approx 44.2°$$

$$\tan 44.2° \approx \frac{50 + l}{75}$$

$$75 \tan 44.2° \approx 50 + l$$

$$72.9 \approx 50 + l$$

$$23 \approx l$$

The length of the antenna is about 23 ft.

33. Sine: $[0, 1)$; cosine: $[0, 1)$; tangent: $[0, \infty)$

34. 1) Use the tangent ratio for 14°.

2) Use the cotangent ratio for 14°.

3) Find the measure of the other acute angle, 76°, and then use the tangent ratio for 76°.

4) Find the measure of the other acute angle, 76°, and then use the cotangent ratio for 76°.

5) Find the length of the hypotenuse. (This could be done using the cosine or secant ratio for 14°or the sine or cosecant ratio for 76°.) Then use the Pythagorean theorem to find c.

Many would probably agree that method (1) is most efficient.

35. $d = \sqrt{(x_1 - x_2)^2 + (y_1 - y_2)^2}$

$$d = \sqrt{[8 - (-6)]^2 + [-2 - (-4)]^2}$$

$$= \sqrt{14^2 + 2^2}$$

$$= \sqrt{200} = 10\sqrt{2}$$

$$\approx 14.142$$

36. $d = \sqrt{(-9 - 0)^2 + (3 - 0)^2} = \sqrt{90} = 3\sqrt{10} \approx 9.487$

37. $\log 0.001 = -3$ is equivalent to $\log_{10} 0.001 = -3$. Remember that the base remains the same, and the logarithm is the exponent. We have $10^{-3} = 0.001$.

38. $\log_e t = 4$, or $\ln t = 4$

39.

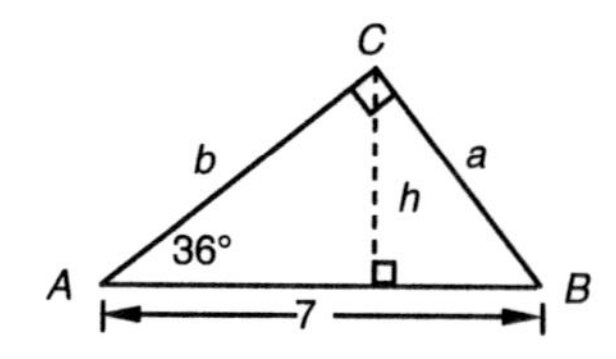

$$\cos 36° = \frac{b}{7}$$

$$b = 7 \cos 36° \approx 5.66$$

$$\sin 36° = \frac{h}{b}$$

$$\sin 36° \approx \frac{h}{5.66}$$

$$h \approx 5.66 \sin 36° \approx 3.3$$

40.

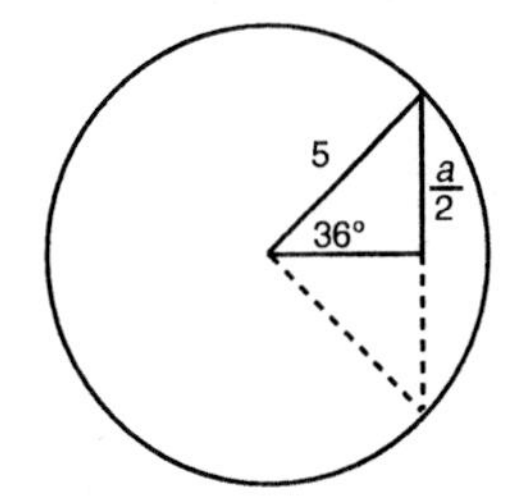

$$\sin 36° = \frac{\frac{a}{2}}{5}$$

$$\sin 36° = \frac{a}{10}$$

$$a = 10 \sin 36° \approx 5.9$$

41. $\tan\theta = \dfrac{8}{1.5}$

$$\theta \approx 79.38°$$

The rafters should be cut so that $\theta = 79.38°$.

42.

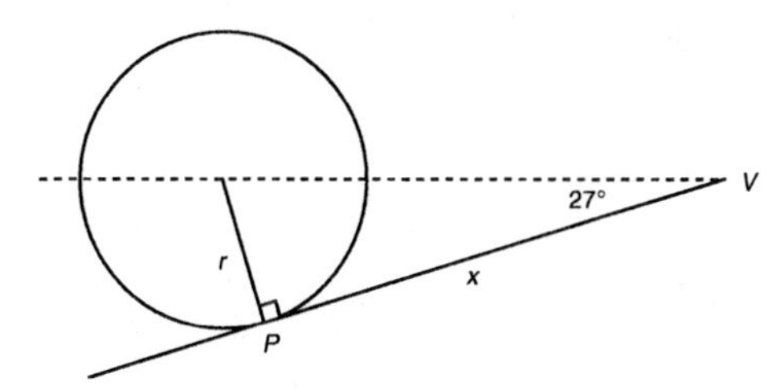

Let $x = \overline{VP}$ and $r = \dfrac{d}{2}$.

a) $r = \dfrac{2 \text{ cm}}{2} = 1 \text{ cm}$

$$\tan 27° = \frac{1 \text{ cm}}{x}$$

$$x = \frac{1 \text{ cm}}{\tan 27°} \approx 1.96 \text{ cm}$$

b) $\tan 27° = \dfrac{\frac{d}{2}}{3.93 \text{ cm}}$

$$\tan 27° = \frac{d}{2(3.93 \text{ cm})}$$

$$d = 2(3.93 \text{ cm}) \tan 27° \approx 4.00 \text{ cm}$$

c) $\tan 27° = \dfrac{\frac{d}{2}}{\overline{VP}}$

$$\tan 27° = \frac{d}{2\overline{VP}}$$

$$d = 2\overline{VP} \tan 27°$$

$$d = 1.02\overline{VP}$$

d) $d = 1.02\overline{VP}$ (See part (c).)

$$\overline{VP} = \frac{d}{1.02}$$

$$\overline{VP} = 0.98d$$

43.

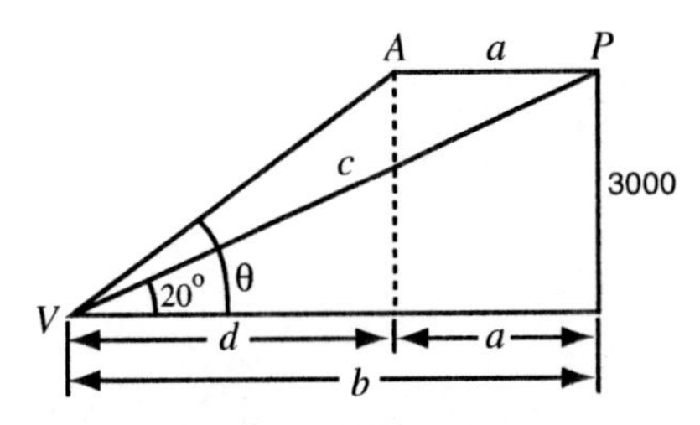

H is the perceived location of the plane;

A is the actual location of the plane when heard.

Plane's speed, 200 mph $\approx$ 293 ft/sec.

$$\csc 20° = \frac{c}{3000} \qquad \cot 20° = \frac{b}{3000}$$

$$c = 3000 \csc 20° \qquad b = 3000 \cot 20°$$

$$c \approx 8771 \qquad b \approx 8242$$

The time it takes the plane to fly from P to A is approximately the same time it takes the sound to travel from P to V. We use the formula distance = rate $\times$ time to find this time.

$$d = rt$$

$$8771 = 1100t$$

$$8 \approx t$$

Then the distance from P to A is given by $a = 293 \cdot 8 = 2344$ ft. Then $d = b - a = 8242 - 2344 = 5898$ ft.

$$\tan\theta = \frac{3000}{d} = \frac{3000}{5898}$$

$$\theta \approx 27°$$

44. $14{,}162 \text{ ft} \approx 2.682197 \text{ mi}$

$\sin 87°53' = \dfrac{R}{R + 2.682197}$ gives

$$R = \frac{2.682197 \sin 87°53'}{1 - \sin 87°53'} \approx 3298 \text{ mi}$$

Exercise Set 5.3

1.

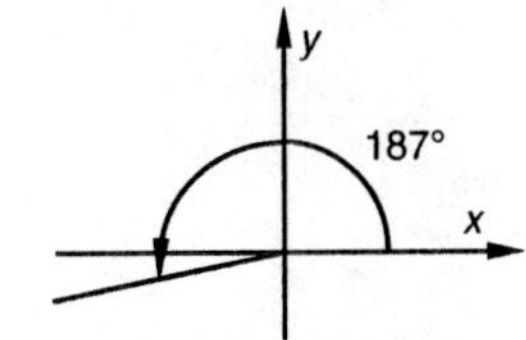

The terminal side lies in quadrant III.

2. IV

3.

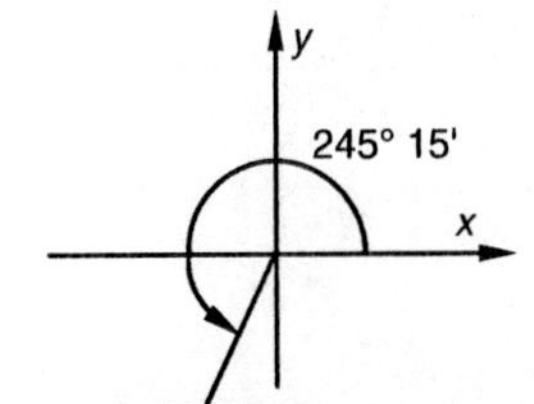

The terminal side lies in quadrant III.

4. III

5.

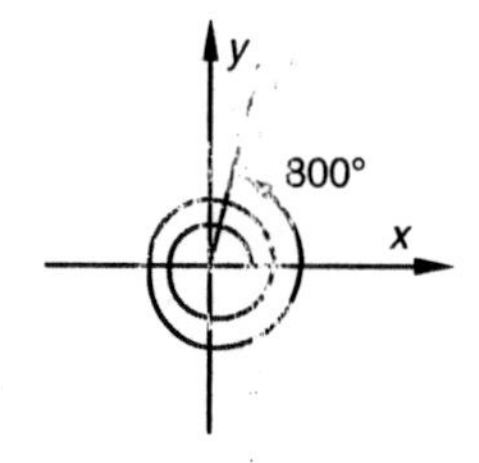

The terminal side lies in quadrant I.

6. IV

7.

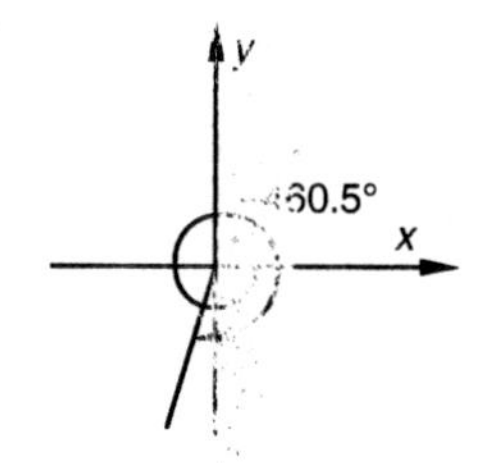

The terminal side lies in quadrant III.

8. IV

9.

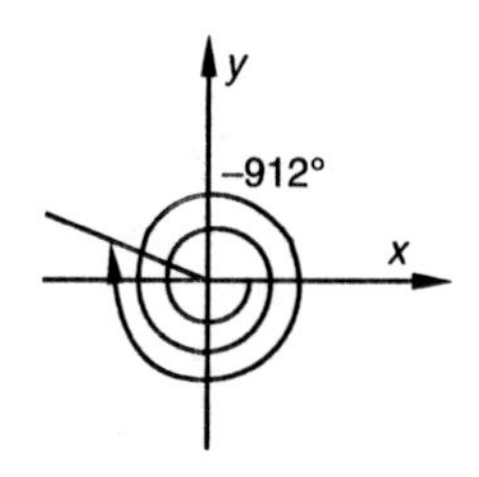

The terminal side lies in quadrant II.

10. I

11.

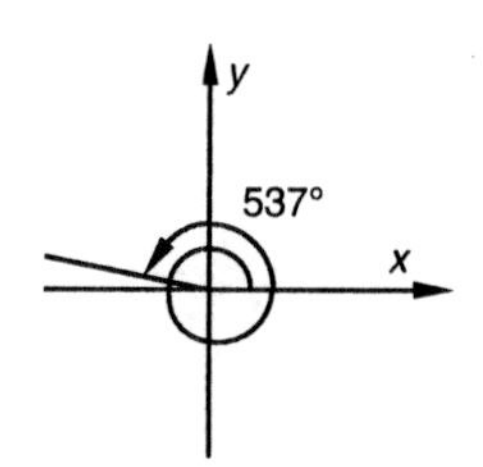

The terminal side lies in quadrant II.

12. I

13. We add and subtract multiples of 360°. Many answers are possible.

$74^\circ + 360^\circ = 434^\circ;$

$74^\circ + 2(360^\circ) = 794^\circ;$

$74^\circ - 360^\circ = -286^\circ;$

$74^\circ - 2(360^\circ) = -646^\circ$

14. Answers may vary.

$-81^\circ + 360^\circ = 279^\circ;$

$-81^\circ + 2(360^\circ) = 639^\circ;$

$-81^\circ - 360^\circ = -441^\circ;$

$-81^\circ - 2(360^\circ) = -801^\circ$

15. We add and subtract multiples of 360°. Many answers are possible.

$115.3^\circ + 360^\circ = 475.3^\circ;$

$115.3^\circ + 2(360^\circ) = 835.3^\circ;$

$115.3^\circ - 360^\circ = -244.7^\circ;$

$115.3^\circ - 2(360^\circ) = -604.7^\circ$

16. Answers may vary.

$275^\circ 10' + 360^\circ = 635^\circ 10';$

$275^\circ 10' + 2(360^\circ) = 995^\circ 10';$

$275^\circ 10' - 360^\circ = -84^\circ 50';$

$275^\circ 10' - 2(360^\circ) = -444^\circ 50'$

17. We add and subtract multiples of 360°. Many answers are possible.

$-180^\circ + 360^\circ = 180^\circ;$

$-180^\circ + 2(360^\circ) = 540^\circ;$

$-180^\circ - 360^\circ = -540^\circ;$

$-180^\circ - 2(360^\circ) = -900^\circ$

18. Answers may vary.

$-310^\circ + 360^\circ = 50^\circ;$

$-310^\circ + 2(360^\circ) = 410^\circ;$

$-310^\circ - 360^\circ = -670^\circ;$

$-310^\circ - 2(360^\circ) = -1030^\circ$

19. $90^\circ - 17.11^\circ = 72.89^\circ$

$180^\circ - 17.11^\circ = 162.89^\circ$

The complement of 17.11° is 72.89° and the supplement is 162.89°.

20. Complement: $90^\circ = \begin{array}{r} 89^\circ 60' \\ -47^\circ 38' \\ \hline 42^\circ 22' \end{array}$

Supplement: $180^\circ = \begin{array}{r} 179^\circ 60' \\ -47^\circ 38' \\ \hline 132^\circ 22' \end{array}$

21. $90^\circ = \begin{array}{r} 89^\circ 59' 60'' \\ -12^\circ\ 3' 14'' \\ \hline 77^\circ 56' 46'' \end{array}$

$180^\circ = \begin{array}{r} 179^\circ 59' 60'' \\ -12^\circ\ 3' 14'' \\ \hline 167^\circ 56' 46'' \end{array}$

The complement of $12^\circ 3' 14''$ is $77^\circ 56' 46''$ and the supplement is $167^\circ 56' 46''$.

22. Complement: $90^\circ - 9.038^\circ = 80.962^\circ$

Supplement: $180^\circ - 9.038^\circ = 170.962^\circ$

23. $90^\circ - 45.2^\circ = 44.8^\circ$

$180^\circ - 45.2^\circ = 134.8^\circ$

The complement of 45.2°is 44.8°and the supplement is 134.8°.

24. Complement: $90° - 67.31° = 22.69°$

Supplement: $180° - 67.31° = 112.69°$

25. We first determine r.

$r = \sqrt{x^2 + y^2}$

$r = \sqrt{(-12)^2 + 5^2}$

$= \sqrt{144 + 25} = \sqrt{169} = 13$

Substituting -12 for x, 5 for y, and 13 for r, the trigonometric function values of θ are

$\sin\beta = \frac{y}{r} = \frac{5}{13}$

$\cos\beta = \frac{x}{r} = \frac{-12}{13} = -\frac{12}{13}$

$\tan\beta = \frac{y}{x} = \frac{5}{-12} = -\frac{5}{12}$

$\csc\beta = \frac{r}{y} = \frac{13}{5}$

$\sec\beta = \frac{r}{x} = \frac{13}{-12} = -\frac{13}{12}$

$\cot\beta = \frac{x}{y} = \frac{-12}{5} = -\frac{12}{5}$

26. $r = \sqrt{(\sqrt{7})^2 + (-3)^2} = \sqrt{16} = 4$

$\sin\theta = -\frac{3}{4}$

$\cos\theta = \frac{\sqrt{7}}{4}$

$\tan\theta = -\frac{3}{\sqrt{7}}$, or $-\frac{3\sqrt{7}}{7}$

$\csc\theta = -\frac{4}{3}$

$\sec\theta = \frac{4}{\sqrt{7}}$, or $\frac{4\sqrt{7}}{7}$

$\cot\theta = -\frac{\sqrt{7}}{3}$

27. We first determine r.

$r = \sqrt{x^2 + y^2}$

$r = \sqrt{(-2\sqrt{3})^2 + (-4)^2} = \sqrt{4 \cdot 3 + 16}$

$= \sqrt{12 + 16} = \sqrt{28} = 2\sqrt{7}$

Substituting $-2\sqrt{3}$ for x, -4 for y and $2\sqrt{7}$ for r, the trigonometric function values are

$\sin\phi = \frac{y}{r} = \frac{-4}{2\sqrt{7}} = -\frac{2}{\sqrt{7}}$, or $-\frac{2\sqrt{7}}{7}$

$\cos\phi = \frac{x}{r} = \frac{-2\sqrt{3}}{2\sqrt{7}} = -\frac{\sqrt{3}}{\sqrt{7}}$, or $-\frac{\sqrt{21}}{7}$

$\tan\phi = \frac{y}{x} = \frac{-4}{-2\sqrt{3}} = \frac{2}{\sqrt{3}}$, or $\frac{2\sqrt{3}}{3}$

$\csc\phi = \frac{r}{y} = \frac{2\sqrt{7}}{-4} = -\frac{\sqrt{7}}{2}$

$\sec\phi = \frac{r}{x} = \frac{2\sqrt{7}}{-2\sqrt{3}} = -\frac{\sqrt{7}}{\sqrt{3}}$, or $-\frac{\sqrt{21}}{3}$

$\cot\phi = \frac{x}{y} = \frac{-2\sqrt{3}}{-4} = \frac{\sqrt{3}}{2}$

28. $r = \sqrt{9^2 + 1^2} = \sqrt{82}$

$\sin\alpha = \frac{1}{\sqrt{82}}$, or $\frac{\sqrt{82}}{82}$

$\cos\alpha = \frac{9}{\sqrt{82}}$, or $\frac{9\sqrt{82}}{82}$

$\tan\alpha = \frac{1}{9}$

$\csc\alpha = \frac{\sqrt{82}}{1} = \sqrt{82}$

$\sec\alpha = \frac{\sqrt{82}}{9}$

$\cot\alpha = \frac{9}{1} = 9$

29. First we draw the graph of $2x + 3y = 0$ and determine a quadrant IV solution of the equation.

We let $x = 3$ and find the corresponding y-value.

$2x + 3y = 0$

$2 \cdot 3 + 3y = 0$ Substituting

$3y = -6$

$y = -2$

Thus, $(3, -2)$ is a point on the terminal side of the angle θ.

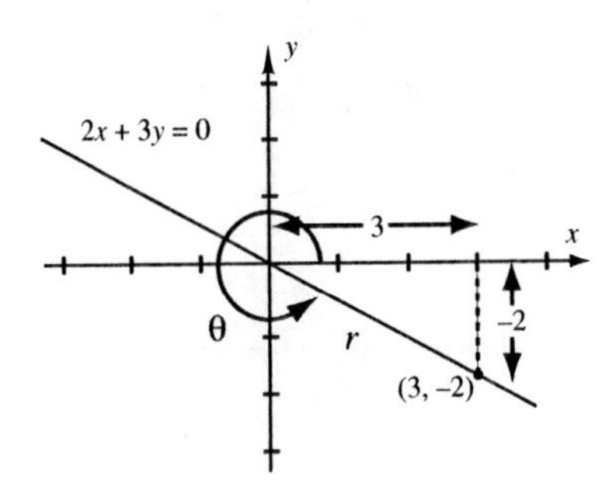

Using $(3, -2)$, we determine r:

$r = \sqrt{3^2 + (-2)^2} = \sqrt{13}$

Then using $x = 3$, $y = -2$, and $r = \sqrt{13}$, we find

$\sin\theta = \frac{-2}{\sqrt{13}}$, or $-\frac{2\sqrt{13}}{13}$,

$\cos\theta = \frac{3}{\sqrt{13}}$, or $\frac{3\sqrt{13}}{13}$,

$\tan\theta = \frac{-2}{3} = -\frac{2}{3}$.

30. A quadrant II solution of $4x + y = 0$ is $(-1, 4)$. Then $r = \sqrt{(-1)^2 + 4^2} = \sqrt{17}$.

$\sin\theta = \frac{4}{\sqrt{17}}$, or $\frac{4\sqrt{17}}{17}$

$\cos\theta = \frac{-1}{\sqrt{17}}$, or $-\frac{\sqrt{17}}{17}$

$\tan\theta = \frac{4}{-1} = -4$

31. First we draw the graph of $5x - 4y = 0$ and determine a quadrant I solution of the equation.

We let $x = 4$ and find the corresponding y-value.

$$5x - 4y = 0$$
$$5 \cdot 4 - 4y = 0 \quad \text{Substituting}$$
$$-4y = -20$$
$$y = 5$$

Thus, $(4, 5)$ is a point on the terminal side of the angle θ.

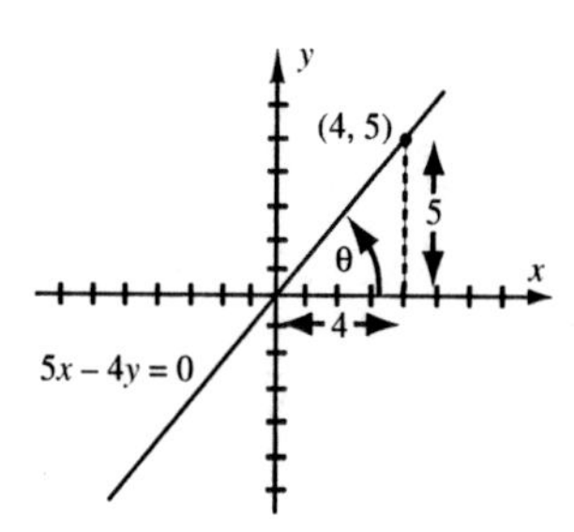

Using $(4, 5)$, we determine r:

$$r = \sqrt{4^2 + 5^2} = \sqrt{41}$$

Then using $x = 4$, $y = 5$, and $r = \sqrt{41}$, we find

$$\sin\theta = \frac{5}{\sqrt{41}}, \text{ or } \frac{5\sqrt{41}}{41},$$

$$\cos\theta = \frac{4}{\sqrt{41}}, \text{ or } \frac{4\sqrt{41}}{41},$$

$$\tan\theta = \frac{5}{4}.$$

32. A quadrant III solution of $y = 0.8x$ is $(-10, -8)$. Then

$$r = \sqrt{(-10)^2 + (-8)^2} = \sqrt{164} = 2\sqrt{41}.$$

$$\sin\theta = \frac{-8}{2\sqrt{41}}, \text{ or } -\frac{4\sqrt{41}}{41}$$

$$\cos\theta = \frac{-10}{2\sqrt{41}}, \text{ or } -\frac{5\sqrt{41}}{41}$$

$$\tan\theta = \frac{-8}{-10} = \frac{4}{5}$$

33. First we sketch a third-quadrant angle and a reference triangle. Since $\sin\theta = -\frac{1}{3} = \frac{-1}{3}$ the length of the vertical leg is 1 and the length of the hypotenuse is 3. The other leg must then have length $\sqrt{8}$, or $2\sqrt{2}$.

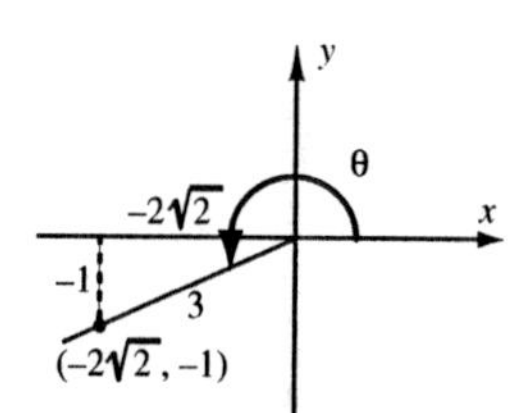

Now we can read off the appropriate ratios:

$$\cos\theta = \frac{-2\sqrt{2}}{3}, \text{ or } -\frac{2\sqrt{2}}{3}$$

$$\tan\theta = \frac{-1}{-2\sqrt{2}} = \frac{1}{2\sqrt{2}}, \text{ or } \frac{\sqrt{2}}{4}$$

$$\csc\theta = \frac{3}{-1} = -3$$

$$\sec\theta = \frac{3}{-2\sqrt{2}} = -\frac{3}{2\sqrt{2}}, \text{ or } -\frac{3\sqrt{2}}{4}$$

$$\cot\theta = \frac{-2\sqrt{2}}{-1} = 2\sqrt{2}$$

34. $\tan\beta = 5 = \frac{5}{1}$

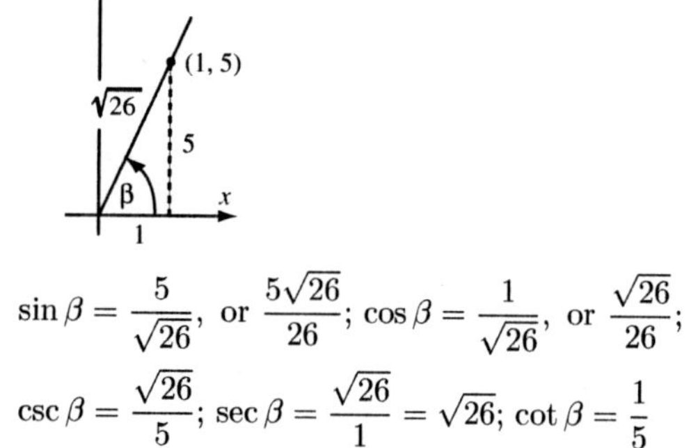

$$\sin\beta = \frac{5}{\sqrt{26}}, \text{ or } \frac{5\sqrt{26}}{26}; \ \cos\beta = \frac{1}{\sqrt{26}}, \text{ or } \frac{\sqrt{26}}{26};$$

$$\csc\beta = \frac{\sqrt{26}}{5}; \ \sec\beta = \frac{\sqrt{26}}{1} = \sqrt{26}; \ \cot\beta = \frac{1}{5}$$

35. Since θ is in quadrant IV, we have $\cot\theta = -2 = \frac{2}{-1}$.

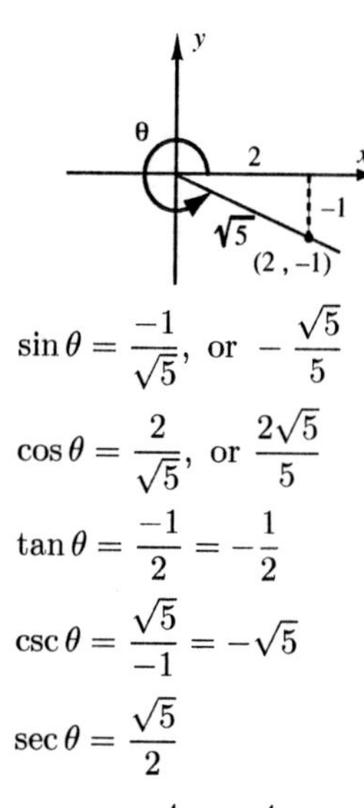

$$\sin\theta = \frac{-1}{\sqrt{5}}, \text{ or } -\frac{\sqrt{5}}{5}$$

$$\cos\theta = \frac{2}{\sqrt{5}}, \text{ or } \frac{2\sqrt{5}}{5}$$

$$\tan\theta = \frac{-1}{2} = -\frac{1}{2}$$

$$\csc\theta = \frac{\sqrt{5}}{-1} = -\sqrt{5}$$

$$\sec\theta = \frac{\sqrt{5}}{2}$$

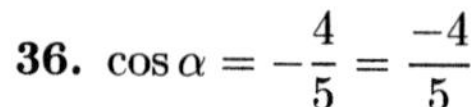

36. $\cos\alpha = -\frac{4}{5} = \frac{-4}{5}$

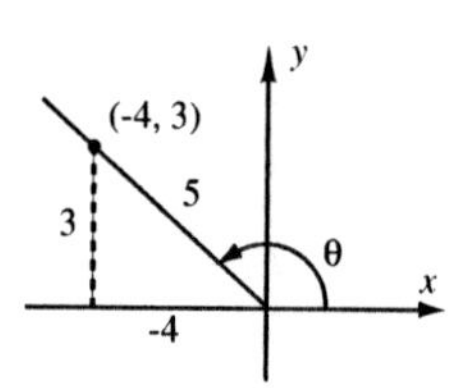

$$\sin\alpha = \frac{3}{5}, \ \tan\alpha = -\frac{3}{4}, \ \csc\alpha = \frac{5}{3}, \ \sec\alpha = -\frac{5}{4},$$

$$\cot\alpha = -\frac{4}{3}$$

37. $\cos\phi = \dfrac{3}{5}$

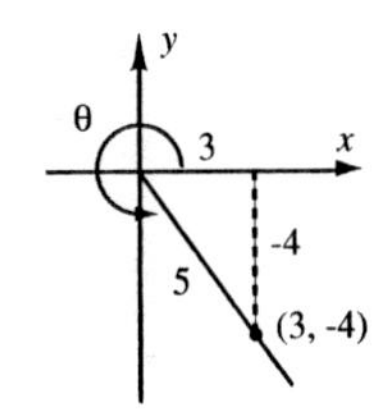

$\sin\theta = -\dfrac{4}{5}$

$\tan\theta = -\dfrac{4}{3}$

$\csc\theta = -\dfrac{5}{4}$

$\sec\theta = \dfrac{5}{3}$

$\cot\theta = -\dfrac{3}{4}$

38. $\sin\theta = -\dfrac{5}{13}$

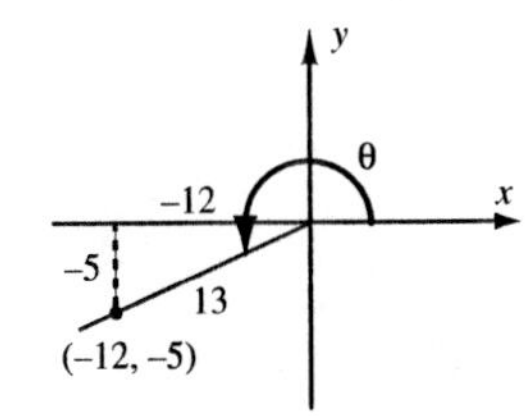

$\cos\theta = \dfrac{-12}{13} = -\dfrac{12}{13}$, $\tan\theta = \dfrac{-5}{-12} = \dfrac{5}{12}$,
$\csc\theta = \dfrac{13}{-5} = -\dfrac{13}{5}$, $\sec\theta = \dfrac{13}{-12} = -\dfrac{13}{12}$,
$\cot\theta = \dfrac{-12}{-5} = \dfrac{12}{5}$

39. Since $180° - 150° = 30°$, the reference angle is $30°$. Note that $150°$is a second-quadrant angle, so the cosine is negative. Recalling that $\cos 30° = \sqrt{3}/2$, we have

$$\cos 150° = -\frac{\sqrt{3}}{2}.$$

40. $-225° + 360° = 135°$, so $-225°$ and $135°$ are coterminal. The reference angle is then $45°$. Note that $-225°$ is a second-quadrant angle, so the secant is negative. Recalling that $\sec 45° = \sqrt{2}$, we have

$$\sec(-225°) = -\sqrt{2}.$$

41. $-135° + 360° = 225°$, so $-135°$ and $225°$ are coterminal. Since $180° + 45° = 225°$, the reference angle is $45°$. Note that $-135°$ is a third-quadrant angle, so the tangent is positive. Recalling that $\tan 45° = 1$, we have

$$\tan(-135°) = 1.$$

42. The reference angle is $45°$. Note that $-45°$ is a fourth-quadrant angle, so the sine is negative. Recalling that $\sin 45° = \sqrt{2}/2$, we have

$$\sin(-45°) = -\frac{\sqrt{2}}{2}.$$

43. Since $7560° = 21\cdot 360°$, or $0° + 21\cdot 360°$, then $7560°$ and $0°$ are coterminal. A point on the terminal side of $0°$is $(1, 0)$. Thus,

$$\sin 7560° = \sin 0° = \frac{0}{1} = 0.$$

44. A point on the terminal side of $270°$ is $(0, -1)$. Thus,

$$\tan 270° = \frac{-1}{0}, \text{ which is not defined.}$$

45. $495° - 360° = 135°$, so $495°$ and $135°$ are coterminal. Since $180° - 135° = 45°$, the reference angle is $45°$. Note that $495°$ is a second-quadrant angle, so the cosine is negative. Recalling that $\cos 45° = \sqrt{2}/2$ we have

$$\cos 495° = -\frac{\sqrt{2}}{2}.$$

46. $675° - 360° = 315°$, so $675°$and $315°$are coterminal. The reference angle is $45°$. Note that $675°$ is a fourth-quadrant angle, so the tangent is negative. Recalling that $\tan 45° = 1$, we have

$$\tan 675° = -1.$$

47. $-210° + 360° = 150°$, so $-210°$ and $150°$ are coterminal. Since $180° - 150° = 30°$, the reference angle is $30°$. Note that $-210°$ is a second-quadrant angle, so the cosecant is positive. Recalling that $\csc 30° = 2$, we have

$$\csc(-210°) = 2.$$

48. $360° - 300° = 60°$, so the reference angle is $60°$. Note that $300°$ is a fourth-quadrant angle, so the sine is negative. Recalling that $\sin 60° = \sqrt{3}/2$, we have

$$\sin 300° = -\frac{\sqrt{3}}{2}.$$

49. $570° - 360° = 210°$, so $570°$ and $210°$ are coterminal. Since $180° + 30° = 210°$, the reference angle is $30°$. Note that $570°$ is a third-quadrant angle, so the cotangent is positive. Recalling that $\cot 30° = \sqrt{3}$, we have

$$\cot 570° = \sqrt{3}.$$

50. $-120° + 360° = 240°$, so $-120°$ and $240°$ are coterminal. Since $180° + 60° = 240°$, the reference angle is $60°$. Note that $-120°$ is a third-quadrant angle, so the cosine is negative. Recalling that $\cos 60° = 1/2$, we have

$$\cos(-120°) = -\frac{1}{2}.$$

51. $360° - 330° = 30°$, so the reference angle is $30°$. Note that $330°$ is a fourth-quadrant angle, so the tangent is negative. Recalling that $\tan 30° = \sqrt{3}/3$, we have

$$\tan 330° = -\frac{\sqrt{3}}{3}.$$

52. $855° - 2\cdot 360° = 135°$, so $855°$ and $135°$ are coterminal. Since $180° - 135° = 45°$, the reference angle is $45°$. Note that $855°$ is a second-quadrant angle, so the cotangent is negative. Recalling that $\cot 45° = 1$, we have

$$\cot 855° = -1.$$

53. A point on the terminal side of $-90°$ is $(0, -1)$. Thus,

$$\sec(-90°) = \frac{1}{0}, \text{ which is not defined.}$$

54. A point on the terminal side of $90°$ is $(0,1)$. Thus,

$$\sin 90° = \frac{1}{1} = 1.$$

55. A point on the terminal side of $-180°$ is $(-1,0)$. Thus,

$$\cos(-180°) = \frac{-1}{1} = -1.$$

56. A point on the terminal side of $90°$ is $(0,1)$. Thus,

$$\csc 90° = \frac{1}{1} = 1.$$

57. $240° - 180° = 60°$, so the reference angle is $60°$. Note that $240°$ is a third-quadrant angle, so the tangent is positive. Recalling that $\tan 60° = \sqrt{3}$, we have

$$\tan 240° = \sqrt{3}.$$

58. A point on the terminal side of $-180°$ is $(-1,0)$. Thus,

$$\cot(-180°) = \frac{-1}{0}, \text{ which is not defined.}$$

59. $495° - 360° = 135°$, so $495°$ and $135°$ are coterminal. Since $180° - 135° = 45°$, the reference angle is $45°$. Note that $495°$ is a second-quadrant angle, so the sine is positive. Recalling that $\sin 45° = \sqrt{2}/2$ we have

$$\sin 495° = \frac{\sqrt{2}}{2}.$$

60. $1050° - 2 \cdot 360° = 330°$, so $1050°$ and $330°$ are coterminal. Since $360° - 330° = 30°$, the reference angle is $30°$. Note that $1050°$ is a fourth-quadrant angle, so the sine is negative. Recalling that $\sin 30° = 1/2$, we have

$$\sin 1050° = -\frac{1}{2}.$$

61. $225° - 180° = 45°$, so the reference angle is $45°$. Note that $225°$ is a third-quadrant angle, so the cosecant is negative. Recalling that $\csc 45° = \sqrt{2}$, we have

$$\csc 225° = -\sqrt{2}.$$

62. $-450° + 360° = -90°$; a point on the terminal side of $-90°$is $(0,-1)$. Thus,

$$\sin(-450°) = \frac{-1}{1} = -1.$$

63. A point on the terminal side of $0°$ is $(1,0)$. Thus,

$$\cos 0° = \frac{1}{1} = 1.$$

64. $480° - 360° = 120°$, so $480°$ and $120°$ are coterminal. Since $180° - 120° = 60°$, the reference angle is $60°$. Note that $480°$ is a second-quadrant angle, so the tangent is negative. Recalling that $\tan 60° = \sqrt{3}$, we have

$$\tan 480° = -\sqrt{3}.$$

65. A point on the terminal side of $-90°$ is $(0,-1)$. Thus,

$$\cot(-90°) = \frac{0}{-1} = 0.$$

66. $360° - 315° = 45°$, so the reference angle is $45°$. Note that $315°$ is a fourth-quadrant angle, so the secant is positive. Recalling that $\sec 45° = \sqrt{2}$, we have

$$\sec 315° = \sqrt{2}.$$

67. A point on the terminal side of $90°$ is $(0,1)$. Thus,

$$\cos 90° = \frac{0}{1} = 0.$$

68. $-135° + 360° = 225°$, so $-135°$ and $225°$ are coterminal. Since $180° + 45° = 225°$, the reference angle is $45°$. Note that $-135°$ is a third-quadrant angle, so the sine is negative. Recalling that $\sin 45° = \sqrt{2}/2$, we have

$$\sin(-135°) = -\frac{\sqrt{2}}{2}.$$

69. A point on the terminal side of $270°$ is $(0,-1)$. Thus,

$$\cos 270° = \frac{0}{1} = 0.$$

70. A point on the terminal side of $0°$ is $(1,0)$. Thus,

$$\tan 0° = \frac{0}{1} = 0.$$

71. $319°$ is a fourth-quadrant angle, so the cosine and secant function values are positive and the sine, cosecant, tangent, and cotangent are negative.

72. $-57°$ is a fourth-quadrant angle, so the cosine and secant function values are positive and the sine, cosecant, tangent, and cotangent are negative.

73. $194°$ is a third-quadrant angle, so the tangent and cotangent function values are positive and the sine, cosecant, cosine, and secant are negative.

74. $-620°$ is a second-quadrant angle, so the sine and cosecant function values are positive and the cosine, secant, tangent, and cotangent are negative.

75. $-215°$ is a second-quadrant angle, so the sine and cosecant function values are positive and the cosine, secant, tangent, and cotangent are negative.

76. $290°$ is a fourth-quadrant angle, so the cosine and secant function values are positive and the sine, cosecant, tangent, and cotangent are negative.

77. $-272°$ is a first-quadrant angle, so all of the trigonometric function values are positive.

78. $91°$ is a second-quadrant angle, so the sine and cosecant function values are positive and the cosine, secant, tangent, and cotangent are negative.

79. $360° - 319° = 41°$, so $41°$ is the reference angle for $319°$. Note that $319°$ is a fourth-quadrant angle. Use the given trigonometric function values for $41°$ to find the trigonometric function values for $319°$. In the fourth quadrant, the cosine and secant functions are positive and the other four are negative.

$$\sin 319° = -\sin 41° = -0.6561$$

$$\cos 319° = \cos 41° = 0.7547$$

$$\tan 319° = -\tan 41° = -0.8693$$

$$\csc 319° = \frac{1}{\sin 319°} = \frac{1}{-0.6561} \approx -1.5242$$

$$\sec 319° = \frac{1}{\cos 319°} = \frac{1}{0.7547} \approx 1.3250$$

$$\cot 319° = \frac{1}{\tan 319°} = \frac{1}{-0.8693} \approx -1.1504$$

80. Reference angle: 27°

Terminal side is in quadrant IV.

$\sin 333° = -\sin 27° = -0.4540$

$\cos 333° = \cos 27° = 0.8910$

$\tan 333° = -\tan 27° = -0.5095$

$\csc 333° = \frac{1}{\sin 333°} = \frac{1}{-0.4540} \approx -2.2026$

$\sec 333° = \frac{1}{\cos 333°} = \frac{1}{0.8910} \approx 1.1223$

$\cot 333° = \frac{1}{\tan 333°} = \frac{1}{-0.5095} \approx -1.9627$

81. $180° - 115° = 65°$, so 65° is the reference angle for 115°. Note that 115° is a second-quadrant angle. Use the given trigonometric function values for 65° to find the trigonometric function values for 115°. In the second quadrant, the sine and cosecant functions are positive and the other four are negative.

$\sin 115° = \sin 65° = 0.9063$

$\cos 115° = -\cos 65° = -0.4226$

$\tan 115° = -\tan 65° = -2.1445$

$\csc 115° = \frac{1}{\sin 115°} = \frac{1}{0.9063} \approx 1.1034$

$\sec 115° = \frac{1}{\cos 115°} = \frac{1}{-0.4226} \approx -2.3663$

$\cot 115° = \frac{1}{\tan 115°} = \frac{1}{-2.1445} \approx -0.4663$

82. Reference angle: 35°

Terminal side is in quadrant III.

$\sin 215° = -\sin 35° = -0.5736$

$\cos 215° = -\cos 35° = -0.8192$

$\tan 215° = \tan 35° = 0.7002$

$\csc 215° = \frac{1}{\sin 215°} = \frac{1}{-0.5736} \approx -1.7434$

$\sec 215° = \frac{1}{\cos 215°} = \frac{1}{-0.8192} \approx -1.2207$

$\cot 215° = \frac{1}{\tan 215°} = \frac{1}{0.7002} \approx 1.4282$

83.

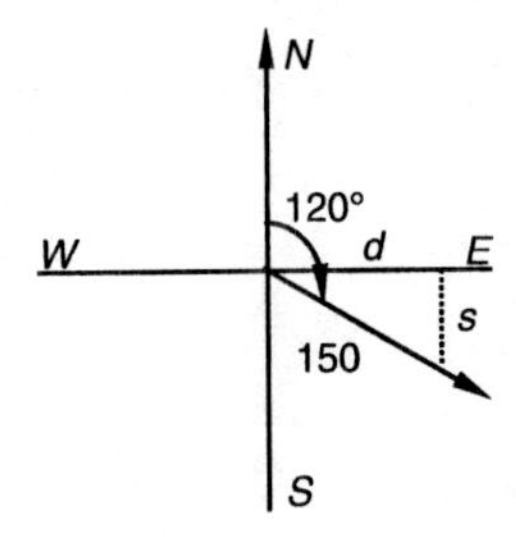

The reference angle is 30°. First find d, the airplane's distance east of the airport.

$$\cos 30° = \frac{d}{150}$$
$$d = 150 \cos 30°$$
$$d \approx 130$$

The airplane is about 130 km east of the airport.

Now find s, the airplane's distance south of the airport.

$$\sin 30° = \frac{s}{150}$$
$$s = 150 \sin 30°$$
$$s = 75$$

The airplane is 75 km south of the airport.

84.

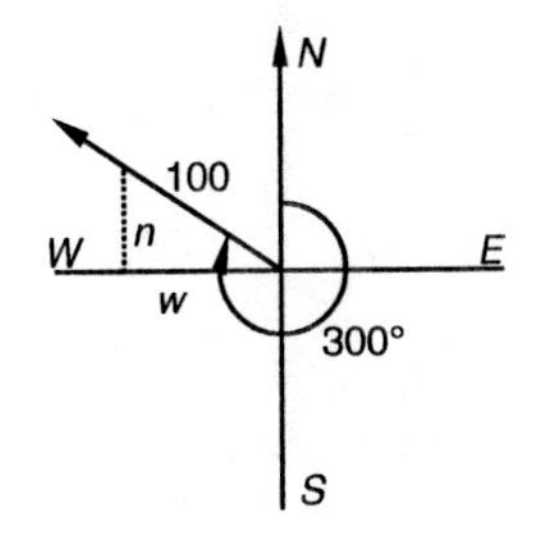

The reference angle is 30°. First find n, the airplane's distance north of the airport.

$$\sin 30° = \frac{n}{100}$$
$$n = 100 \sin 30°$$
$$n = 50 \text{ mi}$$

Now find w, the airplane's distance west of the airport.

$$\cos 30° = \frac{w}{100}$$
$$w = 100 \cos 30°$$
$$w \approx 87 \text{ mi}$$

85.

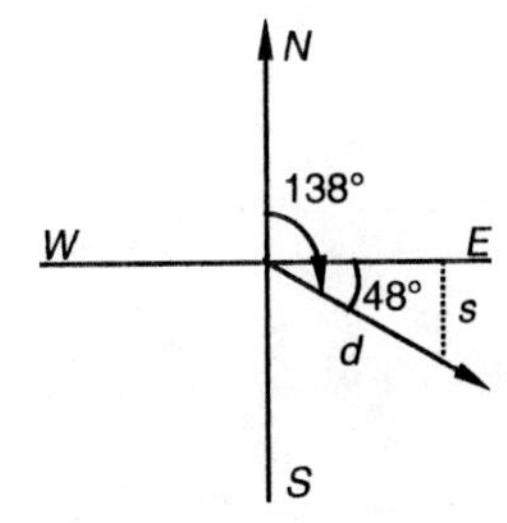

The reference angle is 48°. Use the formula $d = rt$ to find the airplane's distance d from Omaha at the end of 2 hr:

$$d = 150 \frac{\text{km}}{\text{h}} \cdot 2 \text{ hr} = 300 \text{ km}$$

Then find s, the airplane's distance south of Omaha.

$$\sin 48° = \frac{s}{300}$$
$$s = 300 \sin 48°$$
$$s \approx 223$$

The plane is about 223 km south of Omaha.

86.

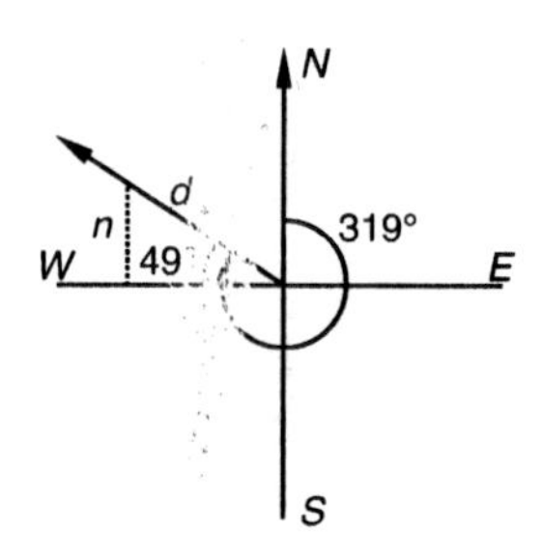

The reference angle is 49°. Use the formula $d = rt$ to find the airplane's distance d from Chicago at the end of 2 hr:

$$d = 120\frac{\text{km}}{\text{h}} \cdot 2 \text{ hr} = 240 \text{ km}$$

Then find n, the airplane's distance north of Chicago.

$$\sin 49° = \frac{n}{240}$$
$$n = 240 \sin 49°$$
$$n \approx 181 \text{ km}$$

87. Use a calculator set in degree mode.
$\tan 310.8° \approx -1.1585$

88. $\cos 205.5° \approx -0.9026$

89. Use a calculator set in degree mode.
$\cot 146.15° = \frac{1}{\tan 146.15°} = -1.4910$

90. $\sin(-16.4°) \approx -0.2823$

91. Use a calculator set in degree mode.
$\sin 118°42' \approx 0.8771$

92. $\cos 273°45' \approx 0.0654$

93. Use a calculator set in degree mode.
$\cos(-295.8°) \approx 0.4352$

94. $\tan 1086.2° \approx 0.1086$

95. Use a calculator set in degree mode.
$\cos 5417° \approx 0.9563$

96. $\sec 240°55' = \frac{1}{\cos 240°55'} \approx -2.0573$

97. Use a calculator set in degree mode.
$\csc 520° = \frac{1}{\sin 520°} \approx 2.9238$

98. $\sin 3824° \approx -0.6947$

99. $\sin\theta = -0.9956$, $270° < \theta < 360°$

We ignore the fact that $\sin\theta$ is negative and use a calculator to find that the reference angle is approximately 84.6°. Since θ is a fourth-quadrant angle, we find θ by subtracting 84.6° from 360°:

$360° - 84.6° = 275.4°$.

Thus, $\theta \approx 275.4°$.

100. $\tan\theta = 0.2460$, $180° < \theta < 270°$

Use a calculator to find that the reference angle is approximately 13.8°. Then $\theta \approx 180° + 13.8° \approx 193.8°$.

101. $\cos\theta = -0.9388$, $180° < \theta < 270°$

We ignore the fact that $\cos\theta$ is negative and use a calculator to find that the reference angle is approximately 20.1°. Since θ is a third-quadrant angle, we find θ by adding 180° and 20.1°:

$180° + 20.1° = 200.1°$.

Thus, $\theta \approx 200.1°$.

102. $\sec\theta = -1.0485$, $90° < \theta < 180°$

$$\cos\theta = \frac{1}{\sec\theta} = \frac{1}{-1.0485} \approx -0.9537$$

Ignore the fact that $\cos\theta$ is negative and use a calculator to find that the reference angle is approximately 17.5°. Then $\theta \approx 180° - 17.5° \approx 162.5°$.

103. $\tan\theta = -3.054$, $270° < \theta < 360°$

We ignore the fact that $\tan\theta$ is negative and use a calculator to find that the reference angle is approximately 71.9°. Since θ is a fourth-quadrant angle, we find θ by subtracting 71.9° from 360°:

$360° - 71.9° = 288.1°$.

Thus, $\theta \approx 288.1°$.

104. $\sin\theta = -0.4313$, $180° < \theta < 270°$

Ignore the fact that $\sin\theta$ is negative and use a calculator to find that the reference angle is approximately 25.6°. Then $\theta \approx 180° + 25.6° \approx 205.6°$.

105. $\csc\theta = 1.0480$, $0 < \theta < 90°$

$$\sin\theta = \frac{1}{\csc\theta} = \frac{1}{1.0480} \approx 0.9542$$

Since θ is a first-quadrant angle, use a calculator to find that $\theta \approx 72.6°$.

106. $\cos\theta = -0.0990$, $90° < \theta < 180°$

Ignore the fact that $\cos\theta$ is negative and use a calculator to find that the reference angle is approximately 84.3°. Then $\theta \approx 180° - 84.3° \approx 95.7°$.

107. Given points P and Q on the terminal side of an angle θ, the reference triangles determined by them are similar. Thus, corresponding sides are proportional and the trigonometric ratios are the same. See the specific example that begins at the bottom of page 363 of the text.

108. Since $\sin\theta = y/r$ and $\cos\theta = x/r$ and $r > 0$ for all angles θ, the domain of the sine and of the cosine functions is the set of all angles θ. However, $\tan\theta = y/x$ and $x = 0$ for all angles which are odd multiples of 90°. Thus, the domain of the tangent function must be restricted to avoid division by 0.

109. $f(x) = \frac{1}{x^2 - 25}$

1. The zeros of the denominator are -5 and 5, so $x = -5$ and $x = 5$ are vertical asymptotes.
2. Because the degree of the numerator is less than the degree of the denominator, the x-axis is the horizontal asymptote. There is no oblique asymptote.

3. The numerator has no zeros, so there is no x-intercept.

4. $f(0) = \dfrac{1}{0^2 - 25} = -\dfrac{1}{25}$, so $\left(0, -\dfrac{1}{25}\right)$ is the y-intercept.

5. Find other function values as needed to determine the general shape and then draw the graph.

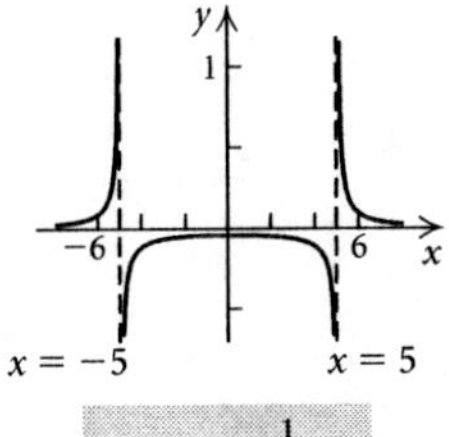

110.

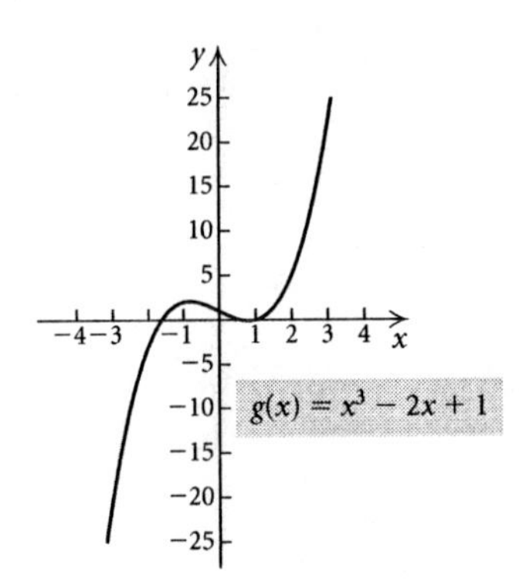

111. $f(x) = \dfrac{x-4}{x+2}$

The denominator is zero for $x = -2$, so the domain is $\{x|x \neq -2\}$.

Examining the graph of the function, we see that the range is $\{x|x \neq 1\}$.

112. $g(x) = \dfrac{x^2 - 9}{2x^2 - 7x - 15}$

$2x^2 - 7x - 15 = (2x+3)(x-5)$; the denominator is zero for $x = -\dfrac{3}{2}$ or $x = 5$, so the domain is $\left\{x \middle| x \neq -\dfrac{3}{2} \text{ and } x \neq 5\right\}$.

Examining the graph of the function, we see that the range is the set of all real numbers.

113. $f(x) = 12 - x$

Solve: $0 = 12 - x$

$x = 12$

The zero of the function is 12.

114. $x^2 - x - 6 = 0$

$(x-3)(x+2) = 0$

$x = 3$ *or* $x = -2$

The zeros are -2 and 3.

115. The first coordinate of the x-intercept is the zero of the function. Thus from Exercise 113 we know that the x-intercept is $(12, 0)$.

116. From Exercise 114 we know that the x-intercepts are $(-2, 0)$ and $(3, 0)$.

117.

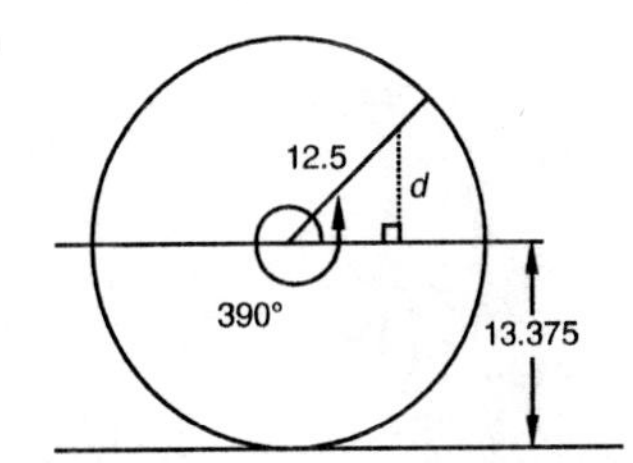

The reference angle is 30°.

Let d be the vertical distance of the valve cap above the center of the wheel.

$\sin 30° = \dfrac{d}{12.5}$

$d = 12.5 \sin 30°$

$d = 6.25$

The distance above the ground is 6.25 in.+ 13.375 in. = 19.625 in.

118.

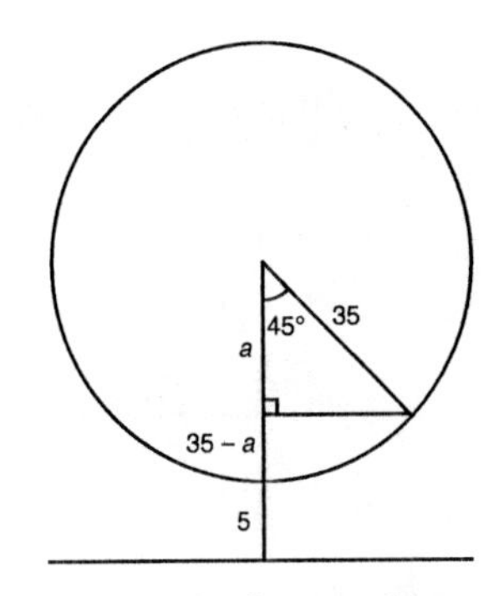

765°is in the first quadrant, and the reference angle is 45°.

$\cos 45° = \dfrac{a}{35}$

$a = 35 \cos 45°$

$a \approx 24.7$

The height above the ground is $(35-a)+5 \approx 35-24.7+5 \approx 15.3$ ft.

Exercise Set 5.4

1. See the answer section in the text.

2.

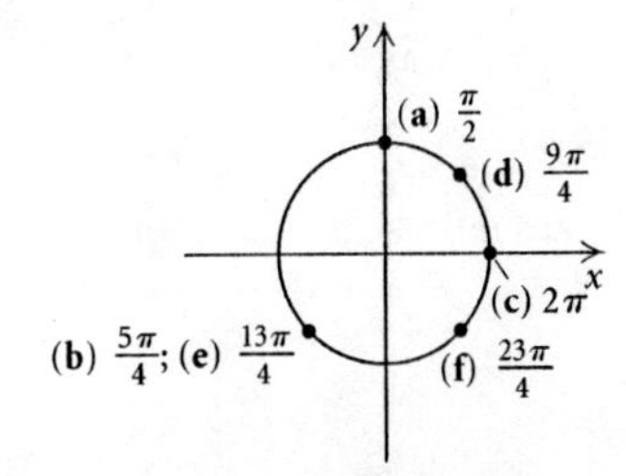

3. See the answer section in the text.

4.

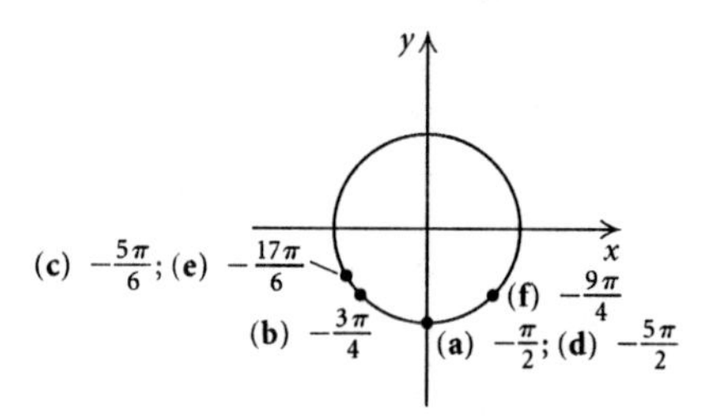

5. Clockwise M is at $\frac{2}{3}\cdot\pi$, or $\frac{2\pi}{3}$. $\left(\text{We could also say that } M \text{ is at } \frac{1}{3}\cdot 2\pi.\right)$ Counterclockwise M is at $\frac{2}{3}\cdot 2\pi$, or $\frac{4\pi}{3}$ so the number $-\frac{4\pi}{3}$ determines M. $\left(\text{We could also say that } M \text{ is at } \frac{8}{6}\cdot\pi \text{ moving counterclockwise, so again the result is } -\frac{4\pi}{3}.\right)$

Clockwise N is at $\frac{3}{4}\cdot 2\pi$, or $\frac{3\pi}{2}$.

Counterclockwise N is at $\frac{1}{4}\cdot 2\pi$, or $\frac{\pi}{2}$, so the number $-\frac{\pi}{2}$ determines N.

Clockwise P is at $\frac{5}{8}\cdot 2\pi$, or $\frac{5\pi}{4}$. $\left(\text{We could also say that } P \text{ is at } \frac{5}{4}\cdot\pi.\right)$

Counterclockwise P is at $\frac{3}{4}\cdot\pi$, or $\frac{3\pi}{4}$, so the number $-\frac{3\pi}{4}$ determines P. $\left(\text{We could also say that } P \text{ is at } \frac{3}{8}\cdot 2\pi \text{ moving counterclockwise, so again the result is } -\frac{3\pi}{4}.\right)$

Clockwise Q is at $\frac{11}{6}\cdot\pi$, or $\frac{11\pi}{6}$. $\left(\text{We could also say that } Q \text{ is at } \frac{11}{12}\cdot 2\pi.\right)$

Counterclockwise Q is at $\frac{1}{6}\cdot\pi$, or $\frac{\pi}{6}$ so the number $-\frac{\pi}{6}$ determines Q. $\left(\text{We could also say that } Q \text{ is at } \frac{1}{12}\cdot 2\pi \text{ moving counterclockwise, so again the result is } -\frac{\pi}{6}.\right)$

6. M: $\pi, -\pi$; N: $\frac{\pi}{4}, -\frac{7\pi}{4}$; P: $\frac{4\pi}{3}, -\frac{2\pi}{3}$; Q: $\frac{7\pi}{6}, -\frac{5\pi}{6}$

7. a) $\frac{2.4}{2\pi} \approx 0.38$, or $2.4 \approx 0.38(2\pi)$, so we move counterclockwise about 0.38 of the way around the circle.

b) $\frac{7.5}{2\pi} \approx 1.19$, or $7.5 \approx 1.19(2\pi)$, or $1(2\pi)+0.19(2\pi)$, so we move completely around the circle counterclockwise once and then continue about another 0.19 of the way around.

c) $\frac{32}{2\pi} \approx 5.09$, or $32 \approx 5.09(2\pi)$, or $5(2\pi)+0.09(2\pi)$, so we move completely around the circle counterclockwise 5 times and then continue about another 0.09 of the way around.

d) $\frac{320}{2\pi} \approx 50.93$, or $320 \approx 50.93(2\pi)$, or $50(2\pi)+0.93(2\pi)$, so we move completely around the circle counterclockwise 50 times and then continue about another 0.93 of the way around.

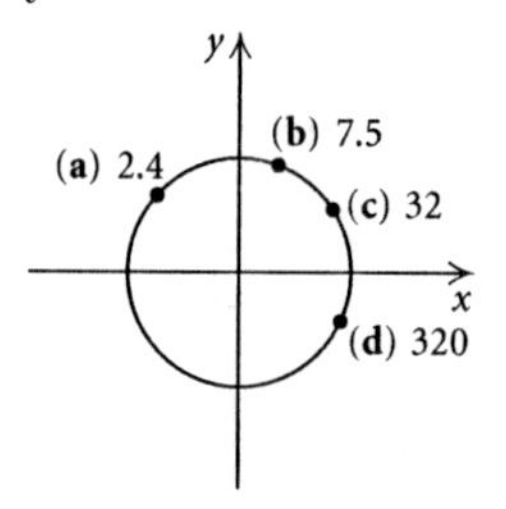

8. a) $0.25 \approx 0.04(2\pi)$, so we move counterclockwise about 0.04 of the way around the circle.

b) $1.8 \approx 0.29(2\pi)$, so we move counterclockwise about 0.29 of the way around the circle.

c) $47 \approx 7.48(2\pi)$, or $7(2\pi)+0.48(2\pi)$, so we move completely around the circle counterclockwise 7 times and then continue about another 0.48 of the way around.

d) $500 \approx 79.58(2\pi)$, or $79(2\pi)+0.58(2\pi)$, so we move completely around the circle counterclockwise 79 times and then continue about another 0.58 of the way around.

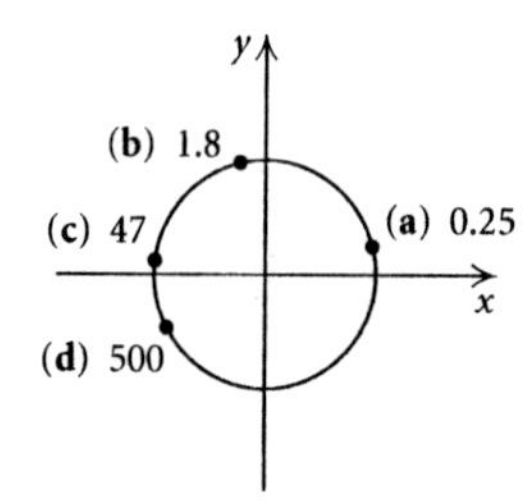

9. We add and subtract multiples of 2π. Answers may vary.

Positive angle: $\frac{\pi}{4}+2\pi = \frac{\pi}{4}+\frac{8\pi}{4} = \frac{9\pi}{4}$

Negative angle: $\frac{\pi}{4}-2\pi = \frac{\pi}{4}-\frac{8\pi}{4} = -\frac{7\pi}{4}$

10. Answers may vary.

Positive angle: $\frac{5\pi}{3}+2\pi = \frac{11\pi}{3}$

Negative angle: $\frac{5\pi}{3}-2\pi = -\frac{\pi}{3}$

11. We add and subtract multiples of 2π. Answers may vary.

Positive angle: $\frac{7\pi}{6}+2\pi = \frac{7\pi}{6}+\frac{12\pi}{6} = \frac{19\pi}{6}$

Negative angle: $\frac{7\pi}{6}-2\pi = \frac{7\pi}{6}-\frac{12\pi}{6} = -\frac{5\pi}{6}$

12. Answers may vary.

Positive angle: $\pi + 2\pi = 3\pi$

Negative angle: $\pi - 2\pi = -\pi$

13. We add and subtract multiples of 2π. Answers may vary.

Positive angle: $-\frac{2\pi}{3} + 2\pi = -\frac{2\pi}{3} + \frac{6\pi}{3} = \frac{4\pi}{3}$

Negative angle: $-\frac{2\pi}{3} - 2\pi = -\frac{2\pi}{3} - \frac{6\pi}{3} = -\frac{8\pi}{3}$

14. Answers may vary.

Positive angle: $-\frac{3\pi}{4} + 2\pi = \frac{5\pi}{4}$

Negative angle: $-\frac{3\pi}{4} - 2\pi = -\frac{11\pi}{4}$

15. Complement: $\frac{\pi}{2} - \frac{\pi}{3} = \frac{3\pi}{6} - \frac{2\pi}{6} = \frac{\pi}{6}$

Supplement: $\pi - \frac{\pi}{3} = \frac{3\pi}{3} - \frac{\pi}{3} = \frac{2\pi}{3}$

16. Complement: $\frac{\pi}{2} - \frac{5\pi}{12} = \frac{\pi}{12}$

Supplement: $\pi - \frac{5\pi}{12} = \frac{7\pi}{12}$

17. Complement: $\frac{\pi}{2} - \frac{3\pi}{8} = \frac{4\pi}{8} - \frac{3\pi}{8} = \frac{\pi}{8}$

Supplement: $\pi - \frac{3\pi}{8} = \frac{8\pi}{8} - \frac{3\pi}{8} = \frac{5\pi}{8}$

18. Complement: $\frac{\pi}{2} - \frac{\pi}{4} = \frac{\pi}{4}$

Supplement: $\pi - \frac{\pi}{4} = \frac{3\pi}{4}$

19. Complement: $\frac{\pi}{2} - \frac{\pi}{12} = \frac{6\pi}{12} - \frac{\pi}{12} = \frac{5\pi}{12}$

Supplement: $\pi - \frac{\pi}{12} = \frac{12\pi}{12} - \frac{\pi}{12} = \frac{11\pi}{12}$

20. Complement: $\frac{\pi}{2} - \frac{\pi}{6} = \frac{3\pi}{6} - \frac{\pi}{6} = \frac{2\pi}{6} = \frac{\pi}{3}$

Supplement: $\pi - \frac{\pi}{6} = \frac{6\pi}{6} - \frac{\pi}{6} = \frac{5\pi}{6}$

21. $75° = 75° \cdot \frac{\pi \text{ radians}}{180°} = \frac{5\pi}{12}$ radians

22. $30° = 30° \cdot \frac{\pi \text{ radians}}{180°} = \frac{\pi}{6}$ radians

23. $200° = 200° \cdot \frac{\pi \text{ radians}}{180°} = \frac{10\pi}{9}$ radians

24. $-135° = -135° \cdot \frac{\pi \text{ radians}}{180°} = -\frac{3\pi}{4}$ radians

25. $-214.6° = -214.6° \cdot \frac{\pi \text{ radians}}{180°} = -\frac{214.6\pi}{180}$ radians

If we multiply by 5/5 to clear the decimal, we have $-\frac{1073\pi}{900}$ radians.

26. $37.71° = 37.71° \cdot \frac{\pi \text{ radians}}{180°} = \frac{37.71\pi}{180}$ radians, or $\frac{419\pi}{2000}$ radians

27. $-180° = -180° \cdot \frac{\pi \text{ radians}}{180°} = -\pi$ radians

28. $90° = 90° \cdot \frac{\pi \text{ radians}}{180°} = \frac{\pi}{2}$ radians

29. $12.5° = 12.5° \cdot \frac{\pi \text{ radians}}{180°} = \frac{12.5\pi}{180}$ radians

If we multiply by 2/2 to clear the decimal and then simplify we have $\frac{5\pi}{72}$ radians.

30. $6.3° = 6.3° \cdot \frac{\pi \text{ radians}}{180°} = \frac{6.3\pi}{180}$ radians, or $\frac{7\pi}{200}$ radians

31. $-340° = -340° \cdot \frac{\pi \text{ radians}}{180°} = -\frac{17\pi}{9}$ radians

32. $-60° = -60° \cdot \frac{\pi \text{ radians}}{180°} = -\frac{\pi}{3}$ radians

33. $240° = 240° \cdot \frac{\pi \text{ radians}}{180°} = \frac{240\pi}{180}$ radians $\approx$ 4.19 radians

34. $15° = 15° \cdot \frac{\pi \text{ radians}}{180°} \approx 0.26$ radians

35. $-60° = -60° \cdot \frac{\pi \text{ radians}}{180°} = -\frac{60\pi}{180}$ radians $\approx$ -1.05 radians

36. $145° = 145° \cdot \frac{\pi \text{ radians}}{180°} \approx 2.53$ radians

37. $117.8° = 117.8° \cdot \frac{\pi \text{ radians}}{180°} = \frac{117.8\pi}{180}$ radians $\approx$ 2.06 radians

38. $-231.2° = -231.2° \cdot \frac{\pi \text{ radians}}{180°} \approx -4.04$ radians

39. $1.354° = 1.354° \cdot \frac{\pi \text{ radians}}{180°} = \frac{1.354\pi}{180}$ radians $\approx$ 0.02 radians

40. $584° = 584° \cdot \frac{\pi \text{ radians}}{180°} \approx 10.19$ radians

41. $345° = 345° \cdot \frac{\pi \text{ radians}}{180°} = \frac{345\pi}{180}$ radians $\approx$ 6.02 radians

42. $-75° = -75° \cdot \frac{\pi \text{ radians}}{180°} \approx -1.31$ radians

43. $95° = 95° \cdot \frac{\pi \text{ radians}}{180°} = \frac{95\pi}{180}$ radians $\approx$ 1.66 radians

44. $24.8° = 24.8° \cdot \frac{\pi \text{ radians}}{180°} \approx 0.43$ radians

45. $-\frac{3\pi}{4} = -\frac{3\pi}{4} \text{ radians} \cdot \frac{180°}{\pi \text{ radians}} = -\frac{3}{4} \cdot 180° =$ $-135°$

46. $\frac{7\pi}{6} = \frac{7\pi}{6} \text{ radians} \cdot \frac{180°}{\pi \text{ radians}} = 210°$

47. $8\pi = 8\pi \text{ radians} \cdot \dfrac{180°}{\pi \text{ radians}} = 8 \cdot 180° = 1440°$

48. $-\dfrac{\pi}{3} = -\dfrac{\pi}{3} \text{ radians} \cdot \dfrac{180°}{\pi \text{ radians}} = -60°$

49. $1 = 1 \text{ radian} \cdot \dfrac{180°}{\pi \text{ radians}} \approx \dfrac{180°}{\pi} \approx 57.30°$

50. $-17.6 = -17.6 \text{ radians} \cdot \dfrac{180°}{\pi \text{ radians}} =$

$-\dfrac{17.6(180°)}{\pi} \approx -1008.41°$

51. $2.347 = 2.347 \text{ radians} \cdot \dfrac{180°}{\pi \text{ radians}} =$

$\dfrac{2.347(180°)}{\pi} \approx 134.47°$

52. $25 = 25 \text{ radians} \cdot \dfrac{180°}{\pi \text{ radians}} \approx 1432.39°$

53. $\dfrac{5\pi}{4} = \dfrac{5\pi}{4} \text{ radians} \cdot \dfrac{180°}{\pi \text{ radians}} = \dfrac{5}{4} \cdot 180° = 225°$

54. $-6\pi = -6\pi \text{ radians} \cdot \dfrac{180°}{\pi \text{ radians}} = -1080°$

55. $-90 = -90 \text{ radians} \cdot \dfrac{180°}{\pi \text{ radians}} = \dfrac{-90(180°)}{\pi} \approx$

$-5156.62°$

56. $37.12 = 37.12 \text{ radians} \cdot \dfrac{180°}{\pi \text{ radians}} \approx 2126.82°$

57. $\dfrac{2\pi}{7} = \dfrac{2\pi}{7} \text{ radians} \cdot \dfrac{180°}{\pi \text{ radians}} = \dfrac{2}{7} \cdot 180° \approx 51.43°$

58. $\dfrac{\pi}{9} = \dfrac{\pi}{9} \text{ radians} \cdot \dfrac{180°}{\pi \text{ radians}} = 20°$

59. $0° = 0° \cdot \dfrac{\pi \text{ radians}}{180°} = 0;$

$30° = 30° \cdot \dfrac{\pi \text{ radians}}{180°} = \dfrac{\pi}{6};$

$45° = 45° \cdot \dfrac{\pi \text{ radians}}{180°} = \dfrac{\pi}{4};$

$60° = 60° \cdot \dfrac{\pi \text{ radians}}{180°} = \dfrac{\pi}{3};$

$90° = 90° \cdot \dfrac{\pi \text{ radians}}{180°} = \dfrac{\pi}{2};$

$135° = 135° \cdot \dfrac{\pi \text{ radians}}{180°} = \dfrac{3\pi}{4};$

$180° = 180° \cdot \dfrac{\pi \text{ radians}}{180°} = \pi;$

$225° = 225° \cdot \dfrac{\pi \text{ radians}}{180°} = \dfrac{5\pi}{4};$

$270° = 270° \cdot \dfrac{\pi \text{ radians}}{180°} = \dfrac{3\pi}{2};$

$315° = 315° \cdot \dfrac{\pi \text{ radians}}{180°} = \dfrac{7\pi}{4};$

$360° = 360° \cdot \dfrac{\pi \text{ radians}}{180°} = 2\pi$

60. (See Exercise 59.) $0° = 0$; $-30° = -\dfrac{\pi}{6}$; $-45° = -\dfrac{\pi}{4}$;

$-60° = -\dfrac{\pi}{3}$; $-90° = -\dfrac{\pi}{2}$; $-135° = -\dfrac{3\pi}{4}$;

$-180° = -\pi$; $-225° = -\dfrac{5\pi}{4}$; $-270° = -\dfrac{3\pi}{2}$;

$-315° = -\dfrac{7\pi}{4}$; $-360° = -2\pi$

61. $\theta = \dfrac{s}{r} = \dfrac{8}{3.5} \approx 2.29$

62. Note that $45° = \dfrac{\pi}{4}$ radians.

$\theta = \dfrac{s}{r}$

$\dfrac{\pi}{4} = \dfrac{200 \text{ cm}}{r}$

$r = \dfrac{4}{\pi} \cdot 200 \text{ cm} \approx 254.65 \text{ cm}$

63. $\theta = \dfrac{s}{r}$

$5 = \dfrac{16 \text{ yd}}{r}$

$r = \dfrac{16 \text{ yd}}{5} = 3.2 \text{ yd}$

64. $\theta = \dfrac{s}{r}$, or $s = r\theta$

$s = 4.2 \text{ in.} \cdot \dfrac{5\pi}{12} \approx 5.50 \text{ in.}$

65. $\theta = \dfrac{s}{r}$

θ is the radian measure of the central angle, s is arc length, and r is radius length.

$\theta = \dfrac{132 \text{ cm}}{120 \text{ cm}}$ Substituting 132 cm for s and 120 cm for r

$\theta = \dfrac{11}{10}$, or 1.1 The unit is understood to be radians.

$1.1 = 1.1 \text{ radians} \cdot \dfrac{180°}{\pi \text{ radians}} \approx 63°$

66. Radius $= 10 \text{ ft}/2 = 5 \text{ ft}$

$\theta = \dfrac{s}{r} = \dfrac{20 \text{ ft}}{5 \text{ ft}} = 4 \text{ radians}$

$4 \text{ radians} \cdot \dfrac{180°}{\pi \text{ radians}} \approx 229°$

67. We use the formula $\theta = \dfrac{s}{r}$, or $s = r\theta$.

$s = r\theta = (2 \text{ yd})(1.6) = 3.2 \text{ yd}$

68. $s = r\theta = (5 \text{ m})(2.1) = 10.5 \text{ m}$

69. Let r = the length of the minute hand. In 60 minutes the minute hand travels the circumference of a circle with radius r, or $2\pi r$. Then in 50 minutes the minute hand travels $\dfrac{50}{60} \cdot 2\pi r$, or $\dfrac{5\pi r}{3}$.

Now find the angle through which the minute hand rotates in 50 minutes.

$\theta = \dfrac{s}{r}$

$$\theta = \frac{\frac{5\pi r}{3}}{r}$$
$$\theta = \frac{5\pi r}{3} \cdot \frac{1}{r}$$
$$\theta = \frac{5\pi}{3} \approx 5.24$$
The minute hand rotates through about 5.24 radians.

70. $1 \text{ mi} = 1 \text{ mi} \cdot \frac{5280 \text{ ft}}{1 \text{ mi}} \cdot \frac{12 \text{ in.}}{1 \text{ ft}} = 63,360 \text{ in.}$
$$r = \frac{d}{2} = \frac{24.877 \text{ in.}}{2} = 12.4385 \text{ in.}$$
$$\theta = \frac{s}{r} = \frac{63,360 \text{ in.}}{12.4385 \text{ in.}} \approx 5094$$

71. Since the linear speed must be in cm/min, the given angular speed, 7 radians/sec, must be changed to radians/min.
$$\omega = \frac{7 \text{ radians}}{1 \text{ sec}} \cdot \frac{60 \text{ sec}}{1 \text{ min}} = \frac{420 \text{ radians}}{1 \text{ min}}$$
$$r = \frac{d}{2} = \frac{15 \text{ cm}}{2} = 7.5 \text{ cm}$$
Using $v = r\omega$, we have:
$$v = 7.5 \text{ cm} \cdot \frac{420}{1 \text{ min}} \quad \text{Substituting and omitting the word radians}$$
$$= 3150 \frac{\text{cm}}{\text{min}}$$
The linear speed of a point on the rim is 3150 cm/min.

72. $r = 30 \text{ cm} \cdot \frac{1 \text{ m}}{100 \text{ cm}} = 0.3 \text{ m}$
$$\omega = \frac{3 \text{ radians}}{1 \text{ sec}} \cdot \frac{60 \text{ sec}}{1 \text{ min}} = \frac{180 \text{ radians}}{1 \text{ min}}$$
$$v = r\omega = 0.3 \text{ m} \cdot \frac{180}{1 \text{ min}} = 54 \text{ m/min}$$

73. First convert 18.33 ft/sec to in./hr.
$$18.33 \text{ ft/sec} = \frac{18.33 \text{ ft}}{1 \text{ sec}} \cdot \frac{12 \text{ in.}}{1 \text{ ft}} \cdot \frac{60 \text{ sec}}{1 \text{ min}} \cdot \frac{60 \text{ min}}{1 \text{ hr}} =$$
791,856 in./hr
$$r = \frac{21 \text{ in.}}{2} = 10.5 \text{ in.}$$
Using $v = r\omega$, we have
$$791,856 \text{ in./hr} = 10.5 \text{ in.} \cdot \omega$$
$$75,415 \text{ radians/hr} \approx \omega$$
Now convert 75,415 radians/hr to revolutions/hr.
$$75,415 \text{ radians/hr} = \frac{75,415 \text{ radians}}{1 \text{ hr}} \cdot \frac{1 \text{ revolution}}{2\pi \text{ radians}} \approx$$
12,003 revolutions/hr

The angular speed of the cylinder is about 12,003 revolutions/hr (or about 12,003 IPH).

74. $19 \text{ ft } 3 \text{ in.} = 19.25 \text{ ft} \cdot \frac{1 \text{ mi}}{5280 \text{ ft}} = \frac{19.25}{5280} \text{ mi}$
$$13 \text{ ft } 11 \text{ in.} = 13\frac{11}{12} \text{ ft} = \frac{167}{12} \text{ ft} = \frac{167}{12} \text{ ft} \cdot \frac{1 \text{ mi}}{5280 \text{ ft}} =$$
$$\frac{167}{63,360} \text{ mi}$$
$$\frac{2.4(2\pi)}{1 \text{ min}} = \frac{2.4(2\pi)}{1 \text{ min}} \cdot \frac{60 \text{ min}}{1 \text{ hr}} = \frac{288\pi}{1 \text{ hr}}$$
Find Alicia's linear speed:
$$v = r\omega = \frac{19.25}{5280} \text{mi} \cdot \frac{288\pi}{1 \text{ hr}} \approx 3.30 \text{ mph}$$
Find Zoe's linear speed:
$$v = r\omega = \frac{167}{63,360} \text{mi} \cdot \frac{288\pi}{1 \text{ hr}} \approx 2.38 \text{ mph}$$
The difference in the linear speeds is about 3.30 − 2.38, or 0.92 mph. (Answers may vary slightly due to rounding differences.)

75. First find ω in radians per hour.
$$\omega = \frac{2\pi}{24 \text{ hr}} = \frac{\pi}{12 \text{ hr}}$$
Using $v = r\omega$, we have
$$v = 4000 \text{ mi} \cdot \frac{\pi}{12 \text{ hr}} \approx 1047 \text{ mph.}$$
The linear speed of a point on the equator is about 1047 mph.

76. $\omega = \frac{2\pi}{365.25 \text{ days}} \cdot \frac{1 \text{ day}}{24 \text{ hr}} = \frac{\pi}{4383 \text{ hr}}$
$$v = r\omega = 93,000,000 \text{ mi} \cdot \frac{\pi}{4383 \text{ hr}} \approx 66,659 \text{ mph}$$

77. First find ω in radians per hour.
$$\omega = \frac{14 \cdot 2\pi}{1 \text{ min}} \cdot \frac{60 \text{ min}}{1 \text{ hr}} = \frac{1680\pi}{\text{hr}}$$
Next find r in miles.
$$r = 10 \text{ ft} \cdot \frac{1 \text{ mi}}{5280 \text{ ft}} = \frac{1}{528} \text{ mi}$$
Using $v = r\omega$, we have
$$v = \frac{1}{528} \text{ mi} \cdot \frac{1680\pi}{1 \text{ hr}} \approx 10 \text{ mph.}$$
The speed of the river is about 10 mph.

78. $67 \text{ cm} = 67 \text{ cm} \cdot \frac{1 \text{ m}}{100 \text{ cm}} \cdot \frac{1 \text{ km}}{1000 \text{ m}} = 0.00067 \text{ km}$
$$r = \frac{0.00067 \text{ km}}{2} = 0.000335 \text{ km}$$
$$\omega = \frac{v}{r} = \frac{40.940 \text{ km/h}}{0.000335 \text{ km}} \approx \frac{122,208.955}{1 \text{ hr}}$$
Convert ω from radians/hr to revolutions/hr:
$$\frac{122,208.955 \text{ radians}}{1 \text{ hr}} \cdot \frac{1 \text{ revolution}}{2\pi \text{ radians}} \approx$$
19,450 revolutions/hr

79. First convert 22 mph to inches/second.
$$v = \frac{22 \text{ mi}}{1 \text{ hr}} \cdot \frac{5280 \text{ ft}}{1 \text{ mi}} \cdot \frac{12 \text{ in.}}{1 \text{ ft}} \cdot \frac{1 \text{ hr}}{60 \text{ min}} \cdot \frac{1 \text{ min}}{60 \text{ sec}} =$$
387.2 in./sec

Using $v = r\omega$, we have
$$387.2 \frac{\text{in.}}{\text{sec}} = 23 \text{ in.} \cdot \omega$$
$$\text{so } \omega = \frac{387.2 \text{ in./sec}}{23 \text{ in.}} \approx \frac{16.835}{1 \text{ sec}}.$$
Then in 12 sec,
$$\theta = \omega t = \frac{16.835}{1 \text{ sec}} \cdot 12 \text{ sec} \approx 202.$$
The wheel rotates through an angle of 202 radians.

80. Left to the student

81. Left to the student

82. Consider Example 7. The division 4 m/2 m simplifies to the real number 2 since m/m = 1. From this point of view it would seem preferable to omit "radians." Also consider Example 8. If "radians" were used throughout the solution, the answer would have been 5.24 cm - radians. Since we are finding a distance the correct unit is centimeters. Thus we omit "radians." Since a measure in radians is simply a real number, it is usually preferable to omit the word "radians."

83. For a point at a distance r from the center of rotation with a fixed angular speed k, the linear speed is given by $v = r \cdot k$, or $r = \frac{1}{k}v$. Thus, the length of the radius is directly proportional to the linear speed.

84. We see from the formula $\theta = \frac{s}{r}$ that the tire with the 15 in. diameter will rotate through a larger angle than the tire with the 16 in. diameter. Thus the car with the 15 in. tires will probably need new tires first.

85. one-to-one

86. cosine of θ

87. exponential function

88. horizontal asymptote

89. odd function

90. natural

91. horizontal line; inverse

92. logarithm

93. One degree of latitude is $\frac{1}{360}$ of the circumference of the earth.

$c = \pi d$, or $2\pi r$

When $r = 6400$ km, $C = 2\pi \cdot 6400 = 12,800\pi$ km.

Thus $1°$ of latitude is $\frac{1}{360} \cdot 12,800\pi \text{ km} = 111.7$ km.

When $r = 4000$ mi, $C = 2\pi \cdot 4000 \text{ mi} = 8000\pi$ mi.

Thus $1°$ of latitude is $\frac{1}{360} \cdot 8000\pi \text{ mi} \approx 69.8$ mi.

94.
$$x^2 + \left(-\frac{\sqrt{21}}{5}\right)^2 = 1$$
$$x^2 = \frac{4}{25}$$
$$x = \pm\frac{2}{5}$$

95. a)
$$\begin{aligned} &100 \text{ mils} \\ &= 100 \text{ mils} \cdot \frac{90°}{1600 \text{ mils}} \\ &= 5.625° \\ &= 5°37'30'' \quad \text{Using the DMS feature on a calculator} \end{aligned}$$

b)
$$\begin{aligned} &350 \text{ mils} \\ &= 350 \text{ mils} \cdot \frac{90°}{1600 \text{ mils}} \\ &= 19.6875° \\ &= 19°41'15'' \quad \text{Using the DMS feature on a calculator} \end{aligned}$$

96. a) $48° = 48° \cdot \frac{100 \text{ grads}}{90°}$

$= \frac{48}{90} \cdot 100 \text{ grads} \approx 53.33$ grads

b) $\frac{5\pi}{7} = \frac{5\pi}{7} \text{ radians} \cdot \frac{100 \text{ grads}}{\frac{\pi}{2} \text{ radians}} = \frac{1000}{7} \text{ grads} \approx$

142.86 grads

97. Let ω_1 = the angular speed of the smaller wheel and ω_2 = the angular speed of the larger wheel. The wheels have the same linear speed, so we have $v = 40\omega_1 = 50\omega_2$.

Convert the angular speed of the smaller wheel, ω_1, to radians per second.

$$\omega_1 = 20 \text{ rpm} = \frac{20 \cdot 2\pi}{1 \text{ min}} \cdot \frac{1 \text{ min}}{60 \text{ sec}} = \frac{2}{3}\pi/\text{ sec}$$

Then

$$\begin{aligned} 40\omega_1 &= 50\omega_2 \\ 40 \cdot \frac{2}{3}\pi/\text{sec} &= 50\omega_2 \\ \frac{8\pi}{15}/\text{sec} &= \omega_2 \\ 1.676/\text{sec} &\approx \omega_2. \end{aligned}$$

The angular speed of the larger wheel is about 1.676 radians/sec.

98. The radii of the pulleys are 25 cm and 15 cm. Let ω_1 = the angular speed of the larger pulley and ω_2 = the angular speed of the smaller pulley. Since they are connected by the same belt, their linear speed v, will be the same.

$v = 25\omega_1 = 15\omega_2$.

Find ω_1, the angular speed of the larger pulley, in radians/sec. The larger pulley makes 12 revolutions per minute.

$$\omega_1 = \frac{12 \cdot 2\pi}{1 \text{ min}} \cdot \frac{1 \text{ min}}{60 \text{ sec}} = 0.4\pi/\text{sec}$$

Then

$$25 \cdot 0.4\pi/\text{sec} = 15\omega_2$$
$$\omega_2 = \frac{2}{3}\pi/\text{sec} \approx 2.094/\text{sec}.$$

99.

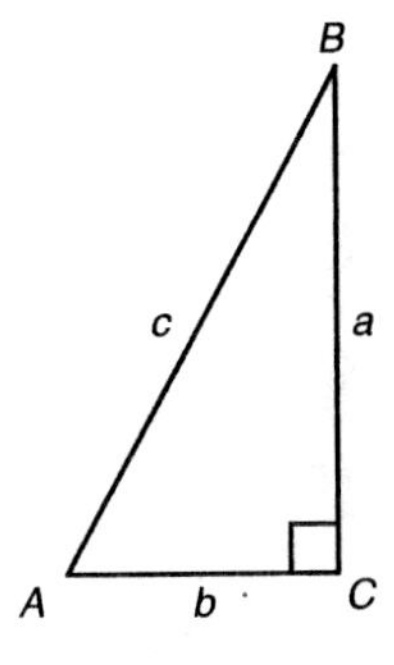

$a = 38°28'45'' - 38°27'30'' = 1'15'' = 1.25' =$
1.25 nautical miles.

$b = 82°57'15'' - 82°56'30'' = 45'' = 0.75' =$
0.75 nautical miles.

$c = \sqrt{a^2 + b^2} = \sqrt{(1.25)^2 + (0.75)^2} \approx$
1.46 nautical miles

100. Let x = the number of minutes after noon when the hands are first perpendicular. The minute hand moves through $\frac{1}{60} \cdot 2\pi$, or $\frac{\pi}{30}$ radians in 1 minute; the hour hand moves through $\frac{1}{12}$ of this rotation, or $\frac{1}{12} \cdot \frac{\pi}{30}$, or $\frac{\pi}{360}$ radians in 1 minute. Then in x minutes the minute hand and hour hand move through $\frac{\pi x}{30}$ and $\frac{\pi x}{360}$ radians, respectively. We want to find x such that the difference in the rotations is $\frac{\pi}{2}$.

$$\frac{\pi x}{30} - \frac{\pi x}{360} = \frac{\pi}{2}$$
$$\frac{11\pi x}{360} = \frac{\pi}{2}$$
$$x = \frac{180}{11} \approx 16.3636$$

Thus the hands are perpendicular about 16.3636 minutes after 12:00 noon, or at about 12:16:22 P.M.

Exercise Set 5.5

1.

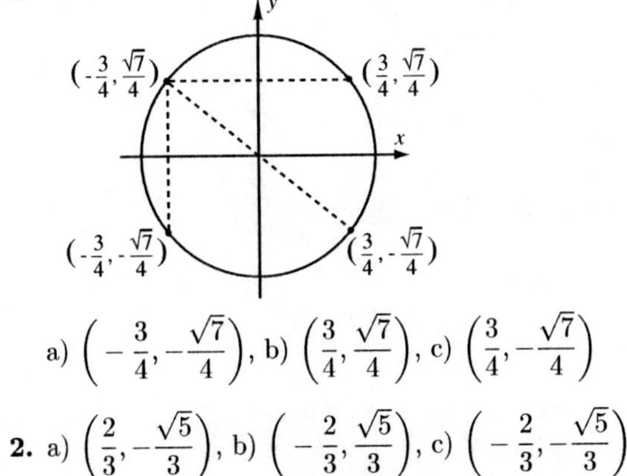

a) $\left(-\frac{3}{4}, -\frac{\sqrt{7}}{4}\right)$, b) $\left(\frac{3}{4}, \frac{\sqrt{7}}{4}\right)$, c) $\left(\frac{3}{4}, -\frac{\sqrt{7}}{4}\right)$

2. a) $\left(\frac{2}{3}, -\frac{\sqrt{5}}{3}\right)$, b) $\left(-\frac{2}{3}, \frac{\sqrt{5}}{3}\right)$, c) $\left(-\frac{2}{3}, -\frac{\sqrt{5}}{3}\right)$

3.

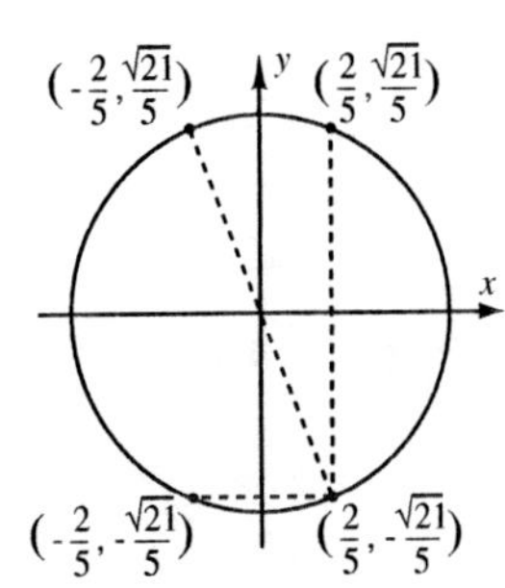

a) $\left(\frac{2}{5}, \frac{\sqrt{21}}{5}\right)$, b) $\left(-\frac{2}{5}, -\frac{\sqrt{21}}{5}\right)$, c) $\left(-\frac{2}{5}, \frac{\sqrt{21}}{5}\right)$

4. a) $\left(-\frac{\sqrt{3}}{2}, \frac{1}{2}\right)$, b) $\left(\frac{\sqrt{3}}{2}, -\frac{1}{2}\right)$, c) $\left(\frac{\sqrt{3}}{2}, \frac{1}{2}\right)$

5. The point determined by $-\pi/4$ is a reflection across the x-axis of the point determined by $\pi/4$, $\left(\frac{\sqrt{2}}{2}, \frac{\sqrt{2}}{2}\right)$. Thus, the coordinates of the point determined by $-\pi/4$ are $\left(\frac{\sqrt{2}}{2}, -\frac{\sqrt{2}}{2}\right)$.

6. The point determined by $-\beta$ is a reflection across the x-axis of the point determined by β. Its coordinates are $\left(-\frac{2}{3}, -\frac{\sqrt{5}}{3}\right)$.

7. The coordinates of the point determined by π are $(-1, 0)$. Thus,

$$\sin \pi = y = 0.$$

8. The point determined by $-\pi/3$ is a reflection across the x-axis of the point determined by $\pi/3$, $\left(\frac{1}{2}, \frac{\sqrt{3}}{2}\right)$. Its coordinates are $\left(\frac{1}{2}, -\frac{\sqrt{3}}{2}\right)$. Thus,

$$\cos\left(-\frac{\pi}{3}\right) = x = \frac{1}{2}.$$

9. The point determined by $\frac{7\pi}{6}$ is a reflection across the origin of the point determined by $\pi/6$, $\left(\frac{\sqrt{3}}{2}, \frac{1}{2}\right)$. Its coordinates are $\left(-\frac{\sqrt{3}}{2}, -\frac{1}{2}\right)$. Thus,

$$\cot\frac{7\pi}{6} = \frac{x}{y} = \frac{-\frac{\sqrt{3}}{2}}{-\frac{1}{2}} = \sqrt{3}.$$

10. The point determined by $11\pi/4$ is a reflection across the y-axis of the point determined by $\pi/4$, $\left(\frac{\sqrt{2}}{2}, \frac{\sqrt{2}}{2}\right)$. Its coordinates are $\left(-\frac{\sqrt{2}}{2}, \frac{\sqrt{2}}{2}\right)$. Thus,

$$\tan\frac{11\pi}{4} = \frac{y}{x} = \frac{\frac{\sqrt{2}}{2}}{-\frac{\sqrt{2}}{2}} = -1.$$

11. The coordinates of the point determined by -3π are $(-1, 0)$. Thus,

$$\sin(-3\pi) = y = 0.$$

12. The point determined by $3\pi/4$ is a reflection across the y-axis of the point determined by $\pi/4$, $\left(\frac{\sqrt{2}}{2}, \frac{\sqrt{2}}{2}\right)$. Its coordinates are $\left(-\frac{\sqrt{2}}{2}, \frac{\sqrt{2}}{2}\right)$. Thus,

$$\csc\frac{3\pi}{4} = \frac{1}{y} = \frac{1}{\frac{\sqrt{2}}{2}} = \frac{2}{\sqrt{2}} = \sqrt{2}.$$

13. The point determined by $5\pi/6$ is a reflection across the y-axis of the point determined by $\pi/6$, $\left(\frac{\sqrt{3}}{2}, \frac{1}{2}\right)$. Its coordinates are $\left(-\frac{\sqrt{3}}{2}, \frac{1}{2}\right)$. Thus,

$$\cos\frac{5\pi}{6} = x = -\frac{\sqrt{3}}{2}.$$

14. The point determined by $-\pi/4$ is a reflection across the x-axis of the point determined by $\pi/4$, $\left(\frac{\sqrt{2}}{2}, \frac{\sqrt{2}}{2}\right)$. Its coordinates are $\left(\frac{\sqrt{2}}{2}, -\frac{\sqrt{2}}{2}\right)$.
Thus,

$$\tan\left(-\frac{\pi}{4}\right) = \frac{y}{x} = \frac{-\frac{\sqrt{2}}{2}}{\frac{\sqrt{2}}{2}} = -1.$$

15. The coordinates of the point determined by $\pi/2$ are $(0, 1)$. Thus,

$$\sec\frac{\pi}{2} = \frac{1}{x} = \frac{1}{0}, \text{ which is not defined.}$$

16. The coordinates of the point determined by 10π are $(1, 0)$. Thus,

$$\cos 10\pi = x = 1.$$

17. The coordinates of the point determined by $\pi/6$ are $\left(\frac{\sqrt{3}}{2}, \frac{1}{2}\right)$. Thus,

$$\cos\frac{\pi}{6} = x = \frac{\sqrt{3}}{2}.$$

18. The point determined by $2\pi/3$ is a reflection across the y-axis of the point determined by $\pi/3$, $\left(\frac{1}{2}, \frac{\sqrt{3}}{2}\right)$. Its coordinates are $\left(-\frac{1}{2}, \frac{\sqrt{3}}{2}\right)$. Thus,

$$\sin\frac{2\pi}{3} = y = \frac{\sqrt{3}}{2}.$$

19. The point determined by $5\pi/4$ is a reflection across the origin of the point determined by $\pi/4$, $\left(\frac{\sqrt{2}}{2}, \frac{\sqrt{2}}{2}\right)$. Its coordinates are $\left(-\frac{\sqrt{2}}{2}, -\frac{\sqrt{2}}{2}\right)$. Thus,

$$\sin\frac{5\pi}{4} = y = -\frac{\sqrt{2}}{2}.$$

20. The point determined by $11\pi/6$ is a reflection across the x-axis of the point determined by $\pi/6$, $\left(\frac{\sqrt{3}}{2}, \frac{1}{2}\right)$. Its coordinates are $\left(\frac{\sqrt{3}}{2}, -\frac{1}{2}\right)$. Thus,

$$\cos\frac{11\pi}{6} = x = \frac{\sqrt{3}}{2}.$$

21. The coordinates of the point determined by -5π are $(-1, 0)$. Thus,

$$\sin(-5\pi) = y = 0.$$

22. The coordinates of the point determined by $3\pi/2$ are $(0, -1)$. Thus,

$$\tan\frac{3\pi}{2} = \frac{y}{x} = \frac{-1}{0}, \text{ which is not defined.}$$

23. The coordinates of the point determined by $5\pi/2$ are $(0, 1)$. Thus,

$$\cot\frac{5\pi}{2} = \frac{x}{y} = \frac{0}{1} = 0.$$

24. The point determined by $5\pi/3$ is a reflection across the x-axis of the point determined by $\pi/3$, $\left(\frac{1}{2}, \frac{\sqrt{3}}{2}\right)$. Its coordinates are $\left(\frac{1}{2}, -\frac{\sqrt{3}}{2}\right)$. Thus,

$$\tan\frac{5\pi}{3} = \frac{y}{x} = \frac{-\frac{\sqrt{3}}{2}}{\frac{1}{2}} = -\sqrt{3}.$$

25. Use a calculator set in radian mode.
$\tan\frac{\pi}{7} \approx 0.4816$

26. Use a calculator set in radian mode.
$\cos\left(-\frac{2\pi}{5}\right) = 0.3090$

27. Use a calculator set in radian mode.
$\sec 37 = \frac{1}{\cos 37} \approx 1.3065$

28. Use a calculator set in radian mode.
$\sin 11.7 \approx -0.7620$

29. Use a calculator set in radian mode.
$\cot 342 = \frac{1}{\tan 342} \approx -2.1599$

30. Use a calculator set in radian mode.
$\tan 1.3 \approx 3.6021$

31. Use a calculator set in radian mode.
$\cos 6\pi = 1$

32. Use a calculator set in radian mode.
$\sin\frac{\pi}{10} \approx 0.3090$

33. Use a calculator set in radian mode.

$$\csc 4.16 = \frac{1}{\sin 4.16} \approx -1.1747$$

34. Use a calculator set in radian mode.

$$\sec \frac{10\pi}{7} = \frac{1}{\cos \frac{10\pi}{7}} \approx -4.4940$$

35. Use a calculator set in radian mode.

$$\tan \frac{7\pi}{4} = -1$$

36. Use a calculator set in radian mode.

$\cos 2000 \approx -0.3675$

37. Use a calculator set in radian mode.

$$\sin\left(-\frac{\pi}{4}\right) \approx -0.7071$$

38. Use a calculator set in radian mode.

$$\cot 7\pi = \frac{1}{\tan 7\pi} \text{ is not defined.}$$

39. Use a calculator set in radian mode.

$\sin 0 = 0$

40. Use a calculator set in radian mode.

$\cos(-29) \approx -0.7481$

41. Use a calculator set in radian mode.

$$\tan \frac{2\pi}{9} \approx 0.8391$$

42. Use a calculator set in radian mode.

$$\sin \frac{8\pi}{3} \approx 0.8660$$

43. a)

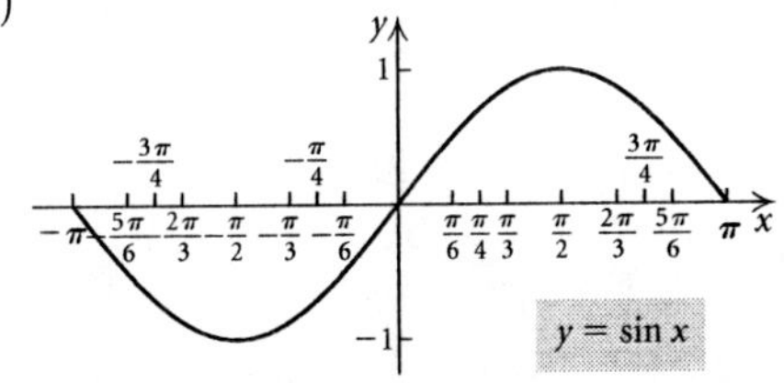

b) Reflect the graph of $y = \sin x$ across the y-axis.

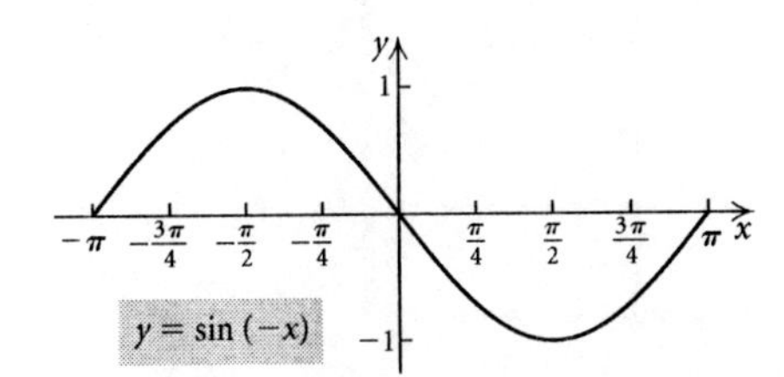

c) Reflect the graph of $y = \sin x$ across the x-axis. The graph is the same as the graph in part (b).

d) They are the same.

44. a)

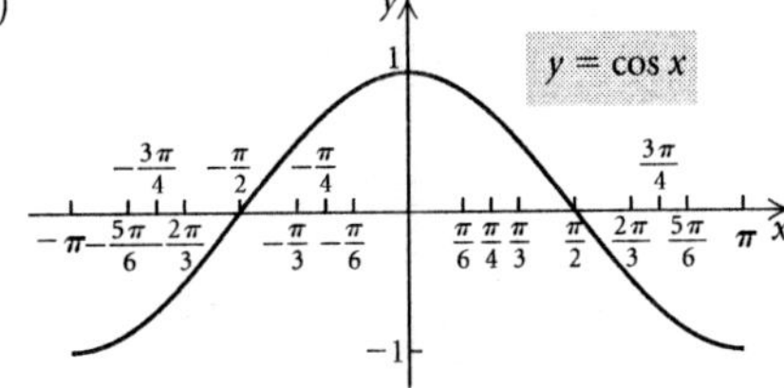

b) Reflect the graph of $y = \cos x$ across the y-axis. The graph is the same as the graph in part (a).

c) Reflect the graph of $y = \cos x$ across the x-axis.

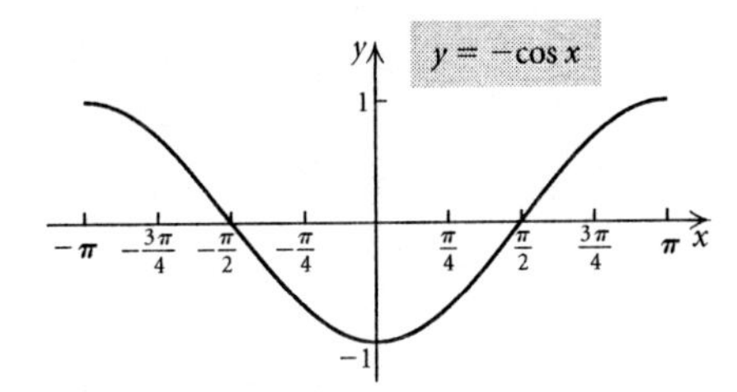

d) They are the same.

45. a) See Exercise 43(a).

b) Shift the graph of $y = \sin x$ left π units.

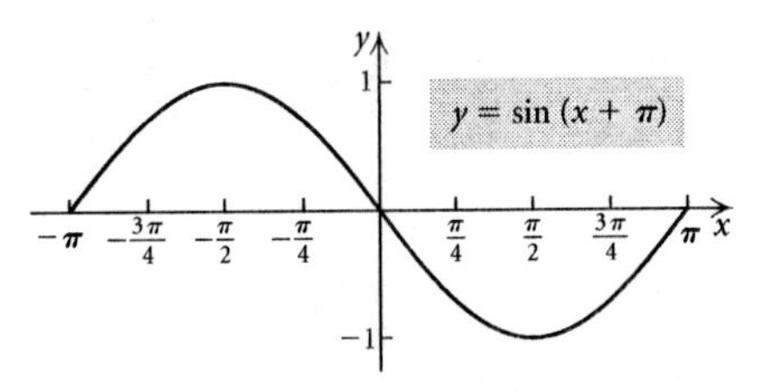

c) Reflect the graph of $y = \sin x$ across the x-axis. The graph is the same as the graph in part (b).

d) They are the same.

46. a) See Exercise 43(a).

b) Shift the graph of $y = \sin x$ right π units.

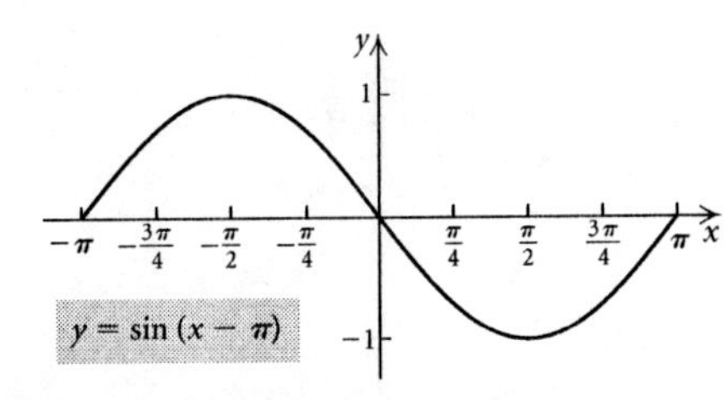

c) Reflect the graph of $y = \sin x$ across the x-axis. The graph is the same as the graph in part (b).

d) They are the same.

47. a)

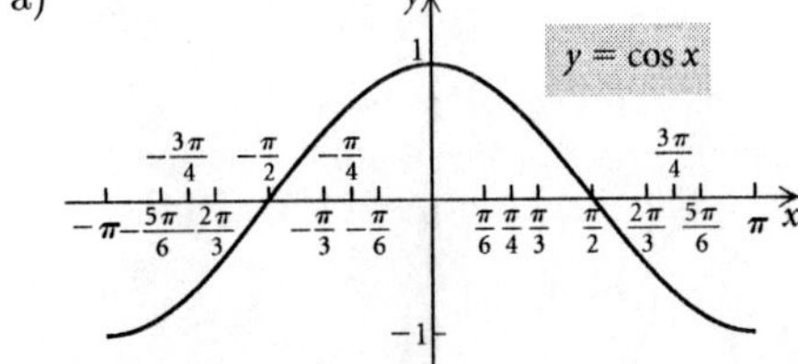

b) Shift the graph of $y = \cos x$ left π units.

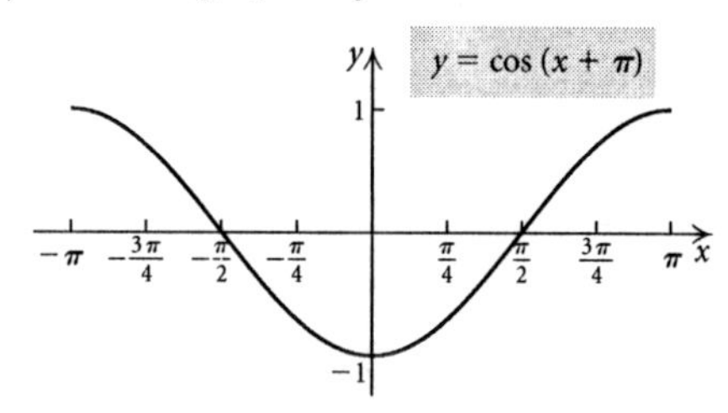

c) Reflect the graph of $y = \cos x$ across the x-axis. The graph is the same as the graph in part (b).

d) They are the same.

48. a) See Exercise 47(a).

b) Shift the graph of $y = \cos x$ right π units.

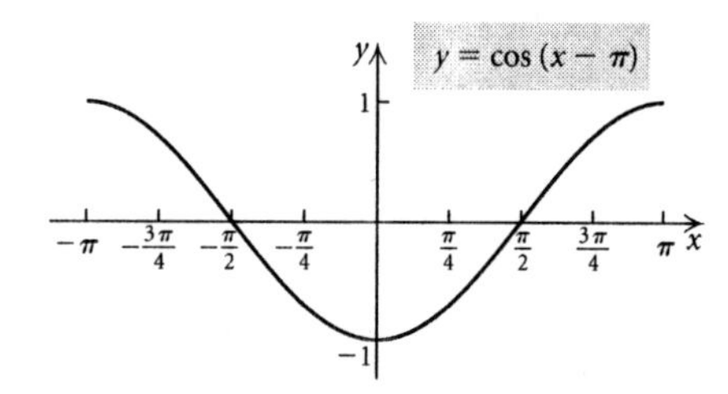

c) Reflect the graph of $y = \cos x$ across the x-axis. The graph is the same as the graph in part (b).

d) They are the same.

49. $\sin(-x) = -\sin x$

The sine function is an odd function.

$\cos(-x) = \cos x$

The cosine function is an even function.

$$\tan(-x) = \frac{\sin(-x)}{\cos(-x)} = \frac{-\sin x}{\cos x} = -\frac{\sin x}{\cos x} = -\tan x$$

The tangent function is an odd function.

$$\csc(-x) = \frac{1}{\sin(-x)} = \frac{1}{-\sin x} = -\frac{1}{\sin x} = -\csc x$$

The cosecant function is an odd function.

$$\sec(-x) = \frac{1}{\cos(-x)} = \frac{1}{\cos x} = \sec x$$

The secant function is an even function.

$$\cot(-x) = \frac{\cos(-x)}{\sin(-x)} = \frac{\cos x}{-\sin x} = -\frac{\cos x}{\sin x} = -\cot x$$

The cotangent function is an odd function.

Thus the cosine and secant functions are even; the sine, tangent, cosecant, and cotangent functions are odd.

50. From the graphs and lists of properties in the text we see that the tangent and cotangent functions have period π and the sine, cosine, cosecant, and secant functions have period 2π.

51. Consider a point (x, y) on the unit circle determined by an angle s. Then $\tan s = \frac{y}{x}$. The coordinates x and y have the same sign in quadrants I and III, so the tangent function is positive in these quadrants; x and y have opposite signs in quadrants II and IV, so the tangent function is negative in these quadrants.

52. Consider a point (x, y) on the unit circle determined by an angle s. Then $\sin s = y$. Since y is positive in quadrants I and II, the sine function is positive in these quadrants; y is negative in quadrants III and IV, so the sine function is negative in these quadrants.

53. Consider a point (x, y) on the unit circle determined by an angle s. Then $\cos s = x$. Since x is positive in quadrants I and IV, the cosine function is positive in these quadrants; x is negative in quadrants II and III, so the cosine function is negative in these quadrants.

54. Consider a point (x, y) on the unit circle determined by an angle s. Then $\csc s = 1/\sin s = 1/y$. From Exercise 52 we see that the cosecant function is positive in quadrants I and II and negative in quadrants III and IV.

55. Graph $y = (\sin x)^2$.

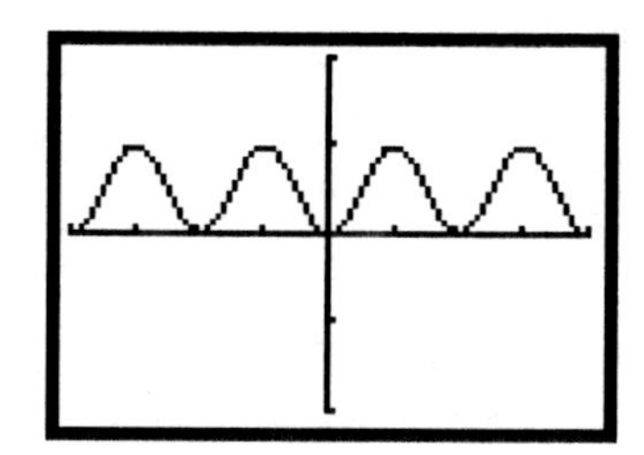

From the graph we find the following information.

Domain: $(-\infty, \infty)$

Range: $[0, 1]$

Period: π

Amplitude: $\frac{1}{2}(1 - 0)$, or $\frac{1}{2}$

56. Graph $y = |\cos x| + 1$.

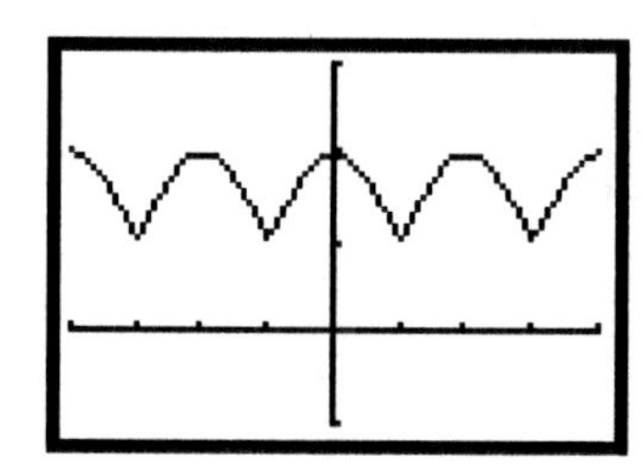

Domain: $(-\infty, \infty)$

Range: $[1, 2]$

Period: π

Amplitude: $\frac{1}{2}(2 - 1)$, or $\frac{1}{2}$

57. $f(x) = \dfrac{\sin x}{x}$, when $0 < x < \dfrac{\pi}{2}$

$\dfrac{\sin \pi/2}{\pi/2} \approx 0.6369$

$\dfrac{\sin 3\pi/8}{3\pi/8} \approx 0.7846$

$\dfrac{\sin \pi/4}{\pi/4} \approx 0.9008$

$\dfrac{\sin \pi/8}{\pi/8} \approx 0.9750$

The limiting value of $\dfrac{\sin x}{x}$ as x approaches 0 is 1.

58. A graph shows that intervals on which $\sin x < \cos x$ are $\left(-\dfrac{11\pi}{4}, -\dfrac{7\pi}{4}\right)$, $\left(-\dfrac{3\pi}{4}, \dfrac{\pi}{4}\right)$, and $\left(\dfrac{5\pi}{4}, \dfrac{9\pi}{4}\right)$.

In general $\sin x < \cos x$ on the intervals $\left(-\dfrac{3\pi}{4} + 2k\pi, \dfrac{\pi}{4} + 2k\pi\right)$, k an integer.

59. The graph of the cosine function is the graph of the sine function shifted left $\pi/2$ units. (Or we can say that the graph of the sine function is the graph of the cosine function shifted right $\pi/2$ units.)

60. The numbers for which the value of the cosine function is 0 are not in the domain of the tangent function.

61. See the answer section in the text.

62.

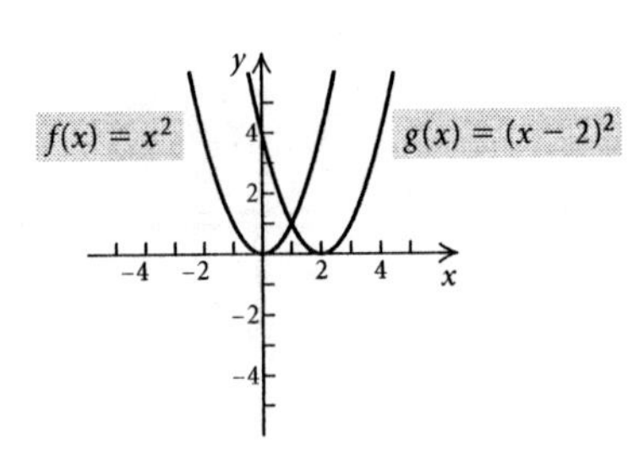

Shift the graph of f right 2 units.

63. See the answer section in the text.

64.

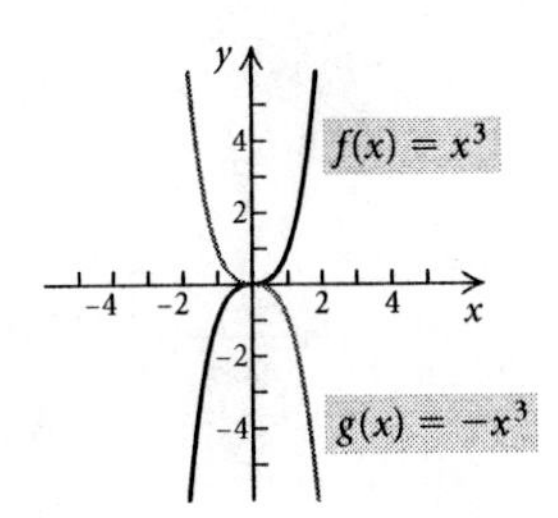

Reflect the graph of f across the x-axis.

65. Start with $y = x^3$.

Reflect it across the x-axis: $y = -x^3$

Shift it right 2 units: $y = -(x-2)^3$

Shift it down 1 unit: $y = -(x-2)^3 - 1$

66. Start with $y = \dfrac{1}{x}$.

Shrink it vertically by a factor of $\dfrac{1}{4}$: $y = \dfrac{1}{4} \cdot \dfrac{1}{x}$, or $y = \dfrac{1}{4x}$

Shift it up 3 units: $y = \dfrac{1}{4x} + 3$

67. Any real numbers x and $-x$ determine points on the unit circle that are symmetric with respect to the x-axis. Their first coordinates are the same, so $\cos(-x) = \cos x$.

68. Any real numbers x and $-x$ determine points on the unit circle that are symmetric with respect to the x-axis. Their second coordinates are opposites, so $\sin(-x) = -\sin x$.

69. Any real numbers x and $x + 2k\pi$ determine the same point on the unit circle, so $\sin(x + 2k\pi) = \sin x$.

70. Any real numbers x and $x + 2k\pi$ determine the same point on the unit circle, so $\cos(x + 2k\pi) = \cos x$.

71. Any real numbers x and $\pi - x$ determine points on the unit circle that are symmetric with respect to the y-axis. Their second coordinates are the same, so $\sin(\pi - x) = \sin x$.

72. Any real numbers x and $\pi - x$ determine points on the unit circle that are symmetric with respect to the y-axis. Their first coordinates are opposites, so $\cos(\pi - x) = -\cos x$.

73. Any real numbers x and $x - \pi$ determine points on the unit circle that are symmetric with respect to the origin. Their first coordinates are opposites, so $\cos(x - \pi) = -\cos x$.

74. Any real number x and $x + \pi$ determine points on the unit circle that are symmetric with respect to the origin. Their first coordinates are opposites, so $\cos(x + \pi) = -\cos x$.

75. Any real number x and $x + \pi$ determine points on the unit circle that are symmetric with respect to the origin. Their second coordinates are opposites, so $\sin(x + \pi) = -\sin x$.

76. Any real numbers x and $x - \pi$ determine points on the unit circle that are symmetric with respect to the origin. Their second coordinates are opposites, so $\sin(x - \pi) = -\sin x$.

77. a) $\sin \dfrac{\pi}{2} = 1$

$\sin\left(\dfrac{\pi}{2} + 2\pi\right) = 1 \qquad \sin\left(\dfrac{\pi}{2} - 2\pi\right) = 1$

$\sin\left(\dfrac{\pi}{2} + 2 \cdot 2\pi\right) = 1 \qquad \sin\left(\dfrac{\pi}{2} - 2 \cdot 2\pi\right) = 1$

$\sin\left(\dfrac{\pi}{2} + 3 \cdot 2\pi\right) = 1 \qquad \sin\left(\dfrac{\pi}{2} - 3 \cdot 2\pi\right) = 1$

$\sin\left(\dfrac{\pi}{2} + k \cdot 2\pi\right) = 1$, k an integer

Thus $x = \dfrac{\pi}{2} + 2k\pi$, k an integer.

b) $\cos \pi = -1$

$\cos(\pi + 2\pi) = -1 \qquad \cos(\pi - 2\pi) = -1$

$\cos(\pi + 2 \cdot 2\pi) = -1 \qquad \cos(\pi - 2 \cdot 2\pi) = -1$

$\cos(\pi + 3 \cdot 2\pi) = -1 \qquad \cos(\pi - 3 \cdot 2\pi) = -1$

$\cos(\pi + k \cdot 2\pi) = 1$, k an integer

Thus $x = \pi + 2k\pi$, or $x = (2k+1)\pi$, k an integer.

c) $\sin 0 = 0$

$\sin \pi = 0 \qquad \sin(-\pi) = 0$

$\sin 2\pi = 0 \qquad \sin(-2\pi) = 0$

$\sin 3\pi = 0 \qquad \sin(-3\pi) = 0$

$\sin k\pi = 0$, k an integer

Thus $x = k\pi$, k an integer.

78. $f(x) = x^2 + 2x, \;\; g(x) = \cos x$

$$f \circ g(x) = f(g(x)) = f(\cos x)$$
$$= (\cos x)^2 + 2\cos x$$
$$= \cos^2 x + 2\cos x$$
$$g \circ f(x) = g(f(x)) = g(x^2 + 2x)$$
$$= \cos(x^2 + 2x)$$

79. $f(x) = \sqrt{\cos x}$

The domain consists of the values of x for which $\cos x \geq 0$. From the graph of the cosine function we see that the domain consists of the intervals

$\left[-\frac{\pi}{2} + 2k\pi, \; \frac{\pi}{2} + 2k\pi\right]$, k an integer.

80. $f(x) = \dfrac{1}{\sin x}$

The domain consists of the values of x for which $\sin x \neq 0$. From the graph of the sine function we see that the domain consists of

$\{x | x \neq k\pi, \; k \text{ an integer}\}$.

81. $f(x) = \dfrac{\sin x}{\cos x}$

The domain consists of the values of x for which $\cos x \neq 0$. From the graph of the cosine function we see that the domain is

$\left\{x \middle| x \neq \frac{\pi}{2} + k\pi, \; k \text{ an integer}\right\}$.

82. $g(x) = \log(\sin x)$

The domain consists of the values of x for which $\sin x > 0$. From the graph of the sine function we see that the domain consists of the intervals $(2k\pi, \; (2k+1)\pi)$, k an integer.

83. See the answer section in the text.

84.

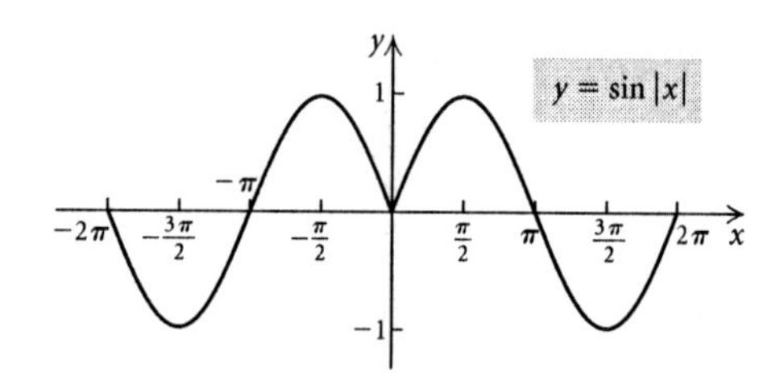

85. See the answer section in the text.

86.

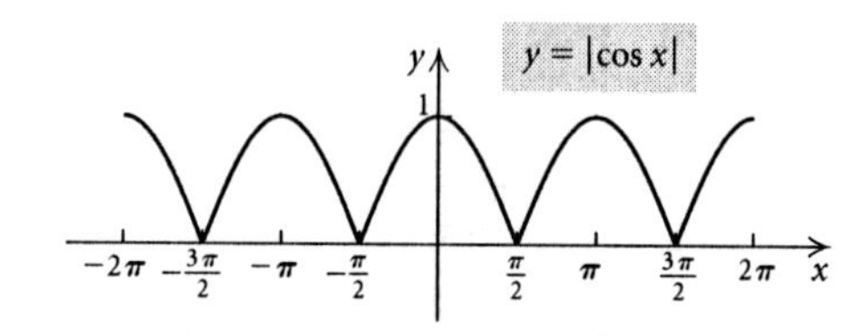

87. a) $\Delta OPA \sim \Delta ODB$

Thus, $\dfrac{AP}{OA} = \dfrac{BD}{OB}$

$\dfrac{\sin\theta}{\cos\theta} = \dfrac{BD}{1}$

$\tan\theta = BD$

b) $\Delta OPA \sim \Delta ODB$

$\dfrac{OD}{OP} = \dfrac{OB}{OA}$

$\dfrac{OD}{1} = \dfrac{1}{\cos\theta}$

$OD = \sec\theta$

c) $\Delta OAP \sim \Delta ECO$

$\dfrac{OE}{PO} = \dfrac{CO}{AP}$

$\dfrac{OE}{1} = \dfrac{1}{\sin\theta}$

$OE = \csc\theta$

d) $\Delta OAP \sim \Delta ECO$

$\dfrac{CE}{AO} = \dfrac{CO}{AP}$

$\dfrac{CE}{\cos\theta} = \dfrac{1}{\sin\theta}$

$CE = \dfrac{\cos\theta}{\sin\theta}$

$CE = \cot\theta$

Exercise Set 5.6

1. $y = \sin x + 1$

$A = 1$, $B = 1$, $C = 0$, $D = 1$

Amplitude: $|A| = |1| = 1$

Period: $\left|\dfrac{2\pi}{B}\right| = \left|\dfrac{2\pi}{1}\right| = 2\pi$

Phase shift: $\dfrac{C}{B} = \dfrac{0}{1} = 0$

Translate the graph of $y = \sin x$ up 1 unit.

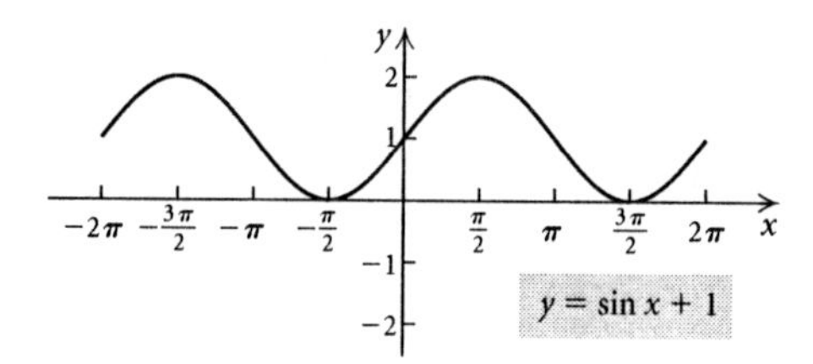

2. $y = \dfrac{1}{4}\cos x$

$A = \dfrac{1}{4}$, $B = 1$, $C = 0$, $D = 0$

Amplitude: $|A| = \left|\dfrac{1}{4}\right| = \dfrac{1}{4}$

Period: $\left|\dfrac{2\pi}{B}\right| = \left|\dfrac{2\pi}{1}\right| = 2\pi$

Phase shift: $\dfrac{C}{B} = \dfrac{0}{1} = 0$

Shrink the graph of $y = \cos x$ vertically by a factor of 4.

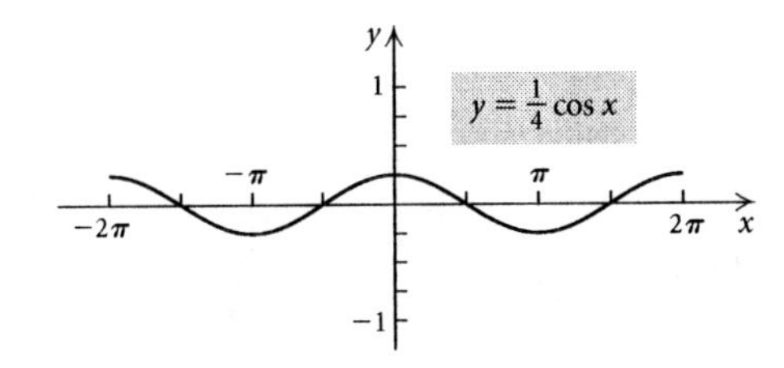

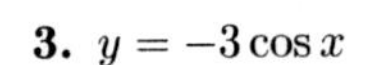

3. $y = -3\cos x$

$A = -3,\ B = 1,\ C = 0,\ D = 0$

Amplitude: $|A| = |-3| = 3$

Period: $\left|\dfrac{2\pi}{B}\right| = \left|\dfrac{2\pi}{1}\right| = 2\pi$

Phase shift: $\dfrac{C}{B} = \dfrac{0}{1} = 0$

Stretch the graph of $y = \cos x$ vertically by a factor of 3 and reflect the graph across the x-axis.

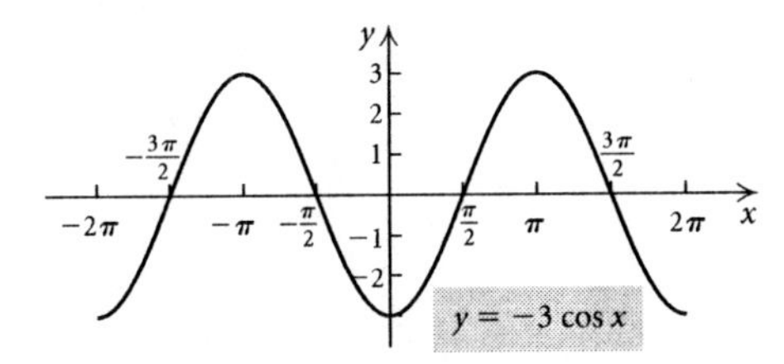

4. $y = \sin(-2x)$

$A = 1,\ B = -2,\ C = 0,\ D = 0$

Amplitude: $|A| = |1| = 1$

Period: $\left|\dfrac{2\pi}{B}\right| = \left|\dfrac{2\pi}{-2}\right| = \pi$

Phase shift: $\dfrac{C}{B} = \dfrac{0}{-2} = 0$

Shrink the graph of $y = \sin x$ horizontally by a factor of 2.

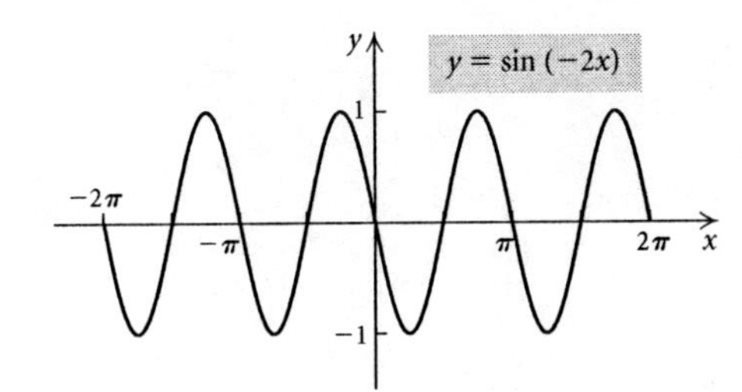

5. $y = \dfrac{1}{2}\cos x$

$A = \dfrac{1}{2},\ B = 1,\ C = 0,\ D = 0$

Amplitude: $|A| = \left|\dfrac{1}{2}\right| = \dfrac{1}{2}$

Period: $\left|\dfrac{2\pi}{B}\right| = \left|\dfrac{2\pi}{1}\right| = 2\pi$

Phase shift: $\dfrac{C}{B} = \dfrac{0}{1} = 0$

Shrink the graph of $y = \cos x$ by a factor of 2.

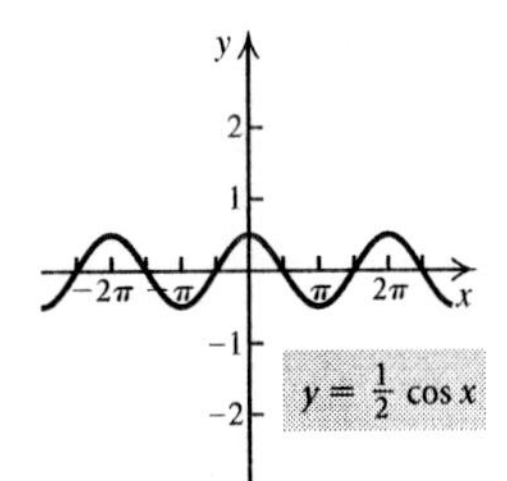

6. $y = \sin\left(\dfrac{1}{2}x\right)$

$A = 1,\ B = \dfrac{1}{2},\ C = 0,\ D = 0$

Amplitude: $|A| = |1| = 1$

Period: $\left|\dfrac{2\pi}{B}\right| = \left|\dfrac{2\pi}{\frac{1}{2}}\right| = 4\pi$

Phase shift: $\dfrac{C}{B} = \dfrac{0}{\frac{1}{2}} = 0$

Stretch the graph of $y = \sin x$ horizontally by a factor of 2.

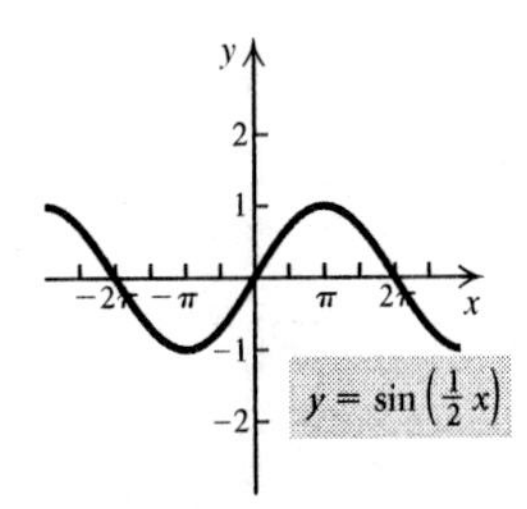

7. $y = \sin(2x)$

$A = 1,\ B = 2,\ C = 0,\ D = 0$

Amplitude: $|A| = |1| = 1$

Period: $\left|\dfrac{2\pi}{B}\right| = \left|\dfrac{2\pi}{2}\right| = \pi$

Phase shift: $\dfrac{C}{B} = \dfrac{0}{2} = 0$

Shrink the graph of $y = \sin x$ horizontally by a factor of 2.

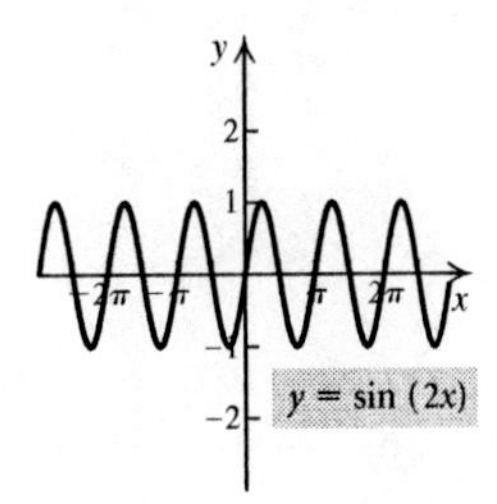

8. $y = \cos x - 1$

$A = 1,\ B = 1,\ C = 0,\ D = -1$

Amplitude: $|A| = |1| = 1$

Period: $\left|\dfrac{2\pi}{B}\right| = \left|\dfrac{2\pi}{1}\right| = 2\pi$

Phase shift: $\dfrac{C}{B} = \dfrac{0}{1} = 0$

Translate the graph of $y = \cos x$ down 1 unit.

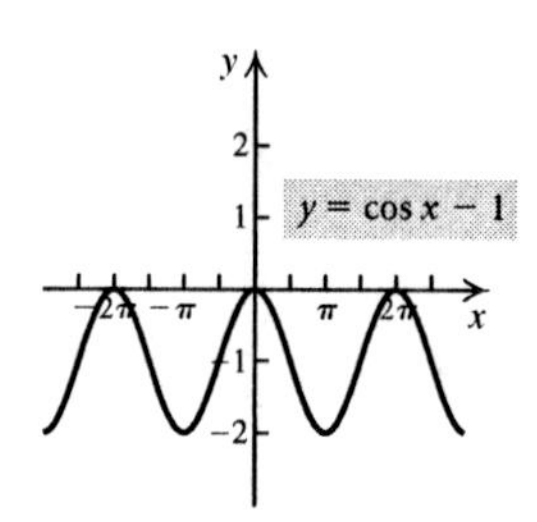

9. $y = 2\sin\left(\dfrac{1}{2}x\right)$

$A = 2,\ B = \dfrac{1}{2},\ C = 0,\ D = 0$

Amplitude: $|A| = |2| = 2$

Period: $\left|\dfrac{2\pi}{B}\right| = \left|\dfrac{2\pi}{\frac{1}{2}}\right| = 4\pi$

Phase shift: $\dfrac{C}{B} = \dfrac{0}{\frac{1}{2}} = 0$

Stretch the graph of $y = \sin x$ horizontally by a factor of 2 and stretch it vertically, also by a factor of 2.

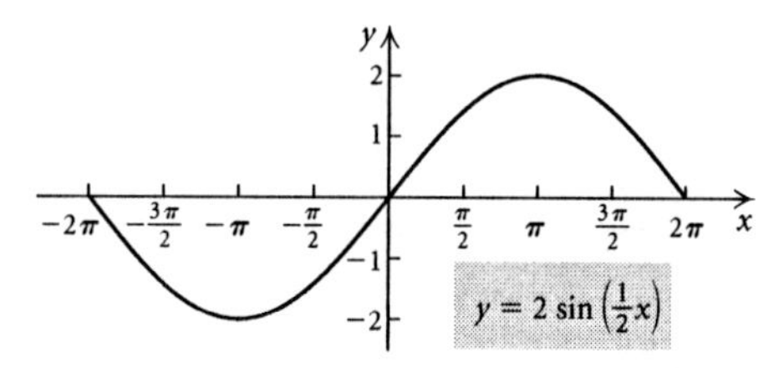

10. $y = \cos\left(x - \dfrac{\pi}{2}\right)$

$A = 1,\ B = 1,\ C = \dfrac{\pi}{2},\ D = 0$

Amplitude: $|A| = |1| = 1$

Period: $\left|\dfrac{2\pi}{B}\right| = \left|\dfrac{2\pi}{1}\right| = 2\pi$

Phase shift: $\dfrac{C}{B} = \dfrac{\frac{\pi}{2}}{1} = \dfrac{\pi}{2}$

Translate the graph of $y = \cos x$ to the right $\dfrac{\pi}{2}$ units.

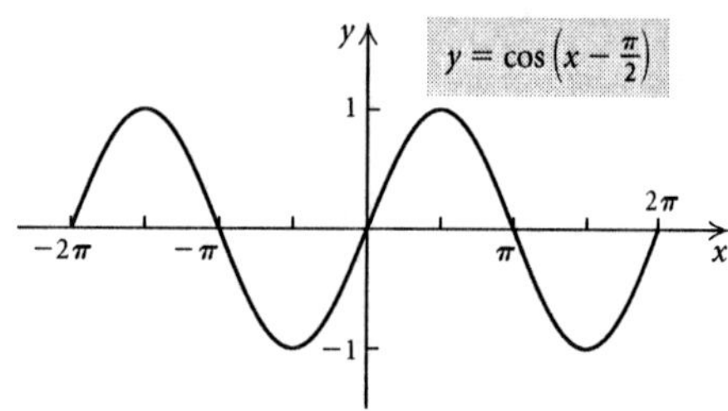

11. $y = \dfrac{1}{2}\sin\left(x + \dfrac{\pi}{2}\right) = \dfrac{1}{2}\sin\left[x - \left(-\dfrac{\pi}{2}\right)\right]$

$A = \dfrac{1}{2},\ B = 1,\ C = -\dfrac{\pi}{2},\ D = 0$

Amplitude: $|A| = \left|\dfrac{1}{2}\right| = \dfrac{1}{2}$

Period: $\left|\dfrac{2\pi}{B}\right| = \left|\dfrac{2\pi}{1}\right| = 2\pi$

Phase shift: $\dfrac{C}{B} = \dfrac{-\frac{\pi}{2}}{1} = -\dfrac{\pi}{2}$

Shrink the graph of $y = \sin x$ vertically by a factor of 2 and translate it to the left $\dfrac{\pi}{2}$ units.

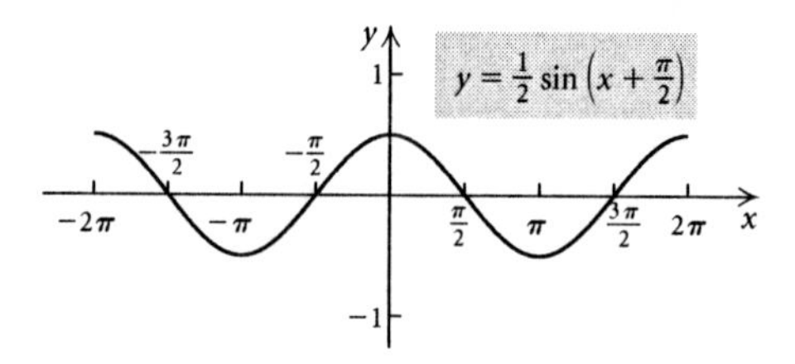

12. $y = \cos x - \dfrac{1}{2}$

$A = 1,\ B = 1,\ C = 0,\ D = -\dfrac{1}{2}$

Amplitude: $|A| = |1| = 1$

Period: $\left|\dfrac{2\pi}{B}\right| = \left|\dfrac{2\pi}{1}\right| = 2\pi$

Phase shift: $\dfrac{C}{B} = \dfrac{0}{1} = 0$

Translate the graph of $y = \cos x$ down $\dfrac{1}{2}$ unit.

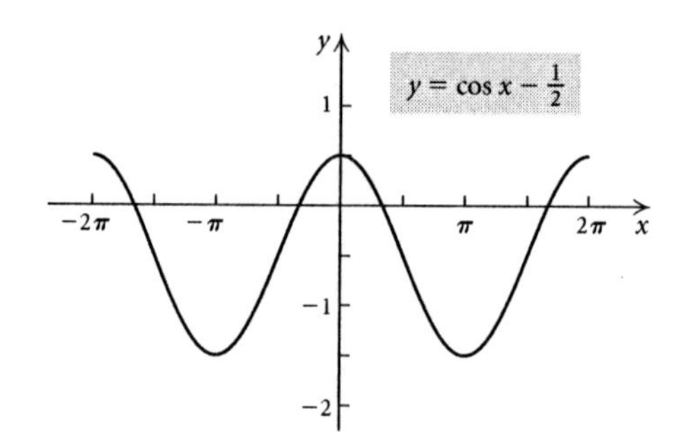

13. $y = 3\cos(x - \pi)$

$A = 3,\ B = 1,\ C = \pi,\ D = 0$

Amplitude: $|A| = |3| = 3$

Period: $\left|\dfrac{2\pi}{B}\right| = \left|\dfrac{2\pi}{1}\right| = 2\pi$

Phase shift: $\dfrac{C}{B} = \dfrac{\pi}{1} = \pi$

Stretch the graph of $y = \cos x$ vertically by a factor of 3 and shift it π units to the right.

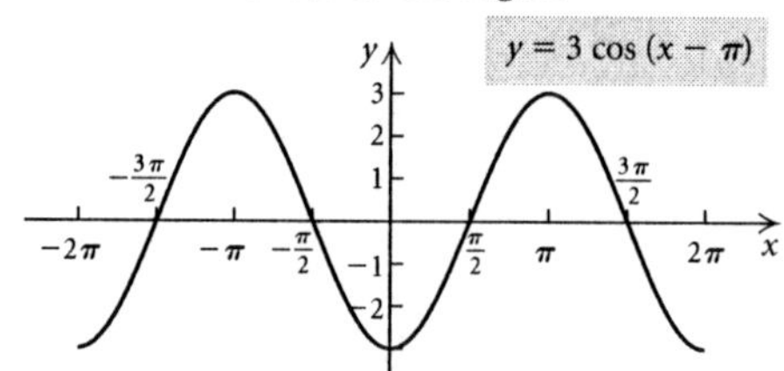

14. $y = -\sin\left(\frac{1}{4}x\right) + 1$

$A = -1,\ B = \frac{1}{4},\ C = 0,\ D = 1$

Amplitude: $|A| = |-1| = 1$

Period: $\left|\frac{2\pi}{B}\right| = \left|\frac{2\pi}{\frac{1}{4}}\right| = 8\pi$

Phase shift: $\frac{C}{B} = \frac{0}{\frac{1}{4}} = 0$

Stretch the graph of $y = \sin x$ horizontally by a factor of 4, reflect it across the x-axis, and then translate it up 1 unit.

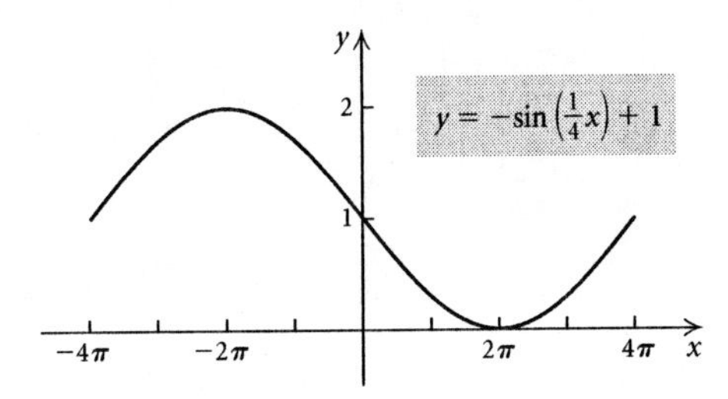

15. $y = \frac{1}{3}\sin x - 4$

$A = \frac{1}{3},\ B = 1,\ C = 0,\ D = -4$

Amplitude: $|A| = \left|\frac{1}{3}\right| = \frac{1}{3}$

Period: $\left|\frac{2\pi}{B}\right| = \left|\frac{2\pi}{1}\right| = 2\pi$

Phase shift: $\frac{C}{B} = \frac{0}{1} = 0$

Shrink the graph of $y = \sin x$ vertically by a factor of 3 and shift it down 4 units.

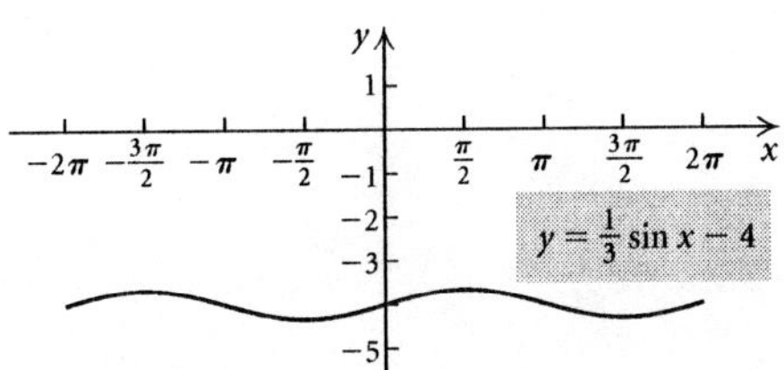

16. $y = \cos\left(\frac{1}{2}x + \frac{\pi}{2}\right) = \cos\left[\frac{1}{2}(x - (-\pi))\right]$

$A = 1,\ B = \frac{1}{2},\ C = -\frac{\pi}{2},\ D = \frac{1}{2}$

Amplitude: $|A| = |1| = 1$

Period: $\left|\frac{2\pi}{B}\right| = \left|\frac{2\pi}{\frac{1}{2}}\right| = 4\pi$

Phase shift: $\frac{C}{B} = \frac{-\frac{\pi}{2}}{\frac{1}{2}} = -\pi$

Stretch the graph of $y = \cos x$ horizontally by a factor of 2 and shift it left π units.

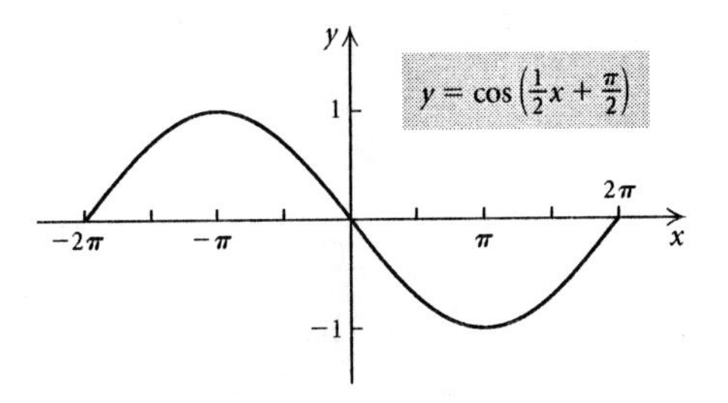

17. $y = -\cos(-x) + 2$

$A = -1,\ B = -1,\ C = 0,\ D = 2$

Amplitude: $|A| = |-1| = 1$

Period: $\left|\frac{2\pi}{B}\right| = \left|\frac{2\pi}{-1}\right| = 2\pi$

Phase shift: $\frac{C}{B} = \frac{0}{-1} = 0$

Reflect the graph of $y = \cos x$ across the y-axis and across the x-axis and translate it up 2 units.

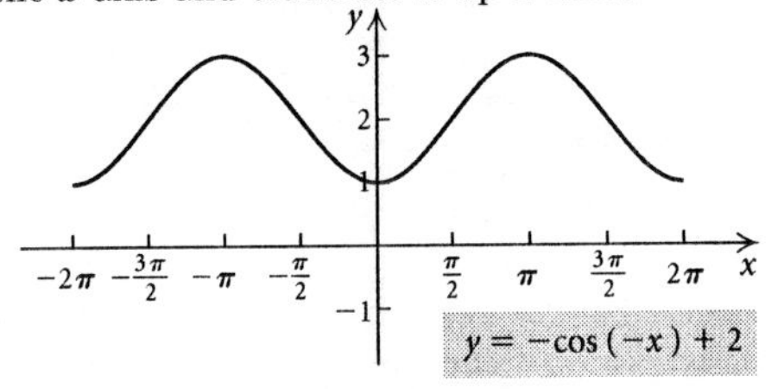

18. $y = \frac{1}{2}\sin\left(2x - \frac{\pi}{4}\right) = \frac{1}{2}\sin\left[2\left(x - \frac{\pi}{8}\right)\right]$

$A = \frac{1}{2},\ B = 2,\ C = \frac{\pi}{4},\ D = 0$

Amplitude: $|A| = \left|\frac{1}{2}\right| = \frac{1}{2}$

Period: $\left|\frac{2\pi}{B}\right| = \left|\frac{2\pi}{2}\right| = \pi$

Phase shift: $\frac{C}{B} = \frac{\frac{\pi}{4}}{2} = \frac{\pi}{8}$

Shrink the graph of $y = \sin x$ horizontally by a factor of 2; shrink it vertically, also by a factor of 2; and shift it $\frac{\pi}{8}$ units to the right.

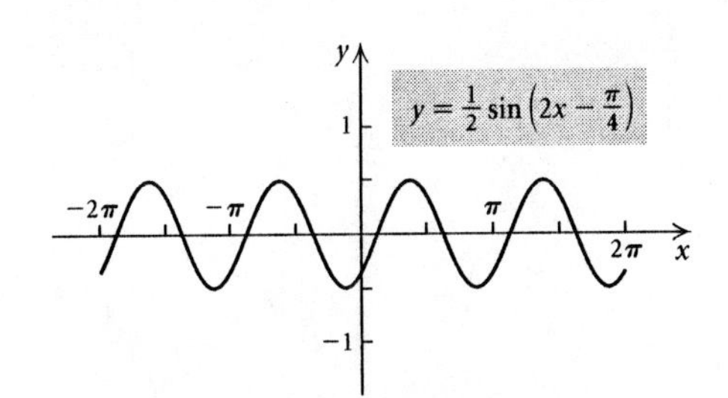

19. $y = 2\cos\left(\frac{1}{2}x - \frac{\pi}{2}\right) = 2\cos\left[\frac{1}{2}(x - \pi)\right]$

$A = 2,\ B = \frac{1}{2},\ C = \frac{\pi}{2},\ D = 0$

Amplitude: $|A| = |2| = 2$

Period: $\left|\frac{2\pi}{B}\right| = \left|\frac{2\pi}{\frac{1}{2}}\right| = 4\pi$

Phase shift: $\dfrac{C}{B} = \dfrac{\frac{\pi}{2}}{\frac{1}{2}} = \pi$

20. $y = 4\sin\left(\frac{1}{4}x + \frac{\pi}{8}\right) = 4\sin\left[\frac{1}{4}\left(x - \left(-\frac{\pi}{2}\right)\right)\right]$

$A = 4,\ B = \frac{1}{4},\ C = -\frac{\pi}{8},\ D = 0$

Amplitude: $|A| = |4| = 4$

Period: $\left|\dfrac{2\pi}{B}\right| = \left|\dfrac{2\pi}{\frac{1}{4}}\right| = 8\pi$

Phase shift: $\dfrac{C}{B} = \dfrac{-\frac{\pi}{8}}{\frac{1}{4}} = -\dfrac{\pi}{2}$

21. $y = -\frac{1}{2}\sin\left(2x + \frac{\pi}{2}\right) = -\frac{1}{2}\sin\left[2\left(x - \left(-\frac{\pi}{4}\right)\right)\right]$

$A = -\frac{1}{2},\ B = 2,\ C = -\frac{\pi}{2},\ D = 0$

Amplitude: $|A| = \left|-\frac{1}{2}\right| = \frac{1}{2}$

Period: $\left|\dfrac{2\pi}{B}\right| = \left|\dfrac{2\pi}{2}\right| = \pi$

Phase shift: $\dfrac{C}{B} = \dfrac{-\frac{\pi}{2}}{2} = -\dfrac{\pi}{4}$

22. $y = -3\cos(4x - \pi) + 2 = -3\cos\left[4\left(x - \frac{\pi}{4}\right)\right] + 2$

$A = -3,\ B = 4,\ C = \pi,\ D = 2$

Amplitude: $|A| = |-3| = 3$

Period: $\left|\dfrac{2\pi}{B}\right| = \left|\dfrac{2\pi}{4}\right| = \dfrac{\pi}{2}$

Phase shift: $\dfrac{C}{B} = \dfrac{\pi}{4}$

23. $y = 2 + 3\cos(\pi x - 3) = 3\cos\left[\pi\left(x - \frac{3}{\pi}\right)\right] + 2$

$A = 3,\ B = \pi,\ C = 3,\ D = 2$

Amplitude: $|A| = |3| = 3$

Period: $\left|\dfrac{2\pi}{B}\right| = \left|\dfrac{2\pi}{\pi}\right| = 2$

Phase shift: $\dfrac{C}{B} = \dfrac{3}{\pi}$

24. $y = 5 - 2\cos\left(\frac{\pi}{2}x + \frac{\pi}{2}\right) = -2\cos\left[\frac{\pi}{2}(x - (-1))\right] + 5$

$A = -2,\ B = \frac{\pi}{2},\ C = -\frac{\pi}{2},\ D = 5$

Amplitude: $|A| = |-2| = 2$

Period: $\left|\dfrac{2\pi}{B}\right| = \left|\dfrac{2\pi}{\frac{\pi}{2}}\right| = 4$

Phase shift: $\dfrac{C}{B} = \dfrac{-\frac{\pi}{2}}{\frac{\pi}{2}} = -1$

25. $y = -\frac{1}{2}\cos(2\pi x) + 2$

$A = -\frac{1}{2},\ B = 2\pi,\ C = 0,\ D = 2$

Amplitude: $|A| = \left|-\frac{1}{2}\right| = \frac{1}{2}$

Period: $\left|\dfrac{2\pi}{B}\right| = \left|\dfrac{2\pi}{2\pi}\right| = 1$

Phase shift: $\dfrac{C}{B} = \dfrac{0}{2\pi} = 0$

26. $y = -2\sin(-2x + \pi) - 2 = -2\sin\left[-2\left(x - \frac{\pi}{2}\right)\right] - 2$

$A = -2,\ B = -2,\ C = -\pi,\ D = -2$

Amplitude: $|A| = |-2| = 2$

Period: $\left|\dfrac{2\pi}{B}\right| = \left|\dfrac{2\pi}{-2}\right| = \pi$

Phase shift: $\dfrac{C}{B} = \dfrac{-\pi}{-2} = \dfrac{\pi}{2}$

27. $y = -\sin\left(\frac{1}{2}x - \frac{\pi}{2}\right) + \frac{1}{2} = -\sin\left[\frac{1}{2}(x - \pi)\right] + \frac{1}{2}$

$A = -1,\ B = \frac{1}{2},\ C = \frac{\pi}{2},\ D = \frac{1}{2}$

Amplitude: $|A| = |-1| = 1$

Period: $\left|\dfrac{2\pi}{B}\right| = \left|\dfrac{2\pi}{\frac{1}{2}}\right| = 4\pi$

Phase shift: $\dfrac{C}{B} = \dfrac{\frac{\pi}{2}}{\frac{1}{2}} = \pi$

28. $y = \frac{1}{3}\cos(-3x) + 1$

$A = \frac{1}{3},\ B = -3,\ C = 0,\ D = 1$

Amplitude: $|A| = \left|\frac{1}{3}\right| = \frac{1}{3}$

Period: $\left|\dfrac{2\pi}{B}\right| = \left|\dfrac{2\pi}{-3}\right| = \dfrac{2\pi}{3}$

Phase shift: $\dfrac{C}{B} = \dfrac{0}{-3} = 0$

29. $y = \cos(-2\pi x) + 2$

$A = 1,\ B = -2\pi,\ C = 0,\ D = 2$

Amplitude: $|A| = |1| = 1$

Period: $\left|\dfrac{2\pi}{B}\right| = \left|\dfrac{2\pi}{-2\pi}\right| = 1$

Phase shift: $\dfrac{C}{B} = \dfrac{0}{-2\pi} = 0$

30. $y = \frac{1}{2}\sin(2\pi x + \pi) = \frac{1}{2}\sin\left[2\pi\left(x - \left(-\frac{1}{2}\right)\right)\right]$

$A = \frac{1}{2},\ B = 2\pi,\ C = -\pi,\ D = 0$

Amplitude: $|A| = \left|\frac{1}{2}\right| = \frac{1}{2}$

Period: $\left|\frac{2\pi}{B}\right| = \left|\frac{2\pi}{2\pi}\right| = 1$

Phase shift: $\frac{C}{B} = \frac{-\pi}{2\pi} = -\frac{1}{2}$

31. $y = -\frac{1}{4}\cos(\pi x - 4) = -\frac{1}{4}\cos\left[\pi\left(x - \frac{4}{\pi}\right)\right]$

$A = -\frac{1}{4}$, $B = \pi$, $C = 4$, $D = 0$

Amplitude: $|A| = \left|-\frac{1}{4}\right| = \frac{1}{4}$

Period: $\left|\frac{2\pi}{B}\right| = \left|\frac{2\pi}{\pi}\right| = 2$

Phase shift: $\frac{C}{B} = \frac{4}{\pi}$

32. $y = 2\sin(2\pi x + 1) = 2\sin\left[2\pi\left(x - \left(-\frac{1}{2\pi}\right)\right)\right]$

$A = 2$, $B = 2\pi$, $C = -1$, $D = 0$

Amplitude: $|A| = |2| = 2$

Period: $\left|\frac{2\pi}{B}\right| = \left|\frac{2\pi}{2\pi}\right| = 1$

Phase shift: $\frac{C}{B} = \frac{-1}{2\pi} = -\frac{1}{2\pi}$

33. $y = -\cos 2x$

Shrink the graph of $y = \cos x$ horizontally by a factor of 2 and reflect it across the x-axis. Graph (b) is the correct choice.

34. $y = \frac{1}{2}\sin x - 2$

Shrink the graph of $y = \sin x$ vertically by a factor of 2 and translate it down 2 units. Graph (e) is the correct choice.

35. $y = 2\cos\left(x + \frac{\pi}{2}\right) = 2\cos\left[x - \left(-\frac{\pi}{2}\right)\right]$

Stretch the graph of $y = \cos x$ vertically by a factor of 2 and translate it left $\pi/2$ units. Graph (h) is the correct choice.

36. $y = -3\sin\frac{1}{2}x - 1$

Stretch the graph of $y = \sin x$ horizontally by a factor of 2, stretch it vertically by a factor of 3, reflect it across the x-axis, and translate it down 1 unit. Graph (d) is the correct choice.

37. $y = \sin(x - \pi) - 2$

Translate the graph of $y = \sin x$ right π units and down 2 units. Graph (a) is the correct choice.

38. $y = -\frac{1}{2}\cos\left(x - \frac{\pi}{4}\right)$

Shrink the graph of $y = \cos x$ vertically by a factor of 2, reflect it across the x-axis, and translate it to the right $\pi/4$ units. Graph (g) is the correct choice.

39. $y = \frac{1}{3}\sin 3x$

Shrink the graph of $y = \sin x$ horizontally and vertically, both by a factor of 3. Graph (f) is the correct choice.

40. $y = \cos\left(x - \frac{\pi}{2}\right)$

Translate the graph of $y = \cos x$ to the right $\pi/2$ units. Graph (c) is the correct choice.

41. This graph has the same shape as $y = \cos x$ but with an amplitude of $\frac{1}{2}$ and shifted up one unit. The equation is $y = \frac{1}{2}\cos x + 1$.

42. This graph has the same shape as $y = \sin x$ but with a phase shift of $\frac{\pi}{2}$ and amplitude of 2. The equation is $y = 2\sin\left(x - \frac{\pi}{2}\right)$.

43. This graph has the same shape as $y = \cos x$ but with a phase shift of $-\frac{\pi}{2}$ and also a shift of 2 units down. The equation is $y = \cos\left(x + \frac{\pi}{2}\right) - 2$.

44. This graph has the same shape as $y = \sin x$ but it is shrunk horizontally by a factor of 2 and reflected across the x-axis. The equation is $y = -\sin 2x$.

45. $y = 2\cos x + \cos 2x$

Graph $y = 2\cos x$ and $y = \cos 2x$ on the same set of axes. Then graphically add some ordinates to obtain points on the graph of $y = 2\cos x + \cos 2x$.

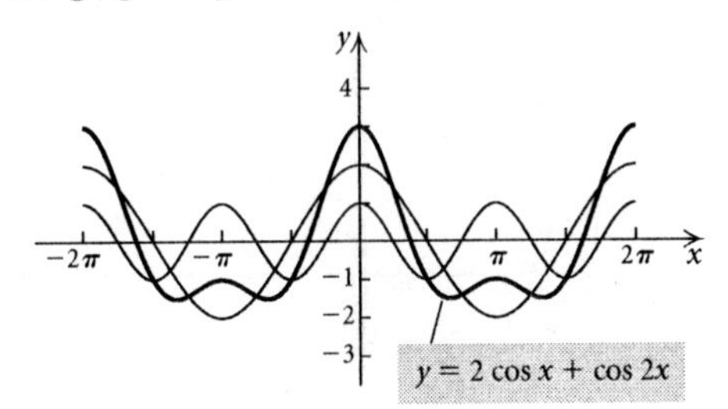

46.

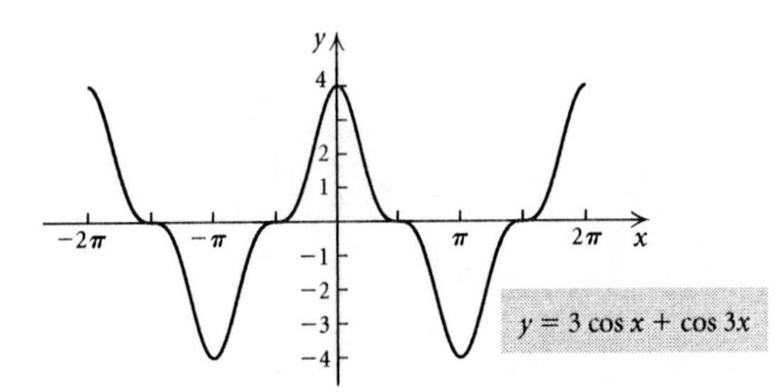

47. $y = \sin x + \cos 2x$

Graph $y = \sin x$ and $y = \cos 2x$ on the same set of axes. Then graphically add some ordinates to obtain points on the graph of $y = \sin x + \cos 2x$.

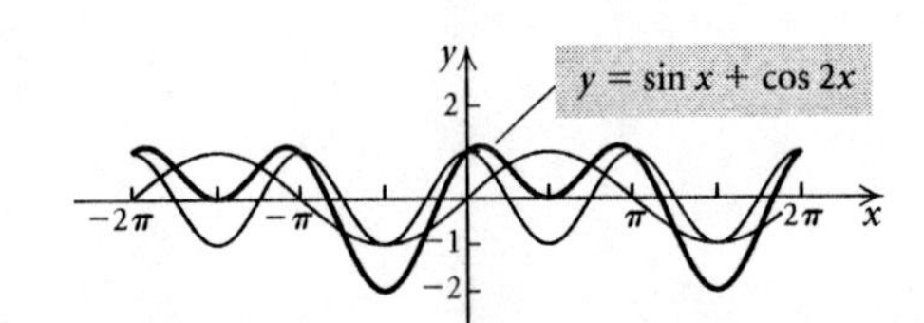

48.

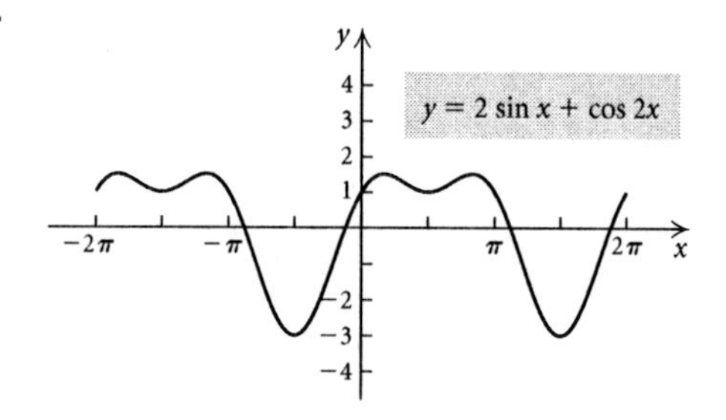

49. $y = \sin x - \cos x$

Graph $y = \sin x$ and $y = -\cos x$ on the same set of axes. Then graphically add some ordinates to obtain points on the graph of $y = \sin x - \cos x$.

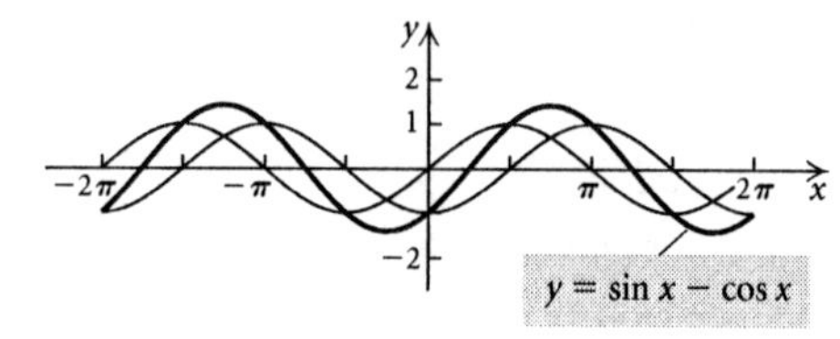

50.

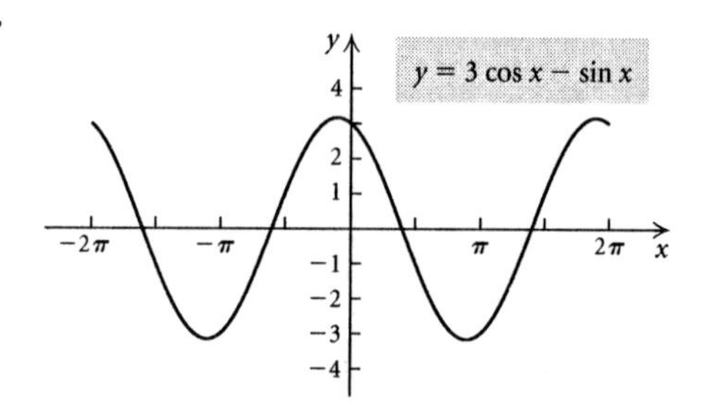

51. $y = 3\cos x + \sin 2x$

Graph $y = 3\cos x$ and $y = \sin 2x$ on the same set of axes. Then graphically add some ordinates to obtain points on the graph of $y = 3\cos x + \sin 2x$.

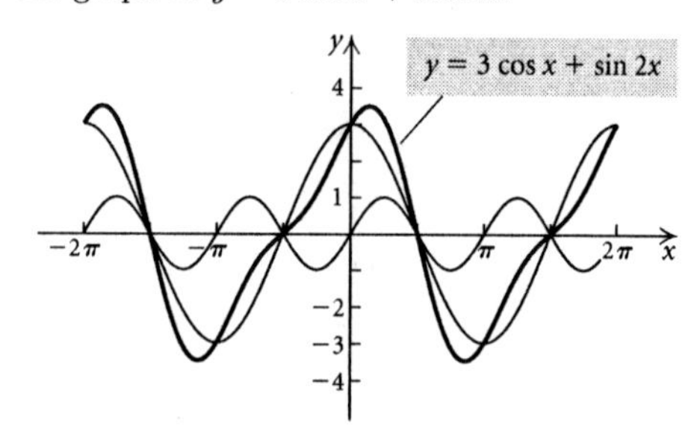

52.

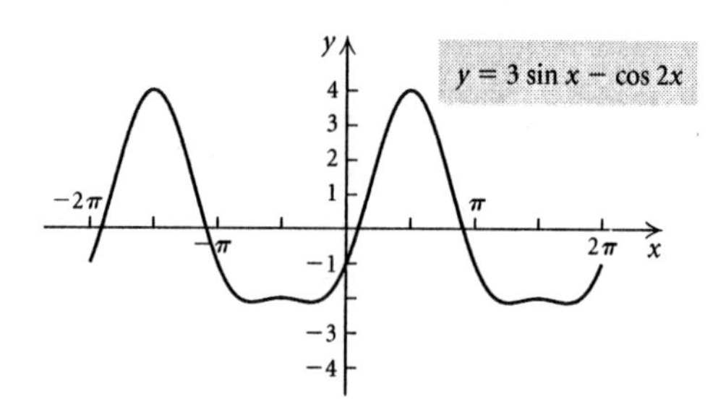

53. See the answer section in the text.

54.

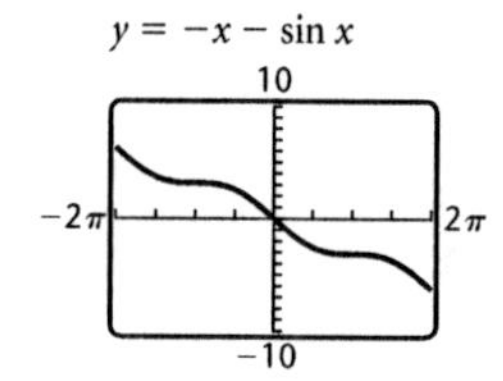

55. See the answer section in the text.

56.

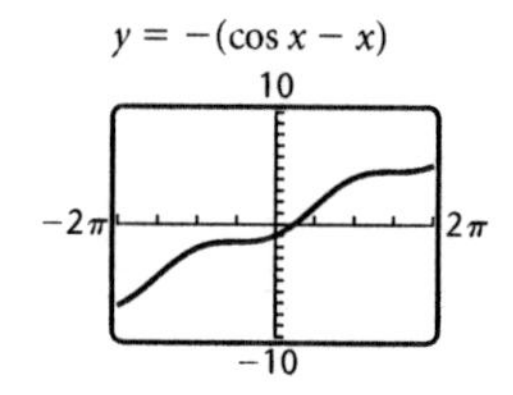

57. See the answer section in the text.

58.

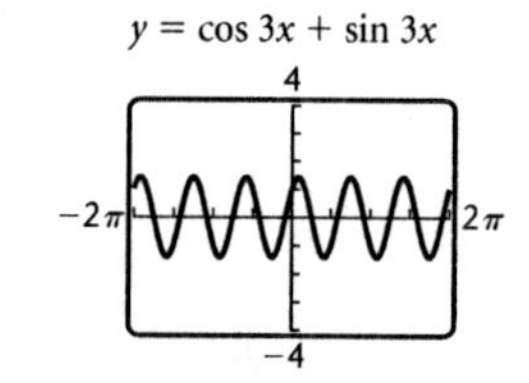

59. See the answer section in the text.

60.

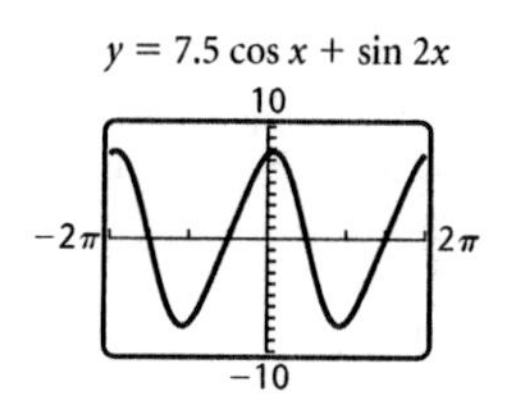

61. Graph $y = \dfrac{\sin x}{x}$ and use the Zero feature to find the zeros in the interval $[-12, 12]$. They are approximately -9.42, -6.28, -3.14, 3.14, 6.28, and 9.42.

62. Graph $y = \dfrac{\cos x - 1}{x}$ and use the Zero feature to find the zeros in the interval $[-12, 12]$. They are approximately -6.28 and 6.28. (Note that although from the graph it might appear that there is also a zero at $x = 0$, the function is not defined for this value of x.)

63. Graph $y = x^3 \sin x$ and use the Zero feature to find the zeros in the interval $[-12, 12]$. They are approximately -3.14, 0, and 3.14.

64. Graph $y = \dfrac{\sin^2 x}{x}$ and use the Zero feature to find the zeros in the interval $[-12, 12]$. They are approximately -3.14 and 3.14.

65. a) $y = 101.6 + 3\sin\left(\frac{\pi}{8}x\right)$

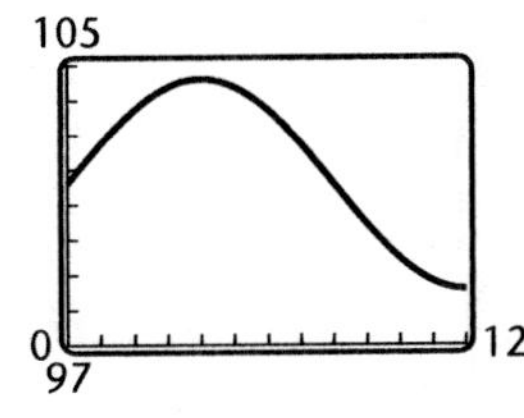

b) The maximum value occurs on day 4 when the sine function takes its maximum value, 1. It is $101.6° + 3° \cdot 1$, or $104.6°$.

The minimum value occurs on day 12 when the sine function takes its minimum value, -1. It is $101.6° + 3°(-1)$, or $98.6°$.

66. a) $y = 10\left(1 - \cos\frac{\pi}{6}x\right)$

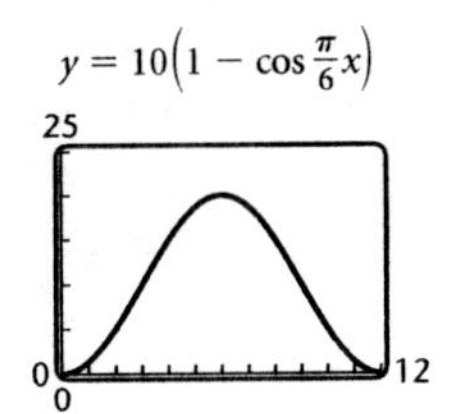

b) $\dfrac{2\pi}{\frac{\pi}{6}} = 12$

c) $0 in July (at $t = 0$ and $t = 12$)

d) $20,000 in January (at $t = 6$)

67. The constants B, C, and D translate the graphs and the constants A and B stretch or shrink the graphs. See the chart on page 491 of the text for a complete description of the effect of each constant.

68. The denominator B in the phase shift C/B serves to shrink or stretch the translation of C units by the same factor as the horizontal shrinking or stretching of the period. Thus, the translation must be done after the horizontal shrinking or stretching. For example, consider $y = \sin(2x - \pi)$. The phase shift of this function is $\pi/2$. First translate the graph of $y = \sin x$ to the right $\pi/2$ units and then shrink it horizontally by a factor of 2. Compare this graph with the one formed by first shrinking the graph of $y = \sin x$ horizontally by a factor of 2 and then translating it to the right $\pi/2$ units. The graphs differ; the second one is correct.

69. $f(x) = \dfrac{x+4}{4} = \dfrac{1}{4}x + 1$

The function can be written in the form $f(x) = mx + b$, so it is linear.

70. The variable in the function is in a logarithm, so the function is logarithmic.

71. This is a polynomial function with degree 4, so it is quartic.

72. $\dfrac{3}{4}x + \dfrac{1}{2}y = -5$, can be written in the equivalent form $y = -\dfrac{3}{2}x - 10$, so it is linear.

73. The variable in $f(x) = \sin x - 3$ is in a trigonometric function, so this is a trigonometric function.

74. The variable is in an exponent, so the function is exponential.

75. $y = \dfrac{2}{5}$ is equivalent to $y = 0x + \dfrac{2}{5}$. Since the function can be written in the form $y = mx + b$, it is linear.

76. The variable is in trigonometric functions, so the function is trigonometric.

77. This is a polynomial function with degree 3, so it is cubic.

78. The variable is in an exponent, so the function is exponential.

79. $y = 2\cos\left[3\left(x - \dfrac{\pi}{2}\right)\right] + 6$

The maximum value of the cosine function is 1, so the maximum value of the given function is $2 \cdot 1 + 6$, or 8.

The minimum value of the cosine function is -1, so the minimum value of the given function is $2(-1) + 6$, or 4.

80. $y = \dfrac{1}{2}\sin(2x - 6\pi) - 4$

The maximum value of the sine function is 1, so the maximum value of the given function is $\dfrac{1}{2} \cdot 1 - 4$, or $-\dfrac{7}{2}$, or $-3\dfrac{1}{2}$.

The minimum value of the sine function is -1, so the minimum value of the given function is $\dfrac{1}{2}(-1) - 4$, or $-\dfrac{9}{2}$, or $-4\dfrac{1}{2}$.

81. $y = -\tan x$

Reflect the graph of $y = \tan x$ across the x-axis.

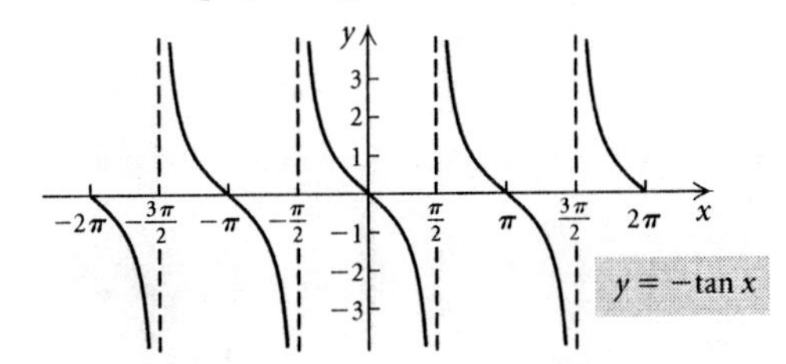

82. $y = \tan(-x)$

Reflect the graph of $y = \tan x$ across the y-axis.

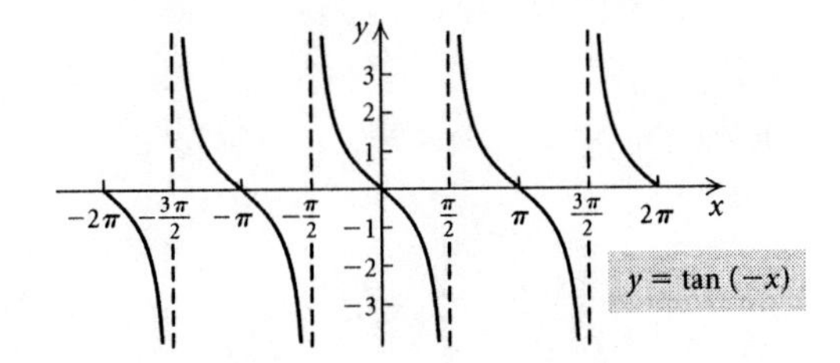

83. $y = -2 + \cot x$

Translate the graph of $y = \cot x$ down 2 units.

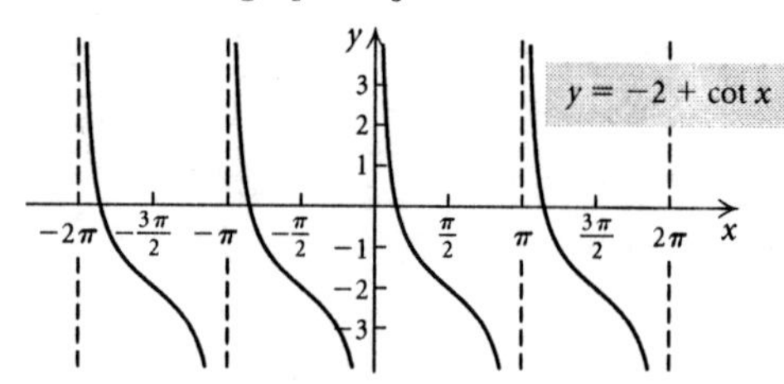

84. $y = -\dfrac{3}{2}\csc x$

Stretch the graph of $y = \csc x$ by a factor of $\dfrac{3}{2}$ and reflect it across the x-axis.

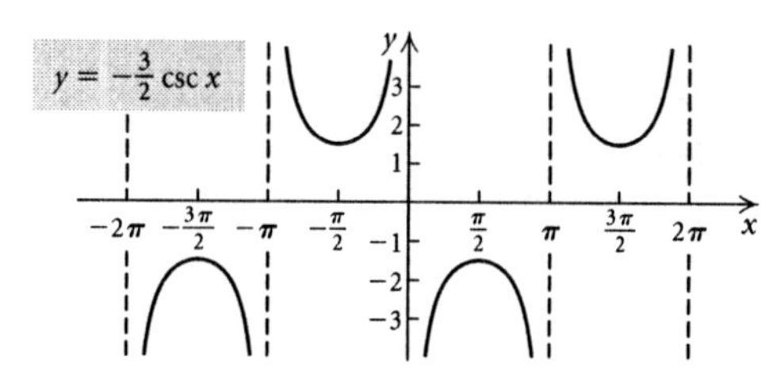

85. $y = 2\tan\dfrac{1}{2}x$

Stretch the graph of $y = \tan x$ horizontally and vertically, both by a factor of 2.

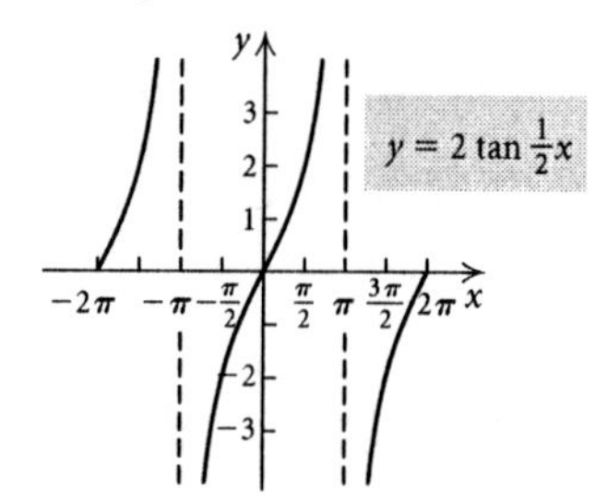

86. $y = \cot 2x$

Shrink the graph of $y = \cot x$ horizontally by a factor of 2.

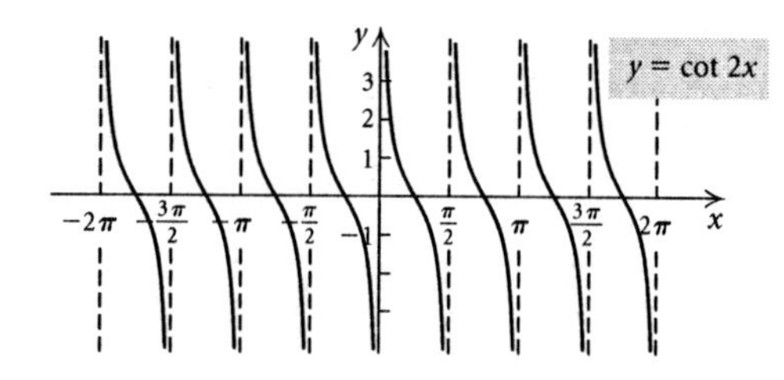

87. $y = 2\sec(x - \pi)$

Stretch the graph of $y = \sec x$ vertically by a factor of 2 and translate it to the right π units.

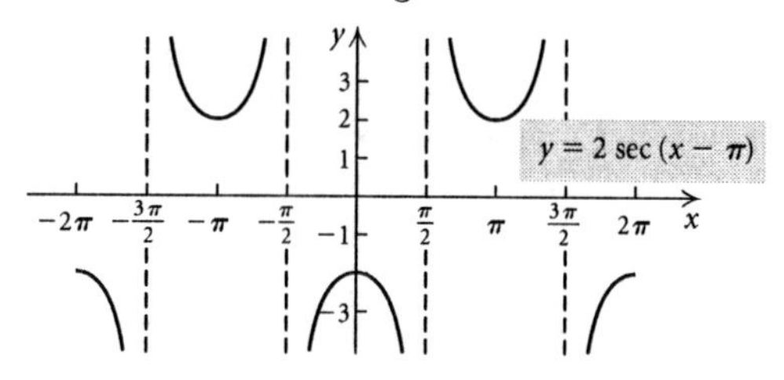

88. $y = 4\tan\left(\dfrac{1}{4}x + \dfrac{\pi}{8}\right) = 4\tan\left[\dfrac{1}{4}\left(x - \left(-\dfrac{\pi}{2}\right)\right)\right]$

Stretch the graph of $y = \tan x$ horizontally and vertically, both by a factor of 4. Then translate it $\pi/2$ units to the left.

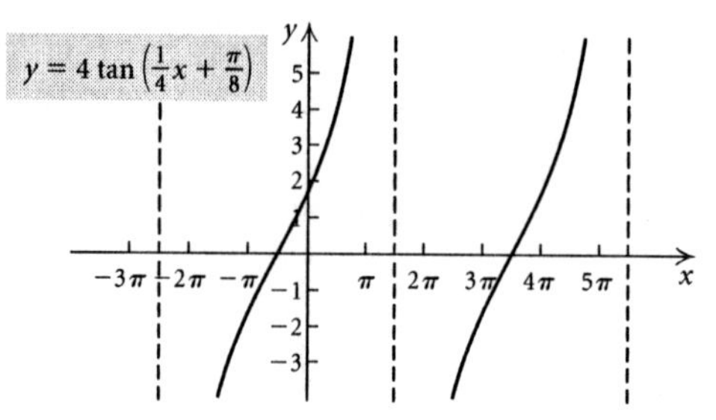

89. $y = 2\csc\left(\dfrac{1}{2}x - \dfrac{3\pi}{4}\right) = 2\csc\left[\dfrac{1}{2}\left(x - \dfrac{3\pi}{2}\right)\right]$

Stretch the graph of $y = \csc x$ horizontally and vertically, both by a factor of 2. Then translate it to the right $3\pi/2$ units.

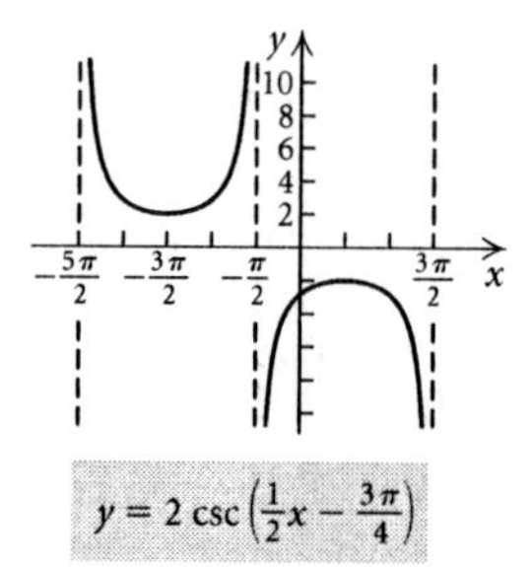

90. $y = 4\sec(2x - \pi) = 4\sec\left[2\left(x - \dfrac{\pi}{2}\right)\right]$

Shrink the graph of $y = \sec x$ horizontally by a factor of 2; stretch is vertically by a factor of 4; translate it to the right $\pi/2$ units.

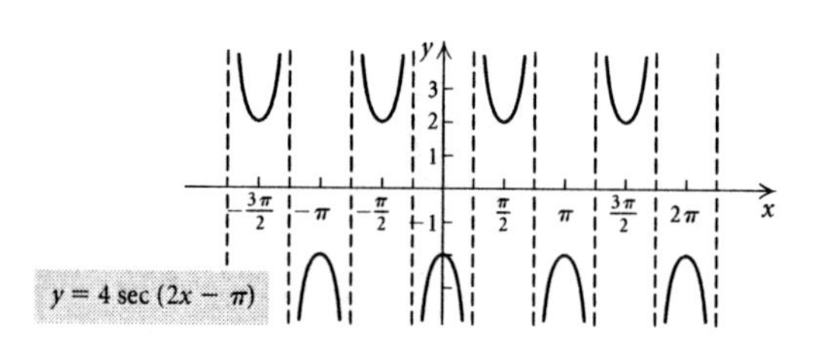

91. Amplitude: $|A| = |3000| = 3000$

Period: $\left|\dfrac{2\pi}{B}\right| = \left|\dfrac{2\pi}{\frac{\pi}{45}}\right| = 90$

Phase shift: $\dfrac{C}{B} = 10$

92. Write the function as $y = 3\sin\left[\dfrac{\pi}{4}(x - (-1))\right]$.

Amplitude: $|A| = |3| = 3$

Period: $\left|\dfrac{2\pi}{B}\right| = \left|\dfrac{2\pi}{\frac{\pi}{4}}\right| = 8$

Phase shift: $\dfrac{C}{B} = -1$

93. As t increases, $e^{-0.8t}$ decreases, and hence $6e^{-0.8t}\cos(6\pi t)$ decreases in the long run and approaches 0. Then the spring would be 4 inches from the ceiling when it stops bobbing.

94. The values of t for which the function is not defined are those values for which the beam is parallel to the wall.

Chapter 6

Trigonometric Identities, Inverse Functions, and Equations

Exercise Set 6.1

1. $(\sin x - \cos x)(\sin x + \cos x)$
$= \sin^2 x - \cos^2 x$

2. $\tan x(\cos x - \csc x)$
$= \dfrac{\sin x}{\cos x}\left(\cos x - \dfrac{1}{\sin x}\right)$
$= \sin x - \dfrac{1}{\cos x}$
$= \sin x - \sec x$

3. $\cos y \sin y(\sec y + \csc y)$
$= \cos y \sin y\left(\dfrac{1}{\cos y} + \dfrac{1}{\sin y}\right)$
$= \sin y + \cos y$

4. $(\sin x + \cos x)(\sec x + \csc x)$
$= \sin x \sec x + \sin x \csc x + \cos x \sec x +$
$\cos x \csc x$
$= \sin x \cdot \dfrac{1}{\cos x} + \sin x \cdot \dfrac{1}{\sin x} + \cos x \cdot x \cdot \dfrac{1}{\cos x} +$
$\cos x \cdot \dfrac{1}{\sin x}$
$= \tan x + 1 + 1 + \cot x$
$= 2 + \tan x + \cot x$

5. $(\sin\phi - \cos\phi)^2$
$= \sin^2\phi - 2\sin\phi\cos\phi + \cos^2\phi$
$= 1 - 2\sin\phi\cos\phi \qquad (\sin^2\phi + \cos^2\phi = 1)$

6. $(1 + \tan x)^2$
$= 1 + 2\tan x + \tan^2 x$
$= \sec^2 x + 2\tan x \qquad (1 + \tan^2 x = \sec^2 x)$

7. $(\sin x + \csc x)(\sin^2 x + \csc^2 x - 1)$
$= \sin^3 x + \sin x \csc^2 x - \sin x + \csc x \sin^2 x +$
$\csc^3 x - \csc x$
$= \sin^3 x + \sin x \cdot \dfrac{1}{\sin^2 x} - \sin x +$
$\dfrac{1}{\sin x} \cdot \sin^2 x + \csc^3 x - \dfrac{1}{\sin x}$
$= \sin^3 x + \dfrac{1}{\sin x} - \sin x + \sin x + \csc^3 x - \dfrac{1}{\sin x}$
$= \sin^3 x + \csc^3 x$

8. $(1 - \sin t)(1 + \sin t)$
$= 1 - \sin^2 t$
$= \cos^2 t$

9. $\sin x \cos x + \cos^2 x$
$= \cos x(\sin x + \cos x)$

10. $\tan^2\theta - \cot^2\theta$
$= (\tan\theta - \cot\theta)(\tan\theta + \cot\theta)$

11. $\sin^4 x - \cos^4 x$
$= (\sin^2 x + \cos^2 x)(\sin^2 x - \cos^2 x)$
$= \sin^2 x - \cos^2 x$
$= (\sin x + \cos x)(\sin x - \cos x)$

12. $4\sin^2 y + 8\sin y + 4$
$= 4(\sin^2 y + 2\sin y + 1)$
$= 4(\sin y + 1)^2$

13. $2\cos^2 x + \cos x - 3$
$= (2\cos x + 3)(\cos x - 1)$

14. $3\cot^2\beta + 6\cot\beta + 3$
$= 3(\cot^2\beta + 2\cot\beta + 1)$
$= 3(\cot\beta + 1)^2$

15. $\sin^3 x + 27$
$= (\sin x)^3 + 3^3$
$= (\sin x + 3)(\sin^2 x - 3\sin x + 9)$

16. $1 - 125\tan^3 s$
$= (1 - 5\tan s)(1 + 5\tan s + 25\tan^2 s)$

17. $\dfrac{\sin^2 x \cos x}{\cos^2 x \sin x}$
$= \dfrac{\sin x}{\cos x} \cdot \dfrac{\sin x \cos x}{\sin x \cos x}$
$= \dfrac{\sin x}{\cos x}$
$= \tan x$

18. $\dfrac{30\sin^3 x \cos x}{6\cos^2 x \sin x}$
$= \dfrac{5\sin^2 x}{\cos x}$, or $5\tan x \sin x$

19. $\dfrac{\sin^2 x + 2\sin x + 1}{\sin x + 1}$

$= \dfrac{(\sin x + 1)^2}{\sin x + 1}$

$= \sin x + 1$

20. $\dfrac{\cos^2 \alpha - 1}{\cos \alpha + 1}$

$= \dfrac{(\cos \alpha + 1)(\cos \alpha - 1)}{\cos \alpha + 1}$

$= \cos \alpha - 1$

21. $\dfrac{4\tan t \sec t + 2\sec t}{6\tan t \sec t + 2\sec t}$

$= \dfrac{2\sec t(2\tan t + 1)}{2\sec t(3\tan t + 1)}$

$= \dfrac{2\tan t + 1}{3\tan t + 1}$

22. $\dfrac{\csc(-x)}{\cot(-x)} = \dfrac{-\csc x}{-\cot x}$

$= \dfrac{\frac{1}{\sin x}}{\frac{\cos x}{\sin x}}$

$= \dfrac{1}{\sin x} \cdot \dfrac{\sin x}{\cos x}$

$= \dfrac{1}{\cos x}$

$= \sec x$

23. $\dfrac{\sin^4 x - \cos^4 x}{\sin^2 x - \cos^2 x}$

$= \dfrac{(\sin^2 x + \cos^2 x)(\sin^2 x - \cos^2 x)}{\sin^2 x - \cos^2 x}$

$= 1 \qquad (\sin^2 x + \cos^2 x = 1)$

24. $\dfrac{4\cos^3 x}{\sin^2 x} \cdot \left(\dfrac{\sin x}{4\cos x}\right)^2$

$= \dfrac{4\cos^3 x \sin^2 x}{16\sin^2 x \cos^2 x}$

$= \dfrac{\cos x}{4}$

25. $\dfrac{5\cos \phi}{\sin^2 \phi} \cdot \dfrac{\sin^2 \phi - \sin \phi \cos \phi}{\sin^2 \phi - \cos^2 \phi}$

$= \dfrac{5\cos \phi \sin \phi(\sin \phi - \cos \phi)}{\sin^2 \phi(\sin \phi + \cos \phi)(\sin \phi - \cos \phi)}$

$= \dfrac{5\cos \phi}{\sin \phi(\sin \phi + \cos \phi)}$

$= \dfrac{5\cot \phi}{\sin \phi + \cos \phi}$

26. $\dfrac{\tan^2 y}{\sec y} \div \dfrac{3\tan^3 y}{\sec y}$

$= \dfrac{\tan^2 y}{\sec y} \cdot \dfrac{\sec y}{3\tan^3 y}$

$= \dfrac{1}{3\tan y} \cdot \dfrac{\sec y \tan^2 y}{\sec y \tan^2 y}$

$= \dfrac{1}{3}\cot y$

27. $\dfrac{1}{\sin^2 s - \cos^2 s} - \dfrac{2}{\cos s - \sin s}$

$= \dfrac{1}{\sin^2 s - \cos^2 s} - \dfrac{2}{-(-\cos s + \sin s)}$

$= \dfrac{1}{\sin^2 s - \cos^2 s} + \dfrac{2}{\sin s - \cos x}$

$= \dfrac{1}{\sin^2 s - \cos^2 s} + \dfrac{2}{\sin s - \cos s} \cdot \dfrac{\sin s + \cos s}{\sin s + \cos s}$

$= \dfrac{1 + 2\sin s + 2\cos s}{\sin^2 s - \cos^2 s}$

28. $\left(\dfrac{\sin x}{\cos x}\right)^2 - \dfrac{1}{\cos^2 x}$

$= \tan^2 x - \sec^2 x$

$= -1 \qquad (1 + \tan^2 x = \sec^2 x)$

29. $\dfrac{\sin^2 \theta - 9}{2\cos \theta + 1} \cdot \dfrac{10\cos \theta + 5}{3\sin \theta + 9}$

$= \dfrac{(\sin \theta + 3)(\sin \theta - 3)}{2\cos \theta + 1} \cdot \dfrac{5(2\cos \theta + 1)}{3(\sin \theta + 3)}$

$= \dfrac{5(\sin \theta - 3)}{3}$

30. $\dfrac{9\cos^2 \alpha - 25}{2\cos \alpha - 2} \cdot \dfrac{\cos^2 \alpha - 1}{6\cos \alpha - 10}$

$= \dfrac{(3\cos \alpha + 5)(3\cos \alpha - 5)(\cos \alpha + 1)(\cos \alpha - 1)}{2(\cos \alpha - 1)(2)(3\cos \alpha - 5)}$

$= \dfrac{(3\cos \alpha + 5)(\cos \alpha + 1)}{4}$

31. $\sqrt{\sin^2 x \cos x} \cdot \sqrt{\cos x}$

$= \sqrt{\sin^2 x \cos^2 x}$

$= \sin x \cos x$

32. $\sqrt{\cos^2 x \sin x} \cdot \sqrt{\sin x}$

$= \sqrt{\cos^2 x \sin^2 x}$

$= \cos x \sin x$

33. $\sqrt{\cos \alpha \sin^2 \alpha} - \sqrt{\cos^3 \alpha}$

$= \sin \alpha \sqrt{\cos \alpha} - \cos \alpha \sqrt{\cos \alpha}$

$= \sqrt{\cos \alpha}(\sin \alpha - \cos \alpha)$

34. $\sqrt{\tan^2 x - 2\tan x \sin x + \sin^2 x}$

$= \sqrt{(\tan x - \sin x)^2}$

$= \tan x - \sin x$

35. $(1-\sqrt{\sin y})(\sqrt{\sin y}+1)$
$= (1-\sqrt{\sin y})(1+\sqrt{\sin y})$
$= 1-\sin y \qquad [(\sqrt{\sin y})^2 = \sin y]$

36. $\sqrt{\cos\theta}(\sqrt{2\cos\theta}+\sqrt{\sin\theta\cos\theta})$
$= \sqrt{2\cos^2\theta}+\sqrt{\sin\theta\cos^2\theta}$
$= \cos\theta\sqrt{2}+\cos\theta\sqrt{\sin\theta}$
$= \cos\theta(\sqrt{2}+\sqrt{\sin\theta})$

37. $\sqrt{\dfrac{\sin x}{\cos x}}$
$= \sqrt{\dfrac{\sin x}{\cos x}\cdot\dfrac{\cos x}{\cos x}}$
$= \sqrt{\dfrac{\sin x\cos x}{\cos^2 x}}$
$= \dfrac{\sqrt{\sin x\cos x}}{\cos x}$

(Note that $\sqrt{\dfrac{\sin x}{\cos x}}$ could also be expressed as $\sqrt{\tan x}$.)

38. $\sqrt{\dfrac{\cos x}{\tan x}} = \sqrt{\dfrac{\cos x\tan x}{\tan^2 x}}$
$= \dfrac{\sqrt{\cos x\cdot\dfrac{\sin x}{\cos x}}}{\tan x}$
$= \dfrac{\sqrt{\sin x}}{\tan x}$

39. $\sqrt{\dfrac{\cos^2 y}{2\sin^2 y}} = \sqrt{\dfrac{\cot^2 y}{2}\cdot\dfrac{2}{2}}$
$= \dfrac{\sqrt{2}\cot y}{2}$

40. $\sqrt{\dfrac{1-\cos\beta}{1+\cos\beta}} = \sqrt{\dfrac{1-\cos^2\beta}{(1+\cos\beta)^2}}$
$= \dfrac{\sqrt{\sin^2\beta}}{1+\cos\beta}$
$= \dfrac{\sin\beta}{1+\cos\beta}$

41. $\sqrt{\dfrac{\cos x}{\sin x}} = \sqrt{\dfrac{\cos x}{\sin x}\cdot\dfrac{\cos x}{\cos x}}$
$= \sqrt{\dfrac{\cos^2 x}{\sin x\cos x}}$
$= \dfrac{\cos x}{\sqrt{\sin x\cos x}}$

42. $\sqrt{\dfrac{\sin x}{\cot x}} = \sqrt{\dfrac{\sin x}{\cot x}\cdot\dfrac{\sin x}{\sin x}}$
$= \sqrt{\dfrac{\sin^2 x}{\cos x}} \qquad \left(\cot x\cdot\sin x = \dfrac{\cos x}{\sin x}\cdot\sin x = \cos x\right)$
$= \dfrac{\sin x}{\sqrt{\cos x}}$

43. $\sqrt{\dfrac{1+\sin y}{1-\sin y}} = \sqrt{\dfrac{1+\sin y}{1-\sin y}\cdot\dfrac{1+\sin y}{1+\sin y}}$
$= \sqrt{\dfrac{(1+\sin y)^2}{1-\sin^2 y}}$
$= \dfrac{1+\sin y}{\sqrt{\cos^2 y}}$
$= \dfrac{1+\sin y}{\cos y}$

44. $\sqrt{\dfrac{\cos^2 x}{2\sin^2 x}} = \sqrt{\dfrac{\cot^2 x}{2}} = \dfrac{\cot x}{\sqrt{2}}$

45. $\sqrt{a^2-x^2} = \sqrt{a^2-(a\sin\theta)^2}$ Substituting
$= \sqrt{a^2-a^2\sin^2\theta}$
$= \sqrt{a^2(1-\sin^2\theta)}$
$= \sqrt{a^2\cos^2\theta}$
$= a\cos\theta \qquad \left(a>0 \text{ and } \cos\theta>0 \text{ for } 0<\theta<\dfrac{\pi}{2}\right)$

Then $\cos\theta = \dfrac{\sqrt{a^2-x^2}}{a}$.

Also $x = a\sin\theta$, so $\sin\theta = \dfrac{x}{a}$. Then

$\tan\theta = \dfrac{\sin\theta}{\cos\theta} = \dfrac{\frac{x}{a}}{\frac{\sqrt{a^2-x^2}}{a}} = \dfrac{x}{a}\cdot\dfrac{a}{\sqrt{a^2-x^2}}$
$= \dfrac{x}{\sqrt{a^2-x^2}}$.

46. $\sqrt{4+x^2} = \sqrt{4+4\tan^2\theta} = \sqrt{4(1+\tan^2\theta)} = 2\sqrt{\sec^2\theta} = 2\sec\theta$

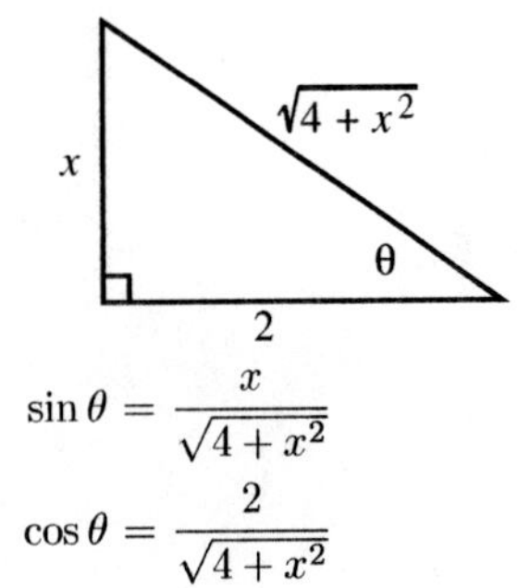

$\sin\theta = \dfrac{x}{\sqrt{4+x^2}}$

$\cos\theta = \dfrac{2}{\sqrt{4+x^2}}$

47. $x = 3\sec\theta$
$x = \dfrac{3}{\cos\theta} \qquad \left(\sec\theta = \dfrac{1}{\cos\theta}\right)$
$\cos\theta = \dfrac{3}{x}$
$\sqrt{x^2-9} = \sqrt{(3\sec\theta)^2-9}$ Substituting
$= \sqrt{9\sec^2\theta-9}$
$= \sqrt{9(\sec^2\theta-1)}$
$= \sqrt{9\tan^2\theta}$
$= 3\tan\theta \qquad \left(\tan\theta>0 \text{ for } 0<\theta<\dfrac{\pi}{2}\right)$

Then $\tan\theta = \dfrac{\sqrt{x^2-9}}{3}$

$$\frac{\sin\theta}{\cos\theta} = \frac{\sqrt{x^2-9}}{3}$$

$$\frac{\sin\theta}{\frac{3}{x}} = \frac{\sqrt{x^2-9}}{3} \quad \left(\cos\theta = \frac{3}{x}\right)$$

$$\sin\theta = \frac{3}{x}\cdot\frac{\sqrt{x^2-9}}{3}$$

$$\sin\theta = \frac{\sqrt{x^2-9}}{x}.$$

48. $\sqrt{x^2-a^2} = \sqrt{a^2\sec^2\theta - a^2} = \sqrt{a^2(\sec^2\theta-1)} = a\sqrt{\tan^2\theta} = a\tan\theta$

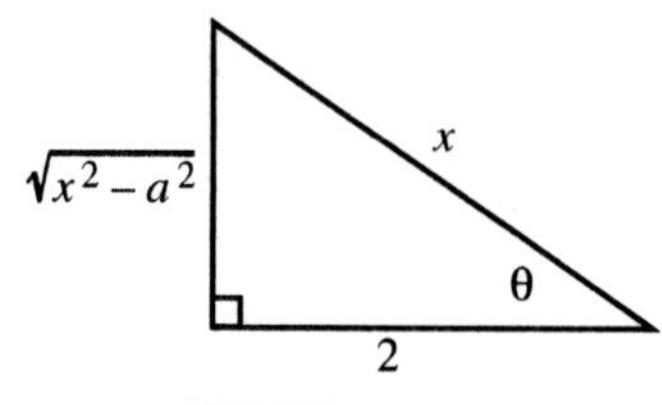

$$\sin\theta = \frac{\sqrt{x^2-a^2}}{x}$$

$$\cos\theta = \frac{a}{x}$$

49. $\dfrac{x^2}{\sqrt{1-x^2}} = \dfrac{\sin^2\theta}{\sqrt{1-\sin^2\theta}}$ Substituting

$$= \frac{\sin^2\theta}{\sqrt{\cos^2\theta}}$$

$$= \frac{\sin^2\theta}{\cos\theta} = \sin\theta\cdot\frac{\sin\theta}{\cos\theta}$$

$$= \sin\theta\tan\theta$$

50. $\dfrac{\sqrt{x^2-16}}{x^2} = \dfrac{\sqrt{16\sec^2\theta-16}}{16\sec^2\theta} = \dfrac{\sqrt{16(\sec^2\theta-1)}}{16\sec^2\theta} = \dfrac{4\sqrt{\tan^2\theta}}{16\sec^2\theta} = \dfrac{\tan\theta}{4\sec^2\theta} = \dfrac{1}{4}\cdot\dfrac{\sin\theta}{\cos\theta}\cdot\cos^2\theta = \dfrac{\sin\theta\cos\theta}{4}$

51. $\dfrac{\pi}{12} = \dfrac{3\pi}{12} - \dfrac{2\pi}{12} = \dfrac{\pi}{4} - \dfrac{\pi}{6}$

$$\sin\frac{\pi}{12} = \sin\left(\frac{\pi}{4}-\frac{\pi}{6}\right)$$

$$= \sin\frac{\pi}{4}\cos\frac{\pi}{6} - \cos\frac{\pi}{4}\sin\frac{\pi}{6}$$

$$= \frac{\sqrt{2}}{2}\cdot\frac{\sqrt{3}}{2} - \frac{\sqrt{2}}{2}\cdot\frac{1}{2}$$

$$= \frac{\sqrt{6}-\sqrt{2}}{4}$$

52. $\cos 75° = \cos(30° + 45°)$

$$= \cos 30°\cos 45° - \sin 30°\sin 45°$$

$$= \frac{\sqrt{3}}{2}\cdot\frac{\sqrt{2}}{2} - \frac{1}{2}\cdot\frac{\sqrt{2}}{2}$$

$$= \frac{\sqrt{6}-\sqrt{2}}{4}$$

53. $\tan 105° = \tan(45° + 60°)$

$$= \frac{\tan 45° + \tan 60°}{1 - \tan 45°\tan 60°}$$

$$= \frac{1+\sqrt{3}}{1-1\cdot\sqrt{3}}$$

$$= \frac{1+\sqrt{3}}{1-\sqrt{3}}$$

(This is equivalent to $\dfrac{\sqrt{6}+\sqrt{2}}{\sqrt{2}-\sqrt{6}}$ and $-2-\sqrt{3}$.)

54. $\dfrac{5\pi}{12} = \dfrac{9\pi}{12} - \dfrac{4\pi}{12} = \dfrac{3\pi}{4} - \dfrac{\pi}{3}$

$$\tan\frac{5\pi}{12} = \tan\left(\frac{3\pi}{4}-\frac{\pi}{3}\right)$$

$$= \frac{\tan\frac{3\pi}{4} - \tan\frac{\pi}{3}}{1+\tan\frac{3\pi}{4}\tan\frac{\pi}{3}}$$

$$= \frac{-1-\sqrt{3}}{1+(-1)\cdot\sqrt{3}}$$

$$= \frac{-1-\sqrt{3}}{1-\sqrt{3}}, \text{ or } \frac{\sqrt{3}+1}{\sqrt{3}-1}$$

(This is equivalent to $\dfrac{3+\sqrt{3}}{3-\sqrt{3}}$.)

55. $\cos 15° = \cos(45° - 30°)$

$$= \cos 45°\cos 30° + \sin 45°\sin 30°$$

$$= \frac{\sqrt{2}}{2}\cdot\frac{\sqrt{3}}{2} + \frac{\sqrt{2}}{2}\cdot\frac{1}{2}$$

$$= \frac{\sqrt{6}+\sqrt{2}}{4}$$

56. $\dfrac{7\pi}{12} = \dfrac{9\pi}{12} - \dfrac{2\pi}{12} = \dfrac{3\pi}{4} - \dfrac{\pi}{6}$

$$\sin\frac{7\pi}{12} = \sin\left(\frac{3\pi}{4}-\frac{\pi}{6}\right)$$

$$= \sin\frac{3\pi}{4}\cos\frac{\pi}{6} - \cos\frac{3\pi}{4}\sin\frac{\pi}{6}$$

$$= \frac{\sqrt{2}}{2}\cdot\frac{\sqrt{3}}{2} - \left(-\frac{\sqrt{2}}{2}\right)\left(\frac{1}{2}\right)$$

$$= \frac{\sqrt{6}+\sqrt{2}}{4}$$

57. $\sin 37°\cos 22° + \cos 37°\sin 22°$

$$= \sin(37° + 22°)$$

$$= \sin 59° \approx 0.8572$$

58. $\cos 83° \cos 53° + \sin 83° \sin 53°$
$= \cos(83° - 53°)$
$= \cos 30° \approx 0.8660$

59. $\cos 19° \cos 5° - \sin 19° \sin 5°$
$= \cos(19° + 5°)$
$= \cos 24° \approx 0.9135$

60. $\sin 40° \cos 15° - \cos 40° \sin 15°$
$= \sin(40° - 15°)$
$= \sin 25° \approx 0.4226$

61. $\dfrac{\tan 20° + \tan 32°}{1 - \tan 20° \tan 32°} = \tan(20° + 32°)$
$= \tan 52° \approx 1.2799$

62. $\dfrac{\tan 35° - \tan 12°}{1 + \tan 35° \tan 12°} = \tan(35° - 12°)$
$= \tan 23° \approx 0.4245$

63. See the answer section in the text.

64. $\tan(u - v)$

$$= \frac{\sin(u-v)}{\cos(u-v)}$$

$$= \frac{\sin u \cos v - \cos u \sin v}{\cos u \cos v + \sin u \sin v}$$

$$= \frac{\sin u \cos v - \cos u \sin v}{\cos u \cos v + \sin u \sin v} \cdot \frac{\frac{1}{\cos u \cos v}}{\frac{1}{\cos u \cos v}}$$

$$= \frac{\frac{\sin u}{\cos u} - \frac{\sin u}{\cos v}}{1 + \frac{\sin u \sin v}{\cos u \cos v}}$$

$$= \frac{\tan u - \tan u}{1 + \tan u \tan v}$$

Use the figures and function values below for Exercises 65-68.

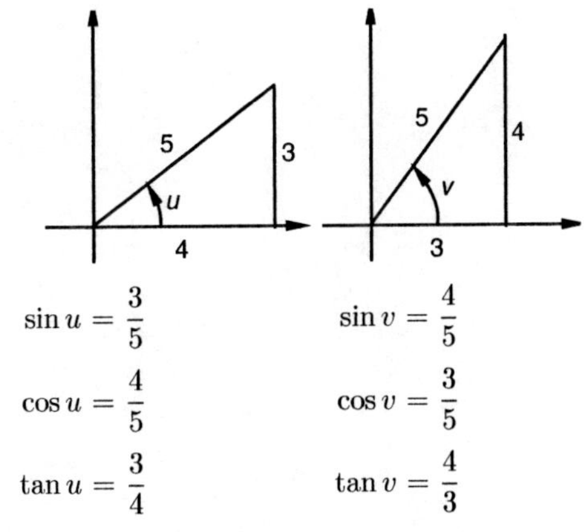

$\sin u = \dfrac{3}{5}$ $\sin v = \dfrac{4}{5}$

$\cos u = \dfrac{4}{5}$ $\cos v = \dfrac{3}{5}$

$\tan u = \dfrac{3}{4}$ $\tan v = \dfrac{4}{3}$

65. See the figures above.

$$\cos(u+v) = \cos u \cos v - \sin u \sin v$$
$$= \frac{4}{5} \cdot \frac{3}{5} - \frac{3}{5} \cdot \frac{4}{5}$$
$$= \frac{12}{25} - \frac{12}{25} = 0$$

66. See the figures before Exercise 65.

$$\tan(u-v) = \frac{\tan u - \tan v}{1 + \tan u \tan v}$$
$$= \frac{\frac{3}{4} - \frac{4}{3}}{1 + \frac{3}{4} \cdot \frac{4}{3}}$$
$$= -\frac{7}{24}$$

67. See the figures before Exercise 65.

$$\sin(u-v) = \sin u \cos v - \cos u \sin v$$
$$= \frac{3}{5} \cdot \frac{3}{5} - \frac{4}{5} \cdot \frac{4}{5}$$
$$= \frac{9}{25} - \frac{16}{25} = -\frac{7}{25}$$

68. See the figures before Exercise 65.

$$\cos(u-v) = \cos u \cos v + \sin u \sin v$$
$$= \frac{4}{5} \cdot \frac{3}{5} + \frac{3}{5} \cdot \frac{4}{5}$$
$$= \frac{12}{25} + \frac{12}{25} = \frac{24}{25}$$

Use the figures and function values below for Exercises 69-72.

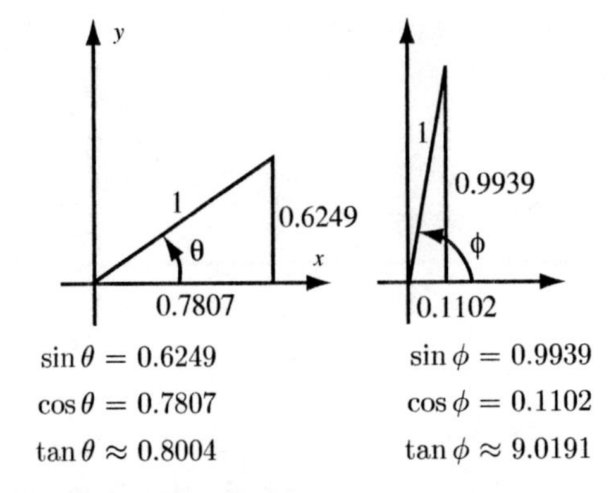

$\sin \theta = 0.6249$ $\sin \phi = 0.9939$

$\cos \theta = 0.7807$ $\cos \phi = 0.1102$

$\tan \theta \approx 0.8004$ $\tan \phi \approx 9.0191$

69. See the figures above.

$$\tan(\theta + \phi) = \frac{\tan \theta + \tan \phi}{1 - \tan \theta \tan \phi}$$
$$= \frac{0.8004 + 9.0191}{1 - 0.8004(9.0191)}$$
$$\approx -1.5790$$

(Answer may vary slightly due to rounding differences.)

70. See the figures before Exercise 69.

$$\sin(\theta - \phi) = \sin \theta \cos \phi - \cos \theta \sin \phi$$
$$= 0.6249(0.1102) - 0.7807(0.9939)$$
$$\approx -0.7071$$

71. See figures before Exercise 69.

$$\cos(\theta - \phi) = \cos \theta \cos \phi + \sin \theta \sin \phi$$
$$= 0.7807(0.1102) + 0.6249(0.9939)$$
$$\approx 0.7071$$

72. See the figures before Exercise 69.

$$\cos(\theta+\phi) = \cos\theta\cos\phi - \sin\theta\sin\phi$$
$$= 0.7807(0.1102) - 0.6249(0.9939)$$
$$\approx -0.5351$$

73. $\sin(\alpha+\beta)+\sin(\alpha-\beta)$

$= (\sin\alpha\cos\beta+\cos\alpha\sin\beta)+$
$(\sin\alpha\cos\beta-\cos\alpha\sin\beta)$

$= 2\sin\alpha\cos\beta$

74. $\cos(\alpha+\beta)-\cos(\alpha-\beta)$

$= (\cos\alpha\cos\beta-\sin\alpha\sin\beta)-$
$(\cos\alpha\cos\beta+\sin\alpha\sin\beta)$

$= -2\sin\alpha\sin\beta$

75. $\cos(u+v)\cos v+\sin(u+v)\sin v$

$= (\cos u\cos v-\sin u\sin v)\cos v+$
$(\sin u\cos v+\cos u\sin v)\sin v$

$= \cos u\cos^2 v-\sin u\sin v\cos v+$
$\sin u\cos v\sin v+\cos u\sin^2 v$

$= \cos u(\cos^2 v+sin^2 v)$

$= \cos u$

76. $\sin(u-v)\cos v+\cos(u-v)\sin v$

$= (\sin u\cos v-\cos u\sin v)\cos v+$
$(\cos u\cos v+\sin u\sin v)\sin v$

$= \sin u\cos^2 v-\cos u\sin v\cos v+$
$(\cos u\cos v\sin v+\sin u\sin^2 v$

$= \sin u\cos^2 v+\sin u\sin^2 v$

$= \sin u(\cos^2 v+\sin^2 v)$

$= \sin u$

77. Left to the student

78. Left to the student

79. A trigonometric equation that is an identity is true for all possible replacements of the variables. A trigonometric equation that is not true for all possible replacements is not an identity. The equation $\sin^2 x+\cos^2 x=1$ is an identity while $\sin^2 x=1$ is not.

80. It is not possible to see the entire graph, so the graphs of the two sides of the identity could differ at a point that hasn't been observed. However, if the two graphs are observed to be different, then we know that the equation is not an identity.

81. $2x-3=2\left(x-\dfrac{3}{2}\right)$

$2x-3=2x-3$

$-3=-3$ Subtracting $2x$ on both sides

The equation is true for all real numbers, so the solution set is the set of all real numbers.

82. $x-7=x+3.4$

$-7=3.4$ Subtracting x on both sides

The equation is false for all values of x. There is no solution.

83. $59°$ and $31°$ are complementary angles. Then $\cos 59° = \sin 31° = 0.5150$, so

$$\sec 59° = \frac{1}{\cos 59°} = \frac{1}{0.5150} \approx 1.9417.$$

84. $\tan 59° = \cot 31° = \dfrac{\cos 31°}{\sin 31°} = \dfrac{0.8572}{0.5150} \approx 1.6645$

85. Solve each equation for y to determine the slope of each line.

$l_1:\ 2x=3-2y$

$y=-x+\dfrac{3}{2}$

Thus $m_1=-1$ and the y-intercept is $\dfrac{3}{2}$.

$l_2:\ x+y=5$

$y=-x+5$

Thus $m_2=-1$ and the y-intercept is 5.

The lines are parallel, so they do not form an angle. When the formula is used, the result is $0°$.

86. l_1 has $m_1=\dfrac{\sqrt{3}}{3}$; l_2 has $m_2=\sqrt{3}$.

Then $\tan\phi = \dfrac{\sqrt{3}-\dfrac{\sqrt{3}}{3}}{1+\sqrt{3}\cdot\dfrac{\sqrt{3}}{3}} = \dfrac{\sqrt{3}}{3}$,

and $\phi=\dfrac{\pi}{6}$, or $30°$

87. Find the slope of each line.

l_1: $y=3$ $(y=0x+3)$

Thus $m=0$.

$l_2:\ x+y=5$

$y=-x+5$

Thus $m=-1$.

Let ϕ be the smallest angle from l_1 to l_2.

$$\tan\phi = \frac{-1-0}{1+(-1)(0)} \qquad \left(\tan\phi = \frac{m_2-m_1}{1+m_2m_1}\right)$$
$$= \frac{-1}{1} = -1$$

Since $\tan\phi$ is negative, we know that ϕ is obtuse. Thus $\phi=\dfrac{3\pi}{4}$, or $135°$.

88. l_1 has $m_1=-2$; l_2 has $m_2=2$.

Then $\tan\phi = \dfrac{2-(-2)}{1+2(-2)} = -\dfrac{4}{3}$,

and $\phi\approx 126.87°$.

89.

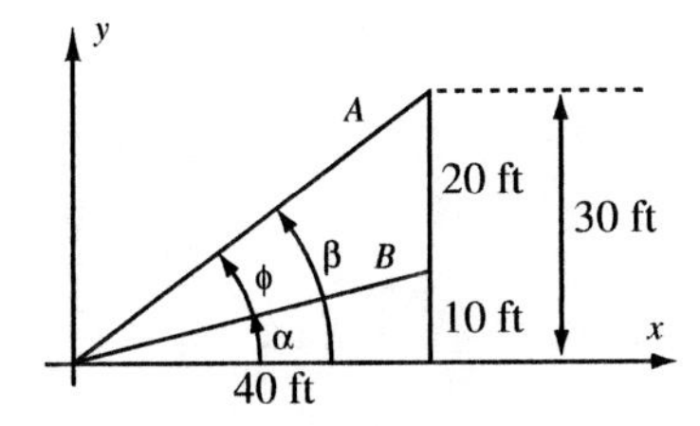

Position a set of coordinate axes as shown above. Let l_1 be the line containing the points $(0,0)$ and $(40,10)$. Find its slope.

$$m_1 = \tan\alpha = \frac{10}{40} = \frac{1}{4}$$

Let l_2 be the line containing the points $(0,0)$ and $(40,30)$. Find its slope.

$$m_2 = \tan\beta = \frac{30}{40} = \frac{3}{4}$$

Now ϕ is the smallest positive angle from l_1 to l_2.

$$\tan\phi = \frac{\frac{3}{4} - \frac{1}{4}}{1 + \frac{3}{4}\cdot\frac{1}{4}} = \frac{\frac{1}{2}}{\frac{19}{16}} = \frac{8}{19} \approx 0.4211$$

$$\phi \approx 22.83°$$

90. $\dfrac{\sin(x+h) - \sin x}{h}$

$$= \frac{\sin x\cos h + \cos x \sin h - \sin x}{h}$$

$$= \frac{\sin x\cos h - \sin x}{h} + \frac{\cos x\sin h}{h}$$

$$= \sin x\left(\frac{\cos h - 1}{h}\right) + \cos h\left(\frac{\sin h}{h}\right)$$

91. See the answer section in the text.

92. Let $\theta = \dfrac{3\pi}{2}$. Then $\sin\theta = -1$, but $\sqrt{\sin^2\theta} = \sqrt{(-1)^2} = 1$. Answers may vary.

93. See the answer section in the text.

94. Let $x = \dfrac{\pi}{2}$. Then $\sin(-x) = \sin\left(-\dfrac{\pi}{2}\right) = -1$, but $\sin x = \sin\dfrac{\pi}{2} = 1$.

Answers may vary.

95. See the answer section in the text.

96. Let $\theta = \dfrac{\pi}{4}$. Then $\tan^2\theta + \cot^2\theta = 1^2 + 1^2 = 2 \neq 1$. Answers may vary.

97. See the answer section in the text.

98.

$$\tan 45° = \frac{\frac{4}{3} - m_1}{1 + \frac{4}{3}\cdot m_1}$$

$$1 = \frac{4 - 3m_1}{3 + 4m_1}$$

$$3 + 4m_1 = 4 - 3m_1$$

$$7m_1 = 1$$

$$m_1 = \frac{1}{7}$$

99.

$$\tan\phi = \frac{m_2 - m_1}{1 + m_2 m_1}$$

$$\tan 30° = \frac{\frac{2}{3} - m_1}{1 + \frac{2}{3}m_1}$$

$$\frac{\sqrt{3}}{3} = \frac{\frac{2}{3} - m_1}{1 + \frac{2}{3}m_1}$$

$$\frac{\sqrt{3}}{3} + \frac{2\sqrt{3}}{9}m_1 = \frac{2}{3} - m_1$$

$$\frac{2\sqrt{3}}{9}m_1 + m_1 = \frac{2}{3} - \frac{\sqrt{3}}{3}$$

$$m_1\left(\frac{2\sqrt{3}}{9} + 1\right) = \frac{2 - \sqrt{3}}{3}$$

$$m_1\left(\frac{2\sqrt{3} + 9}{9}\right) = \frac{2 - \sqrt{3}}{3}$$

$$m_1 = \frac{2 - \sqrt{3}}{3}\cdot\frac{9}{2\sqrt{3} + 9}$$

$$m_1 = \frac{3(2 - \sqrt{3})}{2\sqrt{3} + 9} = \frac{6 - 3\sqrt{3}}{2\sqrt{3} + 9}$$

$$m_1 \approx 0.0645$$

100. Find the slope of l_1.

$$m_1 = \frac{-1 - 4}{5 - (-2)} = \frac{-5}{7} = -\frac{5}{7}$$

Use the following formula to determine the slope of l_2.

$$\tan\phi = \frac{m_2 - m_1}{1 + m_2 m_1}$$

$$\tan 45° = \frac{m_2 - \left(-\frac{5}{7}\right)}{1 + m_2\left(-\frac{5}{7}\right)}$$

$$1 = \frac{7m_2 + 5}{7 - 5m_2}$$

$$7 - 5m_2 = 7m_2 + 5$$

$$2 = 12m_2$$

$$\frac{1}{6} = m_2$$

101. Find the slope of l_1.

$m_1 = \frac{-2-7}{-3-(-3)} = \frac{-9}{0}$, which is undefined so l_1 is vertical.

Find the slope of l_2.

$m_2 = \frac{6-(-4)}{2-0} = \frac{10}{2} = 5$

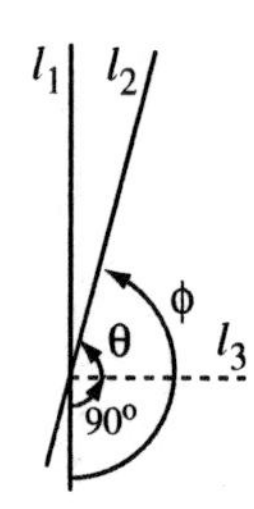

From the drawing we see that $\phi = 90° + \theta$, where θ is the smallest positive angle from the horizontal line l_3 to l_2. We find θ, recalling that the slope of a horizontal line is 0, so $m_3 = 0$.

$$\tan\theta = \frac{m_2 - m_3}{1 + m_2 m_3}$$
$$\tan\theta = \frac{5-0}{1+5\cdot 0} = 5$$
$$\theta \approx 78.7°$$

Then $\phi \approx 90° + 78.7° \approx 168.7°$.

102. $\sin 2\theta = \sin(\theta+\theta)$
$$= \sin\theta\cos\theta + \cos\theta\sin\theta$$
$$= 2\sin\theta\cos\theta$$

103. $\cos 2\theta = \cos(\theta+\theta)$
$$= \cos\theta\cos\theta - \sin\theta\sin\theta$$
$$= \cos^2 - \sin^2\theta,$$
$$\text{or } 1 - 2\sin^2\theta, \qquad (\cos^2\theta = 1-\sin^2\theta)$$
$$\text{or } 2\cos^2\theta - 1 \qquad (\sin^2\theta = 1-\cos^2\theta)$$

104. $\sin\left(x - \frac{3\pi}{2}\right) = \sin x\cos\frac{3\pi}{2} - \cos x\sin\frac{3\pi}{2}$
$$= (\sin x)(0) - (\cos x)(-1)$$
$$= \cos x$$

105. See the answer section in the text.

106.
$$\frac{\tan\alpha+\tan\beta}{1+\tan\alpha\tan\beta} = \frac{\frac{\sin\alpha}{\cos\alpha}+\frac{\sin\beta}{\cos\beta}}{1+\frac{\sin\alpha}{\cos\alpha}\cdot\frac{\sin\beta}{\cos\beta}}$$
$$= \frac{\frac{\sin\alpha\cos\beta+\cos\alpha\sin\beta}{\cos\alpha\cos\beta}}{\frac{\cos\alpha\cos\beta+\sin\alpha\sin\beta}{\cos\alpha\cos\beta}}$$
$$= \frac{\sin\alpha\cos\beta+\cos\alpha\sin\beta}{\cos\alpha\cos\beta+\sin\alpha\sin\beta}$$
$$= \frac{\sin(\alpha+\beta)}{\cos(\alpha-\beta)}$$

107. See the answer section in the text.

Exercise Set 6.2

1. $\sin(3\pi/10) \approx 0.8090$, $\cos(3\pi/10) \approx 0.5878$

a) $\tan\frac{3\pi}{10} = \frac{\sin(3\pi/10)}{\cos(3\pi/10)} \approx \frac{0.8090}{0.5878} \approx 1.3763$,

$\csc\frac{3\pi}{10} = \frac{1}{\sin(3\pi/10)} \approx \frac{1}{0.8090} \approx 1.2361$,

$\sec\frac{3\pi}{10} = \frac{1}{\cos(3\pi/10)} \approx \frac{1}{0.5878} \approx 1.7013$,

$\cot\frac{3\pi}{10} = \frac{1}{\tan(3\pi/10)} \approx \frac{1}{1.3763} \approx 0.7266$

b) $\frac{\pi}{2} - \frac{3\pi}{10} = \frac{2\pi}{10} = \frac{\pi}{5}$, so $\frac{3\pi}{10}$ and $\frac{\pi}{5}$ are complements.

$\sin\frac{\pi}{5} = \cos\frac{3\pi}{10} \approx 0.5878$,

$\cos\frac{\pi}{5} = \sin\frac{3\pi}{10} \approx 0.8090$,

$\tan\frac{\pi}{5} = \cot\frac{3\pi}{10} \approx 0.7266$,

$\csc\frac{\pi}{5} = \sec\frac{3\pi}{10} \approx 1.7013$,

$\sec\frac{\pi}{5} = \csc\frac{3\pi}{10} \approx 1.2361$,

$\cot\frac{\pi}{5} = \tan\frac{3\pi}{10} \approx 1.3763$

2. a) $\tan\frac{\pi}{12} = \frac{\sin(\pi/12)}{\cos(\pi/12)} = \frac{\sqrt{2-\sqrt{3}}}{\sqrt{2+\sqrt{3}}}$,

$\sec\pi/12 = \frac{1}{\sin(\pi/12)} = \frac{2}{\sqrt{2+\sqrt{3}}}$,

$\csc\frac{\pi}{12} = \frac{1}{\sin(\pi/12)} = \frac{2}{\sqrt{2-\sqrt{3}}}$,

$\cot\frac{\pi}{12} = \frac{1}{\tan(\pi/12)} = \frac{\sqrt{2+\sqrt{3}}}{\sqrt{2-\sqrt{3}}}$,

b) Since $\frac{\pi}{12}$ and $\frac{5\pi}{12}$ are complements, we have

$\sin\frac{5\pi}{12} = \frac{\sqrt{2+\sqrt{3}}}{2}$, $\cos\frac{5\pi}{12} = \frac{\sqrt{2-\sqrt{3}}}{2}$,

$\tan\frac{5\pi}{12} = \frac{\sqrt{2+\sqrt{3}}}{\sqrt{2-\sqrt{3}}}$, $\sec\frac{5\pi}{12} = \frac{2}{\sqrt{2-\sqrt{3}}}$,

$\csc\frac{5\pi}{12} = \frac{2}{\sqrt{2+\sqrt{3}}}$, $\cot\frac{5\pi}{12} = \frac{\sqrt{2-\sqrt{3}}}{\sqrt{2+\sqrt{3}}}$.

3. We sketch a second quadrant triangle.

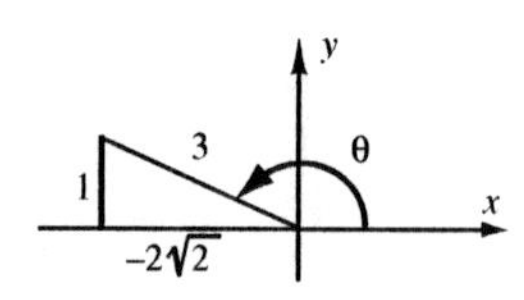

a) $\cos\theta = \dfrac{-2\sqrt{2}}{3} = -\dfrac{2\sqrt{2}}{3}$,

$\tan\theta = \dfrac{1}{-2\sqrt{2}} = -\dfrac{\sqrt{2}}{4}$,

$\csc\theta = \dfrac{3}{1} = 3$,

$\sec\theta = \dfrac{3}{-2\sqrt{2}} = -\dfrac{3\sqrt{2}}{4}$,

$\cot\theta = \dfrac{-2\sqrt{2}}{1} = -2\sqrt{2}$

b) Since θ and $\dfrac{\pi}{2} - \theta$ are complements, we have

$\sin\left(\dfrac{\pi}{2} - \theta\right) = \cos\theta = -\dfrac{2\sqrt{2}}{3}$,

$\cos\left(\dfrac{\pi}{2} - \theta\right) = \sin\theta = \dfrac{1}{3}$,

$\tan\left(\dfrac{\pi}{2} - \theta\right) = \cot\theta = -2\sqrt{2}$,

$\csc\left(\dfrac{\pi}{2} - \theta\right) = \sec\theta = -\dfrac{3\sqrt{2}}{4}$,

$\sec\left(\dfrac{\pi}{2} - \theta\right) = \csc\theta = 3$,

$\cot\left(\dfrac{\pi}{2} - \theta\right) = \tan\theta = -\dfrac{\sqrt{2}}{4}$

c) $\sin\left(\theta - \dfrac{\pi}{2}\right) = \sin\left[-\left(\dfrac{\pi}{2} - \theta\right)\right] =$

$-\sin\left(\dfrac{\pi}{2} - \theta\right) = -\left(-\dfrac{2\sqrt{2}}{3}\right) = \dfrac{2\sqrt{2}}{3}$,

$\cos\left(\theta - \dfrac{\pi}{2}\right) = \cos\left[-\left(\dfrac{\pi}{2} - \theta\right)\right] = \cos\left(\dfrac{\pi}{2} - \theta\right)$

$= \dfrac{1}{3}$,

$\tan\left(\theta - \dfrac{\pi}{2}\right) = \tan\left[-\left(\dfrac{\pi}{2} - \theta\right)\right] =$

$-\tan\left(\frac{\pi}{2} - \theta\right) = -(-2\sqrt{2}) = 2\sqrt{2}$,

$\csc\left(\theta - \dfrac{\pi}{2}\right) = \dfrac{1}{\sin\left(\theta - \dfrac{\pi}{2}\right)} = \dfrac{1}{\dfrac{2\sqrt{2}}{3}} = \dfrac{3}{2\sqrt{2}} =$

$\dfrac{3\sqrt{2}}{4}$,

$\sec\left(\theta - \dfrac{\pi}{2}\right) = \dfrac{1}{\cos\left(\theta - \dfrac{\pi}{2}\right)} = \dfrac{1}{\dfrac{1}{3}} = 3$,

$\cot\left(\theta - \dfrac{\pi}{2}\right) = \dfrac{1}{\tan\left(\theta - \dfrac{\pi}{2}\right)} = \dfrac{1}{2\sqrt{2}} = \dfrac{\sqrt{2}}{4}$

4.

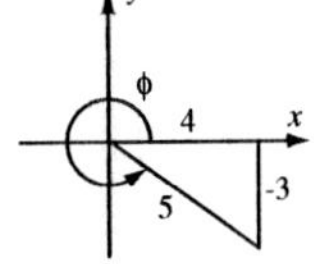

a) $\sin\phi = -\dfrac{3}{5}$, $\tan\phi = -\dfrac{3}{4}$, $\csc\phi = -\dfrac{5}{3}$,

$\sec\phi = \dfrac{5}{4}$, $\cot\phi = -\dfrac{4}{3}$

b) Since ϕ and $\dfrac{\pi}{2} - \phi$ are complements we have

$\sin\left(\dfrac{\pi}{2} - \phi\right) = \dfrac{4}{5}$, $\cos\left(\dfrac{\pi}{2} - \phi\right) = -\dfrac{3}{5}$,

$\tan\left(\dfrac{\pi}{2} - \phi\right) = -\dfrac{4}{3}$, $\csc\left(\dfrac{\pi}{2} - \phi\right) = \dfrac{5}{4}$,

$\sec\left(\dfrac{\pi}{2} - \phi\right) = -\dfrac{5}{3}$, $\cot\left(\dfrac{\pi}{2} - \phi\right) = -\dfrac{3}{4}$.

c) Since $\sin\left(\dfrac{\pi}{2} + \phi\right) = \cos\phi$ and $\cos\left(\dfrac{\pi}{2} + \phi\right) =$

$-\sin\phi$, we have

$\sin\left(\dfrac{\pi}{2} + \phi\right) = \dfrac{4}{5}$, $\cos\left(\dfrac{\pi}{2} + \phi\right) = \dfrac{3}{5}$,

$\tan\left(\dfrac{\pi}{2} + \phi\right) = \dfrac{4}{3}$, $\csc\left(\dfrac{\pi}{2} + \phi\right) = \dfrac{5}{4}$,

$\sec\left(\dfrac{\pi}{2} + \phi\right) = \dfrac{5}{3}$, $\cot\left(\dfrac{\pi}{2} + \phi\right) = \dfrac{3}{4}$.

5. $\sec\left(x + \dfrac{\pi}{2}\right) = \dfrac{1}{\cos\left(x + \dfrac{\pi}{2}\right)} = \dfrac{1}{-\sin x} = -\csc x$

6. $\cot\left(x - \dfrac{\pi}{2}\right) = \dfrac{\cos\left(x - \dfrac{\pi}{2}\right)}{\sin\left(x - \dfrac{\pi}{2}\right)} = \dfrac{\cos\left[-\left(\dfrac{\pi}{2} - x\right)\right]}{\sin\left[-\left(\dfrac{\pi}{2} - x\right)\right]} =$

$\dfrac{\cos\left(\dfrac{\pi}{2} - x\right)}{-\sin\left(\dfrac{\pi}{2} - x\right)} = \dfrac{\sin x}{-\cos x} = -\tan x$

7. $\tan\left(x - \dfrac{\pi}{2}\right) = \tan\left[-\left(\dfrac{\pi}{2} - x\right)\right] =$

$\dfrac{\sin\left[-\left(\dfrac{\pi}{2} - x\right)\right]}{\cos\left[-\left(\dfrac{\pi}{2} - x\right)\right]} = \dfrac{-\sin\left(\dfrac{\pi}{2} - x\right)}{\cos\left(\dfrac{\pi}{2} - x\right)} =$

$-\tan\left(\dfrac{\pi}{2} - x\right) = -\cot x$

8. $\csc\left(x + \dfrac{\pi}{2}\right) = \dfrac{1}{\sin\left(x + \dfrac{\pi}{2}\right)} = \dfrac{1}{\cos x} = \sec x$

9. Make a drawing.

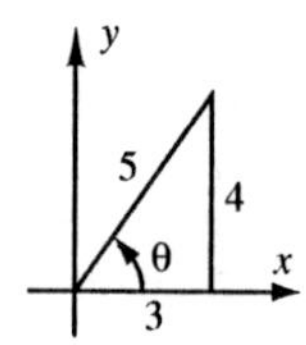

From this drawing we find that $\cos\theta = \frac{3}{5}$ and $\tan\theta = \frac{4}{3}$.

$$\sin 2\theta = 2\sin\theta\cos\theta = 2\cdot\frac{4}{5}\cdot\frac{3}{5} = \frac{24}{25}$$

$$\cos 2\theta = \cos^2\theta - \sin^2\theta = \left(\frac{3}{5}\right)^2 - \left(\frac{4}{5}\right)^2 = \frac{9}{25} - \frac{16}{25} = -\frac{7}{25}$$

$$\tan 2\theta = \frac{2\tan\theta}{1-\tan^2\theta} = \frac{2\cdot\frac{4}{3}}{1-\left(\frac{4}{3}\right)^2} = \frac{\frac{8}{3}}{-\frac{7}{9}} = -\frac{24}{7}$$

(We could have found $\tan 2\theta$ by dividing: $\tan 2\theta = \frac{\sin 2\theta}{\cos 2\theta}$.)

Since $\sin 2\theta$ is positive and $\cos 2\theta$ is negative, 2θ is in quadrant II.

10.

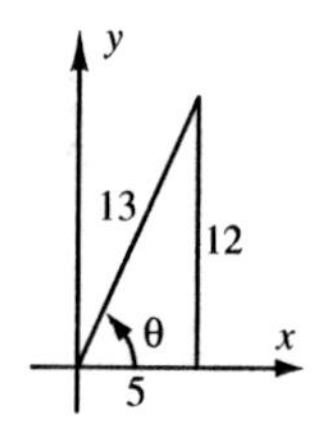

$$\sin 2\theta = 2\sin\theta\cos\theta = 2\cdot\frac{12}{13}\cdot\frac{5}{13} = \frac{120}{169}$$

$$\cos 2\theta = \cos^2\theta - \sin^2\theta = \left(\frac{5}{13}\right)^2 - \left(\frac{12}{13}\right)^2 = -\frac{119}{169}$$

$$\tan 2\theta = \frac{\sin 2\theta}{\cos 2\theta} = \frac{\frac{120}{169}}{-\frac{119}{169}} = -\frac{120}{119}$$

Since $\sin 2\theta$ is positive and $\cos 2\theta$ is negative, 2θ is in quadrant II.

11. Make a drawing.

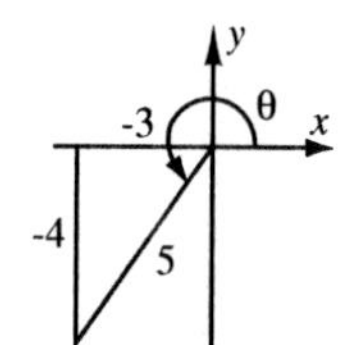

From this drawing we find that $\sin\theta = -\frac{4}{5}$ and $\tan\theta = \frac{4}{3}$.

$$\sin 2\theta = 2\sin\theta\cos\theta = 2\left(-\frac{4}{5}\right)\left(-\frac{3}{5}\right) = \frac{24}{25}$$

$$\cos 2\theta = \cos^2\theta - \sin^2\theta = \left(-\frac{3}{5}\right)^2 - \left(-\frac{4}{5}\right)^2 = -\frac{7}{25}$$

$$\tan 2\theta = \frac{\sin 2\theta}{\cos 2\theta} = -\frac{24}{7}$$

(We could have found $\tan 2\theta$ using a double-angle identity.)

Since $\sin 2\theta$ is positive and $\cos 2\theta$ is negative, 2θ is in quadrant II.

12.

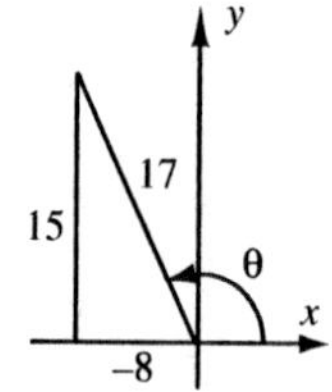

$$\sin 2\theta = 2\sin\theta\cos\theta = 2\cdot\frac{15}{17}\left(-\frac{8}{17}\right) = -\frac{240}{289}$$

$$\cos 2\theta = \cos^2\theta - \sin^2\theta = \left(-\frac{8}{17}\right)^2 - \left(\frac{15}{17}\right)^2 = -\frac{161}{289}$$

$$\tan 2\theta = \frac{\sin 2\theta}{\cos 2\theta} = \frac{-\frac{240}{289}}{-\frac{161}{289}} = \frac{240}{161}$$

Since $\sin 2\theta$ and $\cos 2\theta$ are both negative, 2θ is in quadrant III.

13. Make a drawing.

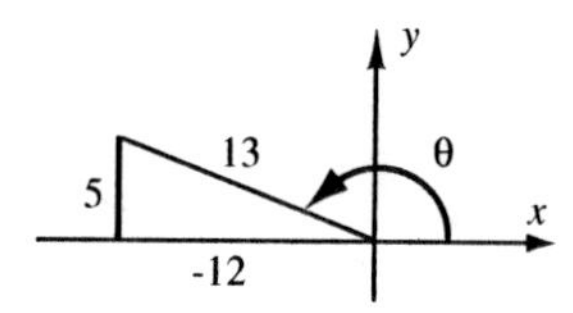

From this drawing we find that $\sin\theta = \frac{5}{13}$ and $\cos\theta = -\frac{12}{13}$.

$$\sin 2\theta = 2\sin\theta\cos\theta = 2\cdot\frac{5}{13}\left(-\frac{12}{13}\right) = -\frac{120}{169}$$

$$\cos 2\theta = \cos^2\theta - \sin^2\theta = \left(-\frac{12}{13}\right)^2 - \left(\frac{5}{13}\right)^2 = \frac{119}{169}$$

$$\tan 2\theta = \frac{\sin 2\theta}{\cos 2\theta} = \frac{-\frac{120}{169}}{\frac{119}{169}} = -\frac{120}{119}$$

(We could have found $\tan 2\theta$ using a double-angle identity.)

Since $\sin 2\theta$ is negative and $\cos 2\theta$ is positive, 2θ is in quadrant IV.

14.

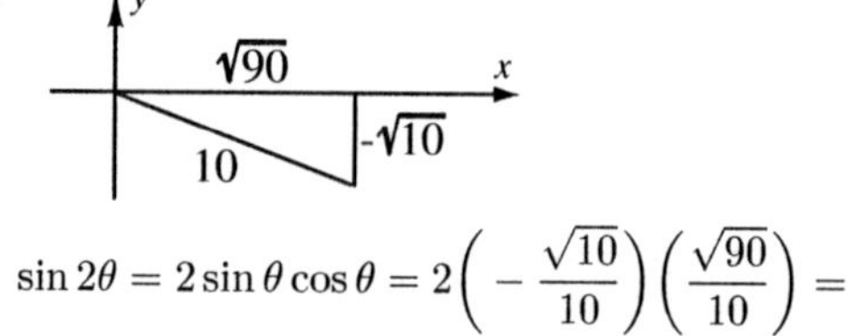

$$\sin 2\theta = 2\sin\theta\cos\theta = 2\left(-\frac{\sqrt{10}}{10}\right)\left(\frac{\sqrt{90}}{10}\right) =$$

$-\frac{2\sqrt{900}}{100} = -\frac{2\cdot 30}{100} = -\frac{3}{5}$

$\cos 2\theta = \cos^2\theta - \sin^2\theta = \left(\frac{\sqrt{90}}{10}\right)^2 - \left(-\frac{\sqrt{10}}{10}\right)^2 =$

$\frac{90}{100} - \frac{10}{100} = \frac{80}{100} = \frac{4}{5}$

$\tan 2\theta = \frac{\sin 2\theta}{\cos 2\theta} = \frac{-\frac{3}{5}}{\frac{4}{5}} = -\frac{3}{4}$

Since $\sin 2\theta$ is negative and $\cos 2\theta$ is positive, 2θ is in quadrant IV.

15. $\cos 4x$

$= \cos[2(2x)]$

$= 1 - 2\sin^2 2x$

$= 1 - 2(2\sin x\cos x)^2$

$= 1 - 8\sin^2 x\cos^2 x$

or

$\cos 4x$

$= \cos[2(2x)]$

$= \cos^2 2x - \sin^2 2x$

$= (\cos^2 x - \sin^2 x)^2 - (2\sin x\cos x)^2$

$= \cos^4 x - 6\sin^2 x\cos^2 x + \sin^4 x$

or

$\cos 4x$

$= \cos[2(2x)]$

$= 2\cos^2 2x - 1$

$= 2(2\cos^2 x - 1)^2 - 1$

$= 8\cos^4 x - 8\cos^2 x + 1$

16. $\sin^4\theta = \sin^2\theta\cdot\sin^2\theta$

$= \frac{1-\cos 2\theta}{2}\cdot\frac{1-\cos 2\theta}{2}$

$= \frac{1 - 2\cos 2\theta + \cos^2 2\theta}{4}$

$= \frac{1 - 2\cos 2\theta + \frac{1+\cos 4\theta}{2}}{4}$

$= \frac{2 - 4\cos 2\theta + 1 + \cos 4\theta}{8}$

$= \frac{3 - 4\cos 2\theta + \cos 4\theta}{8}$

17. $\cos 15° = \cos\frac{30°}{2} = \sqrt{\frac{1+\cos 30°}{2}} = \sqrt{\frac{1+\sqrt{3}/2}{2}} =$

$\sqrt{\frac{2+\sqrt{3}}{4}} = \frac{\sqrt{2+\sqrt{3}}}{2}$

(We choose the positive square root since 15°is in quadrant I where the cosine function is positive.)

18. $\tan 67.5° = \tan\frac{135°}{2} = \frac{1-\cos 135°}{\sin 135°} =$

$\frac{1-\left(-\frac{\sqrt{2}}{2}\right)}{\frac{\sqrt{2}}{2}} = \frac{2+\sqrt{2}}{\sqrt{2}} = \frac{2\sqrt{2}+2}{2} = \sqrt{2}+1$

19. $\sin 112.5° = \sin\frac{225°}{2} = \sqrt{\frac{1-\cos 225°}{2}} =$

$\sqrt{\frac{1-(-\sqrt{2}/2)}{2}} = \sqrt{\frac{2+\sqrt{2}}{4}} = \frac{\sqrt{2+\sqrt{2}}}{2}$

(We choose the positive square root since 112.5°is in quadrant II where the sine function is positive.)

20. $\cos\frac{\pi}{8} = \cos\frac{\frac{\pi}{4}}{2} = \sqrt{\frac{1+\cos(\pi/4)}{2}} =$

$\sqrt{\frac{1+(\sqrt{2}/2)}{2}} = \sqrt{\frac{2+\sqrt{2}}{4}} = \frac{\sqrt{2+\sqrt{2}}}{2}$

(We choose the positive square root since $\pi/8$ is in quadrant I where the cosine function is positive.)

21. $\tan 75° = \tan\frac{150°}{2} = \frac{\sin 150°}{1+\cos 150°} =$

$\frac{\frac{1}{2}}{1+(-\sqrt{3}/2)} = \frac{1}{2-\sqrt{3}}$, or $2+\sqrt{3}$

22. $\sin\frac{5\pi}{2} = \sin\frac{\frac{5\pi}{6}}{2} = \sqrt{\frac{1-\cos(5\pi/6)}{2}} =$

$\sqrt{\frac{1-(-\sqrt{3}/2)}{2}} = \sqrt{\frac{2+\sqrt{3}}{4}} = \frac{\sqrt{2+\sqrt{3}}}{2}$

(We choose the positive square root since $5\pi/12$ is in quadrant I where the sine function is positive.)

23. First find $\cos\theta$.

$\sin^2\theta + \cos^2\theta = 1$

$(0.3416)^2 + \cos^2\theta = 1$

$\cos^2\theta = 1 - (0.3416)^2$

$\cos\theta \approx 0.9398$ (θ is in quadrant I.)

$\sin 2\theta = 2\sin\theta\cos\theta$

$\approx 2(0.3416)(0.9398)$

≈ 0.6421

24. $\cos\frac{\theta}{2} = \sqrt{\frac{1+\cos\theta}{2}}$ (θ is in quadrant I.)

$\approx \sqrt{\frac{1+0.9398}{2}}$ We found $\cos\theta$ in Exercise 23.

≈ 0.9848

25. $\sin\frac{\theta}{2} = \sqrt{\frac{1-\cos\theta}{2}}$ $\left(\frac{\theta}{2}\text{ is in quadrant I.}\right)$

$\approx \sqrt{\frac{1-0.9398}{2}}$ We found $\cos\theta$ in Exercise 23.

≈ 0.1735

26. $\sin 4\theta = 2 \sin 2\theta \cos 2\theta$

We found $\sin 2\theta \approx 0.6421$ in Exercise 23.

Now we find $\cos 2\theta$:

$\cos 2\theta = 1 - 2\sin^2\theta = 1 - 2(0.3416)^2 \approx 0.7666$

Then $\sin 4\theta \approx 2(0.6421)(0.7666) \approx 0.9845$.

27. $2\cos^2\frac{x}{2} - 1 = \cos\left(2\cdot\frac{x}{2}\right) = \cos x$

28.
$$\begin{aligned}&\cos^4 x - \sin^4 x\\ &= (\cos^2 x + \sin^2 x)(\cos^2 x - \sin^2 x)\\ &= (1)\cdot(\cos 2x)\\ &= \cos 2x\end{aligned}$$

29.
$$\begin{aligned}&(\sin x - \cos x)^2 + (\sin 2x)\\ &= (\sin^2 x - 2\sin x\cos x + \cos^2 x) + (2\sin x\cos x)\\ &= \sin^2 x + \cos^2 x\\ &= 1\end{aligned}$$

30.
$$\begin{aligned}&(\sin x + \cos x)^2\\ &= \sin^2 x + 2\sin x\cos x + \cos^2 x\\ &= 1 + \sin 2x\end{aligned}$$

31. $\dfrac{2 - \sec^2 x}{\sec^2 x} = \dfrac{2}{\sec^2 x} - 1 = 2\cos^2 x - 1 = \cos 2x$

32.
$$\begin{aligned}&\frac{1 + \sin 2x + \cos 2x}{1 + \sin 2x - \cos 2x}\\ &= \frac{1 + (2\sin x\cos x) + (2\cos^2 x - 1)}{1 + (2\sin x\cos x) - (1 - 2\sin^2 x)}\\ &= \frac{2\cos x(\sin x + \cos x)}{2\sin x(\cos x + \sin x)}\\ &= \cot x\end{aligned}$$

33.
$$\begin{aligned}&(-4\cos x\sin x + 2\cos 2x)^2 + (2\cos 2x + 4\sin x\cos x)^2\\ &= (16\cos^2 x\sin^2 x - 16\sin x\cos x\cos 2x + 4\cos^2 2x) + (4\cos^2 2x + 16\sin x\cos x\cos 2x + 16\sin^2 x\cos^2 x)\\ &= 8\cos^2 2x + 32\cos^2 x\sin^2 x\\ &= 8(\cos^2 x - \sin^2 x)^2 + 32\cos^2 x\sin^2 x\\ &= 8(\cos^4 x - 2\cos^2 x\sin^2 x + \sin^4 x) + 32\cos^2 x\sin^2 x\\ &= 8\cos^4 x - 16\cos^2 x\sin^2 x + 8\sin^4 x + 32\cos^2 x\sin^2 x\\ &= 8\cos^4 x + 16\cos^2 x\sin^2 x + 8\sin^4 x\\ &= 8(\cos^4 x + 2\cos^2 x\sin^2 x + \sin^4 x)\\ &= 8(\cos^2 x + \sin^2 x)^2\\ &= 8\end{aligned}$$

34.
$$\begin{aligned}&2\sin x\cos^3 x - 2\sin^3 x\cos x\\ &= (2\sin x\cos x)(\cos^2 x - \sin^2 x)\\ &= (\sin 2x)(\cos 2x)\\ &= \frac{1}{2}\sin 4x\end{aligned}$$

35. Using a graphing calculator we see that the graphs of $y_1 = \dfrac{\cos 2x}{\cos x - \sin x}$ and $y_2 = \cos x + \sin x$ are the same on $[-2\pi, 2\pi]$. The correct choice is (d). See the answer section in the text for the algebraic proof.

36. Using a graphing calculator we see that the graphs of $y_1 = 2\cos^2\frac{x}{2}$ and $y_2 = 1 + \cos x$ are the same on $[-2\pi, 2\pi]$. The correct choice is (d).

$2\cos^2\frac{x}{2} = 2\left(\frac{1 + \cos x}{2}\right) = 1 + \cos x$

37. Using a graphing calculator we see that the graphs of $y_1 = \dfrac{\sin 2x}{2\cos x}$ and $y_2 = \sin x$ are the same on $[-2\pi, 2\pi]$. The correct choice is (d). See the answer section in the text for the algebraic proof.

38. Using a graphing calculator we see that the graphs of $y_1 = 2\sin\frac{\theta}{2}\cos\frac{\theta}{2}$ and $y_2 = \sin\theta$ are the same on $[-2\pi, 2\pi]$. The correct choice is (c).

$2\sin\frac{\theta}{2}\cos\frac{\theta}{2} = \sin\left(2\cdot\frac{\theta}{2}\right) = \sin\theta$

39. Each has amplitude 1 and is periodic. The period of $y_1 = \sin x$ is 2π, of $y_2 = \sin 2x$ is π, and of $y_3 = \sin\frac{x}{2}$ is 4π.

40. In the first line, $\cos 4x \neq 2\cos 2x$. In the second line, $\cos 2x \neq \cos^2 x + \sin^2 x$. If the second line had been correct, the third line would have been correct also.

41. $1 - \cos^2 x = \sin^2 x$ $\quad(\sin^2 x + \cos^2 x = 1)$

42. $\sec^2 x - \tan^2 x = 1$ $\quad(1 + \tan^2 x = \sec^2 x)$

43. $\sin^2 x - 1 = -\cos^2 x$ $\quad(\sin^2 x + \cos^2 x = 1)$

44. $1 + \cot^2 x = \csc^2 x$

45. $\csc^2 x - \cot^2 x = 1$ $\quad(1 + \cot^2 x = \csc^2 x)$

46. $1 + \tan^2 x = \sec^2 x$

47. $1 - \sin^2 x = \cos^2 x$ $\quad(\sin^2 x + \cos^2 x = 1)$

48. $\sec^2 x - 1 = \tan^2 x$ $\quad(1 + \tan^2 x = \sec^2 x)$

49. The functions $f(x) = A\sin(Bx - C) + D$, or $f(x) = A\cos(Bx - C) + D$ for which $|A| = 2$ are (a) and (e).

50. The functions $f(x) = A\sin(Bx - C) + D$, or $f(x) = A\cos(Bx - C) + D$ for which $\left|\dfrac{2\pi}{B}\right| = \pi$ are (b), (c), and (f).

51. The function $f(x) = A\sin(Bx - C) + D$, or $f(x) = A\cos(Bx - C) + D$ for which $\left|\dfrac{2\pi}{B}\right| = 2\pi$ is (d).

52. The function $f(x) = A\sin(Bx - C) + D$, or $f(x) = A\cos(Bx - C) + D$ for which $\frac{C}{B} = \frac{\pi}{4}$ is (e).

53. Observe that $141° = 51° + 90°$.

Find $\sin 51°$:

$$\sin^2 51° + \cos^2 51° = 1$$
$$\sin^2 51° + (0.6293)^2 = 1$$
$$\sin^2 51° = 1 - (0.6293)^2$$
$$\sin 51° \approx 0.7772 \quad (51° \text{ is in quadrant I.})$$

$$\sin 141° = \cos 51° \approx 0.6293$$
$$\cos 141° = -\sin 51° \approx -0.7772$$
$$\tan 141° = \frac{\sin 141°}{\cos 141°} \approx \frac{0.6293}{-0.7772} \approx -0.8097$$
$$\csc 141° = \frac{1}{\sin 141°} \approx \frac{1}{0.6293} \approx 1.5891$$
$$\sec 141° = \frac{1}{\cos 141°} \approx \frac{1}{-0.7772} \approx -1.2867$$
$$\cot 141° = \frac{1}{\tan 141°} \approx \frac{1}{-0.8097} \approx -1.2350$$

54.
$$\sin\left(\frac{\pi}{2} - x\right)[\sec x - \cos x]$$
$$= -\sin\left(x - \frac{\pi}{2}\right)\left[\frac{1}{\cos x} - \cos x\right]$$
$$\left(\sin\left(\frac{\pi}{2} - x\right) = \sin\left[-\left(x - \frac{\pi}{2}\right)\right] = -\sin\left(x - \frac{\pi}{2}\right)\right)$$
$$= -(-\cos x)\left(\frac{1 - \cos^2 x}{\cos x}\right) \quad \left[\sin\left(x - \frac{\pi}{2}\right) = -\cos x\right]$$
$$= 1 - \cos^2 x$$
$$= \sin^2 x$$

55.
$$\cos(\pi - x) + \cot x \sin\left(x - \frac{\pi}{2}\right)$$
$$= \cos\pi\cos x + \sin\pi\sin x +$$
$$\cot x(-\cos x) \quad \left[\sin\left(x - \frac{\pi}{2}\right) = -\cos x\right]$$
$$= -\cos x + 0 - \cot x \cos x$$
$$= -\cos x(1 + \cot x)$$

56.
$$\frac{\cos x - \sin\left(\frac{\pi}{2} - x\right)\sin x}{\cos x - \cos(\pi - x)\tan x}$$
$$= \frac{\cos x - \sin\left[-\left(x - \frac{\pi}{2}\right)\right]\sin x}{\cos x - (\cos\pi\cos x + \sin\pi\sin x)\tan x}$$
$$= \frac{\cos x + \sin\left(x - \frac{\pi}{2}\right)\sin x}{\cos x - (-1\cdot\cos x)\tan x}$$
$$= \frac{\cos x - \cos x\sin x}{\cos x + \cos x\tan x}$$
$$= \frac{\cos x(1 - \sin x)}{\cos x(1 + \tan x)}$$
$$= \frac{1 - \sin x}{1 + \tan x}$$

57.
$$\frac{\cos^2 y\sin\left(y + \frac{\pi}{2}\right)}{\sin^2 y\sin\left(\frac{\pi}{2} - y\right)} = \frac{\cos^2 y\cos y}{\sin^2 y\cos y}$$
$$= \frac{\cos^2 y}{\sin^2 y}$$
$$= \cot^2 y$$

58. $\frac{3\pi}{2} \le 2\theta \le 2\pi$, so $\frac{3\pi}{4} \le \theta \le \pi$; $\sin\theta$ is positive; $\cos\theta$ and $\tan\theta$ are negative.

$$\cos 2\theta = 1 - 2\sin^2\theta$$
$$\frac{7}{12} = 1 - 2\sin^2\theta$$
$$2\sin^2\theta = \frac{5}{12}$$
$$\sin^2\theta = \frac{5}{24}$$
$$\sin\theta = \sqrt{\frac{5}{24}} = \frac{\sqrt{30}}{12}$$
$$\cos 2\theta = 2\cos^2\theta - 1$$
$$\frac{7}{12} = 2\cos^2\theta - 1$$
$$\frac{19}{12} = 2\cos^2\theta$$
$$\frac{19}{24} = \cos^2\theta$$
$$-\sqrt{\frac{19}{24}} = \cos\theta, \text{ or}$$
$$-\frac{\sqrt{114}}{12} = \cos\theta$$

$$\tan\theta = \frac{\sin\theta}{\cos\theta} = \frac{\frac{\sqrt{30}}{12}}{-\frac{\sqrt{114}}{12}} = -\sqrt{\frac{5}{19}} = -\frac{\sqrt{95}}{19}$$

59. Since $\pi < \theta \leq \frac{3\pi}{2}$, $\tan\theta$ is positive and $\sin\theta$ and $\cos\theta$ are negative.

$$\tan\frac{\theta}{2} = -\sqrt{\frac{1-\cos\theta}{1+\cos\theta}}$$

$$-\frac{5}{3} = -\sqrt{\frac{1-\cos\theta}{1+\cos\theta}}$$

$$\frac{25}{9} = \frac{1-\cos\theta}{1+\cos\theta}$$

$$25 + 25\cos\theta = 9 - 9\cos\theta$$

$$34\cos\theta = -16$$

$$\cos\theta = -\frac{16}{34} = -\frac{8}{17}$$

$$\sin\theta = -\sqrt{1-\cos^2\theta} = -\sqrt{1-\left(-\frac{8}{17}\right)^2} = -\frac{15}{17}$$

$$\tan\theta = \frac{\sin\theta}{\cos\theta} = \frac{-\frac{15}{17}}{-\frac{8}{17}} = \frac{15}{8}$$

60. a) We substitute 42° for ϕ.

$$N(42^\circ) = 6066 - 31\cos(2\cdot 42^\circ)$$
$$= 6066 - 31\cos 84^\circ$$
$$\approx 6062.76 \text{ ft}$$

b) We substitute 90° for ϕ.

$$N(90^\circ) = 6066 - 31\cos(2\cdot 90^\circ)$$
$$= 6066 - 31\cos 180^\circ$$
$$= 6066 - 31(-1) = 6066 + 31$$
$$= 6097 \text{ ft}$$

c) We substitute $2\cos^2\phi - 1$ for $\cos 2\phi$.

$N(\phi) = 6066 - 31(2\cos^2\phi - 1)$, or

$N(\phi) = 6097 - 62\cos^2\phi$

61. a) Substitute 42° for ϕ.

$$g = 9.78049[1 + 0.005288\sin^2 42^\circ - 0.000006\sin^2(2\cdot 42^\circ)]$$
$$\approx 9.80359 \text{ m/sec}^2$$

b) Substitute 40° for ϕ.

$$g = 9.78049[1 + 0.005288\sin^2 40^\circ - 0.000006\sin^2(2\cdot 40^\circ)]$$
$$\approx 9.80180 \text{ m/sec}^2$$

c) $g = 9.78049[1 + 0.005288\sin^2\phi - 0.000006(2\sin\phi\cos\phi)^2]$

$g = 9.78049(1 + 0.005288\sin^2\phi - 0.000024\sin^2\phi\cos^2\phi)$

$g = 9.78049[1 + 0.005288\sin^2\phi - 0.000024\sin^2\phi(1-\sin^2\phi)]$

$g = 9.78049(1 + 0.005264\sin^2\phi + 0.000024\sin^4\phi)$

Exercise Set 6.3

Note: Answers for the odd-numbered exercises 1-29 are in the answer section in the text.

2. $\dfrac{1+\cos\theta}{\sin\theta} + \dfrac{\sin\theta}{\cos\theta} = \dfrac{\cos\theta + 1}{\sin\theta\cos\theta}$

We start with the left side.

$$\frac{1+\cos\theta}{\sin\theta} + \frac{\sin\theta}{\cos\theta}$$
$$= \frac{1+\cos\theta}{\sin\theta}\cdot\frac{\cos\theta}{\cos\theta} + \frac{\sin\theta}{\cos\theta}\cdot\frac{\sin\theta}{\sin\theta}$$
$$= \frac{\cos\theta + \cos^2\theta + \sin^2\theta}{\sin\theta\cos\theta}$$
$$= \frac{\cos\theta + 1}{\sin\theta\cos\theta}$$

We started with the left side and deduced the right side, so the proof is complete.

4. $\dfrac{1+\tan y}{1+\cot y} = \dfrac{\sec y}{\csc y}$

We start with the left side.

$$\frac{1+\tan y}{1+\cot y}$$
$$= \frac{1+\frac{\sin y}{\cos y}}{1+\frac{\cos y}{\sin y}}$$
$$= \frac{\frac{\cos y + \sin y}{\cos y}}{\frac{\sin y + \cos y}{\sin y}}$$
$$= \frac{\cos y + \sin y}{\cos y}\cdot\frac{\sin y}{\sin y + \cos y}$$
$$= \frac{\sin y}{\cos y}$$
$$= \tan y$$

Now we stop and work with the right side.

$$\frac{\sec y}{\csc y} = \frac{\sin y}{\cos y}$$
$$= \tan y$$

We have deduced the same expression from each side, so the proof is complete.

6. $\dfrac{\sin x + \cos x}{\sec x + \csc x} = \dfrac{\sin x}{\sec x}$

We start with the left side.

$$\frac{\sin x + \cos x}{\sec x + \csc x}$$
$$= \frac{\sin x + \cos x}{\frac{1}{\cos x} + \frac{1}{\sin x}}$$
$$= \frac{\sin x + \cos x}{\frac{\sin x + \cos x}{\sin x \cos x}}$$
$$= (\sin x + \cos x) \cdot \frac{\sin x \cos x}{\sin x + \cos x}$$
$$= \sin x \cos x$$
$$= \sin x \cdot \frac{1}{\sec x}$$
$$= \frac{\sin x}{\sec x}$$

We started with the left side and deduced the right side, so the proof is complete.

8. $\sec 2\theta = \dfrac{\sec^2\theta}{2 - \sec^2\theta}$

We begin with the left side.

$$\sec 2\theta = \frac{1}{\cos 2\theta}$$
$$= \frac{1}{\cos^2\theta - \sin^2\theta}$$

Now we stop and work with the right side.

$$\frac{\sec^2\theta}{2 - \sec^2\theta} = \frac{1 + \tan^2\theta}{2 - 1 - \tan^2\theta}$$
$$= \frac{1 + \tan^2\theta}{1 - \tan^2\theta}$$
$$= \frac{1 + \frac{\sin^2\theta}{\cos^2\theta}}{1 - \frac{\sin^2\theta}{\cos^2\theta}}$$
$$= \frac{\cos^2\theta + \sin^2\theta}{\cos^2\theta - \sin^2\theta}$$
$$= \frac{1}{\cos^2\theta - \sin^2\theta}$$

We have deduced the same expression from each side, so the proof is complete.

10. $\dfrac{\cos(u - v)}{\sin u \sin v} = \tan u + \cot v$

We start with the left side.

$$\frac{\cos(u - v)}{\cos u \sin v}$$
$$= \frac{\cos u \cos v + \sin u \sin v}{\cos u \sin v}$$
$$= \frac{\cos u \cos v}{\cos u \sin v} + \frac{\sin u \sin v}{\cos u \sin v}$$
$$= \frac{\cos v}{\sin v} + \frac{\sin u}{\cos u}$$
$$= \cot v + \tan u$$
$$= \tan u + \cot v$$

We started with the left side and deduced the right side, so the proof is complete.

12. $\cos^4 x - \sin^4 x = \cos 2x$

We start with the left side.

$$\cos^4 x - \sin^4 x$$
$$= (\cos^2 x - \sin^2 x)(\cos^2 x + \sin^2 x)$$
$$= \cos^2 x - \sin^2 x$$
$$= \cos 2x$$

We started with the left side and deduced the right side, so the proof is complete.

14. $\dfrac{\tan 3t - \tan t}{1 + \tan 3t \tan t} = \dfrac{2\tan t}{1 - \tan^2 t}$

We start with the left side.

$$\frac{\tan 3t - \tan t}{1 + \tan 3t \tan t} = \tan(3t - t)$$
$$= \tan 2t$$
$$= \frac{2\tan t}{1 - \tan^2 t}$$

We started with the left side and deduced the right side, so the proof is complete.

16. $\dfrac{\cos^3\beta - \sin^3\beta}{\cos\beta - \sin\beta} = \dfrac{2 + \sin 2\beta}{2}$

We start with the left side.

$$\frac{\cos^3\beta - \sin^3\beta}{\cos\beta - \sin\beta}$$
$$= \frac{(\cos\beta - \sin\beta)(\cos^2\beta + \cos\beta\sin\beta + \sin^2\beta)}{\cos\beta - \sin\beta}$$
$$= 1 + \cos\beta\sin\beta$$

Now we stop and work with the right side.

$$\frac{2 + \sin 2\beta}{2} = \frac{2 + 2\sin\beta\cos\beta}{2}$$
$$= 1 + \sin\beta\cos\beta$$

We have deduced the same expression from each side, so the proof is complete.

18. $\cos^2 x(1 - \sec^2 x) = -\sin^2 x$

We start with the left side.

$$\cos^2 x(1 - \sec^2 x) = \cos^2 x(-\tan^2 x) = \cos^2 x\left(-\frac{\sin^2 x}{\cos^2 x}\right) = -\sin^2 x$$

We started with the left side and deduced the right side, so the proof is complete.

20. $\dfrac{\cos\theta + \sin\theta}{\cos\theta} = 1 + \tan\theta$

We start with the left side.

$$\frac{\cos\theta + \sin\theta}{\cos\theta} = \frac{\cos\theta}{\cos\theta} + \frac{\sin\theta}{\cos\theta} = 1 + \tan\theta$$

We started with the left side and deduced the right side, so the proof is complete.

22. $\dfrac{\tan y + \cot y}{\csc y} = \sec y$

We start with the left side.

$$\frac{\tan y + \cot y}{\csc y} = \frac{\dfrac{\sin y}{\cos y} + \dfrac{\cos y}{\sin y}}{\dfrac{1}{\sin y}} = \frac{\sin^2 y + \cos^2 y}{\cos y \sin y} \cdot \frac{\sin y}{1} = \frac{\sin^2 y + \cos^2 y}{\cos y} = \frac{1}{\cos y} = \sec y$$

We started with the left side and deduced the right side, so the proof is complete.

24. $\tan\theta - \cot\theta = (\sec\theta - \csc\theta)(\sin\theta + \cos\theta)$

We start with the left side.

$$\tan\theta - \cot\theta = \frac{\sin\theta}{\cos\theta} - \frac{\cos\theta}{\sin\theta} = \frac{\sin^2\theta - \cos^2\theta}{\cos\theta\sin\theta}$$

Now we stop and work with the right side.

$$(\sec\theta - \csc\theta)(\sin\theta + \cos\theta) = \left(\frac{1}{\cos\theta} - \frac{1}{\sin\theta}\right)(\sin\theta + \cos\theta) = \left(\frac{\sin\theta - \cos\theta}{\sin\theta\cos\theta}\right)\left(\frac{\sin\theta + \cos\theta}{1}\right) = \frac{\sin^2\theta - \cos^2\theta}{\sin\theta\cos\theta}$$

We have deduced the same expression from each side, so the proof is complete.

26. $\dfrac{\tan x + \cot x}{\sec x + \csc x} = \dfrac{1}{\cos x + \sin x}$

We start with the left side.

$$\frac{\tan x + \cot x}{\sec x + \csc x} = \frac{\dfrac{\sin x}{\cos x} + \dfrac{\cos x}{\sin x}}{\dfrac{1}{\cos x} + \dfrac{1}{\sin x}} = \frac{\dfrac{\sin^2 x + \cos^2 x}{\cos x \sin x}}{\dfrac{\sin x + \cos x}{\cos x \sin x}} = \frac{\sin^2 x + \cos^2 x}{\sin x + \cos x} = \frac{1}{\cos x + \sin x}$$

We started with the left side and deduced the right side, so the proof is complete.

28. $\dfrac{\cot\theta}{\csc\theta - 1} = \dfrac{\csc\theta + 1}{\cot\theta}$

We start with the left side.

$$\frac{\cot\theta}{\csc\theta - 1} = \frac{\cot\theta}{\csc\theta - 1} \cdot \frac{\csc\theta + 1}{\csc\theta + 1} = \frac{\cot\theta(\csc\theta + 1)}{\csc^2\theta - 1} = \frac{\cot\theta(\csc\theta + 1)}{\cot^2\theta} = \frac{\csc\theta + 1}{\cot\theta}$$

We started with the left side and deduced the right side, so the proof is complete.

30. $\sec^4 s - \tan^2 s = \tan^4 s + \sec^2 s$

We start with the left side.

$$\sec^4 s - \tan^2 s = \sec^4 s - (\sec^2 s - 1) = \sec^4 s - \sec^2 s + 1$$

Now we stop and work with the right side.

$$\tan^4 s + \sec^2 s = (\tan^2 s)^2 + \sec^2 s = (\sec^2 s - 1)^2 + \sec^2 s = \sec^4 s - 2\sec^2 s + 1 + \sec^2 s = \sec^4 s - \sec^2 s + 1$$

We have deduced the same expression from each side, so the proof is complete.

31. Expression B completes the identity $\dfrac{\cos x + \cot x}{1 + \csc x} = \cos x$. The proof is in the answer section in the text.

32. Expression E completes the identity $\cot x + \csc x = \dfrac{\sin x}{1 - \cos x}$. To prove it we start with the left side.

$$\cot x + \csc x = \frac{\cos x}{\sin x} + \frac{1}{\sin x} = \frac{\cos x + 1}{\sin x}$$

Now we stop and work with the right side.

$$\frac{\sin x}{1-\cos x} = \frac{\sin x}{1-\cos x}\cdot\frac{1+\cos x}{1+\cos x}$$
$$= \frac{\sin x(1+\cos x)}{1-\cos^2 x}$$
$$= \frac{\sin x(1+\cos x)}{\sin^2 x}$$
$$= \frac{1+\cos x}{\sin x}$$

We have deduced the same expression from each side, so the proof is complete.

33. Expression A completes the identity $\sin x\cos x + 1 = \dfrac{\sin^3 x - \cos^3 x}{\sin x - \cos x}$. The proof is in the answer section in the text.

34. Expression F completes the identity $2\cos^2 x - 1 = \cos^4 x - \sin^4 x$. To prove it we start with the left side.

$2\cos^2 x - 1$

$= 2\cos^2 x - (\sin^2 x + \cos^2 x)$

$= \cos^2 x - \sin^2 x$

Now we stop and work with the right side.

$\cos^4 x - \sin^4 x$

$= (\cos^2 x + \sin^2 x)(\cos^2 x - \sin^2 x)$

$= \cos^2 x - \sin^2 x$

We have deduced the same expression from each side, so the proof is complete.

35. Expression C completes the identity $\dfrac{1}{\cot x\sin^2 x} = \tan x + \cot x$. The proof is in the answer section in the text.

36. Expression D completes the identity $(\cos x + \sin x)(1 - \sin x\cos x) = \cos^3 x + \sin^3 x$. To prove it we start with the right side.

$\cos^3 x + \sin^3 x$

$= (\cos x + \sin x)(\cos^2 x - \cos x\sin x + \sin^2 x)$

$= (\cos x + \sin x)(1 - \sin x\cos x)$

We started with the right side and deduced the left side, so the proof is complete.

37. a) $x \neq k\pi$, k an integer; the tangent function is not defined for these values of x.

b) $\sin x = 0$ for $x = k\pi$, k an integer; $\cos x = -1$ for $x = k\pi$, k an odd integer; thus the restriction $x \neq k\pi$, k an integer applies.

c) The sine and cosine functions are defined for all real numbers so there are no restrictions.

38. The expression $\tan(x + 450°)$ can be simplified using the sine and cosine sum formulas but cannot be simplified using the tangent sum formula because while sin 450° and cos 450° are both defined, tan 450° is undefined.

39. $f(x) = 3x - 2$

a) Find some ordered pairs.

When $x = 0$, $f(0) = 3\cdot 0 - 2 = 2$.

When $x = 2$, $f(2) = 3\cdot 2 - 2 = 4$.

Plot these points and draw the graph.

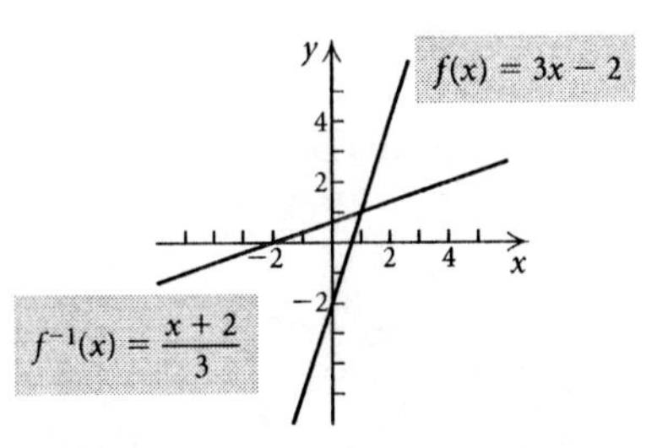

b) Since there is no horizontal line that intersects the graph more than once, the function is one-to-one.

c) Replace $f(x)$ with y: $y = 3x - 2$

Interchange x and y: $x = 3y - 2$

Solve for y: $y = \dfrac{x+2}{3}$

Replace y with $f^{-1}(x)$: $f^{-1}(x) = \dfrac{x+2}{3}$

d) Find some ordered pairs or reflect the graph of $f(x)$ across the line $y = x$. The graph is shown in part (a) above.

40. $f(x) = x^3 + 1$

a)

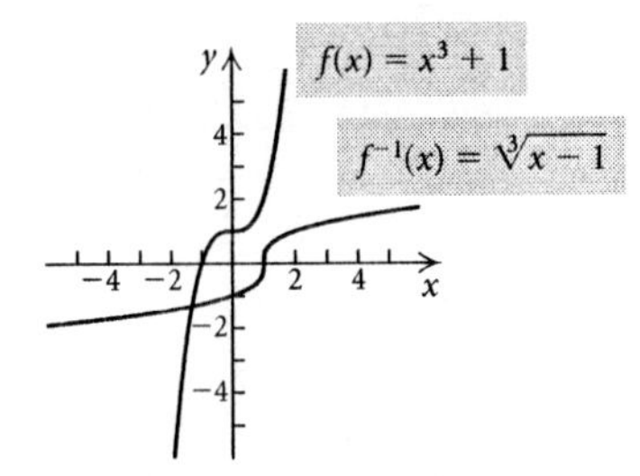

b) Since there is no horizontal line that intersects the graph more than once, the function is one-to-one.

c) Replace $f(x)$ with y: $y = x^3 + 1$

Interchange x and y: $x = y^3 + 1$

Solve for y: $y = \sqrt[3]{x-1}$

Replace y with $f^{-1}(x)$: $f^{-1}(x) = \sqrt[3]{x-1}$

d) See the graph in part (a).

41. $f(x) = x^2 - 4$, $x \geq 0$

a) Find some ordered pairs.

When $x = 0$, $f(0) = 0^2 - 4 = -4$.

When $x = 1$, $f(1) = 1^2 - 4 = -3$.

When $x = 2$, $f(2) = 2^2 - 4 = 0$.

When $x = 3$, $f(3) = 3^2 - 4 = 5$.

Plot these points and draw the graph.

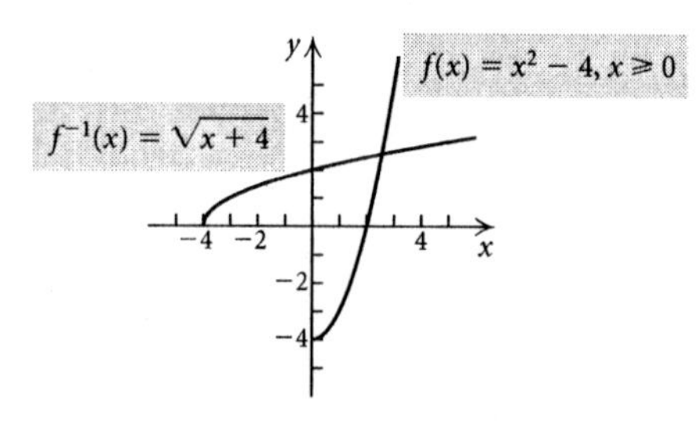

b) Since there is no horizontal line that intersects the graph more than once, the function is one-to-one.

c) Replace $f(x)$ with y: $y = x^2 - 4$

Interchange x and y: $x = y^2 - 4$

Solve for y: $y = \sqrt{x+4}$ (We choose the positive square root since $y \geq 0$.)

Replace y with $f^{-1}(x)$: $f^{-1}(x) = \sqrt{x+4}$

d) Find some ordered pairs or reflect the graph of $f(x)$ across the line $y = x$. The graph is shown in part (a) above.

42. $f(x) = \sqrt{x+2}$

a)

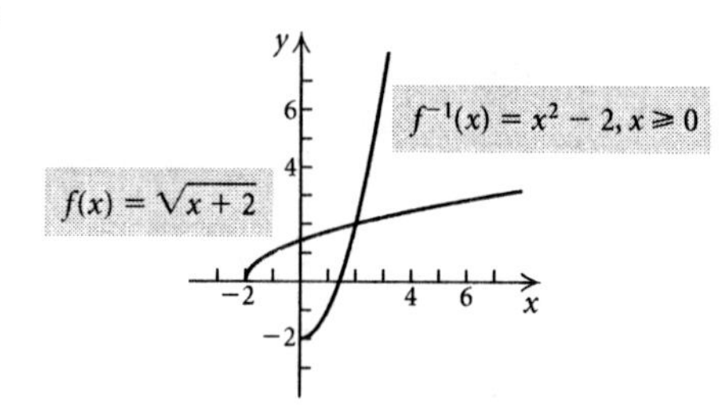

b) Since there is no horizontal line that intersects the graph more than once, the function is one-to-one.

c) Replace $f(x)$ with y: $y = \sqrt{x+2}$

Interchange x and y: $x = \sqrt{y+2}$

Solve for y: $y = x^2 - 2$, $x \geq 0$

Replace y with $f^{-1}(x)$: $f^{-1}(x) = x^2 - 2$, $x \geq 0$

d) See the graph in part (a).

43.
$$2x^2 = 5x$$
$$2x^2 - 5x = 0$$
$$x(2x-5) = 0$$
$$x = 0 \quad or \quad 2x - 5 = 0$$
$$x = 0 \quad or \quad x = \frac{5}{2}$$

The solutions are 0 and $\frac{5}{2}$.

44.
$$3x^2 + 5x - 10 = 18$$
$$3x^2 + 5x - 28 = 0$$
$$(3x-7)(x+4) = 0$$
$$x = \frac{7}{3} \quad or \quad x = -4$$

The solutions are -4 and $\frac{7}{3}$.

45. $x^4 + 5x^2 - 36 = 0$

Let $u = x^2$ and substitute.
$$u^2 + 5u - 36 = 0$$
$$(u+9)(u-4) = 0$$
$$u + 9 = 0 \quad or \quad u - 4 = 0$$
$$u = -9 \quad or \quad u = 4$$
$$x^2 = -9 \quad or \quad x^2 = 4$$
$$x = \pm 3i \quad or \quad x = \pm 2$$

The solutions are $\pm 3i$ and ± 2.

46. $x^2 - 10x + 1 = 0$
$$x = \frac{-(-10) \pm \sqrt{(-10)^2 - 4 \cdot 1 \cdot 1}}{2 \cdot 1}$$
$$= \frac{10 \pm \sqrt{96}}{2} = \frac{10 \pm 4\sqrt{6}}{2}$$
$$= 5 \pm 2\sqrt{6}$$

The solutions are $5 \pm 2\sqrt{6}$.

47. $\sqrt{x-2} = 5$
$$x - 2 = 25 \quad \text{Squaring both sides}$$
$$x = 27$$

This answer checks. The solution is 27.

48.
$$x = \sqrt{x+7} + 5$$
$$x - 5 = \sqrt{x+7}$$
$$x^2 - 10x + 25 = x + 7$$
$$x^2 - 11x + 18 = 0$$
$$(x-2)(x-9) = 0$$
$$x = 2 \quad or \quad x = 9$$

Only 9 checks.

The solution is 9.

49. See the answer section in the text.

50. $\ln|\sec\theta + \tan\theta| = -\ln|\sec\theta - \tan\theta|$

We start with the left side.
$$\ln|\sec\theta + \tan\theta$$
$$= \ln\left|\sec\theta + \tan\theta \cdot \frac{\sec\theta - \tan\theta}{\sec\theta - \tan\theta}\right|$$
$$= \ln\left|\frac{\sec^2\theta - \tan^2\theta}{\sec\theta - \tan\theta}\right|$$
$$= \ln\left|\frac{1}{\sec\theta - \tan\theta}\right|$$
$$= \ln|1| - \ln|\sec\theta - \tan\theta|$$
$$= 0 - \ln|\sec\theta - \tan\theta|$$
$$= -\ln|\sec\theta - \tan\theta|$$

We started with the left side and deduced the right side, so the proof is complete.

51. See the answer section in the text.

52. $\sin\theta = \dfrac{I_1\cos\theta}{\sqrt{(I_1\cos\phi)^2+(I_2\sin\phi)^2}}$

$\sin\theta = \dfrac{I_1\cos\phi}{\sqrt{(I_1\cos\phi)^2+(I_1\sin\phi)^2}} \quad (I_1 = I_2)$

$\sin\theta = \dfrac{I_1\cos\phi}{\sqrt{I_1^2(\cos^2\phi+\sin^2\phi)}}$

$\sin\theta = \dfrac{I_1\cos\phi}{I_1}$

$\sin\theta = \cos\phi.$

53. See the answer section in the text.

54. $\dfrac{E_1+E_2}{2}$

$= \dfrac{\sqrt{2}E_t\cos\left(\theta+\frac{\pi}{P}\right)+\sqrt{2}E_t\cos\left(\theta-\frac{\pi}{P}\right)}{2}$

$= \sqrt{2}E_t\Big[\left(\cos\theta\cos\frac{\pi}{P}-\sin\theta\sin\frac{\pi}{P}\right)+$

$\left(\cos\theta\cos\frac{\pi}{P}+\sin\theta\sin\frac{\pi}{P}\right)\Big]/2$

$= \dfrac{\sqrt{2}E_t\left(2\cos\theta\cos\frac{\pi}{P}\right)}{2}$

$= \sqrt{2}E_t\cos\theta\cos\frac{\pi}{P}$

$\dfrac{E_1-E_2}{2}$

$= \dfrac{\sqrt{2}E_t\cos\left(\theta+\frac{\pi}{P}\right)-\sqrt{2}E_t\cos\left(\theta-\frac{\pi}{P}\right)}{2}$

$= \sqrt{2}E_t\Big[\left(\cos\theta\cos\frac{\pi}{P}-\sin\theta\sin\frac{\pi}{P}\right)-$

$\left(\cos\theta\cos\frac{\pi}{P}+\sin\theta\sin\frac{\pi}{P}\right)\Big]/2$

$= \dfrac{\sqrt{2}E_t\left(-2\sin\theta\sin\frac{\pi}{P}\right)}{2}$

$= -\sqrt{2}E_t\sin\theta\sin\frac{\pi}{P}$

Exercise Set 6.4

1. The only number in the restricted range $[-\pi/2,\pi/2]$ with a sine of $-\sqrt{3}/2$ is $-\pi/3$. Thus, $\sin^{-1}(-\sqrt{3}/2) = -\pi/3$, or $-60°$.

2. $\cos^{-1}\frac{1}{2} = \frac{\pi}{3}$, or $60°$

3. The only number in the restricted range $(-\pi/2,\pi/2)$ with a tangent of 1 is $\pi/4$. Thus, $\tan^{-1} = \pi/4$, or $45°$.

4. $\sin^{-1} = 0$, or $0°$

5. The only number in the restricted range $[0,\pi]$ with a cosine of $\sqrt{2}/2$ is $\pi/4$. Thus, $\cos^{-1}(\sqrt{2}/2) = \pi/4$, or $45°$.

6. $\sec^{-1}\sqrt{2} = \pi/4$, or $45°$

7. The only number in the restricted range $(-\pi/2,\pi/2)$ with a tangent of 0 is 0. Thus, $\tan^{-1}0 = 0$, or $0°$.

8. $\tan^{-1}\frac{\sqrt{3}}{3} = \frac{\pi}{6}$, or $30°$

9. The only number in the restricted range $[0,\pi]$ with a cosine of $\sqrt{3}/2$ is $\pi/6$. Thus, $\cos^{-1}(\sqrt{3}/2) = \pi/6$, or $30°$.

10. $\cot^{-1}(-\sqrt{3}/3) = -\pi/3$, or $-60°$

11. The only number in the restricted range $[-\pi/2,0)\cup(0,\pi/2]$ with a cosecant of 2 is $\pi/6$. Thus, $\csc^{-1}2 = \pi/6$, or $30°$.

12. $\sin^{-1}\frac{1}{2} = \frac{\pi}{6}$, or $30°$

13. The only number in the restricted range $[-\pi/2,0)\cup(0,\pi/2]$ with a cotangent of $-\sqrt{3}$ is $-\pi/6$. Thus, $\cot^{-1}(-\sqrt{3}) = -\pi/6$, or $-30°$.

14. $\tan^{-1}(-1) = -\frac{\pi}{4}$, or $-45°$

15. The only number in the restricted range $[-\pi/2,\pi/2]$ with a sine of $-\frac{1}{2}$ is $-\frac{\pi}{6}$. Thus, $\sin^{-1}\left(-\frac{1}{2}\right) = -\frac{\pi}{6}$, or $-30°$.

16. $\cos^{-1}\left(-\frac{\sqrt{2}}{2}\right) = \frac{3\pi}{4}$, or $135°$.

17. The only number in the restricted range $[0,\pi]$ with a cosine of 0 is $\pi/2$. Thus, $\cos^{-1}0 = \pi/2$, or $90°$.

18. $\sin^{-1}\frac{\sqrt{3}}{2} = \frac{\pi}{3}$, or $60°$

19. The only number in the restricted range $[0,\pi/2)\cup(\pi/2,\pi]$ with a secant of 2 is $\pi/3$. Thus, $\sec^{-1}2 = \pi/3$, or $60°$.

20. $\csc^{-1}(-1) = -\pi/2$, or $-90°$

21. $\tan^{-1}0.3673 \approx 0.3520$, or $20.2°$

22. $\cos^{-1}(-0.2935) \approx 1.8687$, or $107.1°$

23. $\sin^{-1}0.9613 \approx 1.2917$, or $74.0°$

24. $\sin^{-1}(-0.6199) \approx -0.6686$, or $-38.3°$

25. $\cos^{-1}(-0.9810) \approx 2.9463$, or $168.8°$

26. $\tan^{-1}158 \approx 1.5645$, or $89.6°$

27. $\csc^{-1}(-6.2774) = \sin^{-1}\left(\frac{1}{-6.2774}\right) \approx -0.1600$, or $-9.2°$

28. $\sec^{-1}1.1677 = \cos^{-1}\left(\frac{1}{1.1677}\right) \approx 0.5426$, or $31.1°$

29. $\tan^{-1}1.091 \approx 0.8289$, or $47.5°$

30. $\cot^{-1}1.265 = \tan^{-1}\left(\frac{1}{1.265}\right) \approx 0.6689$, or $38.3°$

31. $\sin^{-1}(-0.8192) \approx -0.9600$, or $-55.0°$

32. $\cos^{-1}(-0.2716) \approx 1.8459$, or $105.8°$

33. $\sin^{-1}$: $[-1, 1]$; $\cos^{-1}$: $[-1, 1]$; $\tan^{-1}$: $(-\infty, \infty)$

34. $\sin^{-1}$: $[-\pi/2, \pi/2]$; $\cos^{-1}$: $[0, \pi]$; $\tan^{-1}$: $(-\pi/2, \pi/2)$

35.

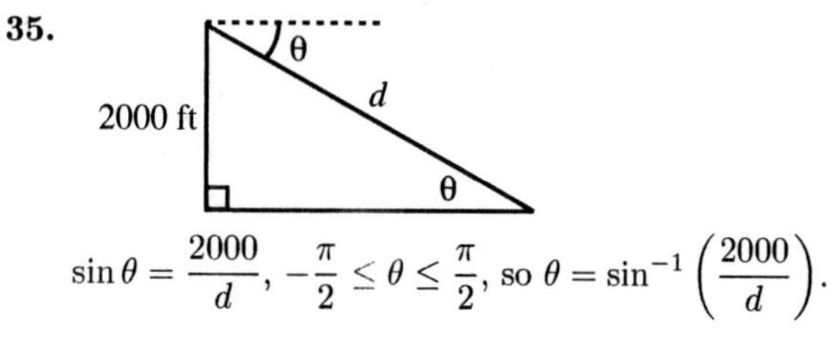

$\sin\theta = \dfrac{2000}{d}$, $-\dfrac{\pi}{2} \leq \theta \leq \dfrac{\pi}{2}$, so $\theta = \sin^{-1}\left(\dfrac{2000}{d}\right)$.

36. $\tan\beta = \dfrac{50}{d}$, $-\dfrac{\pi}{2} < \beta < \dfrac{\pi}{2}$, so $\beta = \tan^{-1}\left(\dfrac{50}{d}\right)$.

37. Since 0.3 is in the interval $[-1, 1]$, $\sin(\sin^{-1} 0.3) = 0.3$.

38. $\tan[\tan^{-1}(-4.2)] = -4.2$

39. $\cos^{-1}\left[\cos\left(-\dfrac{\pi}{4}\right)\right] = \cos^{-1}\left(\dfrac{\sqrt{2}}{2}\right) = \dfrac{\pi}{4}$

40. $\sin^{-1}\left(\sin\dfrac{2\pi}{3}\right) = \sin^{-1}\dfrac{\sqrt{3}}{2} = \dfrac{\pi}{3}$

41. $\sin^{-1}\left(\sin\dfrac{\pi}{5}\right) = \dfrac{\pi}{5}$ because $\dfrac{\pi}{5}$ is in the range of the arcsine function.

42. $\cot^{-1}\left(\cot\dfrac{2\pi}{3}\right) = \cot^{-1}\left(-\dfrac{\sqrt{3}}{3}\right) = -\dfrac{\pi}{3}$

43. $\tan^{-1}\left(\tan\dfrac{2\pi}{3}\right) = \tan^{-1}(-\sqrt{3}) = -\dfrac{\pi}{3}$

44. $\cos^{-1}\left(\cos\dfrac{\pi}{7}\right) = \dfrac{\pi}{7}$ because $\dfrac{\pi}{7}$ is in the range of the arccosine function.

45. $\sin\left(\tan^{-1}\dfrac{\sqrt{3}}{3}\right) = \sin\dfrac{\pi}{6} = \dfrac{1}{2}$

46. $\cos\left(\sin^{-1}\dfrac{\sqrt{3}}{2}\right) = \cos\dfrac{\pi}{3} = \dfrac{1}{2}$

47. $\tan\left(\cos^{-1}\dfrac{\sqrt{2}}{2}\right) = \tan\dfrac{\pi}{4} = 1$

48. $\cos^{-1}(\sin\pi) = \cos^{-1}(0) = \dfrac{\pi}{2}$

49. $\sin^{-1}\left(\cos\dfrac{\pi}{6}\right) = \sin^{-1}\dfrac{\sqrt{3}}{2} = \dfrac{\pi}{3}$

50. $\sin^{-1}\left[\tan\left(-\dfrac{\pi}{4}\right)\right] = \sin^{-1}(-1) = -\dfrac{\pi}{2}$

51. Find $\tan(\arcsin 0.1)$

We wish to find the tangent of an angle whose sine is 0.1, or $\dfrac{1}{10}$.

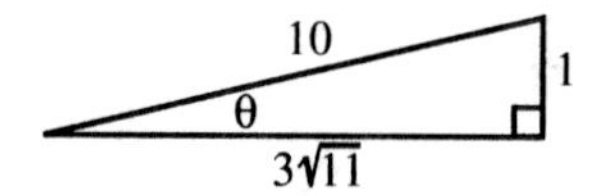

The length of the other leg is $3\sqrt{11}$.

Thus, $\tan(\arcsin 0.1) = \dfrac{1}{3\sqrt{11}}$, or $\dfrac{\sqrt{11}}{33}$.

52.

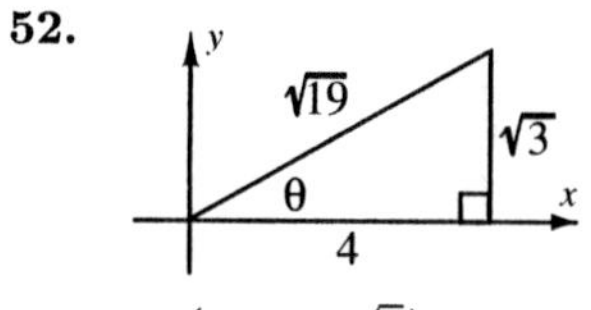

$\cos\left(\tan^{-1}\dfrac{\sqrt{3}}{4}\right) = \dfrac{4}{\sqrt{19}}$, or $\dfrac{4\sqrt{19}}{19}$.

53. $\sin^{-1}\left(\sin\dfrac{7\pi}{6}\right) = \sin^{-1}\left(-\dfrac{\sqrt{3}}{2}\right) = -\dfrac{\pi}{6}$

54. $\tan^{-1}\left(\tan -\dfrac{3\pi}{4}\right) = \tan^{-1} 1 = \dfrac{\pi}{4}$

55. Find $\sin\left(\tan^{-1}\dfrac{a}{3}\right)$.

We draw right triangles whose legs have lengths $|a|$ and 3 so that $\tan\theta = \dfrac{a}{3}$.

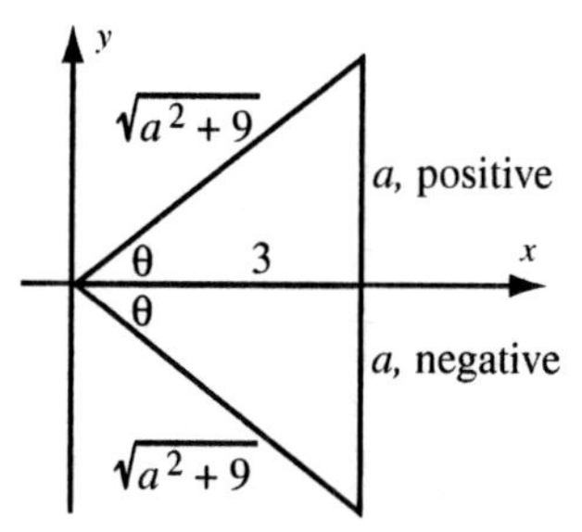

$\sin\left(\tan^{-1}\dfrac{a}{3}\right) = \dfrac{a}{\sqrt{a^2 + 9}}$

56.

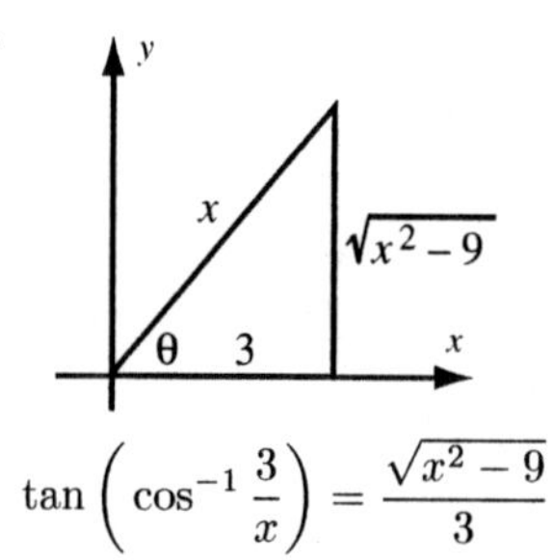

$\tan\left(\cos^{-1}\dfrac{3}{x}\right) = \dfrac{\sqrt{x^2 - 9}}{3}$

57. Find $\cot\left(\sin^{-1}\frac{p}{q}\right)$.

We draw right triangles with one length of length $|p|$ and hypotenuse q so that $\sin\theta = \frac{p}{q}$.

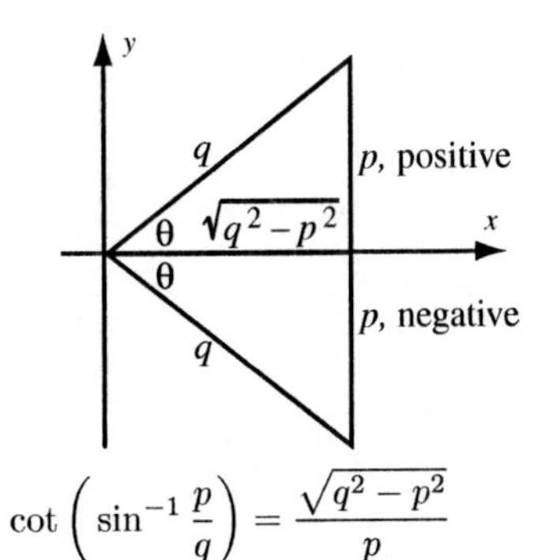

$$\cot\left(\sin^{-1}\frac{p}{q}\right) = \frac{\sqrt{q^2-p^2}}{p}$$

58.

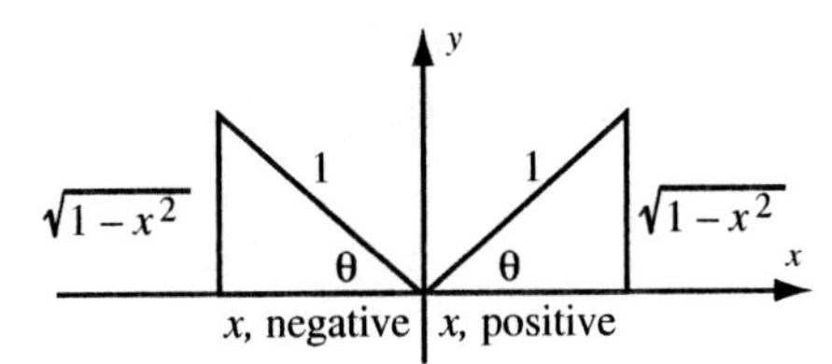

$\sin(\cos^{-1}x) = \frac{\sqrt{1-x^2}}{1}$, or $\sqrt{1-x^2}$.

59. Find $\tan\left(\sin^{-1}\frac{p}{\sqrt{p^2+9}}\right)$.

We draw the right triangle with one leg of length $|p|$ and hypotenuse $\sqrt{p^2+9}$.

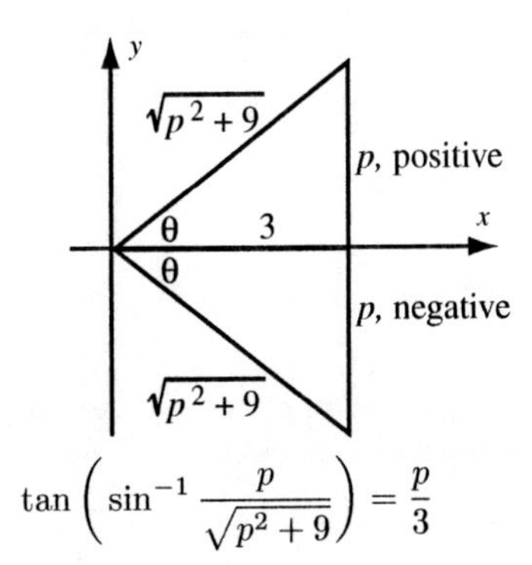

$$\tan\left(\sin^{-1}\frac{p}{\sqrt{p^2+9}}\right) = \frac{p}{3}$$

60.

$$\tan\left(\frac{1}{2}\sin^{-1}\frac{1}{2}\right)$$

$$= \frac{\sin\left(\sin^{-1}\frac{1}{2}\right)}{1+\cos\left(\sin^{-1}\frac{1}{2}\right)} \quad \text{Using } \tan\frac{x}{2} = \frac{\sin x}{1+\cos x}$$

$$= \frac{\frac{1}{2}}{1+\cos\frac{\pi}{6}}$$

$$= \frac{\frac{1}{2}}{1+\frac{\sqrt{3}}{2}}$$

$$= \frac{1}{2+\sqrt{3}}, \text{ or } 2-\sqrt{3}$$

61. $\cos\left(\frac{1}{2}\sin^{-1}\frac{\sqrt{3}}{2}\right) = \cos\left(\frac{1}{2}\cdot 60^\circ\right) = \frac{\sqrt{3}}{2}$

We could also have used a half-angle identity:

$$\cos\left(\frac{1}{2}\sin^{-1}\frac{\sqrt{3}}{2}\right) = \sqrt{\frac{1+\cos[\sin^{-1}(\sqrt{3}/2)]}{2}}$$

$$= \sqrt{\frac{1+\cos\frac{\pi}{3}}{2}}$$

$$= \sqrt{\frac{1+\frac{1}{2}}{2}}$$

$$= \frac{\sqrt{3}}{2}$$

62. We use the identity $\sin 2\theta = 2\sin\theta\cos\theta$.

$$\sin\left(2\cos^{-1}\frac{3}{5}\right)$$

$$= 2\sin\left(\cos^{-1}\frac{3}{5}\right)\cos\left(\cos^{-1}\frac{3}{5}\right)$$

$$= 2\cdot\frac{4}{5}\cdot\frac{3}{5} \quad \text{See the triangle in Exercise 61.}$$

$$= \frac{24}{25}$$

63. Evaluate $\cos\left(\sin^{-1}\frac{\sqrt{2}}{2}+\cos^{-1}\frac{3}{5}\right)$.

This is the cosine of a sum so we use the identity $\cos(u+v) = \cos u\cos v - \sin u\sin v$.

$$\cos\left(\sin^{-1}\frac{\sqrt{2}}{2}+\cos^{-1}\frac{3}{5}\right)$$

$$= \cos\left(\sin^{-1}\frac{\sqrt{2}}{2}\right)\cos\left(\cos^{-1}\frac{3}{5}\right) - \sin\left(\sin^{-1}\frac{\sqrt{2}}{2}\right)\sin\left(\cos^{-1}\frac{3}{5}\right)$$

$$= \left(\cos\frac{\pi}{4}\right)\left(\frac{3}{5}\right) - \frac{\sqrt{2}}{2}\cdot\sin\left(\cos^{-1}\frac{3}{5}\right)$$

$$= \frac{\sqrt{2}}{2}\cdot\frac{3}{5} - \frac{\sqrt{2}}{2}\cdot\sin\left(\cos^{-1}\frac{3}{5}\right)$$

We draw a triangle in order to find $\sin\left(\cos^{-1}\frac{3}{5}\right)$.

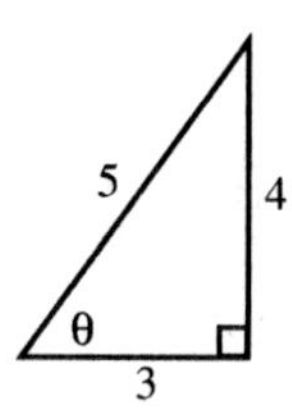

Our expression simplifies to

$\frac{\sqrt{2}}{2}\cdot\frac{3}{5}-\frac{\sqrt{2}}{2}\cdot\frac{4}{5}=\frac{3\sqrt{2}-4\sqrt{2}}{10}=-\frac{\sqrt{2}}{10}$.

64. We use the identity $\sin(u+v)=\sin u\cos v+\cos u\sin v$.

$$\sin\left(\sin^{-1}\frac{1}{2}+\cos^{-1}\frac{3}{5}\right)$$
$$=\sin\left(\sin^{-1}\frac{1}{2}\right)\cos\left(\cos^{-1}\frac{3}{5}\right)+$$
$$\cos\left(\sin^{-1}\frac{1}{2}\right)\sin\left(\cos^{-1}\frac{3}{5}\right)$$
$$=\frac{1}{2}\cdot\frac{3}{5}+\cos\frac{\pi}{6}\cdot\frac{4}{5}$$ See the triangle in Exercise 61.
$$=\frac{1}{2}\cdot\frac{3}{5}+\frac{\sqrt{3}}{2}\cdot\frac{4}{5}$$
$$=\frac{3+4\sqrt{3}}{10}$$

65. Evaluate $\sin(\sin^{-1}x+\cos^{-1}y)$.

We will use the identity $\sin(u+v)=\sin u\cos v+\cos u\sin v$. We draw a triangle with an angle whose sine is x and another with an angle whose cosine is y.

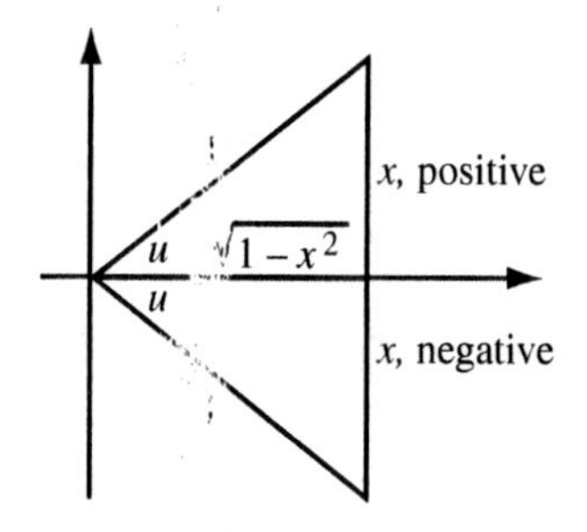

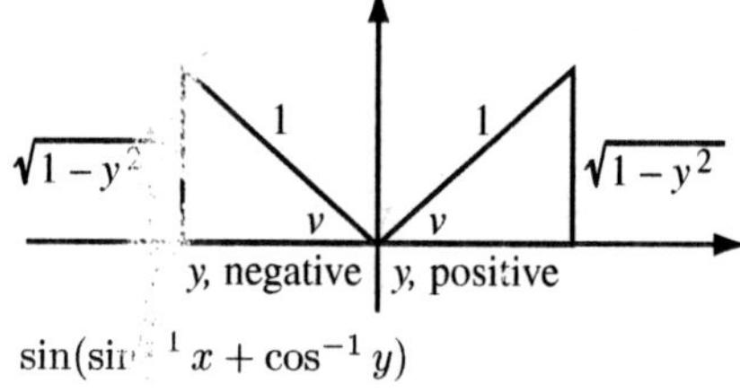

$$\sin(\sin^{-1}x+\cos^{-1}y)$$
$$=\sin(\sin^{-1}x)\cos(\cos^{-1}y)+\cos(\sin^{-1}x)\sin(\cos^{-1}y)$$
$$=x\cdot y+\frac{\sqrt{1-x^2}}{1}\cdot\frac{\sqrt{1-y^2}}{1}$$
$$=xy+\sqrt{(1-x^2)(1-y^2)}$$

66. We use the identity $\cos(u-v)=\cos u\cos v+\sin u\sin v$. We will also use the triangles in Exercise 63.

$$\cos(\sin^{-1}x-\cos^{-1}y)$$
$$=\cos(\sin^{-1}x)\cos(\cos^{-1}y)+\sin(\sin^{-1}x)\sin(\cos^{-1}y)$$
$$=\frac{\sqrt{1-x^2}}{1}\cdot y+x\cdot\frac{\sqrt{1-y^2}}{1}$$
$$=y\sqrt{1-x^2}+x\sqrt{1-y^2}$$

67. Evaluate $\sin(\sin^{-1}0.6032+\cos^{-1}0.4621)$. We will use the identity $\sin(u+v)=\sin u\cos v+\cos u\sin v$. We draw a triangle with an angle whose sine is 0.6032 and another with an angle whose cosine is 0.4621.

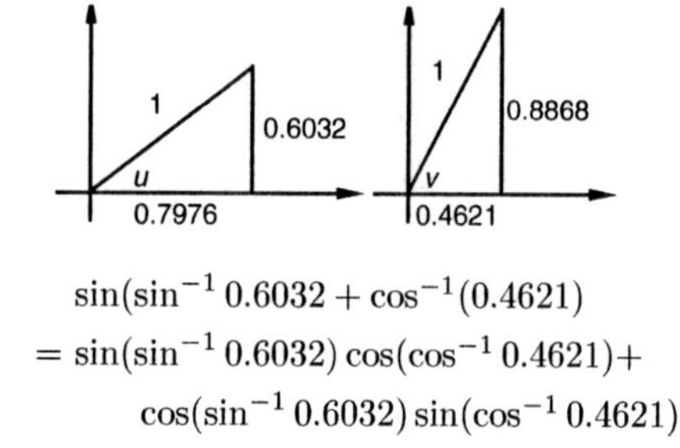

$$\sin(\sin^{-1}0.6032+\cos^{-1}(0.4621)$$
$$=\sin(\sin^{-1}0.6032)\cos(\cos^{-1}0.4621)+$$
$$\cos(\sin^{-1}0.6032)\sin(\cos^{-1}0.4621)$$
$$=0.6032(0.4621)+0.7976(0.8868)$$
$$\approx 0.9861$$

68. We will use the identity $\cos(u-v)=\cos u\cos v+\sin u\sin v$ and the triangles below.

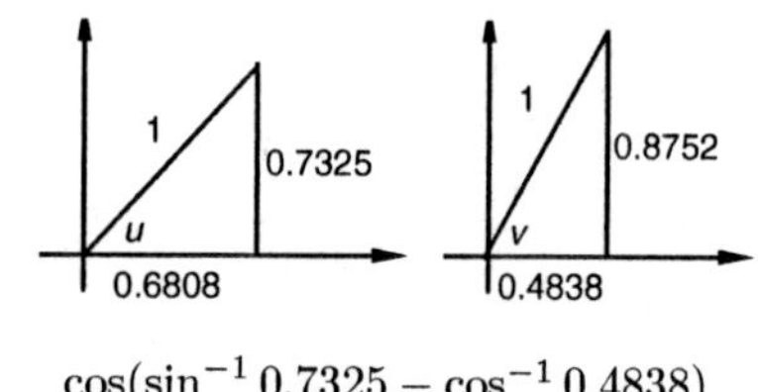

$$\cos(\sin^{-1}0.7325-\cos^{-1}0.4838)$$
$$=\cos(\sin^{-1}0.7325)\cos(\cos^{-1}0.4838)+$$
$$\sin(\sin^{-1}0.7325)\sin(\cos^{-1}0.4838)$$
$$=0.6808(0.4838)+0.7325(0.8752)$$
$$\approx 0.9705$$

Answers will vary slightly depending on when rounding is done.

69. The ranges are restricted so that the inverses of the trigonometric functions are also functions.

70. The graphs have different domains and ranges. The graph of $y=\sin^{-1}x$ is the reflection of the portion of the graph of $y=\sin x$ for $-\frac{\pi}{2}\le x\le\frac{\pi}{2}$, across the line $y=x$.

71. The range of the arcsine function does not include $\frac{5\pi}{6}$. It is $\left[-\frac{\pi}{2},\frac{\pi}{2}\right]$.

72. periodic

73. radian measure

74. similar

75. angle of depression

76. angular speed

77. supplementary

78. amplitude

79. acute

80. circular

81. See the answer section in the text.

82.

$\tan^{-1} x + \cot^{-1} x$	$\frac{\pi}{2}$
$\sin(\tan^{-1} x + \cot^{-1} x)$	$\sin \frac{\pi}{2}$
$\sin(\tan^{-1} x)\cos(\cot^{-1} x) +$ $\cos(\tan^{-1} x)\sin(\cot^{-1} x)$	1
$\frac{x}{\sqrt{1+x^2}} \cdot \frac{x}{\sqrt{1+x^2}} + \frac{1}{\sqrt{1+x^2}} \cdot \frac{1}{\sqrt{1+x^2}}$	
$\frac{x^2+1}{x^2+1}$	
1	

83. See the answer section in the text.

84.

$\tan^{-1} x$	$\sin^{-1} \frac{x}{\sqrt{x^2+1}}$
$\sin(\tan^{-1} x)$	$\sin\left(\sin^{-1} \frac{x}{\sqrt{x^2+1}}\right)$
$\frac{x}{\sqrt{x^2+1}}$	$\frac{x}{\sqrt{x^2+1}}$

85. See the answer section in the text.

86.

$\cos^{-1} x$	$\tan^{-1} \frac{\sqrt{1-x^2}}{x}$
$\cos(\cos^{-1} x)$	$\cos\left(\tan^{-1} \frac{\sqrt{1-x^2}}{x}\right)$
x	x

87.

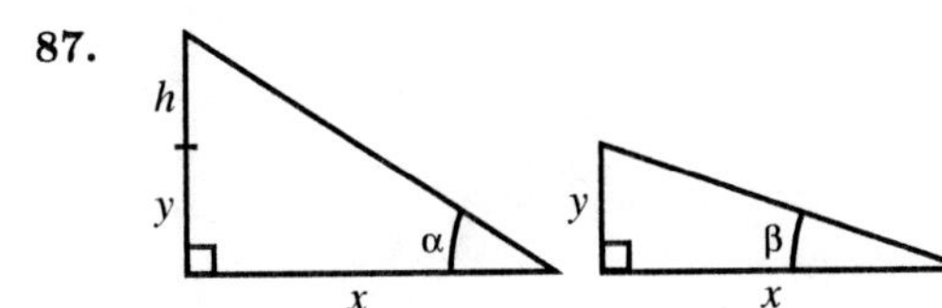

Let $\theta = \alpha - \beta$

$\tan \alpha = \frac{h+y}{x}$, $\qquad \alpha = \tan^{-1} \frac{h+y}{x}$

$\tan \beta = \frac{y}{x}$, $\qquad \beta = \tan^{-1} \frac{y}{x}$

Thus, $\theta = \tan^{-1} \frac{h+y}{x} - \tan^{-1} \frac{y}{x}$.

When $x = 20$ ft, $y = 7$ ft, and $h = 25$ ft we have

$$\theta = \tan^{-1} \frac{25+7}{20} - \tan^{-1} \frac{7}{20}$$
$$\approx 57.99^\circ - 19.29^\circ$$
$$= 38.7^\circ.$$

88. $16 \tan^{-1} \frac{1}{5} - 4 \tan^{-1} \frac{1}{239} \approx 3.141592654$

This expression seems to approximate π.

Exercise Set 6.5

1. $\cos x = \frac{\sqrt{3}}{2}$

Since $\cos x$ is positive the solutions are in quadrants I and IV. They are $\frac{\pi}{6} + 2k\pi$ or $\frac{11\pi}{6} + 2k\pi$, where k is any integer. The solutions can also be expressed as $30^\circ + k \cdot 360^\circ$ or $330^\circ + k \cdot 360^\circ$, where k is any integer.

2. $\sin x = -\frac{\sqrt{2}}{2}$

Since $\sin x$ is negative the solutions are in quadrants III and IV. They are $\frac{5\pi}{4} + 2k\pi$ or $\frac{7\pi}{4} + 2k\pi$, where k is any integer. The solutions can also be expressed as $225^\circ + k \cdot 360^\circ$ or $315^\circ + k \cdot 360^\circ$, where k is any integer.

3. $\tan x = -\sqrt{3}$

Since $\tan x$ is negative the solutions are in quadrants II and IV. They are $\frac{2\pi}{3} + 2k\pi$ or $\frac{5\pi}{3} + 2k\pi$. This can be condensed as $\frac{2\pi}{3} + k\pi$, where k is any integer. The solutions can also be expressed as $120^\circ + k \cdot 180^\circ$, where k is any integer.

4. $\cos x = -\frac{1}{2}$

Since $\cos x$ is negative the solutions are in quadrants II and III. They are $\frac{2\pi}{3} + 2k\pi$ or $\frac{4\pi}{3} + 2k\pi$, where k is any integer. The solutions can also be expressed as $120^\circ + k \cdot 360^\circ$ or $240^\circ + k \cdot 360^\circ$, where k is any integer.

5. $\sin x = \frac{1}{2}$

Since $\sin x$ is positive the solutions are in quadrants I and II. They are $\frac{\pi}{6} + 2k\pi$ or $\frac{5\pi}{6} + 2k\pi$, where k is any integer. The solutions can also be expressed as $30^\circ + k \cdot 360^\circ$ or $150^\circ + k \cdot 360^\circ$, where k is any integer.

6. $\tan x = -1$

Since $\tan x$ is negative the solutions are in quadrants II and IV. They are $\frac{3\pi}{4} + 2k\pi$ or $\frac{7\pi}{4} + 2k\pi$, where k is any integer. This can be condensed as $\frac{3\pi}{4} + k\pi$, where k is any integer. The solutions can also be expressed as $135^\circ + k \cdot 180^\circ$, where k is any integer.

7. $\cos x = -\frac{\sqrt{2}}{2}$

Since $\cos x$ is negative the solutions are in quadrants II and III. They are $\frac{3\pi}{4} + 2k\pi$ or $\frac{5\pi}{4} + 2k\pi$, where k is any integer. The solutions can also be expressed as $135^\circ + k \cdot 360^\circ$ or $225^\circ + k \cdot 360^\circ$, where k is any integer.

8. $\sin x = \dfrac{\sqrt{3}}{2}$

Since $\sin x$ is positive the solutions are in quadrants I and II. They are $\dfrac{\pi}{3} + 2k\pi$ or $\dfrac{2\pi}{3} + 2k\pi$, where k is any integer. The solutions can also be expressed as $60° + k \cdot 360°$ or $120° + k \cdot 360°$, where k is any integer.

9. $2\cos x - 1 = -1.2814$

$2\cos x = -0.2814$

$\cos x = -0.1407$

Using a calculator we find that the reference angle, $\arccos(-0.1407)$ is $x \approx 98.09°$. Since $\cos x$ is negative, the solutions are in quadrants II and III. Thus, one solution is $98.09°$. The reference angle for $98.09°$ is $180° - 98.09°$, or $81.91°$, so the other solution in $[0°, 360°)$ is $180° + 81.91°$, or $261.91°$.

10. $\sin x + 3 = 2.0816$

$\sin x = -0.9184$

Now $x = \arcsin(-0.9184) \approx -66.69°$. Since $\sin x$ is negative the solutions are in quadrants III and IV. The solutions in $[0, 360°)$ are $180° + 66.69°$, or $246.69°$, and $360° - 66.69°$, or $293.31°$.

11. $2\sin x + \sqrt{3} = 0$

$2\sin x = -\sqrt{3}$

$\sin x = -\dfrac{\sqrt{3}}{2}$

The solutions in $[0, 2\pi)$ are $\dfrac{4\pi}{3}$ and $\dfrac{5\pi}{3}$.

12. $2\tan x - 4 = 1$

$2\tan x = 5$

$\tan x = 2.5$

Now $x = \arctan 2.5 \approx 68.20°$. Since $\tan x$ is positive, the solutions are in quadrants I and III. The solutions in $[0, 360°)$ are $68.20°$ and $180° + 68.20°$, or $248.20°$.

13. $2\cos^2 x = 1$

$\cos^2 x = \dfrac{1}{2}$

$\cos x = \pm\dfrac{1}{\sqrt{2}}$, or $\pm\dfrac{\sqrt{2}}{2}$

The solutions in $[0, 2\pi)$ are $\dfrac{\pi}{4}, \dfrac{3\pi}{4}, \dfrac{5\pi}{4}$, and $\dfrac{7\pi}{4}$.

14. $\csc^2 x - 4 = 0$

$\dfrac{1}{\sin^2 x} - 4 = 0$

$\sin^2 x = \dfrac{1}{4}$

$\sin x = \pm\dfrac{1}{2}$

$x = \dfrac{\pi}{6}, \dfrac{5\pi}{6}, \dfrac{7\pi}{6}, \dfrac{11\pi}{6}$

15.

$$\begin{aligned} 2\sin^2 x + \sin x &= 1 \\ 2\sin^2 x + \sin x - 1 &= 0 \\ (2\sin x - 1)(\sin x + 1) &= 0 \\ 2\sin x - 1 = 0 \quad &or \quad \sin x + 1 = 0 \\ 2\sin x = 1 \quad &or \quad \sin x = -1 \\ \sin x = \frac{1}{2} \quad &or \quad \sin x = -1 \end{aligned}$$

The solutions in $[0, 2\pi)$ are $\dfrac{\pi}{6}, \dfrac{5\pi}{6}$, and $\dfrac{3\pi}{2}$.

16.

$$\begin{aligned} \sin^2 x + 2\cos x &= 3 \\ \cos^2 x + 2\cos x - 3 &= 0 \\ (\cos x + 3)(\cos x - 1) &= 0 \\ \cos x + 3 = 0 \quad &or \quad \cos x - 1 = 0 \\ \cos x = -3 \quad &or \quad \cos x = 1 \end{aligned}$$

Since cosine values are never less than -1, $\cos x = -3$ has no solution. Using $\cos x = 1$, we find that the solution in $[0, 2\pi)$ is 0.

17.

$$\begin{aligned} 2\cos^2 x - \sqrt{3}\cos x &= 0 \\ \cos x(2\cos x - \sqrt{3}) &= 0 \\ \cos x = 0 \quad or \quad 2\cos - \sqrt{3} &= 0 \\ \cos x = 0 \quad or \quad \cos x &= \frac{\sqrt{3}}{2} \end{aligned}$$

The solutions in $[0, 2\pi)$ are $\dfrac{\pi}{2}, \dfrac{3\pi}{2}, \dfrac{\pi}{6}$, and $\dfrac{11\pi}{6}$.

18.

$$\begin{aligned} 2\sin^2\theta + 7\sin\theta &= 4 \\ 2\sin^2\theta + 7\sin\theta - 4 &= 0 \\ (2\sin\theta - 1)(\sin\theta + 4) &= 0 \\ 2\sin\theta - 1 = 0 \quad &or \quad \sin\theta + 4 = 0 \\ \sin\theta = \frac{1}{2} \quad &or \quad \sin\theta = -4 \end{aligned}$$

Since sine values are never less than -1, $\sin\theta = -4$ has no solution. Using $\sin\theta = \dfrac{1}{2}$, we find that the solutions in $[0, 2\pi)$ are $\dfrac{\pi}{6}$ and $\dfrac{5\pi}{6}$.

19.

$$\begin{aligned} 6\cos^2\phi + 5\cos\phi + 1 &= 0 \\ (3\cos\phi + 1)(2\cos\phi + 1) &= 0 \\ 3\cos\phi + 1 = 0 \quad &or \quad 2\cos\phi + 1 = 0 \\ \cos\phi = -\frac{1}{3} \quad &or \quad \cos\phi = -\frac{1}{2} \end{aligned}$$

Using $\cos\phi = -\dfrac{1}{3}$, we find that $\phi = \arccos\left(-\dfrac{1}{3}\right) \approx 109.47°$, so one solution in $[0, 360°)$ is $109.47°$. The reference angle for this angle is $180° - 109.47°$, or $70.53°$. Thus, another solution is $180° + 70.53°$, or $250.53°$.

Using $\cos\phi = -\dfrac{1}{2}$, we find that the other solutions in $[0, 360°)$ are $120°$and $240°$.

20.

$$\begin{aligned} 2\sin t\cos t + 2\sin t - \cos t - 1 &= 0 \\ 2\sin t(\cos t + 1) - (\cot t + 1) &= 0 \\ (2\sin t - 1)(\cos t + 1) &= 0 \end{aligned}$$

$$2\sin t - 1 = 0 \quad or \quad \cos t + 1 = 0$$
$$\sin t = \frac{1}{2} \quad or \quad \cos t = -1$$
$$t = \frac{\pi}{6}, \frac{5\pi}{6} \quad or \quad t = \pi$$

21.
$$\sin 2x \cos x - \sin x = 0$$
$$(2\sin x \cos x)\cos - \sin x = 0$$
$$2\sin x \cos^2 x - \sin x = 0$$
$$2\sin x(2\cos^2 x - 1) = 0$$
$$\sin x = 0 \quad or \quad 2\cos^2 x - 1 = 0$$
$$\sin x = 0 \quad or \quad \cos^2 x = \frac{1}{2}$$
$$\sin x = 0 \quad or \quad \cos x = \pm\frac{1}{\sqrt{2}}, \text{ or } \pm\frac{\sqrt{2}}{2}$$

The solutions in $[0, 2\pi)$ are 0, π, $\frac{\pi}{4}$, $\frac{3\pi}{4}$, $\frac{5\pi}{4}$, and $\frac{7\pi}{4}$.

22.
$$5\sin^2 x - 8\sin x = 3$$
$$5\sin^2 x - 8\sin x - 3 = 0$$
$$\sin x = \frac{8 \pm \sqrt{64+60}}{10}$$
$$\sin x = \frac{8 \pm \sqrt{124}}{10}$$
$$\sin x = \frac{8 \pm \sqrt{124}}{10}$$
$$\sin x \approx -0.3136 \text{ or } \sin x \approx 1.9136$$

Since sine values are never greater than 1, $\sin x \approx 1.9136$ has no solution. Using $\sin x \approx -0.3136$, we find that $x = \arcsin(-0.3136) \approx -18.28°$. Thus, the solutions in $[0, 360°)$ are $180° + 18.28°$, or $198.28°$, and $360° - 18.28°$, or $341.72°$. (Answers may vary slightly due to rounding differences.)

23. $\cos^2 x + 6\cos x + 4 = 0$
$$\cos x = \frac{-6 \pm \sqrt{36-16}}{2} = \frac{-6 \pm \sqrt{20}}{2}$$
$$\cos x \approx -0.7639 \quad or \quad \cos x \approx -5.2361$$

Since cosine values are never less than -1, $\cos x \approx -5.2361$ has no solution. Using $\cos x = -0.7639$, we find that $x = \arccos(-0.7639) \approx 139.81°$. Thus, one solution in $[0, 360°)$ is $139.81°$. The reference angle for this angle is $180° - 139.81° = 40.19°$. Then the other solution in $[0, 360°)$ is $180° + 40.19° = 220.19°$.

24.
$$2\tan^2 x = 3\tan x + 7$$
$$2\tan^2 x - 3\tan x - 7 = 0$$
$$\tan x = \frac{3 \pm \sqrt{9+56}}{4}$$
$$\tan x = \frac{3 \pm \sqrt{65}}{4}$$
$$\tan x \approx -1.2656 \quad or \quad \tan x \approx 2.7656$$

Using $\tan x \approx -1.2656$, we find that $x = \arctan(-1.2656) \approx -51.70°$. Thus, two solutions in $[0, 360°)$ are $180° - 51.70°$, or $128.30°$, and $360° - 51.70°$, or $308.30°$.

Using $\tan x \approx 2.7656$, we find that $x = \arctan 2.7656 \approx 70.12°$. Then the other solutions in $[0, 360°)$ are $70.12°$and $180° + 70.12°$, or $250.12°$. (Answers may vary slightly due to rounding differences.)

25. $7 = \cot^2 x + 4\cot x$
$$0 = \cot^2 x + 4\cot x - 7$$
$$\cot x = \frac{-4 \pm \sqrt{16+28}}{2} \quad \text{Using the quadratic formula}$$
$$\cot x = \frac{-4 \pm \sqrt{44}}{2}$$
$$\cot x \approx 1.3166 \quad or \quad \cot x \approx -5.3166$$

Using $\cot x \approx 1.3166$, we find that $x = \text{arccot } 1.3166 \approx 37.22°$. Thus, two solutions in $[0, 360°)$ are $37.22°$ and $180° + 37.22°$, or $217.22°$.

Using $\cot x \approx -5.3166$ we find that $x = \text{arccot } (-5.3166) \approx -10.65°$. Then the other solutions in $[0, 360°)$ are $180° - 10.65°$, or $169.35°$, and $360° - 10.65°$, or $349.35°$.

26.
$$3\sin^2 x = 3\sin x + 2$$
$$3\sin^2 x - 3\sin x - 2 = 0$$
$$\sin x = \frac{3 \pm \sqrt{33}}{6}$$
$$\sin x \approx 1.4574 \quad or \quad \sin x \approx -0.4574$$

Since sine values are never greater than 1, $\sin x \approx 1.4574$ has no solution. Using $\sin x \approx -0.4574$, we find that $x = \arcsin(-0.4574) \approx -27.22°$. Thus, the solutions in $[0, 360°)$ are $180° + 27.22°$, or $207.22°$, and $360° - 27.22°$, or $332.78°$.

27.
$$\cos 2x - \sin x = 1$$
$$1 - 2\sin^2 x - \sin x = 1$$
$$0 = 2\sin^2 x + \sin x$$
$$0 = \sin x(2\sin x + 1)$$
$$\sin x = 0 \quad or \quad 2\sin x + 1 = 0$$
$$\sin x = 0 \quad or \quad \sin x = -\frac{1}{2}$$
$$x = 0, \pi \quad or \quad x = \frac{7\pi}{6}, \frac{11\pi}{6}$$

All values check. The solutions in $[0, 2\pi)$ are 0, π, $\frac{7\pi}{6}$, and $\frac{11\pi}{6}$.

28. $2\sin x \cos x + \sin x = 0$
$$\sin x(2\cos x + 1) = 0$$
$$\sin x = 0 \quad or \quad 2\cos x + 1 = 0$$
$$\sin x = 0 \quad or \quad \cos x = -\frac{1}{2}$$
$$x = 0, \pi \quad or \quad x = \frac{2\pi}{3}, \frac{4\pi}{3}$$

All values check.

29.
$$\sin 4x - 2\sin 2x = 0$$
$$\sin[2(2x)] - 2\sin 2x = 0$$
$$2\sin 2x\cos 2x - 2\sin 2x = 0$$
$$2\sin 2x(\cos 2x - 1) = 0$$

$\sin 2x = 0$ *or* $\cos 2x - 1 = 0$

$\sin 2x = 0$ *or* $\cos 2x = 1$

$2x = 0, \pi, 2\pi, 3\pi$ *or* $2x = 0, 2\pi$

$x = 0, \frac{\pi}{2}, \pi, \frac{3\pi}{2}$ *or* $x = 0, \pi$

All values check. The solutions in $[0, 2\pi)$ are 0, $\frac{\pi}{2}$, π, and $\frac{3\pi}{2}$.

30. $\tan x \sin x - \tan x = 0$

$\tan x(\sin x - 1) = 0$

$\tan x = 0$ *or* $\sin x - 1 = 0$

$\tan x = 0$ *or* $\sin x = 1$

$x = 0, \pi$ $\quad x = \frac{\pi}{2}$

The value $\frac{\pi}{2}$ does not check, but the other values do. Thus the solutions in $[0, 2\pi)$ are 0 and π.

31.
$$\sin 2x\cos x + \sin x = 0$$
$$(2\sin x\cos x)\cos x + \sin x = 0$$
$$2\sin x\cos^2 x + \sin x = 0$$
$$\sin x(2\cos^2 x + 1) = 0$$

$\sin x = 0$ *or* $2\cos^2 x + 1 = 0$

$\sin x = 0$ *or* $\cos^2 x = -\frac{1}{2}$

$x = 0, \pi$ *or* No solution

Both values check. The solutions are 0 and π.

32. $\cos 2x\sin x + \sin x = 0$

$\sin x(\cos 2x + 1) = 0$

$\sin x = 0$ *or* $\cos 2x + 1 = 0$

$\sin x = 0$ *or* $\cos 2x = -1$

$x = 0, \pi$ *or* $2x = \pi, 3\pi$

$x = 0, \pi$ *or* $x = \frac{\pi}{2}, \frac{3\pi}{2}$

All values check.

33. $2\sec x\tan x + 2\sec x + \tan x + 1 = 0$

$2\sec x(\tan x + 1) + (\tan x + 1) = 0$

$(2\sec x + 1)(\tan x + 1) = 0$

$2\sec x + 1 = 0$ *or* $\tan x + 1 = 0$

$\sec x = -\frac{1}{2}$ *or* $\tan x = -1$

No solution $\quad x = \frac{3\pi}{4}, \frac{7\pi}{4}$

Both values check. The solutions in $[0, 2\pi)$ are $\frac{3\pi}{4}$ and $\frac{7\pi}{4}$.

34.
$$\sin 2x\sin x - \cos 2x\cos x = -\cos x$$
$$(2\sin x\cos x)\sin x - (1 - 2\sin^2 x)\cos x = -\cos x$$
$$2\sin^2 x\cos x - \cos x + 2\sin^2 x\cos x = -\cos x$$
$$4\sin^2 x\cos x = 0$$

$\sin^2 x = 0$ *or* $\cos x = 0$

$\sin x = 0$ *or* $\cos x = 0$

$x = 0, \pi$ *or* $x = \frac{\pi}{2}, \frac{3\pi}{2}$

All values check.

35.
$$\sin 2x + \sin x + 2\cos x + 1 = 0$$
$$2\sin x\cos x + \sin x + 2\cos x + 1 = 0$$
$$\sin x(2\cos x + 1) + 2\cos x + 1 = 0$$
$$(\sin x + 1)(2\cos x + 1) = 0$$

$\sin x + 1 = 0$ *or* $2\cos x + 1 = 0$

$\sin x = -1$ *or* $\cos x = -\frac{1}{2}$

$x = \frac{3\pi}{2}$ *or* $x = \frac{2\pi}{3}, \frac{4\pi}{3}$

All values check. The solutions in $[0, 2\pi)$ are $\frac{2\pi}{3}$, $\frac{4\pi}{3}$, and $\frac{3\pi}{2}$.

36. $\tan^2 x + 4 = 2\sec^2 x + \tan x$

$\tan^2 x + 4 = 2(1 + \tan^2 x) + \tan x$

$\tan^2 x + 4 = 2 + 2\tan^2 x + \tan x$

$0 = \tan^2 x + \tan x - 2$

$0 = (\tan x + 2)(\tan x - 1)$

$\tan x + 2 = 0$ *or* $\tan x - 1 = 0$

$\tan x = -2$ *or* $\tan x = 1$

$x \approx 2.034, 5.176$ *or* $x = \frac{\pi}{4}, \frac{5\pi}{4}$

All values check.

37.
$$\sec^2 x - 2\tan^2 x = 0$$
$$1 + \tan^2 x - 2\tan^2 x = 0$$
$$1 - \tan^2 x = 0$$
$$\tan^2 x = 1$$
$$\tan x = \pm 1$$
$$x = \frac{\pi}{4}, \frac{3\pi}{4}, \frac{5\pi}{4}, \frac{7\pi}{4}$$

All values check. The solutions in $[0, 2\pi)$ are $\frac{\pi}{4}$, $\frac{3\pi}{4}$, $\frac{5\pi}{4}$, $\frac{7\pi}{4}$.

38.

$$\cot x = \tan(2x - 3\pi)$$

$$\cot x = \frac{\tan 2x - \tan 3\pi}{1 + \tan 2x \tan 3\pi}$$

$$\cot x = \frac{\tan 2x - 0}{1 + \tan 2x \cdot 0}$$

$$\cot x = \tan 2x$$

$$\frac{1}{\tan x} = \frac{2\tan x}{1 - \tan^2 x}$$

In this form of the equation, x cannot be $\pi/2$ or $3\pi/2$. Therefore $\pi/2$ and $3\pi/2$ must also be checked in the original equation.

$$1 - \tan^2 x = 2\tan^2 x$$

$$1 = 3\tan^2 x$$

$$\frac{1}{3} = \tan^2 x$$

$$\pm\frac{\sqrt{3}}{3} = \tan x$$

$$\frac{\pi}{6}, \frac{5\pi}{6}, \frac{7\pi}{6}, \frac{11\pi}{6} = x$$

These values and also $\frac{\pi}{2}$ and $\frac{3\pi}{2}$ check. The solutions in $[0, 2\pi)$ are $\frac{\pi}{6}, \frac{\pi}{2}, \frac{5\pi}{6}, \frac{7\pi}{6}, \frac{3\pi}{2}$, and $\frac{11\pi}{6}$.

39.

$$2\cos x + 2\sin x = \sqrt{6}$$

$$\cos x + \sin x = \frac{\sqrt{6}}{2}$$

$$\cos^2 x + 2\sin x\cos x + \sin^2 x = \frac{6}{4} \quad \text{Squaring both sides}$$

$$\sin 2x + 1 = \frac{3}{2}$$

$$\sin 2x = \frac{1}{2}$$

$$2x = \frac{\pi}{6}, \frac{5\pi}{6}, \frac{13\pi}{6}, \frac{17\pi}{6}$$

$$x = \frac{\pi}{12}, \frac{5\pi}{12}, \frac{13\pi}{12}, \frac{17\pi}{12}$$

The values $\frac{13\pi}{12}$ and $\frac{17\pi}{12}$ do not check, but the other values do. The solutions in $[0, 2\pi)$ are $\frac{\pi}{12}$ and $\frac{5\pi}{12}$.

40. $\sqrt{3}\cos x - \sin x = 1$

$$\sqrt{3}\cos x = 1 + \sin x$$

$$3\cos^2 x = 1 + 2\sin x + \sin^2 x$$

$$3(1 - \sin^2 x) = 1 + 2\sin x + \sin^2 x$$

$$3 - 3\sin^2 x = 1 + 2\sin x + \sin^2 x$$

$$0 = 4\sin^2 x + 2\sin x - 2$$

$$0 = 2\sin^2 x + \sin x - 1$$

$$0 = (2\sin x - 1)(\sin x + 1)$$

$$2\sin x - 1 = 0 \quad \textit{or} \quad \sin x + 1 = 0$$

$$\sin x = \frac{1}{2} \quad \textit{or} \quad \sin x = -1$$

$$x = \frac{\pi}{6}, \frac{5\pi}{6} \quad \textit{or} \quad x = \frac{3\pi}{2}$$

The value $\frac{5\pi}{6}$ does not check, but the other values do check. The solutions in $[0, 2\pi)$ are $\frac{\pi}{6}$ and $\frac{3\pi}{2}$.

41.

$$\sec^2 x + 2\tan x = 6$$

$$1 + \tan^2 x + 2\tan x = 6$$

$$\tan^2 x + 2\tan x - 5 = 0$$

$$\tan x = \frac{-2 \pm \sqrt{4 + 20}}{2} = \frac{-2 \pm \sqrt{24}}{2}$$

$$\tan x \approx 1.4495 \quad \textit{or} \quad \tan x \approx -3.4495$$

Using $\tan x \approx 1.4495$, we find that $x = \arctan 1.4495 \approx 0.967$. Then two possible solutions in $[0, 2\pi)$ are 0.967 and $\pi + 0.967$, or 4.109.

Using $\tan x \approx -3.4495$, we find that $x = \arctan(-3.4495) \approx -1.289$. Thus, the other two possible solutions in $[0, 2\pi)$ are $\pi - 1.289$, or 1.853, and $2\pi - 1.289$, or 4.994.

All values check. The solutions in $[0, 2\pi)$ are 0.967, 1.853, 4.109, and 4.994. (Answers may vary slightly due to rounding differences.)

42.

$$5\cos 2x + \sin x = 4$$

$$5(1 - 2\sin^2 x) + \sin x = 4$$

$$0 = 10\sin^2 x - \sin x - 1$$

$$\sin x = \frac{1 \pm \sqrt{41}}{20}$$

$$\sin x \approx 0.3702 \quad \textit{or} \quad \sin x \approx -0.2702$$

Using $\sin x \approx 0.3702$, we find that $x = \arcsin 0.3702 \approx 0.379$. Then two possible solutions in $[0, 2\pi)$ are 0.379 and $\pi - 0.379$, or 2.763.

Using $\sin x \approx -0.2702$, we find that $x = \arcsin(-0.2702) \approx -0.274$. Then the other two possible solutions in $[0, 2\pi)$ are $\pi + 0.274$, or 3.416, and $2\pi - 0.274$, or 6.009.

(Answers may vary slightly due to rounding differences.)

43.

$$\cos(\pi - x) + \sin\left(x - \frac{\pi}{2}\right) = 1$$

$$(-\cos x) + (-\cos x) = 1$$

$$-2\cos x = 1$$

$$\cos x = -\frac{1}{2}$$

$$x = \frac{2\pi}{3}, \frac{4\pi}{3}$$

Both values check. The solutions in $[0, 2\pi)$ are $\frac{2\pi}{3}$ and $\frac{4\pi}{3}$.

44.
$$\frac{\sin^2 x - 1}{\cos\left(\frac{\pi}{2} - x\right) + 1} = \frac{\sqrt{2}}{2} - 1$$
$$\frac{\sin^2 x - 1}{\sin x + 1} = \frac{\sqrt{2}}{2} - 1$$
$$\frac{(\sin x + 1)(\sin x - 1)}{\sin x + 1} = \frac{\sqrt{2}}{2} - 1$$
$$\sin x - 1 = \frac{\sqrt{2}}{2} - 1$$
$$\sin x = \frac{\sqrt{2}}{2}$$
$$x = \frac{\pi}{4}, \frac{3\pi}{4}$$
Both values check.

45. Left to the student

46. Left to the student

47. Find the points of intersection of $y_1 = x \sin x$ and $y_2 = 1$ or find the zeros of $y_1 = x \sin x - 1$. The solutions in $[0, 2\pi)$ are 1.114 and 2.773.

48. Find the points of intersection of $y_1 = x^2 + 2$ and $y_2 = \sin x$ or find the zeros of $y_1 = x^2 + 2 - \sin x$. There are no solutions in $[0, 2\pi)$.

49. Find the point of intersection of $y_1 = 2\cos^2 x$ and $y_2 = x + 1$ or find the zero of $y_1 = 2\cos^2 x - x - 1$. The only solution in $[0, 2\pi)$ is 0.515.

50. Find the zero of $y_1 = x \cos x - 2$ The only solution in $[0, 2\pi)$ is 5.114.

51. Find the points of intersection of $y_1 = \cos x - 2$ and $y_2 = x^2 - 3x$ or find the zeros of $y_1 = \cos x - 2 - x^2 + 3x$. The solutions in $[0, 2\pi)$ are 0.422 and 1.756.

52. Find the points of intersection of $y_1 = \sin x$ and $y_2 = \tan \frac{x}{2}$ or find the zeros of $y_1 = \sin x - \tan \frac{x}{2}$. The solutions in $[0, 2\pi)$ are 0, 1.571, and 4.712 or 0, $\frac{\pi}{2}$, and $\frac{3\pi}{2}$.

53. a) Using the sine regression feature on a graphing calculator, we get $y = 7\sin(-2.6180x + 0.5236) + 7$

b) In December, $x = 12$; when $x = 12$, $y \approx 10.5$. Thus, total sales in December are about $10,500.

In July, $x = 7$; when $x = 7$, $y \approx 13.062$. Thus, total sales in July are about $13,062.

54. a) Using the sine regression feature on a graphing calculator, we get $y = 7.8787 \sin(0.0166x - 1.2723) + 12.1840$.

b) On April 22, $x = 112$; when $x = 112$, $y \approx 16.5$. Thus, there are approximately 16.5 hours of daylight on April 22.

On July 4, $x = 185$; when $x = 185$, $y \approx 19.9$. Thus, there are approximately 19.9 hours of daylight on July 4.

On December 15, $x = 349$; when $x = 349$, $y \approx 4.5$. Thus, there are approximately 4.5 hours of daylight on December 15.

55. Yes; first note that $7\pi/6 = \pi/6 + \pi$. Since $\pi/6 + k\pi$ includes both odd and even multiples of π it is equivalent to $\pi/6 + 2k\pi$ and $7\pi/6 + 2k\pi$.

56. "Possible" replacements for the variables are those for which the expressions in the identity are defined.

57. $B = 90° - 55° = 35°$
$$\tan 55° = \frac{201}{b}$$
$$b = \frac{201}{\tan 55°}$$
$$b \approx 140.7$$
$$\sin 55° = \frac{201}{c}$$
$$c = \frac{201}{\sin 55°}$$
$$c \approx 245.4$$

58.
$$\sin R = \frac{3.8}{14.2}$$
$$R \approx 15.5°$$
$$T \approx 90° - 15.5° \approx 74.5°$$
$$t^2 + (3.8)^2 = (14.2)^2$$
$$t^2 = (14.2)^2 - (3.8)^2 = 187.2$$
$$t \approx 13.7$$

59.
$$\frac{x}{27} = \frac{4}{3}$$
$x = 36$ Multiplying by 27

60.
$$\frac{0.01}{0.7} = \frac{0.2}{h}$$
$$h = \frac{0.2(0.7)}{0.01} = 14$$

61.
$$|\sin x| = \frac{\sqrt{3}}{2}$$
$$\sin x = \frac{\sqrt{3}}{2} \quad \textit{or} \quad \sin x = -\frac{\sqrt{3}}{2}$$
$$x = \frac{\pi}{3}, \frac{2\pi}{3} \quad \textit{or} \quad x = \frac{4\pi}{3}, \frac{5\pi}{3}$$
All values check. The solutions in $[0, 2\pi)$ are $\frac{\pi}{3}$, $\frac{2\pi}{3}$, $\frac{4\pi}{3}$, and $\frac{5\pi}{3}$.

62.
$$|\cos x| = \frac{1}{2}$$
$$\cos x = \frac{1}{2} \quad \textit{or} \quad \cos x = -\frac{1}{2}$$
$$x = \frac{\pi}{3}, \frac{5\pi}{3} \quad \textit{or} \quad x = \frac{2\pi}{3}, \frac{4\pi}{3}$$
All values check.

63. $\sqrt{\tan x} = \sqrt[4]{3}$

$(\sqrt{\tan x})^4 = (\sqrt[4]{3})^4$

$\tan^2 x = 3$

$\tan x = \pm\sqrt{3}$

$x = \frac{\pi}{3}, \frac{2\pi}{3}, \frac{4\pi}{3}, \frac{5\pi}{3}$

Only $\frac{\pi}{3}$ and $\frac{4\pi}{3}$ check. They are the solutions in $[0, 2\pi)$.

64. $12\sin x - 7\sqrt{\sin x} + 1 = 0$

Let $u = \sqrt{\sin x}$.

$12u^2 - 7u + 1 = 0$

$(4u-1)(3u-1) = 0$

$4u - 1 = 0$ *or* $3u - 1 = 0$

$u = \frac{1}{4}$ *or* $u = \frac{1}{3}$

$\sqrt{\sin x} = \frac{1}{4}$ *or* $\sqrt{\sin x} = \frac{1}{3}$

$\sin x = \frac{1}{16}$ *or* $\sin x = \frac{1}{9}$

$x \approx 0.063, 3.079$ *or* $x \approx 0.111, 3.031$

All values check. (Answers may vary slightly due to rounding differences.)

65. $\ln(\cos x) = 0$

$\cos x = 1$

$x = 0$

This value checks.

66. $e^{\sin x} = 1$

$\ln e^{\sin x} = \ln 1$

$\sin x = 0$

$x = 0, \pi$

Both values check.

67. $\sin(\ln x) = -1$

$\ln x = \frac{3\pi}{2} + 2k\pi, k$ an integer

$x = e^{3\pi/2+2k\pi}, k$ an integer

x is in the interval $[0, 2\pi)$ when $k \le -1$. Thus, the possible solutions are $e^{3\pi/2+2k\pi}$, k an integer, $k \le -1$. These values check and are the solutions.

68. $e^{\ln \sin x} = 1$

$\sin x = 1$

$x = \frac{\pi}{2}$

This value checks.

69. $T(t) = 101.6° + 3° \sin\left(\frac{\pi}{8}t\right), 0 \le t \le 12$

$103° = 101.6° + 3° \sin\left(\frac{\pi}{8}t\right)$ Substituting

$1.4° = 3° \sin\left(\frac{\pi}{8}t\right)$

$0.4667 \approx \sin\left(\frac{\pi}{8}t\right)$

$\frac{\pi}{8}t \approx 0.4855, 2.6561$ $\left(0 \le t \le 12, \text{ so } 0 \le \frac{\pi}{8}t \le \frac{3\pi}{2}\right)$

$t \approx 1.24, 6.76$

Both values check.

The patient's temperature was 103° at $t \approx 1.24$ days and $t \approx 6.76$ days.

70. Note that $0 \le t \le 240$, so

$-10 \le t - 10 \le 230$ and

$\frac{\pi}{45}(-10) \le \frac{\pi}{45}(t-10) \le \frac{\pi}{45}(230)$, or

$-\frac{2\pi}{9} \le \frac{\pi}{45}(t-10) \le \frac{46\pi}{9}$.

$3000 = 5000\left[\cos\frac{\pi}{45}(t-10)\right]$

$0.6 = \cos\frac{\pi}{45}(t-10)$

We will consider only first quadrant solutions. Solutions in the fourth quadrant represent positions south of the equator.

$\frac{\pi}{45}(t-10) \approx 0.9273, 7.2105, 13.4937$

$t \approx 23.28$ min, 113.28 min, 203.28 min

71. $N(\phi) = 6066 - 31\cos 2\phi$

We consider ϕ in the interval $[0°, 90°]$ since we want latitude north.

$6040 = 6066 - 31\cos 2\phi$ Substituting

$-26 = -31\cos 2\phi$

$0.8387 \approx \cos 2\phi$ $(0° \le 2\phi \le 180°)$

$2\phi \approx 33.0°$

$\phi \approx 16.5°$

The value checks.

At about 16.5°N the length of a British nautical mile is found to be 6040 ft.

72. Using the result of Exercise 53, Exercise Set 6.2, we have $9.8 = 9.78049(1 + 0.0005264\sin^2\phi + 0.000024\sin^4\phi)$.

We solve for $0° \le \phi \le 90°$.

$\phi \approx 37.95615°$N

73. Sketch a triangle having an angle θ whose cosine is $\frac{3}{5}$. (See the triangle in Exercise 63, Exercise Set 6.4) Then $\sin\theta = \frac{4}{5}$. Thus, $\arccos\frac{3}{5} = \arcsin\frac{4}{5}$.

$$\arccos x = \arccos\frac{3}{5} - \arcsin\frac{4}{5}$$
$$\arccos x = 0$$
$$x = 1$$

74. $\sin^{-1} x = \tan^{-1}\frac{1}{3} + \tan^{-1}\frac{1}{2}$, or

$\sin^{-1} x = \alpha + \beta$ where $\alpha = \tan^{-1}\frac{1}{3}$, and $\beta = \tan^{-1}\frac{1}{2}$.

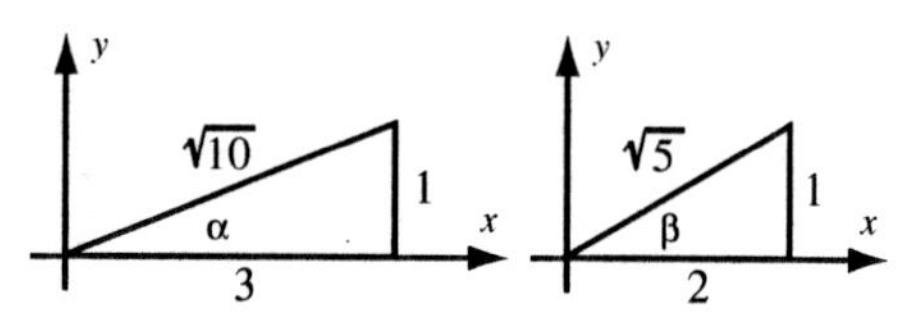

$$\begin{aligned} \text{Then } x &= \sin(\alpha+\beta) \\ &= \sin\alpha\cos\beta + \cos\alpha\sin\beta \\ &= \frac{1}{\sqrt{10}}\cdot\frac{2}{\sqrt{5}} + \frac{3}{\sqrt{10}}\cdot\frac{1}{\sqrt{5}} \\ &= \frac{\sqrt{2}}{2}. \end{aligned}$$

75.
$$\sin x = 5\cos x$$
$$\frac{\sin x}{\cos x} = 5$$
$$\tan x = 5$$
$$x \approx 1.3734, 4.5150$$

Then $\sin x\cos x = \sin(1.3734)\cos(1.3734) \approx 0.1923$.

(The result is the same if $x \approx 4.5150$ is used.)

Chapter 7

Applications of Trigonometry

Exercise Set 7.1

1.

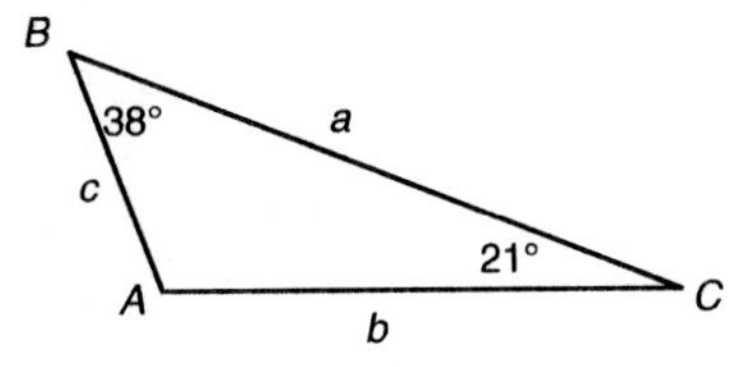

To solve this triangle find A, a, and c.

$A = 180° - (38° + 21°) = 121°$

Use the law of sines to find a and c.

Find a:

$$\frac{a}{\sin A} = \frac{b}{\sin B}$$

$$\frac{a}{\sin 121°} = \frac{24}{\sin 38°}$$

$$a = \frac{24 \sin 121°}{\sin 38°} \approx 33$$

Find c:

$$\frac{c}{\sin C} = \frac{b}{\sin B}$$

$$\frac{c}{\sin 21°} = \frac{24}{\sin 38°}$$

$$c = \frac{24 \sin 21°}{\sin 38°} \approx 14$$

2. $B = 180° - (131° + 23°) = 26°$

$$\frac{a}{\sin A} = \frac{b}{\sin B}, \text{ or } \frac{a}{\sin 131°} = \frac{10}{\sin 26°}$$

$a \approx 17$

$$\frac{c}{\sin C} = \frac{b}{\sin B}, \text{ or } \frac{c}{\sin 23°} = \frac{10}{\sin 26°}$$

$c \approx 9$

3.

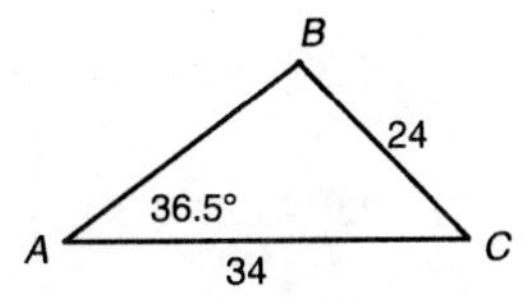

To solve this triangle find A, a, and c.

Find B:

$$\frac{b}{\sin B} = \frac{a}{\sin A}$$

$$\frac{34}{\sin B} = \frac{24}{\sin 36.5°}$$

$$\sin B = \frac{34 \sin 36.5°}{24} \approx 0.8427$$

There are two angles less than 180°having a sine of 0.8427. They are 57.4°and 122.6°. This gives us two possible solutions.

Solution I

If $B \approx 57.4°$, then

$C \approx 180° - (36.5° + 57.4°) \approx 86.1°$.

Find c:

$$\frac{c}{\sin C} = \frac{a}{\sin A}$$

$$\frac{c}{\sin 86.1°} = \frac{24}{\sin 36.5°}$$

$$c = \frac{24 \sin 86.1°}{\sin 36.5°} \approx 40$$

Solution II

If $B \approx 122.6°$, then

$C \approx 180° - (36.5° + 122.6°) \approx 20.9°$.

Find c:

$$\frac{c}{\sin C} = \frac{a}{\sin A}$$

$$\frac{c}{\sin 20.9°} = \frac{24}{\sin 36.5°}$$

$$c = \frac{24 \sin 20.9°}{\sin 36.5°} \approx 14$$

4. $A = 180° - (118.3° + 45.6°) = 16.1°$

$$\frac{a}{\sin A} = \frac{b}{\sin B}, \text{ or } \frac{a}{\sin 16.1°} = \frac{42.1}{\sin 118.3°}$$

$a \approx 13.3$

$$\frac{c}{\sin C} = \frac{b}{\sin B}, \text{ or } \frac{c}{\sin 45.6°} = \frac{42.1}{\sin 118.3°}$$

$c \approx 34.2$

5. Find B:

$$\frac{b}{\sin B} = \frac{c}{\sin C}$$

$$\frac{24.2}{\sin B} = \frac{30.3}{\sin 61°10'}$$

$$\sin B = \frac{24.2 \sin 61°10'}{30.3} \approx 0.6996$$

Then $B \approx 44°24'$ or $B \approx 135°36'$. An angle of $135°36'$ cannot be an angle of this triangle because it already has an angle of $61°10'$ and the two would total more than 180°. Thus $B \approx 44°24'$.

Find A:

$A \approx 180° - (61°10' + 44°24') \approx 74°26'$

Find a:

$$\frac{a}{\sin A} = \frac{c}{\sin C}$$

$$\frac{a}{\sin 74°26'} = \frac{30.3}{\sin 61°10'}$$

$$a = \frac{30.3 \sin 74°26'}{\sin 61°10'} \approx 33.3$$

6. $\dfrac{c}{\sin C} = \dfrac{a}{\sin A}$, or $\dfrac{13.5}{\sin C} = \dfrac{17.2}{\sin 126.5°}$

$\sin C \approx 0.6309$

Then $C \approx 39.1°$ or $C \approx 140.9°$. An angle of 140.9°cannot be an angle of this triangle because it already has an angle of 126.5°and these two would total more than 180°. Thus $C \approx 39.1°$.

$B \approx 180° - (126.5° + 39.1°) \approx 14.4°$

$\dfrac{b}{\sin B} = \dfrac{a}{\sin A}$, or $\dfrac{b}{\sin 14.4°} = \dfrac{17.2}{\sin 126.5°}$

$b \approx 5.3$

7. Find A:

$A = 180° - (37.48° + 32.16°) = 110.36°$

Find a:

$$\frac{a}{\sin A} = \frac{c}{\sin C}$$

$$\frac{a}{\sin 110.36°} = \frac{3}{\sin 32.16°}$$

$$a = \frac{3 \sin 110.36°}{\sin 32.16°} \approx 5 \text{ mi}$$

Find b:

$$\frac{b}{\sin B} = \frac{c}{\sin C}$$

$$\frac{b}{\sin 37.48°} = \frac{3}{\sin 32.16°}$$

$$b = \frac{3 \sin 37.48°}{\sin 32.16°} \approx 3 \text{ mi}$$

8. $\dfrac{b}{\sin B} = \dfrac{a}{\sin A}$, or $\dfrac{2345}{\sin B} = \dfrac{2345}{\sin 124.67°}$

$\sin B \approx 0.8224$

Then $B \approx 55.33°$ or $B \approx 124.67°$. An angle of 55.33°cannot be an angle of this triangle because it already has an angle of 124.67°and the two would total 180°. An angle of 124.67°cannot be an angle of this triangle either because the two 124.67°angles would total more than 180°.

There is no solution.

9. Find B:

$$\frac{b}{\sin B} = \frac{c}{\sin C}$$

$$\frac{56.78}{\sin B} = \frac{56.78}{\sin 83.78°}$$

$$\sin B = \frac{56.78 \sin 83.78°}{56.78} \approx 0.9941$$

Then $B \approx 83.78°$ or $B \approx 96.22°$. An angle of 96.22°cannot be an angle of this triangle because it already has an angle of 83.78°and the two would total 180°. Thus, $B \approx 83.78°$.

Find A:

$A \approx 180° - (83.78° + 83.78°) \approx 12.44°$

Find a:

$$\frac{a}{\sin A} = \frac{c}{\sin C}$$

$$\frac{a}{\sin 12.44°} = \frac{56.78}{\sin 83.78°}$$

$$a = \frac{56.78 \sin 12.44°}{\sin 83.78} \approx 12.30 \text{ yd}$$

10. $B = 180° - (129°32' + 18°28') = 32°$

$\dfrac{a}{\sin A} = \dfrac{b}{\sin B}$, or $\dfrac{a}{\sin 129°32'} = \dfrac{1204}{\sin 32°}$

$a \approx 1752$ in.

$\dfrac{c}{\sin C} = \dfrac{b}{\sin B}$, or $\dfrac{c}{\sin 18°28'} = \dfrac{1204}{\sin 32°}$

$c \approx 720$ in.

11. Find B:

$$\frac{b}{\sin B} = \frac{a}{\sin A}$$

$$\frac{10.07}{\sin B} = \frac{20.01}{\sin 30.3°}$$

$$\sin B = \frac{10.07 \sin 30.3°}{20.01} \approx 0.2539$$

Then $B \approx 14.7°$ or $B \approx 165.3°$. An angle of 165.3°cannot be an angle of this triangle because it already has an angle of 30.3°and the two would total 180°. Thus, $B \approx 14.7°$.

Find C:

$C \approx 180° - (30.3° + 14.7°) \approx 135.0°$

Find c:

$$\frac{c}{\sin C} = \frac{a}{\sin A}$$

$$\frac{c}{\sin 135.0°} = \frac{20.01}{\sin 30.3°}$$

$$c = \frac{20.01 \sin 135.0°}{\sin 30.3°} \approx 28.04 \text{ cm}$$

12. $\dfrac{b}{\sin B} = \dfrac{c}{\sin C}$, or $\dfrac{4.157}{\sin B} = \dfrac{3.446}{\sin 51°48'}$

$\sin B \approx 0.9480$

Then $B \approx 71°26'$ *or* $B \approx 108°34'$

Solution I

If $B \approx 71°26'$, then

$A \approx 180° - (71°26' + 51°48') \approx 56°46'$.

$\dfrac{a}{\sin A} = \dfrac{c}{\sin C}$, or $\dfrac{a}{\sin 56°46'} = \dfrac{3.446}{\sin 51°48'}$

$a \approx 3.668$ km

Solution II

If $B \approx 108°34'$, then

$A \approx 180° - (108°34' + 51°48') \approx 19°38'$.

$\dfrac{a}{\sin A} = \dfrac{c}{\sin C}$, or $\dfrac{a}{\sin 19°38'} = \dfrac{3.446}{\sin 51°48'}$

$a \approx 1.473$ km

13. Find B:

$$\frac{b}{\sin B} = \frac{a}{\sin A}$$

$$\frac{18.4}{\sin B} = \frac{15.6}{\sin 89°}$$

$$\sin B = \frac{18.4 \sin 89°}{15.6} \approx 1.1793$$

Since there is no angle having a sine greater than 1, there is no solution.

14. $\frac{a}{\sin A} = \frac{c}{\sin C}$, or $\frac{56.2}{\sin A} = \frac{22.1}{\sin 46°32'}$

$\sin A \approx 1.8456$

Since there is no angle having a sine greater than 1, there is no solution.

15. Find B:

$B = 180° - (32.76° + 21.97°) = 125.27°$

Find b:

$$\frac{b}{\sin B} = \frac{a}{\sin A}$$

$$\frac{b}{\sin 125.27°} = \frac{200}{\sin 32.76°}$$

$$b = \frac{200 \sin 125.27°}{\sin 32.76°} \approx 302 \text{ m}$$

Find c:

$$\frac{c}{\sin C} = \frac{a}{\sin A}$$

$$\frac{c}{\sin 21.97°} = \frac{200}{\sin 32.76°}$$

$$c = \frac{200 \sin 21.97°}{\sin 32.76°} \approx 138 \text{ m}$$

16. $\frac{c}{\sin C} = \frac{a}{\sin A}$, or $\frac{45.6}{\sin C} = \frac{23.8}{\sin 115°}$

$\sin C \approx 1.7365$

Since there is no angle having a sine greater than 1, there is no solution.

17. $K = \frac{1}{2}ac \sin B$

$K = \frac{1}{2}(7.2)(3.4) \sin 42°$ Substituting

$K \approx 8.2 \text{ ft}^2$

18. $K = \frac{1}{2}bc \sin A$

$K = \frac{1}{2}(10)(13) \sin 17°12' \approx 19 \text{ in}^2$

19. $K = \frac{1}{2}ab \sin C$

$K = \frac{1}{2} \cdot 4 \cdot 6 \cdot \sin 82°54'$ Substituting

$K \approx 12 \text{ yd}^2$

20. $K = \frac{1}{2}ab \sin C$

$K = \frac{1}{2}(1.5)(2.1) \sin 75.16° \approx 1.5 \text{ m}^2$

21. $K = \frac{1}{2}ac \sin B$

$K = \frac{1}{2}(46.12)(36.74) \sin 135.2°$

$K \approx 596.98 \text{ ft}^2$

22. $K = \frac{1}{2}bc \sin A$

$K = \frac{1}{2}(18.2)(23.7) \sin 113° \approx 198.5 \text{ cm}^2$

23. $K = \frac{1}{2}bc \sin A$

$K = \frac{1}{2} \cdot 42 \cdot 53 \sin 135°$

$K \approx 787 \text{ ft}^2$

24. The amount of rope available for side s of triangle STQ is 38 ft − 21 ft − $2\left(4\frac{1}{2} \text{ ft}\right)$, or 8 ft. To determine if it is possible to have a triangle STQ with $S = 35°$, $s = 8$ ft, and $q = 21$ ft, we use the law of sines.

$$\frac{q}{\sin Q} = \frac{s}{\sin S}, \text{ or } \frac{21}{\sin Q} = \frac{8}{\sin 35°}$$

$\sin Q \approx 1.5056$

Since there is no angle whose sine is greater than 1, the triangle described above does not exist. Thus, the rancher does not have enough rope.

25.

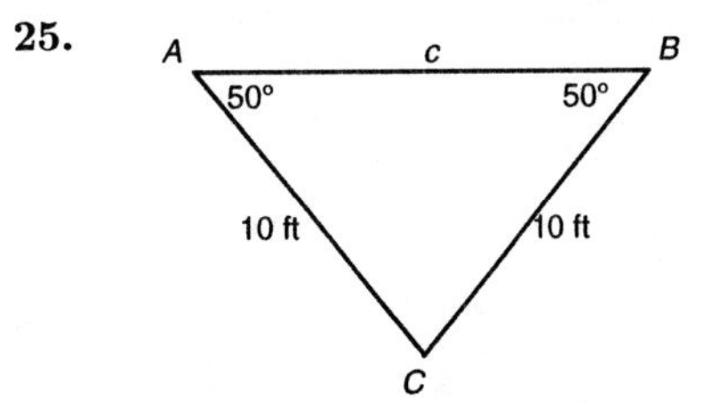

To solve this triangle find A, a, and c.

First we find C:

$C = 180° - (50° + 50°) = 80°$

Now we use the law of sines to find c:

$$\frac{c}{\sin C} = \frac{a}{\sin A}$$

$$\frac{c}{\sin 80°} = \frac{10}{\sin 50°}$$

$$c = \frac{10 \sin 80°}{\sin 50°} \approx 12.86$$

The speakers should be placed about 12.86 ft apart, or about 12 ft 10 in. apart.

26.

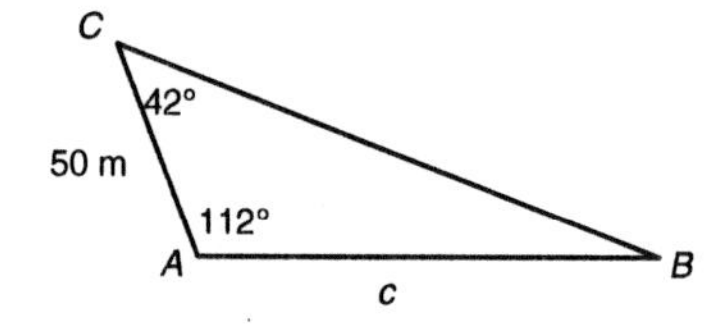

To solve this triangle find A, a, and c.

Let $c =$ the width of the crater.

First find B.

$B = 180° - (112° + 42°) = 26°$

Now we find c:

$\frac{c}{\sin C} = \frac{b}{\sin B}$, or $\frac{c}{\sin 42°} = \frac{50}{\sin 26°}$

$c \approx 76.3$ m

27.

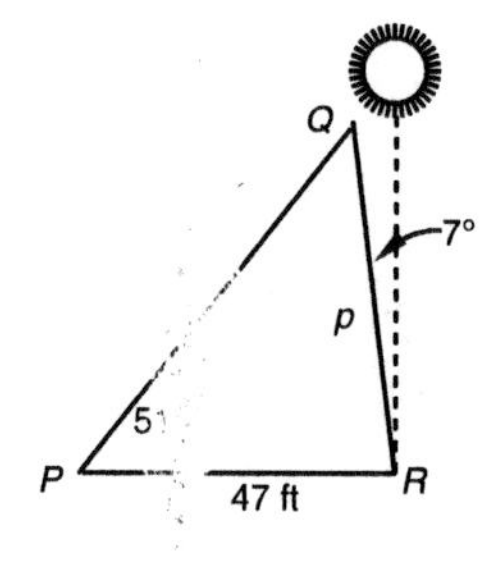

To solve this triangle find A, a, and c.

Find R: $R = 90° - 7° = 83°$

Find Q: $Q = 180° - (51° + 83°) = 46°$

Find p: (p is the length of the pole.)

$$\frac{p}{\sin P} = \frac{Q}{\sin Q}$$

$$\frac{p}{\sin 51°} = \frac{47}{\sin 46°}$$

$$p = \frac{47 \sin 51°}{\sin 46°} \approx 51$$

The pole is about 51 ft long.

28.

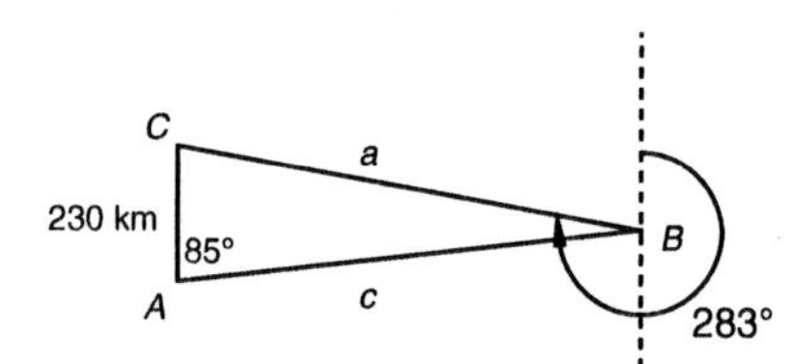

To solve this triangle find A, a, and c.

Find B: $B = 283° - 180° - 85° = 18°$

Find C: $C = 180° - (85° + 18°) = 77°$

Find a:

$\frac{a}{\sin A} = \frac{b}{\sin B}$, or $\frac{a}{\sin 85°} = \frac{230}{\sin 18°}$

$a \approx 742$

Find c:

$\frac{c}{\sin C} = \frac{b}{\sin B}$, or $\frac{c}{\sin 77°} = \frac{230}{\sin 18°}$

$c \approx 725$

The plane flew a total of $742 + 725$, or 1467 km.

29.

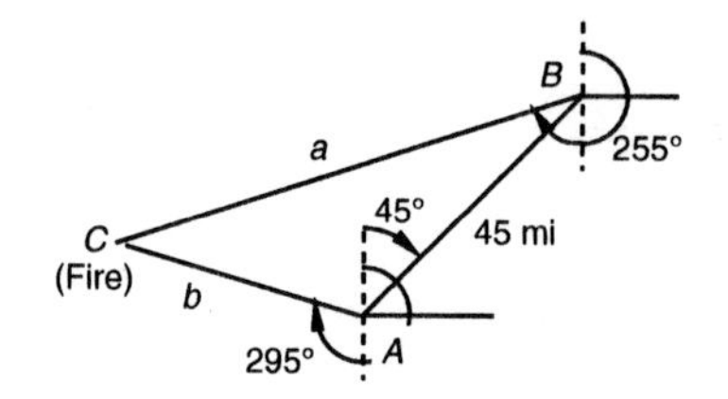

To solve this triangle find A, a, and c.

$A = 45° + (360° - 295°) = 110°$

$B = 255° - 180° - 45° = 30°$

$C = 180° - (110° + 30°) = 40°$

The distance from Tower A to the fire is b:

$$\frac{b}{\sin B} = \frac{c}{\sin C}$$

$$\frac{b}{\sin 30°} = \frac{45}{\sin 40°}$$

$$b = \frac{45 \sin 30°}{\sin 40°} \approx 35 \text{ mi}$$

The distance from Tower B to the fire is a:

$$\frac{a}{\sin A} = \frac{c}{\sin C}$$

$$\frac{a}{\sin 110°} = \frac{45}{\sin 40°}$$

$$a = \frac{45 \sin 110°}{\sin 40°} \approx 66 \text{ mi}$$

30.

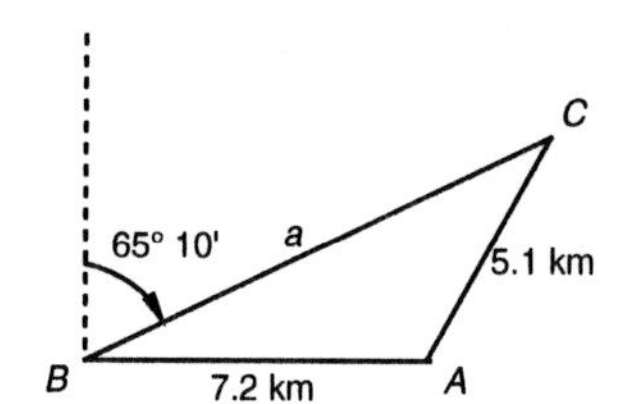

To solve this triangle find A, a, and c.

Find B: $B = 90° - 65°10' = 24°50'$

Find C:

$\frac{c}{\sin C} = \frac{b}{\sin B}$, or $\frac{7.2}{\sin C} = \frac{5.1}{\sin 24°50'}$

$\sin C \approx 0.5929$

Then $C \approx 36°20'$ or $C \approx 143°40'$.

If $C \approx 143°40'$, then

$A \approx 180° - (24°50' + 143°40') \approx 11°30'$.

Find a:

$\frac{a}{\sin A} = \frac{b}{\sin B}$, or $\frac{a}{\sin 11°30'} = \frac{5.1}{\sin 24°50'}$

$a \approx 2.4$

If $C \approx 36°20'$, then

$A \approx 180° - (24°50' + 36°20') \approx 118°50'$.

Find a:

$\frac{a}{\sin A} = \frac{b}{\sin B}$, or $\frac{a}{\sin 118°50'} = \frac{5.1}{\sin 24°50'}$

$a \approx 10.6$

The boat is either 2.4 km or 10.6 km from lighthouse B.

31.

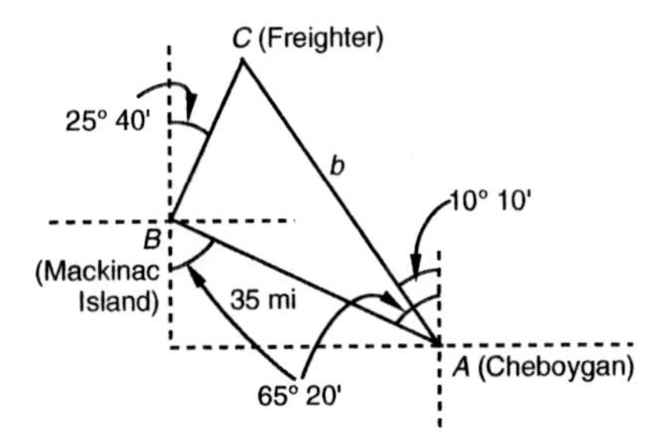

To solve this triangle find A, a, and c.

Find A: $A = 65°20' - 10°10' = 55°10'$

Find B: $B = 180° - 25°40' - 65°20' = 89°$

Find C: $C = 180° - (55°10' + 89°) = 35°50'$

Find b:

$$\frac{b}{\sin B} = \frac{c}{\sin C}$$

$$\frac{b}{\sin 89°} = \frac{35}{\sin 35°50'}$$

$$b = \frac{35 \sin 89°}{\sin 35°50'} \approx 60$$

The freighter is about 60 mi from Cheboygan.

32.

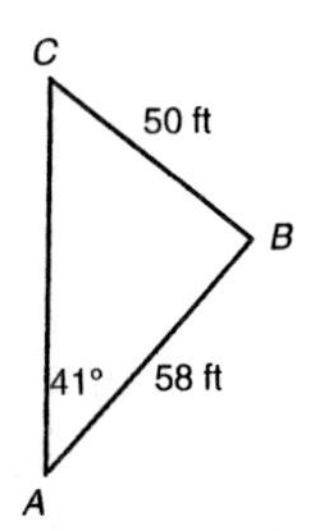

To solve this triangle find A, a, and c.

$a = 28 \text{ ft} + 22 \text{ ft} = 50 \text{ ft}$

$b = 22 \text{ ft} + 36 \text{ ft} = 58 \text{ ft}$

$$\frac{c}{\sin C} = \frac{a}{\sin A}, \text{ or } \frac{58}{\sin C} = \frac{50}{\sin 41°}$$

$\sin C \approx 0.7610$

Then $C \approx 50°$ or $C \approx 130°$.

If $C \approx 50°$, then $\phi \approx 180° - (41° + 50°) \approx 89°$.

If $C \approx 130°$, then $\phi \approx 180° - (41° + 130°) \approx 9°$.

Using the law of sines, we find that $b \approx 76$ ft when $\phi \approx 89°$ and $b \approx 12$ ft when $\phi \approx 9°$. Since b must be at least the sum of the two radii 36 ft and 28 ft, only $\phi \approx 89°$ can be a solution.

33. The law of sines involves two angles of a triangle and the sides opposite them. Three of these four values must be known in order to find the fourth. Thus, we must know the measure of one angle in order to use the law of sines.

34. In Cases 1 and 6, $b < h$; in Case 2, $b = h$, in Cases 3, 4, 5, 7, and 8, $b > h$.

35. $\cos A = 0.2213$

Using a calculator set in Radian mode, press .2213 [SHIFT] [COS], or [2nd] [COS] .2213 [ENTER]. The calculator returns 1.347649005, so $A \approx 1.348$ radians. Set the calculator in Degree mode and repeat the keystrokes above. The calculator returns 77.21460028, so $A \approx 77.2°$.

36. Since there is no angle whose cosine is greater than 1, there is no angle A for which $\cos A = 1.5612$.

37. With a calculator set in Degree mode, enter $18°14'20''$ as $18'14'20'$. We find that $18°14'20'' \approx 18.24°$.

38. With a calculator set in Degree mode, enter $125°3'42''$ as $125'3'42'$. We find that $125°3'42'' \approx 125.06°$.

39. The distance of -5 from 0 is 5, so $|-5| = 5$.

40. $\cos \frac{\pi}{6} = \frac{\sqrt{3}}{2}$

41. $\sin 45° = \frac{1}{\sqrt{2}}$, or $\frac{\sqrt{2}}{2}$

42. $\sin 300° = -\sin 60° = -\frac{\sqrt{3}}{2}$

43. $\cos\left(-\frac{2\pi}{3}\right) = -\cos\frac{\pi}{3} = -\frac{1}{2}$

44. $(1 - i)(1 + i) = 1 - i^2 = 1 - (-1) = 2$

45. See the answer section in the text.

46.

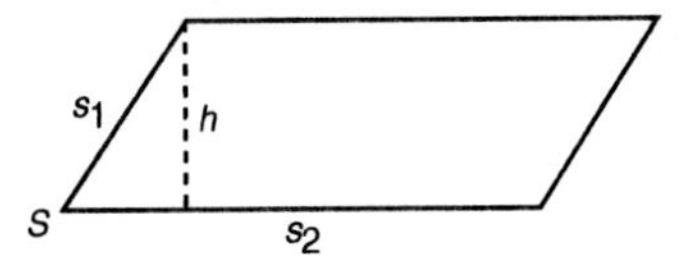

To solve this triangle find A, a, and c.

$A = s_2 h$, $h = s_1 \sin S$, so $A = s_1 s_2 \sin S$.

47. See the answer section in the text.

48. First find the length x of the diagonal of the 11 in. by 12 in. base:

$$x^2 = 11^2 + 12^2$$

$$x^2 = 265$$

$$x \approx 16.2788$$

Now consider the triangle formed by x, d, and the 15 in. side of the figure:

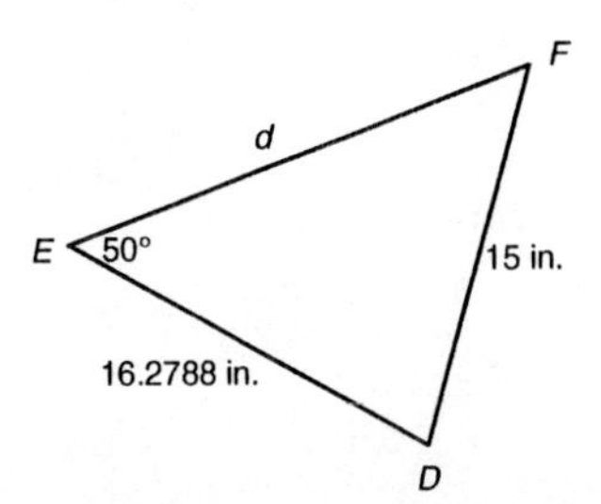

To solve this triangle find A, a, and c.

Find F:

$$\frac{16.2788}{\sin F} = \frac{15}{\sin 50°}$$
$$\sin F \approx 0.8314$$

Then $F \approx 56.24°$ or $F \approx 123.76°$.

The drawing in the text indicates that F is an acute angle, so we consider $F \approx 56.24°$.

Find D:

$D = 180° - (50° + 56.24°) = 73.76°$

Find d:

$$\frac{d}{\sin 73.76°} = \frac{15}{\sin 50°}$$
$$d \approx 18.8 \text{ in.}$$

Exercise Set 7.2

1.

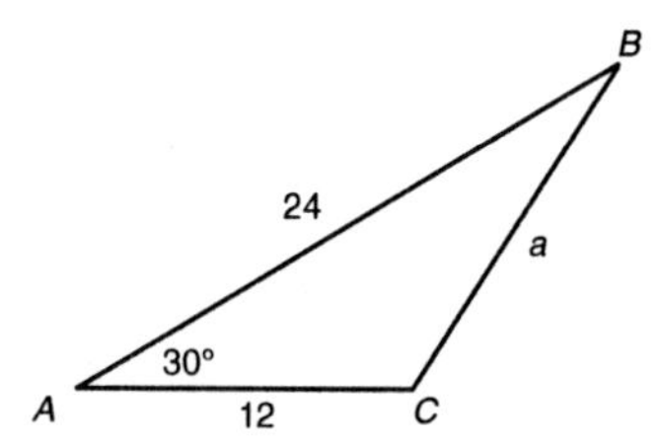

To solve this triangle find A, a, and c.

To solve this triangle find a, B, and C.

From the law of cosines,

$a^2 = b^2 + c^2 - 2bc\cos A$

$a^2 = 12^2 + 24^2 - 2 \cdot 12 \cdot 24\cos 30°$

$a^2 \approx 221$

$a \approx 15$

Next we use the law of cosines again to find a second angle.

$b^2 = a^2 + c^2 - 2ac\cos B$

$12^2 = 15^2 + 24^2 - 2(15)(24)\cos B$

$144 = 225 + 576 - 720\cos B$

$-657 = -720\cos B$

$\cos B \approx 0.9125$

$B \approx 24°$

Now we find the third angle.

$C \approx 180° - (30° + 24°) \approx 126°$

2. $b^2 = a^2 + c^2 - 2ac\cos B$, or

$b^2 = 12^2 + 15^2 - 2 \cdot 12 \cdot 15\cos 133°$

$b \approx 25$

$a^2 = b^2 + c^2 - 2bc\cos A$, or

$12^2 = 25^2 + 15^2 - 2 \cdot 25 \cdot 15\cos A$

$-0.9413 \approx \cos A$

$20° \approx A$

$C \approx 180° - (20° + 133°) \approx 27°$

3.

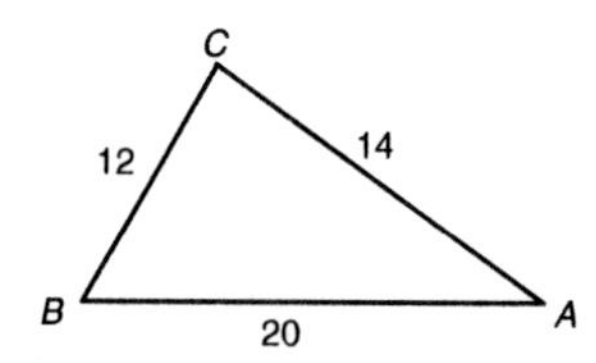

To solve this triangle find A, B, and C.

Find A:

$a^2 = b^2 + c^2 - 2bc\cos A$

$12^2 = 14^2 + 20^2 - 2 \cdot 14 \cdot 20\cos A$

$144 = 196 + 400 - 560\cos A$

$-452 = -560\cos A$

$\cos A \approx 0.8071$

Thus $A \approx 36.18°$.

Find B:

$b^2 = a^2 + c^2 - 2ac\cos B$

$14^2 = 12^2 + 20^2 - 2 \cdot 12 \cdot 20\cos B$

$196 = 144 + 400 - 480\cos B$

$-348 = -480\cos B$

$\cos B \approx 0.7250$

Thus $B \approx 43.53°$.

Then $C \approx 180° - (36.18° + 43.53°) \approx 100.29°$.

4. $a^2 = b^2 + c^2 - 2bc\cos A$, or

$(22.3)^2 = (22.3)^2 + (36.1)^2 - 2(22.3)(36.1)\cos A$

$0.8094 \approx \cos A$

$35.96° \approx A$

$b^2 = a^2 + c^2 - 2ac\cos B$, or

$(22.3)^2 = (22.3)^2 + (36.1)^2 - 2(22.3)(36.1)\cos B$

$0.8094 \approx \cos B$

$35.96° \approx B$

$C \approx 180° - (35.96° + 35.96°) \approx 108.08°$

5. To solve this triangle find b, A, and C.

Find b:

$b^2 = a^2 + c^2 - 2ac\cos B$

$b^2 = 78^2 + 16^2 - 2 \cdot 78 \cdot 16\cos 72°40'$

$b^2 \approx 5596$

$b \approx 75$ m

Find A:

$a^2 = b^2 + c^2 - 2bc\cos A$

$78^2 = 75^2 + 16^2 - 2 \cdot 75 \cdot 16\cos A$

$6084 = 5625 + 256 - 2400\cos A$

$203 = -2400\cos A$

$\cos A \approx -0.0846$

$a \approx 94°51'$

Then $C = 180° - (94°51' + 72°40') = 12°29'$

6. $c^2 = a^2 + b^2 - 2ab\cos C$, or

$c^2 = (25.4)^2 + (73.8)^2 - 2(25.4)(73.8)\cos 22.28°$

$c \approx 51.2$ cm

$a^2 = b^2 + c^2 - 2bc\cos A$

$(25.4)^2 = (73.8)^2 + (51.2)^2 - 2(73.8)(51.2)\cos A$

$\cos A \approx 0.9822$

$A \approx 10.82°$

$B \approx 180° - (10.82° + 22.28°) \approx 146.90°$.

7. Find A:

$a^2 = b^2 + c^2 - 2bc\cos A$

$16^2 = 20^2 + 32^2 - 2 \cdot 20 \cdot 32\cos A$

$256 = 400 + 1024 - 1280\cos A$

$-1168 = -1280\cos A$

$\cos A \approx 0.9125$

$A \approx 24.15°$.

Find B:

$b^2 = a^2 + c^2 - 2ac\cos B$

$20^2 = 16^2 + 32^2 - 2 \cdot 16 \cdot 32\cos B$

$400 = 256 + 1024 - 1024\cos B$

$-880 = -1024\cos B$

$\cos B \approx 0.8594$

$B \approx 30.75°$.

Then $C \approx 180° - (24.15° + 30.75°) \approx 125.10°$.

8. $b^2 = a^2 + c^2 - 2ac\cos B$, or

$b^2 = (23.78)^2 + (25.74)^2 - 2(23.78)(25.74)\cos 72.66°$

$b \approx 29.38$ km

$a^2 = b^2 + c^2 - 2bc\cos A$, or

$(23.78)^2 = (29.38)^2 + (25.74)^2 - 2(29.38)(25.74)\cos A$

$0.6349 \approx \cos A$

$50.59° \approx A$

$C \approx 180° - (50.59° - 72.66°) \approx 56.75°$

9. Find A:

$a^2 = b^2 + c^2 - 2bc\cos A$

$2^2 = 3^2 + 8^2 - 2 \cdot 3 \cdot 8\cos A$

$4 = 9 + 64 - 48\cos A$

$-69 = -48\cos A$

$\cos A \approx 1.4375$

Since there is no angle whose cosine is greater than 1, there is no solution.

10. $a^2 = b^2 + c^2 - 2bc\cos A$, or

$a^2 = (15.8)^2 + (18.4)^2 - 2(15.8)(18.4)\cos 96°13'$

$a \approx 25.5$ yd

$b^2 = a^2 + c^2 - 2ac\cos B$

$(15.8)^2 = (25.5)^2 + (18.4)^2 - 2(25.5)(18.4)\cos B$

$\cos B \approx 0.7877$

$B \approx 38°2'$

$C \approx 180° - (96°13' + 38°2') \approx 45°45'$

11. Find A:

$a^2 = b^2 + c^2 - 2bc\cos A$

$(26.12)^2 = (21.34)^2 + (19.25)^2 - 2(21.34)(19.25)\cos A$

$682.2544 = 455.3956 + 370.5625 - 821.59\cos A$

$-143.7037 = -821.59\cos A$

$\cos A \approx 0.1749$

$A \approx 79.93°$

Find B:

$b^2 = a^2 + c^2 - 2ac\cos B$

$(21.34)^2 = (26.12)^2 + (19.25)^2 - 2(26.12)(19.25)\cos B$

$455.3956 = 682.2544 + 370.5625 - 1005.62\cos B$

$-597.4213 = -1005.62\cos B$

$\cos B \approx 0.5941$

$B \approx 53.55°$

$C \approx 180° - (79.93° + 53.55°) \approx 46.52°$

12. $c^2 = a^2 + b^2 - 2ab\cos C$, or

$c^2 = 6^2 + 9^2 - 2 \cdot 6 \cdot 9(\cos 28°43')$

$c \approx 5$ mm

$a^2 = b^2 + c^2 = 2bc\cos A$, or

$6^2 = 9^2 + 5^2 - 2 \cdot 9 \cdot 5\cos A$

$0.7778 \approx \cos A$

$38°57' \approx A$

$B \approx 180° - (38°57' + 28°43') = 112°20'$

13. Find c:

$c^2 = a^2 + b^2 - 2ab\cos C$

$c^2 = (60.12)^2 + (40.23)^2 - 2(60.12)(40.23)\cos 48.7°$

$c^2 \approx 2040$

$c \approx 45.17$ mi

Find A:

$a^2 = b^2 + c^2 - 2bc\cos A$

$(60.12)^2 = (40.23)^2 + (45.17)^2 - 2(40.23)(45.17)\cos A$

$3614.4144 = 1618.4529 + 2040.3289 - 3634.3782\cos A$

$-44.3674 = -3634.3782\cos A$

$\cos A \approx 0.0122$

$A \approx 89.3°$

$B \approx 180° - (89.3° + 48.7°) \approx 42.0°$

14. $a^2 = b^2 + c^2 - 2bc\cos A$, or

$$(11.2)^2 = (5.4)^2 + 7^2 - 2(5.4)(7)\cos A$$
$$\cos A \approx -0.6254$$
$$A \approx 128.71°$$
$$b^2 = a^2 + c^2 - 2ac\cos B, \text{ or}$$
$$(5.4)^2 = (11.2)^2 + 7^2 - 2(11.2)(7)\cos B$$
$$\cos B \approx 0.9265$$
$$B \approx 22.10°$$
$$C \approx 180° - (128.71° + 22.10°) \approx 29.19°$$

15. Find a:

$$a^2 = b^2 + c^2 - 2bc\cos A$$
$$a^2 = (10.2)^2 + (17.3)^2 - 2(10.2)(17.3)\cos 53.456°$$
$$a^2 \approx 193.19$$
$$a \approx 13.9 \text{ in.}$$

Find B:

$$b^2 = a^2 + c^2 - 2ac\cos B$$
$$(10.2)^2 = (13.9)^2 + (17.3)^2 - 2(13.9)(17.3)\cos B$$
$$104.04 = 193.21 + 299.29 - 480.94\cos B$$
$$-388.46 = -480.94\cos B$$
$$\cos B \approx 0.8077$$
$$B \approx 36.127°$$

Find C:

$$C \approx 180° - (53.456° + 36.127°) \approx 90.417°$$

16. $a^2 = b^2 + c^2 - 2bc\cos A$, or

$$17^2 = (15.4)^2 + (1.5)^2 - 2(15.4)(1.5)\cos A$$
$$\cos A \approx -1.0734$$

Since there is no angle whose cosine is less than -1, there is no solution.

17. We are given two sides and the angle opposite one of them. The law of sines applies.

$$C = 180° - (70° + 12°) = 98°$$
$$\frac{a}{\sin A} = \frac{b}{\sin B}$$
$$\frac{a}{\sin 70°} = \frac{21.4}{\sin 12°}$$
$$a = \frac{21.4\sin 70°}{\sin 12°} \approx 96.7$$
$$\frac{c}{\sin C} = \frac{b}{\sin B}$$
$$\frac{c}{\sin 98°} = \frac{21.4}{\sin 12°}$$
$$c = \frac{21.4\sin 98°}{\sin 12°} \approx 101.9$$

18. We are given two sides and the included angle. The law of cosines applies.

$$b^2 = a^2 + c^2 - 2ac\cos B$$
$$b^2 = 15^2 + 7^2 - 2\cdot 15\cdot 7\cos 62°$$
$$b \approx 13$$
$$a^2 = b^2 + c^2 - 2bc\cos A$$
$$15^2 = 13^2 + 7^2 - 2\cdot 13\cdot 7\cos A$$
$$A \approx 92°$$
$$C \approx 180° - (92° + 62°) \approx 26°$$

19. We are given all three sides of the triangle. The law of cosines applies.

$$a^2 = b^2 + c^2 - 2bc\cos A$$
$$(3.3)^2 = (2.7)^2 + (2.8)^2 - 2(2.7)(2.8)\cos A$$
$$10.89 = 7.29 + 7.84 - 15.12\cos A$$
$$-4.24 = -15.12\cos A$$
$$\cos A \approx 0.2804$$
$$A \approx 73.71°$$
$$b^2 = a^2 + c^2 - 2ac\cos B$$
$$(2.7)^2 = (3.3)^2 + (2.8)^2 - 2(3.3)(2.8)\cos B$$
$$7.29 = 10.89 + 7.84 - 18.48\cos B$$
$$-11.44 = -18.48\cos B$$
$$\cos B \approx 0.6190$$
$$B \approx 51.75°$$
$$C \approx 180° - (73.71° + 51.75°) \approx 54.54°$$

20. We are given two sides and the angle opposite one of them. The law of sines applies.

$$\frac{a}{\sin A} = \frac{b}{\sin B}, \text{ or } \frac{1.5}{\sin 58°} = \frac{2.5}{\sin B}$$

$\sin B \approx 1.4134$, so there is no solution.

21. We are given the three angles of a triangle. Neither law applies. This triangle cannot be solved using the given information.

22. We are given two sides of the triangle and the included angle. The law of cosines applies.

$$c^2 = a^2 + b^2 - 2ab\cos C$$
$$c^2 = 60^2 + 40^2 - 2(60)(40)\cos 47°$$
$$c \approx 44$$
$$a^2 = b^2 + c^2 = 2bc\cos A$$
$$60^2 = 40^2 + 44^2 - 2(40)(44)\cos A$$
$$\cos A \approx -0.0182$$
$$A \approx 91°$$
$$B \approx 180° - (47° + 91°) \approx 42°$$

23. We are given all three sides of the triangle. The law of cosines applies.

$$a^2 = b^2 + c^2 - 2bc\cos A$$
$$(3.6)^2 = (6.2)^2 + (4.1)^2 - 2(6.2)(4.1)\cos A$$
$$12.96 = 38.44 + 16.81 - 50.84\cos A$$
$$-42.29 = -50.84\cos A$$
$$\cos A \approx 0.8318$$
$$A \approx 33.71°$$

$$b^2 = a^2 + c^2 - 2ac\cos B$$
$$(6.2)^2 = (3.6)^2 + (4.1)^2 - 2(3.6)(4.1)\cos B$$
$$38.44 = 12.96 + 16.81 - 29.52\cos B$$
$$8.67 = -29.52\cos B$$
$$\cos B \approx -0.2937$$
$$B \approx 107.08°$$
$$C \approx 180° - (33.71° + 107.08°) \approx 39.21°$$

24. We are given two angles and the side opposite one of them. The law of sines applies.

$$A = 180° - (110°30' + 8°10') = 61°20'$$
$$\frac{a}{\sin A} = \frac{c}{\sin C}, \text{ or } \frac{a}{\sin 61°10'} = \frac{0.912}{\sin 8°10'}$$
$$a \approx 5.633$$
$$\frac{b}{\sin B} = \frac{c}{\sin C}, \text{ or } \frac{b}{\sin 110°30'} = \frac{0.912}{\sin 8°10'}$$
$$b \approx 6.014$$

25.

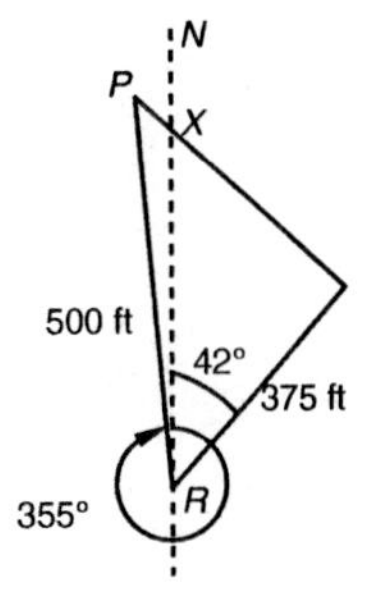

Angle $XRP = 360° - 355° = 5°$, so $R = 42° + 5° = 47°$. Use the law of cosines to find x, the distance from P to G.

$$x^2 = 500^2 + 375^2 - 2 \cdot 500 \cdot 375\cos 47°$$
$$x^2 \approx 134,876$$
$$x \approx 367$$

The distance between the poachers and the game is about 367 ft.

26.

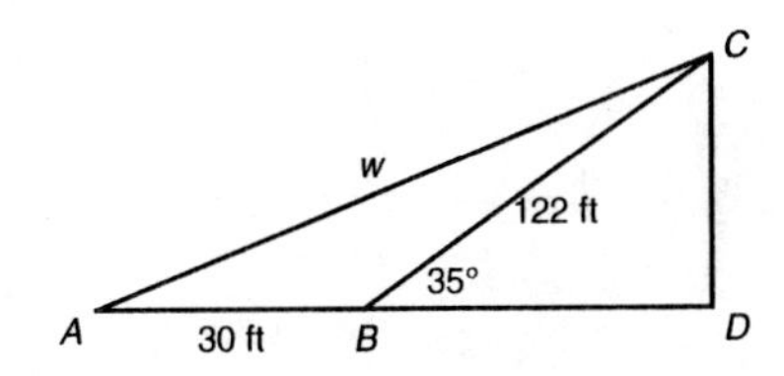

Angle $ABC = 180° - 35° = 145°$. Then

$$w^2 = 30^2 + 122^2 - 2 \cdot 30 \cdot 122\cos 145°$$
$$w \approx 148 \text{ ft}$$

148 ft − 122 ft = 26 ft, so the aerialists will have to walk about 26 ft farther.

Find the new approach angle A:

$$\frac{122}{\sin A} = \frac{148}{\sin 145°}$$
$$\sin A \approx 0.4728$$
$$A \approx 28°$$

The new approach angle is about 28°.

27. Use the law of cosine to find the distance d from A to B.

$$d^2 = (0.5)^2 + (1.3)^2 - 2(0.5)(1.3)\cos 110°$$
$$d^2 \approx 2.38$$
$$d \approx 1.5$$

He skated about 1.5 mi.

28.

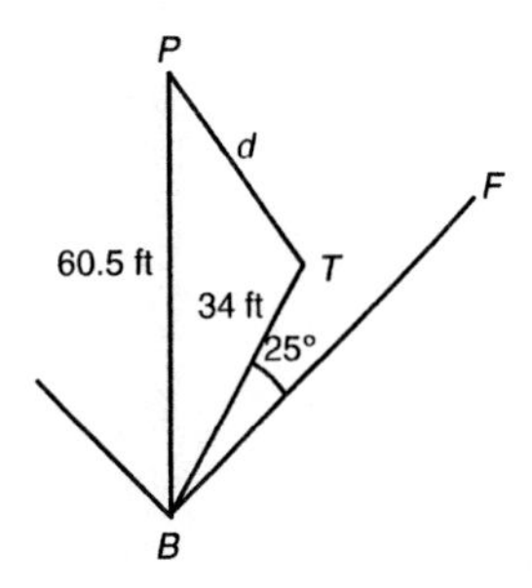

Since the baseball diamond is a square, angle $FBP = \frac{1}{2} \cdot 90°$, or 45°, and angle $TBP = 45° - 25°$, or 20°. Use the law of cosines to find the distance d that the pitcher travels.

$$d^2 = (60.5)^2 + 34^2 - 2(60.5)(34)\cos 20°$$
$$d \approx 30.8 \text{ ft}$$

29. 25 knots × 2 hr = 50 nautical mi

20 knots × 2 hr = 40 nautical mi

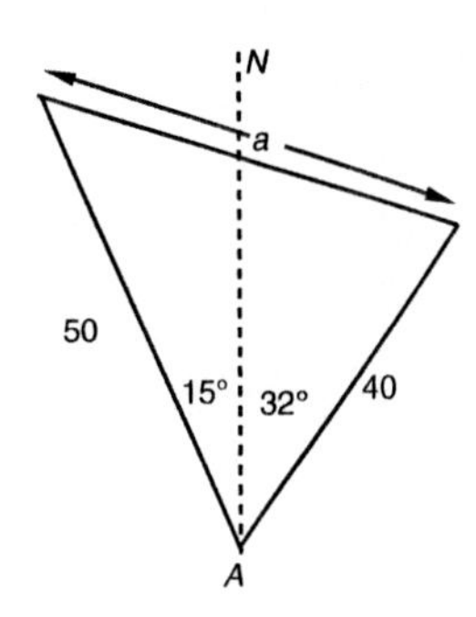

Find A: $A = 15° + 32° = 47°$

Find a:

$$a^2 = b^2 + c^2 - 2bc\cos A$$
$$a^2 = 50^2 + 40^2 - 2 \cdot 50 \cdot 40\cos 47°$$
$$a^2 \approx 1372$$
$$a \approx 37$$

The ships are about 37 nautical mi apart.

30. Find S:

$$(45.2)^2 = (31.6)^2 + (22.4)^2 - 2(31.6)(22.4)\cos S$$
$$-0.3834 \approx \cos S$$
$$112.5° \approx S$$

Find T:

$$\frac{22.4}{\sin T} = \frac{45.2}{\sin 112.5°}$$
$$\sin T \approx 0.4579$$
$$T \approx 27.2°$$

Find U:

$$U \approx 180° - (112.5° + 27.2°) \approx 40.3°$$

31. 150 km/h × 3 hr = 450 km

200 km/h × 3 hr = 600 km

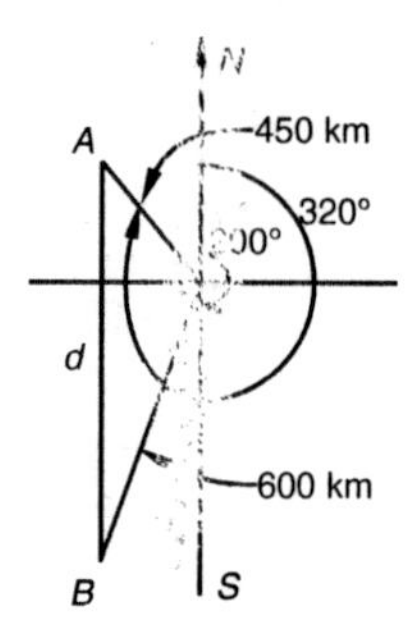

The angle opposite d is $320° - 200°$, or $120°$. Use the law of cosines to find d:

$d^2 = 450^2 + 600^2 - 2 \cdot 450 \cdot 600 \cos 120°$

$d \approx 912$

The planes are about 912 km apart.

32.

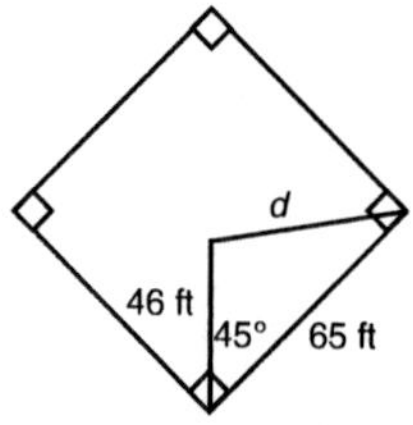

$d^2 = 65^2 + 46^2 - 2 \cdot 65 \cdot 46 \cos 45°$

$d \approx 46$ ft

33.

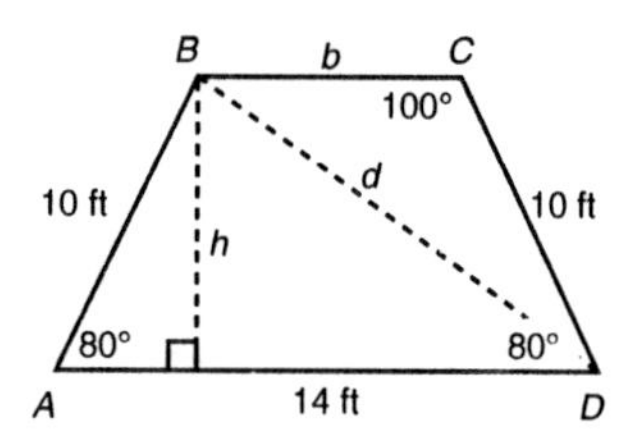

Let b represent the other base, d represent the diagonal, and h the height.

a) Find d:

Using the law of cosines,

$d^2 = 10^2 + 14^2 - 2 \cdot 10 \cdot 14 \cos 80°$

$d^2 \approx 247$

$d \approx 16$

The length of the diagonal is about 16 ft.

b) Find h:

$\frac{h}{10} = \sin 80°$

$h = 10 \sin 80°$

$h \approx 9.85$

Find $\angle CBD$:

Using the law of sines,

$$\frac{10}{\sin \angle CBD} = \frac{16}{\sin 100°}$$

$$\sin \angle CBD = \frac{10 \sin 100°}{16} \approx 0.6155$$

Thus $\angle CBD \approx 38°$.

Then $\angle CDB \approx 180° - (100° + 38°) \approx 42°$.

Find b:

Using the law of sines,

$$\frac{b}{\sin 42°} = \frac{16}{\sin 100°}$$

$$b = \frac{16 \sin 42°}{\sin 100°} \approx 10.87$$

$$\text{Area} = \frac{1}{2}h(b_1 + b_2)$$

$$\text{Area} \approx \frac{1}{2}(9.85)(10.87 + 14)$$

$$\approx 122$$

The area is about 122 ft^2. (Answers may vary due to rounding differences.)

34.

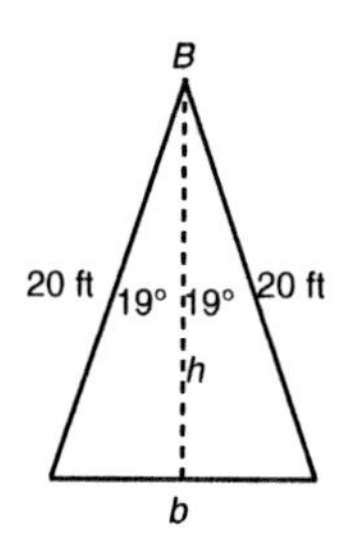

$b^2 = a^2 + c^2 - 2ac \cos B$, or

$b^2 = 20^2 + 20^2 - 2 \cdot 20 \cdot 20 \cos 38°$

$b \approx 13$

$\cos 19° = \frac{h}{20}$, so $h \approx 19$

$\text{Area} = \frac{1}{2}bh \approx \frac{1}{2} \cdot 13 \cdot 19 \approx 124$ ft^2

(Answers may vary due to rounding differences.)

35. Place the figure on a coordinate system as shown.

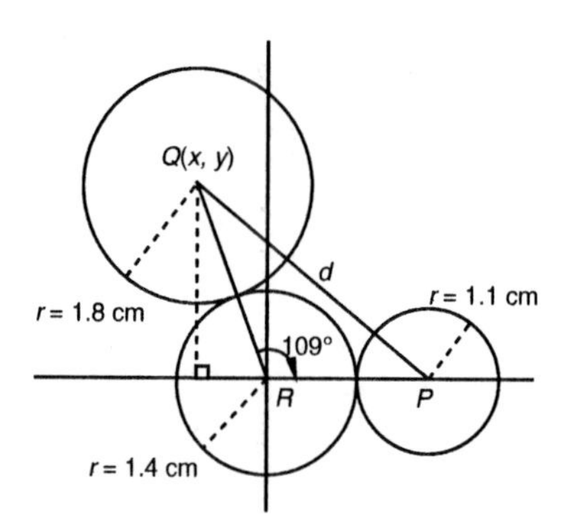

Let (x, y) be the coordinates of point Q. The coordinates of point P are $(1.4 + 1.1, 0)$, or $(2.5, 0)$. The length of QR is $1.8 + 1.4$, or 3.2.

We use the law of cosines to find d.

$d^2 = (2.5)^2 + (3.2)^2 - 2(2.5)(3.2) \cos 109°$

$d^2 \approx 21.70$

$d \approx 4.7$

The length PQ is about 4.7 cm.

36.

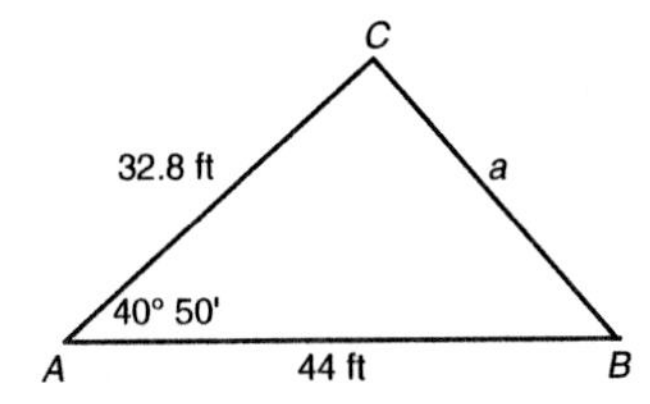

$a^2 = b^2 + c^2 - 2bc \cos A$

$a^2 = (32.8)^2 + 44^2 - 2(32.8)(44) \cos 40.8°$

$a \approx 28.8$ ft

37. Using the law of cosines it is necessary to solve the quadratic equation

$(11.1)^2 = a^2 + (28.5)^2 - 2a(2.85) \cos 19°$, or

$0 = a^2 - [2(2.85) \cos 19°]a + [(28.5)^2 - (11.1)^2]$.

The law of sines requires less complicated computations.

38. The law of sines involves two angles of a triangle and the sides opposite them. Three of these four values must be known in order to find the fourth. Given SAS, only two of these four values are known.

39. This is a polynomial function with degree 4, so it is a quartic function.

40. $y - 3 = 17x$, or $y = 17x + 3$

This function can be written in the form $y = mx + b$, so it is linear.

41. The variable is in a trigonometric function, so this is trigonometric.

42. The variable is in an exponent, so this is an exponential function.

43. This is the quotient of two polynomials, so it is a rational function.

44. This is a polynomial function with degree 3, so it is a cubic function.

45. The variable is in an exponent, so this is an exponential function.

46. The variable is in a logarithmic function, so it is logarithmic.

47. The variable is in a trigonometric function, so this is trigonometric.

48. This is a polynomial function with degree 2, so it is a quadratic function.

49.

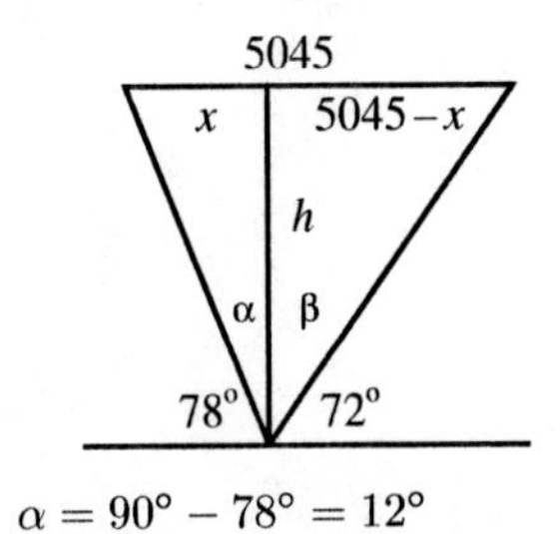

$\alpha = 90° - 78° = 12°$

$\beta = 90 - 72° = 18°$

$\tan 12° = \frac{x}{h}$, so $x = h \tan 12°$

$\tan 18° = \frac{5045 - x}{h}$, or

$\tan 18° = \frac{5045 - h \tan 12°}{h}$ Substituting for x

$h \tan 18° = 5045 - h \tan 12°$

$h \tan 18° + h \tan 12° = 5045$

$h(\tan 18° + \tan 12°) = 5045$

$h = \frac{5045}{\tan 18° + \tan 12°}$

$h \approx 9386$

The canyon is about 9386 ft deep.

50. $K = \frac{1}{2} bc \sin A$

$K^2 = \frac{1}{4} b^2 c^2 \sin^2 A$ Squaring both sides

$K^2 = \frac{1}{4} b^2 c^2 (1 - \cos^2 A)$

$K^2 = \frac{1}{4} b^2 c^2 - \frac{1}{4} b^2 c^2 \cos^2 A$

Now, the law of cosines gives us

$a^2 = b^2 + c^2 - 2bc \cos A$ so

$bc \cos A = \frac{-a^2 + b^2 + c^2}{2}$ and

$b^2 c^2 \cos^2 A = \left(\frac{-a^2 + b^2 + c^2}{2}\right)^2$.

We substitute in the expression for K^2:

$$K^2 = \frac{1}{4} b^2 c^2 - \frac{1}{4}\left(\frac{-a^2 + b^2 + c^2}{2}\right)^2$$
$$= \frac{1}{4} b^2 c^2 - \frac{1}{4} \cdot \frac{1}{4}(-a^2 + b^2 + c^2)^2$$
$$= \frac{1}{16}[4b^2c^2 - (-a^2 + b^2 + c^2)^2]$$
$$= \frac{1}{16}[2bc + (-a^2 + b^2 + c^2)][2bc - (-a^2 + b^2 + c^2)]$$
$$= \frac{1}{16}[(b^2 + 2bc + c^2) - a^2][a^2 - (b^2 - 2bc + c^2)]$$
$$= \frac{1}{16}[(b + c)^2 - a^2][a^2 - (b - c)^2]$$
$$= \frac{1}{16}(b + c + a)(b + c - a)(a + b - c)(a - b + c)$$
$$= \frac{1}{2}(a + b + c) \cdot \left[\frac{1}{2}(a + b + c) - a\right] \cdot \left[\frac{1}{2}(a + b + c) - b\right] \cdot \left[\frac{1}{2}(a + b + c) - c\right]$$

Thus, $K^2 = s(s - a)(s - b)(s - c)$, where

$s = \frac{1}{2}(a + b + c)$.

Then $K = \sqrt{s(s-a)(s-b)(s-c)}$, where
$s = \frac{1}{2}(a+b+c)$.

In Exercise 36, $a = 44$ ft, $b = 32.8$ ft, and $c \approx 28.8$ ft.

$$s = \frac{1}{2}(a+b+c) \approx \frac{1}{2}(44+32.8+28.8) \approx 52.8$$

$$K \approx \sqrt{52.8(52.8-44)(52.8-32.8)(52.8-28.8)}$$

$$K \approx 472.3 \text{ ft}^2$$

51.

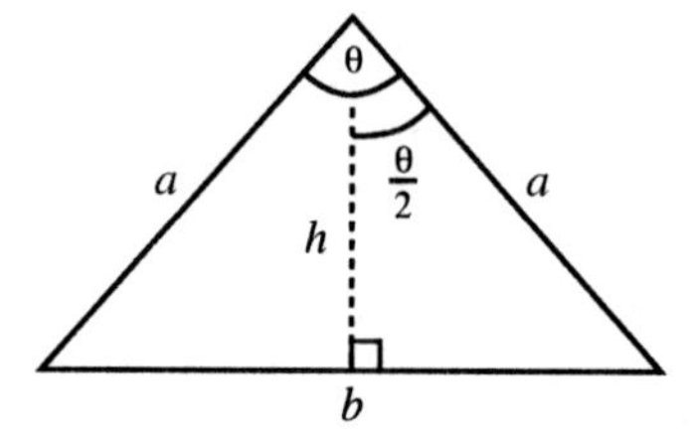

Let b represent the base, h the height, and θ the included angle.

Find h:

$$\frac{h}{a} = \cos\frac{\theta}{2}$$

$$h = a\cos\frac{\theta}{2}$$

Find b:

$$b^2 = a^2 + a^2 - 2 \cdot a \cdot a\cos\theta$$

$$b^2 = 2a^2 - 2a^2\cos\theta$$

$$b^2 = 2a^2(1-\cos\theta)$$

$$b = \sqrt{2}a\sqrt{1-\cos\theta}$$

Find A (area):

$$A = \frac{1}{2}bh$$

$$A = \frac{1}{2}(\sqrt{2}a\sqrt{1-\cos\theta})\left(a\cos\frac{\theta}{2}\right)$$

$$A = \left(a\sqrt{\frac{1-\cos\theta}{2}}\right)\left(a\sqrt{\frac{1+\cos\theta}{2}}\right)$$

$$A = \frac{1}{2}a^2\sqrt{1-\cos^2\theta} = \frac{1}{2}a^2\sqrt{\sin^2\theta}$$

$$A = \frac{1}{2}a^2\sin\theta$$

The maximum area occurs when θ is 90°, because the sine function takes its maximum value, 1, at $\theta = 90°$.

52. We add labels to the drawing in the text.

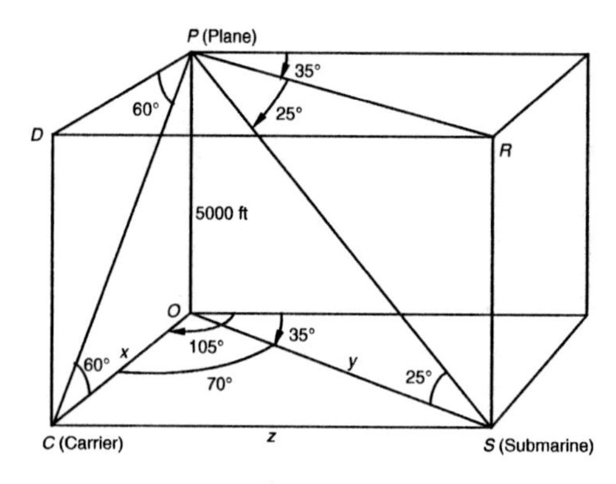

We find x using ΔPCO:

$$\cot 60° = \frac{x}{5000}$$

$$x = 5000\cot 60°$$

$$x = 2886.75$$

We find y using ΔPSO:

$$\cot 25° = \frac{y}{5000}$$

$$y = 5000\cot 25°$$

$$y \approx 10,722.53$$

Finally, we use ΔOCS and the law of cosines to find z, the distance from the submarine to the carrier.

$$z^2 = x^2 + y^2 - 2xy\cos 70°$$

$$z^2 = (2886.75)^2 + (10,722.53)^2 - 2(2886.75)(10,722.53)\cos 70°$$

$$z \approx 10,106 \text{ ft}$$

Exercise Set 7.3

1.

Imaginary, Real axes; point $4 + 3i$ plotted.

$$|4+3i| = \sqrt{4^2+3^2} = \sqrt{16+9} = \sqrt{25} = 5$$

2.

$$|-2-3i| = \sqrt{(-2)^2+(-3)^2} = \sqrt{13}$$

3.

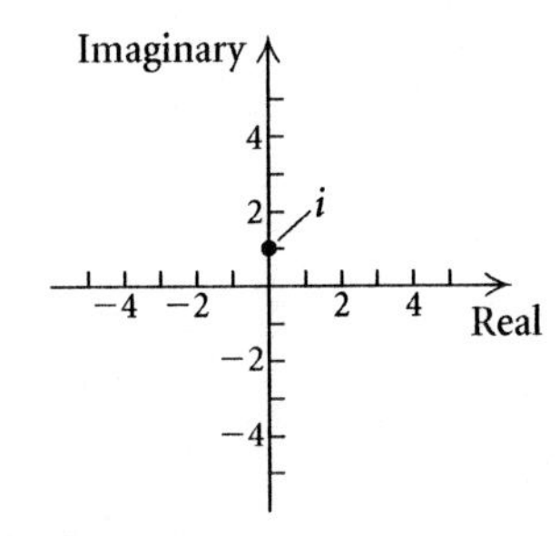

$|i| = |0 + 1 \cdot i| = \sqrt{1^2} = 1$

4.

$|-5 - 2i| = \sqrt{(-5)^2 + (-2)^2} = \sqrt{29}$

5.

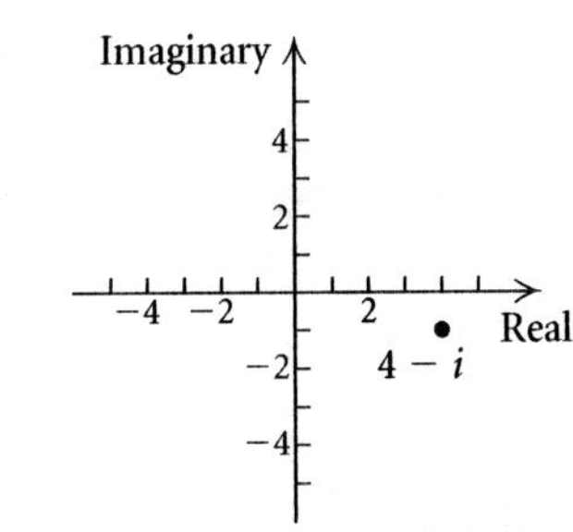

$|4 - i| = \sqrt{4^2 + (-1)^2} = \sqrt{16 + 1} = \sqrt{17}$

6.

$|6 + 3i| = \sqrt{6^2 + 3^2} = \sqrt{45} = 3\sqrt{5}$

7.

$|3| = |3 + 0i| = \sqrt{3^2} = 3$

8.

$|-2i| = \sqrt{(-2)^2} = 2$

9. From the graph we see that standard notation for the number is $3 - 3i$.

Find trigonometric notation:

$r = \sqrt{3^2 + (-3)^2} = \sqrt{18} = 3\sqrt{2}$

$\sin\theta = \dfrac{-3}{3\sqrt{2}} = -\dfrac{1}{\sqrt{2}}$, or $-\dfrac{\sqrt{2}}{2}$

$\cos\theta = \dfrac{3}{3\sqrt{2}} = \dfrac{1}{\sqrt{2}}$, or $\dfrac{\sqrt{2}}{2}$

Thus, $\theta = \dfrac{7\pi}{4}$, or $315°$, and we have

$3 - 3i = 3\sqrt{2}\left(\cos\dfrac{7\pi}{4} + i\sin\dfrac{7\pi}{4}\right)$, or

$3 - 3i = 3\sqrt{2}(\cos 315° + i\sin 315°)$.

10. Standard notation is -1.

Find trigonometric notation:

$r = \sqrt{(-1)^2} = 1$

$\sin\theta = \dfrac{0}{1} = 0$, $\cos\theta = \dfrac{-1}{1} = -1$. Thus, $\theta = \pi$, or $180°$, and we have

$-1 = 1(\cos\pi + i\sin\pi)$, or

$-1 = 1(\cos 180° + i\sin 180°)$.

11. From the graph we see that standard notation for the number is $0 + 4i$, or $4i$.

Find trigonometric notation:

$r = \sqrt{4^2} = 4$

$\sin\theta = \dfrac{4}{4} = 1$, $\cos\theta = \dfrac{0}{4} = 0$. Thus, $\theta = \dfrac{\pi}{2}$, or $90°$, and we have

$4i = 4\left(\cos\dfrac{\pi}{2} + i\sin\dfrac{\pi}{2}\right)$, or

$4i = 4(\cos 90° + i\sin 90°)$.

12. Standard notation: $-2 + 2i$

Find trigonometric notation:

$r = \sqrt{(-2)^2 + 2^2} = \sqrt{8} = 2\sqrt{2}$

$\sin\theta = \dfrac{2}{2\sqrt{2}} = \dfrac{1}{\sqrt{2}}$, or $\dfrac{\sqrt{2}}{2}$

$\cos\theta = \dfrac{-2}{2\sqrt{2}} = -\dfrac{1}{\sqrt{2}}$, or $-\dfrac{\sqrt{2}}{2}$.

Thus, $\theta = \dfrac{3\pi}{4}$, or $135°$, and we have

$-2 + 2i = 2\sqrt{2}\left(\cos\dfrac{3\pi}{4} + i\sin\dfrac{3\pi}{4}\right)$, or

$-2 + 2i = 2\sqrt{2}(\cos 135° + i\sin 135°)$.

13. Find trigonometric notation for $1 - i$.

$r = \sqrt{1^2 + (-1)^2} = \sqrt{2}$

$\sin\theta = \frac{-1}{\sqrt{2}}$, or $-\frac{\sqrt{2}}{2}$, $\cos\theta = \frac{1}{\sqrt{2}}$, or $\frac{\sqrt{2}}{2}$

Thus, $\theta = \frac{7\pi}{4}$, or 315°, and we have

$1 - i = \sqrt{2}\left(\cos\frac{7\pi}{4} + i\sin\frac{7\pi}{4}\right)$, or

$1 - i = \sqrt{2}(\cos 315° + i\sin 315°)$.

14. Find trigonometric notation for $-10\sqrt{3} + 10i$.

$r = \sqrt{(-10\sqrt{3})^2 + 10^2} = \sqrt{400} = 20$

$\sin\theta = \frac{10}{20} = \frac{1}{2}$, $\cos\theta = \frac{-10\sqrt{3}}{20} = -\frac{\sqrt{3}}{2}$

Thus, $\theta = \frac{5\pi}{6}$, or 150°, so we have

$-10\sqrt{3} + 10i = 20\left(\cos\frac{5\pi}{6} + i\sin\frac{5\pi}{6}\right)$, or

$-10\sqrt{3} + 10i = 20(\cos 150° + i\sin 150°)$.

15. Find trigonometric notation for $-3i$.

$r = \sqrt{(-3)^2} = 3$

$\sin\theta = \frac{-3}{3} = -1$, $\cos\theta = \frac{0}{3} = 0$

Thus, $\theta = \frac{3\pi}{2}$, or 270°, and we have

$-3i = 3\left(\cos\frac{3\pi}{2} + i\sin\frac{3\pi}{2}\right)$, or

$-3i = 3(\cos 270° + i\sin 270°)$.

16. Find trigonometric notation for $-5 + 5i$.

$r = \sqrt{(-5)^2 + 5^2} = \sqrt{50} = 5\sqrt{2}$

$\sin\theta = \frac{5}{5\sqrt{2}} = \frac{1}{\sqrt{2}}$, or $\frac{\sqrt{2}}{2}$

$\cos\theta = \frac{-5}{5\sqrt{2}} = -\frac{1}{\sqrt{2}}$, or $-\frac{\sqrt{2}}{2}$

Thus, $\theta = \frac{3\pi}{4}$, or 135°, and we have

$-5 + 5i = 5\sqrt{2}\left(\cos\frac{3\pi}{4} + i\sin\frac{3\pi}{4}\right)$, or

$-5 + 5i = 5\sqrt{2}(\cos 135° + i\sin 135°)$.

17. Find trigonometric notation for $\sqrt{3} + i$.

$r = \sqrt{(\sqrt{3})^2 + 1^2} = \sqrt{4} = 2$

$\sin\theta = \frac{1}{2}$, $\cos\theta = \frac{\sqrt{3}}{2}$. Thus, $\theta = \frac{\pi}{6}$, or 30°, and we have

$\sqrt{3} + i = 2\left(\cos\frac{\pi}{6} + i\sin\frac{\pi}{6}\right)$, or

$\sqrt{3} + i = 2(\cos 30° + i\sin 30°)$.

18. Find trigonometric notation for 4.

$r = \sqrt{4^2} = 4$

$\sin\theta = \frac{0}{4} = 0$, $\cos\theta = \frac{4}{4} = 1$. Thus, $0 = 0$, or 0°and we have

$4 = 4(\cos 0 + i\sin 0)$, or

$4 = 4(\cos 0° + i\sin 0°)$.

19. Find trigonometric notation for $\frac{2}{5}$.

$r = \sqrt{\left(\frac{2}{5}\right)^2} = \frac{2}{5}$

$\sin\theta = \frac{0}{\frac{2}{5}} = 0$, $\cos\theta = \frac{\frac{2}{5}}{\frac{2}{5}} = 1$. Thus, $\theta = 0$, or 0°, and we have

$r = \frac{2}{5}(\cos 0 + i\sin 0)$, or

$r = \frac{2}{5}(\cos 0° + i\sin 0°)$.

20. Find trigonometric notation for $7.5i$.

$r = \sqrt{(7.5)^2} = 7.5$

$\sin\theta = \frac{7.5}{7.5} = 1$, $\cos\theta = \frac{0}{7.5} = 0$. Thus, $\theta = \frac{\pi}{2}$, or 90°, and we have

$7.5i = 7.5\left(\cos\frac{\pi}{2} + i\sin\frac{\pi}{2}\right)$, or

$7.5i = 7.5(\cos 90° + i\sin 90°)$.

21. Find trigonometric notation for $-3\sqrt{2} - 3\sqrt{2}i$.

$r = \sqrt{(-3\sqrt{2})^2 + (-3\sqrt{2})^2} = \sqrt{36} = 6$

$\sin\theta = \frac{-3\sqrt{2}}{6} = -\frac{\sqrt{2}}{2}$, $\cos\theta = \frac{-3\sqrt{2}}{6} = -\frac{\sqrt{2}}{2}$. Thus $\theta = \frac{5\pi}{4}$, or 225°, and we have $-3\sqrt{2} - 3\sqrt{2}i = 6\left(\cos\frac{5\pi}{4} + i\sin\frac{5\pi}{4}\right)$, or $6(\cos 225° + i\sin 225°)$.

22. Find trigonometric notation for $-\frac{9}{2} - \frac{9\sqrt{3}}{2}i$.

$r = \sqrt{\left(-\frac{9}{2}\right)^2 + \left(-\frac{9\sqrt{3}}{2}\right)^2} = \sqrt{\frac{81}{4} + \frac{243}{4}} = \sqrt{\frac{324}{4}} = \sqrt{81} = 9$

$\sin\theta = \frac{-\frac{9\sqrt{3}}{2}}{9} = -\frac{\sqrt{3}}{2}$, $\cos\theta = \frac{-\frac{9}{2}}{9} = -\frac{1}{2}$. Thus $\theta = \frac{4\pi}{3}$, or 240°, and we have

$-\frac{9}{2} - \frac{9\sqrt{3}}{2}i = 9\left(\cos\frac{4\pi}{3} + i\sin\frac{4\pi}{3}\right)$, or $9(\cos 240° + i\sin 240°)$.

23. $3(\cos 30° + i\sin 30°) = 3\cos 30° + (3\sin 30°)i$

$a = 3\cos 30° = 3\cdot\frac{\sqrt{3}}{2} = \frac{3\sqrt{3}}{2}$

$b = 3\sin 30° = 3 \cdot \frac{1}{2} = \frac{3}{2}$

Thus $3(\cos 30° + i\sin 30°) = \frac{3\sqrt{3}}{2} + \frac{3}{2}i$.

24. $6(\cos 120° + i\sin 120°) = 6\cos 120° + (6\sin 120°)i$

$a = 6\cos 120° = 6\left(-\frac{1}{2}\right) = -3$

$b = 6\sin 120° = 6 \cdot \frac{\sqrt{3}}{2} = 3\sqrt{3}$

Thus $6(\cos 120° + i\sin 120°) = -3 + 3\sqrt{3}i$.

25. $10(\cos 270° + i\sin 270°) = 10\cos 270° + (10\sin 270°)i$

$a = 10\cos 270° = 10 \cdot 0 = 0$

$b = 10\sin 270° = 10(-1) = -10$

Thus $10(\cos 270° + i\sin 270°) = 0 + (-10)i = -10i$.

26. $3(\cos 0° + i\sin 0°) = 3\cos 0° + (3\sin 0°)i$

$a = 3\cos 0° = 3 \cdot 1 = 3$

$b = 3\sin 0° = 3 \cdot 0 = 0$

Thus $3(\cos 0° + i\sin 0°) = 3 + 0i = 3$.

27. $\sqrt{8}\left(\cos\frac{\pi}{4} + i\sin\frac{\pi}{4}\right) = \sqrt{8}\cos\frac{\pi}{4} + \left(\sqrt{8}\cos\frac{\pi}{4}\right)i$

$a = \sqrt{8}\cos\frac{\pi}{4} = \sqrt{8} \cdot \frac{\sqrt{2}}{2} = 2$

$b = \sqrt{8}\sin\frac{\pi}{4} = \sqrt{8} \cdot \frac{\sqrt{2}}{2} = 2$

Thus $\sqrt{8}\left(\cos\frac{\pi}{4} + i\sin\frac{\pi}{4}\right) = 2 + 2i$.

28. $5\left(\cos\frac{\pi}{3} + i\sin\frac{\pi}{3}\right) = 5\cos\frac{\pi}{3} + \left(5\sin\frac{\pi}{3}\right)i$

$a = 5\cos\frac{\pi}{3} = 5 \cdot \frac{1}{2} = \frac{5}{2}$

$b = 5\sin\frac{\pi}{3} = 5 \cdot \frac{\sqrt{3}}{2} = \frac{5\sqrt{3}}{2}$

Thus $5\left(\cos\frac{\pi}{3} + i\sin\frac{\pi}{3}\right) = \frac{5}{2} + \frac{5\sqrt{3}}{2}i$.

29. $2\left(\cos\frac{\pi}{2} + i\sin\frac{\pi}{2}\right) = 2\cos\frac{\pi}{2} + \left(2\sin\frac{\pi}{2}\right)i$

$a = 2\cos\frac{\pi}{2} = 2 \cdot 0 = 0$

$b = 2\sin\frac{\pi}{2} = 2 \cdot 1 = 2$

Thus $2\left(\cos\frac{\pi}{2} + i\sin\frac{\pi}{2}\right) = 0 + 2i = 2i$.

30. $3\left[\cos\left(-\frac{3\pi}{4}\right) + i\sin\left(-\frac{3\pi}{4}\right)\right] =$

$3\cos\left(-\frac{3\pi}{4}\right) + \left[\sin\left(-\frac{3\pi}{4}\right)\right]i$

$a = 3\cos\left(-\frac{3\pi}{4}\right) = 3\left(-\frac{\sqrt{2}}{2}\right) = -\frac{3\sqrt{2}}{2}$

$b = 3\sin\left(-\frac{3\pi}{4}\right) = 3\left(-\frac{\sqrt{2}}{2}\right) = -\frac{3\sqrt{2}}{2}$

Thus $3\left[\cos\left(-\frac{3\pi}{4}\right) + i\sin\left(-\frac{3\pi}{4}\right)\right] =$

$-\frac{3\sqrt{2}}{2} - \frac{3\sqrt{2}}{2}i$.

31. $\sqrt{2}[\cos(-60°) + i\ \sin(-60°)] =$

$\sqrt{2}\cos(-60°) + [\sqrt{2}\sin(-60°)]i$

$a = \sqrt{2}\cos(-60°) = \sqrt{2} \cdot \frac{1}{2} = \frac{\sqrt{2}}{2}$

$b = \sqrt{2}\sin(-60°) = \sqrt{2}\left(-\frac{\sqrt{3}}{2}\right) = -\frac{\sqrt{6}}{2}$

Thus $\sqrt{2}[\cos(-60°) + i\ \sin(-60°)] = \frac{\sqrt{2}}{2} - \frac{\sqrt{6}}{2}i$.

32. $4(\cos\ 135° + i\ \sin\ 135°) = 4\cos\ 135° + (4\sin\ 135°)i$

$a = 4\cos\ 135° = 4\left(-\frac{\sqrt{2}}{2}\right) = -2\sqrt{2}$

$a = 4\sin\ 135° = 4 \cdot \frac{\sqrt{2}}{2} = 2\sqrt{2}$

Thus $4(\cos\ 135° + i\ \sin\ 135°) = -2\sqrt{2} + 2\sqrt{2}i$.

33. $\dfrac{12(\cos 48° + i\sin 48°)}{3(\cos 6° + i\sin 6°)}$

$= \frac{12}{3}[\cos(48° - 6°) + i\sin(48° - 6°)]$

$= 4(\cos 42° + i\sin 42°)$

34. $5\left(\cos\frac{\pi}{3} + i\sin\frac{\pi}{3}\right) \cdot 2\left(\cos\frac{\pi}{4} + i\sin\frac{\pi}{4}\right)$

$= 5 \cdot 2\left[\cos\left(\frac{\pi}{3} + \frac{\pi}{4}\right) + i\sin\left(\frac{\pi}{3} + \frac{\pi}{4}\right)\right]$

$= 10\left(\cos\frac{7\pi}{12} + i\sin\frac{7\pi}{12}\right)$

35. $2.5(\cos 35° + i\sin 35°) \cdot 4.5(\cos 21° + i\sin 21°)$

$= 2.5(4.5)[\cos(35° + 21°) + i\sin(35° + 21°)]$

$= 11.25(\cos 56° + i\sin 56°)$

36. $\dfrac{\frac{1}{2}\left(\cos\frac{2\pi}{3} + i\sin\frac{2\pi}{3}\right)}{\frac{3}{8}\left(\cos\frac{\pi}{6} + i\sin\frac{\pi}{6}\right)}$

$= \dfrac{\frac{1}{2}}{\frac{3}{8}}\left[\cos\left(\frac{2\pi}{3} - \frac{\pi}{6}\right) + i\sin\left(\frac{2\pi}{3} - \frac{\pi}{6}\right)\right]$

$= \frac{4}{3}\left(\cos\frac{\pi}{2} + i\sin\frac{\pi}{2}\right)$

37. $(1 - i)(2 + 2i)$

Find trigonometric notation for $1 - i$.

$r = \sqrt{1^2 + (-1)^2} = \sqrt{2}$

$\sin\theta = \frac{-1}{\sqrt{2}} = -\frac{\sqrt{2}}{2}$, $\cos\theta = \frac{1}{\sqrt{2}} = \frac{\sqrt{2}}{2}$

Thus $\theta = \frac{7\pi}{4}$, or 315°, and $1 - i = \sqrt{2}(\cos 315° + i\sin 315°)$.

Find trigonometric notation for $2+2i$.

$r = \sqrt{2^2+2^2} = \sqrt{8} = 2\sqrt{2}$

$\sin\theta = \frac{2}{2\sqrt{2}} = \frac{\sqrt{2}}{2}$, $\cos\theta = \frac{2}{2\sqrt{2}} = \frac{\sqrt{2}}{2}$

Thus $\theta = \frac{\pi}{4}$, or $45°$, and $2+2i = 2\sqrt{2}(\cos 45° + i\sin 45°)$.

$(1-i)(2+2i)$

$= \sqrt{2}(\cos 315° + i\sin 315°)\cdot 2\sqrt{2}(\cos 45° + \sin 45°)$

$= \sqrt{2}\cdot 2\sqrt{2}[\cos(315°+45°) + i\sin(315°+45°)]$

$= 4(\cos 360° + i\sin 360°)$

$= 4(\cos 0° + i\sin 0°)$, or 4

38. For $1+i\sqrt{3}$: $r = \sqrt{1^2+(\sqrt{3})^2} = 2$

$\sin\theta = \frac{\sqrt{3}}{2}$ and $\cos\theta = \frac{1}{2}$, so $\theta = 60°$.

$1+i\sqrt{3} = 2(\cos 60° + i\sin 60°)$

From Example 3(a) we know that $1+i = \sqrt{2}(\cos 45° + i\sin 45°)$.

$1+i = \sqrt{2}(\cos 45° + i\sin 45°)$

$2(\cos 60° + i\sin 60°)\cdot\sqrt{2}(\cos 45° + i\sin 45°) = 2\sqrt{2}(\cos 105° + i\sin 105°)$

Using identities for the sum and difference of angles we find that $\cos 105° = \frac{\sqrt{2}-\sqrt{6}}{4}$ and $\sin 105° = \frac{\sqrt{6}+\sqrt{2}}{4}$. Thus,

$2\sqrt{2}(\cos 105° + i\sin 105°)$

$= 2\sqrt{2}\left(\frac{\sqrt{2}-\sqrt{6}}{4} + \frac{\sqrt{6}+\sqrt{2}}{4}i\right)$

$= 1-\sqrt{3}+(\sqrt{3}+1)i$

39. $\frac{1-i}{1+i}$

Find trigonometric notation for $1-i$.

$r = \sqrt{1^2+(-1)^2} = \sqrt{2}$

$\sin\theta = \frac{-1}{\sqrt{2}} = -\frac{\sqrt{2}}{2}$, $\cos\theta = \frac{1}{\sqrt{2}} = \frac{\sqrt{2}}{2}$

Thus $\theta = \frac{7\pi}{4}$, or $315°$, and $1-i = \sqrt{2}(\cos 315° + i\sin 315°)$

From Example 3(a) we know that $1+i = \sqrt{2}(\cos 45° + i\sin 45°)$.

$\frac{1-i}{1+i}$

$= \frac{\sqrt{2}(\cos 315° + i\sin 315°)}{\sqrt{2}(\cos 45° + i\sin 45°)}$

$= \frac{\sqrt{2}}{\sqrt{2}}[\cos(315°-45°) + i\sin(315°-45°)]$

$= 1(\cos 270° + i\sin 270°)$

$= 1[0+i(-1)]$

$= -i$

40. From Exercise 35 we know that $1-i = \sqrt{2}(\cos 315° + i\sin 315°)$.

From Example 3(b) we know that $\sqrt{3}-i = 2(\cos 330° + i\sin 330°)$.

$\frac{1-i}{\sqrt{3}-i}$

$= \frac{\sqrt{2}(\cos 315° + i\sin 315°)}{2(\cos 330° + i\sin 330°)}$

$= \frac{\sqrt{2}}{2}[\cos(-15°) + i\sin(-15°)]$

Using identities for the sum and difference of angles we find that $\cos(-15°) = \frac{\sqrt{2}+\sqrt{6}}{4}$ and $\sin(-15°) = \frac{\sqrt{2}-\sqrt{6}}{4}$. Thus,

$\frac{\sqrt{2}}{2}[\cos(-15°) + i\sin(-15°)]$

$= \frac{\sqrt{2}}{2}\left(\frac{\sqrt{2}+\sqrt{6}}{4} + \frac{\sqrt{2}-\sqrt{6}}{4}i\right)$

$= \frac{1+\sqrt{3}}{4} + \frac{1-\sqrt{3}}{4}i$

41. $(3\sqrt{3}-3i)(2i)$

Find trigonometric notation for $3\sqrt{3}-3i$.

$r = \sqrt{(3\sqrt{3})^2+(-3)^2} = \sqrt{36} = 6$

$\sin\theta = \frac{-3}{6} = -\frac{1}{2}$, $\cos\theta = \frac{3\sqrt{3}}{6} = \frac{\sqrt{3}}{2}$

Thus $\theta = 330°$, and $3\sqrt{3}-3i = 6(\cos 330° + i\sin 330°)$.

Find trigonometric notation for $2i$.

$r = \sqrt{2^2} = 2$

$\sin\theta = \frac{2}{2} = 1$, $\cos\theta = \frac{0}{2} = 0$

Thus $\theta = 90°$, and $2i = 2(\cos 90° + i\sin 90°)$.

$(3\sqrt{3}-3i)(2i)$

$= 6(\cos 330° + i\sin 330°)\cdot 2(\cos 90° + i\sin 90°)$

$= 6\cdot 2[\cos(330°+90°) + i\sin(330°+90°)]$

$= 12(\cos 420° + i\sin 420°)$

$= 12(\cos 60° + i\sin 60°)$

$= 12\left(\frac{1}{2} + i\cdot\frac{\sqrt{3}}{2}\right)$

$= 6+6\sqrt{3}i$

42. For $2\sqrt{3}+2i$: $r = \sqrt{(2\sqrt{3})^2+2^2} = 4$

$\sin\theta = \frac{2}{4} = \frac{1}{2}$ and $\cos\theta = \frac{2\sqrt{3}}{4} = \frac{\sqrt{3}}{2}$, so $\theta = 30°$.

$2\sqrt{3}+2i = 4(\cos 30° + i\sin 30°)$

From Exercise 41 we know that $2i = 2(\cos 90° + i\sin 90°)$.

$$(2\sqrt{3}+2i)(2i)$$
$$= 4(\cos 30° + i\sin 30°)\cdot 2(\cos 90° + i\sin 90°)$$
$$= 8(\cos 120° + i\sin 120°)$$
$$= 8\left(-\frac{1}{2} + i\cdot\frac{\sqrt{3}}{2}\right)$$
$$= -4 + 4\sqrt{3}i$$

43. $\dfrac{2\sqrt{3}-2i}{1+\sqrt{3}i}$

Find trigonometric notation for $2\sqrt{3}-2i$.

$r = \sqrt{(2\sqrt{3})^2 + (-2)^2} = \sqrt{16} = 4$

$\sin\theta = \dfrac{-2}{4} = -\dfrac{1}{2}$, $\cos\theta = \dfrac{2\sqrt{3}}{4} = \dfrac{\sqrt{3}}{2}$

Thus $\theta = 330°$, and $2\sqrt{3}-2i = 4(\cos 330° + i\sin 330°)$

Find trigonometric notation for $1+\sqrt{3}i$.

$r = \sqrt{1^2 + (\sqrt{3})^2} = \sqrt{4} = 2$

$\sin\theta = \dfrac{\sqrt{3}}{2}$, $\cos\theta = \dfrac{1}{2}$

Thus $\theta = 60°$, and $1+\sqrt{3}i = 2(\cos 60° + i\sin 60°)$

$$\frac{2\sqrt{3}-2i}{1+\sqrt{3}i}$$
$$= \frac{4(\cos 330° + i\sin 330°)}{2(\cos 60° + i\sin 60°)}$$
$$= \frac{4}{2}[\cos(330° - 60°) + i\sin(330° - 60°)]$$
$$= 2(\cos 270° + i\sin 270°)$$
$$= 2[0 + i\cdot(-1)]$$
$$= -2i$$

44. For $3 - 3\sqrt{3}i$: $r = \sqrt{3^2 + (-3\sqrt{3})^2} = 6$

$\sin\theta = \dfrac{-3\sqrt{3}}{6} = -\dfrac{\sqrt{3}}{2}$ and $\cos\theta = \dfrac{3}{6} = \dfrac{1}{2}$, so $\theta = 300°$.

$3 - 3\sqrt{3}i = 6(\cos 300° + i\sin 300°)$

From Example 3(b) we know that

$\sqrt{3} - i = 2(\cos 330° + i\sin 330°)$.

$$\frac{3-3\sqrt{3}i}{\sqrt{3}-i}$$
$$= \frac{6(\cos 300° + i\sin 300°)}{2(\cos 330° + i\sin 330°)}$$
$$= 3[\cos(-30°) + i\sin(-30°)]$$
$$= 3\left[\frac{\sqrt{3}}{2} + i\cdot\left(-\frac{1}{2}\right)\right]$$
$$= \frac{3\sqrt{3}}{2} - \frac{3}{2}i$$

45. $\left[2\left(\cos\dfrac{\pi}{3} + i\sin\dfrac{\pi}{3}\right)\right]^3$

$$= 2^3\left[\cos\left(3\cdot\frac{\pi}{3}\right) + i\sin\left(3\cdot\frac{\pi}{3}\right)\right]$$
$$= 8(\cos\pi + i\sin\pi)$$

46. $[2(\cos 120° + i\sin 120°)]^4$

$$= 2^4[\cos(4\cdot 120°) + i\sin(4\cdot 120°)]$$
$$= 16(\cos 480° + i\sin 480°)$$
$$= 16(\cos 120° + i\sin 120°)$$

47. From Exercise 39 we know that $1+i = \sqrt{2}(\cos 45° + i\sin 45°)$.

$$(1+i)^6 = [\sqrt{2}(\cos 45° + i\sin 45°)]^6$$
$$= (\sqrt{2})^6[\cos(6\cdot 45°) + i\sin(6\cdot 45°)]$$
$$= 8(\cos 270° + i\sin 270°),\ \text{or}$$
$$8\left(\cos\frac{3\pi}{2} + i\sin\frac{3\pi}{2}\right)$$

48. $(-\sqrt{3}+i)^5 = [2(\cos 150° + i\sin 150°)]^5$

$$= 32(\cos 750° + i\sin 750°)$$
$$= 32(\cos 30° + i\sin 30°),\ \text{or}$$
$$32\left(\cos\frac{\pi}{6} + i\sin\frac{\pi}{6}\right)$$

49. $[3(\cos 20° + i\sin 20°)]^3$

$$= 27(\cos 60° + i\sin 60°)$$
$$= 27\left(\frac{1}{2} + i\cdot\frac{\sqrt{3}}{2}\right)$$
$$= \frac{27}{2} + \frac{27\sqrt{3}}{2}i$$

50. $[2(\cos 10° + i\sin 10°)]^9$

$$= 512(\cos 90° + i\sin 90°)$$
$$= 512(0 + i\cdot 1)$$
$$= 512i$$

51. From Exercise 13 we know that $1-i = \sqrt{2}(\cos 315° + i\sin 315°)$.

$$(1-i)^5 = [\sqrt{2}(\cos 315° + i\sin 315°)]^5$$
$$= (\sqrt{2})^5(\cos 1575° + i\sin 1575°)$$
$$= 2^{5/2}(\cos 135° + i\sin 135°)$$
$$= 4\sqrt{2}\left(-\frac{\sqrt{2}}{2} + i\cdot\frac{\sqrt{2}}{2}\right)$$
$$= -4 + 4i$$

52. $(2+2i)^4 = [2\sqrt{2}(\cos 45° + i\sin 45°)]^4$

$$= 64(\cos 180° + i\sin 180°)$$
$$= 64(-1 + i\cdot 0)$$
$$= -64$$

53. Find trigonometric notation for $\frac{1}{\sqrt{2}} - \frac{1}{\sqrt{2}}i$:

$$r = \sqrt{\left(\frac{1}{\sqrt{2}}\right)^2 + \left(-\frac{1}{\sqrt{2}}\right)^2} = \sqrt{1} = 1$$

$$\sin\theta = \frac{-\frac{1}{\sqrt{2}}}{1} = -\frac{1}{\sqrt{2}}, \text{ or } -\frac{\sqrt{2}}{2}$$

$$\cos\theta = \frac{\frac{1}{\sqrt{2}}}{1} = \frac{1}{\sqrt{2}}, \text{ or } \frac{\sqrt{2}}{2}$$

Thus $\theta = 315°$, so $\frac{1}{\sqrt{2}} - \frac{1}{\sqrt{2}}i = 1(\cos 315° + i\sin 315°)$.

$$\begin{aligned}\left(\frac{1}{\sqrt{2}} - \frac{1}{\sqrt{2}}i\right)^{12} &= [1(\cos 315° + i\sin 315°)]^{12} \\ &= 1(\cos 3780° + i\sin 3780°) \\ &= 1(\cos 180° + i\sin 180°) \\ &= 1(-1 + i\cdot 0) \\ &= -1\end{aligned}$$

54.
$$\begin{aligned}\left(\frac{\sqrt{3}}{2} + \frac{1}{2}i\right)^{10} &= [1(\cos 30° + i\sin 30°)]^{10} \\ &= 1(\cos 300° + i\sin 300°) \\ &= 1\left[\frac{1}{2} + i\cdot\left(-\frac{\sqrt{3}}{2}\right)\right] \\ &= \frac{1}{2} - \frac{\sqrt{3}}{2}i\end{aligned}$$

55. $-i = 1(\cos 270° + i\sin 270°)$

$$\begin{aligned}&(-i)^{1/2} \\ &= [1(\cos 270° + i\sin 270)]^{1/2} \\ &= 1^{1/2}\left[\cos\left(\frac{270°}{2} + k\cdot\frac{360°}{2}\right) + i\sin\left(\frac{270°}{2} + k\cdot\frac{360°}{2}\right)\right], k = 0, 1 \\ &= 1[\cos(135° + k\cdot 180°) + i\sin(135° + k\cdot 180°)], \quad k = 0, 1\end{aligned}$$

The roots are

$1(\cos 135° + i\sin 135°)$, for $k = 0$

and

$1(\cos 315° + i\sin 315°)$, for $k = 1$,

or $-\frac{\sqrt{2}}{2} + \frac{\sqrt{2}}{2}i$ and $\frac{\sqrt{2}}{2} - \frac{\sqrt{2}}{2}i$.

56.
$$\begin{aligned}&(1+i)^{1/2} \\ &= [\sqrt{2}(\cos 45° + i\sin 45°)]^{1/2} \\ &= (2^{1/2})^{1/2}\left[\cos\left(\frac{45°}{2} + k\cdot\frac{360°}{2}\right) + i\sin\left(\frac{45°}{2} + k\cdot\frac{360°}{2}\right)\right], k = 0, 1\end{aligned}$$

The roots are

$\sqrt[4]{2}(\cos 22.5° + i\sin 22.5°)$

and

$\sqrt[4]{2}(\cos 202.5° + i\sin 202.5°)$.

57. $2\sqrt{2} - 2\sqrt{2}i = 4(\cos 315° + i\sin 315°)$

$$\begin{aligned}&(2\sqrt{2} - 2\sqrt{2}i)^{1/2} \\ &= [4(\cos 315° + i\sin 315°)]^{1/2} \\ &= 4^{1/2}\left[\cos\left(\frac{315°}{2} + k\cdot\frac{360°}{2}\right) + i\sin\left(\frac{315°}{2} + k\cdot\frac{360°}{2}\right)\right], k = 0, 1\end{aligned}$$

The roots are

$2(\cos 157.5° + i\sin 157.5°)$, for $k = 0$

and

$2(\cos 337.5° + i\sin 337.5°)$, for $k = 1$.

58.
$$\begin{aligned}&(-\sqrt{3} - i)^{1/2} \\ &= [2(\cos 210° + i\sin 210°)]^{1/2} \\ &= 2^{1/2}\left[\cos\left(\frac{210°}{2} + k\cdot\frac{360°}{2}\right) + i\sin\left(\frac{210°}{2} + k\cdot\frac{360°}{2}\right)\right], k = 0, 1\end{aligned}$$

The roots are

$\sqrt{2}(\cos 105° + i\sin 105°)$

and

$\sqrt{2}(\cos 285° + i\sin 285°)$.

59. $i = 1(\cos 90° + i\sin 90°)$

$$\begin{aligned}&i^{1/3} \\ &= [1(\cos 90° + i\sin 90°)]^{1/3} \\ &= 1\left[\cos\left(\frac{90°}{3} + k\cdot\frac{360°}{3}\right) + i\sin\left(\frac{90°}{3} + k\cdot\frac{360°}{3}\right)\right], \\ &k = 0, 1, 2\end{aligned}$$

The roots are $1(\cos 30° + i\sin 30°)$, $1(\cos 150° + i\sin 150°)$, and $1(\cos 270° + i\sin 270°)$, or $\frac{\sqrt{3}}{2} + \frac{1}{2}i$, $-\frac{\sqrt{3}}{2} + \frac{1}{2}i$, and $-i$.

60.
$$\begin{aligned}&(-64i)^{1/3} \\ &= [64(\cos 270° + i\sin 270°)]^{1/3} \\ &= 4\left[\cos\left(\frac{270°}{3} + k\cdot\frac{360°}{3}\right) + i\sin\left(\frac{270°}{3} + k\cdot\frac{360°}{3}\right)\right]^{1/3}, k = 0, 1, 2\end{aligned}$$

The roots are $4(\cos 90° + i\sin 90°)$, $4(\cos 210° + i\sin 210°)$, and $4(\cos 330° + i\sin 330°)$, or $4i$, $-2\sqrt{3} - 2i$, and $2\sqrt{3} - 2i$.

61.
$$\begin{aligned}&(2\sqrt{3} - 2i)^{1/3} \\ &= 4(\cos 330° + i\sin 330°)^{1/3} \\ &= 4^{1/3}\left[\cos\left(\frac{330°}{3} + k\cdot\frac{360°}{3}\right) + i\sin\left(\frac{330°}{3} + k\cdot\frac{360°}{3}\right)\right], k = 0, 1, 2\end{aligned}$$

The roots are $\sqrt[3]{4}(\cos 110° + i\sin 110°)$, $\sqrt[3]{4}(\cos 230° + i\sin 230°)$, and $\sqrt[3]{4}(\cos 350° + i\sin 350°)$.

62. $(1-\sqrt{3}i)^{1/3}$

$$= [2(\cos 300° + i\sin 300°)]^{1/3}$$

$$= 2^{1/3}\left[\cos\left(\frac{300°}{3} + k\cdot\frac{360°}{3}\right) + i\sin\left(\frac{300°}{3} + k\cdot\frac{360°}{3}\right)\right], k = 0, 1, 2$$

The roots are $\sqrt[3]{2}(\cos 100° + i\sin 100°)$, $\sqrt[3]{2}(\cos 220° + i\sin 220°)$, and $\sqrt[3]{2}(\cos 340° + i\sin 340°)$.

63. $16^{1/4}$

$$= [16(\cos 0° + i\sin 0°)]^{1/4}$$

$$= 2\left[\cos\left(\frac{0°}{4} + k\cdot\frac{360°}{4}\right) + i\sin\left(0° + k\cdot\frac{360°}{4}\right)\right],$$

$k = 0, 1, 2, 3$

The roots are $2(\cos 0° + i\sin 0°)$, $2(\cos 90° + i\sin 90°)$, $2(\cos 180° + i\sin 180°)$, and $2(\cos 270° + i\sin 270°)$, or 2, $2i$, -2, and $-2i$.

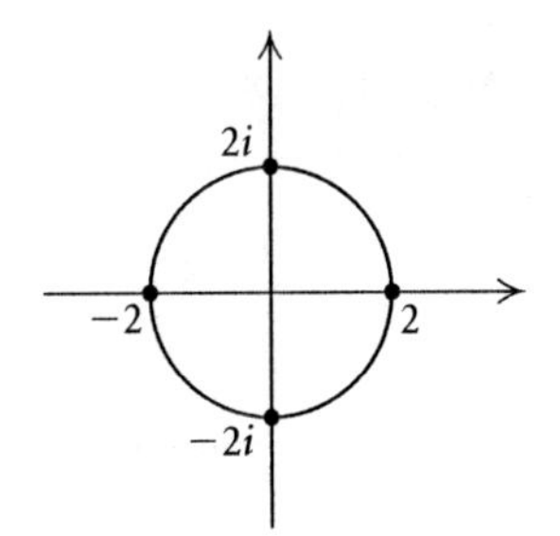

64. $i^{1/4}$

$$= [1(\cos 90° + i\sin 90°)]^{1/4}$$

$$= 1\left[\cos\left(\frac{90°}{4} + k\cdot\frac{360°}{4}\right) + i\sin\left(\frac{90°}{4} + k\cdot\frac{360°}{4}\right)\right],$$

$k = 0, 1, 2, 3$

The roots are $\cos 22.5° + i\sin 22.5°$, $\cos 112.5° + i\sin 112.5°$, $\cos 202.5° + i\sin 202.5°$, and $\cos 292.5° + i\sin 292.5°$.

65. $(-1)^{1/5}$

$$= [1(\cos 180° + i\sin 180°)]^{1/5}$$

$$= 1\left[\cos\left(\frac{180°}{5} + k\cdot\frac{360°}{5}\right) + i\sin\left(\frac{180°}{5} + k\cdot\frac{360°}{5}\right)\right],$$

$k = 0, 1, 2, 3, 4$

The roots are $\cos 36° + i\sin 36°$; $\cos 108° + i\sin 108°$; $\cos 180° + i\sin 180°$, or -1; $\cos 252° + i\sin 252°$; and $\cos 324° + i\sin 324°$.

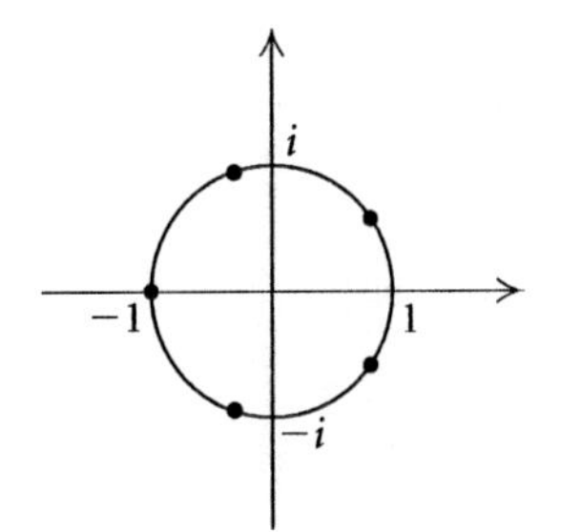

66. $1^{1/6}$

$$= [1(\cos 0° + i\sin 0°)]^{1/6}$$

$$= 1\left[\cos\left(\frac{0°}{6} + k\cdot\frac{360°}{6}\right) + i\sin\left(\frac{0°}{6} + k\cdot\frac{360°}{6}\right)\right],$$

$k = 0, 1, 2, 3, 4, 5$

The roots are $\cos 0° + i\sin 0°$, $\cos 60° + i\sin 60°$, $\cos 120° + i\sin 120°$, $\cos 180° + i\sin 180°$, $\cos 240° + i\sin 240°$, and $\cos 300° + i\sin 300°$, or 1, $\frac{1}{2} + \frac{\sqrt{3}}{2}i$, $-\frac{1}{2} + \frac{\sqrt{3}}{2}i$, -1, $-\frac{1}{2} - \frac{\sqrt{3}}{2}i$, and $\frac{1}{2} - \frac{\sqrt{3}}{2}i$.

67. $8^{1/10}$

$$= [8(\cos 0° + i\sin 0°)]^{1/10}$$

$$= \left[8^{1/10}\left(\cos\frac{0°}{10} + k\cdot\frac{360°}{10}\right) + i\sin\left(\frac{0°}{10} + k\cdot\frac{360°}{10}\right)\right],$$

$k = 0, 1, 2, 3, 4, 5, 6, 7, 8, 9$

The roots are $\sqrt[10]{8}(\cos 0° + i\sin 0°)$, or $\sqrt[10]{8}$; $\sqrt[10]{8}(\cos 36° + i\sin 36°)$; $\sqrt[10]{8}(\cos 72° + i\sin 72°)$, $\sqrt[10]{8}(\cos 108° + i\sin 108°)$; $\sqrt[10]{8}(\cos 144° + i\sin 144°)$; $\sqrt[10]{8}(\cos 180° + i\sin 180°)$, or $-\sqrt[10]{8}$; $\sqrt[10]{8}(\cos 216° + i\sin 216°)$; $\sqrt[10]{8}(\cos 252° + i\sin 252°)$; $\sqrt[10]{8}(\cos 288° + i\sin 288°)$; and $\sqrt[10]{8}(\cos 324° + i\sin 324°)$.

68. $(-4)^{1/9}$

$$= [4(\cos 180° + i\sin 180°)]^{1/9}$$

$$= 4^{1/9}\left[\cos\left(\frac{180°}{9} + k\cdot\frac{360°}{9}\right) + i\sin\left(\frac{180°}{9} + k\cdot\frac{360°}{9}\right)\right],$$

$k = 0, 1, 2, 3, 4, 5, 6, 7, 8$

The roots are $\sqrt[9]{4}(\cos 20° + i \sin 20°)$;

$\sqrt[9]{4}(\cos 60° + i \sin 60°)$, or $\dfrac{\sqrt[9]{4}}{2} + \dfrac{\sqrt[9]{4}\sqrt{3}}{2}i$;

$\sqrt[9]{4}(\cos 100° + i \sin 100°)$; $\sqrt[9]{4}(\cos 140° + i \sin 140°)$;

$\sqrt[9]{4}(\cos 180° + i \sin 180°)$, or $-\sqrt[9]{4}$;

$\sqrt[9]{4}(\cos 220° + i \sin 220°)$; $\sqrt[9]{4}(\cos 260° + i \sin 260°)$;

$\sqrt[9]{4}(\cos 300° + i \sin 300°)$, or $\dfrac{\sqrt[9]{4}}{2} - \dfrac{\sqrt[9]{4}\sqrt{3}}{2}i$;

and $\sqrt[9]{4}(\cos 340° + i \sin 340°)$.

69. $(-1)^{1/6}$

$= [1(\cos 180° + i \sin 180°)]^{1/6}$

$$= 1\left[\cos\left(\frac{180°}{6} + k \cdot \frac{360°}{6}\right) + i \sin\left(\frac{180°}{6} + k \cdot \frac{360°}{6}\right)\right], k = 0, 1, 2, 3, 4, 5$$

The roots are $\cos 30° + i \sin 30°$,

$\cos 90° + i \sin 90°$, $\cos 150° + i \sin 150°$,

$\cos 210° + i \sin 210°$, $\cos 270° + i \sin 270°$,

and $\cos 330° + i \sin 330°$, or $\dfrac{\sqrt{3}}{2} + \dfrac{1}{2}i$,

i, $-\dfrac{\sqrt{3}}{2} + \dfrac{1}{2}i$, $-\dfrac{\sqrt{3}}{2} - \dfrac{1}{2}i$, $-i$, and $\dfrac{\sqrt{3}}{2} - \dfrac{1}{2}i$.

70. $12^{1/4}$

$= [12(\cos 0° + i \sin 0°)]^{1/4}$

$$= 12^{1/4}\left[\cos\left(\frac{0°}{4} + k \cdot \frac{360°}{4}\right) + i \sin\left(\frac{0°}{4} + k \cdot \frac{360°}{4}\right)\right],$$

$k = 0, 1, 2, 3$

The roots are $\sqrt[4]{12}(\cos 0° + i \sin 0°)$,

$\sqrt[4]{12}(\cos 90° + i \sin 90°)$, $\sqrt[4]{12}(\cos 180° + i \sin 180°)$,

and $\sqrt[4]{12}(\cos 270° + i \sin 270°)$, or $\sqrt[4]{12}$, $\sqrt[4]{12}i$,

$-\sqrt[4]{12}$, and $-\sqrt[4]{12}i$.

71. $x^3 = 1$

The solutions of this equation are the cube roots of 1. These were found in Example 11 in the text. They are 1, $-\dfrac{1}{2} + \dfrac{\sqrt{3}}{2}i$, and $-\dfrac{1}{2} - \dfrac{\sqrt{3}}{2}i$.

72. $x^5 - 1 = 0$

$x^5 = 1$

Find the fifth roots of 1.

$1^{1/5}$

$= [1(\cos 0° + i \sin 0°)]^{1/5}$

$$= 1\left[\cos\left(\frac{0°}{5} + k \cdot \frac{360°}{5}\right) + i \sin\left(\frac{0°}{5} + k \cdot \frac{360°}{5}\right)\right],$$

$k = 0, 1, 2, 3, 4$

The solutions are $\cos 0° + i \sin 0°$, or 1;

$\cos 72° + i \sin 72°$; $\cos 144° + i \sin 144°$;

$\cos 216° + i \sin 216°$; and $\cos 288° + i \sin 288°$.

73. $x^4 + i = 0$

$x^4 = -i$

Find the fourth roots of $-i$.

$(-i)^{1/4}$

$= [1(\cos 270° + i \sin 270°)]^{1/4}$

$$= 1\left[\cos\left(\frac{270°}{4} + k \cdot \frac{360°}{4}\right) + i \sin\left(\frac{270°}{4} + k \cdot \frac{360°}{4}\right)\right], k = 0, 1, 2, 3$$

The solutions are $\cos 67.5° + i \sin 67.5°$, $\cos 157.5° + i \sin 157.5°$, $\cos 247.5° + i \sin 247.5°$, and $\cos 337.5° + i \sin 337.5°$.

74. $x^4 + 81 = 0$

$x^4 = -81$

Find the fourth roots of -81.

$(-81)^{1/4}$

$= [81(\cos 180° + i \sin 180°)]^{1/4}$

$$= 3\left[\cos\left(\frac{180°}{4} + k \cdot \frac{360°}{4}\right) + i \sin\left(\frac{180°}{4} + k \cdot \frac{360°}{4}\right)\right], k = 0, 1, 2, 3$$

The solutions are $3(\cos 45° + i \sin 45°)$, $3(\cos 135° + i \sin 135°)$, $3(\cos 225° + i \sin 225°)$,

and $3(\cos 315° + i \sin 315°)$, or $\dfrac{3\sqrt{2}}{2} + \dfrac{3\sqrt{2}}{2}i$,

$-\dfrac{3\sqrt{2}}{2} + \dfrac{3\sqrt{2}}{2}i$, $-\dfrac{3\sqrt{2}}{2} - \dfrac{3\sqrt{2}}{2}i$, and $\dfrac{3\sqrt{2}}{2} - \dfrac{3\sqrt{2}}{2}i$.

75. $x^6 + 64 = 0$

$x^6 = -64$

Find the sixth roots of -64.

$(-64)^{1/6}$

$= [64(\cos 180° + i \sin 180°)]^{1/6}$

$$= 2\left[\cos\left(\frac{180°}{6} + k \cdot \frac{360°}{6}\right) + i \sin\left(\frac{180°}{6} + k \cdot \frac{360°}{6}\right)\right], k = 0, 1, 2, 3, 4, 5$$

The solutions are $2(\cos 30° + i \sin 30°)$, $2(\cos 90° + i \sin 90°)$, $2(\cos 150° + i \sin 150°)$, $2(\cos 210° + i \sin 210°)$, $2(\cos 270° + i \sin 270°)$, and $2(\cos 330° + i \sin 330°)$, or $\sqrt{3} + i$, $2i$, $-\sqrt{3} + i$, $-\sqrt{3} - i$, $-2i$, and $\sqrt{3} - i$.

76. $x^5 + \sqrt{3} + i = 0$

$x^5 = -\sqrt{3} - i$

Find the fifth roots of $-\sqrt{3} - i$.

$(-\sqrt{3} - i)^{1/5}$

$= [2(\cos 210° + i \sin 210°)]^{1/5}$

$$= \sqrt[5]{2}\left[\cos\left(\frac{210°}{5} + k \cdot \frac{360°}{5}\right) + i \sin\left(\frac{210°}{5} + k \cdot \frac{360°}{5}\right)\right], k = 0, 1, 2, 3, 4$$

The solutions are $\sqrt[5]{2}(\cos 42° + i\sin 42°)$;
$\sqrt[5]{2}(\cos 114° + i\sin 114°)$; $\sqrt[5]{2}(\cos 186° + i\sin 186°)$;
$\sqrt[5]{2}(\cos 258° + i\sin 258°)$; and
$\sqrt[5]{2}(\cos 330° + i\sin 330°)$, or $\dfrac{\sqrt[5]{2}\sqrt{3}}{2} - \dfrac{\sqrt[5]{2}}{2}i$.

77. Left to the student

78. Left to the student

79. The square roots of $1 - i$ are
$\sqrt[4]{2}(\cos 157.5° + i\sin 157.5°)$ and
$\sqrt[4]{2}(\cos 337.5° + i\sin 337.5°)$, or approximately
$1.0987 + 0.4551i$ and $-1.0987 - 0.4551i$.

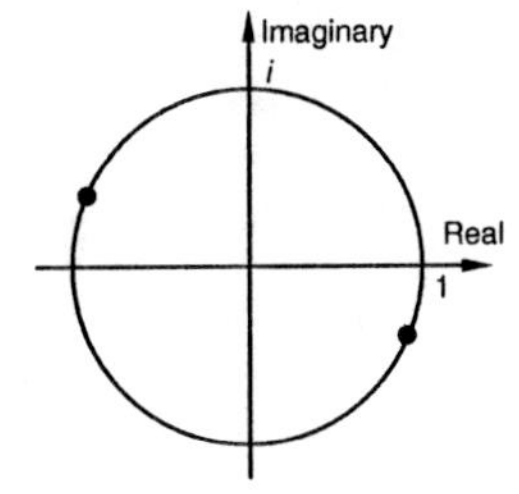

The roots are opposites because they are reflections of each other across the origin.

80. Trigonometric notation is not unique because there are infinitely many angles coterminal with a given angle. Standard notation is unique because any point has a unique ordered pair (a, b) associated with it.

81. $$\frac{\pi}{12}\text{ radians} = \frac{\pi}{12}\text{ radians}\cdot\frac{180°}{\pi\text{ radians}} = \frac{\pi(180°)}{12\pi} = 15°$$

82. $$3\pi\text{ radians} = 3\pi\text{ radians}\cdot\frac{180°}{\pi\text{ radians}} = 540°$$

83. $$330° = 330°\cdot\frac{\pi\text{ radians}}{180°} = \frac{330\pi}{180}\text{ radians} = \frac{11\pi}{6}\text{ radians}$$

84. $$-225° = -225°\cdot\frac{\pi\text{ radians}}{180°} = -\frac{5\pi}{4}\text{ radians}$$

85. Use the Pythagorean theorem.
$$r^2 = 3^2 + 6^2$$
$$r^2 = 9 + 36 = 45$$
$$r = \sqrt{45} = 3\sqrt{5}$$

86.

87. The point determined by $\frac{2\pi}{3}$ is a reflection across the y-axis of the point determined by $\frac{\pi}{3}$. The coordinates of the point determined by $\frac{\pi}{3}$ are $\left(\frac{1}{2}, \frac{\sqrt{3}}{2}\right)$, so the coordinates of the point determined by $\frac{2\pi}{3}$ are $\left(-\frac{1}{2}, \frac{\sqrt{3}}{2}\right)$. Thus,
$$\sin\frac{2\pi}{3} = y = \frac{\sqrt{3}}{2}.$$

88. The coordinates of the point determined by $\frac{\pi}{6}$ are $\left(\frac{\sqrt{3}}{2}, \frac{1}{2}\right)$. Thus,
$$\cos\frac{\pi}{6} = x = \frac{\sqrt{3}}{2}.$$

89. The coordinates of the point determined by $\frac{\pi}{4}$ are $\left(\frac{\sqrt{2}}{2}, \frac{\sqrt{2}}{2}\right)$. Thus,
$$\cos\frac{\pi}{4} = x = \frac{\sqrt{2}}{2}.$$

90. The point determined by $\frac{5\pi}{6}$ is a reflection across the y-axis of the point determined by $\frac{\pi}{6}$. The coordinates of the point determined by $\frac{\pi}{6}$ are $\left(\frac{\sqrt{3}}{2}, \frac{1}{2}\right)$, so the coordinates of the point determined by $\frac{5\pi}{6}$ are $\left(-\frac{\sqrt{3}}{2}, \frac{1}{2}\right)$. Thus,
$$\sin\frac{5\pi}{6} = y = \frac{1}{2}.$$

91. $x^2 + (1 - i)x + i = 0$
Use the quadratic formula.
$a = 1$, $b = 1 - i$, $c = i$
$$x = \frac{-(1-i) \pm \sqrt{(1-i)^2 - 4\cdot 1\cdot i}}{2\cdot 1} = \frac{-1 + i \pm \sqrt{1 - 2i + i^2 - 4i}}{2} = \frac{-1 + i \pm \sqrt{-6i}}{2}$$
Now we find the square roots of $-6i$.
$$(-6i)^{1/2}$$
$$= [6(\cos 270° + i\sin 270°)]^{1/2}$$
$$= 6\left[\cos\left(\frac{270°}{2} + k\cdot\frac{360°}{2}\right) + i\sin\left(\frac{270°}{2} + k\cdot\frac{360°}{5}\right)\right],$$
$k = 0, 1$

The roots are $\sqrt{6}(\cos 135° + i\sin 135°)$ and $\sqrt{6}(\cos 315° + i\sin 315°)$, or $-\sqrt{3}+\sqrt{3}i$ and $\sqrt{3}-\sqrt{3}i$.

Then the solutions of the original equation are

$$x = \frac{-1+i-\sqrt{3}+\sqrt{3}i}{2} = \frac{-1-\sqrt{3}}{2} + \frac{1+\sqrt{3}}{2}i, \text{ or}$$

$$-\frac{1+\sqrt{3}}{2} + \frac{1+\sqrt{3}}{2}i$$

and

$$x = \frac{-1+i+\sqrt{3}-\sqrt{3}i}{2} = \frac{-1+\sqrt{3}}{2} + \frac{1-\sqrt{3}}{2}i, \text{ or}$$

$$-\frac{1-\sqrt{3}}{2} + \frac{1-\sqrt{3}}{2}i.$$

92. $3x^2 + (1+2i)x + (1-i) = 0$

$$x = \frac{-(1+2i) \pm \sqrt{(1+2i)^2 - 4\cdot 3\cdot(1-i)}}{2\cdot 3}$$

$$= \frac{-1-2i \pm \sqrt{-15+16i}}{6}$$

Find the square roots of $-15+16i$.

$$(-15+16i)^{1/2}$$

$$\approx \left[\sqrt{481}(\cos 133.15° + i\sin 133.15°)\right]^{1/2}$$

$$\approx \sqrt[4]{481}\left[\cos\left(\frac{133.15°}{2} + k\cdot\frac{360°}{2}\right) + i\sin\left(\frac{133.15°}{2} + k\cdot\frac{360°}{2}\right)\right],\ k = 0, 1$$

The square roots are approximately $\sqrt[4]{481}(\cos 66.575° + i\sin 66.575°)$ and $\sqrt[4]{481}(\cos 246.575° + i\sin 246.575°)$, or $1.8618 + 4.2972i$ and $-1.8618 - 4.2972i$.

Then the solutions of the original equation are

$$x \approx \frac{-1-2i+1.8618+4.2972i}{6} \approx 0.1436 + 0.3829i$$

and

$$x \approx \frac{-1-2i-1.8618-4.2972i}{6} \approx -0.4770-1.0495i.$$

93. $(\cos\theta + i\sin\theta)^{-1}$

$$= \frac{1}{\cos\theta + i\sin\theta}\cdot\frac{\cos\theta - i\sin\theta}{\cos\theta - i\sin\theta}$$

$$= \frac{\cos\theta - i\sin\theta}{\cos^2\theta + \sin^2\theta}$$

$$= \cos\theta - i\sin\theta$$

94. $z = a+bi$, $|z| = \sqrt{a^2+b^2}$;
$-z = -a-bi$, $|-z| = \sqrt{(-a)^2+(-b)^2} = \sqrt{a^2+b^2}$;
so $|z| = |-z|$.

95. See the answer section in the text.

96. $|z\bar{z}| = |(a+bi)(a-bi)| = |a^2+b^2| = a^2+b^2$;
$|z^2| = |(a+bi)^2| = |a^2+2abi-b^2| =$
$\sqrt{(a^2-b^2)^2+(2ab)^2} = \sqrt{a^4-2a^2b^2+b^4+4a^2b^2} =$
$\sqrt{a^4+2a^2b^2+b^4} = \sqrt{(a^2+b^2)^2} = a^2+b^2$;
so $|z\bar{z}| = |z^2|$.

97. See the answer section in the text.

98. $|z\cdot w| = |r_1(\cos\theta_1 + i\sin\theta_1)\cdot r_2(\cos\theta_2 + i\sin\theta_2)| =$
$|r_1r_2[\cos(\theta_1+\theta_2) + i\sin(\theta_1+\theta_2)]| =$
$\sqrt{r_1^2r_2^2\cos^2(\theta_1+\theta_2) + r_1^2r_2^2\sin^2(\theta_1+\theta_2)} =$
$\sqrt{r_1^2r_2^2[\cos^2(\theta_1+\theta_2) + \sin^2(\theta_1+\theta_2)]} =$
$\sqrt{r_1^2r_2^2} = r_1r_2$;
$|z|\cdot|w| =$
$\sqrt{r_1^2\cos^2\theta_1 + r_1^2\sin^2\theta_1}\sqrt{r_2^2\cos^2\theta_2 + r_2^2\sin^2\theta_2} =$
$\sqrt{r_1^2(\cos^2\theta_1 + \sin^2\theta_1)}\sqrt{r_2^2(\cos^2\theta_2 + \sin^2\theta_2)} =$
$\sqrt{r_1^2}\sqrt{r_2^2} = r_1r_2$;
so $|z\cdot w| = |z|\cdot|w|$.

99. See the answer section in the text.

100. Graph: $|z| = 1$

Let $z = a+bi$

Then $|a+bi| = 1$

$$\sqrt{a^2+b^2} = 1$$

$$a^2+b^2 = 1$$

The graph of $a^2+b^2 = 1$ is a circle whose radius is 1 and whose center is $(0,0)$.

101.

$$z + \bar{z} = 3$$

$$(a+bi) + (a-bi) = 3$$

$$2a = 3$$

$$a = \frac{3}{2}$$

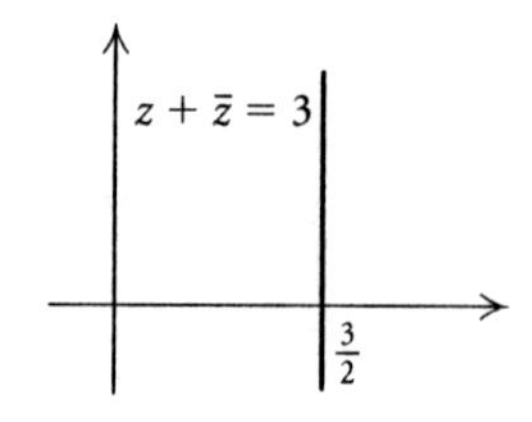

Exercise Set 7.4

See the text answer section for odd exercise answers 1-11.
Even exercises 2-12 are shown below.

13. Answers may vary.

A: $(4, 30°)$, $(4, 390°)$, $(-4, 210°)$;

B: $(5, 300°)$, $(5, -60°)$, $(-5, 120°)$;

C: $(2, 150°)$, $(2, -210°)$, $(-2, 330°)$;

D: $(3, 225°)$, $(3, -135°)$, $(-3, 45°)$

14. Answers may vary.

A: $\left(2, \frac{\pi}{4}\right)$, $\left(2, \frac{9\pi}{4}\right)$, $\left(-2, \frac{5\pi}{4}\right)$;

B: $\left(5, \frac{3\pi}{2}\right)$, $\left(5, -\frac{\pi}{2}\right)$, $\left(-5, \frac{\pi}{2}\right)$;

C: $\left(4, \frac{11\pi}{12}\right)$, $\left(4, -\frac{13\pi}{12}\right)$, $\left(-4, -\frac{\pi}{12}\right)$;

D: $\left(3, \frac{5\pi}{3}\right)$, $\left(3, -\frac{\pi}{3}\right)$, $\left(-3, \frac{2\pi}{3}\right)$

15. $(0, -3)$

$r = \sqrt{0^2 + (-3)^2} = \sqrt{9} = 3$

$\tan\theta = \frac{-3}{0}$, which is not defined; therefore, since $(0, -3)$ lies on the negative y-axis, $\theta = 270°$, or $\frac{3\pi}{2}$.

Thus, $(0, -3) = (3, 270°)$, or $\left(3, \frac{3\pi}{2}\right)$.

16. $(-4, 4)$

$r = \sqrt{(-4)^2 + 4^2} = \sqrt{32} = 4\sqrt{2}$

$\tan\theta = \frac{4}{-4} = -1$, so $\theta = 135°$, or $\frac{3\pi}{4}$. (The point is in quadrant II.)

Thus, $(-4, 4) = (4\sqrt{2}, 135°)$, or $\left(4\sqrt{2}, \frac{3\pi}{4}\right)$.

17. $(3, -3\sqrt{3})$

$r = \sqrt{3^2 + (-3\sqrt{3})^2} = \sqrt{9 + 27} = \sqrt{36} = 6$

$\tan\theta = \frac{-3\sqrt{3}}{3} = -\sqrt{3}$, so $\theta = 300°$, or $\frac{5\pi}{3}$. (The point is in quadrant IV.)

Thus, $(3, -3\sqrt{3}) = (6, 300°)$, or $\left(6, \frac{5\pi}{3}\right)$.

18. $(-\sqrt{3}, 1)$

$r = \sqrt{(-\sqrt{3})^2 + 1^2} = \sqrt{4} = 2$

$\tan\theta = \frac{1}{-\sqrt{3}}$, or $-\frac{\sqrt{3}}{3}$, so $\theta = 150°$, or $\frac{5\pi}{6}$. (The point is in quadrant II.)

Thus, $(-\sqrt{3}, 1) = (2, 150°)$, or $\left(2, \frac{5\pi}{6}\right)$.

19. $(4\sqrt{3}, -4)$

$r = \sqrt{(4\sqrt{3})^2 + (-4)^2} = \sqrt{48 + 16} = \sqrt{64} = 8$

$\tan\theta = \frac{-4}{4\sqrt{3}} = -\frac{1}{\sqrt{3}}$, or $-\frac{\sqrt{3}}{3}$, so $\theta = 330°$, or $\frac{11\pi}{6}$. (The point is in quadrant IV.)

Thus, $(4\sqrt{3}, -4) = (8, 330°)$, or $\left(8, \frac{11\pi}{6}\right)$.

20. $(2\sqrt{3}, 2)$

$r = \sqrt{(2\sqrt{3})^2 + 2^2} = \sqrt{16} = 4$

$\tan\theta = \frac{2}{2\sqrt{3}} = \frac{1}{\sqrt{3}}$, or $\frac{\sqrt{3}}{3}$, so $\theta = 30°$, or $\frac{\pi}{6}$. (The point is in quadrant I.)

Thus, $(2\sqrt{3}, 2) = (4, 30°)$, or $\left(4, \frac{\pi}{6}\right)$.

21. $(-\sqrt{2}, -\sqrt{2})$

$r = \sqrt{(-\sqrt{2})^2 + (-\sqrt{2})^2} = \sqrt{2 + 2} = \sqrt{4} = 2$

$\tan\theta = \frac{-\sqrt{2}}{-\sqrt{2}} = 1$, so $\theta = 225°$, or $\frac{5\pi}{4}$. (The point is in quadrant III.)

Thus, $(-\sqrt{2}, -\sqrt{2}) = (2, 225°)$, or $\left(2, \frac{5\pi}{4}\right)$.

22. $(-3, 3\sqrt{3})$

$r = \sqrt{(-3)^2 + (3\sqrt{3})^2} = \sqrt{36} = 6$

$\tan\theta = \frac{3\sqrt{3}}{-3} = -\sqrt{3}$, so $\theta = 120°$, or $\frac{2\pi}{3}$. (The point is in quadrant II.)

Thus, $(-3, 3\sqrt{3}) = (6, 120°)$, or $\left(6, \frac{2\pi}{3}\right)$.

23. $(1, \sqrt{3})$

$r = \sqrt{1^2 + (\sqrt{3})^2} = \sqrt{4} = 2$

$\tan\theta = \frac{\sqrt{3}}{1} = \sqrt{3}$, so $\theta = 60°$, or $\frac{\pi}{3}$. (The point is in quadrant I.)

Thus, $(1, \sqrt{3}) = (2, 60°)$, or $\left(2, \frac{\pi}{3}\right)$.

24. $(0, -1)$

$r = \sqrt{0^2 + (-1)^2} = 1$

$\tan\theta = \dfrac{-1}{0}$, which is not defined; therefore, since $(0, -1)$ lies on the negative y-axis, $\theta = 270°$, or $\dfrac{3\pi}{2}$. Thus, $(0, -1) = (1, 270°)$, or $\left(1, \dfrac{3\pi}{2}\right)$.

25. $\left(\dfrac{5\sqrt{2}}{2}, -\dfrac{5\sqrt{2}}{2}\right)$

$$r = \sqrt{\left(\frac{5\sqrt{2}}{2}\right)^2 + \left(-\frac{5\sqrt{2}}{2}\right)^2} = \sqrt{\frac{25}{2} + \frac{25}{2}} = \sqrt{25} = 5$$

$\tan\theta = \dfrac{-\frac{5\sqrt{2}}{2}}{\frac{5\sqrt{2}}{2}} = -1$, so $\theta = 315°$, or $\dfrac{7\pi}{4}$. (The point is in quadrant IV.)

Thus, $\left(\dfrac{5\sqrt{2}}{2}, -\dfrac{5\sqrt{2}}{2}\right) = (5, 315°)$, or $\left(5, \dfrac{7\pi}{4}\right)$.

26. $\left(-\dfrac{3}{2}, -\dfrac{3\sqrt{3}}{2}\right)$

$$r = \sqrt{\left(-\frac{3}{2}\right)^2 + \left(-\frac{3\sqrt{3}}{2}\right)^2} = \sqrt{\frac{9}{4} + \frac{27}{4}} = \sqrt{9} = 3$$

$\tan\theta = \dfrac{-\frac{3\sqrt{3}}{2}}{-\frac{3}{2}} = \sqrt{3}$, so $\theta = 240°$, or $\dfrac{4\pi}{3}$. (The point is in quadrant III.)

Thus, $\left(-\dfrac{3}{2}, -\dfrac{3\sqrt{3}}{2}\right) = (3, 240°)$, or $\left(3, \dfrac{4\pi}{3}\right)$.

27. $(5, 60°)$

$x = r\cos\theta = 5\cos 60° = 5 \cdot \dfrac{1}{2} = \dfrac{5}{2}$

$y = r\sin\theta = 5\sin 60° = 5 \cdot \dfrac{\sqrt{3}}{2} = \dfrac{5\sqrt{3}}{2}$

$(5, 60°) = \left(\dfrac{5}{2}, \dfrac{5\sqrt{3}}{2}\right)$

28. $(0, -23°)$

$x = 0\cos(-23°) = 0$

$y = 0\cos(-23°) = 0$

$(0, -23°) = (0, 0)$

29. $(-3, 45°)$

$x = r\cos\theta = -3\cos 45° = -3 \cdot \dfrac{\sqrt{2}}{2} = -\dfrac{3\sqrt{2}}{2}$

$y = r\sin\theta = -3\sin 45° = -3 \cdot \dfrac{\sqrt{2}}{2} = -\dfrac{3\sqrt{2}}{2}$

$(-3, 45°) = \left(-\dfrac{3\sqrt{2}}{2}, -\dfrac{3\sqrt{2}}{2}\right)$

30. $(6, 30°)$

$x = 6\cos 30° = 6 \cdot \dfrac{\sqrt{3}}{2} = 3\sqrt{3}$

$y = 6\sin 30° = 6 \cdot \dfrac{1}{2} = 3$

$(6, 30°) = (3\sqrt{3}, 3)$

31. $(3, -120°)$

$x = 3\cos(-120°) = 3\left(-\dfrac{1}{2}\right) = -\dfrac{3}{2}$

$y = 3\sin(-120°) = 3\left(-\dfrac{\sqrt{3}}{2}\right) = -\dfrac{3\sqrt{3}}{2}$

$(3, -120°) = \left(-\dfrac{3}{2}, -\dfrac{3\sqrt{3}}{2}\right)$

32. $\left(7, \dfrac{\pi}{6}\right)$

$x = 7\cos\dfrac{\pi}{6} = 7 \cdot \dfrac{\sqrt{3}}{2} = \dfrac{7\sqrt{3}}{2}$

$y = 7\sin\dfrac{\pi}{6} = 7 \cdot \dfrac{1}{2} = \dfrac{7}{2}$

$\left(7, \dfrac{\pi}{6}\right) = \left(\dfrac{7\sqrt{3}}{2}, \dfrac{7}{2}\right)$

33. $\left(-2, \dfrac{5\pi}{3}\right)$

$x = -2\cos\dfrac{5\pi}{3} = -2 \cdot \dfrac{1}{2} = -1$

$y = -2\sin\dfrac{5\pi}{3} = -2\left(-\dfrac{\sqrt{3}}{2}\right) = \sqrt{3}$

$\left(-2, \dfrac{5\pi}{3}\right) = (-1, \sqrt{3})$

34. $(1.4, 225°)$

$x = 1.4\cos 225° = 1.4\left(-\dfrac{\sqrt{2}}{2}\right) = -0.7\sqrt{2}$

$y = 1.4\sin 225° = 1.4\left(-\dfrac{\sqrt{2}}{2}\right) = -0.7\sqrt{2}$

$(1.4, 225°) = (-0.7\sqrt{2}, -0.7\sqrt{2})$

35. $(2, 210°)$

$x = 2\cos 210° = 2\left(-\dfrac{\sqrt{3}}{2}\right) = -\sqrt{3}$

$y = 2\sin 210° = 2\left(-\dfrac{1}{2}\right) = -1$

$(2, 210°) = (-\sqrt{3}, -1)$

36. $\left(1, \dfrac{7\pi}{4}\right)$

$x = 1 \cdot \cos\dfrac{7\pi}{4} = \dfrac{\sqrt{2}}{2}$

$y = 1 \cdot \sin\dfrac{7\pi}{4} = -\dfrac{\sqrt{2}}{2}$

$\left(1, \dfrac{7\pi}{4}\right) = \left(\dfrac{\sqrt{2}}{2}, -\dfrac{\sqrt{2}}{2}\right)$

37. $\left(-6, \frac{5\pi}{6}\right)$

$x = -6\cos\frac{5\pi}{6} = -6\left(-\frac{\sqrt{3}}{2}\right) = 3\sqrt{3}$

$x = -6\sin\frac{5\pi}{6} = -6\cdot\frac{1}{2} = -3$

$\left(-6, \frac{5\pi}{6}\right) = (3\sqrt{3}, -3)$

38. $(4, 180°)$

$x = 4\cos 180° = 4(-1) = -4$

$y = 4\sin 180° = 4\cdot 0 = 0$

$(4, 180°) = (-4, 0)$

39.

$$\begin{aligned} 3x + 4y &= 5 \\ 3r\cos\theta + 4r\sin\theta &= 5 \quad (x = r\cos\theta,\ y = r\sin\theta) \\ r(3\cos\theta + 4\sin\theta) &= 5 \end{aligned}$$

40.

$$\begin{aligned} 5x + 3y &= 4 \\ 5r\cos\theta + 3r\sin\theta &= 4 \\ r(5\cos\theta + 3\sin\theta) &= 4 \end{aligned}$$

41.

$$\begin{aligned} x &= 5 \\ r\cos\theta &= 5 \quad (x = r\cos\theta) \end{aligned}$$

42.

$$\begin{aligned} y &= 4 \\ r\sin\theta &= 4 \end{aligned}$$

43.

$$\begin{aligned} x^2 + y^2 &= 36 \\ (r\cos\theta)^2 + (r\sin\theta)^2 &= 36 \quad (x = r\cos\theta,\ y = r\sin\theta) \\ r^2\cos^2\theta + r^2\sin^2\theta &= 36 \\ r^2(\cos^2\theta + \sin^2\theta) &= 36 \\ r^2 &= 36 \quad (\cos^2\theta + \sin^2\theta = 1) \\ r &= 6 \end{aligned}$$

44.

$$\begin{aligned} x^2 - 4y^2 &= 4 \\ (r\cos\theta)^2 - 4(r\sin\theta)^2 &= 4 \\ r^2\cos^2\theta - 4r^2\sin^2\theta &= 4 \\ r^2(\cos^2\theta - 4\sin^2\theta) &= 4 \end{aligned}$$

45.

$$\begin{aligned} x^2 &= 25y \\ (r\cos\theta)^2 &= 25r\sin\theta \quad \text{Substituting for } x \text{ and } y \\ r^2\cos^2\theta &= 25r\sin\theta \end{aligned}$$

46.

$$\begin{aligned} 2x - 9y + 3 &= 0 \\ 2r\cos\theta - 9r\sin\theta + 3 &= 0 \\ r(2\cos\theta - 9\sin\theta) + 3 &= 0 \end{aligned}$$

47.

$$\begin{aligned} y^2 - 5x - 25 &= 0 \\ (r\sin\theta)^2 - 5r\cos\theta - 25 &= 0 \quad \text{Substituting for } x \text{ and } y \\ r^2\sin^2\theta - 5r\cos\theta - 25 &= 0 \end{aligned}$$

48.

$$\begin{aligned} x^2 + y^2 &= 8y \\ (r\cos\theta)^2 + (r\sin\theta)^2 &= 8r\sin\theta \\ r^2\cos^2\theta + r^2\sin^2\theta &= 8r\sin\theta \\ r^2(\cos^2\theta + \sin^2\theta) &= 8r\sin\theta \\ r^2 &= 8r\sin\theta \end{aligned}$$

49.

$$\begin{aligned} x^2 - 2x + y^2 &= 0 \\ (r\cos\theta)^2 - 2r\cos\theta + (r\sin\theta)^2 &= 0 \\ r^2\cos^2\theta - 2r\cos\theta + r^2\sin^2\theta &= 0 \\ r^2(\cos^2\theta + \sin^2\theta) - 2r\cos\theta &= 0 \\ r^2 - 2r\cos\theta &= 0, \text{ or} \\ r^2 &= 2r\cos\theta \end{aligned}$$

50.

$$\begin{aligned} 3x^2y &= 81 \\ 3(r\cos\theta)^2(r\sin\theta) &= 81 \\ 3(r^2\cos^2\theta)(r\sin\theta) &= 81 \\ 3r^3\cos^2\theta\sin\theta &= 81 \\ r^3\cos^2\theta\sin\theta &= 27 \end{aligned}$$

51.

$$\begin{aligned} r &= 5 \\ \pm\sqrt{x^2 + y^2} &= 5 \quad \text{Substituting for } r \\ x^2 + y^2 &= 25 \quad \text{Squaring} \end{aligned}$$

52.

$$\begin{aligned} \tan\theta &= \frac{y}{x} \\ \tan\frac{3\pi}{4} &= \frac{y}{x} \quad \left(\theta = \frac{3\pi}{4}\right) \\ -1 &= \frac{y}{x} \\ y &= -x \end{aligned}$$

53.

$$\begin{aligned} r\sin\theta &= 2 \\ y &= 2 \quad (y = r\sin\theta) \end{aligned}$$

54.

$$\begin{aligned} r &= -3\sin\theta \\ r^2 &= -3(r\sin\theta) \\ x^2 + y^2 &= -3y \end{aligned}$$

55.

$$\begin{aligned} r + r\cos\theta &= 3 \\ \pm\sqrt{x^2 + y^2} + x &= 3 \\ \pm\sqrt{x^2 + y^2} &= 3 - x \\ x^2 + y^2 &= (3 - x)^2 \quad \text{Squaring both sides} \\ x^2 + y^2 &= 9 - 6x + x^2 \\ y^2 &= -6x + 9 \end{aligned}$$

56.

$$\begin{aligned} r &= \frac{2}{1 - \sin\theta} \\ r - r\sin\theta &= 2 \\ r - y &= 2 \\ r &= 2 + y \\ r^2 &= 4 + 4y + y^2 \\ x^2 + y^2 &= 4 + 4y + y^2 \\ x^2 &= 4y + 4 \end{aligned}$$

57.

$$\begin{aligned} r - 9\cos\theta &= 7\sin\theta \\ r^2 - 9r\cos\theta &= 7r\sin\theta \quad \text{Multiplying both sides by } r \\ x^2 + y^2 - 9x &= 7y \\ x^2 - 9x + y^2 - 7y &= 0 \end{aligned}$$

58.
$$\begin{aligned} r + 5\sin\theta &= 7\cos\theta \\ r^2 + 5(r\sin\theta) &= 7(r\cos\theta) \\ (x^2+y^2) + 5y &= 7x \\ x^2 - 7x + y^2 + 5y &= 0 \end{aligned}$$

59.
$$\begin{aligned} r &= 5\sec\theta \\ r &= 5\cdot\frac{1}{\cos\theta} \quad \left(\sec\theta = \frac{1}{\cos\theta}\right) \\ r\cos\theta &= 5 \\ x &= 5 \end{aligned}$$

60.
$$\begin{aligned} r &= 3\cos\theta \\ r^2 &= 3r\cos\theta \\ x^2+y^2 &= 3x \end{aligned}$$

61.
$$\begin{aligned} \tan\theta &= \frac{y}{x} \\ \tan\frac{5\pi}{3} &= \frac{y}{x} \quad \text{Substituting } \frac{5\pi}{3} \text{ for } \theta \\ -\sqrt{3} &= \frac{y}{x} \\ -\sqrt{3}x &= y, \text{ or} \\ y &= -\sqrt{3}x \end{aligned}$$

62.
$$\begin{aligned} r &= \cos\theta - \sin\theta \\ r^2 &= r\cos\theta - r\sin\theta \\ x^2+y^2 &= x - y \end{aligned}$$

63. $r = \sin\theta$

Make a table of values. Note that the points begin to repeat at $\theta = 360°$. Plot the points and draw the graph.

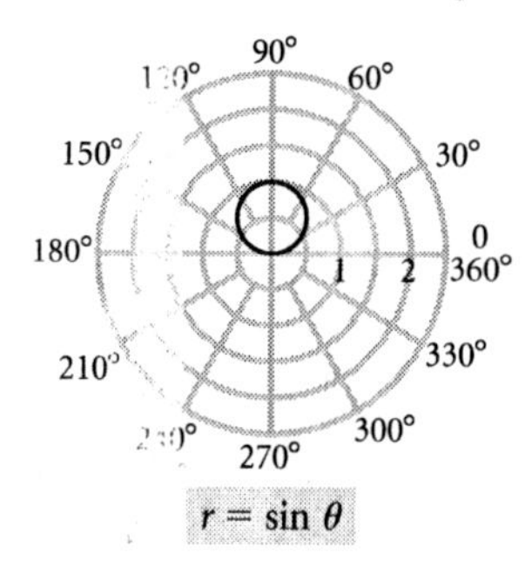

$r = \sin\theta$

64.

65. $r = 4\cos 2\theta$

Make a table of values. Note that the points begin to repeat at $\theta = 180°$. Plot the points and draw the graph.

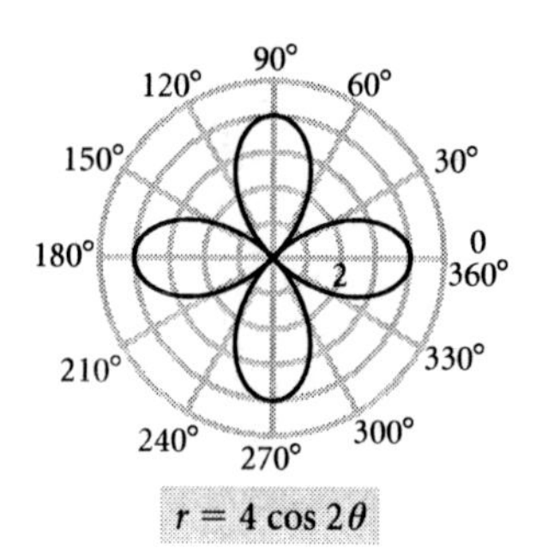

$r = 4\cos 2\theta$

66.

67. $r = \cos\theta$

Make a table of values. Note that the points begin to repeat at $\theta = 360^0$. Plot the points and draw the graph.

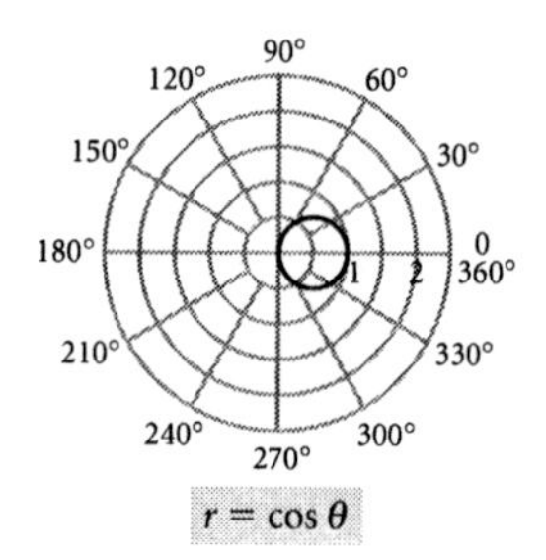

$r = \cos\theta$

68.

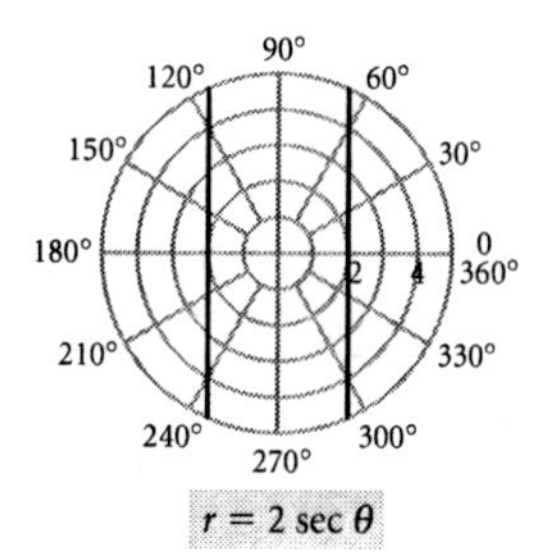

$r = 2\sec\theta$

69. $r = 2 - \cos 3\theta$

Make a table of values. Note that the points begin to repeat at $\theta = 120°$. Plot the points and draw the graph.

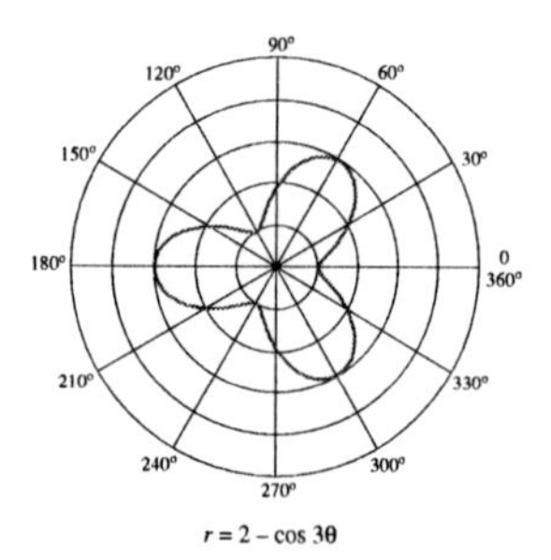

$r = 2 - \cos 3\theta$

70.

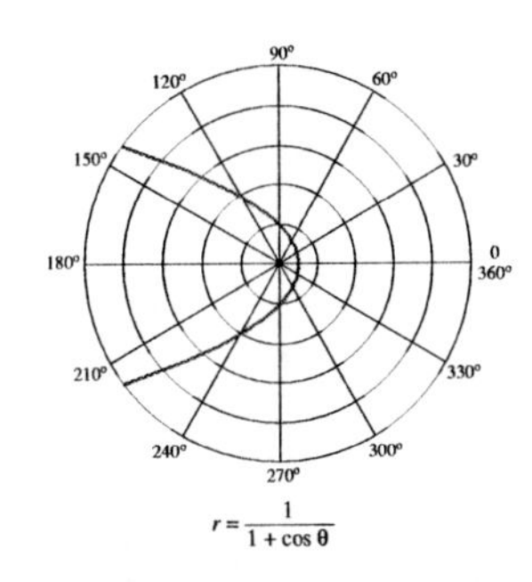

71. Use the ANGLE feature on a graphing calculator.

$(3, 7) = (7.616, 66.8°)$, or $(7.616, 1.166)$

72. $(-2, -\sqrt{5}) = (3, 228.2°)$, or $(3, 3.983)$

(The calculator might return a negative angle. Convert to the smallest positive angle by adding 360°or 2π.)

73. Use the ANGLE feature on a graphing calculator.

$(-\sqrt{10}, 3.4) = (4.643, 132.9°)$, or $(4.643, 2.230)$

74. $(0.9, -6) = (6.067, 278.5°)$, or $(6.067, 4.861)$

(The calculator might return a negative angle. Convert to the smallest positive angle by adding 360°or 2π.)

75. Use the ANGLE feature on a graphing calculator.

$(3, -43°) = (2.19, -2.05)$

76. $\left(-5, \frac{\pi}{7}\right) = (-4.50, -2.17)$

77. Use the ANGLE feature on a graphing calculator.

$\left(-4.2, \frac{3\pi}{5}\right) = (1.30, -3.99)$

78. $(2.8, 166°) = (-2.72, 0.68)$

79. Graph (d) is the graph of $r = 3\sin 2\theta$.

80. Graph (i) is the graph of $r = 4\cos\theta$.

81. Graph (g) is the graph of $r = \theta$.

82. Graph (c) is the graph of $r^2 = \sin 2\theta$.

83. Graph (j) is the graph of $r = \dfrac{5}{1 + \cos\theta}$.

84. Graph (f) is the graph of $r = 1 + 2\sin\theta$.

85. Graph (b) is the graph of $r = 3\cos 2\theta$.

86. Graph (a) is the graph of $r = 3\sec\theta$.

87. Graph (e) is the graph of $r = 3\sin\theta$.

88. Graph (h) is the graph of $r = 4\cos 5\theta$.

89. Graph (k) is the graph of $r = 2\sin 3\theta$.

90. Graph (l) is the graph of $r\sin\theta = 6$.

91. See the answer section in the text.

92.

93. See the answer section in the text.

94.

95. See the answer section in the text.

96.

97. See the answer section in the text.

98.

99. Rectangular coordinates are unique because any point has a unique ordered pair (x, y) associated with it. Polar coordinates are not unique because there are infinitely many angles coterminal with a given angle and also because r can be positive or negative depending on the angle used.

100. One example is the equation of a circle centered at the origin, $r = c$, c a constant. In particular, the polar form of the equation is a function while the rectangular form, $x^2 + y^2 = c^2$, is not.

101. $2x - 4 = x + 8$

$x = 12$ Adding 4 and subtracting x

The solution is 12.

102. $4 - 5y = 3$

$-5y = -1$

$y = \frac{1}{5}$

The solution is $\frac{1}{5}$.

103. $y = 2x - 5$

Make a table of values by choosing values for x and finding the corresponding y-values. Plot points and draw the graph.

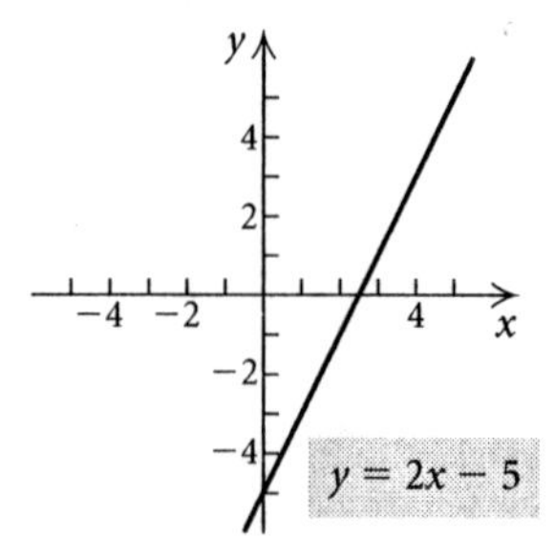

104.

105. $x = -3$

Note that any point on the graph has -3 for its first coordinate. Thus, the graph is a vertical line 3 units left of the y-axis.

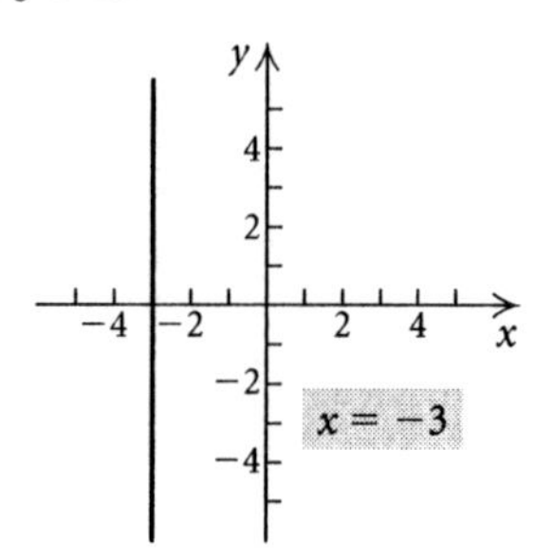

106.

107.

$$r = \sec^2 \frac{\theta}{2}$$

$$r = \frac{1}{\cos^2 \frac{\theta}{2}}$$

$$r = \frac{1}{\frac{1+\cos\theta}{2}}$$

$$r = \frac{2}{1+\cos\theta}$$

$$r + r\cos\theta = 2$$

$$\pm\sqrt{x^2+y^2} + x = 2$$

$$x^2 + y^2 = (2-x)^2$$

$$x^2 + y^2 = 4 - 4x + x^2$$

$$y^2 = -4x + 4$$

108.

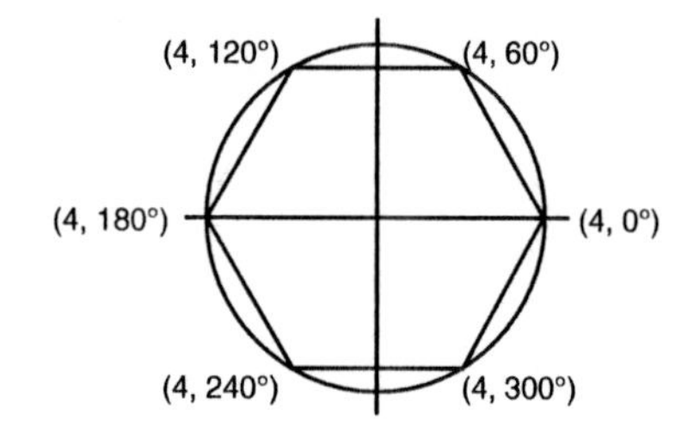

Exercise Set 7.5

1.

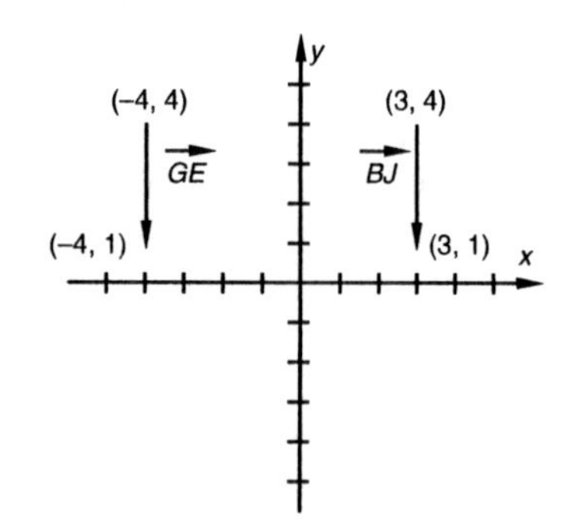

First we find the length of each vector using the distance formula.

$|\overrightarrow{GE}| = \sqrt{[-4-(-4)]^2 + (4-1)^2} = \sqrt{9} = 3$

$|\overrightarrow{BJ}| = \sqrt{(3-3)^2 + (4-1)^2} = \sqrt{9} = 3$

Thus, $|\overrightarrow{GE}| = |\overrightarrow{BJ}|$.

Both vectors point down. To verify that they have the same direction we calculate the slopes of the lines that they lie on.

Slope of $\overrightarrow{GE} = \dfrac{4-1}{-4-(-4)} = \dfrac{3}{0}$ (undefined)

Slope of $\overrightarrow{BJ} = \dfrac{4-1}{3-3} = \dfrac{3}{0}$ (undefined)

Both vectors have undefined slope, so they lie on vertical lines and hence have the same direction.

Since $\overrightarrow{GE}$ and $\overrightarrow{BJ}$ have the same magnitude and same direction, they are equivalent.

2.

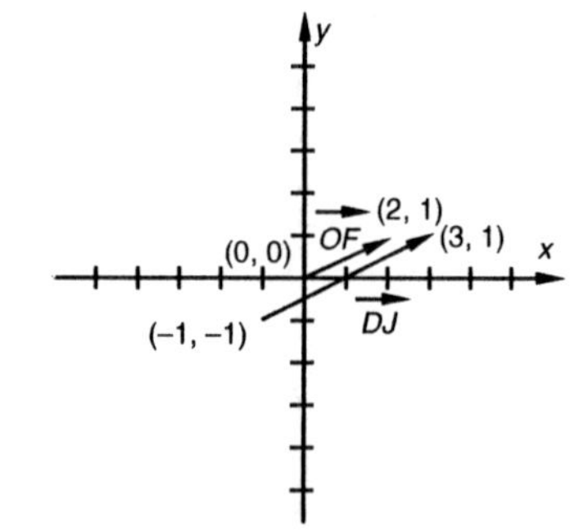

$|\overrightarrow{DJ}| = \sqrt{[3-(-1)]^2 + [1-(-1)]^2} = \sqrt{16+4} = \sqrt{20}$

$|\overrightarrow{OF}| = \sqrt{(2-0)^2 + (1-0)^2} = \sqrt{4+1} = \sqrt{5}$

Since $|\overrightarrow{DJ}| \neq |\overrightarrow{OF}|$, the vectors are not equivalent.

3.

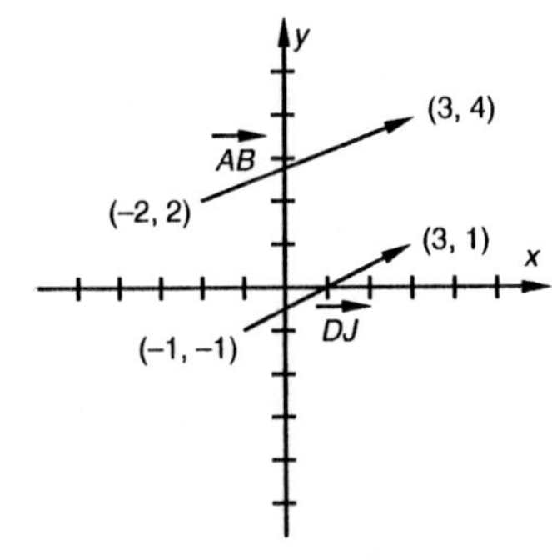

First we find the length of each vector using the distance formula.

$|\overrightarrow{DJ}| = \sqrt{[3-(-1)]^2 + [1-(-1)]^2} = \sqrt{16+4} = \sqrt{20}$

$|\overrightarrow{AB}| = \sqrt{[3-(-2)]^2 + (4-2)^2} = \sqrt{25+4} = \sqrt{29}$

Since $|\overrightarrow{DJ}| \neq |\overrightarrow{AB}|$, the vectors are not equivalent.

4.

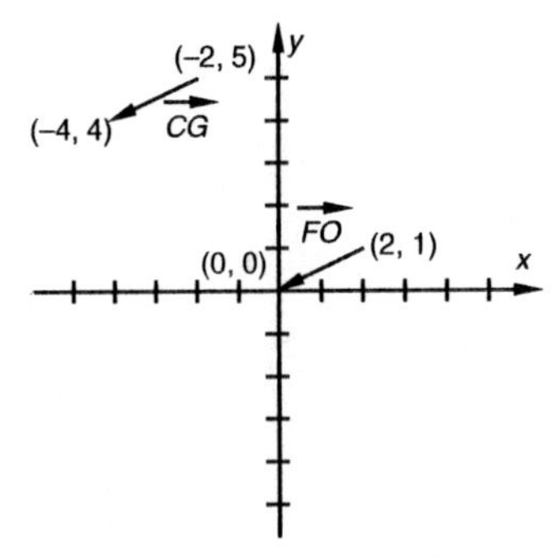

$|\overrightarrow{CG}| = \sqrt{[-2-(-4)]^2 + (5-4)^2} = \sqrt{4+1} = \sqrt{5}$

$|\overrightarrow{FO}| = \sqrt{(2-0)^2 + (1-0)^2} = \sqrt{4+1} = \sqrt{5}$

Thus, $|\overrightarrow{CG}| = |\overrightarrow{FO}|$.

Slope of $\overrightarrow{CG} = \dfrac{5-4}{-2-(-4)} = \dfrac{1}{2}$

Slope of $\overrightarrow{FO} = \dfrac{1-0}{2-0} = \dfrac{1}{2}$

The slopes are the same.

Since $\overrightarrow{CG}$ and $\overrightarrow{FO}$ have the same magnitude and same direction, they are equivalent.

5.

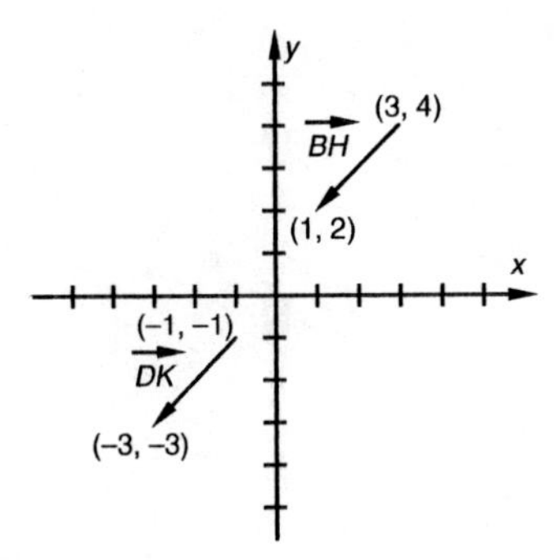

First we find the length of each vector using the distance formula.

$|\overrightarrow{DK}| = \sqrt{[-3-(-1)]^2 + [-3-(-1)]^2} = \sqrt{4+4} = \sqrt{8}$

$|\overrightarrow{BH}| = \sqrt{(3-1)^2 + (4-2)^2} = \sqrt{4+4} = \sqrt{8}$

Thus, $|\overrightarrow{DK}| = |\overrightarrow{BH}|$.

Both vectors point down and to the left. We calculate the slopes of the lines that they lie on.

Slope of $\overrightarrow{DK} = \dfrac{-3-(-1)}{-3-(-1)} = \dfrac{-2}{-2} = 1$

Slope of $\overrightarrow{BH} = \dfrac{4-2}{3-1} = \dfrac{2}{2} = 1$

The slopes are the same.

Since $\overrightarrow{DK}$ and $\overrightarrow{BH}$ have the same magnitude and same direction, they are equivalent.

6.

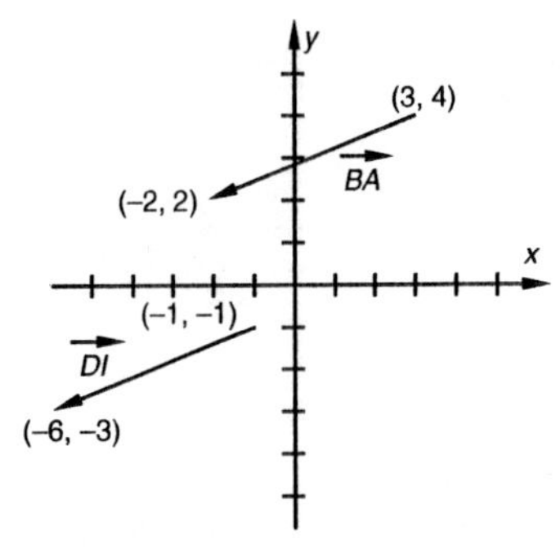

$|\overrightarrow{BA}| = \sqrt{(-2-3)^2 + (2-4)^2} = \sqrt{25+4} = \sqrt{29}$

$|\overrightarrow{DI}| = \sqrt{[-6-(-1)]^2 + [-3-(-1)]^2} = \sqrt{25+4} = \sqrt{29}$

Thus, $|\overrightarrow{BA}| = |\overrightarrow{DI}|$.

Slope of $\overrightarrow{BA} = \dfrac{2-4}{-2-3} = \dfrac{-2}{-5} = \dfrac{2}{5}$

Slope of $\overrightarrow{DI} = \dfrac{-3-(-1)}{-6-(-1)} = \dfrac{-2}{-5} = \dfrac{2}{5}$

The slopes are the same.

Since $\overrightarrow{BA}$ and $\overrightarrow{DI}$ have the same magnitude and same direction, they are equivalent.

7.

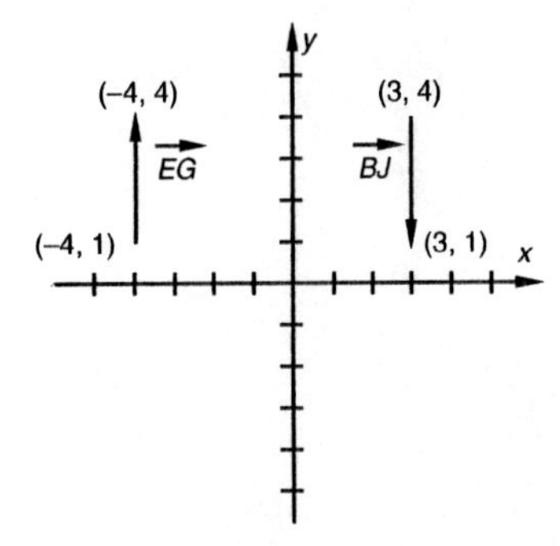

The vectors clearly have different directions, so they are not equivalent.

8.

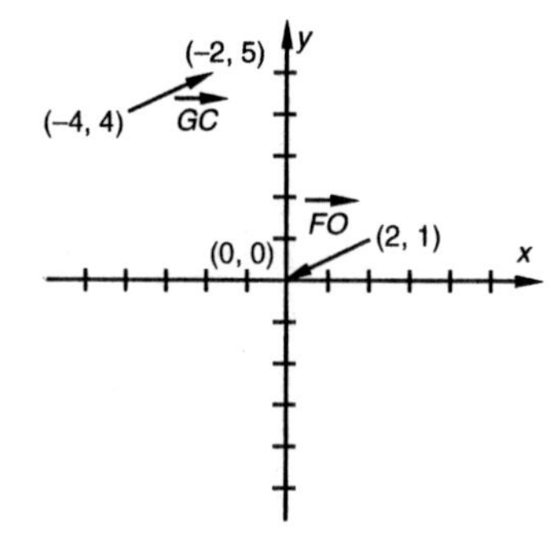

The vectors clearly have different directions, so they are not equivalent.

9.

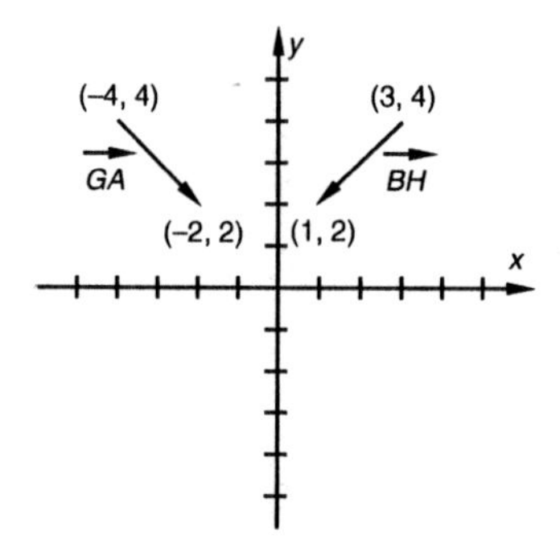

The vectors clearly have different directions, so they are not equivalent.

10.

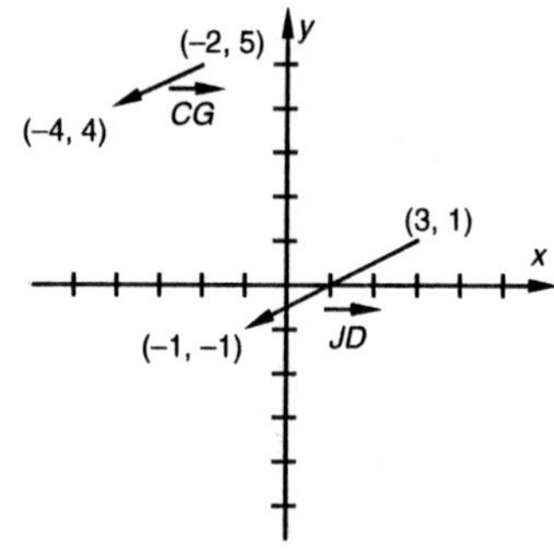

$|\overrightarrow{JD}| = \sqrt{(-1-3)^2 + (-1-1)^2} = \sqrt{16+4} = \sqrt{20}$

$|\overrightarrow{CG}| = \sqrt{[-4-(-2)]^2 + (4-5)^2} = \sqrt{4+1} = \sqrt{5}$

Since $|\overrightarrow{JD}| \neq |\overrightarrow{CG}|$, the vectors are not equivalent.

11.

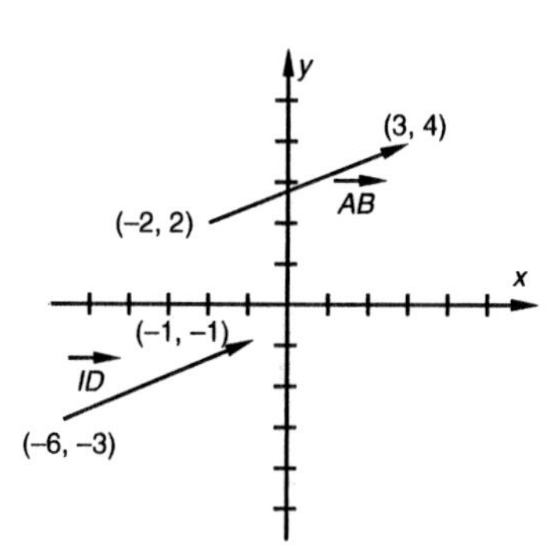

First we find the length of each vector using the distance formula.

$|\overrightarrow{AB}| = \sqrt{[3-(-2)]^2 + (4-2)^2} = \sqrt{25+4} = \sqrt{29}$

$|\overrightarrow{ID}| = \sqrt{[-1-(-6)]^2 + [-1-(-3)]^2} = \sqrt{25+4} = \sqrt{29}$

Thus, $|\overrightarrow{AB}| = |\overrightarrow{ID}|$.

Both vectors point up and to the right. We calculate the slopes of the lines that they lie on.

Slope of $\overrightarrow{AB} = \dfrac{2-4}{-2-3} = \dfrac{-2}{-5} = \dfrac{2}{5}$

Slope of $\overrightarrow{ID} = \dfrac{-3-(-1)}{-6-(-1)} = \dfrac{-2}{-5} = \dfrac{2}{5}$

The slopes are the same.

Since $\overrightarrow{AB}$ and $\overrightarrow{ID}$ have the same magnitude and same direction, they are equivalent.

12.

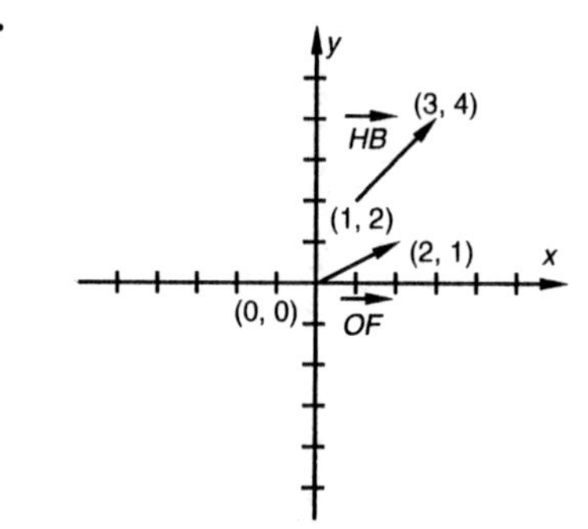

$|\overrightarrow{OF}| = \sqrt{(2-0)^2 + (1-0)^2} = \sqrt{4+1} = \sqrt{5}$

$|\overrightarrow{HB}| = \sqrt{(3-1)^2 + (4-2)^2} = \sqrt{4+4} = \sqrt{8}$

Since $|\overrightarrow{OF}| \neq |\overrightarrow{HB}|$, the vectors are not equivalent.

13.

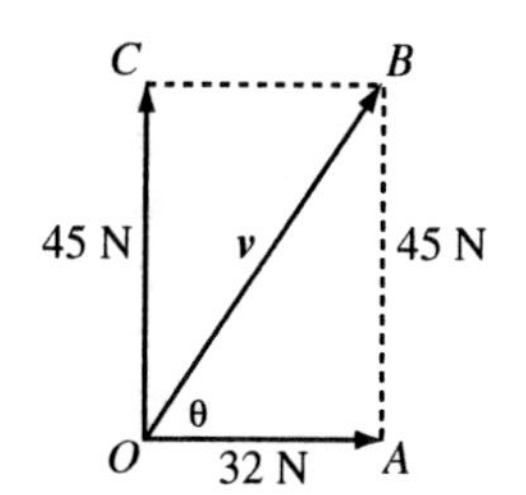

Use the Pythagorean theorem to find the magnitude of the resultant **v**.

$|\mathbf{v}|^2 = 32^2 + 45^2$

$|\mathbf{v}| = \sqrt{32^2 + 45^2} \approx 55$ N

To find θ use the fact that triangle OAB is a right triangle.

$\tan\theta = \dfrac{45}{32}$

$\theta \approx 55°$

14.

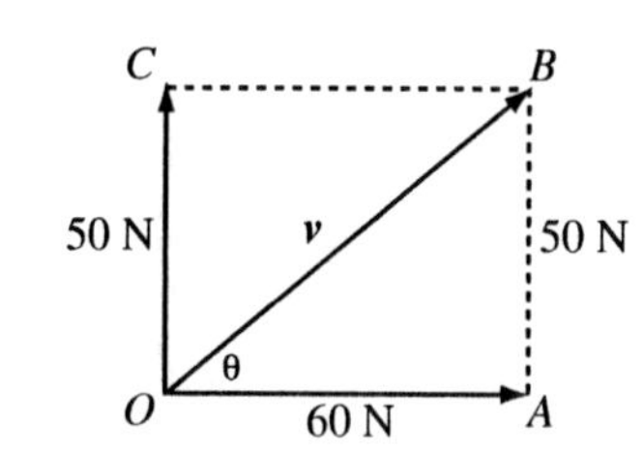

$|\mathbf{v}|^2 = 60^2 + 50^2$

$|\mathbf{v}| = \sqrt{60^2 + 50^2} \approx 78$ N

$\tan\theta = \dfrac{50}{60}$

$\theta \approx 40°$

15.

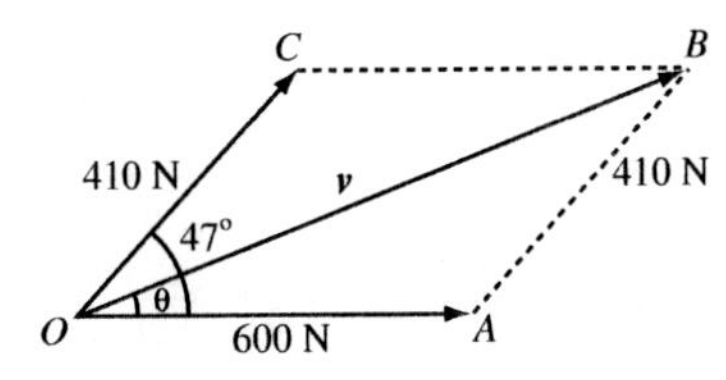

$\angle A = 180° - 47° = 133°$

Use the law of cosines to find the magnitude of the resultant $\mathbf{v}$.

$|\mathbf{v}|^2 = 600^2 + 410^2 - 2 \cdot 600 \cdot 410 \cos 133°$

$|\mathbf{v}| \approx \sqrt{863{,}643} \approx 929$ N

Use the law of sines to find θ.

$\dfrac{410}{\sin\theta} = \dfrac{929}{\sin 133°}$

$\sin\theta = \dfrac{410 \sin 133°}{929} \approx 0.3228$

$\theta \approx 19°$

16.

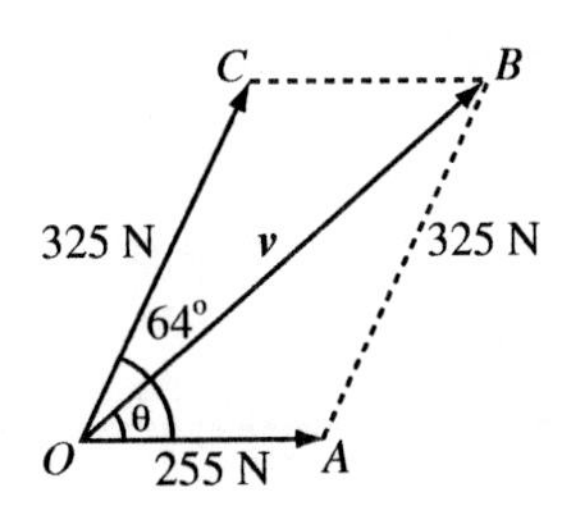

$\angle A = 180° - 64° = 116°$

$|\mathbf{v}|^2 = 255^2 + 325^2 - 2 \cdot 255 \cdot 325 \cos 116°$

$|\mathbf{v}| \approx \sqrt{243{,}310} \approx 493$ N

$\dfrac{325}{\sin\theta} = \dfrac{493}{\sin 116°}$

$\sin\theta = \dfrac{325 \sin 116°}{493} \approx 0.5925$

$\theta \approx 36°$

17.

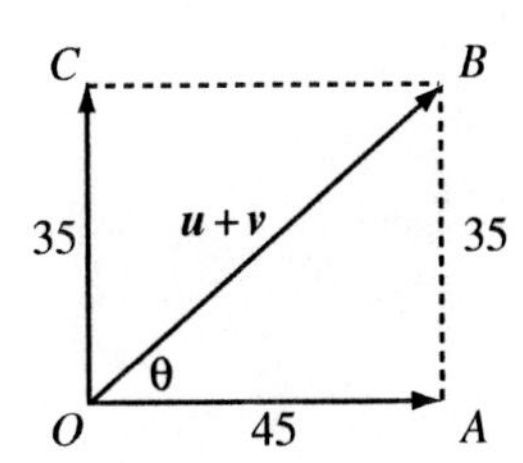

We use the Pythagorean theorem to find the magnitude of $\mathbf{u} + \mathbf{v}$.

$|\mathbf{u} + \mathbf{v}|^2 = 35^2 + 45^2$

$|\mathbf{u} + \mathbf{v}| = \sqrt{35^2 + 45^2}$

$|\mathbf{u} + \mathbf{v}| \approx 57.0$

To find the direction of $\mathbf{u} + \mathbf{v}$ we note that since OAB is a right triangle

$\tan\theta = \dfrac{35}{45}$

$\theta \approx 38°$.

18.

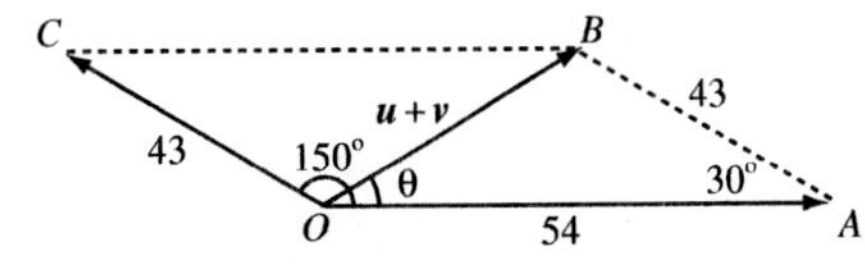

$\angle A = 180° - 150° = 30°$

$|\mathbf{u} + \mathbf{v}|^2 = 54^2 + 43^2 - 2 \cdot 54 \cdot 43 \cos 30°$

$|\mathbf{u} + \mathbf{v}| \approx \sqrt{743.18} \approx 27.3$

$\dfrac{43}{\sin\theta} = \dfrac{27.3}{\sin 30°}$

$\sin\theta = \dfrac{43 \sin 30°}{27.3} \approx 0.7875$

$\theta \approx 52°$

19.

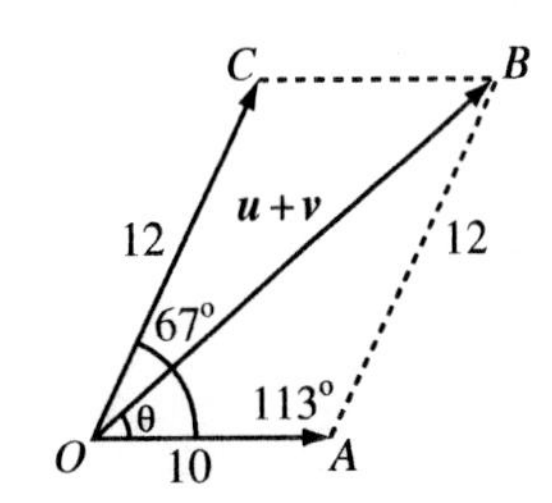

$A = 180° - 67° = 113°$

Use the law of cosines to find $|\mathbf{u} + \mathbf{v}|$.

$|\mathbf{u} + \mathbf{v}|^2 = 10^2 + 12^2 - 2 \cdot 10 \cdot 12 \cos 113°$

$|\mathbf{u} + \mathbf{v}| \approx \sqrt{337.78} \approx 18.4$

Use the law of sines to find θ.

$\dfrac{12}{\sin\theta} = \dfrac{18.4}{\sin 113°}$

$\sin\theta = \dfrac{12 \sin 113°}{18.4} \approx 0.6003$

$\theta \approx 37°$

20.

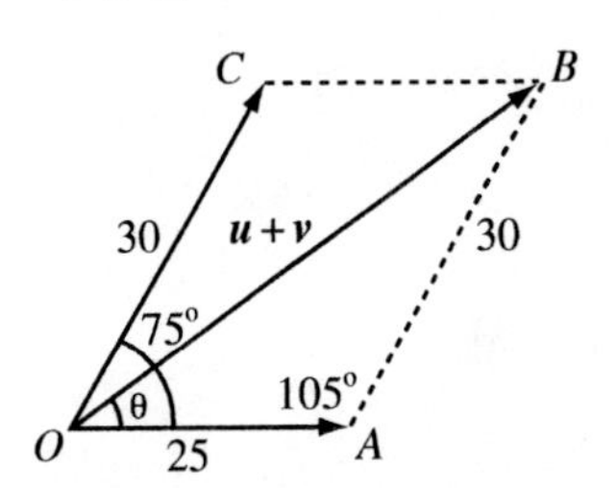

$A = 180° - 75° = 105°$

$|\mathbf{u} + \mathbf{v}|^2 = 25^2 + 30^2 - 2 \cdot 25 \cdot 30 \cos 105°$

$|\mathbf{u} + \mathbf{v}| \approx \sqrt{1913.23} \approx 43.7$

$\dfrac{30}{\sin\theta} = \dfrac{43.7}{\sin 105°}$

$\sin\theta = \dfrac{30 \sin 105°}{43.7} \approx 0.6631$

$\theta \approx 42°$

(Answers may vary slightly due to rounding differences.)

21.

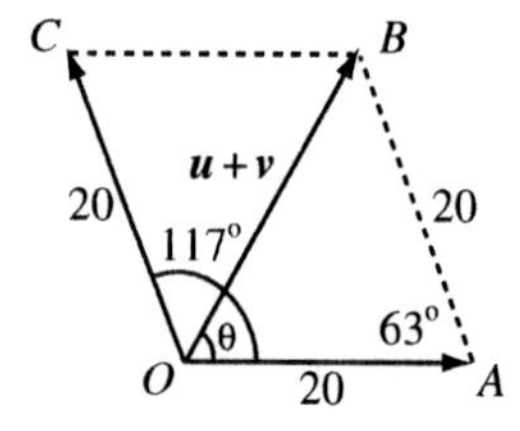

$A = 180° - 117° = 63°$

Use the law of cosines to find $|\mathbf{u} + \mathbf{v}|$.

$$|\mathbf{u} + \mathbf{v}|^2 = 20^2 + 20^2 - 2 \cdot 20 \cdot 20 \cos 63°$$

$$|\mathbf{u} + \mathbf{v}| \approx \sqrt{436.81} \approx 20.9$$

Triangle OAB is isosceles so θ and angle OBA have the same measure.

Thus, $\theta = \frac{1}{2}(180° - 63°) = \frac{1}{2}(117°) = 58.5° \approx 59°$.

(If we use the law of sines and $|\mathbf{u} + \mathbf{v}| \approx 20.9$ to find θ, we get $\theta \approx 58°$.)

22.

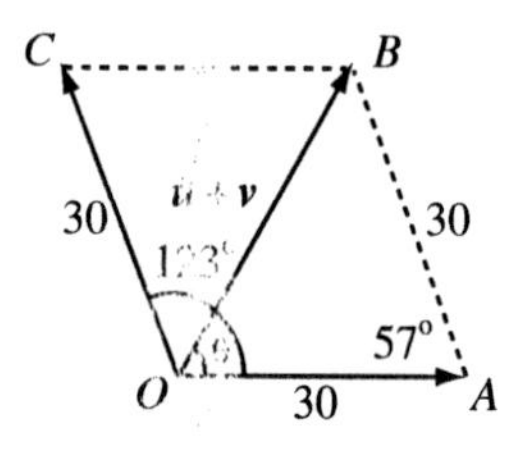

$A = 180° - 123° = 57°$

$$|\mathbf{u} + \mathbf{v}|^2 = 30^2 + 30^2 - 2 \cdot 30 \cdot 30 \cos 57°$$

$$|\mathbf{u} + \mathbf{v}| \approx \sqrt{819.65} \approx 28.6$$

Triangle OAB is isosceles so θ and angle OBA have the same measure.

Thus, $\theta = \frac{1}{2}(180° - 57°) = \frac{1}{2}(123°) = 61.5° \approx 62°$.

23.

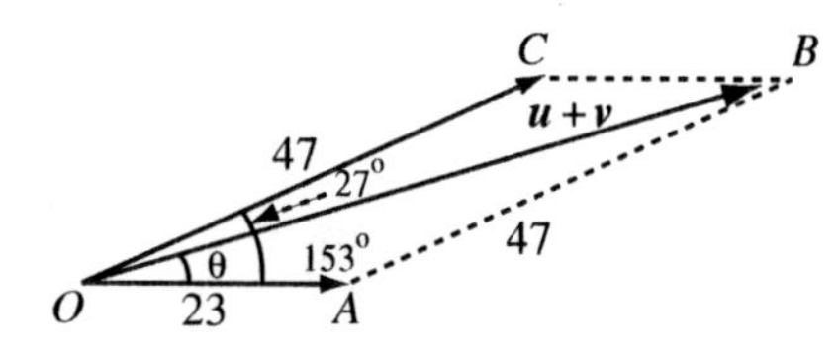

$A = 180° - 27° = 153°$

Use the law of cosines to find $|\mathbf{u} + \mathbf{v}|$.

$$|\mathbf{u} + \mathbf{v}|^2 = 23^2 + 47^2 - 2 \cdot 23 \cdot 47 \cos 153°$$

$$|\mathbf{u} + \mathbf{v}| \approx \sqrt{4664.36} \approx 68.3$$

Use the law of sines to find θ.

$$\frac{47}{\sin\theta} = \frac{68.3}{\sin 153°}$$

$$\sin\theta = \frac{47 \sin 153°}{68.3} \approx 0.3124$$

$$\theta \approx 18°$$

24.

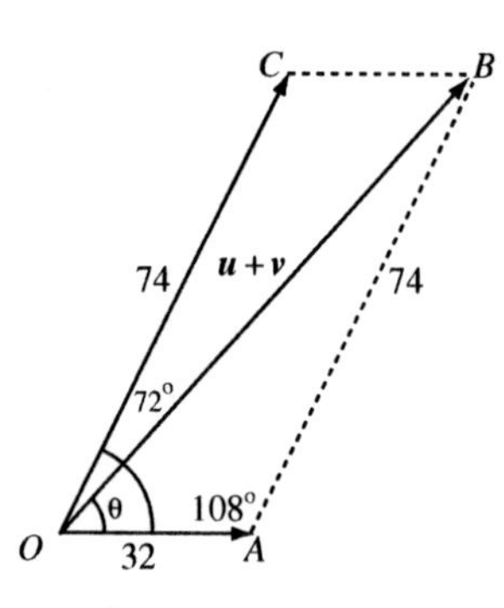

$A = 180° - 72° = 108°$

$$|\mathbf{u} + \mathbf{v}|^2 = 32^2 + 74^2 - 2 \cdot 32 \cdot 74 \cos 108°$$

$$|\mathbf{u} + \mathbf{v}| \approx \sqrt{7963.50} \approx 89.2$$

$$\frac{74}{\sin\theta} = \frac{89.2}{\sin 108°}$$

$$\sin\theta = \frac{74 \sin 108°}{89.2} \approx 0.7890$$

$$\theta \approx 52°$$

25.

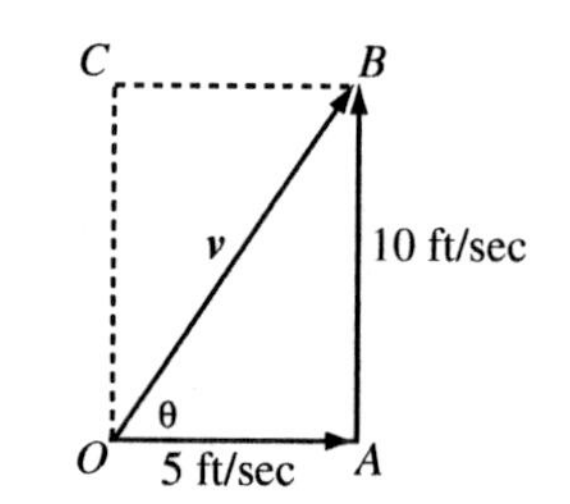

Use the Pythagorean theorem to find the speed of the balloon, $|\mathbf{v}|$.

$$|\mathbf{v}|^2 = 5^2 + 10^2$$

$$|\mathbf{v}| = \sqrt{5^2 + 10^2} \approx 11 \text{ ft/sec}$$

Use the fact that triangle OAB is a right triangle to find θ.

$$\tan\theta = \frac{10}{5} = 2$$

$$\theta \approx 63°$$

26.

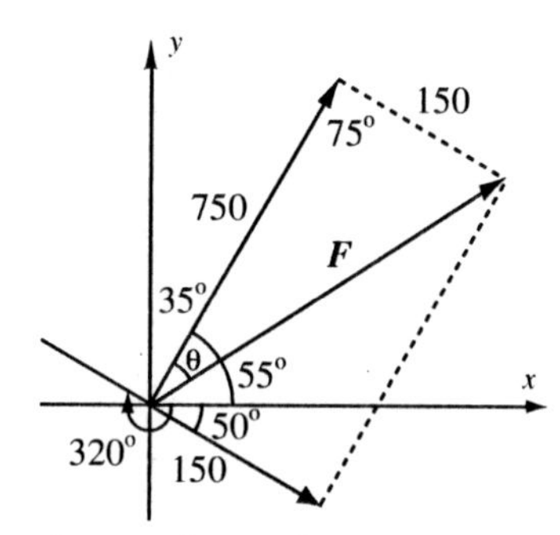

$$|\mathbf{F}|^2 = 750^2 + 150^2 - 2 \cdot 750 \cdot 150 \cos 75°$$

$$|\mathbf{F}| = \sqrt{526,766} \approx 726 \text{ lb}$$

$$\frac{150}{\sin\theta} = \frac{726}{\sin 75°}$$

$$\sin\theta = \frac{150 \sin 75°}{726} \approx 0.1996$$

$$\theta \approx 12°$$

The boat is moving in the direction $35° + 12°$, or $47°$.

27.

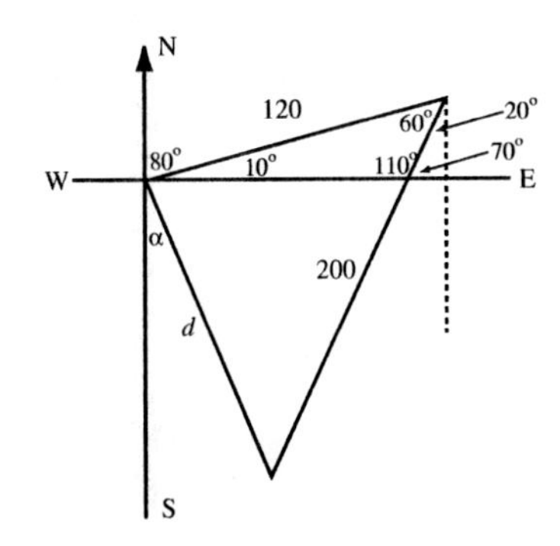

Note:

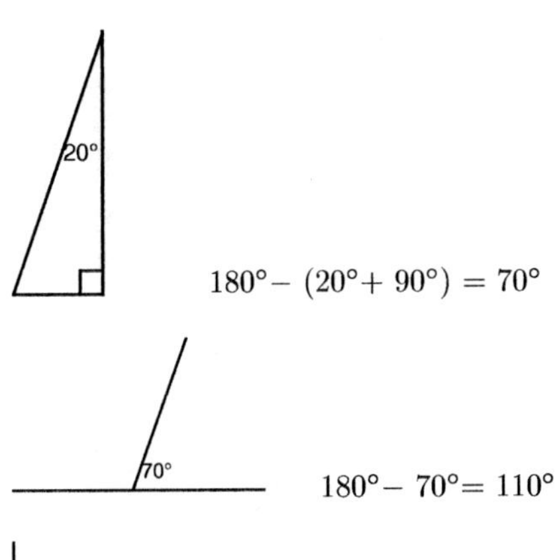

$180° - (20° + 90°) = 70°$

$180° - 70° = 110°$

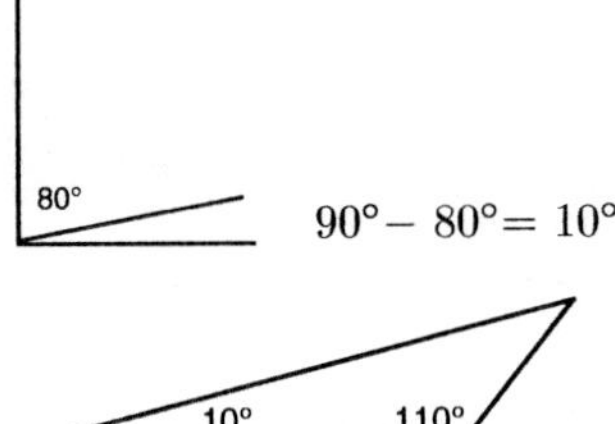

$90° - 80° = 10°$

$180° - (10° + 110°) = 60°$

Use the law of cosines to find $|\mathbf{d}|$.

$$|\mathbf{d}|^2 = 120^2 + 200^2 - 2 \cdot 120 \cdot 200 \cos 60°$$
$$|\mathbf{d}| = \sqrt{30,400} \approx 174$$

Use the law of sines to find θ in this triangle.

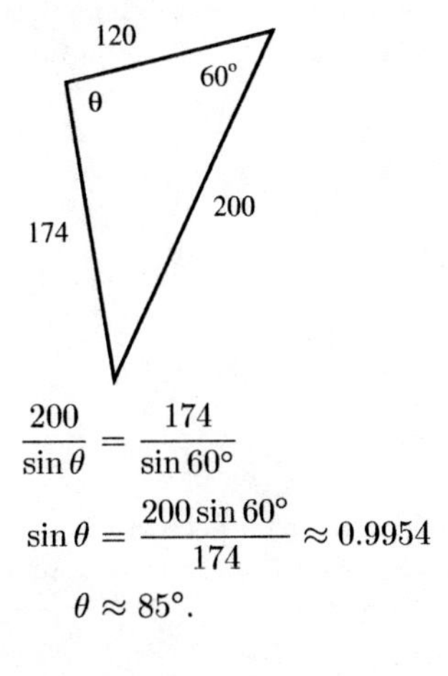

$$\frac{200}{\sin\theta} = \frac{174}{\sin 60°}$$
$$\sin\theta = \frac{200\sin 60°}{174} \approx 0.9954$$
$$\theta \approx 85°.$$

Now we can find α.

$\alpha = 180° - (80° + 85°) = 15°$.

The ship is 174 nautical miles from the starting point in the direction S 15° E. (Answers may vary due to rounding differences.)

28.

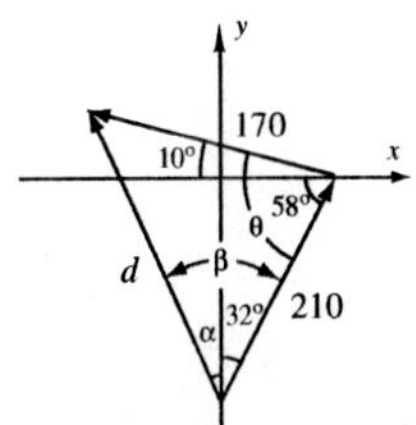

$$|\mathbf{d}|^2 = 170^2 + 210^2 - 2 \cdot 170 \cdot 210 \cos 68°$$
$$|\mathbf{d}| = \sqrt{46,253} \approx 215 \text{ km}$$
$$\frac{170}{\sin\beta} = \frac{215}{\sin 68°}$$
$$\sin\beta = \frac{170\sin 68°}{215} \approx 0.7331$$
$$\beta \approx 47°$$

Then $\alpha = 47° - 32 = 15°$, so the direction of the plane is $360° - 15°$, or $345°$.

29.

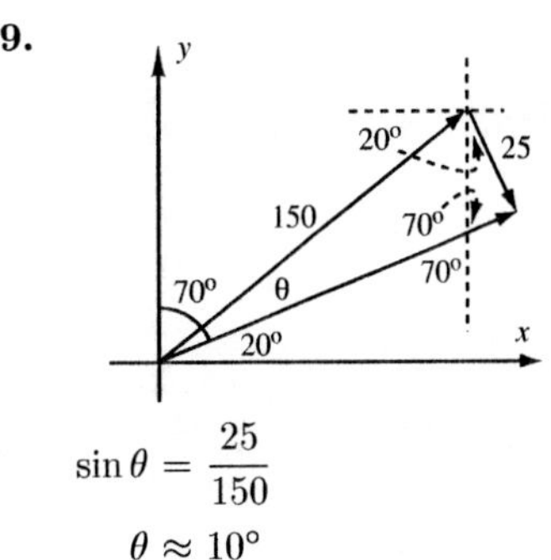

$$\sin\theta = \frac{25}{150}$$
$$\theta \approx 10°$$

Then the airplane's actual heading will be about $90° - (10° + 20°)$, or $60°$.

30.

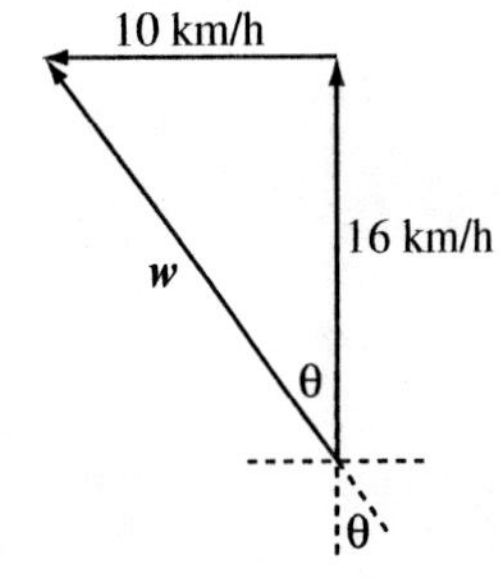

$$|\mathbf{w}| = \sqrt{10^2 + 16^2} \approx 18.9 \text{ km/h}$$
$$\tan\theta = \frac{10}{16}$$
$$\theta \approx 32°$$

The wind is blowing from the direction

$180° - \theta = 180° - 32° = 148°$.

31.

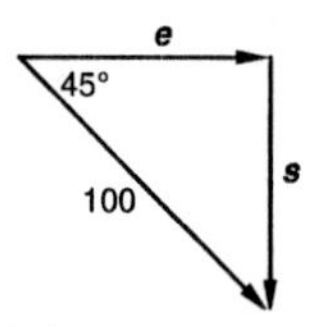

$$\frac{|\mathbf{e}|}{100} = \cos 45°$$
$$|\mathbf{e}| = 100 \cos 45° \approx 70.7$$

$$\frac{|\mathbf{s}|}{100} = \sin 45°$$
$$|\mathbf{s}| = 100 \sin 45° \approx 70.7$$

The easterly and southerly components are both 70.7.

32.

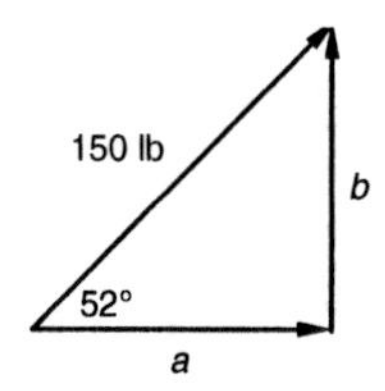

$$|\mathbf{a}| = 150 \cos 52° \approx 92$$
$$|\mathbf{b}| = 150 \cos 52° \approx 118$$

The horizontal component is about 92 lb, and the vertical component is about 118 lb.

33.

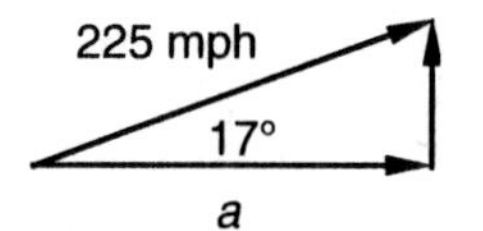

$$|\mathbf{a}| = 225 \cos 17° \approx 215.17$$
$$|\mathbf{b}| = 225 \sin 17° \approx 65.78$$

The horizontal component is about 215.17 mph forward, and the vertical component is about 65.78 mph up.

34.

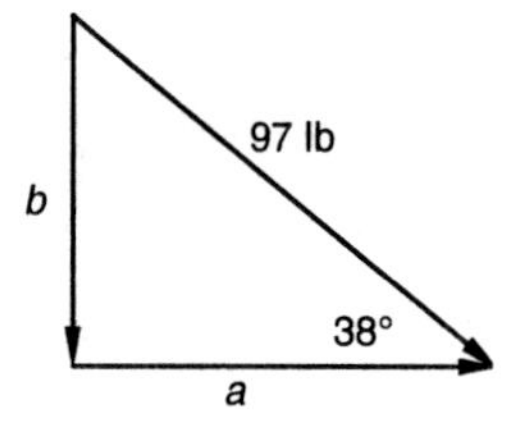

$$|\mathbf{a}| = 97 \cos 38° \approx 76.44$$
$$|\mathbf{b}| = 97 \sin 38° \approx 59.72$$

The horizontal component is about 76.44 lb forward, and the vertical component is about 59.72 lb down.

35.

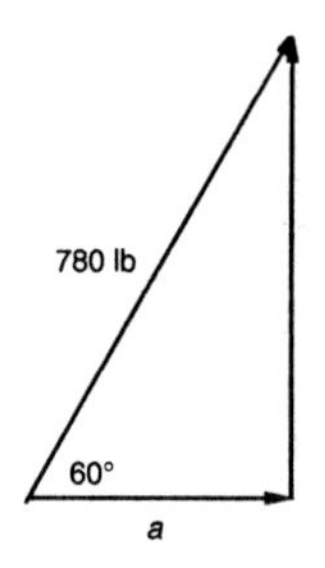

$$|\mathbf{a}| = 780 \cos 60° = 390$$
$$|\mathbf{b}| = 780 \sin 60° \approx 675.5$$

The horizontal component is about 390 lb forward, and the vertical component is about 675.5 lb up.

36.

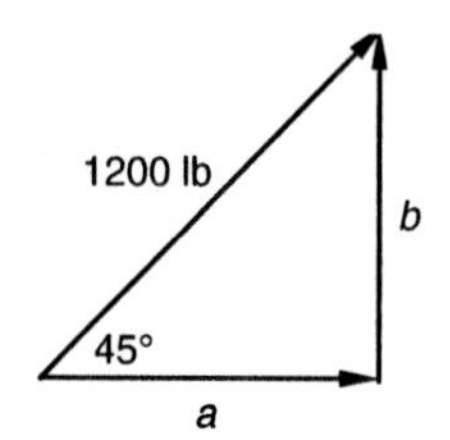

$$|\mathbf{a}| = 1200 \cos 45° \approx 848.5$$
$$|\mathbf{b}| = 1200 \sin 45° \approx 848.5$$

The horizontal component is about 848.5 lb toward the balloon, and the vertical component is about 848.5 lb up.

37.

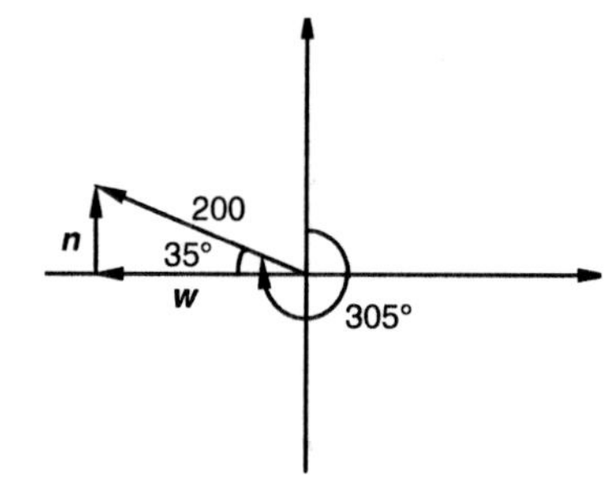

$$|\mathbf{w}| = 200 \cos 35° \approx 164$$
$$|\mathbf{n}| = 200 \sin 35° \approx 115$$

The westerly component is about 164 km/h, and the northerly component is about 115 km/h.

38.

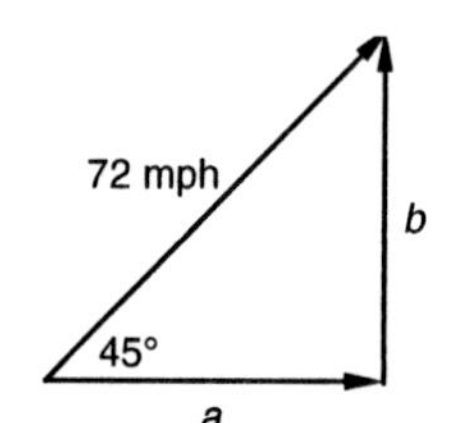

$$|\mathbf{a}| = 72 \cos 45° \approx 50.91$$
$$|\mathbf{b}| = 72 \sin 45° \approx 50.91$$

The horizontal component is about 50.91 mph forward, and the vertical component is about 50.91 mph up.

39.

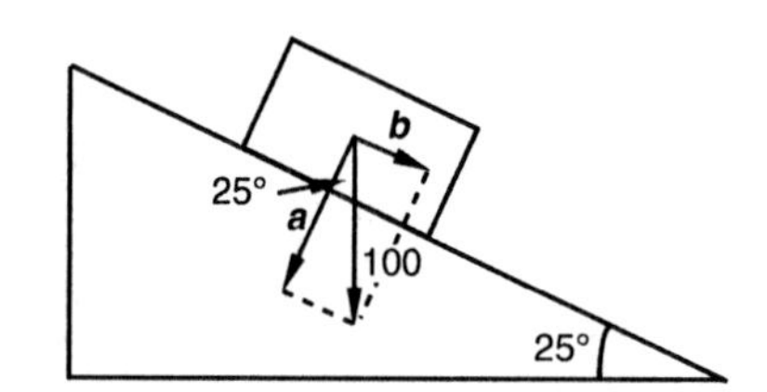

$$|\mathbf{a}| = 100 \cos 25° \approx 90.6$$
$$|\mathbf{b}| = 100 \sin 25° \approx 42.3$$

The magnitude of the component perpendicular to the incline is about 90.6 lb; the magnitude of the component parallel to the incline is about 42.3 lb.

40.

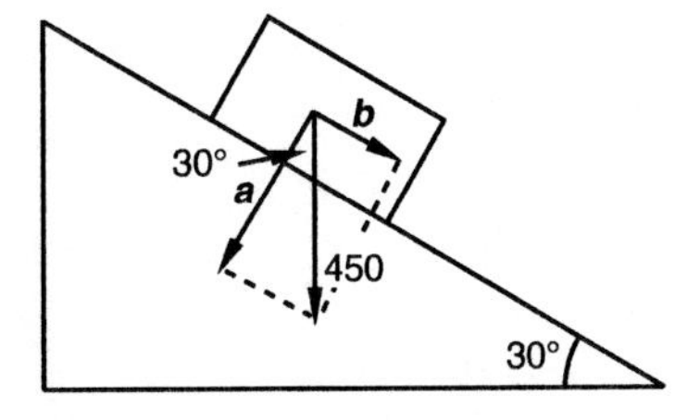

$|\mathbf{a}| = 450 \cos 30° \approx 389.7$ kg

$|\mathbf{b}| = 450 \sin 30° = 225$ kg

41.

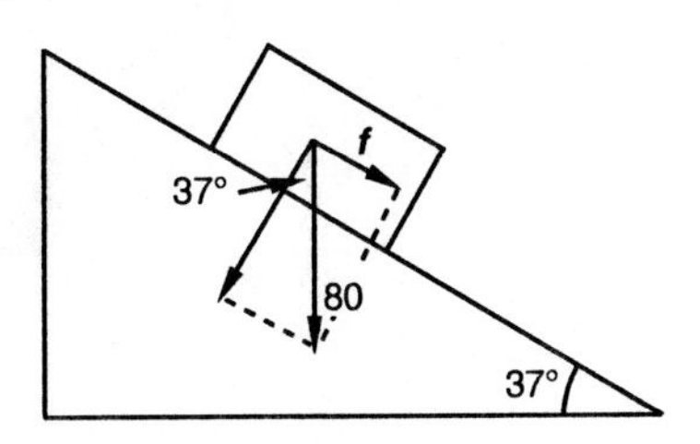

$|\mathbf{f}| = 80 \sin 37° \approx 48.1$ lb

42.

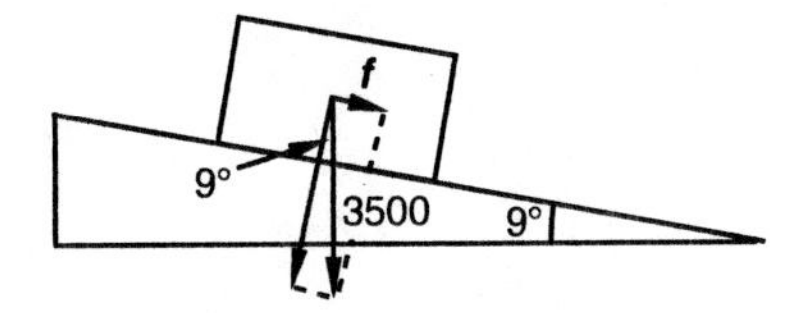

$|\mathbf{f}| = 3500 \sin 9° \approx 547.5$ lb

43. A vector is a quantity that has both magnitude and direction.

44. Vectors $\overrightarrow{QR}$ and $\overrightarrow{RQ}$ have opposite directions, so they are not equivalent.

45. natural

46. half-angle

47. linear speed

48. cosine

49. identity

50. cotangent of θ

51. coterminal

52. sines

53. horizontal line; inverse

54. reference angle; acute

55.

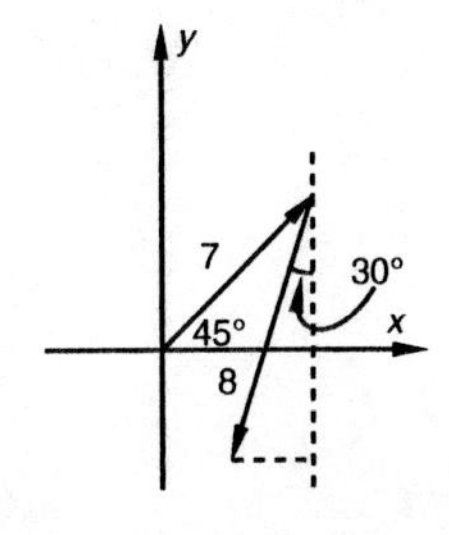

a)

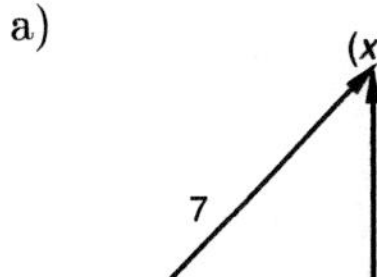

$x = 7 \cos 45° \approx 4.950$

$y = 7 \sin 45° \approx 4.950$

The cliff is located at $(4.950, 4.950)$.

b)

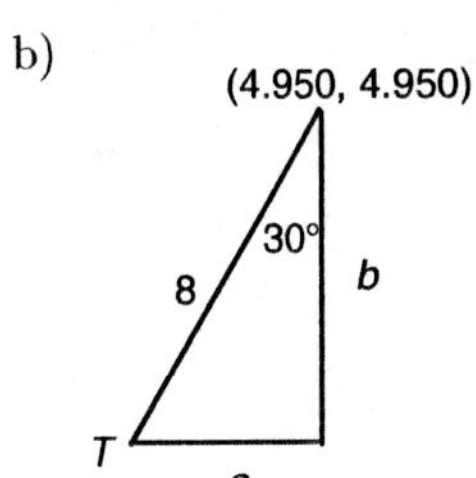

$a = 8 \sin 30° = 4$

$b = 8 \cos 30° \approx 6.928$

Then the coordinates of T are $(4.950 - 4, 4.950 - 6.928)$, or $(0.950, -1.978)$.

Exercise Set 7.6

1. $\overrightarrow{MN} = \langle -3 - 6, -2 - (-7) \rangle = \langle -9, 5 \rangle$

$|\overrightarrow{MN}| = \sqrt{(-9)^2 + 5^2} = \sqrt{81 + 25} = \sqrt{106}$

2. $\overrightarrow{CD} = \langle 5 - 1, 7 - 5 \rangle = \langle 4, 2 \rangle$

$|\overrightarrow{CD}| = \sqrt{4^2 + 2^2} = \sqrt{16 + 4} = \sqrt{20}$, or $2\sqrt{5}$

3. $\overrightarrow{FE} = \langle 8 - 11, 4 - (-2) \rangle = \langle -3, 6 \rangle$

$|\overrightarrow{FE}| = \sqrt{(-3)^2 + 6^2} = \sqrt{9 + 36} = \sqrt{45}$, or $3\sqrt{5}$

4. $\overrightarrow{BA} = \langle 9 - 9, 0 - 7) \rangle = \langle 0, -7 \rangle$

$|\overrightarrow{BA}| = \sqrt{0^2 + (-7)^2} = \sqrt{49} = 7$

5. $\overrightarrow{KL} = \langle 8 - 4, -3 - (-3) \rangle = \langle 4, 0 \rangle$

$|\overrightarrow{KL}| = \sqrt{4^2 + 0^2} = \sqrt{16} = 4$

6. $\overrightarrow{GH} = \langle -3 - (-6), 2 - 10 \rangle = \langle 3, -8 \rangle$

$|\overrightarrow{GH}| = \sqrt{3^2 + (-8)^2} = \sqrt{9 + 64} = \sqrt{73}$

7. $|\mathbf{u}| = \sqrt{(-1)^2 + 6^2} = \sqrt{1 + 36} = \sqrt{37}$

8. $|\overrightarrow{ST}| = \sqrt{(-12)^2 + 5^2} = \sqrt{169} = 13$

9. $\mathbf{u} + \mathbf{w} = \langle 5 + (-1), -2 + (-3) \rangle = \langle 4, -5 \rangle$

10. $\mathbf{w} + \mathbf{u} = \langle -1 + 5, -3 + (-2) \rangle = \langle 4, -5 \rangle$

11. $|3\mathbf{w}-\mathbf{v}| = |3\langle -1,-3\rangle - \langle -4,7\rangle|$
$= |\langle -3,-9\rangle - \langle -4,7\rangle|$
$= |\langle -3,-(-4),-9-7\rangle|$
$= |\langle 1,-16\rangle|$
$= \sqrt{1^2+(-16)^2}$
$= \sqrt{257}$

12. $|6\mathbf{v}+5\mathbf{u}| = 6\langle -4,7\rangle + 5\langle 5,-2\rangle$
$= \langle -24,42\rangle + \langle 25,-10\rangle$
$= \langle -24+25, 42-10\rangle$
$= \langle 1,32\rangle$

13. $\mathbf{v}-\mathbf{u} = \langle -4,7\rangle - \langle 5,-2\rangle = \langle -4-5, 7-(-2)\rangle = \langle -9,9\rangle$

14. $|2\mathbf{w}| = |2\langle -1,-3\rangle|$
$= |\langle -2,-6\rangle|$
$= \sqrt{(-2)^2+(-6)^2}$
$= \sqrt{40}$, or $2\sqrt{10}$

15. $5\mathbf{u}-4\mathbf{v} = 5\langle 5,-2\rangle - 4\langle -4,7\rangle$
$= \langle 25,-10\rangle - \langle -16,28\rangle$
$= \langle 25-(-16), -10-28\rangle$
$= \langle 41,-38\rangle$

16. $-5\mathbf{v} = -5\langle -4,7\rangle = \langle 20,-35\rangle$

17. $|3\mathbf{u}| - |\mathbf{v}| = |3\langle 5,-2\rangle| - |\langle -4,7\rangle|$
$= |\langle 15,-6\rangle| - |\langle -4,7\rangle|$
$= \sqrt{15^2+(-6)^2} - \sqrt{(-4)^2+7^2}$
$= \sqrt{261} - \sqrt{65}$

18. $|\mathbf{v}| + |\mathbf{u}| = |\langle -4,7\rangle| + |\langle 5,-2\rangle|$
$= \sqrt{(-4)^2+7^2} + \sqrt{5^2+(-2)^2}$
$= \sqrt{65} + \sqrt{29}$

19. $\mathbf{v}+\mathbf{u}+2\mathbf{w} = \langle -4,7\rangle + \langle 5,-2\rangle + 2\langle -1,-3\rangle$
$= \langle -4,7\rangle + \langle 5,-2\rangle + \langle -2,-6\rangle$
$= \langle -4+5+(-2), 7+(-2)+(-6)\rangle$
$= \langle -1,-1\rangle$

20. $\mathbf{w}-(\mathbf{u}+4\mathbf{v}) = \langle -1,-3\rangle - (\langle 5,-2\rangle + 4\langle -4,7\rangle)$
$= \langle -1,-3\rangle - (\langle 5,-2\rangle + \langle -16,28\rangle)$
$= \langle -1,-3\rangle - \langle -11,26\rangle$
$= \langle -1-(-11), -3-26\rangle$
$= \langle 10,-29\rangle$

21. $2\mathbf{v}+\mathbf{O} = 2\mathbf{v} = 2\langle -4,7\rangle = \langle -8,14\rangle$

22. $10|7\mathbf{w}-3\mathbf{u}| = 10|7\langle -1,-3\rangle - 3\langle 5,-2\rangle|$
$= 10|\langle -7,-21\rangle - \langle 15,-6\rangle|$
$= 10|\langle -7-15, -21-(-6)\rangle|$
$= 10|\langle -22,-15\rangle|$
$= 10\sqrt{(-22)^2+(-15)^2}$
$= 10\sqrt{709}$

23. $\mathbf{u}\cdot\mathbf{w} = 5(-1)+(-2)(-3) = -5+6 = 1$

24. $\mathbf{w}\cdot\mathbf{u} = -1\cdot 5+(-3)(-2) = -5+6 = 1$

25. $\mathbf{u}\cdot\mathbf{v} = 5(-4)+(-2)(7) = -20-14 = -34$

26. $\mathbf{v}\cdot\mathbf{w} = -4(-1)+7(-3) = 4-21 = -17$

27. See the answer section in the text.

28.

29. See the answer section in the text.

30.

31. (a) $\mathbf{w} = \mathbf{u}+\mathbf{v}$

(b) $\mathbf{v} = \mathbf{w}-\mathbf{u}$

32. Since the diagonals of a parallelogram intersect at their midpoints, P is the midpoint of $\mathbf{u}+\mathbf{w}$. Thus

$\mathbf{v} = \frac{1}{2}(\mathbf{u}+\mathbf{w})$.

33. $|\langle -5,12\rangle| = \sqrt{(-5)^2+12^2} = \sqrt{169} = 13$

$\frac{1}{13}\langle -5,12\rangle = \left\langle -\frac{5}{13}, \frac{12}{13}\right\rangle$

34. $|\langle 3,4\rangle| = \sqrt{3^2+4^2} = \sqrt{25} = 5$

$\frac{1}{5}\langle 3,4\rangle = \left\langle \frac{3}{5}, \frac{4}{5}\right\rangle$

35. $|\langle 1,-10\rangle| = \sqrt{1^2+(-10)^2} = \sqrt{101}$

$\frac{1}{\sqrt{101}}\langle 1,-10\rangle = \left\langle \frac{1}{\sqrt{101}}, -\frac{10}{\sqrt{101}}\right\rangle$

36. $|\langle 6,-7\rangle| = \sqrt{6^2+(-7)^2} = \sqrt{85}$

$\frac{1}{\sqrt{85}}\langle 6,-7\rangle = \left\langle \frac{6}{\sqrt{85}}, -\frac{7}{\sqrt{85}}\right\rangle$

37. $|\langle -2,-8\rangle| = \sqrt{(-2)^2+(-8)^2} = \sqrt{68} = 2\sqrt{17}$

$\frac{1}{2\sqrt{17}}\langle -2,-8\rangle = \left\langle -\frac{2}{2\sqrt{17}}, -\frac{8}{2\sqrt{17}}\right\rangle =$
$\left\langle -\frac{1}{\sqrt{17}}, -\frac{4}{\sqrt{17}}\right\rangle$

38. $|\langle -3,-3\rangle| = \sqrt{(-3)^2+(-3)^2} = \sqrt{18} = 3\sqrt{2}$

$\frac{1}{3\sqrt{2}}\langle -3,-3\rangle = \left\langle -\frac{1}{\sqrt{2}}, -\frac{1}{\sqrt{2}}\right\rangle$

39. $\langle -4,6\rangle = -4\mathbf{i}+6\mathbf{j}$

40. $-15\mathbf{i}+9\mathbf{j}$

41. $\langle 2,5\rangle = 2\mathbf{i}+5\mathbf{j}$

42. $2\mathbf{i}-\mathbf{j}$

43. Horizontal component: $-3 - 4 = -7$

Vertical component: $3 - (-2) = 5$

We write the vector as $-7\mathbf{i} + 5\mathbf{j}$.

44. Horizontal component: $1 - (-3) = 4$

Vertical component: $4 - (-4) = 8$

We write the vector as $4\mathbf{i} + 8\mathbf{j}$.

45. (a) $$\begin{aligned}4\mathbf{u} - 5\mathbf{w} &= 4(2\mathbf{i} + \mathbf{j}) - 5(\mathbf{i} - 5\mathbf{j})\\ &= 8\mathbf{i} + 4\mathbf{j} - 5\mathbf{i} + 25\mathbf{j}\\ &= 3\mathbf{i} + 29\mathbf{j}\end{aligned}$$

(b) $3\mathbf{i} + 29\mathbf{j} = \langle 3, 29\rangle$

46. (a) $$\begin{aligned}\mathbf{v} + 3\mathbf{w} &= -3\mathbf{i} - 10\mathbf{j} + 3(\mathbf{i} - 5\mathbf{j})\\ &= -3\mathbf{i} - 10\mathbf{j} + 3\mathbf{i} - 15\mathbf{j}\\ &= -25\mathbf{j}\end{aligned}$$

(b) $-25\mathbf{j} = 0\mathbf{i} - 25\mathbf{j} = \langle 0, -25\rangle$

47. (a) $$\begin{aligned}\mathbf{u} - (\mathbf{v} + \mathbf{w}) &= 2\mathbf{i} + \mathbf{j} - (-3\mathbf{i} - 10\mathbf{j} + \mathbf{i} - 5\mathbf{j})\\ &= 2\mathbf{i} + \mathbf{j} - (-2\mathbf{i} - 15\mathbf{j})\\ &= 2\mathbf{i} + \mathbf{j} + 2\mathbf{i} + 15\mathbf{j}\\ &= 4\mathbf{i} + 16\mathbf{j}\end{aligned}$$

(b) $4\mathbf{i} + 16\mathbf{j} = \langle 4, 16\rangle$

48. (a) $$\begin{aligned}(\mathbf{u} - \mathbf{v}) + \mathbf{w} &= (2\mathbf{i} + \mathbf{j} - (-3\mathbf{i} - 10\mathbf{j})) + \mathbf{i} - 5\mathbf{j}\\ &= (2\mathbf{i} + \mathbf{j} + 3\mathbf{i} + 10\mathbf{j}) + \mathbf{i} - 5\mathbf{j}\\ &= 5\mathbf{i} + 11\mathbf{j} + \mathbf{i} - 5\mathbf{j}\\ &= 6\mathbf{i} + 6\mathbf{j}\end{aligned}$$

(b) $\langle 6, 6\rangle$

49.

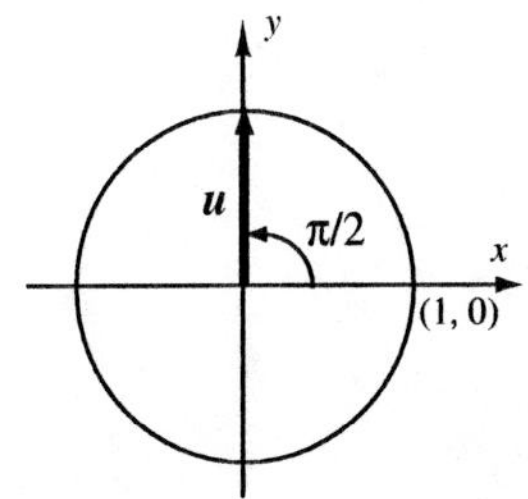

$$\mathbf{u} = \left(\cos\frac{\pi}{2}\right)\mathbf{i} + \left(\sin\frac{\pi}{2}\right)\mathbf{j} = 0\mathbf{i} + 1\mathbf{j} = \mathbf{j}, \text{ or } \langle 0, 1\rangle$$

50.

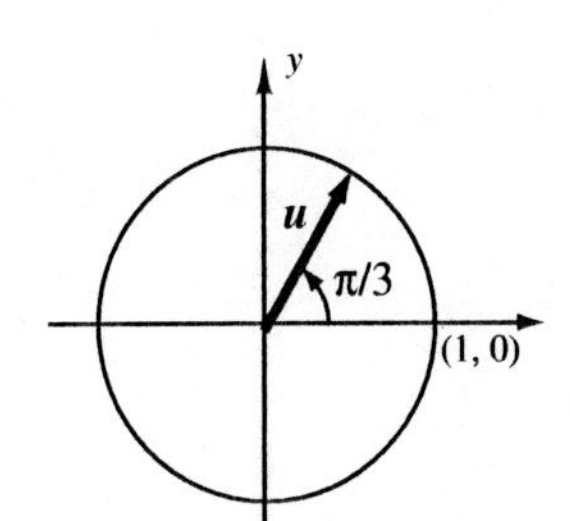

$$\mathbf{u} = \left(\cos\frac{\pi}{3}\right)\mathbf{i} + \left(\sin\frac{\pi}{3}\right)\mathbf{j} = \frac{1}{2}\mathbf{i} + \frac{\sqrt{3}}{2}\mathbf{j}, \text{ or } \left\langle\frac{1}{2}, \frac{\sqrt{3}}{2}\right\rangle$$

51.

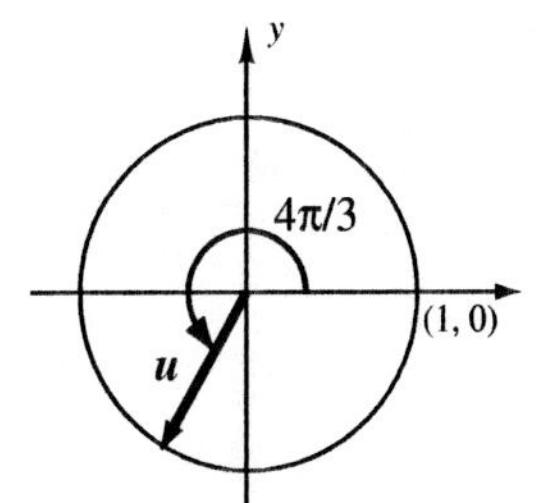

$$\mathbf{u} = \left(\cos\frac{4\pi}{3}\right)\mathbf{i} + \left(\sin\frac{4\pi}{3}\right)\mathbf{j} = -\frac{1}{2}\mathbf{i} - \frac{\sqrt{3}}{2}\mathbf{j}, \text{ or}$$
$$\left\langle -\frac{1}{2}, -\frac{\sqrt{3}}{2}\right\rangle$$

52.

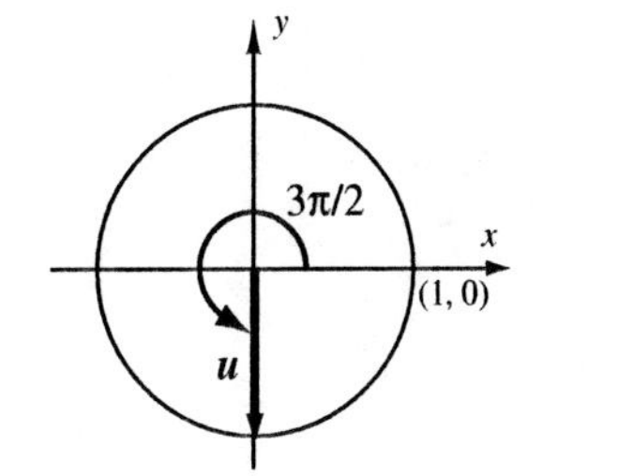

$$\mathbf{u} = \left(\cos\frac{3\pi}{2}\right)\mathbf{i} + \left(\sin\frac{3\pi}{2}\right)\mathbf{j} = 0\mathbf{i} - 1\cdot\mathbf{j} = -\mathbf{j}, \text{ or } \langle 0, -1\rangle$$

53. $\tan\theta = \dfrac{-5}{-2} = \dfrac{5}{2}$

$\theta = \tan^{-1}\dfrac{5}{2}$

The vector is in the third quadrant, so θ is a third-quadrant angle. The reference angle is

$\tan^{-1}\dfrac{5}{2} \approx 68°$

and $\theta \approx 180° + 68°$, or $248°$.

54. $\tan\theta = \dfrac{-3}{4} = -\dfrac{3}{4}$

$\theta = \tan^{-1}\left(-\dfrac{3}{4}\right)$

θ is a fourth-quadrant angle. The reference angle is

$\tan^{-1}\dfrac{3}{4} \approx 37°$

and $\theta \approx 360° - 37°$, or $323°$.

55. $\tan\theta = \dfrac{2}{1} = 2$

$\theta = \tan^{-1} 2$

The vector is in the first quadrant, so θ is a first-quadrant angle. Then

$\theta = \tan^{-1} 2 \approx 63°$.

56. $\tan\theta = \dfrac{-1}{5} = -\dfrac{1}{5}$

$\theta = \tan^{-1}\left(-\dfrac{1}{5}\right)$

θ is a fourth-quadrant angle. The reference angle is

$\tan^{-1}\dfrac{1}{5} \approx 11°$

and $\theta \approx 360° - 11°$, or $349°$.

57. $\tan\theta = \dfrac{6}{5}$

$\theta = \tan^{-1}\dfrac{6}{5}$

The vector is in the first quadrant, so θ is a first-quadrant angle. Then

$\theta = \tan^{-1}\dfrac{6}{5} \approx 50°.$

58. $\tan\theta = \dfrac{-4}{-8} = \dfrac{1}{2}$

$\theta = \tan^{-1}\dfrac{1}{2}$

θ is a third-quadrant angle. The reference angle is

$\tan^{-1}\dfrac{1}{2} \approx 27°$

and $\theta \approx 180° + 27°$, or $207°$.

59. $|\mathbf{u}| = \sqrt{(3\cos 45°)^2 + (3\sin 45°)^2}$

$= \sqrt{9\cos^2 45° + 9\sin^2 45°}$

$= \sqrt{9(\cos^2 45° + \sin^2 45°)}$

$= \sqrt{9 \cdot 1} = \sqrt{9} = 3$

The vector is given in terms of the direction angle, 45°.

60. $|\mathbf{w}| = \sqrt{(6\cos 150°)^2 + (6\sin 150°)^2}$

$= \sqrt{36\cos^2 150° + 36\sin^2 150°}$

$= \sqrt{36(\cos^2 150° + \sin^2 150°)}$

$= \sqrt{36 \cdot 1} = 6$

The vector is given in terms of the direction angle, 150°.

61. $|\mathbf{v}| = \sqrt{\left(-\dfrac{1}{2}\right)^2 + \left(\dfrac{\sqrt{3}}{2}\right)^2} = \sqrt{\dfrac{1}{4} + \dfrac{3}{4}} = \sqrt{1} = 1$

$\tan\theta = \dfrac{\frac{\sqrt{3}}{2}}{-\frac{1}{2}} = \dfrac{\sqrt{3}}{2}\left(-\dfrac{2}{1}\right) = -\sqrt{3}$

$\theta = \tan^{-1}(-\sqrt{3})$

The vector is in the second quadrant, so θ is a second-quadrant angle. The reference angle is

$\tan^{-1}\sqrt{3} = 60°$

and $\theta = 180° - 60°$, or $120°$.

62. $|\mathbf{u}| = \sqrt{(-1)^2 + (-1)^2} = \sqrt{2}$

$\tan\theta = \dfrac{-1}{-1} = 1$

$\theta = \tan^{-1} 1$

θ is a third-quadrant angle. The reference angle is

$\tan^{-1} 1 = 45°$

and $\theta = 180° + 45°$, or $225°$.

63. $\mathbf{u} = \langle 2, -5\rangle,\ \mathbf{v} = \langle 1, 4\rangle$

$\mathbf{u}\cdot\mathbf{v} = 2\cdot 1 + (-5)(4) = -18$

$|\mathbf{u}| = \sqrt{2^2 + (-5)^2} = \sqrt{29}$

$|\mathbf{v}| = \sqrt{1^2 + 4^2} = \sqrt{17}$

$\cos\alpha = \dfrac{\mathbf{u}\cdot\mathbf{v}}{|\mathbf{u}|\,|\mathbf{v}|} = \dfrac{-18}{\sqrt{29}\sqrt{17}}$

$\alpha = \cos^{-1}\dfrac{-18}{\sqrt{29}\sqrt{17}}$

$\alpha \approx 144.2°$

64. $\mathbf{a} = \langle -3, -3\rangle,\ \mathbf{b} = \langle -5, 2\rangle$

$\mathbf{a}\cdot\mathbf{b} = -3(-5) + (-3)(2) = 9$

$|\mathbf{a}| = \sqrt{(-3)^2 + (-3)^2} = \sqrt{18}$

$|\mathbf{b}| = \sqrt{(-5)^2 + 2^2} = \sqrt{29}$

$\cos\alpha = \dfrac{\mathbf{a}\cdot\mathbf{b}}{|\mathbf{a}|\,|\mathbf{b}|} = \dfrac{9}{\sqrt{18}\sqrt{29}}$

$\alpha = \cos^{-1}\dfrac{9}{\sqrt{18}\sqrt{29}}$

$\alpha \approx 66.8°$

65. $\mathbf{w} = \langle 3, 5\rangle,\ \mathbf{r} = \langle 5, 5\rangle$

$\mathbf{w}\cdot\mathbf{r} = 3\cdot 5 + 5\cdot 5 = 40$

$|\mathbf{w}| = \sqrt{3^2 + 5^2} = \sqrt{34}$

$|\mathbf{r}| = \sqrt{5^2 + 5^2} = \sqrt{50}$

$\cos\alpha = \dfrac{\mathbf{w}\cdot\mathbf{r}}{|\mathbf{w}|\,|\mathbf{r}|} = \dfrac{40}{\sqrt{34}\sqrt{50}}$

$\alpha = \cos^{-1}\dfrac{40}{\sqrt{34}\sqrt{50}}$

$\alpha \approx 14.0°$

66. $\mathbf{v} = \langle -4, 2\rangle,\ \mathbf{t} = \langle 1, -4\rangle$

$\mathbf{v}\cdot\mathbf{t} = -4\cdot 1 + 2(-4) = -12$

$|\mathbf{v}| = \sqrt{(-4)^2 + 2^2} = \sqrt{20}$

$|\mathbf{t}| = \sqrt{1^2 + (-4)^2} = \sqrt{17}$

$\cos\alpha = \dfrac{\mathbf{v}\cdot\mathbf{t}}{|\mathbf{v}|\,|\mathbf{t}|} = \dfrac{-12}{\sqrt{20}\sqrt{17}}$

$\alpha = \cos^{-1}\dfrac{-12}{\sqrt{20}\sqrt{17}}$

$\alpha \approx 130.6°$

67. $\mathbf{a} = \mathbf{i} + \mathbf{j},\ \mathbf{t} = 2\mathbf{i} - 3\mathbf{j}$

$\mathbf{a}\cdot\mathbf{b} = 1\cdot 2 + 1(-3) = -1$

$|\mathbf{a}| = \sqrt{1^2 + 1^2} = \sqrt{2}$

$|\mathbf{b}| = \sqrt{2^2 + (-3)^2} = \sqrt{13}$

$\cos\alpha = \dfrac{\mathbf{a}\cdot\mathbf{b}}{|\mathbf{a}|\,|\mathbf{b}|} = \dfrac{-1}{\sqrt{2}\sqrt{13}}$

$\alpha = \cos^{-1}\dfrac{-1}{\sqrt{2}\sqrt{13}}$

$\alpha \approx 101.3°$

68. $\mathbf{u} = 3\mathbf{i} + 2\mathbf{j},\ \mathbf{t} = -\mathbf{i} + 4\mathbf{j}$

$\mathbf{u} \cdot \mathbf{v} = 3(-1) + 2 \cdot 4 = 5$

$|\mathbf{u}| = \sqrt{3^2 + 2^2} = \sqrt{13}$

$|\mathbf{v}| = \sqrt{(-1)^2 + 4^2} = \sqrt{17}$

$$\cos\alpha = \frac{\mathbf{u} \cdot \mathbf{v}}{|\mathbf{u}|\,|\mathbf{v}|} = \frac{5}{\sqrt{13}\sqrt{17}}$$

$$\alpha = \cos^{-1}\frac{5}{\sqrt{13}\sqrt{17}}$$

$$\alpha \approx 70.3°$$

69. For $\theta = \frac{\pi}{6}$, $\mathbf{u} = \left(\cos\frac{\pi}{6}\right)\mathbf{i} + \left(\sin\frac{\pi}{6}\right)\mathbf{j} = \frac{\sqrt{3}}{2}\mathbf{i} + \frac{1}{2}\mathbf{j}$.

For $\theta = \frac{3\pi}{4}$,

$\mathbf{u} = \left(\cos\frac{3\pi}{4}\right)\mathbf{i} + \left(\sin\frac{3\pi}{4}\right)\mathbf{j} = -\frac{\sqrt{2}}{2}\mathbf{i} + \frac{\sqrt{2}}{2}\mathbf{j}$.

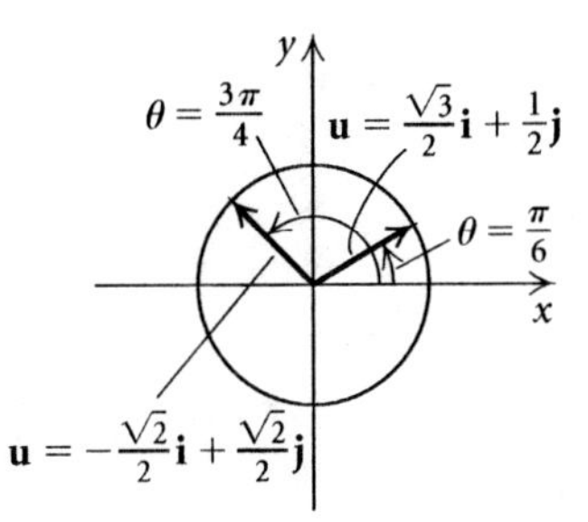

70. For $\theta = -\frac{\pi}{4}$,

$\mathbf{u} = \left[\cos\left(-\frac{\pi}{4}\right)\right]\mathbf{i} + \left[\sin\left(-\frac{\pi}{4}\right)\right]\mathbf{j} =$

$\frac{\sqrt{2}}{2}\mathbf{i} - \frac{\sqrt{2}}{2}\mathbf{j}$.

For $\theta = -\frac{3\pi}{4}$,

$\mathbf{u} = \left[\cos\left(-\frac{3\pi}{4}\right)\right]\mathbf{i} + \left[\sin\left(-\frac{3\pi}{4}\right)\right]\mathbf{j} =$

$-\frac{\sqrt{2}}{2}\mathbf{i} - \frac{\sqrt{2}}{2}\mathbf{j}$.

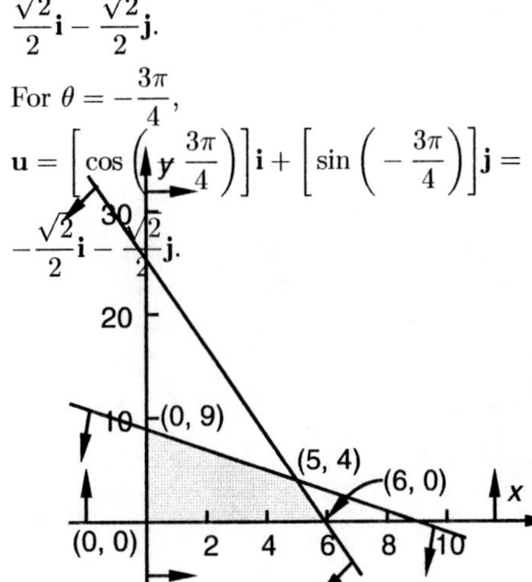

71. $\mathbf{u} = (\cos\theta)\mathbf{i} + (\sin\theta)\mathbf{j}$ where $\theta = \frac{\pi}{2} + \frac{3\pi}{4}$, or $\frac{5\pi}{4}$. Then

$\mathbf{u} = \left(\cos\frac{5\pi}{4}\right)\mathbf{i} + \left(\sin\frac{5\pi}{4}\right)\mathbf{j} = -\frac{\sqrt{2}}{2}\mathbf{i} - \frac{\sqrt{2}}{2}\mathbf{j}$.

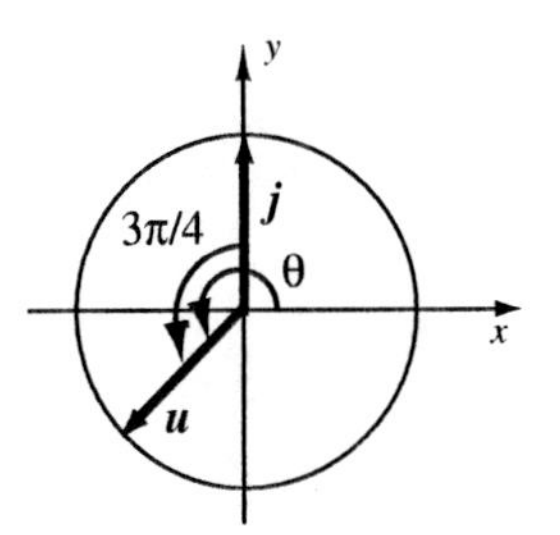

72. $\mathbf{u} = (\cos\theta)\mathbf{i} + (\sin\theta)\mathbf{j}$ where $\theta = 2\pi - \frac{\pi}{6}$, or $\frac{11\pi}{6}$. Then

$\mathbf{u} = \left(\cos\frac{11\pi}{6}\right)\mathbf{i} + \left(\sin\frac{11\pi}{6}\right)\mathbf{j} = \frac{\sqrt{3}}{2}\mathbf{i} - \frac{1}{2}\mathbf{j}$.

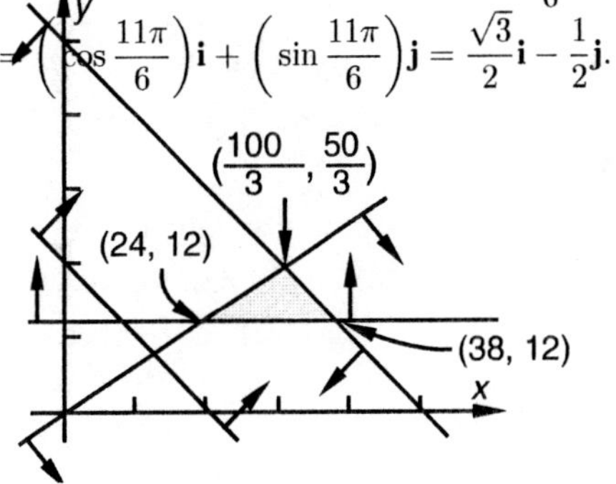

73. Find the magnitude of $-\mathbf{i} + 3\mathbf{j}$.

$\sqrt{(-1)^2 + 3^2} = \sqrt{10}$

Then the desired unit vector is

$\frac{-\mathbf{i} + 3\mathbf{j}}{\sqrt{10}} = -\frac{1}{\sqrt{10}}\mathbf{i} + \frac{3}{\sqrt{10}}\mathbf{j}$, or $-\frac{\sqrt{10}}{10}\mathbf{i} + \frac{3\sqrt{10}}{10}\mathbf{j}$.

74. Find the magnitude of $6\mathbf{i} - 8\mathbf{j}$.

$\sqrt{6^2 + (-8)^2} = \sqrt{100} = 10$

Then the desired unit vector is $\frac{6\mathbf{i} - 8\mathbf{j}}{10} = \frac{3}{5}\mathbf{i} - \frac{4}{5}\mathbf{j}$.

75. Find the magnitude of $2\mathbf{i} - 3\mathbf{j}$.

$\sqrt{2^2 + (-3)^2} = \sqrt{13}$

Then the desired vector is $\sqrt{13}\left(\frac{2\mathbf{i} - 3\mathbf{j}}{\sqrt{13}}\right) =$

$\sqrt{13}\left(\frac{2}{\sqrt{13}}\mathbf{i} - \frac{3}{\sqrt{13}}\mathbf{j}\right)$, or $\sqrt{13}\left(\frac{2\sqrt{13}}{13}\mathbf{i} - \frac{3\sqrt{13}}{13}\mathbf{j}\right)$.

76. Find the magnitude of $5\mathbf{i} + 12\mathbf{j}$.

$\sqrt{5^2 + 13^2} = \sqrt{169} = 13$

Then the desired vector is $13\left(\frac{5\mathbf{i} + 12\mathbf{j}}{13}\right) =$

$13\left(\frac{5}{13}\mathbf{i} + \frac{12}{13}\mathbf{j}\right)$.

77. See the answer section in the text.

78.

79. Refer to the drawing in this manual accompanying the solution for Exercise 27, Exercise Set 7.5. The vector representing the first part of the ship's trip can be given by

$120(\cos 10°)\mathbf{i} + 120(\sin 10°)\mathbf{j}$.

The vector representing the second part of the trip can be given by

$200(\cos 250°)\mathbf{i} + 200(\sin 250°)\mathbf{j}$.

Then the resultant is

$$120(\cos 10°)\mathbf{i} + 120(\sin 10°)\mathbf{j} + 200(\cos 250°)\mathbf{i} + 200(\sin 250°)\mathbf{j}$$

$$= [120(\cos 10°) + 200(\cos 250°)]\mathbf{i} + [120(\sin 10°) + 200(\sin 250°)]\mathbf{j}$$

$$\approx 49.77\mathbf{i} - 167.10\mathbf{j}$$

Then the distance from the starting point is

$\sqrt{(49.77)^2 + (-167.10)^2} \approx 174$ nautical miles.

Now we find the direction angle of the resultant.

$$\tan\theta = \frac{-167.10}{49.77}$$

$$\theta = \tan^{-1}\frac{-167.10}{49.77}$$

θ is a fourth-quadrant angle. The reference angle is

$$\tan^{-1}\frac{167.10}{49.77} \approx 73°$$

and $\theta \approx 360° - 73°$, or $287°$.

Thus, the ship's bearing is about S17°E. (This answer differs from the answer found in Section 7.5 due to rounding differences.)

80. Refer to the drawing in this manual accompanying the solution for Exercise 26, Exercise Set 7.5. The vector representing the boat's velocity can be given by

$750(\cos 55°)\mathbf{i} + 750(\sin 55°)\mathbf{j}$.

The vector representing the wind can be given by

$150(\cos 310°)\mathbf{i} + 150(\sin 310°)\mathbf{j}$.

Then the resultant is the sum of these vectors:

$526.60\mathbf{i} + 499.46\mathbf{j}$.

Then magnitude of the resultant force is

$\sqrt{(526.60)^2 + (499.46)^2} \approx 726$ lb.

Now we find the direction angle of the resultant.

$$\tan\theta = \frac{499.46}{526.60}$$

$$\theta = \tan^{-1}\frac{499.46}{526.60} \approx 43°$$

Then the direction from north is $90° - 43°$, or $47°$.

81. Refer to the drawing in this manual accompanying the solution for Exercise 29, Exercise Set 7.5. We can use the Pythagorean theorem to find the magnitude of the vector representing the desired velocity of the airplane:

$\sqrt{150^2 - 25^2} \approx 148$

Then the vector representing the desired velocity can be given by

$148(\cos 20°)\mathbf{i} + 148(\sin 20°)\mathbf{j}$.

The vector representing the wind can be given by

$25(\cos 290°)\mathbf{i} + 25(\sin 290°)\mathbf{j}$.

The vector representing the desired velocity is the resultant of the vectors representing the airplane's actual velocity and the wind, so we subtract to find the vector representing the airplane's actual velocity.

$$[148(\cos 20°)\mathbf{i} + 148(\sin 20°)\mathbf{j}] - [25(\cos 290°)\mathbf{i} + 25(\sin 290°)\mathbf{j}]$$

$$= [148(\cos 20°) - 25(\cos 290°)]\mathbf{i} + [148(\sin 20°) - 25(\sin 290°)]\mathbf{j}$$

$$\approx 130.52\mathbf{i} + 74.11\mathbf{j}$$

Now we find the direction angle of this vector. Note that α is a first-quadrant angle.

$$\tan\alpha = \frac{74.11}{130.52}$$

$$\alpha = \tan^{-1}\frac{74.11}{130.52} \approx 30°$$

Thus, the airplane's actual bearing is 60°.

82. Refer to the drawing in this manual accompanying the solution for Exercise 28, Exercise Set 7.5. The vector representing the first part of the flight can be given by

$210(\cos 58°)\mathbf{i} + 210(\sin 58°)\mathbf{j}$.

The vector representing the second part of the flight can be given by

$170(\cos 170°)\mathbf{i} + 170(\sin 170°)\mathbf{j}$.

The resultant is the sum of these vectors:

$-56.13\mathbf{i} + 207.61\mathbf{j}$

The magnitude of the resultant is

$\sqrt{(-56.13)^2 + (207.61)^2} \approx 215$ mi.

Now we find the direction angle of the resultant.

$$\tan\phi = \frac{207.61}{-56.13}$$

$$\phi = \tan^{-1}\frac{207.61}{-56.13}$$

ϕ is a second-quadrant angle. The reference angle is

$$\tan^{-1}\frac{207.61}{56.13} \approx 75°$$

and $\phi \approx 180° - 75° = 105°$.

Then the direction of the airplane is 345°.

83. We draw a force diagram with the initial point of each vector at the origin.

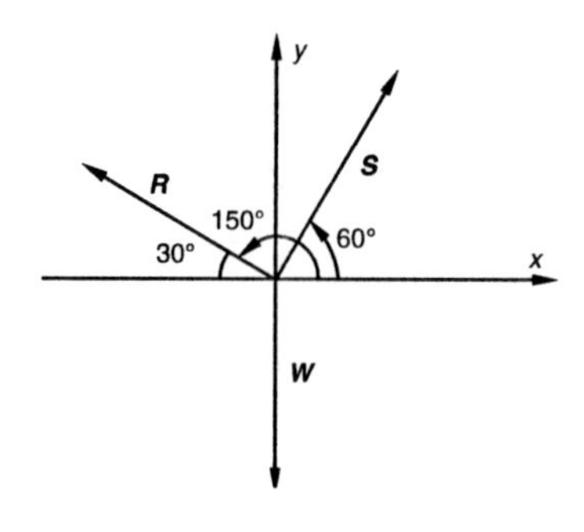

We express each vector in terms of its magnitude and direction angle.

$\mathbf{R} = |\mathbf{R}|[(\cos 150°)\mathbf{i} + (\sin 150°)\mathbf{j}]$

$\mathbf{S} = |\mathbf{S}|[(\cos 60°)\mathbf{i} + (\sin 60°)\mathbf{j}]$

$\mathbf{W} = 1000(\cos 270°)\mathbf{i} + 1000(\sin 270°)\mathbf{j} = -1000\mathbf{j}$

Substituting for **R**, **S**, and **W** in $\mathbf{R}+\mathbf{S}+\mathbf{W} = \mathbf{O}$, we have

$|\mathbf{R}|[(\cos 150°)\mathbf{i} + (\sin 150°)\mathbf{j}]+$
$|\mathbf{S}|[(\cos 60°)\mathbf{i} + (\sin 60°)\mathbf{j}] - 1000\mathbf{j} = 0\mathbf{i} + 0\mathbf{j}$.

This gives us a system of equations.

$$|\mathbf{R}|(\cos 150°) + |\mathbf{S}|(\cos 60°) = 0,$$
$$|\mathbf{R}|(\sin 150°) + |\mathbf{S}|(\sin 60°) = 1000$$

Solving this system, we get

$|\mathbf{R}| = 500,\ |\mathbf{S}| \approx 866.$

The tension in the cable on the left is 500 lb and in the cable on the right is about 866 lb.

84.

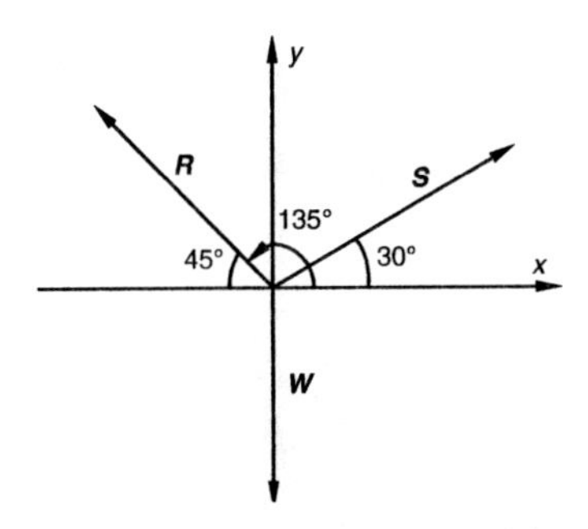

$\mathbf{R} = |\mathbf{R}|[(\cos 135°)\mathbf{i} + (\sin 135°)\mathbf{j}]$

$\mathbf{S} = |\mathbf{S}|[(\cos 30°)\mathbf{i} + (\sin 30°)\mathbf{j}]$

$\mathbf{W} = 2500(\cos 270°)\mathbf{i} + 2500(\sin 270°)\mathbf{j} = -2500\mathbf{j}$

Then we have

$|\mathbf{R}|[(\cos 135°)\mathbf{i} + (\sin 135°)\mathbf{j}]+$
$|\mathbf{S}|[(\cos 30°)\mathbf{i} + (\sin 30°)\mathbf{j}] - 2500\mathbf{j} = 0\mathbf{i} + 0\mathbf{j}$.

This gives us a system of equations.

$$|\mathbf{R}|(\cos 135°) + |\mathbf{S}|(\cos 30°) = 0,$$
$$|\mathbf{R}|(\sin 135°) + |\mathbf{S}|(\sin 30°) = 2500$$

Solving this system, we get

$|\mathbf{R}| \approx 2241$ kg, $|\mathbf{S}| \approx 1830$ kg.

85. We draw a force diagram with the initial point of each vector at the origin.

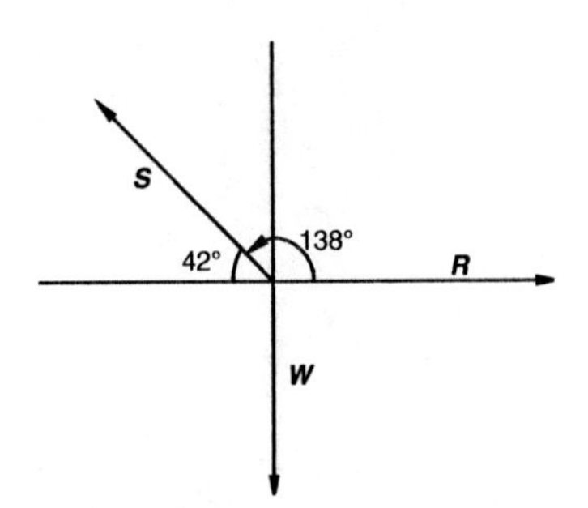

We express each vector in terms of its magnitude and direction angle.

$\mathbf{R} = |\mathbf{R}|[(\cos 0°)\mathbf{i} + (\sin 0°)\mathbf{j}] = |\mathbf{R}|\mathbf{i}$

$\mathbf{S} = |\mathbf{S}|[(\cos 138°)\mathbf{i} + (\sin 138°)\mathbf{j}]$

$\mathbf{W} = 150(\cos 270°)\mathbf{i} + 150(\sin 270°)\mathbf{j} = -150\mathbf{j}$

Substituting for **R**, **S**, and **W** in $\mathbf{R}+\mathbf{S}+\mathbf{W} = \mathbf{O}$, we have

$|\mathbf{R}|\mathbf{i} + |\mathbf{S}|[(\cos 138°)\mathbf{i} + (\sin 138°)\mathbf{j}] - 150\mathbf{j} = 0\mathbf{i} + 0\mathbf{j}$.

This gives us a system of equations.

$$|\mathbf{R}| + |\mathbf{S}|(\cos 138°) = 0,$$
$$|\mathbf{S}|(\sin 138°) = 150$$

Solving this system, we get

$|\mathbf{R}| \approx 167,\ |\mathbf{S}| \approx 224.$

The tension in the cable is about 224 lb, and the compression in the boom is about 167 lb. (Answers may vary slightly due to rounding differences.)

86.

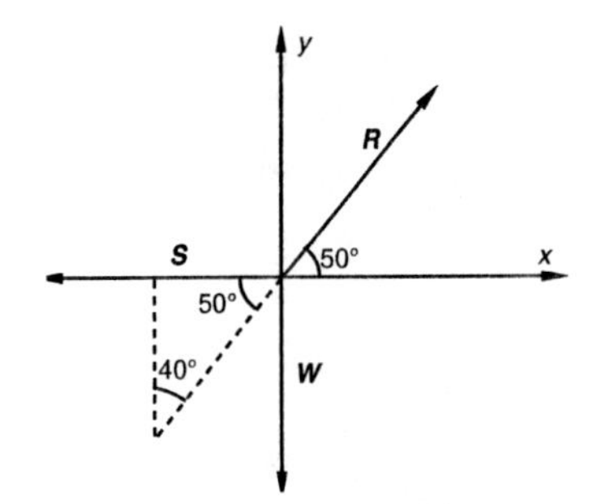

$\mathbf{R} = |\mathbf{R}|[(\cos 50°)\mathbf{i} + (\sin 50°)\mathbf{j}]$

$\mathbf{S} = |\mathbf{S}|[(\cos 180°)\mathbf{i} + (\sin 180°)\mathbf{j}] = -|\mathbf{S}|\mathbf{i}$

$\mathbf{W} = 200(\cos 270°)\mathbf{i} + 200(\sin 270°)\mathbf{j} = -200\mathbf{j}$

Then we have

$|\mathbf{R}|[(\cos 50°)\mathbf{i} + (\sin 50°)\mathbf{j}] - |\mathbf{S}|\mathbf{i} - 200\mathbf{j} = 0\mathbf{i} + 0\mathbf{j}$.

This gives us a system of equations.

$$|\mathbf{R}|(\cos 50°) - |\mathbf{S}| = 0,$$
$$|\mathbf{R}|(\sin 50°) = 200$$

Solving this system, we get

$|\mathbf{R}| \approx 261$ lb, $|\mathbf{S}| \approx 168$ lb.

87.
$$\begin{aligned}\mathbf{u}+\mathbf{v} &= \langle u_1, u_2\rangle + \langle v_1, v_2\rangle \\ &= \langle u_1+v_1, u_2+v_2\rangle \\ &= \langle v_1+u_1, v_2+u_2\rangle \\ &= \langle v_1, v_2\rangle + \langle u_1, u_2\rangle \\ &= \mathbf{v}+\mathbf{u}\end{aligned}$$

88.
$$\begin{aligned}\mathbf{u}\cdot\mathbf{v} &= u_1v_1 + u_2v_2 \\ &= v_1u_1 + v_2u_2 \\ &= \mathbf{v}\cdot\mathbf{u}\end{aligned}$$

89. Vector components are two *vectors* whose sum is a third vector. Scalar components are *numbers* that define a vector.

90. Answers may vary. For $\mathbf{u} = 3\mathbf{i} - 4\mathbf{j}$ and $\mathbf{w} = 2\mathbf{i} - 4\mathbf{j}$ find **v** where $\mathbf{v} = \mathbf{u} + \mathbf{w}$

91. $-\frac{1}{5}x - y = 15$

$-y = \frac{1}{5}x + 15$

$y = -\frac{1}{5}x - 15$ Multiplying by -1

With the equation written in the form $y = mx + b$ we see that the slope is $-\frac{1}{5}$ and the y-intercept is $(0, -15)$.

92. $y = 7 = 0x + 7$

Slope: 0; y-intercept: $(0, 7)$

93. $x^3 - 4x^2 = 0$

$x^2(x - 4) = 0$

$x^2 = 0$ *or* $x - 4 = 0$

$x = 0$ *or* $x = 4$

The zeros are 0 and 4.

94. $6x^2 + 7x = 55$

$6x^2 + 7x - 55 = 0$

$(2x - 5)(3x + 11) = 0$

$2x - 5 = 0$ *or* $3x + 11 = 0$

$2x = 5$ *or* $3x = -11$

$x = \frac{5}{2}$ *or* $x = -\frac{11}{3}$

The zeros are $\frac{5}{2}$ and $-\frac{11}{3}$.

95. (a) Assume neither $|\mathbf{u}|$ nor $|\mathbf{v}|$ is zero.

If $\mathbf{u} \cdot \mathbf{v} = |\mathbf{u}||\mathbf{v}|\cos\theta = 0$, then $\cos\theta = 0$, or $\theta = 90°$ and the vectors are perpendicular.

(b) Answers may vary.

Let $\mathbf{u} = \mathbf{i}$ and $\mathbf{v} = \mathbf{j}$. Then $\mathbf{u} \cdot \mathbf{v} = 1 \cdot 0 + 0 \cdot 1 = 0$.

96. Let $\overrightarrow{PQ} = \langle x, y \rangle$. Then $\overrightarrow{QP} = \langle -x, -y \rangle$ and $PQ + QP = \langle x, y \rangle + \langle -x, -y \rangle = \langle 0, 0 \rangle = \mathbf{0}$

97. Find the magnitude of $\mathbf{u} = \langle 3, -4 \rangle$:

$|\mathbf{u}| = \sqrt{3^2 + (-4)^2} = \sqrt{25} = 5$

The unit vector in the same direction as $\mathbf{u}$ is

$\frac{3\mathbf{i} - 4\mathbf{j}}{5} = \frac{3}{5}\mathbf{i} - \frac{4}{5}\mathbf{j}$.

The unit vector in the opposite direction to $\mathbf{u}$ is

$-\left(\frac{3\mathbf{i} - 4\mathbf{j}}{5}\right)$, or $-\frac{3}{5}\mathbf{i} + \frac{4}{5}\mathbf{j}$.

These are the only unit vectors parallel to $\langle 3, -4 \rangle$.

98. $|\mathbf{v}| = \sqrt{(-1)^2 + 2^2} = \sqrt{5}$

Then to find a vector of length 2 whose direction is the opposite of the direction of $\mathbf{v}$, we multiply $\mathbf{v}$ by $-\frac{2}{\sqrt{5}}$, or $-\frac{2\sqrt{5}}{5}$. We have

$-\frac{2\sqrt{5}}{5}\left(-\mathbf{i} + 2\mathbf{j}\right) = \frac{2\sqrt{5}}{5}\mathbf{i} - \frac{4\sqrt{5}}{5}\mathbf{j}$.

This is the only vector that satisfies the given criteria.

99. B has coordinates (x, y) such that

$x - 2 = 3$, or $x = 5$

and

$y - 9 = -1$, or $y = 8$.

Thus point B is (5,8).

100. A has coordinates (x, y) such that

$-2 - x = 4$, or $x = -6$

and

$5 - y = -2$, or $y = 7$.

Thus point A is $(-6, 7)$. Then the vector from A to the origin is $\langle 0 - (-6), 0 - 7 \rangle$, or $\langle 6, -7 \rangle$, or $6\mathbf{i} - 7\mathbf{j}$.

Chapter 8

Systems of Equations and Matrices

Exercise Set 8.1

1. Graph (c) is the graph of this system.

2. Graph (e) is the graph of this system.

3. Graph (f) is the graph of this system.

4. Graph (a) is the graph of this system.

5. Graph (b) is the graph of this system.

6. Graph (d) is the graph of this system.

7. Graph $x + y = 2$ and $3x + y = 0$ and find the coordinates of the point of intersection.

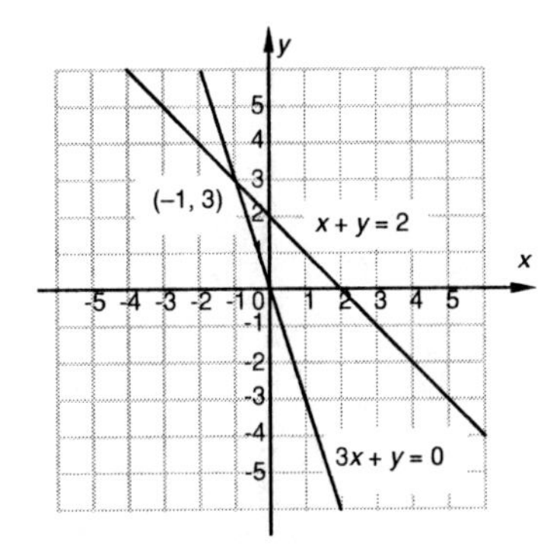

The solution is $(-1, 3)$.

8. Graph $x + y = 1$ and $3x + y = 7$ and find the coordinates of the point of intersection.

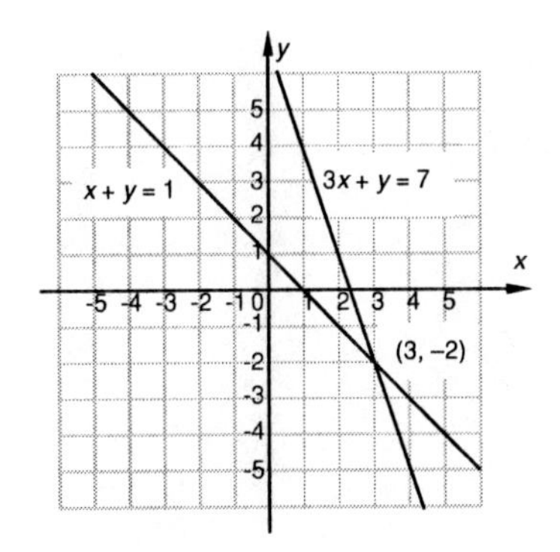

The solution is $(3, -2)$.

9. Graph $x + 2y = 1$ and $x + 4y = 3$ and find the coordinates of the point of intersection.

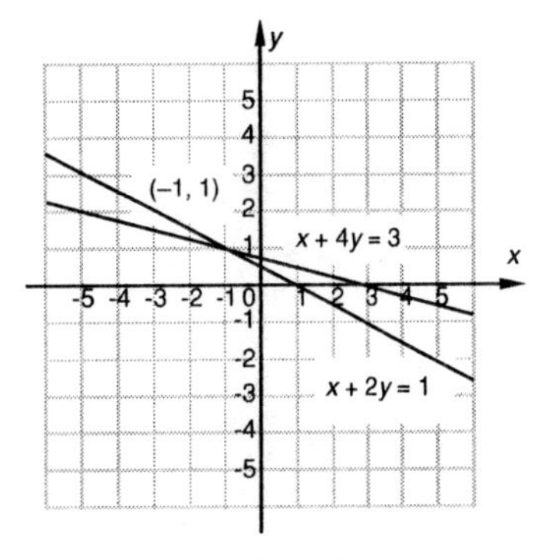

The solution is $(-1, 1)$.

10. Graph $3x + 4y = 5$ and $x - 2y = 5$ and find the coordinates of the point of intersection.

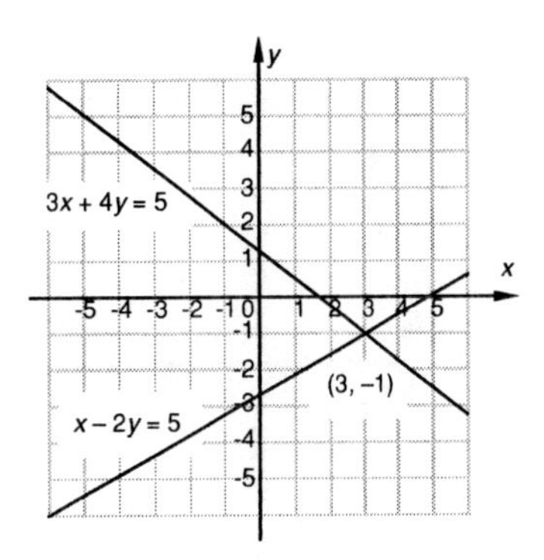

The solution is $(3, -1)$.

11. Graph $y + 1 = 2x$ and $y - 1 = 2x$ and find the coordinates of the point of intersection.

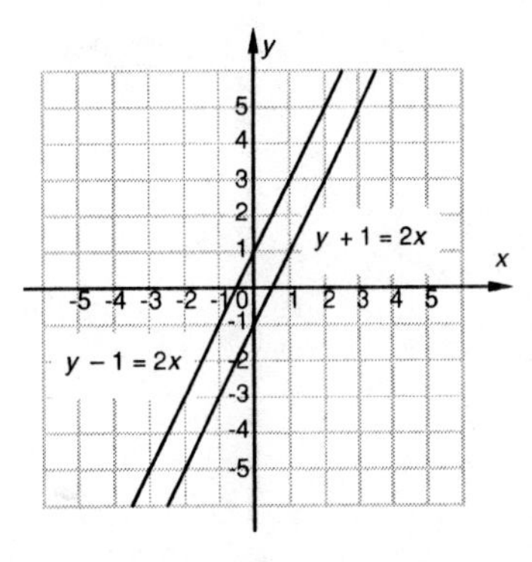

The graphs do not intersect, so there is no solution.

12. Graph $2x - y = 1$ and $3y = 6x - 3$ and find the coordinates of the point of intersection.

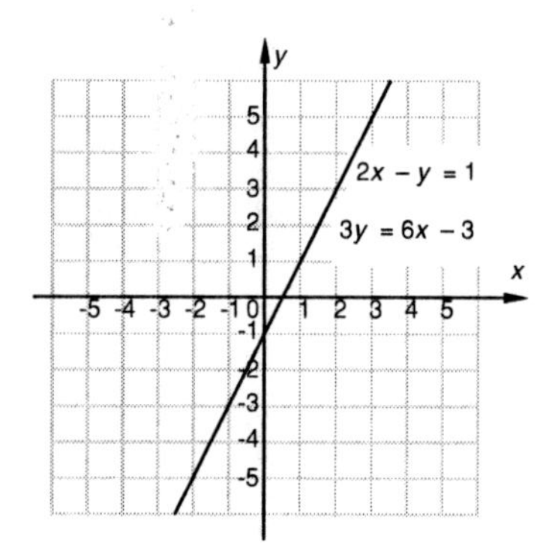

The graphs coincide so there are infinitely many solutions. Solving either equation for y, we have $y = 2x - 1$, so the solutions can be expressed as $(x, 2x-1)$. Similarly, solving either equation for x, we get $x = \dfrac{y+1}{2}$, so the solutions can also be expressed as $\left(\dfrac{y+1}{2}, y\right)$.

13. Graph $x - y = -6$ and $y = -2x$ and find the coordinates of the point of intersection.

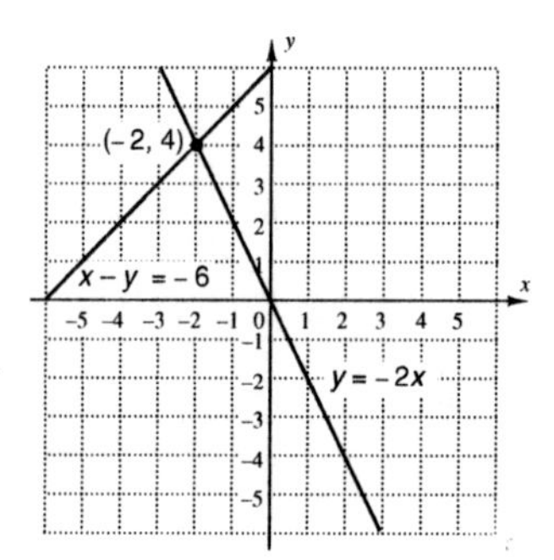

The solution is $(-2, 4)$.

14. Graph $2x + y = 5$ and $x = -3y$ and find the coordinates of the point of intersection.

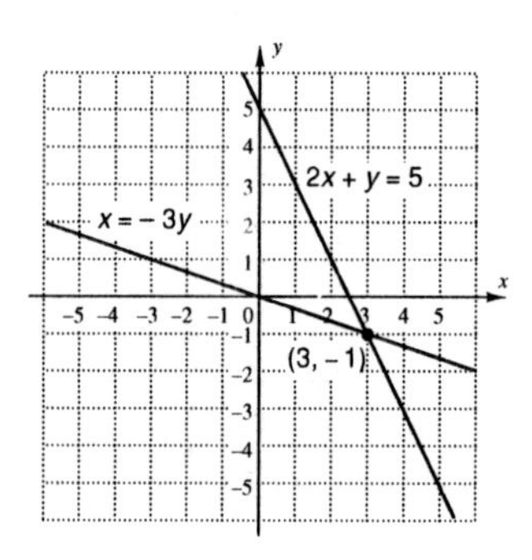

The solution is $(3, -1)$.

15. Graph $2y = x - 1$ and $3x = 6y + 3$ and find the coordinates of the point of intersection.

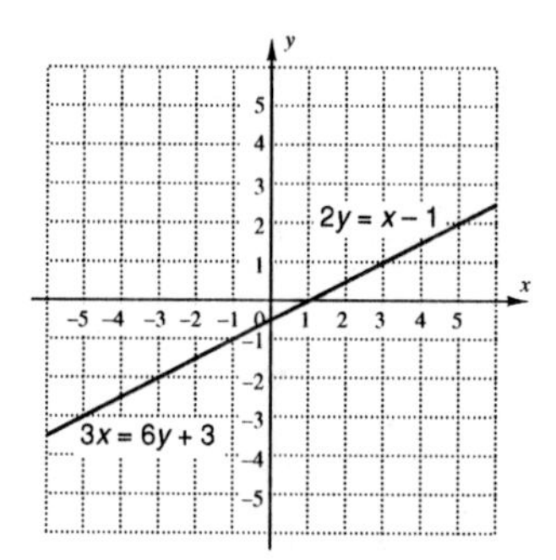

The graphs coincide so there are infinitely many solutions. Solving either equation for y, we get $y = \dfrac{x-1}{2}$, so the solutions can bae expressed as $\left(x, \dfrac{x-1}{2}\right)$. Similarly, solving either equation for x, we get $x = 2y + 1$, so the solutions can also be expressed as $(2y + 1, y)$.

16. Graph $y = 3x + 2$ and $3x - y = -3$ and find the coordinates of the point of intersection.

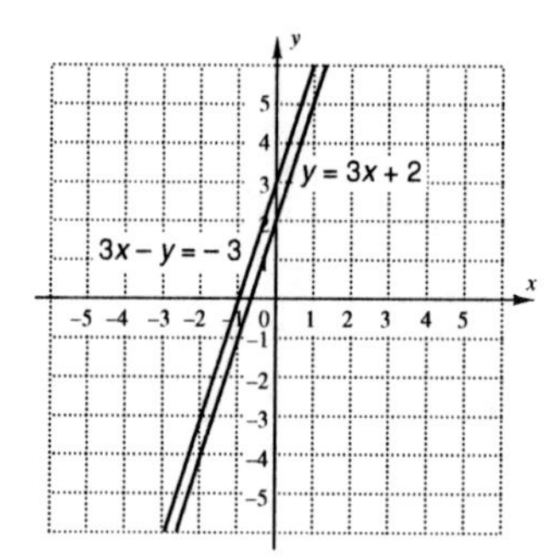

The graphs do not intersect, so there is no solution.

17.

$$x + y = 9, \quad (1)$$
$$2x - 3y = -2 \quad (2)$$

Solve equation (1) for either x or y. We choose to solve for y.

$$y = 9 - x$$

Then substitute $9 - x$ for y in equation (2) and solve the resulting equation.

$$2x - 3(9 - x) = -2$$
$$2x - 27 + 3x = -2$$
$$5x - 27 = -2$$
$$5x = 25$$
$$x = 5$$

Now substitute 5 for x in either equation (1) or (2) and solve for y.

$5 + y = 9$ Using equation (1)

$y = 4$

The solution is $(5, 4)$.

18. $3x - y = 5,\quad (1)$

$x + y = \frac{1}{2}\quad (2)$

Solve equation (2) for y.

$$y = \frac{1}{2} - x$$

Substitute in equation (1) and solve for x.

$$3x - \left(\frac{1}{2} - x\right) = 5$$
$$3x - \frac{1}{2} + x = 5$$
$$4x = \frac{11}{2}$$
$$x = \frac{11}{8}$$

Back-substitute to find y.

$$\frac{11}{8} + y = \frac{1}{2}\quad \text{Using equation (2)}$$
$$y = -\frac{7}{8}$$

The solution is $\left(\frac{11}{8}, -\frac{7}{8}\right)$.

19. $x - 2y = 7,\quad (1)$

$x = y + 4\quad (2)$

Use equation (2) and substitute $y+4$ for x in equation (1). Then solve for y.

$$y + 4 - 2y = 7$$
$$-y + 4 = 7$$
$$-y = 3$$
$$y = -3$$

Substitute -3 for y in equation (2) to find x.

$$x = -3 + 4 = 1$$

The solution is $(1, -3)$.

20. $x + 4y = 6,\quad (1)$

$x = -3y + 3\quad (2)$

Substitute $-3y + 3$ for x in equation (1) and solve for y.

$$-3y + 3 + 4y = 6$$
$$y = 3$$

Back-substitute to find x.

$$x = -3 \cdot 3 + 3 = -6$$

The solution is $(-6, 3)$.

21. $y = 2x - 6,\quad (1)$

$5x - 3y = 16\quad (2)$

Use equation (1) and substitute $2x - 6$ for y in equation (2). Then solve for x.

$$5x - 3(2x - 6) = 16$$
$$5x - 6x + 18 = 16$$
$$-x + 18 = 16$$
$$-x = -2$$
$$x = 2$$

Substitute 2 for x in equation (1) to find y.

$$y = 2 \cdot 2 - 6 = 4 - 6 = -2$$

The solution is $(2, -2)$.

22. $3x + 5y = 2,\quad (1)$

$2x - y = -3\quad (2)$

Solve equation (2) for y.

$$y = 2x + 3$$

Substitute $2x + 3$ for y in equation (1) and solve for x.

$$3x + 5(2x + 3) = 2$$
$$3x + 10x + 15 = 2$$
$$13x = -13$$
$$x = -1$$

Back-substitute to find y.

$$2(-1) - y = -3\quad \text{Using equation (2)}$$
$$-2 - y = -3$$
$$1 = y$$

The solution is $(-1, 1)$.

23. $x - 5y = 4,\quad (1)$

$y = 7 - 2x\quad (2)$

Use equation (2) and substitute $7 - 2x$ for y in equation (1). Then solve for x.

$$x - 5(7 - 2x) = 4$$
$$x - 35 + 10x = 4$$
$$11x - 35 = 4$$
$$11x = 39$$
$$x = \frac{39}{11}$$

Substitute $\frac{39}{11}$ for x in equation (2) to find y.

$$y = 7 - 2 \cdot \frac{39}{11} = 7 - \frac{78}{11} = -\frac{1}{11}$$

The solution is $\left(\frac{39}{11}, -\frac{1}{11}\right)$.

24. $5x + 3y = -1,\quad (1)$

$x + y = 1\quad (2)$

Solve equation (2) for either x or y. We choose to solve for x.

$$x = 1 - y$$

Substitute $1 - y$ for x in equation (1) and solve for y.

$$5(1 - y) + 3y = -1$$
$$5 - 5y + 3y = -1$$
$$-2y = -6$$
$$y = 3$$

Back-substitute to find x.

$$x + 3 = 1\quad \text{Using equation (2)}$$
$$x = -2$$

The solution is $(-2, 3)$.

25. $2x - 3y = 5,$ (1)

$5x + 4y = 1$ (2)

Solve one equation for either x or y. We choose to solve equation (1) for x.

$$2x - 3y = 5$$
$$2x = 3y + 5$$
$$x = \frac{3}{2}y + \frac{5}{2}$$

Substitute $\frac{3}{2}y + \frac{5}{2}$ for x in equation (2) and solve for y.

$$5\left(\frac{3}{2}y + \frac{5}{2}\right) + 4y = 1$$
$$\frac{15}{2}y + \frac{25}{2} + 4y = 1$$
$$\frac{23}{2}y + \frac{25}{2} = 1$$
$$\frac{23}{2}y = -\frac{23}{2}$$
$$y = -1$$

Substitute -1 for y in either equation (1) or (2) and solve for x.

$$2x - 3(-1) = 5 \quad \text{Using equation (1)}$$
$$2x + 3 = 5$$
$$2x = 2$$
$$x = 1$$

The solution is $(1, -1)$.

26. $3x + 4y = 6,$ (1)

$2x + 3y = 5$ (2)

Solve one equation for either x or y. We choose to solve equation (2) for x.

$$x = -\frac{3}{2}y + \frac{5}{2}$$

Substitute $-\frac{3}{2}y + \frac{5}{2}$ for x in equation (1) and solve for y.

$$3\left(-\frac{3}{2}y + \frac{5}{2}\right) + 4y = 6$$
$$-\frac{9}{2}y + \frac{15}{2} + 4y = 6$$
$$-\frac{1}{2}y = -\frac{3}{2}$$
$$y = 3$$

Back-substitute to find x.

$$3x + 4 \cdot 3 = 6 \quad \text{Using equation (1)}$$
$$3x = -6$$
$$x = -2$$

The solution is $(-2, 3)$.

27. $x + 2y = 2,$ (1)

$4x + 4y = 5$ (2)

Solve one equation for either x or y. We choose to solve equation (1) for x since x has a coefficient of 1 in that equation.

$$x + 2y = 2$$
$$x = -2y + 2$$

Substitute $-2y + 2$ for x in equation (2) and solve for y.

$$4(-2y + 2) + 4y = 5$$
$$-8y + 8 + 4y = 5$$
$$-4y + 8 = 5$$
$$-4y = -3$$
$$y = \frac{3}{4}$$

Substitute $\frac{3}{4}$ for y in either equation (1) or equation (2) and solve for x.

$$x + 2y = 2 \quad \text{Using equation (1)}$$
$$x + 2 \cdot \frac{3}{4} = 2$$
$$x + \frac{3}{2} = 2$$
$$x = \frac{1}{2}$$

The solution is $\left(\frac{1}{2}, \frac{3}{4}\right)$.

28. $2x - y = 2,$ (1)

$4x + y = 3$ (2)

Solve one equation for either x or y. We choose to solve equation (2) for y since y has a coefficient of 1 in that equation.

$$y = -4x + 3$$

Substitute $-4x + 3$ for y in equation (1) and solve for x.

$$2x - (-4x + 3) = 2$$
$$2x + 4x - 3 = 2$$
$$6x = 5$$
$$x = \frac{5}{6}$$

Back-substitute to find y.

$$4 \cdot \frac{5}{6} + y = 3$$
$$\frac{10}{3} + y = 3$$
$$y = -\frac{1}{3}$$

The solution is $\left(\frac{5}{6}, -\frac{1}{3}\right)$.

29. $x + 2y = 7,$ (1)

$x - 2y = -5$ (2)

We add the equations to eliminate y.

$$\begin{array}{l} x + 2y = 7 \\ \underline{x - 2y = -5} \\ 2x \quad\;\; = 2 \quad \text{Adding} \end{array}$$
$$x = 1$$

Back-substitute in either equation and solve for y.

$1 + 2y = 7$ Using equation (1)

$2y = 6$

$y = 3$

The solution is $(1, 3)$. Since the system of equations has exactly one solution it is consistent and independent.

30. $3x + 4y = -2$ (1)

$-3x - 5y = 1$ (2)

$-y = -1$ Adding

$y = 1$

Back-substitute to find x.

$3x + 4 \cdot 1 = -2$ Using equation (1)

$3x = -6$

$x = -2$

The solution is $(-2, 1)$. Since the system of equations has exactly one solution it is consistent and independent.

31. $x - 3y = 2,$ (1)

$6x + 5y = -34$ (2)

Multiply equation (1) by -6 and add it to equation (2) to eliminate x.

$-6x + 18y = -12$

$6x + 5y = -34$

$23y = -46$

$y = -2$

Back-substitute to find x.

$x - 3(-2) = 2$ Using equation (1)

$x + 6 = 2$

$x = -4$

The solution is $(-4, -2)$. Since the system of equations has exactly one solution it is consistent and independent.

32. $x + 3y = 0,$ (1)

$20x - 15y = 75$ (2)

Multiply equation (1) by 5 and add.

$5x + 15y = 0$

$20x - 15y = 75$

$25x = 75$

$x = 3$

Back-substitute to find y.

$3 + 3y = 0$ Using equation (1)

$3y = -3$

$y = -1$

The solution is $(3, -1)$. Since the system of equations has exactly one solution it is consistent and independent.

33. $3x - 12y = 6,$ (1)

$2x - 8y = 4$ (2)

Multiply equation (1) by 2 and equation (2) by -3 and add.

$6x - 24y = 12$

$-6x + 24y = -12$

$0 = 0$

The equation $0 = 0$ is true for all values of x and y. Thus, the system of equations has infinitely many solutions. Solving either equation for y, we can write $y = \frac{1}{4}x - \frac{1}{2}$ so the solutions are ordered pairs of the form $\left(x, \frac{1}{4}x - \frac{1}{2}\right)$. Equivalently, if we solve either equation for x we get $x = 4y + 2$ so the solutions can also be expressed as $(4y + 2, y)$. Since there are infinitely many solutions, the system of equations is consistent and dependent.

34. $2x + 6y = 7,$ (1)

$3x + 9y = 10$ (2)

Multiply equation (1) by 3 and equation (2) by -2 and add.

$6x + 18y = 21$

$-6x - 18y = -20$

$0 = 1$

We get a false equation so there is no solution. Since there is no solution the system of equations is inconsistent and independent.

35. $2x = 5 - 3y,$ (1)

$4x = 11 - 7y$ (2)

We rewrite the equations.

$2x + 3y = 5,$ (1a)

$4x + 7y = 11$ (2a)

Multiply equation (2a) by -2 and add to eliminate x.

$-4x - 6y = -10$

$4x + 7y = 11$

$y = 1$

Back-substitute to find x.

$2x = 5 - 3 \cdot 1$ Using equation (1)

$2x = 2$

$x = 1$

The solution is $(1, 1)$. Since the system of equations has exactly one solution it is consistent and independent.

36. $7(x - y) = 14,$ (1)

$2x = y + 5$ (2)

$x - y = 2,$ (1a) Dividing equation (1) by 7

$2x - y = 5$ (2a) Rewriting equation (2)

Multiply equation (1a) by -1 and add.

$-x + y = -2$

$2x - y = 5$

$x = 3$

Back-substitute to find y.

$3 - y = 2$ Using equation (1a)

$1 = y$

The solution is $(3, 1)$. Since the system of equations has exactly one solution it is consistent and independent.

37. $0.3x - 0.2y = -0.9,$

$0.2x - 0.3y = -0.6$

First, multiply each equation by 10 to clear the decimals.

$3x - 2y = -9 \quad (1)$

$2x - 3y = -6 \quad (2)$

Now multiply equation (1) by 3 and equation (2) by -2 and add to eliminate y.

$$\begin{aligned} 9x - 6y &= -27 \\ -4x + 6y &= 12 \\ \hline 5x \quad &= -15 \\ x &= -3 \end{aligned}$$

Back-substitute to find y.

$$\begin{aligned} 3(-3) - 2y &= -9 \quad \text{Using equation (1)} \\ -9 - 2y &= -9 \\ -2y &= 0 \\ y &= 0 \end{aligned}$$

The solution is $(-3, 0)$. Since the system of equations has exactly one solution it is consistent and independent.

38. $0.2x - 0.3y = 0.3,$

$0.4x + 0.6y = -0.2$

First, multiply each equation by 10 to clear the decimals.

$2x - 3y = 3 \quad (1)$

$4x + 6y = -2 \quad (2)$

Now multiply equation (1) by 2 and add.

$$\begin{aligned} 4x - 6y &= 6 \\ 4x + 6y &= -2 \\ \hline 8x \quad &= 4 \\ x &= \frac{1}{2} \end{aligned}$$

Back-substitute to find y.

$$\begin{aligned} 4 \cdot \frac{1}{2} + 6y &= -2 \quad \text{Using equation (2)} \\ 2 + 6y &= -2 \\ 6y &= -4 \\ y &= -\frac{2}{3} \end{aligned}$$

The solution is $\left(\frac{1}{2}, -\frac{2}{3}\right)$. Since the system of equations has exactly one solution it is consistent and independent.

39. $\frac{1}{5}x + \frac{1}{2}y = 6, \quad (1)$

$\frac{3}{5}x - \frac{1}{2}y = 2 \quad (2)$

We could multiply both equations by 10 to clear fractions, but since the y-coefficients differ only by sign we will just add to eliminate y.

$$\begin{aligned} \frac{1}{5}x + \frac{1}{2}y &= 6 \\ \frac{3}{5}x - \frac{1}{2}y &= 2 \\ \hline \frac{4}{5}x \quad &= 8 \\ x &= 10 \end{aligned}$$

Back-substitute to find y.

$$\begin{aligned} \frac{1}{5} \cdot 10 + \frac{1}{2}y &= 6 \quad \text{Using equation (1)} \\ 2 + \frac{1}{2}y &= 6 \\ \frac{1}{2}y &= 4 \\ y &= 8 \end{aligned}$$

The solution is $(10, 8)$. Since the system of equations has exactly one solution it is consistent and independent.

40. $\frac{2}{3}x + \frac{3}{5}y = -17,$

$\frac{1}{2}x - \frac{1}{3}y = -1$

Multiply the first equation by 15 and the second by 6 to clear fractions.

$10x + 9y = -255 \quad (1)$

$3x - 2y = -6 \quad (2)$

Now multiply equation (1) by 2 and equation (2) by 9 and add.

$$\begin{aligned} 20x + 18y &= -510 \\ 27x - 18y &= -54 \\ \hline 47x \quad &= -564 \\ x &= -12 \end{aligned}$$

Back-substitute to find y.

$$\begin{aligned} 3(-12) - 2y &= -6 \quad \text{Using equation (2)} \\ -36 - 2y &= -6 \\ -2y &= 30 \\ y &= -15 \end{aligned}$$

The solution is $(-12, -15)$. Since the system of equations has exactly one solution it is consistent and independent.

41. ***Familiarize***. Let x = the number of tons of ice consumed and y = the number of tons of hot dogs consumed.

Translate. The amount of ice consumed was 50 times the amount of hot dogs consumed, by weight, so we have one equation:

$x = 50y.$

A second equation comes from the fact that a total of 204 tons of ice and hot dogs were consumed:

$x + y = 204.$

Carry out. We solve the system of equations

$x = 50y, \quad (1)$

$x + y = 204. \quad (2)$

Substitute $50y$ for x in equation (2) and solve for y.

$$50y + y = 204$$
$$51y = 204$$
$$y = 4$$

Back-substitute in equation (1) to find x.

$$x = 50 \cdot 4 = 200$$

Check. 200 tons is 50 times 4 tons. Also, 4 tons + 200 tons = 204 tons, so the answer checks.

State. 200 tons of ice and 4 tons of hot dogs were consumed.

42. Let $x =$ the amount spent on textbooks and course materials and $y =$ the amount spent on computer equipment.

Solve: $x = y + 183,$

$x + y = 819.$

$x = \$501,\ y = \$318$

43. ***Familiarize.*** Let $a =$ the cost of each adult's admission and $c =$ the cost of each child's admission.

Translate. One equation comes from the fact that an adult's admission is \$5.50 more than a child's:

$$a = c + 5.50.$$

A second equation comes from the fact that the total cost of 2 adult's admissions and 5 child's admissions is \$39:

$$2a + 5c = 39.$$

Carry out. We solve the system of equations

$$a = c + 5.50, \quad (1)$$
$$2a + 5c = 39. \quad (2)$$

Substitute $c + 5.50$ for a in equation (2) and solve for c.

$$2(c + 5.50) + 5c = 39$$
$$2c + 11 + 5c = 39$$
$$7c + 11 = 39$$
$$7c = 28$$
$$c = 4$$

Back-substitute in equation (1) to find a.

$$a = 4 + 5.50 = 9.50$$

Check. \$9.50 is \$5.50 more than \$4. Also, 2 adult's admissions cost 2(\$9.50), or \$19, and 5 child's admissions cost $5 \cdot \$4$, or \$20. Then the total admission cost for 2 adults and 5 children is \$19 + \$20, or \$39. The answer checks.

State. An adult's admission costs \$9.50, and a child's admission costs \$4.

44. Let $x =$ the number of standard-delivery packages and $y =$ the number of express-delivery packages.

Solve: $x + y = 120,$

$3.5x + 7.5y = 596.$

$x = 76,\ y = 44$

45. ***Familiarize.*** Let $x =$ the number of video rentals and $y =$ the number of boxes of popcorn given away. Then the total cost of the rentals is $1 \cdot x$, or x, and the total cost of the popcorn is $2y$.

Translate. The number of new members is the same as the number of incentives so we have one equation:

$$x + y = 48.$$

The total cost of the incentive was \$86. This gives us another equation:

$$x + 2y = 86.$$

Carry out. We solve the system of equations

$$x + y = 48, \quad (1)$$
$$x + 2y = 86. \quad (2)$$

Multiply equation (1) by -1 and add.

$$\begin{array}{r} -x - y = -48 \\ x + 2y = 86 \\ \hline y = 38 \end{array}$$

Back-substitute to find x.

$x + 38 = 48$ Using equation (1)

$x = 10$

Check. When 10 rentals and 38 boxes of popcorn are given away, a total of 48 new members have signed up. Ten rentals cost the store \$10 and 38 boxes of popcorn cost $2 \cdot 38$, or \$76, so the total cost of the incentives was \$10 + \$76, or \$86. The solution checks.

State. Ten rentals and 38 boxes of popcorn were given away.

46. Let $x =$ the number of tickets sold for pavilion seats and $y =$ the number sold for lawn seats.

Solve: $x + y = 1500,$

$25x + 15y = 28,500.$

$x = 600,\ y = 900$

47. ***Familiarize and Translate.*** We use the system of equations given in the problem.

$$y = 70 + 2x \quad (1)$$
$$y = 175 - 5x, \quad (2)$$

Carry out. Substitute $175 - 5x$ for y in equation (1) and solve for x.

$175 - 5x = 70 + 2x$

$105 = 7x$ Adding $5x$ and subtracting 70

$15 = x$

Back-substitute in either equation to find y. We choose equation (1).

$$y = 70 + 2 \cdot 15 = 70 + 30 = 100$$

Check. Substituting 15 for x and 100 for y in both of the original equations yields true equations, so the solution checks. We could also check graphically by finding the point of intersection of equations (1) and (2).

State. The equilibrium point is $(15, \$100)$.

48. Solve: $y = 240 + 40x,$

$y = 500 - 25x.$

$x = 4,\ y = 400$, so the equilibrium point is $(4, \$400)$.

49. ***Familiarize and Translate.*** We find the value of x for which $C = R$, where

$$C = 14x + 350,$$
$$R = 16.5x.$$

Carry out. When $C = R$ we have:

$$14x + 350 = 16.5x$$
$$350 = 2.5x$$
$$140 = x$$

Check. When $x = 140$, $C = 14 \cdot 140 + 350$, or 2310 and $R = 16.5(140)$, or 2310. Since $C = R$, the solution checks.

State. 140 units must be produced and sold in order to break even.

50. Solve $C = R$, where

$$C = 8.5x + 75,$$
$$R = 10x.$$

$x = 50$

51. ***Familiarize and Translate.*** We find the value of x for which $C = R$, where

$$C = 15x + 12,000,$$
$$R = 18x - 6000.$$

Carry out. When $C = R$ we have:

$$15x + 12,000 = 18x - 6000$$
$$18,000 = 3x \quad \text{Subtracting } 15x \text{ and adding } 6000$$
$$6000 = x$$

Check. When $x = 6000$, $C = 15 \cdot 6000 + 12,000$, or 102,000 and $R = 18 \cdot 6000 - 6000$, or 102,000. Since $C = R$, the solution checks.

State. 6000 units must be produced and sold in order to break even.

52. Solve $C = R$, where

$$C = 3x + 400,$$
$$R = 7x - 600.$$

$x = 250$

53. ***Familiarize.*** Let $x =$ the number of servings of spaghetti and meatballs required and $y =$ the number of servings of iceberg lettuce required. Then x servings of spaghetti contain $260x$ Cal and $32x$ g of carbohydrates; y servings of lettuce contain $5y$ Cal and $1 \cdot y$ or y, g of carbohydrates.

Translate. One equation comes from the fact that 400 Cal are desired:

$$260x + 5y = 400.$$

A second equation comes from the fact that 50g of carbohydrates are required:

$$32x + y = 50.$$

Carry out. We solve the system

$$260x + 5y = 400, \quad (1)$$
$$32x + y = 50. \quad (2)$$

Multiply equation (2) by -5 and add.

$$260x + 5y = 400$$
$$-160x - 5y = -250$$
$$100x \quad = 150$$
$$x = 1.5$$

Back-substitute to find y.

$$32(1.5) + y = 50 \quad \text{Using equation (2)}$$
$$48 + y = 50$$
$$y = 2$$

Check. 1.5 servings of spaghetti contain 260(1.5), or 390 Cal and 32(1.5), or 48 g of carbohydrates; 2 servings of lettuce contain $5 \cdot 2$, or 10 Cal and $1 \cdot 2$, or 2 g of carbohydrates. Together they contain $390 + 10$, or 400 Cal and $48 + 2$, or 50 g of carbohydrates. The solution checks.

State. 1.5 servings of spaghetti and meatballs and 2 servings of iceberg lettuce are required.

54. Let $x =$ the number of servings of tomato soup and $y =$ the number of slices of whole wheat bread required.

Solve: $100x + 70y = 230$,

$$18x + 13y = 42.$$

$x = 1.25$, $y = 1.5$

55. ***Familiarize.*** It helps to make a drawing. Then organize the information in a table. Let $x =$ the speed of the boat and $y =$ the speed of the stream. The speed upstream is $x - y$. The speed downstream is $x + y$.

46 km 2 hr $(x + y)$ km/h
Downstream

51 km 3 hr $(x - y)$ km/h
Upstream

	Distance	Speed	Time
Downstream	46	$x + y$	2
Upstream	51	$x - y$	3

Translate. Using $d = rt$ in each row of the table, we get a system of equations.

$$46 = (x + y)2 \qquad x + y = 23, \quad (1)$$
or
$$51 = (x - y)3 \qquad x - y = 17 \quad (2)$$

Carry out. We begin by adding equations (1) and (2).

$$x + y = 23$$
$$x - y = 17$$
$$2x \quad = 40$$
$$x = 20$$

Back-substitute to find y.

$$20 + y = 23 \quad \text{Using equation (1)}$$
$$y = 23$$

Check. The speed downstream is $20 + 3$, or 23 km/h. The distance traveled downstream in 2 hr is $23 \cdot 2$, or 46 km. The speed upstream is $20 - 3$, or 17 km/h. The distance traveled upstream in 3 hr is $17 \cdot 3$, or 51 km. The solution checks.

State. The speed of the boat is 20 km/h. The speed of the stream is 3 km/h.

56. Let x = the speed of the plane and y = the speed of the wind.

	Speed	Time	Distance
Downwind	$x+y$	3	3000
Upwind	$x-y$	4	3000

Solve: $(x+y)3 = 3000,$

$(x-y)4 = 3000.$

$x = 875$ km/h, $y = 125$ km/h

57. ***Familiarize.*** Let x = the amount invested at 7% and y = the amount invested at 9%. Then the interest from the investments is $7\%x$ and $9\%y$, or $0.07x$ and $0.09y$.

Translate.

The total investment is \$15,000.

$x + y = 15,000$

The total interest is \$1230.

$0.07x + 0.09y = 1230$

We have a system of equations:

$x + y = 15,000,$

$0.07x + 0.09y = 1230$

Multiplying the second equation by 100 to clear the decimals, we have:

$x + y = 15,000,$ (1)

$7x + 9y = 123,000.$ (2)

Carry out. We begin by multiplying equation (1) by -7 and adding.

$$\begin{aligned} -7x - 7y &= -105,000 \\ 7x + 9y &= 123,000 \\ \hline 2y &= 18,000 \\ y &= 9000 \end{aligned}$$

Back-substitute to find x.

$x + 9000 = 15,000$ Using equation (1)

$x = 6000$

Check. The total investment is \$6000+\$9000, or \$15,000. The total interest is 0.07(\$6000) + 0.09(\$9000), or \$420 + \$810, or \$1230. The solution checks.

State. \$6000 was invested at 7% and \$9000 was invested at 9%.

58. Let x = the number of short-sleeved shirts sold and y = the number of long-sleeved shirts sold.

Solve: $x + y = 36,$

$12x + 18y = 522.$

$x = 21$, $y = 15$

59. ***Familiarize.*** Let x = the number of pounds of French roast coffee used and y = the number of pounds of Kenyan coffee. We organize the information in a table.

	French roast	Kenyan	Mixture
Amount	x	y	10 lb
Price per pound	\$9.00	\$7.50	\$8.40
Total cost	\$9x	\$7.50y	\$8.40(10), or \$84

Translate. The first and third rows of the table give us a system of equations.

$x + y = 10,$

$9x + 7.5y = 84$

Multiply the second equation by 10 to clear the decimals.

$x + y = 10,$ (1)

$90x + 75y = 840$ (2)

Carry out. Begin by multiplying equation (1) by -75 and adding.

$$\begin{aligned} -75x - 75y &= -750 \\ 90x + 75y &= 840 \\ \hline 15x \quad &= 90 \\ x &= 6 \end{aligned}$$

Back-substitute to find y.

$6 + y = 10$ Using equation (1)

$y = 4$

Check. The total amount of coffee in the mixture is $6+4$, or 10 lb. The total value of the mixture is 6(\$9)+4(\$7.50), or \$54 + \$30, or \$84. The solution checks.

State. 6 lb of French roast coffee and 4 lb of Kenyan coffee should be used.

60. Let x = the speed of the plane and y = the speed of the wind.

	Distance	Speed	Time
LA to NY	3000	$x+y$	5
NY to LA	3000	$x-y$	6

Solve: $3000 = 5(x+y),$

$3000 = 6(x-y).$

$x = 550$ mph, $y = 50$ mph

61. ***Familiarize.*** Let x = the monthly sales, C = the earnings with the straight-commission plan, and S = the earnings with the salary-plus-commission plan. Then 8% of sales is represented by $8\%x$, or $0.08x$, and 1% of sales is represented by $1\%x$, or $0.01x$.

Translate.

Straight commission pays 8% of sales.

$C = 0.08x$

Salary plus commission pays \$1500 plus 1% of sales.

$S = 1500 + 0.01x$

Carry out. We find the value of x for which $C = S$. We have:

$$0.08x = 1500 + 0.01x$$
$$0.07x = 1500$$
$$x \approx 21,428.57$$

Check. When $x \approx 21,428.57$, $C \approx 0.08(21,428.57) \approx 1714.29$ and $S \approx 1500 + 0.01(21,428.57) \approx 1714.29$. Since $C = S$, the solution checks.

State. The two plans pay the same amount for monthly sales of about $21,428.57.

62. Let d = the distance traveled by the slower plane and t = the time the planes travel.

	Distance	Speed	Time
Slower plane	d	190	t
Faster plane	$780 - d$	200	t

Solve: $d = 190t,$
$780 - d = 200t.$

$t = 2$ hr

63. Left to the student

64. Left to the student

65. a) $b(x) = 0.1714285714x + 63.48571429;$
$c(x) = 1.007142857x + 47.92857143$

b) Graph $y_1 = b(x)$ and $y_2 = c(x)$ and find the first coordinate of the point of intersection of the graphs. It is approximately 18.6, so we estimate that chicken consumption will equal beef consumption about 19 yr after 1995.

66. a) $m(x) = -0.086746988x + 75.0253012;$
$w(x) = 0.221686747x + 58.9686747$

b) Graph $y_1 = m(x)$ and $y_2 = w(x)$ and find the first coordinate or the point of intersection of the graphs. It is approximately 52, so the percentages of men and women in the labor force will be equal about 52 years after 1995.

67. When a variable is not alone on one side of an equation or when solving for a variable is difficult or produces an expression containing fractions, the elimination method is preferable to the substitution method.

68. The solution of the equation $2x + 5 = 3x - 7$ is the first coordinate of the point of intersection of the graphs of $y_1 = 2x + 5$ and $y_2 = 3x - 7$. The solution of the system of equations $y = 2x + 5$, $y = 3x - 7$ is the ordered pair that is the point of intersection of y_1 and y_2.

69. ***Familiarize.*** Let d = the number of DVDs shipped to retailers in the first quarter of 2002, in millions. Then a 93% increase over this number is $d + 0.93d$, or $1.93d$.

Translate.

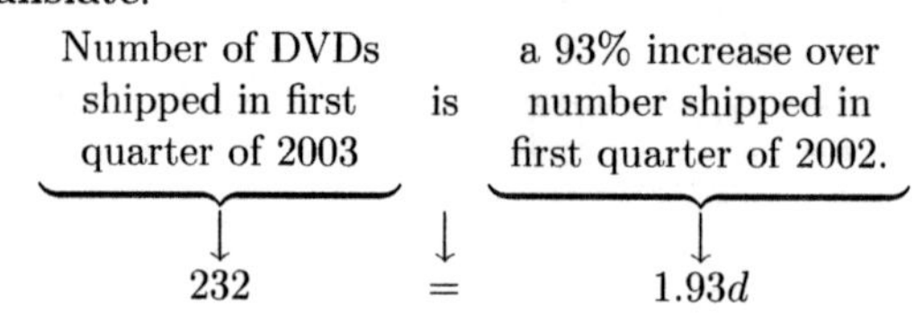

Carry out. We solve the equation.

$$232 = 1.93d$$
$$120 \approx d \quad \text{Dividing by 1.93}$$

Check. 93% of 120 million is about 112 million, and 120 million + 112 million = 232 million. The answer checks.

State. About 120 million DVDs were shipped to retailers in the first quarter of 2002.

70. Let t = the gross ticket sales for *Star Wars* in 2002, in millions of dollars.

Solve: $404 = \frac{4}{3}t$

$t = \$303$ million

71. Substituting 15 for $f(x)$, we solve the following equation.

$$15 = x^2 - 4x + 3$$
$$0 = x^2 - 4x - 12$$
$$0 = (x - 6)(x + 2)$$
$$x - 6 = 0 \quad or \quad x + 2 = 0$$
$$x = 6 \quad or \quad x = -2$$

If the output is 15, the input is 6 or -2.

72. $f(-2) = (-2)^2 - 4(-2) + 3 = 15$

73. Substituting 8 for $f(x)$, we solve the following equation.

$$8 = x^2 - 4x + 3$$
$$0 = x^2 - 4x - 5$$
$$0 = (x - 5)(x + 1)$$
$$x - 5 = 0 \quad or \quad x + 1 = 0$$
$$x = 5 \quad or \quad x = -1$$

Given an output of 8, the corresponding inputs are 5 and -1.

74.
$$x^2 - 4x + 3 = 0$$
$$(x - 1)(x - 3) = 0$$
$$x = 1 \quad or \quad x = 3$$

75. ***Familiarize.*** Let x = the time spent jogging and y = the time spent walking. Then Nancy jogs $8x$ km and walks $4y$ km. We organize the information in a table.

Translate.

The total time is 1 hr.

$$x + y = 1$$

The total distance is 6 km.

$$8x + 4y = 6$$

Carry out. Solve the system

$x + y = 1, \quad (1)$
$8x + 4y = 6. \quad (2)$

Multiply equation (1) by -4 and add.

$$\begin{aligned} -4x - 4y &= -4 \\ 8x + 4y &= 6 \\ \hline 4x \qquad &= 2 \\ x &= \frac{1}{2} \end{aligned}$$

This is the time we need to find the distance spent jogging, so we could stop here. However, we will not be able to check the solution unless we find y also so we continue. We back-substitute.

$\frac{1}{2} + y = 1 \quad$ Using equation (1)

$y = \frac{1}{2}$

Then the distance jogged is $8 \cdot \frac{1}{2}$, or 4 km.

Check. The total time is $\frac{1}{2}$ hr $+ \frac{1}{2}$ hr, or 1 hr. The total distance is 4 km + 2 km, or 6 km. The solution checks.

State. Nancy jogged 4 km on each trip.

76. Let $x =$ the number of one-turtleneck orders and $y =$ the number of two-turtleneck orders.

Solve: $\quad x + 2y = 1250,$
$\qquad 15x + 25y = 16,750.$

$y = 400$

77. ***Familiarize and Translate.*** We let x and y represent the speeds of the trains. Organize the information in a table. Using $d = rt$, we let $3x$, $2y$, $1.5x$, and $3y$ represent the distances the trains travel.

First situation:

3 hours $\quad x$ km/h $\qquad y$ km/h $\quad$ 2 hours
Union $\qquad\qquad\qquad\qquad$ Central
$\vdash$———— 216 km ————$\dashv$

Second situation:

1.5 hours $\quad x$ km/h $\qquad y$ km/h $\quad$ 3 hours
Union $\qquad\qquad\qquad\qquad$ Central
$\vdash$———— 216 km ————$\dashv$

	Distance traveled in first situation	Distance traveled in second situation
Train_1 (from Union to Central)	$3x$	$1.5x$
Train_2 (from Central to Union)	$2y$	$3y$
Total	216	216

The total distance in each situation is 216 km. Thus, we have a system of equations.

$3x + 2y = 216, \quad (1)$
$1.5x + 3y = 216 \quad (2)$

Carry out. Multiply equation (2) by -2 and add.

$$\begin{aligned} 3x + 2y &= 216 \\ -3x - 6y &= -432 \\ \hline -4y &= -216 \\ y &= 54 \end{aligned}$$

Back-substitute to find x.

$3x + 2 \cdot 54 = 216 \quad$ Using equation (1)
$3x + 108 = 216$
$3x = 108$
$x = 36$

Check. If $x = 36$ and $y = 54$, the total distance the trains travel in the first situation is $3 \cdot 36 + 2 \cdot 54$, or 216 km. The total distance they travel in the second situation is $1.5 \cdot 36 + 3 \cdot 54$, or 216 km. The solution checks.

State. The speed of the first train is 36 km/h. The speed of the second train is 54 km/h.

78. Let $x =$ the amount of mixture replaced by 100% antifreeze and $y =$ the amount of 30% mixture retained.

	Replaced	Retained	Total
Amount	x	y	16 L
Percent of antifreeze	100%	30%	50%
Amount of antifreeze	$100\%x$	$30\%y$	$50\% \times 16$, or 8 L

Solve: $\quad x + y = 16,$
$\qquad x + 0.3y = 8.$

$x = 4\frac{4}{7}$ L

79. Substitute the given solutions in the equation $Ax + By = 1$ to get a system of equations.

$3A - B = 1, \quad (1)$
$-4A - 2B = 1 \quad (2)$

Multiply equation (1) by -2 and add.

$$\begin{aligned} -6A + 2B &= -2 \\ -4A - 2B &= 1 \\ \hline -10A \qquad &= -1 \\ A \qquad &= \frac{1}{10} \end{aligned}$$

Back-substitute to find B.

$3\left(\frac{1}{10}\right) - B = 1 \qquad$ Using equation (1)

$\frac{3}{10} - B = 1$

$-B = \frac{7}{10}$

$B = -\frac{7}{10}$

We have $A = \frac{1}{10}$ and $B = -\frac{7}{10}$.

80. Let x = the number of people ahead of you and y = the number of people behind you.

Solve: $x = 2 + y$,

$x + 1 + y = 3y$.

$x = 5$

81. ***Familiarize.*** Let x and y represent the number of gallons of gasoline used in city driving and in highway driving, respectively. Then $46x$ and $51y$ represent the number of miles driven in the city and on the highway, respectively.

Translate. The fact that 13 gal of gasoline were used gives us one equation:

$x + y = 13$.

A second equation comes from the fact that the car is driven 621 mile:

$46x + 51y = 621$.

Carry out. We solve the system of equations

$$x + y = 13, \quad (1)$$
$$46x + 51y = 621. \quad (2)$$

Multiply equation (1) by -46 and add.

$$-46x - 46y = -598$$
$$46x + 51y = 621$$
$$5y = 23$$
$$y = 4.6$$

Back-substitute to find x.

$$x + 4.6 = 13$$
$$x = 8.4$$

Then in the city the car is driven 46(8.4), or 386.4 mi; on the highway it is driven 51(4.6), or 234.6 mi.

Check. The number of gallons of gasoline used is 8.4+4.6, or 13. The number of miles driven is 386.4+234.6, or 621. The answer checks.

State. The car was driven 386.4 mi in the city and 234.6 mi on the highway.

82. First we convert the given distances to miles:

$$300 \text{ ft} = \frac{300}{5280} \text{ mi} = \frac{5}{88} \text{ mi},$$

$$500 \text{ ft} = \frac{500}{5280} \text{ mi} = \frac{25}{264} \text{ mi}$$

Then at 10 mph, Heather can run to point P in $\frac{\frac{5}{88}}{10}$, or $\frac{1}{176}$ hr, and she can run to point Q in $\frac{\frac{25}{264}}{10}$, or $\frac{5}{528}$ hr (using $d = rt$, or $\frac{d}{r} = t$).

Let d = the distance, in miles, from the train to point P in the drawing in the text, and let r the speed of the train, in miles per hour.

	Distance	Speed	Time
Going to P	d	r	$\frac{1}{176}$
Going to Q	$d + \frac{5}{88} + \frac{25}{264}$	r	$\frac{5}{528}$

Solve: $d = r\left(\frac{1}{176}\right)$,

$$d + \frac{5}{88} + \frac{25}{264} = r\left(\frac{5}{528}\right).$$

$r = 40$ mph

Exercise Set 8.2

1.
$$x + y + z = 2, \quad (1)$$
$$6x - 4y + 5z = 31, \quad (2)$$
$$5x + 2y + 2z = 13 \quad (3)$$

Multiply equation (1) by -6 and add it to equation (2). We also multiply equation (1) by -5 and add it to equation (3).

$$x + y + z = 2 \quad (1)$$
$$-10y - z = 19 \quad (4)$$
$$-3y - 3z = 3 \quad (5)$$

Multiply the last equation by 10 to make the y-coefficient a multiple of the y-coefficient in equation (4).

$$x + y + z = 2 \quad (1)$$
$$-10y - z = 19 \quad (4)$$
$$-30y - 30z = 30 \quad (6)$$

Multiply equation (4) by -3 and add it to equation (6).

$$x + y + z = 2 \quad (1)$$
$$-10y - z = 19 \quad (4)$$
$$-27z = -27 \quad (7)$$

Solve equation (7) for z.

$$-27z = -27$$
$$z = 1$$

Back-substitute 1 for z in equation (4) and solve for y.

$$-10y - 1 = 19$$
$$-10y = 20$$
$$y = -2$$

Back-substitute 1 for z for -2 and y in equation (1) and solve for x.

$$x + (-2) + 1 = 2$$
$$x - 1 = 2$$
$$x = 3$$

The solution is $(3, -2, 1)$.

2.
$$x + 6y + 3z = 4, \quad (1)$$
$$2x + y + 2z = 3, \quad (2)$$
$$3x - 2y + z = 0 \quad (3)$$

Multiply equation (1) by -2 and add it to equation (2). We also multiply equation (1) by -3 and add it to equation (3).

$$\begin{aligned} x + 6y + 3z &= 4, & (1)\\ -11y - 4z &= -5, & (4)\\ -20y - 8z &= -12 & (5)\end{aligned}$$

Multiply equation (5) by 11.

$$\begin{aligned} x + 6y + 3z &= 4, & (1)\\ -11y - 4z &= -5, & (4)\\ -220y - 88z &= -132 & (6)\end{aligned}$$

Multiply equation (4) by -20 and add it to equation (6).

$$\begin{aligned} x + 6y + 3z &= 4, & (1)\\ -11y - 4z &= -5, & (4)\\ -8z &= -32 & (7)\end{aligned}$$

Complete the solution.

$$\begin{aligned} -8z &= -32\\ z &= 4\\ -11y - 4 \cdot 4 &= -5\\ -11y &= 11\\ y &= -1\\ x + 6(-1) + 3 \cdot 4 &= 4\\ x &= -2\end{aligned}$$

The solution is $(-2, -1, 4)$.

3.

$$\begin{aligned} x - y + 2z &= -3 & (1)\\ x + 2y + 3z &= 4 & (2)\\ 2x + y + z &= -3 & (3)\end{aligned}$$

Multiply equation (1) by -1 and add it to equation (2). We also multiply equation (1) by -2 and add it to equation (3).

$$\begin{aligned} x - y + 2z &= -3 & (1)\\ 3y + z &= 7 & (4)\\ 3y - 3z &= 3 & (5)\end{aligned}$$

Multiply equation (4) by -1 and add it to equation (5).

$$\begin{aligned} x - y + 2z &= -3 & (1)\\ 3y + z &= 7 & (4)\\ -4z &= -4 & (6)\end{aligned}$$

Solve equation (6) for z.

$$\begin{aligned} -4z &= -4\\ z &= 1\end{aligned}$$

Back-substitute 1 for z in equation (4) and solve for y.

$$\begin{aligned} 3y + 1 &= 7\\ 3y &= 6\\ y &= 2\end{aligned}$$

Back-substitute 1 for z and 2 for y in equation (1) and solve for x.

$$\begin{aligned} x - 2 + 2 \cdot 1 &= -3\\ x &= -3\end{aligned}$$

The solution is $(-3, 2, 1)$.

4.

$$\begin{aligned} x + y + z &= 6, & (1)\\ 2x - y - z &= -3, & (2)\\ x - 2y + 3z &= 6 & (3)\end{aligned}$$

Multiply equation (1) by -2 and add it to equation (2). Also, multiply equation (1) by -1 and add it to equation (3).

$$\begin{aligned} x + y + z &= 6, & (1)\\ -3y - 3z &= -15, & (4)\\ -3y + 2z &= 0 & (5)\end{aligned}$$

Multiply equation (4) by -1 and add it to equation (5).

$$\begin{aligned} x + y + z &= 6, & (1)\\ -3y - 3z &= -15, & (4)\\ 5z &= 15 & (6)\end{aligned}$$

Complete the solution.

$$\begin{aligned} 5z &= 15\\ z &= 3\\ -3y - 3 \cdot 3 &= -15\\ -3y &= -6\\ y &= 2\\ x + 2 + 3 &= 6\\ x &= 1\end{aligned}$$

The solution is $(1, 2, 3)$.

5.

$$\begin{aligned} x + 2y - z &= 5, & (1)\\ 2x - 4y + z &= 0, & (2)\\ 3x + 2y + 2z &= 3 & (3)\end{aligned}$$

Multiply equation (1) by -2 and add it to equation (2). Also, multiply equation (1) by -3 and add it to equation (3).

$$\begin{aligned} x + 2y - z &= 5, & (1)\\ -8y + 3z &= -10, & (4)\\ -4y + 5z &= -12 & (5)\end{aligned}$$

Multiply equation (5) by 2 to make the y-coefficient a multiple of the y-coefficient of equation (4).

$$\begin{aligned} x + 2y - z &= 5, & (1)\\ -8y + 3z &= -10, & (4)\\ -8y + 10z &= -24 & (6)\end{aligned}$$

Multiply equation (4) by -1 and add it to equation (6).

$$\begin{aligned} x + 2y - z &= 5, & (1)\\ -8y + 3z &= -10, & (4)\\ 7z &= -14 & (7)\end{aligned}$$

Solve equation (7) for z.

$$\begin{aligned} 7z &= -14\\ z &= -2\end{aligned}$$

Back-substitute -2 for z in equation (4) and solve for y.

$$\begin{aligned} -8y + 3(-2) &= -10\\ -8y - 6 &= -10\\ -8y &= -4\\ y &= \frac{1}{2}\end{aligned}$$

Back-substitute $\frac{1}{2}$ for y and -2 for z in equation (1) and solve for x.

$$x + 2\cdot\frac{1}{2} - (-2) = 5$$
$$x + 1 + 2 = 5$$
$$x = 2$$

The solution is $\left(2, \frac{1}{2}, -2\right)$.

6. $2x + 3y - z = 1,\quad (1)$
$x + 2y + 5z = 4,\quad (2)$
$3x - y - 8z = -7\quad (3)$

Interchange equations (1) and (2).

$x + 2y + 5z = 4,\quad (2)$
$2x + 3y - z = 1,\quad (1)$
$3x - y - 8z = -7\quad (3)$

Multiply equation (2) by -2 and add it to equation (1). Multiply equation (2) by -3 and add it to equation (3).

$x + 2y + 5z = 4,\quad (2)$
$-y - 11z = -7,\quad (4)$
$-7y - 23z = -19\quad (5)$

Multiply equation (4) by -7 and add it to equation (5).

$x + 2y + 5z = 4,\quad (2)$
$-y - 11z = -7\quad (4)$
$54z = 30$

Complete the solution.

$$54z = 30$$
$$z = \frac{5}{9}$$
$$-y - 11\cdot\frac{5}{9} = -7$$
$$-y = -\frac{8}{9}$$
$$y = \frac{8}{9}$$
$$x + 2\cdot\frac{8}{9} + 5\cdot\frac{5}{9} = 4$$
$$x = -\frac{5}{9}$$

The solution is $\left(-\frac{5}{9}, \frac{8}{9}, \frac{5}{9}\right)$.

7. $x + 2y - z = -8,\quad (1)$
$2x - y + z = 4,\quad (2)$
$8x + y + z = 2\quad (3)$

Multiply equation (1) by -2 and add it to equation (2). Also, multiply equation (1) by -8 and add it to equation (3).

$x + 2y - z = -8,\quad (1)$
$-5y + 3z = 20,\quad (4)$
$-15y + 9z = 66\quad (5)$

Multiply equation (4) by -3 and add it to equation (5).

$x + 2y - z = -8,\quad (1)$
$-5y + 3z = 20,\quad (4)$
$0 = 6\quad (6)$

Equation (6) is false, so the system of equations has no solution.

8. $x + 2y - z = 4,\quad (1)$
$4x - 3y + z = 8,\quad (2)$
$5x - y = 12\quad (3)$

Multiply equation (1) by -4 and add it to equation (2). Also, multiply equation (1) by -5 and add it to equation (3).

$x + 2y - z = 4,\quad (1)$
$-11y + 5z = -8,\quad (4)$
$-11y + 5z = -8\quad (5)$

Multiply equation (4) by -1 and add it to equation (5).

$x + 2y - z = 4,\quad (1)$
$-11y + 5z = -8,\quad (4)$
$0 = 0\quad (6)$

The equation $0 = 0$ tells us that equation (3) of the original system is dependent on the first two equations. The system of equations has infinitely many solutions and is equivalent to

$x + 2y - z = 4,\quad (1)$
$4x - 3y + z = 8.\quad (2)$

To find an expression for the solutions, we first solve equation (4) for either y or z. We choose to solve for y.

$$-11y + 5z = -8$$
$$-11y = -5z - 8$$
$$y = \frac{5z + 8}{11}$$

Back-substitute in equation (1) to find an expression for x in terms of z.

$$x + 2\left(\frac{5z + 8}{11}\right) - z = 4$$
$$x + \frac{10z}{11} + \frac{16}{11} - z = 4$$
$$x = \frac{z}{11} + \frac{28}{11} = \frac{z + 28}{11}$$

The solutions are given by $\left(\frac{z + 28}{11}, \frac{5z + 8}{11}, z\right)$, where z is any real number.

9. $2x + y - 3z = 1,\quad (1)$
$x - 4y + z = 6,\quad (2)$
$4x - 7y - z = 13\quad (3)$

Interchange equations (1) and (2).

$x - 4y + z = 6,\quad (2)$
$2x + y - 3z = 1,\quad (1)$
$4x - 7y - z = 13\quad (3)$

Multiply equation (2) by -2 and add it to equation (1). Also, multiply equation (2) by -4 and add it to equation (3).

$$x - 4y + z = 6, \quad (2)$$
$$9y - 5z = -11, \quad (4)$$
$$9y - 5z = -11 \quad (5)$$

Multiply equation (4) by -1 and add it to equation (5).

$$x - 4y + z = 6, \quad (1)$$
$$9y - 5z = -11, \quad (4)$$
$$0 = 0 \quad (6)$$

The equation $0 = 0$ tells us that equation (3) of the original system is dependent on the first two equations. The system of equations has infinitely many solutions and is equivalent to

$$2x + y - 3z = 1, \quad (1)$$
$$x - 4y + z = 6. \quad (2)$$

To find an expression for the solutions, we first solve equation (4) for either y or z. We choose to solve for z.

$$9y - 5z = -11$$
$$-5z = -9y - 11$$
$$z = \frac{9y + 11}{5}$$

Back-substitute in equation (2) to find an expression for x in terms of y.

$$x - 4y + \frac{9y + 11}{5} = 6$$
$$x - 4y + \frac{9}{5}y + \frac{11}{5} = 6$$
$$x = \frac{11}{5}y + \frac{19}{5} = \frac{11y + 19}{5}$$

The solutions are given by $\left(\frac{11y + 19}{5}, y, \frac{9y + 11}{5}\right)$, where y is any real number.

10. $x + 3y + 4z = 1, \quad (1)$
$3x + 4y + 5z = 3, \quad (2)$
$x + 8y + 11z = 2 \quad (3)$

Multiply equation (1) by -3 and add it to equation (2). Also, multiply equation (1) by -1 and add it to equation (3).

$$x + 3y + 4z = 1, \quad (1)$$
$$-5y - 7z = 0, \quad (4)$$
$$5y + 7z = 1 \quad (5)$$

Add equation (4) to equation (5).

$$x + 3y + 4z = 1, \quad (1)$$
$$-5y - 7z = 0, \quad (4)$$
$$0 = 1 \quad (6)$$

Equation (6) is false, so the system of equations has no solution.

11. $4a + 9b = 8, \quad (1)$
$8a + 6c = -1, \quad (2)$
$6b + 6c = -1 \quad (3)$

Multiply equation (1) by -2 and add it to equation (2).

$$4a + 9b = 8, \quad (1)$$
$$-18b + 6c = -17, \quad (4)$$
$$6b + 6c = -1 \quad (3)$$

Multiply equation (3) by 3 to make the b-coefficient a multiple of the b-coefficient in equation (4).

$$4a + 9b = 8, \quad (1)$$
$$-18b + 6c = -17, \quad (4)$$
$$18b + 18c = -3 \quad (5)$$

Add equation (4) to equation (5).

$$4a + 9b = 8, \quad (1)$$
$$-18b + 6c = -17, \quad (4)$$
$$24c = -20 \quad (6)$$

Solve equation (6) for c.

$$24c = -20$$
$$c = -\frac{20}{24} = -\frac{5}{6}$$

Back-substitute $-\frac{5}{6}$ for c in equation (4) and solve for b.

$$-18b + 6c = -17$$
$$-18b + 6\left(-\frac{5}{6}\right) = -17$$
$$-18b - 5 = -17$$
$$-18b = -12$$
$$b = \frac{12}{18} = \frac{2}{3}$$

Back-substitute $\frac{2}{3}$ for b in equation (1) and solve for a.

$$4a + 9b = 8$$
$$4a + 9 \cdot \frac{2}{3} = 8$$
$$4a + 6 = 8$$
$$4a = 2$$
$$a = \frac{1}{2}$$

The solution is $\left(\frac{1}{2}, \frac{2}{3}, -\frac{5}{6}\right)$.

12. $3p + 2r = 11, \quad (1)$
$q - 7r = 4, \quad (2)$
$p - 6q = 1 \quad (3)$

Interchange equations (1) and (3).

$$p - 6q = 1, \quad (3)$$
$$q - 7r = 4, \quad (2)$$
$$3p + 2r = 11 \quad (1)$$

Multiply equation (3) by -3 and add it to equation (1).

$$p - 6q = 1, \quad (3)$$
$$q - 7r = 4, \quad (2)$$
$$18q + 2r = 8 \quad (4)$$

Multiply equation (2) by -18 and add it to equation (4).

$$\begin{aligned} p - 6q \quad\quad &= 1, \quad (3)\\ q - \quad 7r &= 4, \quad (2)\\ 128r &= -64 \quad (5) \end{aligned}$$

Complete the solution.

$$128r = -64$$
$$r = -\frac{1}{2}$$
$$q - 7\left(-\frac{1}{2}\right) = 4$$
$$q = \frac{1}{2}$$
$$p - 6\cdot\frac{1}{2} = 1$$
$$p = 4$$

The solution is $\left(4, \frac{1}{2}, -\frac{1}{2}\right)$.

13.
$$\begin{aligned} 2x \quad\quad + z &= 1, \quad (1)\\ 3y - 2z &= 6, \quad (2)\\ x - 2y \quad\quad &= -9 \quad (3) \end{aligned}$$

Interchange equations (1) and (3).

$$\begin{aligned} x - 2y \quad\quad &= -9, \quad (3)\\ 3y - 2z &= 6, \quad (2)\\ 2x \quad\quad + z &= 1 \quad (1) \end{aligned}$$

Multiply equation (3) by -2 and add it to equation (1).

$$\begin{aligned} x - 2y \quad\quad &= -9, \quad (3)\\ 3y - 2z &= 6, \quad (2)\\ 4y + z &= 19 \quad (4) \end{aligned}$$

Multiply equation (4) by 3 to make the y-coefficient a multiple of the y-coefficient in equation (2).

$$\begin{aligned} x - 2y \quad\quad &= -9, \quad (3)\\ 3y - 2z &= 6, \quad (2)\\ 12y + 3z &= 57 \quad (5) \end{aligned}$$

Multiply equation (2) by -4 and add it to equation (5).

$$\begin{aligned} x - 2y \quad\quad &= -9, \quad (3)\\ 3y - 2z &= 6, \quad (2)\\ 11z &= 33 \quad (6) \end{aligned}$$

Solve equation (6) for z.

$$11z = 33$$
$$z = 3$$

Back-substitute 3 for z in equation (2) and solve for y.

$$3y - 2z = 6$$
$$3y - 2\cdot 3 = 6$$
$$3y - 6 = 6$$
$$3y = 12$$
$$y = 4$$

Back-substitute 4 for y in equation (3) and solve for x.

$$x - 2y = -9$$
$$x - 2\cdot 4 = -9$$
$$x - 8 = -9$$
$$x = -1$$

The solution is $(-1, 4, 3)$.

14.
$$\begin{aligned} 3x \quad\quad + 4z &= -11, \quad (1)\\ x - 2y \quad\quad &= 5, \quad (2)\\ 4y - z &= -10 \quad (3) \end{aligned}$$

Interchange equations (1) and (2).

$$\begin{aligned} x - 2y \quad\quad &= 5, \quad (2)\\ 3x \quad\quad + 4z &= -11, \quad (1)\\ 4y - z &= -10 \quad (3) \end{aligned}$$

Multiply equation (2) by -3 and add it to equation (1).

$$\begin{aligned} x - 2y \quad\quad &= 5, \quad (2)\\ 6y + 4z &= -26, \quad (4)\\ 4y - z &= -10 \quad (3) \end{aligned}$$

Multiply equation (3) by 3 to make the y-coefficient a multiple of the y-coefficient in equation (4).

$$\begin{aligned} x - 2y \quad\quad &= 5, \quad (2)\\ 6y + 4z &= -26, \quad (4)\\ 12y - 3z &= -30 \quad (5) \end{aligned}$$

Multiply equation (4) by -2 and add it to equation (5).

$$\begin{aligned} x - 2y \quad\quad &= 5, \quad (2)\\ 6y + 4z &= -26, \quad (4)\\ -11z &= 22 \quad (6) \end{aligned}$$

Solve equation (6) for z.

$$-11z = 22$$
$$z = -2$$

Back-substitute -2 for z in equation (4) and solve for y.

$$6y + 4z = -26$$
$$6y + 4(-2) = -26$$
$$6y - 8 = -26$$
$$6y = -18$$
$$y = -3$$

Back-substitute -3 for y in equation (2) and solve for x.

$$x - 2y = 5$$
$$x - 2(-3) = 5$$
$$x + 6 = 5$$
$$x = -1$$

The solution is $(-1, -3, -2)$.

15.
$$\begin{aligned} w + x + y + z &= 2 \quad (1)\\ w + 2x + 2y + 4z &= 1 \quad (2)\\ -w + x - y - z &= -6 \quad (3)\\ -w + 3x + y - z &= -2 \quad (4)\end{aligned}$$

Multiply equation (1) by -1 and add to equation (2). Add equation (1) to equation (3) and to equation (4).

$$\begin{aligned} w + x + y + z &= 2 \quad (1)\\ x + y + 3z &= -1 \quad (5)\\ 2x &= -4 \quad (6)\\ 4x + 2y &= 0 \quad (7)\end{aligned}$$

Solve equation (6) for x.

$$2x = -4$$
$$x = -2$$

Back-substitute -2 for x in equation (7) and solve for y.

$$4(-2) + 2y = 0$$
$$-8 + 2y = 0$$
$$2y = 8$$
$$y = 4$$

Back-substitute -2 for x and 4 for y in equation (5) and solve for z.

$$-2 + 4 + 3z = -1$$
$$3z = -3$$
$$z = -1$$

Back-substitute -2 for x, 4 for y, and -1 for z in equation (1) and solve for w.

$$w - 2 + 4 - 1 = 2$$
$$w = 1$$

The solution is $(1, -2, 4, -1)$.

16.
$$\begin{aligned} w + x - y + z &= 0, \quad (1)\\ -w + 2x + 2y + z &= 5, \quad (2)\\ -w + 3x + y - z &= -4, \quad (3)\\ -2w + x + y - 3z &= -7 \quad (4)\end{aligned}$$

Add equation (1) to equation (2) and equation (3). Also, multiply equation (1) by 2 and add it to equation (4).

$$\begin{aligned} w + x - y + z &= 0, \quad (1)\\ 3x + y + 2z &= 5, \quad (5)\\ 4x &= -4, \quad (6)\\ 3x - y - z &= -7 \quad (7)\end{aligned}$$

Multiply equation (6) by 3.

$$\begin{aligned} w + x - y + z &= 0, \quad (1)\\ 3x + y + 2z &= 5, \quad (5)\\ 12x &= -12, \quad (8)\\ 3x - y - z &= -7 \quad (7)\end{aligned}$$

Multiply equation (5) by -4 and add it to equation (8). Also, multiply equation (5) by -1 and add it to equation (7).

$$\begin{aligned} w + x - y + z &= 0, \quad (1)\\ 3x + y + 2z &= 5, \quad (5)\\ -4y - 8z &= -32, \quad (9)\\ -2y - 3z &= -12 \quad (10)\end{aligned}$$

Multiply equation (10) by 2.

$$\begin{aligned} w + x - y + z &= 0, \quad (1)\\ 3x + y + 2z &= 5, \quad (5)\\ -4y - 8z &= -32, \quad (9)\\ -4y - 6z &= -24 \quad (11)\end{aligned}$$

Multiply equation (9) by -1 and add it to equation (11).

$$\begin{aligned} w + x - y + z &= 0, \quad (1)\\ 3x + y + 2z &= 5, \quad (5)\\ -4y - 8z &= -32, \quad (9)\\ 2z &= 8 \quad (12)\end{aligned}$$

Complete the solution.

$$2z = 8$$
$$z = 4$$
$$-4y - 8 \cdot 4 = -32$$
$$-4y = 0$$
$$y = 0$$
$$3x + 0 + 2 \cdot 4 = 5$$
$$3x = -3$$
$$x = -1$$
$$w - 1 - 0 + 4 = 0$$
$$w = -3$$

The solution is $(-3, -1, 0, 4)$.

17. ***Familiarize.*** Let x = the number of orders under 10 lb, y = the number of orders from 10 lb up to 15 lb, and z = the number of orders of 15 lb or more. Then the total shipping charges for each category of order are $\$3x$, $\$5y$, and $\$7.50z$.

Translate.

The total number of orders was 150.

$$x + y + z = 150$$

Total shipping charges were \$680.

$$3x + 5y + 7.5z = 680$$

The number of orders under 10 lb was three times the number of orders weighing 15 lb or more.

$$x = 3z$$

We have a system of equations

$$\begin{aligned} x + y + z &= 150, & x + y + z &= 150,\\ 3x + 5y + 7.5z &= 680, \quad \text{or} & 30x + 50y + 75z &= 6800,\\ x &= 3z & x - 3z &= 0\end{aligned}$$

Carry out. Solving the system of equations, we get $(60, 70, 20)$.

Check. The total number of orders is $60 + 70 + 20$, or 150. The total shipping charges are $\$3 \cdot 60 + \$5 \cdot 70 + \$7.50(20)$, or \$680. The number of orders under 10 lb, 60, is three times the number of orders weighing over 15 lb, 20. The solution checks.

State. There were 60 packages under 10 lb, 70 packages from 10 lb up to 15 lb, and 20 packages weighing 15 lb or more.

18. Let x, y, and z represent the number of orders of \$25 or less, from \$25.01 to \$75, and over \$75, respectively.

Solve: $x+y+z=600$,

$4x+6y+7z=3340$,

$x=z+80$.

$x=180$, $y=320$, $z=100$

19. ***Familiarize.*** Let x, y, and z represent the number of servings of ground beef, baked potato, and strawberries required, respectively. One serving of ground beef contains $245x$ Cal, $0x$ or 0 g of carbohydrates, and $9x$ mg of calcium. One baked potato contains $145y$ Cal, $34y$ g of carbohydrates, and $8y$ mg of calcium. One serving of strawberries contains $45z$ Cal, $10z$ g of carbohydrates, and $21z$ mg of calcium.

Translate.

The total number of calories is 485.

$$245x+145y+45z=485$$

A total of 41.5 g of carbohydrates is required.

$$34y+10z=41.5$$

A total of 35 mg of calcium is required.

$$9x+8y+21z=35$$

We have a system of equations.

$$\begin{aligned} 245x+145y+45z&=485,\\ 34y+10z&=41.5,\\ 9x+8y+21z&=35 \end{aligned}$$

Carry out. Solving the system of equations, we get $(1.25, 1, 0.75)$.

Check. 1.25 servings of ground beef contains 306.25 Cal, no carbohydrates, and 11.25 mg of calcium; 1 baked potato contains 145 Cal, 34 g of carbohydrates, and 8 mg of calcium; 0.75 servings of strawberries contains 33.75 Cal, 7.5 g of carbohydrates, and 15.75 mg of calcium. Thus, there are a total of $306.25+145+33.75$, or 485 Cal, $34+7.5$, or 41.5 g of carbohydrates, and $11.25+8+15.75$, or 35 mg of calcium. The solution checks.

State. 1.25 servings of ground beef, 1 baked potato, and 0.75 serving of strawberries are required.

20. Let x, y, and z represent the number of servings of chicken, mashed potatoes, and peas to be used, respectively.

Solve: $140x+160y+125z=415$,

$27x+4y+8z=50.5$,

$64x+636y+139z=553$.

$x=1.5$, $y=0.5$, $z=1$

21. ***Familiarize.*** Let x, y, and z represent the amounts invested at 3%, 4%, and 6%, respectively. Then the annual interest from the investments is $3\%x$, $4\%y$, and $6\%z$, or $0.03x$, $0.04y$, and $0.06z$.

Translate.

A total of \$5000 was invested.

$$x+y+z=5000$$

The total interest is \$243.

$$0.03x+0.04y+0.06z=243$$

The amount invested at 6% is \$1500 more than the amount invested at 3%.

$$z=x+1500$$

We have a system of equations.

$$\begin{aligned} x+y+z&=5000,\\ 0.03x+0.04y+0.06z&=243,\\ z&=x+1500 \end{aligned}$$

or

$$\begin{aligned} x+y+z&=5000,\\ 3x+4y+6z&=24,300,\\ -x+z&=1500 \end{aligned}$$

Carry out. Solving the system of equations, we get $(1300, 900, 2800)$.

Check. The total investment was \$1300+\$900+\$2800, or \$5000. The total interest was 0.03(\$1300) + 0.04(\$900) + 0.06(\$2800) = \$39 + \$36 + \$168, or \$243. The amount invested at 6%, \$2800, is \$1500 more than the amount invested at 3%, \$1300. The solution checks.

State. \$1300 was invested at 3%, \$900 at 4%, and \$2800 at 6%.

22. Let x, y, and z represent the amounts invested at 2%, 3%, and 4%, respectively.

Solve: $0.02x+0.03y+0.04z=126$,

$y=x+500$,

$z=3y$.

$x=\$300$, $y=\$800$, $z=\$2400$

23. ***Familiarize.*** Let x, y, and z represent the prices of orange juice, a raisin bagel, and a cup of coffee, respectively. The new price for orange juice is $x+50\%x$, or $x+0.5x$, or $1.5x$; the new price of a bagel is $y+20\%y$, or $y+0.2y$, or $1.2y$.

Translate.

Orange juice, a raisin bagel, and a cup of coffee cost \$3.

$$x+y+x=3$$

After the price increase, orange juice, a raisin bagel, and a cup of coffee will cost \$3.75.

$$1.5x+1.2y+z=3.75$$

After the price increases, orange juice will cost twice as much as coffee.

$$1.5x=2z$$

We have a system of equations.

$$\begin{aligned} x+y+z&=3, & x+y+z&=3,\\ 1.5x+1.2y+z&=3.75, \text{ or } & 150x+120y+100z&=375,\\ 1.5x&=2z & 15x-20z&=0 \end{aligned}$$

Carry out. Solving the system of equations, we get $(1, 1.25, 0.75)$.

Check. If orange juice costs \$1, a bagel costs \$1.25, and a cup of coffee costs \$0.75, then together they cost \$1 + \$1.25 + \$0.75, or \$3. After the price increases orange juice will cost 1.5(\$1), or \$1.50 and a bagel will cost 1.2(\$1.25) or \$1.50. Then orange juice, a bagel, and coffee will cost \$1.50 + \$1.50 + \$0.75, or \$3.75. After the price increase the price of orange juice, \$1.50, will be twice the price of coffee, \$0.75. The solution checks.

State. Before the increase orange juice costs \$1, a raisin bagel cost \$1.25, and a cup of coffee cost \$0.75.

24. Let x, y, and z represent the prices of a carton of milk, a donut, and a cup of coffee, respectively.

Solve: $x + 2y + z = 5,$

$3y + 2z = 5.50,$

$x + y + 2z = 5.25.$

$x = \$1.75$, $y = \$1$, $z = \$1.25$

Then 2 cartons of milk and 2 donuts will cost 2(\$1.75) + 2(\$1), or \$5.50. They will not have enough money. They need 5¢ more.

25. ***Familiarize.*** Let x, y, and z represent the volume of passenger traffic by private automobile, by bus, and by railroads, respectively, in billions of passenger-miles.

Translate.

The total volume of passenger traffic was 1899 billion passenger-miles.

$x + y + z = 1899$

The volume of bus traffic was 21 billion passenger-miles more than the volume of railroad traffic.

$y = z + 21$

The volume of bus traffic was 1815 billion passenger-miles less than the volume of traffic by private automobile.

$y = x - 1815$

We have a system of equations.

$$\begin{aligned} x + y + z &= 1899, \\ y &= z + 21, \\ y &= x - 1815 \end{aligned} \quad \text{or} \quad \begin{aligned} x + y + z &= 1899, \\ y - z &= 21, \\ -x + y &= -1815 \end{aligned}$$

Carry out. Solving the system of equations, we get $(1850, 35, 14)$.

Check. The total volume is $1850 + 35 + 14$, or 1899 billion passenger-miles. The volume of bus traffic, 35 billion passenger-miles, is 21 billion passenger-miles more than the volume of railroad traffic, 14 billion passenger-miles. The volume of bus traffic, 35 billion passenger-miles, is 1815 billion passenger-miles less than the volume of traffic by private automobile, 1850 billion passenger-miles. The solution checks.

State. The volume of traffic by private automobile was 1850 billion passenger-miles, by bus was 35 billion passenger-miles, and by railroad was 14 billion passenger-miles.

26. Let x, y, and z represent the per capita consumption of cheddar, mozzarella, and Swiss cheese, in pounds.

Solve: $x + y + z = 20.4,$

$y + z = x + 0.2,$

$y = z + 8.1.$

$x = 10.1$ lb, $y = 9.2$ lb, $z = 1.1$ lb

27. ***Familiarize.*** Let x, y, and z represent the number of par-3, par-4, and par-5 holes, respectively. A golfer who shoots par on every hole has a score of $3x$ from the par-3 holes, $4y$ from the par-4 holes, and $5z$ from the par-5 holes.

Translate.

The total number of holes is 18.

$x + y + z = 18$

A golfer who shoots par on every hole has a score of 72.

$3x + 4y + 5z = 72$

The sum of the number of par-3 holes and the number of par-5 holes is 8.

$x + z = 8$

We have a system of equations.

$$\begin{aligned} x + y + z &= 18, \\ 3x + 4y + 5z &= 72, \\ x \qquad + z &= 8 \end{aligned}$$

Carry out. We solve the system. The solution is $(4, 10, 4)$.

Check. The total number of holes is $4 + 10 + 4$, or 18. A golfer who shoots par on every hole has a score of $3 \cdot 4 + 4 \cdot 10 + 5 \cdot 4$, or 72. The sum of the number of par-3 holes and the number of par-5 holes is $4 + 4$, or 8. The solution checks.

State. There are 4 par-3 holes, 10 par-4 holes, and 4 par-5 holes.

28. Let x, y, and z represent the number of par-3, par-4, and par-5 holes, respectively.

Solve: $x + y + z = 18,$

$3x + 4y + 5z = 70,$

$y = 2z.$

$x = 6$, $y = 8$, $z = 4$

29. a) Substitute the data points $(0, 237)$, $(5, 200)$, and $(11, 224)$ in the function $f(x) = ax^2 + bx + c$.

$237 = a \cdot 0^2 + b \cdot 0 + c$

$200 = a \cdot 5^2 + b \cdot 5 + c$

$224 = a \cdot 11^2 + b \cdot 11 + c$

We have a system of equations.

$$\begin{aligned} c &= 237, \\ 25a + 5b + c &= 200, \\ 121a + 11b + c &= 224 \end{aligned}$$

Solving the system of equations, we get $\left(\frac{57}{55}, -\frac{692}{55}, 237\right)$, so $f(x) = \frac{57}{55}x^2 - \frac{692}{55}x + 237$, where $f(x)$ is in thousands and x is the number of years after 1990.

b) In 2008, $x = 2008 - 1990$, or 18.

$f(18) = \frac{57}{55}(18)^2 - \frac{692}{55}(18) + 237 \approx 346$

There will be about 346 thousand marriages in California in 2008.

30. a) Solve: $112 = a \cdot 0^2 + b \cdot 0 + c,$

$115 = a \cdot 9^2 + b \cdot 9 + c,$

$114 = a \cdot 10^2 + b \cdot 10 + c,$

or

$c = 112,$

$81a + 9b + c = 115,$

$100a + 10b + c = 114.$

$a = -\frac{2}{15}$, $b = \frac{23}{15}$, $c = 112$, so

$f(x) = -\frac{2}{15}x^2 + \frac{23}{15}x + 112$, where x is the number of years after 1990.

b) In 2010, $x = 2010 - 1990$, or 20.

$f(20) = -\frac{2}{15}(20)^2 + \frac{23}{15}(20) + 112 \approx 89.$

The per capita consumption of red meat in 2010 will be about 89 lb.

31. a) Substitute the data points $(0, 594)$, $(10, 577)$, and $(15, 593)$ in the function $f(x) = ax^2 + bx + c$.

$594 = a \cdot 0^2 + b \cdot 0 + c$

$577 = a \cdot 10^2 + b \cdot 10 + c$

$593 = a \cdot 15^2 + b \cdot 15 + c$

We have a system of equations.

$c = 594,$

$100a + 10b + c = 577,$

$225a + 15b + c = 593$

Solving the system of equations, we get $\left(\frac{49}{150}, -\frac{149}{30}, 594\right)$, so $f(x) = \frac{49}{150}x^2 - \frac{149}{30}x + 594$, where x is the number of years after 1985.

b) In 2007, $x = 2007 - 1985$, or 22.

$f(22) = \frac{49}{150}(22)^2 - \frac{149}{30}(22) + 594 \approx 643.$

The per capita consumption of dairy products in 2007 will be about 643 lb.

32. a) Solve: $773 = a \cdot 0^2 + b \cdot 0 + c,$

$686 = a \cdot 5^2 + b \cdot 5 + c,$

$739 = a \cdot 10^2 + b \cdot 10 + c,$

or

$c = 773,$

$25a + 5b + c = 686,$

$100a + 10b + c = 739.$

$a = \frac{14}{5}$, $b = -\frac{157}{5}$, $c = 773$, so

$f(x) = \frac{14}{5}x^2 - \frac{157}{5}x + 773$, where x is the number of years after 1990 and $f(x)$ is in millions of metric tons.

b) In 2009, $x = 2009 - 1990$, or 19.

$f(19) = \frac{14}{5}(19)^2 - \frac{157}{5}(19) + 773 \approx 1187$

The world product of crude steel in 2009 will be about 1187 million metric tons.

33. a) Using the quadratic regression feature on a graphing calculator, we have $f(x) = 0.1793514633x^2 - 11.15232206x + 468.8939939$, where x is the number of years after 1920.

b) In 2006, $x = 2006 - 1920$, or 86; $f(86) \approx 836$, so we estimate that there will be about 836 morning newspapers in 2006.

In 2009, $x = 2009 - 1920$, or 89; $f(89) \approx 897$, so we estimate that there will be about 897 morning newspapers in 2009.

34. a) $f(x) = -0.0196633073x^2 + 0.2971010746x + 3.370971717$, where x is the number of years after 1990 and $f(x)$ is in millions.

b) $f(17) \approx 2.739$, so we estimate that there will be about 2.739 million children enrolled in nursery school in 2007.

$f(20) \approx 1.448$, so we estimate that there will be about 1.448 million children enrolled in nursery school in 2010.

35. Add a non-zero multiple of one equation to a non-zero multiple of the other equation, where the multiples are not opposites.

36. Answers will vary.

37. Perpendicular

38. The leading-term test

39. A vertical line

40. A one-to-one function

41. A rational function

42. Inverse variation

43. A vertical asymptote

44. A horizontal asymptote

45. $\frac{2}{x} - \frac{1}{y} - \frac{3}{z} = -1,$

$\frac{2}{x} - \frac{1}{y} + \frac{1}{z} = -9,$

$\frac{1}{x} + \frac{2}{y} - \frac{4}{z} = 17$

First substitute u for $\frac{1}{x}$, v for $\frac{1}{y}$, and w for $\frac{1}{z}$ and solve for u, v, and w.

$2u - v - 3w = -1,$

$2u - v + w = -9,$

$u + 2v - 4w = 17$

Solving this system we get $(-1, 5, -2)$.

If $u = -1$, and $u = \frac{1}{x}$, then $-1 = \frac{1}{x}$, or $x = -1$.

If $v = 5$ and $v = \frac{1}{y}$, then $5 = \frac{1}{y}$, or $y = \frac{1}{5}$.

If $w = -2$ and $w = \frac{1}{z}$, then $-2 = \frac{1}{z}$, or $z = -\frac{1}{2}$.

The solution of the original system is $\left(-1, \frac{1}{5}, -\frac{1}{2}\right)$.

46. $\frac{2}{x} + \frac{2}{y} - \frac{3}{z} = 3,$

$\frac{1}{x} - \frac{2}{y} - \frac{3}{z} = 9,$

$\frac{7}{x} - \frac{2}{y} + \frac{9}{z} = -39$

Substitute u for $\frac{1}{x}$, v for $\frac{1}{y}$, and w for $\frac{1}{z}$ and solve for u, v, and w.

$2u + 2v - 3w = 3,$

$u - 2v - 3w = 9,$

$7u - 2v + 9w = -39$

Solving this system we get $(-2, -1, -3)$.

Then $\frac{1}{x} = -2$, or $x = -\frac{1}{2}$; $\frac{1}{y} = -1$, or $y = -1$; and $\frac{1}{z} = -3$, or $z = -\frac{1}{3}$. The solution of the original system is $\left(-\frac{1}{2}, -1, -\frac{1}{3}\right)$.

47. Label the angle measures at the tips of the stars a, b, c, d, and e. Also label the angles of the pentagon p, q, r, s, and t.

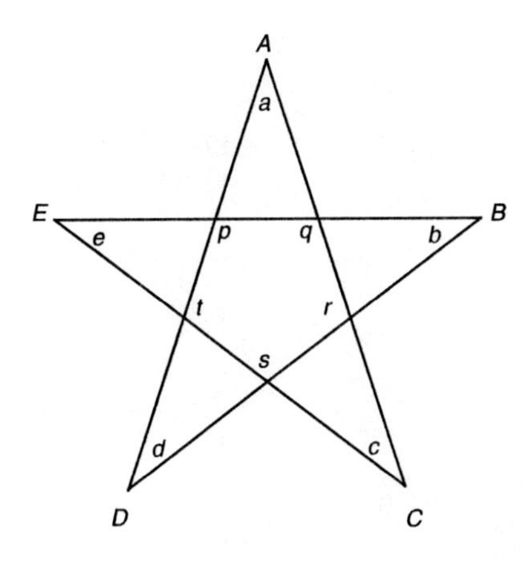

Using the geometric fact that the sum of the angle measures of a triangle is 180°, we get 5 equations.

$p + b + d = 180$

$q + c + e = 180$

$r + a + d = 180$

$s + b + e = 180$

$t + a + c = 180$

Adding these equations, we get

$(p + q + r + s + t) + 2a + 2b + 2c + 2d + 2e = 5(180)$.

The sum of the angle measures of any convex polygon with n sides is given by the formula $S = (n - 2)180$. Thus $p + q + r + s + t = (5 - 2)180$, or 540. We substitute and solve for $a + b + c + d + e$.

$540 + 2(a + b + c + d + e) = 900$

$2(a + b + c + d + e) = 360$

$a + b + c + d + e = 180$

The sum of the angle measures at the tips of the star is 180°.

48. Let h, t, and u represent the hundred's, ten's, and unit's digit of the year, respectively. (We know the thousand's digit is 1.) Using the given information we know the following:

$1 + h + t + u = 24,$

$u = 1 + h,$

$t = k \cdot 3$ and $u = m \cdot 3$

where k, m are positive integers.

We know $h > 5$ (there was no transcontinental railroad before 1600); and, since

$u = 1 + h = m \cdot 3$, $h \neq 6$, $h \neq 7$, $h \neq 9$.

Thus $h = 8$ and $u = 9$. Then $1 + 8 + t + 9 = 24$, or $t = 6$.

The year is 1869.

49. Substituting, we get

$A + \frac{3}{4}B + 3C = 12,$

$\frac{4}{3}A + B + 2C = 12,$

$2A + B + C = 12,$ or

$4A + 3B + 12C = 48,$

$4A + 3B + 6C = 36,$ Clearing fractions

$2A + B + C = 12.$

Solving the system of equations, we get $(3, 4, 2)$. The equation is $3x + 4y + 2z = 12$.

50. Solve: $1 = B - M - 2N,$

$2 = B - 3M + 6N,$

$1 = B - \frac{3}{2}M - N$

$B = 2$, $M = \frac{1}{2}$, $N = \frac{1}{4}$, so $y = 2 - \frac{1}{2}x - \frac{1}{4}z$.

51. Substituting, we get

$$59 = a(-2)^3 + b(-2)^2 + c(-2) + d,$$
$$13 = a(-1)^3 + b(-1)^2 + c(-1) + d,$$
$$-1 = a \cdot 1^3 + b \cdot 1^2 + c \cdot 1 + d,$$
$$-17 = a \cdot 2^3 + b \cdot 2^2 + c \cdot 2 + d, \text{ or}$$

$$-8a + 4b - 2c + d = 59,$$
$$-a + b - c + d = 13,$$
$$a + b + c + d = -1,$$
$$8a + 4b + 2c + d = -17.$$

Solving the system of equations, we get $(-4, 5, -3, 1)$, so $y = -4x^3 + 5x^2 - 3x + 1$.

52. Solve: $-39 = a(-2)^3 + b(-2)^2 + c(-2) + d,$

$$-12 = a(-1)^3 + b(-1)^2 + c(-1) + d,$$
$$-6 = a \cdot 1^3 + b \cdot 1^2 + c \cdot 1 + d,$$
$$16 = a \cdot 3^3 + b \cdot 3^2 + c \cdot 3 + d, \text{ or}$$

$$-8a + 4b - 2c + d = -39,$$
$$-a + b - c + d = -12,$$
$$a + b + c + d = -6,$$
$$27a + 9b + 3c + d = 16.$$

$a = 2$, $b = -4$, $c = 1$, $d = -5$, so $y = 2x^3 - 4x^2 + x - 5$.

53. ***Familiarize and Translate.*** Let a, s, and c represent the number of adults, students, and children in attendance, respectively.

The total attendance was 100.

$$a + s + c = 100$$

The total amount of money taken in was \$100.

(Express 50 cents as $\frac{1}{2}$ dollar.)

$$10a + 3s + \frac{1}{2}c = 100$$

The resulting system is

$$a + s + c = 100,$$
$$10a + 3s + \frac{1}{2}c = 100.$$

Carry out. Multiply the first equation by -3 and add it to the second equation to obtain $7a - \frac{5}{2}c = -200$ or $a = \frac{5}{14}(c - 80)$ where $(c - 80)$ is a positive multiple of 14 (because a must be a positive integer). That is $(c - 80) = k \cdot 14$ or $c = 80 + k \cdot 14$, where k is a positive integer. If $k > 1$, then $c > 100$. This is impossible since the total attendance is 100. Thus $k = 1$, so $c = 80 + 1 \cdot 14 = 94$. Then $a = \frac{5}{14}(94 - 80) = \frac{5}{14} \cdot 14 = 5$, and $5 + s + 94 = 10$, or $s = 1$.

Check. The total attendance is $5 + 1 + 94$, or 100.

The total amount of money taken in was $\$10 \cdot 5 + \$3 \cdot 1 + \$\frac{1}{2} \cdot 94 = \$100$. The result checks.

State. There were 5 adults, 1 student, and 94 children in attendance.

Exercise Set 8.3

1. The matrix has 3 rows and 2 columns, so its order is 3×2.

2. The matrix has 4 rows and 1 column, so its order is 4×1.

3. The matrix has 1 row and 4 columns, so its order is 1×4.

4. The matrix has 1 row and 1 column, so its order is 1×1.

5. The matrix has 3 rows and 3 columns, so its order is 3×3.

6. The matrix has 2 rows and 4 columns, so its order is 2×4.

7. We omit the variables and replace the equals signs with a vertical line.

$$\left[\begin{array}{rr|r} 2 & -1 & 7 \\ 1 & 4 & -5 \end{array}\right]$$

8. $$\left[\begin{array}{rr|r} 3 & 2 & 8 \\ 2 & -3 & 15 \end{array}\right]$$

9. We omit the variables, writing zeros for the missing terms, and replace the equals signs with a vertical line.

$$\left[\begin{array}{rrr|r} 1 & -2 & 3 & 12 \\ 2 & 0 & -4 & 8 \\ 0 & 3 & 1 & 7 \end{array}\right]$$

10. $$\left[\begin{array}{rrr|r} 1 & 1 & -1 & 7 \\ 0 & 3 & 2 & 1 \\ -2 & -5 & 0 & 6 \end{array}\right]$$

11. Insert variables and replace the vertical line with equals signs.

$$3x - 5y = 1,$$
$$x + 4y = -2$$

12. $$x + 2y = -6,$$
$$4x + y = -3$$

13. Insert variables and replace the vertical line with equals signs.

$$2x + y - 4z = 12,$$
$$3x + 5z = -1,$$
$$x - y + z = 2$$

14. $$-x - 2y + 3z = 6,$$
$$4y + z = 2,$$
$$2x - y = 9$$

15. $4x + 2y = 11,$

$3x - \quad y = 2$

Write the augmented matrix. We will use Gaussian elimination.

$$\left[\begin{array}{rr|r} 4 & 2 & 11 \\ 3 & -1 & 2 \end{array}\right]$$

Multiply row 2 by 4 to make the first number in row 2 a multiple of 4.

$$\left[\begin{array}{rr|r} 4 & 2 & 11 \\ 12 & -4 & 8 \end{array}\right]$$

Multiply row 1 by -3 and add it to row 2.

$$\left[\begin{array}{rr|r} 4 & 2 & 11 \\ 0 & -10 & -25 \end{array}\right]$$

Multiply row 1 by $\frac{1}{4}$ and row 2 by $-\frac{1}{10}$.

$$\left[\begin{array}{rr|r} 1 & \frac{1}{2} & \frac{11}{4} \\ 0 & 1 & \frac{5}{2} \end{array}\right]$$

Write the system of equations that corresponds to the last matrix.

$$x + \frac{1}{2}y = \frac{11}{4}, \quad (1)$$

$$y = \frac{5}{2} \quad (2)$$

Back-substitute in equation (1) and solve for x.

$$x + \frac{1}{2} \cdot \frac{5}{2} = \frac{11}{4}$$

$$x + \frac{5}{4} = \frac{11}{4}$$

$$x = \frac{6}{4} = \frac{3}{2}$$

The solution is $\left(\frac{3}{2}, \frac{5}{2}\right)$.

16. $2x + \quad y = 1,$

$3x + 2y = -2$

Write the augmented matrix. We will use Gauss-Jordan elimination.

$$\left[\begin{array}{rr|r} 2 & 1 & 1 \\ 3 & 2 & -2 \end{array}\right]$$

Multiply row 2 by 2.

$$\left[\begin{array}{rr|r} 2 & 1 & 1 \\ 6 & 4 & -4 \end{array}\right]$$

Multiply row 1 by -3 and add it to row 2.

$$\left[\begin{array}{rr|r} 2 & 1 & 1 \\ 0 & 1 & -7 \end{array}\right]$$

Multiply row 2 by -1 and add it to row 1.

$$\left[\begin{array}{rr|r} 2 & 0 & 8 \\ 0 & 1 & -7 \end{array}\right]$$

Multiply row 1 by $\frac{1}{2}$.

$$\left[\begin{array}{rr|r} 1 & 0 & 4 \\ 0 & 1 & -7 \end{array}\right]$$

The solution is $(4, -7)$.

17. $5x - 2y = -3,$

$2x + 5y = -24$

Write the augmented matrix. We will use Gaussian elimination.

$$\left[\begin{array}{rr|r} 5 & -2 & -3 \\ 2 & 5 & -24 \end{array}\right]$$

Multiply row 2 by 5 to make the first number in row 2 a multiple of 5.

$$\left[\begin{array}{rr|r} 5 & -2 & -3 \\ 10 & 25 & -120 \end{array}\right]$$

Multiply row 1 by -2 and add it to row 2.

$$\left[\begin{array}{rr|r} 5 & -2 & -3 \\ 0 & 29 & -114 \end{array}\right]$$

Multiply row 1 by $\frac{1}{5}$ and row 2 by $\frac{1}{29}$.

$$\left[\begin{array}{rr|r} 1 & -\frac{2}{5} & -\frac{3}{5} \\ 0 & 1 & -\frac{114}{29} \end{array}\right]$$

Write the system of equations that corresponds to the last matrix.

$$x - \frac{2}{5}y = -\frac{3}{5}, \quad (1)$$

$$y = -\frac{114}{29} \quad (2)$$

Back-substitute in equation (1) and solve for x.

$$x - \frac{2}{5}\left(-\frac{114}{29}\right) = -\frac{3}{5}$$

$$x + \frac{228}{145} = -\frac{3}{5}$$

$$x = -\frac{315}{145} = -\frac{63}{29}$$

The solution is $\left(-\frac{63}{29}, -\frac{114}{29}\right)$.

18. $2x + y = 1,$
$3x - 6y = 4$

Write the augmented matrix. We will use Gaussian elimination.

$$\left[\begin{array}{cc|c} 2 & 1 & 1 \\ 3 & -6 & 4 \end{array}\right]$$

Multiply row 2 by 2.

$$\left[\begin{array}{cc|c} 2 & 1 & 1 \\ 6 & -12 & 8 \end{array}\right]$$

Multiply row 1 by -3 and add it to row 2.

$$\left[\begin{array}{cc|c} 2 & 1 & 1 \\ 0 & -15 & 5 \end{array}\right]$$

Multiply row 1 by $\frac{1}{2}$ and row 2 by $-\frac{1}{15}$.

$$\left[\begin{array}{cc|c} 1 & \frac{1}{2} & \frac{1}{2} \\ 0 & 1 & -\frac{1}{3} \end{array}\right]$$

We have:

$$x + \frac{1}{2}y = 1, \quad (1)$$

$$y = -\frac{1}{3} \quad (2)$$

Back-substitute in equation (1) and solve for x.

$$x + \frac{1}{2}\left(-\frac{1}{3}\right) = \frac{1}{2}$$

$$x = \frac{2}{3}$$

The solution is $\left(\frac{2}{3}, -\frac{1}{3}\right)$.

19. $3x + 4y = 7,$
$-5x + 2y = 10$

Write the augmented matrix. We will use Gaussian elimination.

$$\left[\begin{array}{cc|c} 3 & 4 & 7 \\ -5 & 2 & 10 \end{array}\right]$$

Multiply row 2 by 3 to make the first number in row 2 a multiple of 3.

$$\left[\begin{array}{cc|c} 3 & 4 & 7 \\ -15 & 6 & 30 \end{array}\right]$$

Multiply row 1 by 5 and add it to row 2.

$$\left[\begin{array}{cc|c} 3 & 4 & 7 \\ 0 & 26 & 65 \end{array}\right]$$

Multiply row 1 by $\frac{1}{3}$ and row 2 by $\frac{1}{26}$.

$$\left[\begin{array}{cc|c} 1 & \frac{4}{3} & \frac{7}{3} \\ 0 & 1 & \frac{5}{2} \end{array}\right]$$

Write the system of equations that corresponds to the last matrix.

$$x + \frac{4}{3}y = \frac{7}{3}, \quad (1)$$

$$y = \frac{5}{2} \quad (2)$$

Back-substitute in equation (1) and solve for x.

$$x + \frac{4}{3} \cdot \frac{5}{2} = \frac{7}{3}$$

$$x + \frac{10}{3} = \frac{7}{3}$$

$$x = -\frac{3}{3} = -1$$

The solution is $\left(-1, \frac{5}{2}\right)$.

20. $5x - 3y = -2,$
$4x + 2y = 5$

Write the augmented matrix. We will use Gaussian elimination.

$$\left[\begin{array}{cc|c} 5 & -3 & -2 \\ 4 & 2 & 5 \end{array}\right]$$

Multiply row 2 by 5.

$$\left[\begin{array}{cc|c} 5 & -3 & -2 \\ 20 & 10 & 25 \end{array}\right]$$

Multiply row 1 by -4 and add it to row 2.

$$\left[\begin{array}{cc|c} 5 & -3 & -2 \\ 0 & 22 & 33 \end{array}\right]$$

Multiply row 1 by $\frac{1}{5}$ and row 2 by $\frac{1}{22}$.

$$\left[\begin{array}{cc|c} 1 & -\frac{3}{5} & -\frac{2}{5} \\ 0 & 1 & \frac{3}{2} \end{array}\right]$$

We have:

$$x - \frac{3}{5}y = -\frac{2}{5}, \quad (1)$$

$$y = \frac{3}{2} \quad (2)$$

Back-substitute in (1) and solve for x.

$$x - \frac{3}{5} \cdot \frac{3}{2} = -\frac{2}{5}$$

$$x - \frac{9}{10} = -\frac{2}{5}$$

$$x = \frac{1}{2}$$

The solution is $\left(\frac{1}{2}, \frac{3}{2}\right)$.

21. $3x + 2y = 6,$

$2x - 3y = -9$

Write the augmented matrix. We will use Gauss-Jordan elimination.

$$\left[\begin{array}{cc|c} 3 & 2 & 6 \\ 2 & -3 & -9 \end{array}\right]$$

Multiply row 2 by 3 to make the first number in row 2 a multiple of 3.

$$\left[\begin{array}{cc|c} 3 & 2 & 6 \\ 6 & -9 & -27 \end{array}\right]$$

Multiply row 1 by -2 and add it to row 2.

$$\left[\begin{array}{cc|c} 3 & 2 & 6 \\ 0 & -13 & -39 \end{array}\right]$$

Multiply row 2 by $-\frac{1}{13}$.

$$\left[\begin{array}{cc|c} 3 & 2 & 6 \\ 0 & 1 & 3 \end{array}\right]$$

Multiply row 2 by -2 and add it to row 1.

$$\left[\begin{array}{cc|c} 3 & 0 & 0 \\ 0 & 1 & 3 \end{array}\right]$$

Multiply row 1 by $\frac{1}{3}$.

$$\left[\begin{array}{cc|c} 1 & 0 & 0 \\ 0 & 1 & 3 \end{array}\right]$$

We have $x = 0$, $y = 3$. The solution is $(0, 3)$.

22. $x - 4y = 9,$

$2x + 5y = 5$

Write the augmented matrix. We will use Gauss-Jordan elimination.

$$\left[\begin{array}{cc|c} 1 & -4 & 9 \\ 2 & 5 & 5 \end{array}\right]$$

Multiply row 1 by -2 and add it to row 2.

$$\left[\begin{array}{cc|c} 1 & -4 & 9 \\ 0 & 13 & -13 \end{array}\right]$$

Multiply row 2 by $\frac{1}{13}$.

$$\left[\begin{array}{cc|c} 1 & -4 & 9 \\ 0 & 1 & -1 \end{array}\right]$$

Multiply row 2 by 4 and add it to row 1.

$$\left[\begin{array}{cc|c} 1 & 0 & 5 \\ 0 & 1 & -1 \end{array}\right]$$

The solution is $(5, -1)$.

23. $x - 3y = 8,$

$2x - 6y = 3$

Write the augmented matrix.

$$\left[\begin{array}{cc|c} 1 & -3 & 8 \\ 2 & -6 & 3 \end{array}\right]$$

Multiply row 1 by -2 and add it to row 2.

$$\left[\begin{array}{cc|c} 1 & -3 & 8 \\ 0 & 0 & -13 \end{array}\right]$$

The last row corresponds to the false equation $0 = -13$, so there is no solution.

24. $4x - 8y = 12,$

$-x + 2y = -3$

Write the augmented matrix.

$$\left[\begin{array}{cc|c} 4 & -8 & 12 \\ -1 & 2 & -3 \end{array}\right]$$

Interchange the rows.

$$\left[\begin{array}{cc|c} -1 & 2 & -3 \\ 4 & -8 & 12 \end{array}\right]$$

Multiply row 1 by 4 and add it to row 2.

$$\left[\begin{array}{cc|c} -1 & 2 & -3 \\ 0 & 0 & 0 \end{array}\right]$$

The last row corresponds to $0 = 0$, so the system is dependent and equivalent to $-x + 2y = -3$. Solving this system for x gives us $x = 2y + 3$. Then the solutions are of the form $(2y + 3, y)$, where y is any real number.

25. $-2x + 6y = 4,$
$3x - 9y = -6$

Write the augmented matrix.

$$\left[\begin{array}{rr|r} -2 & 6 & 4 \\ 3 & -9 & -6 \end{array}\right]$$

Multiply row 1 by $-\frac{1}{2}$.

$$\left[\begin{array}{rr|r} 1 & -3 & -2 \\ 3 & -9 & -6 \end{array}\right]$$

Multiply row 1 by -3 and add it to row 2.

$$\left[\begin{array}{rr|r} 1 & -3 & -2 \\ 0 & 0 & 0 \end{array}\right]$$

The last row corresponds to the equation $0 = 0$ which is true for all values of x and y. Thus, the system of equations is dependent and is equivalent to the first equation $-2x + 6y = 4$, or $x - 3y = -2$. Solving for x, we get $x = 3y - 2$. Then the solutions are of the form $(3y - 2, y)$, where y is any real number.

26. $6x + 2y = -10,$
$-3x - y = 6$

Write the augmented matrix.

$$\left[\begin{array}{rr|r} 6 & 2 & -10 \\ -3 & -1 & 6 \end{array}\right]$$

Interchange the rows.

$$\left[\begin{array}{rr|r} -3 & -1 & 6 \\ 6 & 2 & -10 \end{array}\right]$$

Multiply row 1 by 2 and add it to row 2.

$$\left[\begin{array}{rr|r} 3 & -1 & 6 \\ 0 & 0 & 2 \end{array}\right]$$

The last row corresponds to the false equation $0 = 2$, so there is no solution.

27. $x + 2y - 3z = 9,$
$2x - y + 2z = -8,$
$3x - y - 4z = 3$

Write the augmented matrix. We will use Gauss-Jordan elimination.

$$\left[\begin{array}{rrr|r} 1 & 2 & -3 & 9 \\ 2 & -1 & 2 & -8 \\ 3 & -1 & -4 & 3 \end{array}\right]$$

Multiply row 1 by -2 and add it to row 2. Also, multiply row 1 by -3 and add it to row 3.

$$\left[\begin{array}{rrr|r} 1 & 2 & -3 & 9 \\ 0 & -5 & 8 & -26 \\ 0 & -7 & 5 & -24 \end{array}\right]$$

Multiply row 2 by $-\frac{1}{5}$ to get a 1 in the second row, second column.

$$\left[\begin{array}{rrr|r} 1 & 2 & -3 & 9 \\ 0 & 1 & -\frac{8}{5} & \frac{26}{5} \\ 0 & -7 & 5 & -24 \end{array}\right]$$

Multiply row 2 by -2 and add it to row 1. Also, multiply row 2 by 7 and add it to row 3.

$$\left[\begin{array}{rrr|r} 1 & 0 & \frac{1}{5} & -\frac{7}{5} \\ 0 & 1 & -\frac{8}{5} & \frac{26}{5} \\ 0 & 0 & -\frac{31}{5} & \frac{62}{5} \end{array}\right]$$

Multiply row 3 by $-\frac{5}{31}$ to get a 1 in the third row, third column.

$$\left[\begin{array}{rrr|r} 1 & 0 & \frac{1}{5} & -\frac{7}{5} \\ 0 & 1 & -\frac{8}{5} & \frac{26}{5} \\ 0 & 0 & 1 & -2 \end{array}\right]$$

Multiply row 3 by $-\frac{1}{5}$ and add it to row 1. Also, multiply row 3 by $\frac{8}{5}$ and add it to row 2.

$$\left[\begin{array}{rrr|r} 1 & 0 & 0 & -1 \\ 0 & 1 & 0 & 2 \\ 0 & 0 & 1 & -2 \end{array}\right]$$

We have $x = -1$, $y = 2$, $z = -2$. The solution is $(-1, 2, -2)$.

28. $x - y + 2z = 0,$
$x - 2y + 3z = -1,$
$2x - 2y + z = -3$

Write the augmented matrix. We will use Gauss-Jordan elimination.

$$\left[\begin{array}{rrr|r} 1 & -1 & 2 & 0 \\ 1 & -2 & 3 & -1 \\ 2 & -2 & 1 & -3 \end{array}\right]$$

Multiply row 1 by -1 and add it to row 2. Also, multiply row 1 by -2 and add it to row 3.

$$\left[\begin{array}{rrr|r} 1 & -1 & 2 & 0 \\ 0 & -1 & 1 & -1 \\ 0 & 0 & -3 & -3 \end{array}\right]$$

Multiply row 2 by -1.

$$\left[\begin{array}{rrr|r} 1 & -1 & 2 & 0 \\ 0 & 1 & -1 & 1 \\ 0 & 0 & -3 & -3 \end{array}\right]$$

Add row 2 to row 1.

$$\left[\begin{array}{rrr|r} 1 & 0 & 1 & 1 \\ 0 & 1 & -1 & 1 \\ 0 & 0 & -3 & -3 \end{array}\right]$$

Multiply row 3 by $-\frac{1}{3}$.

$$\left[\begin{array}{rrr|r} 1 & 0 & 1 & 1 \\ 0 & 1 & -1 & 1 \\ 0 & 0 & 1 & 1 \end{array}\right]$$

Multiply row 3 by -1 and add it to row 1. Also, add row 3 to row 2.

$$\left[\begin{array}{rrr|r} 1 & 0 & 0 & 0 \\ 0 & 1 & 0 & 2 \\ 0 & 0 & 1 & 1 \end{array}\right]$$

The solution is $(0, 2, 1)$.

29. $4x - y - 3z = 1$,
$8x + y - z = 5$,
$2x + y + 2z = 5$

Write the augmented matrix. We will use Gauss-Jordan elimination.

$$\left[\begin{array}{rrr|r} 4 & -1 & -3 & 1 \\ 8 & 1 & -1 & 5 \\ 2 & 1 & 2 & 5 \end{array}\right]$$

First interchange rows 1 and 3 so that each number below the first number in the first row is a multiple of that number.

$$\left[\begin{array}{rrr|r} 2 & 1 & 2 & 5 \\ 8 & 1 & -1 & 5 \\ 4 & -1 & -3 & 1 \end{array}\right]$$

Multiply row 1 by -4 and add it to row 2. Also, multiply row 1 by -2 and add it to row 3.

$$\left[\begin{array}{rrr|r} 2 & 1 & 2 & 5 \\ 0 & -3 & -9 & -15 \\ 0 & -3 & -7 & -9 \end{array}\right]$$

Multiply row 2 by -1 and add it to row 3.

$$\left[\begin{array}{rrr|r} 2 & 1 & 2 & 5 \\ 0 & -3 & -9 & -15 \\ 0 & 0 & 2 & 6 \end{array}\right]$$

Multiply row 2 by $-\frac{1}{3}$ to get a 1 in the second row, second column.

$$\left[\begin{array}{rrr|r} 2 & 1 & 2 & 5 \\ 0 & 1 & 3 & 5 \\ 0 & 0 & 2 & 6 \end{array}\right]$$

Multiply row 2 by -1 and add it to row 1.

$$\left[\begin{array}{rrr|r} 2 & 0 & -1 & 0 \\ 0 & 1 & 3 & 5 \\ 0 & 0 & 2 & 6 \end{array}\right]$$

Multiply row 3 by $\frac{1}{2}$ to get a 1 in the third row, third column.

$$\left[\begin{array}{rrr|r} 2 & 0 & -1 & 0 \\ 0 & 1 & 3 & 5 \\ 0 & 0 & 1 & 3 \end{array}\right]$$

Add row 3 to row 1. Also multiply row 3 by -3 and add it to row 2.

$$\left[\begin{array}{rrr|r} 2 & 0 & 0 & 3 \\ 0 & 1 & 0 & -4 \\ 0 & 0 & 1 & 3 \end{array}\right]$$

Finally, multiply row 1 by $\frac{1}{2}$.

$$\left[\begin{array}{rrr|r} 1 & 0 & 0 & \frac{3}{2} \\ 0 & 1 & 0 & -4 \\ 0 & 0 & 1 & 3 \end{array}\right]$$

We have $x = \frac{3}{2}$, $y = -4$, $z = 3$. The solution is $\left(\frac{3}{2}, -4, 3\right)$.

30. $3x + 2y + 2z = 3$,
$x + 2y - z = 5$,
$2x - 4y + z = 0$

Write the augmented matrix. We will use Gauss-Jordan elimination.

$$\left[\begin{array}{rrr|r} 3 & 2 & 2 & 3 \\ 1 & 2 & -1 & 5 \\ 2 & -4 & 1 & 0 \end{array}\right]$$

Interchange the first two rows.

$$\left[\begin{array}{rrr|r} 1 & 2 & -1 & 5 \\ 3 & 2 & 2 & 3 \\ 2 & -4 & 1 & 0 \end{array}\right]$$

Multiply row 1 by -3 and add it to row 2. Also, multiply row 1 by -2 and add it to row 3.

$$\left[\begin{array}{rrr|r} 1 & 2 & -1 & 5 \\ 0 & -4 & 5 & -12 \\ 0 & -8 & 3 & -10 \end{array}\right]$$

Multiply row 2 by $-\frac{1}{4}$.

$$\left[\begin{array}{ccc|c} 1 & 2 & -1 & 5 \\ 0 & 1 & -\frac{5}{4} & 3 \\ 0 & -8 & 3 & -10 \end{array}\right]$$

Multiply row 2 by -2 and add it to row 1. Also, multiply row 2 by 8 and add it to row 3.

$$\left[\begin{array}{ccc|c} 1 & 0 & \frac{3}{2} & -1 \\ 0 & 1 & -\frac{5}{4} & 3 \\ 0 & 0 & -7 & 14 \end{array}\right]$$

Multiply row 3 by $-\frac{1}{7}$.

$$\left[\begin{array}{ccc|c} 1 & 0 & \frac{3}{2} & -1 \\ 0 & 1 & -\frac{5}{4} & 3 \\ 0 & 0 & 1 & -2 \end{array}\right]$$

Multiply row 3 by $-\frac{3}{2}$ and add it to row 1. Also, multiply row 3 by $\frac{5}{4}$ and add it to row 2.

$$\left[\begin{array}{ccc|c} 1 & 0 & 0 & 2 \\ 0 & 1 & 0 & \frac{1}{2} \\ 0 & 0 & 1 & -2 \end{array}\right]$$

The solution is $\left(2, \frac{1}{2}, -2\right)$.

31. $x - 2y + 3z = 4,$
$3x + y - z = 0,$
$2x + 3y - 5z = 1$

Write the augmented matrix. We will use Gaussian elimination.

$$\left[\begin{array}{ccc|c} 1 & -2 & 3 & -4 \\ 3 & 1 & -1 & 0 \\ 2 & 3 & -5 & 1 \end{array}\right]$$

Multiply row 1 by -3 and add it to row 2. Also, multiply row 1 by -2 and add it to row 3.

$$\left[\begin{array}{ccc|c} 1 & -2 & 3 & -4 \\ 0 & 7 & -10 & 12 \\ 0 & 7 & -11 & 9 \end{array}\right]$$

Multiply row 2 by -1 and add it to row 3.

$$\left[\begin{array}{ccc|c} 1 & -2 & 3 & -4 \\ 0 & 7 & -10 & 12 \\ 0 & 0 & -1 & -3 \end{array}\right]$$

Multiply row 2 by $\frac{1}{7}$ and multiply row 3 by -1.

$$\left[\begin{array}{ccc|c} 1 & -2 & 3 & -4 \\ 0 & 1 & -\frac{10}{7} & \frac{12}{7} \\ 0 & 0 & 1 & 3 \end{array}\right]$$

Now write the system of equations that corresponds to the last matrix.

$$x - 2y + 3z = -4, \quad (1)$$
$$y - -\frac{10}{7}z = \frac{12}{7}, \quad (2)$$
$$z = 3 \quad (3)$$

Back-substitute 3 for z in equation (2) and solve for y.

$$y - \frac{10}{7} \cdot 3 = \frac{12}{7}$$
$$y - \frac{30}{7} = \frac{12}{7}$$
$$y = \frac{42}{7} = 6$$

Back-substitute 6 for y and 3 for z in equation (1) and solve for x.

$$x - 2 \cdot 6 + 3 \cdot 3 = -4$$
$$x - 3 = -4$$
$$x = -1$$

The solution is $(-1, 6, 3)$.

32. $2x - 3y + 2z = 2,$
$x + 4y - z = 9,$
$-3x + y - 5z = 5$

Write the augmented matrix. We will use Gaussian elimination.

$$\left[\begin{array}{ccc|c} 2 & -3 & 2 & 2 \\ 1 & 4 & -1 & 9 \\ -3 & 1 & -5 & 5 \end{array}\right]$$

Interchange the first two rows.

$$\left[\begin{array}{ccc|c} 1 & 4 & -1 & 9 \\ 2 & -3 & 2 & 2 \\ -3 & 1 & -5 & 5 \end{array}\right]$$

Multiply row 1 by -2 and add it to row 2. Also, multiply row 1 by 3 and add it to row 3.

$$\left[\begin{array}{ccc|c} 1 & 4 & -1 & 9 \\ 0 & -11 & 4 & -16 \\ 0 & 13 & -8 & 32 \end{array}\right]$$

Multiply row 3 by 11.

$$\left[\begin{array}{ccc|c} 1 & 4 & -1 & 9 \\ 0 & -11 & 4 & -16 \\ 0 & 143 & -88 & 352 \end{array}\right]$$

Multiply row 2 by 13 and add it to row 3.

$$\left[\begin{array}{ccc|c} 1 & 4 & -1 & 9 \\ 0 & -11 & 4 & -16 \\ 0 & 0 & -36 & 144 \end{array}\right]$$

Multiply row 2 by $-\frac{1}{11}$ and multiply row 3 by $-\frac{1}{36}$.

$$\left[\begin{array}{ccc|c} 1 & 4 & -1 & 9 \\ 0 & 1 & -\frac{4}{11} & \frac{16}{11} \\ 0 & 0 & 1 & -4 \end{array}\right]$$

We have

$$x + 4y - z = 9, \quad (1)$$
$$y - \frac{4}{11}z = \frac{16}{11}, \quad (2)$$
$$z = -4. \quad (3)$$

Back-substitute in (2) to find y.

$$y - \frac{4}{11}(-4) = \frac{16}{11}$$
$$y + \frac{16}{11} = \frac{16}{11}$$
$$y = 0$$

Back-substitute in (1) to find x.

$$x - 4 \cdot 0 - (-4) = 9$$
$$x = 5$$

The solution is $(5, 0, -4)$.

33. $2x - 4y - 3z = 3,$
$x + 3y + z = -1,$
$5x + y - 2z = 2$

Write the augmented matrix.

$$\left[\begin{array}{ccc|c} 2 & -4 & -3 & 3 \\ 1 & 3 & 1 & -1 \\ 5 & 1 & -2 & 2 \end{array}\right]$$

Interchange the first two rows to get a 1 in the first row, first column.

$$\left[\begin{array}{ccc|c} 1 & 3 & 1 & -1 \\ 2 & -4 & -3 & 3 \\ 5 & 1 & -2 & 2 \end{array}\right]$$

Multiply row 1 by -2 and add it to row 2. Also, multiply row 1 by -5 and add it to row 3.

$$\left[\begin{array}{ccc|c} 1 & 3 & 1 & -1 \\ 0 & -10 & -5 & 5 \\ 0 & -14 & -7 & 7 \end{array}\right]$$

Multiply row 2 by $-\frac{1}{10}$ to get a 1 in the second row, second column.

$$\left[\begin{array}{ccc|c} 1 & 3 & 1 & -1 \\ 0 & 1 & \frac{1}{2} & -\frac{1}{2} \\ 0 & -14 & -7 & 7 \end{array}\right]$$

Multiply row 2 by 14 and add it to row 3.

$$\left[\begin{array}{ccc|c} 1 & 3 & 1 & -1 \\ 0 & 1 & \frac{1}{2} & -\frac{1}{2} \\ 0 & 0 & 0 & 0 \end{array}\right]$$

The last row corresponds to the equation $0 = 0$. This indicates that the system of equations is dependent. It is equivalent to

$$x + 3y + z = -1,$$
$$y + \frac{1}{2}z = -\frac{1}{2}$$

We solve the second equation for y.

$$y = -\frac{1}{2}z - \frac{1}{2}$$

Substitute for y in the first equation and solve for x.

$$x + 3\left(-\frac{1}{2}z - \frac{1}{2}\right) + z = -1$$
$$x - \frac{3}{2}z - \frac{3}{2} + z = -1$$
$$x = \frac{1}{2}z + \frac{1}{2}$$

The solution is $\left(\frac{1}{2}z + \frac{1}{2}, -\frac{1}{2}z - \frac{1}{2}, z\right)$, where z is any real number.

34. $x + y - 3z = 4,$
$4x + 5y + z = 1,$
$2x + 3y + 7z = -7$

Write the augmented matrix.

$$\left[\begin{array}{ccc|c} 1 & 1 & -3 & 4 \\ 4 & 5 & 1 & 1 \\ 2 & 3 & 7 & -7 \end{array}\right]$$

Multiply row 1 by -4 and add it to row 2. Also, multiply row 1 by -2 and add it to row 3.

$$\left[\begin{array}{ccc|c} 1 & 1 & -3 & 4 \\ 0 & 1 & 13 & -15 \\ 0 & 1 & 13 & -15 \end{array}\right]$$

Multiply row 2 by -1 and add it to row 3.

$$\left[\begin{array}{ccc|c} 1 & 1 & -3 & 4 \\ 0 & 1 & 13 & -15 \\ 0 & 0 & 0 & 0 \end{array}\right]$$

We have a dependent system of equations that is equivalent to

$$x + y - 3z = 4,$$
$$y + 13z = -15.$$

Solve the second equation for y.

$$y = -13z - 15$$

Substitute for y in the first equation and solve for x.

$$x - 13z - 15 - 3z = 4$$
$$x = 16z + 19$$

The solution is $(16z+19, -13z-15, z)$, where z is any real number.

35. $p + q + r = 1,$
$p + 2q + 3r = 4,$
$4p + 5q + 6r = 7$

Write the augmented matrix.

$$\left[\begin{array}{ccc|c} 1 & 1 & 1 & 1 \\ 1 & 2 & 3 & 4 \\ 4 & 5 & 6 & 7 \end{array}\right]$$

Multiply row 1 by -1 and add it to row 2. Also, multiply row 1 by -4 and add it to row 3.

$$\left[\begin{array}{ccc|c} 1 & 1 & 1 & 1 \\ 0 & 1 & 2 & 3 \\ 0 & 1 & 2 & 3 \end{array}\right]$$

Multiply row 2 by -1 and add it to row 3.

$$\left[\begin{array}{ccc|c} 1 & 1 & 1 & 1 \\ 0 & 1 & 2 & 3 \\ 0 & 0 & 0 & 0 \end{array}\right]$$

The last row corresponds to the equation $0 = 0$. This indicates that the system of equations is dependent. It is equivalent to

$$p + q + r = 1,$$
$$q + 2r = 3.$$

We solve the second equation for q.

$$q = -2r + 3$$

Substitute for y in the first equation and solve for p.

$$p - 2r + 3 + r = 1$$
$$p - r + 3 = 1$$
$$p = r - 2$$

The solution is $(r - 2, -2r + 3, r)$, where r is any real number.

36. $m + n + t = 9,$
$m - n - t = -15,$
$3m + n + t = 2$

Write the augmented matrix.

$$\left[\begin{array}{ccc|c} 1 & 1 & 1 & 9 \\ 1 & -1 & -1 & -15 \\ 3 & 1 & 1 & 2 \end{array}\right]$$

Multiply row 1 by -1 and add it to row 2. Also, multiply row 1 by -3 and add it to row 3.

$$\left[\begin{array}{ccc|c} 1 & 1 & 1 & 9 \\ 0 & -2 & -2 & -24 \\ 0 & -2 & -2 & -25 \end{array}\right]$$

Multiply row 2 by -1 and add it to row 3.

$$\left[\begin{array}{ccc|c} 1 & 1 & 1 & 9 \\ 0 & -2 & -2 & -24 \\ 0 & 0 & 0 & -1 \end{array}\right]$$

The last row corresponds to the false equation $0 = -1$. Thus, the system of equations has no solution.

37. $a + b - c = 7,$
$a - b + c = 5,$
$3a + b - c = -1$

Write the augmented matrix.

$$\left[\begin{array}{ccc|c} 1 & 1 & -1 & 7 \\ 1 & -1 & 1 & 5 \\ 3 & 1 & -1 & -1 \end{array}\right]$$

Multiply row 1 by -1 and add it to row 2. Also, multiply row 1 by -3 and add it to row 3.

$$\left[\begin{array}{ccc|c} 1 & 1 & -1 & 7 \\ 0 & -2 & 2 & -2 \\ 0 & -2 & 2 & -22 \end{array}\right]$$

Multiply row 2 by -1 and add it to row 3.

$$\left[\begin{array}{ccc|c} 1 & 1 & -1 & 7 \\ 0 & -2 & 2 & -2 \\ 0 & 0 & 0 & -20 \end{array}\right]$$

The last row corresponds to the false equation $0 = -20$. Thus, the system of equations has no solution.

38. $a - b + c = 3,$
$2a + b - 3c = 5,$
$4a + b - c = 11$

Write the augmented matrix. We will use Gaussian elimination.

$$\left[\begin{array}{ccc|c} 1 & -1 & 1 & 3 \\ 2 & 1 & -3 & 5 \\ 4 & 1 & -1 & 11 \end{array}\right]$$

Multiply row 1 by -2 and add it to row 2. Also, multiply row 1 by -4 and add it to row 3.

$$\left[\begin{array}{ccc|c} 1 & -1 & 1 & 3 \\ 0 & 3 & -5 & -1 \\ 0 & 5 & -5 & -1 \end{array}\right]$$

Multiply row 3 by 3.

$$\left[\begin{array}{ccc|c} 1 & -1 & 1 & 3 \\ 0 & 3 & -5 & -1 \\ 0 & 15 & -15 & -3 \end{array}\right]$$

Multiply row 2 by -5 and add it to row 3.

$$\left[\begin{array}{ccc|c} 1 & -1 & 1 & 3 \\ 0 & 3 & -5 & -1 \\ 0 & 0 & 10 & 2 \end{array}\right]$$

Multiply row 2 by $\frac{1}{3}$ and multiply row 3 by $\frac{1}{10}$.

$$\left[\begin{array}{ccc|c} 1 & -1 & 1 & 3 \\ 0 & 1 & -\frac{5}{3} & -\frac{1}{3} \\ 0 & 0 & 1 & \frac{1}{5} \end{array}\right]$$

We have

$$\begin{aligned} x - y + z &= 3, & (1) \\ y - \frac{5}{3}z &= -\frac{1}{3}, & (2) \\ z &= \frac{1}{5}. & (3) \end{aligned}$$

Back-substitute in equation (2) to find y.

$$y - \frac{5}{3}\cdot\frac{1}{5} = -\frac{1}{3}$$
$$y = 0$$

Back-substitute in equation (1) to find x.

$$x - 0 + \frac{1}{5} = 3$$
$$x = \frac{14}{5}$$

The solution is $\left(\frac{14}{5}, 0, \frac{1}{5}\right)$.

39. $-2w + 2x + 2y - 2z = -10,$

$w + x + y + z = -5,$

$3w + x - y + 4z = -2,$

$w + 3x - 2y + 2z = -6$

Write the augmented matrix. We will use Gaussian elimination.

$$\left[\begin{array}{cccc|c} -2 & 2 & 2 & -2 & -10 \\ 1 & 1 & 1 & 1 & -5 \\ 3 & 1 & -1 & 4 & -2 \\ 1 & 3 & -2 & 2 & -6 \end{array}\right]$$

Interchange rows 1 and 2.

$$\left[\begin{array}{cccc|c} 1 & 1 & 1 & 1 & -5 \\ -2 & 2 & 2 & -2 & -10 \\ 3 & 1 & -1 & 4 & -2 \\ 1 & 3 & -2 & 2 & -6 \end{array}\right]$$

Multiply row 1 by 2 and add it to row 2. Multiply row 1 by -3 and add it to row 3. Multiply row 1 by -1 and add it to row 4.

$$\left[\begin{array}{cccc|c} 1 & 1 & 1 & 1 & -5 \\ 0 & 4 & 4 & 0 & -20 \\ 0 & -2 & -4 & 1 & 13 \\ 0 & 2 & -3 & 1 & -1 \end{array}\right]$$

Interchange rows 2 and 3.

$$\left[\begin{array}{cccc|c} 1 & 1 & 1 & 1 & -5 \\ 0 & -2 & -4 & 1 & 13 \\ 0 & 4 & 4 & 0 & -20 \\ 0 & 2 & -3 & 1 & -1 \end{array}\right]$$

Multiply row 2 by 2 and add it to row 3. Add row 2 to row 4.

$$\left[\begin{array}{cccc|c} 1 & 1 & 1 & 1 & -5 \\ 0 & -2 & -4 & 1 & 13 \\ 0 & 0 & -4 & 2 & 6 \\ 0 & 0 & -7 & 2 & 12 \end{array}\right]$$

Multiply row 4 by 4.

$$\left[\begin{array}{cccc|c} 1 & 1 & 1 & 1 & -5 \\ 0 & -2 & -4 & 1 & 13 \\ 0 & 0 & -4 & 2 & 6 \\ 0 & 0 & -28 & 8 & 48 \end{array}\right]$$

Multiply row 3 by -7 and add it to row 4.

$$\left[\begin{array}{cccc|c} 1 & 1 & 1 & 1 & -5 \\ 0 & -2 & -4 & 1 & 13 \\ 0 & 0 & -4 & 2 & 6 \\ 0 & 0 & 0 & -6 & 6 \end{array}\right]$$

Multiply row 2 by $-\frac{1}{2}$, row 3 by $-\frac{1}{4}$, and row 6 by $-\frac{1}{6}$.

$$\left[\begin{array}{cccc|c} 1 & 1 & 1 & 1 & -5 \\ 0 & 1 & 2 & -\frac{1}{2} & -\frac{13}{2} \\ 0 & 0 & 1 & -\frac{1}{2} & -\frac{3}{2} \\ 0 & 0 & 0 & 1 & -1 \end{array}\right]$$

Write the system of equations that corresponds to the last matrix.

$$\begin{aligned} w + x + y + z &= -5, & (1) \\ x + 2y - \frac{1}{2}z &= -\frac{13}{2}, & (2) \\ y - \frac{1}{2}z &= -\frac{3}{2}, & (3) \\ z &= -1 & (4) \end{aligned}$$

Back-substitute in equation (3) and solve for y.

$$y - \frac{1}{2}(-1) = -\frac{3}{2}$$
$$y + \frac{1}{2} = -\frac{3}{2}$$
$$y = -2$$

Back-substitute in equation (2) and solve for x.

$$x + 2(-2) - \frac{1}{2}(-1) = -\frac{13}{2}$$
$$x - 4 + \frac{1}{2} = -\frac{13}{2}$$
$$x = -3$$

Back-substitute in equation (1) and solve for w.

$$w - 3 - 2 - 1 = -5$$
$$w = 1$$

The solution is $(1, -3, -2, -1)$.

40. $-w + 2x - 3y + z = -8,$
$-w + x + y - z = -4,$
$w + x + y + z = 22,$
$-w + x - y - z = -14$

Write the augmented matrix. We will use Gauss-Jordan elimination.

$$\left[\begin{array}{rrrr|r} -1 & 2 & -3 & 1 & -8 \\ -1 & 1 & 1 & -1 & -4 \\ 1 & 1 & 1 & 1 & 22 \\ -1 & 1 & -1 & -1 & -14 \end{array}\right]$$

Multiply row 1 by -1.

$$\left[\begin{array}{rrrr|r} 1 & -2 & 3 & -1 & 8 \\ -1 & 1 & 1 & -1 & -4 \\ 1 & 1 & 1 & 1 & 22 \\ -1 & 1 & -1 & -1 & -14 \end{array}\right]$$

Add row 1 to row 2 and to row 4. Multiply row 1 by -1 and add it to row 3.

$$\left[\begin{array}{rrrr|r} 1 & -2 & 3 & -1 & 8 \\ 0 & -1 & 4 & -2 & 4 \\ 0 & 3 & -2 & 2 & 14 \\ 0 & -1 & 2 & -2 & -6 \end{array}\right]$$

Multiply row 2 by -1.

$$\left[\begin{array}{rrrr|r} 1 & -2 & 3 & -1 & 8 \\ 0 & 1 & -4 & 2 & -4 \\ 0 & 3 & -2 & 2 & 14 \\ 0 & -1 & 2 & -2 & -6 \end{array}\right]$$

Multiply row 2 by -3 and add it to row 3. Also, add row 2 to row 4.

$$\left[\begin{array}{rrrr|r} 1 & -2 & 3 & -1 & 8 \\ 0 & 1 & -4 & 2 & -4 \\ 0 & 0 & 10 & -4 & 26 \\ 0 & 0 & -2 & 0 & -10 \end{array}\right]$$

Interchange row 3 and row 4.

$$\left[\begin{array}{rrrr|r} 1 & -2 & 3 & -1 & 8 \\ 0 & 1 & -4 & 2 & -4 \\ 0 & 0 & -2 & 0 & -10 \\ 0 & 0 & 10 & -4 & 26 \end{array}\right]$$

Multiply row 3 by $-\frac{1}{2}$.

$$\left[\begin{array}{rrrr|r} 1 & -2 & 3 & -1 & 8 \\ 0 & 1 & -4 & 2 & -4 \\ 0 & 0 & 1 & 0 & 5 \\ 0 & 0 & 10 & -4 & 26 \end{array}\right]$$

Multiply row 3 by -10 and add it to row 4.

$$\left[\begin{array}{rrrr|r} 1 & -2 & 3 & -1 & 8 \\ 0 & 1 & -4 & 2 & -4 \\ 0 & 0 & 1 & 0 & 5 \\ 0 & 0 & 0 & -4 & -24 \end{array}\right]$$

Multiply row 4 by $-\frac{1}{4}$.

$$\left[\begin{array}{rrrr|r} 1 & -2 & 3 & -1 & 8 \\ 0 & 1 & -4 & 2 & -4 \\ 0 & 0 & 1 & 0 & 5 \\ 0 & 0 & 0 & 1 & 6 \end{array}\right]$$

Add row 4 to row 1. Also, multiply row 4 by -2 and add it to row 2.

$$\left[\begin{array}{rrrr|r} 1 & -2 & 3 & 0 & 14 \\ 0 & 1 & -4 & 0 & -16 \\ 0 & 0 & 1 & 0 & 5 \\ 0 & 0 & 0 & 1 & 6 \end{array}\right]$$

Multiply row 3 by -3 and add it to row 1. Also, multiply row 3 by 4 and add it to row 2.

$$\left[\begin{array}{rrrr|r} 1 & -2 & 0 & 0 & -1 \\ 0 & 1 & 0 & 0 & 4 \\ 0 & 0 & 1 & 0 & 5 \\ 0 & 0 & 0 & 1 & 6 \end{array}\right]$$

Multiply row 2 by 2 and add it to row 1.

$$\left[\begin{array}{rrrr|r} 1 & 0 & 0 & 0 & 7 \\ 0 & 1 & 0 & 0 & 4 \\ 0 & 0 & 1 & 0 & 5 \\ 0 & 0 & 0 & 1 & 6 \end{array}\right]$$

The solution is $(7, 4, 5, 6)$.

41. ***Familiarize***. Let x = the number of hours the Houlihans were out before 11 P.M. and y = the number of hours after 11 P.M. Then they pay the babysitter $\$5x$ before 11 P.M. and $\$7.50y$ after 11 P.M.

Translate.

The Houlihans were out for a total of 5 hr.

$$x + y = 5$$

They paid the sitter a total of \$30.

$$5x + 7.5y = 30$$

Carry out. Use Gaussian elimination or Gauss-Jordan elimination to solve the system of equations.

$$x + y = 5,$$
$$5x + 7.5y = 30.$$

The solution is $(3, 2)$. The coordinate $y = 2$ indicates that the Houlihans were out 2 hr after 11 P.M., so they came home at 1 A.M.

Check. The total time is $3 + 2$, or 5 hr. The total pay is $\$5 \cdot 3 + \$7.50(2)$, or $\$15 + \$15$, or \$30. The solution checks.

State. The Houlihans came home at 1 A.M.

42. Let x, y, and z represent the amount spent on advertising in fiscal years 2001, 2002, and 2003, respectively, in millions of dollars.

Solve: $x + y + z = 11,$
$$z = 3x,$$
$$y = z - 3$$

$x = \$2$ million, $y = \$3$ million, $z = \$6$ million

43. ***Familiarize.*** Let x, y, and z represent the amounts borrowed at 8%, 10%, and 12%, respectively. Then the annual interest is $8\%x$, $10\%y$, and $12\%z$, or $0.08x$, $0.1y$, and $0.12z$.

Translate.

The total amount borrowed was \$30,000.

$$x + y + z = 30,000$$

The total annual interest was \$3040.

$$0.08x + 0.1y + 0.12z = 3040$$

The total amount borrowed at 8% and 10% was twice the amount borrowed at 12%.

$$x + y = 2z$$

We have a system of equations.

$$x + y + z = 30,000,$$
$$0.08x + 0.1y + 0.12z = 3040,$$
$$x + y = 2z, \text{ or}$$

$$x + y + z = 30,000,$$
$$0.08x + 0.1y + 0.12z = 3040,$$
$$x + y - 2z = 0$$

Carry out. Using Gaussian elimination or Gauss-Jordan elimination, we find that the solution is $(8000, 12,000, 10,000)$.

Check. The total amount borrowed was $\$8000 + \$12,000 + \$10,000$, or \$30,000. The total annual interest was $0.08(\$8000) + 0.1(\$12,000) + 0.12(\$10,000)$, or $\$640 + \$1200 + \$1200$, or \$3040. The total amount borrowed at 8% and 10%, $\$8000 + \$12,000$ or \$20,000, was twice the amount borrowed at 12%, \$10,000. The solution checks.

State. The amounts borrowed at 8%, 10%, and 12% were \$8000, \$12,000 and \$10,000, respectively.

44. Let x and y represent the number of 37¢ and 23¢ stamps purchased, respectively.

Solve: $x + y = 60,$
$$0.37x + 0.23y = 20.10$$

$x = 45$, $y = 15$

45. Answers will vary.

46. See Example 4 in this section of the text.

47. The function has a variable in the exponent, so it is an exponential function.

48. The function is of the form $f(x) = mx + b$, so it is linear.

49. The function is the quotient of two polynomials, so it is a rational function.

50. This is a polynomial function of degree 4, so it is a quartic function.

51. The function is of the form $f(x) = \log_a x$, so it is logarithmic.

52. This is a polynomial function of degree 3, so it is a cubic function.

53. The function is of the form $f(x) = mx + b$, so it is linear.

54. This is a polynomial function of degree 2, so it is quadratic.

55. Substitute to find three equations.

$$12 = a(-3)^2 + b(-3) + c$$
$$-7 = a(-1)^2 + b(-1) + c$$
$$-2 = a \cdot 1^2 + b \cdot 1 + c$$

We have a system of equations.

$$9a - 3b + c = 12,$$
$$a - b + c = -7,$$
$$a + b + c = -2$$

Write the augmented matrix. We will use Gaussian elimination.

$$\left[\begin{array}{rrr|r} 9 & -3 & 1 & 12 \\ 1 & -1 & 1 & -7 \\ 1 & 1 & 1 & -2 \end{array}\right]$$

Interchange the first two rows.

$$\left[\begin{array}{rrr|r} 1 & -1 & 1 & -7 \\ 9 & -3 & 1 & 12 \\ 1 & 1 & 1 & -2 \end{array}\right]$$

Multiply row 1 by -9 and add it to row 2. Also, multiply row 1 by -1 and add it to row 3.

$$\left[\begin{array}{rrr|r} 1 & -1 & 1 & -7 \\ 0 & 6 & -8 & 75 \\ 0 & 2 & 0 & 5 \end{array}\right]$$

Interchange row 2 and row 3.

$$\left[\begin{array}{rrr|r} 1 & -1 & 1 & -7 \\ 0 & 2 & 0 & 5 \\ 0 & 6 & -8 & 75 \end{array}\right]$$

Multiply row 2 by -3 and add it to row 3.

$$\left[\begin{array}{ccc|c} 1 & -1 & 1 & -7 \\ 0 & 2 & 0 & 5 \\ 0 & 0 & -8 & 60 \end{array}\right]$$

Multiply row 2 by $\frac{1}{2}$ and row 3 by $-\frac{1}{8}$.

$$\left[\begin{array}{ccc|c} 1 & -1 & 1 & -7 \\ 0 & 1 & 0 & \frac{5}{2} \\ 0 & 0 & 1 & -\frac{15}{2} \end{array}\right]$$

Write the system of equations that corresponds to the last matrix.

$$x - y + z = -7,$$
$$y = \frac{5}{2},$$
$$z = -\frac{15}{2}$$

Back-substitute $\frac{5}{2}$ for y and $-\frac{15}{2}$ for z in the first equation and solve for x.

$$x - \frac{5}{2} - \frac{15}{2} = -7$$
$$x - 10 = -7$$
$$x = 3$$

The solution is $\left(3, \frac{5}{2}, -\frac{15}{2}\right)$, so the equation is

$y = 3x^2 + \frac{5}{2}x - \frac{15}{2}$.

56. Solve: $0 = a(-1)^2 + b(-1) + c,$

$$-3 = a \cdot 1^2 + b \cdot 1 + c,$$
$$-22 = a \cdot 3^2 + b \cdot 3 + c, \text{ or}$$
$$a - b + c = 0,$$
$$a + b + c = -3,$$
$$9a + 3b + c = -22$$

$a = -2$, $b = -\frac{3}{2}$, $c = \frac{1}{2}$, so $y = -2x^2 - \frac{3}{2}x + \frac{1}{2}$.

57. $\begin{bmatrix} 1 & 5 \\ 3 & 2 \end{bmatrix}$

Multiply row 1 by -3 and add it to row 2.

$$\begin{bmatrix} 1 & 5 \\ 0 & -13 \end{bmatrix}$$

Multiply row 2 by $-\frac{1}{13}$.

$$\begin{bmatrix} 1 & 5 \\ 0 & 1 \end{bmatrix}$$ Row-echelon form

Multiply row 2 by -5 and add it to row 1.

$$\begin{bmatrix} 1 & 0 \\ 0 & 1 \end{bmatrix}$$ Reduced row-echelon form

58. $x - y + 3z = -8,$

$$2x + 3y - z = 5,$$
$$3x + 2y + 2kz = -3k$$

Write the augmented matrix.

$$\left[\begin{array}{ccc|c} 1 & -1 & 3 & -8 \\ 2 & 3 & -1 & 5 \\ 3 & 2 & 2k & -3k \end{array}\right]$$

Multiply row 1 by -2 and add it to row 2. Also, multiply row 1 by -3 and add it to row 3.

$$\left[\begin{array}{ccc|c} 1 & -1 & 3 & -8 \\ 0 & 5 & -7 & 21 \\ 0 & 5 & 2k-9 & -3k+24 \end{array}\right]$$

Multiply row 2 by -1 and add it to row 3.

$$\left[\begin{array}{ccc|c} 1 & -1 & 3 & -8 \\ 0 & 5 & -7 & 21 \\ 0 & 0 & 2k-2 & -3k+3 \end{array}\right]$$

a) If the system has no solution we have:

$$2k - 2 = 0 \text{ and } -3k + 3 \neq 0$$
$$k = 1 \text{ and } k \neq 1$$

This is impossible, so there is no value of k for which the system has no solution.

b) If the system has exactly one solution, we have:

$$2k - 2 \neq -3k + 3$$
$$5k \neq 5$$
$$k \neq 1$$

c) If the system has infinitely many solutions, we have:

$$2k - 2 = 0 \text{ and } -3k + 3 = 0$$
$$k = 1 \text{ and } k = 1, \text{ or}$$
$$k = 1$$

59. $y = x + z,$

$$3y + 5z = 4,$$
$$x + 4 = y + 3z, \text{ or}$$
$$x - y + z = 0,$$
$$3y + 5z = 4,$$
$$x - y - 3z = -4$$

Write the augmented matrix. We will use Gauss-Jordan elimination.

$$\left[\begin{array}{ccc|c} 1 & -1 & 1 & 0 \\ 0 & 3 & 5 & 4 \\ 1 & -1 & -3 & -4 \end{array}\right]$$

Multiply row 1 by -1 and add it to row 3.

$$\left[\begin{array}{rrr|r} 1 & -1 & 1 & 0 \\ 0 & 3 & 5 & 4 \\ 0 & 0 & -4 & -4 \end{array}\right]$$

Multiply row 3 by $-\frac{1}{4}$.

$$\left[\begin{array}{rrr|r} 1 & -1 & 1 & 0 \\ 0 & 3 & 5 & 4 \\ 0 & 0 & 1 & 1 \end{array}\right]$$

Multiply row 3 by -1 and add it to row 1. Also, multiply row 3 by -5 and add it to row 2.

$$\left[\begin{array}{rrr|r} 1 & -1 & 0 & -1 \\ 0 & 3 & 0 & -1 \\ 0 & 0 & 1 & 1 \end{array}\right]$$

Multiply row 2 by $\frac{1}{3}$.

$$\left[\begin{array}{rrr|r} 1 & -1 & 0 & -1 \\ 0 & 1 & 0 & -\frac{1}{3} \\ 0 & 0 & 1 & 1 \end{array}\right]$$

Add row 2 to row 1.

$$\left[\begin{array}{rrr|r} 1 & 0 & 0 & -\frac{4}{3} \\ 0 & 1 & 0 & -\frac{1}{3} \\ 0 & 0 & 1 & 1 \end{array}\right]$$

Read the solution from the last matrix. It is $\left(-\frac{4}{3}, -\frac{1}{3}, 1\right)$.

60. $x + y = 2z,$

$2x - 5z = 4,$

$x - z = y + 8,$ or

$x + y - 2z = 0,$

$2x \quad - 5z = 4,$

$x - y - \quad z = 8$

Write the augmented matrix. We will use Gauss-Jordan elimination.

$$\left[\begin{array}{rrr|r} 1 & 1 & -2 & 0 \\ 2 & 0 & -5 & 4 \\ 1 & -1 & -1 & 8 \end{array}\right]$$

Multiply row 1 by -2 and add it to row 2. Also, multiply row 1 by -1 and add it to row 3.

$$\left[\begin{array}{rrr|r} 1 & 1 & -2 & 0 \\ 0 & -2 & -1 & 4 \\ 0 & -2 & 1 & 8 \end{array}\right]$$

Multiply row 2 by $-\frac{1}{2}$.

$$\left[\begin{array}{rrr|r} 1 & 1 & -2 & 0 \\ 0 & 1 & \frac{1}{2} & -2 \\ 0 & -2 & 1 & 8 \end{array}\right]$$

Multiply row 2 by -1 and add it to row 1. Also, multiply row 2 by 2 and add it to row 3.

$$\left[\begin{array}{rrr|r} 1 & 0 & -\frac{5}{2} & 2 \\ 0 & 1 & \frac{1}{2} & -2 \\ 0 & 0 & 2 & 4 \end{array}\right]$$

Multiply row 3 by $\frac{1}{2}$.

$$\left[\begin{array}{rrr|r} 1 & 0 & -\frac{5}{2} & 2 \\ 0 & 1 & \frac{1}{2} & -2 \\ 0 & 0 & 1 & 2 \end{array}\right]$$

Multiply row 3 by $\frac{5}{2}$ and add it to row 1. Also, multiply row 3 by $-\frac{1}{2}$ and add it to row 2.

$$\left[\begin{array}{rrr|r} 1 & 0 & 0 & 7 \\ 0 & 1 & 0 & -3 \\ 0 & 0 & 1 & 2 \end{array}\right]$$

The solution is $(7, -3, 2)$.

61. $x - 4y + 2z = 7,$

$3x + \quad y + 3z = -5$

Write the augmented matrix.

$$\left[\begin{array}{rrr|r} 1 & -4 & 2 & 7 \\ 3 & 1 & 3 & -5 \end{array}\right]$$

Multiply row 1 by -3 and add it to row 2.

$$\left[\begin{array}{rrr|r} 1 & -4 & 2 & 7 \\ 0 & 13 & -3 & -26 \end{array}\right]$$

Multiply row 2 by $\frac{1}{13}$.

$$\left[\begin{array}{rrr|r} 1 & -4 & 2 & 7 \\ 0 & 1 & -\frac{3}{13} & -2 \end{array}\right]$$

Write the system of equations that corresponds to the last matrix.

$$x - 4y + 2z = 7,$$

$$y - \frac{3}{13}z = -2$$

Solve the second equation for y.

$$y = \frac{3}{13}z - 2$$

Substitute in the first equation and solve for x.

$$x - 4\left(\frac{3}{13}z - 2\right) + 2z = 7$$

$$x - \frac{12}{13}z + 8 + 2z = 7$$

$$x = -\frac{14}{13}z - 1$$

The solution is $\left(-\frac{14}{13}z - 1, \frac{3}{13}z - 2, z\right)$, where z is any real number.

62. $x - y - 3z = 3,$
$-x + 3y + z = -7$

Write the augmented matrix.

$$\left[\begin{array}{rrr|r} 1 & -1 & -3 & 3 \\ -1 & 3 & 1 & -7 \end{array}\right]$$

Add row 1 to row 2.

$$\left[\begin{array}{rrr|r} 1 & -1 & -3 & 3 \\ 0 & 2 & -2 & -4 \end{array}\right]$$

Multiply row 2 by $\frac{1}{2}$.

$$\left[\begin{array}{rrr|r} 1 & -1 & -3 & 3 \\ 0 & 1 & -1 & -2 \end{array}\right]$$

We have

$x - y - 3z = 3,$
$y - z = -2.$

Then $y = z - 2$. Substitute in the first equation and solve for x.

$$x - (z - 2) - 3z = 3$$

$$x - z + 2 - 3z = 3$$

$$x = 4z + 1$$

The solution is $(4z+1, z-2, z)$, where z is any real number.

63. $4x + 5y = 3,$
$-2x + y = 9,$
$3x - 2y = -15$

Write the augmented matrix.

$$\left[\begin{array}{rr|r} 4 & 5 & 3 \\ -2 & 1 & 9 \\ 3 & -2 & -15 \end{array}\right]$$

Multiply row 2 by 2 and row 3 by 4.

$$\left[\begin{array}{rr|r} 4 & 5 & 3 \\ -4 & 2 & 18 \\ 12 & -8 & -60 \end{array}\right]$$

Add row 1 to row 2. Also, multiply row 1 by -3 and add it to row 3.

$$\left[\begin{array}{rr|r} 4 & 5 & 3 \\ 0 & 7 & 21 \\ 0 & -23 & -69 \end{array}\right]$$

Multiply row 2 by $\frac{1}{7}$ and row 3 by $-\frac{1}{23}$.

$$\left[\begin{array}{rr|r} 4 & 5 & 3 \\ 0 & 1 & 3 \\ 0 & 1 & 3 \end{array}\right]$$

Multiply row 2 by -1 and add it to row 3.

$$\left[\begin{array}{rr|r} 4 & 5 & 3 \\ 0 & 1 & 3 \\ 0 & 0 & 0 \end{array}\right]$$

The last row corresponds to the equation $0 = 0$. Thus we have a dependent system that is equivalent to

$4x + 5y = 3,$ (1)
$y = 3.$ (2)

Back-substitute in equation (1) to find x.

$$4x + 5 \cdot 3 = 3$$

$$4x + 15 = 3$$

$$4x = -12$$

$$x = -3$$

The solution is $(-3, 3)$.

64. $2x - 3y = -1,$
$-x + 2y = -2,$
$3x - 5y = 1$

Write the augmented matrix.

$$\left[\begin{array}{rr|r} 2 & -3 & -1 \\ -1 & 2 & -2 \\ 3 & -5 & 1 \end{array}\right]$$

Interchange the first two rows.

$$\left[\begin{array}{rr|r} -1 & 2 & -2 \\ 2 & -3 & -1 \\ 3 & -5 & 1 \end{array}\right]$$

Multiply row 1 by 2 and add it to row 2. Also, multiply row 1 by 3 and add it to row 3.

$$\left[\begin{array}{rr|r} -1 & 2 & -2 \\ 0 & 1 & -5 \\ 0 & 1 & -5 \end{array}\right]$$

Multiply row 2 by -1 and add it to row 3.

$$\left[\begin{array}{rr|r} -1 & 2 & -2 \\ 0 & 1 & -5 \\ 0 & 0 & 0 \end{array}\right]$$

We have a dependent system that is equivalent to

$$-x+2y=-2, \quad (1)$$
$$y=-5. \quad (2)$$

Back-substitute in equation (1) to find x.

$$-x+2(-5)=-2$$
$$-x-10=-2$$
$$-x=8$$
$$x=-8$$

The solution is $(-8,-5)$.

Exercise Set 8.4

1. $\begin{bmatrix} 5 & x \end{bmatrix} = \begin{bmatrix} y & -3 \end{bmatrix}$

Corresponding entries of the two matrices must be equal. Thus we have $5=y$ and $x=-3$.

2. $\begin{bmatrix} 6x \\ 25 \end{bmatrix} = \begin{bmatrix} -9 \\ 5y \end{bmatrix}$

$$6x=-9 \quad \text{and} \quad 25=5y$$
$$x=-\frac{3}{2} \quad \text{and} \quad 5=y$$

3. $\begin{bmatrix} 3 & 2x \\ y & -8 \end{bmatrix} = \begin{bmatrix} 3 & -2 \\ 1 & -8 \end{bmatrix}$

Corresponding entries of the two matrices must be equal. Thus, we have:

$$2x=-2 \quad \text{and} \quad y=1$$
$$x=-1 \quad \text{and} \quad y=1$$

4. $\begin{bmatrix} x-1 & 4 \\ y+3 & -7 \end{bmatrix} = \begin{bmatrix} 0 & 4 \\ -2 & -7 \end{bmatrix}$

$$x-1=0 \quad \text{and} \quad y+3=-2$$
$$x=1 \quad \text{and} \quad y=-5$$

5. $\mathbf{A}+\mathbf{B} = \begin{bmatrix} 1 & 2 \\ 4 & 3 \end{bmatrix} + \begin{bmatrix} -3 & 5 \\ 2 & -1 \end{bmatrix}$

$$= \begin{bmatrix} 1+(-3) & 2+5 \\ 4+2 & 3+(-1) \end{bmatrix}$$
$$= \begin{bmatrix} -2 & 7 \\ 6 & 2 \end{bmatrix}$$

6. $\mathbf{B}+\mathbf{A} = \mathbf{A}+\mathbf{B} = \begin{bmatrix} -2 & 7 \\ 6 & 2 \end{bmatrix}$ (See Exercise 5.)

7. $\mathbf{E}+\mathbf{O} = \begin{bmatrix} 1 & 3 \\ 2 & 6 \end{bmatrix} + \begin{bmatrix} 0 & 0 \\ 0 & 0 \end{bmatrix}$

$$= \begin{bmatrix} 1+0 & 3+0 \\ 2+0 & 6+0 \end{bmatrix}$$
$$= \begin{bmatrix} 1 & 3 \\ 2 & 6 \end{bmatrix}$$

8. $2\mathbf{A} = \begin{bmatrix} 2\cdot 1 & 2\cdot 2 \\ 2\cdot 4 & 2\cdot 3 \end{bmatrix} = \begin{bmatrix} 2 & 4 \\ 8 & 6 \end{bmatrix}$

9. $3\mathbf{F} = 3\begin{bmatrix} 3 & 3 \\ -1 & -1 \end{bmatrix}$

$$= \begin{bmatrix} 3\cdot 3 & 3\cdot 3 \\ 3\cdot(-1) & 3\cdot(-1) \end{bmatrix}$$
$$= \begin{bmatrix} 9 & 9 \\ -3 & -3 \end{bmatrix}$$

10. $(-1)\mathbf{D} = \begin{bmatrix} -1\cdot 1 & -1\cdot 1 \\ -1\cdot 1 & -1\cdot 1 \end{bmatrix} = \begin{bmatrix} -1 & -1 \\ -1 & -1 \end{bmatrix}$

11. $3\mathbf{F} = 3\begin{bmatrix} 3 & 3 \\ -1 & -1 \end{bmatrix} = \begin{bmatrix} 9 & 9 \\ -3 & -3 \end{bmatrix}$,

$$2\mathbf{A} = 2\begin{bmatrix} 1 & 2 \\ 4 & 3 \end{bmatrix} = \begin{bmatrix} 2 & 4 \\ 8 & 6 \end{bmatrix}$$

$$3\mathbf{F}+2\mathbf{A} = \begin{bmatrix} 9 & 9 \\ -3 & -3 \end{bmatrix} + \begin{bmatrix} 2 & 4 \\ 8 & 6 \end{bmatrix}$$
$$= \begin{bmatrix} 9+2 & 9+4 \\ -3+8 & -3+6 \end{bmatrix}$$
$$= \begin{bmatrix} 11 & 13 \\ 5 & 3 \end{bmatrix}$$

12. $\mathbf{A}-\mathbf{B} = \begin{bmatrix} 1 & 2 \\ 4 & 3 \end{bmatrix} - \begin{bmatrix} -3 & 5 \\ 2 & -1 \end{bmatrix} = \begin{bmatrix} 4 & -3 \\ 2 & 4 \end{bmatrix}$

13. $\mathbf{B}-\mathbf{A} = \begin{bmatrix} -3 & 5 \\ 2 & -1 \end{bmatrix} - \begin{bmatrix} 1 & 2 \\ 4 & 3 \end{bmatrix}$

$$= \begin{bmatrix} -3 & 5 \\ 2 & -1 \end{bmatrix} + \begin{bmatrix} -1 & -2 \\ -4 & -3 \end{bmatrix}$$
$$[\mathbf{B}-\mathbf{A} = \mathbf{B}+(-\mathbf{A})]$$
$$= \begin{bmatrix} -3+(-1) & 5+(-2) \\ 2+(-4) & -1+(-3) \end{bmatrix}$$
$$= \begin{bmatrix} -4 & 3 \\ -2 & -4 \end{bmatrix}$$

14. $\mathbf{AB} = \begin{bmatrix} 1 & 2 \\ 4 & 3 \end{bmatrix}\begin{bmatrix} -3 & 5 \\ 2 & -1 \end{bmatrix}$

$$= \begin{bmatrix} 1(-3)+2\cdot 2 & 1\cdot 5+2(-1) \\ 4(-3)+3\cdot 2 & 4\cdot 5+3(-1) \end{bmatrix}$$
$$= \begin{bmatrix} 1 & 3 \\ -6 & 17 \end{bmatrix}$$

15. $\mathbf{BA} = \begin{bmatrix} -3 & 5 \\ 2 & -1 \end{bmatrix}\begin{bmatrix} 1 & 2 \\ 4 & 3 \end{bmatrix}$

$$= \begin{bmatrix} -3\cdot 1+5\cdot 4 & -3\cdot 2+5\cdot 3 \\ 2\cdot 1+(-1)4 & 2\cdot 2+(-1)3 \end{bmatrix}$$
$$= \begin{bmatrix} 17 & 9 \\ -2 & 1 \end{bmatrix}$$

16. $\mathbf{OF} = \mathbf{O} = \begin{bmatrix} 0 & 0 \\ 0 & 0 \end{bmatrix}$

17. $\mathbf{CD} = \begin{bmatrix} 1 & -1 \\ -1 & 1 \end{bmatrix}\begin{bmatrix} 1 & 1 \\ 1 & 1 \end{bmatrix}$

$= \begin{bmatrix} 1\cdot1+(-1)\cdot1 & 1\cdot1+(-1)\cdot1 \\ -1\cdot1+1\cdot1 & -1\cdot1+1\cdot1 \end{bmatrix}$

$= \begin{bmatrix} 0 & 0 \\ 0 & 0 \end{bmatrix}$

18. $\mathbf{EF} = \begin{bmatrix} 1 & 3 \\ 2 & 6 \end{bmatrix}\begin{bmatrix} 3 & 3 \\ -1 & -1 \end{bmatrix}$

$= \begin{bmatrix} 1\cdot3+3(-1) & 1\cdot3+3(-1) \\ 2\cdot3+6(-1) & 2\cdot3+6(-1) \end{bmatrix}$

$= \begin{bmatrix} 0 & 0 \\ 0 & 0 \end{bmatrix}$

19. $\mathbf{AI} = \begin{bmatrix} 1 & 2 \\ 4 & 3 \end{bmatrix}\begin{bmatrix} 1 & 0 \\ 0 & 1 \end{bmatrix}$

$= \begin{bmatrix} 1\cdot1+2\cdot0 & 1\cdot0+2\cdot1 \\ 4\cdot1+3\cdot0 & 4\cdot0+3\cdot1 \end{bmatrix}$

$= \begin{bmatrix} 1 & 2 \\ 4 & 3 \end{bmatrix}$

20. $\mathbf{IA} = \mathbf{A} = \begin{bmatrix} 1 & 2 \\ 4 & 3 \end{bmatrix}$

21. $\begin{bmatrix} -1 & 0 & 7 \\ 3 & -5 & 2 \end{bmatrix}\begin{bmatrix} 6 \\ -4 \\ 1 \end{bmatrix}$

$= \begin{bmatrix} -1\cdot6+0(-4)+7\cdot1 \\ 3\cdot6+(-5)(-4)+2\cdot1 \end{bmatrix}$

$= \begin{bmatrix} 1 \\ 40 \end{bmatrix}$

22. $\begin{bmatrix} 6 & -1 & 2 \end{bmatrix}\begin{bmatrix} 1 & 4 \\ -2 & 0 \\ 5 & -3 \end{bmatrix}$

$= \begin{bmatrix} 6\cdot1+(-1)(-2)+2\cdot5 & 6\cdot4+(-1)\cdot0+2(-3) \end{bmatrix}$

$= \begin{bmatrix} 18 & 18 \end{bmatrix}$

23. $\begin{bmatrix} -2 & 4 \\ 5 & 1 \\ -1 & -3 \end{bmatrix}\begin{bmatrix} 3 & -6 \\ -1 & 4 \end{bmatrix}$

$= \begin{bmatrix} -2\cdot3+4(-1) & -2(-6)+4\cdot4 \\ 5\cdot3+1(-1) & 5(-6)+1\cdot4 \\ -1\cdot3+(-3)(-1) & -1(-6)+(-3)\cdot4 \end{bmatrix}$

$= \begin{bmatrix} -10 & 28 \\ 14 & -26 \\ 0 & -6 \end{bmatrix}$

24. $\begin{bmatrix} 2 & -1 & 0 \\ 0 & 5 & 4 \end{bmatrix}\begin{bmatrix} -3 & 1 & 0 \\ 0 & 2 & -1 \\ 5 & 0 & 4 \end{bmatrix}$

$= \begin{bmatrix} -6+0+0 & 2-2+0 & 0+1+0 \\ 0+0+20 & 0+10+0 & 0-5+16 \end{bmatrix}$

$= \begin{bmatrix} -6 & 0 & 1 \\ 20 & 10 & 11 \end{bmatrix}$

25. $\begin{bmatrix} 1 \\ -5 \\ 3 \end{bmatrix}\begin{bmatrix} -6 & 5 & 8 \\ 0 & 4 & -1 \end{bmatrix}$

This product is not defined because the number of columns of the first matrix, 1, is not equal to the number of rows of the second matrix, 2.

26. $\begin{bmatrix} 2 & 0 & 0 \\ 0 & -1 & 0 \\ 0 & 0 & 3 \end{bmatrix}\begin{bmatrix} 0 & -4 & 3 \\ 2 & 1 & 0 \\ -1 & 0 & 6 \end{bmatrix}$

$= \begin{bmatrix} 0+0+0 & -8+0+0 & 6+0+0 \\ 0-2+0 & 0-1+0 & 0+0+0 \\ 0+0-3 & 0+0+0 & 0+0+18 \end{bmatrix}$

$= \begin{bmatrix} 0 & -8 & 6 \\ -2 & -1 & 0 \\ -3 & 0 & 18 \end{bmatrix}$

27. $\begin{bmatrix} 1 & -4 & 3 \\ 0 & 8 & 0 \\ -2 & -1 & 5 \end{bmatrix}\begin{bmatrix} 3 & 0 & 0 \\ 0 & -4 & 0 \\ 0 & 0 & 1 \end{bmatrix}$

$= \begin{bmatrix} 3+0+0 & 0+16+0 & 0+0+3 \\ 0+0+0 & 0-32+0 & 0+0+0 \\ -6+0+0 & 0+4+0 & 0+0+5 \end{bmatrix}$

$= \begin{bmatrix} 3 & 16 & 3 \\ 0 & -32 & 0 \\ -6 & 4 & 5 \end{bmatrix}$

28. $\begin{bmatrix} 4 \\ -5 \end{bmatrix}\begin{bmatrix} 2 & 0 \\ 6 & -7 \\ 0 & -3 \end{bmatrix}$

This product is not defined because the number of columns of the first matrix, 1, is not equal to the number of rows of the second matrix, 3.

29. a) $\mathbf{B} = \begin{bmatrix} 150 & 80 & 40 \end{bmatrix}$

b) $\$150 + 5\% \cdot \$150 = 1.05(\$150) = \$157.50$

$\$80 + 5\% \cdot \$80 = 1.05(\$80) = \$84$

$\$40 + 5\% \cdot \$40 = 1.05(\$40) = \$42$

We write the matrix that corresponds to these amounts.

$\mathbf{R} = \begin{bmatrix} 157.5 & 84 & 42 \end{bmatrix}$

c) $\mathbf{B}+\mathbf{R} = \begin{bmatrix} 150 & 80 & 40 \end{bmatrix} + \begin{bmatrix} 157.5 & 84 & 42 \end{bmatrix}$

$= \begin{bmatrix} 307.5 & 164 & 82 \end{bmatrix}$

The entries represent the total budget for each type of expenditure for June and July.

30. a) $\mathbf{A} = \begin{bmatrix} 40 & 20 & 30 \end{bmatrix}$

b) $1.1(40) = 44$, $1.1(20) = 22$, $1.1(30) = 33$

$\mathbf{B} = \begin{bmatrix} 44 & 22 & 33 \end{bmatrix}$

c) $\mathbf{A}+\mathbf{B} = \begin{bmatrix} 40 & 20 & 30 \end{bmatrix} + \begin{bmatrix} 44 & 22 & 33 \end{bmatrix}$

$= \begin{bmatrix} 84 & 42 & 63 \end{bmatrix}$

The entries represent the total amount of each type of produce ordered for both weeks.

31. a) $\mathbf{C} = \begin{bmatrix} 140 & 27 & 3 & 13 & 64 \end{bmatrix}$

$\mathbf{P} = \begin{bmatrix} 180 & 4 & 11 & 24 & 662 \end{bmatrix}$

$\mathbf{B} = \begin{bmatrix} 50 & 5 & 1 & 82 & 20 \end{bmatrix}$

b) $\mathbf{C} + 2\mathbf{P} + 3\mathbf{B}$

$= \begin{bmatrix} 140 & 27 & 3 & 13 & 64 \end{bmatrix} + \begin{bmatrix} 360 & 8 & 22 & 48 & 1324 \end{bmatrix} + \begin{bmatrix} 150 & 15 & 3 & 246 & 60 \end{bmatrix}$

$= \begin{bmatrix} 650 & 50 & 28 & 307 & 1448 \end{bmatrix}$

The entries represent the total nutritional value of one serving of chicken, 1 cup of potato salad, and 3 broccoli spears.

32. a) $\mathbf{P} = \begin{bmatrix} 290 & 15 & 9 & 39 \end{bmatrix}$

$\mathbf{G} = \begin{bmatrix} 70 & 2 & 0 & 17 \end{bmatrix}$

$\mathbf{M} = \begin{bmatrix} 150 & 8 & 8 & 11 \end{bmatrix}$

b) $3\mathbf{P} + 2\mathbf{G} + 2\mathbf{M}$

$= \begin{bmatrix} 870 & 45 & 27 & 117 \end{bmatrix} + \begin{bmatrix} 140 & 4 & 0 & 34 \end{bmatrix} + \begin{bmatrix} 300 & 16 & 16 & 22 \end{bmatrix}$

$= \begin{bmatrix} 1310 & 65 & 43 & 173 \end{bmatrix}$

The entries represent the total nutritional value of 3 slices of pizza, 1 cup of gelatin, and 2 cups of whole milk.

33. a) $\mathbf{M} = \begin{bmatrix} 45.29 & 6.63 & 10.94 & 7.42 & 8.01 \\ 53.78 & 4.95 & 9.83 & 6.16 & 12.56 \\ 47.13 & 8.47 & 12.66 & 8.29 & 9.43 \\ 51.64 & 7.12 & 11.57 & 9.35 & 10.72 \end{bmatrix}$

b) $\mathbf{N} = \begin{bmatrix} 65 & 48 & 93 & 57 \end{bmatrix}$

c) $\mathbf{NM} =$

$\begin{bmatrix} 12{,}851.86 & 1862.1 & 3019.81 & 2081.9 & 2611.56 \end{bmatrix}$

d) The entries of **NM** represent the total cost, in cents, of each item for the day's meals.

34. a) $\mathbf{M} = \begin{bmatrix} 1 & 2.5 & 0.75 & 0.5 \\ 0 & 0.5 & 0.25 & 0 \\ 0.75 & 0.25 & 0.5 & 0.5 \\ 0.5 & 0 & 0.5 & 1 \end{bmatrix}$

b) $\mathbf{C} = \begin{bmatrix} 15 & 28 & 54 & 83 \end{bmatrix}$

c) $\mathbf{CM} = \begin{bmatrix} 97 & 65 & 86.75 & 117.5 \end{bmatrix}$

d) The entries of **CM** represent the total cost, in cents, of each menu item.

35. a) $\mathbf{S} = \begin{bmatrix} 8 & 15 \\ 6 & 10 \\ 4 & 3 \end{bmatrix}$

b) $\mathbf{C} = \begin{bmatrix} 3 & 1.5 & 2 \end{bmatrix}$

c) $\mathbf{CS} = \begin{bmatrix} 41 & 66 \end{bmatrix}$

d) The entries of **CS** represent the total cost, in dollars, of ingredients for each coffee shop.

36. a) $\mathbf{M} = \begin{bmatrix} 900 & 500 \\ 450 & 1000 \\ 600 & 700 \end{bmatrix}$

b) $\mathbf{P} = \begin{bmatrix} 5 & 8 & 4 \end{bmatrix}$

c) $\mathbf{PM} = \begin{bmatrix} 10{,}500 & 13{,}300 \end{bmatrix}$

d) The entries of **PM** represent the total profit from each distributor.

37. a) $\mathbf{P} = \begin{bmatrix} 6 & 4.5 & 5.2 \end{bmatrix}$

b) $\mathbf{PS} = \begin{bmatrix} 6 & 4.5 & 5.2 \end{bmatrix} \begin{bmatrix} 8 & 15 \\ 6 & 10 \\ 4 & 3 \end{bmatrix}$

$= \begin{bmatrix} 95.8 & 150.6 \end{bmatrix}$

The profit from Mugsey's Coffee Shop is \$95.80, and the profit from The Coffee Club is \$150.60.

38. a) $\mathbf{C} = \begin{bmatrix} 20 & 25 & 15 \end{bmatrix}$

b) $\mathbf{CM} = \begin{bmatrix} 20 & 25 & 15 \end{bmatrix} \begin{bmatrix} 900 & 500 \\ 450 & 1000 \\ 600 & 700 \end{bmatrix}$

$= \begin{bmatrix} 38{,}250 & 45{,}500 \end{bmatrix}$

The total production costs for the products shipped to Distributors 1 and 2 are \$38,250 and \$45,500, respectively.

39. $2x - 3y = 7,$

$x + 5y = -6$

Write the coefficients on the left in a matrix. Then write the product of that matrix and the column matrix containing the variables, and set the result equal to the column matrix containing the constants on the right.

$$\begin{bmatrix} 2 & -3 \\ 1 & 5 \end{bmatrix} \begin{bmatrix} x \\ y \end{bmatrix} = \begin{bmatrix} 7 \\ -6 \end{bmatrix}$$

40. $\begin{bmatrix} -1 & 1 \\ 5 & -4 \end{bmatrix} \begin{bmatrix} x \\ y \end{bmatrix} = \begin{bmatrix} 3 \\ 16 \end{bmatrix}$

41. $x + y - 2z = 6,$

$3x - y + z = 7,$

$2x + 5y - 3z = 8$

Write the coefficients on the left in a matrix. Then write the product of that matrix and the column matrix containing the variables, and set the result equal to the column matrix containing the constants on the right.

$$\begin{bmatrix} 1 & 1 & -2 \\ 3 & -1 & 1 \\ 2 & 5 & -3 \end{bmatrix} \begin{bmatrix} x \\ y \\ z \end{bmatrix} = \begin{bmatrix} 6 \\ 7 \\ 8 \end{bmatrix}$$

42. $\begin{bmatrix} 3 & -1 & 1 \\ 1 & 2 & -1 \\ 4 & 3 & -2 \end{bmatrix} \begin{bmatrix} x \\ y \\ z \end{bmatrix} = \begin{bmatrix} 1 \\ 3 \\ 11 \end{bmatrix}$

43. $3x - 2y + 4z = 17,$

$2x + y - 5z = 13$

Write the coefficients on the left in a matrix. Then write the product of that matrix and the column matrix containing the variables, and set the result equal to the column matrix containing the constants on the right.

$$\begin{bmatrix} 3 & -2 & 4 \\ 2 & 1 & -5 \end{bmatrix} \begin{bmatrix} x \\ y \\ z \end{bmatrix} = \begin{bmatrix} 17 \\ 13 \end{bmatrix}$$

44. $$\begin{bmatrix} 3 & 2 & 5 \\ 4 & -3 & 2 \end{bmatrix} \begin{bmatrix} x \\ y \\ z \end{bmatrix} = \begin{bmatrix} 9 \\ 10 \end{bmatrix}$$

45. $-4w + x - y + 2z = 12,$

$w + 2x - y - z = 0,$

$-w + x + 4y - 3z = 1,$

$2w + 3x + 5y - 7z = 9$

Write the coefficients on the left in a matrix. Then write the product of that matrix and the column matrix containing the variables, and set the result equal to the column matrix containing the constants on the right.

$$\begin{bmatrix} -4 & 1 & -1 & 2 \\ 1 & 2 & -1 & -1 \\ -1 & 1 & 4 & -3 \\ 2 & 3 & 5 & -7 \end{bmatrix} \begin{bmatrix} w \\ x \\ y \\ z \end{bmatrix} = \begin{bmatrix} 12 \\ 0 \\ 1 \\ 9 \end{bmatrix}$$

46. $$\begin{bmatrix} 12 & 2 & 4 & -5 \\ -1 & 4 & -1 & 12 \\ 2 & -1 & 4 & 0 \\ 0 & 2 & 10 & 1 \end{bmatrix} \begin{bmatrix} w \\ x \\ y \\ z \end{bmatrix} = \begin{bmatrix} 2 \\ 5 \\ 13 \\ 5 \end{bmatrix}$$

47. Left to the student

48. Left to the student

49. No; see Exercise 17, for example.

50. She could make decisions regarding cost and profit, including how many of each product to prepare and distribute to each shop.

51. $f(x) = x^2 - x - 6$

a) $-\dfrac{b}{2a} = -\dfrac{-1}{2 \cdot 1} = \dfrac{1}{2}$

$f\left(\dfrac{1}{2}\right) = \left(\dfrac{1}{2}\right)^2 - \dfrac{1}{2} - 6 = -\dfrac{25}{4}$

The vertex is $\left(\dfrac{1}{2}, -\dfrac{25}{4}\right)$.

b) The axis of symmetry is $x = \dfrac{1}{2}$.

c) Since the coefficient of x^2 is positive, the function has a minimum value. It is the second coordinate of the vertex, $-\dfrac{25}{4}$.

d) Plot some points and draw the graph of the function.

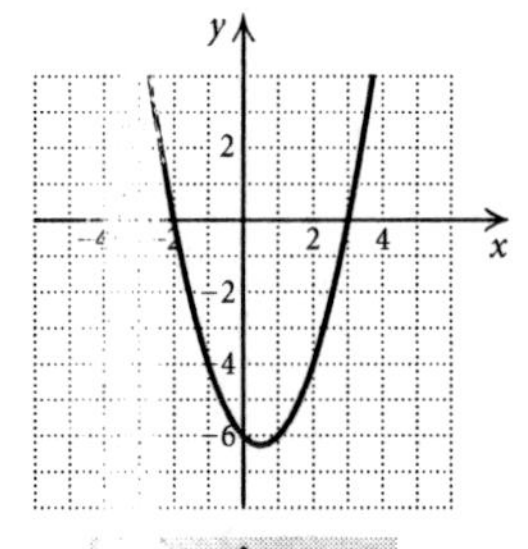

$f(x) = x^2 - x - 6$

52. $f(x) = 2x^2 - 5x - 3$

a) $-\dfrac{b}{2a} = -\dfrac{-5}{2 \cdot 2} = \dfrac{5}{4}$

$f\left(\dfrac{5}{4}\right) = 2\left(\dfrac{5}{4}\right)^2 - 5\left(\dfrac{5}{4}\right) - 3 = -\dfrac{49}{8}$

The vertex is $\left(\dfrac{5}{4}, -\dfrac{49}{8}\right)$.

b) $x = \dfrac{5}{4}$

c) Minimum: $-\dfrac{49}{8}$

d)

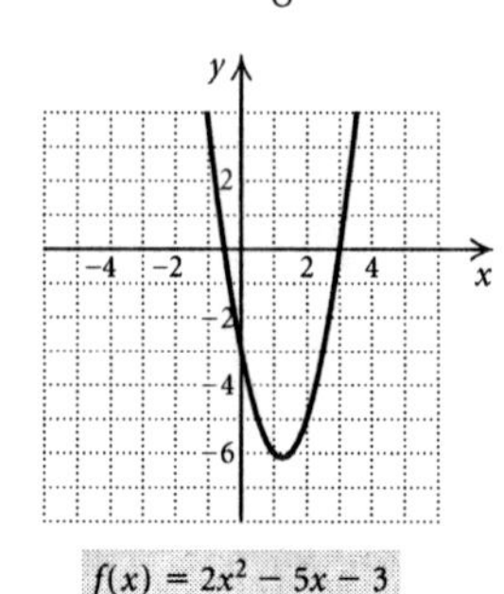

$f(x) = 2x^2 - 5x - 3$

53. $f(x) = -x^2 - 3x + 2$

a) $-\dfrac{b}{2a} = -\dfrac{-3}{2(-1)} = -\dfrac{3}{2}$

$f\left(-\dfrac{3}{2}\right) = -\left(-\dfrac{3}{2}\right)^2 - 3\left(-\dfrac{3}{2}\right) + 2 = \dfrac{17}{4}$

The vertex is $\left(-\dfrac{3}{2}, \dfrac{17}{4}\right)$.

b) The axis of symmetry is $x = -\dfrac{3}{2}$.

c) Since the coefficient of x^2 is negative, the function has a maximum value. It is the second coordinate of the vertex, $\dfrac{17}{4}$.

d) Plot some points and draw the graph of the function.

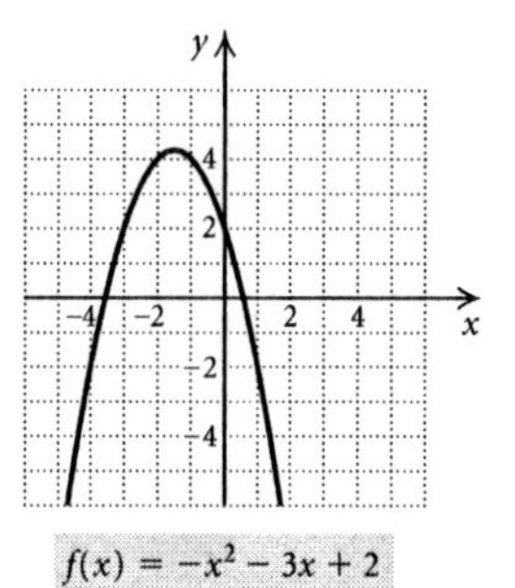

$f(x) = -x^2 - 3x + 2$

54. $f(x) = -3x^2 + 4x + 4$

a) $-\dfrac{b}{2a} = -\dfrac{4}{2(-3)} = \dfrac{2}{3}$

$f\left(\dfrac{2}{3}\right) = -3\left(\dfrac{2}{3}\right)^2 + 4\left(\dfrac{2}{3}\right) + 4 = \dfrac{16}{3}$

The vertex is $\left(\dfrac{2}{3}, \dfrac{16}{3}\right)$.

b) $x = \frac{2}{3}$

c) Maximum: $\frac{16}{3}$

d)

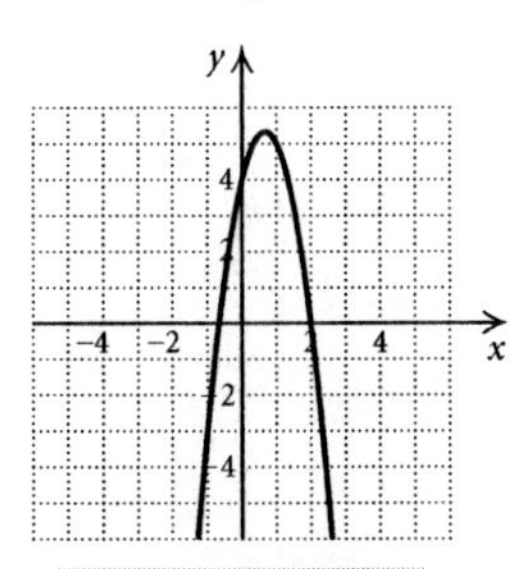

55. $\mathbf{A} = \begin{bmatrix} -1 & 0 \\ 2 & 1 \end{bmatrix}, \mathbf{B} = \begin{bmatrix} 1 & -1 \\ 0 & 2 \end{bmatrix}$

$(\mathbf{A}+\mathbf{B})(\mathbf{A}-\mathbf{B}) = \begin{bmatrix} 0 & -1 \\ 2 & 3 \end{bmatrix}\begin{bmatrix} -2 & 1 \\ 2 & -1 \end{bmatrix}$

$= \begin{bmatrix} -2 & 1 \\ 2 & -1 \end{bmatrix}$

$\mathbf{A}^2 - \mathbf{B}^2$

$= \begin{bmatrix} -1 & 0 \\ 2 & 1 \end{bmatrix}\begin{bmatrix} -1 & 0 \\ 2 & 1 \end{bmatrix} - \begin{bmatrix} 1 & -1 \\ 0 & 2 \end{bmatrix}\begin{bmatrix} 1 & -1 \\ 0 & 2 \end{bmatrix}$

$= \begin{bmatrix} 1 & 0 \\ 0 & 1 \end{bmatrix} - \begin{bmatrix} 1 & -3 \\ 0 & 4 \end{bmatrix}$

$= \begin{bmatrix} 0 & 3 \\ 0 & -3 \end{bmatrix}$

Thus $(\mathbf{A}+\mathbf{B})(\mathbf{A}-\mathbf{B}) \neq \mathbf{A}^2 - \mathbf{B}^2$.

56. $\mathbf{A} = \begin{bmatrix} -1 & 0 \\ 2 & 1 \end{bmatrix}, \mathbf{B} = \begin{bmatrix} 1 & -1 \\ 0 & 2 \end{bmatrix}$

$(\mathbf{A}+\mathbf{B})(\mathbf{A}+\mathbf{B}) = \begin{bmatrix} 0 & -1 \\ 2 & 3 \end{bmatrix}\begin{bmatrix} 0 & -1 \\ 2 & 3 \end{bmatrix}$

$= \begin{bmatrix} -2 & -3 \\ 6 & 7 \end{bmatrix}$

We found $\mathbf{A}^2$ and $\mathbf{B}^2$ in Exercise 55.

$2\mathbf{AB} = 2\begin{bmatrix} -1 & 0 \\ 2 & 1 \end{bmatrix}\begin{bmatrix} 1 & -1 \\ 0 & 2 \end{bmatrix} = \begin{bmatrix} -2 & 2 \\ 4 & 0 \end{bmatrix}$

$\mathbf{A}^2 + 2\mathbf{AB} + \mathbf{B}^2$

$= \begin{bmatrix} 1 & 0 \\ 0 & 1 \end{bmatrix} + \begin{bmatrix} -2 & 2 \\ 4 & 0 \end{bmatrix} + \begin{bmatrix} 1 & -3 \\ 0 & 4 \end{bmatrix}$

$= \begin{bmatrix} 0 & -1 \\ 4 & 5 \end{bmatrix}$

Thus $(\mathbf{A}+\mathbf{B})(\mathbf{A}+\mathbf{B}) \neq \mathbf{A}^2 + 2\mathbf{AB} + \mathbf{B}^2$.

57. In Exercise 55 we found that $(\mathbf{A}+\mathbf{B})(\mathbf{A}-\mathbf{B}) = \begin{bmatrix} -2 & 1 \\ 2 & -1 \end{bmatrix}$

and we also found $\mathbf{A}^2$ and $\mathbf{B}^2$.

$\mathbf{BA} = \begin{bmatrix} 1 & -1 \\ 0 & 2 \end{bmatrix}\begin{bmatrix} -1 & 0 \\ 2 & 1 \end{bmatrix} = \begin{bmatrix} -3 & -1 \\ 4 & 2 \end{bmatrix}$

$\mathbf{AB} = \begin{bmatrix} -1 & 0 \\ 2 & 1 \end{bmatrix}\begin{bmatrix} 1 & -1 \\ 0 & 2 \end{bmatrix} = \begin{bmatrix} -1 & 1 \\ 2 & 0 \end{bmatrix}$

$\mathbf{A}^2 + \mathbf{BA} - \mathbf{AB} - \mathbf{B}^2$

$= \begin{bmatrix} 1 & 0 \\ 0 & 1 \end{bmatrix} + \begin{bmatrix} -3 & -1 \\ 4 & 2 \end{bmatrix} - \begin{bmatrix} -1 & 1 \\ 2 & 0 \end{bmatrix} - \begin{bmatrix} 1 & -3 \\ 0 & 4 \end{bmatrix}$

$= \begin{bmatrix} -2 & 1 \\ 2 & -1 \end{bmatrix}$

Thus $(\mathbf{A}+\mathbf{B})(\mathbf{A}-\mathbf{B}) = \mathbf{A}^2 + \mathbf{BA} - \mathbf{AB} - \mathbf{B}^2$.

58. In Exercise 56 we found that

$(\mathbf{A}+\mathbf{B})(\mathbf{A}+\mathbf{B}) = \begin{bmatrix} -2 & -3 \\ 6 & 7 \end{bmatrix}$.

We found $\mathbf{A}^2$ and $\mathbf{B}^2$ in Exercise 55, and we found $\mathbf{BA}$ and $\mathbf{AB}$ in Exercise 55.

$\mathbf{A}^2 + \mathbf{BA} + \mathbf{AB} + \mathbf{B}^2$

$= \begin{bmatrix} 1 & 0 \\ 0 & 1 \end{bmatrix} + \begin{bmatrix} -3 & -1 \\ 4 & 2 \end{bmatrix} + \begin{bmatrix} -1 & 1 \\ 2 & 0 \end{bmatrix} + \begin{bmatrix} 1 & -3 \\ 0 & 4 \end{bmatrix}$

$= \begin{bmatrix} -2 & -3 \\ 6 & 7 \end{bmatrix}$

Thus $(\mathbf{A}+\mathbf{B})(\mathbf{A}+\mathbf{B}) = \mathbf{A}^2 + \mathbf{BA} + \mathbf{AB} + \mathbf{B}^2$.

59. See the answer section in the text.

60. $\mathbf{A} + (\mathbf{B} + \mathbf{C})$

$= \begin{bmatrix} a_{11}+(b_{11}+c_{11}) & \cdots & a_{1n}+(b_{1n}+c_{1n}) \\ a_{21}+(b_{21}+c_{21}) & \cdots & a_{2n}+(b_{2n}+c_{2n}) \\ \vdots & \vdots & \vdots \\ a_{m1}+(b_{m1}+c_{m1}) & \cdots & a_{mn}+(b_{mn}+c_{mn}) \end{bmatrix}$

$= \begin{bmatrix} (a_{11}+b_{11})+c_{11} & \cdots & (a_{1n}+b_{1n})+c_{1n} \\ (a_{21}+b_{21})+c_{21} & \cdots & (a_{2n}+b_{2n})+c_{2n} \\ \vdots & \vdots & \vdots \\ (a_{m1}+b_{m1})+c_{m1} & \cdots & (a_{mn}+b_{mn})+c_{mn} \end{bmatrix}$

$= (\mathbf{A} + \mathbf{B}) + \mathbf{C}$

61. See the answer section in the text.

62. $k(\mathbf{A}+\mathbf{B})$

$$=\begin{bmatrix} k(a_{11}+b_{11}) & \cdots & k(a_{1n}+b_{1n}) \\ k(a_{21}+b_{21}) & \cdots & k(a_{2n}+b_{2n}) \\ \vdots & & \vdots \\ k(a_{m1}+b_{m1}) & \cdots & k(a_{mn}+b_{mn}) \end{bmatrix}$$

$$=\begin{bmatrix} ka_{11}+kb_{11} & \cdots & ka_{1n}+kb_{1n} \\ ka_{21}+kb_{21} & \cdots & ka_{2n}+kb_{2n} \\ \vdots & & \vdots \\ ka_{m1}+kb_{m1} & \cdots & ka_{mn}+kb_{mn} \end{bmatrix}$$

$=k\mathbf{A}+k\mathbf{B}$

63. See the answer section in the text.

Exercise Set 8.5

1. $\mathbf{BA}=\begin{bmatrix} 7 & 3 \\ 2 & 1 \end{bmatrix}\begin{bmatrix} 1 & -3 \\ -2 & 7 \end{bmatrix}=\begin{bmatrix} 1 & 0 \\ 0 & 1 \end{bmatrix}$

$\mathbf{AB}=\begin{bmatrix} 1 & -3 \\ -2 & 7 \end{bmatrix}\begin{bmatrix} 7 & 3 \\ 2 & 1 \end{bmatrix}=\begin{bmatrix} 1 & 0 \\ 0 & 1 \end{bmatrix}$

Since $\mathbf{BA}=\mathbf{I}=\mathbf{AB}$, $\mathbf{B}$ is the inverse of $\mathbf{A}$.

2. $\mathbf{BA}=\mathbf{I}=\mathbf{AB}$, so $\mathbf{B}$ is the inverse of $\mathbf{A}$.

3. $\mathbf{BA}=\begin{bmatrix} 2 & 3 & 2 \\ 3 & 3 & 4 \\ 1 & 1 & 1 \end{bmatrix}\begin{bmatrix} -1 & -1 & 6 \\ 1 & 0 & -2 \\ 1 & 0 & -3 \end{bmatrix}=$

$\begin{bmatrix} 3 & -2 & 0 \\ 4 & -3 & 0 \\ 1 & -1 & 1 \end{bmatrix}$

Since $\mathbf{BA}\neq\mathbf{I}$, $\mathbf{B}$ is not the inverse of $\mathbf{A}$.

4. $\mathbf{BA}=\begin{bmatrix} 1 & 0 & -24 \\ 0 & 1 & 8 \\ 0 & 0 & 17 \end{bmatrix}\neq\mathbf{I}$, so $\mathbf{B}$ is not the inverse of $\mathbf{A}$.

5. $\mathbf{A}=\begin{bmatrix} 3 & 2 \\ 5 & 3 \end{bmatrix}$

Write the augmented matrix.

$\left[\begin{array}{cc|cc} 3 & 2 & 1 & 0 \\ 5 & 3 & 0 & 1 \end{array}\right]$

Multiply row 2 by 3.

$\left[\begin{array}{cc|cc} 3 & 2 & 1 & 0 \\ 15 & 9 & 0 & 3 \end{array}\right]$

Multiply row 1 by -5 and add it to row 2.

$\left[\begin{array}{cc|cc} 3 & 2 & 1 & 0 \\ 0 & -1 & -5 & 3 \end{array}\right]$

Multiply row 2 by 2 and add it to row 1.

$\left[\begin{array}{cc|cc} 3 & 0 & -9 & 6 \\ 0 & -1 & -5 & 3 \end{array}\right]$

Multiply row 1 by $\dfrac{1}{3}$ and row 2 by -1.

$\left[\begin{array}{cc|cc} 1 & 0 & -3 & 2 \\ 0 & 1 & 5 & -3 \end{array}\right]$

Then $\mathbf{A}^{-1}=\begin{bmatrix} -3 & 2 \\ 5 & -3 \end{bmatrix}$.

6. $\mathbf{A}=\begin{bmatrix} 3 & 5 \\ 1 & 2 \end{bmatrix}$

Write the augmented matrix.

$\left[\begin{array}{cc|cc} 3 & 5 & 1 & 0 \\ 1 & 2 & 0 & 1 \end{array}\right]$

Interchange the rows.

$\left[\begin{array}{cc|cc} 1 & 2 & 0 & 1 \\ 3 & 5 & 1 & 0 \end{array}\right]$

Multiply row 1 by -3 and add it to row 2.

$\left[\begin{array}{cc|cc} 1 & 2 & 0 & 1 \\ 0 & -1 & 1 & -3 \end{array}\right]$

Multiply row 2 by 2 and add it to row 1.

$\left[\begin{array}{cc|cc} 1 & 0 & 2 & -5 \\ 0 & -1 & 1 & -3 \end{array}\right]$

Multiply row 2 by -1.

$\left[\begin{array}{cc|cc} 1 & 0 & 2 & -5 \\ 0 & 1 & -1 & 3 \end{array}\right]$

Then $\mathbf{A}^{-1}=\begin{bmatrix} 2 & -5 \\ -1 & 3 \end{bmatrix}$.

7. $\mathbf{A}=\begin{bmatrix} 6 & 9 \\ 4 & 6 \end{bmatrix}$

Write the augmented matrix.

$\left[\begin{array}{cc|cc} 6 & 9 & 1 & 0 \\ 4 & 6 & 0 & 1 \end{array}\right]$

Multiply row 2 by 3.

$\left[\begin{array}{cc|cc} 6 & 9 & 1 & 0 \\ 12 & 18 & 0 & 3 \end{array}\right]$

Multiply row 1 by -2 and add it to row 2.

$\left[\begin{array}{cc|cc} 6 & 9 & 1 & 0 \\ 0 & 0 & -2 & 3 \end{array}\right]$

We cannot obtain the identity matrix on the left since the second row contains only zeros to the left of the vertical line. Thus, $\mathbf{A}^{-1}$ does not exist.

8. $\mathbf{A}=\begin{bmatrix} -4 & -6 \\ 2 & 3 \end{bmatrix}$

Write the augmented matrix.

$\left[\begin{array}{cc|cc} -4 & -6 & 1 & 0 \\ 2 & 3 & 0 & 1 \end{array}\right]$

Multiply row 2 by 2.

$\left[\begin{array}{cc|cc} -4 & -6 & 1 & 0 \\ 4 & 6 & 0 & 2 \end{array}\right]$

Add row 1 to row 2.

$\left[\begin{array}{cc|cc} -4 & -6 & 1 & 0 \\ 0 & 0 & 1 & 2 \end{array}\right]$

We cannot obtain the identity matrix on the left since the second row contains only zeros to the left of the vertical line. Thus, $\mathbf{A}^{-1}$ does not exist.

9. $\mathbf{A}=\begin{bmatrix} 4 & -3 \\ 1 & -2 \end{bmatrix}$

Write the augmented matrix.

$\left[\begin{array}{cc|cc} 4 & -3 & 1 & 0 \\ 1 & -2 & 0 & 1 \end{array}\right]$

Interchange the rows.

$$\left[\begin{array}{rr|rr} 1 & -2 & 0 & 1 \\ 4 & -3 & 1 & 0 \end{array}\right]$$

Multiply row 1 by -4 and add it to row 2.

$$\left[\begin{array}{rr|rr} 1 & -2 & 0 & 1 \\ 0 & 5 & 1 & -4 \end{array}\right]$$

Multiply row 2 by $\frac{1}{5}$.

$$\left[\begin{array}{rr|rr} 1 & -2 & 0 & 1 \\ 0 & 1 & \frac{1}{5} & -\frac{4}{5} \end{array}\right]$$

Multiply row 2 by 2 and add it to row 1.

$$\left[\begin{array}{rr|rr} 1 & 0 & \frac{2}{5} & -\frac{3}{5} \\ 0 & 1 & \frac{1}{5} & -\frac{4}{5} \end{array}\right]$$

Then $\mathbf{A}^{-1} = \begin{bmatrix} \frac{2}{5} & -\frac{3}{5} \\ \frac{1}{5} & -\frac{4}{5} \end{bmatrix}$.

10. $\mathbf{A} = \begin{bmatrix} 0 & -1 \\ 1 & 0 \end{bmatrix}$

Write the augmented matrix.

$$\left[\begin{array}{rr|rr} 0 & -1 & 1 & 0 \\ 1 & 0 & 0 & 1 \end{array}\right]$$

Interchange the rows.

$$\left[\begin{array}{rr|rr} 1 & 0 & 0 & 1 \\ 0 & -1 & 1 & 0 \end{array}\right]$$

Multiply row 2 by -1.

$$\left[\begin{array}{rr|rr} 1 & 0 & 0 & 1 \\ 0 & 1 & -1 & 0 \end{array}\right]$$

Then $\mathbf{A}^{-1} = \begin{bmatrix} 0 & 1 \\ -1 & 0 \end{bmatrix}$.

11. $\mathbf{A} = \begin{bmatrix} 3 & 1 & 0 \\ 1 & 1 & 1 \\ 1 & -1 & 2 \end{bmatrix}$

Write the augmented matrix.

$$\left[\begin{array}{rrr|rrr} 3 & 1 & 0 & 1 & 0 & 0 \\ 1 & 1 & 1 & 0 & 1 & 0 \\ 1 & -1 & 2 & 0 & 0 & 1 \end{array}\right]$$

Interchange the first two rows.

$$\left[\begin{array}{rrr|rrr} 1 & 1 & 1 & 0 & 1 & 0 \\ 3 & 1 & 0 & 1 & 0 & 0 \\ 1 & -1 & 2 & 0 & 0 & 1 \end{array}\right]$$

Multiply row 1 by -3 and add it to row 2. Also, multiply row 1 by -1 and add it to row 3.

$$\left[\begin{array}{rrr|rrr} 1 & 1 & 1 & 0 & 1 & 0 \\ 0 & -2 & -3 & 1 & -3 & 0 \\ 0 & -2 & 1 & 0 & -1 & 1 \end{array}\right]$$

Multiply row 2 by $-\frac{1}{2}$.

$$\left[\begin{array}{rrr|rrr} 1 & 1 & 1 & 0 & 1 & 0 \\ 0 & 1 & \frac{3}{2} & -\frac{1}{2} & \frac{3}{2} & 0 \\ 0 & -2 & 1 & 0 & -1 & 1 \end{array}\right]$$

Multiply row 2 by -1 and add it to row 1. Also, multiply row 2 by 2 and add it to row 3.

$$\left[\begin{array}{rrr|rrr} 1 & 0 & -\frac{1}{2} & \frac{1}{2} & -\frac{1}{2} & 0 \\ 0 & 1 & \frac{3}{2} & -\frac{1}{2} & \frac{3}{2} & 0 \\ 0 & 0 & 4 & -1 & 2 & 1 \end{array}\right]$$

Multiply row 3 by $\frac{1}{4}$.

$$\left[\begin{array}{rrr|rrr} 1 & 0 & -\frac{1}{2} & \frac{1}{2} & -\frac{1}{2} & 0 \\ 0 & 1 & \frac{3}{2} & -\frac{1}{2} & \frac{3}{2} & 0 \\ 0 & 0 & 1 & -\frac{1}{4} & \frac{1}{2} & \frac{1}{4} \end{array}\right]$$

Multiply row 3 by $\frac{1}{2}$ and add it to row 1. Also, multiply row 3 by $-\frac{3}{2}$ and add it to row 2.

$$\left[\begin{array}{rrr|rrr} 1 & 0 & 0 & \frac{3}{8} & -\frac{1}{4} & \frac{1}{8} \\ 0 & 1 & 0 & -\frac{1}{8} & \frac{3}{4} & -\frac{3}{8} \\ 0 & 0 & 1 & -\frac{1}{4} & \frac{1}{2} & \frac{1}{4} \end{array}\right]$$

Then $\mathbf{A}^{-1} = \begin{bmatrix} \frac{3}{8} & -\frac{1}{4} & \frac{1}{8} \\ -\frac{1}{8} & \frac{3}{4} & -\frac{3}{8} \\ -\frac{1}{4} & \frac{1}{2} & \frac{1}{4} \end{bmatrix}$.

12. $\mathbf{A} = \begin{bmatrix} 1 & 0 & 1 \\ 2 & 1 & 0 \\ 1 & -1 & 1 \end{bmatrix}$

Write the augmented matrix.

$$\left[\begin{array}{rrr|rrr} 1 & 0 & 1 & 1 & 0 & 0 \\ 2 & 1 & 0 & 0 & 1 & 0 \\ 1 & -1 & 1 & 0 & 0 & 1 \end{array}\right]$$

Multiply row 1 by -2 and add it to row 2. Also, multiply row 1 by -1 and add it to row 3.

$$\left[\begin{array}{rrr|rrr} 1 & 0 & 1 & 1 & 0 & 0 \\ 0 & 1 & -2 & -2 & 1 & 0 \\ 0 & -1 & 0 & -1 & 0 & 1 \end{array}\right]$$

Add row 2 to row 3.

$$\left[\begin{array}{rrr|rrr} 1 & 0 & 1 & 1 & 0 & 0 \\ 0 & 1 & -2 & -2 & 1 & 0 \\ 0 & 0 & -2 & -3 & 1 & 1 \end{array}\right]$$

Multiply row 3 by $-\frac{1}{2}$.

$$\left[\begin{array}{rrr|rrr} 1 & 0 & 1 & 1 & 0 & 0 \\ 0 & 1 & -2 & -2 & 1 & 0 \\ 0 & 0 & 1 & \frac{3}{2} & -\frac{1}{2} & -\frac{1}{2} \end{array}\right]$$

Multiply row 3 by -1 and add it to row 1. Also, multiply row 3 by 2 and add it to row 2.

$$\left[\begin{array}{ccc|ccc} 1 & 0 & 0 & -\frac{1}{2} & \frac{1}{2} & \frac{1}{2} \\ 0 & 1 & 0 & 1 & 0 & -1 \\ 0 & 0 & 1 & \frac{3}{2} & -\frac{1}{2} & -\frac{1}{2} \end{array}\right]$$

Then $\mathbf{A}^{-1} = \begin{bmatrix} -\frac{1}{2} & \frac{1}{2} & \frac{1}{2} \\ 1 & 0 & -1 \\ \frac{3}{2} & -\frac{1}{2} & -\frac{1}{2} \end{bmatrix}$.

13. $\mathbf{A} = \begin{bmatrix} 1 & -4 & 8 \\ 1 & -3 & 2 \\ 2 & -7 & 10 \end{bmatrix}$

Write the augmented matrix.

$$\left[\begin{array}{ccc|ccc} 1 & -4 & 8 & 1 & 0 & 0 \\ 1 & -3 & 2 & 0 & 1 & 0 \\ 2 & -7 & 10 & 0 & 0 & 1 \end{array}\right]$$

Multiply row 1 by -1 and add it to row 2. Also, multiply row 1 by -2 and add it to row 3.

$$\left[\begin{array}{ccc|ccc} 1 & -4 & 8 & 1 & 0 & 0 \\ 0 & 1 & -6 & -1 & 1 & 0 \\ 0 & 1 & -6 & -2 & 0 & 1 \end{array}\right]$$

Since the second and third rows are identical left of the vertical line, it will not be possible to obtain the identity matrix on the left side. Thus, $\mathbf{A}^{-1}$ does not exist.

14. $\mathbf{A} = \begin{bmatrix} -2 & 5 & 3 \\ 4 & -1 & 3 \\ 7 & -2 & 5 \end{bmatrix}$

Write the augmented matrix.

$$\left[\begin{array}{ccc|ccc} -2 & 5 & 3 & 1 & 0 & 0 \\ 4 & -1 & 3 & 0 & 1 & 0 \\ 7 & -2 & 5 & 0 & 0 & 1 \end{array}\right]$$

Multiply row 3 by 2.

$$\left[\begin{array}{ccc|ccc} -2 & 5 & 3 & 1 & 0 & 0 \\ 4 & -1 & 3 & 0 & 1 & 0 \\ 14 & -4 & 10 & 0 & 0 & 2 \end{array}\right]$$

Multiply row 1 by 2 and add it to row 2. Also, multiply row 1 by 7 and add it to row 3.

$$\left[\begin{array}{ccc|ccc} -2 & 5 & 3 & 1 & 0 & 0 \\ 0 & 9 & 9 & 2 & 1 & 0 \\ 0 & 31 & 31 & 7 & 0 & 2 \end{array}\right]$$

To the left of the vertical line, row 3 is a multiple of row 2 so it will not be possible to obtain the identity matrix on the left. Thus, $\mathbf{A}^{-1}$ does not exist.

15. $\mathbf{A} = \begin{bmatrix} 2 & 3 & 2 \\ 3 & 3 & 4 \\ -1 & -1 & -1 \end{bmatrix}$

Write the augmented matrix.

$$\left[\begin{array}{ccc|ccc} 2 & 3 & 2 & 1 & 0 & 0 \\ 3 & 3 & 4 & 0 & 1 & 0 \\ -1 & -1 & -1 & 0 & 0 & 1 \end{array}\right]$$

Interchange rows 1 and 3.

$$\left[\begin{array}{ccc|ccc} -1 & -1 & -1 & 0 & 0 & 1 \\ 3 & 3 & 4 & 0 & 1 & 0 \\ 2 & 3 & 2 & 1 & 0 & 0 \end{array}\right]$$

Multiply row 1 by 3 and add it to row 2. Also, multiply row 1 by 2 and add it to row 3.

$$\left[\begin{array}{ccc|ccc} -1 & -1 & -1 & 0 & 0 & 1 \\ 0 & 0 & 1 & 0 & 1 & 3 \\ 0 & 1 & 0 & 1 & 0 & 2 \end{array}\right]$$

Multiply row 1 by -1.

$$\left[\begin{array}{ccc|ccc} 1 & 1 & 1 & 0 & 0 & -1 \\ 0 & 0 & 1 & 0 & 1 & 3 \\ 0 & 1 & 0 & 1 & 0 & 2 \end{array}\right]$$

Interchange rows 2 and 3.

$$\left[\begin{array}{ccc|ccc} 1 & 1 & 1 & 0 & 0 & -1 \\ 0 & 1 & 0 & 1 & 0 & 2 \\ 0 & 0 & 1 & 0 & 1 & 3 \end{array}\right]$$

Multiply row 2 by -1 and add it to row 1.

$$\left[\begin{array}{ccc|ccc} 1 & 0 & 1 & -1 & 0 & -3 \\ 0 & 1 & 0 & 1 & 0 & 2 \\ 0 & 0 & 1 & 0 & 1 & 3 \end{array}\right]$$

Multiply row 3 by -1 and add it to row 1.

$$\left[\begin{array}{ccc|ccc} 1 & 0 & 0 & -1 & -1 & -6 \\ 0 & 1 & 0 & 1 & 0 & 2 \\ 0 & 0 & 1 & 0 & 1 & 3 \end{array}\right]$$

Then $\mathbf{A}^{-1} = \begin{bmatrix} -1 & -1 & -6 \\ 1 & 0 & 2 \\ 0 & 1 & 3 \end{bmatrix}$.

16. $\mathbf{A} = \begin{bmatrix} 1 & 2 & 3 \\ 2 & -1 & -2 \\ -1 & 3 & 3 \end{bmatrix}$

Write the augmented matrix.

$$\left[\begin{array}{ccc|ccc} 1 & 2 & 3 & 1 & 0 & 0 \\ 2 & -1 & -2 & 0 & 1 & 0 \\ -1 & 3 & 3 & 0 & 0 & 1 \end{array}\right]$$

Multiply row 1 by -2 and add it to row 2. Also, add row 1 to row 3.

$$\left[\begin{array}{ccc|ccc} 1 & 2 & 3 & 1 & 0 & 0 \\ 0 & -5 & -8 & -2 & 1 & 0 \\ 0 & 5 & 6 & 1 & 0 & 1 \end{array}\right]$$

Multiply row 2 by $-\frac{1}{5}$.

$$\left[\begin{array}{ccc|ccc} 1 & 2 & 3 & 1 & 0 & 0 \\ 0 & 1 & \frac{8}{5} & \frac{2}{5} & -\frac{1}{5} & 0 \\ 0 & 5 & 6 & 1 & 0 & 1 \end{array}\right]$$

Multiply row 2 by -2 and add it to row 1. Also, multiply row 2 by -5 and add it to row 3.

$$\left[\begin{array}{ccc|ccc} 1 & 0 & -\frac{1}{5} & \frac{1}{5} & \frac{2}{5} & 0 \\ 0 & 1 & \frac{8}{5} & \frac{2}{5} & -\frac{1}{5} & 0 \\ 0 & 0 & -2 & -1 & 1 & 1 \end{array}\right]$$

Multiply row 3 by $-\frac{1}{2}$.

$$\left[\begin{array}{ccc|ccc} 1 & 0 & -\frac{1}{5} & \frac{1}{5} & \frac{2}{5} & 0 \\ 0 & 1 & \frac{8}{5} & \frac{2}{5} & -\frac{1}{5} & 0 \\ 0 & 0 & 1 & \frac{1}{2} & -\frac{1}{2} & -\frac{1}{2} \end{array}\right]$$

Multiply row 3 by $\frac{1}{5}$ and add it to row 1. Also, multiply row 3 by $-\frac{8}{5}$ and add it to row 2.

$$\left[\begin{array}{ccc|ccc} 1 & 0 & 0 & \frac{3}{10} & \frac{3}{10} & -\frac{1}{10} \\ 0 & 1 & 0 & -\frac{2}{5} & \frac{3}{5} & \frac{4}{5} \\ 0 & 0 & 1 & \frac{1}{2} & -\frac{1}{2} & -\frac{1}{2} \end{array}\right]$$

Then $\mathbf{A}^{-1} = \begin{bmatrix} \frac{3}{10} & \frac{3}{10} & -\frac{1}{10} \\ -\frac{2}{5} & \frac{3}{5} & \frac{4}{5} \\ \frac{1}{2} & -\frac{1}{2} & -\frac{1}{2} \end{bmatrix}$, or

$$\begin{bmatrix} 0.3 & 0.3 & -0.1 \\ -0.4 & 0.6 & 0.8 \\ 0.5 & -0.5 & -0.5 \end{bmatrix}.$$

17. $\mathbf{A} = \begin{bmatrix} 1 & 2 & -1 \\ -2 & 0 & 1 \\ 1 & -1 & 0 \end{bmatrix}$

Write the augmented matrix.

$$\left[\begin{array}{ccc|ccc} 1 & 2 & -1 & 1 & 0 & 0 \\ -2 & 0 & 1 & 0 & 1 & 0 \\ 1 & -1 & 0 & 0 & 0 & 1 \end{array}\right]$$

Multiply row 1 by 2 and add it to row 2. Also, multiply row 1 by -1 and add it to row 3.

$$\left[\begin{array}{ccc|ccc} 1 & 2 & -1 & 1 & 0 & 0 \\ 0 & 4 & -1 & 2 & 1 & 0 \\ 0 & -3 & 1 & -1 & 0 & 1 \end{array}\right]$$

Add row 3 to row 1 and also to row 2.

$$\left[\begin{array}{ccc|ccc} 1 & -1 & 0 & 0 & 0 & 1 \\ 0 & 1 & 0 & 1 & 1 & 1 \\ 0 & -3 & 1 & -1 & 0 & 1 \end{array}\right]$$

Add row 2 to row 1. Also, multiply row 2 by 3 and add it to row 3.

$$\left[\begin{array}{ccc|ccc} 1 & 0 & 0 & 1 & 1 & 2 \\ 0 & 1 & 0 & 1 & 1 & 1 \\ 0 & 0 & 1 & 2 & 3 & 4 \end{array}\right]$$

Then $\mathbf{A}^{-1} = \begin{bmatrix} 1 & 1 & 2 \\ 1 & 1 & 1 \\ 2 & 3 & 4 \end{bmatrix}$.

18. $\mathbf{A} = \begin{bmatrix} 7 & -1 & -9 \\ 2 & 0 & -4 \\ -4 & 0 & 6 \end{bmatrix}$

Write the augmented matrix.

$$\left[\begin{array}{ccc|ccc} 7 & -1 & -9 & 1 & 0 & 0 \\ 2 & 0 & -4 & 0 & 1 & 0 \\ -4 & 0 & 6 & 0 & 0 & 1 \end{array}\right]$$

Interchange row 1 and row 2.

$$\left[\begin{array}{ccc|ccc} 2 & 0 & -4 & 0 & 1 & 0 \\ 7 & -1 & -9 & 1 & 0 & 0 \\ -4 & 0 & 6 & 0 & 0 & 1 \end{array}\right]$$

Multiply row 2 by 2.

$$\left[\begin{array}{ccc|ccc} 2 & 0 & -4 & 0 & 1 & 0 \\ 14 & -2 & -18 & 2 & 0 & 0 \\ -4 & 0 & 6 & 0 & 0 & 1 \end{array}\right]$$

Multiply row 1 by -7 and add it to row 2. Also, multiply row 1 by 2 and add it to row 3.

$$\left[\begin{array}{ccc|ccc} 2 & 0 & -4 & 0 & 1 & 0 \\ 0 & -2 & 10 & 2 & -7 & 0 \\ 0 & 0 & -2 & 0 & 2 & 1 \end{array}\right]$$

Multiply row 3 by -2 and add it to row 1. Also, multiply row 3 by 5 and add it to row 2.

$$\left[\begin{array}{ccc|ccc} 2 & 0 & 0 & 0 & -3 & -2 \\ 0 & -2 & 0 & 2 & 3 & 5 \\ 0 & 0 & -2 & 0 & 2 & 1 \end{array}\right]$$

Multiply row 1 by $\frac{1}{2}$. Also, multiply rows 2 and 3 by $-\frac{1}{2}$.

$$\left[\begin{array}{ccc|ccc} 1 & 0 & 0 & 0 & -\frac{3}{2} & -1 \\ 0 & 1 & 0 & -1 & -\frac{3}{2} & -\frac{5}{2} \\ 0 & 0 & 1 & 0 & -1 & -\frac{1}{2} \end{array}\right]$$

Then $\mathbf{A}^{-1} = \begin{bmatrix} 0 & -\frac{3}{2} & -1 \\ -1 & -\frac{3}{2} & -\frac{5}{2} \\ 0 & -1 & -\frac{1}{2} \end{bmatrix}$, or

$$\begin{bmatrix} 0 & -1.5 & -1 \\ -1 & -1.5 & -2.5 \\ 0 & -1 & -0.5 \end{bmatrix}.$$

19. $\mathbf{A} = \begin{bmatrix} 1 & 3 & -1 \\ 0 & 2 & -1 \\ 1 & 1 & 0 \end{bmatrix}$

Write the augmented matrix.

$$\left[\begin{array}{ccc|ccc} 1 & 3 & -1 & 1 & 0 & 0 \\ 0 & 2 & -1 & 0 & 1 & 0 \\ 1 & 1 & 0 & 0 & 0 & 1 \end{array}\right]$$

Multiply row 1 by -1 and add it to row 3.

$$\left[\begin{array}{ccc|ccc} 1 & 3 & -1 & 1 & 0 & 0 \\ 0 & 2 & -1 & 0 & 1 & 0 \\ 0 & -2 & 1 & -1 & 0 & 1 \end{array}\right]$$

Add row 3 to row 1 and also to row 2.

$$\left[\begin{array}{ccc|ccc} 1 & 1 & 0 & 0 & 0 & 0 \\ 0 & 0 & 0 & -1 & 1 & 1 \\ 0 & -2 & 1 & -1 & 0 & 1 \end{array}\right]$$

Since the second row consists only of zeros to the left of the vertical line, it will not be possible to obtain the identity matrix on the left side. Thus, $\mathbf{A}^{-1}$ does not exist.

20. $\mathbf{A} = \begin{bmatrix} -1 & 0 & -1 \\ -1 & 1 & 0 \\ 0 & 1 & 1 \end{bmatrix}$

Write the augmented matrix.

$$\left[\begin{array}{ccc|ccc} -1 & 0 & -1 & 1 & 0 & 0 \\ -1 & 1 & 0 & 0 & 1 & 0 \\ 0 & 1 & 1 & 0 & 0 & 1 \end{array}\right]$$

Multiply row 1 by -1.

$$\left[\begin{array}{ccc|ccc} 1 & 0 & 1 & -1 & 0 & 0 \\ -1 & 1 & 0 & 0 & 1 & 0 \\ 0 & 1 & 1 & 0 & 0 & 1 \end{array}\right]$$

Add row 1 to row 2.

$$\left[\begin{array}{ccc|ccc} 1 & 0 & 1 & -1 & 0 & 0 \\ 0 & 1 & 1 & -1 & 1 & 0 \\ 0 & 1 & 1 & 0 & 0 & 1 \end{array}\right]$$

Since the second and third rows are identical to the left of the vertical line, it will not be possible to obtain the identity matrix on the left side. Thus, $\mathbf{A}^{-1}$ does not exist.

21. $\mathbf{A} = \begin{bmatrix} 1 & 2 & 3 & 4 \\ 0 & 1 & 3 & -5 \\ 0 & 0 & 1 & -2 \\ 0 & 0 & 0 & -1 \end{bmatrix}$

Write the augmented matrix.

$$\left[\begin{array}{cccc|cccc} 1 & 2 & 3 & 4 & 1 & 0 & 0 & 0 \\ 0 & 1 & 3 & -5 & 0 & 1 & 0 & 0 \\ 0 & 0 & 1 & -2 & 0 & 0 & 1 & 0 \\ 0 & 0 & 0 & -1 & 0 & 0 & 0 & 1 \end{array}\right]$$

Multiply row 4 by -1.

$$\left[\begin{array}{cccc|cccc} 1 & 2 & 3 & 4 & 1 & 0 & 0 & 0 \\ 0 & 1 & 3 & -5 & 0 & 1 & 0 & 0 \\ 0 & 0 & 1 & -2 & 0 & 0 & 1 & 0 \\ 0 & 0 & 0 & 1 & 0 & 0 & 0 & -1 \end{array}\right]$$

Multiply row 4 by -4 and add it to row 1. Multiply row 4 by 5 and add it to row 2. Also, multiply row 4 by 2 and add it to row 3.

$$\left[\begin{array}{cccc|cccc} 1 & 2 & 3 & 0 & 1 & 0 & 0 & 4 \\ 0 & 1 & 3 & 0 & 0 & 1 & 0 & -5 \\ 0 & 0 & 1 & 0 & 0 & 0 & 1 & -2 \\ 0 & 0 & 0 & 1 & 0 & 0 & 0 & -1 \end{array}\right]$$

Multiply row 3 by -3 and add it to row 1 and to row 2.

$$\left[\begin{array}{cccc|cccc} 1 & 2 & 0 & 0 & 1 & 0 & -3 & 10 \\ 0 & 1 & 0 & 0 & 0 & 1 & -3 & 1 \\ 0 & 0 & 1 & 0 & 0 & 0 & 1 & -2 \\ 0 & 0 & 0 & 1 & 0 & 0 & 0 & -1 \end{array}\right]$$

Multiply row 2 by -2 and add it to row 1.

$$\left[\begin{array}{cccc|cccc} 1 & 0 & 0 & 0 & 1 & -2 & 3 & 8 \\ 0 & 1 & 0 & 0 & 0 & 1 & -3 & 1 \\ 0 & 0 & 1 & 0 & 0 & 0 & 1 & -2 \\ 0 & 0 & 0 & 1 & 0 & 0 & 0 & -1 \end{array}\right]$$

Then $\mathbf{A}^{-1} = \begin{bmatrix} 1 & -2 & 3 & 8 \\ 0 & 1 & -3 & 1 \\ 0 & 0 & 1 & -2 \\ 0 & 0 & 0 & -1 \end{bmatrix}$.

22. $\mathbf{A} = \begin{bmatrix} -2 & -3 & 4 & 1 \\ 0 & 1 & 1 & 0 \\ 0 & 4 & -6 & 1 \\ -2 & -2 & 5 & 1 \end{bmatrix}$

Write the augmented matrix.

$$\left[\begin{array}{cccc|cccc} -2 & -3 & 4 & 1 & 1 & 0 & 0 & 0 \\ 0 & 1 & 1 & 0 & 0 & 1 & 0 & 0 \\ 0 & 4 & -6 & 1 & 0 & 0 & 1 & 0 \\ -2 & -2 & 5 & 1 & 0 & 0 & 0 & 1 \end{array}\right]$$

Multiply row 1 by -1 and add it to row 4.

$$\left[\begin{array}{cccc|cccc} -2 & -3 & 4 & 1 & 1 & 0 & 0 & 0 \\ 0 & 1 & 1 & 0 & 0 & 1 & 0 & 0 \\ 0 & 4 & -6 & 1 & 0 & 0 & 1 & 0 \\ 0 & 1 & 1 & 0 & -1 & 0 & 0 & 1 \end{array}\right]$$

Since the second and fourth rows are identical to the left of the vertical line, it will not be possible to get the identity matrix on the left side. Thus, $\mathbf{A}^{-1}$ does not exist.

23. $\mathbf{A} = \begin{bmatrix} 1 & -14 & 7 & 38 \\ -1 & 2 & 1 & -2 \\ 1 & 2 & -1 & -6 \\ 1 & -2 & 3 & 6 \end{bmatrix}$

Write the augmented matrix.

$$\left[\begin{array}{cccc|cccc} 1 & -14 & 7 & 38 & 1 & 0 & 0 & 0 \\ -1 & 2 & 1 & -2 & 0 & 1 & 0 & 0 \\ 1 & 2 & -1 & -6 & 0 & 0 & 1 & 0 \\ 1 & -2 & 3 & 6 & 0 & 0 & 0 & 1 \end{array}\right]$$

Add row 1 to row 2. Also, multiply row 1 by -1 and add it to row 3 and to row 4.

$$\left[\begin{array}{cccc|cccc} 1 & -14 & 7 & 38 & 1 & 0 & 0 & 0 \\ 0 & -12 & 8 & 36 & 1 & 1 & 0 & 0 \\ 0 & 16 & -8 & -44 & -1 & 0 & 1 & 0 \\ 0 & 12 & -4 & -32 & -1 & 0 & 0 & 1 \end{array}\right]$$

Add row 2 to row 4.

$$\left[\begin{array}{cccc|cccc} 1 & -14 & 7 & 38 & 1 & 0 & 0 & 0 \\ 0 & -12 & 8 & 36 & 1 & 1 & 0 & 0 \\ 0 & 16 & -8 & -44 & -1 & 0 & 1 & 0 \\ 0 & 0 & 4 & 4 & 0 & 1 & 0 & 1 \end{array}\right]$$

Multiply row 4 by $\frac{1}{4}$.

$$\left[\begin{array}{cccc|cccc} 1 & -14 & 7 & 38 & 1 & 0 & 0 & 0 \\ 0 & -12 & 8 & 36 & 1 & 1 & 0 & 0 \\ 0 & 16 & -8 & -44 & -1 & 0 & 1 & 0 \\ 0 & 0 & 1 & 1 & 0 & \frac{1}{4} & 0 & \frac{1}{4} \end{array}\right]$$

Multiply row 4 by -38 and add it to row 1. Multiply row 4 by -36 and add it to row 2. Also, multiply row 4 by 44 and add it to row 3.

$$\left[\begin{array}{cccc|cccc} 1 & -14 & -31 & 0 & 1 & -\frac{19}{2} & 0 & -\frac{19}{2} \\ 0 & -12 & -28 & 0 & 1 & -8 & 0 & -9 \\ 0 & 16 & 36 & 0 & -1 & 11 & 1 & 11 \\ 0 & 0 & 1 & 1 & 0 & \frac{1}{4} & 0 & \frac{1}{4} \end{array}\right]$$

Multiply row 3 by $\frac{1}{36}$.

$$\left[\begin{array}{cccc|cccc} 1 & -14 & -31 & 0 & 1 & -\frac{19}{2} & 0 & -\frac{19}{2} \\ 0 & -12 & -28 & 0 & 1 & -8 & 0 & -9 \\ 0 & \frac{4}{9} & 1 & 0 & -\frac{1}{36} & \frac{11}{36} & \frac{1}{36} & \frac{11}{36} \\ 0 & 0 & 1 & 1 & 0 & \frac{1}{4} & 0 & \frac{1}{4} \end{array}\right]$$

Multiply row 3 by 31 and add it to row 1. Multiply row 3 by 28 and add it to row 2. Also, multiply row 3 by -1 and add it to row 4.

$$\left[\begin{array}{cccc|cccc} 1 & -\frac{2}{9} & 0 & 0 & \frac{5}{36} & -\frac{1}{36} & \frac{31}{36} & -\frac{1}{36} \\ 0 & \frac{4}{9} & 0 & 0 & \frac{2}{9} & \frac{5}{9} & \frac{7}{9} & -\frac{4}{9} \\ 0 & \frac{4}{9} & 1 & 0 & -\frac{1}{36} & \frac{11}{36} & \frac{1}{36} & \frac{11}{36} \\ 0 & -\frac{4}{9} & 0 & 1 & \frac{1}{36} & -\frac{1}{18} & -\frac{1}{36} & -\frac{1}{18} \end{array}\right]$$

Multiply row 2 by $\frac{1}{2}$ and add it to row 1. Also, multiply row 2 by -1 and add it to row 3. Add row 2 to row 4.

$$\left[\begin{array}{cccc|cccc} 1 & 0 & 0 & 0 & \frac{1}{4} & \frac{1}{4} & \frac{5}{4} & -\frac{1}{4} \\ 0 & \frac{4}{9} & 0 & 0 & \frac{2}{9} & \frac{5}{9} & \frac{7}{9} & -\frac{4}{9} \\ 0 & 0 & 1 & 0 & -\frac{1}{4} & -\frac{1}{4} & -\frac{3}{4} & \frac{3}{4} \\ 0 & 0 & 0 & 1 & \frac{1}{4} & \frac{1}{2} & \frac{3}{4} & -\frac{1}{2} \end{array}\right]$$

Multiply row 2 by $\frac{9}{4}$.

$$\left[\begin{array}{cccc|cccc} 1 & 0 & 0 & 0 & \frac{1}{4} & \frac{1}{4} & \frac{5}{4} & -\frac{1}{4} \\ 0 & 1 & 0 & 0 & \frac{1}{2} & \frac{5}{4} & \frac{7}{4} & -1 \\ 0 & 0 & 1 & 0 & -\frac{1}{4} & -\frac{1}{4} & -\frac{3}{4} & \frac{3}{4} \\ 0 & 0 & 0 & 1 & \frac{1}{4} & \frac{1}{2} & \frac{3}{4} & -\frac{1}{2} \end{array}\right]$$

Then $\mathbf{A}^{-1} = \begin{bmatrix} \frac{1}{4} & \frac{1}{4} & \frac{5}{4} & -\frac{1}{4} \\ \frac{1}{2} & \frac{5}{4} & \frac{7}{4} & -1 \\ -\frac{1}{4} & -\frac{1}{4} & -\frac{3}{4} & \frac{3}{4} \\ \frac{1}{4} & \frac{1}{2} & \frac{3}{4} & -\frac{1}{2} \end{bmatrix}$, or

$$\begin{bmatrix} 0.25 & 0.25 & 1.25 & -0.25 \\ 0.5 & 1.25 & 1.75 & -1 \\ -0.25 & -0.25 & -0.75 & 0.75 \\ 0.25 & 0.5 & 0.75 & -0.5 \end{bmatrix}.$$

24. $\mathbf{A} = \begin{bmatrix} 10 & 20 & -30 & 15 \\ 3 & -7 & 14 & -8 \\ -7 & -2 & -1 & 2 \\ 4 & 4 & -3 & 1 \end{bmatrix}$

Write the augmented matrix.

$$\left[\begin{array}{cccc|cccc} 10 & 20 & -30 & 15 & 1 & 0 & 0 & 0 \\ 3 & -7 & 14 & -8 & 0 & 1 & 0 & 0 \\ -7 & -2 & -1 & 2 & 0 & 0 & 1 & 0 \\ 4 & 4 & -3 & 1 & 0 & 0 & 0 & 1 \end{array}\right]$$

Multiply rows 2 and 3 by 10. Also, multiply row 4 by 5.

$$\left[\begin{array}{cccc|cccc} 10 & 20 & -30 & 15 & 1 & 0 & 0 & 0 \\ 30 & -70 & 140 & -80 & 0 & 10 & 0 & 0 \\ -70 & -20 & -10 & 20 & 0 & 0 & 10 & 0 \\ 20 & 20 & -15 & 5 & 0 & 0 & 0 & 5 \end{array}\right]$$

Multiply row 1 by -3 and add it to row 2. Multiply row 1 by 7 and add it to row 3. Also, multiply row 1 by -2 and add it to row 4.

$$\left[\begin{array}{cccc|cccc} 10 & 20 & -30 & 15 & 1 & 0 & 0 & 0 \\ 0 & -130 & 230 & -125 & -3 & 10 & 0 & 0 \\ 0 & 120 & -220 & 125 & 7 & 0 & 10 & 0 \\ 0 & -20 & 45 & -25 & -2 & 0 & 0 & 5 \end{array}\right]$$

Interchange rows 2 and 4.

$$\left[\begin{array}{cccc|cccc} 10 & 20 & -30 & 15 & 1 & 0 & 0 & 0 \\ 0 & -20 & 45 & -25 & -2 & 0 & 0 & 5 \\ 0 & 120 & -220 & 125 & 7 & 0 & 10 & 0 \\ 0 & -130 & 230 & -125 & -3 & 10 & 0 & 0 \end{array}\right]$$

Multiply row 4 by 2.

$$\left[\begin{array}{cccc|cccc} 10 & 20 & -30 & 15 & 1 & 0 & 0 & 0 \\ 0 & -20 & 45 & -25 & -2 & 0 & 0 & 5 \\ 0 & 120 & -220 & 125 & 7 & 0 & 10 & 0 \\ 0 & -260 & 460 & -250 & -6 & 20 & 0 & 0 \end{array}\right]$$

Add row 2 to row 1. Multiply row 2 by 6 and add it to row 3. Also, multiply row 2 by -13 and add it to row 4.

$$\left[\begin{array}{cccc|cccc} 10 & 0 & 15 & -10 & -1 & 0 & 0 & 5 \\ 0 & -20 & 45 & -25 & -2 & 0 & 0 & 5 \\ 0 & 0 & 50 & -25 & -5 & 0 & 10 & 30 \\ 0 & 0 & -125 & 75 & 20 & 20 & 0 & -65 \end{array}\right]$$

Multiply rows 1 and 2 by 10 and multiply row 4 by 2.

$$\left[\begin{array}{cccc|cccc} 100 & 0 & 150 & -100 & -10 & 0 & 0 & 50 \\ 0 & -200 & 450 & -250 & -20 & 0 & 0 & 50 \\ 0 & 0 & 50 & -25 & -5 & 0 & 10 & 30 \\ 0 & 0 & -250 & 150 & 40 & 40 & 0 & -130 \end{array}\right]$$

Multiply row 3 by -3 and add it to row 1. Multiply row 3 by -9 and add it to row 2. Also, multiply row 3 by 5 and add it to row 4.

$$\left[\begin{array}{cccc|cccc} 100 & 0 & 0 & -25 & 5 & 0 & -30 & -40 \\ 0 & -200 & 0 & -25 & 25 & 0 & -90 & -220 \\ 0 & 0 & 50 & -25 & -5 & 0 & 10 & 30 \\ 0 & 0 & 0 & 25 & 15 & 40 & 50 & 20 \end{array}\right]$$

Add row 4 to rows 1, 2, and 3.

$$\left[\begin{array}{cccc|cccc} 100 & 0 & 0 & 0 & 20 & 40 & 20 & -20 \\ 0 & -200 & 0 & 0 & 40 & 40 & -40 & -200 \\ 0 & 0 & 50 & 0 & 10 & 40 & 60 & 50 \\ 0 & 0 & 0 & 25 & 15 & 40 & 50 & 20 \end{array}\right]$$

Multiply row 1 by $\frac{1}{100}$, multiply row 2 by $-\frac{1}{200}$, multiply row 3 by $\frac{1}{50}$, and multiply row 4 by $\frac{1}{25}$.

$$\left[\begin{array}{cccc|cccc} 1 & 0 & 0 & 0 & \frac{1}{5} & \frac{2}{5} & \frac{1}{5} & -\frac{1}{5} \\ 0 & 1 & 0 & 0 & -\frac{1}{5} & -\frac{1}{5} & \frac{1}{5} & 1 \\ 0 & 0 & 1 & 0 & \frac{1}{5} & \frac{4}{5} & \frac{6}{5} & 1 \\ 0 & 0 & 0 & 1 & \frac{3}{5} & \frac{8}{5} & 2 & \frac{4}{5} \end{array}\right]$$

Then $\mathbf{A}^{-1} = \begin{bmatrix} \frac{1}{5} & \frac{2}{5} & \frac{1}{5} & -\frac{1}{5} \\ -\frac{1}{5} & -\frac{1}{5} & \frac{1}{5} & 1 \\ \frac{1}{5} & \frac{4}{5} & \frac{6}{5} & 1 \\ \frac{3}{5} & \frac{8}{5} & 2 & \frac{4}{5} \end{bmatrix}$, or

$$\begin{bmatrix} 0.2 & 0.4 & 0.2 & -0.2 \\ -0.2 & -0.2 & 0.2 & 1 \\ 0.2 & 0.8 & 1.2 & 1 \\ 0.6 & 1.6 & 2 & 0.8 \end{bmatrix}.$$

25. Write an equivalent matrix equation, $\mathbf{AX} = \mathbf{B}$.

$$\begin{bmatrix} 11 & 3 \\ 7 & 2 \end{bmatrix}\begin{bmatrix} x \\ y \end{bmatrix} = \begin{bmatrix} -4 \\ 5 \end{bmatrix}$$

Then we have $\mathbf{X} = \mathbf{A}^{-1}\mathbf{B}$.

$$\begin{bmatrix} x \\ y \end{bmatrix} = \begin{bmatrix} 2 & -3 \\ -7 & 11 \end{bmatrix}\begin{bmatrix} -4 \\ 5 \end{bmatrix} = \begin{bmatrix} -23 \\ 83 \end{bmatrix}$$

The solution is $(-23, 83)$.

26. $\begin{bmatrix} x \\ y \end{bmatrix} = \begin{bmatrix} -3 & 5 \\ 5 & -8 \end{bmatrix}\begin{bmatrix} -6 \\ 2 \end{bmatrix} = \begin{bmatrix} 28 \\ -46 \end{bmatrix}$

The solution is $(28, -46)$.

27. Write an equivalent matrix equation, $\mathbf{AX} = \mathbf{B}$.

$$\begin{bmatrix} 3 & 1 & 0 \\ 2 & -1 & 2 \\ 1 & 1 & 1 \end{bmatrix}\begin{bmatrix} x \\ y \\ z \end{bmatrix} = \begin{bmatrix} 2 \\ -5 \\ 5 \end{bmatrix}$$

Then we have $\mathbf{X} = \mathbf{A}^{-1}\mathbf{B}$.

$$\begin{bmatrix} x \\ y \\ z \end{bmatrix} = \frac{1}{9}\begin{bmatrix} 3 & 1 & -2 \\ 0 & -3 & 6 \\ -3 & 2 & 5 \end{bmatrix}\begin{bmatrix} 2 \\ -5 \\ 5 \end{bmatrix} = \frac{1}{9}\begin{bmatrix} -9 \\ 45 \\ 9 \end{bmatrix} = \begin{bmatrix} -1 \\ 5 \\ 1 \end{bmatrix}$$

The solution is $(-1, 5, 1)$.

28. $\begin{bmatrix} x \\ y \\ z \end{bmatrix} = \frac{1}{5}\begin{bmatrix} -3 & 2 & -1 \\ 12 & -3 & 4 \\ 7 & -3 & 4 \end{bmatrix}\begin{bmatrix} -4 \\ -3 \\ 1 \end{bmatrix} = \frac{1}{5}\begin{bmatrix} 5 \\ -35 \\ -15 \end{bmatrix} = \begin{bmatrix} 1 \\ -7 \\ -3 \end{bmatrix}$

The solution is $(1, -7, -3)$.

29. $4x + 3y = 2,$
$x - 2y = 6$

Write an equivalent matrix equation, $\mathbf{AX} = \mathbf{B}$.

$$\begin{bmatrix} 4 & 3 \\ 1 & -2 \end{bmatrix}\begin{bmatrix} x \\ y \end{bmatrix} = \begin{bmatrix} 2 \\ 6 \end{bmatrix}$$

Then $\mathbf{X} = \mathbf{A}^{-1}\mathbf{B} = \begin{bmatrix} \frac{2}{11} & \frac{3}{11} \\ \frac{1}{11} & -\frac{4}{11} \end{bmatrix}\begin{bmatrix} 2 \\ 6 \end{bmatrix} = \begin{bmatrix} 2 \\ -2 \end{bmatrix}$.

The solution is $(2, -2)$.

30. $2x - 3y = 7,$
$4x + y = -7$

Write an equivalent matrix equation, $\mathbf{AX} = \mathbf{B}$.

$$\begin{bmatrix} 2 & -3 \\ 4 & 1 \end{bmatrix}\begin{bmatrix} x \\ y \end{bmatrix} = \begin{bmatrix} 7 \\ -7 \end{bmatrix}$$

Then $\mathbf{X} = \mathbf{A}^{-1}\mathbf{B} = \begin{bmatrix} \frac{1}{14} & \frac{3}{14} \\ -\frac{2}{7} & \frac{1}{7} \end{bmatrix}\begin{bmatrix} 7 \\ -7 \end{bmatrix} = \begin{bmatrix} -1 \\ -3 \end{bmatrix}$.

The solution is $(-1, -3)$.

31. $5x + y = 2,$
$3x - 2y = -4$

Write an equivalent matrix equation, $\mathbf{AX} = \mathbf{B}$.

$$\begin{bmatrix} 5 & 1 \\ 3 & -2 \end{bmatrix}\begin{bmatrix} x \\ y \end{bmatrix} = \begin{bmatrix} \frac{2}{13} & \frac{1}{13} \\ \frac{3}{13} & -\frac{5}{13} \end{bmatrix}\begin{bmatrix} 2 \\ -4 \end{bmatrix} = \begin{bmatrix} 2 \\ -4 \end{bmatrix}$$

Then $\mathbf{X} = \mathbf{A}^{-1}\mathbf{B} = \begin{bmatrix} 0 \\ 2 \end{bmatrix}$.

The solution is $(0, 2)$.

32. $x - 6y = 5,$
$-x + 4y = -5$

Write an equivalent matrix equation, $\mathbf{AX} = \mathbf{B}$.

$$\begin{bmatrix} 1 & -6 \\ -1 & 4 \end{bmatrix}\begin{bmatrix} x \\ y \end{bmatrix} = \begin{bmatrix} 5 \\ -5 \end{bmatrix}$$

Then $\mathbf{X} = \mathbf{A}^{-1}\mathbf{B} = \begin{bmatrix} -2 & -3 \\ -\frac{1}{2} & -\frac{1}{2} \end{bmatrix}\begin{bmatrix} 5 \\ -5 \end{bmatrix} = \begin{bmatrix} 5 \\ 0 \end{bmatrix}$.

The solution is $(5, 0)$.

33. $x + z = 1,$
$2x + y = 3,$
$x - y + z = 4$

Write an equivalent matrix equation, $\mathbf{AX} = \mathbf{B}$.

$$\begin{bmatrix} 1 & 0 & 1 \\ 2 & 1 & 0 \\ 1 & -1 & 1 \end{bmatrix}\begin{bmatrix} x \\ y \\ z \end{bmatrix} = \begin{bmatrix} 1 \\ 3 \\ 4 \end{bmatrix}$$

Then $\mathbf{X} = \mathbf{A}^{-1}\mathbf{B} = \begin{bmatrix} -\frac{1}{2} & \frac{1}{2} & \frac{1}{2} \\ 1 & 0 & -1 \\ \frac{3}{2} & -\frac{1}{2} & -\frac{1}{2} \end{bmatrix}\begin{bmatrix} 1 \\ 3 \\ 4 \end{bmatrix} = \begin{bmatrix} 3 \\ -3 \\ -2 \end{bmatrix}$.

The solution is $(3, -3, -2)$.

34. $x + 2y + 3z = -1,$
$2x - 3y + 4z = 2,$
$-3x + 5y - 6z = 4$

Write an equivalent matrix equation, $\mathbf{AX} = \mathbf{B}$.

$$\begin{bmatrix} 1 & 2 & 3 \\ 2 & -3 & 4 \\ -3 & 5 & -6 \end{bmatrix}\begin{bmatrix} x \\ y \\ z \end{bmatrix} = \begin{bmatrix} -1 \\ 2 \\ 4 \end{bmatrix}$$

Then $\mathbf{X} = \mathbf{A}^{-1}\mathbf{B} = \begin{bmatrix} -2 & 27 & 17 \\ 0 & 3 & 2 \\ 1 & -11 & -7 \end{bmatrix}\begin{bmatrix} -1 \\ 2 \\ 4 \end{bmatrix} = \begin{bmatrix} 124 \\ 14 \\ -51 \end{bmatrix}$.

The solution is $(124, 14, -51)$.

35. $2x + 3y + 4z = 2,$
$x - 4y + 3z = 2,$
$5x + y + z = -4$

Write an equivalent matrix equation, $\mathbf{AX} = \mathbf{B}$.

$$\begin{bmatrix} 2 & 3 & 4 \\ 1 & -4 & 3 \\ 5 & 1 & 1 \end{bmatrix}\begin{bmatrix} x \\ y \\ z \end{bmatrix} = \begin{bmatrix} 2 \\ 2 \\ -4 \end{bmatrix}$$

Then $\mathbf{X} = \mathbf{A}^{-1}\mathbf{B} = \begin{bmatrix} -\frac{1}{16} & \frac{1}{112} & \frac{25}{112} \\ \frac{1}{8} & -\frac{9}{56} & -\frac{1}{56} \\ \frac{3}{16} & \frac{13}{112} & -\frac{11}{112} \end{bmatrix}\begin{bmatrix} 2 \\ 2 \\ -4 \end{bmatrix} = \begin{bmatrix} -1 \\ 0 \\ 1 \end{bmatrix}$.

The solution is $(-1, 0, 1)$.

36. $x + y = 2,$
$3x + 2z = 5,$
$2x + 3y - 3z = 9$

Write an equivalent matrix equation, $\mathbf{AX} = \mathbf{B}$.

$$\begin{bmatrix} 1 & 1 & 0 \\ 3 & 0 & 2 \\ 2 & 3 & -3 \end{bmatrix}\begin{bmatrix} x \\ y \\ z \end{bmatrix} = \begin{bmatrix} 2 \\ 5 \\ 9 \end{bmatrix}$$

Then $\mathbf{X} = \mathbf{A}^{-1}\mathbf{B} = \begin{bmatrix} -\frac{6}{7} & \frac{3}{7} & \frac{2}{7} \\ \frac{13}{7} & -\frac{3}{7} & -\frac{2}{7} \\ \frac{9}{7} & -\frac{1}{7} & -\frac{3}{7} \end{bmatrix}\begin{bmatrix} 2 \\ 5 \\ 9 \end{bmatrix} = \begin{bmatrix} 3 \\ -1 \\ -2 \end{bmatrix}$.

The solution is $(3, -1, -2)$.

37. $2w - 3x + 4y - 5z = 0,$
$3w - 2x + 7y - 3z = 2,$
$w + x - y + z = 1,$
$-w - 3x - 6y + 4z = 6$

Write an equivalent matrix equation, $\mathbf{AX} = \mathbf{B}$.

$$\begin{bmatrix} 2 & -3 & 4 & -5 \\ 3 & -2 & 7 & -3 \\ 1 & 1 & -1 & 1 \\ -1 & -3 & -6 & 4 \end{bmatrix}\begin{bmatrix} w \\ x \\ y \\ z \end{bmatrix} = \begin{bmatrix} 0 \\ 2 \\ 1 \\ 6 \end{bmatrix}$$

Then $\mathbf{X} = \mathbf{A}^{-1}\mathbf{B} = \frac{1}{203}\begin{bmatrix} 26 & 11 & 127 & 9 \\ -8 & -19 & 39 & -34 \\ -37 & 39 & -48 & -5 \\ -55 & 47 & -11 & 20 \end{bmatrix}\begin{bmatrix} 0 \\ 2 \\ 1 \\ 6 \end{bmatrix} = \begin{bmatrix} 1 \\ -1 \\ 0 \\ 1 \end{bmatrix}$.

The solution is $(1, -1, 0, 1)$.

38. $5w - 4x + 3y - 2z = -6,$
$w + 4x - 2y + 3z = -5,$
$2w - 3x + 6y - 9z = 14,$
$3w - 5x + 2y - 4z = -3$

Write an equivalent matrix equation, $\mathbf{AX} = \mathbf{B}$.

$$\begin{bmatrix} 5 & -4 & 3 & -2 \\ 1 & 4 & -2 & 3 \\ 2 & -3 & 6 & -9 \\ 3 & -5 & 2 & -4 \end{bmatrix}\begin{bmatrix} w \\ x \\ y \\ z \end{bmatrix} = \begin{bmatrix} -6 \\ -5 \\ 14 \\ -3 \end{bmatrix}$$

Then

$$\mathbf{X} = \mathbf{A}^{-1}\mathbf{B} = \frac{1}{302}\begin{bmatrix} 18 & 75 & 1 & 45 \\ -10 & 59 & 33 & -25 \\ 112 & -87 & 23 & -173 \\ 82 & -61 & -29 & -97 \end{bmatrix}\begin{bmatrix} -6 \\ -5 \\ 14 \\ -3 \end{bmatrix} = \begin{bmatrix} -2 \\ 1 \\ 2 \\ -1 \end{bmatrix}.$$

The solution is $(-2, 1, 2, -1)$.

39. ***Familiarize.*** Let x = the number of hot dogs sold and y = the number of sausages.

Translate.

The total number of items sold was 145.

$x + y = 145$

The number of hot dogs sold is 45 more than the number of sausages.

$x = y + 45$

We have a system of equations:

$x + y = 145,$ $\quad$ $x + y = 145,$
$\quad$ or
$x = y + 45,$ $\quad$ $x - y = 45.$

Carry out. Write an equivalent matrix equation, $\mathbf{AX} = \mathbf{B}$.

$$\begin{bmatrix} 1 & 1 \\ 1 & -1 \end{bmatrix}\begin{bmatrix} x \\ y \end{bmatrix} = \begin{bmatrix} 145 \\ 45 \end{bmatrix}$$

Then $\mathbf{X} = \mathbf{A}^{-1}\mathbf{B} = \begin{bmatrix} \frac{1}{2} & \frac{1}{2} \\ \frac{1}{2} & -\frac{1}{2} \end{bmatrix}\begin{bmatrix} 145 \\ 45 \end{bmatrix} = \begin{bmatrix} 95 \\ 50 \end{bmatrix}$, so the solution is $(95, 50)$.

Check. The total number of items is $95 + 50$, or 145. The number of hot dogs, 95, is 45 more than the number of sausages. The solution checks.

State. Stefan sold 95 hot dogs and 50 Italian sausages.

40. Let x = the price of a lab record book and y = the price of a highlighter.

Solve: $4x + 3y = 13.93,$
$3x + 2y = 10.25$

Writing $\begin{bmatrix} 4 & 3 \\ 3 & 2 \end{bmatrix} \begin{bmatrix} x \\ y \end{bmatrix} = \begin{bmatrix} 13.93 \\ 10.25 \end{bmatrix}$, we find that $x =$ \$2.89, $y =$ \$0.79.

41. ***Familiarize***. Let x, y, and z represent the prices of one ton of topsoil, mulch, and pea gravel, respectively.

Translate.

Four tons of topsoil, 3 tons of mulch, and 6 tons of pea gravel costs \$2825.

$$4x + 3y + 6z = 2825$$

Five tons of topsoil, 2 tons of mulch, and 5 tons of pea gravel costs \$2663.

$$5x + 2y + 5z = 2663$$

Pea gravel costs \$17 less per ton than topsoil.

$$z = x - 17$$

We have a system of equations.

$$4x + 3y + 6z = 2825,$$
$$5x + 2y + 5z = 2663,$$
$$z = x - 17, \text{ or}$$

$$4x + 3y + 6z = 2825,$$
$$5x + 2y + 5z = 2663,$$
$$x \quad - z = 17$$

Carry out. Write an equivalent matrix equation, $\mathbf{AX} = \mathbf{B}$.

$$\begin{bmatrix} 4 & 3 & 6 \\ 5 & 2 & 5 \\ 1 & 0 & -1 \end{bmatrix} \begin{bmatrix} x \\ y \\ z \end{bmatrix} = \begin{bmatrix} 2825 \\ 2663 \\ 17 \end{bmatrix}$$

Then $\mathbf{X} = \mathbf{A}^{-1}\mathbf{B} = \begin{bmatrix} -\frac{1}{5} & \frac{3}{10} & \frac{3}{10} \\ 1 & -1 & 1 \\ -\frac{1}{5} & \frac{3}{10} & -\frac{7}{10} \end{bmatrix} \begin{bmatrix} 2825 \\ 2663 \\ 17 \end{bmatrix} =$

$\begin{bmatrix} 239 \\ 179 \\ 222 \end{bmatrix}$, so the solution is $(239, 179, 222)$.

Check. Four tons of topsoil, 3 tons of mulch, and 6 tons of pea gravel costs $4 \cdot \$239 + 3 \cdot \$179 + 6 \cdot \$222$, or $\$956 + \$537 + \$1332$, or \$2825. Five tons of topsoil, 2 tons of mulch, and 5 tons of pea gravel costs $5 \cdot \$239 + 2 \cdot \$179 + 5 \cdot \$222$, or $\$1195 + \$358 + \$1110$, or \$2663. The price of pea gravel, \$222, is \$17 less than the price of topsoil, \$239. The solution checks.

State. The price of topsoil is \$239 per ton, of mulch is \$179 per ton, and of pea gravel is \$222 per ton.

42. Let x, y, and z represent the amounts invested at 2.2%, 2.65%, and 3.05%, respectively.

Solve: $x + y + z = 8500$,

$0.022x + 0.0265y + 0.0305z = 230$,

$z = x + 1500$

Writing

$$\begin{bmatrix} 1 & 1 & 1 \\ 0.022 & 0.0265 & 0.0305 \\ -1 & 0 & 1 \end{bmatrix} \begin{bmatrix} x \\ y \\ z \end{bmatrix} = \begin{bmatrix} 8500 \\ 230 \\ 1500 \end{bmatrix},$$

we find that $x = \$2500$, $y = \$2000$, $z = \$4000$.

43. Left to the student

44. left to the student

45. Left to the student

46. Left to the student

47. No; for example, let $\mathbf{A} = \mathbf{B} = \begin{bmatrix} 1 & 0 \\ 0 & 1 \end{bmatrix}$.

Then $\mathbf{A} + \mathbf{B} = \begin{bmatrix} 2 & 0 \\ 0 & 2 \end{bmatrix}$ and $(\mathbf{A} + \mathbf{B})^{-1} = \begin{bmatrix} 0.5 & 0 \\ 0 & 0.5 \end{bmatrix}$,

but $\mathbf{A}^{-1} + \mathbf{B}^{-1} = \begin{bmatrix} 1 & 0 \\ 0 & 1 \end{bmatrix} + \begin{bmatrix} 1 & 0 \\ 0 & 1 \end{bmatrix} = \begin{bmatrix} 2 & 0 \\ 0 & 2 \end{bmatrix}$.

48. No; for example, let $\mathbf{A} = \begin{bmatrix} 3 & 2 \\ 5 & 3 \end{bmatrix}$ and $\mathbf{B} = \begin{bmatrix} 11 & 3 \\ 7 & 2 \end{bmatrix}$.

Then $\mathbf{AB} = \begin{bmatrix} 47 & 13 \\ 76 & 21 \end{bmatrix}$ and $(\mathbf{AB})^{-1} =$

$\begin{bmatrix} -21 & 13 \\ 76 & -47 \end{bmatrix}$,

but

$\mathbf{A}^{-1}\mathbf{B}^{-1} = \begin{bmatrix} -3 & 2 \\ 5 & -3 \end{bmatrix} \begin{bmatrix} 2 & -3 \\ -7 & 11 \end{bmatrix} =$

$\begin{bmatrix} -20 & 31 \\ 31 & -48 \end{bmatrix}$.

49.
$$\begin{array}{r|rrrr} -2 & 1 & -6 & 4 & -8 \\ & & -2 & 16 & -40 \\ \hline & 1 & -8 & 20 & -48 \end{array}$$

$f(-2) = -48$

50.
$$\begin{array}{r|rrrrr} 3 & 2 & -1 & 5 & 6 & -4 \\ & & 6 & 15 & 60 & 198 \\ \hline & 2 & 5 & 20 & 66 & 194 \end{array}$$

$f(3) = 194$

51.
$$2x^2 + x = 7$$
$$2x^2 + x - 7 = 0$$
$$a = 2,\ b = 1,\ c = -7$$
$$x = \frac{-b \pm \sqrt{b^2 - 4ac}}{2a}$$
$$= \frac{-1 \pm \sqrt{1^2 - 4 \cdot 2 \cdot (-7)}}{2 \cdot 2} = \frac{-1 \pm \sqrt{1 + 56}}{4}$$
$$= \frac{-1 \pm \sqrt{57}}{4}$$

The solutions are $\frac{-1 + \sqrt{57}}{4}$ and $\frac{-1 - \sqrt{57}}{4}$, or $\frac{-1 \pm \sqrt{57}}{4}$.

52. $\dfrac{1}{x+1} - \dfrac{6}{x-1} = 1$, LCD is $(x+1)(x-1)$

$$(x+1)(x-1)\left(\frac{1}{x+1} - \frac{6}{x-1}\right) = (x+1)(x-1)\cdot 1$$
$$x-1-6(x+1) = x^2-1$$
$$x-1-6x-6 = x^2-1$$
$$0 = x^2+5x+6$$
$$0 = (x+3)(x+2)$$

$x=-3$ *or* $x=-2$

Both numbers check.

53.
$$\sqrt{2x+1}-1 = \sqrt{2x-4}$$
$$(\sqrt{2x+1}-1)^2 = (\sqrt{2x-4})^2 \quad \text{Squaring both sides}$$
$$2x+1-2\sqrt{2x+1}+1 = 2x-4$$
$$2x+2-2\sqrt{2x+1} = 2x-4$$
$$2-2\sqrt{2x+1} = -4 \quad \text{Subtracting } 2x$$
$$-2\sqrt{2x+1} = -6 \quad \text{Subtracting 2}$$
$$\sqrt{2x+1} = 3 \quad \text{Dividing by } -2$$
$$(\sqrt{2x+1})^2 = 3^2 \quad \text{Squaring both sides}$$
$$2x+1 = 9$$
$$2x = 8$$
$$x = 4$$

The number 4 checks. It is the solution.

54. $x-\sqrt{x}-6=0$

Let $u=\sqrt{x}$.

$$u^2-u-6=0$$
$$(u-3)(u+2)=0$$

$u=3$ *or* $u=-2$

$\sqrt{x}=3$ *or* $\sqrt{x}=-2$

$x=9$ No solution

The number 9 checks. It is the solution.

55. $f(x)=x^3-3x^2-6x+8$

We use synthetic division to find one factor. We first try $x-1$.

$$\begin{array}{r|rrrr} 1 & 1 & -3 & -6 & 8 \\ & & 1 & -2 & -8 \\ \hline & 1 & -2 & -8 & 0 \end{array}$$

Since $f(1)=0$, $x-1$ is a factor of $f(x)$. We have $f(x)=(x-1)(x^2-2x-8)$. Factoring the trinomial we get $f(x)=(x-1)(x-4)(x+2)$.

56. $f(x)=x^4+2x^3-16x^2-2x+15$

We try $x-1$.

$$\begin{array}{r|rrrrr} 1 & 1 & 2 & -16 & -2 & 15 \\ & & 1 & 3 & -13 & -15 \\ \hline & 1 & 3 & -13 & -15 & 0 \end{array}$$

We have $f(x)=(x-1)(x^3+3x^2-13x-15)$. Now we factor the cubic polynomial. We try $x+1$.

$$\begin{array}{r|rrrr} -1 & 1 & 3 & -13 & -15 \\ & & -1 & -2 & 15 \\ \hline & 1 & 2 & -15 & 0 \end{array}$$

Now we have $f(x)=(x-1)(x+1)(x^2+2x-15)=(x-1)(x+1)(x+5)(x-3)$.

57. $\mathbf{A}=[x]$

Write the augmented matrix.

$$\left[\begin{array}{c|c} x & 1 \end{array}\right]$$

Multiply by $\dfrac{1}{x}$.

$$\left[\begin{array}{c|c} 1 & \frac{1}{x} \end{array}\right]$$

Then $\mathbf{A}^{-1}$ exists if and only if $x \neq 0$. $\mathbf{A}^{-1}=\left[\dfrac{1}{x}\right]$.

58. $\mathbf{A}=\begin{bmatrix} x & 0 \\ 0 & y \end{bmatrix}$

Write the augmented matrix.

$$\left[\begin{array}{cc|cc} x & 0 & 1 & 0 \\ 0 & y & 0 & 1 \end{array}\right]$$

Multiply row 1 by $\dfrac{1}{x}$ and row 2 by $\dfrac{1}{y}$.

$$\left[\begin{array}{cc|cc} 1 & 0 & \frac{1}{x} & 0 \\ 0 & 1 & 0 & \frac{1}{y} \end{array}\right]$$

Then $\mathbf{A}^{-1}$ exists if and only if $x \neq 0$ and $y \neq 0$, or if and only if $xy \neq 0$.

$$\mathbf{A}^{-1}=\begin{bmatrix} \frac{1}{x} & 0 \\ 0 & \frac{1}{y} \end{bmatrix}$$

59. $\mathbf{A}=\begin{bmatrix} 0 & 0 & x \\ 0 & y & 0 \\ z & 0 & 0 \end{bmatrix}$

Write the augmented matrix.

$$\left[\begin{array}{ccc|ccc} 0 & 0 & x & 1 & 0 & 0 \\ 0 & y & 0 & 0 & 1 & 0 \\ z & 0 & 0 & 0 & 0 & 1 \end{array}\right]$$

Interchange row 1 and row 3.

$$\left[\begin{array}{ccc|ccc} z & 0 & 0 & 0 & 0 & 1 \\ 0 & y & 0 & 0 & 1 & 0 \\ 0 & 0 & x & 1 & 0 & 0 \end{array}\right]$$

Multiply row 1 by $\dfrac{1}{z}$, row 2 by $\dfrac{1}{y}$, and row 3 by $\dfrac{1}{x}$.

$$\left[\begin{array}{ccc|ccc} 1 & 0 & 0 & 0 & 0 & \frac{1}{z} \\ 0 & 1 & 0 & 0 & \frac{1}{y} & 0 \\ 0 & 0 & 1 & \frac{1}{x} & 0 & 0 \end{array}\right]$$

Then $\mathbf{A}^{-1}$ exists if and only if $x \neq 0$ and $y \neq 0$ and $z \neq 0$, or if and only if $xyz \neq 0$.

$$\mathbf{A}^{-1} = \begin{bmatrix} 0 & 0 & \frac{1}{z} \\ 0 & \frac{1}{y} & 0 \\ \frac{1}{x} & 0 & 0 \end{bmatrix}$$

60. $\mathbf{A} = \begin{bmatrix} x & 1 & 1 & 1 \\ 0 & y & 0 & 0 \\ 0 & 0 & z & 0 \\ 0 & 0 & 0 & w \end{bmatrix}$

Write the augmented matrix.

$$\left[\begin{array}{cccc|cccc} x & 1 & 1 & 1 & 1 & 0 & 0 & 0 \\ 0 & y & 0 & 0 & 0 & 1 & 0 & 0 \\ 0 & 0 & z & 0 & 0 & 0 & 1 & 0 \\ 0 & 0 & 0 & w & 0 & 0 & 0 & 1 \end{array}\right]$$

Multiply row 4 by $-\frac{1}{w}$ and add it to row 1.

$$\left[\begin{array}{cccc|cccc} x & 1 & 1 & 0 & 1 & 0 & 0 & -\frac{1}{w} \\ 0 & y & 0 & 0 & 0 & 1 & 0 & 0 \\ 0 & 0 & z & 0 & 0 & 0 & 1 & 0 \\ 0 & 0 & 0 & w & 0 & 0 & 0 & 1 \end{array}\right]$$

Multiply row 3 by $-\frac{1}{z}$ and add it to row 1.

$$\left[\begin{array}{cccc|cccc} x & 1 & 0 & 0 & 1 & 0 & -\frac{1}{z} & -\frac{1}{w} \\ 0 & y & 0 & 0 & 0 & 1 & 0 & 0 \\ 0 & 0 & z & 0 & 0 & 0 & 1 & 0 \\ 0 & 0 & 0 & w & 0 & 0 & 0 & 1 \end{array}\right]$$

Multiply row 2 by $-\frac{1}{y}$ and add it to row 1.

$$\left[\begin{array}{cccc|cccc} x & 0 & 0 & 0 & 1 & -\frac{1}{y} & -\frac{1}{z} & -\frac{1}{w} \\ 0 & y & 0 & 0 & 0 & 1 & 0 & 0 \\ 0 & 0 & z & 0 & 0 & 0 & 1 & 0 \\ 0 & 0 & 0 & w & 0 & 0 & 0 & 1 \end{array}\right]$$

Multiply row 1 by $\frac{1}{x}$, row 2 by $\frac{1}{y}$, row 3 by $\frac{1}{z}$, and row 4 by $\frac{1}{w}$.

$$\left[\begin{array}{cccc|cccc} 1 & 0 & 0 & 0 & \frac{1}{x} & -\frac{1}{xy} & -\frac{1}{xz} & -\frac{1}{xw} \\ 0 & 1 & 0 & 0 & 0 & \frac{1}{y} & 0 & 0 \\ 0 & 0 & 1 & 0 & 0 & 0 & \frac{1}{z} & 0 \\ 0 & 0 & 0 & 1 & 0 & 0 & 0 & \frac{1}{w} \end{array}\right]$$

Then $\mathbf{A}^{-1}$ exists if and only if $w \neq 0$ and $x \neq 0$ and $y \neq 0$ and $z \neq 0$, or if and only if $wxyz \neq 0$.

$$\mathbf{A}^{-1} = \begin{bmatrix} \frac{1}{x} & -\frac{1}{xy} & -\frac{1}{xz} & -\frac{1}{xw} \\ 0 & \frac{1}{y} & 0 & 0 \\ 0 & 0 & \frac{1}{z} & 0 \\ 0 & 0 & 0 & \frac{1}{w} \end{bmatrix}$$

Exercise Set 8.6

1. $\begin{vmatrix} -2 & -\sqrt{5} \\ -\sqrt{5} & 3 \end{vmatrix} = -2\cdot 3 - (-\sqrt{5})(-\sqrt{5}) = -6 - 5 = -11$

2. $\begin{vmatrix} \sqrt{5} & -3 \\ 4 & 2 \end{vmatrix} = \sqrt{5}(2) - 4(-3) = 2\sqrt{5} + 12$

3. $\begin{vmatrix} x & 4 \\ x & x^2 \end{vmatrix} = x\cdot x^2 - x\cdot 4 = x^3 - 4x$

4. $\begin{vmatrix} y^2 & -2 \\ y & 3 \end{vmatrix} = y^2\cdot 3 - y(-2) = 3y^2 + 2y$

5. We will expand across the first row. We could have chosen any other row or column just as well.

$$\begin{vmatrix} 3 & 1 & 2 \\ -2 & 3 & 1 \\ 3 & 4 & -6 \end{vmatrix}$$

$$= 3(-1)^{1+1}\begin{vmatrix} 3 & 1 \\ 4 & -6 \end{vmatrix} + 1\cdot(-1)^{1+2}\begin{vmatrix} -2 & 1 \\ 3 & -6 \end{vmatrix} + 2(-1)^{1+3}\begin{vmatrix} -2 & 3 \\ 3 & 4 \end{vmatrix}$$

$$= 3\cdot 1[3(-6) - 4\cdot 1] + 1(-1)[-2(-6) - 3\cdot 1] + 2\cdot 1(-2\cdot 4 - 3\cdot 3)$$

$$= 3(-22) - (9) + 2(-17)$$

$$= -109$$

6. We will expand down the third column.

$$\begin{vmatrix} 3 & -2 & 1 \\ 2 & 4 & 3 \\ -1 & 5 & 1 \end{vmatrix}$$

$$= 1(-1)^{1+3}\begin{vmatrix} 2 & 4 \\ -1 & 5 \end{vmatrix} + 3(-1)^{2+3}\begin{vmatrix} 3 & -2 \\ -1 & 5 \end{vmatrix} + 1(-1)^{3+3}\begin{vmatrix} 3 & -2 \\ 2 & 4 \end{vmatrix}$$

$$= 1\cdot 1[2\cdot 5 - (-1)\cdot 4] + 3(-1)[3\cdot 5 - (-1)(-2)] + 1\cdot 1[3\cdot 4 - 2(-2)]$$

$$= 1\cdot 14 - 3\cdot 13 + 1\cdot 16$$

$$= -9$$

7. We will expand down the second column. We could have chosen any other row or column just as well.

$$\begin{vmatrix} x & 0 & -1 \\ 2 & x & x^2 \\ -3 & x & 1 \end{vmatrix}$$

$$= 0(-1)^{1+2}\begin{vmatrix} 2 & x^2 \\ -3 & 1 \end{vmatrix} + x(-1)^{2+2}\begin{vmatrix} x & -1 \\ -3 & 1 \end{vmatrix} +$$

$$x(-1)^{3+2}\begin{vmatrix} x & -1 \\ 2 & x^2 \end{vmatrix}$$

$$= 0(-1)[2\cdot 1-(-3)x^2]+x\cdot 1[x\cdot 1-(-3)(-1)]+$$
$$x(-1)[x\cdot x^2-2(-1)]$$
$$= 0+x(x-3)-x(x^3+2)$$
$$= x^2-3x-x^4-2x = -x^4+x^2-5x$$

8. We will expand across the third row.

$$\begin{vmatrix} x & 1 & -1 \\ x^2 & x & x \\ 0 & x & 1 \end{vmatrix}$$

$$= 0(-1)^{3+1}\begin{vmatrix} 1 & -1 \\ x & x \end{vmatrix} + x(-1)^{3+2}\begin{vmatrix} x & -1 \\ x^2 & x \end{vmatrix} +$$

$$1(-1)^{3+3}\begin{vmatrix} x & 1 \\ x^2 & x \end{vmatrix}$$

$$= 0+x(-1)[x\cdot x-x^2(-1)]+1\cdot 1(x\cdot x-x^2\cdot 1)$$
$$= -x(x^2+x^2)+1(x^2-x^2)$$
$$= -x(2x^2)+1(0) = -2x^3$$

9. $\mathbf{A} = \begin{bmatrix} 7 & -4 & -6 \\ 2 & 0 & -3 \\ 1 & 2 & -5 \end{bmatrix}$

M_{11} is the determinant of the matrix formed by deleting the first row and first column of $\mathbf{A}$:

$$M_{11} = \begin{vmatrix} 0 & -3 \\ 2 & -5 \end{vmatrix} = 0(-5)-2(-3) = 0+6 = 6$$

M_{32} is the determinant of the matrix formed by deleting the third row and second column of $\mathbf{A}$:

$$M_{32} = \begin{vmatrix} 7 & -6 \\ 2 & -3 \end{vmatrix} = 7(-3)-2(-6) = -21+12 = -9$$

M_{22} is the determinant of the matrix formed by deleting the second row and second column of $\mathbf{A}$:

$$M_{22} = \begin{vmatrix} 7 & -6 \\ 1 & -5 \end{vmatrix} = 7(-5)-1(-6) = -35+6 = -29$$

10. See matrix $\mathbf{A}$ in Exercise 9 above:

$$M_{13} = \begin{vmatrix} 2 & 0 \\ 1 & 2 \end{vmatrix} = 2\cdot 2-1\cdot 0 = 4-0 = 4$$

$$M_{31} = \begin{vmatrix} -4 & -6 \\ 0 & -3 \end{vmatrix} = -4(-3)-0(-6) = 12-0 = 12$$

$$M_{23} = \begin{vmatrix} 7 & -4 \\ 1 & 2 \end{vmatrix} = 7\cdot 2-1(-4) = 14+4 = 18$$

11. In Exercise 9 we found that $M_{11} = 6$.

$A_{11} = (-1)^{1+1}M_{11} = 1\cdot 6 = 6$

In Exercise 9 we found that $M_{32} = -9$.

$A_{32} = (-1)^{3+2}M_{32} = -1(-9) = 9$

In Exercise 9 we found that $M_{22} = -29$.

$A_{22} = (-1)^{2+2}(-29) = 1(-29) = -29$

12. In Exercise 10 we found that $M_{13} = 4$.

$A_{13} = (-1)^{1+3}M_{13} = 1\cdot 4 = 4$

In Exercise 10 we found that $M_{31} = 12$.

$A_{31} = (-1)^{3+1}M_{31} = 1\cdot 12 = 12$

In Exercise 10 we found that $M_{23} = 18$.

$A_{23} = (-1)^{2+3}M_{23} = -1\cdot 18 = -18$

13. $\mathbf{A} = \begin{bmatrix} 7 & -4 & -6 \\ 2 & 0 & -3 \\ 1 & 2 & -5 \end{bmatrix}$

$$|\mathbf{A}|$$
$$= 2A_{21}+0A_{22}+(-3)A_{23}$$
$$= 2(-1)^{2+1}\begin{vmatrix} -4 & -6 \\ 2 & -5 \end{vmatrix} + 0 + (-3)(-1)^{2+3}\begin{vmatrix} 7 & -4 \\ 1 & 2 \end{vmatrix}$$
$$= 2(-1)[-4(-5)-2(-6)]+0+$$
$$(-3)(-1)[7\cdot 2-1(-4)]$$
$$= -2(32)+0+3(18) = -64+0+54$$
$$= -10$$

14. See matrix $\mathbf{A}$ in Exercise 13 above.

$$|\mathbf{A}|$$
$$= -4A_{12}+0A_{22}+2A_{32}$$
$$= -4(-1)^{1+2}\begin{vmatrix} 2 & -3 \\ 1 & -5 \end{vmatrix} + 0 + 2(-1)^{3+2}\begin{vmatrix} 7 & -6 \\ 2 & -3 \end{vmatrix}$$
$$= -4(-1)[2(-5)-1(-3)]+0+$$
$$2(-1)[7(-3)-2(-6)]$$
$$= 4(-7)+0-2(-9) = -28+0+18$$
$$= -10$$

15. $\mathbf{A} = \begin{bmatrix} 7 & -4 & -6 \\ 2 & 0 & -3 \\ 1 & 2 & -5 \end{bmatrix}$

$$|\mathbf{A}|$$
$$= -6A_{13}+(-3)A_{23}+(-5)A_{33}$$
$$= -6(-1)^{1+3}\begin{vmatrix} 2 & 0 \\ 1 & 2 \end{vmatrix} + (-3)(-1)^{2+3}\begin{vmatrix} 7 & -4 \\ 1 & 2 \end{vmatrix} +$$
$$(-5)(-1)^{3+3}\begin{vmatrix} 7 & -4 \\ 2 & 0 \end{vmatrix}$$
$$= -6\cdot 1(2\cdot 2-1\cdot 0)+(-3)(-1)[7\cdot 2-1(-4)]+$$
$$-5\cdot 1(7\cdot 0-2(-4))$$
$$= -6(4)+3(18)-5(8) = -24+54-40$$
$$= -10$$

16. See matrix $\mathbf{A}$ in Exercise 15 above.

$|\mathbf{A}|$

$= 7A_{11} - 4A_{12} - 6A_{13}$

$= 7(-1)^{1+1}\begin{vmatrix} 0 & -3 \\ 2 & -5 \end{vmatrix} + (-4)(-1)^{1+2}\begin{vmatrix} 2 & -3 \\ 1 & -5 \end{vmatrix} +$

$(-6)(-1)^{1+3}\begin{vmatrix} 2 & 0 \\ 1 & 2 \end{vmatrix}$

$= 7\cdot 1[0(-5)-2(-3)]+(-4)(-1)[2(-5)-1(-3)]+$
$(-6)(1)(2\cdot 2 - 1\cdot 0)$

$= 7(6) + 4(-7) - 6(4) = 42 - 28 - 24$

$= -10$

17. $\mathbf{A} = \begin{bmatrix} 1 & 0 & 0 & -2 \\ 4 & 1 & 0 & 0 \\ 5 & 6 & 7 & 8 \\ -2 & -3 & -1 & 0 \end{bmatrix}$

M_{41} is the determinant of the matrix formed by deleting the fourth row and the first column of $\mathbf{A}$.

$M_{41} = \begin{bmatrix} 0 & 0 & -2 \\ 1 & 0 & 0 \\ 6 & 7 & 8 \end{bmatrix}$

We will expand M_{41} across the first row.

$M_{41} = 0(-1)^{1+1}\begin{vmatrix} 0 & 0 \\ 7 & 8 \end{vmatrix} + 0(-1)^{1+2}\begin{vmatrix} 1 & 0 \\ 6 & 8 \end{vmatrix} +$

$(-2)(-1)^{3+1}\begin{vmatrix} 1 & 0 \\ 6 & 7 \end{vmatrix}$

$= 0 + 0 + (-2)(1)(1\cdot 7 - 6\cdot 0)$

$= 0 + 0 - 2(7) = -14$

M_{33} is the determinant of the matrix formed by deleting the third row and the third column of $\mathbf{A}$.

$M_{33} = \begin{bmatrix} 1 & 0 & -2 \\ 4 & 1 & 0 \\ -2 & -3 & 0 \end{bmatrix}$

We will expand M_{33} down the third column.

$M_{33} = -2(-1)^{1+3}\begin{vmatrix} 4 & 1 \\ -2 & -3 \end{vmatrix} + 0(-1)^{2+3}\begin{vmatrix} 1 & 0 \\ -2 & -3 \end{vmatrix} +$

$0(-1)^{3+3}\begin{vmatrix} 1 & 0 \\ 4 & 1 \end{vmatrix}$

$= -2(1)[4(-3) - (-2)(1)] + 0 + 0$

$= -2(-10) + 0 + 0 = 20$

18. See matrix $\mathbf{A}$ in Exercise 17 above.

$M_{12} = \begin{bmatrix} 4 & 0 & 0 \\ 5 & 7 & 8 \\ -2 & -1 & 0 \end{bmatrix}$

We will expand M_{12} across the first row.

$M_{12} = 4(-1)^{1+1}\begin{vmatrix} 7 & 8 \\ -1 & 0 \end{vmatrix} + 0(-1)^{1+2}\begin{vmatrix} 5 & 8 \\ -2 & 0 \end{vmatrix} +$

$0(-1)^{1+3}\begin{vmatrix} 5 & 7 \\ -2 & -1 \end{vmatrix}$

$= 4\cdot 1[7\cdot 0 - (-1)8] + 0 + 0$

$= 4(8) = 32$

$M_{44} = \begin{bmatrix} 1 & 0 & 0 \\ 4 & 1 & 0 \\ 5 & 6 & 7 \end{bmatrix}$

We will expand M_{44} across the first row.

$M_{44} = 1(-1)^{1+1}\begin{vmatrix} 1 & 0 \\ 6 & 7 \end{vmatrix} + 0(-1)^{1+2}\begin{vmatrix} 4 & 0 \\ 5 & 7 \end{vmatrix} +$

$0(-1)^{1+3}\begin{vmatrix} 4 & 1 \\ 5 & 6 \end{vmatrix}$

$= 1\cdot 1(1\cdot 7 - 6\cdot 0) + 0 + 0$

$= 1(7) = 7$

19. $\mathbf{A} = \begin{bmatrix} 1 & 0 & 0 & -2 \\ 4 & 1 & 0 & 0 \\ 5 & 6 & 7 & 8 \\ -2 & -3 & -1 & 0 \end{bmatrix}$

$A_{24} = (-1)^{2+4}M_{24} = 1\cdot M_{24} = M_{24}$

$= \begin{vmatrix} 1 & 0 & 0 \\ 5 & 6 & 7 \\ -2 & -3 & -1 \end{vmatrix}$

We will expand across the first row.

$\begin{vmatrix} 1 & 0 & 0 \\ 5 & 6 & 7 \\ -2 & -3 & -1 \end{vmatrix}$

$= 1(-1)^{1+1}\begin{vmatrix} 6 & 7 \\ -3 & -1 \end{vmatrix} + 0(-1)^{1+2}\begin{vmatrix} 5 & 7 \\ -2 & -1 \end{vmatrix} +$

$0(-1)^{1+3}\begin{vmatrix} 5 & 6 \\ -2 & -3 \end{vmatrix}$

$= 1\cdot 1[6(-1) - (-3)(7)] + 0 + 0$

$= 1(15) = 15$

$A_{43} = (-1)^{4+3}M_{43} = -1\cdot M_{43}$

$= -1\cdot\begin{vmatrix} 1 & 0 & -2 \\ 4 & 1 & 0 \\ 5 & 6 & 8 \end{vmatrix}$

We will expand across the first row.

$-1\cdot\begin{vmatrix} 1 & 0 & -2 \\ 4 & 1 & 0 \\ 5 & 6 & 8 \end{vmatrix}$

$= -1\Bigg[1(-1)^{1+1}\begin{vmatrix} 1 & 0 \\ 6 & 8 \end{vmatrix} + 0(-1)^{1+2}\begin{vmatrix} 4 & 0 \\ 5 & 8 \end{vmatrix} +$

$(-2)(-1)^{1+3}\begin{vmatrix} 4 & 1 \\ 5 & 6 \end{vmatrix}\Bigg]$

$= -1[1\cdot 1(1\cdot 8 - 6\cdot 0) + 0 + (-2)\cdot 1(4\cdot 6 - 5\cdot 1)]$

$= -1[1(8) + 0 - 2(19)]$

$= -1(8 - 38) = -1(-30)$

$= 30$

20. See matrix $\mathbf{A}$ in Exercise 19 above.

$A_{22} = (-1)^{2+2}M_{22} = M_{22}$

$= \begin{vmatrix} 1 & 0 & -2 \\ 5 & 7 & 8 \\ -2 & -1 & 0 \end{vmatrix}$

We will expand across the first row.

$$\begin{vmatrix} 1 & 0 & -2 \\ 5 & 7 & 8 \\ -2 & -1 & 0 \end{vmatrix}$$

$$= 1(-1)^{1+1}\begin{vmatrix} 7 & 8 \\ -1 & 0 \end{vmatrix} + 0(-1)^{1+2}\begin{vmatrix} 5 & 8 \\ -2 & 0 \end{vmatrix} +$$

$$(-2)(-1)^{1+3}\begin{vmatrix} 5 & 7 \\ -2 & -1 \end{vmatrix}$$

$$= 1\cdot 1[7\cdot 0-(-1)8] + 0 + (-2)\cdot 1[5(-1)-(-2)7]$$

$$= 1(8) + 0 - 2(9) = 8 + 0 - 18$$

$$= -10$$

$$A_{34} = (-1)^{3+4}M_{34} = -1\cdot M_{34}$$

$$= -1\cdot\begin{vmatrix} 1 & 0 & 0 \\ 4 & 1 & 0 \\ -2 & -3 & -1 \end{vmatrix}$$

We will expand across the first row.

$$-1\cdot\begin{vmatrix} 1 & 0 & 0 \\ 4 & 1 & 0 \\ -2 & -3 & -1 \end{vmatrix}$$

$$= -1\left[1(-1)^{1+1}\begin{vmatrix} 1 & 0 \\ -3 & -1 \end{vmatrix} + 0(-1)^{1+2}\begin{vmatrix} 4 & 0 \\ -2 & -1 \end{vmatrix} +\right.$$

$$\left. 0(-1)^{1+3}\begin{vmatrix} 4 & 1 \\ -2 & -3 \end{vmatrix}\right]$$

$$= -1[1\cdot 1(1(-1)-(-3)\cdot 0) + 0 + 0]$$

$$= -1[1(-1)] = 1$$

21. $\mathbf{A} = \begin{bmatrix} 1 & 0 & 0 & -2 \\ 4 & 1 & 0 & 0 \\ 5 & 6 & 7 & 8 \\ -2 & -3 & -1 & 0 \end{bmatrix}$

$|\mathbf{A}|$

$$= 1\cdot A_{11} + 0\cdot A_{12} + 0\cdot A_{13} + (-2)A_{14}$$

$$= A_{11} + (-2)A_{14}$$

$$= (-1)^{1+1}\begin{vmatrix} 1 & 0 & 0 \\ 6 & 7 & 8 \\ -3 & -1 & 0 \end{vmatrix} +$$

$$(-2)(-1)^{1+4}\begin{vmatrix} 4 & 1 & 0 \\ 5 & 6 & 7 \\ -2 & -3 & -1 \end{vmatrix}$$

$$= \begin{vmatrix} 1 & 0 & 0 \\ 6 & 7 & 8 \\ -3 & -1 & 0 \end{vmatrix} + 2\begin{vmatrix} 4 & 1 & 0 \\ 5 & 6 & 7 \\ -2 & -3 & -1 \end{vmatrix}$$

We will expand each determinant across the first row. We have:

$$1(-1)^{1+1}\begin{vmatrix} 7 & 8 \\ -1 & 0 \end{vmatrix} + 0 + 0 +$$

$$2\left[4(-1)^{1+1}\begin{vmatrix} 6 & 7 \\ -3 & -1 \end{vmatrix} + 1(-1)^{1+2}\begin{vmatrix} 5 & 7 \\ -2 & -1 \end{vmatrix} + 0\right]$$

$$= 1\cdot 1[7\cdot 0-(-1)8] + 2[4\cdot 1[6(-1)-(-3)\cdot 7] +$$

$$1(-1)[5(-1)-(-2)\cdot 7]$$

$$= 1(8) + 2[4(15) - 1(9)] = 8 + 2(51)$$

$$= 8 + 102 = 110$$

22. See matrix **A** in Exercise 21 above.

$$|\mathbf{A}| = 0\cdot A_{13} + 0\cdot A_{23} + 7A_{33} + (-1)A_{43}$$

$$= 7A_{33} - A_{43}$$

$$= 7(-1)^{3+3}\begin{vmatrix} 1 & 0 & -2 \\ 4 & 1 & 0 \\ -2 & -3 & 0 \end{vmatrix} -$$

$$(-1)^{4+3}\begin{vmatrix} 1 & 0 & -2 \\ 4 & 1 & 0 \\ 5 & 6 & 8 \end{vmatrix}$$

We will expand each determinant down the third column. We have:

$$7\left[-2(-1)^{1+3}\begin{vmatrix} 4 & 1 \\ -2 & 3 \end{vmatrix} + 0 + 0\right] +$$

$$\left[-2(-1)^{1+3}\begin{vmatrix} 4 & 1 \\ 5 & 6 \end{vmatrix} + 0 + 8(-1)^{3+3}\begin{vmatrix} 1 & 0 \\ 4 & 1 \end{vmatrix}\right]$$

$$= 7[-2(4(-3)-(-2)\cdot 1)] + [-2(4\cdot 6 - 5\cdot 1) +$$

$$8(1\cdot 1 - 4\cdot 0)]$$

$$= 7[-2(-10)] + [-2(19) + 8(1)]$$

$$= 7(20) + (-30) = 140 - 30$$

$$= 110$$

23. $-2x + 4y = 3,$

$3x - 7y = 1$

$$D = \begin{vmatrix} -2 & 4 \\ 3 & -7 \end{vmatrix} = -2(-7) - 3\cdot 4 = 14 - 12 = 2$$

$$D_x = \begin{vmatrix} 3 & 4 \\ 1 & -7 \end{vmatrix} = 3(-7) - 1\cdot 4 = -21 - 4 = -25$$

$$D_y = \begin{vmatrix} -2 & 3 \\ 3 & 1 \end{vmatrix} = -2\cdot 1 - 3\cdot 3 = -2 - 9 = -11$$

$$x = \frac{D_x}{D} = \frac{-25}{2} = -\frac{25}{2}$$

$$y = \frac{D_y}{D} = \frac{-11}{2} = -\frac{11}{2}$$

The solution is $\left(-\frac{25}{2}, -\frac{11}{2}\right)$.

24. $5x - 4y = -3,$
$7x + 2y = 6$

$$D = \begin{vmatrix} 5 & -4 \\ 7 & 2 \end{vmatrix} = 5 \cdot 2 - 7(-4) = 10 + 28 = 38$$

$$D_x = \begin{vmatrix} -3 & -4 \\ 6 & 2 \end{vmatrix} = -3 \cdot 2 - 6(-4) = -6 + 24 = 18$$

$$D_y = \begin{vmatrix} 5 & -3 \\ 7 & 6 \end{vmatrix} = 5 \cdot 6 - 7(-3) = 30 + 21 = 51$$

$$x = \frac{D_x}{D} = \frac{18}{38} = \frac{9}{19}$$
$$y = \frac{D_y}{D} = \frac{51}{38}$$

The solution is $\left(\frac{9}{19}, \frac{51}{38}\right)$.

25. $2x - y = 5,$
$x - 2y = 1$

$$D = \begin{vmatrix} 2 & -1 \\ 1 & -2 \end{vmatrix} = 2(-2) - 1(-1) = -4 + 1 = -3$$

$$D_x = \begin{vmatrix} 5 & -1 \\ 1 & -2 \end{vmatrix} = 5(-2) - 1(-1) = -10 + 1 = -9$$

$$D_y = \begin{vmatrix} 2 & 5 \\ 1 & 1 \end{vmatrix} = 2 \cdot 1 - 1 \cdot 5 = 2 - 5 = -3$$

$$x = \frac{D_x}{D} = \frac{-9}{-3} = 3$$
$$y = \frac{D_y}{D} = \frac{-3}{-3} = 1$$

The solution is $(3, 1)$.

26. $3x + 4y = -2,$
$5x - 7y = 1$

$$D = \begin{vmatrix} 3 & 4 \\ 5 & -7 \end{vmatrix} = 3(-7) - 5 \cdot 4 = -21 - 20 = -41$$

$$D_x = \begin{vmatrix} -2 & 4 \\ 1 & -7 \end{vmatrix} = -2(-7) - 1 \cdot 4 = 14 - 4 = 10$$

$$D_y = \begin{vmatrix} 3 & -2 \\ 5 & 1 \end{vmatrix} = 3 \cdot 1 - 5(-2) = 3 + 10 = 13$$

$$x = \frac{D_x}{D} = \frac{10}{-41} = -\frac{10}{41}$$
$$y = \frac{D_y}{D} = \frac{13}{-41} = -\frac{13}{41}$$

The solution is $\left(-\frac{10}{41}, -\frac{13}{41}\right)$.

27. $2x + 9y = -2,$
$4x - 3y = 3$

$$D = \begin{vmatrix} 2 & 9 \\ 4 & -3 \end{vmatrix} = 2(-3) - 4 \cdot 9 = -6 - 36 = -42$$

$$D_x = \begin{vmatrix} -2 & 9 \\ 3 & -3 \end{vmatrix} = -2(-3) - 3 \cdot 9 = 6 - 27 = -21$$

$$D_y = \begin{vmatrix} 2 & -2 \\ 4 & 3 \end{vmatrix} = 2 \cdot 3 - 4(-2) = 6 + 8 = 14$$

$$x = \frac{D_x}{D} = \frac{-21}{-42} = \frac{1}{2}$$
$$y = \frac{D_y}{D} = \frac{14}{-42} = -\frac{1}{3}$$

The solution is $\left(\frac{1}{2}, -\frac{1}{3}\right)$.

28. $2x + 3y = -1,$
$3x + 6y = -0.5$

$$D = \begin{vmatrix} 2 & 3 \\ 3 & 6 \end{vmatrix} = 2 \cdot 6 - 3 \cdot 3 = 12 - 9 = 3$$

$$D_x = \begin{vmatrix} -1 & 3 \\ -0.5 & 6 \end{vmatrix} = -1 \cdot 6 - (-0.5)3 =$$
$$-6 + 1.5 = -4.5$$

$$D_y = \begin{vmatrix} 2 & -1 \\ 3 & -0.5 \end{vmatrix} = 2(-0.5) - 3(-1) = -1 + 3 = 2$$

$$x = \frac{D_x}{D} = \frac{-4.5}{3} = -1.5, \text{ or } -\frac{3}{2}$$
$$y = \frac{D_y}{D} = \frac{2}{3}$$

The solution is $\left(-\frac{3}{2}, \frac{2}{3}\right)$.

29. $2x + 5y = 7,$
$3x - 2y = 1$

$$D = \begin{vmatrix} 2 & 5 \\ 3 & -2 \end{vmatrix} = 2(-2) - 3 \cdot 5 = -4 - 15 = -19$$

$$D_x = \begin{vmatrix} 7 & 5 \\ 1 & -2 \end{vmatrix} = 7(-2) - 1 \cdot 5 = -14 - 5 = -19$$

$$D_y = \begin{vmatrix} 2 & 7 \\ 3 & 1 \end{vmatrix} = 2 \cdot 1 - 3 \cdot 7 = 2 - 21 = -19$$

$$x = \frac{D_x}{D} = \frac{-19}{-19} = 1$$
$$y = \frac{D_y}{D} = \frac{-19}{-19} = 1$$

The solution is $(1, 1)$.

30. $3x + 2y = 7,$

$2x + 3y = -2$

$$D = \begin{vmatrix} 3 & 2 \\ 2 & 3 \end{vmatrix} = 3 \cdot 3 - 2 \cdot 2 = 9 - 4 = 5$$

$$D_x = \begin{vmatrix} 7 & 2 \\ -2 & 3 \end{vmatrix} = 7 \cdot 3 - (-2) \cdot 2 = 21 + 4 = 25$$

$$D_y = \begin{vmatrix} 3 & 7 \\ 2 & -2 \end{vmatrix} = 3(-2) - 2 \cdot 7 = -6 - 14 = -20$$

$$x = \frac{D_x}{D} = \frac{25}{5} = 5$$

$$y = \frac{D_y}{D} = \frac{-20}{5} = -4$$

The solution is $(5, -4)$.

31. $3x + 2y - z = 4,$

$3x - 2y + z = 5,$

$4x - 5y - z = -1$

$$D = \begin{vmatrix} 3 & 2 & -1 \\ 3 & -2 & 1 \\ 4 & -5 & -1 \end{vmatrix} = 42$$

$$D_x = \begin{vmatrix} 4 & 2 & -1 \\ 5 & -2 & 1 \\ -1 & -5 & -1 \end{vmatrix} = 63$$

$$D_y = \begin{vmatrix} 3 & 4 & -1 \\ 3 & 5 & 1 \\ 4 & -1 & -1 \end{vmatrix} = 39$$

$$D_z = \begin{vmatrix} 3 & 2 & 4 \\ 3 & -2 & 5 \\ 4 & -5 & -1 \end{vmatrix} = 99$$

$$x = \frac{D_x}{D} = \frac{63}{42} = \frac{3}{2}$$

$$y = \frac{D_y}{D} = \frac{39}{42} = \frac{13}{14}$$

$$z = \frac{D_z}{D} = \frac{99}{42} = \frac{33}{14}$$

The solution is $\left(\frac{3}{2}, \frac{13}{14}, \frac{33}{14}\right)$.

(Note that we could have used Cramer's rule to find only two of the values and then used substitution to find the remaining value.)

32. $3x - y + 2z = 1,$

$x - y + 2z = 3,$

$-2x + 3y + z = 1$

$$D = \begin{vmatrix} 3 & -1 & 2 \\ 1 & -1 & 2 \\ -2 & 3 & 1 \end{vmatrix} = -14$$

$$D_x = \begin{vmatrix} 1 & -1 & 2 \\ 3 & -1 & 2 \\ 1 & 3 & 1 \end{vmatrix} = 14$$

$$D_y = \begin{vmatrix} 3 & 1 & 2 \\ 1 & 3 & 2 \\ -2 & 1 & 1 \end{vmatrix} = 12$$

$$D_z = \begin{vmatrix} 3 & -1 & 1 \\ 1 & -1 & 3 \\ -2 & 3 & 1 \end{vmatrix} = -22$$

$$x = \frac{D_x}{D} = \frac{14}{-14} = -1$$

$$y = \frac{D_y}{D} = \frac{12}{-14} = -\frac{6}{7}$$

$$z = \frac{D_z}{D} = \frac{-22}{-14} = \frac{11}{7}$$

The solution is $\left(-1, -\frac{6}{7}, \frac{11}{7}\right)$.

33. $3x + 5y - z = -2,$

$x - 4y + 2z = 13,$

$2x + 4y + 3z = 1$

$$D = \begin{vmatrix} 3 & 5 & -1 \\ 1 & -4 & 2 \\ 2 & 4 & 3 \end{vmatrix} = -67$$

$$D_x = \begin{vmatrix} -2 & 5 & -1 \\ 13 & -4 & 2 \\ 1 & 4 & 3 \end{vmatrix} = -201$$

$$D_y = \begin{vmatrix} 3 & -2 & -1 \\ 1 & 13 & 2 \\ 2 & 1 & 3 \end{vmatrix} = 134$$

$$D_z = \begin{vmatrix} 3 & 5 & -2 \\ 1 & -4 & 13 \\ 2 & 4 & 1 \end{vmatrix} = -67$$

$$x = \frac{D_x}{D} = \frac{-201}{-67} = 3$$

$$y = \frac{D_y}{D} = \frac{134}{-67} = -2$$

$$z = \frac{D_z}{D} = \frac{-67}{-67} = 1$$

The solution is $(3, -2, 1)$.

(Note that we could have used Cramer's rule to find only two of the values and then used substitution to find the remaining value.)

34. $3x + 2y + 2z = 1,$
$5x - y - 6z = 3,$
$2x + 3y + 3z = 4$

$$D = \begin{vmatrix} 3 & 2 & 2 \\ 5 & -1 & -6 \\ 2 & 3 & 3 \end{vmatrix} = 25$$

$$D_x = \begin{vmatrix} 1 & 2 & 2 \\ 3 & -1 & -6 \\ 4 & 3 & 3 \end{vmatrix} = -25$$

$$D_y = \begin{vmatrix} 3 & 1 & 2 \\ 5 & 3 & -6 \\ 2 & 4 & 3 \end{vmatrix} = 100$$

$$D_z = \begin{vmatrix} 3 & 2 & 1 \\ 5 & -1 & 3 \\ 2 & 3 & 4 \end{vmatrix} = -50$$

$$x = \frac{D_x}{D} = \frac{-25}{25} = -1$$

$$y = \frac{D_y}{D} = \frac{100}{25} = 4$$

$$z = \frac{D_z}{D} = \frac{-50}{25} = -2$$

The solution is $(-1, 4, -2)$.

35. $x - 3y - 7z = 6,$
$2x + 3y + z = 9,$
$4x + y = 7$

$$D = \begin{vmatrix} 1 & -3 & -7 \\ 2 & 3 & 1 \\ 4 & 1 & 0 \end{vmatrix} = 57$$

$$D_x = \begin{vmatrix} 6 & -3 & -7 \\ 9 & 3 & 1 \\ 7 & 1 & 0 \end{vmatrix} = 57$$

$$D_y = \begin{vmatrix} 1 & 6 & -7 \\ 2 & 9 & 1 \\ 4 & 7 & 0 \end{vmatrix} = 171$$

$$D_z = \begin{vmatrix} 1 & -3 & 6 \\ 2 & 3 & 9 \\ 4 & 1 & 7 \end{vmatrix} = -114$$

$$x = \frac{D_x}{D} = \frac{57}{57} = 1$$

$$y = \frac{D_y}{D} = \frac{171}{57} = 3$$

$$z = \frac{D_z}{D} = \frac{-114}{57} = -2$$

The solution is $(1, 3, -2)$.

(Note that we could have used Cramer's rule to find only two of the values and then used substitution to find the remaining value.)

36. $x - 2y - 3z = 4,$
$3x - 2z = 8,$
$2x + y + 4z = 13$

$$D = \begin{vmatrix} 1 & -2 & -3 \\ 3 & 0 & -2 \\ 2 & 1 & 4 \end{vmatrix} = 25$$

$$D_x = \begin{vmatrix} 4 & -2 & -3 \\ 8 & 0 & -2 \\ 13 & 1 & 4 \end{vmatrix} = 100$$

$$D_y = \begin{vmatrix} 1 & 4 & -3 \\ 3 & 8 & -2 \\ 2 & 13 & 4 \end{vmatrix} = -75$$

$$D_z = \begin{vmatrix} 1 & -2 & 4 \\ 3 & 0 & 8 \\ 2 & 1 & 13 \end{vmatrix} = 50$$

$$x = \frac{D_x}{D} = \frac{100}{25} = 4$$

$$y = \frac{D_y}{D} = \frac{-75}{25} = -3$$

$$z = \frac{D_z}{D} = \frac{50}{25} = -2$$

The solution is $(4, -3, 2)$.

37. $6y + 6z = -1,$
$8x + 6z = -1,$
$4x + 9y = 8$

$$D = \begin{vmatrix} 0 & 6 & 6 \\ 8 & 0 & 6 \\ 4 & 9 & 0 \end{vmatrix} = 576$$

$$D_x = \begin{vmatrix} -1 & 6 & 6 \\ -1 & 0 & 6 \\ 8 & 9 & 0 \end{vmatrix} = 288$$

$$D_y = \begin{vmatrix} 0 & -1 & 6 \\ 8 & -1 & 6 \\ 4 & 8 & 0 \end{vmatrix} = 384$$

$$D_z = \begin{vmatrix} 0 & 6 & -1 \\ 8 & 0 & -1 \\ 4 & 9 & 8 \end{vmatrix} = -480$$

$$x = \frac{D_x}{D} = \frac{288}{576} = \frac{1}{2}$$

$$y = \frac{D_y}{D} = \frac{384}{576} = \frac{2}{3}$$

$$z = \frac{D_z}{D} = \frac{-480}{576} = -\frac{5}{6}$$

The solution is $\left(\frac{1}{2}, \frac{2}{3}, -\frac{5}{6}\right)$.

(Note that we could have used Cramer's rule to find only two of the values and then used substitution to find the remaining value.)

38. $3x + 5y \quad = 2,$

$2x \quad - 3z = 7,$

$4y + 2z = -1$

$$D = \begin{vmatrix} 3 & 5 & 0 \\ 2 & 0 & -3 \\ 0 & 4 & 2 \end{vmatrix} = 16$$

$$D_x = \begin{vmatrix} 2 & 5 & 0 \\ 7 & 0 & -3 \\ -1 & 4 & 2 \end{vmatrix} = -31$$

$$D_y = \begin{vmatrix} 3 & 2 & 0 \\ 2 & 7 & -3 \\ 0 & -1 & 2 \end{vmatrix} = 25$$

$$D_z = \begin{vmatrix} 3 & 5 & 2 \\ 2 & 0 & 7 \\ 0 & 4 & -1 \end{vmatrix} = -58$$

$$x = \frac{D_x}{D} = \frac{-31}{16} = -\frac{31}{16}$$

$$y = \frac{D_y}{D} = \frac{25}{16}$$

$$z = \frac{D_z}{D} = \frac{-58}{16} = -\frac{29}{8}$$

The solution is $\left(-\frac{31}{16}, \frac{25}{16}, -\frac{29}{8}\right)$.

39. Left to the student

40. Left to the student

41. Enter matrix **A** and use the "det(" option from the MATRIX MATH menu.

$|\mathbf{A}| = 110$

42. Enter matrix **A** and use the "det(" option from the MATRIX MATH menu.

$|\mathbf{A}| = -195$

43. Left to the student

44. Left to the student

45. If $\begin{vmatrix} a_1 & b_1 \\ a_2 & b_2 \end{vmatrix} = 0$, then $a_1 = ka_2$ and $b_1 = kb_2$ for some number k. This means that the equations $a_1x + b_1y = c$, and $a_2x + b_2y = c_2$ are either dependent, if $c_1 = kc_2$, or inconsistent, if $c_1 \neq kc_2$.

46. If $a_1x + b_1y = c_1$ and $a_2x + b_2y = c_2$ are parallel lines, then $a_1 = ka_2$, $b_1 = kb_2$, and $c_1 \neq kc_2$, for some number k. Then $\begin{vmatrix} a_1 & b_1 \\ a_2 & b_2 \end{vmatrix} = 0$, $\begin{vmatrix} c_1 & b_1 \\ c_2 & b_2 \end{vmatrix} \neq 0$, and $\begin{vmatrix} a_1 & c_1 \\ a_2 & c_2 \end{vmatrix} \neq 0$.

47. The graph of $f(x) = 3x+2$ is shown below. Since it passes the horizontal-line test, the function is one-to-one.

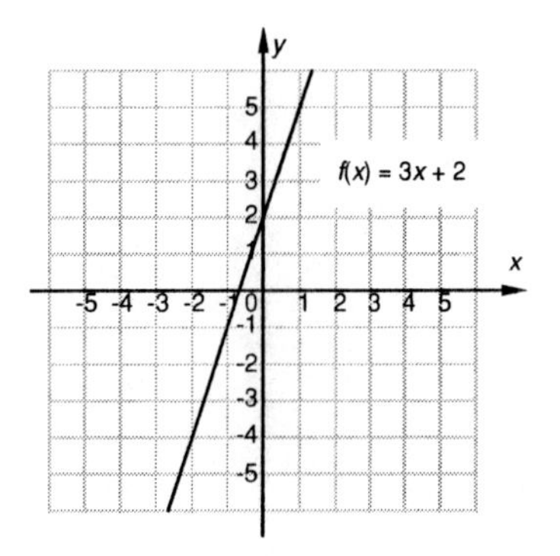

We find a formula for $f^{-1}(x)$.

Replace $f(x)$ with y: $y = 3x + 2$

Interchange x and y: $x = 3y + 2$

Solve for y: $y = \dfrac{x-2}{3}$

Replace y with $f^{-1}(x)$: $f^{-1}(x) = \dfrac{x-2}{3}$

48. The graph of $f(x) = x^2 - 4$ is shown below. It fails the horizontal-line test, so it is not one-to-one.

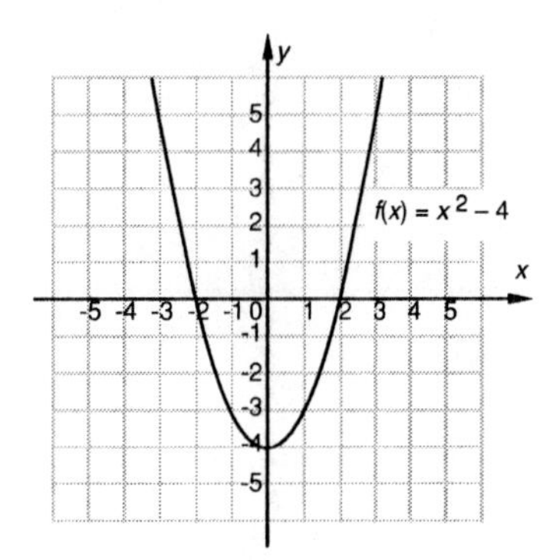

49. The graph of $f(x) = |x| + 3$ is shown below. It fails the horizontal-line test, so it is not one-to-one.

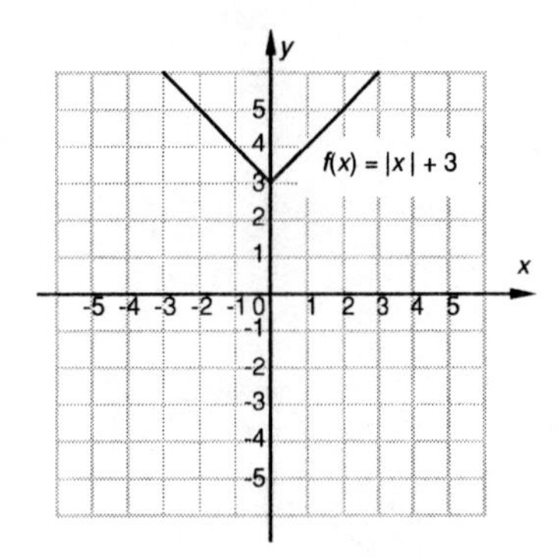

50. The graph of $f(x) = \sqrt[3]{x}+1$ is show below. Since it passes the horizontal-line test, the function is one-to-one.

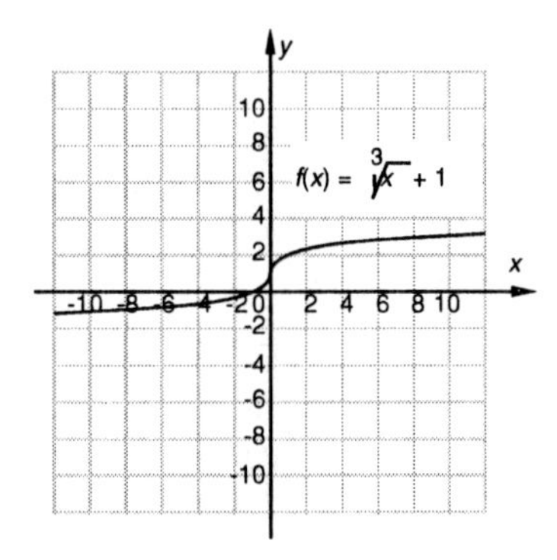

We find a formula for $f^{-1}(x)$.

Replace $f(x)$ with y: $y = \sqrt[3]{x}+1$

Interchange x and y: $x = \sqrt[3]{y}+1$

Solve for y: $y = (x-1)^3$

Replace y with $f^{-1}(x)$: $f^{-1}(x) = (x-1)^3$

51. $(3-4i)-(-2-i) = 3-4i+2+i =$
$(3+2)+(-4+1)i = 5-3i$

52. $(5+2i)+(1-4i) = 6-2i$

53. $(1-2i)(6+2i) = 6+2i-12i-4i^2 =$
$6+2i-12i+4 = 10-10i$

54. $\dfrac{3+i}{4-3i} = \dfrac{3+i}{4-3i}\cdot\dfrac{4+3i}{4+3i} = \dfrac{12+9i+4i+3i^2}{16-9i^2} =$
$\dfrac{12+9i+4i-3}{16+9} = \dfrac{9+13i}{25} = \dfrac{9}{25}+\dfrac{13}{25}i$

55. $\begin{vmatrix} x & 5 \\ -4 & x \end{vmatrix} = 24$

$x\cdot x-(-4)(5) = 24$ Evaluating the determinant

$x^2+20 = 24$

$x^2 = 4$

$x = \pm 2$

The solutions are -2 and 2.

56. $\begin{vmatrix} y & 2 \\ 3 & y \end{vmatrix} = y$

$y^2-6 = y$

$y^2-y-6 = 0$

$(y-3)(y+2) = 0$

$y-3 = 0$ *or* $y+2 = 0$

$y = 3$ *or* $y = -2$

The solutions are 3 and -2.

57. $\begin{vmatrix} x & -3 \\ -1 & x \end{vmatrix} \geq 0$

$x\cdot x-(-1)(-3) \geq 0$

$x^2-3 \geq 0$

We solve the related equation.

$x^2-3 = 0$

$x^2 = 3$

$x = \pm\sqrt{3}$

The numbers $-\sqrt{3}$ and $\sqrt{3}$ divide the x-axis into three intervals. We let $f(x) = x^2-3$ and test a value in each interval.

$(-\infty,-\sqrt{3}):\ f(-2) = (-2)^2-3 = 1 > 0$

$(-\sqrt{3},\sqrt{3}):\ f(0) = 0^2-3 = -3 < 0$

$(\sqrt{3},\infty):\ f(2) = 2^2-3 = 1 > 0$

The function is positive in $(-\infty,-\sqrt{3})$, and $(\sqrt{3},\infty)$. We also include the endpoints of the intervals since the inequality symbol is $\geq$. The solution set is $\{x|x \leq -\sqrt{3}$ *or* $x \geq \sqrt{3}\}$, or $(-\infty,-\sqrt{3}]\cup[\sqrt{3},\infty)$.

58. $\begin{vmatrix} y & -5 \\ -2 & y \end{vmatrix} < 0$

$y^2-10 < 0$

Solve the related equation.

$y^2-10 = 0$

$y^2 = 10$

$y = \pm\sqrt{10}$

The number $-\sqrt{10}$ and $\sqrt{10}$ divide the axis into three intervals. We let $f(y) = y^2-10$ and test a value in each interval.

$(-\infty,-\sqrt{10}):\ f(-4) = (-4)^2-10 = 6 > 0$

$(-\sqrt{10},\sqrt{10}):\ f(0) = 0^2-10 = -10 < 0$

$(\sqrt{10},\infty):\ f(4) = 4^2-10 = 6 > 0$

The solution set is $\{y|-\sqrt{10} < y < \sqrt{10}\}$, or $(-\sqrt{10},\sqrt{10})$.

59. $\begin{vmatrix} x+3 & 4 \\ x-3 & 5 \end{vmatrix} = -7$

$(x+3)(5)-(x-3)(4) = -7$

$5x+15-4x+12 = -7$

$x+27 = -7$

$x = -34$

The solution is -34.

60. $\begin{vmatrix} m+2 & -3 \\ m+5 & -4 \end{vmatrix} = 3m-5$

$-4m-8+3m+15 = 3m-5$

$-m+7 = 3m-5$

$-4m = -12$

$m = 3$

61. $\begin{vmatrix} 2 & x & 1 \\ 1 & 2 & -1 \\ 3 & 4 & -2 \end{vmatrix} = -6$

$-x-2 = -6$ Evaluating the determinant

$-x = -4$

$x = 4$

The solution is 4.

62. $\begin{vmatrix} x & 2 & x \\ 3 & -1 & 1 \\ 1 & -2 & 2 \end{vmatrix} = -10$

$$-5x - 10 = -10$$
$$-5x = 0$$
$$x = 0$$

63. Answers may vary.

$\begin{vmatrix} L & -W \\ 2 & 2 \end{vmatrix}$

64. Answers may vary.

$\begin{vmatrix} \pi & -\pi \\ h & r \end{vmatrix}$

65. Answers may vary.

$\begin{vmatrix} a & b \\ -b & a \end{vmatrix}$

66. Answers may vary.

$\begin{vmatrix} \frac{h}{2} & -\frac{h}{2} \\ b & a \end{vmatrix}$

67. Answers may vary.

$\begin{vmatrix} 2\pi r & 2\pi r \\ -h & r \end{vmatrix}$

68. Answers may vary.

$\begin{vmatrix} xy & Q \\ Q & xy \end{vmatrix}$

Exercise Set 8.7

1. Graph (f) is the graph of $y > x$.
2. Graph (c) is the graph of $y < -2x$.
3. Graph (h) is the graph of $y \leq x - 3$.
4. Graph (a) is the graph of $y \geq x + 5$.
5. Graph (g) is the graph of $2x + y < 4$.
6. Graph (d) is the graph of $3x + y < -6$.
7. Graph (b) is the graph of $2x - 5y > 10$.
8. Graph (e) is the graph of $3x - 9y < 9$.
9. Graph: $y > 2x$

   1. We first graph the related equation $y = 2x$. We draw the line dashed since the inequality symbol is $>$.
   2. To determine which half-plane to shade, test a point not on the line. We try $(1, 1)$ and substitute:

   $$\begin{array}{c|c} y & > 2x \\ \hline 1 & ?\ 2 \cdot 1 \\ 1 & 2 \quad \text{FALSE} \end{array}$$

   Since $1 > 2$ is false, $(1, 1)$ is not a solution, nor are any points in the half-plane containing $(1, 1)$. The points in the opposite half-plane are solutions, so we shade that half-plane and obtain the graph.

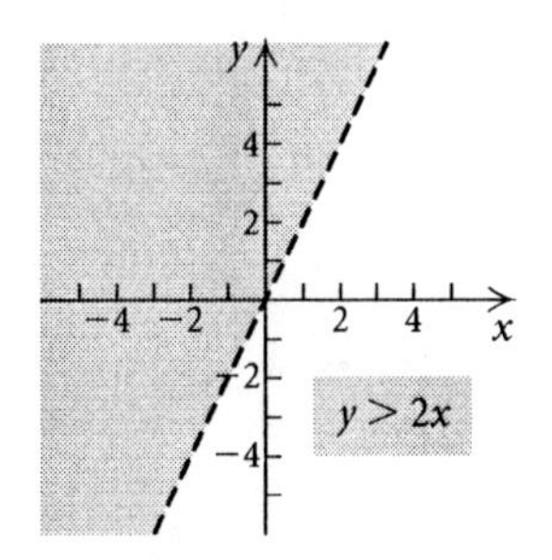

10.

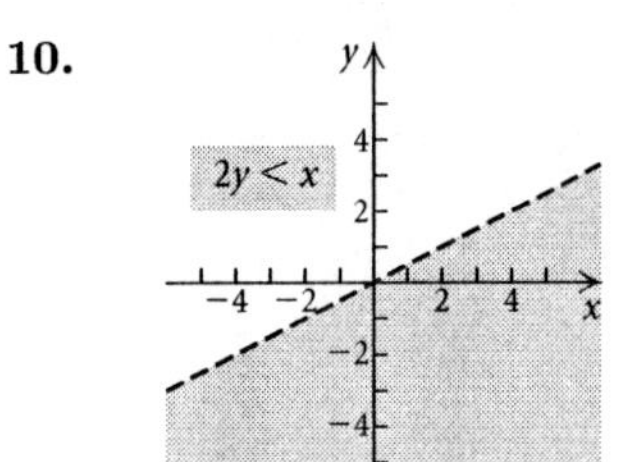

11. Graph: $y + x \geq 0$

1. First graph the related equation $y + x = 0$. Draw the line solid since the inequality is $\geq$.
2. Next determine which half-plane to shade by testing a point not on the line. Here we use $(2, 2)$ as a check.

$$\begin{array}{c|c} y + x & \geq 0 \\ \hline 2 + 2 & ?\ 0 \\ 4 & 0 \quad \text{TRUE} \end{array}$$

Since $4 \geq 0$ is true, $(2, 2)$ is a solution. Thus shade the half-plane containing $(2, 2)$.

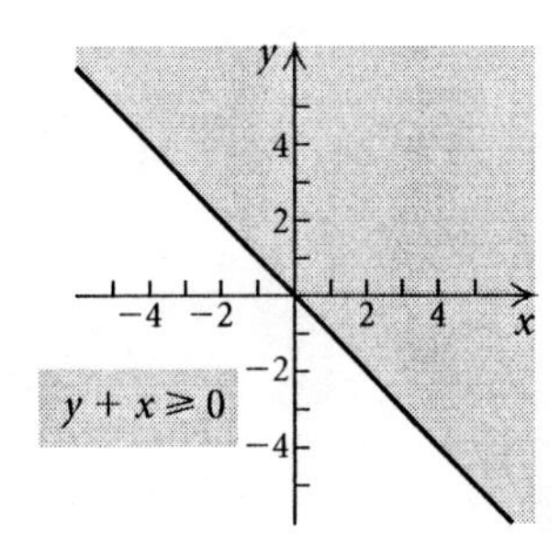

12.

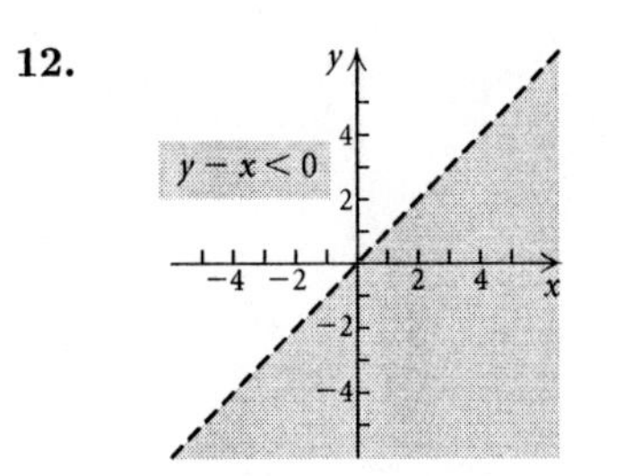

13. Graph: $y > x - 3$

1. We first graph the related equation $y = x - 3$. Draw the line dashed since the inequality symbol is $>$.

2. To determine which half-plane to shade, test a point not on the line. We try $(0,0)$.

$$\begin{array}{c|l} y > x - 3 & \\ \hline 0 \; ? \; 0-3 & \\ 0 & -3 \quad \text{TRUE} \end{array}$$

Since $0 > -3$ is true, $(0,0)$ is a solution. Thus we shade the half-plane containing $(0,0)$.

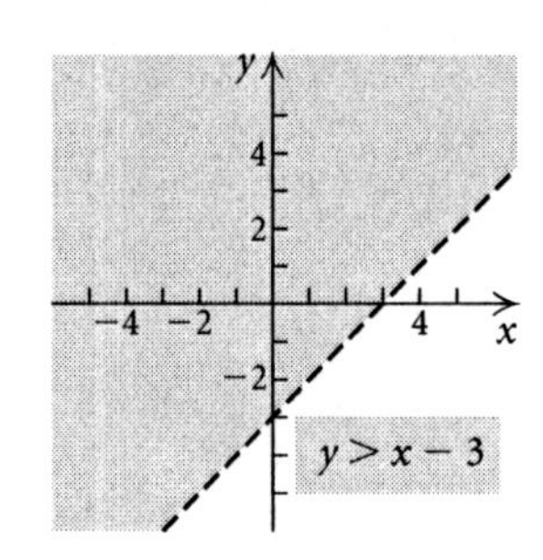

14.

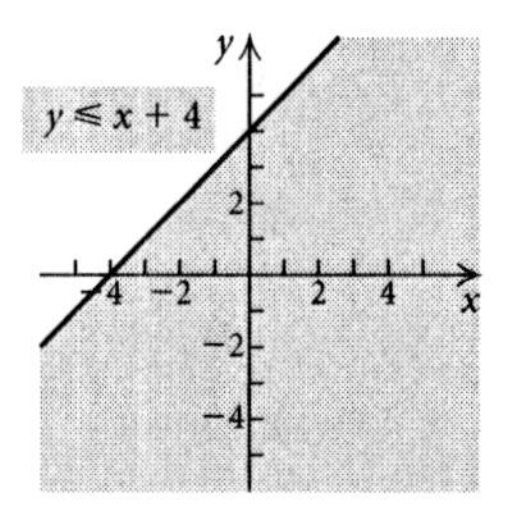

15. Graph: $x + y < 4$

1. First graph the related equation $x + y = 4$. Draw the line dashed since the inequality is $<$.
2. To determine which half-plane to shade, test a point not on the line. We try $(0,0)$.

$$\begin{array}{c|l} x + y < 4 & \\ \hline 0+0 \; ? \; 4 & \\ 0 & 4 \quad \text{TRUE} \end{array}$$

Since $0 < 4$ is true, $(0,0)$ is a solution. Thus shade the half-plane containing $(0,0)$.

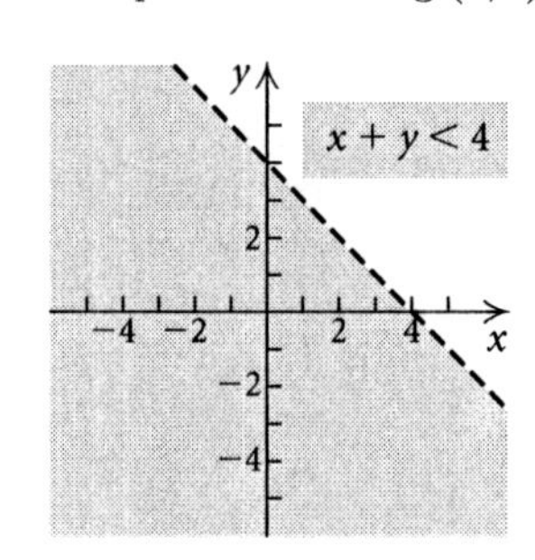

16.

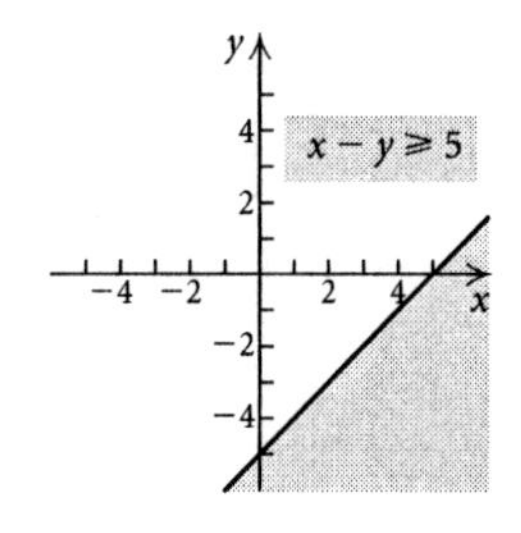

17. Graph: $3x - 2y \leq 6$

1. First graph the related equation $3x - 2y = 6$. Draw the line solid since the inequality is $\leq$.
2. To determine which half-plane to shade, test a point not on the line. We try $(0,0)$.

$$\begin{array}{c|l} 3x - 2y \leq 6 & \\ \hline 3(0) - 2(0) \; ? \; 6 & \\ 0 & 6 \quad \text{TRUE} \end{array}$$

Since $0 \leq 6$ is true, $(0,0)$ is a solution. Thus shade the half-plane containing $(0,0)$.

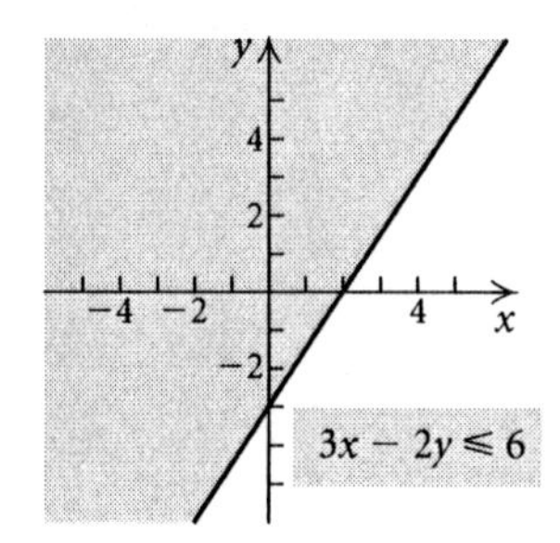

18.

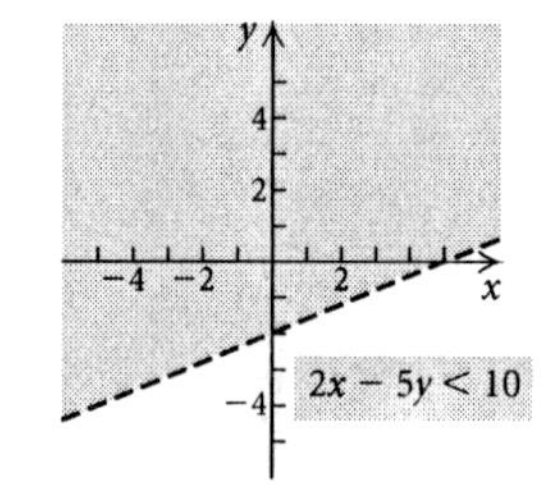

19. Graph: $3y + 2x \geq 6$

1. First graph the related equation $3y + 2x = 6$. Draw the line solid since the inequality is $\geq$.
2. To determine which half-plane to shade, test a point not on the line. We try $(0,0)$.

$$\begin{array}{c|l} 3y + 2x \geq 6 & \\ \hline 3 \cdot 0 + 2 \cdot 0 \; ? \; 6 & \\ 0 & 6 \quad \text{FALSE} \end{array}$$

Since $0 \geq 6$ is false, $(0,0)$ is not a solution. We shade the half-plane which does not contain $(0,0)$.

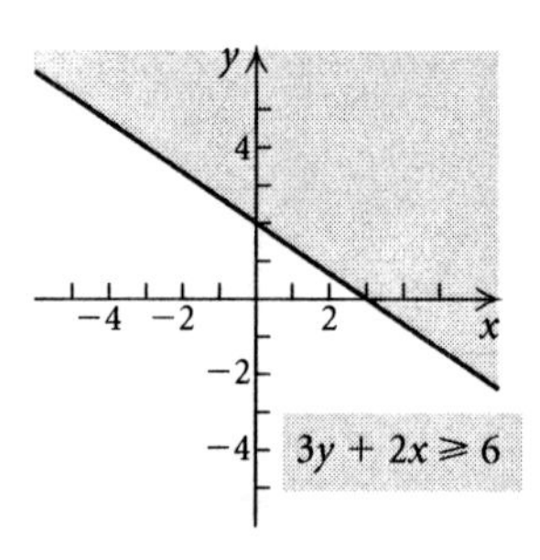

20.

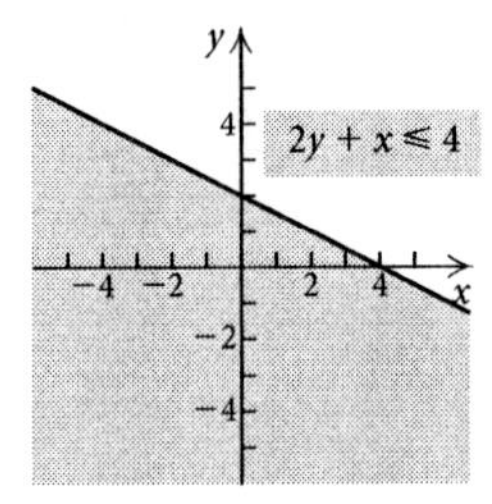

21. Graph: $3x - 2 \leq 5x + y$

$$-2 \leq 2x + y \quad \text{Adding } -3x$$

1. First graph the related equation $2x+y = -2$. Draw the line solid since the inequality is $\leq$.
2. To determine which half-plane to shade, test a point not on the line. We try $(0,0)$.

$$\begin{array}{c|c} 2x + y \geq & -2 \\ \hline 2(0) + 0 \ ? & -2 \\ 0 & -2 \quad \text{TRUE} \end{array}$$

Since $0 \geq -2$ is true, $(0,0)$ is a solution. Thus shade the half-plane containing the origin.

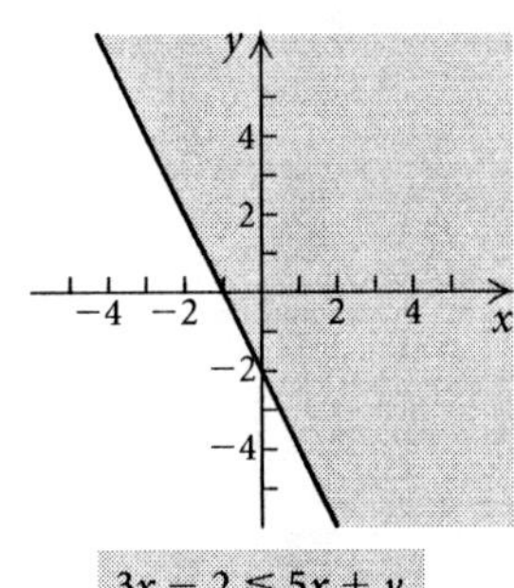

22.

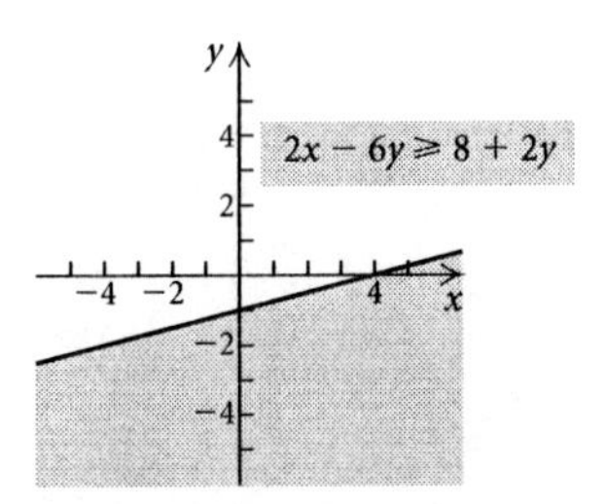

23. Graph: $x < -4$

1. We first graph the related equation $x = -4$. Draw the line dashed since the inequality is $<$.
2. To determine which half-plane to shade, test a point not on the line. We try $(0,0)$.

$$\begin{array}{c} x < -4 \\ \hline 0 \ ? \ -4 \quad \text{FALSE} \end{array}$$

Since $0 < -4$ is false, $(0,0)$ is not a solution. Thus, we shade the half-plane which does not contain the origin.

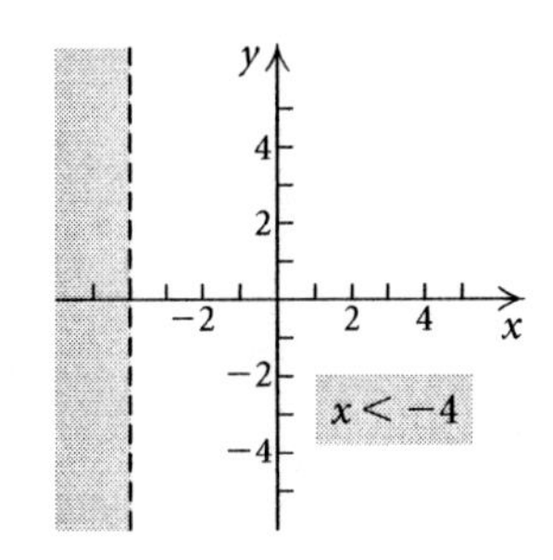

24.

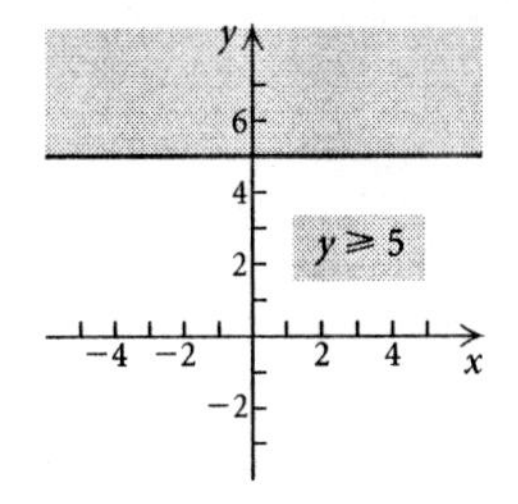

25. Graph: $y > -3$

1. First we graph the related equation $y = -3$. Draw the line dashed since the inequality is $>$.
2. To determine which half-plane to shade we test a point not on the line. We try $(0,0)$.

$$\begin{array}{c} y > -3 \\ \hline 0 \ ? \ -3 \quad \text{TRUE} \end{array}$$

Since $0 > -3$ is true, $(0,0)$ is a solution. We shade the half-plane containing $(0,0)$.

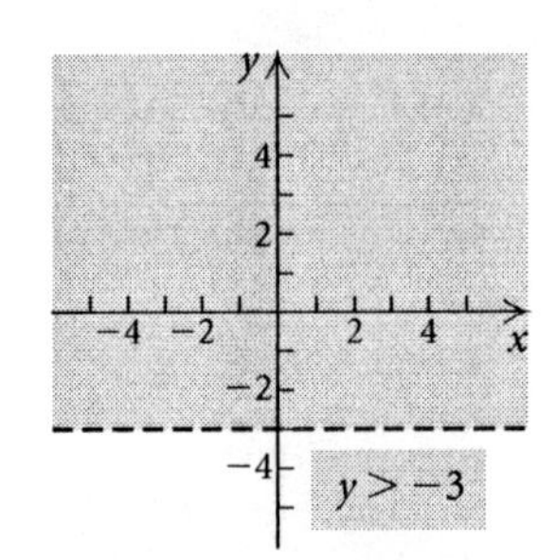

26.

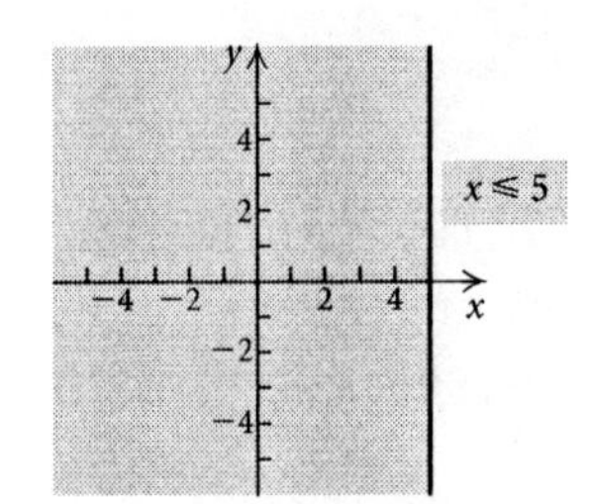

27. Graph: $-4 < y < -1$

This is a conjunction of two inequalities

$4 < y$ *and* $y < -1$.

We can graph $-4 < y$ and $y < -1$ separately and then graph the intersection, or region in both solution sets.

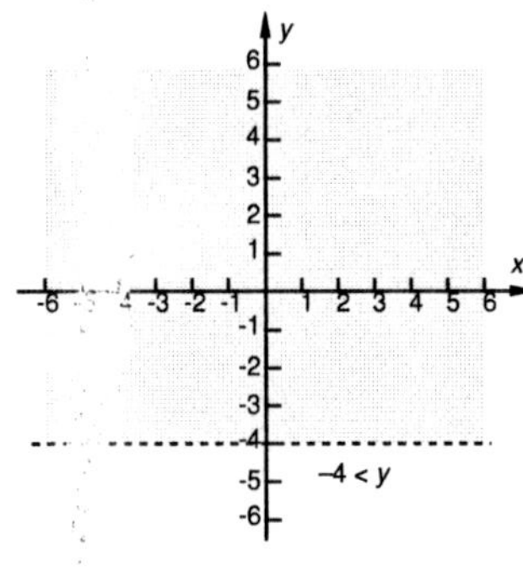

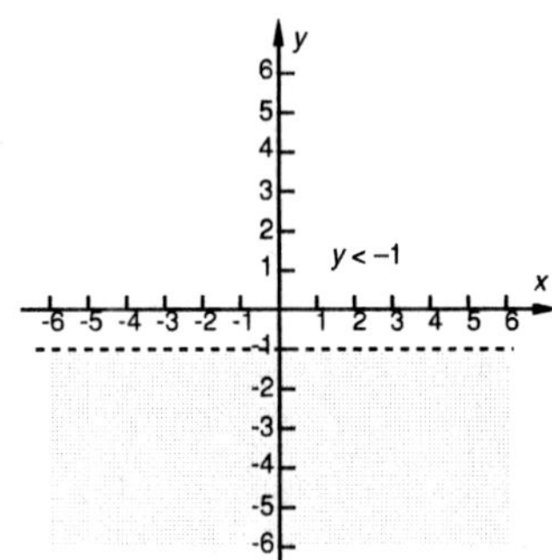

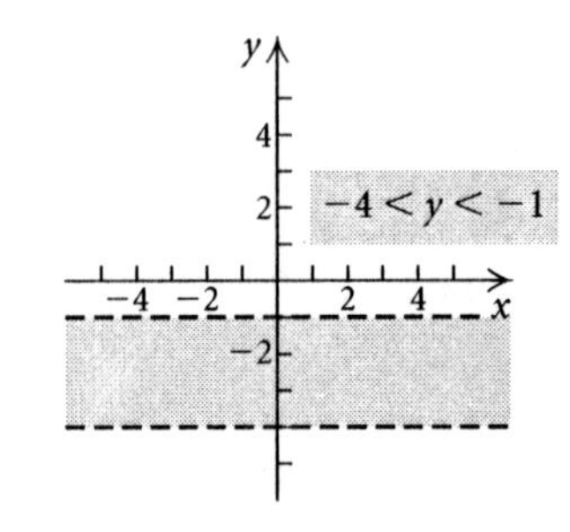

28.

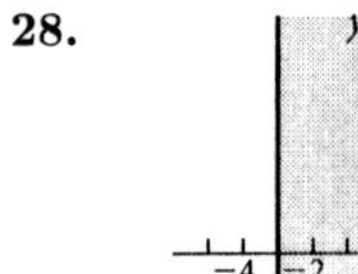

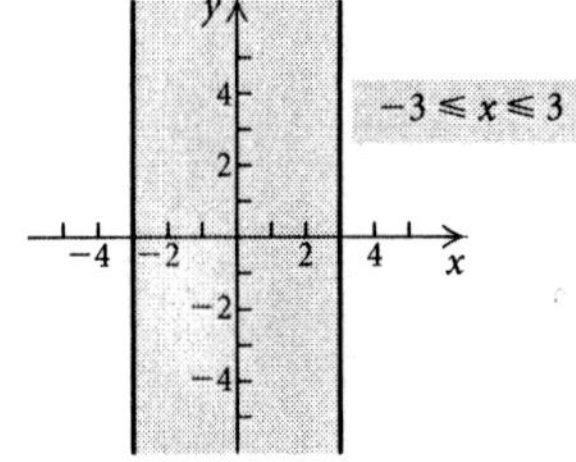

29. Graph: $y \geq |x|$

1. Graph the related equation $y = |x|$. Draw the line solid since the inequality symbol is $\geq$.

2. To determine the region to shade, observe that the solution set consists of all ordered pairs (x, y) where the second coordinate is greater than or equal to the absolute value of the first coordinate. We see that the solutions are the points on or above the graph of $y = |x|$.

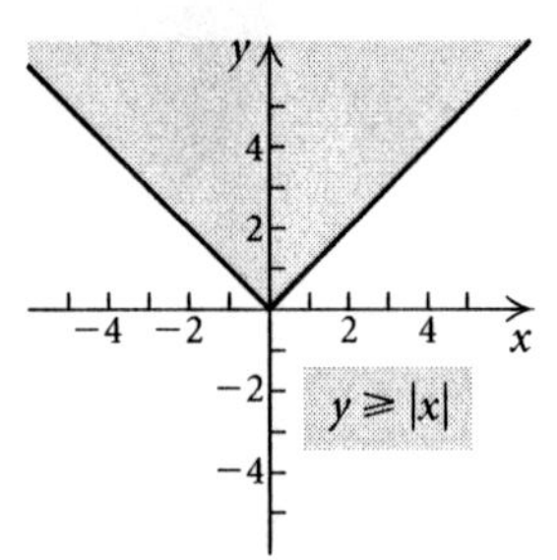

30.

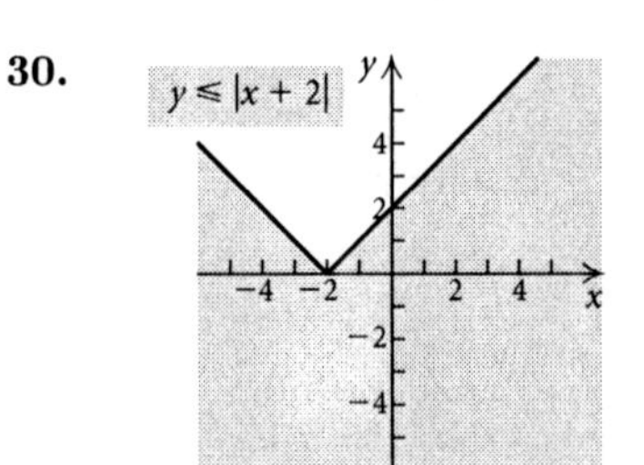

31. Graph (f) is the correct graph.

32. Graph (c) is the correct graph.

33. Graph (a) is the correct graph.

34. Graph (d) is the correct graph.

35. Graph (b) is the correct graph.

36. Graph (e) is the correct graph.

37. First we find the related equations. One line goes through $(0, 4)$ and $(4, 0)$. We find its slope:

$$m = \frac{0-4}{4-0} = \frac{-4}{4} = -1$$

This line has slope -1 and y-intercept $(0, 4)$, so its equation is $y = -x + 4$.

The other line goes through $(0, 0)$ and $(1, 3)$. We find the slope.

$$m = \frac{3-0}{1-0} = 3$$

This line has slope 3 and y-intercept $(0, 0)$, so its equation is $y = 3x + 0$, or $y = 3x$.

Observing the shading on the graph and the fact that the lines are solid, we can write the system of inqualities as

$$y \leq -x + 4,$$
$$y \leq 3x.$$

Answers may vary.

38. First we find the related equations. One line goes through $(-1, 2)$ and $(0, 0)$. We find its slope:

$$m = \frac{0-2}{0-(-1)} = -2$$

Then the equation of the line is $y = -2x + 0$, or $y = -2x$.

The other line goes through $(0, -3)$ and $(3, 0)$. We find its slope:

$$m = \frac{0-(-3)}{3-0} = \frac{3}{3} = 1$$

Then the equation of this line is $y = x - 3$.

Observing the shading on the graph and the fact that the lines are solid, we can write the system of inequalities as

$y \leq -2x,$

$y \geq x - 3.$

Answers may vary.

39. The equation of the vertical line is $x = 2$ and the equation of the horizontal line is $y = -1$. The lines are dashed and the shaded area is to the left of the vertical line and above the horizontal line, so the system of inequalities can be written

$x < 2,$

$y > -1.$

40. The equation of the vertical line is $x = -1$ and the equation of the horizontal line is $y = 2$. The lines are dashed and the shaded area is to the right of the vertical line and below the horizontal line, so the system of inequalities can be written

$x > -1,$

$y < 2.$

41. First we find the related equations. One line goes through $(0, 3)$ and $(3, 0)$. We find its slope:

$$m = \frac{0-3}{3-0} = \frac{-3}{3} = -1$$

This line has slope -1 and y-intercept $(0, 3)$, so its equation is $y = -x + 3$.

The other line goes through $(0, 1)$ and $(1, 2)$. We find its slope:

$$m = \frac{2-1}{1-0} = \frac{1}{1} = 1$$

This line has slope 1 and y-intercept $(0, 1)$, so its equation is $y = x + 1$.

Observe that both lines are solid and that the shading lies below both lines, to the right of the y-axis, and above the x-axis. We can write this system of inequalities as

$y \leq -x + 3,$

$y \leq x + 1,$

$x \geq 0,$

$y \geq 0.$

42. One line goes through $(-2, 0)$ and $(0, 2)$. We find its slope:

$$m = \frac{2-0}{0-(-2)} = \frac{2}{2} = 1$$

Then the equation of this line is $y = x + 2$. The other line is the vertical line $x = 2$. Observe that both lines are solid and that the shading lies below $y = x + 2$, to the left of $x = 2$, to the right of the y-axis, and above the x-axis. We can write this system of inequalities as

$y \leq x + 2,$

$x \leq 2,$

$x \geq 0,$

$y \geq 0.$

43. Graph: $y \leq x,$

$y \geq 3 - x$

We graph the related equations $y = x$ and $y = 3 - x$ using solid lines. The arrows at the ends of the lines indicate the half-plane containing the solution set for each inequality. Shade the region common to both solution sets.

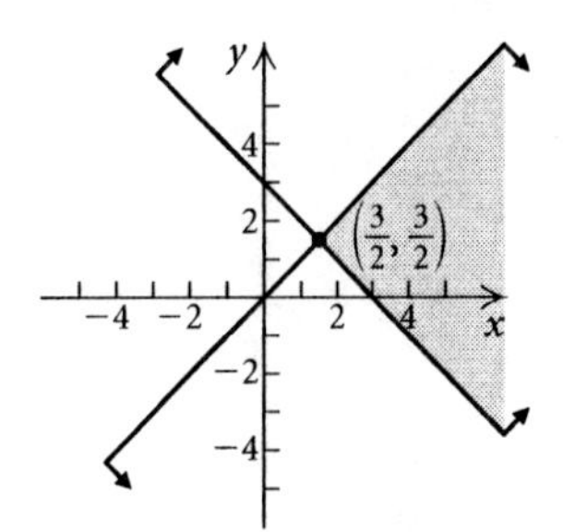

We find the vertex $\left(\frac{3}{2}, \frac{3}{2}\right)$ by solving the system

$y = x,$

$y = 3 - x.$

44. Graph: $y \geq x,$

$y \leq x - 5$

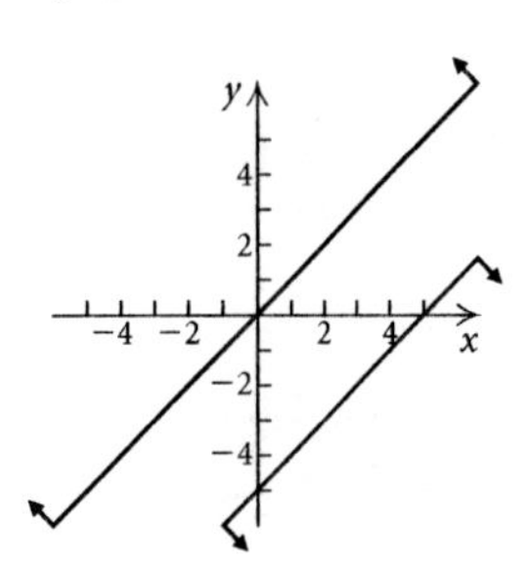

45. Graph: $y \geq x,$

$y \leq x - 4$

We graph the related equations $y = x$ and $y = x - 4$ using solid lines. The arrows at the ends of the lines indicate the half-plane containing the solution set for each inequality. There is not a region common to both solution sets, so there are no vertices.

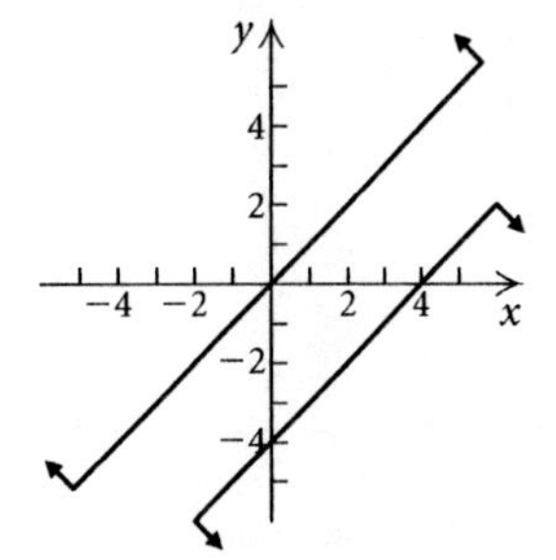

46. Graph: $y \geq x$,
$y \leq 2 - x$

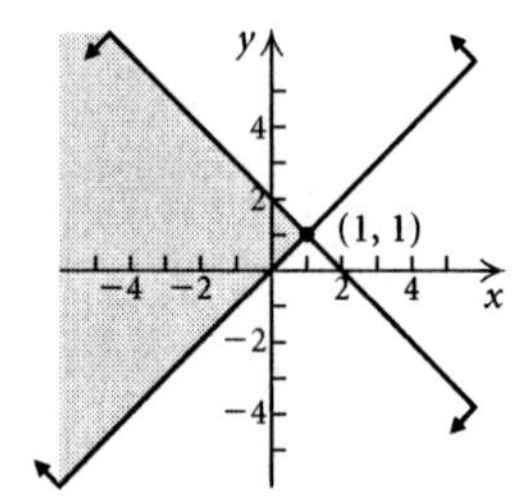

47. Graph: $y \geq -3$,
$x \geq 1$

We graph the related equations $y = -3$ and $x = 1$ using solid lines. The arrows at the ends of the lines indicate the half-plane containing the solution set for each inequality. Shade the region common to both solution sets.

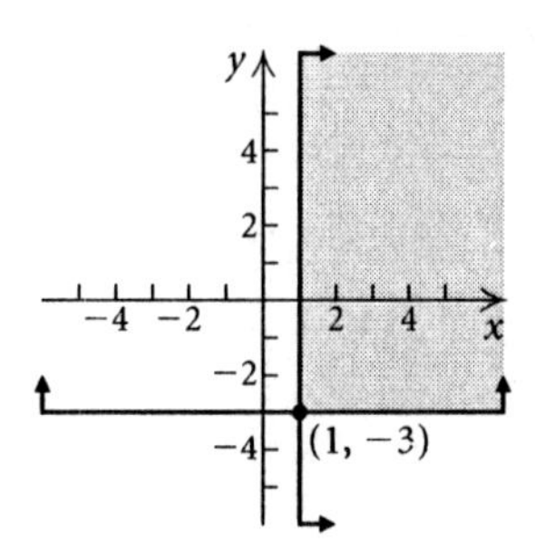

We find the vertex $(1, -3)$ by solving the system

$y = -3$,
$x = 1$.

48. Graph: $y \leq -2$,
$x \geq 2$

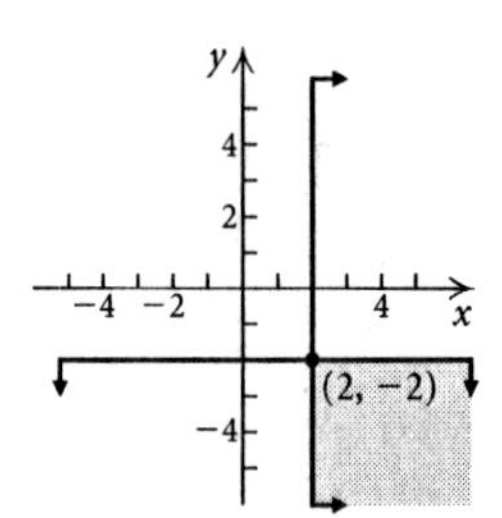

49. Graph: $x \leq 3$,
$y \geq 2 - 3x$

We graph the related equations $x = 3$ and $y = 2 - 3x$ using solid lines. The arrows at the ends of the lines indicate the half-plane containing the solution set for each inequality. Shade the region common to both solution sets.

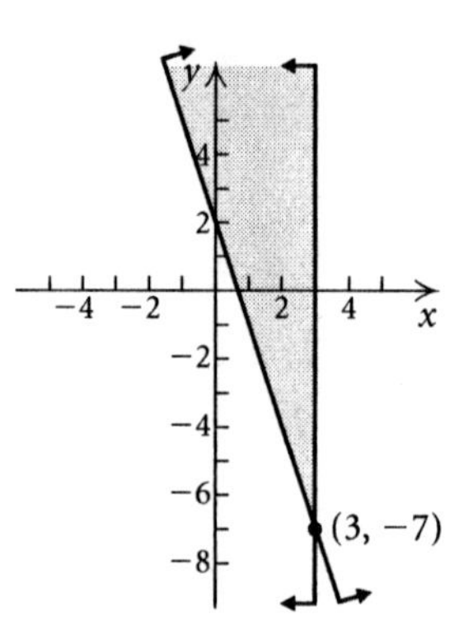

We find the vertex $(3, -7)$ by solving the system

$x = 3$,
$y = 2 - 3x$.

50. Graph: $x \geq -2$,
$y \leq 3 - 2x$

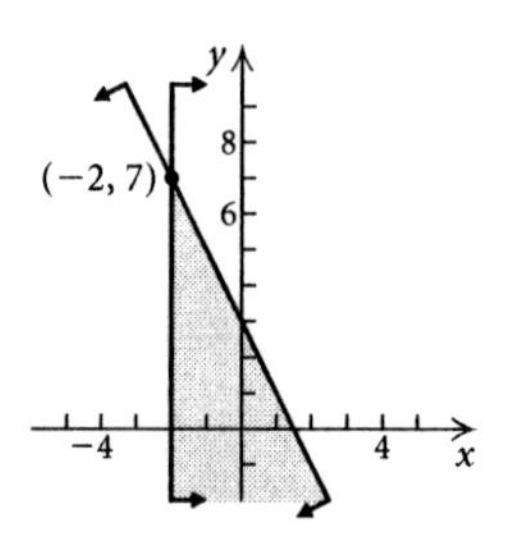

51. Graph: $x + y \leq 1$,
$x - y \leq 2$

We graph the related equations $x + y = 1$ and $x - y = 2$ using solid lines. The arrows at the ends of the lines indicate the half-plane containing the solution set for each inequality. Shade the region common to both solution sets.

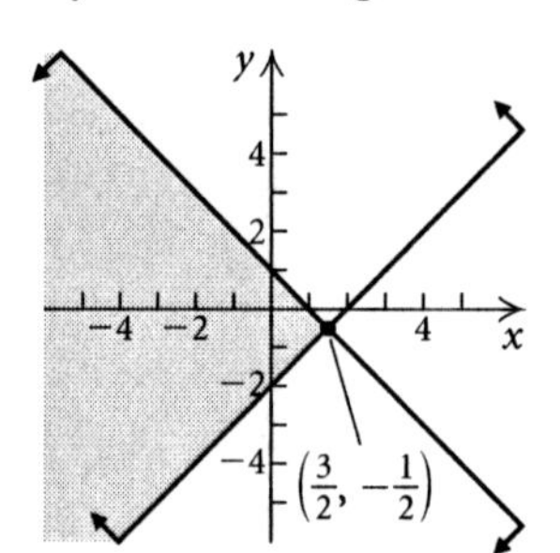

We find the vertex $\left(\frac{3}{2}, -\frac{1}{2}\right)$ by solving the system

$x + y = 1$,
$x - y = 2$.

52. Graph: $y + 3x \geq 0,$
$y + 3x \leq 2$

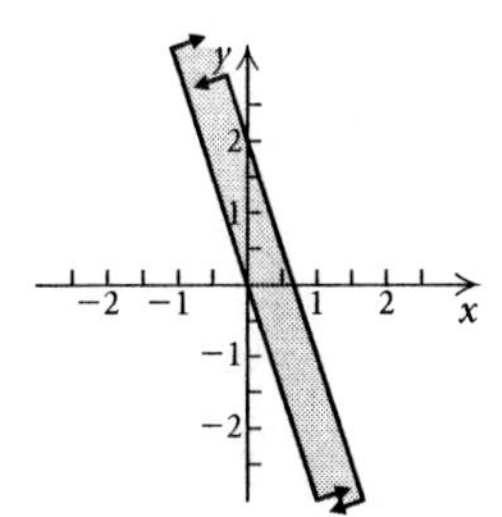

53. Graph: $2y - x \leq 2,$
$y + 3x \geq -1$

We graph the related equations $2y - x = 2$ and $y + 3x = -1$ using solid lines. The arrows at the ends of the lines indicate the half-plane containing the solution set for each inequality. Shade the region common to both solution sets.

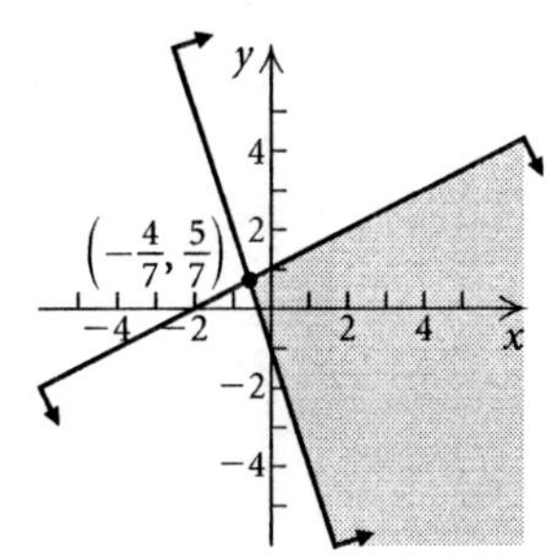

We find the vertex $\left(-\frac{4}{7}, \frac{5}{7}\right)$ by solving the system

$2y - x = 2,$
$y + 3x = -1.$

54. Graph: $y \leq 2x + 1,$
$x \geq -2x + 1,$
$x \leq 2$

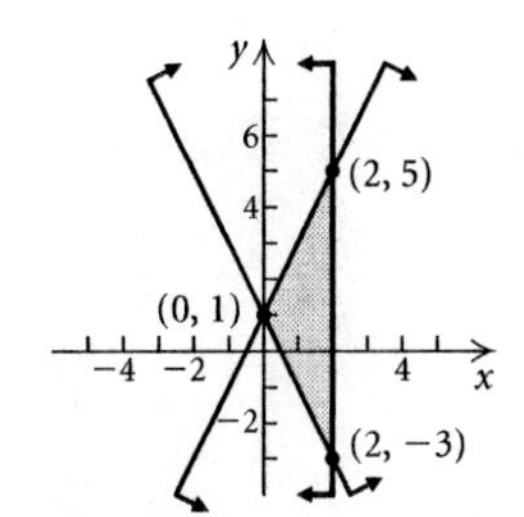

55. Graph: $x - y \leq 2,$
$x + 2y \geq 8,$
$y \leq 4$

We graph the related equations $x - y = 2$, $x + 2y = 8$, and $y = 4$ using solid lines. The arrows at the ends of the lines indicate the half-plane containing the solution set for each inequality. Shade the region common to all three solution sets.

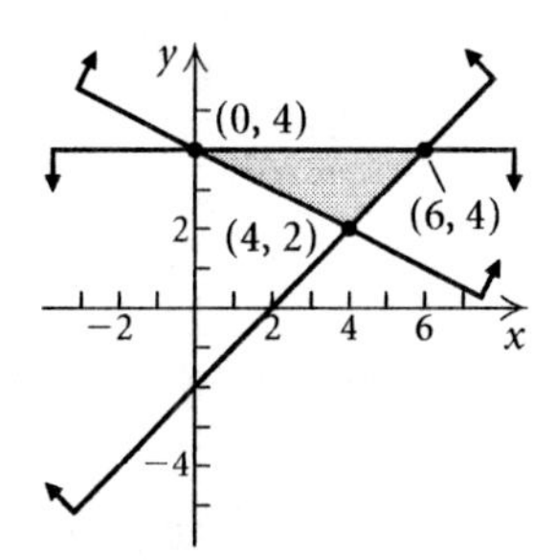

We find the vertex $(0, 4)$ by solving the system

$x + 2y = 8,$
$y = 4.$

We find the vertex $(6, 4)$ by solving the system

$x - y = 2,$
$y = 4.$

We find the vertex $(4, 2)$ by solving the system

$x - y = 2,$
$x + 2y = 8.$

56. Graph: $x + 2y \leq 12,$
$2x + y \leq 12,$
$x \geq 0,$
$y \geq 0$

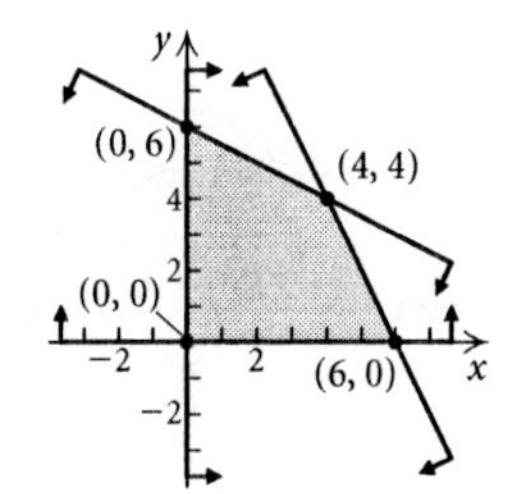

57. Graph: $4x - 3y \geq -12,$
$4x + 3y \geq -36,$
$y \leq 0,$
$x \leq 0$

Shade the intersection of the graphs of the four inequalities.

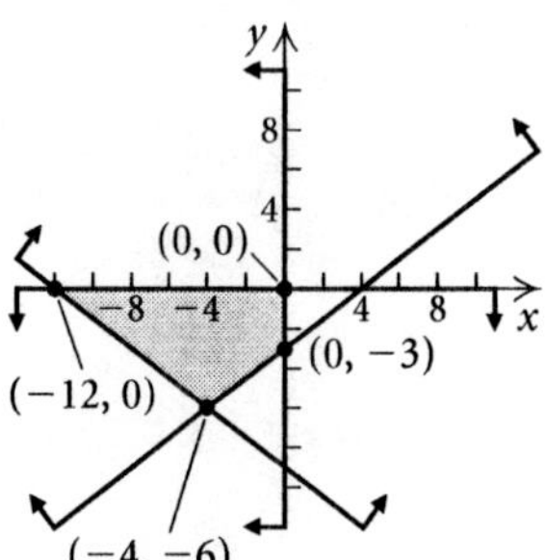

We find the vertex $(-12, 0)$ by solving the system

$4y + 3x = -36,$
$y = 0.$

We find the vertex $(0, 0)$ by solving the system

$y = 0,$

$x = 0.$

We find the vertex $(0, -3)$ by solving the system

$4y - 3x = -12,$

$x = 0.$

We find the vertex $(-4, -6)$ by solving the system

$4y - 3x = -12,$

$4y + 3x = -36.$

58. Graph: $8x + 5y \leq 40,$

$x + 2y \leq 8,$

$x \geq 0,$

$y \geq 0$

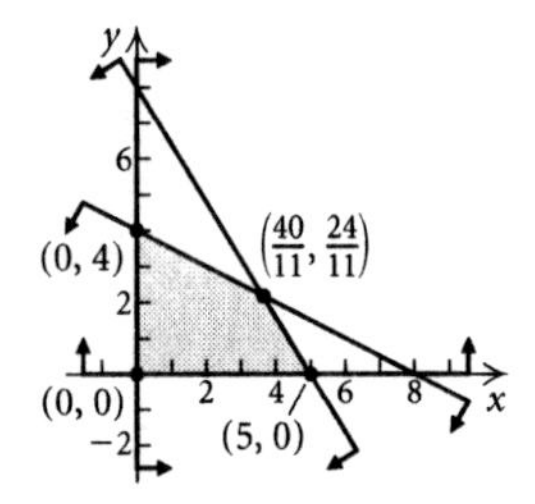

59. Graph: $3x + 4y \geq 12,$

$5x + 6y \leq 30,$

$1 \leq x \leq 3$

Shade the intersection of the graphs of the given inequalities.

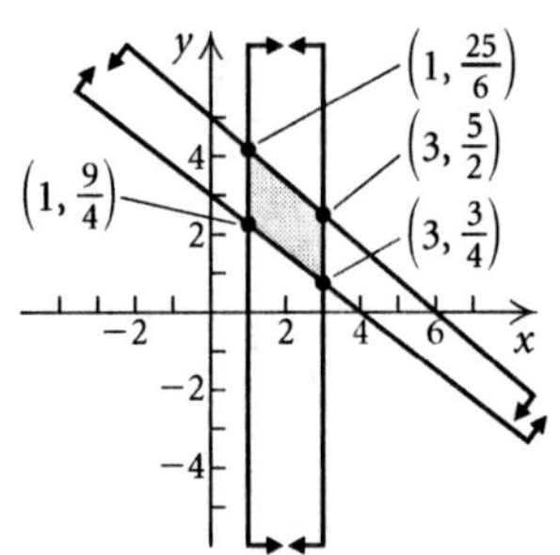

We find the vertex $\left(1, \dfrac{25}{6}\right)$ by solving the system

$5x + 6y = 30,$

$x = 1.$

We find the vertex $\left(3, \dfrac{5}{2}\right)$ by solving the system

$5x + 6y = 30,$

$x = 3.$

We find the vertex $\left(3, \dfrac{3}{4}\right)$ by solving the system

$3x + 4y = 12,$

$x = 3.$

We find the vertex $\left(1, \dfrac{9}{4}\right)$ by solving the system

$3x + 4y = 12,$

$x = 1.$

60. Graph: $y - x \geq 1,$

$y - x \leq 3,$

$2 \leq x \leq 5$

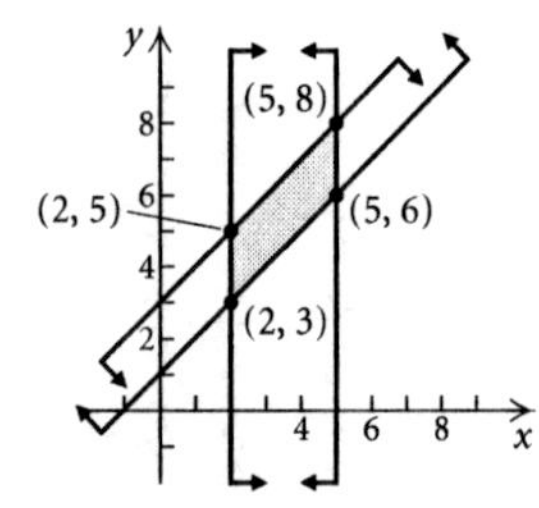

61. Find the maximum and minimum values of $P = 17x - 3y + 60$, subject to

$6x + 8y \leq 48,$

$0 \leq y \leq 4,$

$0 \leq x \leq 3.$

Graph the system of inequalities and determine the vertices.

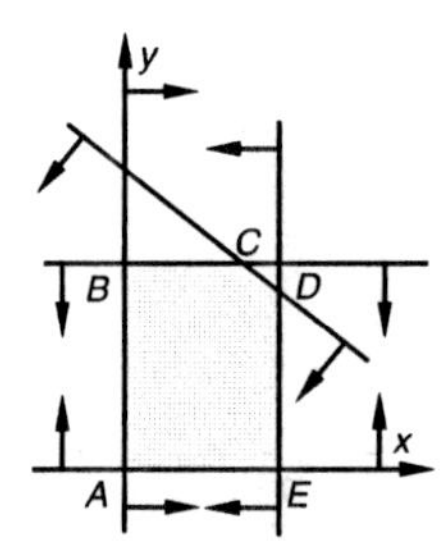

Vertex A: $(0, 0)$

Vertex B:
We solve the system $x = 0$ and $y = 4$. The coordinates of point B are $(0, 4)$.

Vertex C:
We solve the system $6x + 8y = 48$ and $y = 4$. The coordinates of point C are $\left(\dfrac{8}{3}, 4\right)$.

Vertex D:
We solve the system $6x + 8y = 48$ and $y = 7$. The coordinates of point D are $\left(7, \dfrac{3}{4}\right)$.

Vertex E:
We solve the system $x = 7$ and $y = 0$. The coordinates of point E are $(7, 0)$.

Evaluate the objective function P at each vertex.

Vertex	$P = 17x - 3y + 60$
$A(0, 0)$	$17 \cdot 0 - 3 \cdot 0 + 60 = 60$
$B(0, 4)$	$17 \cdot 0 - 3 \cdot 4 + 60 = 48$
$C\left(\frac{8}{3}, 3\right)$	$17 \cdot \frac{8}{3} - 3 \cdot 4 + 60 = 66\frac{2}{3}$
$D\left(7, \frac{3}{4}\right)$	$17 \cdot 7 - 3 \cdot \frac{3}{4} + 60 = 176\frac{3}{4}$
$E(7, 0)$	$17 \cdot 7 - 3 \cdot 0 + 60 = 179$

The maximum value of P is 179 when $x = 7$ and $y = 0$.

The minimum value of P is 48 when $x = 0$ and $y = 4$.

62. We graph the system of inequalities and find the vertices:

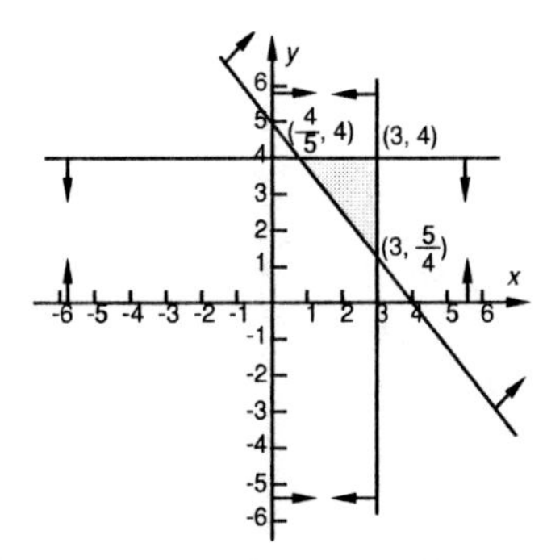

Vertex	$Q = 28x - 4y + 72$
$\left(\frac{4}{5}, 4\right)$	$78\frac{2}{5}$ ⟵ Minimum
$(3, 4)$	140
$\left(3, \frac{5}{4}\right)$	151 ⟵ Maximum

63. Find the maximum and minimum values of

$F = 5x + 36y$, subject to

$$5x + 3y \le 34,$$
$$3x + 5y \le 30,$$
$$x \ge 0,$$
$$y \ge 0.$$

Graph the system of inequalities and find the vertices.

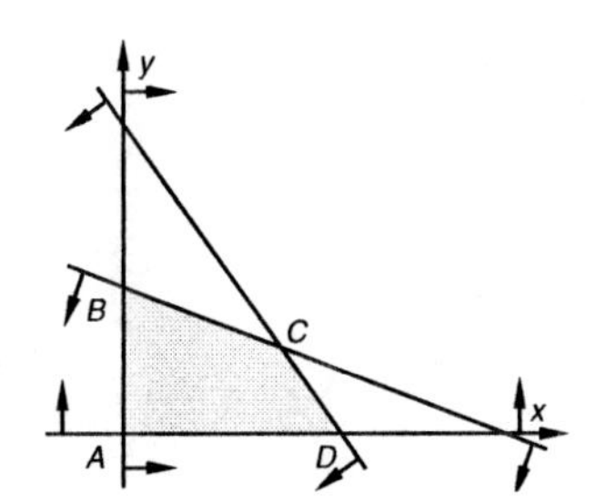

Vertex A: $(0, 0)$

Vertex B:

We solve the system $3x + 5y = 30$ and $x = 0$. The coordinates of point B are $(0, 6)$.

Vertex C:

We solve the system $5x + 3y = 34$ and $3x + 5y = 30$. The coordinates of point C are $(5, 3)$.

Vertex D:

We solve the system $5x + 3y = 34$ and $y = 0$. The coordinates of point D are $\left(\frac{34}{5}, 0\right)$.

Evaluate the objective function F at each vertex.

Vertex	$F = 5x + 36y$
$A(0, 0)$	$5 \cdot 0 + 36 \cdot 0 + = 0$
$B(0, 6)$	$5 \cdot 0 + 36 \cdot 6 = 216$
$C(5, 3)$	$5 \cdot 5 + 39 \cdot 3 = 133$
$D\left(\frac{34}{5}, 0\right)$	$5 \cdot \frac{34}{5} + 36 \cdot 0 = 34$

The maximum value of F is 216 when $x = 0$ and $y = 6$.

The minimum value of F is 0 when $x = 0$ and $y = 0$.

64. We graph the system of inequalities and find the vertices:

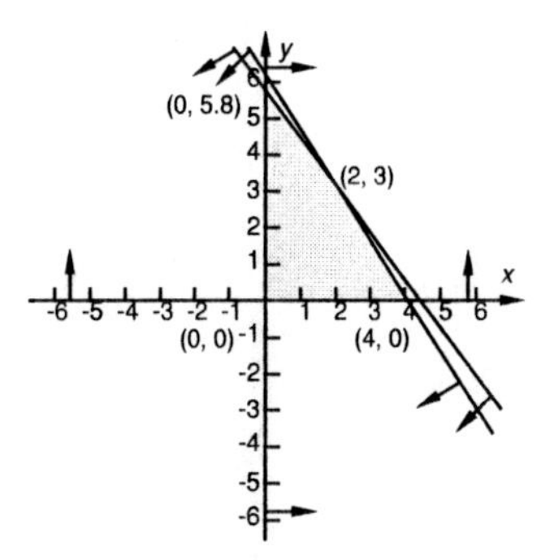

Vertex	$G = 16x + 14y$
$(0, 0)$	0 ⟵ Minimum
$(0, 5.8)$	81.2 ⟵ Maximum
$(2, 3)$	74
$(4, 0)$	64

65. Let $x =$ the number of jumbo biscuits and $y =$ the number of regular biscuits to be made per day. The income I is given by

$$I = 0.10x + 0.08y$$

subject to the constrains

$$x + y \le 200,$$
$$2x + y \le 300,$$
$$x \ge 0,$$
$$y \ge 0.$$

We graph the system of inequalities, determine the vertices, and find the value if I at each vertex.

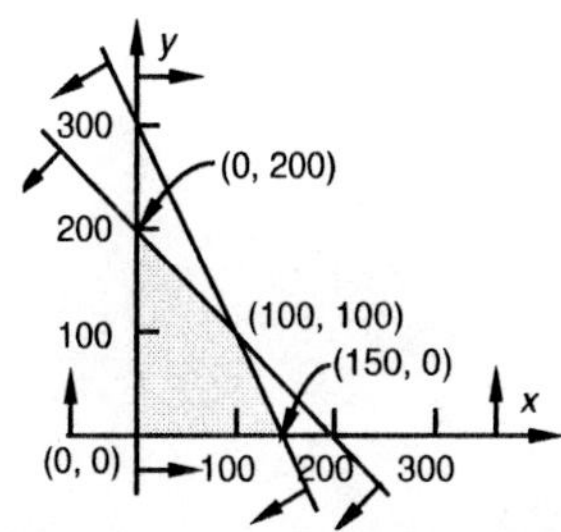

Vertex	$I = 0.10x + 0.08y$
$(0, 0)$	$0.10(0) + 0.08(0) = 0$
$(0, 200)$	$0.10(0) + 0.08(200) = 16$
$(100, 100)$	$0.10(100) + 0.08(100) = 18$
$(150, 0)$	$0.10(150) + 0.08(0) = 15$

The company will have a maximum income of \$18 when 100 of each type of biscuit is made.

66. Let $x =$ the number of gallons the car uses and $y =$ the number of gallons the moped uses. Find the maximum value of

$$M = 20x + 100y$$

subject to

$$x + y \leq 12,$$
$$0 \leq x \leq 10,$$
$$0 \leq y \leq 3.$$

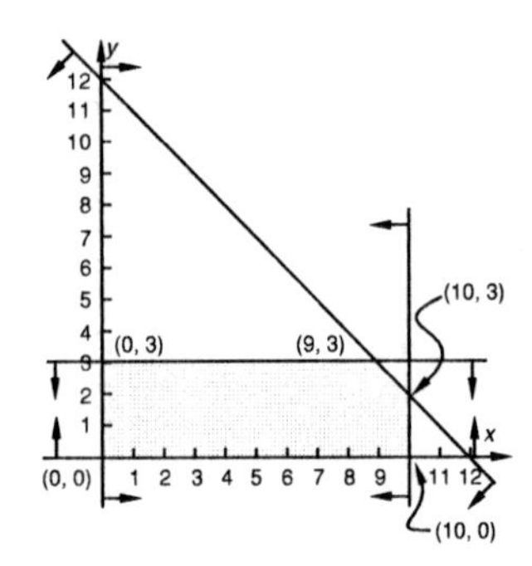

Vertex	$M = 20x + 100y$
$(0, 0)$	0
$(0, 3)$	300
$(9, 3)$	480
$(10, 2)$	400
$(10, 0)$	200

The maximum number of miles is 480 when the car uses 9 gal and the moped uses 3 gal.

67. Let x = the number of units of lumber and y = the number of units of plywood produced per week. The profit P is given by

$$P = 20x + 30y$$

subject to the constraints

$$x + y \leq 400,$$
$$x \geq 100,$$
$$y \geq 150.$$

We graph the system of inequalities, determine the vertices and find the value of P at each vertex.

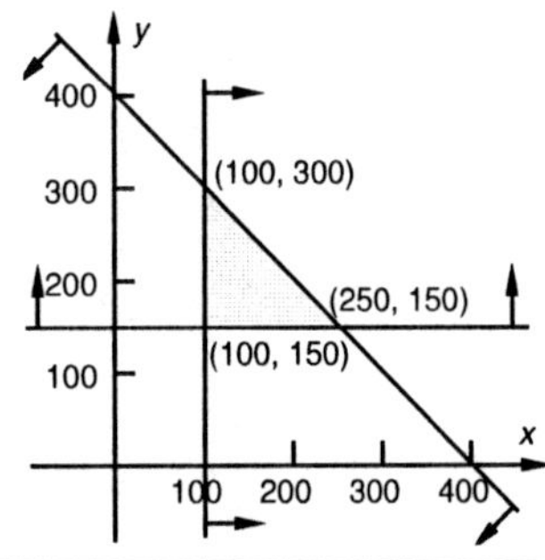

Vertex	$P = 20x + 30y$
$(100, 150)$	$20 \cdot 100 + 30 \cdot 150 = 6500$
$(100, 300)$	$20 \cdot 100 + 30 \cdot 300 = 11,000$
$(250, 150)$	$20 \cdot 250 + 30 \cdot 150 = 9500$

The maximum profit of \$11,000 is achieved by producing 100 units of lumber and 300 units of plywood.

68. Let x = the corn acreage and y = the oats acreage. Find the maximum value of

$$P = 40x + 30y$$

subject to

$$x + y \leq 240,$$
$$2x + y \leq 320,$$
$$x \geq 0,$$
$$y \geq 0.$$

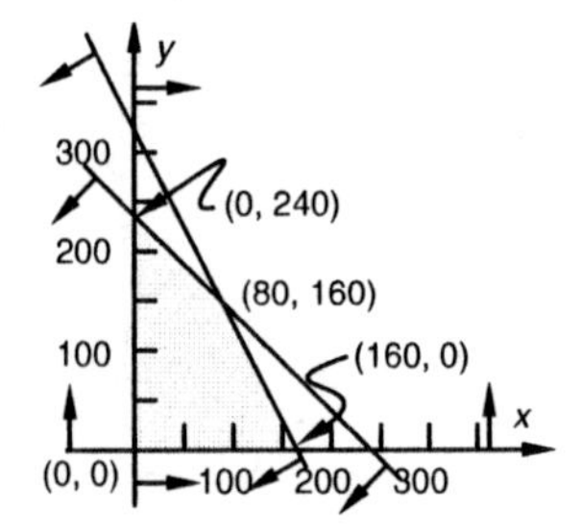

Vertex	$P = 40x + 30y$
$(0, 0)$	0
$(0, 240)$	7200
$(80, 160)$	8000
$(160, 0)$	6400

The maximum profit of \$8000 occurs when 80 acres of corn and 160 acres of oats are planted.

69. Let x = the number of sacks of soybean meal to be used and y = the number of sacks of oats. The minimum cost is given by

$$C = 15x + 5y$$

subject to the constraints

$$50x + 15y \geq 120,$$
$$8x + 5y \geq 24,$$
$$5x + y \geq 10,$$
$$x \geq 0,$$
$$y \geq 0.$$

Graph the system of inequalities, determine the vertices, and find the value of C at each vertex.

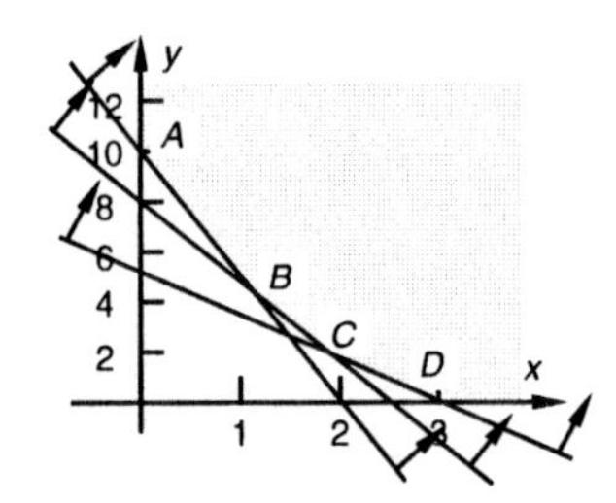

Vertex	$C = 15x + 5y$
$A(0, 10)$	$15 \cdot 0 + 5 \cdot 10 = 50$
$B\left(\frac{6}{5}, 4\right)$	$15 \cdot \frac{6}{5} + 5 \cdot 4 = 38$
$C\left(\frac{24}{13}, \frac{24}{13}\right)$	$15 \cdot \frac{24}{13} + 5 \cdot \frac{24}{13} = 36\frac{12}{13}$
$D(3, 0)$	$15 \cdot 3 + 5 \cdot 0 = 45$

The minimum cost of $\$36\frac{12}{13}$ is achieved by using $\frac{24}{13}$, or $1\frac{11}{13}$ sacks of soybean meal and $\frac{24}{13}$, or $1\frac{11}{13}$ sacks of oats.

70. Let x = the number of sacks of soybean meal to be used and y = the number of sacks of alfalfa. Find the minimum value of

$C = 15x + 8y$

subject to the constraints

$$50x + 20y \geq 120,$$
$$8x + 6y \geq 24,$$
$$5x + 8y \geq 10,$$
$$x \geq 0,$$
$$y \geq 0.$$

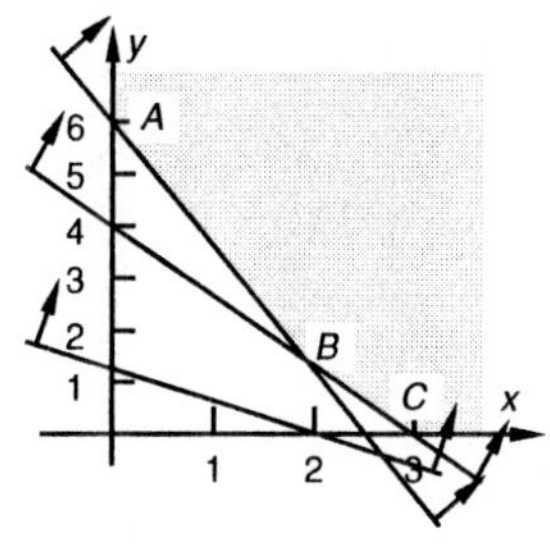

Vertex	$C = 15x + 8y$
$A(0, 6)$	$15 \cdot 0 + 8 \cdot 6 = 48$
$B\left(\frac{12}{7}, \frac{12}{7}\right)$	$15 \cdot \frac{12}{7} + 8 \cdot \frac{12}{7} = 39\frac{3}{7}$
$C(3, 0)$	$15 \cdot 3 + 8 \cdot 0 = 45$

The minimum cost of $\$39\frac{3}{7}$ is achieved by using $\frac{12}{7}$, or $1\frac{5}{7}$ sacks of soybean meal and $\frac{12}{7}$, or $1\frac{5}{7}$ sacks of alfalfa.

71. Let x = the amount invested in corporate bonds and y = the amount invested in municipal bonds. The income I is given by

$I = 0.08x + 0.075y$

subject to the constrains

$$x + y \leq 40,000,$$
$$6000 \leq x \leq 22,000,$$
$$y \leq 30,000.$$

We graph the system of inequalities, determine the vertices, and find the value of I at each vertex.

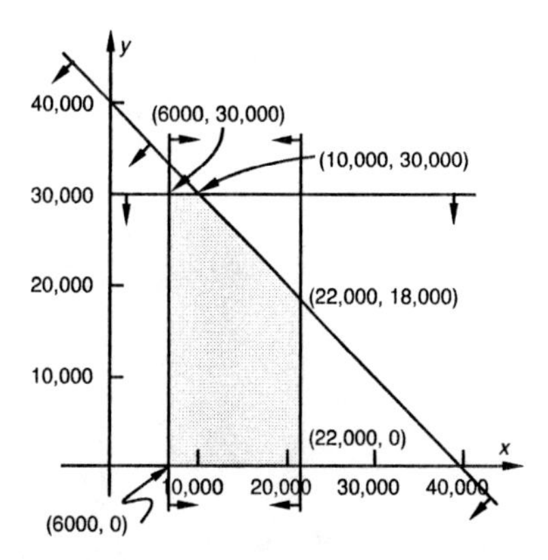

Vertex	$I = 0.08x + 0.075y$
$(6000, 0)$	480
$(6000, 30,000)$	2730
$(10,000, 30,000)$	3050
$(22,000, 18,000)$	3110
$(22,000, 0)$	1760

The maximum income of \$3110 occurs when \$22,000 is invested in corporate bonds and \$18,000 is invested in municipal bonds.

72. Let x = the amount invested in City Bank and y = the amount invested in People's Bank. Find the maximum value of

$I = 0.06x + 0.065y$

subject to

$$x + y \leq 22,000,$$
$$2000 \leq x \leq 14,000$$
$$0 \leq y \leq 15,000.$$

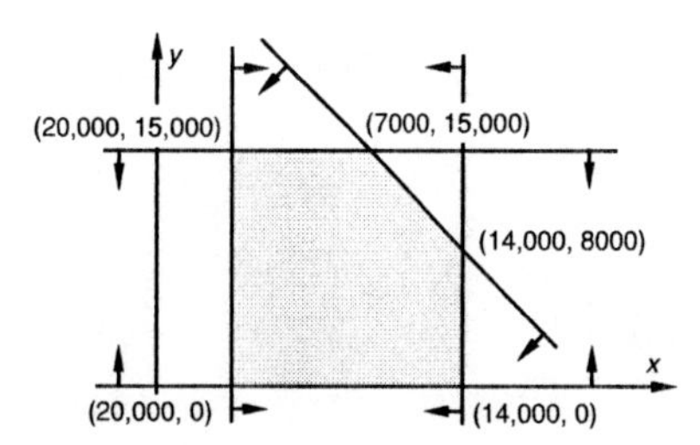

Vertex	$I = 0.06x + 0.065y$
$(2000, 0)$	120
$(2000, 15,000)$	1095
$(7000, 15,000)$	1395
$(14,000, 8,000)$	1360
$(14,000, 0)$	840

The maximum interest income is \$1395 when \$7000 is invested in City Bank and \$15,000 is invested in People's Bank.

73. Let x = the number of P_1 airplanes and y = the number of P_2 airplanes to be used. The operating cost C, in thousands of dollars, is given by

$$C = 12x + 10y$$

subject to the constraints

$$40x + 80y \geq 2000,$$
$$40x + 30y \geq 1500,$$
$$120x + 40y \geq 2400,$$
$$x \geq 0,$$
$$y \geq 0.$$

Graph the system of inequalities, determine the vertices, and find the value of C at each vertex.

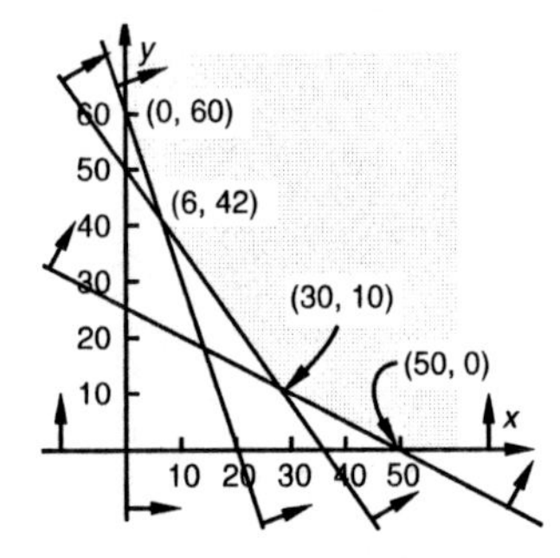

Vertex	$C = 12x + 10y$
$(0, 60)$	$12 \cdot 0 + 10 \cdot 60 = 600$
$(6, 42)$	$12 \cdot 6 + 10 \cdot 42 = 492$
$(30, 10)$	$12 \cdot 30 + 10 \cdot 10 = 460$
$(50, 0)$	$12 \cdot 50 + 10 \cdot 0 = 600$

The minimum cost of \$460 thousand is achieved using 30 P_1's and 10 P_2's.

74. Let x = the number of P_2 airplanes and y = the number of P_3 airplanes to be used. Find the minimum value of

$$C = 10x + 15y$$

subject to

$$80x + 40y \geq 2000,$$
$$30x + 40y \geq 1500,$$
$$40x + 80y \geq 2400,$$
$$x \geq 0,$$
$$y \geq 0.$$

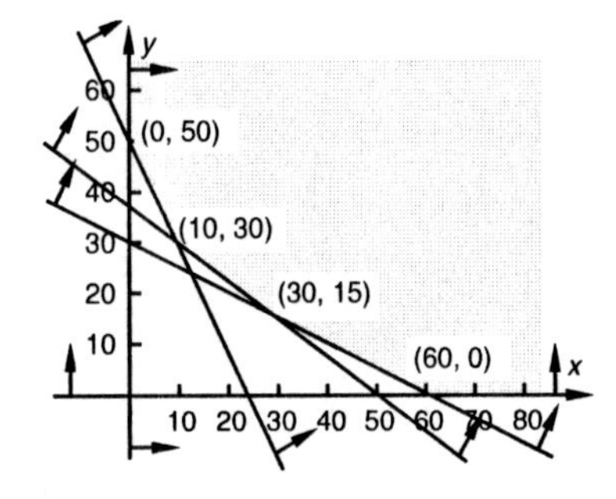

Vertex	$C = 10x + 15y$
$(0, 50)$	$10 \cdot 0 + 15 \cdot 50 = 750$
$(10, 30)$	$10 \cdot 10 + 15 \cdot 30 = 550$
$(30, 15)$	$10 \cdot 30 + 15 \cdot 15 = 525$
$(60, 0)$	$10 \cdot 60 + 15 \cdot 0 = 600$

The minimum cost of \$525 thousand is achieved using 30 P_2's and 15 P_3's.

75. Let x = the number of knit suits and y = the number of worsted suits made. The profit is given by

$$P = 34x + 31y$$

subject to

$$2x + 4y \leq 20,$$
$$4x + 2y \leq 16,$$
$$x \geq 0,$$
$$y \geq 0.$$

Graph the system of inequalities, determine the vertices, and find the value of P at each vertex.

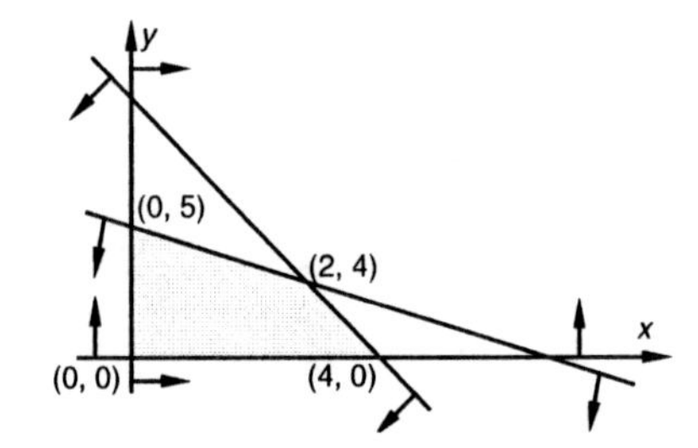

Vertex	$P = 34x + 31y$
$(0, 0)$	$34 \cdot 0 + 31 \cdot 0 = 0$
$(0, 5)$	$34 \cdot 0 + 31 \cdot 5 = 155$
$(2, 4)$	$34 \cdot 2 + 31 \cdot 4 = 192$
$(4, 0)$	$34 \cdot 4 + 31 \cdot 0 = 136$

The maximum profit per day is \$192 when 2 knit suits and 4 worsted suits are made.

76. Let x = the number of smaller gears and y = the number of larger gears produced each day. Find the maximum value of

$$P = 25x + 10y$$

subject to

$$4x + y \leq 24,$$
$$x + y \leq 9,$$
$$x \geq 0,$$
$$y \geq 0.$$

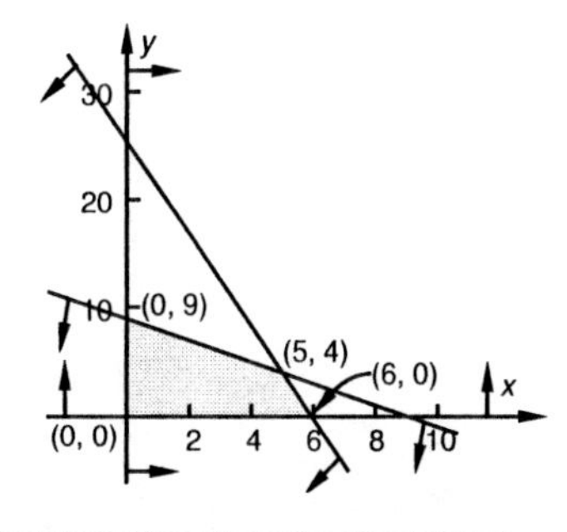

Vertex	$P = 25x + 10y$
$(0, 0)$	$25 \cdot 0 + 10 \cdot 0 = 0$
$(0, 9)$	$25 \cdot 0 + 10 \cdot 9 = 90$
$(5, 4)$	$25 \cdot 5 + 10 \cdot 4 = 165$
$(6, 0)$	$25 \cdot 6 + 10 \cdot 0 = 150$

The maximum profit of $165 per day is achieved when 5 smaller gears and 4 larger gears are produced.

77. Let x = the number of pounds of meat and y = the number of pounds of cheese in the diet in a week. The cost is given by

$$C = 3.50x + 4.60y$$

subject to

$$2x + 3y \geq 12,$$
$$2x + y \geq 6,$$
$$x \geq 0,$$
$$y \geq 0.$$

Graph the system of inequalities, determine the vertices, and find the value of C at each vertex.

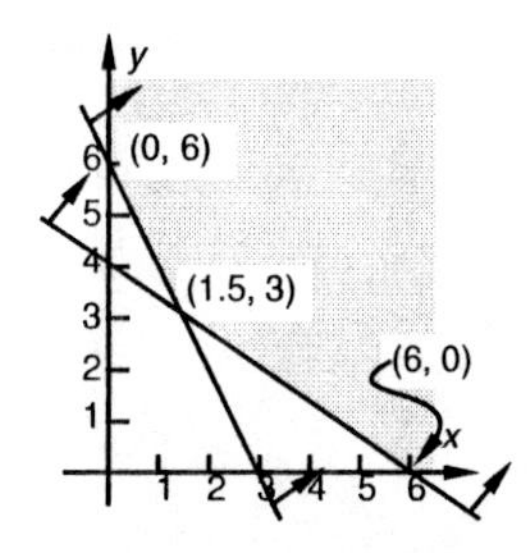

Vertex	$C = 3.50x + 4.60y$
$(0, 6)$	$3.50(0) + 4.60(6) = 27.60$
$(1.5, 3)$	$3.50(1.5) + 4.60(3) = 19.05$
$(6, 0)$	$3.50(6) + 4.60(0) = 21.00$

The minimum weekly cost of $19.05 is achieved when 1.5 lb of meat and 3 lb of cheese are used.

78. Let x = the number of teachers and y = the number of teacher's aides. Find the minimum value of

$$C = 35,000x + 18,000y$$

subject to

$$x + y \leq 50,$$
$$x + y \geq 20,$$
$$y \geq 12,$$
$$x \geq 2y.$$

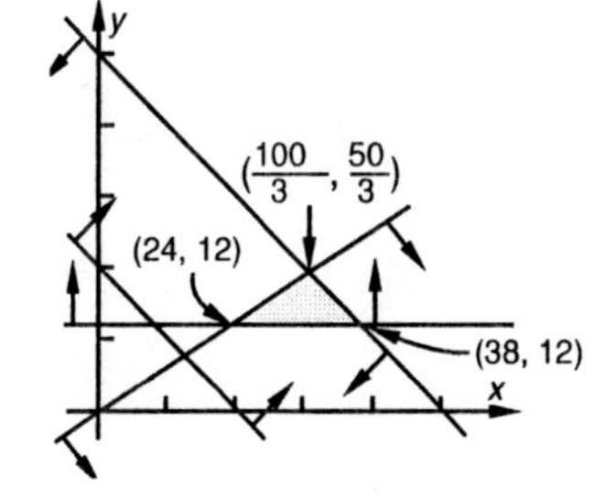

Vertex	$C = 35,000x + 18,000y$
$(24, 12)$	$35,000 \cdot 24 + 18,000 \cdot 12 = 1,056,000$
$\left(\frac{100}{3}, \frac{50}{3}\right)$	$35,000 \cdot \frac{100}{3} + 18,000 \cdot \frac{50}{3} \approx 1,466,667$
$(38, 12)$	$35,000 \cdot 38 + 18,000 \cdot 12 = 1,546,000$

The minimum cost of $1,056,000 is achieved when 24 teachers and 12 teacher's aides are hired.

79. Let x = the number of animal A and y = the number of animal B. The total number of animals is given by

$$T = x + y$$

subject to

$$x + 0.2y \leq 600,$$
$$0.5x + y \leq 525,$$
$$x \geq 0,$$
$$y \geq 0.$$

Graph the system of inequalities, determine the vertices, and find the value of T at each vertex.

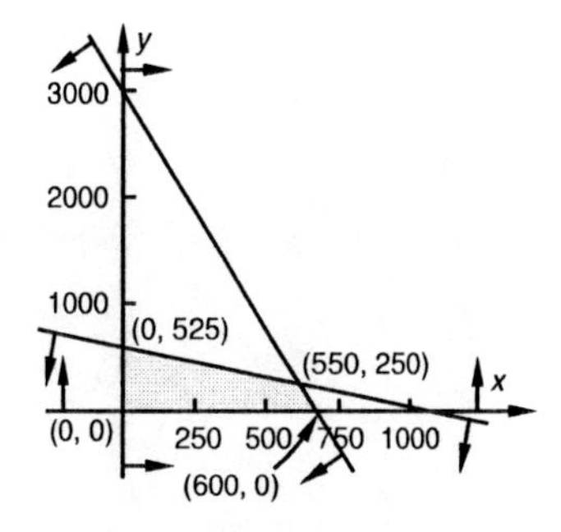

Vertex	$T = x + y$
$(0, 0)$	$0 + 0 = 0$
$(0, 525)$	$0 + 525 = 525$
$(550, 250)$	$550 + 250 = 800$
$(600, 0)$	$600 + 0 = 600$

The maximum total number of 800 is achieved when there are 550 of A and 250 of B.

80. Let x = the number of animal A and y = the number of animal B. Find the maximum value of

$$T = x + y$$

subject to

$$x + 0.2y \le 1080,$$
$$0.5x + y \le 810,$$
$$x \ge 0,$$
$$y \ge 0.$$

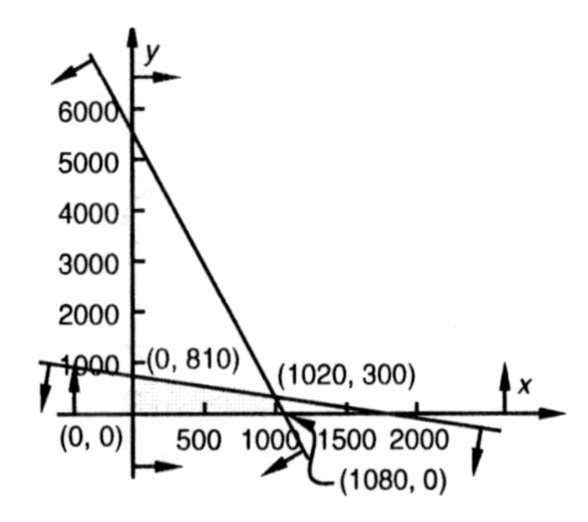

Vertex	$T = x + y$
$(0, 0)$	$0 + 0 = 0$
$(0, 810)$	$0 + 810 = 810$
$(1020, 300)$	$1020 + 300 = 1320$
$(1080, 0)$	$1080 + 0 = 1080$

The maximum total number of 1320 is achieved when there are 1020 of A and 300 of B.

81. Left to the student

82. Left to the student

83. Answers will vary. Exercise 59 can be used as a model, if desired.

84. The graph of a linear equation consists of a set of points on a line. The graph of a linear inequality consists of the set of points in a half-plane and might also include the points on the line that is the boundary of the half-plane.

85. $-5 \le x + 2 < 4$

$-7 \le x < 2$ Subtracting 2

The solution set is $\{x| - 7 \le x < 2\}$, or $[-7, 2)$.

86. $|x - 3| \ge 2$

$x - 3 \le -2$ *or* $x - 3 \ge 2$

$x \le 1$ *or* $x \ge 5$

The solution set is $\{x|x \le 1$ *or* $x \ge 5\}$, or $(-\infty, 1] \cup [5, \infty)$.

87. $x^2 - 2x \le 3$ Polynomial inequality

$x^2 - 2x - 3 \le 0$

$x^2 - 2x - 3 = 0$ Related equation

$(x + 1)(x - 3) = 0$ Factoring

Using the principle of zero products or by observing the graph of $y = x^2 - 2x - 3$, we see that the solutions of the related equation are -1 and 3. These numbers divide the x-axis into the intervals $(-\infty, -1)$, $(-1, 3)$, and $(3, \infty)$. We let $f(x) = x^2 - 2x - 3$ and test a value in each interval.

$(-\infty, -1)$: $f(-2) = 5 > 0$

$(-1, 3)$: $f(0) = -3 < 0$

$(3, \infty)$: $f(4) = 5 > 0$

Function values are negative on $(-1, 3)$. This can also be determined from the graph of $y = x^2 - 2x - 3$. Since the inequality symbol is $\le$, the endpoints of the interval must be included in the solution set. It is $\{x| - 1 \le x \le 3\}$ or $[-1, 3]$.

88. $\dfrac{x-1}{x+2} > 4$ Rational inequality

$\dfrac{x-1}{x+2} - 4 > 0$

$\dfrac{x-1}{x+2} - 4 = 0$ Related equation

The denominator of $f(x) = \dfrac{x-1}{x+2} - 4$ is 0 when $x = -2$, so the function is not defined for $x = -2$. We solve the related equation $f(x) = 0$.

$$\frac{x-1}{x+2} - 4 = 0$$

$x - 1 - 4(x + 2) = 0$ Multiplying by $x + 2$

$$x - 1 - 4x - 8 = 0$$
$$-3x - 9 = 0$$
$$-3x = 9$$
$$x = -3$$

Thus, the critical values are -3 and -2. They divide the x-axis into the intervals $(-\infty, -3)$, $(-3, -2)$, and $(-2, \infty)$. We test a value in each interval.

$(-\infty, -3)$: $f(-4) = -\dfrac{3}{2} < 0$

$(-3, -2)$: $f(-2.5) = 3 > 0$

$(-2, \infty)$: $f(0) = -\dfrac{9}{2} < 0$

Function values are positive on $(-3, -2)$. This can also be determined from the graph of $y = \dfrac{x-1}{x+2} - 4$. The solution set is $\{x| - 3 < x < 2\}$, or $(-3, -2)$.

89. Graph: $y \ge x^2 - 2$,

$y \le 2 - x^2$

First graph the related equations $y = x^2 - 2$ and $y = 2 - x^2$ using solid lines. The solution set consists of the region above the graph of $y = x^2 - 2$ and below the graph of $y = 2 - x^2$.

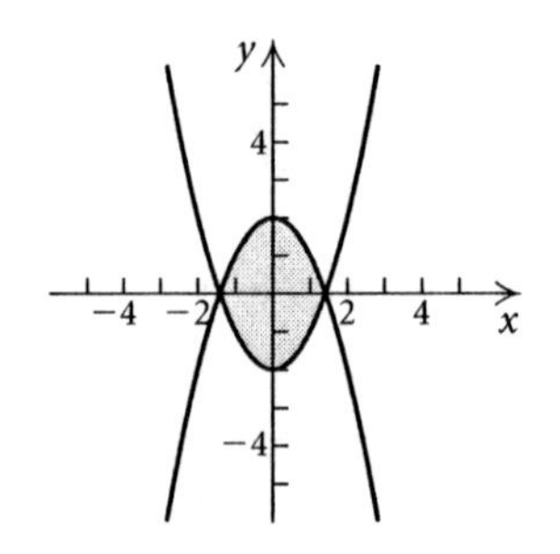

90. $y < x + 1$
$y \geq x^2$

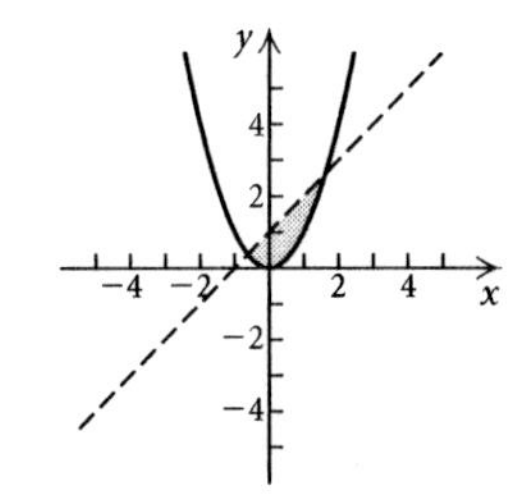

91. See the answer section in the text.

92.

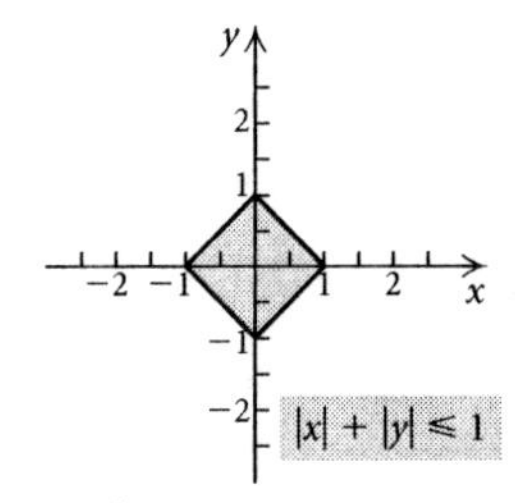

93. See the answer section in the text.

94.

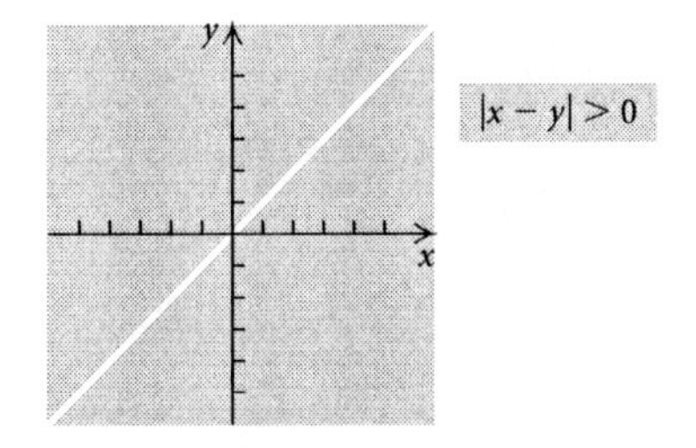

95. Let x = the number of less expensive speaker assemblies and y = the number of more expensive assemblies. The income is given by

$$I = 350x + 600y$$

subject to

$$y \leq 44$$
$$x + y \leq 60,$$
$$x + 2y \leq 90,$$
$$x \geq 0,$$
$$y \geq 0.$$

Graph the system of inequalities, determine the vertices, and find the value of I at each vertex.

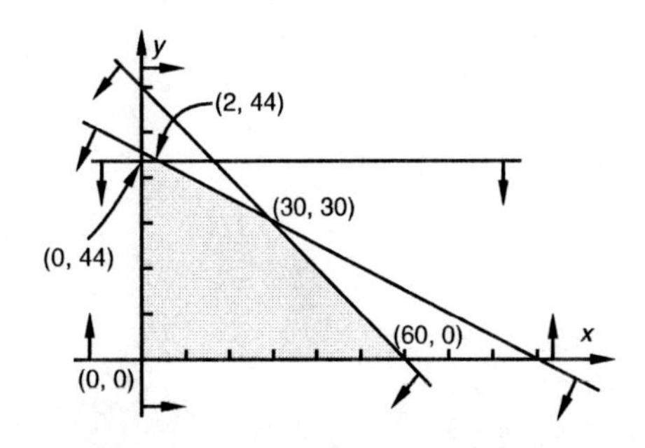

Vertex	$I = 350x + 600y$
$(0, 0)$	$350 \cdot 0 + 600 \cdot 0 = 0$
$(0, 44)$	$350 \cdot 0 + 600 \cdot 44 = 26,400$
$(2, 44)$	$350 \cdot 2 + 600 \cdot 44 = 27,100$
$(30, 30)$	$350 \cdot 30 + 600 \cdot 30 = 28,500$
$(60, 0)$	$350 \cdot 60 + 600 \cdot 0 = 21,000$

The maximum income of \$28,500 is achieved when 30 less expensive and 30 more expensive assemblies are made.

96. Let x = the number of chairs and y = the number of sofas produced. Find the maximum value of

$$I = 80x + 300y$$

subject to

$$20x + 100y \leq 1900,$$
$$x + 50y \leq 500,$$
$$2x + 20y \leq 240,$$
$$x \geq 0,$$
$$y \geq 0.$$

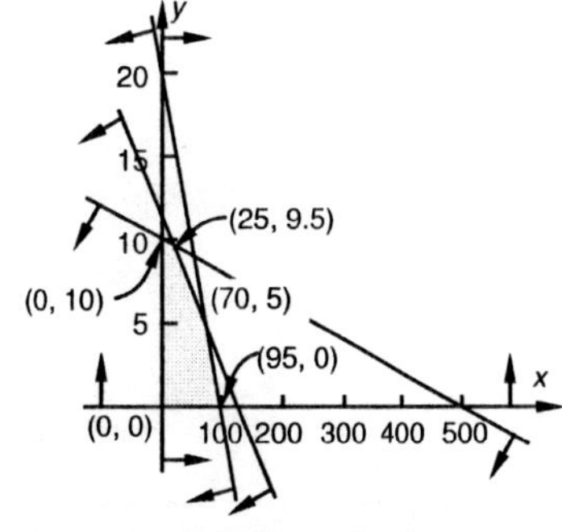

Vertex	$I = 80x + 300y$
$(0, 0)$	$80 \cdot 0 + 300 \cdot 0 = 0$
$(0, 10)$	$80 \cdot 0 + 300 \cdot 10 = 3000$
$(25, 9.5)$	$80 \cdot 25 + 300(9.5) = 4850$
$(70, 5)$	$80 \cdot 70 + 300 \cdot 5 = 7100$
$(95, 0)$	$80 \cdot 95 + 300 \cdot 0 = 7600$

The maximum income of \$7600 is achieved by making 95 chairs and 0 sofas.

Exercise Set 8.8

1. $$\frac{x+7}{(x-3)(x+2)} = \frac{A}{x-3} + \frac{B}{x+2}$$

$$\frac{x+7}{(x-3)(x+2)} = \frac{A(x+2) + B(x-3)}{(x-3)(x+2)} \quad \text{Adding}$$

Equate the numerators:

$x + 7 = A(x+2) + B(x-3)$

Let $x + 2 = 0$, or $x = -2$. Then we get

$-2 + 7 = 0 + B(-2-3)$

$5 = -5B$

$-1 = B$

Next let $x - 3 = 0$, or $x = 3$. Then we get

$$3+7 = A(3+2)+0$$
$$10 = 5A$$
$$2 = A$$

The decomposition is as follows:

$$\frac{2}{x-3} - \frac{1}{x+2}$$

2. $$\frac{2x}{(x+1)(x-1)} = \frac{A}{x+1} + \frac{B}{x-1}$$
$$\frac{2x}{(x+1)(x-1)} = \frac{A(x-1)+B(x+1)}{(x+1)(x-1)}$$
$$2x = A(x-1)+B(x+1)$$

Let $x=1$: $2\cdot 1 = 0 + B(1+1)$
$$2 = 2B$$
$$1 = B$$

Let $x=-1$: $2(-1) = A(-1-1)+0$
$$-2 = -2A$$
$$1 = A$$

The decomposition is $\frac{1}{x+1} + \frac{1}{x-1}$.

3. $$\frac{7x-1}{6x^2-5x+1}$$
$$= \frac{7x-1}{(3x-1)(2x-1)} \quad \text{Factoring the denominator}$$
$$= \frac{A}{3x-1} + \frac{B}{2x-1}$$
$$= \frac{A(2x-1)+B(3x-1)}{(3x-1)(2x-1)} \quad \text{Adding}$$

Equate the numerators:

$7x-1 = A(2x-1)+B(3x-1)$

Let $2x-1=0$, or $x=\frac{1}{2}$. Then we get

$$7\left(\frac{1}{2}\right) - 1 = 0 + B\left(3\cdot\frac{1}{2}-1\right)$$
$$\frac{5}{2} = \frac{1}{2}B$$
$$5 = B$$

Next let $3x-1=0$, or $x=\frac{1}{3}$. We get

$$7\left(\frac{1}{3}\right) - 1 = A\left(2\cdot\frac{1}{3}-1\right)$$
$$\frac{7}{3} - 1 = A\left(\frac{2}{3}-1\right)$$
$$\frac{4}{3} = -\frac{1}{3}A$$
$$-4 = A$$

The decomposition is as follows:

$$-\frac{4}{3x-1} + \frac{5}{2x-1}$$

4. $$\frac{13x+46}{12x^2-11x-15} = \frac{13x+46}{(4x+3)(3x-5)}$$
$$= \frac{A}{4x+3} + \frac{B}{3x-5}$$
$$= \frac{A(3x-5)+B(4x+3)}{(4x+3)(3x-5)}$$

Let $x=\frac{5}{3}$: $13\cdot\frac{5}{3}+46 = 0 + B\left(4\cdot\frac{5}{3}+3\right)$
$$\frac{203}{3} = \frac{29}{3}B$$
$$7 = B$$

Let $x=-\frac{3}{4}$:

$$13\left(-\frac{3}{4}\right)+46 = A\left[3\left(-\frac{3}{4}\right)-5\right]+0$$
$$\frac{145}{4} = -\frac{29}{4}A$$
$$-5 = A$$

The decomposition is $-\frac{5}{4x+3} + \frac{7}{3x-5}$.

5. $$\frac{3x^2-11x-26}{(x^2-4)(x+1)}$$
$$= \frac{3x^2-11x-26}{(x+2)(x-2)(x+1)} \quad \text{Factoring the denominator}$$
$$= \frac{A}{x+2} + \frac{B}{x-2} + \frac{C}{x+1}$$
$$= \frac{A(x-2)(x+1)+B(x+2)(x+1)+C(x+2)(x-2)}{(x+2)(x-2)(x+1)} \quad \text{Adding}$$

Equate the numerators:

$$3x^2-11x-26 = A(x-2)(x+1) + B(x+2)(x+1) + C(x+2)(x-2)$$

Let $x+2=0$ or $x=-2$. Then we get

$$3(-2)^2-11(-2)-26 = A(-2-2)(-2+1)+0+0$$
$$12+22-26 = A(-4)(-1)$$
$$8 = 4A$$
$$2 = A$$

Next let $x-2=0$, or $x=2$. Then, we get

$$3\cdot 2^2 - 11\cdot 2 - 26 = 0 + B(2+2)(2+1)+0$$
$$12-22-26 = B\cdot 4\cdot 3$$
$$-36 = 12B$$
$$-3 = B$$

Finally let $x+1=0$, or $x=-1$. We get

$$3(-1)^2-11(-1)-26 = 0+0+C(-1+2)(-1-2)$$
$$3+11-26 = C(1)(-3)$$
$$-12 = -3C$$
$$4 = C$$

The decomposition is as follows:

$$\frac{2}{x+2} - \frac{3}{x-2} + \frac{4}{x+1}$$

6. $\dfrac{5x^2+9x-56}{(x-4)(x-2)(x+1)}$

$= \dfrac{A}{x-4}+\dfrac{B}{x-2}+\dfrac{C}{x+1}$

$= \dfrac{A(x-2)(x+1)+B(x-4)(x+1)+C(x-4)(x-2)}{(x-4)(x-2)(x+1)}$

$5x^2+9x-56 = A(x-2)(x+1)+$
$B(x-4)(x+1)+C(x-4)(x-2)$

Let $x=4$:

$5\cdot 4^2+9\cdot 4-56 = A(4-2)(4+1)+0+0$

$60 = 10A$

$6 = A$

Let $x=2$:

$5\cdot 2^2+9\cdot 2-56 = 0+B(2-4)(2+1)+0$

$-18 = -6B$

$3 = B$

Let $x=-1$:

$5(-1)^2+9(-1)-56 = 0+0+C(-1-4)(-1-2)$

$-60 = 15C$

$-4 = C$

The decomposition is $\dfrac{6}{x-4}+\dfrac{3}{x-2}-\dfrac{4}{x+1}$.

7. $\dfrac{9}{(x+2)^2(x-1)}$

$= \dfrac{A}{x+2}+\dfrac{B}{(x+2)^2}+\dfrac{C}{x-1}$

$= \dfrac{A(x+2)(x-1)+B(x-1)+C(x+2)^2}{(x+2)^2(x-1)}$ Adding

Equate the numerators:

$9 = A(x+2)(x-1)+B(x-1)+C(x+2)^2 \quad (1)$

Let $x-1=0$, or $x=1$. Then, we get

$9 = 0+0+C(1+2)^2$

$9 = 9C$

$1 = C$

Next let $x+2=0$, or $x=-2$. Then, we get

$9 = 0+B(-2-1)+0$

$9 = -3B$

$-3 = B$

To find A we first simplify equation (1).

$9 = A(x^2+x-2)+B(x-1)+C(x^2+4x+4)$

$= Ax^2+Ax-2A+Bx-B+Cx^2+4Cx+4C$

$= (A+C)x^2+(A+B+4C)x+(-2A-B+4C)$

Then we equate the coefficients of x^2.

$0 = A+C$

$0 = A+1$ Substituting 1 for C

$-1 = A$

The decomposition is as follows:

$-\dfrac{1}{x+2}-\dfrac{3}{(x+2)^2}+\dfrac{1}{x-1}$

8. $\dfrac{x^2-x-4}{(x-2)^3}$

$= \dfrac{A}{x-2}+\dfrac{B}{(x-2)^2}+\dfrac{C}{(x-2)^3}$

$= \dfrac{A(x-2)^2+B(x-2)+C}{(x-2)^3}$

$x^2-x-4 = A(x-2)^2+B(x-2)+C \quad (1)$

Let $x=2$: $2^2-2-4 = 0+0+C$

$-2 = C$

Simplify equation (1).

$x^2-x-4 = Ax^2-4Ax+4A+Bx-2B+C$

$= Ax^2+(-4A+B)x+(4A-2B+C)$

Then $1 = A$ and

$-1 = -4A+B$, so $B=3$.

The decomposition is $\dfrac{1}{x-2}+\dfrac{3}{(x-2)^2}-\dfrac{2}{(x-2)^3}$.

9. $\dfrac{2x^2+3x+1}{(x^2-1)(2x-1)}$

$= \dfrac{2x^2+3x+1}{(x+1)(x-1)(2x-1)}$ Factoring the denominator

$= \dfrac{A}{x+1}+\dfrac{B}{x-1}+\dfrac{C}{2x-1}$

$= \dfrac{A(x-1)(2x-1)+B(x+1)(2x-1)+C(x+1)(x-1)}{(x+1)(x-1)(2x-1)}$ Adding

Equate the numerators:

$2x^2+3x+1 = A(x-1)(2x-1)+$
$B(x+1)(2x-1)+C(x+1)(x-1)$

Let $x+1=0$, or $x=-1$. Then, we get

$2(-1)^2+3(-1)+1 = A(-1-1)[2(-1)-1]+0+0$

$2-3+1 = A(-2)(-3)$

$0 = 6A$

$0 = A$

Next let $x-1=0$, or $x=1$. Then, we get

$2\cdot 1^2+3\cdot 1+1 = 0+B(1+1)(2\cdot 1-1)+0$

$2+3+1 = B\cdot 2\cdot 1$

$6 = 2B$

$3 = B$

Finally we let $2x-1=0$, or $x=\dfrac{1}{2}$. We get

$2\left(\dfrac{1}{2}\right)^2+3\left(\dfrac{1}{2}\right)+1 = 0+0+C\left(\dfrac{1}{2}+1\right)\left(\dfrac{1}{2}-1\right)$

$\dfrac{1}{2}+\dfrac{3}{2}+1 = C\cdot\dfrac{3}{2}\cdot\left(-\dfrac{1}{2}\right)$

$3 = -\dfrac{3}{4}C$

$-4 = C$

The decomposition is as follows:

$\dfrac{3}{x-1}-\dfrac{4}{2x-1}$

10. $\dfrac{x^2-10x+13}{(x^2-5x+6)(x-1)}$

$= \dfrac{x^2-10x+13}{(x-3)(x-2)(x-1)}$

$= \dfrac{A}{x-3}+\dfrac{B}{x-2}+\dfrac{C}{x-1}$

$= \dfrac{A(x-2)(x-1)+B(x-3)(x-1)+C(x-3)(x-2)}{(x-3)(x-2)(x-1)}$

$x^2-10x+13 = A(x-2)(x-1)+B(x-3)(x-1)+C(x-3)(x-2)$

Let $x=3$:

$3^2-10\cdot 3+13 = A(3-2)(3-1)+0+0$

$-8 = 2A$

$-4 = A$

Let $x=2$:

$2^2-10\cdot 2+13 = 0+B(2-3)(2-1)+0$

$-3 = -B$

$3 = B$

Let $x=1$:

$1^2-10\cdot 1+13 = 0+0+C(1-3)(1-2)$

$4 = 2C$

$2 = C$

The decomposition is $-\dfrac{4}{x-3}+\dfrac{3}{x-2}+\dfrac{2}{x-1}$.

11. $\dfrac{x^4-3x^3-3x^2+10}{(x+1)^2(x-3)}$

$= \dfrac{x^4-3x^3-3x^2+10}{x^3-x^2-5x-3}$ Multiplying the denominator

Since the degree of the numerator is greater than the degree of the denominator, we divide.

$$
\begin{array}{r}
x-2 \\
x^3-x^2-5x-3\,\overline{)\,x^4-3x^3-3x^2+\ 0x+10} \\
\underline{x^4-\ x^3-5x^2-\ 3x} \\
-2x^3+2x^2+\ 3x+10 \\
\underline{-2x^3+2x^2+10x+\ 6} \\
-\ 7x+\ 4
\end{array}
$$

The original expression is thus equivalent to the following:

$x-2+\dfrac{-7x+4}{x^3-x^2-5x-3}$

We proceed to decompose the fraction.

$\dfrac{-7x+4}{(x+1)^2(x-3)}$

$= \dfrac{A}{x+1}+\dfrac{B}{(x+1)^2}+\dfrac{C}{x-3}$

$= \dfrac{A(x+1)(x-3)+B(x-3)+C(x+1)^2}{(x+1)^2(x-3)}$ Adding

Equate the numerators:

$-7x+4 = A(x+1)(x-3)+B(x-3)+C(x+1)^2 \quad (1)$

Let $x-3=0$, or $x=3$. Then, we get

$-7\cdot 3+4 = 0+0+C(3+1)^2$

$-17 = 16C$

$-\dfrac{17}{16} = C$

Let $x+1=0$, or $x=-1$. Then, we get

$-7(-1)+4 = 0+B(-1-3)+0$

$11 = -4B$

$-\dfrac{11}{4} = B$

To find A we first simplify equation (1).

$-7x+4$

$= A(x^2-2x-3)+B(x-3)+C(x^2+2x+1)$

$= Ax^2-2Ax-3A+Bx-3B+Cx^2-2Cx+C$

$= (A+C)x^2+(-2A+B-2C)x+(-3A-3B+C)$

Then equate the coefficients of x^2.

$0 = A+C$

Substituting $-\dfrac{17}{16}$ for C, we get $A=\dfrac{17}{16}$.

The decomposition is as follows:

$\dfrac{17/16}{x+1}-\dfrac{11/4}{(x+1)^2}-\dfrac{17/16}{x-3}$

The original expression is equivalent to the following:

$x-2+\dfrac{17/16}{x+1}-\dfrac{11/4}{(x+1)^2}-\dfrac{17/16}{x-3}$

12. $\dfrac{10x^3-15x^2-35x}{x^2-x-6} = 10x-5+\dfrac{20x-30}{x^2-x-6}$ Dividing

$\dfrac{20x-30}{x^2-x-6} = \dfrac{20x-30}{(x-3)(x+2)}$

$= \dfrac{A}{x-3}+\dfrac{B}{x+2}$

$= \dfrac{A(x+2)+B(x-3)}{(x-3)(x+2)}$

$20x-30 = A(x+2)+B(x-3)$

Let $x=3$: $20\cdot 3-30 = A(3+2)+0$

$30 = 5A$

$6 = A$

Let $x=-2$: $20(-2)-30 = 0+B(-2-3)$

$-70 = -5B$

$14 = B$

The decomposition is $10x-5+\dfrac{6}{x-3}+\dfrac{14}{x+2}$.

13. $\dfrac{-x^2+2x-13}{(x^2+2)(x-1)}$

$= \dfrac{Ax+B}{x^2+2}+\dfrac{C}{x-1}$

$= \dfrac{(Ax+B)(x-1)+C(x^2+2)}{(x^2+2)(x-1)}$ Adding

Equate the numerators:

$-x^2+2x-13 = (Ax+B)(x-1)+C(x^2+2) \quad (1)$

Let $x-1=0$, or $x=1$. Then we get

$$-1^2+2\cdot 1-13=0+C(1^2+2)$$
$$-1+2-13=C(1+2)$$
$$-12=3C$$
$$-4=C$$

To find A and B we first simplify equation (1).

$$-x^2+2x-13$$
$$=Ax^2-Ax+Bx-B+Cx^2+2C$$
$$=(A+C)x^2+(-A+B)x+(-B+2C)$$

Equate the coefficients of x^2:

$-1=A+C$

Substituting -4 for C, we get $A=3$.

Equate the constant terms:

$-13=-B+2C$

Substituting -4 for C, we get $B=5$.

The decomposition is as follows:

$$\frac{3x+5}{x^2+2}-\frac{4}{x-1}$$

14. $$\frac{26x^2+208x}{(x^2+1)(x+5)}$$
$$=\frac{Ax+B}{x^2+1}+\frac{C}{x+5}$$
$$=\frac{(Ax+B)(x+5)+C(x^2+1)}{(x^2+1)(x+5)}$$

$$26x^2+208x=(Ax+B)(x+5)+C(x^2+1) \qquad (1)$$

Let $x=-5$:

$$26(-5)^2+208(-5)=0+C[(-5)^2+1]$$
$$-390=26C$$
$$-15=C$$

Simplify equation (1).

$$26x^2+208x$$
$$=Ax^2+5Ax+Bx+5B+Cx^2+C$$
$$=(A+C)x^2+(5A+B)x+(5B+C)$$

$26=A+C$

$26=A-15$

$41=A$

$0=5B+C$

$0=5B-15$

$3=B$

The decomposition is $\dfrac{41x+3}{x^2+1}-\dfrac{15}{x+5}$.

15. $$\frac{6+26x-x^2}{(2x-1)(x+2)^2}$$
$$=\frac{A}{2x-1}+\frac{B}{x+2}+\frac{C}{(x+2)^2}$$
$$=\frac{A(x+2)^2+B(2x-1)(x+2)+C(2x-1)}{(2x-1)(x+2)^2}$$

Adding

Equate the numerators:

$$6+26x-x^2=A(x+2)^2+B(2x-1)(x+2)+ C(2x-1) \qquad (1)$$

Let $2x-1=0$, or $x=\dfrac{1}{2}$. Then, we get

$$6+26\cdot\frac{1}{2}-\left(\frac{1}{2}\right)^2=A\left(\frac{1}{2}+2\right)^2+0+0$$
$$6+13-\frac{1}{4}=A\left(\frac{5}{2}\right)^2$$
$$\frac{75}{4}=\frac{25}{4}A$$
$$3=A$$

Let $x+2=0$, or $x=-2$. We get

$$6+26(-2)-(-2)^2=0+0+C[2(-2)-1]$$
$$6-52-4=-5C$$
$$-50=-5C$$
$$10=C$$

To find B we first simplify equation (1).

$$6+26x-x^2$$
$$=A(x^2+4x+4)+B(2x^2+3x-2)+C(2x-1)$$
$$=Ax^2+4Ax+4A+2Bx^2+3Bx-2B+2Cx-C$$
$$=(A+2B)x^2+(4A+3B+2C)x+(4A-2B-C)$$

Equate the coefficients of x^2:

$-1=A+2B$

Substituting 3 for A, we obtain $B=-2$.

The decomposition is as follows:

$$\frac{3}{2x-1}-\frac{2}{x+2}+\frac{10}{(x+2)^2}$$

16. $$\frac{5x^3+6x^2+5x}{(x^2-1)(x+1)^3}$$
$$=\frac{5x^3+6x^2+5x}{(x-1)(x+1)^4}$$
$$=\frac{A}{x-1}+\frac{B}{x+1}+\frac{C}{(x+1)^2}+\frac{D}{(x+1)^3}+\frac{E}{(x+1)^4}$$
$$=[A(x+1)^4+B(x-1)(x+1)^3+C(x-1)(x+1)^2+ D(x-1)(x+1)+E(x-1)]/[(x-1)(x+1)^4]$$

$$5x^3+6x^2+5x=$$
$$A(x+1)^4+B(x-1)(x+1)^3+C(x-1)(x+1)^2+ D(x-1)(x+1)+E(x-1) \qquad (1)$$

Let $x=1$: $5\cdot 1^3+6\cdot 1^2+5\cdot 1=A(1+1)^4$

$$16=16A$$
$$A=1$$

Let $x=-1$:

$$5(-1)^3+6(-1)^2+5(-1)=E(-1-1)$$
$$-4=-2E$$
$$2=E$$

Simplify equation (1).

$$5x^3+6x^2+5x=(A+B)x^4+(4A+2B+C)x^3+$$
$$(6A+C+D)x^2+(4A-2B-C+E)x+$$
$$(A-B-C-D-E)$$
$$0=A+B$$
$$0=1+B$$
$$-1=B$$
$$5=4A+2B+C$$
$$5=4\cdot 1+2(-1)+C$$
$$3=C$$
$$0=A-B-C-D-E$$
$$0=1-(-1)-3-D-2$$
$$D=-3$$

The decomposition is

$$\frac{1}{x-1}-\frac{1}{x+1}+\frac{3}{(x+1)^2}-\frac{3}{(x+1)^3}+\frac{2}{(x+1)^4}.$$

17. $\dfrac{6x^3+5x^2+6x-2}{2x^2+x-1}$

Since the degree of the numerator is greater than the degree of the denominator, we divide.

$$\begin{array}{r} 3x+\ \ 1 \\ 2x^2+x-1 \overline{\big)\,6x^3+5x^2+6x-2} \\ \underline{6x^3+3x^2-3x} \\ 2x^2+9x-2 \\ \underline{2x^2+\ x-1} \\ 8x-1 \end{array}$$

The original expression is equivalent to

$$3x+1+\frac{8x-1}{2x^2+x-1}$$

We proceed to decompose the fraction.

$$\frac{8x-1}{2x^2+x-1}=\frac{8x-1}{(2x-1)(x+1)} \quad \text{Factoring the denominator}$$
$$=\frac{A}{2x-1}+\frac{B}{x+1}$$
$$=\frac{A(x+1)+B(2x-1)}{(2x-1)(x+1)} \quad \text{Adding}$$

Equate the numerators:

$$8x-1=A(x+1)+B(2x-1)$$

Let $x+1=0$, or $x=-1$. Then we get

$$8(-1)-1=0+B[2(-1)-1]$$
$$-8-1=B(-2-1)$$
$$-9=-3B$$
$$3=B$$

Next let $2x-1=0$, or $x=\dfrac{1}{2}$. We get

$$8\left(\frac{1}{2}\right)-1=A\left(\frac{1}{2}+1\right)+0$$
$$4-1=A\left(\frac{3}{2}\right)$$
$$3=\frac{3}{2}A$$
$$2=A$$

The decomposition is

$$\frac{2}{2x-1}+\frac{3}{x+1}.$$

The original expression is equivalent to

$$3x+1+\frac{2}{2x-1}+\frac{3}{x+1}.$$

18. $\dfrac{2x^3+3x^2-11x-10}{x^2+2x-3}=2x-1+\dfrac{-3x-13}{x^2+2x-3}$ Dividing

$$\frac{-3x-13}{x^2+2x-3}=\frac{A}{x+3}+\frac{B}{x-1}$$
$$=\frac{A(x-1)+B(x+3)}{(x+3)(x-1)}$$
$$-3x-13=A(x-1)+B(x+3)$$

Let $x=-3$: $-3(-3)-13=A(-3-1)+0$

$$-4=-4A$$
$$1=A$$

Let $x=1$: $-3\cdot 1-13=0+B(1+3)$

$$-16=4B$$
$$-4=B$$

The decomposition is $2x-1+\dfrac{1}{x+3}-\dfrac{4}{x-1}$.

19. $\dfrac{2x^2-11x+5}{(x-3)(x^2+2x-5)}$

$$=\frac{A}{x-3}+\frac{Bx+C}{x^2+2x-5}$$
$$=\frac{A(x^2+2x-5)+(Bx+C)(x-3)}{(x-3)(x^2+2x-5)} \quad \text{Adding}$$

Equate the numerators:

$$2x^2-11x+5=A(x^2+2x-5)+(Bx+C)(x-3) \quad (1)$$

Let $x-3=0$, or $x=3$. Then, we get

$$2\cdot 3^2-11\cdot 3+5=A(3^2+2\cdot 3-5)+0$$
$$18-33+5=A(9+6-5)$$
$$-10=10A$$
$$-1=A$$

To find B and C, we first simplify equation (1).

$$2x^2-11x+5=Ax^2+2Ax-5A+Bx^2-3Bx+Cx-3C$$
$$=(A+B)x^2+(2A-3B+C)x+(-5A-3C)$$

Equate the coefficients of x^2:

$2=A+B$

Substituting -1 for A, we get $B=3$.

Equate the constant terms:

$5=-5A-3C$

Substituting -1 for A, we get $C=0$.

The decomposition is as follows:

$$-\frac{1}{x-3}+\frac{3x}{x^2+2x-5}$$

20. $\dfrac{3x^2-3x-8}{(x-5)(x^2+x-4)}$

$= \dfrac{A}{x-5} + \dfrac{Bx+C}{x^2+x-4}$

$= \dfrac{A(x^2+x-4)+(Bx+C)(x-5)}{(x-5)(x^2+x-4)}$

$3x^2-3x-8 = A(x^2+x-4)+$
$(Bx+C)(x-5) \qquad (1)$

Let $x=5$: $3\cdot 5^2 - 3\cdot 5 - 8 = A(5^2+5-4)+0$
$52 = 26A$
$2 = A$

Simplify equation (1).

$3x^2-3x-8$
$= Ax^2+Ax-4A+Bx^2-5Bx+Cx-5C$
$= (A+B)x^2+(A-5B+C)x+(-4A-5C)$

$3 = A+B$
$3 = 2+B$
$1 = B$

$-8 = -4A-5C$
$-8 = -4\cdot 2-5C$
$0 = C$

The decomposition is $\dfrac{2}{x-5}+\dfrac{x}{x^2+x-4}$.

21. $\dfrac{-4x^2-2x+10}{(3x+5)(x+1)^2}$

The decomposition looks like

$\dfrac{A}{3x+5}+\dfrac{B}{x+1}+\dfrac{C}{(x+1)^2}$.

Add and equate the numerators.

$-4x^2-2x+10$
$= A(x+1)^2+B(3x+5)(x+1)+C(3x+5)$
$= A(x^2+2x+1)+B(3x^2+8x+5)+C(3x+5)$

or

$-4x^2-2x+10$
$= (A+3B)x^2+(2A+8B+3C)x+(A+5B+5C)$

Then equate corresponding coefficients.

$-4 = A+3B$ Coefficients of x^2-terms
$-2 = 2A+8B+3C$ Coefficients of x-terms
$10 = A+5B+5C$ Constant terms

We solve this system of three equations and find $A=5$, $B=-3$, $C=4$.

The decomposition is

$\dfrac{5}{3x+5}-\dfrac{3}{x+1}+\dfrac{4}{(x+1)^2}$.

22. $\dfrac{26x^2-36x+22}{(x-4)(2x-1)^2} = \dfrac{A}{x-4}+\dfrac{B}{2x-1}+\dfrac{C}{(2x-1)^2}$

Add and equate numerators.

$26x^2-36x+22$
$= A(2x-1)^2+B(x-4)(2x-1)+C(x-4)$
$= A(4x^2-4x+1)+B(2x^2-9x+4)+C(x-4)$

or

$26x^2-36x+22 = (4A+2B)x^2+(-4A-9B+C)x+$
$(A+4B-4C)$

Solving the system of equations

$26 = 4A+2B,$
$-36 = -4A-9B+C,$
$22 = A+4B-4C$

we get $A=6$, $B=1$, and $C=-3$.

Then the decomposition is

$\dfrac{6}{x-4}+\dfrac{1}{2x-1}-\dfrac{3}{(2x-1)^2}$.

23. $\dfrac{36x+1}{12x^2-7x-10} = \dfrac{36x+1}{(4x-5)(3x+2)}$

The decomposition looks like

$\dfrac{A}{4x-5}+\dfrac{B}{3x+2}$.

Add and equate the numerators.

$36x+1 = A(3x+2)+B(4x-5)$

or $36x+1 = (3A+4B)x+(2A-5B)$

Then equate corresponding coefficients.

$36 = 3A+4B$ Coefficients of x-terms
$1 = 2A-5B$ Constant terms

We solve this system of equations and find $A=8$ and $B=3$.

The decomposition is

$\dfrac{8}{4x-5}+\dfrac{3}{3x+2}$.

24. $\dfrac{-17x+61}{6x^2+39x-21} = \dfrac{A}{6x-3}+\dfrac{B}{x+7}$

$-17x+61 = (A+6B)x+(7A-3B)$
$-17 = A+6B,$
$61 = 7A-3B$

Then $A=7$ and $B=-4$.

The decomposition is

$\dfrac{7}{6x-3}-\dfrac{4}{x+7}$.

25. $\dfrac{-4x^2-9x+8}{(3x^2+1)(x-2)}$

The decomposition looks like

$\dfrac{Ax+B}{3x^2+1}+\dfrac{C}{x-2}$.

Add and equate the numerators.

$-4x^2-9x+8$
$= (Ax+B)(x-2)+C(3x^2+1)$
$= Ax^2-2Ax+Bx-2B+3Cx^2+C$

or

$-4x^2-9x+8$
$= (A+3C)x^2+(-2A+B)x+(-2B+C)$

Then equate corresponding coefficients.

$-4 = A + 3C$ Coefficients of x^2-terms

$-9 = -2A + B$ Coefficients of x-terms

$8 = -2B + C$ Constant terms

We solve this system of equations and find

$A = 2,\ B = -5,\ C = -2.$

The decomposition is

$$\frac{2x-5}{3x^2+1} - \frac{2}{x-2}.$$

26. $\dfrac{11x^2-39x+16}{(x^2+4)(x-8)} = \dfrac{Ax+B}{x^2+4} + \dfrac{C}{x-8}$

$11x^2 - 39x + 16 = (A+C)x^2 + (-8A+B)x + (-8B+4C)$

$11 = A + C,$

$-39 = -8A + B,$

$16 = -8B + 4C$

Then $A = 5$, $B = 1$, and $C = 6$.

The decomposition is

$$\frac{5x+1}{x^2+4} + \frac{6}{x-8}.$$

27. See the procedure on page 721 of the text. One of the algebraic methods referred to in Step 5 involves substituting values for the variable that allow us to find the constants. The other method involves equating numerators, equating coefficients of the like terms, and then using a system of equations to find the constants. Answers will vary regarding the method preferred.

28. The denominator of the second fraction, x^2-5x+6, can be factored into linear factors with real coefficients: $(x-3)(x-2)$. Thus, the given expression is not a partial fraction decomposition.

29. The degree of the numerator is equal to the degree of the denominator, so the first step should be to divide the numerator by the denominator in order to express the fraction as a quotient + remainder/denominator.

30. $x^3 - 3x^2 + x - 3 = 0$

$x^2(x-3) + (x-3) = 0$

$(x-3)(x^2+1) = 0$

$x - 3 = 0$ *or* $x^2 + 1 = 0$

$x = 3$ *or* $x^2 = -1$

$x = 3$ *or* $x = \pm i$

The solutions are 3, i, and $-i$.

31. $f(x) = x^3 + x^2 - 3x - 2$

We use synthetic division to factor the polynomial. Using the possibilities found by the rational zeros theorem we find that $x+2$ is a factor:

```
-2 |  1   1  -3  -2
   |     -2   2   2
   ----------------
      1  -1  -1 | 0
```

We have $x^3 + x^2 - 3x - 2 = (x+2)(x^2 - x - 1)$.

$x^3 + x^2 - 3x - 2 = 0$

$(x+2)(x^2 - x - 1) = 0$

$x + 2 = 0$ *or* $x^2 - x - 1 = 0$

The solution of the first equation is -2. We use the quadratic formula to solve the second equation.

$$x = \frac{-b \pm \sqrt{b^2-4ac}}{2a} = \frac{-(-1) \pm \sqrt{(-1)^2 - 4\cdot 1\cdot(-1)}}{2\cdot 1} = \frac{1 \pm \sqrt{5}}{2}$$

The solutions are -2, $\dfrac{1+\sqrt{5}}{2}$ and $\dfrac{1-\sqrt{5}}{2}$.

32. $f(x) = x^4 - x^3 - 5x^2 - x - 6$

```
-2 |  1  -1  -5  -1  -6
   |     -2   6  -2   6
   --------------------
      1  -3   1  -3 | 0
```

$x^4 - x^3 - 5x^2 - x - 6 = 0$

$(x+2)(x^3 - 3x^2 + x - 3) = 0$

$(x+2)[x^2(x-3) + (x-3)] = 0$

$(x+2)(x-3)(x^2+1) = 0$

$x + 2 = 0$ *or* $x - 3 = 0$ *or* $x^2 + 1 = 0$

$x = -2$ *or* $x = 3$ *or* $x = \pm i$

The solutions are -2, 3, i, and $-i$.

33. $f(x) = x^3 + 5x^2 + 5x - 3$

```
-3 |  1   5   5  -3
   |     -3  -6   3
   ----------------
      1   2  -1 | 0
```

$x^3 + 5x^2 + 5x - 3 = 0$

$(x+3)(x^2 + 2x - 1) = 0$

$x + 3 = 0$ *or* $x^2 + 2x - 1 = 0$

The solution of the first equation is -3. We use the quadratic formula to solve the second equation.

$$x = \frac{-b \pm \sqrt{b^2-4ac}}{2a} = \frac{-2 \pm \sqrt{2^2 - 4\cdot 1\cdot(-1)}}{2\cdot 1} = \frac{-2 \pm \sqrt{8}}{2} = \frac{-2 \pm 2\sqrt{2}}{2} = \frac{2(-1 \pm \sqrt{2})}{2} = -1 \pm \sqrt{2}$$

The solutions are -3, $-1+\sqrt{2}$, and $-1-\sqrt{2}$.

34. $\dfrac{9x^3 - 24x^2 + 48x}{(x-2)^4(x+1)} = \dfrac{A}{x+1} + \dfrac{P(x)}{(x-2)^4}$

Add and equate numerators.

$9x^3 - 24x^2 + 48x = A(x-2)^4 + P(x)(x+1)$

Let $x = -1$:

$9(-1)^3 - 24(-1)^2 + 48(-1) = A(-1-2)^4 + 0$

$-81 = 81A$

$-1 = A$

Then

$$9x^3 - 24x^2 + 48x = -(x-2)^4 + P(x)(x+1)$$

$$9x^3 - 24x^2 + 48x = -x^4 + 8x^3 - 24x^2 + 32x - 16 + P(x)(x+1)$$

$$x^4 + x^3 + 16x + 16 = P(x)(x+1)$$

$$x^3 + 16 = P(x) \quad \text{Dividing by } x+1$$

Now decompose $\frac{x^3+16}{(x-2)^4}$.

$$\frac{x^3+16}{(x-2)^4} = \frac{B}{x-2} + \frac{C}{(x-2)^2} + \frac{D}{(x-2)^3} + \frac{E}{(x-2)^4}$$

Add and equate numerators.

$$x^3 + 16 = B(x-2)^3 + C(x+2)^2 + D(x-2) + E \qquad (1)$$

Let $x = 2$: $2^3 + 16 = 0 + 0 + 0 + E$

$$24 = E$$

Simplify equation (1).

$$x^3 + 16 = Bx^3 + (-6B+C)x^2 + (12B-4C+D)x + (-8B+4C-2D+E)$$

$$1 = B$$

$$0 = -6B + C$$

$$0 = -6 \cdot 1 + C$$

$$6 = C$$

$$0 = 12B - 4C + D$$

$$0 = 12 \cdot 1 - 4 \cdot 6 + D$$

$$12 = D$$

The decomposition is

$$-\frac{1}{x+1} + \frac{1}{x-2} + \frac{6}{(x-2)^2} + \frac{12}{(x-2)^3} + \frac{24}{(x-2)^4}.$$

35. $\frac{x}{x^4 - a^4}$

$$= \frac{x}{(x^2+a^2)(x+a)(x-a)} \quad \text{Factoring the denominator}$$

$$= \frac{Ax+B}{x^2+a^2} + \frac{C}{x+a} + \frac{D}{x-a}$$

$$= [(Ax+B)(x+a)(x-a) + C(x^2+a^2)(x-a) + D(x^2+a^2)(x+a)]/[(x^2+a^2)(x+a)(x-a)]$$

Equate the numerators:

$$x = (Ax+B)(x+a)(x-a) + C(x^2+a^2)(x-a) + D(x^2+a^2)(x+a)$$

Let $x - a = 0$, or $x = a$. Then, we get

$$a = 0 + 0 + D(a^2+a^2)(a+a)$$

$$a = D(2a^2)(2a)$$

$$a = 4a^3 D$$

$$\frac{1}{4a^2} = D$$

Let $x + a = 0$, or $x = -a$. We get

$$-a = 0 + C[(-a)^2 + a^2](-a-a) + 0$$

$$-a = C(2a^2)(-2a)$$

$$-a = -4a^3 C$$

$$\frac{1}{4a^2} = C$$

Equate the coefficients of x^3:

$$0 = A + C + D$$

Substituting $\frac{1}{4a^2}$ for C and for D, we get

$$A = -\frac{1}{2a^2}.$$

Equate the constant terms:

$$0 = -Ba^2 - Ca^3 + Da^3$$

Substitute $\frac{1}{4a^2}$ for C and for D. Then solve for B.

$$0 = -Ba^2 - \frac{1}{4a^2} \cdot a^3 + \frac{1}{4a^2} \cdot a^3$$

$$0 = -Ba^2$$

$$0 = B$$

The decomposition is as follows:

$$-\frac{\frac{1}{2a^2}x}{x^2+a^2} + \frac{\frac{1}{4a^2}}{x+a} + \frac{\frac{1}{4a^2}}{x-a}$$

36. $\frac{1}{e^{-x} + 3 + 2e^x} = \frac{e^x}{1 + 3e^x + 2e^{2x}}$

Multiplying by e^x/e^x

Let $y = e^x$, decompose $\frac{y}{2y^2+3y+1}$, and then substitute e^x for y. The result is $\frac{1}{e^x+1} - \frac{1}{2e^x+1}$.

37. $\frac{1 + \ln x^2}{(\ln x + 2)(\ln x - 3)^2} = \frac{1 + 2\ln x}{(\ln x + 2)(\ln x - 3)^2}$

Let $u = \ln x$. Then we have:

$$\frac{1+2u}{(u+2)(u-3)^2}$$

$$= \frac{A}{u+2} + \frac{B}{u-3} + \frac{C}{(u-3)^2}$$

$$= \frac{A(u3)^2 + B(u+2)(u-3) + C(u+2)}{(u+2)(u-3)^2}$$

Equate the numerators:

$$1 + 2u = A(u-3)^2 + B(u+2)(u-3) + C(u+2)$$

Let $u - 3 = 0$, or $u = 3$.

$$1 + 2 \cdot 3 = 0 + 0 + C(5)$$

$$7 = 5C$$

$$\frac{7}{5} = C$$

Let $u + 2 = 0$, or $u = -2$.

$$1 + 2(-2) = A(-2-3)^2 + 0 + 0$$

$$-3 = 25A$$

$$-\frac{3}{25} = A$$

To find B, we equate the coefficients of u^2:

$$0 = A + B$$

Substituting $-\frac{3}{25}$ for A and solving for B, we get $B = \frac{3}{25}$.

The decomposition of $\frac{1+2u}{(u+2)(u-3)^2}$ is as follows:

$$-\frac{3}{25(u+2)} + \frac{3}{25(u-3)} + \frac{7}{5(u-3)^2}$$

Substituting $\ln x$ for u we get

$$-\frac{3}{25(\ln x+2)} + \frac{3}{25(\ln x-3)} + \frac{7}{5(\ln x-3)^2}.$$

Chapter 9

Conic Sections

Exercise Set 9.1

1. Graph (f) is the graph of $x^2 = 8y$.

2. Graph (c) is the graph of $y^2 = -10x$.

3. Graph (b) is the graph of $(y-2)^2 = -3(x+4)$.

4. Graph (e) is the graph of $(x+1)^2 = 5(y-2)$.

5. Graph (d) is the graph of $13x^2 - 8y - 9 = 0$.

6. Graph (a) is the graph of $41x + 6y^2 = 12$.

7. $x^2 = 20y$

$x^2 = 4 \cdot 5 \cdot y$ Writing $x^2 = 4py$

Vertex: $(0,0)$

Focus: $(0,5)$ $[(0,p)]$

Directrix: $y = -5$ $(y = -p)$

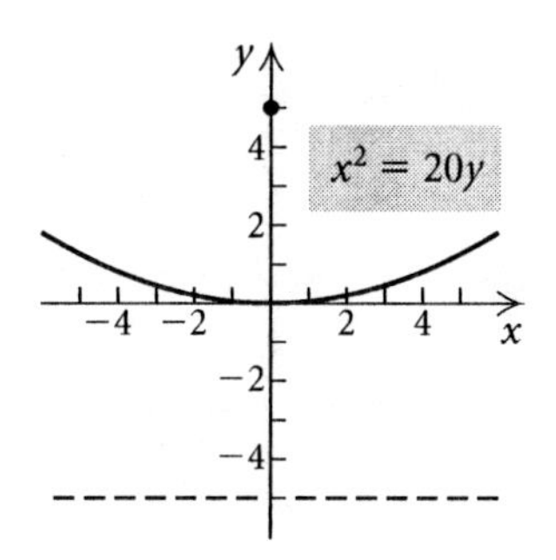

8. $x^2 = 16y$

$x^2 = 4 \cdot 4 \cdot y$

$V: (0,0),\ F: (0,4),\ D: y = -4$

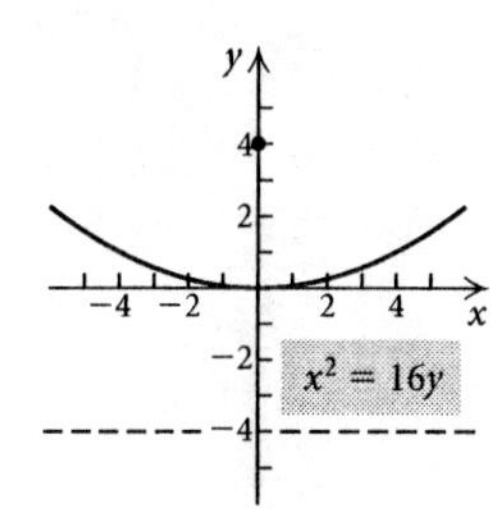

9. $y^2 = -6x$

$y^2 = 4\left(-\frac{3}{2}\right)x$ Writing $y^2 = 4px$

Vertex: $(0,0)$

Focus: $\left(-\frac{3}{2}, 0\right)$ $[(p,0)]$

Directrix: $x = -\left(-\frac{3}{2}\right) = \frac{3}{2}$ $(x = -p)$

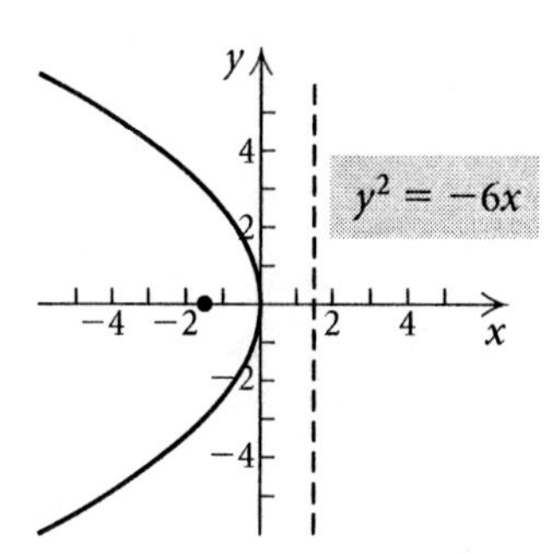

10. $y^2 = -2x$

$y^2 = 4\left(-\frac{1}{2}\right)x$

$V: (0,0), F: \left(-\frac{1}{2}, 0\right), D: x = \frac{1}{2}$

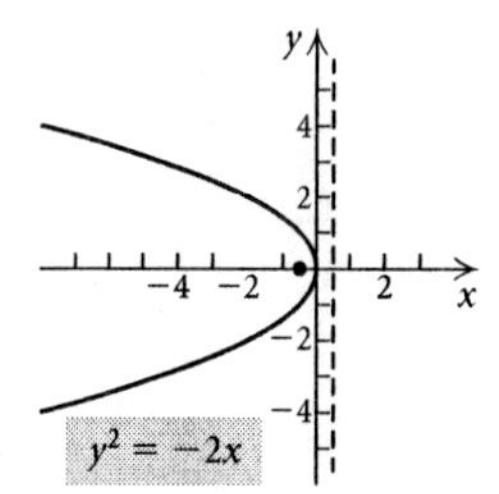

11. $x^2 - 4y = 0$

$x^2 = 4y$

$x^2 = 4 \cdot 1 \cdot y$ Writing $x^2 = 4py$

Vertex: $(0,0)$

Focus: $(0,1)$ $[(0,p)]$

Directrix: $y = -1$ $(y = -p)$

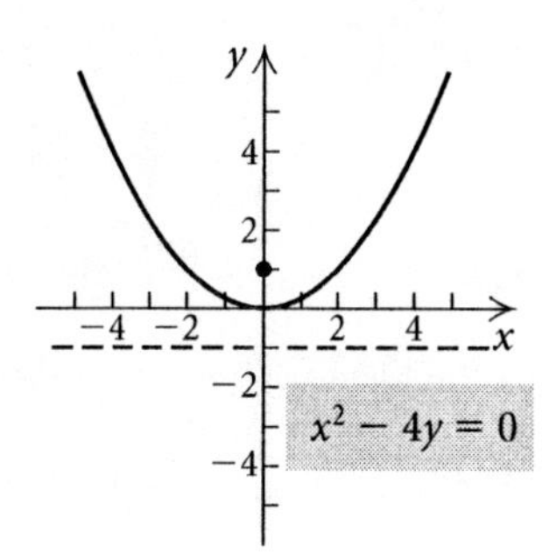

12. $y^2 + 4x = 0$

$y^2 = -4x$

$y^2 = 4(-1)x$

$V: (0,0), F: (-1,0), D: x = 1$

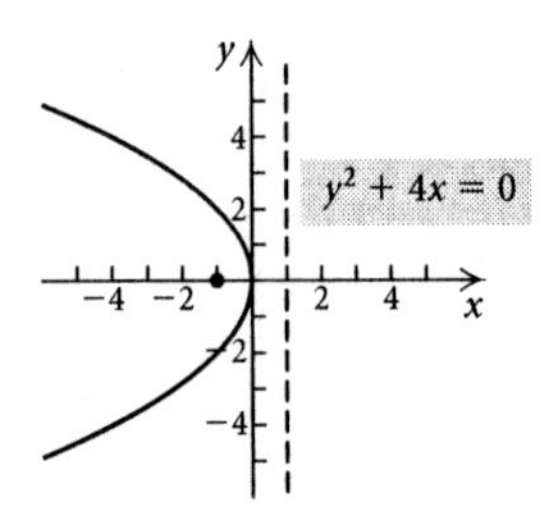

13. $x = 2y^2$

$y^2 = \frac{1}{2}x$

$y^2 = 4 \cdot \frac{1}{8} \cdot x$ Writing $y^2 = 4px$

Vertex: $(0, 0)$

Focus: $\left(\frac{1}{8}, 0\right)$

Directrix: $x = -\frac{1}{8}$

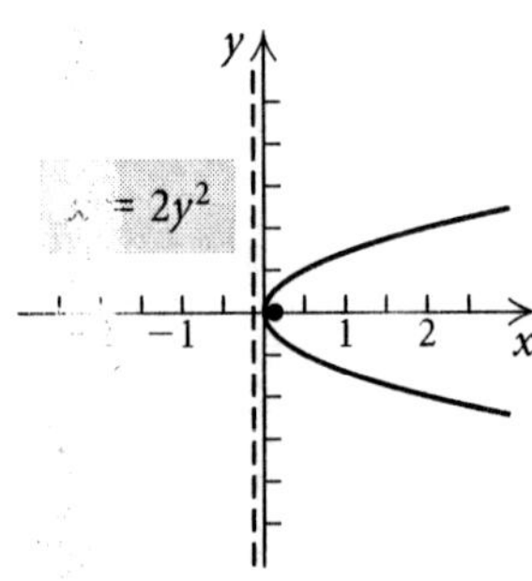

14. $y = \frac{1}{2}x^2$

$V: (0,0),\ F: \left(0, \frac{1}{2}\right),\ D: y = -\frac{1}{2}$

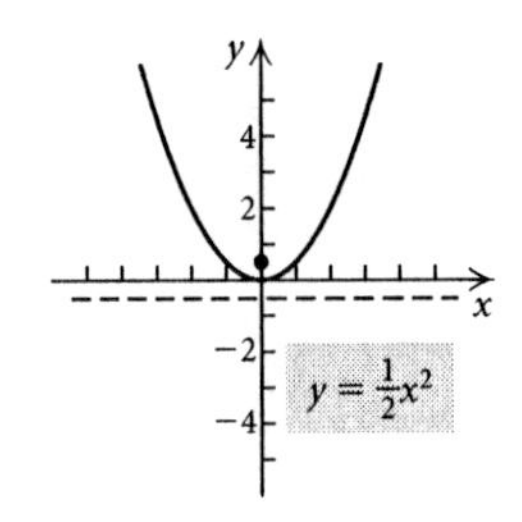

15. Since the directrix, $x = -4$, is a vertical line, the equation is of the form $(y - k)^2 = 4p(x - h)$. The focus, $(4, 0)$, is on the x-axis so the line of symmetry is the x-axis and $p = 4$. The vertex, (h, k), is the point on the x-axis midway between the directrix and the focus. Thus, it is $(0, 0)$. We have

$$(y - k)^2 = 4p(x - h)$$
$$(y - 0)^2 = 4 \cdot 4(x - 0) \quad \text{Substituting}$$
$$y^2 = 16x.$$

16. $(x - h)^2 = 4p(y - k)$

$(x - 0)^2 = 4 \cdot \frac{1}{4}(y - 0)$

$x^2 = y$

17. Since the directrix, $y = \pi$, is a horizontal line, the equation is of the form $(x-h)^2 = 4p(y-k)$. The focus, $(0, -\pi)$, is on the y-axis so the line of symmetry is the y-axis and $p = -\pi$. The vertex (h, k) is the point on the y-axis midway between the directrix and the focus. Thus, it is $(0, 0)$. We have

$$(x - h)^2 = 4p(y - k)$$
$$(x - 0)^2 = 4(-\pi)(y - 0) \quad \text{Substituting}$$
$$x^2 = -4\pi y$$

18. $(y - k)^2 = 4p(x - h)$

$(y - 0)^2 = 4(-\sqrt{2})(x - 0)$

$y^2 = -4\sqrt{2}x$

19. Since the directrix, $x = -4$, is a vertical line, the equation is of the form $(y - k)^2 = 4p(x - h)$. The focus, $(3, 2)$, is on the horizontal line $y = 2$, so the line of symmetry is $y = 2$. The vertex is the point on the line $y = 2$ that is midway between the directrix and the focus. That is, it is the midpoint of the segment from $(-4, 2)$ to $(3, 2)$: $\left(\frac{-4+3}{2}, \frac{2+2}{2}\right)$, or $\left(-\frac{1}{2}, 2\right)$. Then $h = -\frac{1}{2}$ and the directrix is $x = h - p$, so we have

$$x = h - p$$
$$-4 = -\frac{1}{2} - p$$
$$-\frac{7}{2} = -p$$
$$\frac{7}{2} = p.$$

Now we find the equation of the parabola.

$$(y - k)^2 = 4p(x - h)$$
$$(y - 2)^2 = 4\left(\frac{7}{2}\right)\left[x - \left(-\frac{1}{2}\right)\right]$$
$$(y - 2)^2 = 14\left(x + \frac{1}{2}\right)$$

20. Since the directrix, $y = -3$, is a horizontal line, the equation is of the form $(x - h)^2 = 4p(y - k)$. The focus $(-2, 3)$, is on the vertical line $x = -2$, so the line of symmetry is $x = -2$. The vertex is the point on the line $x = -2$ that is midway between the directrix and the focus. That is, it is the midpoint of the segment from $(-2, 3)$ to $(-2, -3)$: $\left(\frac{-2-2}{3}, \frac{3-3}{2}\right)$, or $(-2, 0)$. Then $k = 0$ and the directrix is $y = k - p$, so we have

$$y = k - p$$
$$-3 = 0 - p$$
$$3 = p$$

Now we find the equation of the parabola.

$$(x - h)^2 = 4p(y - k)$$
$$[x - (-2)]^2 = 4 \cdot 3(y - 0)$$
$$(x + 2)^2 = 12y$$

21. $(x + 2)^2 = -6(y - 1)$

$[x - (-2)]^2 = 4\left(-\frac{3}{2}\right)(y - 1) \quad [(x-h)^2 = 4p(y-k)]$

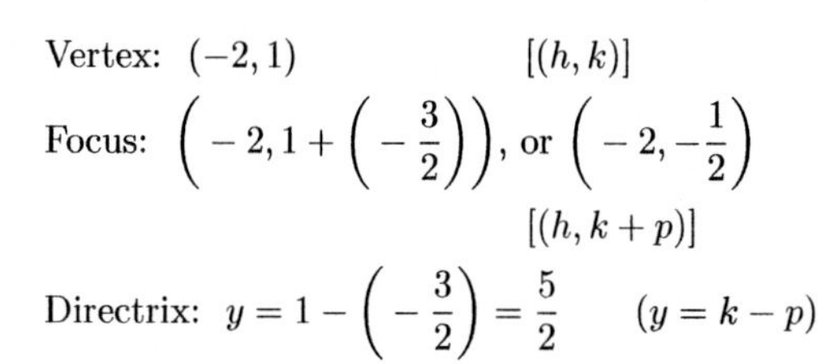

Vertex: $(-2,1)$ $\qquad [(h,k)]$

Focus: $\left(-2, 1+\left(-\frac{3}{2}\right)\right)$, or $\left(-2,-\frac{1}{2}\right)$

$[(h, k+p)]$

Directrix: $y = 1-\left(-\frac{3}{2}\right) = \frac{5}{2} \qquad (y = k-p)$

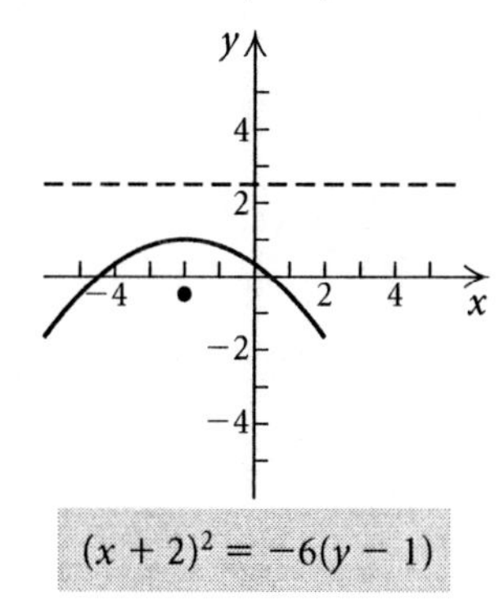

22. $(y-3)^2 = -20(x+2)$

$(y-3)^2 = 4(-5)[x-(-2)]$

$V:\ (-2,3)$

$F:\ (-2-5,3)$, or $(-7,3)$

$D:\ x = -2-(-5) = 3$

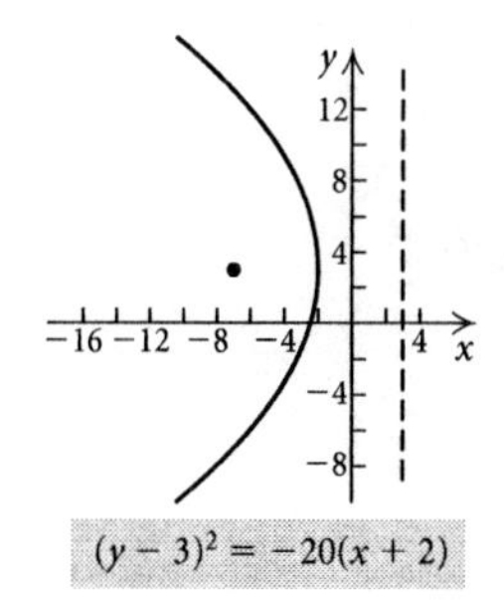

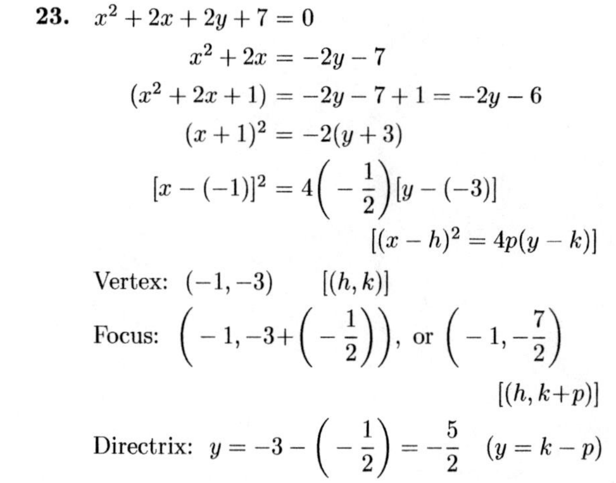

23. $x^2+2x+2y+7=0$

$$x^2+2x = -2y-7$$
$$(x^2+2x+1) = -2y-7+1 = -2y-6$$
$$(x+1)^2 = -2(y+3)$$
$$[x-(-1)]^2 = 4\left(-\frac{1}{2}\right)[y-(-3)]$$

$[(x-h)^2 = 4p(y-k)]$

Vertex: $(-1,-3) \qquad [(h,k)]$

Focus: $\left(-1,-3+\left(-\frac{1}{2}\right)\right)$, or $\left(-1,-\frac{7}{2}\right)$

$[(h,k+p)]$

Directrix: $y = -3-\left(-\frac{1}{2}\right) = -\frac{5}{2} \qquad (y = k-p)$

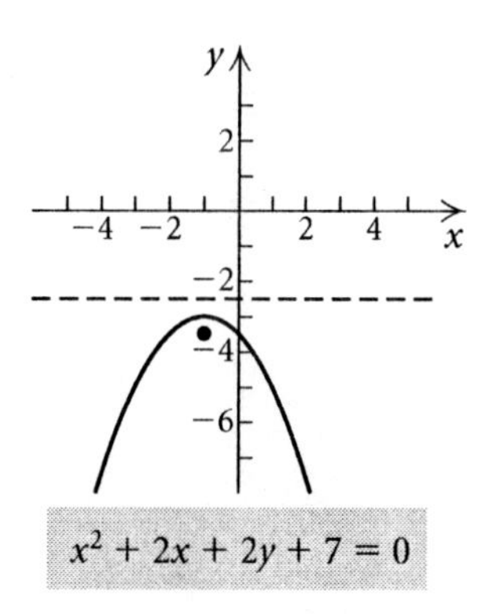

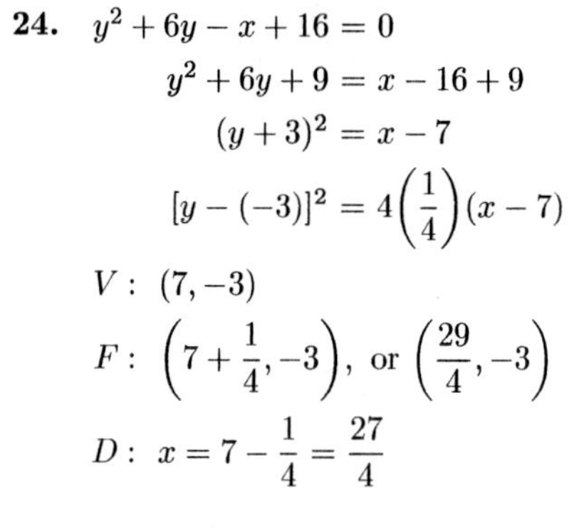

24. $y^2+6y-x+16=0$

$$y^2+6y+9 = x-16+9$$
$$(y+3)^2 = x-7$$
$$[y-(-3)]^2 = 4\left(\frac{1}{4}\right)(x-7)$$

$V:\ (7,-3)$

$F:\ \left(7+\frac{1}{4},-3\right)$, or $\left(\frac{29}{4},-3\right)$

$D:\ x = 7-\frac{1}{4} = \frac{27}{4}$

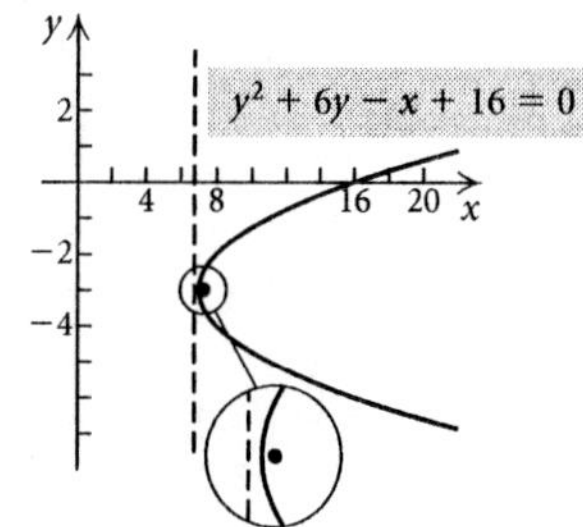

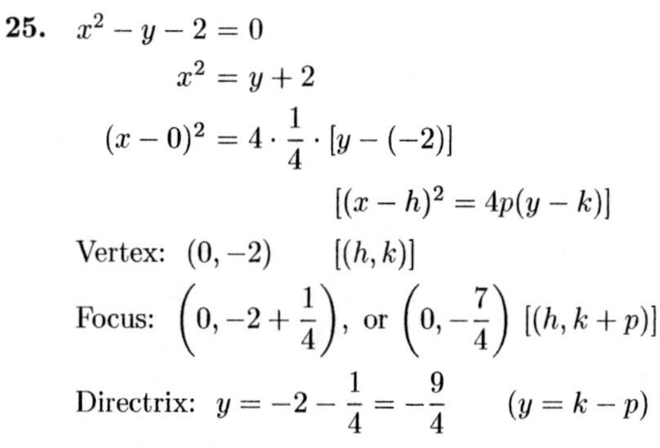

25. $x^2-y-2=0$

$$x^2 = y+2$$
$$(x-0)^2 = 4\cdot\frac{1}{4}\cdot[y-(-2)]$$

$[(x-h)^2 = 4p(y-k)]$

Vertex: $(0,-2) \qquad [(h,k)]$

Focus: $\left(0,-2+\frac{1}{4}\right)$, or $\left(0,-\frac{7}{4}\right) \quad [(h,k+p)]$

Directrix: $y = -2-\frac{1}{4} = -\frac{9}{4} \qquad (y = k-p)$

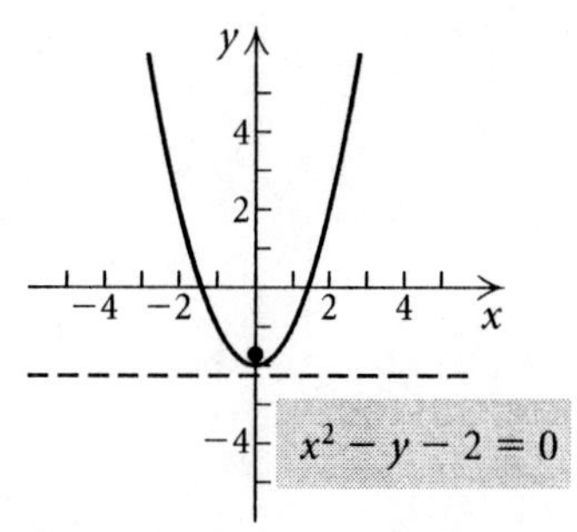

26. $x^2 - 4x - 2y = 0$

$$x^2 - 4x + 4 = 2y + 4$$
$$(x-2)^2 = 2(y+2)$$
$$(x-2)^2 = 4\left(\frac{1}{2}\right)[y-(-2)]$$

$V:\ (2,-2)$

$F:\ \left(2, -2+\frac{1}{2}\right)$, or $\left(2, -\frac{3}{2}\right)$

$D:\ y = -2 - \frac{1}{2} = -\frac{5}{2}$

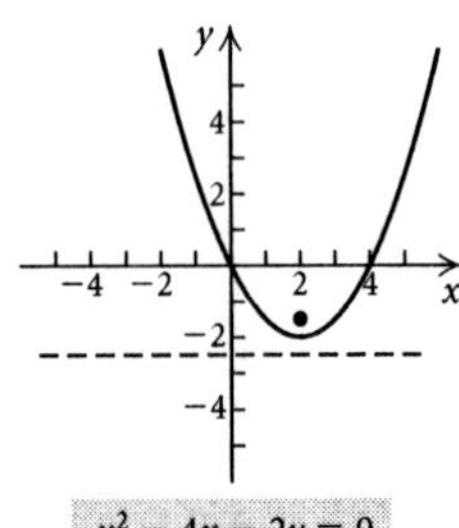

$x^2 - 4x - 2y = 0$

27.

$$y = x^2 + 4x + 3$$
$$y - 3 = x^2 + 4x$$
$$y - 3 + 4 = x^2 + 4x + 4$$
$$y + 1 = (x+2)^2$$
$$4 \cdot \frac{1}{4} \cdot [y-(-1)] = [x-(-2)]^2$$

$[(x-h)^2 = 4p(y-k)]$

Vertex: $(-2,-1)$ $[(h,k)]$

Focus: $\left(-2, -1+\frac{1}{4}\right)$, or $\left(-2, -\frac{3}{4}\right)$ $[(h, k+p)]$

Directrix: $y = -1 - \frac{1}{4} = -\frac{5}{4}$ $(y = k - p)$

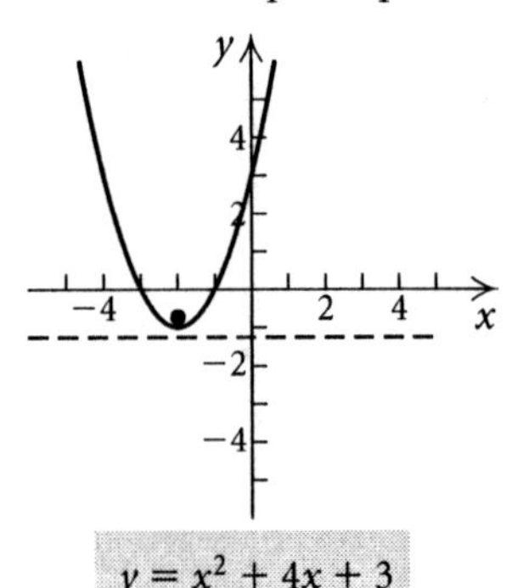

$y = x^2 + 4x + 3$

28.

$$y = x^2 + 6x + 10$$
$$y - 10 + 9 = x^2 + 6x + 9$$
$$y - 1 = (x+3)^2$$
$$4\left(\frac{1}{4}\right)(y-1) = [x-(-3)]^2$$

$V:\ (-3,1)$

$F:\ \left(-3, 1+\frac{1}{4}\right)$, or $\left(-3, \frac{5}{4}\right)$

$D:\ y = 1 - \frac{1}{4} = \frac{3}{4}$

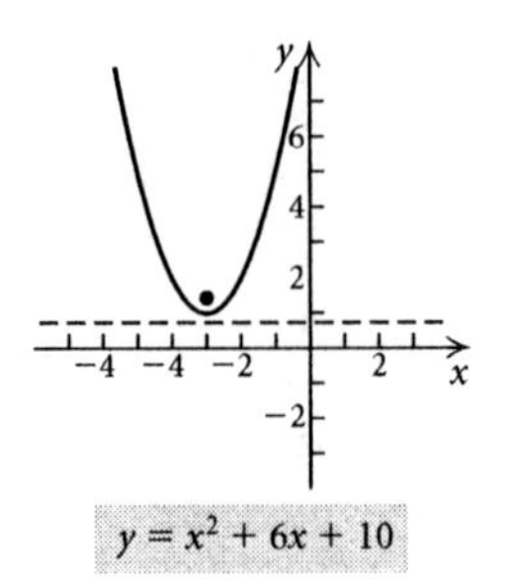

$y = x^2 + 6x + 10$

29. $y^2 - y - x + 6 = 0$

$$y^2 - y = x - 6$$
$$y^2 - y + \frac{1}{4} = x - 6 + \frac{1}{4}$$
$$\left(y - \frac{1}{2}\right)^2 = x - \frac{23}{4}$$
$$\left(y - \frac{1}{2}\right)^2 = 4 \cdot \frac{1}{4}\left(x - \frac{23}{4}\right)$$

$[(y-k)^2 = 4p(x-h)]$

Vertex: $\left(\frac{23}{4}, \frac{1}{2}\right)$ $[(h,k)]$

Focus: $\left(\frac{23}{4} + \frac{1}{4}, \frac{1}{2}\right)$, or $\left(6, \frac{1}{2}\right)$ $[(h+p, k)]$

Directrix: $x = \frac{23}{4} - \frac{1}{4} = \frac{22}{4}$ or $\frac{11}{2}$ $(x = h - p)$

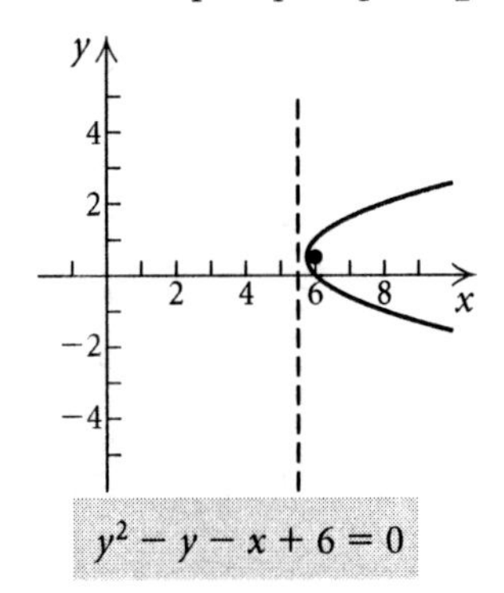

$y^2 - y - x + 6 = 0$

30. $y^2 + y - x - 4 = 0$

$$y^2 + y = x + 4$$
$$y^2 + y + \frac{1}{4} = x + 4 + \frac{1}{4}$$
$$\left(y + \frac{1}{2}\right)^2 = x + \frac{17}{4}$$
$$\left[y - \left(-\frac{1}{2}\right)\right]^2 = 4 \cdot \frac{1}{4}\left[x - \left(-\frac{17}{4}\right)\right]$$

$V: \left(-\frac{17}{4}, -\frac{1}{2}\right)$

$F: \left(-\frac{17}{4} + \frac{1}{4}, -\frac{1}{2}\right)$, or $\left(-4, -\frac{1}{2}\right)$

$D: x = -\frac{17}{4} - \frac{1}{4} = -\frac{9}{2}$

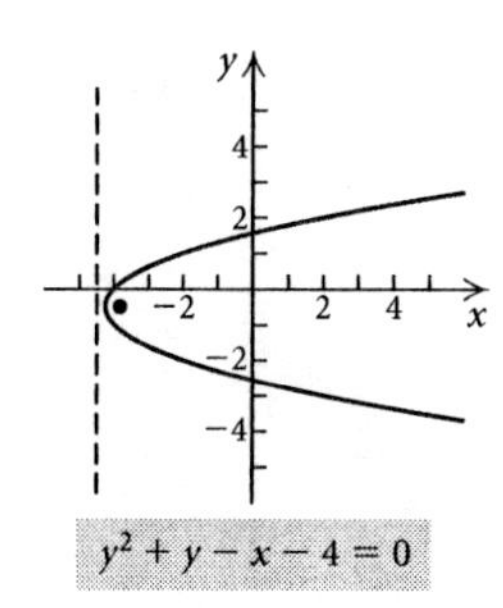

31. a) The vertex is $(0,0)$. The focus is $(4,0)$, so $p = 4$. The parabola has a horizontal axis of symmetry so the equation is of the form $y^2 = 4px$. We have

$$y^2 = 4px$$
$$y^2 = 4 \cdot 4 \cdot x$$
$$y^2 = 16x$$

b) We make a drawing.

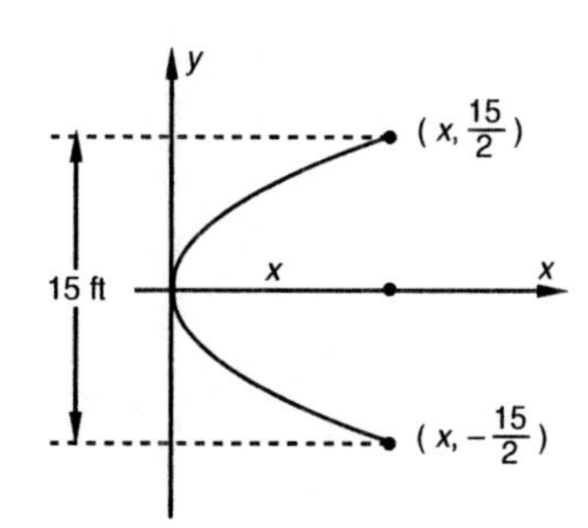

The depth of the satellite dish at the vertex is x where $\left(x, \frac{15}{2}\right)$ is a point on the parabola.

$$y^2 = 16x$$
$$\left(\frac{15}{2}\right)^2 = 16x \quad \text{Substituting } \frac{15}{2} \text{ for } y$$
$$\frac{225}{4} = 16x$$
$$\frac{225}{64} = x, \text{ or}$$
$$3\frac{33}{64} = x$$

The depth of the satellite dish at the vertex is $3\frac{33}{64}$ ft.

32. a) The parabola is of the form $y^2 = 4px$. A point on the parabola is $\left(1, \frac{6}{2}\right)$, or $(1,3)$.

$$y^2 = 4px$$
$$3^2 = 4 \cdot p \cdot 1$$
$$\frac{9}{4} = p$$

Then the equation of the parabola is $y^2 = 4 \cdot \frac{9}{4}x$, or $y^2 = 9x$.

b) The focus is at $(p,0)$, or $\left(\frac{9}{4}, 0\right)$, so the bulb should be placed $\frac{9}{4}$ in., or $2\frac{1}{4}$ in., from the vertex.

33. We position a coordinate system with the origin at the vertex and the x-axis on the parabola's axis of symmetry.

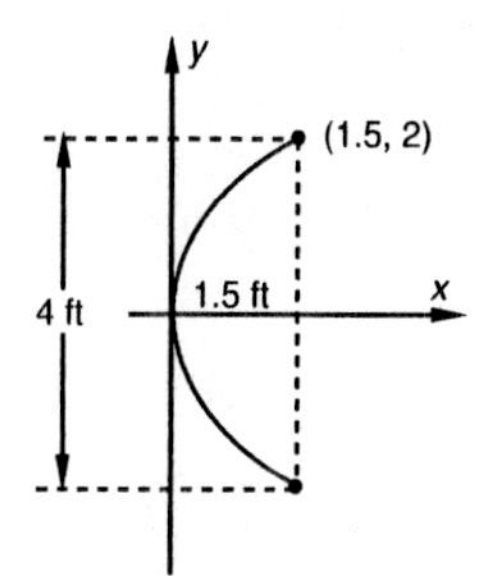

The parabola is of the form $y^2 = 4px$ and a point on the parabola is $\left(1.5, \frac{4}{2}\right)$, or $(1.5, 2)$.

$$y^2 = 4px$$
$$2^2 = 4 \cdot p \cdot (1.5) \quad \text{Substituting}$$
$$4 = 6p$$
$$\frac{4}{6} = p, \text{ or}$$
$$\frac{2}{3} = p$$

Since the focus is at $(p,0)$, or $\left(\frac{2}{3}, 0\right)$, the focus is $\frac{2}{3}$ ft, or 8 in., from the vertex.

34.

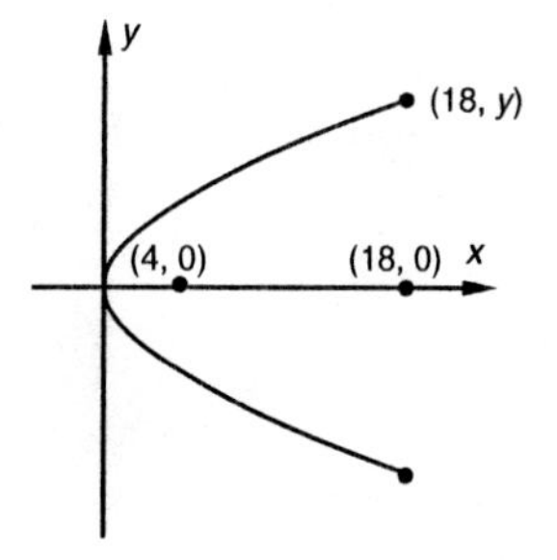

$$y^2 = 4px$$
$$y^2 = 4 \cdot 4 \cdot 18$$
$$y^2 = 288$$
$$y = 12\sqrt{2}$$

The width at the opening is $2 \cdot 12\sqrt{2}$, or $24\sqrt{2}$ in. ≈ 34 in.

35. Vertex: $(0.867, 0.348)$
Focus: $(0.867, -0.190)$
Directrix: $y = 0.887$

36. Vertex: $(7.126, 1.180)$
Focus: $(7.045, 1.180)$
Directrix: $x = 7.207$

37. No; parabolas with a horizontal axis of symmetry fail the vertical-line test.

38. See page 737 of the text.

39. When we let $y = 0$ and solve for x, the only equation for which $x = \frac{2}{3}$ is (h), so only equation (h) has x-intercept $\left(\frac{2}{3}, 0\right)$.

40. Equations (a) - (f) are in the form $y = mx + b$ and $b = 7$ in equation (d). When we solve equations (g) and (h) for y we get $y = 2x - \frac{7}{4}$ and $y = -\frac{1}{2}x + \frac{1}{3}$, respectively. Neither has $b = 7$, so only equation (d) has y-intercept $(0, 7)$.

41. Note that equation (g) is equivalent to $y = 2x - \frac{7}{4}$ and equation (h) is equivalent to $y = -\frac{1}{2}x + \frac{1}{3}$. When we look at the equations in the form $y = mx+b$, we see that $m > 0$ for (a), (b), (f), and (g) so these equations have positive slope, or slant up front left to right.

42. The equation for which $|m|$ is smallest is (b), so it has the least steep slant.

43. When we look at the equations in the form $y = mx+b$ (See Exercise 41.), only (b) has $m = \frac{1}{3}$ so only (b) has slope $\frac{1}{3}$.

44. When we substitute 3 for x in each equation, we see that $y = 7$ only in equation (f) so only (f) contains the point $(3, 7)$.

45. Parallel lines have the same slope and different y-intercepts. When we look at the equations in the form $y = mx + b$ (See Exercise 41.), we see that (a) and (g) represent parallel lines.

46. The pairs of equations for which the product of the slopes is -1 are (a) and (h), (g) and (h), and (b) and (c).

47. A parabola with a vertical axis of symmetry has an equation of the type $(x - h)^2 = 4p(y - k)$.

Solve for p substituting $(-1, 2)$ for (h, k) and $(-3, 1)$ for (x, y).

$$[-3 - (-1)]^2 = 4p(1 - 2)$$
$$4 = -4p$$
$$-1 = p$$

The equation of the parabola is

$$[x - (-1)]^2 = 4(-1)(y - 2), \text{ or}$$
$$(x + 1)^2 = -4(y - 2).$$

48. A parabola with a horizontal axis of symmetry has an equation of the type $(y - k)^2 = 4p(x - h)$.

Find p by substituting $(-2, 1)$ for (h, k) and $(-3, 5)$ for (x, y).

$$(5 - 1)^2 = 4p[-3 - (-2)]$$
$$16 = 4p(-1)$$
$$16 = -4p$$
$$-4 = p$$

The equation of the parabola is

$$(y - 1)^2 = 4(-4)[x - (-2)], \text{ or}$$
$$(y - 1)^2 = -16(x + 2).$$

49. Position a coordinate system as shown below with the y-axis on the parabola's axis of symmetry.

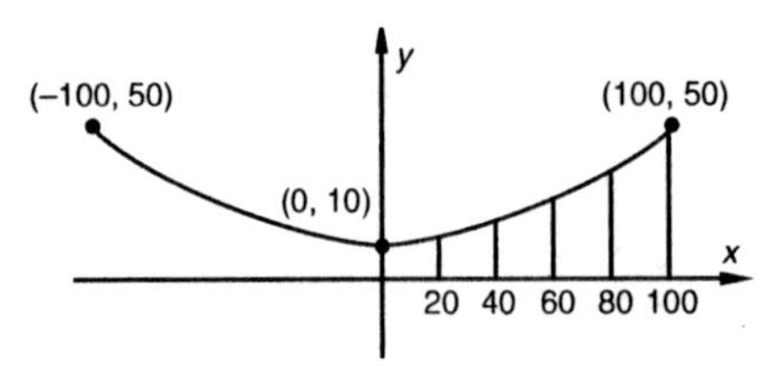

The equation of the parabola is of the form $(x - h)^2 = 4p(y - k)$. Substitute 100 for x, 50 for y, 0 for h, and 10 for k and solve for p.

$$(x - h)^2 = 4p(y - k)$$
$$(100 - 0)^2 = 4p(50 - 10)$$
$$10{,}000 = 160p$$
$$\frac{250}{4} = p$$

Then the equation is

$$x^2 = 4\left(\frac{250}{4}\right)(y - 10), \text{ or}$$
$$x^2 = 250(y - 10).$$

To find the lengths of the vertical cables, find y when $x = 0$, 20, 40, 60, 80, and 100.

When $x = 0$: $0^2 = 250(y - 10)$
$$0 = y - 10$$
$$10 = y$$

When $x = 20$: $20^2 = 250(y - 10)$
$$400 = 250(y - 10)$$
$$1.6 = y - 10$$
$$11.6 = y$$

When $x = 40$: $40^2 = 250(y - 10)$
$$1600 = 250(y - 10)$$
$$6.4 = y - 10$$
$$16.4 = y$$

When $x = 60$: $60^2 = 250(y - 10)$
$$3600 = 250(y - 10)$$
$$14.4 = y - 10$$
$$24.4 = y$$

When $x = 80$: $80^2 = 250(y - 10)$
$$6400 = 250(y - 10)$$
$$25.6 = y - 10$$
$$35.6 = y$$

When $x = 100$, we know from the given information that $y = 50$.

The lengths of the vertical cables are 10 ft, 11.6 ft, 16.4 ft, 24.4 ft, 35.6 ft, and 50 ft.

Exercise Set 9.2

1. Graph (b) is the graph of $x^2 + y^2 = 5$.

2. Graph (f) is the graph of $y^2 = 20 - x^2$.

3. Graph (d) is the graph of $x^2 + y^2 - 6x + 2y = 6$.

4. Graph (c) is the graph of $x^2 + y^2 + 10x - 12y = 3$.

5. Graph (a) is the graph of $x^2 + y^2 - 5x + 3y = 0$.

6. Graph (e) is the graph of $x^2 + 4x - 2 = 6y - y^2 - 6$.

7. Complete the square twice.

$$x^2 + y^2 - 14x + 4y = 11$$
$$x^2 - 14x + y^2 + 4y = 11$$
$$x^2 - 14x + 49 + y^2 + 4y + 4 = 11 + 49 + 4$$
$$(x-7)^2 + (y+2)^2 = 64$$
$$(x-7)^2 + [y-(-2)]^2 = 8^2$$

Center: $(7, -2)$

Radius: 8

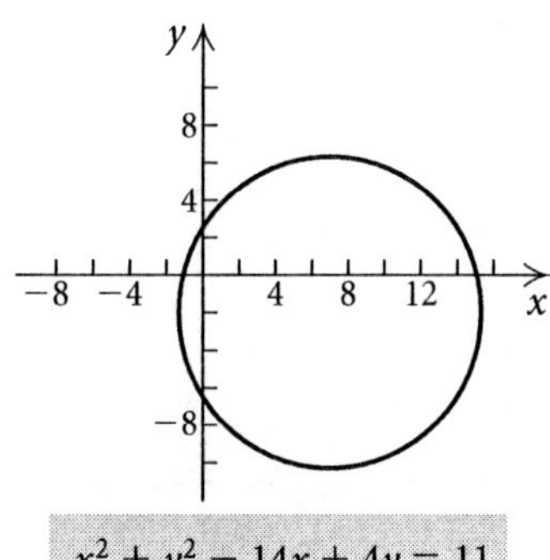

$x^2 + y^2 - 14x + 4y = 11$

8.

$$x^2 + y^2 + 2x - 6y = -6$$
$$x^2 + 2x + 1 + y^2 - 6y + 9 = -6 + 1 + 9$$
$$(x+1)^2 + (y-3)^2 = 4$$
$$[x-(-1)]^2 + (y-3)^2 = 2^2$$

Center: $(-1, 3)$

Radius: 2

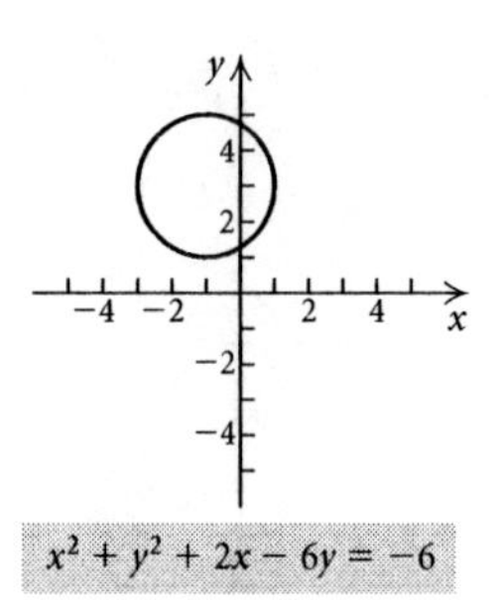

$x^2 + y^2 + 2x - 6y = -6$

9. Complete the square twice.

$$x^2 + y^2 + 6x - 2y = 6$$
$$x^2 + 6x + y^2 - 2y = 6$$
$$x^2 + 6x + 9 + y^2 - 2y + 1 = 6 + 9 + 1$$
$$(x+3)^2 + (y-1)^2 = 16$$
$$[x-(-3)]^2 + (y-1)^2 = 4^2$$

Center: $(-3, 1)$

Radius: 4

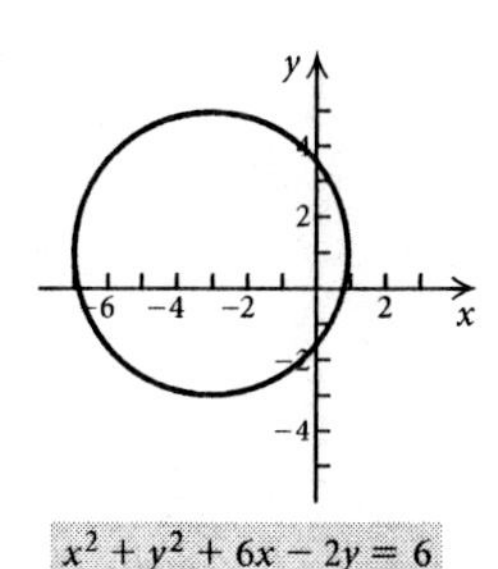

$x^2 + y^2 + 6x - 2y = 6$

10.

$$x^2 + y^2 - 4x + 2y = 4$$
$$x^2 - 4x + 4 + y^2 + 2y + 1 = 4 + 4 + 1$$
$$(x-2)^2 + (y+1)^2 = 9$$
$$(x-2)^2 + [y-(-1)]^2 = 3^2$$

Center: $(2, -1)$

Radius: 3

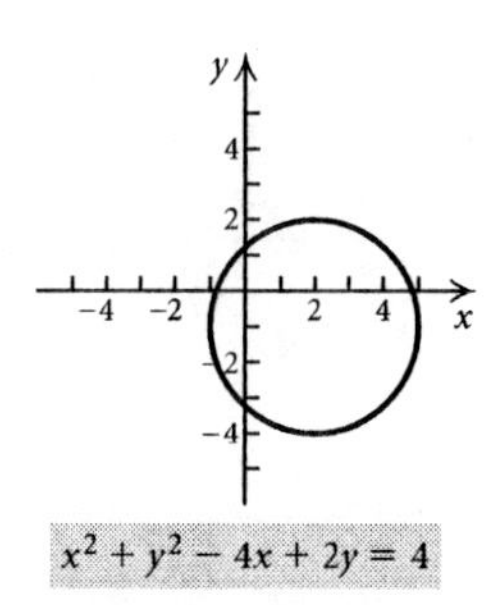

$x^2 + y^2 - 4x + 2y = 4$

11. Complete the square twice.

$$x^2 + y^2 + 4x - 6y - 12 = 0$$
$$x^2 + 4x + y^2 - 6y = 12$$
$$x^2 + 4x + 4 + y^2 - 6y + 9 = 12 + 4 + 9$$
$$(x+2)^2 + (y-3)^2 = 25$$
$$[x-(-2)]^2 + (y-3)^2 = 5^2$$

Center: $(-2, 3)$

Radius: 5

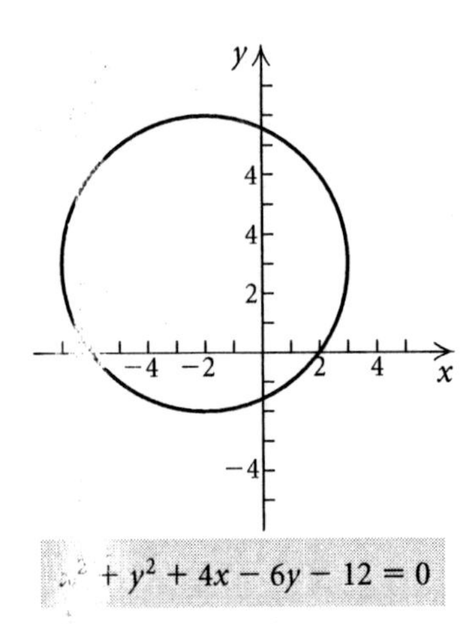

$x^2 + y^2 + 4x - 6y - 12 = 0$

12.
$$\begin{aligned} x^2 + y^2 - 8x - 2y - 19 &= 0 \\ x^2 - 8x + y^2 - 2y &= 19 \\ x^2 - 8x + 16 + y^2 - 2y + 1 &= 19 + 16 + 1 \\ (x-4)^2 + (y-1)^2 &= 36 \\ (x-4)^2 + (y-1)^2 &= 6^2 \end{aligned}$$

Center: $(4, 1)$

Radius: 6

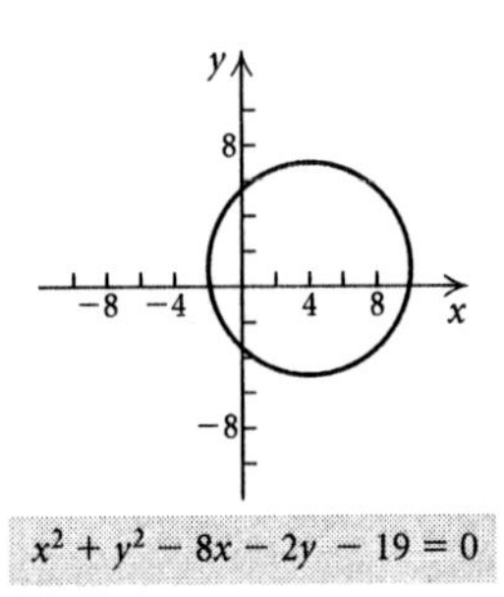

$x^2 + y^2 - 8x - 2y - 19 = 0$

13. Complete the square twice.
$$\begin{aligned} x^2 + y^2 - 6x - 8y + 16 &= 0 \\ x^2 - 6x + y^2 - 8y &= -16 \\ x^2 - 6x + 9 + y^2 - 8y + 16 &= -16 + 9 + 16 \\ (x-3)^2 + (y-4)^2 &= 9 \\ (x-3)^2 + (y-4)^2 &= 3^2 \end{aligned}$$

Center: $(3, 4)$

Radius: 3

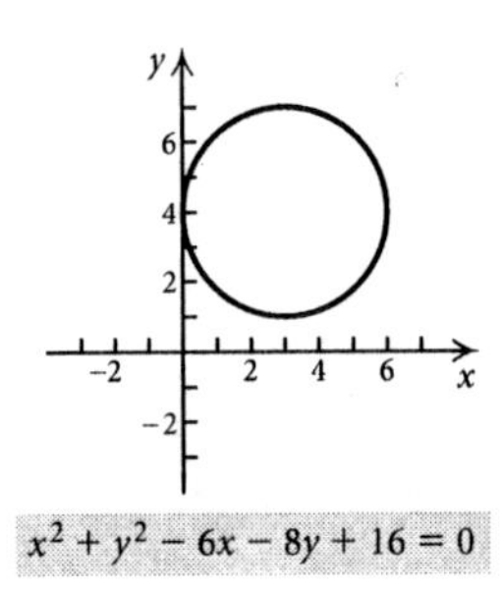

$x^2 + y^2 - 6x - 8y + 16 = 0$

14.
$$\begin{aligned} x^2 + y^2 - 2x + 6y + 1 &= 0 \\ x^2 - 2x + y^2 + 6y &= -1 \\ x^2 - 2x + 1 + y^2 + 6y + 9 &= -1 + 1 + 9 \\ (x-1)^2 + (y+3)^2 &= 9 \\ (x-1)^2 + [y-(-3)]^2 &= 3^2 \end{aligned}$$

Center: $(1, -3)$

Radius: 3

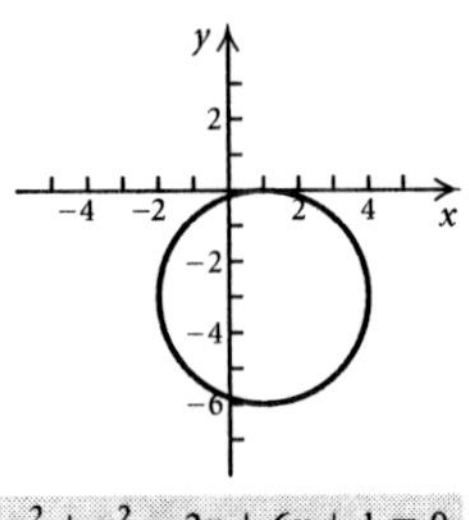

$x^2 + y^2 - 2x + 6y + 1 = 0$

15. Complete the square twice.
$$\begin{aligned} x^2 + y^2 + 6x - 10y &= 0 \\ x^2 + 6x + y^2 - 10y &= 0 \\ x^2 + 6x + 9 + y^2 - 10y + 25 &= 0 + 9 + 25 \\ (x+3)^2 + (y-5)^2 &= 34 \\ [x-(-3)]^2 + (y-5)^2 &= (\sqrt{34})^2 \end{aligned}$$

Center: $(-3, 5)$

Radius: $\sqrt{34}$

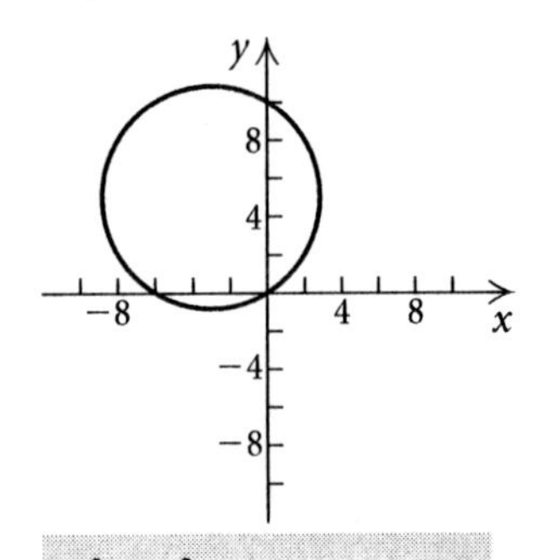

$x^2 + y^2 + 6x - 10y = 0$

16.
$$\begin{aligned} x^2 + y^2 - 7x - 2y &= 0 \\ x^2 - 7x + \frac{49}{4} + y^2 - 2y + 1 &= \frac{49}{4} + 1 \\ \left(x - \frac{7}{2}\right)^2 + (y-1)^2 &= \frac{53}{4} \\ \left(x - \frac{7}{2}\right)^2 + (y-1)^2 &= \left(\frac{\sqrt{53}}{2}\right)^2 \end{aligned}$$

Center: $\left(\frac{7}{2}, 1\right)$

Radius: $\frac{\sqrt{53}}{2}$

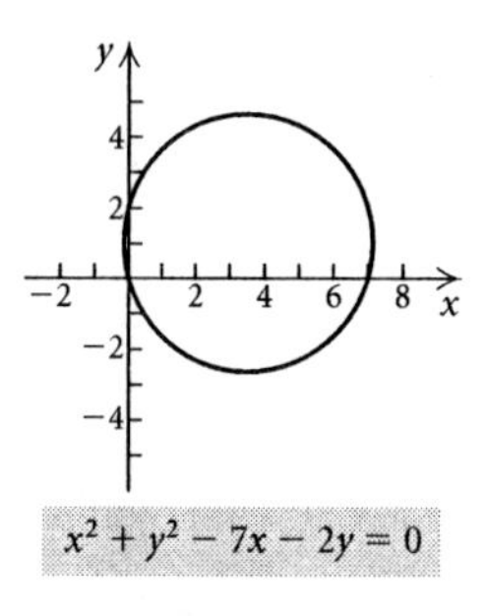

17. Complete the square twice.

$$x^2 + y^2 - 9x = 7 - 4y$$
$$x^2 - 9x + y^2 + 4y = 7$$
$$x^2 - 9x + \frac{81}{4} + y^2 + 4y + 4 = 7 + \frac{81}{4} + 4$$
$$\left(x - \frac{9}{2}\right)^2 + (y+2)^2 = \frac{125}{4}$$
$$\left(x - \frac{9}{2}\right)^2 + [y - (-2)]^2 = \left(\frac{5\sqrt{5}}{2}\right)^2$$

Center: $\left(\frac{9}{2}, -2\right)$

Radius: $\frac{5\sqrt{5}}{2}$

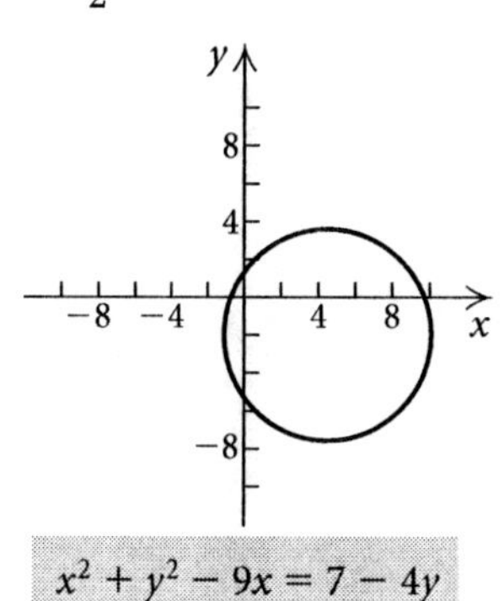

18.

$$y^2 - 6y - 1 = 8x - x^2 + 3$$
$$x^2 - 8x + y^2 - 6y = 4$$
$$x^2 - 8x + 16 + y^2 - 6y + 9 = 4 + 16 + 9$$
$$(x-4)^2 + (y-3)^2 = 29$$
$$(x-4)^2 + (y-3)^2 = (\sqrt{29})^2$$

Center: $(4, 3)$

Radius: $\sqrt{29}$

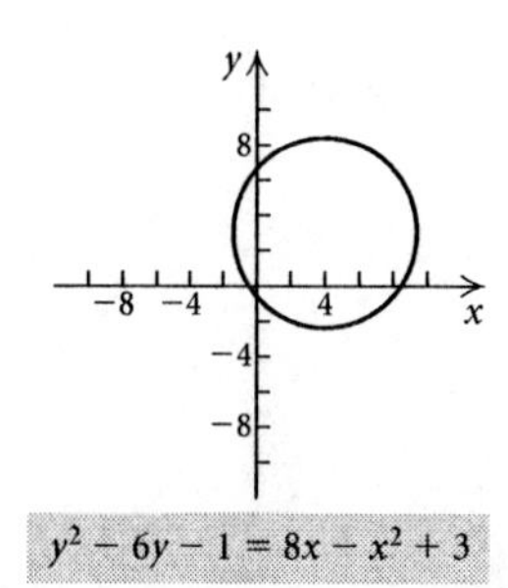

19. Graph (c) is the graph of $16x^2 + 4y^2 = 64$.

20. Graph (a) is the graph of $4x^2 + 5y^2 = 20$.

21. Graph (d) is the graph of $x^2 + 9y^2 - 6x + 90y = -225$.

22. Graph (b) is the graph of $9x^2 + 4y^2 + 18x - 16y = 11$.

23. $\frac{x^2}{4} + \frac{y^2}{1} = 1$

$\frac{x^2}{2^2} + \frac{y^2}{1^2} = 1$ Standard form

$a = 2$, $b = 1$

The major axis is horizontal, so the vertices are $(-2, 0)$ and $(2, 0)$. Since we know that $c^2 = a^2 - b^2$, we have $c^2 = 4 - 1 = 3$, so $c = \sqrt{3}$ and the foci are $(-\sqrt{3}, 0)$ and $(\sqrt{3}, 0)$.

To graph the ellipse, plot the vertices. Note also that since $b = 1$, the y-intercepts are $(0, -1)$ and $(0, 1)$. Plot these points as well and connect the four plotted points with a smooth curve.

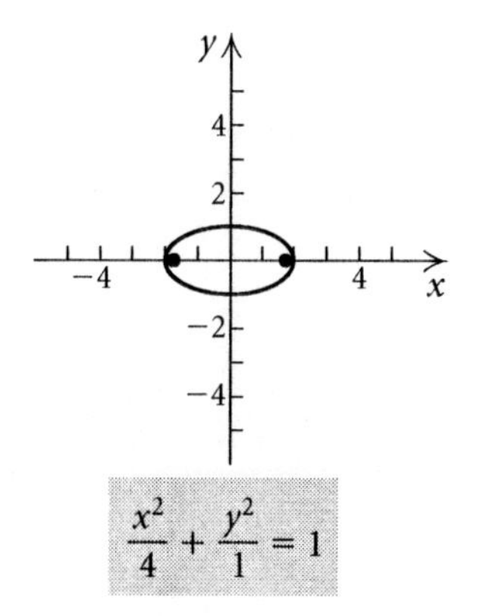

24. $\frac{x^2}{25} + \frac{y^2}{36} = 1$, or $\frac{x^2}{5^2} + \frac{y^2}{6^2} = 1$

$a = 6$, $b = 5$

The major axis is vertical, so the vertices are $(0, -6)$ and $(0, 6)$. Since $c^2 = a^2 - b^2$, we have $c^2 = 36 - 25 = 11$, so $c = \sqrt{11}$ and the foci are $(0, -\sqrt{11})$ and $(0, \sqrt{11})$.

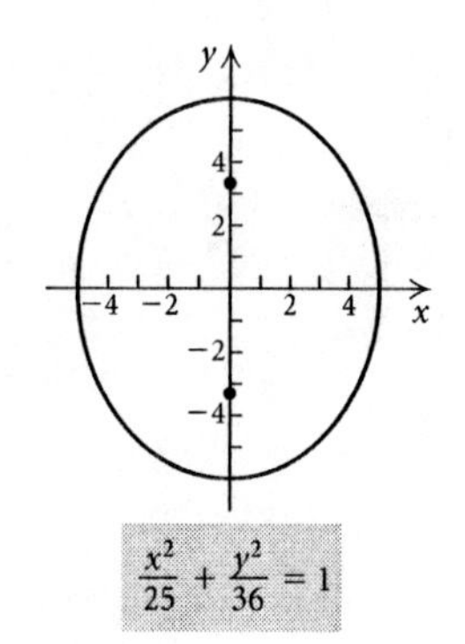

25. $16x^2 + 9y^2 = 144$

$\frac{x^2}{9} + \frac{y^2}{16} = 1$ Dividing by 144

$\frac{x^2}{3^2} + \frac{y^2}{4^2} = 1$ Standard form

$a = 4$, $b = 3$

The major axis is vertical, so the vertices are $(0, -4)$ and $(0, 4)$. Since $c^2 = a^2 - b^2$, we have $c^2 = 16 - 9 = 7$, so $c = \sqrt{7}$ and the foci are $(0, -\sqrt{7})$ and $(0, \sqrt{7})$.

To graph the ellipse, plot the vertices. Note also that since $b = 3$, the x-intercepts are $(-3, 0)$ and $(3, 0)$. Plot these points as well and connect the four plotted points with a smooth curve.

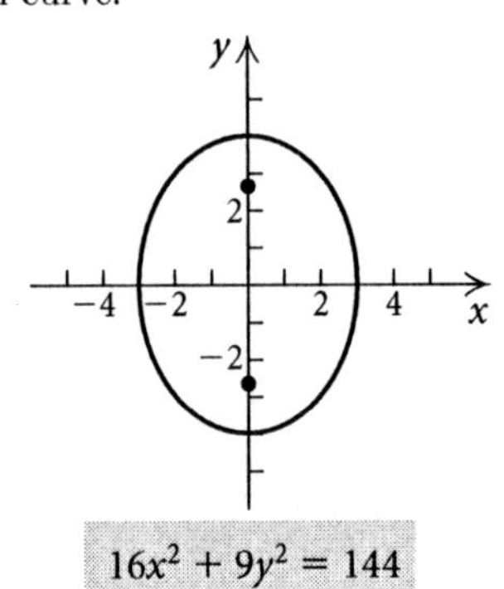

26. $9x^2 + 4y^2 = 36$

$\frac{x^2}{4} + \frac{y^2}{9} = 1$

$\frac{x^2}{2^2} + \frac{y^2}{3^2} = 1$

$a = 3$, $b = 2$

The major axis is vertical, so the vertices are $(0, -3)$ and $(0, 3)$. Since $c^2 = a^2 - b^2$, we have $c^2 = 9 - 4 = 5$, so $c = \sqrt{5}$ and the foci are $(0, -\sqrt{5})$ and $(0, \sqrt{5})$.

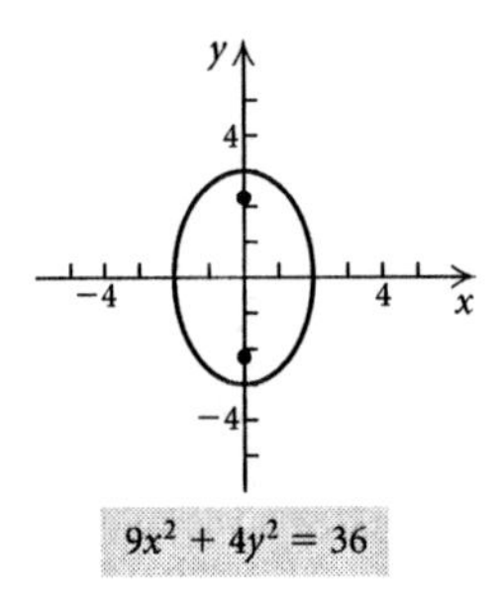

27. $2x^2 + 3y^2 = 6$

$\frac{x^2}{3} + \frac{y^2}{2} = 1$

$\frac{x^2}{(\sqrt{3})^2} + \frac{y^2}{(\sqrt{2})^2} = 1$

$a = \sqrt{3}$, $b = \sqrt{2}$

The major axis is horizontal, so the vertices are $(-\sqrt{3}, 0)$ and $(\sqrt{3}, 0)$. Since $c^2 = a^2 - b^2$, we have $c^2 = 3 - 2 = 1$, so $c = 1$ and the foci are $(-1, 0)$ and $(1, 0)$.

To graph the ellipse, plot the vertices. Note also that since $b = \sqrt{2}$, the y-intercepts are $(0, -\sqrt{2})$ and $(0, \sqrt{2})$. Plot these points as well and connect the four plotted points with a smooth curve.

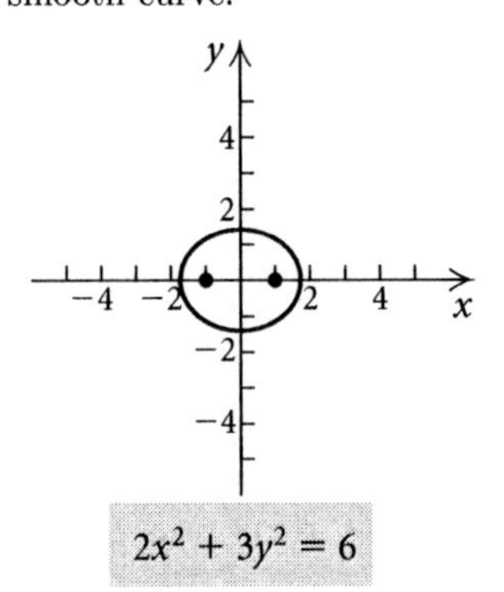

28. $5x^2 + 7y^2 = 35$

$\frac{x^2}{7} + \frac{y^2}{5} = 1$

$\frac{x^2}{(\sqrt{7})^2} + \frac{y^2}{(\sqrt{5})^2} = 1$

$a = \sqrt{7}$, $b = \sqrt{5}$

The major axis is horizontal, so the vertices are $(-\sqrt{7}, 0)$ and $(\sqrt{7}, 0)$. Since $c^2 = a^2 - b^2$, we have $c^2 = 7 - 5 = 2$, so $c = \sqrt{2}$ and the foci are $(-\sqrt{2}, 0)$ and $(\sqrt{2}, 0)$.

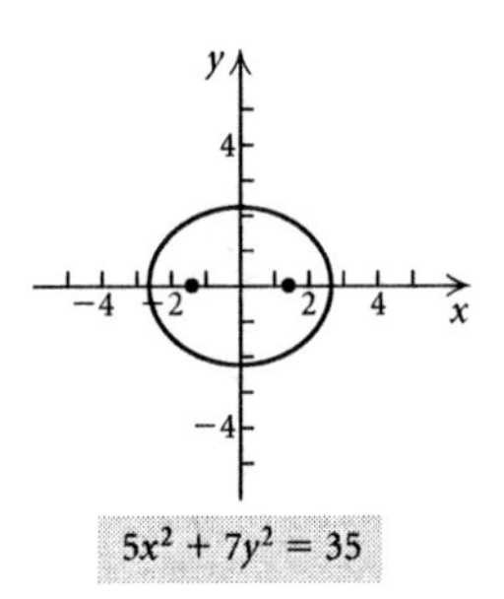

29. $4x^2 + 9y^2 = 1$

$\frac{x^2}{\frac{1}{4}} + \frac{y^2}{\frac{1}{9}} = 1$

$\frac{x^2}{\left(\frac{1}{2}\right)^2} + \frac{y^2}{\left(\frac{1}{3}\right)^2} = 1$

$a = \frac{1}{2}$, $b = \frac{1}{3}$

The major axis is horizontal, so the vertices are $\left(-\frac{1}{2}, 0\right)$ and $\left(\frac{1}{2}, 0\right)$. Since $c^2 = a^2 - b^2$, we have $c^2 = \frac{1}{4} - \frac{1}{9} = \frac{5}{36}$, so $c = \frac{\sqrt{5}}{6}$ and the foci are $\left(-\frac{\sqrt{5}}{6}, 0\right)$ and $\left(\frac{\sqrt{5}}{6}, 0\right)$.

To graph the ellipse, plot the vertices. Note also that since $b = \frac{1}{3}$, the y-intercepts are $\left(0, -\frac{1}{3}\right)$ and $\left(0, \frac{1}{3}\right)$. Plot these points as well and connect the four plotted points with a smooth curve.

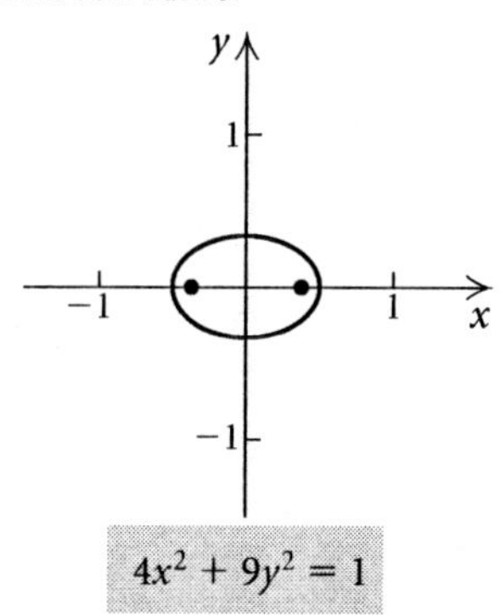

30. $$25x^2 + 16y^2 = 1$$

$$\frac{x^2}{\frac{1}{25}} + \frac{y^2}{\frac{1}{16}} = 1$$

$$\frac{x^2}{\left(\frac{1}{5}\right)^2} + \frac{y^2}{\left(\frac{1}{4}\right)^2} = 1$$

$$a = \frac{1}{4},\ b = \frac{1}{5}$$

The major axis is vertical, so the vertices are $\left(0, -\frac{1}{4}\right)$ and $\left(0, \frac{1}{4}\right)$. Since $c^2 = a^2 - b^2$, we have $c^2 = \frac{1}{16} - \frac{1}{25} = \frac{9}{400}$, so $c = \frac{3}{20}$ and the foci are $\left(0, -\frac{3}{20}\right)$ and $\left(0, \frac{3}{20}\right)$.

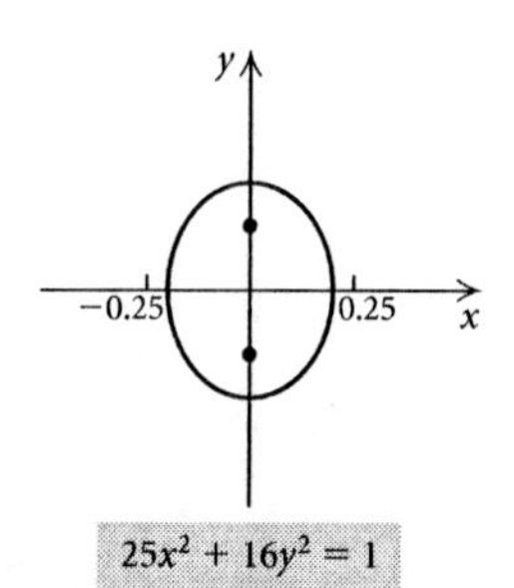

31. The vertices are on the x-axis, so the major axis is horizontal. We have $a = 7$ and $c = 3$, so we can find b^2:

$$c^2 = a^2 - b^2$$
$$3^2 = 7^2 - b^2$$
$$b^2 = 49 - 9 = 40$$

Write the equation:

$$\frac{x^2}{a^2} + \frac{y^2}{b^2} = 1$$
$$\frac{x^2}{49} + \frac{y^2}{40} = 1$$

32. The major axis is vertical; $a = 6$ and $c = 4$.

$$c^2 = a^2 - b^2$$
$$16 = 36 - b^2$$
$$b^2 = 20$$

The equation is $\frac{x^2}{20} + \frac{y^2}{36} = 1$.

33. The vertices, $(0, -8)$ and $(0, 8)$, are on the y-axis, so the major axis is vertical and $a = 8$. Since the vertices are equidistant from the origin, the center of the ellipse is at the origin. The length of the minor axis is 10, so $b = 10/2$, or 5.

Write the equation:

$$\frac{x^2}{b^2} + \frac{y^2}{a^2} = 1$$
$$\frac{x^2}{5^2} + \frac{y^2}{8^2} = 1$$
$$\frac{x^2}{25} + \frac{y^2}{64} = 1$$

34. The vertices, $(-5, 0)$ and $(5, 0)$ are on the x-axis, so the major axis is horizontal and $a = 5$. Since the vertices are equidistant from the origin, the center of the ellipse is at the origin. The length of the minor axis is 6, so $b = 6/2$, or 3. The equation is $\frac{x^2}{25} + \frac{y^2}{9} = 1$.

35. The foci, $(-2, 0)$ and $(2, 0)$ are on the x-axis, so the major axis is horizontal and $c = 2$. Since the foci are equidistant from the origin, the center of the ellipse is at the origin. The length of the major axis is 6, so $a = 6/2$, or 3. Now we find b^2:

$$c^2 = a^2 - b^2$$
$$2^2 = 3^2 - b^2$$
$$4 = 9 - b^2$$
$$b^2 = 5$$

Write the equation:

$$\frac{x^2}{a^2} + \frac{y^2}{b^2} = 1$$
$$\frac{x^2}{9} + \frac{y^2}{5} = 1$$

36. The foci, $(0, -3)$ and $(0, 3)$, are on the y-axis, so the major axis is vertical. The foci are equidistant from the origin, so the center of the ellipse is at the origin. The length of the major axis is 10, so $a = 10/2$, or 5. Find b^2:

$$c^2 = a^2 - b^2$$
$$9 = 25 - b^2$$
$$b^2 = 16$$

The equation is $\frac{x^2}{16} + \frac{y^2}{25} = 1$.

37. $$\frac{(x-1)^2}{9} + \frac{(y-2)^2}{4} = 1$$

$$\frac{(x-1)^2}{3^2} + \frac{(y-2)^2}{2^2} = 1 \quad \text{Standard form}$$

The center is $(1, 2)$. Note that $a = 3$ and $b = 2$. The major axis is horizontal so the vertices are 3 units left and right of the center:

$(1 - 3, 2)$ and $(1 + 3, 2)$, or $(-2, 2)$ and $(4, 2)$.

We know that $c^2 = a^2 - b^2$, so $c^2 = 9 - 4 = 5$ and $c = \sqrt{5}$. Then the foci are $\sqrt{5}$ units left and right of the center:

$$(1 - \sqrt{5}, 2) \text{ and } (1 + \sqrt{5}, 2).$$

To graph the ellipse, plot the vertices. Since $b = 2$, two other points on the graph are 2 units below and above the center:

$(1, 2 - 2)$ and $(1, 2 + 2)$ or $(1, 0)$ and $(1, 4)$

Plot these points also and connect the four plotted points with a smooth curve.

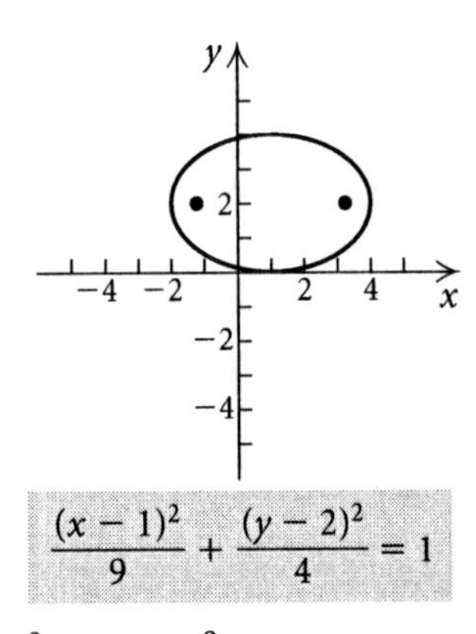

38. $\dfrac{(x-1)^2}{1} + \dfrac{(y-2)^2}{4} = 1$

$\dfrac{(x-1)^2}{1^2} + \dfrac{(y-2)^2}{2^2} = 1$ Standard form

The center is $(1, 2)$. Note that $a = 2$ and $b = 1$. The major axis is vertical so the vertices are 2 units below and above the center:

$(1, 2 - 2)$ and $(1, 2 + 2)$, or $(1, 0)$ and $(1, 4)$.

We know that $c^2 = a^2 - b^2$, so $c^2 = 4 - 1 = 3$ and $c = \sqrt{3}$. Then the foci are $\sqrt{3}$ units below and above the center:

$$(1, 2 - \sqrt{3}) \text{ and } (1, 2 + \sqrt{3}).$$

Since $b = 1$, two points on the graph other than the vertices are $(1 - 1, 2)$ and $(1 + 1, 2)$ or $(0, 2)$ and $(2, 2)$.

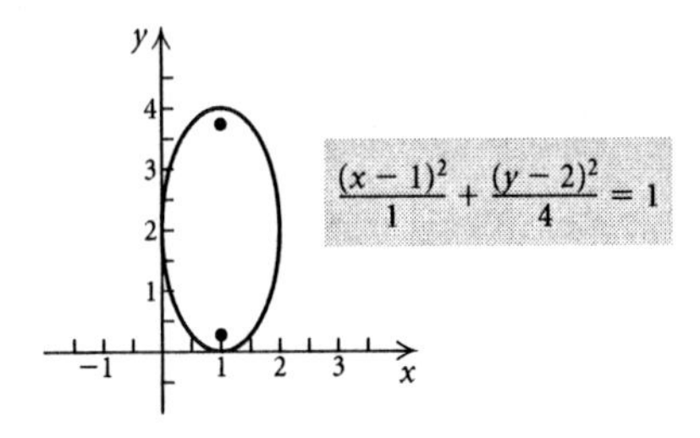

39. $\dfrac{(x+3)^2}{25} + \dfrac{(y-5)^2}{36} = 1$

$\dfrac{[x-(-3)]^2}{5^2} + \dfrac{(y-5)^2}{6^2} = 1$ Standard form

The center is $(-3, 5)$. Note that $a = 6$ and $b = 5$. The major axis is vertical so the vertices are 6 units below and above the center:

$(-3, 5 - 6)$ and $(-3, 5 + 6)$, or $(-3, -1)$ and $(-3, 11)$.

We know that $c^2 = a^2 - b^2$, so $c^2 = 36 - 25 = 11$ and $c = \sqrt{11}$. Then the foci are $\sqrt{11}$ units below and above the vertex:

$$(-3, 5 - \sqrt{11}) \text{ and } (-3, 5 + \sqrt{11}).$$

To graph the ellipse, plot the vertices. Since $b = 5$, two other points on the graph are 5 units left and right of the center:

$(-3 - 5, 5)$ and $(-3 + 5, 5)$, or $(-8, 5)$ and $(2, 5)$

Plot these points also and connect the four plotted points with a smooth curve.

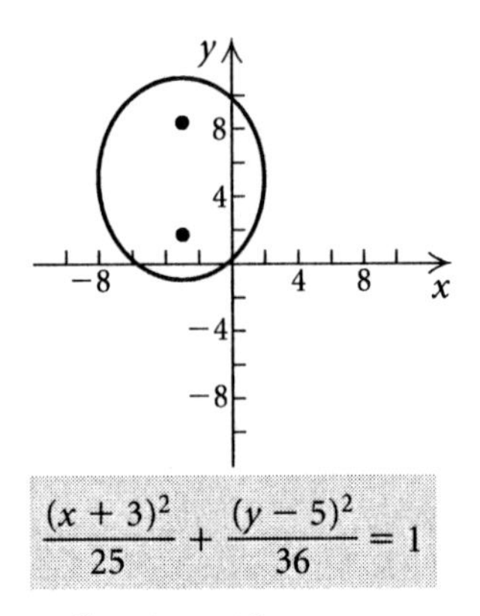

40. $\dfrac{(x-2)^2}{16} + \dfrac{(y+3)^2}{25} = 1$

$\dfrac{(x-2)^2}{4^2} + \dfrac{[y-(-3)]^2}{5^2} = 1$ Standard form

The center is $(2, -3)$. Note that $a = 5$ and $b = 4$. The major axis is vertical so the vertices are 5 units below and above the center:

$(2, -3 - 5)$ and $(2, -3 + 5)$, or $(2, -8)$ and $(2, 2)$.

We know that $c^2 = a^2 - b^2$, so $c^2 = 25 - 16 = 9$ and $c = 3$. Then the foci are 3 units below and above the center:

$(2, -3 - 3)$ and $(2, -3 + 3)$, or $(2, -6)$ and $(2, 0)$.

Since $b = 4$, two points on the graph other than the vertices are $(2 - 4, -3)$ and $(2 + 4, -3)$, or $(-2, -3)$ and $(6, -3)$.

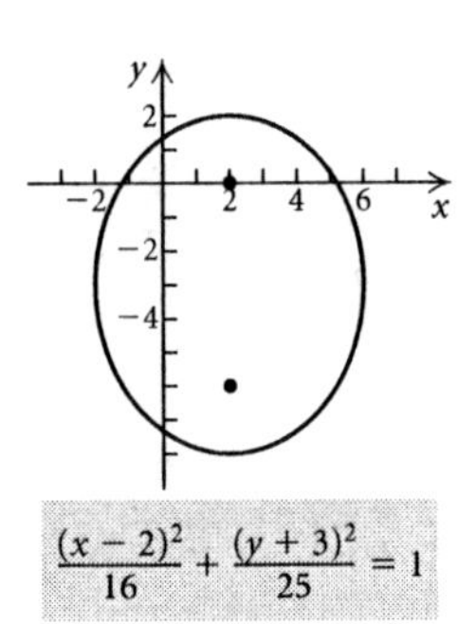

41. $3(x+2)^2 + 4(y-1)^2 = 192$

$\dfrac{(x+2)^2}{64} + \dfrac{(y-1)^2}{48} = 1$ Dividing by 192

$\dfrac{[x-(-2)]^2}{8^2} + \dfrac{(y-1)^2}{(\sqrt{48})^2} = 1$ Standard form

The center is $(-2, 1)$. Note that $a = 8$ and $b = \sqrt{48}$, or $4\sqrt{3}$. The major axis is horizontal so the vertices are 8 units left and right of the center:

$(-2-8, 1)$ and $(-2+8, 1)$, or $(-10, 1)$ and $(6, 1)$.

We know that $c^2 = a^2 - b^2$, so $c^2 = 64-48 = 16$ and $c = 4$. Then the foci are 4 units left and right of the center:

$(-2-4, 1)$ and $(-2+4, 1)$ or $(-6, 1)$ and $(2, 1)$.

To graph the ellipse, plot the vertices. Since $b = 4\sqrt{3} \approx 6.928$, two other points on the graph are about 6.928 units below and above the center:

$(-2, 1-6.928)$ and $(-2, 1+6.928)$, or

$(-2, -5.928)$ and $(-2, 7.928)$.

Plot these points also and connect the four plotted points with a smooth curve.

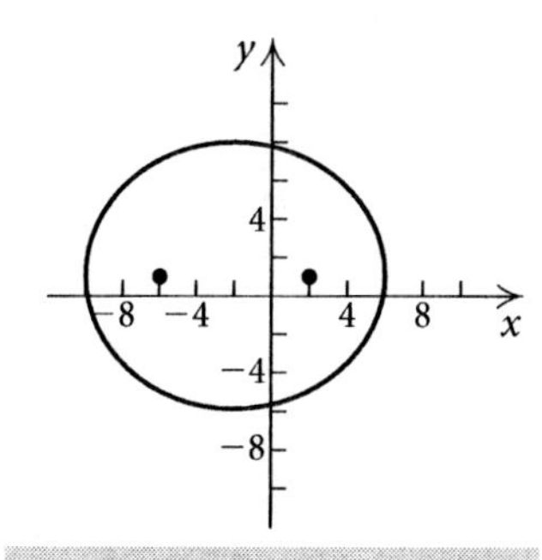

42. $4(x-5)^2 + 3(y-4)^2 = 48$

$$\frac{(x-5)^2}{12} + \frac{(y-4)^2}{16} = 1$$

$$\frac{(x-5)^2}{(\sqrt{12})^2} + \frac{(y-4)^2}{4^2} = 1 \quad \text{Standard form}$$

The center is $(5, 4)$. Note that $a = 16$ and $b = \sqrt{12}$, or $2\sqrt{3}$. The major axis is vertical so the vertices are 4 units below and above the center:

$(5, 4-4)$ and $(5, 4+4)$, or $(5, 0)$ and $(5, 8)$.

We know that $c^2 = a^2 - b^2$, so $c^2 = 16 - 12 = 4$ and $c = 2$. Then the foci are 2 units below and above the center:

$(5, 4-2)$ and $(5, 4+2)$, or $(5, 2)$ and $(5, 6)$.

Since $b = 2\sqrt{3} \approx 3.464$, two points on the graph other than the vertices are about $(5-3.464, 4)$ and $(5+3.464, 4)$, or $(1.536, 4)$ and $(8.464, 4)$.

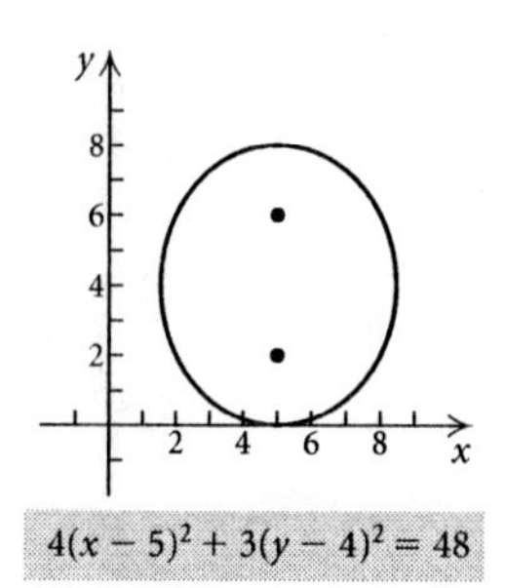

43. Begin by completing the square twice.

$$4x^2 + 9y^2 - 16x + 18y - 11 = 0$$
$$4x^2 - 16x + 9y^2 + 18y = 11$$
$$4(x^2 - 4x) + 9(y^2 + 2y) = 11$$
$$4(x^2 - 4x + 4) + 9(y^2 + 2y + 1) = 11 + 4 \cdot 4 + 9 \cdot 1$$
$$4(x-2)^2 + 9(y+1)^2 = 36$$
$$\frac{(x-2)^2}{9} + \frac{(y+1)^2}{4} = 1$$
$$\frac{(x-2)^2}{3^2} + \frac{[y-(-1)]^2}{2^2} = 1$$

The center is $(2, -1)$. Note that $a = 3$ and $b = 2$. The major axis is horizontal so the vertices are 3 units left and right of the center:

$(2-3, -1)$ and $(2+3, -1)$, or $(-1, -1)$ and $(5, -1)$.

We know that $c^2 = a^2 - b^2$, so $c^2 = 9 - 4 = 5$ and $c = \sqrt{5}$. Then the foci are $\sqrt{5}$ units left and right of the center:

$(2-\sqrt{5}, -1)$ and $(2+\sqrt{5}, -1)$.

To graph the ellipse, plot the vertices. Since $b = 2$, two other points on the graph are 2 units below and above the center:

$(2, -1-2)$ and $(2, -1+2)$, or $(2, -3)$ and $(2, 1)$.

Plot these points also and connect the four plotted points with a smooth curve.

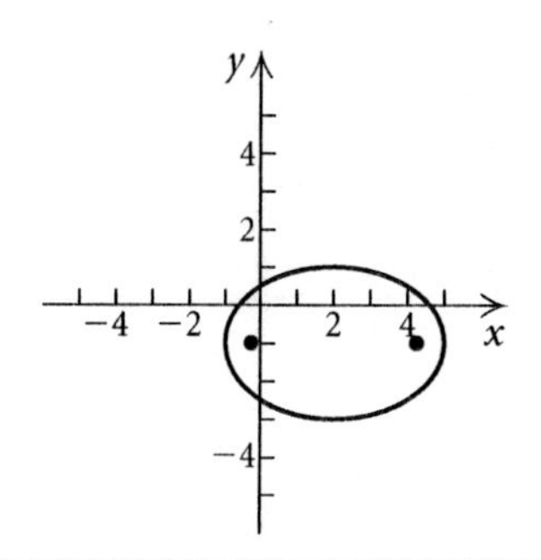

44. Begin by completing the square twice.

$$x^2 + 2y^2 - 10x + 8y + 29 = 0$$
$$x^2 - 10x + 2(y^2 + 4y) = -29$$
$$x^2 - 10x + 25 + 2(y^2 + 4y + 4) = -29 + 25 + 2 \cdot 4$$
$$(x-5)^2 + 2(y+2)^2 = 4$$
$$\frac{(x-5)^2}{4} + \frac{(y+2)^2}{2} = 1$$
$$\frac{(x-5)^2}{2^2} + \frac{[y-(-2)]^2}{(\sqrt{2})^2} = 1$$

The center is $(5, -2)$. Note that $a = 2$ and $b = \sqrt{2}$. The major axis is horizontal so the vertices are 2 units left and right of the center:

$(5-2, -2)$ and $(5+2, -2)$, or $(3, -2)$ and $(7, -2)$.

We know that $c^2 = a^2 - b^2$, so $c^2 = 4 - 2 = 2$ and $c = \sqrt{2}$. Then the foci are $\sqrt{2}$ units left and right of the center:

$(5-\sqrt{2},-2)$ and $(5+\sqrt{2},-2)$.

Since $b=\sqrt{2}\approx 1.414$, two points on the graph other than the vertices are about

$(5,-2-1.414)$ and $(5,-2+1.414)$, or

$(5,-3.414)$ and $(5,-0.586)$.

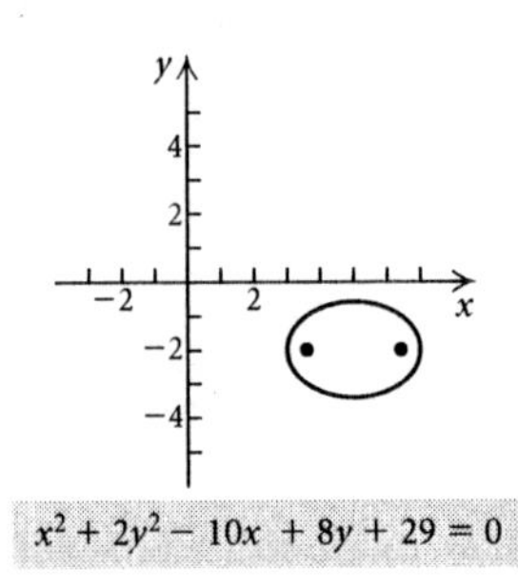

$x^2+2y^2-10x+8y+29=0$

45. Begin by completing the square twice.

$$4x^2+y^2-8x-2y+1=0$$
$$4x^2-8x+y^2-2y=-1$$
$$4(x^2-2x)+y^2-2y=-1$$
$$4(x^2-2x+1)+y^2-2y+1=-1+4\cdot 1+1$$
$$4(x-1)^2+(y-1)^2=4$$
$$\frac{(x-1)^2}{1}+\frac{(y-1)^2}{4}=1$$
$$\frac{(x-1)^2}{1^2}+\frac{(y-1)^2}{2^2}=1$$

The center is $(1,1)$. Note that $a=2$ and $b=1$. The major axis is vertical so the vertices are 2 units below and above the center:

$(1,1-2)$ and $(1,1+2)$, or $(1,-1)$ and $(1,3)$.

We know that $c^2=a^2-b^2$, so $c^2=4-1=3$ and $c=\sqrt{3}$. Then the foci are $\sqrt{3}$ units below and above the center:

$(1,1-\sqrt{3})$ and $(1,1+\sqrt{3})$.

To graph the ellipse, plot the vertices. Since $b=1$, two other points on the graph are 1 unit left and right of the center:

$(1-1,1)$ and $(1+1,1)$ or $(0,1)$ and $(2,1)$.

Plot these points also and connect the four plotted points with a smooth curve.

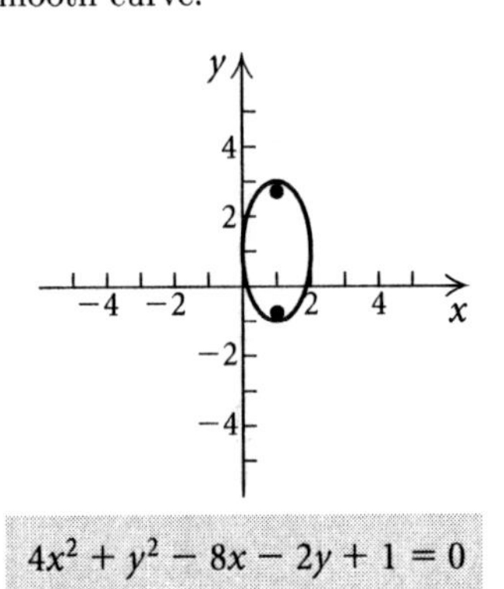

$4x^2+y^2-8x-2y+1=0$

46. Begin by completing the square twice.

$$9x^2+4y^2+54x-8y+49=0$$
$$9(x^2+6x)+4(y^2-2y)=-49$$
$$9(x^2+6x+9)+4(y^2-2y+1)=-49+9\cdot 9+4\cdot 1$$
$$9(x+3)^2+4(y-1)^2=36$$
$$\frac{(x-3)^2}{4}+\frac{(y-1)^2}{9}=1$$
$$\frac{[x-(-3)]^2}{2^2}+\frac{(y-1)^2}{3^2}=1$$

The center is $(-3,1)$. Note that $a=3$ and $b=2$. The major axis is vertical so the vertices are 3 units below and above the center:

$(-3,1-3)$ and $(-3,1+3)$, or $(-3,-2)$ and $(-3,4)$.

We know that $c^2=a^2-b^2$, so $c^2=9-4=5$ and $c=\sqrt{5}$. Then the foci are $\sqrt{5}$ units below and above the center:

$(-3,1-\sqrt{5})$ and $(-3,1+\sqrt{5})$.

Since $b=2$, two points on the graph other than the vertices are

$(-3-2,1)$ and $(-3+2,1)$, or $(-5,1)$ and $(-1,1)$.

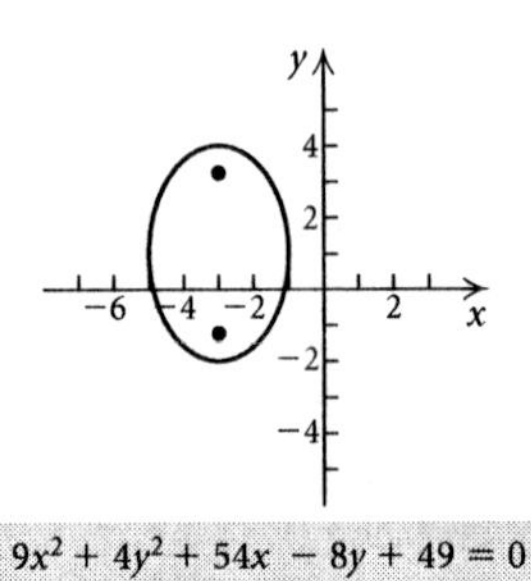

$9x^2+4y^2+54x-8y+49=0$

47. The ellipse in Example 4 is flatter than the one in Example 2, so the ellipse in Example 2 has the smaller eccentricity.

We compute the eccentricities: In Example 2, $c=3$ and $a=5$, so $e=c/a=3/5=0.6$. In Example 4, $c=2\sqrt{3}$ and $a=4$, so $e=c/a=2\sqrt{3}/4\approx 0.866$. These computations confirm that the ellipse in Example 2 has the smaller eccentricity.

48. Ellipse (b) is flatter than ellipse (a), so ellipse (a) has the smaller eccentricity.

49. Since the vertices, $(0,-4)$ and $(0,4)$ are on the y-axis and are equidistant from the origin, we know that the major axis of the ellipse is vertical, its center is at the origin, and $a=4$. Use the information that $e=1/4$ to find c:

$$e=\frac{c}{a}$$
$$\frac{1}{4}=\frac{c}{4}\quad \text{Substituting}$$
$$c=1$$

Now $c^2=a^2-b^2$, so we can find b^2:

$$1^2=4^2-b^2$$
$$1=16-b^2$$
$$b^2=15$$

Write the equation of the ellipse:

$$\frac{x^2}{b^2}+\frac{y^2}{a^2}=1$$

$$\frac{x^2}{15}+\frac{y^2}{16}=1$$

50. Since the vertices, $(-3,0)$ and $(3,0)$, are on the x-axis and are equidistant from the origin, we know that the major axis of the ellipse is horizontal, its center is at the origin, and $a=3$.

Find c:

$$\frac{c}{a}=\frac{7}{10}$$

$$\frac{c}{3}=\frac{7}{10}$$

$$c=\frac{21}{10}$$

Now find b^2:

$$c^2=a^2-b^2$$

$$\left(\frac{21}{10}\right)^2=3^2-b^2$$

$$b^2=\frac{459}{100}$$

The equation of the ellipse is $\frac{x^2}{9}+\frac{y^2}{459/100}=1$.

51. From the figure in the text we see that the center of the ellipse is $(0,0)$, the major axis is horizontal, the vertices are $(-50,0)$ and $(50,0)$, and one y-intercept is $(0,12)$. Then $a=50$ and $b=12$. The equation is

$$\frac{x^2}{a^2}+\frac{y^2}{b^2}=1$$

$$\frac{x^2}{50^2}+\frac{y}{12^2}=1$$

$$\frac{x^2}{2500}+\frac{y^2}{144}=1.$$

52. Find the equation of the ellipse with center $(0,0)$, $a=1048/2=524$, $b=898/2=449$, and a horizontal major axis:

$$\frac{x^2}{524^2}+\frac{y^2}{449^2}=1$$

$$\frac{x^2}{274,576}+\frac{y^2}{201,601}=1$$

53. Position a coordinate system as shown below where 1 unit $=10^7$ mi.

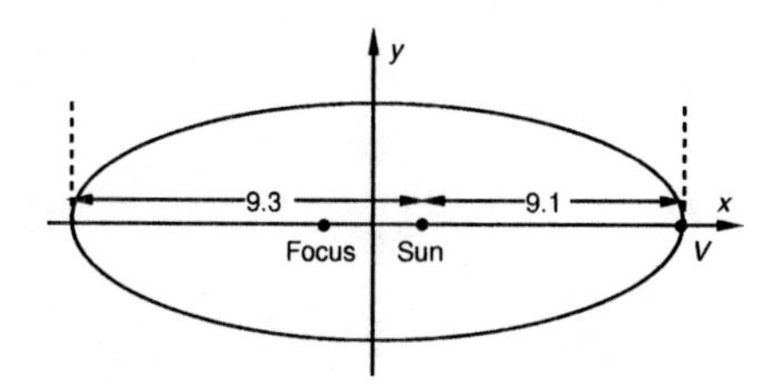

The length of the major axis is $9.3+9.1$, or 18.4. Then the distance from the center of the ellipse (the origin) to V is $18.4/2$, or 9.2. Since the distance from the sun to V is 9.1, the distance from the sun to the center is $9.2-9.1$, or 0.1. Then the distance from the sun to the other focus is twice this distance:

$$2(0.1\times10^7\text{ mi})=0.2\times10^7\text{ mi}$$
$$=2\times10^6\text{ mi}$$

54. Position a coordinate system as shown.

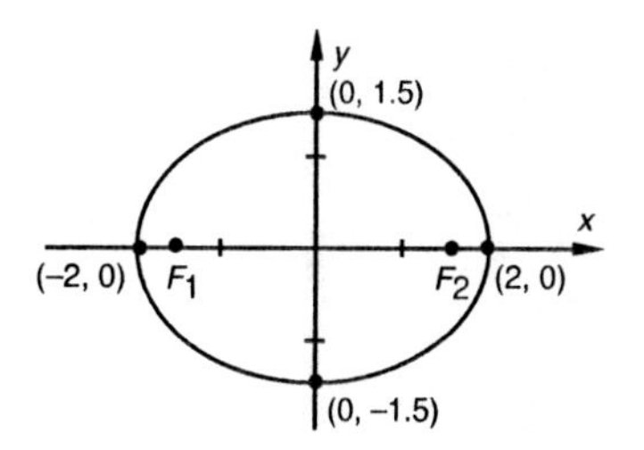

a) We have an ellipse with $a=2$ and $b=1.5$. The foci are at $(\pm c,0)$ where $c^2=a^2-b^2$, so $c^2=4-2.25=1.75$ and $c=\sqrt{1.75}$. Then the string should be attached $2-\sqrt{1.75}$ ft, or about 0.7 ft from the ends of the board.

b) The string should be the length of the major axis, 4 ft.

55. Center: $(2.003,-1.005)$

Vertices: $(-1.017,-1.005)$, $(5.023,-1.005)$

56. Center: $(-3.004,1.002)$

Vertices: $(-3.004,-1.970)$, $(-3.004,3.974)$

57. Circles and ellipses are not functions.

58. No, the center of an ellipse is not part of the graph of the ellipse. Its coordinates do not satisfy the equation of the ellipse.

59. midpoint

60. zero

61. y-intercept

62. two different real-number solutions

63. remainder

64. ellipse

65. parabola

66. circle

67. The center of the ellipse is the midpoint of the segment connecting the vertices:

$$\left(\frac{3+3}{2},\frac{-4+6}{2}\right),\text{ or }(3,1).$$

Now a is the distance from the origin to a vertex. We use the vertex $(3,6)$.

$$a=\sqrt{(3-3)^2+(6-1)^2}=5$$

Also b is one-half the length of the minor axis.

$$b=\frac{\sqrt{(5-1)^2+(1-1)^2}}{2}=\frac{4}{2}=2$$

The vertices lie on the vertical line $x = 3$, so the major axis is vertical. We write the equation of the ellipse.

$$\frac{(x-h)^2}{b^2} + \frac{(y-k)^2}{a^2} = 1$$

$$\frac{(x-3)^2}{4} + \frac{(y-1)^2}{25} = 1$$

68. Center: $\left(\frac{-1-1}{2}, \frac{-1+5}{2}\right)$, or $(-1, 2)$

$$a = \sqrt{[1-(-1)]^2 + (-1-2)^2} = 3$$

$$b = \frac{\sqrt{(-3-1)^2 + (2-2)^2}}{2} = \frac{4}{2} = 2$$

The vertices are on the line $x = -1$, so the major axis is vertical. The equation is

$$\frac{(x+1)^2}{4} + \frac{(y-2)^2}{9} = 1.$$

69. The center is the midpoint of the segment connecting the vertices:

$$\left(\frac{-3+3}{0}, \frac{0+0}{0}\right), \text{ or } (0, 0).$$

Then $a = 3$ and since the vertices are on the x-axis, the major axis is horizontal. The equation is of the form $\frac{x^2}{a^2} + \frac{y^2}{b^2} = 1$.

Substitute 3 for a, 2 for x, and $\frac{22}{3}$ for y and solve for b^2.

$$\frac{4}{9} + \frac{\frac{484}{9}}{b^2} = 1$$

$$\frac{4}{9} + \frac{484}{9b^2} = 1$$

$$4b^2 + 484 = 9b^2$$

$$484 = 5b^2$$

$$\frac{484}{5} = b^2$$

Then the equation is $\frac{x^2}{9} + \frac{y^2}{484/5} = 1$.

70. $a = 4/2 = 2$; $b = 1/2$

The equation is $\frac{(x+2)^2}{1/4} + \frac{(y-3)^2}{4} = 1$.

71. Position a coordinate system as shown.

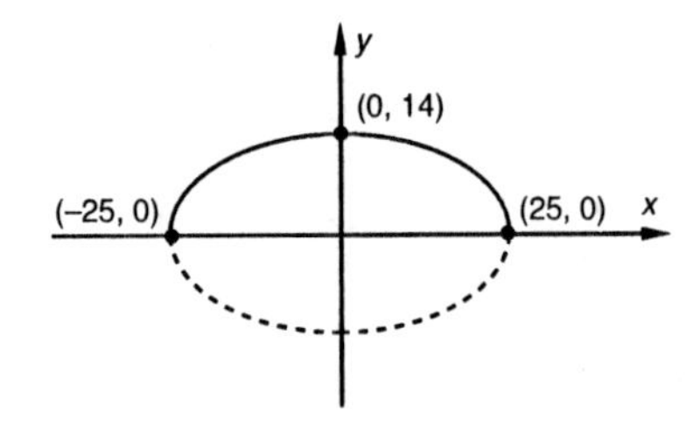

The equation of the ellipse is

$$\frac{x^2}{25^2} + \frac{y^2}{14^2} = 1$$

$$\frac{x^2}{625} + \frac{y^2}{196} = 1.$$

A point 6 ft from the riverbank corresponds to $(25-6, 0)$, or $(19, 0)$ or to $(-25+6, 0)$, or $(-19, 0)$. Substitute either 19 or -19 for x and solve for y, the clearance.

$$\frac{19^2}{625} + \frac{y^2}{196} = 1$$

$$\frac{y^2}{196} = 1 - \frac{361}{625}$$

$$y^2 = 196\left(1 - \frac{361}{625}\right)$$

$$y \approx 9.1$$

The clearance 6 ft from the riverbank is about 9.1 ft.

Exercise Set 9.3

1. Graph (b) is the graph of $\frac{x^2}{25} - \frac{y^2}{9} = 1$.

2. Graph (e) is the graph of $\frac{y^2}{4} - \frac{x^2}{36} = 1$.

3. Graph (c) is the graph of $\frac{(y-1)^2}{16} - \frac{(x+3)^2}{1} = 1$.

4. Graph (f) is the graph of $\frac{(x+4)^2}{100} - \frac{(y-2)^2}{81} = 1$.

5. Graph (a) is the graph of $25x^2 - 16y^2 = 400$.

6. Graph (d) is the graph of $y^2 - x^2 = 9$.

7. The vertices are equidistant from the origin and are on the y-axis, so the center is at the origin and the transverse axis is vertical. Since $c^2 = a^2 + b^2$, we have $5^2 = 3^2 + b^2$ so $b^2 = 16$.

The equation is of the form $\frac{y^2}{a^2} - \frac{x^2}{b^2} = 1$, so we have $\frac{y^2}{9} - \frac{x^2}{16} = 1$.

8. The vertices are equidistant from the origin and are on the x-axis, so the center is at the origin and the transverse axis is horizontal. Since $c^2 = a^2 + b^2$, we have $2^2 = 1^2 + b^2$ so $b^2 = 3$. The equation is $\frac{x^2}{1} - \frac{y^2}{3} = 1$.

9. The asymptotes pass through the origin, so the center is the origin. The given vertex is on the x-axis, so the transverse axis is horizontal. Since $\frac{b}{a}x = \frac{3}{2}x$ and $a = 2$, we have $b = 3$. The equation is of the form $\frac{x^2}{a^2} - \frac{y^2}{b^2} = 1$, so we have $\frac{x^2}{2^2} - \frac{y^2}{3^2} = 1$, or $\frac{x^2}{4} - \frac{y^2}{9} = 1$.

10. The asymptotes pass through the origin, so the center is the origin. The given vertex is on the y-axis, so the transverse axis is vertical. We use the equation of an asymptote to find b.

$$\frac{a}{b}x = \frac{5}{4}x$$

$$\frac{3}{b}x = \frac{5}{4}x \quad \text{Substituting 3 for } a$$

$$\frac{12}{5} = b$$

The equation is $\frac{y^2}{9} - \frac{x^2}{144/25} = 1$.

11. $\frac{x^2}{4} - \frac{y^2}{4} = 1$

$$\frac{x^2}{2^2} - \frac{y^2}{2^2} = 1 \quad \text{Standard form}$$

The center is $(0,0)$; $a = 2$ and $b = 2$. The transverse axis is horizontal so the vertices are $(-2,0)$ and $(2,0)$. Since $c^2 = a^2 + b^2$, we have $c^2 = 4 + 4 = 8$ and $c = \sqrt{8}$, or $2\sqrt{2}$. Then the foci are $(-2\sqrt{2},0)$ and $(2\sqrt{2},0)$.

Find the asymptotes:

$$y = \frac{b}{a}x \text{ and } y = -\frac{b}{a}x$$

$$y = \frac{2}{2}x \text{ and } y = -\frac{2}{2}x$$

$$y = x \quad \text{and} \quad y = -x$$

To draw the graph sketch the asymptotes, plot the vertices, and draw the branches of the hyperbola outward from the vertices toward the asymptotes.

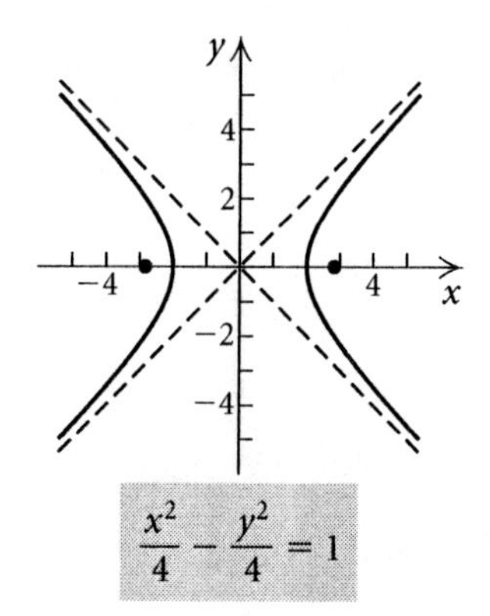

12. $\frac{x^2}{1} - \frac{y^2}{9} = 1$

$$\frac{x^2}{1^2} - \frac{y^2}{3^2} = 1 \quad \text{Standard form}$$

The center is $(0,0)$; $a = 1$ and $b = 3$. The transverse axis is horizontal so the vertices are $(-1,0)$ and $(1,0)$. Since $c^2 = a^2 + b^2$, we have $c^2 = 1 + 9 = 10$ and $c = \sqrt{10}$. Then the foci are $(-\sqrt{10},0)$ and $(\sqrt{10},0)$. Find the asymptotes:

$$y = \frac{b}{a}x \text{ and } y = -\frac{b}{a}x$$

$$y = \frac{3}{1}x \text{ and } y = -\frac{3}{1}x$$

$$y = 3x \quad \text{and} \quad y = -3x$$

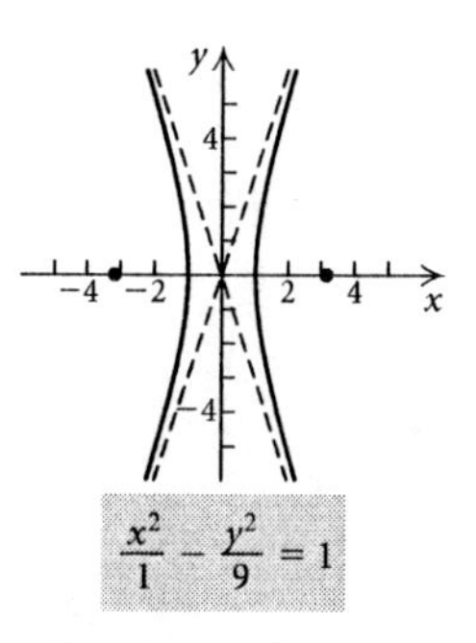

13. $\frac{(x-2)^2}{9} - \frac{(y+5)^2}{1} = 1$

$$\frac{(x-2)^2}{3^2} - \frac{[y-(-5)]^2}{1^2} = 1 \quad \text{Standard form}$$

The center is $(2,-5)$; $a = 3$ and $b = 1$. The transverse axis is horizontal, so the vertices are 3 units left and right of the center:

$(2-3,-5)$ and $(2+3,-5)$, or $(-1,-5)$ and $(5,-5)$.

Since $c^2 = a^2 + b^2$, we have $c^2 = 9 + 1 = 10$ and $c = \sqrt{10}$. Then the foci are $\sqrt{10}$ units left and right of the center:

$$(2-\sqrt{10},-5) \text{ and } (2+\sqrt{10},-5).$$

Find the asymptotes:

$$y-k=\frac{b}{a}(x-h) \text{ and } \quad y-k=-\frac{b}{a}(x-h)$$

$$y-(-5)=\frac{1}{3}(x-2) \text{ and } y-(-5)=-\frac{1}{3}(x-2)$$

$$y+5=\frac{1}{3}(x-2) \text{ and } \quad y+5=-\frac{1}{3}(x-2), \text{ or}$$

$$y=\frac{1}{3}x-\frac{17}{3} \text{ and } \quad y=-\frac{1}{3}x-\frac{13}{3}$$

Sketch the asymptotes, plot the vertices, and draw the graph.

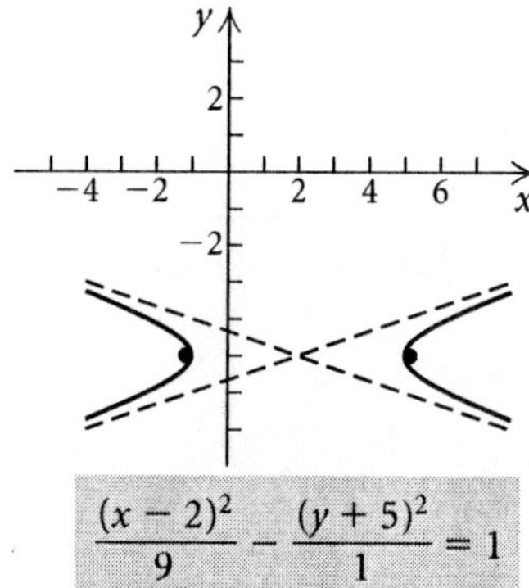

14. $\frac{(x-5)^2}{16} - \frac{(y+2)^2}{9} = 1$

$$\frac{(x-5)^2}{4^2} - \frac{[y-(-2)]^2}{3^2} = 1 \quad \text{Standard form}$$

The center is $(5,-2)$; $a = 4$ and $b = 3$. The transverse axis is horizontal, so the vertices are 4 units left and right of the center:

$(5-4,-2)$ and $(5+4,-2)$, or $(1,-2)$ and $(9,-2)$.

Since $c^2 = a^2 + b^2$, we have $c^2 = 16 + 9 = 25$ and $c = 5$. Then the foci are 5 units left and right of the center:

$(5 - 5, -2)$ and $(5 + 5, -2)$, or $(0, -2)$ and $(10, -2)$.

Find the asymptotes:

$$y - k = \frac{b}{a}(x - h) \text{ and } y - k = -\frac{b}{a}(x - h)$$

$$y + 2 = \frac{3}{4}(x - 5) \text{ and } y + 2 = -\frac{3}{4}(x - 5), \text{ or}$$

$$y = \frac{3}{4}x - \frac{23}{4} \quad \text{and} \quad y = -\frac{3}{4}x + \frac{7}{4}$$

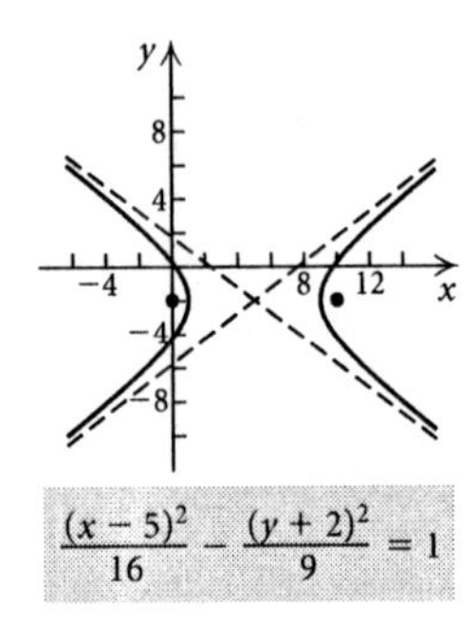

$\frac{(x-5)^2}{16} - \frac{(y+2)^2}{9} = 1$

15. $\dfrac{(y+3)^2}{4} - \dfrac{(x+1)^2}{16} = 1$

$$\frac{[y-(-3)]^2}{2^2} - \frac{[x-(-1)]^2}{4^2} = 1 \quad \text{Standard form}$$

The center is $(-1, -3)$; $a = 2$ and $b = 4$. The transverse axis is vertical, so the vertices are 2 units below and above the center:

$(-1, -3 - 2)$ and $(1, -3 + 2)$, or $(-1, -5)$ and $(-1, -1)$.

Since $c^2 = a^2 + b^2$, we have $c^2 = 4 + 16 = 20$ and $c = \sqrt{20}$, or $2\sqrt{5}$. Then the foci are $2\sqrt{5}$ units below and above of the center:

$$(-1, -3 - 2\sqrt{5}) \text{ and } (-1, -3 + 2\sqrt{5}).$$

Find the asymptotes:

$$y - k = \frac{a}{b}(x - h) \quad \text{and} \quad y - k = -\frac{a}{b}(x - h)$$

$$y - (-3) = \frac{2}{4}(x - (-1)) \text{ and } y - (-3) = -\frac{2}{4}(x - (-1))$$

$$y + 3 = \frac{1}{2}(x + 1) \quad \text{and} \quad y + 3 = -\frac{1}{2}(x + 1), \text{ or}$$

$$y = \frac{1}{2}x - \frac{5}{2} \quad \text{and} \quad y = -\frac{1}{2}x - \frac{7}{2}$$

Sketch the asymptotes, plot the vertices, and draw the graph.

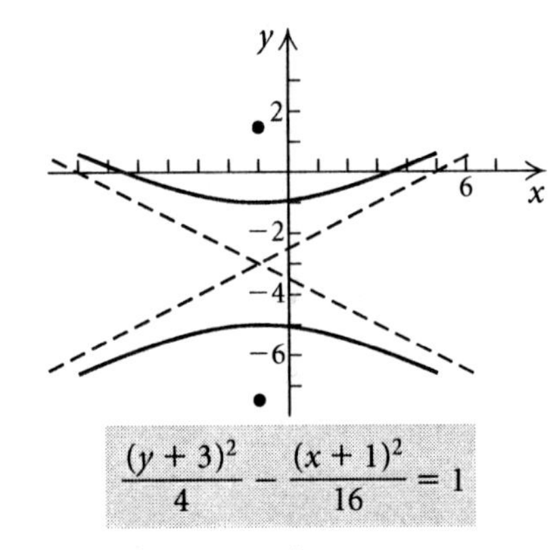

$\frac{(y+3)^2}{4} - \frac{(x+1)^2}{16} = 1$

16. $\dfrac{(y+4)^2}{25} - \dfrac{(x+2)^2}{16} = 1$

$$\frac{[y-(-4)]^2}{5^2} - \frac{[x-(-2)]^2}{4^2} = 1 \quad \text{Standard form}$$

The center is $(-2, -4)$; $a = 5$ and $b = 4$. The transverse axis is vertical, so the vertices are 5 units below and above the center:

$(-2, -4 - 5)$ and $(-2, -4 + 5)$, or $(-2, -9)$ and $(-2, 1)$.

Since $c^2 = a^2 + b^2$, we have $c^2 = 25 + 16 = 41$ and $c = \sqrt{41}$. Then the foci are $\sqrt{41}$ units below and above the center:

$$(-2, -4 - \sqrt{41}) \text{ and } (-2, -4 + \sqrt{41}).$$

Find the asymptotes:

$$y - k = \frac{a}{b}(x - h) \text{ and } y - k = -\frac{a}{b}(x - h)$$

$$y + 4 = \frac{5}{4}(x + 2) \text{ and } y + 4 = -\frac{5}{4}(x + 2), \text{ or}$$

$$y = \frac{5}{4}x - \frac{3}{2} \quad \text{and} \quad y = -\frac{5}{4}x - \frac{13}{2}$$

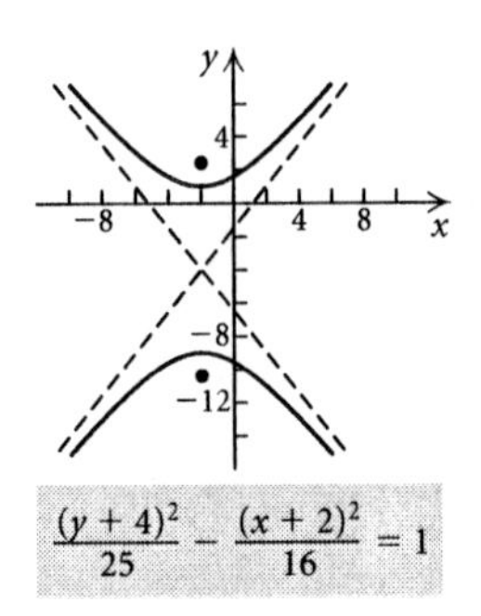

$\frac{(y+4)^2}{25} - \frac{(x+2)^2}{16} = 1$

17. $x^2 - 4y^2 = 4$

$$\frac{x^2}{4} - \frac{y^2}{1} = 1$$

$$\frac{x^2}{2^2} - \frac{y^2}{1^2} = 1 \quad \text{Standard form}$$

The center is $(0, 0)$; $a = 2$ and $b = 1$. The transverse axis is horizontal, so the vertices are $(-2, 0)$ and $(2, 0)$. Since $c^2 = a^2 + b^2$, we have $c^2 = 4 + 1 = 5$ and $c = \sqrt{5}$. Then the foci are $(-\sqrt{5}, 0)$ and $(\sqrt{5}, 0)$.

Find the asymptotes:

$$y = \frac{b}{a}x \text{ and } y = -\frac{b}{a}x$$

$$y = \frac{1}{2}x \text{ and } y = -\frac{1}{2}x$$

Sketch the asymptotes, plot the vertices, and draw the graph.

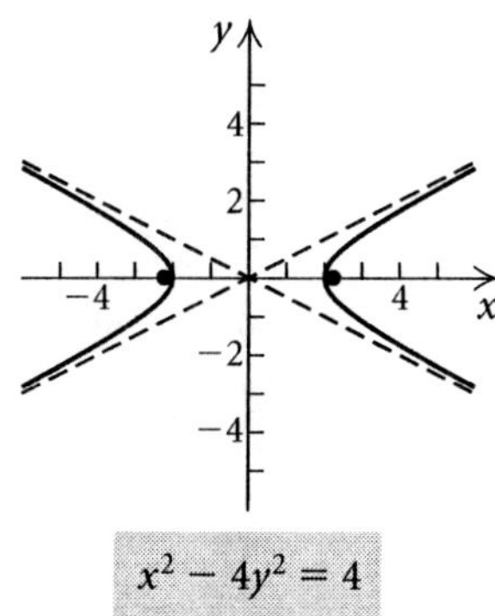

18. $4x^2 - y^2 = 16$

$$\frac{x^2}{4} - \frac{y^2}{16} = 1$$

$$\frac{x^2}{2^2} - \frac{y^2}{4^2} = 1 \quad \text{Standard form}$$

The center is $(0, 0)$; $a = 2$ and $b = 4$. The transverse axis is horizontal, so the vertices are $(-2, 0)$ and $(2, 0)$. Since $c^2 = a^2 + b^2$, we have $c^2 = 4 + 16 = 20$ and $c = \sqrt{20}$, or $2\sqrt{5}$. Then the foci are $(-2\sqrt{5}, 0)$ and $(2\sqrt{5}, 0)$.

Find the asymptotes:

$$y = \frac{b}{a}x \text{ and } y = -\frac{b}{a}x$$

$$y = \frac{4}{2}x \text{ and } y = -\frac{4}{2}x$$

$$y = 2x \text{ and } y = -2x$$

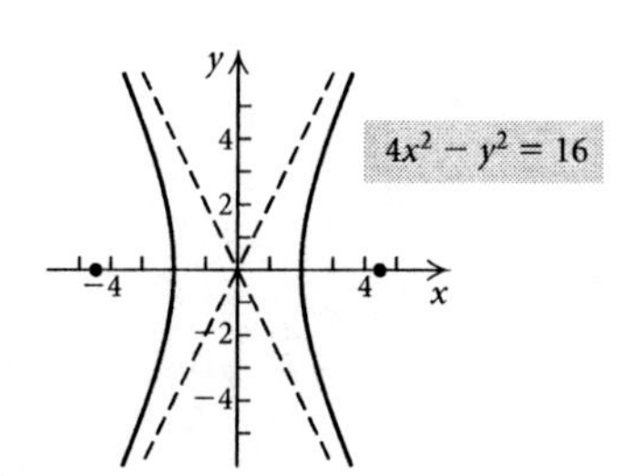

19. $9y^2 - x^2 = 81$

$$\frac{y^2}{9} - \frac{x^2}{81} = 1$$

$$\frac{y^2}{3^2} - \frac{x^2}{9^2} = 1 \quad \text{Standard form}$$

The center is $(0, 0)$; $a = 3$ and $b = 9$. The transverse axis is vertical, so the vertices are $(0, -3)$ and $(0, 3)$. Since $c^2 = a^2 + b^2$, we have $c^2 = 9 + 81 = 90$ and $c = \sqrt{90}$, or $3\sqrt{10}$. Then the foci are $(0, -3\sqrt{10})$ and $(0, 3\sqrt{10})$.

Find the asymptotes:

$$y = \frac{a}{b}x \text{ and } y = -\frac{a}{b}x$$

$$y = \frac{3}{9}x \text{ and } y = -\frac{3}{9}x$$

$$y = \frac{1}{3}x \text{ and } y = -\frac{1}{3}x$$

Sketch the asymptotes, plot the vertices, and draw the graph.

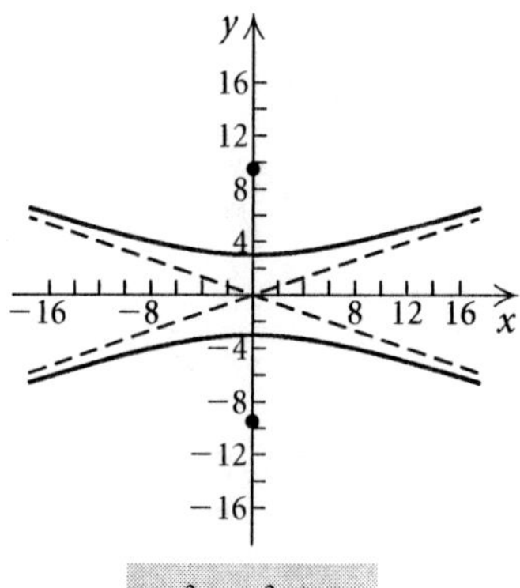

20. $y^2 - 4x^2 = 4$

$$\frac{y^2}{4} - \frac{x^2}{1} = 1$$

$$\frac{y^2}{2^2} - \frac{x^2}{1^2} = 1 \quad \text{Standard form}$$

The center is $(0, 0)$; $a = 2$ and $b = 1$. The transverse axis is vertical, so the vertices are $(0, -2)$ and $(0, 2)$. Since $c^2 = a^2 + b^2$, we have $c^2 = 4 + 1 = 5$ and $c = \sqrt{5}$. Then the foci are $(0, -\sqrt{5})$ and $(0, \sqrt{5})$.

Find the asymptotes:

$$y = \frac{a}{b}x \text{ and } y = -\frac{a}{b}x$$

$$y = \frac{2}{1}x \text{ and } y = -\frac{2}{1}x$$

$$y = 2x \text{ and } y = -2x$$

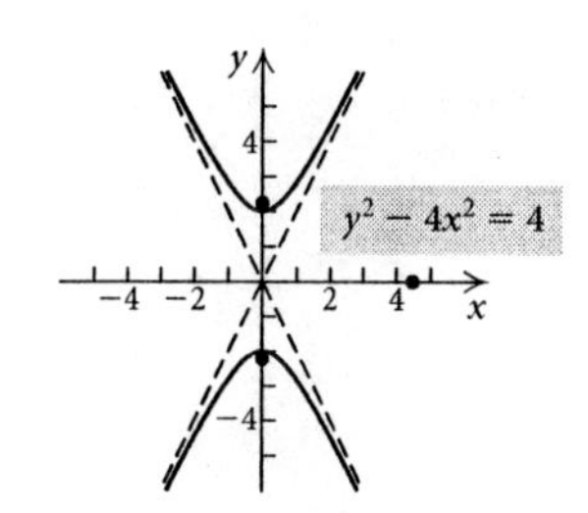

21.
$$x^2 - y^2 = 2$$
$$\frac{x^2}{2} - \frac{y^2}{2} = 1$$
$$\frac{x^2}{(\sqrt{2})^2} - \frac{y^2}{(\sqrt{2})^2} = 1 \quad \text{Standard form}$$

The center is $(0, 0)$; $a = \sqrt{2}$ and $b = \sqrt{2}$. The transverse axis is horizontal, so the vertices are $(-\sqrt{2}, 0)$ and $(\sqrt{2}, 0)$. Since $c^2 = a^2 + b^2$, we have $c^2 = 2 + 2 = 4$ and $c = 2$. Then the foci are $(-2, 0)$ and $(2, 0)$.

Find the asymptotes:
$$y = \frac{b}{a}x \quad \text{and} \quad y = -\frac{b}{a}x$$
$$y = \frac{\sqrt{2}}{\sqrt{2}}x \quad \text{and} \quad y = -\frac{\sqrt{2}}{\sqrt{2}}x$$
$$y = x \quad \text{and} \quad y = -x$$

Sketch the asymptotes, plot the vertices, and draw the graph.

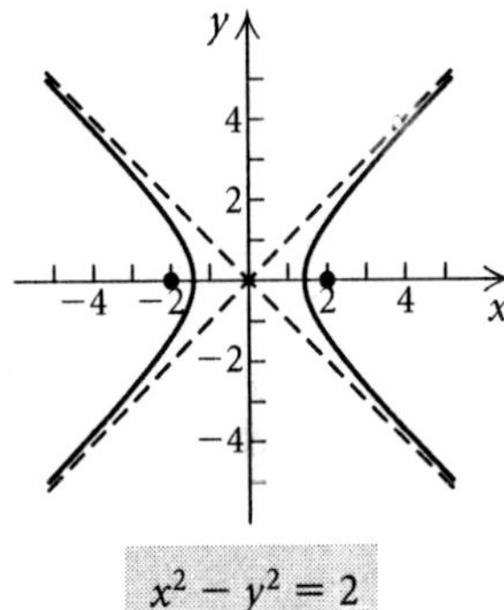

$x^2 - y^2 = 2$

22.
$$x^2 - y^2 = 3$$
$$\frac{x^2}{3} - \frac{y^2}{3} = 1$$
$$\frac{x^2}{(\sqrt{3})^2} - \frac{y^2}{(\sqrt{3})^2} = 1 \quad \text{Standard form}$$

The center is $(0, 0)$; $a = \sqrt{3}$ and $b = \sqrt{3}$. The transverse axis is horizontal, so the vertices are $(-\sqrt{3}, 0)$ and $(\sqrt{3}, 0)$. Since $c^2 = a^2 + b^2$, we have $c^2 = 3 + 3 = 6$ so $c = \sqrt{6}$. Then the foci are $(-\sqrt{6}, 0)$ and $(\sqrt{6}, 0)$.

Find the asymptotes:
$$y = \frac{b}{a}x \quad \text{and} \quad y = -\frac{b}{a}x$$
$$y = \frac{\sqrt{3}}{\sqrt{3}}x \quad \text{and} \quad y = -\frac{\sqrt{3}}{\sqrt{3}}x$$
$$y = x \quad \text{and} \quad y = -x$$

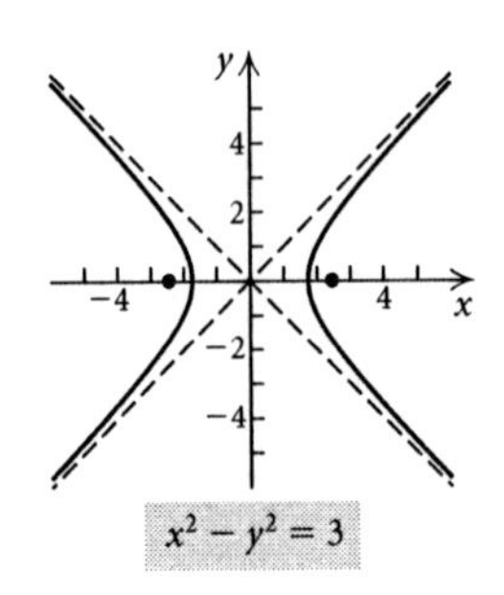

$x^2 - y^2 = 3$

23.
$$y^2 - x^2 = \frac{1}{4}$$
$$\frac{y^2}{1/4} - \frac{x^2}{1/4} = 1$$
$$\frac{y^2}{(1/2)^2} - \frac{x^2}{(1/2)^2} = 1 \quad \text{Standard form}$$

The center is $(0, 0)$; $a = \frac{1}{2}$ and $b = \frac{1}{2}$. The transverse axis is vertical, so the vertices are $\left(0, -\frac{1}{2}\right)$ and $\left(0, \frac{1}{2}\right)$. Since $c^2 = a^2 + b^2$, we have $c^2 = \frac{1}{4} + \frac{1}{4} = \frac{1}{2}$ and $c = \sqrt{\frac{1}{2}}$, or $\frac{\sqrt{2}}{2}$. Then the foci are $\left(0, -\frac{\sqrt{2}}{2}\right)$ and $\left(0, \frac{\sqrt{2}}{2}\right)$.

Find the asymptotes:
$$y = \frac{a}{b}x \quad \text{and} \quad y = -\frac{a}{b}x$$
$$y = \frac{1/2}{1/2}x \quad \text{and} \quad y = -\frac{1/2}{1/2}x$$
$$y = x \quad \text{and} \quad y = -x$$

Sketch the asymptotes, plot the vertices, and draw the graph.

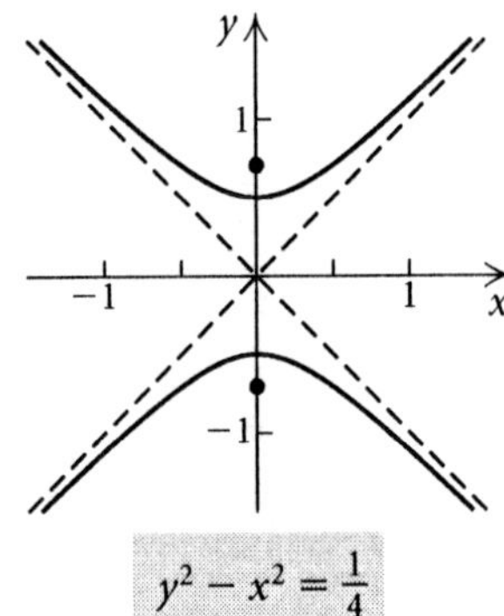

$y^2 - x^2 = \frac{1}{4}$

24.
$$y^2 - x^2 = \frac{1}{9}$$
$$\frac{y^2}{1/9} - \frac{x^2}{1/9} = 1$$
$$\frac{y^2}{(1/3)^2} - \frac{x^2}{(1/3)^2} = 1 \quad \text{Standard form}$$

The center is $(0, 0)$; $a = \frac{1}{3}$ and $b = \frac{1}{3}$. The transverse axis is vertical, so the vertices are $\left(0, -\frac{1}{3}\right)$ and $\left(0, \frac{1}{3}\right)$. Since $c^2 = a^2 + b^2$, we have $c^2 = \frac{1}{9} + \frac{1}{9} = \frac{2}{9}$ and $c = \frac{\sqrt{2}}{3}$. Then the foci are $\left(0, -\frac{\sqrt{2}}{3}\right)$ and $\left(0, \frac{\sqrt{2}}{3}\right)$.

Find the asymptotes:
$$y = \frac{a}{b}x \quad \text{and} \quad y = -\frac{a}{b}x$$
$$y = \frac{1/3}{1/3}x \quad \text{and} \quad y = -\frac{1/3}{1/3}x$$
$$y = x \quad \text{and} \quad y = -x$$

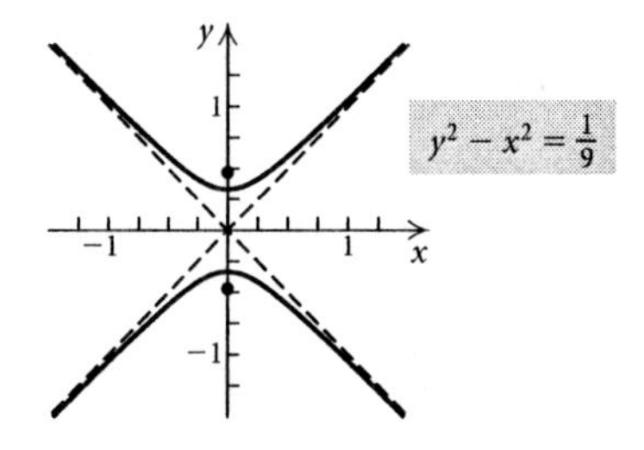

25. Begin by completing the square twice.

$$x^2 - y^2 - 2x - 4y - 4 = 0$$
$$(x^2 - 2x) - (y^2 + 4y) = 4$$
$$(x^2 - 2x + 1) - (y^2 + 4y + 4) = 4 + 1 - 1 \cdot 4$$
$$(x-1)^2 - (y+2)^2 = 1$$
$$\frac{(x-1)^2}{1^2} - \frac{[y-(-2)]^2}{1^2} = 1 \quad \text{Standard form}$$

The center is $(1, -2)$; $a = 1$ and $b = 1$. The transverse axis is horizontal, so the vertices are 1 unit left and right of the center:

$(1-1, -2)$ and $(1+1, -2)$ or $(0, -2)$ and $(2, -2)$

Since $c^2 = a^2 + b^2$, we have $c^2 = 1 + 1 = 2$ and $c = \sqrt{2}$. Then the foci are $\sqrt{2}$ units left and right of the center:

$(1 - \sqrt{2}, -2)$ and $(1 + \sqrt{2}, -2)$.

Find the asymptotes:

$$y - k = \frac{b}{a}(x - h) \text{ and } y - k = -\frac{b}{a}(x - h)$$
$$y - (-2) = \frac{1}{1}(x - 1) \text{ and } y - (-2) = -\frac{1}{1}(x - 1)$$
$$y + 2 = x - 1 \quad \text{and} \quad y + 2 = -(x - 1), \text{ or}$$
$$y = x - 3 \quad \text{and} \quad y = -x - 1$$

Sketch the asymptotes, plot the vertices, and draw the graph.

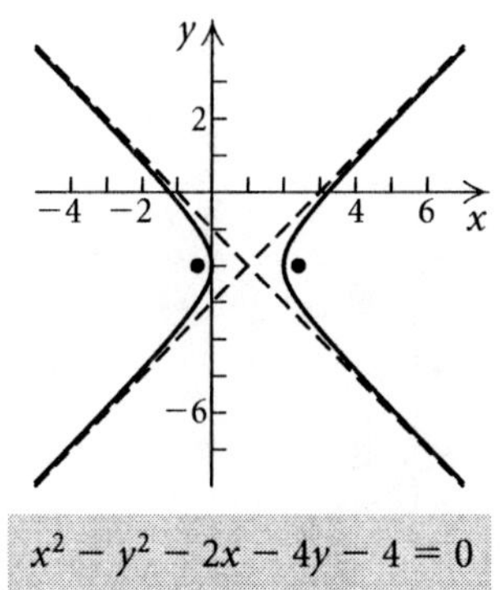

26. Begin by completing the square twice.

$$4x^2 - y^2 + 8x - 4y - 4 = 0$$
$$4(x^2 + 2x) - (y^2 + 4y) = 4$$
$$4(x^2 + 2x + 1) - (y^2 + 4y + 4) = 4 + 4 \cdot 1 - 1 \cdot 4$$
$$4(x+1)^2 - (y+2)^2 = 4$$
$$\frac{(x+1)^2}{1} - \frac{(y+2)^2}{4} = 1$$
$$\frac{[x-(-1)]^2}{1^2} - \frac{[y-(-2)]^2}{2^2} = 1 \quad \text{Standard form}$$

The center is $(-1, -2)$; $a = 1$ and $b = 2$. The transverse axis is horizontal, so the vertices are 1 unit left and right of the center:

$(-1-1, -2)$ and $(-1+1, -2)$ or $(-2, -2)$ and $(0, -2)$

Since $c^2 = a^2 + b^2$, we have $c^2 = 1 + 4 = 5$ and $c = \sqrt{5}$. Then the foci are $\sqrt{5}$ units left and right of the center:

$(-1 - \sqrt{5}, -2)$ and $(1 + \sqrt{5}, -2)$.

Find the asymptotes:

$$y - k = \frac{b}{a}(x - h) \text{ and } y - k = -\frac{b}{a}(x - h)$$
$$y + 2 = \frac{2}{1}(x + 1) \text{ and } y + 2 = -\frac{2}{1}(x + 1)$$
$$y + 2 = 2(x + 1) \text{ and } y + 2 = -2(x + 1), \text{ or}$$
$$y = 2x \quad \text{and} \quad y = -2x - 4$$

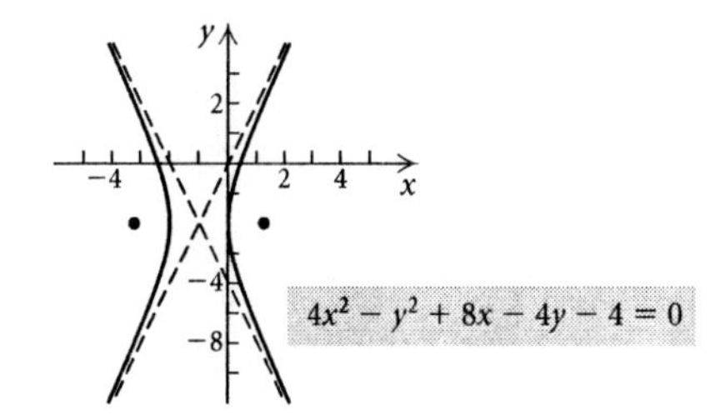

27. Begin by completing the square twice.

$$36x^2 - y^2 - 24x + 6y - 41 = 0$$
$$(36x^2 - 24x) - (y^2 - 6y) = 41$$
$$36\left(x^2 - \frac{2}{3}x\right) - (y^2 - 6y) = 41$$
$$36\left(x^2 - \frac{2}{3}x + \frac{1}{9}\right) - (y^2 - 6y + 9) = 41 + 36 \cdot \frac{1}{9} - 1 \cdot 9$$
$$36\left(x - \frac{1}{3}\right)^2 - (y-3)^2 = 36$$
$$\frac{\left(x - \frac{1}{3}\right)^2}{1} - \frac{(y-3)^2}{36} = 1$$
$$\frac{\left(x - \frac{1}{3}\right)^2}{1^2} - \frac{(y-3)^2}{6^2} = 1 \quad \text{Standard form}$$

The center is $\left(\frac{1}{3}, 3\right)$; $a = 1$ and $b = 6$. The transverse axis is horizontal, so the vertices are 1 unit left and right of the center:

$\left(\frac{1}{3} - 1, 3\right)$ and $\left(\frac{1}{3} + 1, 3\right)$ or $\left(-\frac{2}{3}, 3\right)$ and $\left(\frac{4}{3}, 3\right)$.

Since $c^2 = a^2 + b^2$, we have $c^2 = 1 + 36 = 37$ and $c = \sqrt{37}$. Then the foci are $\sqrt{37}$ units left and right of the center:

$\left(\frac{1}{3} - \sqrt{37}, 3\right)$ and $\left(\frac{1}{3} + \sqrt{37}, 3\right)$.

Find the asymptotes:

$$y - k = \frac{b}{a}(x - h) \quad \text{and} \quad y - k = -\frac{b}{a}(x - h)$$

$$y - 3 = \frac{6}{1}\left(x - \frac{1}{3}\right) \quad \text{and} \quad y - 3 = -\frac{6}{1}\left(x - \frac{1}{3}\right)$$

$$y - 3 = 6\left(x - \frac{1}{3}\right) \quad \text{and} \quad y - 3 = -6\left(x - \frac{1}{3}\right), \text{ or}$$

$$y = 6x + 1 \quad \text{and} \quad y = -6x + 5$$

Sketch the asymptotes, plot the vertices, and draw the graph.

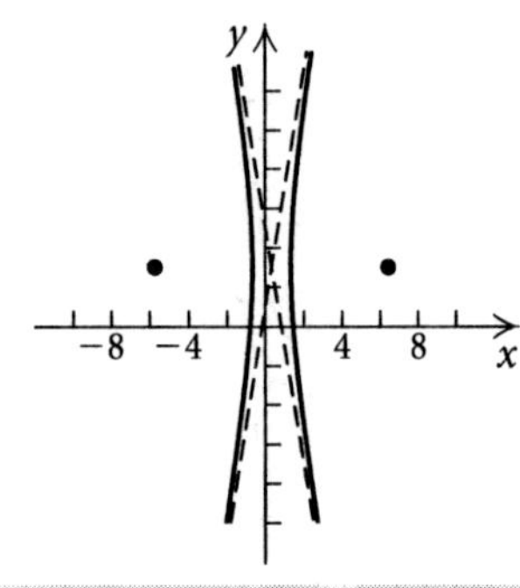

$36x^2 - y^2 - 24x + 6y - 41 = 0$

28. Begin by completing the square twice.

$$9x^2 - 4y^2 + 54x + 8y + 41 = 0$$

$$9(x^2 + 6x) - 4(y^2 - 2y) = -41$$

$$9(x^2 + 6x + 9) - 4(y^2 - 2y + 1) = -41 + 9 \cdot 9 - 4 \cdot 1$$

$$9(x + 3)^2 - 4(y - 1)^2 = 36$$

$$\frac{(x + 3)^2}{4} - \frac{(y - 1)^2}{9} = 1$$

$$\frac{[x - (-3)]^2}{2^2} - \frac{(y - 1)^2}{3^2} = 1 \quad \text{Standard form}$$

The center is $(-3, 1)$; $a = 2$ and $b = 3$. The transverse axis is horizontal, so the vertices are 2 units left and right of the center:

$(-3 - 2, 1)$ and $(-3 + 2, 1)$, or $(-5, 1)$ and $(-1, 1)$.

Since $c^2 = a^2 + b^2$, we have $c^2 = 4 + 9 = 13$ and $c = \sqrt{13}$. Then the foci are $\sqrt{13}$ units left and right of the center:

$(-3 - \sqrt{13}, 1)$ and $(-3 + \sqrt{13}, 1)$.

Find the asymptotes:

$$y - k = \frac{b}{a}(x - h) \text{ and } y - k = -\frac{b}{a}(x - h)$$

$$y - 1 = \frac{3}{2}(x + 3) \text{ and } y - 1 = -\frac{3}{2}(x + 3), \text{ or}$$

$$y = \frac{3}{2}x + \frac{11}{2} \quad \text{and} \quad y = -\frac{3}{2}x - \frac{7}{2}$$

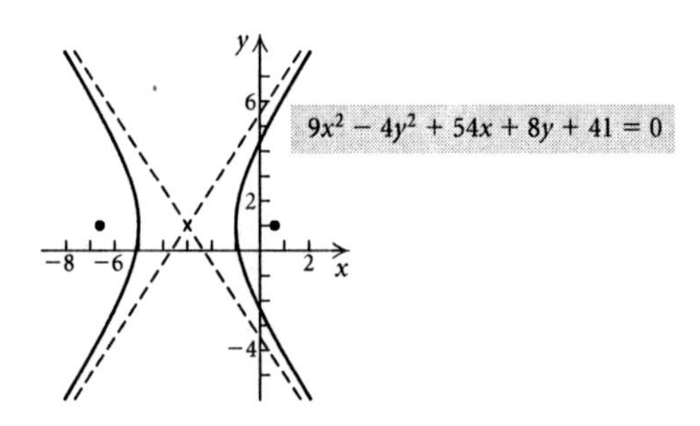

$9x^2 - 4y^2 + 54x + 8y + 41 = 0$

29. Begin by completing the square twice.

$$9y^2 - 4x^2 - 18y + 24x - 63 = 0$$

$$9(y^2 - 2y) - 4(x^2 - 6x) = 63$$

$$9(y^2 - 2y + 1) - 4(x^2 - 6x + 9) = 63 + 9 \cdot 1 - 4 \cdot 9$$

$$9(y - 1)^2 - 4(x - 3)^2 = 36$$

$$\frac{(y - 1)^2}{4} - \frac{(x - 3)^2}{9} = 1$$

$$\frac{(y - 1)^2}{2^2} - \frac{(x - 3)^2}{3^2} = 1 \quad \text{Standard form}$$

The center is $(3, 1)$; $a = 2$ and $b = 3$. The transverse axis is vertical, so the vertices are 2 units below and above the center:

$(3, 1 - 2)$ and $(3, 1 + 2)$, or $(3, -1)$ and $(3, 3)$.

Since $c^2 = a^2 + b^2$, we have $c^2 = 4 + 9 = 13$ and $c = \sqrt{13}$. Then the foci are $\sqrt{13}$ units below and above the center:

$(3, 1 - \sqrt{13})$ and $(3, 1 + \sqrt{13})$.

Find the asymptotes:

$$y - k = \frac{a}{b}(x - h) \text{ and } y - k = -\frac{a}{b}(x - h)$$

$$y - 1 = \frac{2}{3}(x - 3) \text{ and } y - 1 = -\frac{2}{3}(x - 3), \text{ or}$$

$$y = \frac{2}{3}x - 1 \quad \text{and} \quad y = -\frac{2}{3}x + 3$$

Sketch the asymptotes, plot the vertices, and draw the graph.

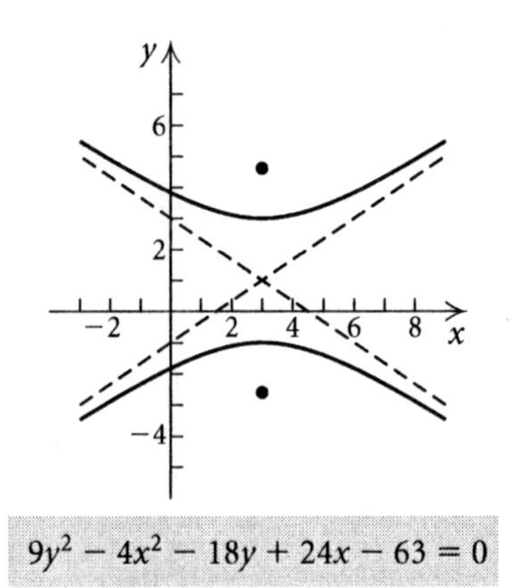

$9y^2 - 4x^2 - 18y + 24x - 63 = 0$

30. Begin by completing the square twice.

$$x^2 - 25y^2 + 6x - 50y = 41$$
$$x^2 + 6x + 9 - 25(y^2 + 2y + 1) = 41 + 9 - 25 \cdot 1$$
$$(x+3)^2 - 25(y+1)^2 = 25$$
$$\frac{(x+3)^2}{25} - \frac{(y+1)^2}{1} = 1$$
$$\frac{[x-(-3)]^2}{5^2} - \frac{[y-(-1)]^2}{1^2} = 1$$

The center is $(-3, -1)$; $a = 5$ and $b = 1$. The transverse axis is horizontal, so the vertices are 5 units left and right of the center:

$(-3-5, -1)$ and $(-3+5, -1)$, or $(-8, -1)$ and $(2, -1)$.

Since $c^2 = a^2 + b^2$, we have $c^2 = 25 + 1 = 26$ and $c = \sqrt{26}$. Then the foci are $\sqrt{26}$ units left and right of the center:

$$(-3 - \sqrt{26}, -1) \text{ and } (-3 + \sqrt{26}, -1).$$

Find the asymptotes:

$$y - k = \frac{b}{a}(x-h) \text{ and } y - k = -\frac{b}{a}(x-h)$$
$$y + 1 = \frac{1}{5}(x+3) \text{ and } y + 1 = -\frac{1}{5}(x+3), \text{ or}$$
$$y = \frac{1}{5}x - \frac{2}{5} \quad \text{and} \quad y = -\frac{1}{5}x - \frac{8}{5}$$

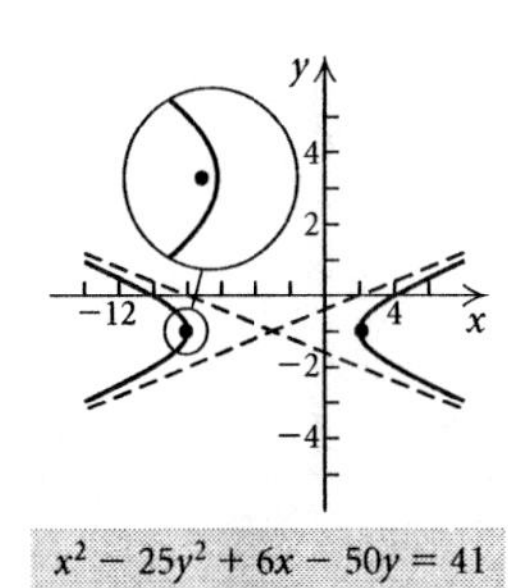

$x^2 - 25y^2 + 6x - 50y = 41$

31. Begin by completing the square twice.

$$x^2 - y^2 - 2x - 4y = 4$$
$$(x^2 - 2x + 1) - (y^2 + 4y + 4) = 4 + 1 - 4$$
$$(x-1)^2 - (y+2)^2 = 1$$
$$\frac{(x-1)^2}{1^2} - \frac{[y-(-2)]^2}{1^2} = 1 \quad \text{Standard form}$$

The center is $(1, -2)$; $a = 1$ and $b = 1$. The transverse axis is horizontal, so the vertices are 1 unit left and right of the center:

$(1-1, -2)$ and $(1+1, -2)$, or $(0, -2)$ and $(2, -2)$.

Since $c^2 = a^2 + b^2$, we have $c^2 = 1 + 1 = 2$ and $c = \sqrt{2}$. Then the foci are $\sqrt{2}$ units left and right of the center:

$$(1 - \sqrt{2}, -2) \text{ and } (1 + \sqrt{2}, -2).$$

Find the asymptotes:

$$y - k = \frac{b}{a}(x-h) \text{ and } \quad y - k = -\frac{b}{a}(x-h)$$
$$y - (-2) = \frac{1}{1}(x-1) \text{ and } y - (-2) = -\frac{1}{1}(x-1)$$
$$y + 2 = x - 1 \quad \text{and} \quad y + 2 = -(x-1), \text{ or}$$
$$y = x - 3 \quad \text{and} \quad y = -x - 1$$

Sketch the asymptotes, plot the vertices, and draw the graph.

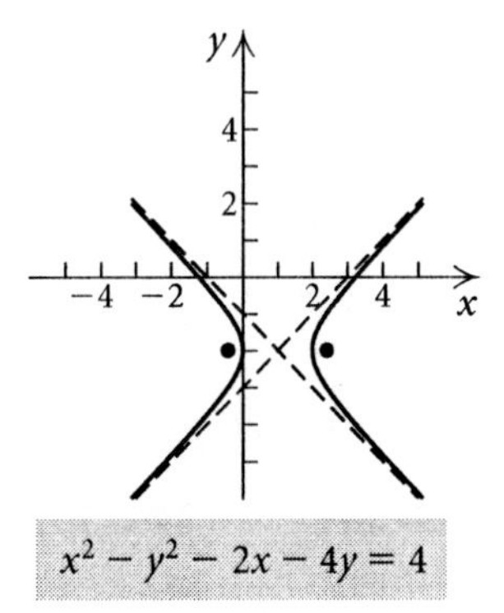

$x^2 - y^2 - 2x - 4y = 4$

32. Begin by completing the square twice.

$$9y^2 - 4x^2 - 54y - 8x + 41 = 0$$
$$9(y^2 - 6y + 9) - 4(x^2 + 2x + 1) = -41 + 9 \cdot 9 - 4 \cdot 1$$
$$9(y-3)^2 - 4(x+1)^2 = 36$$
$$\frac{(y-3)^2}{4} - \frac{(x+1)^2}{9} = 1$$
$$\frac{(y-3)^2}{2^2} - \frac{[x-(-1)]^2}{3^2} = 1 \quad \text{Standard form}$$

The center is $(-1, 3)$; $a = 2$ and $b = 3$. The transverse axis is vertical, so the vertices are 2 units below and above the center:

$(-1, 3-2)$ and $(-1, 3+2)$, or $(-1, 1)$ and $(-1, 5)$.

Since $c^2 = a^2 + b^2$, we have $c^2 = 4 + 9$ and $c = \sqrt{13}$. Then the foci are $\sqrt{13}$ units below and above the center:

$$(-1, 3 - \sqrt{13}) \text{ and } (-1, 3 + \sqrt{13}).$$

Find the asymptotes:

$$y - k = \frac{a}{b}(x-h) \text{ and } y - k = -\frac{a}{b}(x-h)$$
$$y - 3 = \frac{2}{3}(x+1) \text{ and } y - 3 = -\frac{2}{3}(x+1), \text{ or}$$
$$y = \frac{2}{3}x + \frac{11}{3} \quad \text{and} \quad y = -\frac{2}{3}x + \frac{7}{3}$$

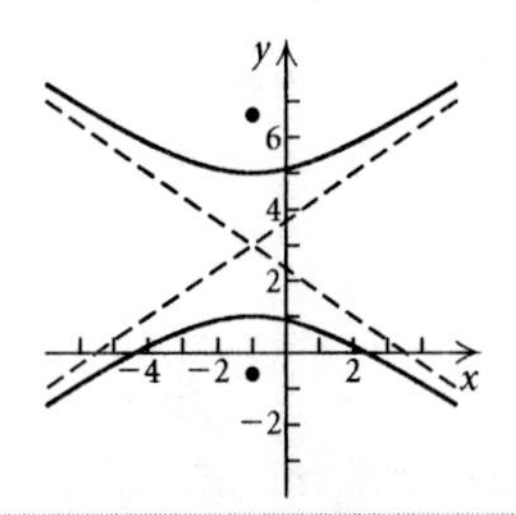

$9y^2 - 4x^2 - 54y - 8x + 41 = 0$

33. Begin by completing the square twice.

$$y^2 - x^2 - 6x - 8y - 29 = 0$$
$$(y^2 - 8y + 16) - (x^2 + 6x + 9) = 29 + 16 - 9$$
$$(y-4)^2 - (x+3)^2 = 36$$
$$\frac{(y-4)^2}{36} - \frac{(x+3)^2}{36} = 1$$
$$\frac{(y-4)^2}{6^2} - \frac{[x-(-3)]^2}{6^2} = 1 \quad \text{Standard form}$$

The center is $(-3, 4)$; $a = 6$ and $b = 6$. The transverse axis is vertical, so the vertices are 6 units below and above the center:

$(-3, 4-6)$ and $(-3, 4+6)$, or $(-3, -2)$ and $(-3, 10)$.

Since $c^2 = a^2 + b^2$, we have $c^2 = 36 + 36 = 72$ and $c = \sqrt{72}$, or $6\sqrt{2}$. Then the foci are $6\sqrt{2}$ units below and above the center:

$(-3, 4 - 6\sqrt{2})$ and $(-3, 4 + 6\sqrt{2})$.

Find the asymptotes:

$$y - k = \frac{a}{b}(x-h) \quad \text{and} \quad y - k = -\frac{a}{b}(x-h)$$
$$y - 4 = \frac{6}{6}(x-(-3)) \quad \text{and} \quad y - 4 = -\frac{6}{6}(x-(-3))$$
$$y - 4 = x + 3 \quad \text{and} \quad y - 4 = -(x+3), \text{ or}$$
$$y = x + 7 \quad \text{and} \quad y = -x + 1$$

Sketch the asymptotes, plot the vertices, and draw the graph.

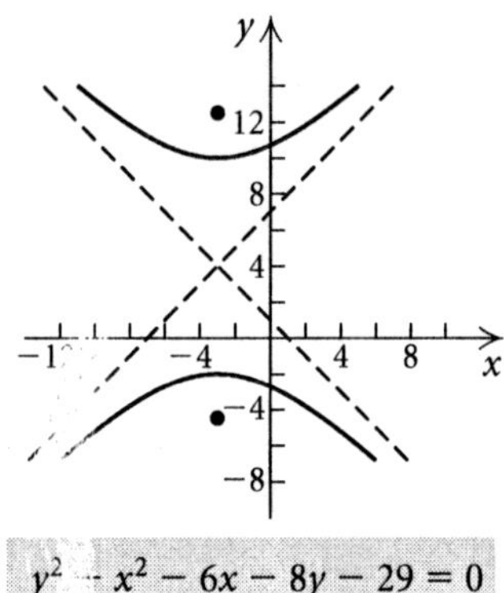

$y^2 - x^2 - 6x - 8y - 29 = 0$

34. Begin by completing the square twice.

$$x^2 - y^2 = 8x - 2y - 13$$
$$x^2 - 8x - y^2 + 2y = 13$$
$$x^2 - 8x + 16 - (y^2 - 2y + 1) = -13 + 16 - 1$$
$$(x-4)^2 - (y-1)^2 = 2$$
$$\frac{(x-4)^2}{(\sqrt{2})^2} - \frac{(y-1)^2}{(\sqrt{2})^2} = 1 \quad \text{Standard form}$$

The center is $(4, 1)$; $a = \sqrt{2}$ and $b = \sqrt{2}$. The transverse axis is horizontal, so the vertices are $\sqrt{2}$ units left and right of the center:

$(4 - \sqrt{2}, 1)$ and $(4 + \sqrt{2}, 1)$.

Since $c^2 = a^2 + b^2$, we have $c^2 = 2 + 2 = 4$ and $c = 2$. Then the foci are 2 units left and right of the center:

$(4-2, 1)$ and $(4+2, 1)$, or $(2, 1)$ and $(6, 1)$.

Find the asymptotes:

$$y - k = \frac{b}{a}(x-h) \quad \text{and} \quad y - k = -\frac{b}{a}(x-h)$$
$$y - 1 = \frac{\sqrt{2}}{\sqrt{2}}(x-4) \quad \text{and} \quad y - 1 = -\frac{\sqrt{2}}{\sqrt{2}}(x-4)$$
$$y - 1 = x - 4 \quad \text{and} \quad y - 1 = -(x-4), \text{ or}$$
$$y = x - 3 \quad \text{and} \quad y = -x + 5$$

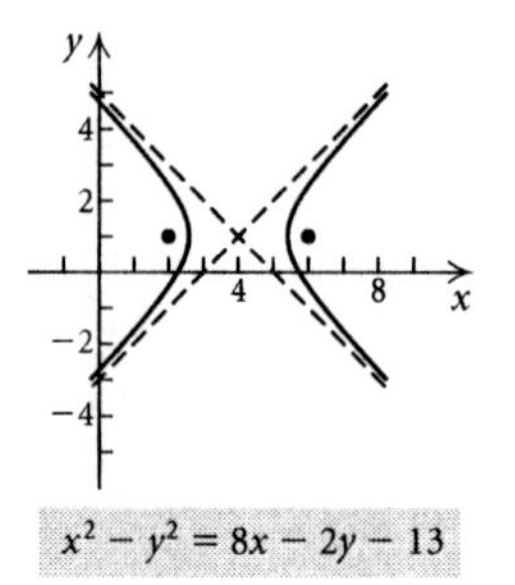

$x^2 - y^2 = 8x - 2y - 13$

35. The hyperbola in Example 3 is wider than the one in Example 2, so the hyperbola in Example 3 has the larger eccentricity.

Compute the eccentricities: In Example 2, $c = 5$ and $a = 4$, so $e = 5/4$, or 1.25. In Example 3, $c = \sqrt{5}$ and $a = 1$, so $e = \sqrt{5}/1 \approx 2.24$. These computations confirm that the hyperbola in Example 3 has the larger eccentricity.

36. Hyperbola (*b*) is wider so it has the larger eccentricity.

37. The center is the midpoint of the segment connecting the vertices:

$$\left(\frac{3-3}{2}, \frac{7+7}{2}\right), \text{ or } (0, 7).$$

The vertices are on the horizontal line $y = 7$, so the transverse axis is horizontal. Since the vertices are 3 units left and right of the center, $a = 3$.

Find c:

$$e = \frac{c}{a} = \frac{5}{3}$$
$$\frac{c}{3} = \frac{5}{3} \quad \text{Substituting 3 for } a$$
$$c = 5$$

Now find b^2:

$$c^2 = a^2 + b^2$$
$$5^2 = 3^2 + b^2$$
$$16 = b^2$$

Write the equation:

$$\frac{(x-h)^2}{a^2} - \frac{(y-k)^2}{b^2} = 1$$
$$\frac{x^2}{9} - \frac{(y-7)^2}{16} = 1$$

38. The center is the midpoint of the segment connecting the vertices:

$$\left(\frac{-1-1}{2}, \frac{3+7}{2}\right), \text{ or } (-1, 5).$$

The vertices are on the vertical line $x = -1$, so the transverse axis is vertical. Since the vertices are 2 units below and above the center, $a = 2$.

Find c:

$$e = \frac{c}{a} = 4$$
$$\frac{c}{2} = 4$$
$$c = 8$$

Now find b^2:

$$c^2 = a^2 + b^2$$
$$64 = 4 + b^2$$
$$60 = b^2$$

The equation is $\frac{(y-5)^2}{4} - \frac{(x+1)^2}{60} = 1$.

39.

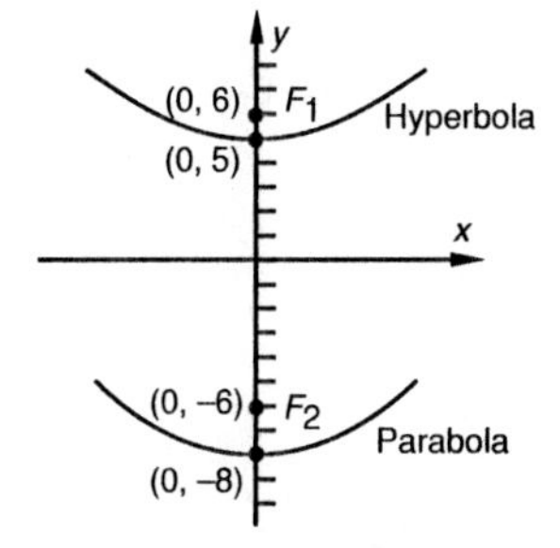

One focus is 6 units above the center of the hyperbola, so $c = 6$. One vertex is 5 units above the center, so $a = 5$. Find b^2:

$$c^2 = a^2 + b^2$$
$$6^2 = 5^2 + b^2$$
$$11 = b^2$$

Write the equation:

$$\frac{y^2}{a^2} - \frac{x^2}{b^2} = 1$$
$$\frac{y^2}{25} - \frac{x^2}{11} = 1$$

40.

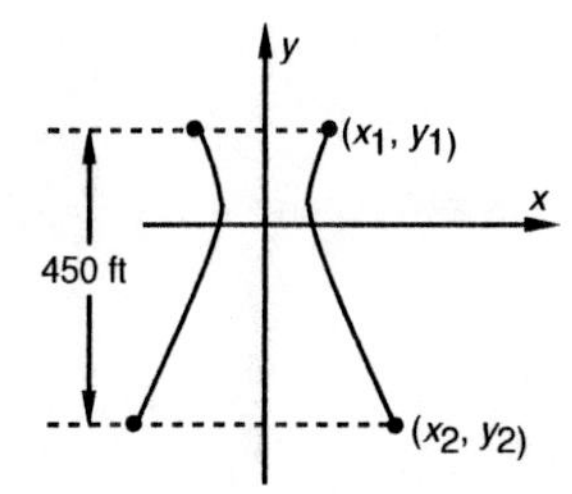

$y_1 = \frac{1}{2}y_2$, so $y_2 + \frac{1}{2}y_2 = 450$. Then $y_2 = 300$ and $y_1 = \frac{1}{2} \cdot 300 = 150$.

Find x_1:

$$\frac{x_1^2}{90^2} - \frac{150^2}{130^2} = 1$$
$$x_1^2 = 90^2\left(1 + \frac{150^2}{130^2}\right)$$
$$x_1 \approx 137.4$$

Then the diameter of the top of the tower is $2x_1 \approx 2(137.4) \approx 275$ ft.

Find x_2:

$$\frac{x_2^2}{90^2} - \frac{300^2}{130^2} = 1$$
$$x_2^2 = 90^2\left(1 + \frac{300^2}{130^2}\right)$$
$$x_2 \approx 226.4$$

Then the diameter of the bottom of the tower is $2x_2 \approx 2(226.4) \approx 453$ ft.

41. Center: $(-1.460, -0.957)$

Vertices: $(-2.360, -0.957)$, $(-0.560, -0.957)$

Asymptotes: $y = -1.429x - 3.043$, $y = 1.429x + 1.129$

42. Center: $(1.023, -2.044)$

Vertices: $(2.07, -2.044)$, $(-0.024, -2.044)$

Asymptotes: $y = x - 3.067$, $y = -x - 1.021$

43. See the figure on page 736 of the text.

44. No, the asymptotes of a hyperbola are not part of the graph of the hyperbola. The coordinates of points on the asymptotes do not satisfy the equation of the hyperbola.

45. a) The graph of $f(x) = 2x - 3$ is shown below.

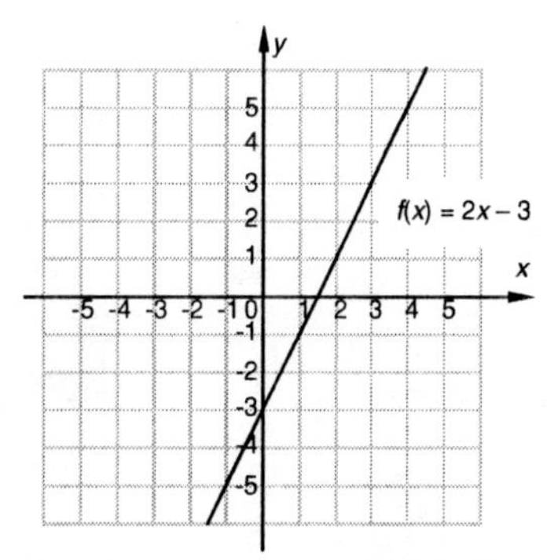

Since there is no horizontal line that crosses the graph more than once, the function is one-to-one.

b) Replace $f(x)$ with y: $y = 2x - 3$

Interchange x and y: $x = 2y - 3$

Solve for y: $x + 3 = 2y$

$$\frac{x+3}{2} = y$$

Replace y with $f^{-1}(x)$: $f^{-1}(x) = \frac{x+3}{2}$

46. a) The graph of $f(x) = x^3 + 2$ is shown below. It passes the horizontal line test, so it is one-to-one.

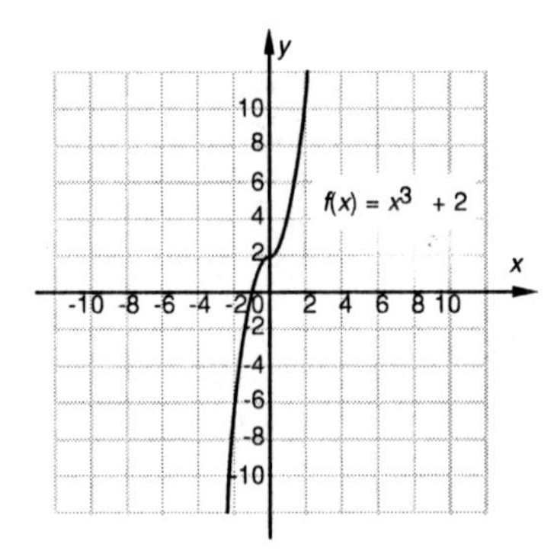

b) Replace $f(x)$ with y: $y = x^3 + 2$

Interchange x and y: $x = y^3 + 2$

Solve for y: $x - 2 = y^3$

$\sqrt[3]{x-2} = y$

Replace y with $f^{-1}(x)$: $f^{-1}(x) = \sqrt[3]{x-2}$

47. a) The graph of $f(x) = \dfrac{5}{x-1}$ is shown below.

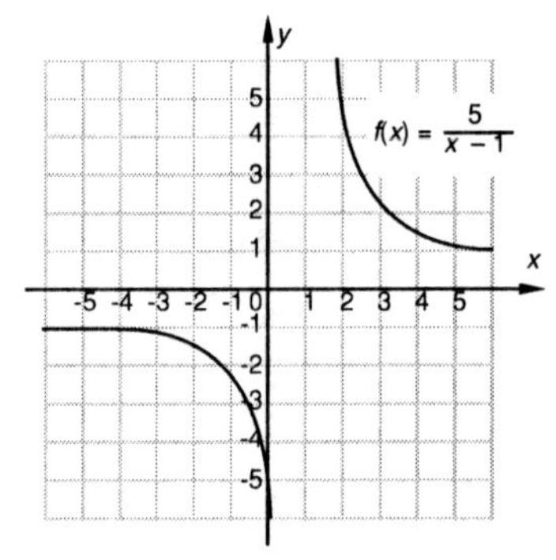

Since there is no horizontal line that crosses the graph more than once, the function is one-to-one.

b) Replace $f(x)$ with y: $y = \dfrac{5}{x-1}$

Interchange x and y: $x = \dfrac{5}{y-1}$

Solve for y: $x(y-1) = 5$

$y - 1 = \dfrac{5}{x}$

$y = \dfrac{5}{x} + 1$

Replace y with $f^{-1}(x)$: $f^{-1} = \dfrac{5}{x} + 1$, or $\dfrac{5+x}{x}$

48. a) The graph of $f(x) = \sqrt{x+4}$ is shown below. It passes the horizontal line test, so it is one-to-one.

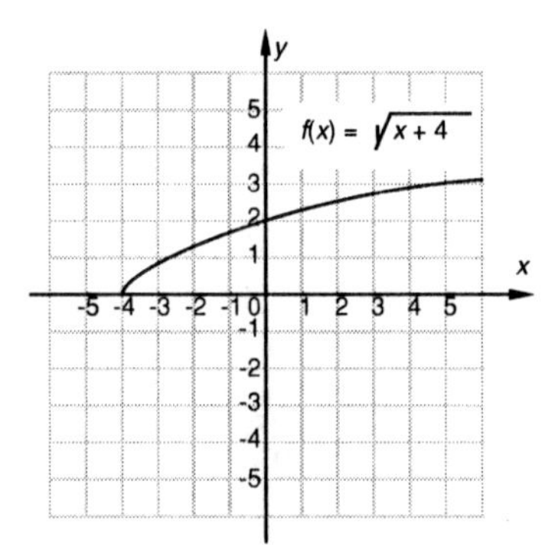

b) Replace $f(x)$ with y: $y = \sqrt{x+4}$

Interchange x and y: $x = \sqrt{y+4}$

Solve for y: $x^2 = y + 4$

$x^2 - 4 = y$

Replace y with $f^{-1}(x)$: $f^{-1}(x) = x^2 - 4$, $x \geq 0$

49.
$$\begin{array}{rl} x + y = 5, & (1) \\ x - y = 7 & (2) \\ \hline 2x \quad = 12 & \text{Adding} \\ x = 6 & \end{array}$$

Back-substitute in either equation (1) or (2) and solve for y. We use equation (1).

$6 + y = 5$

$y = -1$

The solution is $(6, -1)$.

50.
$$\begin{array}{rl} 3x - 2y = 5, & (1) \\ 5x + 2y = 3 & (2) \\ \hline 8x \quad = 8 & \text{Adding} \\ x = 1 & \end{array}$$

Back-substitute and solve for y.

$5 \cdot 1 + 2y = 3$ Using equation (2)

$2y = -2$

$y = -1$

The solution is $(1, -1)$.

51. $2x - 3y = 7$, (1)

$3x + 5y = 1$ (2)

Multiply equation (1) by 5 and equation (2) by 3 and add to eliminate y.

$$\begin{array}{r} 10x - 15y = 35 \\ 9x + 15y = 3 \\ \hline 19x \quad = 38 \\ x = 2 \end{array}$$

Back-substitute and solve for y.

$3 \cdot 2 + 5y = 1$ Using equation (2)

$5y = -5$

$y = -1$

The solution is $(2, -1)$.

52. $3x + 2y = -1$ (1)

$2x + 3y = 6$ (2)

Multiply equation (1) by 3 and equation (2) by -2 and add.

$$\begin{array}{rl} 9x + 6y &= -3 \\ -4x - 6y &= -12 \\ \hline 5x \quad &= -15 \\ x &= -3 \end{array}$$

Back-substitute and solve for y.

$$\begin{aligned} 3(-3) + 2y &= -1 \quad \text{Using equation (1)} \\ 2y &= 8 \\ y &= 4 \end{aligned}$$

The solution is $(-3, 4)$.

53. The center is the midpoint of the segment connecting $(3, -8)$ and $(3, -2)$:

$$\left(\frac{3+3}{2}, \frac{-8-2}{2}\right), \text{ or } (3, -5).$$

The vertices are on the vertical line $x = 3$ and are 3 units above and below the center so the transverse axis is vertical and $a = 3$. Use the equation of an asymptote to find b:

$$\begin{aligned} y - k &= \frac{a}{b}(x - h) \\ y + 5 &= \frac{3}{b}(x - 3) \\ y &= \frac{3}{b}x - \frac{9}{b} - 5 \end{aligned}$$

This equation corresponds to the asymptote $y = 3x - 14$, so $\frac{3}{b} = 3$ and $b = 1$.

Write the equation of the hyperbola:

$$\frac{(y-k)^2}{a^2} - \frac{(x-h)^2}{b^2} = 1$$

$$\frac{(y+5)^2}{9} - \frac{(x-3)^2}{1} = 1$$

54. The center is the midpoint of the segment connecting the vertices:

$$\left(\frac{-9-5}{2}, \frac{4+4}{2}\right), \text{ or } (-7, 4).$$

The vertices are on the horizontal line $y = 4$ and are 2 units left and right of the center, so the transverse axis is horizontal and $a = 2$. Use the equation of an asymptote to find b:

$$\begin{aligned} y - k &= \frac{b}{a}(x - h) \\ y - 4 &= \frac{b}{2}(x + 7) \\ y &= \frac{b}{2}x + \frac{7}{2}b + 4 \end{aligned}$$

This equation corresponds to the asymptote $y = 3x + 25$, so $\frac{b}{2} = 3$ and $b = 6$.

Write the equation of the hyperbola:

$$\frac{(x+7)^2}{4} - \frac{(y-4)^2}{36} = 1$$

55. S and T are the foci of the hyperbola, so $c = 300/2 = 150$.

$200 \text{ microseconds} \cdot \dfrac{0.186 \text{ mi}}{1 \text{ microsecond}} = 37.2$ mi, the difference of the ships' distances from the foci. That is, $2a = 37.2$, so $a = 18.6$.

Find b^2:

$$\begin{aligned} c^2 &= a^2 + b^2 \\ 150^2 &= 18.6^2 + b^2 \\ 22,154.04 &= b^2 \end{aligned}$$

Then the equation of the hyperbola is

$$\frac{x^2}{18.6^2} - \frac{y^2}{22,154.04} = 1, \text{ or } \frac{x^2}{345.96} - \frac{y^2}{22,154.04} = 1.$$

Exercise Set 9.4

1. The correct graph is (e).

2. The correct graph is (a).

3. The correct graph is (c).

4. The correct graph is (f).

5. The correct graph is (b).

6. The correct graph is (d).

7. $x^2 + y^2 = 25,$ (1)

$y - x = 1$ (2)

First solve equation (2) for y.

$y = x + 1$ (3)

Then substitute $x + 1$ for y in equation (1) and solve for x.

$$\begin{aligned} x^2 + y^2 &= 25 \\ x^2 + (x+1)^2 &= 25 \\ x^2 + x^2 + 2x + 1 &= 25 \\ 2x^2 + 2x - 24 &= 0 \\ x^2 + x - 12 &= 0 \quad \text{Multiplying by } \frac{1}{2} \\ (x+4)(x-3) &= 0 \quad \text{Factoring} \end{aligned}$$

$x + 4 = 0$ or $x - 3 = 0$ Principle of zero products

$x = -4$ or $x = 3$

Now substitute these numbers into equation (3) and solve for y.

$y = -4 + 1 = -3$

$y = 3 + 1 = 4$

The pairs $(-4, -3)$ and $(3, 4)$ check, so they are the solutions.

8. $x^2 + y^2 = 100,$
$y - x = 2$

$y = x + 2$

$$\begin{aligned} x^2 + (x+2)^2 &= 100 \\ x^2 + x^2 + 4x + 4 &= 100 \\ 2x^2 + 4x - 96 &= 0 \\ x^2 + 2x - 48 &= 0 \\ (x+8)(x-6) &= 0 \end{aligned}$$

$x = -8$ or $x = 6$

$y = -8 + 2 = -6$

$y = 6 + 2 = 8$

The pairs $(-8, -6)$ and $(6, 8)$ check.

9. $4x^2 + 9y^2 = 36,$ (1)
$3y + 2x = 6$ (2)

First solve equation (2) for y.

$3y = -2x + 6$

$y = -\frac{2}{3}x + 2$ (3)

Then substitute $-\frac{2}{3}x + 2$ for y in equation (1) and solve for x.

$$\begin{aligned} 4x^2 + 9y^2 &= 36 \\ 4x^2 + 9\left(-\frac{2}{3}x + 2\right)^2 &= 36 \\ 4x^2 + 9\left(\frac{4}{9}x^2 - \frac{8}{3}x + 4\right) &= 36 \\ 4x^2 + 4x^2 - 24x + 36 &= 36 \\ 8x^2 - 24x &= 0 \\ x^2 - 3x &= 0 \\ x(x-3) &= 0 \end{aligned}$$

$x = 0$ or $x = 3$

Now substitute these numbers in equation (3) and solve for y.

$y = -\frac{2}{3} \cdot 0 + 2 = 2$

$y = -\frac{2}{3} \cdot 3 + 2 = 0$

The pairs $(0, 2)$ and $(3, 0)$ check, so they are the solutions.

10. $9x^2 + 4y^2 = 36,$
$3x + 2y = 6$

$$\begin{aligned} x &= 2 - \frac{2}{3}y \\ 9\left(2 - \frac{2}{3}y\right)^2 + 4y^2 &= 36 \\ 9\left(4 - \frac{8}{3}y + \frac{4}{9}y^2\right) + 4y^2 &= 36 \\ 36 - 24y + 4y^2 + 4y^2 &= 36 \\ 8y^2 - 24y &= 0 \\ y^2 - 3y &= 0 \\ y(y-3) &= 0 \end{aligned}$$

$y = 0$ or $y = 3$

$x = 2 - \frac{2}{3}(0) = 2$

$x = 2 - \frac{2}{3}(3) = 0$

The pairs $(2, 0)$ and $(0, 3)$ check.

11. $x^2 + y^2 = 25,$ (1)
$y^2 = x + 5$ (2)

We substitute $x + 5$ for y^2 in equation (1) and solve for x.

$$\begin{aligned} x^2 + y^2 &= 25 \\ x^2 + (x+5) &= 25 \\ x^2 + x - 20 &= 0 \\ (x+5)(x-4) &= 0 \end{aligned}$$

$x + 5 = 0$ or $x - 4 = 0$

$x = -5$ or $x = 4$

We substitute these numbers for x in either equation (1) or equation (2) and solve for y. Here we use equation (2).

$y^2 = -5 + 5 = 0$ and $y = 0$.

$y^2 = 4 + 5 = 9$ and $y = \pm 3$.

The pairs $(-5, 0)$, $(4, 3)$ and $(4, -3)$ check. They are the solutions.

12. $y = x^2,$
$x = y^2$

$x = (x^2)^2$

$x = x^4$

$0 = x^4 - x$

$0 = x(x^3 - 1)$

$0 = x(x-1)(x^2 + x + 1)$

$x = 0$ or $x = 1$ or $x = \frac{-1 \pm \sqrt{1^2 - 4 \cdot 1 \cdot 1}}{2}$

$x = 0$ or $x = 1$ or $x = -\frac{1}{2} \pm \frac{\sqrt{3}}{2}i$

$y = 0^2 = 0$

$y = 1^2 = 1$

$y = \left(-\frac{1}{2} + \frac{\sqrt{3}}{2}i\right)^2 = -\frac{1}{2} - \frac{\sqrt{3}}{2}i$

$y = \left(-\frac{1}{2} - \frac{\sqrt{3}}{2}i\right)^2 = -\frac{1}{2} + \frac{\sqrt{3}}{2}i$

The pairs $(0, 0)$, $(1, 1)$, $\left(-\frac{1}{2} + \frac{\sqrt{3}}{2}i, -\frac{1}{2} - \frac{\sqrt{3}}{2}i\right)$, and $\left(-\frac{1}{2} - \frac{\sqrt{3}}{2}i, -\frac{1}{2} + \frac{\sqrt{3}}{2}i\right)$ check.

13. $x^2 + y^2 = 9,$ (1)

$x^2 - y^2 = 9$ (2)

Here we use the elimination method.

$$\begin{array}{rll} x^2 + y^2 = & 9 & (1) \\ x^2 - y^2 = & 9 & (2) \\ \hline 2x^2 \quad = & 18 & \text{Adding} \\ x^2 = & 9 & \\ x = & \pm 3 & \end{array}$$

If $x = 3$, $x^2 = 9$, and if $x = -3$, $x^2 = 9$, so substituting 3 or -3 in equation (1) gives us

$$x^2 + y^2 = 9$$
$$9 + y^2 = 9$$
$$y^2 = 0$$
$$y = 0.$$

The pairs $(3, 0)$ and $(-3, 0)$ check. They are the solutions.

14. $y^2 - 4x^2 = 4$ (1)

$4x^2 + y^2 = 4$ (2)

$$\begin{array}{rll} -4x^2 + y^2 = 4 & (1) \\ 4x^2 + y^2 = 4 & (2) \\ \hline 2y^2 = 8 & \text{Adding} \\ y^2 = 4 & \\ y = \pm 2 & \end{array}$$

Substitute for y in equation (2).

$$4x^2 + 4 = 4$$
$$4x^2 = 0$$
$$x = 0$$

The pairs $(0, 2)$ and $(0, -2)$ check.

15. $y^2 - x^2 = 9$ (1)

$2x - 3 = y$ (2)

Substitute $2x - 3$ for y in equation (1) and solve for x.

$$y^2 - x^2 = 9$$
$$(2x - 3)^2 - x^2 = 9$$
$$4x^2 - 12x + 9 - x^2 = 9$$
$$3x^2 - 12x = 0$$
$$x^2 - 4x = 0$$
$$x(x - 4) = 0$$

$x = 0$ or $x = 4$

Now substitute these numbers into equation (2) and solve for y.

If $x = 0$, $y = 2 \cdot 0 - 3 = -3$.

If $x = 4$, $y = 2 \cdot 4 - 3 = 5$.

The pairs $(0, -3)$ and $(4, 5)$ check. They are the solutions.

16. $x + y = -6,$

$xy = -7$

$$y = -x - 6$$
$$x(-x - 6) = -7$$
$$-x^2 - 6x = -7$$
$$0 = x^2 + 6x - 7$$
$$0 = (x + 7)(x - 1)$$

$x = -7$ or $x = 1$

$y = -(-7) - 6 = 1$

$y = -1 - 6 = -7$

The pairs $(-7, 1)$ and $(1, -7)$ check.

17. $y^2 = x + 3,$ (1)

$2y = x + 4$ (2)

First solve equation (2) for x.

$2y - 4 = x$ (3)

Then substitute $2y - 4$ for x in equation (1) and solve for y.

$$y^2 = x + 3$$
$$y^2 = (2y - 4) + 3$$
$$y^2 = 2y - 1$$
$$y^2 - 2y + 1 = 0$$
$$(y - 1)(y - 1) = 0$$

$y - 1 = 0$ or $y - 1 = 0$

$y = 1$ or $y = 1$

Now substitute 1 for y in equation (3) and solve for x.

$$2 \cdot 1 - 4 = x$$
$$-2 = x$$

The pair $(-2, 1)$ checks. It is the solution.

18. $y = x^2,$

$3x = y + 2$

$y = 3x - 2$

$$3x - 2 = x^2$$
$$0 = x^2 - 3x + 2$$
$$0 = (x - 2)(x - 1)$$

$x = 2$ or $x = 1$

$y = 3 \cdot 2 - 2 = 4$

$y = 3 \cdot 1 - 2 = 1$

The pairs $(2, 4)$ and $(1, 1)$ check.

19. $x^2 + y^2 = 25,$ (1)

$xy = 12$ (2)

First we solve equation (2) for y.

$$xy = 12$$
$$y = \frac{12}{x}$$

Then we substitute $\frac{12}{x}$ for y in equation (1) and solve for x.

$$x^2 + y^2 = 25$$
$$x^2 + \left(\frac{12}{x}\right)^2 = 25$$
$$x^2 + \frac{144}{x^2} = 25$$
$$x^4 + 144 = 25x^2 \quad \text{Multiplying by } x^2$$
$$x^4 - 25x^2 + 144 = 0$$
$$u^2 - 25u + 144 = 0 \quad \text{Letting } u = x^2$$
$$(u - 9)(u - 16) = 0$$
$$u = 9 \quad \text{or} \quad u = 16$$

We now substitute x^2 for u and solve for x.

$$x^2 = 9 \quad \text{or} \quad x^2 = 16$$
$$x = \pm 3 \quad \text{or} \quad x = \pm 4$$

Since $y = 12/x$, if $x = 3$, $y = 4$; if $x = -3$, $y = -4$; if $x = 4$, $y = 3$; and if $x = -4$, $y = -3$. The pairs $(3, 4)$, $(-3, -4)$, $(4, 3)$, and $(-4, -3)$ check. They are the solutions.

20. $x^2 - y^2 = 16$, (1)
$x + y^2 = 4$ (2)

$$\begin{array}{rl} x^2 - y^2 = 16 & (1) \\ x + y^2 = 4 & (2) \\ \hline x^2 + x = 20 & \text{Adding} \end{array}$$

$$x^2 + x - 20 = 0$$
$$(x + 5)(x - 4) = 0$$
$$x = -5 \quad \text{or} \quad x = 4$$
$$y^2 = 4 - x \quad \text{Solving equation (2) for } y^2$$
$$y^2 = 4 - (-5) = 9 \text{ and } y = \pm 3$$
$$y^2 = 4 - 4 = 0 \text{ and } y = 0$$

The pairs $(-5, 3)$, $(-5, -3)$, and $(4, 0)$ check.

21. $x^2 + y^2 = 4$ (1)
$16x^2 + 9y^2 = 144$ (2)

$$\begin{array}{rll} -9x^2 - 9y^2 = & -36 & \text{Multiplying (1) by } -9 \\ 16x^2 + 9y^2 = & 144 & \\ \hline 7x^2 \qquad = & 108 & \text{Adding} \end{array}$$

$$x^2 = \frac{108}{7}$$
$$x = \pm\sqrt{\frac{108}{7}} = \pm 6\sqrt{\frac{3}{7}}$$
$$x = \pm\frac{6\sqrt{21}}{7} \quad \text{Rationalizing the denominator}$$

Substituting $\frac{6\sqrt{21}}{7}$ or $-\frac{6\sqrt{21}}{7}$ for x in equation (1) gives us

$$\frac{36 \cdot 21}{49} + y^2 = 4$$
$$y^2 = 4 - \frac{108}{7}$$
$$y^2 = -\frac{80}{7}$$
$$y = \pm\sqrt{-\frac{80}{7}} = \pm 4i\sqrt{\frac{5}{7}}$$
$$y = \pm\frac{4i\sqrt{35}}{7}. \quad \text{Rationalizing the denominator}$$

The pairs $\left(\frac{6\sqrt{21}}{7}, \frac{4i\sqrt{35}}{7}\right)$, $\left(\frac{6\sqrt{21}}{7}, -\frac{4i\sqrt{35}}{7}\right)$, $\left(-\frac{6\sqrt{21}}{7}, \frac{4i\sqrt{35}}{7}\right)$, and $\left(-\frac{6\sqrt{21}}{7}, -\frac{4i\sqrt{35}}{7}\right)$ check. They are the solutions.

22. $x^2 + y^2 = 25$, (1)
$25x^2 + 16y^2 = 400$ (2)

$$\begin{array}{rll} -16x^2 - 16y^2 = & -400 & \text{Multiplying (1) by } -16 \\ 25x^2 + 16y^2 = & 400 & \\ \hline 9x^2 \qquad = & 0 & \text{Adding} \\ x = & 0 & \end{array}$$

$$0^2 + y^2 = 25 \quad \text{Substituting in (1)}$$
$$y = \pm 5$$

The pairs $(0, 5)$ and $(0, -5)$ check.

23. $x^2 + 4y^2 = 25$, (1)
$x + 2y = 7$ (2)

First solve equation (2) for x.

$x = -2y + 7$ (3)

Then substitute $-2y + 7$ for x in equation (1) and solve for y.

$$x^2 + 4y^2 = 25$$
$$(-2y + 7)^2 + 4y^2 = 25$$
$$4y^2 - 28y + 49 + 4y^2 = 25$$
$$8y^2 - 28y + 24 = 0$$
$$2y^2 - 7y + 6 = 0$$
$$(2y - 3)(y - 2) = 0$$
$$y = \frac{3}{2} \text{ or } y = 2$$

Now substitute these numbers in equation (3) and solve for x.

$$x = -2 \cdot \frac{3}{2} + 7 = 4$$
$$x = -2 \cdot 2 + 7 = 3$$

The pairs $\left(4, \frac{3}{2}\right)$ and $(3, 2)$ check, so they are the solutions.

24. $y^2 - x^2 = 16,$

$2x - y = 1$

$y = 2x - 1$

$$(2x-1)^2 - x^2 = 16$$
$$4x^2 - 4x + 1 - x^2 = 16$$
$$3x^2 - 4x - 15 = 0$$
$$(3x+5)(x-3) = 0$$
$$x = -\frac{5}{3} \text{ or } x = 3$$
$$y = 2\left(-\frac{5}{3}\right) - 1 = -\frac{13}{3}$$
$$y = 2(3) - 1 = 5$$

The pairs $\left(-\frac{5}{3}, -\frac{13}{3}\right)$ and $(3,5)$ check.

25. $x^2 - xy + 3y^2 = 27,$ (1)

$x - y = 2$ (2)

First solve equation (2) for y.

$x - 2 = y$ (3)

Then substitute $x-2$ for y in equation (1) and solve for x.

$$x^2 - xy + 3y^2 = 27$$
$$x^2 - x(x-2) + 3(x-2)^2 = 27$$
$$x^2 - x^2 + 2x + 3x^2 - 12x + 12 = 27$$
$$3x^2 - 10x - 15 = 0$$
$$x = \frac{-(-10) \pm \sqrt{(-10)^2 - 4(3)(-15)}}{2\cdot 3}$$
$$x = \frac{10 \pm \sqrt{100+180}}{6} = \frac{10 \pm \sqrt{280}}{6}$$
$$x = \frac{10 \pm 2\sqrt{70}}{6} = \frac{5 \pm \sqrt{70}}{3}$$

Now substitute these numbers in equation (3) and solve for y.

$$y = \frac{5+\sqrt{70}}{3} - 2 = \frac{-1+\sqrt{70}}{3}$$
$$y = \frac{5-\sqrt{70}}{3} - 2 = \frac{-1-\sqrt{70}}{3}$$

The pairs $\left(\frac{5+\sqrt{70}}{3}, \frac{-1+\sqrt{70}}{3}\right)$ and $\left(\frac{5-\sqrt{70}}{3}, \frac{-1-\sqrt{70}}{3}\right)$ check, so they are the solutions.

26. $2y^2 + xy + x^2 = 7,$

$x - 2y = 5$

$x = 2y + 5$

$$2y^2 + (2y+5)y + (2y+5)^2 = 7$$
$$2y^2 + 2y^2 + 5y + 4y^2 + 20y + 25 = 7$$
$$8y^2 + 25y + 18 = 0$$
$$(8y+9)(y+2) = 0$$
$$y = -\frac{9}{8} \text{ or } y = -2$$
$$x = 2\left(-\frac{9}{8}\right) + 5 = \frac{11}{4}$$
$$x = 2(-2) + 5 = 1$$

The pairs $\left(\frac{11}{4}, -\frac{9}{8}\right)$ and $(1,-2)$ check.

27. $x^2 + y^2 = 16,$ or $x^2 + y^2 = 16,$ (1)

$y^2 - 2x^2 = 10$ $\quad -2x^2 + y^2 = 10$ (2)

Here we use the elimination method.

$$\begin{aligned} 2x^2 + 2y^2 &= 32 \quad \text{Multiplying (1) by 2} \\ -2x^2 + y^2 &= 10 \\ \hline 3y^2 &= 42 \quad \text{Adding} \\ y^2 &= 14 \\ y &= \pm\sqrt{14} \end{aligned}$$

Substituting $\sqrt{14}$ or $-\sqrt{14}$ for y in equation (1) gives us

$$x^2 + 14 = 16$$
$$x^2 = 2$$
$$x = \pm\sqrt{2}$$

The pairs $(-\sqrt{2}, -\sqrt{14})$, $(-\sqrt{2}, \sqrt{14})$, $(\sqrt{2}, -\sqrt{14})$, and $(\sqrt{2}, \sqrt{14})$ check. They are the solutions.

28.

$$\begin{aligned} x^2 + y^2 &= 14, \quad (1) \\ x^2 - y^2 &= 4 \quad (2) \\ \hline 2x^2 &= 18 \quad \text{Adding} \\ x^2 &= 9 \\ x &= \pm 3 \end{aligned}$$

$9 + y^2 = 14$ Substituting in equation (1)

$$y^2 = 5$$
$$y = \pm\sqrt{5}$$

The pairs $(-3, -\sqrt{5})$, $(-3, \sqrt{5})$, $(3, -\sqrt{5})$, and $(3, \sqrt{5})$ check.

29. $x^2 + y^2 = 5,$ (1)

$xy = 2$ (2)

First we solve equation (2) for y.

$$xy = 2$$
$$y = \frac{2}{x}$$

Then we substitute $\frac{2}{x}$ for y in equation (1) and solve for x.

$$x^2 + y^2 = 5$$
$$x^2 + \left(\frac{2}{x}\right)^2 = 5$$
$$x^2 + \frac{4}{x^2} = 5$$
$$x^4 + 4 = 5x^2 \quad \text{Multiplying by } x^2$$
$$x^4 - 5x^2 + 4 = 0$$
$$u^2 - 5u + 4 = 0 \quad \text{Letting } u = x^2$$
$$(u-4)(u-1) = 0$$
$$u = 4 \text{ or } u = 1$$

We now substitute x^2 for u and solve for x.

$x^2 = 4$ or $x^2 = 1$

$x = \pm 2$ $\quad x = \pm 1$

Since $y = 2/x$, if $x = 2$, $y = 1$; if $x = -2$, $y = -1$; if $x = 1$, $y = 2$; and if $x = -1$, $y = -2$. The pairs $(2, 1)$, $(-2, -1)$, $(1, 2)$, and $(-1, -2)$ check. They are the solutions.

30. $x^2 + y^2 = 20$,

$xy = 8$

$y = \frac{8}{x}$

$x^2 + \left(\frac{8}{x}\right)^2 = 20$

$x^2 + \frac{64}{x^2} = 20$

$x^4 + 64 = 20x^2$

$x^4 - 20x^2 + 64 = 0$

$u^2 - 20u + 64 = 0$ $\quad$ Letting $u = x^2$

$(u - 16)(u - 4) = 0$

$u = 16$ or $u = 4$

$x^2 = 16$ or $x^2 = 4$

$x = \pm 4$ or $x = \pm 2$

$y = 8/x$, so if $x = 4$, $y = 2$; if $x = -4$, $y = -2$; if $x = 2$, $y = 4$; if $x = -2$, $y = -4$. The pairs $(4, 2)$, $(-4, -2)$, $(2, 4)$, and $(-2, -4)$ check.

31. $3x + y = 7$ $\quad (1)$

$4x^2 + 5y = 56$ $\quad (2)$

First solve equation (1) for y.

$3x + y = 7$

$y = 7 - 3x$ $\quad (3)$

Next substitute $7 - 3x$ for y in equation (2) and solve for x.

$4x^2 + 5y = 56$

$4x^2 + 5(7 - 3x) = 56$

$4x^2 + 35 - 15x = 56$

$4x^2 - 15x - 21 = 0$

Using the quadratic formula, we find that

$x = \frac{15 - \sqrt{561}}{8}$ or $x = \frac{15 + \sqrt{561}}{8}$.

Now substitute these numbers into equation (3) and solve for y.

If $x = \frac{15 - \sqrt{561}}{8}$, $y = 7 - 3\left(\frac{15 - \sqrt{561}}{8}\right)$, or $\frac{11 + 3\sqrt{561}}{8}$.

If $x = \frac{15 + \sqrt{561}}{8}$, $y = 7 - 3\left(\frac{15 + \sqrt{561}}{8}\right)$, or $\frac{11 - 3\sqrt{561}}{8}$.

The pairs $\left(\frac{15 - \sqrt{561}}{8}, \frac{11 + 3\sqrt{561}}{8}\right)$ and $\left(\frac{15 + \sqrt{561}}{8}, \frac{11 - 3\sqrt{561}}{8}\right)$ check and are the solutions.

32. $2y^2 + xy = 5$,

$4y + x = 7$

$x = -4y + 7$

$2y^2 + (-4y + 7)y = 5$

$2y^2 - 4y^2 + 7y = 5$

$0 = 2y^2 - 7y + 5$

$0 = (2y - 5)(y - 1)$

$y = \frac{5}{2}$ or $y = 1$

$x = -4\left(\frac{5}{2}\right) + 7 = -3$

$x = -4(1) + 7 = 3$

The pairs $\left(-3, \frac{5}{2}\right)$ and $(3, 1)$ check.

33. $a + b = 7$, $\quad (1)$

$ab = 4$ $\quad (2)$

First solve equation (1) for a.

$a = -b + 7$ $\quad (3)$

Then substitute $-b + 7$ for a in equation (2) and solve for b.

$(-b + 7)b = 4$

$-b^2 + 7b = 4$

$0 = b^2 - 7b + 4$

$b = \frac{-(-7) \pm \sqrt{(-7)^2 - 4 \cdot 1 \cdot 4}}{2 \cdot 1}$

$b = \frac{7 \pm \sqrt{33}}{2}$

Now substitute these numbers in equation (3) and solve for a.

$a = -\left(\frac{7 + \sqrt{33}}{2}\right) + 7 = \frac{7 - \sqrt{33}}{2}$

$a = -\left(\frac{7 - \sqrt{33}}{2}\right) + 7 = \frac{7 + \sqrt{33}}{2}$

The pairs $\left(\frac{7 - \sqrt{33}}{2}, \frac{7 + \sqrt{33}}{2}\right)$ and $\left(\frac{7 + \sqrt{33}}{2}, \frac{7 - \sqrt{33}}{2}\right)$ check, so they are the solutions.

34. $p + q = -4$,

$pq = -5$

$p = -q - 4$

$(-q - 4)q = -5$

$-q^2 - 4q = -5$

$0 = q^2 + 4q - 5$

$0 = (q + 5)(q - 1)$

$q = -5$ or $q = 1$

$p = -(-5) - 4 = 1$

$p = -1 - 4 = -5$

The pairs $(1, -5)$ and $(-5, 1)$ check.

35. $x^2 + y^2 = 13, \quad (1)$

$xy = 6 \quad (2)$

First we solve Equation (2) for y.

$$xy = 6$$
$$y = \frac{6}{x}$$

Then we substitute $\frac{6}{x}$ for y in equation (1) and solve for x.

$$x^2 + y^2 = 13$$
$$x^2 + \left(\frac{6}{x}\right)^2 = 13$$
$$x^2 + \frac{36}{x^2} = 13$$
$$x^4 + 36 = 13x^2 \quad \text{Multiplying by } x^2$$
$$x^4 - 13x^2 + 36 = 0$$
$$u^2 - 13u + 36 = 0 \quad \text{Letting } u = x^2$$
$$(u - 9)(u - 4) = 0$$
$$u = 9 \quad \text{or} \quad u = 4$$

We now substitute x^2 for u and solve for x.

$$x^2 = 9 \quad \text{or} \quad x^2 = 4$$
$$x = \pm 3 \quad \text{or} \quad x = \pm 2$$

Since $y = 6/x$, if $x = 3$, $y = 2$; if $x = -3$, $y = -2$; if $x = 2$, $y = 3$; and if $x = -2$, $y = -3$. The pairs $(3, 2)$, $(-3, -2)$, $(2, 3)$, and $(-2, -3)$ check. They are the solutions.

36. $x^2 + 4y^2 = 20,$

$xy = 4$

$$y = \frac{4}{x}$$
$$x^2 + 4\left(\frac{4}{x}\right)^2 = 20$$
$$x^2 + \frac{64}{x^2} = 20$$
$$x^4 + 64 = 20x^2$$
$$x^4 - 20x^2 + 64 = 0$$
$$u^2 - 20u + 64 = 0 \quad \text{Letting } u = x^2$$
$$(u - 16)(u - 4) = 0$$
$$u = 16 \quad \text{or} \quad u = 4$$
$$x^2 = 16 \quad \text{or} \quad x^2 = 4$$
$$x = \pm 4 \quad \text{or} \quad x = \pm 2$$

$y = 4/x$, so if $x = 4$, $y = 1$; if $x = -4$, $y = -1$; if $x = 2$, $y = 2$; and if $x = -2$, $y = -2$. The pairs $(4, 1)$, $(-4, -1)$, $(2, 2)$, and $(-2, -2)$ check.

37. $x^2 + y^2 + 6y + 5 = 0 \quad (1)$

$x^2 + y^2 - 2x - 8 = 0 \quad (2)$

Using the elimination method, multiply equation (2) by -1 and add the result to equation (1).

$$\begin{array}{rl} x^2 + y^2 + 6y + 5 = 0 & (1) \\ -x^2 - y^2 + 2x + 8 = 0 & (2) \\ \hline 2x + 6y + 13 = 0 & (3) \end{array}$$

Solve equation (3) for x.

$$2x + 6y + 13 = 0$$
$$2x = -6y - 13$$
$$x = \frac{-6y - 13}{2}$$

Substitute $\frac{-6y - 13}{2}$ for x in equation (1) and solve for y.

$$x^2 + y^2 + 6y + 5 = 0$$
$$\left(\frac{-6y - 13}{2}\right)^2 + y^2 + 6y + 5 = 0$$
$$\frac{36y^2 + 156y + 169}{4} + y^2 + 6y + 5 = 0$$
$$36y^2 + 156y + 169 + 4y^2 + 24y + 20 = 0$$
$$40y^2 + 180y + 189 = 0$$

Using the quadratic formula, we find that $y = \frac{-45 \pm 3\sqrt{15}}{20}$. Substitute $\frac{-45 \pm 3\sqrt{15}}{20}$ for y in $x = \frac{-6y - 13}{2}$ and solve for x.

If $y = \frac{-45 + 3\sqrt{15}}{20}$, then

$$x = \frac{-6\left(\frac{-45 + 3\sqrt{15}}{20}\right) - 13}{2} = \frac{5 - 9\sqrt{15}}{20}.$$

If $y = \frac{-45 - 3\sqrt{15}}{20}$, then

$$x = \frac{-6\left(\frac{-45 - 3\sqrt{15}}{20}\right) - 13}{2} = \frac{5 + 9\sqrt{15}}{20}.$$

The pairs $\left(\frac{5 + 9\sqrt{15}}{20}, \frac{-45 - 3\sqrt{15}}{20}\right)$ and $\left(\frac{5 - 9\sqrt{15}}{20}, \frac{-45 + 3\sqrt{15}}{20}\right)$ check and are the solutions.

38. $2xy + 3y^2 = 7, \quad (1)$

$3xy - 2y^2 = 4 \quad (2)$

$$\begin{array}{rl} 6xy + 9y^2 = 21 & \text{Multiplying (1) by 3} \\ -6xy + 4y^2 = -8 & \text{Multiplying (2) by } -2 \\ \hline 13y^2 = 13 & \\ y^2 = 1 & \\ y = \pm 1 & \end{array}$$

Substitute for y in equation (1) and solve for x.

When $y = 1$: $\quad 2 \cdot x \cdot 1 + 3 \cdot 1^2 = 7$

$$2x = 4$$
$$x = 2$$

$$\begin{aligned} \text{When } y = -1: \quad 2 \cdot x \cdot (-1) + 3(-1)^2 &= 7 \\ -2x &= 4 \\ x &= -2 \end{aligned}$$

The pairs $(2, 1)$ and $(-2, -1)$ check.

39. $2a + b = 1$, (1)

$b = 4 - a^2$ (2)

Equation (2) is already solved for b. Substitute $4 - a^2$ for b in equation (1) and solve for a.

$$2a + 4 - a^2 = 1$$
$$0 = a^2 - 2a - 3$$
$$0 = (a - 3)(a + 1)$$
$$a = 3 \quad \text{or} \quad a = -1$$

Substitute these numbers in equation (2) and solve for b.

$$b = 4 - 3^2 = -5$$
$$b = 4 - (-1)^2 = 3$$

The pairs $(3, -5)$ and $(-1, 3)$ check. They are the solutions.

40. $4x^2 + 9y^2 = 36$,

$x + 3y = 3$

$$x = -3y + 3$$

$$\begin{aligned} 4(-3y + 3)^2 + 9y^2 &= 36 \\ 4(9y^2 - 18y + 9) + 9y^2 &= 36 \\ 36y^2 - 72y + 36 + 9y^2 &= 36 \\ 45y^2 - 72y &= 0 \\ 5y^2 - 8y &= 0 \\ y(5y - 8) &= 0 \end{aligned}$$

$$y = 0 \text{ or } y = \frac{8}{5}$$
$$x = -3 \cdot 0 + 3 = 3$$
$$x = -3\left(\frac{8}{5}\right) + 3 = -\frac{9}{5}$$

The pairs $(3, 0)$ and $\left(-\frac{9}{5}, \frac{8}{5}\right)$ check.

41. $a^2 + b^2 = 89$, (1)

$a - b = 3$ (2)

First solve equation (2) for a.

$a = b + 3$ (3)

Then substitute $b + 3$ for a in equation (1) and solve for b.

$$\begin{aligned} (b + 3)^2 + b^2 &= 89 \\ b^2 + 6b + 9 + b^2 &= 89 \\ 2b^2 + 6b - 80 &= 0 \\ b^2 + 3b - 40 &= 0 \\ (b + 8)(b - 5) &= 0 \end{aligned}$$

$$b = -8 \text{ or } b = 5$$

Substitute these numbers in equation (3) and solve for a.

$$a = -8 + 3 = -5$$
$$a = 5 + 3 = 8$$

The pairs $(-5, -8)$ and $(8, 5)$ check. They are the solutions.

42. $xy = 4$,

$x + y = 5$

$$x = -y + 5$$
$$(-y + 5)y = 4$$
$$-y^2 + 5y = 4$$
$$0 = y^2 - 5y + 4$$
$$0 = (y - 4)(y - 1)$$
$$y = 4 \quad \text{or} \quad y = 1$$
$$x = -4 + 5 = 1$$
$$x = -1 + 5 = 4$$

The pairs $(1, 4)$ and $(4, 1)$ check.

43. $xy - y^2 = 2$, (1)

$2xy - 3y^2 = 0$ (2)

$$\begin{aligned} -2xy + 2y^2 &= -4 \quad \text{Multiplying (1) by } -2 \\ 2xy - 3y^2 &= 0 \\ \hline -y^2 &= -4 \quad \text{Adding} \\ y^2 &= 4 \\ y &= \pm 2 \end{aligned}$$

We substitute for y in equation (1) and solve for x.

$$\begin{aligned} \text{When } y = 2: \quad x \cdot 2 - 2^2 &= 2 \\ 2x - 4 &= 2 \\ 2x &= 6 \\ x &= 3 \end{aligned}$$

$$\begin{aligned} \text{When } y = -2: \quad x(-2) - (-2)^2 &= 2 \\ -2x - 4 &= 2 \\ -2x &= 6 \\ x &= -3 \end{aligned}$$

The pairs $(3, 2)$ and $(-3, -2)$ check. They are the solutions.

44. $4a^2 - 25b^2 = 0$, (1)

$2a^2 - 10b^2 = 3b + 4$ (2)

$$\begin{aligned} 4a^2 - 25b^2 &= 0 \\ -4a^2 + 20b^2 &= -6b - 8 \quad \text{Multiplying (2) by } -2 \\ \hline -5b^2 &= -6b - 8 \\ 0 &= 5b^2 - 6b - 8 \\ 0 &= (5b + 4)(b - 2) \end{aligned}$$

$$b = -\frac{4}{5} \quad \text{or} \quad b = 2$$

Substitute for b in equation (1) and solve for a.

$$\begin{aligned} \text{When } b = -\frac{4}{5}: \quad 4a^2 - 25\left(-\frac{4}{5}\right)^2 &= 0 \\ 4a^2 &= 16 \\ a^2 &= 4 \\ a &= \pm 2 \end{aligned}$$

When $b = 2$: $4a^2 - 25(2)^2 = 0$

$$4a^2 = 100$$
$$a^2 = 25$$
$$a = \pm 5$$

The pairs $\left(2, -\frac{4}{5}\right)$, $\left(-2, -\frac{4}{5}\right)$, $(5, 2)$ and $(-5, 2)$ check.

45. $m^2 - 3mn + n^2 + 1 = 0$, (1)
$3m^2 - mn + 3n^2 = 13$ (2)

$m^2 - 3mn + n^2 = -1$ (3) Rewriting (1)
$3m^2 - mn + 3n^2 = 13$ (2)

$-3m^2 + 9mn - 3n^2 = 3$ Multiplying (3) by -3
$3m^2 - mn + 3n^2 = 13$

$$8mn = 16$$
$$mn = 2$$
$$n = \frac{2}{m} \quad (4)$$

Substitute $\frac{2}{m}$ for n in equation (1) and solve for m.

$$m^2 - 3m\left(\frac{2}{m}\right) + \left(\frac{2}{m}\right)^2 + 1 = 0$$
$$m^2 - 6 + \frac{4}{m^2} + 1 = 0$$
$$m^2 - 5 + \frac{4}{m^2} = 0$$
$$m^4 - 5m^2 + 4 = 0 \quad \text{Multiplying by } m^2$$

Substitute u for m^2.

$$u^2 - 5u + 4 = 0$$
$$(u - 4)(u - 1) = 0$$
$$u = 4 \quad or \quad u = 1$$
$$m^2 = 4 \quad or \quad m^2 = 1$$
$$m = \pm 2 \quad or \quad m = \pm 1$$

Substitute for m in equation (4) and solve for n.

When $m = 2$, $n = \frac{2}{2} = 1$.

When $m = -2$, $n = \frac{2}{-2} = -1$.

When $m = 1$, $n = \frac{2}{1} = 2$.

When $m = -1$, $n = \frac{2}{-1} = -2$.

The pairs $(2, 1)$, $(-2, -1)$, $(1, 2)$, and $(-1, -2)$ check. They are the solutions.

46. $ab - b^2 = -4$, (1)
$ab - 2b^2 = -6$ (2)

$ab - b^2 = -4$
$-ab + 2b^2 = 6$ Multiplying (2) by -1

$$b^2 = 2$$
$$b = \pm\sqrt{2}$$

Substitute for b in equation (1) and solve for a.

When $b = \sqrt{2}$: $a(\sqrt{2}) - (\sqrt{2})^2 = -4$

$$a\sqrt{2} = -2$$
$$a = -\frac{2}{\sqrt{2}} = -\sqrt{2}$$

When $b = -\sqrt{2}$: $a(-\sqrt{2}) - (-\sqrt{2})^2 = -4$

$$-a\sqrt{2} = -2$$
$$a = \frac{-2}{-\sqrt{2}} = \sqrt{2}$$

The pairs $(-\sqrt{2}, \sqrt{2})$ and $(\sqrt{2}, -\sqrt{2})$ check.

47. $x^2 + y^2 = 5$, (1)
$x - y = 8$ (2)

First solve equation (2) for x.

$x = y + 8$ (3)

Then substitute $y + 8$ for x in equation (1) and solve for y.

$$(y + 8)^2 + y^2 = 5$$
$$y^2 + 16y + 64 + y^2 = 5$$
$$2y^2 + 16y + 59 = 0$$
$$y = \frac{-16 \pm \sqrt{(16)^2 - 4(2)(59)}}{2 \cdot 2}$$
$$y = \frac{-16 \pm \sqrt{-216}}{4}$$
$$y = \frac{-16 \pm 6i\sqrt{6}}{4}$$
$$y = -4 \pm \frac{3}{2}i\sqrt{6}$$

Now substitute these numbers in equation (3) and solve for x.

$$x = -4 + \frac{3}{2}i\sqrt{6} + 8 = 4 + \frac{3}{2}i\sqrt{6}$$
$$x = -4 - \frac{3}{2}i\sqrt{6} + 8 = 4 - \frac{3}{2}i\sqrt{6}$$

The pairs $\left(4 + \frac{3}{2}i\sqrt{6}, -4 + \frac{3}{2}i\sqrt{6}\right)$ and $\left(4 - \frac{3}{2}i\sqrt{6}, -4 - \frac{3}{2}i\sqrt{6}\right)$ check. They are the solutions.

48. $4x^2 + 9y^2 = 36$,
$y - x = 8$

$y = x + 8$

$$4x^2 + 9(x + 8)^2 = 36$$
$$4x^2 + 9(x^2 + 16x + 64) = 36$$
$$4x^2 + 9x^2 + 144x + 576 = 36$$
$$13x^2 + 144x + 540 = 0$$

$$x = \frac{-144 \pm \sqrt{(144)^2 - 4(13)(540)}}{2 \cdot 13}$$

$$x = \frac{-72 \pm 6i\sqrt{51}}{13} = -\frac{72}{13} \pm \frac{6}{13}i\sqrt{51}$$

$$y = -\frac{72}{13} + \frac{6}{13}i\sqrt{51} + 8 = \frac{32}{13} + \frac{6}{13}i\sqrt{51}$$

$$y = -\frac{72}{13} - \frac{6}{13}i\sqrt{51} + 8 = \frac{32}{13} - \frac{6}{13}i\sqrt{51}$$

The pairs $\left(-\frac{72}{13} + \frac{6}{13}i\sqrt{51}, \frac{32}{13} + \frac{6}{13}i\sqrt{51}\right)$ and $\left(-\frac{72}{13} - \frac{6}{13}i\sqrt{51}, \frac{32}{13} - \frac{6}{13}i\sqrt{51}\right)$ check.

49. $a^2 + b^2 = 14,$ (1)

$ab = 3\sqrt{5}$ (2)

Solve equation (2) for b.

$$b = \frac{3\sqrt{5}}{a}$$

Substitute $\frac{3\sqrt{5}}{a}$ for b in equation (1) and solve for a.

$$a^2 + \left(\frac{3\sqrt{5}}{a}\right)^2 = 14$$

$$a^2 + \frac{45}{a^2} = 14$$

$$a^4 + 45 = 14a^2$$

$$a^4 - 14a^2 + 45 = 0$$

$u^2 - 14u + 45 = 0$ Letting $u = a^2$

$$(u-9)(u-5) = 0$$

$u = 9$ *or* $u = 5$

$a^2 = 9$ *or* $a^2 = 5$

$a = \pm 3$ *or* $a = \pm\sqrt{5}$

Since $b = 3\sqrt{5}/a$, if $a = 3$, $b = \sqrt{5}$; if $a = -3$, $b = -\sqrt{5}$; if $a = \sqrt{5}$, $b = 3$; and if $a = -\sqrt{5}$, $b = -3$. The pairs $(3, \sqrt{5})$, $(-3, -\sqrt{5})$, $(\sqrt{5}, 3)$, $(-\sqrt{5}, -3)$ check. They are the solutions.

50. $x^2 + xy = 5,$ (1)

$2x^2 + xy = 2$ (2)

$-x^2 - xy = -5$ Multiplying (1) by -1

$2x^2 + xy = 2$

$x^2 = -3$

$x = \pm i\sqrt{3}$

Substitute for x in equation (1) and solve for y.

When $x = i\sqrt{3}$: $(i\sqrt{3})^2 + (i\sqrt{3})y = 5$

$$i\sqrt{3}y = 8$$

$$y = \frac{8}{i\sqrt{3}}$$

$$y = -\frac{8i\sqrt{3}}{3}$$

When $x = -i\sqrt{3}$: $(-i\sqrt{3})^2 + (-i\sqrt{3})y = 5$

$$-i\sqrt{3}y = 8$$

$$y = -\frac{8}{i\sqrt{3}}$$

$$y = -\frac{8i\sqrt{3}}{3}$$

The pairs $\left(i\sqrt{3}, -\frac{8i\sqrt{3}}{3}\right)$ and $\left(-i\sqrt{3}, \frac{8i\sqrt{3}}{3}\right)$.

51. $x^2 + y^2 = 25,$ (1)

$9x^2 + 4y^2 = 36$ (2)

$-4x^2 - 4y^2 = -100$ Multiplying (1) by -4

$9x^2 + 4y^2 = 36$

$5x^2 = -64$

$$x^2 = -\frac{64}{5}$$

$$x = \pm\sqrt{\frac{-64}{5}} = \pm\frac{8i}{\sqrt{5}}$$

$x = \pm\frac{8i\sqrt{5}}{5}$ Rationalizing the denominator

Substituting $\frac{8i\sqrt{5}}{5}$ or $-\frac{8i\sqrt{5}}{5}$ for x in equation (1) and solving for y gives us

$$-\frac{64}{5} + y^2 = 25$$

$$y^2 = \frac{189}{5}$$

$$y = \pm\sqrt{\frac{189}{5}} = \pm 3\sqrt{\frac{21}{5}}$$

$y = \pm\frac{3\sqrt{105}}{5}.$ Rationalizing the denominator

The pairs $\left(\frac{8i\sqrt{5}}{5}, \frac{3\sqrt{105}}{5}\right)$, $\left(-\frac{8i\sqrt{5}}{5}, \frac{3\sqrt{105}}{5}\right)$, $\left(\frac{8i\sqrt{5}}{5}, -\frac{3\sqrt{105}}{5}\right)$, and $\left(-\frac{8i\sqrt{5}}{5}, -\frac{3\sqrt{105}}{5}\right)$ check.

They are the solutions.

52. $x^2 + y^2 = 1,$ (1)

$9x^2 - 16y^2 = 144$ (2)

$16x^2 + 16y^2 = 16$ Multiplying (1) by 16

$9x^2 - 16y^2 = 144$

$25x^2 = 160$

$$x^2 = \frac{160}{25}$$

$$x = \pm\frac{4\sqrt{10}}{5}$$

Substituting for x in equation (1) and solving for y gives us

$$\frac{160}{25} + y^2 = 1$$
$$y^2 = -\frac{135}{25}$$
$$y = \pm\sqrt{-\frac{135}{25}}$$
$$y = \pm\frac{3i\sqrt{15}}{5}.$$

The pairs $\left(\frac{4\sqrt{10}}{5}, \frac{3i\sqrt{15}}{5}\right)$, $\left(\frac{4\sqrt{10}}{5}, -\frac{3i\sqrt{15}}{5}\right)$, $\left(-\frac{4\sqrt{10}}{5}, \frac{3i\sqrt{15}}{5}\right)$, and $\left(-\frac{4\sqrt{10}}{5}, -\frac{3i\sqrt{15}}{5}\right)$ check.

53. $5y^2 - x^2 = 1,\quad (1)$
$xy = 2 \quad (2)$

Solve equation (2) for x.

$$x = \frac{2}{y}$$

Substitute $\frac{2}{y}$ for x in equation (1) and solve for y.

$$5y^2 - \left(\frac{2}{y}\right)^2 = 1$$
$$5y^2 - \frac{4}{y^2} = 1$$
$$5y^4 - 4 = y^2$$
$$5y^4 - y^2 - 4 = 0$$
$$5u^2 - u - 4 = 0 \quad \text{Letting } u = y^2$$
$$(5u+4)(u-1) = 0$$
$$5u + 4 = 0 \quad \text{or} \quad u - 1 = 0$$
$$u = -\frac{4}{5} \quad \text{or} \quad u = 1$$
$$y^2 = -\frac{4}{5} \quad \text{or} \quad y^2 = 1$$
$$y = \pm\frac{2i}{\sqrt{5}} \quad \text{or} \quad y = \pm 1$$
$$y = \pm\frac{2i\sqrt{5}}{5} \quad \text{or} \quad y = \pm 1$$

Since $x = 2/y$, if $y = \frac{2i\sqrt{5}}{5}$, $x = \frac{2}{\frac{2i\sqrt{5}}{5}} = \frac{5}{i\sqrt{5}} = \frac{5}{i\sqrt{5}} \cdot \frac{-i\sqrt{5}}{-i\sqrt{5}} = -i\sqrt{5}$; if $y = -\frac{2i\sqrt{5}}{5}$, $x = \frac{2}{-\frac{2i\sqrt{5}}{5}} = i\sqrt{5}$; if $y = 1$, $x = 2/1 = 2$; if $y = -1$, $x = 2/-1 = -2$.

The pairs $\left(-i\sqrt{5}, \frac{2i\sqrt{5}}{5}\right)$, $\left(i\sqrt{5}, -\frac{2i\sqrt{5}}{5}\right)$, $(2, 1)$ and $(-2, -1)$ check. They are the solutions.

54. $x^2 - 7y^2 = 6,$
$xy = 1$

$$y = \frac{1}{x}$$
$$x^2 - 7\left(\frac{1}{x}\right)^2 = 6$$
$$x^2 - \frac{7}{x^2} = 6$$
$$x^4 - 7 = 6x^2$$
$$x^4 - 6x^2 - 7 = 0$$
$$u^2 - 6u - 7 = 0 \quad \text{Letting } u = x^2$$
$$(u-7)(u+1) = 0$$
$$u = 7 \quad \text{or} \quad u = -1$$
$$x^2 = 7 \quad \text{or} \quad x^2 = -1$$
$$x = \pm\sqrt{7} \quad \text{or} \quad x = \pm i$$

Since $y = 1/x$, if $x = \sqrt{7}$, $y = 1/\sqrt{7} = \sqrt{7}/7$; if $x = -\sqrt{7}$, $y = 1/(-\sqrt{7}) = -\sqrt{7}/7$; if $x = i$, $y = 1/i = -i$, if $x = -i$, $y = 1/(-i) = i$.

The pairs $\left(\sqrt{7}, \frac{\sqrt{7}}{7}\right)$, $\left(-\sqrt{7}, -\frac{\sqrt{7}}{7}\right)$, $(i, -i)$, and $(-i, i)$ check.

55. ***Familiarize***. We first make a drawing. We let l and w represent the length and width, respectively.

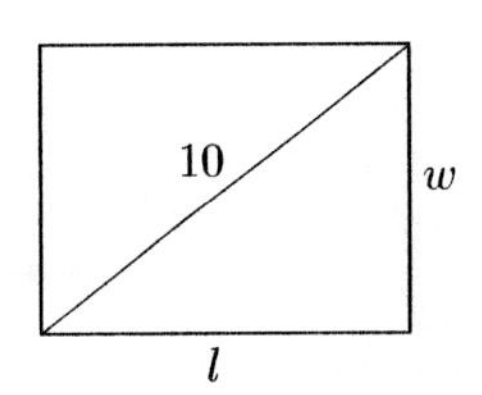

Translate. The perimeter is 28 cm.

$2l + 2w = 28$, or $l + w = 14$

Using the Pythagorean theorem we have another equation.

$l^2 + w^2 = 10^2$, or $l^2 + w^2 = 100$

Carry out. We solve the system:

$l + w = 14,\quad (1)$
$l^2 + w^2 = 100 \quad (2)$

First solve equation (1) for w.

$w = 14 - l \quad (3)$

Then substitute $14 - l$ for w in equation (2) and solve for l.

$$l^2 + w^2 = 100$$
$$l^2 + (14-l)^2 = 100$$
$$l^2 + 196 - 28l + l^2 = 100$$
$$2l^2 - 28l + 96 = 0$$
$$l^2 - 14l + 48 = 0$$
$$(l-8)(l-6) = 0$$
$$l = 8 \quad \text{or} \quad l = 6$$

If $l = 8$, then $w = 14 - 8$, or 6. If $l = 6$, then $w = 14 - 6$, or 8. Since the length is usually considered to be longer than the width, we have the solution $l = 8$ and $w = 6$, or $(8, 6)$.

Check. If $l = 8$ and $w = 6$, then the perimeter is $2\cdot8+2\cdot6$, or 28. The length of a diagonal is $\sqrt{8^2+6^2}$, or $\sqrt{100}$, or 10. The numbers check.

State. The length is 8 cm, and the width is 6 cm.

56. Let l and w represent the length and width, respectively. Solve the system:

$$2l + 2w = 6,$$
$$l^2 + w^2 = (\sqrt{5})^2, \text{ or}$$

$$l + w = 3$$
$$l^2 + w^2 = 5$$

The solutions are $(1, 2)$ and $(2, 1)$. Choosing the larger number as the length, we have the solution. The length is 2 m, and the width is 1 m.

57. ***Familiarize.*** We first make a drawing. Let l = the length and w = the width of the brochure.

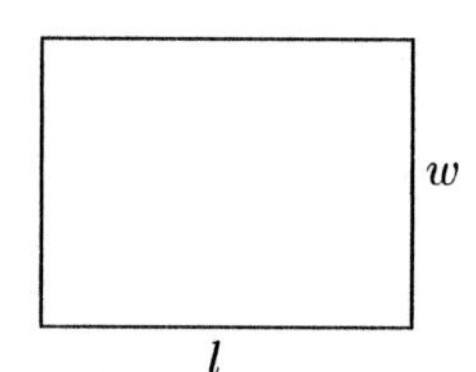

Translate.

Area: $lw = 20$

Perimeter: $2l + 2w = 18$, or $l + w = 9$

Carry out. We solve the system:

Solve the second equation for l: $l = 9 - w$

Substitute $9 - w$ for l in the first equation and solve for w.

$$(9 - w)w = 20$$
$$9w - w^2 = 20$$
$$0 = w^2 - 9w + 20$$
$$0 = (w - 5)(w - 4)$$
$$w = 5 \text{ or } w = 4$$

If $w = 5$, then $l = 9 - w$, or 4. If $w = 4$, then $l = 9 - 4$, or 5. Since length is usually considered to be longer than width, we have the solution $l = 5$ and $w = 4$, or $(5, 4)$.

Check. If $l = 5$ and $w = 4$, the area is $5 \cdot 4$, or 20. The perimeter is $2 \cdot 5 + 2 \cdot 4$, or 18. The numbers check.

State. The length of the brochure is 5 in. and the width is 4 in.

58. Let l and w represent the length and width, respectively. We solve the system:

$$lw = 2,$$
$$2l + 2w = 6$$

The solutions are $(1, 2)$ and $(2, 1)$. We choose the larger number to be the length, so the length is 2 yd and the width is 1 yd.

59. ***Familiarize.*** We first make a drawing. Let l = the length and w = the width.

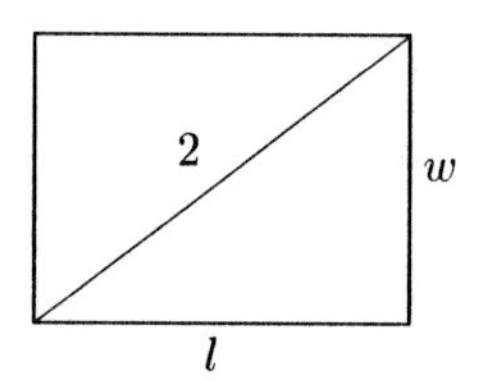

Translate.

Area: $lw = \sqrt{3}$ (1)

From the Pythagorean theorem: $l^2 + w^2 = 2^2$ (2)

Carry out. We solve the system of equations.

We first solve equation (1) for w.

$$lw = \sqrt{3}$$
$$w = \frac{\sqrt{3}}{l}$$

Then we substitute $\frac{\sqrt{3}}{l}$ for w in equation 2 and solve for l.

$$l^2 + \left(\frac{\sqrt{3}}{l}\right)^2 = 4$$
$$l^2 + \frac{3}{l^2} = 4$$
$$l^4 + 3 = 4l^2$$
$$l^4 - 4l^2 + 3 = 0$$
$$u^2 - 4u + 3 = 0 \quad \text{Letting } u = l^2$$
$$(u - 3)(u - 1) = 0$$
$$u = 3 \text{ or } u = 1$$

We now substitute l^2 for u and solve for l.

$$l^2 = 3 \quad \text{or} \quad l^2 = 1$$
$$l = \pm\sqrt{3} \quad \text{or} \quad l = \pm 1$$

Measurements cannot be negative, so we only need to consider $l = \sqrt{3}$ and $l = 1$. Since $w = \sqrt{3}/l$, if $l = \sqrt{3}$, $w = 1$ and if $l = 1$, $w = \sqrt{3}$. Length is usually considered to be longer than width, so we have the solution $l = \sqrt{3}$ and $w = 1$, or $(\sqrt{3}, 1)$.

Check. If $l = \sqrt{3}$ and $w = 1$, the area is $\sqrt{3} \cdot 1 = \sqrt{3}$. Also $(\sqrt{3})^2 + 1^2 = 3 + 1 = 4 = 2^2$. The numbers check.

State. The length is $\sqrt{3}$ m, and the width is 1 m.

60. Let l = the length and w = the width. Solve the system

$$lw = \sqrt{2},$$
$$l^2 + w^2 = (\sqrt{3})^2.$$

The solutions are $(\sqrt{2}, 1)$, $(-\sqrt{2}, -1)$, $(1, \sqrt{2})$, and $(-1, -\sqrt{2})$. Only the pairs $(\sqrt{2}, 1)$ and $(1, \sqrt{2})$ have meaning in this problem. Since length is usually considered to be longer than width, the length is $\sqrt{2}$ m, and the width is 1 m.

61. ***Familiarize.*** We make a drawing of the dog run. Let $l =$ the length and $w =$ the width.

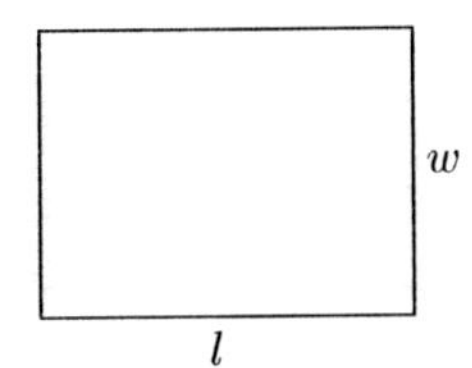

Since it takes 210 yd of fencing to enclose the run, we know that the perimeter is 210 yd.

Translate.

Perimeter: $2l + 2w = 210$, or $l + w = 105$

Area: $lw = 2250$

Carry out. We solve the system:

Solve the first equation for l: $l = 105 - w$

Substitute $105 - w$ for l in the second equation and solve for w.

$$(105 - w)w = 2250$$
$$105w - w^2 = 2250$$
$$0 = w^2 - 105w + 2250$$
$$0 = (w - 30)(w - 75)$$

$w = 30$ or $w = 75$

If $w = 30$, then $l = 105 - 30$, or 75. If $w = 75$, then $l = 105 - 75$, or 30. Since length is usually considered to be longer than width, we have the solution $l = 75$ and $w = 30$, or $(75, 30)$.

Check. If $l = 75$ and $w = 30$, the perimeter is $2 \cdot 75 + 2 \cdot 30$, or 210. The area is $75(30)$, or 2250. The numbers check.

State. The length is 75 yd and the width is 30 yd.

62. Let l and w represent the length and width, respectively. Solve the system:

$$\sqrt{l^2 + w^2} = l + 1,$$
$$\sqrt{l^2 + w^2} = 2w + 3$$

The solutions are $(12, 5)$ and $(0, -1)$. Only $(12, 5)$ has meaning in this problem. It checks. The length is 12 ft and the width is 5 ft.

63. ***Familiarize.*** We let $x =$ the length of a side of one test plot and $y =$ the length of a side of the other plot. Make a drawing.

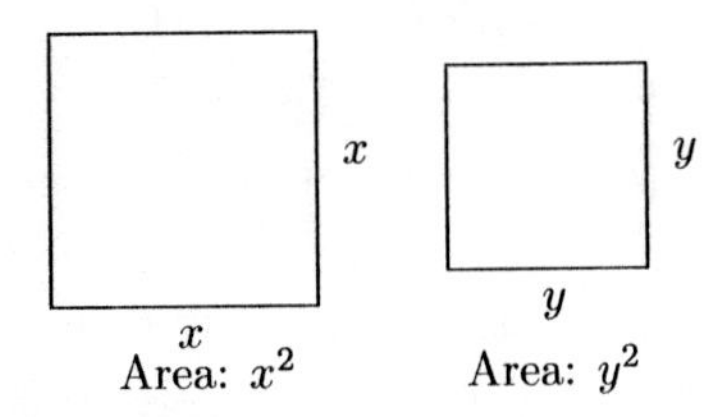

Translate.

The sum of the areas is 832 ft^2.

$$x^2 + y^2 = 832$$

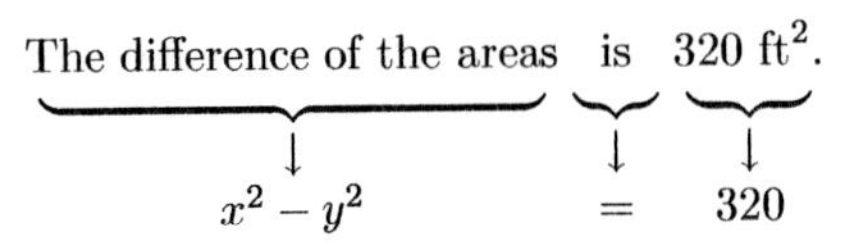

Carry out. We solve the system of equations.

$$x^2 + y^2 = 832$$
$$x^2 - y^2 = 320$$
$$2x^2 = 1152 \quad \text{Adding}$$
$$x^2 = 576$$
$$x = \pm 24$$

Since measurements cannot be negative, we consider only $x = 24$. Substitute 24 for x in the first equation and solve for y.

$$24^2 + y^2 = 832$$
$$576 + y^2 = 832$$
$$y^2 = 256$$
$$y = \pm 16$$

Again, we consider only the positive value, 16. The possible solution is $(24, 16)$.

Check. The areas of the test plots are 24^2, or 576, and 16^2, or 256. The sum of the areas is $576 + 256$, or 832. The difference of the areas is $576 - 256$, or 320. The values check.

State. The lengths of the test plots are 24 ft and 16 ft.

64. Let $p =$ the principal and $r =$ the interest rate. Solve the system:

$$pr = 7.5,$$
$$(p + 25)(r - 0.01) = 7.5$$

The solutions are $(125, 0.06)$ and $(-1.50, -0.05)$. Only $(125, 0.06)$ has meaning in this problem. The principal was \$125 and the interest rate was 0.06, or 6%.

65. Find the points of intersection of $y_1 = \ln x + 2$ and $y_2 = x^2$. They are $(1.564, 2.448)$ and $(0.138, 0.019)$.

66. Graph $y_1 = \ln(x + 4)$, $y_2 = \sqrt{6 - x^2}$, and $y_3 = -\sqrt{6 - x^2}$ and find the points of intersection. They are $(1.720, 1.744)$ and $(-2.405, 0.467)$.

67. Find the point of intersection of $y_1 = e^x - 1$ and $y_2 = -3x + 4$. It is $(0.871, 1.388)$.

68. Find the point of intersection of $y_1 = e^{-x} + 1$ and $y_2 = 2x + 5$. It is $(-0.841, 3.318)$.

69. Find the points of intersection of $y_1 = e^x$ and $y_2 = x + 2$. They are $(1.146, 3.146)$ and $(-1.841, 0.159)$.

70. Find the points of intersection of $y_1 = e^{-x}$ and $y_2 = 3 - x$. They are $(2.948, 0.052)$ and $(-1.505, 4.505)$.

71. Graph $y_1 = \sqrt{19,380,510.36 - x^2}$,

$y_2 = -\sqrt{19,380,510.36 - x^2}$, and

$y_3 = 27,941.25x/6.125$ and find the points of intersection. They are $(0.965, 4402.33)$ and $(-0.965, -4402.33)$.

72. Find the points of intersection of $y = (1660 - 2x)/2$ (or $y = 830 - x$) and $y = 35,325/x$. They are $(785, 45)$ and $(45, 785)$.

73. Graph $y_1 = \sqrt{\dfrac{14.5x^2 - 64.5}{13.5}}$,

$y_2 = -\sqrt{\dfrac{14.5x^2 - 64.5}{13.5}}$, and $y_3 = (5.5x - 12.3)/6.3$ and find the points of intersection. They are $(2.112, -0.109)$ and $(-13.041, -13.337)$.

74. Find the points of intersection of $y = -15.6/(13.5x)$ and $y_2 = (5.6x - 42.3)/6.7$. They are $(7.366, -0.157)$ and $(0.188, -6.157)$.

75. Graph $y_1 = \sqrt{\dfrac{56,548 - 0.319x^2}{2688.7}}$,

$y_2 = -\sqrt{\dfrac{56,548 - 0.319x^2}{2688.7}}$,

$y_3 = \sqrt{\dfrac{0.306x^2 - 43,452}{2688.7}}$,

and $y_4 = -\sqrt{\dfrac{0.306x^2 - 43,452}{2688.7}}$ and find the points of intersection. They are $(400, 1.431)$, $(-400, 1.431)$, $(400, -1.431)$, and $(-400, -1.431)$.

76. Graph $y_1 = \sqrt{\dfrac{6408 - 18.465x^2}{788.723}}$,

$y_2 = \sqrt{\dfrac{6408 - 18.465x^2}{788.723}}$, $y_3 = \sqrt{\dfrac{106.535x^2 - 2692}{788.723}}$,

and $y_4 = \sqrt{\dfrac{106.535x^2 - 2692}{788.723}}$ and find the points of intersection. They are $(8.532, 2.534)$, $(8.532, -2.534)$, $(-8.532, 2.534)$, and $(-8.532, -2.534)$.

77. We can only find the real-number solutions graphically.

78. Answers will vary.

79.
$$\begin{aligned} 2^{3x} &= 64 \\ 2^{3x} &= 2^6 \\ 3x &= 6 \\ x &= 2 \end{aligned}$$

The solution is 2.

80.
$$\begin{aligned} 5^x &= 27 \\ \ln 5^x &= \ln 27 \\ x \ln 5 &= \ln 27 \\ x &= \frac{\ln 27}{\ln 5} \\ x &\approx 2.048 \end{aligned}$$

81.
$$\begin{aligned} \log_3 x &= 4 \\ x &= 3^4 \\ x &= 81 \end{aligned}$$

The solution is 81.

82.
$$\begin{aligned} \log(x-3) + \log x &= 1 \\ \log(x-3)(x) &= 1 \\ x^2 - 3x &= 10 \\ x^2 - 3x - 10 &= 0 \\ (x-5)(x+2) &= 0 \end{aligned}$$

$x = 5$ *or* $x = -2$

Only 5 checks.

83. $(x-h)^2 + (y-k)^2 = r^2$

If $(2, 4)$ is a point on the circle, then

$(2-h)^2 + (4-k)^2 = r^2$.

If $(3, 3)$ is a point on the circle, then

$(3-h)^2 + (3-k)^2 = r^2$.

Thus

$$\begin{aligned} (2-h)^2 + (4-k)^2 &= (3-h)^2 + (3-k)^2 \\ 4 - 4h + h^2 + 16 - 8k + k^2 &= \\ & 9 - 6h + h^2 + 9 - 6k + k^2 \\ -4h - 8k + 20 &= -6h - 6k + 18 \\ 2h - 2k &= -2 \\ h - k &= -1 \end{aligned}$$

If the center (h, k) is on the line $3x - y = 3$, then $3h - k = 3$.

Solving the system

$h - k = -1$,

$3h - k = 3$

we find that $(h, k) = (2, 3)$.

Find r^2, substituting $(2, 3)$ for (h, k) and $(2, 4)$ for (x, y). We could also use $(3, 3)$ for (x, y).

$$\begin{aligned} (x-h)^2 + (y-k)^2 &= r^2 \\ (2-2)^2 + (4-3)^2 &= r^2 \\ 0 + 1 &= r^2 \\ 1 &= r^2 \end{aligned}$$

The equation of the circle is $(x-2)^2 + (y-3)^2 = 1$.

84. Let (h, k) represent the point on the line $5x + 8y = -2$ which is the center of a circle that passes through the points $(-2, 3)$ and $(-4, 1)$. The distance between (h, k) and $(-2, 3)$ is the same as the distance between (h, k) and $(-4, 1)$. This gives us one equation:

$$\begin{aligned} \sqrt{[h-(-2)]^2 + (k-3)^2} &= \sqrt{[h-(-4)]^2 + (k-1)^2} \\ (h+2)^2 + (k-3)^2 &= (h+4)^2 + (k-1)^2 \\ h^2 + 4h + 4 + k^2 - 6k + 9 &= h^2 + 8h + 16 + k^2 - 2k + 1 \\ 4h - 6k + 13 &= 8h - 2k + 17 \\ -4h - 4k &= 4 \\ h + k &= -1 \end{aligned}$$

We get a second equation by substituting (h, k) in $5x + 8y = -2$.

$5h + 8k = -2$

We now solve the following system:

$h + k = -1$,

$5h + 8k = -2$

The solution, which is the center of the circle, is $(-2, 1)$.

Next we find the length of the radius. We can find the distance between either $(-2, 3)$ or $(-4, 1)$ and the center $(-2, 1)$. We use $(-2, 3)$.

$$r = \sqrt{[-2-(-2)]^2 + (1-3)^2}$$
$$r = \sqrt{0^2 + (-2)^2}$$
$$r = \sqrt{4} = 2$$

We can write the equation of the circle with center $(-2, 1)$ and radius 2.

$$(x-h)^2 + (y-k)^2 = r^2$$
$$[x-(-2)]^2 + (y-1)^2 = 2^2$$
$$(x+2)^2 + (y-1)^2 = 4$$

85. The equation of the ellipse is of the form $\dfrac{x^2}{a^2} + \dfrac{y^2}{b^2} = 1$. Substitute $\left(1, \dfrac{\sqrt{3}}{2}\right)$ and $\left(\sqrt{3}, \dfrac{1}{2}\right)$ for (x, y) to get two equations.

$$\frac{1^2}{a^2} + \frac{\left(\frac{\sqrt{3}}{2}\right)^2}{b^2} = 1, \text{ or } \frac{1}{a^2} + \frac{3}{4b^2} = 1$$

$$\frac{(\sqrt{3})^2}{a^2} + \frac{\left(\frac{1}{2}\right)^2}{b^2} = 1, \text{ or } \frac{3}{a^2} + \frac{1}{4b^2} = 1$$

Substitute u for $\dfrac{1}{a^2}$ and v for $\dfrac{1}{b^2}$.

$$u + \frac{3}{4}v = 1, \qquad 4u + 3v = 4,$$
$$\text{or}$$
$$3u + \frac{1}{4}v = 1 \qquad 12u + v = 4$$

Solving for u and v, we get $u = \dfrac{1}{4}$, $v = 1$. Then $u = \dfrac{1}{a^2} = \dfrac{1}{4}$, so $a^2 = 4$; $v = \dfrac{1}{b^2} = 1$, so $b^2 = 1$.

Then the equation of the ellipse is

$$\frac{x^2}{4} + \frac{y^2}{1} = 1, \text{ or } \frac{x^2}{4} + y^2 = 1.$$

86. $\dfrac{x^2}{a^2} - \dfrac{y^2}{b^2} = 1$

Substitute each ordered pair for (x, y).

$$\frac{(-3)^2}{a^2} - \frac{\left(-\frac{3\sqrt{5}}{2}\right)^2}{b^2} = 1,$$

$$\frac{(-3)^2}{a^2} - \frac{\left(\frac{3\sqrt{5}}{b^2}\right)^2}{b^2} = 1,$$

$$\frac{\left(-\frac{3}{2}\right)^2}{a^2} - \frac{0^2}{b^2} = 1$$

$$\frac{9}{a^2} - \frac{45}{4b^2} = 1, \quad (1)$$
$$\frac{9}{a^2} - \frac{45}{4b^2} = 1, \quad (2)$$
$$\frac{9}{4a^2} = 1 \quad (3)$$

Note that equation (1) and equation (2) are identical. Multiply both sides of equation (3) by 4:

$$\frac{9}{a^2} = 4$$

Substitute 4 for $\dfrac{9}{a^2}$ in equation (1) and solve for b^2.

$$4 - \frac{45}{4b^2} = 1$$
$$16b^2 - 45 = 4b^2$$
$$12b^2 = 45$$
$$b^2 = \frac{45}{12}, \text{ or } \frac{15}{4}$$

Solve equation (3) for a^2.

$$\frac{9}{4a^2} = 1$$
$$\frac{9}{4} = a^2$$

The equation of the hyperbola is $\dfrac{x^2}{9/4} - \dfrac{y^2}{15/4} = 1$.

87. $(x-h)^2 + (y-k)^2 = r^2$ Standard form

Substitute $(4, 6)$, $(-6, 2)$, and $(1, -3)$ for (x, y).

$$(4-h)^2 + (6-k)^2 = r^2 \quad (1)$$
$$(-6-h)^2 + (2-k)^2 = r^2 \quad (2)$$
$$(1-h)^2 + (-3-k)^2 = r^2 \quad (3)$$

Thus

$(4-h)^2 + (6-k)^2 = (-6-h)^2 + (2-k)^2$, or

$5h + 2k = 3$

and

$(4-h)^2 + (6-k)^2 = (1-h)^2 + (-3-k)^2$, or

$h + 3k = 7$.

We solve the system

$$5h + 2k = 3,$$
$$h + 3k = 7.$$

Solving we get $h = -\dfrac{5}{13}$ and $k = \dfrac{32}{13}$. Substituting these values in equation (1), (2), or (3), we find that $r^2 = \dfrac{5365}{169}$.

The equation of the circle is

$$\left(x + \frac{5}{13}\right)^2 + \left(y - \frac{32}{13}\right)^2 = \frac{5365}{169}.$$

88. Using $(x-h)^2 + (y-k)^2 = r^2$ and the given points, we have

$$(2-h)^2 + (3-k)^2 = r^2 \quad (1)$$
$$(4-h)^2 + (5-k)^2 = r^2 \quad (2)$$
$$(0-h)^2 + (-3-k)^2 = r^2 \quad (3)$$

Then equation (1) − equation (2) gives $h + k = 7$ and equation (2) − equation (3) gives $h + 2k = 4$. We solve this system:

$$h + k = 7,$$
$$h + 2k = 4.$$

Then $h = 10$, $k = -3$, $r = 10$ and the equation of the circle is $(x-10)^2+[y-(-3)]^2 = 10^2$, or $(x-10)^2+(y+3)^2 = 100$.

89. See the answer section in the text.

90. Let x and y represent the numbers. Solve:

$$xy = 2,$$
$$\frac{1}{x}+\frac{1}{y} = \frac{33}{8}.$$

The solutions are $\left(\frac{1}{4}, 8\right)$ and $\left(8, \frac{1}{4}\right)$. In either case the numbers are $\frac{1}{4}$ and 8.

91. ***Familiarize.*** Let x and y represent the numbers.

Translate.

The square of a certain number exceeds twice the square of another number by $\frac{1}{8}$.

$$x^2 = 2y^2 + \frac{1}{8}$$

The sum of the squares is $\frac{5}{16}$.

$$x^2 + y^2 = \frac{5}{16}$$

Carry out. We solve the system.

$$x^2 - 2y^2 = \frac{1}{8}, \quad (1)$$
$$x^2 + y^2 = \frac{5}{16} \quad (2)$$

$$x^2 - 2y^2 = \frac{1}{8},$$
$$2x^2 + 2y^2 = \frac{5}{8} \quad \text{Multiplying (2) by 2}$$
$$3x^2 = \frac{6}{8}$$
$$x^2 = \frac{1}{4}$$
$$x = \pm\frac{1}{2}$$

Substitute $\pm\frac{1}{2}$ for x in (2) and solve for y.

$$\left(\pm\frac{1}{2}\right)^2 + y^2 = \frac{5}{16}$$
$$\frac{1}{4} + y^2 = \frac{5}{16}$$
$$y^2 = \frac{1}{16}$$
$$y = \pm\frac{1}{4}$$

We get $\left(\frac{1}{2}, \frac{1}{4}\right)$, $\left(-\frac{1}{2}, \frac{1}{4}\right)$, $\left(\frac{1}{2}, -\frac{1}{4}\right)$ and $\left(-\frac{1}{2}, -\frac{1}{4}\right)$.

Check. It is true that $\left(\pm\frac{1}{2}\right)^2$ exceeds twice $\left(\pm\frac{1}{4}\right)^2$ by $\frac{1}{8}$: $\frac{1}{4} = 2\left(\frac{1}{16}\right) + \frac{1}{8}$

Also $\left(\pm\frac{1}{2}\right)^2 + \left(\pm\frac{1}{4}\right)^2 = \frac{5}{16}$. The pairs check.

State. The numbers are $\frac{1}{2}$ and $\frac{1}{4}$ or $-\frac{1}{2}$ and $\frac{1}{4}$ or $\frac{1}{2}$ and $-\frac{1}{4}$ or $-\frac{1}{2}$ and $-\frac{1}{4}$.

92. Make a drawing.

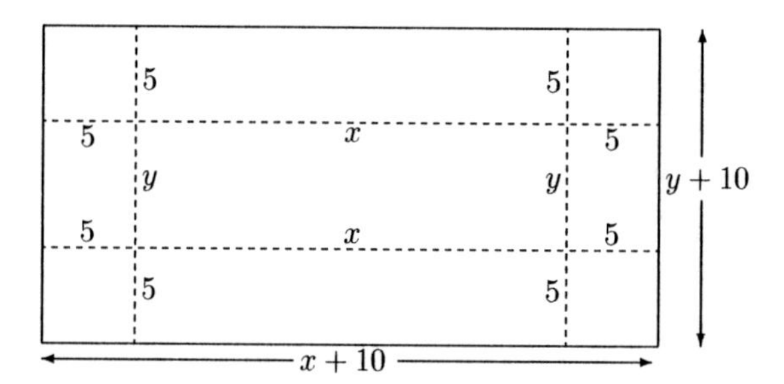

We let x and y represent the length and width of the base of the box, respectively. Then the dimensions of the metal sheet are $x + 10$ and $y + 10$.

Solve the system

$$(x + 10)(y + 10) = 340,$$
$$x \cdot y \cdot 5 = 350.$$

The solutions are $(10, 7)$ and $(7, 10)$. The dimensions of the box are 10 in. by 7 in. by 5 in.

93. See the answer section in the text.

94. $x^2 - y^2 = a^2 - b^2, \quad (1)$

$x - y = a - b \quad (2)$

Solve equation (2) for x.

$x = y + a - b \quad (3)$

Substitute for x in equation (1) and solve for y.

$$(y + a - b)^2 - y^2 = a^2 - b^2$$
$$y^2 + 2ay - 2by + a^2 - 2ab + b^2 - y^2 = a^2 - b^2$$
$$2ay - 2by = 2ab - 2b^2$$
$$2y(a - b) = 2b(a - b)$$
$$y = b$$

Substitute for y in equation (3) and solve for x.

$$x = b + a - b = a$$

The pair (a, b) checks.

95. $x^3 + y^3 = 72,\quad (1)$

$x + y = 6\quad (2)$

Solve equation (2) for y: $y = 6 - x$

Substitute for y in equation (1) and solve for x.

$$x^3 + (6-x)^3 = 72$$
$$x^3 + 216 - 108x + 18x^2 - x^3 = 72$$
$$18x^2 - 108x + 144 = 0$$
$$x^2 - 6x + 8 = 0 \quad \text{Multiplying by } \frac{1}{18}$$
$$(x-4)(x-2) = 0$$

$x = 4$ or $x = 2$

If $x = 4$, then $y = 6 - 4 = 2$.

If $x = 2$, then $y = 6 - 2 = 4$.

The pairs $(4, 2)$ and $(2, 4)$ check.

96. $a + b = \frac{5}{6},\quad (1)$

$\frac{a}{b} + \frac{b}{a} = \frac{13}{6}\quad (2)$

$b = \frac{5}{6} - a = \frac{5-6a}{6}$ Solving equation (1) for b

$$\frac{a}{\frac{5-6a}{6}} + \frac{\frac{5-6a}{6}}{a} = \frac{13}{6} \quad \text{Substituting for } b \text{ in equation (2)}$$
$$\frac{6a}{5-6a} + \frac{5-6a}{6a} = \frac{13}{6}$$
$$36a^2 + 25 - 60a + 36a^2 = 65a - 78a^2$$
$$150a^2 - 125a + 25 = 0$$
$$6a^2 - 5a + 1 = 0$$
$$(3a-1)(2a-1) = 0$$

$a = \frac{1}{3}$ or $a = \frac{1}{2}$

Substitute for a and solve for b.

When $a = \frac{1}{3}$, $b = \dfrac{5 - 6\left(\frac{1}{3}\right)}{6} = \frac{1}{2}$.

When $a = \frac{1}{2}$, $b = \dfrac{5 - 6\left(\frac{1}{2}\right)}{6} = \frac{1}{3}$.

The pairs $\left(\frac{1}{3}, \frac{1}{2}\right)$ and $\left(\frac{1}{2}, \frac{1}{3}\right)$ check. They are the solutions.

97. $p^2 + q^2 = 13,\quad (1)$

$\frac{1}{pq} = -\frac{1}{6}\quad (2)$

Solve equation (2) for p.

$$\frac{1}{q} = -\frac{p}{6}$$
$$-\frac{6}{q} = p$$

Substitute $-6/q$ for p in equation (1) and solve for q.

$$\left(-\frac{6}{q}\right)^2 + q^2 = 13$$
$$\frac{36}{q^2} + q^2 = 13$$
$$36 + q^4 = 13q^2$$
$$q^4 - 13q^2 + 36 = 0$$
$$u^2 - 13u + 36 = 0 \quad \text{Letting } u = q^2$$
$$(u-9)(u-4) = 0$$
$$u = 9 \quad \text{or} \quad u = 4$$
$$x^2 = 9 \quad \text{or} \quad x^2 = 4$$
$$x = \pm 3 \text{ or} \quad x = \pm 2$$

Since $p = -6/q$, if $q = 3$, $p = -2$; if $q = -3$, $p = 2$; if $q = 2$, $p = -3$; and if $q = -2$, $p = 3$. The pairs $(-2, 3)$, $(2, -3)$, $(-3, 2)$, and $(3, -2)$ check. They are the solutions.

98. $x^2 + y^2 = 4,\quad (1)$

$(x-1)^2 + y^2 = 4\quad (2)$

Solve equation (1) for y^2.

$y^2 = 4 - x^2\quad (3)$

Substitute for $4 - x^2$ for y^2 in equation (2) and solve for x.

$$(x-1)^2 + (4 - x^2) = 4$$
$$x^2 - 2x + 1 + 4 - x^2 = 4$$
$$-2x = -1$$
$$x = \frac{1}{2}$$

Substitute $\frac{1}{2}$ for x in equation (3) and solve for y.

$$y^2 = 4 - \left(\frac{1}{2}\right)^2$$
$$y^2 = 4 - \frac{1}{4}$$
$$y^2 = \frac{15}{4}$$
$$y = \pm\frac{\sqrt{15}}{2}$$

The pairs $\left(\frac{1}{2}, \frac{\sqrt{15}}{2}\right)$ and $\left(\frac{1}{2}, -\frac{\sqrt{15}}{2}\right)$ check. They are the solutions.

99. $5^{x+y} = 100,$

$3^{2x-y} = 1000$

$(x+y)\log 5 = 2,$ Taking logarithms and

$(2x-y)\log 3 = 3$ simplifying

$x\log 5 + y\log 5 = 2,\quad (1)$

$2x\log 3 - y\log 3 = 3\quad (2)$

Multiply equation (1) by $\log 3$ and equation (2) by $\log 5$ and add.

$$\begin{array}{rl} x\log 3\cdot\log 5 + y\log 3\cdot\log 5 &= 2\log 3 \\ 2x\log 3\cdot\log 5 - y\log 3\cdot\log 5 &= 3\log 5 \\ \hline 3x\log 3\cdot\log 5 \qquad\qquad &= 2\log 3 + 3\log 5 \end{array}$$

$$x = \frac{2\log 3 + 3\log 5}{3\log 3\cdot\log 5}$$

Substitute in (1) to find y.

$$\frac{2\log 3 + 3\log 5}{3\log 3\cdot\log 5}\cdot\log 5 + y\log 5 = 2$$

$$y\log 5 = 2 - \frac{2\log 3 + 3\log 5}{3\log 3}$$

$$y\log 5 = \frac{6\log 3 - 2\log 3 - 3\log 5}{3\log 3}$$

$$y\log 5 = \frac{4\log 3 - 3\log 5}{3\log 3}$$

$$y = \frac{4\log 3 - 3\log 5}{3\log 3\cdot\log 5}$$

The pair $\left(\frac{2\log 3 + 3\log 5}{3\log 3\cdot\log 5}, \frac{4\log 3 - 3\log 5}{3\log 3\cdot\log 5}\right)$ checks. It is the solution.

100. $e^x - e^{x+y} = 0,\quad (1)$

$e^y - e^{x-y} = 0 \quad (2)$

Factor (1): $e^x(1 - e^y) = 0$

$e^x = 0$ or $1 - e^y = 0$

No solution $\quad y = 0$

Substitute in (2).

$$e^0 - e^{x-0} = 0$$

$$1 - e^x = 0$$

$$x = 0$$

The solution is $(0, 0)$.

Chapter 10

Sequences, Series, and Combinatorics

Exercise Set 10.1

1. $a_n = 4n - 1$

$a_1 = 4 \cdot 1 - 1 = 3,$

$a_2 = 4 \cdot 2 - 1 = 7,$

$a_3 = 4 \cdot 3 - 1 = 11,$

$a_4 = 4 \cdot 4 - 1 = 15;$

$a_{10} = 4 \cdot 10 - 1 = 39;$

$a_{15} = 4 \cdot 15 - 1 = 59$

2. $a_1 = (1-1)(1-2)(1-3) = 0,$

$a_2 = (2-1)(2-2)(2-3) = 0,$

$a_3 = (3-1)(3-2)(3-3) = 0,$

$a_4 = (4-1)(4-2)(4-3) = 3 \cdot 2 \cdot 1 = 6;$

$a_{10} = (10-1)(10-2)(10-3) = 9 \cdot 8 \cdot 7 = 504;$

$a_{15} = (15-1)(15-2)(15-3) = 14 \cdot 13 \cdot 12 = 2184$

3. $a_n = \dfrac{n}{n-1},\ n \geq 2$

The first 4 terms are a_2, a_3, a_4, and a_5:

$a_2 = \dfrac{2}{2-1} = 2,$

$a_3 = \dfrac{3}{3-1} = \dfrac{3}{2},$

$a_4 = \dfrac{4}{4-1} = \dfrac{4}{3},$

$a_5 = \dfrac{5}{5-1} = \dfrac{5}{4};$

$a_{10} = \dfrac{10}{10-1} = \dfrac{10}{9};$

$a_{15} = \dfrac{15}{15-1} = \dfrac{15}{14}$

4. $a_3 = 3^2 - 1 = 8,$

$a_4 = 4^2 - 1 = 15,$

$a_5 = 5^2 - 1 = 24,$

$a_6 = 6^2 - 1 = 35;$

$a_{10} = 10^2 - 1 = 99;$

$a_{15} = 15^2 - 1 = 224$

5. $a_n = \dfrac{n^2 - 1}{n^2 + 1},$

$a_1 = \dfrac{1^2 - 1}{1^2 + 1} = 0,$

$a_2 = \dfrac{2^2 - 1}{2^2 + 1} = \dfrac{3}{5},$

$a_3 = \dfrac{3^2 - 1}{3^2 + 1} = \dfrac{8}{10} = \dfrac{4}{5},$

$a_4 = \dfrac{4^2 - 1}{4^2 + 1} = \dfrac{15}{17};$

$a_{10} = \dfrac{10^2 - 1}{10^2 + 1} = \dfrac{99}{101};$

$a_{15} = \dfrac{15^2 - 1}{15^2 + 1} = \dfrac{224}{226} = \dfrac{112}{113}$

6. $a_1 = \left(-\dfrac{1}{2}\right)^{1-1} = 1,$

$a_2 = \left(-\dfrac{1}{2}\right)^{2-1} = -\dfrac{1}{2},$

$a_3 = \left(-\dfrac{1}{2}\right)^{3-1} = \dfrac{1}{4},$

$a_4 = \left(-\dfrac{1}{2}\right)^{4-1} = -\dfrac{1}{8},$

$a_{10} = \left(-\dfrac{1}{2}\right)^{10-1} = -\dfrac{1}{512};$

$a_{15} = \left(-\dfrac{1}{2}\right)^{15-1} = \dfrac{1}{16,384}$

7. $a_n = (-1)^n n^2$

$a_1 = (-1)^1 1^2 = -1,$

$a_2 = (-1)^2 2^2 = 4,$

$a_3 = (-1)^3 3^2 = -9,$

$a_4 = (-1)^4 4^2 = 16;$

$a_{10} = (-1)^{10} 10^2 = 100;$

$a_{15} = (-1)^{15} 15^2 = -225$

8. $a_1 = (-1)^{1-1}(3 \cdot 1 - 5) = -2,$

$a_2 = (-1)^{2-1}(3 \cdot 2 - 5) = -1,$

$a_3 = (-1)^{3-1}(3 \cdot 3 - 5) = 4,$

$a_4 = (-1)^{4-1}(3 \cdot 4 - 5) = -7;$

$a_{10} = (-1)^{10-1}(3 \cdot 10 - 5) = -25,$

$a_{15} = (-1)^{15-1}(3 \cdot 15 - 5) = 40$

9. $a_n = 5 + \frac{(-2)^{n+1}}{2^n}$

$a_1 = 5 + \frac{(-2)^{1+1}}{2^1} = 5 + \frac{4}{2} = 7,$

$a_2 = 5 + \frac{(-2)^{2+1}}{2^2} = 5 + \frac{-8}{4} = 3,$

$a_3 = 5 + \frac{(-2)^{3+1}}{2^3} = 5 + \frac{16}{8} = 7,$

$a_4 = 5 + \frac{(-2)^{4+1}}{2^4} = 5 + \frac{-32}{16} = 3;$

$a_{10} = 5 + \frac{(-2)^{10+1}}{2^{10}} = 5 + \frac{-1 \cdot 2^{11}}{2^{10}} = 3;$

$a_{15} = 5 + \frac{(-2)^{15+1}}{2^{15}} = 5 + \frac{2^{16}}{2^{15}} = 7$

10. $a_1 = \frac{2 \cdot 1 - 1}{1^2 + 2 \cdot 1} = \frac{1}{3},$

$a_2 = \frac{2 \cdot 2 - 1}{2^2 + 2 \cdot 2} = \frac{3}{8},$

$a_3 = \frac{2 \cdot 3 - 1}{3^2 + 2 \cdot 3} = \frac{5}{15} = \frac{1}{3},$

$a_4 = \frac{2 \cdot 4 - 1}{4^2 + 2 \cdot 4} = \frac{7}{24};$

$a_{10} = \frac{2 \cdot 10 - 1}{10^2 + 2 \cdot 10} = \frac{19}{120};$

$a_{15} = \frac{2 \cdot 15 - 1}{15^2 + 2 \cdot 15} = \frac{29}{255}$

11. $a_n = 5n - 6$

$a_8 = 5 \cdot 8 - 6 = 40 - 6 = 34$

12. $a_7 = (3 \cdot 7 - 4)(2 \cdot 7 + 5) = 17 \cdot 19 = 323$

13. $a_n = (2n + 3)^2$

$a_6 = (2 \cdot 6 + 3)^2 = 225$

14. $a_{12} = (-1)^{12-1}[4.6(12) - 18.3] = -36.9$

15. $a_n = 5n^2(4n - 100)$

$a_{11} = 5(11)^2(4 \cdot 11 - 100) = 5(121)(-56) =$
$-33,880$

16. $a_{80} = \left(1 + \frac{1}{80}\right)^2 = \left(\frac{81}{80}\right)^2 = \frac{6561}{6400}$

17. $a_n = \ln e^n$

$a_{67} = \ln e^{67} = 67$

18. $a_{100} = 2 - \frac{1000}{100} = 2 - 10 = -8$

19. $2, 4, 6, 8, 10, \ldots$

These are the even integers, so the general term might be $2n$.

20. 3^n

21. $-2, 6, -18, 54, \ldots$

We can see a pattern if we write the sequence as

$-1 \cdot 2 \cdot 1, 1 \cdot 2 \cdot 3, -1 \cdot 2 \cdot 9, 1 \cdot 2 \cdot 27, \ldots$

The general term might be $(-1)^n 2(3)^{n-1}$.

22. $5n - 7$

23. $\frac{2}{3}, \frac{3}{4}, \frac{4}{5}, \frac{5}{6}, \frac{6}{7}, \ldots$

These are fractions in which the denominator is 1 greater than the numerator. Also, each numerator is 1 greater than the preceding numerator. The general term might be $\frac{n+1}{n+2}$.

24. $\sqrt{2n}$

25. $1 \cdot 2, 2 \cdot 3, 3 \cdot 4, 4 \cdot 5, \ldots$

These are the products of pairs of consecutive natural numbers. The general term might be $n(n + 1)$.

26. $-1 - 3(n - 1)$, or $-3n + 2$, or $-(3n - 2)$

27. $0, \log 10, \log 100, \log 1000, \ldots$

We can see a pattern if we write the sequence as

$\log 1, \log 10, \log 100, \log 1000, \ldots$

The general term might be $\log 10^{n-1}$. This is equivalent to $n - 1$.

28. $\ln e^{n+1}$, or $n + 1$

29. $1, 2, 3, 4, 5, 6, 7, \ldots$

$S_3 = 1 + 2 + 3 = 6$

$S_7 = 1 + 2 + 3 + 4 + 5 + 6 + 7 = 28$

30. $S_2 = 1 - 3 = -2$

$S_8 = 1 - 3 + 5 - 7 + 9 = 5$

31. $2, 4, 6, 8, \ldots$

$S_4 = 2 + 4 + 6 + 8 = 20$

$S_5 = 2 + 4 + 6 + 8 + 10 = 30$

32. $S_1 = 1$

$S_5 = 1 + \frac{1}{4} + \frac{1}{9} + \frac{1}{16} + \frac{1}{25} = \frac{5269}{3600}$

33.
$$\sum_{k=1}^{5} \frac{1}{2k} = \frac{1}{2 \cdot 1} + \frac{1}{2 \cdot 2} + \frac{1}{2 \cdot 3} + \frac{1}{2 \cdot 4} + \frac{1}{2 \cdot 5}$$
$$= \frac{1}{2} + \frac{1}{4} + \frac{1}{6} + \frac{1}{8} + \frac{1}{10}$$
$$= \frac{60}{120} + \frac{30}{120} + \frac{20}{120} + \frac{15}{120} + \frac{12}{120}$$
$$= \frac{137}{120}$$

34. $\frac{1}{3} + \frac{1}{5} + \frac{1}{7} + \frac{1}{9} + \frac{1}{11} + \frac{1}{13} = \frac{43,024}{45,045}$

35.
$$\sum_{i=0}^{6} 2^i = 2^0 + 2^1 + 2^2 + 2^3 + 2^4 + 2^5 + 2^6$$
$$= 1 + 2 + 4 + 8 + 16 + 32 + 64$$
$$= 127$$

36. $\sqrt{7} + \sqrt{9} + \sqrt{11} + \sqrt{13} \approx 12.5679$

37. $\sum_{k=7}^{10} \ln k = \ln 7 + \ln 8 + \ln 9 + \ln 10 =$

$\ln(7 \cdot 8 \cdot 9 \cdot 10) = \ln 5040 \approx 8.5252$

38. $\pi + 2\pi + 3\pi + 4\pi = 10\pi \approx 31.4159$

39. $$\sum_{k=1}^{8} \frac{k}{k+1} = \frac{1}{1+1} + \frac{2}{2+1} + \frac{3}{3+1} + \frac{4}{4+1} + \frac{5}{5+1} + \frac{6}{6+1} + \frac{7}{7+1} + \frac{8}{8+1}$$
$$= \frac{1}{2} + \frac{2}{3} + \frac{3}{4} + \frac{4}{5} + \frac{5}{6} + \frac{6}{7} + \frac{7}{8} + \frac{8}{9}$$
$$= \frac{15,551}{2520}$$

40. $$\frac{1-1}{1+3} + \frac{2-1}{2+3} + \frac{3-1}{3+3} + \frac{4-1}{4+3} + \frac{5-1}{5+3}$$
$$= 0 + \frac{1}{5} + \frac{2}{6} + \frac{3}{7} + \frac{4}{8}$$
$$= 0 + \frac{1}{5} + \frac{1}{3} + \frac{3}{7} + \frac{1}{2}$$
$$= \frac{307}{210}$$

41. $$\sum_{i=1}^{5} (-1)^i$$
$$= (-1)^1 + (-1)^2 + (-1)^3 + (-1)^4 + (-1)^5$$
$$= -1 + 1 - 1 + 1 - 1$$
$$= -1$$

42. $-1 + 1 - 1 + 1 - 1 + 1 = 0$

43. $$\sum_{k=1}^{8} (-1)^{k+1} 3k$$
$$= (-1)^2 3 \cdot 1 + (-1)^3 3 \cdot 2 + (-1)^4 3 \cdot 3 + (-1)^5 3 \cdot 4 + (-1)^6 3 \cdot 5 + (-1)^7 3 \cdot 6 + (-1)^8 3 \cdot 7 + (-1)^9 3 \cdot 8$$
$$= 3 - 6 + 9 - 12 + 15 - 18 + 21 - 24$$
$$= -12$$

44. $4 - 4^2 + 4^3 - 4^4 + 4^5 - 4^6 + 4^7 - 4^8 = -52,428$

45. $$\sum_{k=0}^{6} \frac{2}{k^2+1} = \frac{2}{0^2+1} + \frac{2}{1^2+1} + \frac{2}{2^2+1} + \frac{2}{3^2+1} + \frac{2}{4^2+1} + \frac{2}{5^2+1} + \frac{2}{6^2+1}$$
$$= 2 + 1 + \frac{2}{5} + \frac{2}{10} + \frac{2}{17} + \frac{2}{26} + \frac{2}{37}$$
$$= 2 + 1 + \frac{2}{5} + \frac{1}{5} + \frac{2}{17} + \frac{1}{13} + \frac{2}{37}$$
$$= \frac{157,351}{40,885}$$

46. $1 \cdot 2 + 2 \cdot 3 + 3 \cdot 4 + 4 \cdot 5 + 5 \cdot 6 + 6 \cdot 7 + 7 \cdot 8 + 8 \cdot 9 + 9 \cdot 10 + 10 \cdot 11 = 440$

47. $$\sum_{k=0}^{5} (k^2 - 2k + 3)$$
$$= (0^2 - 2 \cdot 0 + 3) + (1^2 - 2 \cdot 1 + 3) + (2^2 - 2 \cdot 2 + 3) + (3^2 - 2 \cdot 3 + 3) + (4^2 - 2 \cdot 4 + 3) + (5^2 - 2 \cdot 5 + 3)$$
$$= 3 + 2 + 3 + 6 + 11 + 18$$
$$= 43$$

48. $$\frac{1}{1 \cdot 2} + \frac{1}{2 \cdot 3} + \frac{1}{3 \cdot 4} + \frac{1}{4 \cdot 5} + \frac{1}{5 \cdot 6} + \frac{1}{6 \cdot 7} + \frac{1}{7 \cdot 8} + \frac{1}{8 \cdot 9} + \frac{1}{9 \cdot 10} + \frac{1}{10 \cdot 11}$$
$$= \frac{1}{2} + \frac{1}{6} + \frac{1}{12} + \frac{1}{20} + \frac{1}{30} + \frac{1}{42} + \frac{1}{56} + \frac{1}{72} + \frac{1}{90} + \frac{1}{110}$$
$$= \frac{10}{11}$$

49. $$\sum_{i=0}^{10} \frac{2i}{2^i + 1}$$
$$= \frac{2^0}{2^0+1} + \frac{2^1}{2^1+1} + \frac{2^2}{2^2+1} + \frac{2^3}{2^3+1} + \frac{2^4}{2^4+1} + \frac{2^5}{2^5+1} + \frac{2^6}{2^6+1} + \frac{2^7}{2^7+1} + \frac{2^8}{2^8+1} + \frac{2^9}{2^9+1} + \frac{2^{10}}{2^{10}+1}$$
$$= \frac{1}{2} + \frac{2}{3} + \frac{4}{5} + \frac{8}{9} + \frac{16}{17} + \frac{32}{33} + \frac{64}{65} + \frac{128}{129} + \frac{256}{257} + \frac{512}{513} + \frac{1024}{1025}$$
$$\approx 9.736$$

50. $(-2)^0 + (-2)^2 + (-2)^4 + (-2)^6 = 1 + 4 + 16 + 64 = 85$

51. $5 + 10 + 15 + 20 + 25 + \ldots$

This is a sum of multiples of 5, and it is an infinite series. Sigma notation is

$$\sum_{k=1}^{\infty} 5k.$$

52. $\sum_{k=1}^{\infty} 7k$

53. $2 - 4 + 8 - 16 + 32 - 64$

This is a sum of powers of 2 with alternating signs. Sigma notation is

$$\sum_{k=1}^{6} (-1)^{k+1} 2k, \text{ or } \sum_{k=1}^{6} (-1)^{k-1} 2k$$

54. $\sum_{k=1}^{5} 3k$

55. $-\frac{1}{2} + \frac{2}{3} - \frac{3}{4} + \frac{4}{5} - \frac{5}{6} + \frac{6}{7}$

This is a sum of fractions in which the denominator is one greater than the numerator. Also, each numerator is 1 greater than the preceding numerator and the signs alternate. Sigma notation is

$$\sum_{k=1}^{6} (-1)^k \frac{k}{k+1}.$$

56. $\sum_{k=1}^{5} \frac{1}{k^2}$

57. $4 - 9 + 16 - 25 + \ldots + (-1)^n n^2$

This is a sum of terms of the form $(-1)^k k^2$, beginning with $k = 2$ and continuing through $k = n$. Sigma notation is

$$\sum_{k=2}^{n} (-1)^k k^2.$$

58. $\sum_{k=3}^{n} (-1)^{k+1} k^2$, or $\sum_{k=3}^{n} (-1)^{k-1} k^2$

59. $\frac{1}{1 \cdot 2} + \frac{1}{2 \cdot 3} + \frac{1}{3 \cdot 4} + \frac{1}{4 \cdot 5} + \ldots$

This is a sum of fractions in which the numerator is 1 and the denominator is a product of two consecutive integers. The larger integer in each product is the smaller integer in the succeeding product. It is an infinite series. Sigma notation is

$$\sum_{k=1}^{\infty} \frac{1}{k(k+1)}.$$

60. $\sum_{k=1}^{\infty} \frac{1}{k(k+1)^2}$

61. $a_1 = 4, \quad a_{k+1} = 1 + \frac{1}{a_k}$

$a_2 = 1 + \frac{1}{4} = 1\frac{1}{4}$, or $\frac{5}{4}$

$a_3 = 1 + \frac{1}{\frac{5}{4}} = 1 + \frac{4}{5} = 1\frac{4}{5}$, or $\frac{9}{5}$

$a_4 = 1 + \frac{1}{\frac{9}{5}} = 1 + \frac{5}{9} = 1\frac{5}{9}$, or $\frac{14}{9}$

62. $a_1 = 256,\ a_2 = \sqrt{256} = 16,\ a_3 = \sqrt{16} = 4,$

$a_4 = \sqrt{4} = 2$

63. $a_1 = 6561, \quad a_{k+1} = (-1)^k \sqrt{a_k}$

$a_2 = (-1)^1 \sqrt{6561} = -81$

$a_3 = (-1)^2 \sqrt{-81} = 9i$

$a_4 = (-1)^3 \sqrt{9i} = -3\sqrt{i}$

64. $a_1 = e^Q,\ a_2 = \ln e^Q = Q,\ a_3 = \ln Q,\ a_4 = \ln(\ln Q)$

65. $a_1 = 2, \quad a_{k+1} = a_k + a_{k-1}$

$a_2 = 3$

$a_3 = 3 + 2 = 5$

$a_4 = 5 + 3 = 8$

66. $a_1 = -10,\ a_2 = 8,\ a_3 = 8 - (-10) = 18,$

$a_4 = 18 - 8 = 10$

67. a) $a_1 = \$1000(1.062)^1 = \$1062$

$a_2 = \$1000(1.062)^2 \approx \$1127.84$

$a_3 = \$1000(1.062)^3 \approx \$1197.77$

$a_4 = \$1000(1.062)^4 \approx \$1272.03$

$a_5 = \$1000(1.062)^5 \approx \$1350.90$

$a_6 = \$1000(1.062)^6 \approx \$1434.65$

$a_7 = \$1000(1.062)^7 \approx \$1523.60$

$a_8 = \$1000(1.062)^8 \approx \$1618.07$

$a_9 = \$1000(1.062)^9 \approx \$1718.39$

$a_{10} = \$1000(1.062)^{10} \approx \$1824.93$

b) $a_{20} = \$1000(1.062)^{20} \approx \$3330.35$

68. Find each term by multiplying the preceding term by 0.75:

\$5200, \$3900, \$2925, \$2193.75, \$1645.31,

\$1233.98, \$925.49, \$694.12, \$520.59, \$390.44

69. Find each term by multiplying the preceding term by 2. Find 17 terms, beginning with $a_1 = 1$, since there are 16 fifteen minute periods in 4 hr.

1, 2, 4, 8, 16, 32, 64, 128, 256, 512, 1024,

2048, 4096, 8192, 16,384, 32,768, 65,536

70. Find each term by adding \$0.30 to the preceding term:

\$8.30, \$8.60, \$8.90, \$9.20, \$9.50, \$9.80,

\$10.10, \$10.40, \$10.70, \$11.00

71. $a_1 = 1$ (Given)

$a_2 = 1$ (Given)

$a_3 = a_2 + a_1 = 1 + 1 = 2$

$a_4 = a_3 + a_2 = 2 + 1 = 3$

$a_5 = a_4 + a_3 = 3 + 2 = 5$

$a_6 = a_5 + a_4 = 5 + 3 = 8$

$a_7 = a_6 + a_5 = 8 + 5 = 13$

72.

n	Un
1	.41421
2	.31784
3	.26795
4	.23607
5	.21342
6	.19626
7	.18268
8	.17157
9	.16228
10	.15435

73. See the answer section in the text.

74.

n	Un
1	2
2	1.5
3	1.4167
4	1.4142
5	1.4142
6	1.4142
7	1.4142
8	1.4142
9	1.4142
10	1.4142

75. See the answer section in the text.

76. a) Using the quadratic regression operation on a graphing calculator we get $a_n = -50.27142857n^2 + 4375.891429n - 38,603.23714$.

b) $a_{15} \approx \$15,724$

$a_{25} \approx \$39,374$

$a_{40} \approx \$55,998$

$a_{60} \approx \$42,973$

$a_{75} \approx \$6812$

77. a)

n	Un
15	27.25
16	26.108
17	25.002
18	23.932
19	22.898
20	21.9
21	20.938
22	20.012
23	19.122
24	18.268

b) From the table in part (a), we know that $a_{16} = 26.108 \approx 26$ and $a_{23} = 19.122 \approx 19$. Using a table set in Ask mode we find that $a_{50} = 8.7 \approx 9$, $a_{75} = 22.45 \approx 22$, and $a_{85} = 34.25 \approx 34$.

78. a) $a_1 = 41$, $a_2 = 43$, $a_3 = 47$, $a_4 = 53$; the terms fit the pattern $41, 41+2, 41+6, 41+12, \cdots$, or $41+1\cdot 0$, $41+2\cdot 1, 41+3\cdot 2, 41+4\cdot 3, \cdots$, or $41+n(n-1)$.

b) Yes; $n^2 - n + 41 = n(n-1) + 41 = 41 + n(n-1)$.

79. For a Fibonacci sequence, as $n \to \infty$, $a_{n+1}/a_n \to \tau \approx 1.618033989$ where τ is a ratio called the golden section. It is the ratio that results when a line is divided so that the ratio of the larger part to the smaller part is equal to the ratio of the whole to the larger part.

80. $3x - 2y = 3,\quad (1)$

$2x + 3y = -11\quad (2)$

Multiply equation (1) by 3 and equation (2) by 2 and add.

$$\begin{aligned} 9x - 6y &= 9 \\ 4x + 6y &= -22 \\ \hline 13x \quad &= -13 \\ x &= -1 \end{aligned}$$

Back-substitute to find y.

$$\begin{aligned} 2(-1) + 3y &= -11 \quad \text{Using equation (2)} \\ -2 + 3y &= -11 \\ 3y &= -9 \\ y &= -3 \end{aligned}$$

The solution is $(-1, -3)$.

81. ***Familiarize.*** Let x and y represent the amount of individual income taxes collected by the federal government in 2001 and 2002, respectively, in billions of dollars.

Translate. A total of \$1943.5 billion was collected in 2001 and 2002, so we have one equation:

$$x + y = 1943.5$$

We also know that the amount collected in 2002 was \$45.1 billion less than the amount collected in 2001. This gives us another equation:

$$y = x - 45.1$$

We have a system of equations:

$$x + y = 1943.5,$$
$$y = x - 45.1$$

Carry out. We use the substitution method.

$$\begin{aligned} x + (x - 45.1) &= 1943.5 \\ 2x - 45.1 &= 1943.5 \\ 2x &= 1988.6 \\ x &= 994.3 \end{aligned}$$

Back-substitute to find y.

$$y = 994.3 - 45.1 = 949.2$$

Check. The total amount collected is \$994.3 billion + \$949.2 billion, or \$1943.5 billion. Also, the amount collected in 2002, \$949.2 billion, is \$45.1 billion less than \$994.3 billion, the amount collected in 2001. The answer checks.

State. In 2001 \$994.3 billion was collected, and \$949.2 billion was collected in 2002.

82.
$$\begin{aligned} x^2 + y^2 - 6x + 4y &= 3 \\ x^2 - 6x + 9 + y^2 + 4y + 4 &= 3 + 9 + 4 \\ (x-3)^2 + (y+2)^2 &= 16 \end{aligned}$$

Center: $(3, -2)$; radius: 4

83. We complete the square twice.

$$\begin{aligned} x^2 + y^2 + 5x - 8y &= 2 \\ x^2 + 5x + y^2 - 8y &= 2 \\ x^2 + 5x + \frac{25}{4} + y^2 - 8y + 16 &= 2 + \frac{25}{4} + 16 \\ \left(x + \frac{5}{2}\right)^2 + (y-4)^2 &= \frac{97}{4} \\ \left[x - \left(-\frac{5}{2}\right)\right]^2 + (y-4)^2 &= \left(\frac{\sqrt{97}}{2}\right)^2 \end{aligned}$$

The center is $\left(-\frac{5}{2}, 4\right)$ and the radius is $\frac{\sqrt{97}}{2}$.

84. $a_n = \frac{1}{2^n} \log 1000^n$

$a_1 = \frac{1}{2^1} \log 1000^1 = \frac{1}{2} \log 10^3 = \frac{1}{2} \cdot 3 = \frac{3}{2}$

$a_2 = \frac{1}{2^2} \log 1000^2 = \frac{1}{4} \log (10^3)^2 = \frac{1}{4} \log 10^6 = \frac{1}{4} \cdot 6 = \frac{3}{2}$

$a_3 = \frac{1}{2^3} \log 1000^3 = \frac{1}{8} \log (10^3)^3 = \frac{1}{8} \log 10^9 = \frac{1}{8} \cdot 9 = \frac{9}{8}$

$a_4 = \frac{1}{2^4} \log 1000^4 = \frac{1}{16} \log (10^3)^4 = \frac{1}{16} \log 10^{12} = \frac{1}{16} \cdot 12 = \frac{3}{4}$

$a_5 = \frac{1}{2^5} \log 1000^5 = \frac{1}{32} \log (10^3)^5 = \frac{1}{32} \log 10^{15} = \frac{1}{32} \cdot 15 = \frac{15}{32}$

$S_5 = \frac{3}{2} + \frac{3}{2} + \frac{9}{8} + \frac{3}{4} + \frac{15}{32} = \frac{171}{32}$

85. $a_n = i^n$

$a_1 = i$

$a_2 = i^2 = -1$

$a_3 = i^3 = -i$

$a_4 = i^4 = 1$

$a_5 = i^5 = i^4 \cdot i = i;$

$S_5 = i - 1 - i + 1 + i = i$

86. $a_n = \ln(1 \cdot 2 \cdot 3 \cdots n)$

$a_1 = \ln 1 = 0$

$a_2 = \ln(1 \cdot 2) = \ln 2$

$a_3 = \ln(1 \cdot 2 \cdot 3) = \ln 6$

$a_4 = \ln(1 \cdot 2 \cdot 3 \cdot 4) = \ln 24$

$a_5 = \ln(1 \cdot 2 \cdot 3 \cdot 4 \cdot 5) = \ln 120$

$S_5 = 0 + \ln 2 + \ln 6 + \ln 24 + \ln 120;$

$= \ln(2 \cdot 6 \cdot 24 \cdot 120) = \ln 34,560 \approx 10.450$

87. $S_n = \ln 1 + \ln + 2 \ln 3 + \cdots + \ln n$

$= \ln(1 \cdot 2 \cdot 3 \cdots n)$

88. $S_n = \left(1 - \frac{1}{2}\right) + \left(\frac{1}{2} - \frac{1}{3}\right) + \left(\frac{1}{3} - \frac{1}{4}\right) + \cdots + \left(\frac{1}{n-1} - \frac{1}{n}\right) + \left(\frac{1}{n} - \frac{1}{n+1}\right)$

$= 1 - \frac{1}{n+1} = \frac{n}{n+1}$

Exercise Set 10.2

1. 3, 8, 13, 18, . . .

$a_1 = 3$

$d = 5 \quad (8 - 3 = 5,\ 13 - 8 = 5,\ 18 - 13 = 5)$

2. $a_1 = \$1.08$, $d = \$0.08$, $(\$1.16 - \$1.08 = \$0.08)$

3. 9, 5, 1, −3, . . .

$a_1 = 9$

$d = -4 \quad (5-9 = -4,\ 1-5 = -4,\ -3-1 = -4)$

4. $a_1 = -8$, $d = 3$ $(-5 - (-8) = 3)$

5. $\frac{3}{2}, \frac{9}{4}, 3, \frac{15}{4}, \ldots$

$a_1 = \frac{3}{2}$

$d = \frac{3}{4} \quad \left(\frac{9}{4} - \frac{3}{2} = \frac{3}{4}, 3 - \frac{9}{4} = \frac{3}{4}\right)$

6. $a_1 = \frac{3}{5}$, $d = -\frac{1}{2} \left(\frac{1}{10} - \frac{3}{5} = -\frac{1}{2}\right)$

7. $a_1 = \$316$

$d = -\$3 \quad (\$313 - \$316 = -\$3,\ \$310 - \$313 = -\$3,\ \$307 - \$310 = -\$3)$

8. $a_1 = 0.07$, $d = 0.05$, and $n = 11$

$a_{11} = 0.07 + (11 - 1)(0.05) = 0.07 + 0.5 = 0.57$

9. 2, 6, 10, . . .

$a_1 = 2$, $d = 4$, and $n = 12$

$a_n = a_1 + (n-1)d$

$a_{12} = 2 + (12 - 1)4 = 2 + 11 \cdot 4 = 2 + 44 = 46$

10. $a_1 = 7$, $d = -3$, and $n = 17$

$a_{17} = 7 + (17 - 1)(-3) = 7 + 16(-3) = 7 - 48 = -41$

11. $3, \frac{7}{3}, \frac{5}{3}, \ldots$

$a_1 = 3$, $d = -\frac{2}{3}$, and $n = 14$

$a_n = a_1 + (n-1)d$

$a_{14} = 3 + (14 - 1)\left(-\frac{2}{3}\right) = 3 - \frac{26}{3} = -\frac{17}{3}$

12. $a_1 = \$1200$, $d = \$964.32 - \$1200 = -\$235.68$, and $n = 13$

$a_{13} = \$1200 + (13 - 1)(-\$235.68) = \$1200 + 12(-\$235.68) = \$1200 - \$2828.16 = -\$1628.16$

13. \$2345.78, \$2967.54, \$3589.30, . . .

$a_1 = \$2345.78$, $d = \$621.76$, and $n = 10$

$a_n = a_1 + (n-1)d$

$a_{10} = \$2345.78 + (10 - 1)(\$621.76) = \$7941.62$

14. $106 = 2 + (n-1)(4)$

$106 = 2 + 4n - 4$

$108 = 4n$

$27 = n$

The 27th term is 106.

15. $a_1 = 0.07,\ d = 0.05$

$a_n = a_1 + (n-1)d$

Let $a_n = 1.67$, and solve for n.

$1.67 = 0.07 + (n-1)(0.05)$

$1.67 = 0.07 + 0.05n - 0.05$

$1.65 = 0.05n$

$33 = n$

The 33rd term is 1.67.

16. $-296 = 7 + (n-1)(-3)$

$-296 = 7 - 3n + 3$

$-306 = -3n$

$102 = n$

The 102nd term is -296.

17. $a_1 = 3,\ d = -\frac{2}{3}$

$a_n = a_1 + (n-1)d$

Let $a_n = -27$, and solve for n.

$-27 = 3 + (n-1)\left(-\frac{2}{3}\right)$

$-81 = 9 + (n-1)(-2)$

$-81 = 9 - 2n + 2$

$-92 = -2n$

$46 = n$

The 46th term is -27.

18. $a_{20} = 14 + (20-1)(-3) = 14 + 19(-3) = -43$

19. $a_n = a_1 + (n-1)d$

$33 = a_1 + (8-1)4$ Substituting 33 for a_8, 8 for n, and 4 for d

$33 = a_1 + 28$

$5 = a_1$

(Note that this procedure is equivalent to subtracting d from a_8 seven times to get a_1: $33 - 7(4) = 33 - 28 = 5$)

20. $26 = 8 + (11-1)d$

$26 = 8 + 10d$

$18 = 10d$

$1.8 = d$

21. $a_n = a_1 + (n-1)d$

$-507 = 25 + (n-1)(-14)$

$-507 = 25 - 14n + 14$

$-546 = -14n$

$39 = n$

22. We know that $a_{17} = -40$ and $a_{28} = -73$. We would have to add d eleven times to get from a_{17} to a_{28}. That is,

$-40 + 11d = -73$

$11d = -33$

$d = -3.$

Since $a_{17} = -40$, we subtract d sixteen times to get to a_1.

$a_1 = -40 - 16(-3) = -40 + 48 = 8$

We write the first five terms of the sequence:

$8, 5, 2, -1, -4$

23. $\frac{25}{3} + 15d = \frac{95}{6}$

$15d = \frac{45}{6}$

$d = \frac{1}{2}$

$a_1 = \frac{25}{3} - 16\left(\frac{1}{2}\right) = \frac{25}{3} - 8 = \frac{1}{3}$

The first five terms of the sequence are $\frac{1}{3}, \frac{5}{6}, \frac{4}{3}, \frac{11}{6}, \frac{7}{3}$.

24. $a_{14} = 11 + (14-1)(-4) = 11 + 13(-4) = -41$

$S_{14} = \frac{14}{2}[11 + (-41)] = 7(-30) = -210$

25. $5 + 8 + 11 + 14 + \ldots$

Note that $a_1 = 5$, $d = 3$, and $n = 20$. First we find a_{20}:

$a_n = a_1 + (n-1)d$

$a_{20} = 5 + (20-1)3$

$= 5 + 19 \cdot 3 = 62$

Then

$S_n = \frac{n}{2}(a_1 + a_n)$

$S_{20} = \frac{20}{2}(5 + 62)$

$= 10(67) = 670.$

26. $1 + 2 + 3 + \ldots + 299 + 300.$

$S_{300} = \frac{300}{2}(1 + 300) = 150(301) = 45,150$

27. The sum is $2+4+6+\ldots+798+800$. This is the sum of the arithmetic sequence for which $a_1 = 2$, $a_n = 800$, and $n = 400$.

$S_n = \frac{n}{2}(a_1 + a_n)$

$S_{400} = \frac{400}{2}(2 + 800) = 200(802) = 160,400$

28. $1 + 3 + 5 + \ldots + 197 + 199.$

$S_{100} = \frac{100}{2}(1 + 199) = 50(200) = 10,000$

29. The sum is $7 + 14 + 21 + \ldots + 91 + 98$. This is the sum of the arithmetic sequence for which $a_1 = 7$, $a_n = 98$, and $n = 14$.

$S_n = \frac{n}{2}(a_1 + a_n)$

$S_{14} = \frac{14}{2}(7 + 98) = 7(105) = 735$

30. $16 + 20 + 24 + \ldots + 516 + 520$

$S_{127} = \frac{127}{2}(16 + 520) = 34,036$

31. First we find a_{20}:

$$a_n = a_1 + (n-1)d$$
$$a_{20} = 2 + (20-1)5 = 2 + 19 \cdot 5 = 97$$

Then

$$S_n = \frac{n}{2}(a_1 + a_n)$$
$$S_{20} = \frac{20}{2}(2+97) = 10(99) = 990.$$

32. $a_{32} = 7 + (32-1)(-3) = 7 + (31)(-3) = -86$

$$S_{32} = \frac{32}{2}[7 + (-86)] = 16(-79) = -1264$$

33. $\sum_{k=1}^{40} (2k+3)$

Write a few terms of the sum:

$5 + 7 + 9 + \ldots + 83$

This is a series coming from an arithmetic sequence with $a_1 = 5$, $n = 40$, and $a_{40} = 83$. Then

$$S_n = \frac{n}{2}(a_1 + a_n)$$
$$S_{40} = \frac{40}{2}(5+83) = 20(88) = 1760$$

34. $\sum_{k=5}^{20} 8k$

$40 + 48 + 56 + 64 + \ldots + 160$

This is equivalent to a series coming from an arithmetic sequence with $a_1 = 40$, $n = 16$, and $n_{16} = 160$.

$$S_{16} = \frac{16}{2}(40+160) = 1600$$

35. $\sum_{k=0}^{19} \frac{k-3}{4}$

Write a few terms of the sum:

$-\frac{3}{4} - \frac{1}{2} - \frac{1}{4} + 0 + \frac{1}{4} + \ldots + 4$

Since k goes from 0 through 19, there are 20 terms. Thus, this is equivalent to a series coming from an arithmetic sequence with $a_1 = -\frac{3}{4}$, $n = 20$, and $a_{20} = 4$. Then

$$S_n = \frac{n}{2}(a_1 + a_n)$$
$$S_{20} = \frac{20}{2}\left(-\frac{3}{4} + 4\right) = 10 \cdot \frac{13}{4} = \frac{65}{2}.$$

36. $\sum_{k=2}^{50} (2000 - 3k)$

$1994 + 1991 + 1988 + \ldots + 1850$

This is equivalent to a series coming from an arithmetic sequence with $a_1 = 1994$, $n = 49$, and $n_{49} = 1850$.

$$S_{49} = \frac{49}{2}(1994 + 1850) = 94,178$$

37. $\sum_{k=12}^{57} \frac{7-4k}{13}$

Write a few terms of the sum:

$-\frac{41}{13} - \frac{45}{13} - \frac{49}{13} - \ldots - \frac{221}{13}$

Since k goes from 12 through 57, there are 46 terms. Thus, this is equivalent to a series coming from an arithmetic sequence with $a_1 = -\frac{41}{13}$, $n = 46$, and $a_{46} = -\frac{221}{13}$. Then

$$S_n = \frac{n}{2}(a_1 + a_n)$$
$$S_{46} = \frac{46}{2}\left(-\frac{41}{13} - \frac{221}{13}\right) = 23\left(-\frac{262}{13}\right) = -\frac{6026}{13}.$$

38. First find $\sum_{k=101}^{200} (1.14k - 2.8)$.

$112.34 + 113.48 + \ldots + 225.2$

This is equivalent to a series coming from an arithmetic sequence with $a_1 = 112.34$, $n = 100$, and $a_{100} = 225.2$.

$$S_{100} = \frac{100}{2}(112.34 + 225.2) = 16,877$$

Next find $\sum_{k=1}^{5} \left(\frac{k+4}{10}\right)$.

$$S_5 = \frac{5}{2}\left(\frac{1}{2} + \frac{9}{10}\right) = 3.5$$

Then $16,877 - 3.5 = 16,873.5$.

39. ***Familiarize.*** We go from 50 poles in a row, down to six poles in the top row, so there must be 45 rows. We want the sum $50 + 49 + 48 + \ldots + 6$. Thus we want the sum of an arithmetic sequence. We will use the formula $S_n = \frac{n}{2}(a_1 + a_n)$.

Translate. We want to find the sum of the first 45 terms of an arithmetic sequence with $a_1 = 50$ and $a_{45} = 6$.

Carry out. Substituting into the formula, we have

$$S_{45} = \frac{45}{2}(50+6) = \frac{45}{2} \cdot 56 = 1260$$

Check. We can do the calculation again, or we can do the entire addition:

$50 + 49 + 48 + \ldots + 6$.

State. There will be 1260 poles in the pile.

40. $a_{25} = 5000 + (25-1)(1125) = 5000 + 24 \cdot 1125 = 32,000$

$$S_{25} = \frac{25}{2}(5000 + 32,000) = \frac{25}{2}(37,000) = \$462,500$$

41. We first find how many plants will be in the last row.

Familiarize. The sequence is 35, 31, 27, ... It is an arithmetic sequence with $a_1 = 35$ and $d = -4$. Since each row must contain a positive number of plants, we must determine how many times we can add -4 to 35 and still have a positive result.

Translate. We find the largest integer x for which $35 + x(-4) > 0$. Then we evaluate the expression $35 - 4x$ for that value of x.

Carry out. We solve the inequality.

$$35 - 4x > 0$$
$$35 > 4x$$
$$\frac{35}{4} > x$$
$$8\frac{3}{4} > x$$

The integer we are looking for is 8. Thus $35 - 4x = 35 - 4(8) = 3$.

Check. If we add -4 to 35 eight times we get 3, a positive number, but if we add -4 to 35 more than eight times we get a negative number.

State. There will be 3 plants in the last row.

Next we find how many plants there are altogether.

Familiarize. We want to find the sum $35+31+27+\ldots+3$. We know $a_1 = 35$ $a_n = 3$, and, since we add -4 to 35 eight times, $n = 9$. (There are 8 terms after a_1, for a total of 9 terms.) We will use the formula $S_n = \frac{n}{2}(a_1 + a_n)$.

Translate. We want to find the sum of the first 9 terms of an arithmetic sequence in which $a_1 = 35$ and $a_9 = 3$.

Carry out. Substituting into the formula, we have

$$S_9 = \frac{9}{2}(35 + 3)$$
$$= \frac{9}{2} \cdot 38 = 171$$

Check. We can check the calculations by doing them again. We could also do the entire addition:

$$35 + 31 + 27 + \ldots + 3.$$

State. There are 171 plants altogether.

42. $a_8 = 10 + (8 - 1)(2) = 10 + 7 \cdot 2 = 24$ marchers

$S_8 = \frac{8}{2}(10 + 24) = 4 \cdot 34 = 136$ marchers

43. ***Familiarize.*** We have a sequence 10, 20, 30, . . . It is an arithmetic sequence with $a_1 = 10$, $d = 10$, and $n = 31$.

Translate. We want to find $S_n = \frac{n}{2}(a_1 + a_n)$ where $a_n = a_1 + (n - 1)d$, $a_1 = 10$, $d = 10$, and $n = 31$.

Carry out. First we find a_{31}.

$a_{31} = 10 + (31 - 1)10 = 10 + 30 \cdot 10 = 310$

Then $S_{31} = \frac{31}{2}(10 + 310) = \frac{31}{2} \cdot 320 = 4960$.

Check. We can do the calculation again, or we can do the entire addition:

$$10 + 20 + 30 + \ldots + 310.$$

State. A total of 4960¢, or \$49.60 is saved.

44. Yes; $d = 48 - 16 = 80 - 48 = 112 - 80 = 144 - 112 = 32$.

$a_{10} = 16 + (10 - 1)32 = 304$

$S_{10} = \frac{10}{2}(16 + 304) = 1600$ ft

45. ***Familiarize.*** We have arithmetic sequence with $a_1 = 28$, $d = 4$, and $n = 20$.

Translate. We want to find $S_n = \frac{n}{2}(a_1 + a_n)$ where $a_n = a_1 + (n - 1)d$, $a_1 = 28$, $d = 4$, and $n = 20$.

Carry out. First we find a_{20}.

$a_{20} = 28 + (20 - 1)4 = 104$

Then $S_{20} = \frac{20}{2}(28 + 104) = 10 \cdot 132 = 1320$.

Check. We can do the calculations again, or we can do the entire addition:

$$28 + 32 + 36 + \ldots 104.$$

State. There are 1320 seats in the first balcony.

46. Yes; $d = 0.6080 - 0.5908 = 0.6252 - 0.6080 = \ldots = 0.7112 - 0.6940 = 0.0172$

47. Yes; $d = 6 - 3 = 9 - 6 = 3n - 3(n - 1) = 3$.

48. The first formula can be derived from the second by substituting $a_1 + (n - 1)d$ for a_n. When the first and last terms of the sum are known, the second formula is the better one to use. If the last term is not known, the first formula allows us to compute the sum in one step without first finding a_n.

49.

$$1 + 2 + 3 + \ldots + 100$$
$$= (1 + 100) + (2 + 99) + (3 + 98) + \ldots + (50 + 51)$$
$$= \underbrace{101 + 101 + 101 + \ldots + 101}_{\text{50 addends of 101}}$$
$$= 50 \cdot 101$$
$$= 5050$$

A formula for the first n natural numbers is $\frac{n}{2}(1 + n)$.

50. $7x - 2y = 4$, (1)

$x + 3y = 17$ (2)

Multiply equation (1) by 3 and equation (2) by 2 and add.

$$21x - 6y = 12$$
$$2x + 6y = 34$$
$$23x = 46$$
$$x = 2$$

Back-substitute to find y.

$2 + 3y = 17$ Using equation (2)

$3y = 15$

$y = 5$

The solution is $(2, 5)$.

51. $2x + y + 3z = 12$
$x - 3y - 2z = -1$
$5x + 2y - 4z = -4$

We will use Gauss-Jordan elimination with matrices. First we write the augmented matrix.

$$\left[\begin{array}{ccc|c} 2 & 1 & 3 & 12 \\ 1 & -3 & -2 & -1 \\ 5 & 2 & -4 & -4 \end{array}\right]$$

Next we interchange the first two rows.

$$\left[\begin{array}{ccc|c} 1 & -3 & -2 & -1 \\ 2 & 1 & 3 & 12 \\ 5 & 2 & -4 & -4 \end{array}\right]$$

Now multiply the first row by -2 and add it to the second row. Also multiply the first row by -5 and add it to the third row.

$$\left[\begin{array}{ccc|c} 1 & -3 & -2 & -1 \\ 0 & 7 & 7 & 14 \\ 0 & 17 & 6 & 1 \end{array}\right]$$

Multiply the second row by $\frac{1}{7}$.

$$\left[\begin{array}{ccc|c} 1 & -3 & -2 & -1 \\ 0 & 1 & 1 & 2 \\ 0 & 17 & 6 & 1 \end{array}\right]$$

Multiply the second row by 3 and add it to the first row. Also multiply the second row by -17 and add it to the third row.

$$\left[\begin{array}{ccc|c} 1 & 0 & 1 & 5 \\ 0 & 1 & 1 & 2 \\ 0 & 0 & -11 & -33 \end{array}\right]$$

Multiply the third row by $-\frac{1}{11}$.

$$\left[\begin{array}{ccc|c} 1 & 0 & 1 & 5 \\ 0 & 1 & 1 & 2 \\ 0 & 0 & 1 & 3 \end{array}\right]$$

Multiply the third row by -1 and add it to the first row and also to the second row.

$$\left[\begin{array}{ccc|c} 1 & 0 & 0 & 2 \\ 0 & 1 & 0 & -1 \\ 0 & 0 & 1 & 3 \end{array}\right]$$

Now we can read the solution from the matrix. It is $(2, -1, 3)$.

52. $9x^2 + 16y^2 = 144$

$$\frac{x^2}{16} + \frac{y^2}{9} = 1$$

Vertices: $(-4, 0)$, $(4, 0)$

$c^2 = a^2 - b^2 = 16 - 9 = 7$

$c = \sqrt{7}$

Foci: $(-\sqrt{7}, 0)$, $(\sqrt{7}, 0)$

53. The vertices are on the y-axis, so the transverse axis is vertical and $a = 5$. The length of the minor axis is 4, so $b = 4/2 = 2$. The equation is

$$\frac{x^2}{4} + \frac{y^2}{25} = 1.$$

54. Let $x =$ the first number in the sequence and let $d =$ the common difference. Then the three numbers in the sequence are x, $x + d$, and $x + 2d$. Solve:

$$x + x + 2d = 10,$$
$$x(x + d) = 15.$$

We get $x = 3$ and $d = 2$ so the numbers are 3, $3 + 2$, and $3 + 2 \cdot 2$, or 3, 5, and 7.

55. $S_n = \frac{n}{2}(1 + 2n - 1) = n^2$

56.

$$\begin{aligned}
a_1 &= \$8760 \\
a_2 &= \$8760 + (-\$798.23) = \$7961.77 \\
a_3 &= \$8760 + 2(-\$798.23) = \$7163.54 \\
a_4 &= \$8760 + 3(-\$798.23) = \$6365.31 \\
a_5 &= \$8760 + 4(-\$798.23) = \$5567.08 \\
a_6 &= \$8760 + 5(-\$798.23) = \$4768.85 \\
a_7 &= \$8760 + 6(-\$798.23) = \$3970.62 \\
a_8 &= \$8760 + 7(-\$798.23) = \$3172.39 \\
a_9 &= \$8760 + 8(-\$798.23) = \$2374.16 \\
a_{10} &= \$8760 + 9(-\$798.23) = \$1575.93 \\
S_{10} &= \frac{10}{2}(\$8760 + \$1575.93) = \$51{,}679.65
\end{aligned}$$

57. Let $d =$ the common difference. Then $a_4 = a_2 + 2d$, or

$$\begin{aligned}
10p + q &= 40 - 3q + 2d \\
10p + 4q - 40 &= 2d \\
5p + 2q - 20 &= d.
\end{aligned}$$

Also, $a_1 = a_2 - d$, so we have

$$\begin{aligned}
a_1 &= 40 - 3q - (5p + 2q - 20) \\
&= 40 - 3q - 5p - 2q + 20 \\
&= 60 - 5p - 5q.
\end{aligned}$$

58. $P(x) = x^4 + 4x^3 - 84x^2 - 176x + 640$ has at most 4 zeros because $P(x)$ is of degree 4. By the rational roots theorem, the possible zeros are

$\pm 1, \pm 2, \pm 4, \pm 5, \pm 8, \pm 10, \pm 16, \pm 20, \pm 32, \pm 40, \pm 64, \pm 80, \pm 128, \pm 160, \pm 320, \pm 640$

Using the graph of the function and synthetic division, we find that two zeros are -4 and 2.

Also by synthetic division we determine that ± 1 and -2 are not zeros. Therefore we determine that $d = 6$ in the arithmetic sequence.

Possible arithmetic sequences:

a) $-4,\ 2,\ 8,\ 14$

b) $-4,\ 2,\ 8$

c) $-16,\ -10,\ -4,\ 2$

The solution cannot be (a) because 14 is not a possible zero. Checking -16 by synthetic division we find that -16 is not a zero. Thus (*b*) is the only possible arithmetic sequence which contains all four zeros. Synthetic division confirms that -10 and 8 are also zeros. The zeros are -10, -4, 2, and 8.

59. $4,\ m_1,\ m_2,\ m_3,\ 1$

We look for m_1, m_2, and m_3 such that $4,\ m_1,\ m_2,\ m_3,\ 13$ is an arithmetic sequence. In this case, $a_1 = 4$, $n = 5$, and $a_5 = 12$. First we find d:

$$a_n = a_1 + (n-1)d$$
$$12 = 4 + (5-1)d$$
$$12 = 4 + 4d$$
$$8 = 4d$$
$$2 = d$$

Then we have

$$m_1 = a_1 + d = 4 + 2 = 6$$
$$m_2 = m_1 + d = 6 + 2 = 8$$
$$m_3 = m_2 + d = 8 + 2 = 10$$

60. $-3,\ m_1,\ m_2,\ m_3,\ 5$

We look for m_1, m_2, and m_3 such that $-3,\ m_1,\ m_2,\ m_3,$ 5 is an arithmetic sequence. In this case, $a_1 = -3$, $n = 5$, and $a_5 = 5$. First we find d:

$$a_n = a_1 + (n-1)d$$
$$5 = -3 + (5-1)d$$
$$8 = 4d$$
$$2 = d$$

Then we have

$$m_1 = a_1 + d = -3 + 2 = -1,$$
$$m_2 = m_1 + d = -1 + 2 = 1,$$
$$m_3 = m_2 + d = 1 + 2 = 3.$$

61. $4,\ m_1,\ m_2,\ m_3,\ m_4,\ 13$

We look for m_1, m_2, m_3, and m_4 such that $4,\ m_1,\ m_2,$ $m_3,\ m_4,\ 13$ is an arithmetic sequence. In this case $a_1 = 4$, $n = 6$, and $a_6 = 13$. First we find d.

$$a_n = a_1 + (n-1)d$$
$$13 = 4 + (6-1)d$$
$$9 = 5d$$
$$1\frac{4}{5} = d$$

Then we have

$$m_1 = a_1 + d = 4 + 1\frac{4}{5} = 5\frac{4}{5},$$
$$m_2 = m_1 + d = 5\frac{4}{5} + 1\frac{4}{5} = 6\frac{8}{5} = 7\frac{3}{5},$$
$$m_3 = m_2 + d = 7\frac{3}{5} + 1\frac{4}{5} = 8\frac{7}{5} = 9\frac{2}{5},$$
$$m_4 = m_3 + d = 9\frac{2}{5} + 1\frac{4}{5} = 10\frac{6}{5} = 11\frac{1}{5}.$$

62. $27,\ m_1,\ m_2,\ \ldots,\ m_9,\ m_{10},\ 300$

$$300 = 27 + (12-1)d$$
$$273 = 11d$$
$$\frac{273}{11} = d, \text{ or}$$
$$24\frac{9}{11} = d$$
$$m_1 = 27 + 24\frac{9}{11} = 51\frac{9}{11}$$
$$m_2 = 51\frac{9}{11} + 24\frac{9}{11} = 76\frac{7}{11}$$
$$m_3 = 76\frac{7}{11} + 24\frac{9}{11} = 101\frac{5}{11},$$
$$m_4 = 101\frac{5}{11} + 24\frac{9}{11} = 126\frac{3}{11},$$
$$m_5 = 126\frac{3}{11} + 24\frac{9}{11} = 151\frac{1}{11},$$
$$m_6 = 151\frac{1}{11} + 24\frac{9}{11} = 175\frac{10}{11},$$
$$m_7 = 175\frac{10}{11} + 24\frac{9}{11} = 200\frac{8}{11},$$
$$m_8 = 200\frac{8}{11} + 24\frac{9}{11} = 225\frac{6}{11},$$
$$m_9 = 225\frac{6}{11} + 24\frac{9}{11} = 250\frac{4}{11},$$
$$m_{10} = 250\frac{4}{11} + 24\frac{9}{11} = 275\frac{2}{11}$$

63. $1,\ 1+d,\ 1+2d,\ \ldots,\ 50$ has n terms and $S_n = 459$.

Find n:

$$459 = \frac{n}{2}(1 + 50)$$
$$18 = n$$

Find d:

$$50 = 1 + (18-1)d$$
$$\frac{49}{17} = d$$

The sequence has a total of 18 terms, so we insert 16 arithmetic means between 1 and 50 with $d = \dfrac{49}{17}$.

64. a) $a_t = \$5200 - t\left(\dfrac{\$5200 - \$1100}{8}\right)$

$$a_t = \$5200 - \$512.50t$$

b)
$$a_0 = \$5200 - \$512.50(0) = \$5200$$
$$a_1 = \$5200 - \$512.50(1) = \$4687.50$$
$$a_2 = \$5200 - \$512.50(2) = \$4175$$
$$a_3 = \$5200 - \$512.50(3) = \$3662.50$$
$$a_4 = \$5200 - \$512.50(4) = \$3150$$
$$a_7 = \$5200 - \$512.50(7) = \$1612.50$$
$$a_8 = \$5200 - \$512.50(8) = \$1100$$

65.
$$\begin{array}{l} m = p + d \\ m = q - d \\ \hline 2m = p + q \quad \text{Adding} \\ m = \frac{p+q}{2} \end{array}$$

Exercise Set 10.3

1. 2, 4, 8, 16, . . .

$\frac{4}{2} = 2$, $\frac{8}{4} = 2$, $\frac{16}{8} = 2$

$r = 2$

2. $r = -\frac{6}{18} = -\frac{1}{3}$

3. 1, −1, 1, −1, . . .

$\frac{-1}{1} = -1$, $\frac{1}{-1} = -1$, $\frac{-1}{1} = -1$

$r = -1$

4. $r = \frac{-0.8}{8} = 0.1$

5. $\frac{2}{3}, -\frac{4}{3}, \frac{8}{3}, -\frac{16}{3}, \ldots$

$\frac{-\frac{4}{3}}{\frac{2}{3}} = -2$, $\frac{\frac{8}{3}}{-\frac{4}{3}} = -2$, $\frac{-\frac{16}{3}}{\frac{8}{3}} = -2$

$r = -2$

6. $r = \frac{15}{75} = \frac{1}{5}$

7. $\frac{0.6275}{6.275} = 0.1$, $\frac{0.06275}{0.6275} = 0.1$

$r = 0.1$

8. $r = \frac{\frac{1}{x^2}}{\frac{1}{x}} = \frac{1}{x}$

9. $\frac{\frac{5a}{2}}{5} = \frac{a}{2}$, $\frac{\frac{5a^2}{4}}{\frac{5a}{2}} = \frac{a}{2}$, $\frac{\frac{5a^3}{8}}{\frac{5a}{4}} = \frac{a}{2}$

$r = \frac{a}{2}$

10. $r = \frac{\$858}{\$780} = 1.1$

11. 2, 4, 8, 16, . . .

$a_1 = 2$, $n = 7$, and $r = \frac{4}{2}$, or 2.

We use the formula $a_n = a_1 r^{n-1}$.

$a_7 = 2(2)^{7-1} = 2 \cdot 2^6 = 2 \cdot 64 = 128$

12. $a_9 = 2(-5)^{9-1} = 781{,}250$

13. $2, 2\sqrt{3}, 6, \ldots$

$a_1 = 2$, $n = 9$, and $r = \frac{2\sqrt{3}}{2}$, or $\sqrt{3}$

$a_n = a_1 r^{n-1}$

$a_9 = 2(\sqrt{3})^{9-1} = 2(\sqrt{3})^8 = 2 \cdot 81 = 162$

14. $a_{57} = 1(-1)^{57-1} = 1$

15. $\frac{7}{625}, -\frac{7}{25}, \ldots$

$a_1 = \frac{7}{625}$, $n = 23$, and $r = \frac{-\frac{7}{25}}{\frac{7}{625}} = -25.$

$a_n = a_1 r^{n-1}$

$a_{23} = \frac{7}{625}(-25)^{23-1} = \frac{7}{625}(-25)^{22}$

$= \frac{7}{25^2} \cdot 25^2 \cdot 25^{20} = 7(25)^{20}$, or $7(5)^{40}$

16. $a_5 = \$1000(1.06)^{5-1} \approx \$1262.48$

17. 1, 3, 9, . . .

$a_1 = 1$ and $r = \frac{3}{1}$, or 3

$a_n = a_1 r^{n-1}$

$a_n = 1(3)^{n-1} = 3^{n-1}$

18. $a_n = 25\left(\frac{1}{5}\right)^{n-1} = \frac{5^2}{5^{n-1}} = 5^{3-n}$

19. 1, −1, 1, −1, . . .

$a_1 = 1$ and $r = \frac{-1}{1} = -1$

$a_n = a_1 r^{n-1}$

$a_n = 1(-1)^{n-1} = (-1)^{n-1}$

20. $a_n = (-2)^n$

21. $\frac{1}{x}, \frac{1}{x^2}, \frac{1}{x^2}, \ldots$

$a_1 = \frac{1}{x}$ and $r = \frac{\frac{1}{x^2}}{\frac{1}{x}} = \frac{1}{x}$

$a_n = a_1 r^{n-1}$

$a_n = \frac{1}{x}\left(\frac{1}{x}\right)^{n-1} = \frac{1}{x} \cdot \frac{1}{x^{n-1}} = \frac{1}{x^{1+n-1}} = \frac{1}{x^n}$

22. $a_n = 5\left(\frac{a}{2}\right)^{n-1}$

23. $6 + 12 + 24 + \ldots$

$a_1 = 6$, $n = 7$, and $r = \frac{12}{6}$, or 2

$S_n = \frac{a_1(1 - r^n)}{1 - r}$

$S_7 = \frac{6(1 - 2^7)}{1 - 2} = \frac{6(1 - 128)}{-1} = \frac{6(-127)}{-1} = 762$

24. $S_{10} = \dfrac{16\left[1-\left(-\frac{1}{2}\right)^{10}\right]}{1-\left(-\frac{1}{2}\right)} = \dfrac{16\left(1-\frac{1}{1024}\right)}{\frac{3}{2}} =$

$\dfrac{16\left(\frac{1023}{1024}\right)}{\frac{3}{2}} = \dfrac{341}{32}$, or $10\dfrac{21}{32}$

25. $\dfrac{1}{18} - \dfrac{1}{6} + \dfrac{1}{2} - \ldots$

$a_1 = \dfrac{1}{18}$, $n = 9$, and $r = \dfrac{-\frac{1}{6}}{\frac{1}{18}} = -\dfrac{1}{6} \cdot \dfrac{18}{1} = -3$

$S_n = \dfrac{a_1(1-r^n)}{1-r}$

$S_9 = \dfrac{\frac{1}{18}\left[1-(-3)^9\right]}{1-(-3)} = \dfrac{\frac{1}{18}(1+19,683)}{4}$

$\dfrac{\frac{1}{18}(19,684)}{4} = \dfrac{1}{18}(19,684)\left(\dfrac{1}{4}\right) = \dfrac{4921}{18}$

26. $a_n = a_1 r^{n-1}$

$-\dfrac{1}{32} = (-8)\cdot\left(-\dfrac{1}{2}\right)^{n-1}$

$n = 9$

$S_9 = \dfrac{-8\left[\left(-\frac{1}{2}\right)^9 - 1\right]}{-\frac{1}{2} - 1} = -\dfrac{171}{32}$

27. $4 + 2 + 1 + \ldots$

$|r| = \left|\dfrac{2}{4}\right| = \left|\dfrac{1}{2}\right| = \dfrac{1}{2}$, and since $|r| < 1$, the series does have a sum.

$S_\infty = \dfrac{a_1}{1-r} = \dfrac{4}{1-\frac{1}{2}} = \dfrac{4}{\frac{1}{2}} = 4 \cdot \dfrac{2}{1} = 8$

28. $|r| = \left|\dfrac{3}{7}\right| = \dfrac{3}{7} < 1$, so the series has a sum.

$S_\infty = \dfrac{7}{1-\frac{3}{7}} = \dfrac{7}{\frac{4}{7}} = \dfrac{49}{4}$

29. $25 + 20 + 16 + \ldots$

$|r| = \left|\dfrac{20}{25}\right| = \left|\dfrac{4}{5}\right| = \dfrac{4}{5}$, and since $|r| < 1$, the series does have a sum.

$S_\infty = \dfrac{a_1}{1-r} = \dfrac{25}{1-\frac{4}{5}} = \dfrac{25}{\frac{1}{5}} = 25 \cdot \dfrac{5}{1} = 125$

30. $|r| = \left|\dfrac{-10}{100}\right| = \dfrac{1}{10} < 1$, so the series has a sum.

$S_\infty = \dfrac{100}{1-\left(-\frac{1}{10}\right)} = \dfrac{100}{\frac{11}{10}} = \dfrac{1000}{11}$

31. $8 + 40 + 200 + \ldots$

$|r| = \left|\dfrac{40}{8}\right| = |5| = 5$, and since $|r| > 1$ the series does not have a sum.

32. $|r| = \left|\dfrac{3}{-6}\right| = \left|-\dfrac{1}{2}\right| = \dfrac{1}{2} < 1$, so the series has a sum.

$S_\infty = \dfrac{-6}{1-\left(-\frac{1}{2}\right)} = \dfrac{-6}{\frac{3}{2}} = -6 \cdot \dfrac{2}{3} = -4$

33. $0.6 + 0.06 + 0.006 + \ldots$

$|r| = \left|\dfrac{0.06}{0.6}\right| = |0.1| = 0.1$, and since $|r| < 1$, the series does have a sum.

$S_\infty = \dfrac{a_1}{1-r} = \dfrac{0.6}{1-0.1} = \dfrac{0.6}{0.9} = \dfrac{6}{9} = \dfrac{2}{3}$

34. $\sum_{k=0}^{10} 3^k$

$a_1 = 1$, $|r| = 3$, $n = 11$

$S_{11} = \dfrac{1(1-3^{11})}{1-3} = 88,573$

35. $\sum_{k=1}^{11} 15\left(\dfrac{2}{3}\right)^k$

$a_1 = 15 \cdot \dfrac{2}{3}$ or 10; $|r| = \left|\dfrac{2}{3}\right| = \dfrac{2}{3}$, $n = 11$

$S_{11} = \dfrac{10\left[1-\left(\frac{2}{3}\right)^{11}\right]}{1-\frac{2}{3}} = \dfrac{10\left[1-\frac{2048}{177,147}\right]}{\frac{1}{3}}$

$= 10 \cdot \dfrac{175,099}{177,147} \cdot 3$

$= \dfrac{1,750,990}{59,049}$, or $29\dfrac{38,569}{59,049}$

36. $\sum_{k=0}^{50} 200(1.08)^k$

$a_1 = 200$, $|r| = 1.08$, $n = 51$

$S_{51} = \dfrac{200[1-(1.08)^{51}]}{1-1.08} \approx 124,134.354$

37. $\sum_{k=1}^{\infty} \left(\dfrac{1}{2}\right)^{k-1}$

$a_1 = 1$, $|r| = \left|\dfrac{1}{2}\right| = \dfrac{1}{2}$

$S_\infty = \dfrac{a_1}{1-r} = \dfrac{1}{1-\frac{1}{2}} = \dfrac{1}{\frac{1}{2}} = 2$

38. Since $|r| = |2| > 1$, the sum does not exist.

39. $\sum_{k=1}^{\infty} 12.5^k$

Since $|r| = 12.5 > 1$, the sum does not exist.

40. Since $|r| = 1.0625 > 1$, the sum does not exist.

41. $\sum_{k=1}^{\infty} \$500(1.11)^{-k}$

$a_1 = \$500(1.11)^{-1}$, or $\frac{\$500}{1.11}$; $|r| = |1.11^{-1}| = \frac{1}{1.11}$

$$S_\infty = \frac{a_1}{1-r} = \frac{\frac{\$500}{1.11}}{1-\frac{1}{1.11}} = \frac{\frac{\$500}{1.11}}{\frac{0.11}{1.11}} \approx \$4545.\overline{45}$$

42. $\sum_{k=1}^{\infty} \$1000(1.06)^{-k}$

$a_1 = \frac{\$1000}{1.06}$, $|r| = \frac{1}{1.06}$

$$S_\infty = \frac{\frac{\$1000}{1.06}}{1-\frac{1}{1.06}} = \frac{\frac{\$1000}{1.06}}{\frac{0.06}{1.06}} \approx \$16,666.\overline{66}$$

43. $\sum_{k=1}^{\infty} 16(0.1)^{k-1}$

$a_1 = 16$, $|r| = |0.1| = 0.1$

$$S_\infty = \frac{a_1}{1-r} = \frac{16}{1-0.1} = \frac{16}{0.9} = \frac{160}{9}$$

44. $\sum_{k=1}^{\infty} \frac{8}{3}\left(\frac{1}{2}\right)^{k-1}$

$a_1 = \frac{8}{3}$, $|r| = \frac{1}{2}$

$$S_\infty = \frac{\frac{8}{3}}{1-\frac{1}{2}} = \frac{\frac{8}{3}}{\frac{1}{2}} = \frac{16}{3}$$

45. $0.131313\ldots = 0.13 + 0.0013 + 0.000013 + \ldots$

This is an infinite geometric series with $a_1 = 0.13$.

$|r| = \left|\frac{0.0013}{0.13}\right| = |0.01| = 0.01 < 1$, so the series has a limit.

$$S_\infty = \frac{a_1}{1-r} = \frac{0.13}{1-0.01} = \frac{0.13}{0.99} = \frac{13}{99}$$

46. $0.2222 = 0.2 + 0.02 + 0.002 + 0.0002 + \ldots$

$|r| = \left|\frac{0.02}{0.2}\right| = |0.1| = 0.1$

$$S_\infty = \frac{0.2}{1-0.1} = \frac{0.2}{0.9} = \frac{2}{9}$$

47. We will find fractional notation for $0.999\overline{9}$ and then add 8.

$0.999\overline{9} = 0.9 + 0.09 + 0.009 + 0.0009 + \ldots$

This is an infinite geometric series with $a_1 = 0.9$.

$|r| = \left|\frac{0.09}{0.9}\right| = |0.1| = 0.1 < 1$, so the series has a limit.

$$S_\infty = \frac{a_1}{1-r} = \frac{0.9}{1-0.1} = \frac{0.9}{0.9} = 1$$

Then $8.999\overline{9} = 8 + 1 = 9$.

48. $0.1\overline{6} = 0.16 + 0.0016 + 0.000016 + \ldots$

$|r| = \left|\frac{0.0016}{0.16}\right| = |0.01| = 0.01$

$$S_\infty = \frac{0.16}{1-0.01} = \frac{0.16}{0.99} = \frac{16}{99}$$

Then $6.1\overline{6} = 6 + \frac{16}{99} = \frac{610}{99}$.

49. $3.4125\overline{125} = 3.4 + 0.0125\overline{125}$

We will find fractional notation for $0.0125\overline{125}$ and then add 3.4, or $\frac{34}{10}$, or $\frac{17}{5}$.

$0.0125\overline{125} = 0.0125 + 0.0000125 + \ldots$

This is an infinite geometric series with $a_1 = 0.0125$.

$|r| = \left|\frac{0.0000125}{0.0125}\right| = |0.001| = 0.001 < 1$, so the series has a limit.

$$S_\infty = \frac{a_1}{1-r} = \frac{0.0125}{1-0.001} = \frac{0.0125}{0.999} = \frac{125}{9990}$$

Then $\frac{17}{5} + \frac{125}{9990} = \frac{33,966}{9990} + \frac{125}{9990} = \frac{34,091}{9990}$

50. $12.7809\overline{809} = 12.7 + 0.0809\overline{809}$

$0.0809\overline{809} = 0.0809 + 0.0000809 + \ldots$

$|r| = \left|\frac{0.0000809}{0.0809}\right| = |0.001| = 0.001$

$$S_\infty = \frac{0.0809}{1-0.001} = \frac{0.0809}{0.999} = \frac{809}{9990}$$

Then $12.7 + \frac{809}{9990} = \frac{127}{10} + \frac{809}{9990} = \frac{127,682}{9990} = \frac{63,841}{4995}$

51. a) ***Familiarize.*** The rebound distances form a geometric sequence:

$$\frac{1}{4}\times 16,\quad \left(\frac{1}{4}\right)^2\times 16,\quad \left(\frac{1}{4}\right)^3\times 16,\ \ldots,$$

or $4,\quad \frac{1}{4}\times 4,\quad \left(\frac{1}{4}\right)^2\times 4,\ \ldots$

The height of the 6th rebound is the 6th term of the sequence.

Translate. We will use the formula $a_n = a_1 r^{n-1}$, with $a_1 = 4$, $r = \frac{1}{4}$, and $n = 6$:

$$a_6 = 4\left(\frac{1}{4}\right)^{6-1}$$

Carry out. We calculate to obtain $a_6 = \frac{1}{256}$.

Check. We can do the calculation again.

State. It rebounds $\frac{1}{256}$ ft the 6th time.

b) $S_\infty = \frac{a}{1-r} = \frac{4}{1-\frac{1}{4}} = \frac{4}{\frac{3}{4}} = \frac{16}{3}$ ft, or $5\frac{1}{3}$ ft

52. $a_1 = \$0.01$, $r = 2$, $n = 28$

$$S_{28} = \frac{\$0.01(1-2^{28})}{1-2} \approx \$2,684,355$$

53. a) ***Familiarize.*** The rebound distances form a geometric sequence:

$0.6 \times 200,\ (0.6)^2 \times 200,\ (0.6)^3 \times 200,\ \ldots,$

or $120,\ 0.6 \times 120,\ (0.6)^2 \times 120,\ \ldots$

The total rebound distance after 9 rebounds is the sum of the first 9 terms of this sequence.

Translate. We will use the formula

$S_n = \dfrac{a_1(1-r^n)}{1-r}$ with $a_1 = 120$, $r = 0.6$, and $n = 9$.

Carry out.

$$S_9 = \frac{120[1-(0.6)^9]}{1-0.6} \approx 297$$

Check. We repeat the calculation.

State. The bungee jumper has traveled about 297 ft upward after 9 rebounds.

b) $S_\infty = \dfrac{a_1}{1-r} = \dfrac{120}{1-0.6} = 300$ ft

54. a) $a_1 = 100{,}000$, $r = 1.03$. The population in 15 years will be the 16th term of the sequence $100{,}000,\ (1.03)100{,}000,\ (1.03)^2 100{,}000,\ \ldots$

$$a_{16} = 100{,}000(1.03)^{16-1} \approx 155{,}797$$

b) Solve: $200{,}000 = 100{,}000(1.03)^{n-1}$

$n \approx 24$ yr

55. ***Familiarize.*** The amount of the annuity is the geometric series

$\$1000 + \$1000(1.032) + \$1000(1.032)^2 + \ldots + \$1000(1.032)^{17}$, where $a_1 = \$1000$, $r = 1.032$, and $n = 18$.

Translate. Using the formula

$$S_n = \frac{a_1(1-r^n)}{1-r}$$

we have

$$S_{18} = \frac{\$1000[1-(1.032)^{18}]}{1-1.032}$$

Carry out. We carry out the computation and get $S_{18} \approx \$23{,}841.50$.

Check. Repeat the calculations.

State. The amount of the annuity is \$23,841.50.

56. a) We have a geometric sequence with $a_1 = P$ and $r = 1+i$. Then

$$S_N = V = \frac{P[1-(1+i)^N]}{1-(1+i)} = \frac{P[1-(1+i)^N]}{-i} = \frac{P[(1+i)^N - 1]}{i}$$

b) We have a geometric sequence with $a_1 = P$ and $r = 1 + \dfrac{i}{n}$.

The number of terms is nN. Then

$$S_{nN} = V = \frac{P\left[1-\left(1+\frac{i}{n}\right)^{nN}\right]}{1-\left(1+\frac{i}{n}\right)} = \frac{P\left[1-\left(1+\frac{i}{n}\right)^{nN}\right]}{-\frac{i}{n}} = \frac{P\left[\left(1+\frac{i}{n}\right)^{nN} - 1\right]}{\frac{i}{n}}$$

57. ***Familiarize.*** The amounts owed at the beginning of successive years form a geometric sequence:

$\$120{,}000,\ (1.12)\$120{,}000,\ (1.12)^2\$120{,}000,\ \ldots$

The amount to be repaid at the end of 13 years is the amount owed at the beginning of the 14th year.

Translate. Use the formula $a_n = a_1 r^{n-1}$ with $a_1 = 120{,}000$, $r = 1.12$, and $n = 14$:

$$a_{14} = 120{,}000(1.12)^{14-1}$$

Carry out. We perform the calculation, obtaining $a_{14} \approx \$523{,}619.17$.

Check. Repeat the calculation.

State. At the end of 13 years, \$523,619.17 will be repaid.

58. We have a sequence $0.01,\ 2(0.01),\ 2^2(0.01),\ 2^3(0.01),\ \ldots$. The thickness after 20 folds is given by the 21st term of the sequence.

$$a_{21} = 0.01(2)^{21-1} = 10{,}485.76 \text{ in.}$$

59. ***Familiarize.*** The total effect on the economy is the sum of an infinite geometric series

$\$13{,}000{,}000{,}000 + \$13{,}000{,}000{,}000(0.85) + \$13{,}000{,}000{,}000(0.85)^2 + \ldots$

with $a_1 = \$13{,}000{,}000{,}000$ and $r = 0.85$.

Translate. Using the formula

$$S_\infty = \frac{a_1}{1-r}$$

we have

$$S_\infty = \frac{\$13{,}000{,}000{,}000}{1-0.85}.$$

Carry out. Perform the calculation:

$S_\infty \approx \$86{,}666{,}666{,}667$.

Check. Repeat the calculation.

State. The total effect on the economy is \$86,666,666,667.

60. $S_\infty = \dfrac{5{,}000{,}000(0.3)}{1-0.7} \approx 2{,}142{,}857$

$\dfrac{2{,}142{,}857}{5{,}000{,}000} \approx 0.429$, so this is about 42.9% of the population.

61. Left to the student

62. Left to the student

63. Answers may vary. One possibility is given. Casey invests \$900 at 8% interest, compounded annually. How much will be in the account at the end of 40 years?

64. $S_1 = 2$, $S_2 = 2.5$, $S_3 = 2.\overline{6}$, $S_4 = 2.708\overline{3}$, $S_5 = 2.71\overline{6}$, $S_6 = 2.7180\overline{5}$

$2.718 < S_\infty < 2.719$ since the terms from a_7 on will cause changes in S_n in the fourth decimal place and beyond.

65. $f(x) = x^2$, $g(x) = 4x + 5$

$(f \circ g)(x) = f(g(x)) = f(4x+5) = (4x+5)^2 = 16x^2 + 40x + 25$

$(g \circ f)(x) = g(f(x)) = g(x^2) = 4x^2 + 5$

66. $f(x) = x - 1$, $g(x) = x^2 + x + 3$

$(f \circ g)(x) = f(g(x)) = f(x^2 + x + 3) = x^2 + x + 3 - 1 = x^2 + x + 2$

$(g \circ f)(x) = g(f(x)) = g(x-1) = (x-1)^2 + (x-1) + 3 = x^2 - 2x + 1 + x - 1 + 3 = x^2 - x + 3$

67.
$$5^x = 35$$
$$\ln 5^x = \ln 35$$
$$x \ln 5 = \ln 35$$
$$x = \frac{\ln 35}{\ln 5}$$
$$x \approx 2.209$$

68.
$$\log_2 x = -4$$
$$x = 2^{-4} = \frac{1}{2^4}$$
$$x = \frac{1}{16}$$

69. See the answer section in the text.

70. The sequence is not geometric; $a_4/a_3 \neq a_3/a_2$.

71. a) If the sequence is arithmetic, then $a_2 - a_1 = a_3 - a_2$.

$$x + 7 - (x+3) = 4x - 2 - (x+7)$$
$$x = \frac{13}{3}$$

The three given terms are $\frac{13}{3} + 3 = \frac{22}{3}$, $\frac{13}{3} + 7 = \frac{34}{3}$, and $4 \cdot \frac{13}{3} - 2 = \frac{46}{3}$.

Then $d = \frac{12}{3}$, or 4, so the fourth term is $\frac{46}{3} + \frac{12}{3} = \frac{58}{3}$.

b) If the sequence is geometric, then $a_2/a_1 = a_3/a_2$.

$$\frac{x+7}{x+3} = \frac{4x-2}{x+7}$$
$$x = -\frac{11}{3} \text{ or } x = 5$$

For $x = -\frac{11}{3}$: The three given terms are $-\frac{11}{3} + 3 = -\frac{2}{3}$, $-\frac{11}{3} + 7 = \frac{10}{3}$, and $4\left(-\frac{11}{3}\right) - 2 = -\frac{50}{3}$.

Then $r = -5$, so the fourth term is $-\frac{50}{3}(-5) = \frac{250}{3}$.

For $x = 5$: The three given terms are $5 + 3 = 8$, $5 + 7 = 12$, and $4 \cdot 5 - 2 = 18$. Then $r = \frac{3}{2}$, so the fourth term is $18 \cdot \frac{3}{2} = 27$.

72. $S_n = \frac{1(1-x^n)}{1-x} = \frac{1-x^n}{1-x}$

73. $x^2 - x^3 + x^4 - x^5 + \ldots$

This is a geometric series with $a_1 = x^2$ and $r = -x$.

$$S_n = \frac{a_1(1-r^n)}{1-r} = \frac{x^2(1-(-x)^n)}{1-(-x)} = \frac{x^2(1-(-x)^n)}{1+x}$$

74. $\frac{a_n + 1}{a_n} = r$, so $\frac{(a_n+1)^2}{(a_n)^2} = r^2$; Thus $a_1^2, a_2^2, a_3^2, \ldots$ is a geometric sequence with the common ratio r^2.

75. See the answer section in the text.

76. Let the arithmetic sequence have the common difference $d = a_{n+1} - a_n$. Then for the sequence $5^{a_1}, 5^{a_2}, 5^{a_3}, \ldots$, we have $\frac{5^{a_{n+1}}}{5^{a_n}} = 5^{a_{n+1} - a_n} = 5^d$. Thus, we have a geometric sequence with the common ratio 5^d.

77. ***Familiarize.*** The length of a side of the first square is 16 cm. The length of a side of the next square is the length of the hypotenuse of a right triangle with legs 8 cm and 8 cm, or $8\sqrt{2}$ cm. The length of a side of the next square is the length of the hypotenuse of a right triangle with legs $4\sqrt{2}$ cm and $4\sqrt{2}$ cm, or 8 cm. The areas of the squares form a sequence:

$(16)^2, \ (8\sqrt{2})^2, \ (8)^2, \ldots$, or

$256, \ 128, \ 64, \ldots$.

This is a geometric sequence with $a_1 = 256$ and $r = \frac{1}{2}$.

Translate. We find the sum of the infinite geometric series $256 + 128 + 64 + \ldots$.

$$S_\infty = \frac{a_1}{1-r}$$
$$S_\infty = \frac{256}{1 - \frac{1}{2}}$$

Carry out. We calculate to obtain $S_\infty = 512$.

Check. We can do the calculation again.

State. The sum of the areas is 512 cm^2.

Exercise Set 10.4

1. $n^2 < n^3$

$1^2 < 1^3$, $2^2 < 2^3$, $3^2 < 3^3$, $4^2 < 4^3$, $5^2 < 5^3$

The first statement is false, and the others are true.

2. $1^2 - 1 + 41$ is prime, $2^2 - 2 + 41$ is prime, $3^3 - 3 + 41$ is prime, $4^2 - 4 + 41$ is prime, $5^2 - 5 + 41$ is prime. Each of these statements is true.

The statement is false for $n = 41$; $41^2 - 41 + 41$ is not prime.

3. A polygon of n sides has $\frac{n(n-3)}{2}$ diagonals.

A polygon of 3 sides has $\frac{3(3-3)}{2}$ diagonals.

A polygon of 4 sides has $\frac{4(4-3)}{2}$ diagonals.

A polygon of 5 sides has $\frac{5(5-3)}{2}$ diagonals.

A polygon of 6 sides has $\frac{6(6-3)}{2}$ diagonals.

A polygon of 7 sides has $\frac{7(7-3)}{2}$ diagonals.

Each of these statements is true.

4. The sum of the angles of a polygon of 3 sides is $(3-2)\cdot 180°$.

The sum of the angles of a polygon of 4 sides is $(4-2)\cdot 180°$.

The sum of the angles of a polygon of 5 sides is $(5-2)\cdot 180°$.

The sum of the angles of a polygon of 6 sides is $(6-2)\cdot 180°$.

The sum of the angles of a polygon of 7 sides is $(7-2)\cdot 180°$.

Each of these statements is true.

5. See the answer section in the text.

6. $S_n: \ 4+8+12+...+4n = 2n(n+1)$

$S_1: \ 4 = 2\cdot 1\cdot(1+1)$

$S_k: \ 4+8+12+...+4k = 2k(k+1)$

$S_{k+1}: \ 4+8+12+...+4k+4(k+1) = 2(k+1)(k+2)$

1) *Basis step:* Since $2\cdot 1\cdot(1+1) = 2\cdot 2 = 4$, S_1 is true.

2) *Induction step:* Let k be any natural number. Assume S_k. Deduce S_{k+1}. Starting with the left side of S_{k+1}, we have

$$\underbrace{4+8+12+...+4k}_{} + 4(k+1)$$
$$= 2k(k+1) + 4(k+1) \quad \text{By } S_k$$
$$= (k+1)(2k+4)$$
$$= 2(k+1)(k+2)$$

7. See the answer section in the text.

8. $S_n: \ 3+6+9+...+3n = \frac{3n(n+1)}{2}$

$S_1: \ 3 = \frac{3\cdot 1(1+1)}{2}$

$S_k: \ 3+6+9+...+3k = \frac{3k(k+1)}{2}$

$S_{k+1}: \ 3+6+9+...+3k+3(k+1) = \frac{3(k+1)(k+2)}{2}$

1) *Basis step:* Since $\frac{3\cdot 1(1+1)}{2} = \frac{3\cdot 2}{2} = 3$, S_1 is true.

2) *Induction step:* Let k be any natural number. Assume S_k. Deduce S_{k+1}. Starting with the left side of S_{k+1}, we have

$$\underbrace{3+6+9+...+3k}_{} + 3(k+1)$$
$$= \frac{3k(k+1)}{2} + 3(k+1) \quad \text{By } S_k$$
$$= \frac{3k(k+1)+6(k+1)}{2}$$
$$= \frac{(k+1)(3k+6)}{2}$$
$$= \frac{3(k+1)(k+2)}{2}$$

9. See the answer section in the text.

10. $S_n: \ 2 \le 2^n$

$S_1: \ 2 \le 2^1$

$S_k: \ 2 \le 2^k$

$S_{k+1}: \ 2 \le 2^{k+1}$

1) *Basis step:* Since $2 \le 2$, S_1 is true.

2) *Induction step:* Let k be any natural number. Assume S_k. Deduce S_{k+1}.

$$2 \le 2^k \quad S_k$$
$$2\cdot 2 \le 2^k\cdot 2 \quad \text{Multiplying by 2}$$
$$2 < 2\cdot 2 \le 2^{k+1} \quad (2 < 2\cdot 2)$$
$$2 \le 2^{k+1}$$

11. See the answer section in the text.

12. $S_n: \ 3^n < 3^{n+1}$

$S_1: \ 3^1 < 3^{1+1}$

$S_k: \ 3^k < 3^{k+1}$

$S_{k+1}: \ 3^{k+1} < 3^{k+2}$

1) *Basis step:* Since $3^1 < 3^{1+1}$, or $3 < 9$, S_1 is true.

2) *Induction step:* Let k be any natural number. Assume S_k. Deduce S_{k+1}.

$$3^k < 3^{k+1} \quad S_k$$
$$3^k\cdot 3 < 3^{k+1}\cdot 3 \quad \text{Multiplying by 3}$$
$$3^{k+1} < 3^{k+2}$$

13. See the answer section in the text.

14. $S_n: \ \frac{1}{1\cdot 2} + \frac{1}{2\cdot 3} + ... + \frac{1}{n(n+1)} = \frac{n}{n+1}$

$S_1: \ \frac{1}{1\cdot 2} = \frac{1}{1+1}$

$S_k: \ \frac{1}{1\cdot 2} + \frac{1}{2\cdot 3} + ... + \frac{1}{k(k+1)} = \frac{k}{k+1}$

$S_{k+1}: \ \frac{1}{1\cdot 2} + \frac{1}{2\cdot 3} + ... + \frac{1}{k(k+1)} + \frac{1}{(k+1)(k+2)} = \frac{k+1}{k+2}$

1) *Basis step:* Since $\frac{1}{1+1} = \frac{1}{2} = \frac{1}{1\cdot 2}$, S_1 is true.

2) *Induction step:* Let k be any natural number. Assume S_k. Deduce S_{k+1}.

$$\frac{1}{1\cdot 2}+\frac{1}{2\cdot 3}+\ldots+\frac{1}{k(k+1)}=\frac{k}{k+1} \quad (S_k)$$

$$\frac{1}{1\cdot 2}+\frac{1}{2\cdot 3}+\ldots+\frac{1}{k(k+1)}+\frac{1}{(k+1)(k+2)}=$$

$$\frac{k}{k+1}+\frac{1}{(k+1)(k+2)}$$

$$\left(\text{Adding } \frac{1}{(k+1)(k+2)} \text{ on both sides}\right)$$

$$=\frac{k(k+2)+1}{(k+1)(k+2)}$$

$$=\frac{k^2+2k+1}{(k+1)(k+2)}$$

$$=\frac{(k+1))(k+1)}{(k+1)(k+2)}$$

$$=\frac{k+1}{k+2}$$

15. See the answer section in the text.

16. $S_n:\ x\le x^n$

$S_1:\ x\le x$

$S_k:\ x\le x^k$

$S_{k+1}:\ x\le x^{k+1}$

1) *Basis step:* Since $x=x$, S_1 is true.

2) *Induction step:* Let k be any natural number. Assume S_k. Deduce S_{k+1}.

$$x\le x^k \qquad S_k$$

$$x\cdot x\le x^k\cdot x \qquad \text{Multiplying by } x,\ x>1$$

$$x\le x\cdot x\le x^k\cdot x$$

$$x\le x^{k+1}$$

17. See the answer section in the text.

18. $S_n:\ 1^2+2^2+3^2+\ldots+n^2=\dfrac{n(n+1)(2n+1)}{6}$

$S_1:\ 1^2=\dfrac{1(1+1)(2\cdot 1+1)}{6}$

$S_k:\ 1^2+2^2+3^2+\ldots+k^2=\dfrac{k(k+1)(2k+1)}{6}$

$S_{k+1}:\ 1^2+2^2+\ldots+k^2+(k+1)^2=\dfrac{(k+1)(k+1+1)(2(k+1)+1)}{6}$

1) *Basis step:* $1^2=\dfrac{1(1+1)(2\cdot 1+1)}{6}$ is true.

2) *Induction step:* Let k be any natural number. Assume S_k. Deduce S_{k+1}.

$$1^2+2^2+\ldots+k^2=\frac{k(k+1)(2k+1)}{6}$$

$$1^2+2^2+\ldots+k^2+(k+1)^2$$

$$=\frac{k(k+1)(2k+1)}{6}+(k+1)^2$$

$$=\frac{k(k+1)(2k+1)+6(k+1)^2}{6}$$

$$=\frac{(k+1)(2k^2+7k+6)}{6}$$

$$=\frac{(k+1)(k+2)(2k+3)}{6}$$

$$=\frac{(k+1)(k+1+1)(2(k+1)+1)}{6}$$

19. See the answer section in the text.

20. $S_n:\ 1^4+2^4+3^4+\ldots+n^4=\dfrac{n(n+1)(2n+1)(3n^2+3n-1)}{30}$

$S_1:\ 1^4=\dfrac{1(1+1)(2\cdot 1+1)(3\cdot 1^2+3\cdot 1-1)}{30}$

$S_k:\ 1^4+2^4+3^4+\ldots+k^4=\dfrac{k(k+1)(2k+1)(3k^2+3k-1)}{30}$

$S_{k+1}:\ 1^4+2^4+\ldots+k^4+(k+1)^4=\dfrac{(k+1)(k+1+1)(2(k+1)+1)(3(k+1)^2-3(k+1)-1)}{30}$

1) *Basis step:*

$1^4=\dfrac{1(1+1)(2\cdot 1+1)(3\cdot 1^2+3\cdot 1-1)}{30}$ is true.

2) *Induction step:*

$$1^4+2^4+\ldots+k^4=\frac{k(k+1)(2k+1)(3k^2+3k-1)}{30}$$

$$1^4+2^4+\ldots+k^4+(k+1)^4$$

$$=\frac{k(k+1)(2k+1)(3k^2+3k-1)}{30}+(k+1)^4$$

$$=\frac{k(k+1)(2k+1)(3k^2+3k-1)+30(k+1)^4}{30}$$

$$=\frac{(k+1)(6k^4+39k^3+91k^2+89k+30)}{30}$$

$$=\frac{(k+1)(k+2)(2k+3)(3k^2+9k+5)}{30}$$

$$=[(k+1)(k+1+1)(2(k+1)+1)(3(k+1)^2+3(k+1)-1)]/30$$

21. See the answer section in the text.

22. $S_n:\ 2+5+8+...+3n-1=\dfrac{n(3n+1)}{2}$

$S_1:\ 2=\dfrac{1(3\cdot 1+1)}{2}$

$S_k:\ 2+5+8+...+3k-1=\dfrac{k(3k+1)}{2}$

$S_{k+1}:\ 2+5+...+(3k-1)+(3(k+1)-1)=$
$\dfrac{(k+1)(3(k+1)+1)}{2}$

1) *Basis step:* $2=\dfrac{1(3\cdot 1+1)}{2}$ is true.

2) *Induction step:* Let k be any natural number. Assume S_k. Deduce S_{k+1}.

$$2+5+...+3k-1=\frac{k(3k+1)}{2}$$

$2+5+...+(3k-1)+(3(k+1)-1)=$

$$\frac{k(3k+1)}{2}+(3(k+1)-1)$$
$$=\frac{3k^2+k+6k+6-2}{2}$$
$$=\frac{3k^2+7k+4}{2}$$
$$=\frac{(k+1)(3k+4)}{2}$$
$$=\frac{(k+1)(3(k+1)+1)}{2}$$

23. See the answer section in the text.

24. $S_n\ \left(1+\frac{1}{1}\right)\left(1+\frac{1}{2}\right)\left(1+\frac{1}{3}\right)\cdots\left(1+\frac{1}{n}\right)=$
$n+1$

$S_1:\ 1+\dfrac{1}{1}=1+1$

$S_k\ \left(1+\frac{1}{1}\right)\cdots\left(1+\frac{1}{k}\right)=k+1$

$S_{k+1}\ \left(1+\frac{1}{1}\right)\cdots\left(1+\frac{1}{k}\right)\left(1+\frac{1}{k+1}\right)=(k+1)+1$

1) *Basis step:* $\left(1+\dfrac{1}{1}\right)=1+1$ is true.

2) *Induction step:* Let k be any natural number. Assume S_k. Deduce S_{k+1}.

$$\left(1+\frac{1}{1}\right)\cdots\left(1+\frac{1}{k}\right)=k+1$$

$$\left(1+\frac{1}{1}\right)\cdots\left(1+\frac{1}{k}\right)\left(1+\frac{1}{k+1}\right)=(k+1)\left(1+\frac{1}{k+1}\right)$$

Multiplying by $\left(1+\dfrac{1}{k+1}\right)$

$$=(k+1)\left(\frac{k+1+1}{k+1}\right)$$
$$=(k+1)+1$$

25. See the answer section in the text.

26. We can prove an infinite sequence of statements S_n by showing that a basis statement S_1 is true and then that for all natural numbers k, if S_k is true, then S_{k+1} is true.

27. Two possibilities are $n<n^2$ and $n^2\le 2^n$. The basis step is false for the first and the induction step fails for the second.

28. $2x-3y=1,\quad (1)$
$3x-4y=3\quad (2)$

Multiply equation (1) by 4 and multiply equation (2) by -3 and add.

$$\begin{aligned}8x-12y&=4\\-9x+12y&=-9\\\hline -x\qquad&=-5\\x&=5\end{aligned}$$

Back-substitute to find y. We use equation (1).

$$\begin{aligned}2\cdot 5-3y&=1\\10-3y&=1\\-3y&=-9\\y&=3\end{aligned}$$

The solution is $(5,3)$.

29. $x+y+z=3,\quad (1)$
$2x-3y-2z=5,\quad (2)$
$3x+2y+2z=8\quad (3)$

We will use Gaussian elimination. First multiply equation (1) by -2 and add it to equation (2). Also multiply equation (1) by -3 and add it to equation (3).

$$\begin{aligned}x+\ y+\ z&=\ 3\\-5y-4z&=-1\\-y-\ z&=-1\end{aligned}$$

Now multiply the last equation above by 5 to make the y-coefficient a multiple of the y-coefficient in the equation above it.

$$\begin{aligned}x+y+z&=3\quad(1)\\-5y-4z&=-1\quad(4)\\-5y-5z&=-5\quad(5)\end{aligned}$$

Multiply equation (4) by -1 and add it to equation (3).

$$\begin{aligned}x+y+z&=3\quad(1)\\-5y-4z&=-1\quad(4)\\-z&=-4\quad(6)\end{aligned}$$

Now solve equation (6) for z.

$$\begin{aligned}-z&=-4\\z&=4\end{aligned}$$

Back-substitute 4 for z in equation (4) and solve for y.

$$\begin{aligned}-5y-4\cdot 4&=-1\\-5y-16&=-1\\-5y&=15\\y&=-3\end{aligned}$$

Finally, back-substitute -3 for y and 4 for z in equation (1) and solve for x.

$$x-3+4=3$$
$$x+1=3$$
$$x=2$$

The solution is $(2,-3,4)$.

30. Let h = the number of hardback books sold and p = the number of paperback books sold.

Solve: $h+p=80,$

$24.95h+9.95p=1546$

$h=50$, $p=30$

31. ***Familiarize.*** Let x, y, and z represent the amounts invested at 1.5%, 2%, and 3%, respectively.

Translate. We know that simple interest for one year was \$104. This gives us one equation:

$$0.015x+0.02y+0.03z=104$$

The amount invested at 2% is twice the amount invested at 1.5%:

$$y=2x, \text{ or } -2x+y=0$$

There is \$400 more invested at 3% than at 2%:

$$z=y+400, \text{ or } -y+z=400$$

We have a system of equations:

$$\begin{aligned} 0.015x+0.02y+0.03z&=104,\\ -2x+y\quad&=0\\ y+z&=400\end{aligned}$$

Carry out. Solving the system of equations, we get $(800, 1600, 2000)$.

Check. Simple interest for one year would be 0.015(\$800) + 0.02(\$1600)+0.03(\$2000), or \$12+\$32+\$60, or \$104. The amount invested at 2%, \$1600, is twice \$800, the amount invested at 1.5%. The amount invested at 3%, \$2000, is \$400 more than \$1600, the amount invested at 2%. The answer checks.

State. Martin invested \$800 at 1.5%, \$1600 at 2%, and \$2000 at 3%.

32. $S_n: \ a_1+a_1r+a_1r^2+...+a_1r^{n-1}=\dfrac{a_1-a_1r^n}{1-r}$

$S_1: \ a_1=\dfrac{a_1-a_1r}{1-r}$

$S_k: \ a_1+a_1r+a_2r^2+...+a_1r^{k-1}=\dfrac{a_1-a_1r^k}{1-r}$

$S_{k+1}: \ a_1+a_1r+...+a_1r^{k-1}+a_1r^{(k+1))-1}=\dfrac{a_1-a_1r^{k+1}}{1-r}$

1) *Basis step:* $a_1=\dfrac{a_1(1-r)}{1-r}=\dfrac{a_1-a_1r}{1-r}$ is true.

2) *Induction step:* Let n be any natural number. Assume S_k. Deduce S_{k+1}.

$$a_1+a_1r+...+a_1r^{k-1}=\frac{a_1-a_1r^k}{1-r}$$

$$a_1+a+_1r+...+a_1r^{k-1}+a_1r^k=\frac{a_1-a_1r^k}{1-r}+a_1r^k$$

Adding a_1r^k

$$=\frac{a_1-a_1r^k+a_1r^k-a_1r^{k+1}}{1-r}$$

$$=\frac{a_1-a_1r^{k+1}}{1-r}$$

33. See the answer section in the text.

34. $S_n: \ 2n+1<3^n$

$S_2: \ 2\cdot 2+1<3^2$

$S_k: \ 2k+1<3^k$

$S_{k+1}: \ 2(k+1)+1=3^{k+1}$

1) *Basis step:* $2\cdot 2+1<3^2$ is true.

2) *Induction step:* Let k be any natural number greater than or equal to 2. Assume S_k. Deduce S_{k+1}.

$$2k+1<3^k$$

$$3(2k+1)<3\cdot 3^k \quad \text{Multiplying by 3}$$

$$6k+3<3^{k+1}$$

$$2k+3<6k+3<3^{k+1} \quad (2k<6k)$$

$$2(k+1)+1<3^{k+1}$$

35. See the answer section in the text.

36. $S_1: \ \overline{z^1}=\overline{z}^1$ If $z=a+bi$, $\overline{z}=a-bi$.

$S_k: \ \overline{z^k}=\overline{z}^k$

$\overline{z^k}\cdot\overline{z}=\overline{z}^k\cdot\overline{z}$ Multiplying both sides of S_k by $\overline{z}$

$\overline{z^k}\cdot\overline{z}=\overline{z}^{k+1}$

$\overline{z^{k+1}}=\overline{z}^{k+1}$

37. See the answer section in the text.

38. $S_2: \ \overline{z_1z_2}=\overline{z_1}\cdot\overline{z_2}$

$S_k: \ \overline{z_1z_2\cdots z_k}=\overline{z}_1\overline{z}_2\cdots\overline{z}_k$

Starting with the left side of S_{k+1}, we have

$\overline{z_1z_2\cdots z_kz_{k+1}}=\overline{z_1z_2\cdots z_k}\cdot\overline{z_{k+1}}$ By S_2

$=\overline{z}_1\overline{z}_2\cdots\overline{z}_k\cdot\overline{z}_{k+1}$ By S_k

39. See the answer section in the text.

40. $S_1: 2$ is a factor of 1^2+1.

$S_k: 2$ is a factor of k^2+k.

$$(k+1)^2+(k+1)=k^2+2k+1+k+1$$
$$=k^2+k+2(k+1)$$

By S_k, 2 is a factor of k^2+k; hence 2 is a factor of the right-hand side, so 2 is a factor of $(k+1)^2+(k+1)$.

41. See the answer section in the text.

42. a) The least number of moves for

1 disk(s) is $1 = 2^1 - 1$,

2 disk(s) is $3 = 2^2 - 1$,

3 disk(s) is $7 = 2^3 - 1$,

4 disk(s) is $15 = 2^4 - 1$; etc.

b) Let P_n be the least number of moves for n disks. We conjecture and must show:

$S_n: \; P_n = 2^n - 1$.

1) *Basis step:* S_1 is true by substitution.

2) *Induction step:* Assume S_k for k disks: $P_k = 2^k - 1$. Show: $P_{k+1} = 2^{k+1} - 1$. Now suppose there are $k+1$ disks on one peg. Move k of them to another peg in $2^k - 1$ moves (by S_k) and move the remaining disk to the free peg (1 move). Then move the k disks onto it in (another) $2^k - 1$ moves. Thus the total moves P_{k+1} is $2(2^k - 1) + 1 = 2^{k+1} - 1$: $P_{k+1} = 2^{k+1} - 1$.

Exercise Set 10.5

1. ${}_6P_6 = 6! = 6 \cdot 5 \cdot 4 \cdot 3 \cdot 2 \cdot 1 = 720$

2. ${}_4P_3 = 4 \cdot 3 \cdot 2 = 24$, or

$${}_4P_3 = \frac{4!}{(4-3)!} = \frac{4!}{1!} = \frac{4 \cdot 3 \cdot 2 \cdot 1}{1} = 24$$

3. Using formula (1), we have

${}_{10}P_7 = 10 \cdot 9 \cdot 8 \cdot 7 \cdot 6 \cdot 5 \cdot 4 = 604,800$.

Using formula (2), we have

$${}_{10}P_7 = \frac{10!}{(10-7)!} = \frac{10!}{3!} = \frac{10 \cdot 9 \cdot 8 \cdot 7 \cdot 6 \cdot 5 \cdot 4 \cdot 3!}{3!} =$$

$604,800$.

4. ${}_{10}P_3 = 10 \cdot 9 \cdot 8 = 720$, or

$${}_{10}P_3 = \frac{10!}{(10-3)!} = \frac{10!}{7!} = \frac{10 \cdot 9 \cdot 8 \cdot 7!}{7!} = 720$$

5. $5! = 5 \cdot 4 \cdot 3 \cdot 2 \cdot 1 = 120$

6. $7! = 7 \cdot 6 \cdot 5 \cdot 4 \cdot 3 \cdot 2 \cdot 1 = 5040$

7. $0!$ is defined to be 1.

8. $1! = 1$

9. $\frac{9!}{5!} = \frac{9 \cdot 8 \cdot 7 \cdot 6 \cdot 5!}{5!} = 9 \cdot 8 \cdot 7 \cdot 6 = 3024$

10. $\frac{9!}{4!} = \frac{9 \cdot 8 \cdot 7 \cdot 6 \cdot 5 \cdot 4!}{4!} = 9 \cdot 8 \cdot 7 \cdot 6 \cdot 5 = 15,120$

11. $(8-3)! = 5! = 5 \cdot 4 \cdot 3 \cdot 2 \cdot 1 = 120$

12. $(8-5)! = 3! = 3 \cdot 2 \cdot 1 = 6$

13. $\frac{10!}{7!3!} = \frac{10 \cdot 9 \cdot 8 \cdot 7!}{7!3 \cdot 2 \cdot 1} = \frac{10 \cdot 3 \cdot 3 \cdot 4 \cdot 2}{3 \cdot 2 \cdot 1} =$

$10 \cdot 3 \cdot 4 = 120$

14. $\frac{7!}{(7-2)!} = \frac{7!}{5!} = \frac{7 \cdot 6 \cdot 5!}{5!} = 42$

15. Using formula (2), we have

$${}_8P_0 = \frac{8!}{(8-0)!} = \frac{8!}{8!} = 1.$$

16. ${}_{13}P_1 = 13$ (Using formula (1))

17. Using a calculator, we find ${}_{52}P_4 = 6,497,400$

18. ${}_{52}P_5 = 311,875,200$

19. Using formula (1), we have ${}_nP_3 = n(n-1)(n-2)$.

Using formula (2), we have

$${}_nP_3 = \frac{n!}{(n-3)!} = \frac{n(n-1)(n-2)(n-3)!}{(n-3)!} =$$

$n(n-1)(n-2)$.

20. ${}_nP_2 = n(n-1)$

21. Using formula (1), we have ${}_nP_1 = n$.

Using formula (2), we have

$${}_nP_1 = \frac{n!}{(n-1)!} = \frac{n(n-1)!}{(n-1)!} = n.$$

22. ${}_nP_0 = \frac{n!}{(n-0)!} = \frac{n!}{n!} = 1$

23. ${}_6P_6 = 6! = 720$

24. ${}_4P_4 = 4! = 24$

25. ${}_9P_9 = 9! = 362,880$

26. ${}_8P_8 = 8! = 40,320$

27. ${}_9P_4 = 9 \cdot 8 \cdot 7 \cdot 6 = 3024$

28. ${}_8P_5 = 8 \cdot 7 \cdot 6 \cdot 5 \cdot 4 = 6720$

29. Without repetition: ${}_5P_5 = 5! = 120$

With repetition: $5^5 = 3125$

30. ${}_7P_7 = 7! = 5040$

31. BUSINESS: 1 B, 1 U, 3 S's, 1 I, 1 N, 1 E, a total of 8.

$$= \frac{8!}{1! \cdot 1! \cdot 3! \cdot 1! \cdot 1! \cdot 1!}$$

$$= \frac{8!}{3!} = \frac{8 \cdot 7 \cdot 6 \cdot 5 \cdot 4 \cdot 3!}{3!} = 8 \cdot 7 \cdot 6 \cdot 5 \cdot 4 = 6720$$

BIOLOGY: 1 B, 1 I, 2 O's, 1 L, 1 G, 1 Y, a total of 7.

$$= \frac{7!}{1! \cdot 1! \cdot 2! \cdot 1! \cdot 1! \cdot 1!}$$

$$= \frac{7!}{2!} = \frac{7 \cdot 6 \cdot 5 \cdot 4 \cdot 3 \cdot 2!}{2!} = 7 \cdot 6 \cdot 5 \cdot 4 \cdot 3 = 2520$$

MATHEMATICS: 2 M's, 2 A's, 2 T's, 1 H, 1 E, 1 I, 1 C, 1 S, a total of 11.

$$= \frac{11!}{2! \cdot 2! \cdot 2! \cdot 1! \cdot 1! \cdot 1! \cdot 1! \cdot 1!}$$

$$= \frac{11!}{2! \cdot 2! \cdot 2!} = \frac{11 \cdot 10 \cdot 9 \cdot 8 \cdot 7 \cdot 6 \cdot 5 \cdot 4 \cdot 3 \cdot 2!}{2! \cdot 2! \cdot 2!}$$

$$= \frac{11 \cdot 10 \cdot 9 \cdot 8 \cdot 7 \cdot 6 \cdot 5 \cdot 4 \cdot 3}{2 \cdot 1 \cdot 2 \cdot 1}$$

$= 4,989,600$

32. $\dfrac{24!}{3!5!9!4!3!} = 16,491,024,950,400$

33. The first number can be any of the eight digits other than 0 and 1. The remaining 6 numbers can each be any of the ten digits 0 through 9. We have

$$8 \cdot 10^6 = 8,000,000$$

Accordingly, there can be 8,000,000 telephone numbers within a given area code before the area needs to be split with a new area code.

34. ${}_5P_5 \cdot {}_4P_4 = 5!4! = 2880$

35. $a^2b^3c^4 = a \cdot a \cdot b \cdot b \cdot b \cdot c \cdot c \cdot c \cdot c$

There are 2 a's, 3 b's, and 4 c's, for a total of 9. We have

$$\frac{9!}{2! \cdot 3! \cdot 4!}$$

$$= \frac{9 \cdot 8 \cdot 7 \cdot 6 \cdot 5 \cdot 4!}{2 \cdot 1 \cdot 3 \cdot 2 \cdot 1 \cdot 4!} = \frac{9 \cdot 8 \cdot 7 \cdot 6 \cdot 5}{2 \cdot 3 \cdot 2} = 1260.$$

36. a) ${}_4P_4 = 4! = 24$

b) There are 4 choices for the first coin and 2 possibilities (head or tail) for each choice. This results in a total of 8 choices for the first selection.

Likewise there are 6 choices for the second selection, 4 for the third, and 2 for the fourth. Then the number of ways in which the coins can be lined up is $8 \cdot 6 \cdot 4 \cdot 2$, or 384.

37. a) ${}_6P_5 = 6 \cdot 5 \cdot 4 \cdot 3 \cdot 2 = 720$

b) $6^5 = 7776$

c) The first letter can only be D. The other four letters are chosen from A, B, C, E, F without repetition. We have

$$1 \cdot {}_5P_4 = 1 \cdot 5 \cdot 4 \cdot 3 \cdot 2 = 120.$$

d) The first letter can only be D. The second letter can only be E. The other three letters are chosen from A, B, C, F without repetition. We have

$$1 \cdot 1 \cdot {}_4P_3 = 1 \cdot 1 \cdot 4 \cdot 3 \cdot 2 = 24.$$

38. There are 80 choices for the number of the county, 26 choices for the letter of the alphabet, and 9999 choices for the number that follows the letter. By the fundamental counting principle we know there are $80 \cdot 26 \cdot 9999$, or 20,797,920 possible license plates.

39. a) Since repetition is allowed, each of the 5 digits can be chosen in 10 ways. The number of zip-codes possible is $10 \cdot 10 \cdot 10 \cdot 10 \cdot 10$, or 100,000.

b) Since there are 100,000 possible zip-codes, there could be 100,000 post offices.

40. $10^9 = 1,000,000,000$

41. a) Since repetition is allowed, each digit can be chosen in 10 ways. There can be $10 \cdot 10 \cdot 10 \cdot 10 \cdot 10 \cdot 10 \cdot 10 \cdot 10 \cdot 10$, or 1,000,000,000 social security numbers.

b) Since more than 285 million social security numbers are possible, each person can have a social security number.

42. Put the following in the form of a paragraph.

First find the number of seconds in a year (365 days):

$$365 \text{ days} \cdot \frac{24 \text{ hr}}{1 \text{ day}} \cdot \frac{60 \text{ min}}{1 \text{ hr}} \cdot \frac{60 \text{ sec}}{1 \text{ min}} =$$

31,536,000 sec.

The number of arrangements possible is 15!.

The time is $\dfrac{15!}{31,536,000} \approx 41,466$ yr.

43. For each circular arrangement of the numbers on a clock face there are 12 distinguishable ordered arrangements on a line. The number of arrangements of 12 objects on a line is ${}_{12}P_{12}$, or 12!. Thus, the number of circular permutations is $\dfrac{{}_{12}P_{12}}{12} = \dfrac{12!}{12} = 11! = 39,916,800.$

In general, for each circular arrangement of n objects, there are n distinguishable ordered arrangements on a line. The total number of arrangements of n objects on a line is ${}_nP_n$, or $n!$. Thus, the number of circular permutations is $\dfrac{n!}{n} = \dfrac{n(n-1)!}{n} = (n-1)!$.

44.
$$4x - 9 = 0$$
$$4x = 9$$
$$x = \frac{9}{4}, \text{ or } 2.25$$

The solution is $\dfrac{9}{4}$, or 2.25.

45.
$$x^2 + x - 6 = 0$$
$$(x+3)(x-2) = 0$$
$$x + 3 = 0 \quad or \quad x - 2 = 0$$
$$x = -3 \quad or \quad x = 2$$

The solutions are -3 and 2.

46. $2x^2 - 3x - 1 = 0$

$$x = \frac{-(-3) \pm \sqrt{(-3)^2 - 4 \cdot 2 \cdot (-1)}}{2 \cdot 2}$$
$$= \frac{3 \pm \sqrt{17}}{4}$$

The solutions are $\dfrac{3+\sqrt{17}}{4}$ and $\dfrac{3-\sqrt{17}}{4}$, or $\dfrac{3 \pm \sqrt{17}}{4}$.

47. $f(x) = x^3 - 4x^2 - 7x + 10$

We use synthetic division to find one factor of the polynomial. We try $x - 1$.

$$\begin{array}{r|rrrr} 1 & 1 & -4 & -7 & 10 \\ & & 1 & -3 & -10 \\ \hline & 1 & -3 & -10 & 0 \end{array}$$

$$x^3 - 4x^2 - 7x + 10 = 0$$
$$(x-1)(x^2 - 3x - 10) = 0$$
$$(x-1)(x-5)(x+2) = 0$$
$$x - 1 = 0 \quad or \quad x - 5 = 0 \quad or \quad x + 2 = 0$$
$$x = 1 \quad or \quad x = 5 \quad or \quad x = -2$$

The solutions are -2, 1, and 5.

48.
$$_nP_5 = 7 \cdot {_nP_4}$$
$$\frac{n!}{(n-5)!} = 7 \cdot \frac{n!}{(n-4)!}$$
$$\frac{n!}{7(n-5)!} = \frac{n!}{(n-4)!}$$
$$7(n-5)! = (n-4)! \quad \text{The denominators must be the same.}$$
$$7(n-5)! = (n-4)(n-5)!$$
$$7 = n-4$$
$$11 = n$$

49.
$$_nP_4 = 8 \cdot {_{n-1}P_3}$$
$$\frac{n!}{(n-4)!} = 8 \cdot \frac{(n-1)!}{(n-1-3)!}$$
$$\frac{n!}{(n-4)!} = 8 \cdot \frac{(n-1)!}{(n-4)!}$$
$$n! = 8 \cdot (n-1)! \quad \text{Multiplying by } (n-4)!$$
$$n(n-1)! = 8 \cdot (n-1)!$$
$$n = 8 \quad \text{Dividing by } (n-1)!$$

50.
$$_nP_5 = 9 \cdot {_{n-1}P_4}$$
$$\frac{n!}{(n-5)!} = 9 \cdot \frac{(n-1)!}{(n-1-4)!}$$
$$\frac{n!}{(n-5)!} = 9 \cdot \frac{(n-1)!}{(n-5)!}$$
$$n! = 9(n-1)!$$
$$n(n-1)! = 9(n-1)!$$
$$n = 9$$

51.
$$_nP_4 = 8 \cdot {_nP_3}$$
$$\frac{n!}{(n-4)!} = 8 \cdot \frac{n!}{(n-3)!}$$
$$(n-3)! = 8(n-4)! \quad \text{Multiplying by } \frac{(n-4)!(n-3)!}{n!}$$
$$(n-3)(n-4)! = 8(n-4)!$$
$$n-3 = 8 \quad \text{Dividing by } (n-4)!$$
$$n = 11$$

52. $n! = n(n-1)(n-2)(n-3)(n-4)\cdots 1 =$
$n(n-1)(n-2)[(n-3)(n-4)\cdots 1] =$
$n(n-1)(n-2)(n-3)!$

53. There is one losing team per game. In order to leave one tournament winner there must be $n-1$ losers produced in $n-1$ games.

54. 2 losses for each of $(n-1)$ losing teams means $2n-2$ losses. The tournament winner will have lost at most 1 game; thus at most there are $(2n-2)+1$ or $(2n-1)$ losses requiring $2n-1$ games.

Exercise Set 10.6

1. $_{13}C_2 = \frac{13!}{2!(13-2)!}$
$= \frac{13!}{2!11!} = \frac{13 \cdot 12 \cdot 11!}{2 \cdot 1 \cdot 11!}$
$= \frac{13 \cdot 12}{2 \cdot 1} = \frac{13 \cdot 6 \cdot 2}{2 \cdot 1}$
$= 78$

2. $_9C_6 = \frac{9!}{6!(9-6)!}$
$= \frac{9!}{6!3!} = \frac{9 \cdot 8 \cdot 7 \cdot 6!}{6! \cdot 3 \cdot 2 \cdot 1}$
$= 84$

3. $\binom{13}{11} = \frac{13!}{11!(13-11)!}$
$= \frac{13!}{11!2!}$
$= 78$ (See Exercise 1.)

4. $\binom{9}{3} = \frac{9!}{3!(9-3)!}$
$= \frac{9!}{3!6!}$
$= 84$ (See Exercise 2.)

5. $\binom{7}{1} = \frac{7!}{1!(7-1)!}$
$= \frac{7!}{1!6!} = \frac{7 \cdot 6!}{1 \cdot 6!}$
$= 7$

6. $\binom{8}{8} = \frac{8!}{8!(8-8)!}$
$= \frac{8!}{8!0!} = \frac{8!}{8! \cdot 1}$
$= 1$

7. $\frac{_5P_3}{3!} = \frac{5 \cdot 4 \cdot 3}{3!}$
$= \frac{5 \cdot 4 \cdot 3}{3 \cdot 2 \cdot 1} = \frac{5 \cdot 2 \cdot 2 \cdot 3}{3 \cdot 2 \cdot 1}$
$= 5 \cdot 2 = 10$

8. $\frac{_{10}P_5}{5!} = \frac{10 \cdot 9 \cdot 8 \cdot 7 \cdot 6}{5 \cdot 4 \cdot 3 \cdot 2 \cdot 1} = 252$

9. $\binom{6}{0} = \frac{6!}{0!(6-0)!}$
$= \frac{6!}{0!6!} = \frac{6!}{6! \cdot 1}$
$= 1$

10. $\binom{6}{1} = \frac{6}{1} = 6$

11. $\binom{6}{2} = \frac{6 \cdot 5}{2 \cdot 1} = 15$

12. $\binom{6}{3} = \frac{6!}{3!(6-3)!}$

$= \frac{6!}{3!3!} = \frac{6 \cdot 5 \cdot 4 \cdot 3!}{3! \cdot 3 \cdot 2 \cdot 1}$

$= 20$

13. $\binom{n}{r} = \binom{n}{n-r}$, so

$\binom{7}{0} + \binom{7}{1} + \binom{7}{2} + \binom{7}{3} + \binom{7}{4} +$

$\binom{7}{5} + \binom{7}{6} + \binom{7}{7}$

$= 2\left[\binom{7}{0} + \binom{7}{1} + \binom{7}{2} + \binom{7}{3}\right]$

$= 2\left[\frac{7!}{7!0!} + \frac{7!}{6!1!} + \frac{7!}{5!2!} + \frac{7!}{4!3!}\right]$

$= 2(1 + 7 + 21 + 35) = 2 \cdot 64 = 128$

14. $\binom{6}{0} + \binom{6}{1} + \binom{6}{2} + \binom{6}{3} + \binom{6}{4} +$

$\binom{6}{5} + \binom{6}{6}$

$= 2\left[\binom{6}{0} + \binom{6}{1} + \binom{6}{2}\right] + \binom{6}{3}$

$= 2(1 + 6 + 15) + 20 = 64$

15. We will use form (1).

${}_{52}C_4 = \frac{52!}{4!(52-4)!}$

$= \frac{52 \cdot 51 \cdot 50 \cdot 49 \cdot 48!}{4 \cdot 3 \cdot 2 \cdot 1 \cdot 48!}$

$= \frac{52 \cdot 51 \cdot 50 \cdot 49}{4 \cdot 3 \cdot 2 \cdot 1}$

$= 270,725$

16. We will use form (1).

${}_{52}C_5 = \frac{52!}{5!(52-5)!}$

$= \frac{52 \cdot 51 \cdot 50 \cdot 49 \cdot 48 \cdot 47!}{5 \cdot 4 \cdot 3 \cdot 2 \cdot 1 \cdot 47!}$

$= 2,598,960$

17. We will use form (2).

$\binom{27}{11}$

$= \frac{27 \cdot 26 \cdot 25 \cdot 24 \cdot 23 \cdot 22 \cdot 21 \cdot 20 \cdot 19 \cdot 18 \cdot 17}{11 \cdot 10 \cdot 9 \cdot 8 \cdot 7 \cdot 6 \cdot 5 \cdot 4 \cdot 3 \cdot 2 \cdot 1}$

$= 13,037,895$

18. We will use form (2).

$\binom{37}{8}$

$= \frac{37 \cdot 36 \cdot 35 \cdot 34 \cdot 33 \cdot 32 \cdot 31 \cdot 30}{8 \cdot 7 \cdot 6 \cdot 5 \cdot 4 \cdot 3 \cdot 2 \cdot 1}$

$= 38,608,020$

19. $\binom{n}{1} = \frac{n!}{1!(n-1)!} = \frac{n(n-1)!}{1!(n-1)!} = n$

20. $\binom{n}{3} = \frac{n!}{3!(n-3)!} =$

$\frac{n(n-1)(n-2)(n-3)!}{3 \cdot 2 \cdot 1 \cdot (n-3)!} = \frac{n(n-1)(n-2)}{6}$

21. $\binom{m}{m} = \frac{m!}{m!(m-m)!} = \frac{m!}{m!0!} = 1$

22. $\binom{t}{4} = \frac{t!}{4!(t-4)!} = \frac{t(t-1)(t-2)(t-3)(t-4)!}{4 \cdot 3 \cdot 2 \cdot 1 \cdot (t-4)!} =$

$\frac{t(t-1)(t-2)(t-3)}{12}$

23. ${}_{23}C_4 = \frac{23!}{4!(23-4)!}$

$= \frac{23!}{4!19!} = \frac{23 \cdot 22 \cdot 21 \cdot 20 \cdot 19!}{4 \cdot 3 \cdot 2 \cdot 1 \cdot 19!}$

$= \frac{23 \cdot 22 \cdot 21 \cdot 20}{4 \cdot 3 \cdot 2 \cdot 1} = \frac{23 \cdot 2 \cdot 11 \cdot 3 \cdot 7 \cdot 4 \cdot 5}{4 \cdot 3 \cdot 2 \cdot 1}$

$= 8855$

24. Playing all other teams once: ${}_9C_2 = 36$

Playing all other teams twice: $2 \cdot {}_9C_2 = 72$

25. ${}_{13}C_{10} = \frac{13!}{10!(13-10)!}$

$= \frac{13!}{10!3!} = \frac{13 \cdot 12 \cdot 11 \cdot 10!}{10! \cdot 3 \cdot 2 \cdot 1}$

$= \frac{13 \cdot 12 \cdot 11}{3 \cdot 2 \cdot 1} = \frac{13 \cdot 3 \cdot 2 \cdot 2 \cdot 11}{3 \cdot 2 \cdot 1}$

$= 286$

26. ${}_{10}C_7 \cdot {}_5C_3 = \binom{10}{7} \cdot \binom{5}{3}$ Using the fundamental counting principle

$= \frac{10!}{7!(10-7)!} \cdot \frac{5!}{3!(5-3)!}$

$= \frac{10 \cdot 9 \cdot 8 \cdot 7!}{7! \cdot 3!} \cdot \frac{5 \cdot 4 \cdot 3!}{3! \cdot 2!}$

$= \frac{10 \cdot 9 \cdot 8}{3 \cdot 2 \cdot 1} \cdot \frac{5 \cdot 4}{2 \cdot 1} = 120 \cdot 10 = 1200$

27. Since two points determine a line and no three of these 8 points are colinear, we need to find the number of combinations of 8 points taken 2 at a time, ${}_8C_2$.

${}_8C_2 = \binom{8}{2} = \frac{8!}{2!(8-2)!}$

$= \frac{8 \cdot 7 \cdot 6!}{2 \cdot 1 \cdot 6!} = \frac{4 \cdot 2 \cdot 7}{2 \cdot 1}$

$= 28$

Thus 28 lines are determined.

Since three noncolinear points determine a triangle, we need to find the number of combinations of 8 points taken 3 at a time, ${}_8C_3$.

$$\begin{aligned}{}_8C_3 = \binom{8}{3} &= \frac{8!}{3!(8-3)!} \\ &= \frac{8\cdot 7\cdot 6\cdot 5!}{3\cdot 2\cdot 1\cdot 5!} = \frac{8\cdot 7\cdot 3\cdot 2}{3\cdot 2\cdot 1} \\ &= 56\end{aligned}$$

Thus 56 triangles are determined.

28. Using the fundamental counting principle, we have ${}_{58}C_6 \cdot {}_{42}C_4$.

29. ${}_{52}C_5 = 2,598,960$

30. ${}_{52}C_{13} = 635,013,559,600$

31. a) ${}_{31}P_2 = 930$

b) $31 \cdot 31 = 961$

c) ${}_{31}C_2 = 465$

32. Order is considered in a combination lock.

33. Choosing k objects from a set of n objects is equivalent to not choosing the other $n-k$ objects.

34.
$$\begin{aligned}3x - 7 &= 5x + 10 \\ -17 &= 2x \\ -\frac{17}{2} &= x\end{aligned}$$

The solution is $-\frac{17}{2}$.

35.
$$\begin{aligned}2x^2 - x &= 3 \\ 2x^2 - x - 3 &= 0 \\ (2x-3)(x+1) &= 0 \\ 2x - 3 = 0 \quad &or \quad x + 1 = 0 \\ 2x = 3 \quad &or \quad x = -1 \\ x = \frac{3}{2} \quad &or \quad x = -1\end{aligned}$$

The solutions are $\frac{3}{2}$ and -1.

36. $x^2 + 5x + 1 = 0$

$$\begin{aligned}x &= \frac{-5 \pm \sqrt{5^2 - 4\cdot 1\cdot 1}}{2\cdot 1} \\ &= \frac{-5 \pm \sqrt{21}}{2}\end{aligned}$$

The solutions are $\frac{-5+\sqrt{21}}{2}$ and $\frac{-5-\sqrt{21}}{2}$, or $\frac{-5\pm\sqrt{21}}{2}$.

37.
$$\begin{aligned}x^3 + 3x^2 - 10x &= 24 \\ x^3 + 3x^2 - 10x - 24 &= 0\end{aligned}$$

We use synthetic division to find one factor of the polynomial on the left side of the equation. We try $x - 3$.

$$\begin{array}{r|rrrr} 3 & 1 & 3 & -10 & -24 \\ & & 3 & 18 & 24 \\ \hline & 1 & 6 & 8 & 0 \end{array}$$

Now we have:

$$\begin{aligned}(x-3)(x^2+6x+8) &= 0 \\ (x-3)(x+2)(x+4) &= 0 \\ x - 3 = 0 \quad or \quad x + 2 = 0 \quad &or \quad x + 4 = 0 \\ x = 3 \quad or \quad x = -2 \quad &or \quad x = -4\end{aligned}$$

The solutions are -4, -2, and 3.

38. ${}_4C_3 \cdot {}_4C_2 = 24$

39. There are 13 diamonds, and we choose 5. We have ${}_{13}C_5 = 1287$.

40. ${}_nC_4$

41. Playing once: ${}_nC_2$

Playing twice: $2 \cdot {}_nC_2$

42.
$$\begin{aligned}\binom{n+1}{3} &= 2\cdot\binom{n}{2} \\ \frac{(n+1)!}{(n+1-3)!3!} &= 2\cdot\frac{n!}{(n-2)!2!} \\ \frac{(n+1)!}{(n-2)!3!} &= 2\cdot\frac{n!}{(n-2)!2!} \\ \frac{(n+1)(n)(n-1)(n-2)!}{(n-2)!3\cdot 2\cdot 1} &= 2\cdot\frac{n(n-1)(n-2)!}{(n-2)!\cdot 2\cdot 1} \\ \frac{(n+1)(n)(n-1)}{6} &= n(n-1) \\ \frac{n^3 - n}{6} &= n^2 - n \\ n^3 - n &= 6n^2 - 6n \\ n^3 - 6n^2 + 5n &= 0 \\ n(n^2 - 6n + 5) &= 0 \\ n(n-5)(n-1) &= 0 \\ n = 0 \quad \text{or} \quad n = 5 \quad &\text{or} \quad n = 1\end{aligned}$$

Only 5 checks. The solution is 5.

43.
$$\begin{aligned}\binom{n}{n-2} &= 6 \\ \frac{n!}{(n-(n-2))!(n-2)!} &= 6 \\ \frac{n!}{2!(n-2)!} &= 6 \\ \frac{n(n-1)(n-2)!}{2\cdot 1\cdot (n-2)!} &= 6 \\ \frac{n(n-1)}{2} &= 6 \\ n(n-1) &= 12 \\ n^2 - n &= 12 \\ n^2 - n - 12 &= 0 \\ (n-4)(n+3) &= 0 \\ n = 4 \quad &\text{or} \quad n = -3\end{aligned}$$

Only 4 checks. The solution is 4.

44. $\binom{n}{3} = 2 \cdot \binom{n-1}{2}$

$$\frac{n!}{(n-3)!3!} = 2 \cdot \frac{(n-1)!}{(n-1-2)!2!}$$
$$\frac{n!}{(n-3)!3!} = 2 \cdot \frac{(n-1)!}{(n-3)!2!}$$
$$\frac{n!}{3!} = 2 \cdot \frac{(n-1)!}{2!}$$
$$n! = 3!(n-1)!$$
$$\frac{n(n-1)!}{(n-1)!} = 6$$
$$n = 6$$

This number checks. The solution is 6.

45. $\binom{n+2}{4} = 6 \cdot \binom{n}{2}$

$$\frac{(n+2)!}{(n+2-4)!4!} = 6 \cdot \frac{n!}{(n-2)!2!}$$
$$\frac{(n+2)!}{(n-2)!4!} = 6 \cdot \frac{n!}{(n-2)!2!}$$
$$\frac{(n+2)!}{4!} = 6 \cdot \frac{n!}{2!} \quad \text{Multiplying by } (n-2)!$$
$$4! \cdot \frac{(n+2)!}{4!} = 4! \cdot 6 \cdot \frac{n!}{2!}$$
$$(n+2)! = 72 \cdot n!$$
$$(n+2)(n+1)n! = 72 \cdot n!$$
$$(n+2)(n+1) = 72 \quad \text{Dividing by } n!$$
$$n^2 + 3n + 2 = 72$$
$$n^2 + 3n - 70 = 0$$
$$(n+10)(n-7) = 0$$
$$n = -10 \text{ or } n = 7$$

Only 7 checks. The solution is 7.

46. Line segments: ${}_nC_2 = \frac{n!}{2!(n-2)!} = \frac{n(n-1)(n-2)!}{2 \cdot 1 \cdot (n-2)!} = \frac{n(n-1)}{2}$

Diagonals: The n line segments that form the sides of the n-agon are not diagonals. Thus, the number of diagonals is ${}_nC_2 - n = \frac{n(n-1)}{2} - n = \frac{n^2 - n - 2n}{2} = \frac{n^2 - 3n}{2} = \frac{n(n-3)}{2}$, $n \geq 4$.

Let D_n be the number of diagonals on an n-agon. Prove the result above for diagonals using mathematical induction.

$$S_n: \ D_n = \frac{n(n-3)}{2}, \text{ for } n = 4, 5, 6, ...$$
$$S_4: \ D_4 = \frac{4 \cdot 1}{2}$$
$$S_k: \ D_k = \frac{k(k-3)}{2}$$
$$S_{k+1}: \ D_{k+1} = \frac{(k+1)(k-2)}{2}$$

1) *Basis step:* S_4 is true (a quadrilateral has 2 diagonals).

2) *Induction step:* Assume S_k. Observe that when an additional vertex V_{k+1} is added to the k-gon, we gain k segments, 2 of which are sides of the $(k+1)$-gon], and a former side $\overline{V_1V_k}$ becomes a diagonal. Thus the additional number of diagonals is $k - 2 + 1$, or $k - 1$. Then the new total of diagonals is $D_k + (k-1)$, or

$$D_{k+1} = D_k + (k-1)$$
$$= \frac{k(k-3)}{2} + (k-1) \quad (\text{by } S_k)$$
$$= \frac{(k+1)(k-2)}{2}$$

47. See the answer section in the text.

Exercise Set 10.7

1. Expand: $(x+5)^4$.

We have $a = x$, $b = 5$, and $n = 4$.

Pascal's triangle method: Use the fifth row of Pascal's triangle.

1 4 6 4 1

$$(x+5)^4$$
$$= 1 \cdot x^4 + 4 \cdot x^3 \cdot 5 + 6 \cdot x^2 \cdot 5^2 + 4 \cdot x \cdot 5^3 + 1 \cdot 5^4$$
$$= x^4 + 20x^3 + 150x^2 + 500x + 625$$

Factorial notation method:

$$(x+5)^4$$
$$= \binom{4}{0}x^4 + \binom{4}{1}x^3 \cdot 5 + \binom{4}{2}x^2 \cdot 5^2 + \binom{4}{3}x \cdot 5^3 + \binom{4}{4}5^4$$
$$= \frac{4!}{0!4!}x^4 + \frac{4!}{1!3!}x^3 \cdot 5 + \frac{4!}{2!2!}x^2 \cdot 5^2 + \frac{4!}{3!1!}x \cdot 5^3 + \frac{4!}{4!0!}5^4$$
$$= x^4 + 20x^3 + 150x^2 + 500x + 625$$

2. Expand: $(x-1)^4$.

Pascal's triangle method: Use the 5th row of Pascal's triangle.

1 4 6 4 1

$$(x-1)^4$$
$$= 1 \cdot x^4 + 4 \cdot x^3(-1) + 6x^2(-1)^2 + 4x(-1)^3 + 1 \cdot (-1)^4$$
$$= x^4 - 4x^3 + 6x^2 - 4x + 1$$

Factorial notation method:

$(x-1)^4$

$= \binom{4}{0}x^4 + \binom{4}{1}x^3(-1) + \binom{4}{2}x^2(-1)^2 +$

$\binom{4}{3}x(-1)^3 + \binom{4}{4}(-1)^4$

$= x^4 - 4x^3 + 6x^2 - 4x + 1$

3. Expand: $(x-3)^5$.

We have $a = x$, $b = -3$, and $n = 5$.

Pascal's triangle method: Use the sixth row of Pascal's triangle.

1 5 10 10 5 1

$(x-3)^5$

$= 1\cdot x^5 + 5x^4(-3) + 10x^3(-3)^2 + 10x^2(-3)^3 +$

$5x(-3)^4 + 1\cdot(-3)^5$

$= x^5 - 15x^4 + 90x^3 - 270x^2 + 405x - 243$

Factorial notation method:

$(x-3)^5$

$= \binom{5}{0}x^5 + \binom{5}{1}x^4(-3) + \binom{5}{2}x^3(-3)^2 +$

$\binom{5}{3}x^2(-3)^3 + \binom{5}{4}x(-3)^4 + \binom{5}{5}(-3)^5$

$= \frac{5!}{0!5!}x^5 + \frac{5!}{1!4!}x^4(-3) + \frac{5!}{2!3!}x^3(9) +$

$\frac{5!}{3!2!}x^2(-27) + \frac{5!}{4!1!}x(81) + \frac{5!}{5!0!}(-243)$

$= x^5 - 15x^4 + 90x^3 - 270x^2 + 405x - 243$

4. Expand: $(x+2)^9$.

Pascal's triangle method: Use the 10th row of Pascal's triangle.

1 9 36 84 126 126 84 36 9 1

$(x+2)^9$

$= 1\cdot x^9 + 9x^8\cdot 2 + 36x^7\cdot 2^2 + 84x^6\cdot 2^3 +$

$126x^5\cdot 2^4 + 126x^4\cdot 2^5 + 84x^3\cdot 2^6 + 36x^2\cdot 2^7 +$

$9x\cdot 2^8 + 1\cdot 2^9$

$= x^9 + 18x^8 + 144x^7 + 672x^6 + 2016x^5 + 4032x^4 +$

$5376x^3 + 4608x^2 + 2304x + 512$

Factorial notation method:

$(x+2)^9$

$= \binom{9}{0}x^9 + \binom{9}{1}x^8\cdot 2 + \binom{9}{2}x^7\cdot 2^2 +$

$\binom{9}{3}x^6\cdot 2^3 + \binom{9}{4}x^5\cdot 2^4 + \binom{9}{5}x^4\cdot 2^5 +$

$\binom{9}{6}x^3\cdot 2^6 + \binom{9}{7}x^2\cdot 2^7 + \binom{9}{8}x\cdot 2^8 +$

$\binom{9}{9}2^9$

$= x^9 + 18x^8 + 144x^7 + 672x^6 + 2016x^5 + 4032x^4 +$

$5376x^3 + 4608x^2 + 2304x + 512$

5. Expand: $(x-y)^5$.

We have $a = x$, $b = -y$, and $n = 5$.

Pascal's triangle method: We use the sixth row of Pascal's triangle.

1 5 10 10 5 1

$(x-y)^5$

$= 1\cdot x^5 + 5x^4(-y) + 10x^3(-y)^2 + 10x^2(-y)^3 +$

$5x(-y)^4 + 1\cdot(-y)^5$

$= x^5 - 5x^4y + 10x^3y^2 - 10x^2y^3 + 5xy^4 - y^5$

Factorial notation method:

$(x-y)^5$

$= \binom{5}{0}x^5 + \binom{5}{1}x^4(-y) + \binom{5}{2}x^3(-y)^2 +$

$\binom{5}{3}x^2(-y)^3 + \binom{5}{4}x(-y)^4 + \binom{5}{5}(-y)^5$

$= \frac{5!}{0!5!}x^5 + \frac{5!}{1!4!}x^4(-y) + \frac{5!}{2!3!}x^3(y^2) +$

$\frac{5!}{3!2!}x^2(-y^3) + \frac{5!}{4!1!}x(y^4) + \frac{5!}{5!0!}(-y^5)$

$= x^5 - 5x^4y + 10x^3y^2 - 10x^2y^3 + 5xy^4 - y^5$

6. Expand: $(x+y)^8$.

Pascal's triangle method: Use the ninth row of Pascal's triangle.

1 8 28 56 70 56 28 8 1

$(x+y)^8$

$= x^8 + 8x^7y + 28x^6y^2 + 56x^5y^3 + 70x^4y^4 +$

$56x^3y^5 + 28x^2y^6 + 8xy^7 + y^8$

Factorial notation method:

$(x+y)^8$

$= \binom{8}{0}x^8 + \binom{8}{1}x^7y + \binom{8}{2}x^6y^2 +$

$\binom{8}{3}x^5y^3 + \binom{8}{4}x^4y^4 + \binom{8}{5}x^3y^5 +$

$\binom{8}{6}x^2y^6 + \binom{8}{7}xy^7 + \binom{8}{8}y^8$

$= x^8 + 8x^7y + 28x^6y^2 + 56x^5y^3 + 70x^4y^4 +$

$56x^3y^5 + 28x^2y^6 + 8xy^7 + y^8$

7. Expand: $(5x+4y)^6$.

We have $a = 5x$, $b = 4y$, and $n = 6$.

Pascal's triangle method: Use the seventh row of Pascal's triangle.

1 6 15 20 15 6 1

$(5x+4y)^6$

$= 1\cdot(5x)^6 + 6\cdot(5x)^5(4y) + 15(5x)^4(4y)^2 +$

$20(5x)^3(4y)^3 + 15(5x)^2(4y)^4 + 6(5x)(4y)^5 +$

$1\cdot(4y)^6$

$= 15,625x^6 + 75,000x^5y + 150,000x^4y^2 +$

$160,000x^3y^3 + 96,000x^2y^4 + 30,720xy^5 + 4096y^6$

Factorial notation method:

$(5x+4y)^6$

$= \binom{6}{0}(5x)^6 + \binom{6}{1}(5x)^5(4y) +$

$\binom{6}{2}(5x)^4(4y)^2 + \binom{6}{3}(5x)^3(4y)^3 +$

$\binom{6}{4}(5x)^2(4y)^4 + \binom{6}{5}(5x)(4y)^5 + \binom{6}{6}(4y)^6$

$= \frac{6!}{0!6!}(15,625x^6) + \frac{6!}{1!5!}(3125x^5)(4y) +$

$\frac{6!}{2!4!}(625x^4)(16y^2) + \frac{6!}{3!3!}(125x^3)(64y^3) +$

$\frac{6!}{4!2!}(25x^2)(256y^4) + \frac{6!}{5!1!}(5x)(1024y^5) +$

$\frac{6!}{6!0!}(4096y^6)$

$= 15,625x^6 + 75,000x^5y + 150,000x^4y^2 +$

$160,000x^3y^3 + 96,000x^2y^4 + 30,720xy^5 +$

$4096y^6$

8. Expand: $(2x-3y)^5$.

Pascal's triangle method: Use the sixth row of Pascal's triangle.

1 5 10 10 5 1

$(2x-3y)^5$

$= 1\cdot(2x)^5 + 5(2x)^4(-3y) + 10(2x)^3(-3y)^2 +$

$10(2x)^2(-3y)^3 + 5(2x)(-3y)^4 + 1\cdot(-3y)^5$

$= 32x^5 - 240x^4y + 720x^3y^2 - 1080x^2y^3 +$

$810xy^4 - 243y^5$

Factorial notation method:

$(2x-3y)^5$

$= \binom{5}{0}(2x)^5 + \binom{5}{1}(2x)^4(-3y) +$

$\binom{5}{2}(2x)^3(-3y)^2 + \binom{5}{3}(2x)^2(-3y)^3 +$

$\binom{5}{4}(2x)(-3y)^4 + \binom{5}{5}(-3y)^5$

$= 32x^5 - 240x^4y + 720x^3y^2 - 1080x^2y^3 +$

$810xy^4 - 243y^5$

9. Expand: $\left(2t+\frac{1}{t}\right)^7$.

We have $a = 2t$, $b = \frac{1}{t}$, and $n = 7$.

Pascal's triangle method: Use the eighth row of Pascal's triangle.

1 7 21 35 35 21 7 1

$\left(2t+\frac{1}{t}\right)^7$

$= 1\cdot(2t)^7 + 7(2t)^6\left(\frac{1}{t}\right) + 21(2t)^5\left(\frac{1}{t}\right)^2 +$

$35(2t)^4\left(\frac{1}{t}\right)^3 + 35(2t)^3\left(\frac{1}{t}\right)^4 + 21(2t)^2\left(\frac{1}{t}\right)^5 +$

$7(2t)\left(\frac{1}{t}\right)^6 + 1\cdot\left(\frac{1}{t}\right)^7$

$= 128t^7 + 7\cdot 64t^6\cdot\frac{1}{t} + 21\cdot 32t^5\cdot\frac{1}{t^2} +$

$35\cdot 16t^4\cdot\frac{1}{t^3} + 35\cdot 8t^3\cdot\frac{1}{t^4} + 21\cdot 4t^2\cdot\frac{1}{t^5} +$

$7\cdot 2t\cdot\frac{1}{t^6} + \frac{1}{t^7}$

$= 128t^7 + 448t^5 + 672t^3 + 560t + 280t^{-1} +$

$84t^{-3} + 14t^{-5} + t^{-7}$

Factorial notation method:

$\left(2t+\frac{1}{t}\right)^7$

$= \binom{7}{0}(2t)^7 + \binom{7}{1}(2t)^6\left(\frac{1}{t}\right) +$

$\binom{7}{2}(2t)^5\left(\frac{1}{t}\right)^2 + \binom{7}{3}(2t)^4\left(\frac{1}{t}\right)^3 +$

$\binom{7}{4}(2t)^3\left(\frac{1}{t}\right)^4 + \binom{7}{5}(2t)^2\left(\frac{1}{t}\right)^5 +$

$\binom{7}{6}(2t)\left(\frac{1}{t}\right)^6 + \binom{7}{7}\left(\frac{1}{t}\right)^7$

$= \frac{7!}{0!7!}(128t^7) + \frac{7!}{1!6!}(64t^6)\left(\frac{1}{t}\right) + \frac{7!}{2!5!}(32t^5)\left(\frac{1}{t^2}\right) +$

$\frac{7!}{3!4!}(16t^4)\left(\frac{1}{t^3}\right) + \frac{7!}{4!3!}(8t^3)\left(\frac{1}{t^4}\right) +$

$\frac{7!}{5!2!}(4t^2)\left(\frac{1}{t^5}\right) + \frac{7!}{6!1!}(2t)\left(\frac{1}{t^6}\right) + \frac{7!}{7!0!}\left(\frac{1}{t^7}\right)$

$= 128t^7 + 448t^5 + 672t^3 + 560t + 280t^{-1} +$

$84t^{-3} + 14t^{-5} + t^{-7}$

10. Expand: $\left(3y - \frac{1}{y}\right)^4$.

Pascal's triangle method: Use the fifth row of Pascal's triangle.

1 4 6 4 1

$$\left(3y - \frac{1}{y}\right)^4$$
$$= 1\cdot(3y)^4 + 4(3y)^3\left(-\frac{1}{y}\right) + 6(3y)^2\left(-\frac{1}{y}\right)^2 +$$
$$4(3y)\left(-\frac{1}{y}\right)^3 + 1\cdot\left(-\frac{1}{y}\right)^4$$
$$= 81y^4 - 108y^2 + 54 - 12y^{-2} + y^{-4}$$

Factorial notation method:

$$\left(3y - \frac{1}{y}\right)^4$$
$$= \binom{4}{0}(3y)^4 + \binom{4}{1}(3y)^3\left(-\frac{1}{y}\right) +$$
$$\binom{4}{2}(3y)^2\left(-\frac{1}{y}\right)^2 + \binom{4}{3}(3y)\left(-\frac{1}{y}\right)^3 +$$
$$\binom{4}{4}\left(-\frac{1}{y}\right)^4$$
$$= 81y^4 - 108y^2 + 54 - 12y^{-2} + y^{-4}$$

11. Expand: $(x^2 - 1)^5$.

We have $a = x^2$, $b = -1$, and $n = 5$.

Pascal's triangle method: Use the sixth row of Pascal's triangle.

1 5 10 10 5 1

$$(x^2 - 1)^5$$
$$= 1\cdot(x^2)^5 + 5(x^2)^4(-1) + 10(x^2)^3(-1)^2 +$$
$$10(x^2)^2(-1)^3 + 5(x^2)(-1)^4 + 1\cdot(-1)^5$$
$$= x^{10} - 5x^8 + 10x^6 - 10x^4 + 5x^2 - 1$$

Factorial notation method:

$$(x^2 - 1)^5$$
$$= \binom{5}{0}(x^2)^5 + \binom{5}{1}(x^2)^4(-1) +$$
$$\binom{5}{2}(x^2)^3(-1)^2 + \binom{5}{3}(x^2)^2(-1)^3 +$$
$$\binom{5}{4}(x^2)(-1)^4 + \binom{5}{5}(-1)^5$$
$$= \frac{5!}{0!5!}(x^{10}) + \frac{5!}{1!4!}(x^8)(-1) + \frac{5!}{2!3!}(x^6)(1) +$$
$$\frac{5!}{3!2!}(x^4)(-1) + \frac{5!}{4!1!}(x^2)(1) + \frac{5!}{5!0!}(-1)$$
$$= x^{10} - 5x^8 + 10x^6 - 10x^4 + 5x^2 - 1$$

12. Expand: $(1 + 2q^3)^8$.

Pascal's triangle method: Use the ninth row of Pascal's triangle.

1 8 28 56 70 56 28 8 1

$$(1 + 2q^3)^8$$
$$= 1\cdot 1^8 + 8\cdot 1^7(2q^3) + 28\cdot 1^6(2q^3)^2 + 56\cdot 1^5(2q^3)^3 +$$
$$70\cdot 1^4(2q^3)^4 + 56\cdot 1^3(2q^3)^5 + 28\cdot 1^2(2q^3)^6 +$$
$$8\cdot 1(2q^3)^7 + 1\cdot(2q^3)^8$$
$$= 1 + 16q^3 + 112q^6 + 448q^9 + 1120q^{12} + 1792q^{15} +$$
$$1792q^{18} + 1024q^{21} + 256q^{24}$$

Factorial notation method:

$$(1 + 2q^3)^8$$
$$= \binom{8}{0}(1)^8 + \binom{8}{1}(1)^7(2q^3) + \binom{8}{2}(1)^6(2q^3)^2 +$$
$$\binom{8}{3}(1)^5(2q^3)^3 + \binom{8}{4}(1)^4(2q^3)^4 +$$
$$\binom{8}{5}(1)^3(2q^3)^5 + \binom{8}{6}(1)^2(2q^3)^6 +$$
$$\binom{8}{7}(1)(2q^3)^7 + \binom{8}{8}(2q^3)^8$$
$$= 1 + 16q^3 + 112q^6 + 448q^9 + 1120q^{12} + 1792q^{15} +$$
$$1792q^{18} + 1024q^{21} + 256q^{24}$$

13. Expand: $(\sqrt{5} + t)^6$.

We have $a = \sqrt{5}$, $b = t$, and $n = 6$.

Pascal's triangle method: We use the seventh row of Pascal's triangle:

1 6 15 20 15 6 1

$$(\sqrt{5} + t)^6 = 1\cdot(\sqrt{5})^6 + 6(\sqrt{5})^5(t) +$$
$$15(\sqrt{5})^4(t^2) + 20(\sqrt{5})^3(t^3) +$$
$$15(\sqrt{5})^2(t^4) + 6\sqrt{5}t^5 + 1\cdot t^6$$
$$= 125 + 150\sqrt{5}\,t + 375t^2 + 100\sqrt{5}\,t^3 +$$
$$75t^4 + 6\sqrt{5}\,t^5 + t^6$$

Factorial notation method:

$$(\sqrt{5} + t)^6 = \binom{6}{0}(\sqrt{5})^6 + \binom{6}{1}(\sqrt{5})^5(t) +$$
$$\binom{6}{2}(\sqrt{5})^4(t^2) + \binom{6}{3}(\sqrt{5})^3(t^3) +$$
$$\binom{6}{4}(\sqrt{5})^2(t^4) + \binom{6}{5}(\sqrt{5})(t^5) +$$
$$\binom{6}{6}(t^6)$$
$$= \frac{6!}{0!6!}(125) + \frac{6!}{1!5!}(25\sqrt{5})t + \frac{6!}{2!4!}(25)(t^2) +$$
$$\frac{6!}{3!3!}(5\sqrt{5})(t^3) + \frac{6!}{4!2!}(5)(t^4) +$$
$$\frac{6!}{5!1!}(\sqrt{5})(t^5) + \frac{6!}{6!0!}(t^6)$$
$$= 125 + 150\sqrt{5}\,t + 375t^2 + 100\sqrt{5}\,t^3 +$$
$$75t^4 + 6\sqrt{5}\,t^5 + t^6$$

14. Expand: $(x-\sqrt{2})^6$.

Pascal's triangle method: Use the seventh row of Pascal's triangle.

1 6 15 20 15 6 1

$(x-\sqrt{2})^6$

$= 1\cdot x^6 + 6x^5(-\sqrt{2}) + 15x^4(-\sqrt{2})^2 + 20x^3(-\sqrt{2})^3 + 15x^2(-\sqrt{2})^4 + 6x(-\sqrt{2})^5 + 1\cdot(-\sqrt{2})^6$

$= x^6 - 6\sqrt{2}x^5 + 30x^4 - 40\sqrt{2}x^3 + 60x^2 - 24\sqrt{2}x + 8$

Factorial notation method:

$(x-\sqrt{2})^6$

$= \binom{6}{0}x^6 + \binom{6}{1}(x^5)(-\sqrt{2}) + \binom{6}{2}(x^4)(-\sqrt{2})^2 + \binom{6}{3}(x^3)(-\sqrt{2})^3 + \binom{6}{4}(x^2)(-\sqrt{2})^4 + \binom{6}{5}(x)(-\sqrt{2})^5 + \binom{6}{6}(-\sqrt{2})^6$

$= x^6 - 6\sqrt{2}x^5 + 30x^4 - 40\sqrt{2}x^3 + 60x^2 - 24\sqrt{2}x + 8$

15. Expand: $\left(a-\frac{2}{a}\right)^9$.

We have $a = a$, $b = -\frac{2}{a}$, and $n = 9$.

Pascal's triangle method: Use the tenth row of Pascal's triangle.

1 9 36 84 126 126 84 36 9 1

$\left(a-\frac{2}{a}\right)^9 = 1\cdot a^9 + 9a^8\left(-\frac{2}{a}\right) + 36a^7\left(-\frac{2}{a}\right)^2 + 84a^6\left(-\frac{2}{a}\right)^3 + 126a^5\left(-\frac{2}{a}\right)^4 + 126a^4\left(-\frac{2}{a}\right)^5 + 84a^3\left(-\frac{2}{a}\right)^6 + 36a^2\left(-\frac{2}{a}\right)^7 + 9a\left(-\frac{2}{a}\right)^8 + 1\cdot\left(-\frac{2}{a}\right)^9$

$= a^9 - 18a^7 + 144a^5 - 672a^3 + 2016a - 4032a^{-1} + 5376a^{-3} - 4608a^{-5} + 2304a^{-7} - 512a^{-9}$

Factorial notation method:

$\left(a-\frac{2}{a}\right)^9$

$= \binom{9}{0}a^9 + \binom{9}{1}a^8\left(-\frac{2}{a}\right) + \binom{9}{2}a^7\left(-\frac{2}{a}\right)^2 + \binom{9}{3}a^6\left(-\frac{2}{a}\right)^3 + \binom{9}{4}a^5\left(-\frac{2}{a}\right)^4 + \binom{9}{5}a^4\left(-\frac{2}{a}\right)^5 + \binom{9}{6}a^3\left(-\frac{2}{a}\right)^6 + \binom{9}{7}a^2\left(-\frac{2}{a}\right)^7 + \binom{9}{8}a\left(-\frac{2}{a}\right)^8 + \binom{9}{9}\left(-\frac{2}{a}\right)^9$

$= \frac{9!}{9!0!}a^9 + \frac{9!}{8!1!}a^8\left(-\frac{2}{a}\right) + \frac{9!}{7!2!}a^7\left(\frac{4}{a^2}\right) + \frac{9!}{6!3!}a^6\left(-\frac{8}{a^3}\right) + \frac{9!}{5!4!}a^5\left(\frac{16}{a^4}\right) + \frac{9!}{4!5!}a^4\left(-\frac{32}{a^5}\right) + \frac{9!}{3!6!}a^3\left(\frac{64}{a^6}\right) + \frac{9!}{2!7!}a^2\left(-\frac{128}{a^7}\right) + \frac{9!}{1!8!}a\left(\frac{256}{a^8}\right) + \frac{9!}{0!9!}\left(-\frac{512}{a^9}\right)$

$= a^9 - 9(2a^7) + 36(4a^5) - 84(8a^3) + 126(16a) - 126(32a^{-1}) + 84(64a^{-3}) - 36(128a^{-5}) + 9(256a^{-7}) - 512a^{-9}$

$= a^9 - 18a^7 + 144a^5 - 672a^3 + 2016a - 4032a^{-1} + 5376a^{-3} - 4608a^{-5} + 2304a^{-7} - 512a^{-9}$

16. Expand: $(1+3)^n$

Use the factorial notation method.

$(1+3)^n$

$= \binom{n}{0}(1)^n + \binom{n}{1}(1)^{n-1}3 + \binom{n}{2}(1)^{n-2}3^2 + \binom{n}{3}(1)^{n-3}3^3 + \cdots + \binom{n}{n-2}(1)^2 3^{n-2} + \binom{n}{n-1}(1)3^{n-1} + \binom{n}{n}3^n$

$= 1 + 3n + \binom{n}{2}3^2 + \binom{n}{3}3^3 + \cdots + \binom{n}{n-2}3^{n-2} + 3^{n-1}n + 3^n$

17. $(\sqrt{2}+1)^6-(\sqrt{2}-1)^6$

First, expand $(\sqrt{2}+1)^6$.

$$(\sqrt{2}+1)^6=\binom{6}{0}(\sqrt{2})^6+\binom{6}{1}(\sqrt{2})^5(1)+\binom{6}{2}(\sqrt{2})^4(1)^2+\binom{6}{3}(\sqrt{2})^3(1)^3+\binom{6}{4}(\sqrt{2})^2(1)^4+\binom{6}{5}(\sqrt{2})(1)^5+\binom{6}{6}(1)^6$$

$$=\frac{6!}{6!0!}\cdot 8+\frac{6!}{5!1!}\cdot 4\sqrt{2}+\frac{6!}{4!2!}\cdot 4+\frac{6!}{3!3!}\cdot 2\sqrt{2}+\frac{6!}{2!4!}\cdot 2+\frac{6!}{1!5!}\cdot\sqrt{2}+\frac{6!}{0!6!}$$

$$=8+24\sqrt{2}+60+40\sqrt{2}+30+6\sqrt{2}+1$$

$$=99+70\sqrt{2}$$

Next, expand $(\sqrt{2}-1)^6$.

$$(\sqrt{2}-1)^6$$

$$=\binom{6}{0}(\sqrt{2})^6+\binom{6}{1}(\sqrt{2})^5(-1)+\binom{6}{2}(\sqrt{2})^4(-1)^2+\binom{6}{3}(\sqrt{2})^3(-1)^3+\binom{6}{4}(\sqrt{2})^2(-1)^4+\binom{6}{5}(\sqrt{2})(-1)^5+\binom{6}{6}(-1)^6$$

$$=\frac{6!}{6!0!}\cdot 8-\frac{6!}{5!1!}\cdot 4\sqrt{2}+\frac{6!}{4!2!}\cdot 4-\frac{6!}{3!3!}\cdot 2\sqrt{2}+\frac{6!}{2!4!}\cdot 2-\frac{6!}{1!5!}\cdot\sqrt{2}+\frac{6!}{0!6!}$$

$$=8-24\sqrt{2}+60-40\sqrt{2}+30-6\sqrt{2}+1$$

$$=99-70\sqrt{2}$$

$$(\sqrt{2}+1)^6-(\sqrt{2}-1)^6$$

$$=(99+70\sqrt{2})-(99-70\sqrt{2})$$

$$=99+70\sqrt{2}-99+70\sqrt{2}$$

$$=140\sqrt{2}$$

18. $(1-\sqrt{2})^4=1\cdot 1^4+4\cdot 1^3(-\sqrt{2})+6\cdot 1^2(-\sqrt{2})^2+4\cdot 1\cdot(-\sqrt{2})^3+1\cdot(-\sqrt{2})^4$

$$=1-4\sqrt{2}+12-8\sqrt{2}+4$$

$$=17-12\sqrt{2}$$

$(1+\sqrt{2})^4=1+4\sqrt{2}+12+8\sqrt{2}+4$ Using the result above

$$=17+12\sqrt{2}$$

$$(1-\sqrt{2})^4+(1+\sqrt{2})^4=17-12\sqrt{2}+17+12\sqrt{2}=34$$

19. Expand: $(x^{-2}+x^2)^4$.

We have $a=x^{-2}$, $b=x^2$, and $n=4$.

Pascal's triangle method: Use the fifth row of Pascal's triangle.

1 4 6 4 1.

$$(x^{-2}+x^2)^4$$

$$=1\cdot(x^{-2})^4+4(x^{-2})^3(x^2)+6(x^{-2})^2(x^2)^2+4(x^{-2})(x^2)^3+1\cdot(x^2)^4$$

$$=x^{-8}+4x^{-4}+6+4x^4+x^8$$

Factorial notation method:

$$(x^{-2}+x^2)^4$$

$$=\binom{4}{0}(x^{-2})^4+\binom{4}{1}(x^{-2})^3(x^2)+\binom{4}{2}(x^{-2})^2(x^2)^2+\binom{4}{3}(x^{-2})(x^2)^3+\binom{4}{4}(x^2)^4$$

$$=\frac{4!}{4!0!}(x^{-8})+\frac{4!}{3!1!}(x^{-6})(x^2)+\frac{4!}{2!2!}(x^{-4})(x^4)+\frac{4!}{1!3!}(x^{-2})(x^6)+\frac{4!}{0!4!}(x^8)$$

$$=x^{-8}+4x^{-4}+6+4x^4+x^8$$

20. Expand: $\left(\frac{1}{\sqrt{x}}-\sqrt{x}\right)^6$.

Pascal's triangle method: We use the seventh row of Pascal's triangle:

1 6 15 20 15 6 1

$$\left(\frac{1}{\sqrt{x}}-\sqrt{x}\right)^6$$

$$=1\cdot\left(\frac{1}{\sqrt{x}}\right)^6+6\left(\frac{1}{\sqrt{x}}\right)^5(-\sqrt{x})+15\left(\frac{1}{\sqrt{x}}\right)^4(-\sqrt{x})^2+20\left(\frac{1}{\sqrt{x}}\right)^3(-\sqrt{x})^3+15\left(\frac{1}{\sqrt{x}}\right)^2(-\sqrt{x})^4+6\left(\frac{1}{\sqrt{x}}\right)(-\sqrt{x})^5+1\cdot(\sqrt{x})^6$$

$$=x^{-3}-6x^{-2}+15x^{-1}-20+15x-6x^2+x^3$$

Factorial notation method:

$$\left(\frac{1}{\sqrt{x}}-\sqrt{x}\right)^6$$

$$=\binom{6}{0}\left(\frac{1}{\sqrt{x}}\right)^6+\binom{6}{1}\left(\frac{1}{\sqrt{x}}\right)^5(-\sqrt{x})+\binom{6}{2}\left(\frac{1}{\sqrt{x}}\right)^4(-\sqrt{x})^2+\binom{6}{3}\left(\frac{1}{\sqrt{x}}\right)^3(-\sqrt{x})^3+\binom{6}{4}\left(\frac{1}{\sqrt{x}}\right)^2(-\sqrt{x})^4+\binom{6}{5}\left(\frac{1}{\sqrt{x}}\right)(-\sqrt{x})^5+\binom{6}{6}(-\sqrt{x})^6$$

$$=x^{-3}-6x^{-2}+15x^{-1}-20+15x-6x^2+x^3$$

21. Find the 3rd term of $(a+b)^7$.

First, we note that $3=2+1$, $a=a$, $b=b$, and $n=7$. Then the 3rd term of the expansion of $(a+b)^7$ is

$$\binom{7}{2}a^{7-2}b^2,\text{ or }\frac{7!}{2!5!}a^5b^2,\text{ or }21a^5b^2.$$

22. $\binom{8}{5}x^3y^5 = 56x^3y^5$

23. Find the 6th term of $(x-y)^{10}$.

First, we note that $6 = 5+1$, $a = x$, $b = -y$, and $n = 10$. Then the 6th term of the expansion of $(x-y)^{10}$ is

$$\binom{10}{5}x^5(-y)^5, \text{ or } -252x^5y^5.$$

24. $\binom{9}{4}p^5(-2q)^4 = 2016p^5q^4$

25. Find the 12th term of $(a-2)^{14}$.

First, we note that $12 = 11+1$, $a = a$, $b = -2$, and $n = 14$. Then the 12th term of the expansion of $(a-2)^{14}$ is

$$\begin{aligned}\binom{14}{11}a^{14-11}\cdot(-2)^{11} &= \frac{14!}{3!11!}a^3(-2048)\\ &= 364a^3(-2048)\\ &= -745,472a^3\end{aligned}$$

26. $\binom{12}{10}x^{12-10}(-3)^{10} = 3,897,234x^2$

27. Find the 5th term of $(2x^3-\sqrt{y})^8$.

First, we note that $5 = 4+1$, $a = 2x^3$, $b = -\sqrt{y}$, and $n = 8$. Then the 5th term of the expansion of $(2x^3-\sqrt{y})^8$ is

$$\begin{aligned}&\binom{8}{4}(2x^3)^{8-4}(-\sqrt{y})^4\\ &= \frac{8!}{4!4!}(2x^3)^4(-\sqrt{y})^4\\ &= 70(16x^{12})(y^2)\\ &= 1120x^{12}y^2\end{aligned}$$

28. $\binom{7}{3}\left(\frac{1}{b^2}\right)^{7-3}\left(\frac{b}{3}\right)^3 = \frac{35}{27}b^{-5}$

29. The expansion of $(2u-3v^2)^{10}$ has 11 terms so the 6th term is the middle term. Note that $6 = 5+1$, $a = 2u$, $b = -3v^2$, and $n = 10$. Then the 6th term of the expansion of $(2u-3v^2)^{10}$ is

$$\begin{aligned}&\binom{10}{5}(2u)^{10-5}(-3v^2)^5\\ &= \frac{10!}{5!5!}(2u)^5(-3v^2)^5\\ &= 252(32u^5)(-243v^{10})\\ &= -1,959,552u^5v^{10}\end{aligned}$$

30. 3rd term: $\binom{5}{2}(\sqrt{x})^{5-2}(\sqrt{3})^2 = 30x\sqrt{x}$

4th term: $\binom{5}{3}(\sqrt{x})^{5-3}(\sqrt{3})^3 = 30x\sqrt{3}$

31. The number of subsets is 2^7, or 128

32. 2^6, or 64

33. The number of subsets is 2^{24}, or 16,777,216.

34. 2^{26}, or 67,108,864

35. The term of highest degree of $(x^5+3)^4$ is the first term, or

$$\binom{4}{0}(x^5)^{4-0}3^0 = \frac{4!}{4!0!}x^{20} = x^{20}.$$

Therefore, the degree of $(x^5+3)^4$ is 20.

36. The term of highest degree of $(2-5x^3)^7$ is the last term, or

$$\binom{7}{7}(-5x^3)^7 = -78,125x^{21}.$$

Therefore, the degree of $(2-5x^3)^7$ is 21.

37. We use factorial notation. Note that $a = 3$, $b = i$, and $n = 5$.

$$\begin{aligned}&(3+i)^5\\ &= \binom{5}{0}(3^5)+\binom{5}{1}(3^4)(i)+\binom{5}{2}(3^3)(i^2)+\\ &\quad\binom{5}{3}(3^2)(i^3)+\binom{5}{4}(3)(i^4)+\binom{5}{5}(i^5)\\ &= \frac{5!}{0!5!}(243)+\frac{5!}{1!4!}(81)(i)+\frac{5!}{2!3!}(27)(-1)+\\ &\quad\frac{5!}{3!2!}(9)(-i)+\frac{5!}{4!1!}(3)(1)+\frac{5!}{5!0!}(i)\\ &= 243+405i-270-90i+15+i\\ &= -12+316i\end{aligned}$$

38.

$$\begin{aligned}&(1+i)^6\\ &= \binom{6}{0}1^6+\binom{6}{1}1^5\cdot i+\binom{6}{2}1^4\cdot i^2+\\ &\quad\binom{6}{3}1^3\cdot i^3+\binom{6}{4}1^2\cdot i^4+\binom{6}{5}1\cdot i^5+\\ &\quad\binom{6}{6}i^6\\ &= 1+6i-15-20i+15+6i-1\\ &= -8i\end{aligned}$$

39. We use factorial notation. Note that $a = \sqrt{2}$, $b = -i$, and $n = 4$.

$$\begin{aligned}(\sqrt{2}-i)^4 &= \binom{4}{0}(\sqrt{2})^4+\binom{4}{1}(\sqrt{2})^3(-i)+\\ &\quad\binom{4}{2}(\sqrt{2})^2(-i)^2+\binom{4}{3}(\sqrt{2})(-i)^3+\\ &\quad\binom{4}{4}(-i)^4\\ &= \frac{4!}{0!4!}(4)+\frac{4!}{1!3!}(2\sqrt{2})(-i)+\\ &\quad\frac{4!}{2!2!}(2)(-1)+\frac{4!}{3!1!}(\sqrt{2})(i)+\\ &\quad\frac{4!}{4!0!}(1)\\ &= 4-8\sqrt{2}i-12+4\sqrt{2}i+1\\ &= -7-4\sqrt{2}i\end{aligned}$$

40. $\left(\frac{\sqrt{3}}{2}-\frac{1}{2}i\right)^{11}$

$=\binom{11}{0}\left(\frac{\sqrt{3}}{2}\right)^{11}+\binom{11}{1}\left(\frac{\sqrt{3}}{2}\right)^{10}\left(-\frac{1}{2}i\right)+$

$\binom{11}{2}\left(\frac{\sqrt{3}}{2}\right)^{9}\left(-\frac{1}{2}i\right)^{2}+$

$\binom{11}{3}\left(\frac{\sqrt{3}}{2}\right)^{8}\left(-\frac{1}{2}i\right)^{3}+$

$\binom{11}{4}\left(\frac{\sqrt{3}}{2}\right)^{7}\left(-\frac{1}{2}i\right)^{4}+$

$\binom{11}{5}\left(\frac{\sqrt{3}}{2}\right)^{6}\left(-\frac{1}{2}i\right)^{5}+$

$\binom{11}{6}\left(\frac{\sqrt{3}}{2}\right)^{5}\left(-\frac{1}{2}i\right)^{6}+$

$\binom{11}{7}\left(\frac{\sqrt{3}}{2}\right)^{4}\left(-\frac{1}{2}i\right)^{7}+$

$\binom{11}{8}\left(\frac{\sqrt{3}}{2}\right)^{3}\left(-\frac{1}{2}i\right)^{8}+$

$\binom{11}{9}\left(\frac{\sqrt{3}}{2}\right)^{2}\left(-\frac{1}{2}i\right)^{9}+$

$\binom{11}{10}\left(\frac{\sqrt{3}}{2}\right)\left(-\frac{1}{2}i\right)^{10}+\binom{11}{11}\left(-\frac{1}{2}i\right)^{11}$

$=\frac{243\sqrt{3}}{2048}-\frac{2673}{2048}i-\frac{4455\sqrt{3}}{2048}+\frac{13,365}{2048}i+$

$\frac{8910\sqrt{3}}{2048}-\frac{12,474}{2048}i-\frac{4158\sqrt{3}}{2048}+\frac{2970}{2048}i+$

$\frac{495\sqrt{3}}{2048}-\frac{165}{2048}i-\frac{11\sqrt{3}}{2048}+\frac{1}{2048}i$

$=\frac{1024\sqrt{3}}{2048}+\frac{1024}{2048}i$

$=\frac{\sqrt{3}}{2}+\frac{1}{2}i$

41. $(a-b)^n=\binom{n}{0}a^n(-b)^0+\binom{n}{1}a^{n-1}(-b)^1+$

$\binom{n}{2}a^{n-2}(-b)^2+\cdots+$

$\binom{n}{n-1}a^1(-b)^{n-1}+\binom{n}{n}a^0(-b)^n$

$=\binom{n}{0}(-1)^0a^nb^0+\binom{n}{1}(-1)^1a^{n-1}b^1+$

$\binom{n}{2}(-1)^2a^{n-2}b^2+\cdots+$

$\binom{n}{n-1}(-1)^{n-1}a^1b^{n-1}+$

$\binom{n}{n}(-1)^na^0b^n$

$=\sum_{k=0}^{n}\binom{n}{k}(-1)^ka^{n-k}b^k$

42. $\frac{(x+h)^{13}-x^{13}}{h}$

$=(x^{13}+13x^{12}h+78x^{11}h^2+286x^{10}h^3+$
$715x^9h^4+1287x^8h^5+1716x^7h^6+1716x^6h^7+$
$1287x^5h^8+715x^4h^9+286x^3h^{10}+78x^2h^{11}+$
$13xh^{12}+h^{13}-x^{13})/h$

$=13x^{12}+78x^{11}h+286x^{10}h^2+715x^9h^3+$
$1287x^8h^4+1716x^7h^5+1716x^6h^6+1287x^5h^7+$
$715x^4h^8+286x^3h^9+78x^2h^{10}+13xh^{11}+h^{12}$

43. $\frac{(x+h)^n-x^n}{h}$

$=\frac{\binom{n}{0}x^n+\binom{n}{1}x^{n-1}h+\cdots+\binom{n}{n}h^n-x^n}{h}$

$=\binom{n}{1}x^{n-1}+\binom{n}{2}x^{n-2}h+\cdots+\binom{n}{n}h^{n-1}$

$=\sum_{k=1}^{n}\binom{n}{k}x^{n-k}h^{k-1}$

44. In expanding $(a+b)^n$, it would probably be better to use Pascal's triangle when n is relatively small. When n is large, and many rows of Pascal's triangle must be computed to get to the $(n+1)$st row, it would probably be better to use factorial notation. In addition, factorial notation allows us to write a particular term of the expansion more efficiently than Pascal's triangle.

45. The array of numbers that is known as Pascal's triangle appeared as early as 1303 in a work of the Chinese algebraist Chu Shï-kié. Because Pascal developed many of the triangle's properties and then found many applications of these properties, this array became known as Pascal's triangle.

46. $(f+g)(x)=f(x)+g(x)=(x^2+1)+(2x-3)=$
x^2+2x-2

47. $(fg)(x)=f(x)g(x)=(x^2+1)(2x-3)=$
$2x^3-3x^2+2x-3$

48. $(f\circ g)(x)=f(g(x))=f(2x-3)=(2x-3)^2+1=4x^2-$
$12x+9+1=4x^2-12x+10$

49. $(g\circ f)(x)=g(f(x))=g(x^2+1)=2(x^2+1)-3=$
$2x^2+2-3=2x^2-1$

50. $\sum_{k=0}^{8}\binom{8}{k}x^{8-k}3^k=0$

The left side of the equation is sigma notation for $(x+3)^8$, so we have:

$(x+3)^8=0$

$x+3=0$ Taking the 8th root on both sides

$x=-3$

51. $\sum_{k=0}^{4} \binom{4}{k} 5^{4-k} x^k = 64$

The left side of the equation is sigma notation for $(5+x)^4$, so we have:

$(5+x)^4 = 64$

$5 + x = \pm 2\sqrt{2}$ Taking the 4th root on both sides

$x = -5 \pm 2\sqrt{2}$

The real solutions are $-5 \pm 2\sqrt{2}$.

If we also observe that $(2\sqrt{2}i)^4 = 64$, we also find the imaginary solutions $-5 \pm 2\sqrt{2}i$.

52. $\sum_{k=0}^{5} \binom{5}{k} (-1)^k x^{5-k} 3^k = \sum_{k=0}^{5} \binom{5}{k} x^{5-k}(-3)^k$, so the left side of the equation is sigma notation for $(x-3)^5$. We have:

$(x-3)^5 = 32$

$x - 3 = 2$ Taking the 5th root on both sides

$x = 5$

53. $\sum_{k=0}^{4} \binom{4}{k} (-1)^k x^{4-k} 6^k = \sum_{k=0}^{4} \binom{4}{k} x^{4-k}(-6)^k$, so the left side of the equation is sigma notation for $(x-6)^4$. We have:

$(x-6)^4 = 81$

$x - 6 = \pm 3$ Taking the 4th root on both sides

$x - 6 = 3$ or $x - 6 = -3$

$x = 9$ or $x = 3$

The solutions are 9 and 3.

If we also observe that $(3i)^4 = 81$, we also find the imaginary solutions $6 \pm 3i$.

54. The $(k+1)$st term of $\left(\frac{3x^2}{2} - \frac{1}{3x}\right)^{12}$ is $\binom{12}{k}\left(\frac{3x^2}{2}\right)^{12-k}\left(-\frac{1}{3x}\right)^k$. In the term which does not contain x, the exponent of x in the numerator is equal to the exponent of x in the denominator.

$2(12-k) = k$

$24 - 2k = k$

$24 = 3k$

$8 = k$

Find the $(8+1)$st, or 9th term:

$\binom{12}{8}\left(\frac{3x^2}{2}\right)^4\left(-\frac{1}{3x}\right)^8 = \frac{12!}{4!8!}\left(\frac{3^4x^8}{2^4}\right)\left(\frac{1}{3^8x^8}\right) = \frac{55}{144}$

55. The expansion of $(x^2 - 6y^{3/2})^6$ has 7 terms, so the 4th term is the middle term.

$\binom{6}{3}(x^2)^3(-6y^{3/2})^3 = \frac{6!}{3!3!}(x^6)(-216y^{9/2}) = -4320x^6y^{9/2}$

56. $\dfrac{\binom{5}{3}(p^2)^2\left(-\frac{1}{2}p\sqrt[3]{q}\right)^3}{\binom{5}{2}(p^2)^3\left(-\frac{1}{2}p\sqrt[3]{q}\right)^2} = \dfrac{-\frac{1}{8}p^7q}{\frac{1}{4}p^8\sqrt[3]{q^2}} = -\dfrac{\sqrt[3]{q}}{2p}$

57. The $(k+1)$st term of $\left(\sqrt[3]{x} - \frac{1}{\sqrt{x}}\right)^7$ is $\binom{7}{k}(\sqrt[3]{x})^{7-k}\left(-\frac{1}{\sqrt{x}}\right)^k$. The term containing $\frac{1}{x^{1/6}}$ is the term in which the sum of the exponents is $-1/6$. That is,

$\left(\frac{1}{3}\right)(7-k) + \left(-\frac{1}{2}\right)(k) = -\frac{1}{6}$

$\frac{7}{3} - \frac{k}{3} - \frac{k}{2} = -\frac{1}{6}$

$-\frac{5k}{6} = -\frac{15}{6}$

$k = 3$

Find the $(3+1)$st, or 4th term.

$\binom{7}{3}(\sqrt[3]{x})^4\left(-\frac{1}{\sqrt{x}}\right)^3 = \frac{7!}{4!3!}(x^{4/3})(-x^{-3/2}) = -35x^{-1/6}$, or $-\frac{35}{x^{1/6}}$.

58. The total number of subsets of a set of 7 bills is 2^7. This includes the empty set. Thus, $2^7 - 1$, or 127, different sums of money can be formed.

59. ${}_{100}C_0 + {}_{100}C_1 + \cdots + {}_{100}C_{100}$ is the total number of subsets of a set with 100 members, or 2^{100}.

60. ${}_nC_0 + {}_nC_1 + \ldots + {}_nC_n$ is the total number of subsets of a set with n members, or 2^n.

61. $\sum_{k=0}^{23} \binom{23}{k} (\log_a x)^{23-k}(\log_a t)^k = (\log_a x + \log_a t)^{23} = [\log_a(xt)]^{23}$

62. $\sum_{k=0}^{15} \binom{15}{k} i^{30-2k}$

$= i^{30} + 15i^{28} + 105i^{26} + 455i^{24} + 1365i^{22} + 3003i^{20} + 5005i^{18} + 6435i^{16} + 6435i^{14} + 5005i^{12} + 3003i^{10} + 1365i^8 + 455i^6 + 105i^4 + 15i^2 + 1$

$= -1 + 15 - 105 + 455 - 1365 + 3003 - 5005 + 6435 - 6435 + 5005 - 3003 + 1365 - 455 + 105 - 15 + 1$

$= 0$

63. See the answer section in the text.

Exercise Set 10.8

1. a) We use Principle P.

For 1: $P = \frac{18}{100}$, or 0.18

For 2: $P = \frac{24}{100}$, or 0.24

For 3: $P = \frac{23}{100}$, or 0.23

For 4: $P = \frac{23}{100}$, or 0.23

For 5: $P = \frac{12}{100}$, or 0.12

b) Opinions may vary, but it seems that people tend not to select the first or last numbers.

2. The total number of gumdrops is $7+8+9+4+5+6$, or 39.

Lemon: $\frac{8}{39}$

Lime: $\frac{5}{39}$

Orange: $\frac{9}{39} = \frac{3}{13}$

Grape: $\frac{6}{39} = \frac{2}{13}$

Strawberry: $\frac{7}{39}$

Licorice: $\frac{0}{39} = 0$

3. The company can expect 78% of the 15,000 pieces of advertising to be opened and read. We have:

$78\%(15,000) = 0.78(15,000) = 11,700.$

4. a) B: $136/9136 \approx 1.5\%$

C: $273/9136 \approx 3.0\%$

D: $286/9136 \approx 3.1\%$

E: $1229/9136 \approx 13.5\%$

F: $173/9136 \approx 1.9\%$

G: $190/9136 \approx 2.1\%$

H: $399/9136 \approx 4.4\%$

I: $539/9136 \approx 5.9\%$

J: $21/9136 \approx 0.2\%$

K: $57/9136 \approx 0.6\%$

L: $417/9136 \approx 4.6\%$

M: $231/9136 \approx 2.5\%$

N: $597/9136 \approx 6.5\%$

O: $705/9136 \approx 7.7\%$

P: $238/9136 \approx 2.6\%$

Q: $4/9136 \approx 0.04\%$

R: $609/9136 \approx 6.7\%$

S: $745/9136 \approx 8.2\%$

T: $789/9136 \approx 8.6\%$

U: $240/9136 \approx 2.6\%$

V: $113/9136 \approx 1.2\%$

W: $127/9136 \approx 1.4\%$

X: $20/9136 \approx 0.2\%$

Y: $124/9136 \approx 1.4\%$

b) $\frac{853+1229+539+705+240}{9136} = \frac{3566}{9136} \approx 39\%$

c) In part (b) we found that there are 3566 vowels. Thus, there are $9136 - 3566$, or 5570 consonants.

$$P = \frac{5570}{9136} \approx 61\%$$

(We could also find this by subtracting the probability of a vowel occurring from 100%: $100\% - 39\% = 61\%$)

5. a) The consonants with the 5 greatest numbers of occurrences are the 5 consonants with the greatest probability of occurring. They are T, S, R, N, and L.

b) E is the vowel with the greatest number of occurrences, so E is the vowel with the greatest probability of occurring.

c) Yes

6. a) 52

b) $\frac{4}{52}$, or $\frac{1}{13}$

c) $\frac{13}{52}$, or $\frac{1}{4}$

d) $\frac{4}{52}$, or $\frac{1}{13}$

e) $\frac{26}{52}$, or $\frac{1}{2}$

f) $\frac{4+4}{52}$, or $\frac{2}{13}$

g) $\frac{2}{52}$, or $\frac{1}{26}$

7. a) Since there are 14 equally likely ways of selecting a marble from a bag containing 4 red marbles and 10 green marbles, we have, by Principle P,

$P(\text{selecting a red marble}) = \frac{4}{14} = \frac{2}{7}.$

b) Since there are 14 equally likely ways of selecting a marble from a bag containing 4 red marbles and 10 green marbles, we have, by Principle P,

$P(\text{selecting a green marble}) = \frac{10}{14} = \frac{5}{7}.$

c) Since there are 14 equally likely ways of selecting a marble from a bag containing 4 red marbles and 10 green marbles, we have, by Principle P,

$P(\text{selecting a purple marble}) = \frac{0}{14} = 0.$

d) Since there are 14 equally likely ways of selecting a marble from a bag containing 4 red marbles and 10 green marbles, we have, by Principle P,

P(selecting a red or a green marble) =

$\frac{4+10}{14} = 1.$

8. $\frac{{}_{10}C_2 \cdot {}_{10}C_2}{{}_{20}C_4} = \frac{45 \cdot 45}{4845} = \frac{135}{323}$

9. The total number of coins is $7+5+10$, or 22 and the total number of coins to be drawn is $4+3+1$, or 8. The number of ways of selecting 8 coins from a group of 22 is ${}_{22}C_8$. Four dimes can be selected in ${}_7C_4$ ways, 3 nickels in ${}_5C_3$ ways, and 1 quarter in ${}_{10}C_1$ ways.

P(selecting 4 dimes, 3 nickels, and 1 quarter) =

$\frac{{}_7C_4 \cdot {}_5C_3 \cdot {}_{10}C_1}{{}_{22}C_8}$, or $\frac{350}{31,977}$

10. $\frac{1}{{}_{49}C_6} = \frac{1}{13,983,816} \approx 0.000007\%$

11. The number of ways of selecting 5 cards from a deck of 52 cards is ${}_{52}C_5$. Three sevens can be selected in ${}_4C_3$ ways and 2 kings in ${}_4C_2$ ways.

P(drawing 3 sevens and 2 kings) $= \frac{{}_4C_3 \cdot {}_4C_2}{{}_{52}C_5}$, or

$\frac{1}{108,290}.$

12. Since there are only 4 aces in the deck, $P(5 \text{ aces}) = 0$.

13. The number of ways of selecting 5 cards from a deck of 52 cards is ${}_{52}C_5$. Since 13 of the cards are spades, then 5 spades can be drawn in ${}_{13}C_5$ ways

P(drawing 5 spades) $= \frac{{}_{13}C_5}{{}_{52}C_5} = \frac{1287}{2,598,960} =$

$\frac{33}{66,640}$

14. $\frac{{}_4C_4 \cdot {}_4C_1}{{}_{52}C_5} = \frac{1}{649,740}$

15. a) HHH, HHT, HTH, HTT, THH, THT, TTH, TTT

b) Three of the 8 outcomes have exactly one head. Thus, P(exactly one head) $= \frac{3}{8}$.

c) Seven of the 8 outcomes have exactly 0, 1, or 2 heads. Thus, P(at most two heads) $= \frac{7}{8}$.

d) Seven of the 8 outcomes have 1, 2, or 3 heads. Thus, P(at least one head) $= \frac{7}{8}$.

e) Three of the 8 outcomes have exactly two tails. Thus, P(exactly two tails) $= \frac{3}{8}$.

16. $\frac{18}{38}$, or $\frac{9}{19}$

17. The roulette wheel contains 38 equally likely slots. Eighteen of the 38 slots are colored black. Thus, by Principle P,

P(the ball falls in a black slot) $= \frac{18}{38} = \frac{9}{19}$.

18. $\frac{1}{38}$

19. The roulette wheel contains 38 equally likely slots. Only 1 slot is numbered 0. Then, by Principle P,

P(the ball falls in the 0 slot) $= \frac{1}{38}$.

20. $\frac{2}{38} = \frac{1}{19}$

21. The roulette wheel contains 38 equally likely slots. Thirty-six of the slots are colored red or black. Then, by Principle P,

P(the ball falls in a red or a black slot) $= \frac{36}{38} = \frac{18}{19}$.

22. $\frac{1}{38}$

23. The roulette wheel contains 38 equally likely slots. Eighteen of the slots are odd-numbered. Then, by Principle P,

P(the ball falls in a an odd-numbered slot) =

$\frac{18}{38} = \frac{9}{19}$.

24. The dartboard can be thought of as having 18 areas of equal size. Of these, 6 are red, 4 are green, 3 are blue, and 5 are yellow.

$P(\text{red}) = \frac{6}{18} = \frac{1}{3}$

$P(\text{green}) = \frac{4}{18} = \frac{2}{9}$

$P(\text{blue}) = \frac{3}{18} = \frac{1}{6}$

$P(\text{yellow}) = \frac{5}{18}$

25. a), b) Answers may vary

26. From the choices of the 26 letters of the alphabet or a space, the probability of selecting each of the 50 letters or spaces is $\frac{1}{27}$. Then the probability that the given passage could have been written by a monkey is $\left(\frac{1}{27}\right)^{50} \approx 2.7 \times 10^{-72}$. Thus, it is extremely unlikely that the given passage, much less an entire novel, could be written by a monkey.

27. Answers may vary.

28. zero

29. one-to-one

30. function; domain; range; domain; range

31. zero

32. combination

33. inverse variation

34. factor

35. geometric sequence

36. a) 4

b) $\frac{4}{_{52}C_5} = \frac{4}{2,598,960} \approx 1.54 \times 10^{-6}$

37. Consider a suit

A K Q J 10 9 8 7 6 5 4 3 2

A straight flush can be any of the following combinations in the same suit.

K Q J 10 9
Q J 10 9 8
J 10 9 8 7
10 9 8 7 6
9 8 7 6 5
8 7 6 5 4
7 6 5 4 3
6 5 4 3 2
5 4 3 2 A

Remember a straight flush does not include A K Q J 10 which is a royal flush.

a) Since there are 9 straight flushes per suit, there are 36 straight flushes in all 4 suits.

b) Since 2,598,960, or $_{52}C_5$, poker hands can be dealt from a standard 52-card deck and 36 of those hands are straight flushes, the probability of getting a straight flush is $\frac{36}{2,598,960}$, or about 1.39×10^{-5}.

38. a) There are 13 ways to choose a denomination. Then there are 48 ways to choose one of the 48 cards remaining after 4 cards of the same denomination are chosen. Thus there are $13 \cdot 48$, or 624, four of a kind hands.

b) $\frac{624}{_{52}C_5} = \frac{624}{2,598,960} \approx 2.4 \times 10^{-4}$

39. a) There are 13 ways to select a denomination. Then from that denomination there are $_4C_3$ ways to pick 3 of the 4 cards in that denomination. Now there are 12 ways to select any one of the remaining 12 denominations and $_4C_2$ ways to pick 2 cards from the 4 cards in that denomination. Thus the number of full houses is $(13 \cdot {_4C_3}) \cdot (12 \cdot {_4C_2})$ or 3744.

b) $\frac{3744}{_{52}C_5} = \frac{3744}{2,598,960} \approx 0.00144$

40. a) There are 13 ways to select a denomination and then $\binom{4}{3}$ ways to choose 3 of the 4 cards in that denomination. Now there are $\binom{48}{2}$ ways to choose 2 cards from the 12 remaining denominations $(4 \cdot 12$, or 48 cards). But these combinations include the 3744 hands in a full house like Q-Q-Q-4-4 (Exercise 31), so these must be subtracted. Thus the number of three of a kind hands is $13 \cdot \binom{4}{3} \cdot \binom{48}{2} - 3744$, or 54,912.

b) $\frac{54,912}{_{52}C_5} = \frac{54,912}{2,598,960} \approx 0.0211$

41. a) There are 4 ways to select a suit and then $\binom{13}{5}$ ways to choose 5 cards from that suit. But these combinations include hands with all the cards in sequence, so we subtract the 4 royal flushes (Exercise 28) and the 36 straight flushes (Exercise 29). Thus there are $4 \cdot \binom{13}{5} - 4 - 36$, or 5108 flushes.

b) $\frac{5108}{_{52}C_5} = \frac{5108}{2,598,960} \approx 0.00197$

42. a) There are $\binom{13}{2}$ ways to select 2 denominations from the 13 denominations. Then in each denomination there are $\binom{4}{2}$ ways to choose 2 of the 4 cards.

Finally there are $\binom{44}{1}$ ways to choose the fifth card from the 11 remaining denominations $(4 \cdot 11$, or 44 cards). Thus the number of two pairs hands is

$\binom{13}{2} \cdot \binom{4}{2} \cdot \binom{4}{2} \cdot \binom{44}{1}$, or 123,552.

b) $\frac{123,552}{_{52}C_5} = \frac{123,552}{2,598,960} \approx 0.0475$

43. a) There are 10 sets of 5 consecutive cards:

A K Q J 10
K Q J 10 9
.
.
.
5 4 3 2 A

In each of these 10 sets there are 4 ways to choose (from 4 suits) each of the 5 cards. These combinations include the 4 royal flushes and the 36 straight flushes, both of which consist of 5 cards of the *same suit* in sequence.

Thus there are $10 \cdot 4 \cdot 4 \cdot 4 \cdot 4 \cdot 4 - 4 - 36$, or 10,200 straights.

b) $\frac{10,200}{_{52}C_5} = \frac{10,200}{2,598,960} \approx 0.00392$